Feature Papers in Bioactives and Nutraceuticals-Volume I

Feature Papers in Bioactives and Nutraceuticals-Volume I

Editor

Maurizio Battino

MDPI

Basel • Beijing • Wuhan • Barcelona • Belgrade • Novi Sad • Cluj • Manchester

Editor
Maurizio Battino
Università Politecnica delle Marche
Ancona
Italy

Editorial Office
MDPI
St. Alban-Anlage 66
4052 Basel, Switzerland

This is a reprint of articles from the Topical Collection published online in the open access journal *International Journal of Molecular Sciences* (ISSN 1422-0067) (available at: https://www.mdpi.com/journal/ijms/topical_collections/featurepapers_bioactives).

For citation purposes, cite each article independently as indicated on the article page online and as indicated below:

Lastname, A.A.; Lastname, B.B. Article Title. *Journal Name* **Year**, *Volume Number*, Page Range.

Volume I
ISBN 978-3-7258-0999-8 (Hbk)
ISBN 978-3-7258-1000-0 (PDF)
doi.org/10.3390/books978-3-7258-1000-0

Set
ISBN 978-3-7258-0997-4 (Hbk)
ISBN 978-3-7258-0998-1 (PDF)

Contents

About the Editor

Maurizio Battino

Maurizio Battino is a Professor of Nutrition in the Faculty of Medicine at Università Politecnica delle Marche, Ancona, Italy. He obtained his B.Sc. from the University of Bologna, Italy, in 1984, a Ph.D. from the Italian Minister of Science in 1991, and an M.D. (Hon) from the Faculty of Medicine and Pharmacy, Carol Davila, Bucharest, Romania, in 2008. He also obtained an M.S. in ITC and Telemedicine from the Faculty of Medicine, Università Politecnica delle Marche, Ancona, Italy, in 2010. Currently, he hold the positions of Full Professor at the Faculty of Medicine and Director of various Research Centers in Spain (University of Vigo, Vigo, and Universidad Europea del Atlantico, Santander) and China (Jiangsu University). Battino's current and previous research interests include bioenergetics, nutritional biochemistry, membranes, and periodontology. He have published approximately 500 papers with an h-index of 80 (Scopus) and 72 (WoS). Battino has also been a member of SFRBM, PSE, and other international societies and a highly cited researcher, according to Clarivate-Web of Science, during the last 9 consecutive years.

International Journal of
Molecular Sciences

MDPI

Review

Interactions of Whey Proteins with Metal Ions

Agnieszka Rodzik [1,2], Paweł Pomastowski [2,*], Gulyaim N. Sagandykova [1,2] and Bogusław Buszewski [1,2]

1 Department of Environmental Chemistry and Bioanalysis, Faculty of Chemistry, Nicolaus Copernicus University, Gagarina 7, 87-100 Toruń, Poland; agnieszka.rodzik1@gmail.com (A.R.); sagandykova.gulyaim1@gmail.com (G.N.S.); bbusz@chem.umk.pl (B.B.)

2 Centre for Modern Interdisciplinary Technologies, Nicolaus Copernicus University, Wileńska 4, 87-100 Toruń, Poland

* Correspondence: p.pomastowski@umk.pl; Tel.: +48-56-6114308; Fax: +48-56-6656038

Received: 13 February 2020; Accepted: 18 March 2020; Published: 20 March 2020

Abstract: Whey proteins tend to interact with metal ions, which have implications in different fields related to human life quality. There are two impacts of such interactions: they can provide opportunities for applications in food and nutraceuticals, but may lead to analytical challenges related to their study and outcomes for food processing, storage, and food interactions. Moreover, interactions of whey proteins with metal ions are complicated, requiring deep understanding, leading to consequences, such as metalloproteins, metallocomplexes, nanoparticles, or aggregates, creating a biologically active system. To understand the phenomena of metal–protein interactions, it is important to develop analytical approaches combined with studies of changes in the biological activity and to analyze the impact of such interactions on different fields. The aim of this review was to discuss chemistry of β-lactoglobulin, α-lactalbumin, and lactotransferrin, their interactions with different metal ions, analytical techniques used to study them and the implications for food and nutraceuticals.

Keywords: whey proteins; metal–protein interactions; food storage; food safety; nutraceuticals; metallocomplexes

1. Introduction

The protein fraction of whey is composed of different proteins, including β-lactoglobulin (β-LG), α-lactalbumin (α-LA), lactoferrin (LTF), small amounts of immunoglobulin (IG), bovine serum albumin (BSA), and lactoperoxidase (LP) [1]. Structures of LTF, β-LG, and α-LA are presented in Figure 1.

1.1. *β-Lactoglobulin*

β-Lactoglobulin is the major bovine whey protein, accounting for approximately 10% of the total protein in bovine milk and approximately 50% in ruminants [2], but it is not present in human milk. β-LG contains 162 amino acid residues, which form nine antiparallel β-sheets [3]. It belongs to the lipocalin family and has the ability to bind different hydrophobic molecules [4], which can be useful for reducing allergenicity owing to its covalent conjugation to flavonoids because β-LG is one of the major milk allergens responsible for cow milk allergy [5].

β-LG has eleven genetic variants (A, B, C, D, E, F, G, H, W, I, and J). Genetic variants A and B are most common in bovine milk and differ in positions 64 and 118. Two bovine β-lactoglobulins I and J were isolated from bovine milk by isoelectric focusing by Godovac-Zimmermann et al. [4]. Moreover, various variants of β-LG translate into different metal affinities, e.g., to nickel or cobalt complexes [6].

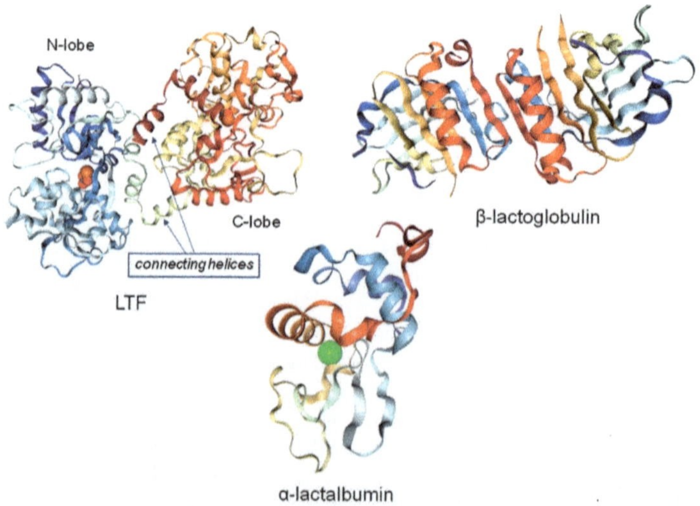

Figure 1. Structures of whey proteins at high resolution extracted from the protein data bank database; iron atoms are denoted in red color for LTF together with carbonate ions; calcium atom is denoted in green color for α-lactalbumin, PDB codes: LTF—1BIY; α-LA—1HFX; β-LG—6RYT.

Additionally, β-LG has two disulphide bonds (Cys-106 to Cys-119; Cys-66 to Cys-160) [7] that maintain the structural integrity during hydrolysis and heat treatment and one free cysteine group (Cys-121) as the binding site for d-block metal ions, such as iron (II/III), copper (II), and silver (I) [8]. Due to the disulphide bonds and free sulfhydryl group in its hydrophobic core, β-LG prevents oxidation by capturing reactive oxygen species (ROS) [9]. β-LG may be modified by phosphorylation [10] or glycation [11], which are examples of post-translational modifications (PTMs) of the protein after its translation by proteolytic cuts or by adding a modifying group to one or more amino acids [12]. The molecular weight of β-LG and of other whey proteins is dependent on post-translational modifications (Table 1). It also can be observed that the number of significant figures in the value of molecular weight can be determined by the precision of the analytical method applied for its analysis.

Heating causes changes in a protein's structure, and subsequently, its properties, thus affecting the quality of food products. De Wit summarized the thermal behavior of β-LG up to 150 °C and concluded that thermal behavior of β-LG is dependent on pH, temperature, time of heating, and concentration [13]. Reversible conformational changes up to 60 °C, which are known as the Tanford transition (negligible between pH 6.5 and 7.8, accounts for 18% at pH 7.0), irreversible denaturation by unfolding and aggregation of monomers between 60 and 70 °C at pH ≥ 7.0 in the presence of OH^- ions) have been reported by several authors [13]. In addition, thiols oxidation between 65 and 75 °C, disulphide/thiol exchange reactions prevailing between 75 and 85 °C and induction of larger aggregates by specific non-covalent aggregation, and unfolding of the residual protein structures above 125 °C have been indicated in a number of publications [13]. Liu et al. reported a loss of antioxidant activity of β-lactoglobulin as a result of cross-linking free thiol groups upon heating (100 °C for 2 min) [14]. Wijayanti and co-authors evaluated the effect of lipoic acid in its acidic and reduced forms on heat-induced unfolding of β-LG and obtained results showed that the reduced form was more effective and its effects were similar to N-ethylmaleimide (NEM) and dithio(bis)-p-nitrobenzoate (DTNB) [15]. In contrast to heating, the antioxidant activity of β-lactoglobulin can be enhanced by ultrasound and enzymatic treatment, which modify its secondary structure and strengthen proteolysis [8].

Table 1. Different whey protein molecular weights associated with post-translational modifications (PTMs).

Protein	Mol. Weight (kDa)	Theoretical mol. Weight (kDa)*	PTM	Method of Isolation/Purification	Identification	Ref.
β-LG	18 18.5	18.277	-	standard of β-LG (protein content > 90%)	SDS-PAGE MALDI-TOF-MS	[5]
β-LG	18.3 36.6	18.277	monomeric and the dimeric forms at pH 7.4 glycated β-lactoglobulin	β-LG was dissolved in 9.1 mM glucose in water, and the pH was adjusted to 7 with 50 mM phosphate buffer	MALDI-TOF-MS	[16]
β-LG	17.4	18.277	-	anion-exchange chromatography (DEAE-Sepharose)	SDS-PAGE	[17]
β-LG	19.9	18.277	proteins appeared as strings of spots, indicating their different isoforms with different charges as a result of PTMs occurring prior to secretion	precipitation via ammonium sulphate fractionation	2-DE MALDI-MS	[18]
α-LA	16.2	16.247				
α-LA	14.1	16.247	small mass differences ruled out PTMs, such as phosphorylation and glycosylation	precipitation by ammonium sulphate	MALDI-TOF-MS	[19]
SA	67.7 (SA) 79.8 (LTF)	69.367	glycosylation of specific milk proteins was shown to vary during lactation; no potential N-glycosylation and O-linked glycans (SA), known N-linked glycoprotein (LTF)	0.5 mL of raw milk was centrifuged at 4 °C for 30 min, fat and cellular layers were removed; residual lipids were removed by addition of three volumes (1.5 mL) of 2:1 chloroform/methanol, agitation, retaining of supernatant; protein was precipitated from supernatant with ethanol overnight at 4 °C, followed by centrifugation; precipitate was re-suspended in 50 mM ammonium bicarbonate buffer (pH 7.5); glycans were separated by SDS-PAGE and extracted for MALDI-MS analysis	MALDI-MS	[20]
LTF	69.0 (SA) 78.0 (LTF)	78.056			LC–MS/MS	
LTF	80.002	78.056	-	milk was defatted by centrifugation, and the pH was then adjusted to 4.6 using hydrochloric acid; precipitated casein was removed by centrifugation	RP-LC–MS/MS	[21]

*values of theoretical molecular weight of the proteins from the Uniprot database (bovine, and for SA-human); Uniprot KB: α-LA—P00711; LTF—P24627; SA—P02768, the value for β-LG accounts for form B from publication of Eigel et al. [22].

Interestingly, Mercadante et al. [23] reported the ability of bovine β-LG to form dimers and studied the dissociation equilibrium and rate constant over the pH range of 2.5–7.5. The equilibrium constant increased with an increase in |pH-pI|, thus indicating the major role of the hydrophobic effect in the stabilization of the dimer and suggesting that electrostatic repulsion destabilizes the dimer, especially at low pH.

1.2. *α-Lactalbumin*

α-Lactalbumin consists of 123 amino acids, except for rat α-lactalbumin, which contains 17 more amino acids and is an extension of the carboxyl end enriched with proline [24]. α-LA constitutes approximately 22% of the total protein of human milk and approximately 36% of whey protein in human milk, and it constitutes approximately 3.5% of the total protein and approximately 17% of the whey protein in bovine milk [25].

The native α-LA consists of two domains: a large α-helical domain and a small β-sheet domain connected by a calcium-binding loop. α-LA possesses a strong calcium-binding site with residues of Lys79, Asp82, Asp84, Asp87, and Asp88 [26]. Calcium-binding has a significant influence on the molecular stability of LA. Moreover, it is required for the refolding and formation of a native disulphide bond in the reduced, denatured protein [27]. The nuclear magnetic resonance (NMR) and circular dichroism (CD) pH titration studies reported by Kim et al. suggested that critical electrostatic interactions concentrated in the calcium-binding region contribute to the denaturation of the protein by determination of the pK_a values of individual functional ionizable groups [28]. When calcium ions dissociate from α-LA at an acidic pH, the protein adopts the molten globule conformation, which has been described as a compact state with a significant degree of secondary structure in the native protein but with a fluctuating tertiary structure [29]. The molten globule has a weakly folded α-helix domain and a domain with a disordered β-sheet domain [30]. The removal of calcium (II) ions resulted in conformational changes, as indicated by spectral (fluorescence and absorbance) changes [31]. Interestingly, the work of Noyelle and co-authors showed that magnesium (II) binding occurred more likely via interactions with the residues belonging to the zinc (II)-binding site in contrast to its expected binding to the calcium (II)-binding site [26]. A study by Wehbi et al. demonstrated that binding of calcium to bovine α-LA increases the resistance of the protein structure to thermal treatment [32].

α-LA is stabilized by four disulphide bonds between the cysteine residues (Cys-6 to Cys-120, Cys-61 to Cys-77, Cys-73 to Cys-91, and Cys-28 to Cys-111) [33]. The active molecular form of α-LA may have various post-translational modifications in contrast to the native form [19]. Moreover, for structural reasons, the α-LA has a metal affinity to ions of s-block elements, such as magnesium (II) and transition metal ions, e.g., zinc (II), which is especially promoted in the reaction with β4-galactosyltransferase, according to immobilized metal-affinity chromatography (IMAC) [34]. In addition, calcium ions increase the stability of α-LA in its native state [35]. Zinc ions may also bind to the calcium-binding site, thus increasing its absorption and bioavailability. In this way, the α-LA complex of zinc can be used as a natural carrier for the supply of zinc in food systems [36].

1.3. *Lactoferrin*

Lactoferrin (LTF) is a highly glycosylated protein of the transferrin family [37] that has a molecular weight of approximately 80 kDa, depending on its post-translational modifications [20], [21]. Wei et al. suggested the presence of five N-glycosylated sites of bovine LTF-a (bLTF-a): -Asn-233, -281, -368, -476, and -545 [38]. The degree of glycosylation may vary and thus determines the rate of resistance to proteases or to very low pH [39].

Lactoferrin consists of a single polypeptide chain with approximately 700 amino acids folded into two symmetrical lobes: a N-lobe and C-lobe. These are homologues with respect to each other (33%–41% homology). Each lobe consists of two domains, such as C_1, C_2, N_1, and N_2 [37]. Both lobes contain approximately 345 residues, and their disposition in each lobe creates an interdomain pocket with a high affinity to iron; the binding is accompanied by synergistic binding of carbonate ions [39]. In more details, in each lobe, a single Fe atom is coordinated by amino acid side chains that are dispersed in each domain and connecting region because of the changes in conformations occur, causing domains to come together. A distorted octahedral coordination sphere is formed by coordinating ligands as carboxylate-O (Asp), two phenolate-O (Tyr), and imidazole-N (His), which is completed by bidentate binding of carbonate or bicarbonate ion. Carbonate is considered as synergistic since its presence is essential for iron binding. The stability constant for iron (III) complex is high (logβ 28 at pH 7.4) [40,41].

LTF can exist in two forms, apo-Lf and holo-Lf, depending on whether it binds iron (III) or not [37]. In addition to iron, LTF is capable of binding other ions, such as aluminum (III), gallium (III), manganese (III), cobalt (III), copper (II), and zinc (II), but with lower affinity [42]. It was reported that LTF releases iron in acidic conditions (pH below 4) [43], and diferric transferrin readily loses iron at pH < 6.7 [44]. It is also very important to consider the iron saturation and concentration at low pH, especially in places of infection and inflammation, where, as a result of metabolic activity of bacteria or stimulated leucocytes, the pH may be lower than 4.5 [45].

2. Interaction of Whey Proteins with Metals

Metal ions interact with proteins, thus affecting their biological activity [46]. The evaluation of these changes is of considerable importance because proteins have many functions in the human body and applications in many industries. The following factors are crucial for the assessment of changes induced by interactions with metals: (a) the creation of new binding sites that determine the interactions of protein with other ligands, (b) changes in the protein structure, (c) interacting groups for studying the nature and thus strength of the interaction, and (d) possible protein aggregation.

The results of metal–protein interactions may include metalloproteins, metallocomplexes, and nanoparticles. Metalloprotein is primarily formed by coordination bonds between metal ions and functional groups of amino acids, for example, carboxyl, of the protein, thus embedding in the protein structure. These protein functional groups form a special binding site in a form of 'cavity' that is determined by protein quaternary structure and its biological activity. In addition, metalloprotein can interact with metal ions, forming either metallocomplexes (as first step) or metal/metal oxide nanoparticles. A metallocomplex is an artificial system defined by weak interactions such as electrostatic, hydrogen bonding, Wan der Waals forces, or donor–acceptor bonds, which are stronger than previously mentioned interactions. The binding affinity of metal ions to protein in the metallocomplex can be defined also by inductive (artificial) binding sites and collective strength of weak interactions. In contrast to metalloproteins, the interactions, leading to formation of metallocomplex, occur mostly via sorption or by intraparticle diffusion (modeled by Weber–Morris), while the metal ion of metalloprotein is embedded or 'buried' in a protein structure forming a natural system as for, e.g., hemoglobin, transferrin, etc. Interactions of a protein and metal ions with active functional groups of amino acids can be reversible and labile and lead to the formation of nanoparticles as was reported in the study of LTF and silver ions [47]. The formed system consisting of metalloprotein, the formed metallocomplex, and nanoparticles can be considered as a nanocomposite [48]. A graphical representation of the possible results of metal–protein interactions is illustrated in Figure 2. Additionally, it is worthy to mention that statistically, different types of interactions of metal ion and different functional groups of the protein can occur and even simultaneously occur from a theoretical point of view, however the final result is dependent on the conditions of such interaction (temperature, pH, etc.) that determines conformation of the protein, its folding/unfolding and thus formation of binding sites for the metal ion, orientation of functional groups of the protein (steric factor for interaction). This also explains strong coordination bonds of the metal ion and protein in the metalloprotein since many factors can lead to formation of 'cavities' as binding sites for metal ion that are quite challenging to reproduce in the artificial system, thus making it possible the synthesis mostly of metallocomplexes, while metalloproteins are formed mostly by a natural way.

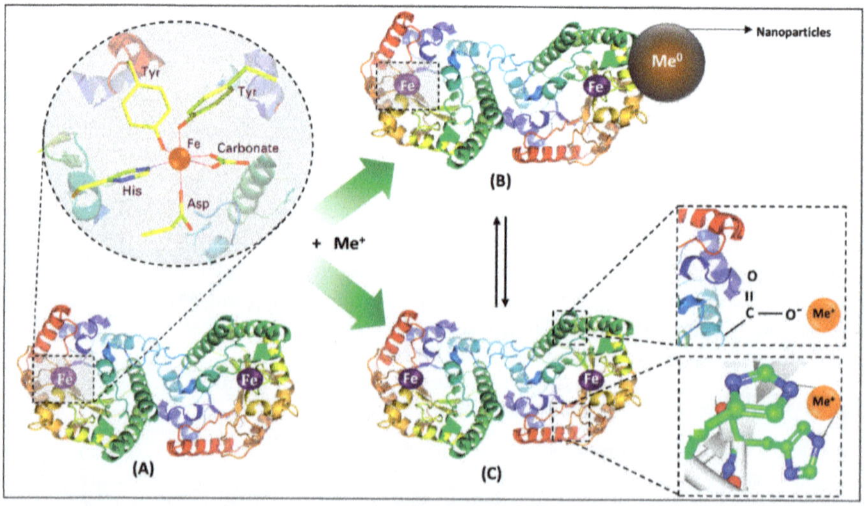

Figure 2. Consequences of metal–protein interactions: (**A**) metalloproteins, (**B**) nanoparticles, and (**C**) metallocomplexes; 2A—carbonate binding site of LTF as a metalloprotein; 2B—formed nanoparticles as a result of interaction of LTF with a metal ion; 2C—metallocomplexes formed by weak electrostatic and sandwich interactions; B and C can form a nanocomposite.

Besides the formation of metallocomplexes and their nanocomposites, metal-induced protein aggregation also occurs, thus causing a loss in biological activity, protein precipitation from solution thus being detrimental for the quality of the product. Chemistry of the metal-induced protein aggregation can be explained by the Derjagin–Landau–Verwey–Overbeek (DLVO) theory of intraparticle interactions, which treats the stability of a biocolloidal system in terms of balance between attractive van der Waals forces and repulsive electrical double-layer forces [49]. The addition of metal ions shifts the attractive forces to increase, thus strengthening the interactions between protein units, causing the formation of large aggregates. An interesting example was shown by Hedberg et al., where the synergistic effects of iron (II/III), chromium (III), and nickel (II) ions were suggested for aggregation of the human serum albumin (HSA) protein. This study also emphasized the importance of considering the safety of metal-based materials upon exposure to the human body and serves as a brilliant example of changes in the biological activity after protein interactions with metals. Changes in the electrostatic forces upon binding were evidenced by changes in electrophoretic mobility [50].

Moreover, such aggregation could cause (a) adsorption of proteins on metal surfaces, creating a problem with cleaning, (b) fouling of the filtration membranes by protein, complicating dairy processing by further disruption of membranes for ultra- and microfiltration, which could be explained by the increase in adhesion of proteins on the metal surface after the addition of metal ions. Interesting studies have been carried out to investigate the factors affecting whey proteins fouling. Yang et al. [51] reported that addition of calcium ions was a more prevailing factor for fouling of β-LG rather than its temperature-induced denaturation. Magens et al. showed the influence of surface type on the fouling performance, deposit structure, and composition in terms of the interactions between whey proteins and surface forces. DLVO theory was applied as mentioned earlier for the analysis of particle–surface forces and surface energy [52]. Such studies serve as direct evidence of the importance of studying whey protein–metal interactions and its implications for daily life.

The current section will be dedicated to interactions between whey proteins and metals, such as copper, zinc, silver, lanthanum, palladium, and ruthenium, as well as to the effects of iron, chromium, nickel, and calcium on the biological activity of whey proteins.

First of all, the major part of metals described in this review are transition metals because the most common inorganic cofactors in biological systems are d-block metals that facilitate various functions of proteins and their complexes [53], thus making the studies of changes in the biological activity of proteins upon their interaction with metals relevant. Moreover, d-block metals have a natural affinity for interacting with proteins owing to the lack of electrons, thus making them able to coordinate to electron-rich moieties in proteins. This makes metalloproteins widespread in nature because metal cofactors are bound to proteins via coordination bonds: it is estimated that more than 50% of all proteins are metalloproteins [46]. Moreover, such metal ions as copper (II), iron (II), manganese (II), and molybdenum (II) have the ability to have the strongest coordination due to their properties such as density and small atomic radius [54]. More importantly, the presence of transition metals is ubiquitous in the environment, and sources and routes of human exposure can greatly vary. In addition, interestingly, metal complexes are applied as drugs for patients with iron-deficient anemia, kidney diseases [55], cancer (complexes of palladium, ruthenium, and platinum [56]), and malnutrition.

Finally, the increased consumption of metal–microelements as supplements to food increases the probability of their interactions with proteins because global demand for protein-rich food has increased with improved living standards [57] along with the prevailing percentage of proteins functions in the human body.

2.1. Nature of the Metal–Protein Interaction

The nature of metal–protein interactions provides insights into the type and strength of the interaction as well as involvement of functional groups of a protein. For example, sorption at the surface of the protein, which is characteristic for metallocomplexes, may occur via weak non-covalent interactions that are unstable and reversible. Coordination bonds that allow embedding of a metal into the structure of protein are stronger.

Polypeptide chains of proteins usually coordinate with the metal ion, and side chains with functional groups can act as an additional binding site for metals, including the imidazole group of histidine, carboxyl group of aspartate and glutamate, the phenol ring of tyrosine [58], and nitrogen of lysine and arginine side chains. Hydrogen, electrostatic, and hydrophobic bonds, and van der Waals interactions are significant for the metal–protein interaction, which has a considerable effect on the stabilization of protein structures [58]. Factors affecting the binding of a metal to proteins include the metal properties, such as the valence state, ionic radius, charge-accepting ability, and free metal concentration in the respective biological compartment [59]. However, studies on the nature of metal–protein interactions and characterization of metal binding to proteins are challenging and carried out by instrumental analytical techniques as well as combinations of techniques, as summarized in Table 2.

Table 2. Interactions of metal ions with proteins and their characterization.

Metal/Conc.	Compound/Conc.	Interaction	Analytical method	Ref.
Zn^{2+}		strong binding affinities:	ITC	[60]
	LTF	2.7×10^5 M^{-1}		
	BSA	2.3×10^5 M^{-1}		
	α-LA	1.5×10^5 M^{-1}		
	β-LG	1.5×10^5 M^{-1}		
Zn^{2+}(6.23 mM)	α-LA (63.9 μM)	two sets of independent binding sites for zinc (II)	ITC	[34]
		two ions bind with the binding constant of 4.53×10^4 M^{-1}	fluorescence	
		four ions bind with the binding constant of 963 M^{-1}	CD	
		electrostatic interactions	DSC	
Zn^{2+}	whey-derived peptides	zinc chelation	FT-IR	[61]
		electrostatic interactions	zinc chelating capacity	
ZnO	WPI	DSC curves allowed to suggest; hydrogen bonding; O–Zn–O bonding; or electrostatic interactions; XRD and UV-Vis allowed to observe evidence for phase structure and crystal quality of ZnO nanoparticles; TEM—image of ZnO-WPI nanocomposite	XRD, TEM, DSC, UV-Vis	[62]
Ag^+	LTF	two stages: (i) internal diffusion and sorption onto the external surface of lactoferrin globules; (ii) internal diffusion and binding to the lactoferrin structure; via electrostatic and hydrophobic interactions	MALDI-TOF/TOF-MS, ICP-MS, FT-IR, SERS, TEM, EDX, electrophoretic techniques	[47]
La (III)-Cys complex		hydrogen bonds, van der Waals interactions	NMR, UV-Vis, FT-IR, TG-DTA, FRET, CD	[63]
	BSA	$K_{BSA\text{-}La}$ 0.11×10^4 M^{-1};		
	β-LG	$K_{\beta\text{-}LG\text{-}La}$ 0.63×10^3 M^{-1}		
La (III)-Trp complex	HSA	hydrophobic interactions: K_b 0.138×10^4 M^{-1} (303 K)	NMR, UV-Vis, FT-IR, TG-DTA	[64]
La (III)-Phe complex	HSA	hydrogen bonds, hydrophobic interactions K_b 0.174×10^4 M^{-1} (303 K)	NMR, UV-Vis, FT-IR	[65]

Table 2. *Cont.*

Metal/Conc.	Compound/Conc.	Interaction	Analytical method	Ref.
Pd (II) complex		hydrogen bonds, van der Waals interactions	NMR, UV-Vis, FT-IR	[66]
	HSA (1×10^5 M)	K_b 0.5×10^4 M^{-1};		
	β-LG (1×10^5 M)	K_b 0.2×10^3 M^{-1}		
Pd (II) complexes (10^{-4} M)	HSA (2 mg/mL)	hydrogen bonds, van der Waals interactions I complex: K_b 0.49×10^4 M^{-1} (293 K); II complex: K_b 0.79×10^4 M^{-1} (293 K)	NMR, UV-Vis, FT-IR, FRET	[67]
Co (II)-Ni (II) complexes		hydrogen bonds, van der Waals interactions	UV-Vis, FT-IR, fluorescence	[68]
	HSA	K_b 3.16×10^6 M^{-1} (303 K);		
	β-LG	K_b 0.54×10^5 M^{-1} (303 K)		
Mn (II)-Co (II) complexes (5×10^{-3} M)		hydrogen bonds, hydrophobic interactions	UV-Vis, FT-IR, FRET	[69]
	HSA (5×10^{-4} M)	I: K_b $7.4 \pm 0.04 \times 10^4$ M^{-1} (303 K); II: K_b $6.08 \pm 0.09 \times 10^3$ M^{-1} (303 K)		
	β-LG (5×10^{-4} M)	I: K_b $7.13 \pm 0.03 \times 10^4$ M^{-1} (303 K); II: K_b $2.62 \pm 0.05 \times 10^3$ M^{-1} (303 K)		

Another factor affecting metal–protein interactions is pH. Magyar et al. reported interesting results showing potential pitfalls during metal–protein interactions studies and discussed how pH, temperature, use of different buffers, and the presence of competing ligands affect the K_d value [70]. The effect of pH on such interactions can be explained by the protonation state of amino acids of the proteins. Firstly, Asp, Glu, and His respond to pH changes leading to association/dissociation of their complexes with metal ions and they are deprotonated at neutral-alkaline pH that leads to the increase in electrostatic attraction and strengthening of the complex, while in acidic pH protonation occurs and weakens the attractive forces [61,71]. Furthermore, cysteine has a thiol group that has to be deprotonated to be involved in metal coordination as well as tyrosine can be deprotonated to produce a phenolate oxygen donor atom, which, e.g., can be a good ligand for Fe (III) [40]. Interesting, that in the case of the effect of pH on binding affinity of iron to LTF, the carbonate ion is involved, since it is essential for iron binding and this ion is unstable at low pH leading to a release of iron from LTF [40]. Another aspect of the effect of pH is related to conformational changes in a protein structure, e.g., it was reported that β-LG at acidic conditions it caused dimerization that involved changes in the exposed β-strands, but in alkaline conditions the denaturation was observed [72].

Tang et al. also showed that glutamate and aspartate should be combined with a nitrogen donor or a sulphur donor to facilitate zinc-binding in peptides or proteins by using isothermal titration calorimetry (ITC). Whey proteins, such as lactoferrin, α-lactalbumin, and β-lactoglobulin showed strong zinc-binding affinities that were similar to each other, even though zinc binding to bovine serum albumin and lactoferrin was exothermic while binding to α-lactalbumin and β-lactoglobulin was slightly endothermic. In addition, authors suggested that zinc binds to the disulphide bonds of oxidized cysteine in LTF and to the thiol group of the cysteine (Cys34) in BSA with significant heat evolution, whereas zinc binds to histidine, aspartate, or glutamate in α-LA and β-LG [60].

Shahraki et al. reported ultraviolet-visible spectroscopy (UV-Vis) results that showed that the interaction of the lanthanum (III)–cysteine complex with β-lactoglobulin and bovine serum albumin induces conformational changes for both proteins. In addition, the lanthanum (III)–cysteine complex strongly quenched the fluorescence of Trp fluorophore in β-LG and BSA in the static quenching mode. Hydrogen bonds and van der Waals forces stabilized the complexes for both proteins [63]. Similar studies on lanthanum–protein interactions were carried out for a lanthanum (III) complex with tryptophan [64] and phenylalanine [65] in relation to human serum albumin. In both cases, spectroscopic techniques indicated the inhibition of protein fluorescence by a static quenching mechanism, whereas data on thermodynamic parameters indicated hydrophobic interactions and hydrogen bonds between the lanthanum (III) complex and protein. In addition, structural studies indicated the conformational changes in proteins in the presence of the lanthanum complex. The lanthanum complex with tryptophan also showed moderate to good antibacterial activity against different bacterial strains. Comparable results were obtained for interactions between human serum albumin and β-lactoglobulin with palladium (II) complexes [66,67]. Spectroscopic studies indicated conformational changes in proteins as a result of the action of the palladium (II) complex. Data on thermodynamic parameters of interaction showed that hydrogen bonds and van der Waals interactions play an important role in HSA/β-LG and palladium (II) complexes associations. In addition, the results of the study showed strong fluorescence quenching of HSA and β-LG by Pd (II) complex via static mechanism [66,67].

2.2. Analytical techniques for separation and analysis of whey proteins

Methods of separation of whey proteins include chromatographic (affinity, anion-, cation-exchange, and reverse-phased), membrane-based (ultrafiltration and microfiltration), and electrophoretic methods. All these methods, except for the electrophoretic, were applied at all stages including isolation, purification, and separation, prior to analysis. Table 3 shows the methods applied for separation of whey proteins for their isolation, analysis, and identification, including real matrixes (different types of cheese and milk).

Table 3. Analytical techniques used for the separation and identification of whey proteins.

Proteins	Matrix	Isolation	Separation	Identification	Ref.
α-LA β-LG A β-LG B	cheese	cheese extracts were desalted and preconcentrated using microcon membranes	CE with fused silica capillaries	DAD	[73]
β-LG α-LA	cow, goat, and ewe cheeses, incl. those of a single animal origin, binary ternary mixtures	desalted, preconcentrated samples were obtained with microcon membranes	CE with fused silica uncoated capillaries	DAD	[74]
α-LA β-LG	raw milk	mixture of standards of purified proteins, separation was achieved by adding SEP and TPS buffers to milk	SDS-PAGE; Microfluidic chip electrophoresis	Fluorescence	[75]
α-LA β-LG caseins	fresh skim milk	mixed protein standards were prepared by combining each of the individual protein solutions (1 mL)	SDS-PAGE; Microfluidic chip electrophoresis	Fluorescence	[76]
β-LG α-LA SA	milk	diluting 200 μL of ultracentrifuged whey with 400 μL of HPLC-grade water	LC, Jupiter C4 column; Microchip electrophoresis	UV, MS; Fluorescence	[77]
LTF β-LG α-LA	milk	samples were centrifuged to remove fat; skim milk was loaded onto lactoferrin immunoaffinity column	LC, Symmetry C4 Column	PDA	[78]
β-LG	buffalo mozzarella	mixtures of cow's milk, water buffalo's milk, mixtures of brine from cow's milk mozzarella, brine from buffalo mozzarella were prepared in diff. vol. ratios for calibration purposes	LC, Supelco Discovery Bio Wide Pore C8 column	MS	[79]
α-LA β-LG BSA	WPC	standard pure proteins to determine ret. times; equilibration buffer Tris-HCl at 20 mM; elution buffer Tris-HCl at 20 mM with 1 M NaCl were used for separation	Mono Q5/50 GL anion-exchange column, FPLC	UV-Vis; SDS-PAGE	[80]
α-LA β-LG BSA	mozzarella cheese whey	different equilibration and elution buffers were prepared	Chromatographic column; packed with SP Sepharose Big; Beads cation exchanger, HPLC	UV-Vis; SDS-PAGE	[81]

Membrane-based methods are very diverse and can be applied for the isolation of whey proteins from real matrices, for the preparation of whey protein concentrate (WPC) and whey protein isolate (WPI), and for separation prior to analysis/detection.

Electrophoretic techniques include capillary (CE), gel (SDS-PAGE, Native PAGE), and microchip electrophoresis (MCE). The microfluidic "lab-on-a-chip" technique for the separation of proteins has been reported as a high-throughput, automated alternative to conventional SDS-PAGE that allows the separation and quantification of many samples within 30 min. Another advantage is the low sample and material volumes required, which are usually less than 0.5 mL of the total volume chip (10 samples) [76]. In comparison, SDS-PAGE requires several liters of materials (acrylamide solutions, running buffers, and staining/destaining solutions). Thus, Anema S.G. applied this technique for the separation of α-LA and β-LG in different forms and compared with traditional SDS-PAGE and concluded that it is a rapid alternative for separation and quantification of milk proteins [76]. Buffoni et al. used MCE together with LC-ESI-MS to characterize the major proteins from milk of Mediterranean water buffalo [77]. However, the main limitation of the capillary and microfluidic system is the relatively low electro-separation reproducibility owing to the adhesion of proteins to the capillary, the denaturation and sensitivity of the system to changes in pH, and the ionic strength of the buffers. However, Costa et al. indicated that addition of buffers (TPS and SEP) to milk prior to separation showed excellent effects on α-LA and β-LG, whereas separation of caseins was better with the SEP buffer, and the results were comparable to those obtained by SDS-PAGE [75].

Chromatographic techniques for separating whey proteins are also diverse. Separation in this case is based on hydrophobic, ionic, and specific (based on affinity) interactions and consists of adsorption of proteins on a solid (column or membrane) that are eluted with the liquid phase [82]. Ion-exchange chromatography has many advantages over column-packed technology, such as a rapid association rate between the target protein and functional groups, short processing times, ease of scale-up, no heat, and chemical pretreatments or pH changes that could affect the protein structure by altering its properties [80]. Moreover, the column-packed chromatographic technique is expensive and thus, not economically viable suitable for industrial scale-up in the food industry [82]. Anion- [80] and cation-exchange [81] membrane columns were exploited for fractionation of whey proteins from WPC and mozzarella cheese whey, followed by fast protein liquid chromatography (FPLC) and high performance liquid chromatography (HPLC) separation, detection with UV-Vis, and identification with SDS-PAGE, respectively. Doultani et al. [81] found that the one ion exchange system can be used for different purposes depending on the elution buffer. The first step included capturing positively charged whey proteins in the cation exchange column, the second step was the removal of unbound contaminants by rinsing, and the last step involved selective desorption of one or more proteins of interest. One advantage of this study was the use of inexpensive, food-grade buffers within one column to yield high-purity proteins. However, chromatographic approaches in comparison with classical electrophoretic techniques, e.g., two-dimensional gel electrophoresis, still have quite a low resolution power.

2.3. *Analytical Techniques for Studies of Interactions of Whey Proteins with Metal Ions*

Instrumental techniques used for the study of the consequences of metal–protein interactions and their potential applications are summarized in Table 4. They include microscopic, spectroscopic, spectrometric, electrophoretic, and even quantum mechanical techniques that provide data regarding the structure, morphology, and chemical composition of metal to protein binding.

Table 4. Instrumental techniques used to study the metal–protein interactions and their potential applications.

Compounds	Form	Application	Analytical methods	Ref.
LTF	metalloprotein	regulation of inflammation and oxidative stress in vertebrates	AFM	[83]
α-LA LF serum albumin	metalloprotein	nutrition of infants in a long breastfeeding stage	Native-PAGE; SEC-ICP-MS; MALDI-TOF/TOF-MS	[84]
LTF	nanoparticles	gene delivery carrier with targeting abilities	TEM	[85]
WPI	nanoparticles	production of antimicrobial cotton fabrics	UV-Vis; TEM; SEM	[86]
LTF	nanoparticles	increased therapeutic efficacy of treatment of malignant melanoma	TEM; SEM; DLS; FT-IR	[87]
LTF	metallocomposites; nanoparticles	in medicine and food industry as an antimicrobial agent	MALDI-TOF/TOF-MS; ICP-MS; FT-IR; SERS; TEM; I-DE; zeta potential measurements	[47]
LTF	nanoparticles	drug delivery strategy against the neurotoxicity in dopaminergic neurons	FE-SEM; AFM; DLS	[88]
LTF	metalloprotein	a therapeutic lead for controlling neutrophil extracellular traps (NETs) release in autoimmune and inflammatory diseases	TEM; SEM; fluorescence microscopy; agarose gel electrophoresis	[89]
IgA	metalloprotein	health and nutrition of breastfed newborns	ESI-MS/MS; FAAS	[90]
LTF	nanoparticles	target specific drug delivery, encapsulation of the drug	FE-SEM; AFM; FT-IR	[91]
LTF BSA	nanoparticles	drug delivery for effective targeting therapy of brain glioma	Particle electrophoresis	[92]
LTF	metalloprotein	antimicrobial biomaterials for dental applications	HPLC; SEM; XPS	[93]
LTF	metalloprotein	inhibition of the attachment of free HIV-1 to epithelial cells	ELISA; flow cytometry	[94]
LTF	Metalloprotein; metallocomplex	immobilized DNA effective for LTF purification	HPCEC; HPIMAC; HPLC; HPSEC; SDS-PAGE	[95]

2.3.1. Mass Spectrometry

Mass spectrometric methods as highly sensitive techniques that have low detection limits were applied for studying metal–protein interactions by measuring the concentrations of metal ions and protein. Inductively coupled plasma mass spectrometry (ICP-MS) is capable of detecting metal concentrations as it was reported for determination of concentration of the silver (I) ion to study the mechanism of binding of silver to LTF [47]. Acosta et al. used size exclusion chromatography (SEC) in combination with ICP-MS to determine the metals, such as manganese, cobalt, copper, and selenium, present in different whey milk protein fractions of human breast milk (HBM) to detect the elements with appropriate sensitivity and accuracy, whereas MALDI-TOF/TOF-MS and nano-LC–MS/MS were used to analyze the protein fraction composition and quantitative profile [84]. Despite the advantages and fast development of MALDI-TOF-MS protocols for analysis of proteins [96], the gold standard for determination of mass, sequence of proteins, their analysis, and characterization is ESI-MS [97]. Limitations of ESI-MS for metal–protein interactions can include redox reactions that may occur during the ionization, atmospheric pressure that can contribute to oxidation of sensitive species, requirements for high purity of the sample and incompatibility with most commonly used non-organic buffers and salts [98]. Moreover, a drift cell was developed for ion mobility mass spectrometry that allowed one to characterize 14 proteins and protein complexes [99]. In addition, Allen et al. studied the effects of polarity on the structures and charge states of native-like proteins and complexes in the gas phase by ESI-MS and ion-mobility mass spectrometry [100]. Additionally, Lermyte et al. studied metal ion binding to the β-amyloid monomer by native FT-ICR mass spectrometry [101] and effects of transition metals in proteinopathies by ESI-MS [98].

2.3.2. Spectroscopic Techniques

In addition to mass spectrometric techniques, other methods used to understand the mechanism of binding silver to LTF have included Fourier transform infrared spectroscopy (FTIR) and Raman spectroscopy (RS). Both FTIR analysis and Raman spectroscopy have shown significant differences between LTF spectra with the addition of silver from native protein in terms of additional peaks. These techniques are complementary for the study of metal ion interactions with active functional groups of proteins. However, FTIR in comparison with RS is less specific and sensitive owing to the presence of water in the system, limiting the participation of the hydroxyl group in the interaction. Alternatively, RS is limited by the fluorescence processes of aromatic residues of LTF and requires surface-enhanced procedures, e.g., by gold or silver nanoparticle sputtering [102]. Additionally, to accurately indicate the location of silver cation binding with LTF, Pomastowski et al. used molecular dynamics (MD) analysis. Moreover, by determining the locations of silver cation bonding with LTF, the reduction of silver ions to elementary silver via density functional theory (DFT) was indicated. On the other hand, X-ray photoelectron spectroscopy (XPS) was used to confirm the attachment of the LTF peptide (hLf1-11) to titanium surfaces by determining the chemical composition of the surface of the system [93]. Another important technique is fluorescence spectroscopy (FS) since it allowed one to determine binding affinity of whey proteins and metal ions by measurements of quenched fluorescence of the protein upon addition of metal. However, this method is suitable mostly for moderate and strong affinities and includes a number of details that are necessary to consider during the measurements [103]. Binding affinities of different metal ions and whey proteins were mostly determined by fluorescence spectroscopy [60–66]. The second method that was used for the determination of binding affinity of the interaction of metal ions and whey proteins was isothermal titration calorimetry (ITC). Obviously, it is not a spectroscopic method, but in contrast to FS, it was used only in two studies (Table 2). ITC is based on measurements of the heat changes during the interaction. One advantage of this method over other techniques is that it is possible to measure thermodynamic parameters of the interaction together with binding affinity, however for very high- or low-affinity complexes it is challenging [104,105].

2.3.3. Microscopic Techniques

Microscopic techniques can serve as indirect methods for study of consequences of interaction and description of the mechanism of metal ions binding to a protein leading to the formation of nanoparticles as it was carried out previously [47]. Kumari and Kondapi used FT-IR spectroscopy to confirm 5-FU entrapment in LTF nanoparticles in their studies on fluorouracil capture in lactoferrin nanoparticles to increase its effectiveness in the treatment of malignant melanoma. Transmission electron microscopy (TEM) and scanning electron microscopy (SEM) analyses were used for physicochemical characterization of LTF nanoparticles formed as a result of an interaction of LTF with silver ions. TEM analysis helped to determine the morphology and size of nanoparticles, while SEM helped to determine that the obtained particles have a spherical shape in the range of 90–110 nm. Additionally, studies based on dynamic light scattering showed that the hydrodynamic diameter of the LTF nanoparticles obtained was 150 ± 20 nm. These differences can be explained by the fact that dynamic light scattering (DLS) analysis takes into account the hydrodynamic size of the solvated protein, but electron microscopy approaches are used for the study of topology, porosity, and metal–organic core size [87]. However, Kumar et al. characterized the morphology of LTF nanoparticles using both SEM and atomic force microscopy (AFM), which showed that the nanoparticles had a spherical shape and a diameter of 50–60 nm [91]. In the study of Bollimpelli et al., DLS analysis of LTF nanoparticles loaded with curcumin was used to determine their size (100 nm), whereas SEM and AFM analyses indicated their sizes were 43–60 nm. In this case, different sizes were related to the surface charge of the particles and their interaction with the water shell [88].

2.3.4. Complementarity of MALDI- and NALDI-TOF-MS for Metal–Protein Interactions Studies

The matrix-assisted laser desorption ionization technique (MALDI) coupled to the time-of-flight mass spectrometry (TOF-MS) has become one of the common methods for protein characterization in addition to other MS-based approaches. The advantages of MALDI-TOF-MS include the simplicity of use, sensitivity, large mass range, and relative resistance to interferences from matrices [96]. Moreover, this technique is reasonable to use not only for protein characterization but also for studies of protein interactions because biological activity and biomolecular recognition are defined by non-covalent interactions. The interactions of various ligands with proteins are interesting to study because they fulfill many functions in nature as well as in the human body. For example, MALDI-TOF-MS analysis of β-lactoglobulin was used to study its influence on human immunity and promotion of cell proliferation [106], and the analysis of carbonic anhydrase IX [107] complexes with potential synthetic inhibitors allowed the identification of only strong binding inhibitors to proteins owing to the nature of MALDI ionization. Authors suggested that the stoichiometry of binding showed a possible second binding site, supporting the hypothesis of the induced-fit model of the interaction. In some cases, in studies of the origins of disease, the role of metal–protein complexes and metal speciation are of big interest, particularly for Alzheimer's disease and its implications. Although MALDI-MS is not able to quantify metals in biological samples as well as in model systems for such studies, it can be used for quantitative determination of proteins and identification of post-translational modifications sites. Despite the challenges of MALDI-MS quantification associated with the reproducibility of results, approaches, such as using isobaric tags for relative and absolute quantization (iTRAQ), are widely used for differentiation of expressed proteins in comparative proteomics owing to large-scale, high throughput, and highly sensitive procedures of different MS-based approaches [108]. This technology applies a 4-plex set of amine reactive isobaric tags to derivatize peptides at the N-terminus and at the lysine side chains, thus labeling all peptides in a digest mixture [109].

Moreover, PTMs are crucial for protein biological activity, and their identification is an important analytical challenge. The MALDI-TOF-MS technique can be applied for this purpose by protein digestion to peptides and then identification of PTM sites. Although MALDI-TOF-MS is widely used for the analysis of digested proteins as well, analysis of LMW peptides can be complicated owing to the matrix interference and thus, signal suppression (<700 Da). Matrix-free approaches include many methods of analysis with the use of materials to replace the matrix. Nano-assisted laser desorption ionization (NALDI) is a promising technique. Nanomaterials assist the ionization that leads to signal enhancement owing to surface plasmon resonance. For example, Shenar et al. reported that a NALDI plate provided results with better sensitivity as compared to a DIOS chip, carbon powder, and porous silica for model peptides in the range of 519–2853 Da [110]. Interestingly, the NALDI approach can be promising for studies of metal–peptide interactions, which is also an important part of metal–protein interactions studies because they can indicate if some peptides can be specifically bound to a metal of interest.

Thus, the MALDI and NALDI techniques can be complementary to each other in terms of characterization of metal–protein interactions when coupled with other instrumental techniques in the following cases: (a) analysis and characterization of intact and digested proteins, analysis of metal–protein adducts by MALDI, (b) analysis of peptides digested from original proteins in the low molecular mass range, peptide adducts with metals and PTM sites located on the peptides with a low MW using NALDI in the mid to high range by MALDI, and (c) quantitative analysis of proteins by MALDI/NALDI.

3. Implications of Whey Protein–Metal Interactions in Food and Nutraceuticals

3.1. Changes in Bioactivity after Metal Interactions

There are many publications that have reported biological claimed activities of whey proteins; however, the changes in these activities upon metal addition have not yet been deeply discussed.

To the best of our knowledge, whey proteins exert antioxidant activity by forming glutathione (GSH) [111], which may be connected with anticancer activity. GSH can be formed due to high sulphur content of whey proteins provided by the presence of cysteine, which forms γ-glutamylcysteine, and this step is rate-limiting in GSH synthesis [112]. In addition, LTF can serve as an antioxidant agent by a binding iron in a form that prevents it to act as a Haber–Weiss catalyst [113]. However, complexes of whey proteins with metals demonstrate anticancer activity, serving as carriers for metal complexes, such as lanthanum (III) with β-LG [63] and cobalt and nickel complexes with HSA and β-LG [68].

Nevertheless, the effect of metals on anticancer activity of whey proteins can be shown by the example of LTF. It has been reported that iron-unsaturated apo-LTF inhibits the growth of cervical cancer (HeLa) cells after 48 h of treatment, whereas diferric-bLf was not effective [114]. Additionally, Gibbons et al. indicated that both iron-unsaturated apo-LTF and iron-saturated b-LTF (> 90% Fe^{3+} sat.) showed antitumorigenic properties; however, the apo-form showed a higher effect in inducing cytotoxicity in both cell lines (MDA-MB-231 and MCF-7) compared to Fe-bLf, which was more effective at inducing apoptosis in MCF-7 cell lines [115]. Studies conducted on mice by Kanwar et al. showed that chemotherapy eradicated EL-4 lymphomas in mice that received iron-saturated LTF for 6 weeks before chemotherapy but not in mice receiving lesser iron-saturated forms. It was concluded that bLf may be a potential natural adjuvant agent for supporting chemotherapy, but when saturated with iron, it can be more effective [116].

LTF possesses antibacterial activity itself along with having immunomodulatory functions [117]. However, in previous research by Pomastowski et al. [47], it was reported that the formed nanocomposite as a result of the interaction of silver ions with LTF and a spontaneous reduction of silver ions to nanoparticles exerted a strong antibacterial activity against *Pseudomonas aeruginosa, Staphylococcus aureus, Escherichia coli,* and *Enterococcus faecalis*. The growth of the drug-resistant *P. aeruginosa* strain was inhibited by more than 97%, which was comparable to the traditional cefotaxime antibiotic. Moreover, Komatsu et al. demonstrated that lactoferrin might be used to reduce the risk of aspiration pneumonia among elderly people for whom oral care remains difficult [118]. Additionally, the antiparasitic effect of LTF is different depending on the species. LTF interferes with iron acquisition by some parasites; however, LTF may act as a specific iron donor for some parasites that use LTF for their growth. More specifically, holo-LTF interacts with cell membrane receptors, and the cell starts to secrete iron reductase complex, which includes NADPH that requires the reductase, and it donates electrons, thus changing the membrane potential and leading to its disintegration [119].

An interesting study by Thawari et al. [120] on β-LG and apo-α-LA was reported on the effect of copper (II) ions on the formation of protein dimers and the catalytic activity of the formed nanobiomaterials, which were inorganic hybrid nanomaterials in its nature. Nanostructures in the form of nanoflowers and nanofibers combined with copper-LG and copper-LA with Cu as the inorganic part exhibited peroxidase-like catalytic activity. Important conclusion of the study was that β-LG produced a pH-dependent protein–protein dimer that formed at up to pH 12 and at a concentration of 1000 equiv. of copper (II), whereas in the case of α-LA, the formation of a dimer was not observed, which was explained by the non-availability of free Cys-SH compared to exposed Cys121 in β-LG.

3.2. *Consequences of Changes in Biological Activity in Food and Nutraceuticals*

Whey proteins as a component of milk and dairy products are a part of the everyday diet of many people. Together with the trend of uncontrolled consumption of supplements by the public, it creates a window for interactions between whey proteins and metals. Clearly, additional supplementation to food in the form of organic complexes that have metals can make up the deficiency of microelements, which has some advantages. Microelement deficiency is a serious problem that is common for developing countries, especially for infants, preschoolers, pregnant and lactating woman, and older adults [121], and in the long-term perspective it can cause chronic diseases and inflammation [122]. Since inorganic complexes with metals (e.g., chloride, etc.) have many side effects, they have been replaced by organic complexes with proteins and peptides that, in addition to minimizing the side effects, are more biologically available [123]. However, accidental interaction between WP and metals can be harmful because it can lead to unpredictable consequences. The formation of nanoparticles as a result of interactions, as it was discussed before, or the enhanced release of metals owing to the instability of the complex in gastric fluid can be toxic to humans. For example, zinc in the form of a dietary supplement is quite popular among modern consumers and has been studied for its chelating ability. It was found [61] that zinc was released from zinc–peptide complexes after simulated gastric digestion, thus confirming the suggestions that such complexes are not stable. However, two types of whey protein hydrolysate (WPH) were studied, the most negatively charged (1) and the least negatively charged (2), according to zeta potential measurements; and in a dispersion stability study, zinc release was found to be much higher for the second hydrolysate. The authors suggested that such differences in stability could be explained by the differences in the surface charge, particle size, and strength of the formed complexes. The molecular size of the peptides could also have an effect because smaller-sized peptides could reach the metal target more easily; however, both hydrolysates have a similar degree of hydrolysis and amount of peptide per mass. Moreover, zinc release increased with pancreatic digestion with a higher level of the second hydrolysate. Although similar studies were performed for casein, silver [124], and zinc-peptide complexes [125], the problem of the stability of metal–organic complexes as well as the unpredictable consequences of such interactions for the production of nutraceuticals has not been well-studied and deserves attention. In the study by Wang et al. [125], complexes of zinc and three peptides were almost unaffected under gastric conditions; however, the release of zinc ions was observed under pancreatic conditions; nevertheless, the zinc-Asn-Cys complex has potential for the improvement of zinc bioavailability.

Another window for whey protein–metal interactions is the contact of milk products with stainless steel surfaces during storage, transport, and processing. Stainless steel alloys are of widespread use in the food industry owing to their high resistance to corrosion, good mechanical properties [126], and relative ease of cleaning. Although a few studies on metal release from stainless steel have been conducted, little attention has been paid to the effects of proteins on metal release, and data on these interactions during food storage and processing is scarce and not systemically investigated. Atapour et al. reported a study on metal release from stainless steel 316L under static and stirring conditions using a whey protein solution, simulated milk solution, and phosphate buffered saline solution for mechanistic comparison, and the results showed the release of iron, chromium, and nickel ion was much more significant for the whey protein solution, causing enhanced rates of protein aggregation and its precipitation from solution. Additionally, the authors remarked that the released concentrations of iron, chromium, and nickel did not exceed the limits estimated by European guidelines; nevertheless, the ratio of the solution volume of milk needs to be considered, and future toxicological assessments need to be performed. Alternative contact materials might not be better from a health perspective, but the effects of contact with milk needs to be investigated [127]. Then, the same research group studied the effect of the grade of corrosion on metal release using electrochemical methods, also confirming the importance of metal complexation by whey proteins on metal release [128].

One more important point is whey protein–metal interactions occurring in the process of purification of whey by magnetic metal nanoparticles that is a good alternative to membrane processes. It is crucial not only for the design of functional foods and its safety but also for utilization of whey because its utilization in the purified form is more safe compared to that of the raw form [129].

4. Conclusion

Whey proteins have a rich chemistry, thus providing interesting properties for many applications. Their natural affinity to metals has advantages as well as disadvantages in the fields of food chemistry and nutraceuticals. Advantages may include their potential applications as nutraceuticals, for the design of functional foods, whey purification, and utilization of whey wastes, while the drawbacks include their interactions with metals in food products, metals as drugs, and dietary supplements that may cause loss of the drug/supplement activity or toxic effects. This may occur with metals contact surfaces during processing, storage, and transport.

Moreover, the effect of metals on their biological activity deserves attention because this also creates opportunities for new applications and is an important factor to consider in all stages of food processing, food safety, quality, and utilization. Formation of complex biocolloidal systems as a result of metal–protein interactions may occur in different ways, such as via (i) metalloproteins, (ii) metallocomplexes, (iii) nanocomposites, and (iv) aggregated systems. This review is focused on a discussion of the biological activities, interactions with metals, and analytical techniques applied for their study.

Author Contributions: Conceptualization, P.P., A.R., G.N.S.; writing—original draft preparation, A.R.; writing—review and editing, G.N.S. and P.P; visualization, A.R. and G.N.S.; supervision, B.B.; project administration, P.P.; funding acquisition, P.P. All authors have read and agreed to the published version of the manuscript.

Funding: This work was financially supported by the National Science Centre in frame of Opus 14 project No. 2017/27/B/ST4/02628 (2018-2021).

Acknowledgments: We acknowledge Oleksandra Pryshchepa for constructive criticism and help with improving a manuscript and Dr. Viorica Railean-Plugaru for help with figure preparation. Paweł Pomastowski and Bogusław Buszewski are members of Toruń Center of Excellence "Towards Personalized Medicine" operating under Excellence Initiative-Research University.

Conflicts of Interest: There are no conflicts of interest to declare.

Abbreviations

5-FU	5-Fluorouracil
α-LA	α-lactalbumin
β-LG	β-lactoglobulin
AFM	Atomic Force Microscopy
BSA	Bovine Serum Albumin
CE	Capillary Electrophoresis
CD	Circular Dichroism
DAD	*Diode Array Detector*
DFT	Density Functional Theory
DLS	Dynamic Light Scattering
DLVO	Derjagin–Landau–Verwey–Overbeek theory of intraparticle interactions
DSC	Differential Scanning Calorimetry
DTNB	Dithio(bis)-p-nitrobenzoate
EDX	Energy Dispersive X-ray Spectroscopy
ELISA	Enzyme-Linked Immunosorbent Assay
ESI-MS/MS	Electrospray Ionization Tandem Mass Spectrometry
FAAS	Flame Atomic Absorption Spectrometry
FE-SEM	Field Emission Scanning Electron Microscopes
FRET	Förster Resonance Energy Transfer
FT-ICR	Fourier transform ion cyclotron resonance
FPLC	Fast Protein Liquid Chromatography
HBM	Human Breast Milk
HPCEC	High Performance Cation Exchange Chromatography
HPIMAC	High Performance Immobilized Metal Ion Affinity Chromatography
HPLC	High Performance Liquid Chromatography
HPSEC	High Performance Size Exclusion Chromatography
HSA	Human Serum Albumin
IG	Immunoglobulin
ICP-MS	Inductively Coupled Plasma Mass Spectrometry
IMAC	Immobilized Metal Affinity Chromatography
ITC	Isothermal Titration Calorimetry
iTRAQ	Isobaric Tags for Relative and Absolute Quantization
LC-ESI-MS	Liquid Chromatography Electrospray Ionization-Mass Spectrometry
LTF	Lactoferrin
LP	Lactoperoxidase
MALDI-TOF MS	Matrix-Assisted Laser Desorption Ionization technique coupled to Time-of-Flight Mass Spectrometry
MCE	Microchip Electrophoresis
MD	Molecular Dynamics
NALDI	Nano-Assisted Laser Desorption Ionization
NEM	N-ethylmaleimide
NMR	Nuclear Magnetic Resonance
PDA	Photodiode Array Detector
PTMs	Posttranslational Modifications
ROS	Reactive Oxygen Species

RS	Raman Spectroscopy
SDS-PAGE	Sodium Dodecyl Sulfate–Polyacrylamide Gel Electrophoresis
SEC	Size Exclusion Chromatography
SEC-ICP-MS	Size Exclusion Chromatography-Inductively Coupled Plasma-Mass Spectrometry
SERS	Surface Enhanced *Raman* Spectroscopy
SEM	Scanning Electron Microscope
SEP buffer	Separating Milk Protein Buffer
TEM	Transmission Electron Microscope
TPS Buffer	Total Protein Solubilization Buffer
TG-DTA	Thermogravimetry/Differential Thermal Analysis
UV-Vis	Ultraviolet/ Visible Spectroscopy
WPC	Whey Protein Concentrate
WPH	Whey Protein Hydrolysate
WPI	Whey Protein Isolate
XPS	X-ray Photoelectron Spectroscopy

References

1. Madureira, A.R.; Pereira, C.I.; Gomes, A.M.P.; Pintado, M.E.; Xavier Malcata, F. Bovine whey proteins—Overview on their main biological properties. *Food Res. Int.* **2007**, *40*, 1197–1211. [CrossRef]
2. Chatterton, D.E.W.; Smithers, G.; Roupas, P.; Brodkorb, A. Bioactivity of b -lactoglobulin and a -lactalbumin — Technological implications for processing. *Int. Dairy J.* **2006**, *16*, 1229–1240. [CrossRef]
3. Corrochano, A.R.; Buckin, V.; Kelly, P.M.; Giblin, L. Invited review: Whey proteins as antioxidants and promoters of cellular antioxidant pathways. *J. Dairy Sci.* **2018**, *101*, 4747–4761. [CrossRef] [PubMed]
4. Godovac-Zimmermann, J.; Krause, I.; Baranyi, M.; Fischer-Frühholz, S.; Juszczak, J.; Erhardt, G.; Buchberger, J.; Klostertneyer, H. Isolation and Rapid Sequence Characterization of Two Novel Bovine 13-Lactoglobulins I and J. *J. Protein Chem.* **1996**, *15*, 743–750. [CrossRef] [PubMed]
5. Wu, X.; Lu, Y.; Xu, H.; Lin, D.; He, Z.; Wu, H.; Liu, L.; Wang, Z. Reducing the allergenic capacity of β -lactoglobulin by covalent conjugation with dietary polyphenols. *Food Chem.* **2018**, *256*, 427–434. [CrossRef] [PubMed]
6. Mensi, A.; Choiset, Y.; Rabesona, H.; Haertlé, T.; Borel, P.; Chobert, J.-M. Interactions of β-Lactoglobulin Variants A and B with Vitamin A. Competitive Binding of Retinoids and Carotenoids. *J. Agric. Food Chem.* **2013**, *61*, 4114–4119. [CrossRef]
7. Qin, B.Y.; Bewley, M.C.; Creamer, L.K.; Baker, H.M.; Baker, E.N.; Jameson, G.B. Structural Basis of the Tanford transition of Bovine beta-Lactoglobulin. *Biochemistry* **1998**, *37*, 14014–14023. [CrossRef]
8. Ma, S.; Wang, C.; Guo, M. Changes in structure and antioxidant activity of β-lactoglobulin by ultrasound and enzymatic treatment. *Ultrason. Sonochem.* **2018**, *43*, 227–236. [CrossRef]
9. Sakai, K.; Sakurai, K.; Sakai, M.; Hoshino, M.; Goto, Y. Conformation and stability of thiol-modified bovine beta-lactoglobulin. *Protein Sci.* **2000**, *9*, 1719–1729.
10. Enomoto, H.; Li, C.P.; Morizane, K.; Ibrahim, H.R.; Sugimoto, Y.; Ohki, S.; Ohtomo, H.; Aoki, T. Glycation and Phosphorylation of β-Lactoglobulin by Dry-Heating: Effect on Protein Structure and Some Properties. *J. Agric. Food Chem.* **2007**, *55*, 2392–2398. [CrossRef]
11. Pinto, M.S.; Léonil, J.; Henry, G.; Cauty, C.; Carvalho, A.F.; Bouhallab, S. Heating and glycation of β -lactoglobulin and β -casein: Aggregation and in vitro digestion. *Food Res. Int.* **2014**, *55*, 70–76. [CrossRef]
12. Mann, M.; Jensen, O.N. Proteomic analysis of post-translational modifications. *Nat. Biotechnol.* **2003**, *21*, 255–261. [CrossRef] [PubMed]
13. de Wit, J.N. Thermal behaviour of bovine β-lactoglobulin at temperatures up to 150 °C. A review. *Trends Food Sci. Technol.* **2009**, *20*, 27–34. [CrossRef]
14. Liu, H.C.; Chen, W.L.; Mao, S.J.T. Antioxidant Nature of Bovine Milk β -Lactoglobulin. *J. Dairy Sci.* **2007**, *90*, 547–555. [CrossRef]
15. Wijayanti, H.B.; Oh, H.E.; Sharma, R.; Deeth, H.C. Reduction of aggregation of β-lactoglobulin during heating by dihydrolipoic acid. *J. Dairy Res.* **2013**, *80*, 383–389. [CrossRef]
16. Medrano, A.; Abirached, C.; Panizzolo, L.; Moyna, P.; Añón, M.C. The effect of glycation on foam and structural properties of β-lactoglobulin. *Food Chem.* **2009**, *113*, 127–133. [CrossRef]

17. Aich, R.; Batabyal, S.; Joardar, S.N. Isolation and purification of beta-lactoglobulin from cow milk. *Vet. World* **2015**, *8*, 621–624. [CrossRef]
18. Hogarth, C.J.; Fitzpatrick, J.L.; Nolan, A.M.; Young, F.J.; Pitt, A.; Eckersall, P.D. Differential protein composition of bovine whey: A comparison of whey from healthy animals and from those with clinical mastitis. *Proteomics* **2004**, *4*, 2094–2100. [CrossRef]
19. Svensson, M.; Sabharwal, H.; Håkansson, A.; Mossberg, A.K.; Lipniunas, P.; Leffler, H.; Svanborg, C.; Linse, S. Molecular characterization of alpha-lactalbumin folding variants that induce apoptosis in tumor cells. *J. Biol. Chem.* **1999**, *274*, 6388–6396. [CrossRef]
20. Froehlich, J.W.; Dodds, E.D.; Barboza, M.; Mcjimpsey, E.L.; Richard, R.; Francis, J.; An, H.J.; Freeman, S.; German, J.B.; Lebrilla, C.B. Glycoprotein Expression in Human Milk During Lactation. *J. Agric. Food Chem.* **2010**, *58*, 6440–6448. [CrossRef]
21. Fong, B.Y.; Norris, C.S.; Palmano, K.P. Fractionation of bovine whey proteins and characterisation by proteomic techniques. *Int. Dairy J.* **2008**, *18*, 23–46. [CrossRef]
22. Eigel, W.N.; Butler, J.E.; Ernstrom, C.A.; Farrell, H.M.; Harwalkar, V.R.; Jenness, R.; Whitney, R.M.L. Nomenclature of Proteins of Cow's Milk: Fifth Revision. *J. Dairy Sci.* **1984**, *67*, 1599–1631. [CrossRef]
23. Mercadante, D.; Melton, L.D.; Norris, G.E.; Loo, T.S.; Williams, M.A.K.; Dobson, R.C.J.; Jameson, G.B. Bovine β-Lactoglobulin is dimeric under imitative physiological conditions: Dissociation equilibrium and rate constants over the pH range of 2.5-7.5. *Biophys. J.* **2012**, *103*, 303–312. [CrossRef] [PubMed]
24. Prasad, R.V.; Butkowski, R.J.; Ebner, K.E.; Hamilton, J.W. Amino Acid Sequence of Rat α-Lactalbumin: A Unique α-Lactalbumin. *Biochemistry* **1982**, *21*, 1479–1482. [CrossRef]
25. Layman, D.K.; Lönnerdal, B.; Fernstrom, J.D. Applications for a-lactalbumin in human nutrition. *Nutr. Rev.* **2018**, *76*, 444–460. [CrossRef]
26. Noyelle, K.; Van Dael, H. Kinetics of conformational changes induced by the binding of various metal ions to bovine α-lactalbumin. *J. Inorg. Biochem.* **2002**, *88*, 69–76. [CrossRef]
27. Chandra, N.; Brew, K.; Ravi, A.K. Structural Evidence for the Presence of a Secondary Calcium Binding Site in Human R-Lactalbumin. *Biochemistry* **1998**, *37*, 4767–4772. [CrossRef]
28. Kim, S.; Baum, J. Electrostatic interactions in the acid denaturation of α-lactalbumin determined by NMR. *Protein Sci.* **1998**, *7*, 1930–1938. [CrossRef]
29. Veprintsev, D.B.; Permyakov, S.E.; Permyakov, E.A.; Rogov, V.V.; Cawthern, K.M.; Berliner, L.J. Cooperative thermal transitions of bovine and human apo-α-lactalbumins: Evidence for a new intermediate state. *FEBS Lett.* **1997**, *412*, 625–628. [CrossRef]
30. Shinozaki, R.; Iwaoka, M. Effects of metal ions, temperature, and a denaturant on the oxidative folding pathways of bovine α-lactalbumin. *Int. J. Mol. Sci.* **2017**, *18*. [CrossRef]
31. Kronman, M.J.; Sinha, S.K.; Brew, K. Characteristics of the binding of Ca2+ and other divalent metal ions to bovine alpha-lactalbumin. *J. Biol. Chem.* **1981**, *256*, 8582–8587. [PubMed]
32. Wehbi, Z.; Pérez, M.D.; Sánchez, L.; Pocoví, C.; Barbana, C.; Calvo, M. Effect of heat treatment on denaturation of bovine α-lactalbumin: Determination of kinetic and thermodynamic parameters. *J. Agric. Food Chem.* **2005**, *53*, 9730–9736. [CrossRef] [PubMed]
33. Permyakov, E.A.; Berliner, L.J. α-Lactalbumin: structure and function. *FEBS Lett.* **2000**, *473*, 269–274. [CrossRef]
34. Fujita-Yamaguchi, Y. Affinity chromatography of native and recombinant proteins from receptors for insulin and IGF-I to recombinant single chain antibodies. *Front. Endocrinol. (Lausanne).* **2015**, *6*. [CrossRef]
35. Bushmarina, N.A.; Blanchet, C.E.; Vernier, G.; Forge, V. Cofactor effects on the protein folding reaction: Acceleration of α-lactalbumin refolding by metal ions. *Protein Sci.* **2006**, *15*, 659–671. [CrossRef]
36. Atri, M.S.; Saboury, A.A.; Moosavi-Movahedi, A.A.; Kavousi, K.; Ariaeenejad, S. Effects of zinc binding on the structure and thermal stability of camel alpha-lactalbumin. *J. Therm. Anal. Calorim.* **2015**, *120*, 481–488. [CrossRef]
37. García-Montoya, I.A.; Cendón, T.S.; Arévalo-Gallegos, S.; Rascón-Cruz, Q. Lactoferrin a multiple bioactive protein: An overview. *Biochim. Biophys. Acta* **2012**, *1820*, 226–236. [CrossRef]
38. Wei, Z.; Nishimura, T.; Yoshida, S. Presence of a Glycan at a Potential N-Glycosylation Site, Asn-281, of Bovine Lactoferrin. *J. Dairy Sci.* **2000**, *83*, 683–689. [CrossRef]
39. Redwan, E.M.; Uversky, V.N.; El-Fakharany, E.M.; Al-Mehdar, H. Potential lactoferrin activity against pathogenic viruses. *Comptes Rendus Biol.* **2014**, *337*, 581–595. [CrossRef]

40. Atkins, P.W.; Overton, T.L.; Rourke, J.P.; Weller, M.T.; Armstrong, F. *Inorganic Chemistry, Fifth Edition*; W. H. Freeman and Company: New York, NY, USA, 2010; ISBN 978-1-42-921820-7.
41. Housecroft, C.E.; Sharpe, A.G. *Inorganic Chemistry Third Edition*; Pearson Education: Edinburgh Gate, UK, 2008; ISBN 978-0-13-175553-6.
42. Baker, E.N. Structure and reactivity of transferrins. *Adv. Inorg. Chem.* **1994**, *41*, 389–463.
43. Mesonjesi, I. Are extrinsic black stains of teeth iron-saturated bovine lactoferrin and a sign of iron deficient anemia or iron overload? *Med. Hypotheses* **2012**, *79*, 219–221. [CrossRef]
44. Lestas, A.N. The Effect of pH upon Human Transferrin: Selective Labelling of the Two Iron-binding Sites. *Br. J. Haematol.* **1976**, *32*, 341–350. [CrossRef]
45. Valenti, P.; Antonini, G. Lactoferrin: An important host defence against microbial and viral attack. *Cell. Mol. Life Sci.* **2005**, *62*, 2576–2587. [CrossRef]
46. Thomson, A.J.; Gray, H.B. Bio-inorganic chemistry. *Curr. Opin. Chem. Biol.* **1998**, *2*, 155–158. [CrossRef]
47. Pomastowski, P.; Sprynsky, M.; Žuvela, P.; Rafińska, K.; Milanowski, M.; Liu, J.J.; Yi, M.; Buszewski, B. Silver-Lactoferrin Nanocomplexes as a Potent Antimicrobial Agent. *J. Am. Chem. Soc.* **2016**, *138*, 7899–7909. [CrossRef]
48. Railean-Plugaru, V.; Pomastowski, P.; Meller, K.; Złoch, M.; Katarzyna, R.; Boguslaw, B. Lactococcus lactis as a safe and inexpensive source of bioactive silver composites. *Appl. Microbiol. Biotechnol.* **2017**, *101*, 7141–7153.
49. Verwey, E.J.W.; Overbeek, J.T.G. Theory of the stability of lyophobic colloids. *J. Colloid Sci.* **1955**, *10*, 224–225. [CrossRef]
50. Hedberg, Y.S.; Dobryden, I.; Chaudhary, H.; Wei, Z.; Claesson, P.M.; Lendel, C. Synergistic effects of metal-induced aggregation of human serum albumin. *Colloids Surf. B Biointerfaces* **2019**, *173*, 751–758. [CrossRef] [PubMed]
51. Yang, W.; Li, D.; Chen, X.D.; Mercadé-Prieto, R. Effect of calcium on the fouling of whey protein isolate on stainless steel using QCM-D. *Chem. Eng. Sci.* **2018**, *177*, 501–508. [CrossRef]
52. Magens, O.M.; Hofmans, J.F.A.; Adriaenssens, Y.; Ian Wilson, D. Comparison of fouling of raw milk and whey protein solution on stainless steel and fluorocarbon coated surfaces: Effects on fouling performance, deposit structure and composition. *Chem. Eng. Sci.* **2019**, *195*, 423–432. [CrossRef]
53. Kluska, K.; Adamczyk, J.; Krężel, A. Metal binding properties, stability and reactivity of zinc fingers. *Coord. Chem. Rev.* **2018**, *367*, 18–64. [CrossRef]
54. Garcia, J.S.; De Magalhães, C.S.; Arruda, M.A.Z. Trends in metal-binding and metalloprotein analysis. *Talanta* **2006**, *69*, 1–15. [CrossRef] [PubMed]
55. Chan, S.; Au, K.; Francis, R.S.; Mudge, D.W.; Johnson, D.W.; Pillans, P.I. Phosphate binders in patients with chronic kidney disease. *Aust. Prescr.* **2017**, *40*, 9–14. [CrossRef] [PubMed]
56. Butler, J.S.; Sadler, P.J. Targeted delivery of platinum-based anticancer complexes. *Curr. Opin. Chem. Biol.* **2013**, *17*, 175–188. [CrossRef]
57. Henchion, M.; Hayes, M.; Mullen, A.; Fenelon, M.; Tiwari, B. Future Protein Supply and Demand: Strategies and Factors Influencing a Sustainable Equilibrium. *Foods* **2017**, *6*, 53. [CrossRef]
58. Yamauchi, O.; Odani, A.; Takani, M. Metal–amino acid chemistry. Weak interactions and related functions of side chain groups. *J. Chem. Soc. Dalt. Trans.* **2002**, 3411–3421. [CrossRef]
59. Dudev, T.; Lim, C. Competition among metal ions for protein binding sites: Determinants of metal ion selectivity in proteins. *Chem. Rev.* **2014**, *114*, 538–556. [CrossRef]
60. Tang, N.; Skibsted, L.H. Zinc bioavailability from whey. Enthalpy-entropy compensation in protein binding. *Food Res. Int.* **2016**, *89*, 749–755. [CrossRef]
61. Udechukwu, M.C.; Downey, B.; Udenigwe, C.C. Influence of structural and surface properties of whey-derived peptides on zinc-chelating capacity, and in vitro gastric stability and bioaccessibility of the zinc-peptide complexes. *Food Chem.* **2018**, *240*, 1227–1232. [CrossRef]
62. Shi, L.; Zhou, J.; Gunasekaran, S. Low temperature fabrication of ZnO-whey protein isolate nanocomposite. *Mater. Lett.* **2008**, *62*, 4383–4385. [CrossRef]
63. Shahraki, S.; Shiri, F.; Majd, M.H.; Dahmardeh, S. Anti-cancer study and whey protein complexation of new lanthanum (III) complex with the aim of achieving bioactive anticancer metal-based drugs. *J. Biomol. Struct. Dyn.* **2018**, *37*, 2072–2085. [CrossRef] [PubMed]

64. Shahraki, S.; Shiri, F.; Beyzaei, H.; Khosravi, F. Synthesis, characterization, protein interaction and antibacterial activity of a lanthanum(iii) complex [La(Trp)3(OH2)2] (Trp = tryptophan) as a new precursor for synthesis of La2O2CO3 nanoparticles. *New J. Chem.* **2017**, *41*, 8413–8421. [CrossRef]
65. Shahraki, S.; Shiri, F.; Saeidifar, M. Evaluation of in silico ADMET analysis and human serum albumin interactions of a new lanthanum (III) complex by spectroscopic and molecular modeling studies. *Inorganica Chim. Acta* **2017**, *463*, 80–87. [CrossRef]
66. Shahraki, S.; Heydari, A. Binding forces between a novel Schiff base palladium (II) complex and two carrier proteins: human serum albumi and β-lactoglobulin. *J. Biomol. Struct. Dyn.* **2018**, *36*, 2807–2821. [CrossRef]
67. Zareian-Jahromi, S.; Mansouri-Torshizi, H. Synthesis, characterization, DNA and HSA binding studies of isomeric Pd (II) antitumor complexes using spectrophotometry techniques. *J. Biomol. Struct. Dyn.* **2018**, *36*, 1329–1350. [CrossRef] [PubMed]
68. Shahraki, S.; Shiri, F.; Majd, M.H.; Razmara, Z. Comparative study on the anticancer activities and binding properties of a hetero metal binuclear complex [Co(dipic)2Ni(OH2)5]·2H2O (dipic = dipicolinate) with two carrier proteins. *J. Pharm. Biomed. Anal.* **2017**, *145*, 273–282. [CrossRef]
69. Shahraki, S.; Shiri, F.; Razmara, Z.; Majd, M.H. A comparative study of the impact of metal complex size on the in vitro biological behavior of hetero di- and poly-nuclear Mn-Co complexes. *J. Mol. Struct.* **2019**, *1178*, 617–629. [CrossRef]
70. Magyar, J.S.; Godwin, H.A. Spectropotentiometric analysis of metal binding to structural zinc-binding sites: Accounting quantitatively for pH and metal ion buffering effects. *Anal. Biochem.* **2003**, *320*, 39–54. [CrossRef]
71. Kállay, C.; Várnagy, K.; Micera, G.; Sanna, D.; Sóvágó, I. Copper(II) complexes of oligopeptides containing aspartyl and glutamyl residues. Potentiometric and spectroscopic studies. *J. Inorg. Biochem.* **2005**, *99*, 1514–1525. [CrossRef]
72. Casal, H.L.; Köhler, U.; Mantsch, H.H. Structural and conformational changes of β-lactoglobulin B: an infrared spectroscopic study of the effect of pH and temperature. *Biochim. Biophys. Acta (BBA)/Protein Struct. Mol.* **1988**, *957*, 11–20. [CrossRef]
73. Herrero-Martínez, J.M.; Simó-Alfonso, E.F.; Ramis-Ramos, G.; Gelfi, C.; Righetti, P.G. Determination of cow's milk in non-bovine and mixed cheeses by capillary electrophoresis of whey proteins in acidic isoelectric buffers. *J. Chromatogr. A* **2000**, *878*, 261–271. [CrossRef]
74. Miralles, B.; Rothbauer, V.; Manso, M.A.; Amigo, L.; Krause, I.; Ramos, M. Improved method for the simultaneous determination of whey proteins, caseins and para-κ-casein in milk and dairy products by capillary electrophoresis. *J. Chromatogr. A* **2001**, *915*, 225–230. [CrossRef]
75. Costa, F.F.; Vasconcelos Paiva Brito, M.A.; Moreira Furtado, M.A.; Martins, M.F.; Leal De Oliveira, M.A.; Mendonça De Castro Barra, P.; Amigo Garrido, L.; De Oliveira Dos Santos, A.S. Microfluidic chip electrophoresis investigation of major milk proteins: Study of buffer effects and quantitative approaching. *Anal. Methods* **2014**, *6*, 1666–1673. [CrossRef]
76. Anema, S.G. The use of "lab-on-a-chip" microfluidic SDS electrophoresis technology for the separation and quantification of milk proteins. *Int. Dairy J.* **2009**, *19*, 198–204. [CrossRef]
77. Buffoni, J.N.; Bonizzi, I.; Pauciullo, A.; Ramunno, L.; Feligini, M. Characterization of the major whey proteins from milk of Mediterranean water buffalo (Bubalus bubalis). *Food Chem.* **2011**, *127*, 1515–1520. [CrossRef]
78. Li, Z.; Wen, F.; Li, Z.; Zheng, N.; Jiang, J.; Xu, D. Simultaneous detection of α-Lactoalbumin, β-Lactoglobulin and Lactoferrin in milk by Visualized Microarray. *BMC Biotechnol.* **2017**, *17*, 1–9. [CrossRef]
79. Czerwenka, C.; Muller, L.; Lindner, W. Detection of the adulteration of water buffalo milk and mozzarella with cow's milk by liquid chromatography-mass spectrometry analysis of β-lactoglobulin variants. *Food Chem.* **2010**, *122*, 901–908. [CrossRef]
80. Santos, M.J.; Teixeira, J.A.; Rodrigues, L.R. Fractionation of the major whey proteins and isolation of β-Lactoglobulin variants by anion exchange chromatography. *Sep. Purif. Technol.* **2012**, *90*, 133–139. [CrossRef]
81. Doultani, S.; Turhan, K.N.; Etzel, M.R. Fractionation of proteins from whey using cation exchange chromatography. *Process Biochem.* **2004**, *39*, 1737–1743. [CrossRef]
82. Nicolás, P.; Ferreira, M.L.; Lassalle, V. A review of magnetic separation of whey proteins and potential application to whey proteins recovery, isolation and utilization. *J. Food Eng.* **2019**, *246*, 7–15. [CrossRef]

83. Barinov, N.A.; Vlasova, I.I.; Sokolov, A.V.; Kostevich, V.A.; Dubrovin, E.V.; Klinov, D.V. High-resolution atomic force microscopy visualization of metalloproteins and their complexes. *Biochim. Biophys. Acta - Gen. Subj.* **2018**, *1862*, 2862–2868. [CrossRef]
84. Acosta, M.; Sabier, T.; Mariño-Repizo, L.; Martinez, L.D.; Gil, R.A. Novel method for metalloproteins determination in human breast milk by size exclusion chromatography coupled to inductively coupled plasma mass spectrometry. *J. Pharm. Biomed. Anal.* **2018**, *158*, 209–213. [CrossRef] [PubMed]
85. Kumari, S.; Kondapi, A.K. Receptor-mediated targeted delivery of DNA using Lactoferrin nanoparticles. *Int. J. Biol. Macromol.* **2018**, *108*, 401–407. [CrossRef] [PubMed]
86. Srisod, S.; Motina, K.; Inprasit, T.; Pisitsak, P. A green and facile approach to durable antimicrobial coating of cotton with silver nanoparticles, whey protein, and natural tannin. *Prog. Org. Coatings* **2018**, *120*, 123–131. [CrossRef]
87. Kumari, S.; Kondapi, A.K. Lactoferrin nanoparticle mediated targeted delivery of 5-fluorouracil for enhanced therapeutic efficacy. *Int. J. Biol. Macromol.* **2017**, *95*, 232–237. [CrossRef]
88. Bollimpelli, V.S.; Kumar, P.; Kumari, S.; Kondapi, A.K. Neuroprotective effect of curcumin-loaded lactoferrin nano particles against rotenone induced neurotoxicity. *Neurochem. Int.* **2016**, *95*, 37–45. [CrossRef]
89. Okubo, K.; Kamiya, M.; Urano, Y.; Nishi, H.; Herter, J.M.; Mayadas, T.; Hirohama, D.; Suzuki, K.; Kawakami, H.; Tanaka, M.; et al. Lactoferrin Suppresses Neutrophil Extracellular Traps Release in Inflammation. *EBioMedicine* **2016**, *10*, 204–215. [CrossRef]
90. Pozzi, C.M.C.; Braga, C.P.; Vieira, J.C.S.; Cavecci, B.; Vitor de Queiroz, J.; de Souza Barbosa, H.; Arruda, M.A.Z.; Gozzo, F.C.; Padillha, P.; de Magalhães Padilha, M. Metal ions bound to the human milk immunoglobulin A: Metalloproteomic approach. *Food Chem.* **2015**, *166*, 492–497. [CrossRef]
91. Kumar, P.; Lakshmi, Y.S.; Bhaskar, C.; Golla, K.; Kondapi, A.K. Improved Safety, Bioavailability and Pharmacokinetics of Zidovudine through Lactoferrin Nanoparticles during Oral Administration in Rats. *PLoS One* **2015**, *10*, 1–17. [CrossRef]
92. Su, Z.; Xing, L.; Chen, Y.; Xu, Y.; Yang, F.; Zhang, C.; Ping, Q.; Xiao, Y. Lactoferrin-Modified Poly(ethylene glycol)-Grafted BSA Nanoparticles as a Dual-Targeting Carrier for Treating Brain Gliomas. *Mol. Pharm.* **2014**, *11*, 1823–1834. [CrossRef]
93. Godoy-Gallardo, M.; Mas-Moruno, C.; Fernández-Calderón, M.C.; Pérez-Giraldo, C.; Manero, J.M.; Albericio, F.; Gil, F.J.; Rodríguez, D. Covalent immobilization of hLf1-11 peptide on a titanium surface reduces bacterial adhesion and biofilm formation. *Acta Biomater.* **2014**, *10*, 3522–3534. [CrossRef] [PubMed]
94. Carthagena, L.; Becquart, P.; Hocini, H.; Kazatchkine, M.D.; Bouhlal, H.; Belec, L. Modulation of HIV Binding to Epithelial Cells and HIV Transfer from Immature Dendritic Cells to CD4 T Lymphocytes by Human Lactoferrin and its Major Exposed LF-33 Peptide. *Open Virol. JournalOpen Virol. J.* **2011**, *5*, 27–34. [CrossRef]
95. Hutchens, T.W.; Magnuson, J.S.; Yip, T. Rapid purification of porcine colostral whey lactoferrin by affinity chromatography on single-stranded DNA-agarose. Characterization, amino acid composition and N-terminal amino acid sequence. *Biochim. Biophys. Acta* **1989**, *999*, 323–329. [CrossRef]
96. Pomastowski, P.; Buszewski, B. Complementarity of matrix-and nanostructure-assisted laser desorption/ionization approaches. *Nanomaterials* **2019**, *9*, 260. [CrossRef] [PubMed]
97. Leney, A.C.; Heck, A.J.R. Native Mass Spectrometry: What is in the Name? *J. Am. Soc. Mass Spectrom.* **2017**, *28*, 5–13. [CrossRef] [PubMed]
98. Lermyte; Everett; Brooks; Bellingeri; Billimoria; Sadler; O'Connor; Telling; Collingwood Emerging Approaches to Investigate the Influence of Transition Metals in the Proteinopathies. *Cells* **2019**, *8*, 1231. [CrossRef] [PubMed]
99. Allen, S.J.; Giles, K.; Gilbert, T.; Bush, M.F. Ion mobility mass spectrometry of peptide, protein, and protein complex ions using a radio-frequency confining drift cell. *Analyst* **2016**, *141*, 884–891. [CrossRef] [PubMed]
100. Allen, S.J.; Schwartz, A.M.; Bush, M.F. Effects of polarity on the structures and charge states of native-like proteins and protein complexes in the gas phase. *Anal. Chem.* **2013**, *85*, 12055–12061. [CrossRef]
101. Lermyte, F.; Everett, J.; Lam, Y.P.Y.; Wootton, C.A.; Brooks, J.; Barrow, M.P.; Telling, N.D.; Sadler, P.J.; O'Connor, P.B.; Collingwood, J.F. Metal Ion Binding to the Amyloid β Monomer Studied by Native Top-Down FTICR Mass Spectrometry. *J. Am. Soc. Mass Spectrom.* **2019**, *30*, 2123–2134. [CrossRef]
102. Lee, C.W.; Tseng, F.G. Surface enhanced Raman scattering (SERS) based biomicrofluidics systems for trace protein analysis. *Biomicrofluidics* **2018**, *12*. [CrossRef]

103. Van De Weert, M.; Stella, L. Fluorescence quenching and ligand binding: A critical discussion of a popular methodology. *J. Mol. Struct.* **2011**, *998*, 144–150. [CrossRef]
104. Kastritis, P.L.; Bonvin, A.M.J.J. On the binding affinity of macromolecular interactions: daring to ask why proteins interact. *J. R. Soc. Interface* **2012**, *10*, 20120835. [CrossRef] [PubMed]
105. Vuignier, K.; Schappler, J.; Veuthey, J.L.; Carrupt, P.A.; Martel, S. Drug-protein binding: A critical review of analytical tools. *Anal. Bioanal. Chem.* **2010**, *398*, 53–66. [CrossRef] [PubMed]
106. Tai, C.S.; Chen, Y.Y.; Chen, W.L. β-Lactoglobulin Influences Human Immunity and Promotes Cell Proliferation. *Biomed Res. Int.* **2016**, *2016*, 1–12. [CrossRef]
107. Žuvela, P.; Liu, J.J.; Yi, M.; Pomastowski, P.P.; Sagandykova, G.; Belka, M.; David, J.; Bączek, T.; Szafrański, K.; Żołnowska, B.; et al. Target-based drug discovery through inversion of quantitative structure-drug-property relationships and molecular simulation: CA IX-sulphonamide complexes. *J. Enzyme Inhib. Med. Chem.* **2018**, *33*, 1430–1443. [CrossRef]
108. Chen, X.; Sun, L.; Yu, Y.; Xue, Y.; Yang, P. Amino acid-coded tagging approaches in quantitative proteomics. *Expert Rev. Proteomics* **2007**, *4*, 25–37. [CrossRef]
109. Ross, P.L.; Huang, Y.N.; Marchese, J.N.; Williamson, B.; Parker, K.; Hattan, S.; Khainovski, N.; Pillai, S.; Dey, S.; Daniels, S.; et al. Multiplexed protein quantitation in Saccharomyces cerevisiae using amine-reactive isobaric tagging reagents. *Mol. Cell. Proteomics* **2004**, *3*, 1154–1169. [CrossRef]
110. Shenar, N.; Cantel, S.; Martinez, J.; Enjalbal, C. Comparison of inert supports in laser desorption/ionization mass spectrometry of peptides: pencil lead, porous silica gel, DIOS-chip and NALDITM target. *Rapid Commun. Mass Spectrom.* **2009**, *23*, 2371–2379. [CrossRef]
111. Zarogoulidis, P.; Tsakiridis, K.; Karapantzou, C.; Lampaki, S.; Kioumis, I.; Pitsiou, G.; Papaiwannou, A.; Hohenforst-Schmidt, W.; Huang, H.; Kesisis, G.; et al. Use of Proteins as Biomarkers and Their Role in Carcinogenesis. *J. Cancer* **2015**, *6*, 9–18. [CrossRef]
112. Shelly, C.; Lu, M.D. Glutathione synthesis. *Biochim. Biophys. Acta* **2013**, *1830*, 3143–3153.
113. Britigan, B.E.; Serody, J.S.; Cohen, M.S. The role of lactoferrin as an anti-inflammatory molecule. *Adv. Exp. Med. Biol.* **1994**, *357*, 143–156. [PubMed]
114. Luzi, C.; Brisdelli, F.; Iorio, R.; Bozzi, A.; Carnicelli, V.; Di Giulio, A.; Lizzi, A.R. Apoptotic effects of bovine apo-lactoferrin on HeLa tumor cells. *Cell Biochem. Funct.* **2017**, *35*, 33–41. [CrossRef] [PubMed]
115. Gibbons, J.A.; Kanwar, J.R.; Kanwar, R.K. Iron-free and iron-saturated bovine lactoferrin inhibit survivin expression and differentially modulate apoptosis in breast cancer. *BMC Cancer* **2015**, *15*, 1–16. [CrossRef] [PubMed]
116. Kanwar, J.R.; Palmano, K.P.; Sun, X.; Kanwar, R.K.; Gupta, R.; Haggarty, N.; Rowan, A.; Ram, S.; Krissansen, G.W. "Iron-saturated" lactoferrin is a potent natural adjuvant for augmenting cancer chemotherapy. *Immunol. Cell Biol.* **2008**, *86*, 277–288. [CrossRef] [PubMed]
117. Sanchez, L.; Calvo, M.; Brock, J.H. Biological role of lactoferrin. *Arch. Dis. Child.* **1992**, *67*, 657–661. [CrossRef] [PubMed]
118. Komatsu, A.; Satoh, T.; Wakabayashi, H.; Ikeda, F. Effects of bovine lactoferrin to oral Candida albicans and Candida glabrata isolates recovered from the saliva in elderly people. *Odontology* **2015**, *103*, 50–55. [CrossRef]
119. Leboffe, L.; Giansanti, F.; Antonini, G. Antifungal and Antiparasitic Activities of Lactoferrin. *Antiinfect. Agents Med. Chem.* **2009**, *8*, 114–127. [CrossRef]
120. Thawari, A.G.; Rao, C.P. Peroxidase-like Catalytic Activity of Copper-Mediated Protein-Inorganic Hybrid Nanoflowers and Nanofibers of β-Lactoglobulin and α-Lactalbumin: Synthesis, Spectral Characterization, Microscopic Features, and Catalytic Activity. *ACS Appl. Mater. Interfaces* **2016**, *8*, 10392–10402. [CrossRef]
121. Wegmüller, R.; Tay, F.; Zeder, C.; Brnić, M.; Hurrell, R.F. Zinc Absorption by Young Adults from Supplemental Zinc Citrate Is Comparable with That from Zinc Gluconate and Higher than from Zinc Oxide. *J. Nutr.* **2014**, *144*, 132–136. [CrossRef]
122. Prasad, A.S. Discovery of human zinc deficiency: 50 years later. *J. Trace Elem. Med. Biol.* **2012**, *26*, 66–69. [CrossRef]
123. Udechukwu, M.C.; Collins, S.A.; Udenigwe, C.C. Prospects of enhancing dietary zinc bioavailability with food-derived zinc-chelating peptides. *Food Funct.* **2016**, *7*, 4137–4144. [CrossRef] [PubMed]
124. Pryshchepa, O.; Sagandykova, G.N.; Pomastowski, P.; Railean-Plugaru, V.; Król, A.; Rogowska, A.; Rodzik, A.; Sprynskyy, M.; Buszewski, B. A New Approach for Spontaneous Silver Ions Immobilization onto Casein. *Int. J. Mol. Sci.* **2019**, *20*, 3864. [CrossRef] [PubMed]

125. Wang, C.; Li, B.; Wang, B.; Xie, N. Degradation and antioxidant activities of peptides and zinc-peptide complexes during in vitro gastrointestinal digestion. *Food Chem.* **2015**, *173*, 733–740. [CrossRef] [PubMed]
126. Jellesen, M.S.; Rasmussen, A.A.; Hilbert, L.R. A review of metal release in the food industry. *Mater. Corros.* **2006**, *57*, 387–393. [CrossRef]
127. Atapour, M.; Wei, Z.; Chaudhary, H.; Lendel, C.; Odnevall Wallinder, I.; Hedberg, Y. Metal release from stainless steel 316L in whey protein - And simulated milk solutions under static and stirring conditions. *Food Control* **2019**, *101*, 163–172. [CrossRef]
128. Atapour, M.; Odnevall Wallinder, I.; Hedberg, Y. Stainless steel in simulated milk and whey protein solutions – Influence of grade on corrosion and metal release. *Electrochim. Acta* **2020**, *331*, 135428. [CrossRef]
129. Nicolás, P.; Ferreira, M.L.; Lassalle, V. Magnetic solid-phase extraction: A nanotechnological strategy for cheese whey protein recovery. *J. Food Eng.* **2019**, *263*, 380–387. [CrossRef]

International Journal of *Molecular Sciences*

MDPI

Article

New Methodology for the Identification of Metabolites of Saccharides and Cyclitols by Off-Line EC-MALDI-TOF-MS

Gulyaim Sagandykova [1,2], Justyna Walczak-Skierska [1], Fernanda Monedeiro [1], Paweł Pomastowski [1] and Bogusław Buszewski [1,2,*]

1 Centre for Modern Interdisciplinary Technologies Nicolaus Copernicus University in Toruń, Wileńska 4, 87-100 Toruń, Poland; sagandykova.gulyaim1@gmail.com (G.S.); walczak-justyna@wp.pl (J.W.-S.); fernandamonedeiro@usp.br (F.M.); pawel_pomastowski@wp.pl (P.P.)

2 Department of Environmental Chemistry and Bioanalytics, Faculty of Chemistry, Nicolaus Copernicus University in Torun, Gagarina 7, 87-100 Toruń, Poland

* Correspondence: bbusz@chem.umk.pl; Tel.: +48-56-6114308

Received: 24 June 2020; Accepted: 22 July 2020; Published: 24 July 2020

Abstract: A combination of electrochemistry (EC) and matrix-assisted laser desorption/ionization time-of-flight mass spectrometry (*off-line* EC-MALDI-TOF-MS) was applied for determination of the studied biologically active compounds (D-glucose, D-fructose, D-galactose, D-pinitol, L-*chiro*-inositol, and *myo*-inositol) and their possible electrochemical metabolites. In this work, boron-doped diamond electrode (BDD) was used as a working electrode. MALDI-TOF-MS experiments were carried out (both in positive and negative ion modes and using two matrices) to identify the structures of electrochemical products. This was one of the first applications of the EC system for the generation of electrochemical products produced from saccharides and cyclitols. Moreover, exploratory data analysis approaches (correlation networks, hierarchical cluster analysis, weighted plots) were used in order to present differences/similarities between the obtained spectra, regarding the class of analyzed compounds, ionization modes, and used matrices. This work presents the investigation and comparison of fragmentation patterns of sugars, cyclitols, and their respective products generated through the electrochemistry (EC) process.

Keywords: biologically active compounds; saccharides; cyclitols; electrochemistry; MALDI-TOF-MS

1. Introduction

Carbohydrates are a complex group of organic compounds occurring in all plants. They are the main source of energy for both humans and plant organisms. They play an important role in the proper functioning of the brain [1,2]. Cyclitols are another group of compounds found in plant material. Cyclitols are responsible for the plant's self-protection against unfavorable environmental conditions, they are involved in signal transduction, in biogenesis and osmoregulation, and in phosphate storage [1,3–6].

Biologically active substances and compounds with potential biological activity of plant origin, i.e., both those with beneficial effects and used as medicines and those that have harmful effects on human health, require monitoring in the environment and in the human body [7]. One of the techniques that allows the monitoring of biologically active compounds is matrix-assisted laser desorption/ionization time-of-flight mass spectrometry (MALDI-TOF-MS) [8]. The MALDI-TOF-MS technique in recent decades became irreplaceable in the analysis of individual compounds and their mixtures and also in complex biological matrices. The development of 'soft' ionization methods, high-throughput techniques and quantitative methods has expanded the use of mass spectrometry

(MS) in the study of the structure, function, and modification of low molecular weight compounds [8,9]. The use of MALDI-TOF-MS in the analysis of metabolites of low molecular weight compounds allows eliminating chromatographic separation [10]. MALDI-MS is widely used for the analysis of large compounds, such as proteins, peptides, and also whole bacteria cells [11–13]. Analysis of low molecular weight compounds by MALDI-MS is a current analytical challenge [14]. However, despite many limitations such as signal suppression, acidity of the matrix solution, and degradation of the analyte, MALDI-MS has found its place in application in the analysis of low molecular weight compounds [14,15]. MALDI-TOF-MS was used in the analysis of anthocyanins, sugars, and organic acids in strawberry [16], the analysis and characterization of oligosaccharides [17], and the identification of flavonoids in berry extracts [18] as well as sugars and alkaloids in food [19].

In recent years, electrochemistry (EC) has been increasingly used in research on the metabolic biotransformation of biologically active compounds. Electrochemical methods allow for real-time measurements often with high resolution in both in vitro and in vivo studies. The electrodes can be used to analyze metabolites released by the cell or to assess the consumption of compounds already in the culture medium, but they can also generate compounds in the immediate vicinity of the cell and then measure how the cell responds to such stimulation [20,21]. The combination of electrochemistry with mass spectrometry allows mimicking the reactions of phase I and II metabolism as well as identifying electrochemical products that can serve as potential metabolites [22,23]. This simple instrumental analytical technique allows the production, detection, and identification of a number of metabolic products, including reactive intermediate metabolites, that are responsible for drug activation. EC/MS also allows the identification of the final products of electrochemical reactions of the tested compounds and is used in redox (oxidation–reduction) studies [24–26].

The study of a compound biotransformation is a complex process. Among the available analytical techniques, EC-MS is a universal tool that has been used to confirm the presence of parent compounds and identify their candidate metabolites, which are most often targeted in a complex biological environment in metabolomics studies. In this work, a new method using an *off-line* EC-MALDI-TOF-MS system for the analysis of possible products of low-molecular biologically active compounds was developed. For the first time, an electrochemical system and matrix-assisted laser desorption/ionization time-of-flight mass spectrometry were used to determine and identify electrochemical products of selected saccharides (D-glucose, D-fructose, and D-galactose) and cyclitols (D-pinitol, L-*chiro*-inositol, and *myo*-inositol). Statistical methods were used to provide data visualization and to highlight the main characteristics of data structure, supporting the discussion of fragmentation pathways. The present goals comprised the elucidation of fragmentation patterns of sugars, cyclitols, and their suggested metabolites, as well as the investigation of correlations between their MS spectra. The intention was to propose a wide range of EC products of these compounds, presenting a methodology that could be used in the monitoring of such species and the modeling of potential metabolic reactions of phase I oxidative cell metabolism.

2. Results and Discussion

2.1. *Effect of Matrix*

MALDI-TOF-MS has been mostly used to identify large biomolecules such as proteins, peptides, nucleic acids, and polymers. Qualitative and quantitative analysis of low molecular weight compounds (LMWC) is challenging due to the large number of signals coming from the ionizing matrix, which consequently prevents identification of the analytes. Despite this limitation, the application of MALDI-MS in the analysis of low molecular weight compounds has been constantly increasing due to the large tolerance of the buffers used during the analysis, formation of mostly single-charged ions, high sensitivity, and throughput [15].

Two matrices, a benzoic acid derivative, 2,5-dihydroxybenzoic acid (DHB), and a cinnamic acid derivative, α-cyano-4-hydroxycinnamic acid (HCCA), were employed. Both matrices belong to the first-generation matrices with molecular weight below 300 Da [15]. The use of these matrices ensured stable ionization, high reproducibility, and a high resolution of recorded MS spectra. MALDI spectra for DHB and HCCA matrices were registered to eliminate the signals coming from matrices from the spectra. Several ions were formed during laser irradiation for the HCCA matrix, as it can be observed from the spectra. Except for a molecular ion $[M + H]^+$ at *m/z* 190.0, a fragment of high intensity was probably formed at *m/z* 172.0 and *m/z* 379.0, which may correspond to $[M\text{-}H_2O + H]^+$ and matrix cluster $[2M + H]^+$, respectively. Less intense fragments also could appear in the spectrum at *m/z* 146.1, *m/z* 164.1, *m/z* 212.0, and *m/z* 335.1, which may correspond to $[M\text{-}CO_2 + H]^+$, $[MH\text{-}CN]^+$, sodium adduct $[M + Na]^+$, and $[2M + H\text{-}CO_2]^+$, respectively [15,27]. In case of the DHB matrix, less ions coming from the matrix were generated. A signal of high intensity at *m/z* 137.0 that may correspond to $[M{+}H\text{-}H_2O]^+$ can be observed from the spectrum as well as less intense signals at *m/z* 155.0 and *m/z* 177.0 that may correspond to the protonated molecular ion from matrix $[M + H]^+$ and the sodium adduct $[M{+}Na]^+$, respectively. It can be assumed that the DHB matrix is found to be more suitable for the analysis of low molecular weight analytes such as sugars and cyclitols, since less ions coming from the matrix were generated during laser irradiation, which can cause interferences during the identification of analytes.

2.2. *Off-Line EC-MALDI-TOF-MS*

In order to generate possible transformation products of cyclitols and saccharides, a system consisting of an off-line electrochemical cell coupled with matrix-assisted laser desorption/ionization was used. The off-line EC-MALDI system was used for the first time in the analysis of metabolites of low molecular weight compounds. The methodology of the experiment included use of the boron-doped diamond electrode (BDD) working electrode and ammonium acetate at an approximate pH of 7.4. Standards of cyclitols and saccharides were introduced to the electrochemical cells, and fractions after conversion were collected manually. Each fraction was evaporated using a Labconco Centri Vap DNA concentrator (Kansas City, USA). In a second step, fractions were analyzed by MALDI-TOF-MS. Figures 1–6 present the suggested fragmentation pathways of cyclitols and saccharides before and after electrochemical analysis. There are very few reports regarding the analysis of cyclitols and saccharides using MALDI-TOF-MS in the literature.

Moreover, representatives of species of sugars and cyclitols were selected in order to observe the possible similarities and differences in their EC products, since these analytes have similar structural features. Such similarity was interesting, since they possess opposite activities (e.g., in *Diabetes Mellitus*) that in particular cases is structure-dependent. In addition, the anti-diabetic activity of cyclitols has different aspects, but the mechanisms are still understood poorly. Besides that, literature on the potential metabolites of cyclitols in human organism is quite scarce [28].

2.2.1. D-Pinitol

Figure 1a shows a suggested fragmentation pathway of D-pinitol (signals before the electrochemical process are denoted in red color, signals after the electrochemical process are denoted in blue). Figure 1b,c show the mass spectra of D-pinitol before and after electrochemical analysis. The signals A and B (in both mass spectra; denoted in black color) belong to the DHB matrix. In the spectrum, the molecular ion of D-pinitol at *m/z* 195.0 (C) can be observed. The signal at *m/z* = 193.0 (D) could be derived from the signal C by the loss of hydrogen (dehydrogenation). The ion at *m/z* = 212.1 (R) could be originated from electrochemical oxidation and the detachment of a methyl group from ion 193.0 (D). The signal *m/z* 228.0 (S) appeared only in the spectrum before the electrochemical process and may correspond to the addition of a hydroxyl group. Ions at *m/z* 230.9 (more intense before the electrochemical process and less intense after the electrochemical process; T) and *m/z* 249.9 (occurred only before electrochemical process; U) may correspond to the addition of two hydrogen atoms (hydrogenation) and the addition of sodium and loss of four hydrogen atoms, respectively. Other products of D-pinitol are E, J, and K (Figure 1a,c). The ions at *m/z* 370.9 (E), 354.9 (J), and 346.9 (K) may correspond to $[C_{14}H_{26}O_{11}]^+$ (dimer of D-pinitol), $[C_{13}H_{22}O_{11}]^+$ (loss of methyl group), and $[C_{13}H_{14}O_{11}]^+$ (loss of eight hydrogen atoms from hydroxyl groups). The fragment that was observed at *m/z* 392.9 (F) may correspond to $[C_{14}H_{25}NaO_{11}]^+$ (dimer of D-pinitol and adduct with sodium). The signals at *m/z* 376.9 $[C_{13}H_{21}NaO_{11}]^+$ (G) and *m/z* 361.9 $[C_{13}H_{23}NaO_{10}]^+$ (I) could result from the ion at *m/z* 392.9 by the loss of a methyl group (14 u) and oxygen atom (16 u), respectively. The ions detected at *m/z* 366.9 (H), 344.0 (L), 328.9 (M), 288.2 (N), 273.0 (O), and 265.9 (P) can be ascribed to $[C_{14}H_{22}O_{11}]^+$, $[C_{13}H_{12}O_{11}]^+$, $[C_{13}H_{12}O_{10}]^+$, $[C_{13}H_{20}O_7]^+$, $[C_{13}H_{21}O_6]^+$, and $[C_{13}H_{13}O_6]^+$, probably corresponding to the formation of a dimer of D-pinitol and the loss of two hydrogen atoms (2 u), loss of one carbon atom and 10 hydrogen atoms (22 u), loss of one oxygen atom (16 u), loss of 3 × O, addition of 8 × H, and the loss of oxygen (16 u), and the loss of five hydrogen atoms (5 u), respectively. The ions at *m/z* 273.0 (O) and *m/z* 265.9 (P) may correspond to the formation of radical cations (Figure 1a,b).

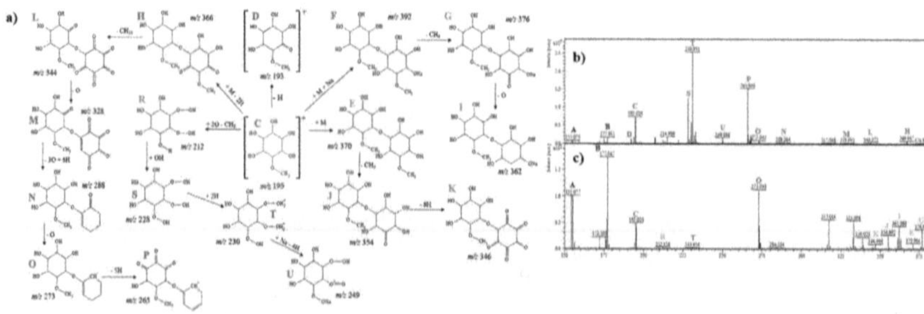

Figure 1. (**a**) Fragmentation pathways of D-pinitol; (**b**) MS spectrum before electrochemical process, (**c**) MS spectrum after electrochemical process.

2.2.2. L-*Chiro*-Inositol

The signal observed at *m/z* 164.1 (D) may correspond to the loss of hydroxyl group (17 u) from L-*chiro*-inositol (Figure 2b). The molecular ion $[M + H]^+$ was not observed in the MS spectra (Figure 2b,c). The ion at *m/z* 193.0 (C) could be formed by the addition of a methyl group to the hydroxyl group. In Figure 2c, L-*chiro*-inositol's derivative at *m/z* 199.0 $[C_6H_8NaO_6]^+$ (E) can be observed. This ion could be formed by replacement of the methyl group to sodium (Figure 2a). The fragment at *m/z* 230.9 $[C_6H_{14}O_9]^+$ (G) could be originated from the ion at *m/z* 214.9 $[C_7H_{11}NaO_6]^+$ (F) by the loss of CH–Na and the addition of three hydroxyl groups (51 u). The dimer of L-*chiro*-inositol can be observed at *m/z* 343.2 $[2M]^+$ (H) presented in Figure 2a,b. Ions at *m/z* 304.2 $[C_{12}H_{16}O_9]^+$ (I), *m/z* 298.3 $[C_{12}H_{10}O_9]^+$ (J), *m/z* 256.2 $[C_{12}H_{16}O_6]^+$ (K), and *m/z* 244.2 $[C_{11}H_{16}O_6]^+$ (L) may correspond to the loss of two oxygen atoms and seven hydrogen atoms from the phenol group (H_7O_2, 39 u), the loss of six hydrogen atoms (dehydrogenation), the loss of three oxygen atoms (oxidation) and addition of six hydrogen atoms (hydrogenation), and the loss of one carbon atom (cleavage of ring), respectively. The ion at *m/z* 339.0 (M) can be ascribed to the adduct of one molecule of the DHB matrix with L-*chiro*-inositol and sodium (Figure 2c). The signal at *m/z* 317.0 (N) is created from the ion M by the loss of sodium and addition of one hydrogen atom. The ion at *m/z* 316.2 (O) may result from the dehydrogenation of a carboxyl group from DHB molecules at *m/z* 317.0. Another signal at *m/z* 288.2 (P) was observed in spectra before the electrochemical process and may correspond to $[C_{13}H_{20}O_7]^+$. The peak at *m/z* 284.3 (R) may correspond to the loss of four hydrogen atoms (dehydrogenation) from ion $[C_{13}H_{20}O_7]^+$. The ion at *m/z* 376.9 (S) could result from the formation of a sodium adduct of L-*chiro*-inositol with one molecule of the DHB matrix and two hydroxyl groups $[C_{13}H_{21}NaO_{11}]^+$. The ions detected at *m/z* 360.9 (T), 354.8 (U), 331.017 (V), and 332.2 (X) can be ascribed to $[C_{13}H_{21}NaO_{10}]^+$, $[C_{13}H_{15}NaO_{10}]^+$, $[C_{13}H_{15}O_{10}]^+$, and $[C_{13}H_{16}O_{10}]^+$, probably corresponding to the loss of one hydroxyl group (17 u) from the DHB ring, dehydrogenation (the loss of four hydrogen atoms), the loss of one sodium atom from the carboxyl group of DHB molecules, and hydrogenation (addition of one hydrogen), respectively. Furthermore, the ion at *m/z* 361.9 (Y) could result from the addition of one hydrogen atom to the ion at *m/z* 360.9. The ion at *m/z* 346.9 (W) may be derived from the ion at *m/z* 354.8 by the loss of eight hydrogen atoms.

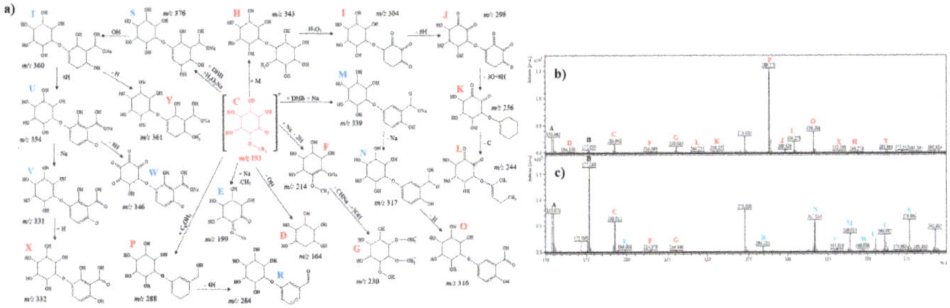

Figure 2. (**a**) Fragmentation pathways of L-*chiro*-inositol; (**b**) MS spectrum before electrochemical process, (**c**) MS spectrum after electrochemical process.

2.2.3. *Myo*-Inositol

In the case of *myo*-inositol, the molecular ion also could not be observed (Figure 3b,c). Signals at *m/z* 174.9 $[C_6H_6O_6]^+$ (D) and 172.0 $[C_6H_4O_6]^+$ (E) could correspond to *myo*-inositol after the loss of three and four hydrogen atoms (dehydrogenation) from hydroxyl groups, respectively. In the spectrum before and after electrochemical analysis, the fragment at *m/z* = 193.0 (C) can be observed, which may correspond to the addition of a methyl group to the hydroxyl group similarly to L-*chiro*-inositol (Figure 3a–c). Figure 3c shows the ions at *m/z* 199.0 (F) and 212.1 (G), which may correspond to the exchange of the methyl group to sodium $[C_7H_{12}O_6 - CH_4 + Na]^+$, the addition of two oxygen atoms, and the loss of sodium $[C_6H_{12}O_6 - Na + O_2]^+$, respectively. The ion at *m/z* 203.0 $[C_6H_{12}NaO_6]^+$ (H) may correspond to the addition of sodium and the loss of carbon and hydrogen atoms from ion C. The signals at *m/z* 219.9 (I), 228.0 (J), 230.9 (K), and 242.2 (L) may correspond to $[C_6H_{12}NaO_7]^+$ (addition of one hydroxyl group, 17 u), $[C_6H_{12}O_9]^+$ (addition of 2 hydroxyl groups, 34 u, and loss of one sodium atom, 23 u), $[C_6H_{14}O_9]^+$ (addition of 2 hydrogen atoms, 2 u) and $[C_7H_{14}O_9]^+$ (addition of methyl group, 12 u), respectively. Ions at *m/z* 288.2 $[C_{13}H_{20}O_7]^+$ (M) and 284.3 $[C_{13}H_{16}O_7]^+$ (N) may correspond to the addition of hydroxymethyl-phenol and the loss of four hydrogen atoms (dehydrogenation), respectively. Ions observed at *m/z* 366.9 (O), 339.0 (P), and 325.2 (R) may be obtained by the addition of C_5HO_7 (173 u), the loss of 2 × O, and the addition of 5 × H (27 u) and loss of the CH_2 group (14 u) from the ring, respectively (Figure 3a). The fragmentation pathway of *myo*-inositol S→T→U→V looks similar to that of L-*chiro*-inositol (Figures 2a and 3a). The ion at *m/z* 328.9 (W) could be derived from the ion at *m/z* 346.9 (V) by the loss of sodium and the addition of five hydrogen atoms. Moreover, the ion at *m/z* 299.0 $[C_{13}H_{15}O_8]^+$ (X) could be originated from the ion at *m/z* 328.9 $[C_{13}H_{12}O_{10}]^+$ and was characteristic for *myo*-inositol after the electrochemical analysis.

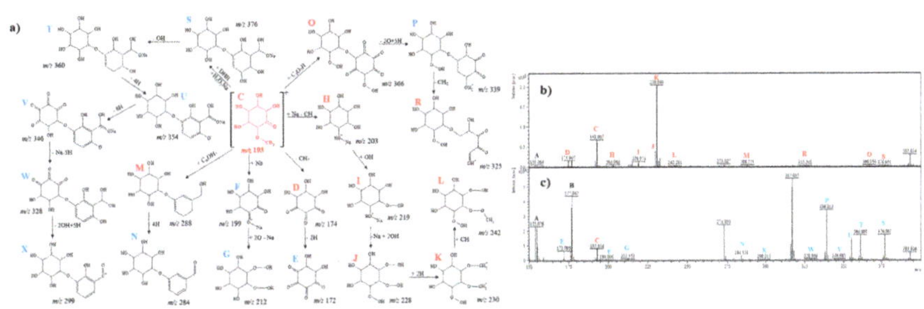

Figure 3. (**a**) Fragmentation pathways of myo-inositol; (**b**) MS spectrum before electrochemical process, (**c**) MS spectrum after electrochemical process.

2.2.4. D-Glucose

The saccharides can be ionized by the MALDI technique in positive and negative ionization modes. In the positive mode, the fragmentation process was preferentially toward ring–ring cleavage. Furthermore, in the negative mode, the cross-ring cleavage mechanism of fragmentation was dominant [29]. In the case of positive ionization, MALDI-TOF-MS mass spectra for sugars were characterized by similar signals to those of cyclitols. Therefore, mass spectra were present in the negative ionization mode for sugars to show a different fragmentation mechanism. Post-source decay (PSD) spectra of molecular ion at *m/z* 179.0 (B) of D-glucose are shown in Figure 4a–c. The ion detected at *m/z* 114.6 (C) may correspond to tetrahydropyran with two carbonyl groups $[C_5H_6O_3]^-$. The fragment at *m/z* 225.0 (D) could be formed by the addition of a carboxyl group (46 u). Ions equivalent to *m/z* 261.6 (E) and 247.7 (F) correspond to the addition of a cyclohexane ring and the loss of CH_2 (cleavage of a cyclohexane ring) (Figure 4a). A dimer of glucose can be observed at *m/z* 343.3 $[C_{12}H_{23}O_{11}]^-$ (G). The fragment ion at *m/z* 345.3 (L) could be formed from the ion at *m/z* 343.3 by the addition of hydrogen (hydrogenation). The fragment ion at *m/z* 327.3 (H) could be the result also from the ion at *m/z* 343.3 by loss of a hydroxyl group. The ion at *m/z* 321.4 $[C_{12}H_{17}O_{10}]^-$ (M) could be formed by the loss of six hydrogen atoms from the ion G, and probably in this case, the ring closes between the free hydroxyl groups (Figure 4a). The ions detected at *m/z* 307.4 (I), 289.5 (J), and 273.6 (K) can be ascribed to $[C_{12}H_{19}O_9]^-$ (loss of a hydroxyl group and three hydrogen atoms, 20 u), $[C_{12}H_{17}O_8]^-$ (loss of water molecules, 18 u), $[C_{12}H_{17}O_7]^-$ (loss of an oxygen atom, 16 u), respectively. The ion at *m/z* 315.4 $[C_{10}H_{19}O_{11}]^-$ (N) could be derived from the molecular ion $[C_6H_{11}O_6]^-$ by the addition of a "glucose ring" (Figure 4a). The loss of hydrogen from the ion at *m/z* 315.4 may produce the ion at *m/z* 314.7 $[C_{10}H_{18}O_{11}]^-$ (O).

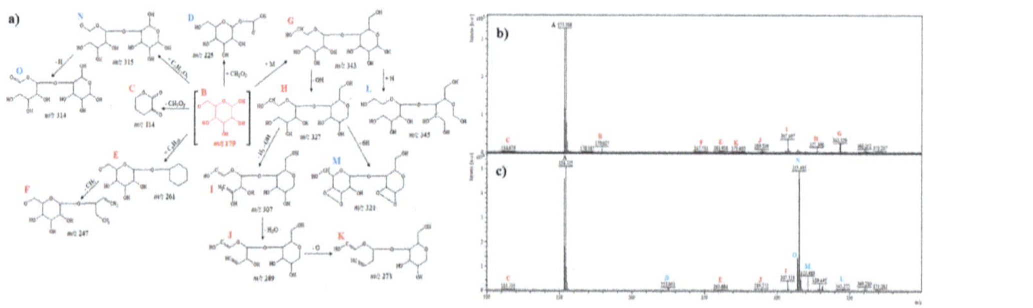

Figure 4. (**a**) Fragmentation pathways of D-glucose; (**b**) MS spectrum before electrochemical process, (**c**) MS spectrum after electrochemical process.

2.2.5. D-Fructose

MALDI-TOF-MS spectra of D-fructose are shown in Figure 5a–c. D-fructose belongs to monohexoses and Figure 5b presents the deprotonated molecule $[M - H]^-$ at *m/z* 179.4 (B). The signal at *m/z* 170.2 (C) may correspond to deprotonated D-fructose $[C_6H_2O_6]^-$. The signal at *m/z* 114.7 (D) could be created from the molecular ion B by the loss of three hydroxyl groups and loss of a CH_2 group. The ion at *m/z* 289.6 (E) can be attributed to the addition of a $C_5H_2O_3$ (110 u) molecule to a molecular ion. The ion at *m/z* 267.7 $[C_9H_{15}O_9]^-$ (F) may be formed due to the cross-ring cleavages and the loss of a carbon atom (Figure 5a). The signal at *m/z* 251.7 $[C_9H_{15}O_9 - O]^-$ (G) could be formed due to the loss of an oxygen atom from the ion at *m/z* 267.7. The sodium adduct with two molecules of D-fructose followed by the loss of two hydrogen atoms may correspond to the ion at *m/z* 361.4 $[C_{12}H_{18}NaO_{11}]^-$ (H). The ion observed at *m/z* 377.3 $[C_{12}H_{18}NaO_{11} + O]^-$ (I) can be formed by the addition of an oxygen atom. The loss of one hydrogen atom (dehydrogenation) from the ion at *m/z* 361.485 may form the ion at *m/z* 360.3 $[C_{12}H_{17}NaO_{11}]^-$ (J). In addition, the loss of an oxygen atom from the ion at *m/z* 361.4 could create the ion at *m/z* 345.4 $[C_{12}H_{18}NaO_{10}]^-$ (K). The fragment formation at *m/z* 329.4 (L) with low abundance can correspond to $[C_{12}H_{18}NaO_9]^-$ (loss of an oxygen atom from the ion at *m/z* 345.4). The addition of four hydrogen atoms (hydrogenation) and cross-ring cleavages from the ion at *m/z* 329.4 can produce the fragment at *m/z* 321.5 $[C_{12}H_{18}NaO_9 - C + 4H]^-$ (M). The signals detected at 315.5 (N) and 307.6 (O) can be ascribed to $[C_{11}H_{22}NaO_9 - 5H]^-$ (the loss of five hydrogen atoms and the formation of a radical anion) and $[C_{11}H_{16}NaO_9 - C + 4H]^-$ (the loss of one carbon atom and addition of four hydrogen atoms), respectively (Figure 5a).

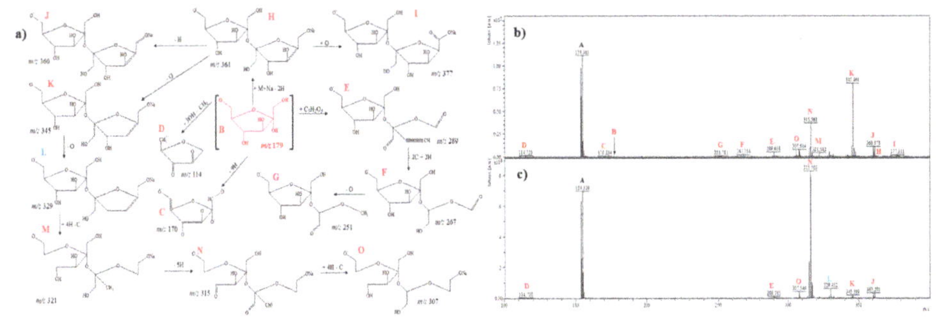

Figure 5. (**a**) Fragmentation pathways of D-fructose; (**b**) MS spectrum before electrochemical process, (**c**) MS spectrum after electrochemical process.

2.2.6. D-Galactose

The ion at *m/z* 179.2 (B) was detected in the MS spectra for D-galactose $[M - H]^-$ (Figure 6a–c). The signals at *m/z* 170.1 (C) and 114.6 (D) may correspond to $[C_6H_2O_6]^-$ (loss of eight hydrogen atoms and formation of a radical anion) and $[C_5H_6O_3]^-$ (loss of three hydroxyl groups and loss of a CH_2 group). The fragments at *m/z* 315.4 $[C_{11}H_{23}O_{10}]^-$ (E) may correspond to $[M - H + C_5H_{11}O_5]^-$ from a molecular ion. The loss of 3 × OH and loss of a CH group from the ion at *m/z* 315.4 could create the ion at *m/z* 251.7 (F). The signal at *m/z* 233.8 $[C_9H_{13}O_7]^-$ (G) can be obtained from the ion at *m/z* 251.7 $[C_{10}H_{19}O_7]^-$ by the loss of CH_6 (18 u). The ion observed at *m/z* 361.1 $[C_{12}H_{18}NaO_{11}]^-$ (N) could be formed by the addition of one molecule of D-galactose and sodium and the loss of two hydrogen atoms (Figure 6a). The addition of a hydroxyl group (17 u) to the ion observed at *m/z* 361.1 could create the ion at *m/z* 377.2 $[C_{12}H_{18}NaO_{11} + OH]^-$ (P). Moreover, the loss of six hydrogen atoms from the ion at *m/z* 377.2 could produce the ion at *m/z* 371.3 $[C_{12}H_{18}NaO_{12} - 6H]^-$ (R). The small intense signal at *m/z* 360.3 (O) could be formed by the dehydrogenation process from the ion $[C_{12}H_{18}NaO_{11}]^-$ (Figure 6c). The ions at *m/z* 345.3 (H), 329.3 (I), 307.4 (K), 289.4 (L), and 273.5 (M) probably corresponding to $[C_{12}H_{25}O_{11}]^-$ (addition of one molecule of D-galactose and four hydrogen atoms with the cross-ring cleavages), $[C_{12}H_{25}O_{10}]^-$ (loss of one hydroxyl group, 17 u), $[C_{12}H_{19}O_9]^-$ (loss of one hydroxyl group, 17 u, and five hydrogen atoms, 5 u), $[C_{12}H_{17}O_8]^-$ (loss of one water molecule, 18 u) and $[C_{12}H_{17}O_7]^-$ (loss of an oxygen atom, 16 u), respectively. Additionally, the fragment at *m/z* 321.3 $[C_{12}H_{17}O_{10}]^-$ (J) could be generated from the ion at *m/z* 329.3 by the loss of eight hydrogen atoms (dehydrogenation).

In general, analysis of the obtained mass spectra (Figures 1–6) has shown that some signals appeared only after the electrochemical process (Figure 1c, Figure 2c, Figure 3c, Figure 4c, Figure 5c, Figure 6c), and some signals had different intensities before and after the electrochemical process. This can be evidence of the occurrence of possible metabolites of the analyzed compounds. Regarding cyclitols, Figure 1b,c have shown that many signals (denoted as E, G, I, J, K, and R) occurred only after the electrochemical process. The ion at *m/z* 265.9 disappeared after the electrochemical oxidation. In addition, the fragments T and S (denoted in red color) had higher intensity before the electrochemical analysis (Figure 1b,c). The electrochemical oxidation for chiro-inositol also seemed to generate many signals since they occurred only after the EC fragmentation. The signals E, N, M, R, S, T, U, V, and W were found to be characteristic for chiro-inositol and thus can be indicators of the simulated bioconversion process. In the spectrum in Figure 2b, it can observed that ions at m/z 244.2 and 256.2 disappeared. In case of myo-inositol, characteristic signals also appeared as well as in the case of L-*chiro*-inositol (Figure 3b,c). Moreover, *chiro*-inositol and *myo*-inositol are supposed to have higher affinity for the formation of DHB and HCCA matrix adducts compared to D-pinitol [30]. Considering saccharides before and after EC assay, fewer products of electrochemical conversion are observed. For D-glucose and D-galactose, only five signals appeared after the electrochemical process (Figures 4c and 6c). In the case of D-fructose (Figure 5c), only one signal can be observed, which appears in the spectrum after the electrochemical process. Moreover, the saccharides have a high ability to form dimers (disaccharides).

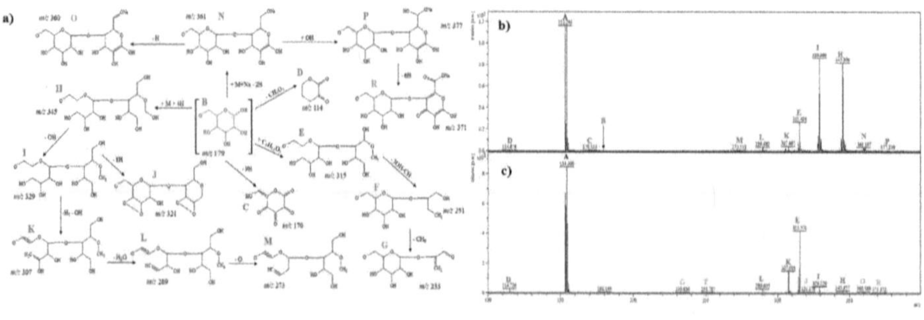

Figure 6. (**a**) Fragmentation pathways of D-galactose; (**b**) MS spectrum before electrochemical process, (**c**) MS spectrum after electrochemical process.

Table 1 presents the fragments of sugars and cyclitols that were suggested to be formed during the electrochemical oxidation and detected by MALDI-TOF-MS. Fragments arising from reactions such as hydrogenation/dehydrogenation and hydroxylation can be considered as possible metabolites, since these are mainly expected metabolic reactions of phase I oxidative cell metabolism. Cross-ring cleavage was also suggested to occur for saccharides. Fragments coming from the association of molecules with sodium (metal adducts), matrix compounds, or units that belong to the core of the matrix's structure (matrix adducts) cannot be considered as potential metabolites that could be produced endogenously. However, such adducts may be valuable for compound identification using the present technique; additionally, metabolic reactions that might occur with the target analyte still can be observed. Therefore, the mentioned fragments were also included in Table 1. Compounds marked with an asterisk have a higher probability to be formed endogenously. Other fragments may refer to less probable candidate metabolites, since MALDI ionization adducts may be formed due to the lower stability of the parent molecule. Once the proposed approach consists of an *off-line* mode, only stable analytes could be analyzed. Nevertheless, the formation of adducts during ionization indicates that these are thermodynamically less stable forms in a gas phase, unlikely acting as effective bioactive compounds in the organism because are prone to further rearrangement. Thus, it is important to consider that candidate metabolites created in the electrochemical cell were identified in this study by their ionization in a gas phase. Subsequently, the study is addressed to model potential metabolic reactions that may occur in phase I of cell metabolism. Enzymatic reactions also should be taken into account to propose metabolites for studied analytes similarly to the application of state-of-the-art *on-line* EC-LC-MS approach.

Table 1. Fragments electrochemically generated for cyclitols and sugars, characterized by different *m/z* values and the proposed involved reaction (* = fragments referring to candidate metabolites once these were assumed to be possibly produced in a biological system).

Compounds	Fragment	*m/z*	Ionization	Proposed Reaction
D-pinitol	$C_7H_6O_5$	172.1	+	OH − 7H loss (dehydroxylation/dehydrogenation)
	* $C_6H_{12}O_8$	212.1	+	CH_3 loss − 2 OH gain (hydroxylation)
	$C_{13}H_{16}O_7$	284.3	+	CH_3 loss − Benzaldehyde gain
	$C_{13}H_{17}O_{10}$	333.0	+	Cyclohexane + H_2O + 3O gain (hydration/reduction)
	$C_{13}H_{16}NaO_9$	339.0	+	CH_3 loss − DHB + Na gain (adduct with matrix and metal)
	$C_{13}H_{14}O_{11}$	346.9	+	Cyclohexane + 5O gain (reduction)
	$C_{13}H_{22}O_{11}$	354.9	+	Cyclohexane + 4OH + O gain (hydroxylation/reduction)
	$C_{13}H_{23}NaO_{10}$	361.9	+	D-pinitol + Na gain − OH − CH_3 loss (dimerization/adduct with metal/dehydroxylation)
	$C_{14}H_{26}O_{11}$	370.9	+	D-pinitol gain (dimerization)
	$C_{15}H_{27}O_{11}$	383.9	+	D-pinitol + $=CH_2$ gain (dimerization)
	$C_{13}H_{22}NaO_{10}$	360.3	−	D-pinitol + Na gain − OH − CH_3 loss (dimerization/adduct with metal/dehydroxylation)
	$C_{20}H_{36}O_{12}$	467.1	−	D-pinitol + phenol gain (dimerization)
	$C_{20}H_{35}NaO_{12}$	490.2	−	D-pinitol + phenol + Na gain (dimerization/adduct with metal)
	$C_{20}H_{35}NaO_{14}$	522.9	−	D-pinitol + phenol + Na + 2OH gain (dimerization/adduct with metal/hydroxylation)
	$C_{27}H_{48}O_{18}$	660.0	−	2D-pinitol + phenol + OH gain (dimerization/hydroxylation)
L-chiro-inositol	$C_6H_4O_6$	172.0	+	8H loss (dehydrogenation)
	$C_6H_8NaO_6$	199.0	+	Na gain − 4H loss (adduct with metal/dehydrogenation)
	$C_{13}H_{16}O_7$	284.3	+	Benzaldehyde gain
	$C_{13}H_{17}O_9$	317.0	+	DHB gain (adduct with matrix)
	$C_{13}H_{15}O_{10}$	331.0	+	DHB + OH gain (adduct with matrix/hydroxylation)
	$C_{13}H_{16}NaO_9$	339.0	+	DHB + Na gain (adduct with matrix and metal)
	$C_{13}H_7NaO_{10}$	346.9	+	DHB + Na + OH gain − 10H loss (adduct with matrix and metal/hydroxylation/ dehydrogenation)
	$C_{13}H_{15}NaO_{10}$	354.8	+	DHB + Na + OH gain (adduct with matrix and metal/hydroxylation)

Table 1. *Cont.*

Compounds	Fragment	m/z	Ionization	Proposed Reaction
L-chiro-inositol	$C_{13}H_{21}NaO_{10}$	360.9	+	DHB + Na + OH gain + 4H (adduct with matrix and metal/hydroxylation/hydrogenation)
	$C_{14}H_{26}O_{11}$	370.9	+	L-chiro-inositol + 2CH_3 gain (dimerization)
	$C_{13}H_{21}NaO_{11}$	376.9	+	DHB + Na + 2OH gain − 4H (adduct with matrix and metal/hydroxylation/ dehydrogenation)
	$C_{14}H_{23}O_{12}$	383.9	+	L-chiro-inositol + C_2H_4 + =O gain (dimerization/reduction)
	* $C_6H_{10}O_6^{\cdot}$	178.8	−	H loss (dehydrogenation)
	$C_{12}H_{15}O_6$	255.9	−	Benzene gain
	$C_{12}H_{15}O_7$	271.9	−	Benzene + OH gain (hydroxylation)
	$C_{12}H_{15}O_{10}$	319.9	−	Benzene + 4OH gain (hydroxylation)
	$C_{18}H_{31}O_{13}$	455.9	−	L-chiro-inositol + phenol + OH gain (dimerization/hydroxylation)
	$C_{24}H_{37}O_{19}$	629.0	−	2L-chiro-inositol + phenol + 2O gain (dimerization/reduction)
	$C_{24}H_{41}O_{20}$	649.0	−	2L-chiro-inositol + phenol + 3OH gain (dimerization/hydroxylation)
Myo-inositol	$C_6H_4O_6$	172.0	+	8H loss (dehydrogenation)
	$C_6H_8NaO_6$	199.0	+	Na gain − 4H loss (adduct with metal/dehydrogenation)
	* $C_6H_{12}O_8$	212.1	+	2OH gain (hydroxylation)
	$C_{13}H_{16}O_7$	284.3	+	Benzaldehyde gain
	$C_{13}H_{15}O_8$	299.0	+	Benzene + OH + C=O gain (hydroxylation/reduction)
	$C_{13}H_{12}O_{10}$	328.9	+	DHB + OH gain − 4H loss (hydroxylation/dehydrogenation)
	$C_{12}H_{19}O_{11}$	339.0	+	Cyclohexane + 2OH + =2O + H_2O gain (hydroxylation/reduction/hydration)
	$C_{13}H_7NaO_{10}$	346.9	+	DHB + Na + OH gain − 10H loss (adduct with matrix and metal/hydroxylation/ dehydrogenation)
	$C_{13}H_{21}NaO_{10}$	360.9	+	DHB + Na + OH gain + 4H (adduct with matrix and metal/hydroxylation/ hydrogenation)
	$C_{12}H_{21}O_6$	261.7	−	Cyclohexane gain
	$C_{12}H_{21}O_{10}$	325.5	−	Cyclohexane + 4OH gain (hydroxylation)
	$C_{18}H_{31}O_{13}$	455.2	−	Myo-inositol + Cyclohexane + 2OH gain (dimerization/hydroxylation)
	$C_{18}H_{31}O_{16}$	503.2	−	2Myo-inositol gain (dimerization)
	$C_{24}H_{40}NaO_{18}$	639.0	−	2Myo-inositol + Cyclohexane + 2OH + Na gain (dimerization/hydroxylation/adduct with metal)
	$C_{24}H_{39}Na_2O_{18}$	661.0	−	2Myo-inositol + Cyclohexane + 2OH + 2Na gain (dimerization/hydroxylation/adduct with metal)

Table 1. *Cont.*

Compounds	Fragment	*m/z*	Ionization	Proposed Reaction
D-glucose	* $C_7H_{13}O_6$	193.0	+	$=CH_2$ gain
	* $C_6H_{12}O_8$	212.9	+	2OH gain (hydroxylation)
	$C_{12}H_{16}O_8$	288.3	+	Benzene + 2OH gain (hydroxylation)
	$C_{12}H_{22}O_9$	310.0	+	Cyclohexane + 3OH gain (hydroxylation)
	$C_{12}H_{22}NaO_8$	317.0	+	Cyclohexane + H_2O + Na + O gain (hydration/adduct with metal/oxidation)
	$C_{12}H_{20}NaO_9$	331.9	+	Cyclohexane + H_2O + Na + 2O gain (hydration/adduct with metal/oxidation)
	$C_{13}H_{23}O_{10}$	339.0	+	Cyclohexane + 3OH + H_2O + CH_2 gain (hydroxylation/hydration)
	* $C_7H_{13}O_8$	225.0	−	CH + 2OH gain (hydroxylation)
	$C_{10}H_{19}O_{11}$	315.4	−	D-glucose gain − $2CH_2$ (dimerization, cross-ring cleavage)
	$C_{12}H_{17}O_{10}$	321.4	−	D-glucose gain − OH − 3H loss (dimerization/dehydroxylation/dehydrogenation)
	$C_{18}H_{31}O_{13}$	455.1	−	D-glucose + Cyclohexane + 2OH gain (dimerization/hydroxylation)
	$C_{24}H_{31}O_{21}$	655.9	−	3D-glucose gain − 10H loss (dimerization/dehydrogenation)
	$C_{24}H_{35}O_{21}$	659.9	−	3D-glucose gain − 6H loss (dimerization/dehydrogenation)
	$C_{24}H_{41}O_{21}$	665.9	−	3D-glucose gain (dimerization)
D-fructose	* $C_6H_8O_3$	128.1	+	3OH − 4H loss (dehydroxylation /dehydrogenation)
	$C_6H_4O_6$	172.0	+	8H loss (dehydrogenation)
	* $C_7H_{13}O_6$	193.0	+	CH_2 gain
	$C_{10}H_{18}O_7$	250.9	+	Tetrahydrofuran gain
	$C_{11}H_{20}O_7$	264.0	+	Tetrahydrofuran + CH_3 gain
	$C_{11}H_{20}O_8$	280.0	+	Tetrahydrofuran + CH_2OH gain
	$C_{11}H_{20}O_9$	296.0	+	Tetrahydrofuran + CH_2OH + OH gain (hydroxylation)
	$C_{12}H_{18}O_{11}$	338.0	+	D-fructose gain − 4H loss (dimerization/dehydrogenation)
	$C_{12}H_{18}O_{13}$	370.9	+	D-fructose + 2OH gain − 4H loss (dimerization/hydroxylation/dehydrogenation)
	$C_{17}H_{29}O_{13}$	441.2	−	3D-fructose gain − 3OH − C loss (dimerization/dehydroxylation)
	$C_{17}H_{29}O_{14}$	457.1	−	3D-fructose gain − 2OH − C loss (dimerization/dehydroxylation)
	$C_{18}H_{27}O_{15}$	483.1	−	3D-fructose gain − OH − 4H loss (dimerization/dehydroxylation/dehydrogenation)
	$C_{23}H_{38}NaO_{18}$	625.0	−	4D-fructose + Na gain − 3OH − CH_2 loss (dimerization/adduct with metal/ dehydroxylation)
	$C_{24}H_{38}NaO_{19}$	653.0	−	4D-fructose + Na gain − 2OH loss (dimerization/adduct with metal/ dehydroxylation)
	$C_{24}H_{37}O_{21}$	661.9	−	4D-fructose gain − 4H loss (dimerization/ dehydrogenation)

Table 1. *Cont.*

Compounds	Fragment	m/z	Ionization	Proposed Reaction
	$C_6H_4O_6$	172.0	+	8H loss (dehydrogenation)
	$C_{12}H_{22}O_6$	262.0	+	Cyclohexane gain
	$C_{12}H_{18}O_8$	284.3	+	Benzene + 2OH gain − 10H loss (hydroxylation/dehydrogenation)
	$C_{12}H_{22}O_8$	294.0	+	Cyclohexane + 2OH gain (hydroxylation)
	$C_{12}H_{22}O_9$	310.0	+	Cyclohexane + 3OH gain (hydroxylation)
	$C_{12}H_{20}NaO_9$	331.0	+	Cyclohexane + H_2O + Na + O gain (hydration/adduct with metal/reduction)
	$C_{13}H_{23}O_{10}$	339.0	+	Cyclohexane + 3OH + H_2O + CH_2 gain (hydroxylation/hydration)
	* $C_9H_{13}O_7$	233.8	−	C_3HOH gain
D-galactose	* $C_{10}H_{19}O_7$	251.7	−	C_4H_7OH gain (cross-ring cleavage)
	$C_{12}H_{17}O_{10}$	321.3	−	D-galactose gain − OH − 4H loss (dimerization/dehydroxylation/dehydrogenation)
	$C_{12}H_{12}NaO_{12}$	371.3	−	D-galactose + Na + O gain − 6H loss (dimerization/adduct with metal/reduction/ dehydrogenation)
	$C_{18}H_{31}O_{13}$	455.2	−	D-galactose + Cyclohexane + 2OH gain (dimerization/hydroxylation)
	$C_{18}H_{32}O_{14}$	472.2	−	D-galactose + Cyclohexane + 3OH gain (dimerization/hydroxylation)
	$C_{24}H_{31}O_{21}$	655.1	−	3D-galactose gain − 10H loss (dimerization/dehydrogenation)
	$C_{24}H_{37}O_{21}$	661.1	−	3D-galactose gain − 4H loss (dimerization/dehydrogenation)
	$C_{24}H_{41}O_{21}$	665.0	−	3D-galactose gain (dimerization)

The application of MALDI-TOF-MS for the analysis of sugar and cyclitol transformations is a new approach in the identification of candidate metabolites. Additionally, little is known regarding the elucidation of the fragmentation pathways of these products or the determination of such compounds in real samples. Only a few papers reporting on the application of the MALDI technique for the determination of sugars and cyclitols can be found in the literature [30–35].

2.3. Data Analysis Approaches

Network analysis (NA) depicted relationships (edges) among groups of studied compounds (nodes), allowing a visualization of the ions (generated MS fragments) concurrent between sugars and cyclitols and their correspondent variants, which were submitted to the electrochemical method (Figure 7). Overlapping ions refer to ions connected to more than one node, while ions specifically incident in the MS spectrum of a given analyte are represented by loose edges coming from a corresponding node. Each of the species (colored blocks) emit lines, which can be connected to a second compound, demonstrating that they have that feature in common. For example, in Figure 7a, *m/z* 268 is present in the MS spectra of D-pinitol, D-galactose, *myo*-inositol, and also the D-fructose variant after EC. On the other hand, *m/z* 228 is a fragment that is exclusively found in D-pinitol, once the edge representing this fragment is emitted by the D-pinitol node and is not connected to any other node. In this way, fragments situated in the center of the net are shared between most of the nodes surrounding it; thus, they are shared between many species. Those located at the periphery of the net are more exclusive fragments that are less recurrent in the MS spectra of analytes. Concurrent fragments were evidenced in the case of all used sets of analysis parameters, suggesting a relevant relationship between the fragmentation patterns of the different investigated compounds. This may be due to the similarities in their original structures and in the processes of adduct formation. Regarding the DHB matrix, the positive ionization mode yielded more correlated spectra between the original and electrochemically processed analytes, whereas for the HCCA matrix, the opposite is observed. Concerning the observation of distinctive ions with greater intensity, these were hindered in the case of EC-treated analytes when the DHB matrix was used. Such an aspect may hamper the differentiation of such compounds in impurified samples.

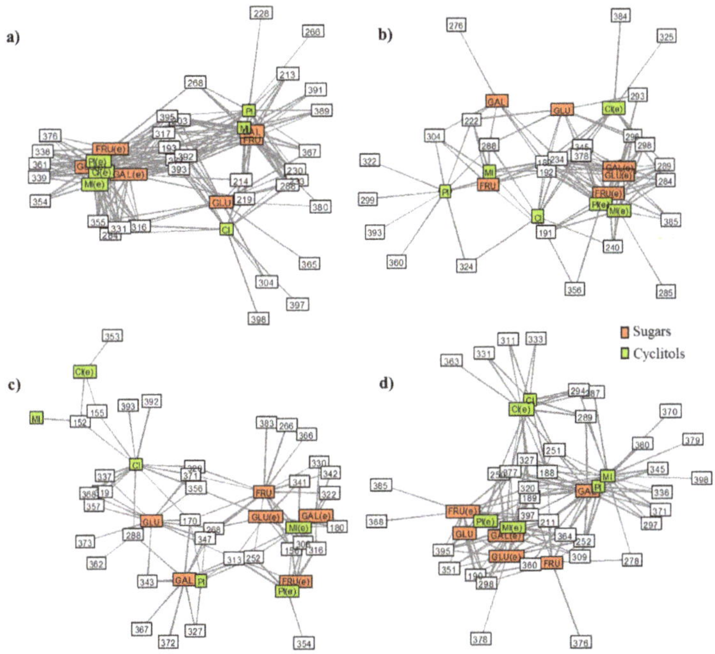

Figure 7. Networks built based on MS spectra obtained using (**a**) 2,5-dihydroxybenzoic acid (DHB) and (**b**) α-cyano-4-hydroxycinnamic acid (HCCA) matrices, in positive ionization mode; (**c**) DHB and (**d**) HCCA matrices, in negative ionization mode. Colored nodes: compounds, white rectangles: MS ions, e—compounds subjected to the electrochemical process, FRU: D-fructose, GAL: D-galactose, GLU: D-glucose, PI: D-pinitol, CI: L-*chiro*-inositol, and MI: *myo*-inositol.

The relative distance between the nodes also indicates their level of correlation based on the number of interconnected variables. For example, more intricate networks are formed by using DHB in the positive mode and HCCA in the negative mode. This shows that in these conditions, less unique MS spectra are obtained for the investigated species in terms of incident ions. Additionally, different patterns of correlation are observed, depending on the used analytical parameters. However, there is a more prominent grouping between species submitted to the electrochemical assay, indicating that the species obtained from this process share more fragmentation products with considerable intensity.

Considering the analysis in the positive mode and using the DHB matrix, the ions shared between saccharides and cyclitols—original and artificially metabolized—were *m*/*z* 377 and *m*/*z* 393, with *m*/*z* 193 being absent solely in the D-glucose spectrum. The fragment *m*/*z* 377 corresponds to the association of a sodium adduct of methylated $[M + H]^+$ ion with a DHB unit (in the case of D-pinitol, it corresponds to the demethylated sodium adduct of the dimerized form); *m*/*z* 393 can be addressed as a precursor of the later fragment, due to an oxygen loss in the structure. The ion *m/z* 193 may be ascribed to $[C_7H_{12}O_6 + H]^+$ or $[C_6H_8O_7 + H]^+$ fragments, the first one preferably generated by cyclitols and the second one by D-fructose and D-galactose.

The fragments *m/z* 361, 338, 376, and 339 were coincident among all species after the electrochemical treatment. The ion *m/z* 343 can be formed by dimerization; in D-pinitol, it refers to the dimer after the loss of 2H and one methyl group; *m/z* 361 can be related to the latter after the gain of one hydroxyl group and one proton (or simply, in the case of L-*chiro*-inositol and *myo*-inositol, to $[2M + H]^+$); *m/z* 376 can be related to the gain of one oxygen and one hydroxyl molecule; *m/z* 339 can be related to the abstraction of 4H (in D-pinitol and *myo*-inositol, *m/z* 339 may also be generated from $[M + H]^+$ by the combined associations: + M + Na + DHB unit). Finally, *m/z* 338 can be produced by the subtraction of oxygen in ion *m/z* 354 in cyclitols and from the subtraction of 4H in the ion *m/z* 342 in sugars (generated by $[M+H]^+$ combined with one more molecule).

Ions such as *m/z* 230, 233, and 288 were coincident solely in both species before electrochemical processing. The fragment *m/z* 203 can be connected with the ion $[M + Na]^+$ (in cyclitols, it can be obtained by the scission of $[2M + H]^+$ and subsequent combination with sodium and abstraction of a hydrogen atom), which in combination with an ethyl group yields *m/z* 230. The same process having as its precursor $[M + H + Na]^+$ may lead to *m/z* 233. The ion *m/z* 288 may refer to the association of a cyclitol molecular ion (or its directed derivative -H, for D-pinitol) together with a benzaldehyde unit. In saccharides, the association of $[M + H]^+$ with the moiety $C_3H_8O_3$ can lead to *m/z* 273 in the spectrum, in which methylation can give rise to *m/z* 288.

Using the HCCA matrix, in the positive mode, the ions *m/z* 192 and *m/z* 189 appear to be shared among both original and treated analytes. $[M + H]^+$(or scission product of cyclitol core in $[2M + H]^+$) suffers water loss and if combined with the formyl group, it produces the fragment *m/z* 192. If associated with the loss of a CN group from the matrix, the ion *m/z* 189 is produced. Ions *m/z* 284 and *m/z* 289 were common for all species subjected to the electrochemical process: the first can be obtained by the addition of moiety $C_4H_7O_3$ to the molecular ion and the second can be obtained by association of the latter with the benzyl alcohol portion of the 4-HCCA molecule. Still regarding this analysis method, *m/z* 288 again appears recurrent in all analytes in their original form. In negative mode analysis, any characteristic ions were found regarding the groups of studied chemical species.

HCA (hierarchical cluster analysis) (Figure 8a,b) provides information regarding sample grouping with its basis on the calculated dissimilarity coefficient. The similarity between two MS spectra is measured according to a parameter, in this case, the Spearman coefficient. The level of correspondence between a pair of profiles that are being compared is related to height of the dendrogram: the shorter the height, the greater the similarity. In this way, in HCA, the assays are spatially organized according to their level of congruence with each other. Correspondence between the obtained MS spectra did not enable a clear discrimination between saccharides and cyclitols species once these did not form an isolated cluster for each type of ionization mode. However, compounds after EC appeared grouped together, indicating that analytes subjected to electrochemical assay generated more coincident fragments, and thus, coincident possible metabolites. Figure 8c presents a chart with ions incident solely in the MS spectrum of one of the analytes, showing that distinctive ions were not detected for most of the performed assays, as expected due to the structural similarity of analytes. The formation of an ion at *m/z* 152 $[C_5H_{11}O_5 + H]^+$ was favored for L-*chiro*-inositol, it consists of a molecular ion after dehydrogenation and decarboxylation. The ion at *m/z* 207 may refer to the gain of –CH and the loss of a proton in the region of D-pinitol's methyl radical, leading to an epoxy structure, while *m/z* 228 can be obtained after the loss of a methyl group, followed by oxidation (the gain of two atoms of oxygen and one hydroxyl group). For *myo*-inositol, a signal at *m/z* 325 appears as a specific fragment that was probably generated by a loss of water followed by the loss of CH_2O in a molecule dimer. D-glucose has associated the ion at *m/z* 256, which may arise from the combination of the original structure with $C_3H_7O_2$. The methylation and dehydrogenation of a glucose dimer can result in the fragment at *m/z* 356; at the same time, *m/z* 330 in the D-fructose MS spectrum can be a product of hydroxymethyl group loss followed by hydroxyl gain. D-galactose ion *m/z* 276 can be generated by the loss of two molecules of water and one $-CH_2O$ group in the structure of the dimer (*m/z* 343). Species that were electrochemically transformed also presented distinguishable fragments. The ion at *m/z* 384 can be a product of L-chiro-inositol dimerization with posterior dehydrogenation and the incorporation of sodium. The fragment *m/z* 274 from D-pinitol may result from *m/z* 387 $[2M-H]^+$, followed by the loss of two $-CH_2O$ groups and three molecules of water. Transformed *myo*-inositol is characterized by the *m/z* 285 ion, which may refer to $[2M-H]^+$ after the loss of four molecules of water. In negative mode, transformed L-chiro-inositol was associated to the formation of a signal at *m/z* 188, which may correspond to $[C_7H_8O_6]^-$ after methylation and successive dehydrogenations. Although investigated analytes are structurally similar and display corresponding molecular weights, the verification of ions specific for some of the studied species suggests that some mechanisms related to molecule rearrangement and fragmentation may be favored to the detriment of others in circumstances of factors such as the steric effect.

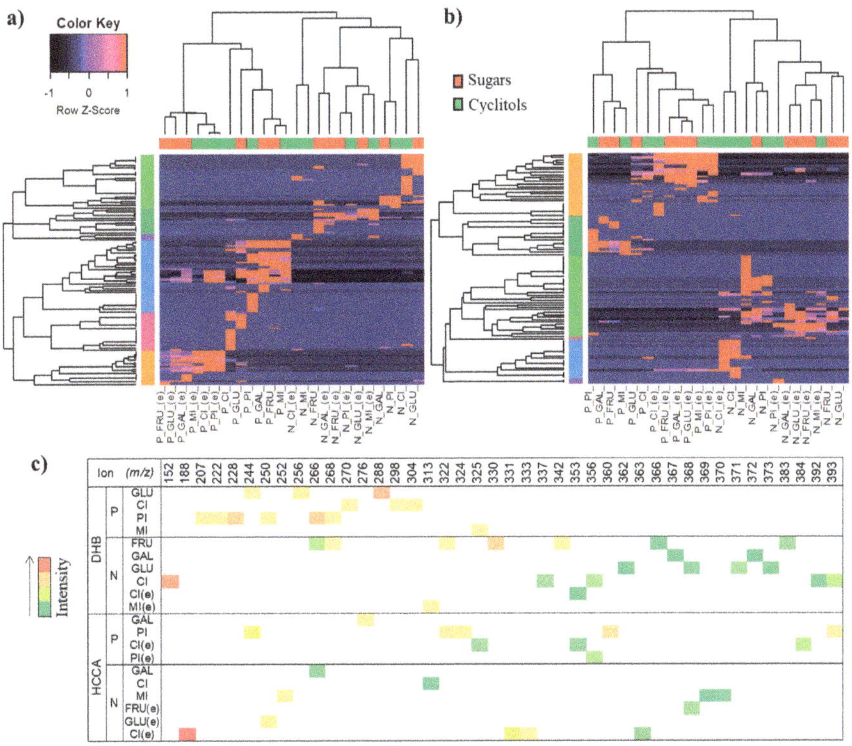

Figure 8. Heatmaps associated to hierarchical cluster analysis (HCA), which were generated for analysis using (**a**) DHB and (**b**) HCCA matrix; (**c**) chart showing ions specifically incident in the spectrum of one of the analytes. P: positive mode, N: negative mode, e—compounds subject to electrochemical process, FRU: D-fructose, GAL: D-galactose, GLU: D-glucose, PI: D-pinitol, CI: L-chiro-inositol, MI: *myo*-inositol.

Figure 9 presents a scatter plot of values of fold change in ion intensity after compounds were electrochemically treated. All the represented ions are those that suffered statistically relevant changes in their intensities in comparison to the EC-treated form. The increasing size of plotted variables indicates their increasing level of significance (expressed in terms of common logarithm of *p* value). Considering this, bigger dots refer to fragments associated to smaller p-values. Colors of features represent four main responses in the spectra obtained after modification of the products: ions that have become absent after the EC process, those that have become incident, and those that increased or decreased their intensities after EC treatment. The graphs allow to observe that, in positive mode, the DHB matrix provided spectra with a greater number of discriminating ions. In negative mode, only a few significant changes between the compounds' spectra before and after treatment were observed. This aspect is probably related to a greater dissimilarity occurring among the spectra of sugars and cyclitols in negative mode analysis. The existence of such discriminating variables denotes the resemblance of MS profiles belonging to the two investigated groups of compounds, a factor that can hinder a proper differentiation between them. However, these features may serve as indicative species that are useful in the monitoring of the metabolization process of the studied compounds. Together with previously presented evidence, the compounds subject to the electrochemistry step tend to present MS spectra containing a greater variety of ions addressed to the total or partial combination of molecular ions with moieties of itself or with matrix compounds. In general, the addressed fragments involve dimerization, alterations, and sodium adduct formation in the dimer structure as well as combination with DHB and 4-HCCA molecules. This suggests that the transformed varieties of analytes tend to be more prone to recombination, due to their increased reactivity. After processing, the MS spectra similarity increases and the number of diverse fragments decreases, pointing to an allowed interconversion of these species into coincident structures, which can refer to ions marked in Figure 9 with "incident" behavior after EC assay. Ions that had their intensities significantly decreased are related to fragments coming from minor transformations in the molecular ion unit. Those related to punctual chemical alterations of molecular ions, or from the fragmentation of original structure itself, tend to appear "absent" after EC assay. Such approach provides an overview of the significance of possible metabolites. The conversion of bioactive compounds as a part of metabolism by cells is important to consider for biological activity studies and thus, new approaches for the simulation of cell metabolism should be proposed. The developed *off-line* EC-MALDI-TOF-MS approach can find its application in the metabolomics studies for the identification of candidate metabolites of potential drugs. In addition, the approach can be interesting for researchers in the field of natural bioactive compounds since, as it was mentioned above, the identification of metabolites can be valuable for the assessment of biological activity.

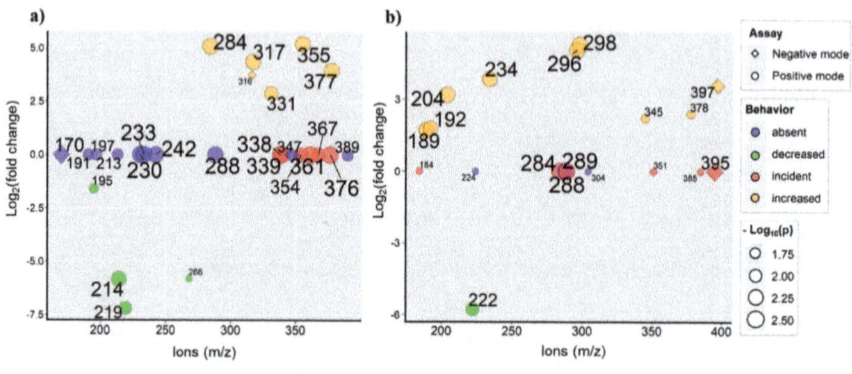

Figure 9. Scatter plots weighted according to the significance of ions (-$\log_{10}$ (p)), referring to a comparison between MS spectra obtained before and after the electrochemical process, using (**a**) DHB and (**b**) the HCCA matrix.

3. Materials and Methods

3.1. Chemicals and Reagents

D-pinitol (95% mass), L-*chiro*-inositol (95% mass), *myo*-inositol (≥98% mass), D-glucose, D-fructose, and D-galactose were purchased from Sigma-Aldrich (Steinheim, Germany). Water, acetonitrile, methanol (all solvents were of LC-MS grade purity), trifluoroacetic acid (≥99% purity), and ammonium acetate (≥99% purity) were purchased from Sigma-Aldrich (Steinheim, Germany). MALDI matrices such as 2,5-dihydroxybenzoic acid (DHB), α-cyano-4-hydroxycinnamic acid (HCCA) as well as cesium triiodide for mass calibration were purchased from Sigma-Aldrich (Steinheim, Germany). Water was obtained with a Milli-Q RG apparatus by Millipore (Millipore Intertech, Bedford, MA, USA).

3.2. Sample Preparation

Stock solutions of standards were prepared by dissolving standards of cyclitols and saccharides in water (Milli-Q). Solutions for subjection to electrochemical conversion were prepared at neutral pH by dissolving 50 μL of standard in 5 mL of 10 mM ammonium acetate and 2 mL of acetonitrile to a final concentration of 10 μg/mL. Control samples that were not subjected to EC conversion were prepared by the same procedure. Two fractions of each compound after EC conversion were collected to 2-mL Eppendorf tubes (the first fraction was collected during the first 10 min and the second was collected during the next 7 min). After collection of the fractions, the solvent was evaporated and frozen in −80 °C. Prior to analysis, samples were re-dissolved in 50 μL of methanol–water (1:1) mixture.

3.3. Instrumentation

The ROXY™EC system (Antec, Zoeterwoude, The Netherlands) was used for the electrochemical conversion of analytes. The system consisted of a potentiostat equipped with a ReactorCell™ with a reference electrode HyREF™ (Pd/H_2) and working electrode (boron-doped diamond electrode, BDD), an infusion pump (Harvard, Holliston, MA, USA), and all necessary capillary tubes. The Dialogue™ version 2.02.145 software (Antec, Zoeterwoude, The Netherlands) was used. The BDD electrode consisted of an ultra-thin layer deposited on top of a silicon substrate in a potential range from 0 to 3000 mV. Additionally, an electrochemical three-electrode arrangement was composed of a Pd counter electrode and a HyREF (Pd/H_2) reference electrode. All working solutions of cyclitols were freshly prepared from stock solutions at final concentrations of 10 μg/mL. Working solutions of cyclitols and saccharides were prepared using 10 mM ammonium acetate (pH 7.4) with the addition of 2 mL of acetonitrile and injected to the system using a 1-mL syringe (Hamilton, Reno, NV, USA). The analysis was performed using a flow rate of 10 μL/min and 37 °C oven temperature. All the fractions and controlled samples were transferred to 2-mL Eppendorf tubes and analyzed by MALDI-TOF-MS (Bruker Daltonics, Bremen, Germany).

3.4. MALDI-TOF-MS

The samples of cyclitols and saccharides were analyzed by an ultra-fleXtreme MALDI-TOF-MS instrument (Bruker Daltonics, Bremen, Germany). The instrument is equipped with a modified neodymium-doped yttrium aluminum garnet (Nd:YAG) laser (Smart beam II™) operating at the wavelength of 355 nm and frequency of 2 kHz, which was used for all measurements. The spectra were analyzed in a reflective positive and negative ionization mode in the 60–1600 *m*/*z* range at 80% of laser power and global attenuator at 50%. Fragment spectra were determined using the LIFT mode in the *m*/*z* range of 50–1000. All mass spectra were acquired and processed using software such as Flex Control and Flex Analysis, respectively (Bruker Daltonics, Bremen, Germany).

The matrices such as α-cyano-4-hydroxycinnamic acid (HCCA) and 2,5-dihydroxybenzoic acid (DHB) at a concentration of 10 mg/mL were prepared by dissolving 10 mg of HCCA and DHB in 1 mL of standard solution (30% acetonitrile 70% H_2O and 0.1% trifluoroacetic acid. The mixtures of 1 μL of each sample and 1 μL of matrix solutions were applied to the spot on a MALDI-TOF-MS MTP AnchorChip

384 plate, in triplicate. All mass spectra were calibrated by using the cesium triiodide-cluster (CsI_3); 10 mg of CsI_3 was dissolved in 1 mL of mixtures of methanol–DHB (1:1).

3.5. Exploratory Data Analysis

The following methods were conducted in R environment, using RStudio console v. 1.1.463 (RStudio, Boston, MA, USA) and employing the packages "gplots", "sna", and "ggplot2". NA aimed to represent coincident ions among the spectra of the investigated compounds. HCA using the Spearman rank correlation method was conducted in order to observe the grouping of performed assays according to the distribution of MS ions. For NA, data concerning only ions with at least 1% of the total spectral abundance were considered, and the dataset was converted into binary entries. For HCA input, Z-score normalization of the data was carried out. Finally, weighted scatter plots were intended to highlight the main significant trends presented by MS ions when submitted to the electrochemical method. These results were expressed in terms of binary logarithm of the fold change in ion intensity. Using IBM SPSS Statistics v. 24 (IBM, Armonk, NY, USA), the Mann–Whitney U test was performed, aiming to point out discriminant features between the group of analytes in original form and when submitted to the electrochemical process; a p value < 0.05 was considered as the significance criterion. All prepared datasets comprehended ions within the range of 150 to 400 *m/z*. Matrix spectra ("blanks") were subtracted from analyses, and the average of intensity values coming from fractions 1 and 2 were used. Exploratory data analysis encompassed the general fragments generated by MALDI-TOF-MS analysis, obeying the aforementioned thresholds of intensity and statistical significance.

4. Conclusions

The study used electrochemical analysis with MALDI-TOF-MS detection for the first time to identify candidate metabolites of biologically active compounds. Using the *off-line* EC-MALDI-TOF-MS method, it was possible to identify and characterize the EC-generated profile of cyclitols (pinitol—*m/z* 346.9, 354.9, 361.9, 370.9, chiro-inositol—*m/z* 317.0, 331.0, 376.9, *myo*-inositol—*m/z* 172.0, 199.0, 212.1, 284.3, 299.0, 328.9, 339.0, 346.9, 360.9) and sugars (glucose—*m/z* 225.0, 315.4, 321.4, 345.3, fructose—*m/z* 329.4, galactose—*m/z* 233.8, 251.7, 321.3, 360.3). In addition, the identified products were characteristic for the sugars and cyclitols tested. The use of electrochemistry is a new approach for the simulation and detection of possible bioconversion products using electrochemical reactions. In addition, the use of statistical analysis showed differences/similarities between MS spectra acquired for compounds before and after the electrochemical process for the two tested matrices. EC assay products provided less distinguishable MS spectra and were mainly characterized by the combination of analytes with more complex structures. The applied method proved to be adequate for the identification and assessment of the analytes and their derived species to model potential metabolic reactions of phase I oxidative cell metabolism for cyclitols and saccharides. Candidate metabolites of the studied substances were indicated, and these results can be introduced in further studies regarding the bioactivity of these compounds, influencing parameters, and in the design of possible pharmacological applications.

Author Contributions: Conceptualization, P.P., J.W.-S., G.S., F.M and B.B.; methodology, J.W.-S., P.P., G.S., F.M.; software, F.M.; validation, F.M., J.W.-S.; formal analysis, F.M.; investigation, J.W.-S., P.P., G.S., F.M.; resources, P.P.; data curation, J.W.-S. and G.S.; writing—original draft preparation, J.W.-S. and F.M.; writing—review and editing, P.P., J.W.-S., G.S., F.M.; visualization, J.W.-S. and F.M.; supervision, B.B.; project administration, P.P. and B.B.; funding acquisition, B.B. All authors have read and agreed to the published version of the manuscript.

Funding: The "Advanced biocomposites for tomorrow's economy BIOG-NET" project (FNP POIR.04.04.00-00-1792/18-00) is carried out within the TEAM-NET programme of the Foundation for Polish Science co-financed by the European Union under the European Regional Development Fund. Additionally, it was supported by Toruń Center of Excellence "Towards Personalized Medicine" operating under Excellence Initiative-Research University (P. Pomastowski, B. Buszewski).

Conflicts of Interest: The authors declare no conflict of interest.

References

1. Ratiu, I.A.; Al-Suod, H.; Ligor, M.; Ligor, T.; Krakowska, A.; Górecki, R.; Buszewski, B. Simultaneous Determination of Cyclitols and Sugars Following a Comprehensive Investigation of 40 Plants. *Food Anal. Methods* **2019**, *12*, 1466–1478. [CrossRef]
2. Evans, C.E.L. Sugars and health: A review of current evidence and future policy. *Proc. Nutr. Soc.* **2017**, *76*, 400–407. [CrossRef]
3. Al-Suod, H.; Ratiu, I.-A.; Górecki, R.; Buszewski, B. Pressurized liquid extraction of cyclitols and sugars: Optimization of extraction parameters and selective separation. *J. Sep. Sci.* **2019**, *42*, 1265–1272. [CrossRef]
4. Ratiu, I.-A.; Al-Suod, H.; Bukowska, M.; Ligor, M.; Buszewski, B. Correlation Study of Honey Regarding their Physicochemical Properties and Sugars and Cyclitols Content. *Molecules* **2019**, *25*, 34. [CrossRef]
5. Al-Suod, H.; Ligor, M.; Ratiu, I.-A.; Rafińska, K.; Górecki, R.; Buszewski, B. A window on cyclitols: Characterization and analytics of inositols. *Phytochem. Lett.* **2017**, *20*, 507–519. [CrossRef]
6. Al-Suod, H.; Ratiu, I.-A.; Ligor, M.; Ligor, T.; Buszewski, B. Determination of sugars and cyclitols isolated from various morphological parts of *Medicago sativa* L. *J. Sep. Sci.* **2018**, *41*, 1118–1128. [CrossRef] [PubMed]
7. Atanasov, A.G.; Waltenberger, B.; Pferschy-Wenzig, E.-M.; Linder, T.; Wawrosch, C.; Uhrin, P.; Temml, V.; Wang, L.; Schwaiger, S.; Heiss, E.H.; et al. Discovery and resupply of pharmacologically active plant-derived natural products: A review. *Biotechnol. Adv.* **2015**, *33*, 1582–1614. [CrossRef]
8. Wang, H.-Y.; Zhao, Z.; Guo, Y. Chemical and Biochemical Applications of MALDI TOF-MS Based on Analyzing the Small Organic Compounds. *Top. Curr. Chem.* **2012**, *331*, 165–192. [CrossRef]
9. Chakraborty, P.; Pradeep, T. The emerging interface of mass spectrometry with materials. *NPG Asia Mater.* **2019**, *11*, 1–22. [CrossRef]
10. Weidner, S.M.; Falkenhagen, J.; Maltsev, S.; Sauerland, V.; Rinken, M. A novel software tool for copolymer characterization by coupling of liquid chromatography with matrix-assisted laser desorption/ionization time-of-flight mass spectrometry. *Rapid Commun. Mass Spectrom.* **2007**, *21*, 2750–2758. [CrossRef]
11. Clark, A.E.; Kaleta, E.J.; Arora, A.; Wolk, D.M. Matrix-Assisted Laser Desorption Ionization-Time of Flight Mass Spectrometry: A Fundamental Shift in the Routine Practice of Clinical Microbiology. *Clin. Microbiol. Rev.* **2013**, *26*, 547–603. [CrossRef] [PubMed]
12. Heap, R.E.; Segarra-Fas, A.; Blain, A.P.; Findlay, G.M.; Trost, M. Profiling embryonic stem cell differentiation by MALDI TOF mass spectrometry: Development of a reproducible and robust sample preparation workflow. *Analyst* **2019**, *144*, 6371–6381. [CrossRef] [PubMed]
13. Clark, C.M.; Costa, M.S.; Sanchez, L.M.; Murphy, B.T. Coupling MALDI-TOF mass spectrometry protein and specialized metabolite analyses to rapidly discriminate bacterial function. *Proc. Natl. Acad. Sci. USA* **2018**, *115*, 4981–4986. [CrossRef] [PubMed]
14. Bronzel, J.L.; Milagre, C.D.; Milagre, H.M.; Jr, J.L.B. Analysis of low molecular weight compounds using MALDI- and LDI-TOF-MS: Direct detection of active pharmaceutical ingredients in different formulations. *J. Mass Spectrom.* **2017**, *52*, 752–758. [CrossRef] [PubMed]
15. Calvano, C.D.; Monopoli, A.; Cataldi, T.R.I.; Palmisano, F. MALDI matrices for low molecular weight compounds: An endless story? *Anal. Bioanal. Chem.* **2018**, *410*, 4015–4038. [CrossRef] [PubMed]
16. Enomoto, H.; Sato, K.; Miyamoto, K.; Ohtsuka, A.; Yamane, H. Distribution Analysis of Anthocyanins, Sugars, and Organic Acids in Strawberry Fruits Using Matrix-Assisted Laser Desorption/Ionization-Imaging Mass Spectrometry. *J. Agric. Food Chem.* **2018**, *66*, 4958–4965. [CrossRef] [PubMed]
17. Mechref, Y.; Novotny, M.V.; Krishnan, C. Structural Characterization of Oligosaccharides Using Maldi-TOF/TOF Tandem Mass Spectrometry. *Anal. Chem.* **2003**, *75*, 4895–4903. [CrossRef]
18. Grant, D.C.; Helleur, R. Rapid screening of anthocyanins in berry samples by surfactant-mediated matrix-assisted laser desorption/ionization time-of-flight mass spectrometry. *Rapid Commun. Mass Spectrom.* **2007**, *22*, 156–164. [CrossRef]
19. Morisasa, M.; Sato, T.; Kimura, K.; Mori, T.; Goto-Inoue, N. Application of Matrix-Assisted Laser Desorption/Ionization Mass Spectrometry Imaging for Food Analysis. *Foods* **2019**, *8*, 633. [CrossRef]
20. Portychová, L.; Schug, K.A. Instrumentation and applications of electrochemistry coupled to mass spectrometry for studying xenobiotic metabolism: A review. *Anal. Chim. Acta* **2017**, *993*, 1–21. [CrossRef]
21. Adams, K.L.; Puchades, M.A.; Ewing, A.G. In Vitro Electrochemistry of Biological Systems. *Annu. Rev. Anal. Chem.* **2008**, *1*, 329. [CrossRef] [PubMed]

22. Szultka-Młyńska, M.; Buszewski, B. Electrochemical oxidation of selected immunosuppressants and identification of their oxidation products by means of liquid chromatography and tandem mass spectrometry (EC-HPLC-MS/MS). *J. Pharm. Biomed. Anal.* **2019**, *176*, 112799. [CrossRef] [PubMed]
23. Szultka-Młyńska, M.; Bajkacz, S.; Kaca, M.; Baranowska, I.; Buszewski, B. Electrochemical simulation of three novel cardiovascular drugs phase I metabolism and development of a new method for determination of them by liquid chromatography coupled with tandem mass spectrometry. *J. Chromatogr. B* **2018**, 100–112. [CrossRef]
24. Bruins, A.P. An overview of electrochemistry combined with mass spectrometry. *TrAC Trends Anal. Chem.* **2015**, *70*, 14–19. [CrossRef]
25. Potęga, A.; Garwolińska, D.; Nowicka, A.M.; Fau, M.; Kot-Wasik, A.; Mazerska, Z. Phase I and phase II metabolism simulation of antitumor-active 2-hydroxyacridinone with electrochemistry coupled on-line with mass spectrometry. *Xenobiotica* **2019**, *49*, 922–934. [CrossRef] [PubMed]
26. Lecours, M.-A.; Eysseric, E.; Yargeau, V.; Lessard, J.; Brisard, G.; Segura, P.A. Electrochemistry-High Resolution Mass Spectrometry to Study Oxidation Products of Trimethoprim. *Environments* **2018**, *5*, 18. [CrossRef]
27. Guo, Z.; Zhang, Q.; Zou, H.; Guo, B.; Ni, J. A Method for the Analysis of Low-Mass Molecules by MALDI-TOF Mass Spectrometry. *Anal. Chem.* **2002**, *74*, 1637–1641. [CrossRef]
28. Thomas, M.P.; Mills, S.J.; Potter, B.V.L. ChemInform Abstract: The "Other" Inositols and Their Phosphates: Synthesis, Biology, and Medicine (with Recent Advances in myo-Inositol Chemistry). *Angew. Chem. Int. Ed.* **2016**, *47*, 1614–1650. [CrossRef]
29. Harvey, D.J. Analysis of carbohydrates and glycoconjugates by matrix-assisted laser desorption/ionization mass spectrometry: An update for the period 2005–2006. *Mass Spectrom. Rev.* **2010**, *30*, 1–100. [CrossRef]
30. Al-Suod, H.; Pomastowski, P.; Ligor, M.; Railean-Plugaru, V.; Buszewski, B. New approach for fast identification of cyclitols by MALDI-TOF mass spectrometry. *Phytochem. Anal.* **2018**, *29*, 528–537. [CrossRef]
31. Gil, J.H.; Seo, J.; Kim, K.-J.; Jung, J.H.; Jung, O.-S.; Kim, M.-S.; Hong, J. Structural determination of cyclitol derivatives by fast-atom bombardment tandem mass spectrometry. *Rapid Commun. Mass Spectrom.* **2006**, *20*, 1253–1256. [CrossRef]
32. Calvano, C.D.; Cataldi, T.R.I.; Kögel, J.F.; Monopoli, A.; Palmisano, F.; Sundermeyer, J. Structural Characterization of Neutral Saccharides by Negative Ion MALDI Mass Spectrometry Using a Superbasic Proton Sponge as Deprotonating Matrix. *J. Am. Soc. Mass Spectrom.* **2017**, *28*, 1666–1675. [CrossRef] [PubMed]
33. He, Q.; Chen, S.; Wang, J.; Hou, J.; Wang, J.; Xiong, S.; Nie, Z. 1-Naphthylhydrazine hydrochloride: A new matrix for the quantification of glucose and homogentisic acid in real samples by MALDI-TOF MS. *Clin. Chim. Acta* **2013**, *420*, 94–98. [CrossRef] [PubMed]
34. Bald, I.; Flosadottir, H.D.; Kopyra, J.; Illenberger, E.; Ingolfsson, O. Fragmentation of deprotonated d-ribose and d-fructose in MALDI—Comparison with dissociative electron attachment. *Int. J. Mass Spectrom.* **2009**, *280*, 190–197. [CrossRef]
35. Erra-Balsells, R.; Nonami, H. UV-MALDI-TOF MS Analysis of Carbohydrates. Reviewing Comparative Studies Performed Using nor-Harmane and Classical UV-MALDI Matrices. *Environ. Control. Biol.* **2008**, *46*, 65–90. [CrossRef]

International Journal of
Molecular Sciences

MDPI

Review

The Beneficial Effects of Morusin, an Isoprene Flavonoid Isolated from the Root Bark of *Morus*

Dong Wook Choi [1,†,‡], Sang Woo Cho [1,†], Seok-Geun Lee [2,3,*] and Cheol Yong Choi [1,*]

1 Department of Biological Sciences, Sungkyunkwan University, Suwon 16419, Korea; dongwookchoi85@gmail.com (D.W.C.); swcho0628@naver.com (S.W.C.)
2 Department of Science in Korean Medicine, Kyung Hee University, Seoul 02447, Korea
3 KHU-KIST Department of Converging Science & Technology, Kyung Hee University, Seoul 02447, Korea
* Correspondence: seokgeun@khu.ac.kr (S.-G.L.); choicy@skku.ac.kr (C.Y.C.); Tel.: +82-2-961-2355 (S.-G.L.); +82-31-290-7010 (C.Y.C.); Fax: +82-2-961-9623 (S.-G.L.); +82-31-290-7015 (C.Y.C.)
† These authors contributed equally to this work.
‡ Current address: Department of Cancer Biology, Dana-Farber Cancer Institute, Harvard Medical School, Boston, MA 02215, USA.

Received: 25 August 2020; Accepted: 5 September 2020; Published: 7 September 2020

Abstract: The root bark of *Morus* has long been appreciated as an antiphlogistic, diuretic and expectorant drug in Chinese herbal medicine, albeit with barely known targets and mechanisms of action. In the 1970s, the development of analytic chemistry allowed for the discovery of morusin as one of 7 different isoprene flavonoid derivatives in the root bark of *Morus*. However, the remarkable antioxidant capacity of morusin with the unexpected potential for health benefits over the other flavonoid derivatives has recently sparked scientific interest in the biochemical identification of target proteins and signaling pathways and further clinical relevance. In this review, we discuss recent advances in the understanding of the functional roles of morusin in multiple biological processes such as inflammation, apoptosis, metabolism and autophagy. We also highlight recent in vivo and in vitro evidence on the clinical potential of morusin treatment for multiple human pathologies including inflammatory diseases, neurological disorders, diabetes, cancer and the underlying mechanisms.

Keywords: morusin; antioxidants; natural products; cell signaling; inflammatory diseases; neurological disorders; diabetes; cancer

1. Introduction

The therapeutic relevance of the root bark of the mulberry tree (genus *Morus*) as an antiphlogistic, diuretic and expectorant drug has long been acknowledged in Chinese herbal medicine [1]. The rapid development of analytic chemistry in the 1970s led researchers to explore the constituents of this mysterious plant with remarkable clinical potential [2–5]. As a result of these efforts, a group at Toho university in Japan achieved the consecutive purification of two 2-arylbenzofuran derivatives and seven isoprene-substituted flavonoid derivatives from the benzene extracts of the root bark of a variety of plants in the genus *Morus* including *Morus alba* L. and *Morus nigra* L. Such discoveries, taken together with the anti-inflammatory properties of flavonoid species, significantly provoked scientific attention to the potential application of *Morus*-derived flavonoids to a myriad of human pathologies beyond its known benefits, leading to subsequent biochemical studies from multiple research groups in which the chemical and bioactive properties of the phytochemicals were tested.

Morusin is one of the *Morus*-derived flavonoids, which has been highlighted for its outstanding antioxidant capacity over the other flavonoids due to unique chemical and bioactive properties (Figure 1) [6]. Its versatile potential against human pathologies including cancer, immune dysfunction and metabolic disorders has been intensively tested in in vitro systems, although the underlying mechanisms and clinical evidence in vivo are yet to be fully explored. In this review, we describe the previously reported physiological processes induced by morusin, as well as in vitro and in vivo evidence connecting its therapeutic potential to relevant human pathologies including cancer. Finally, we highlight our recent discoveries of mechanisms regarding the strategies employed by cancer cells to resist morusin treatment and propose the cotreatment of morusin and relevant inhibitors as a potential tactic for boosting the anti-tumor capacity of the morusin.

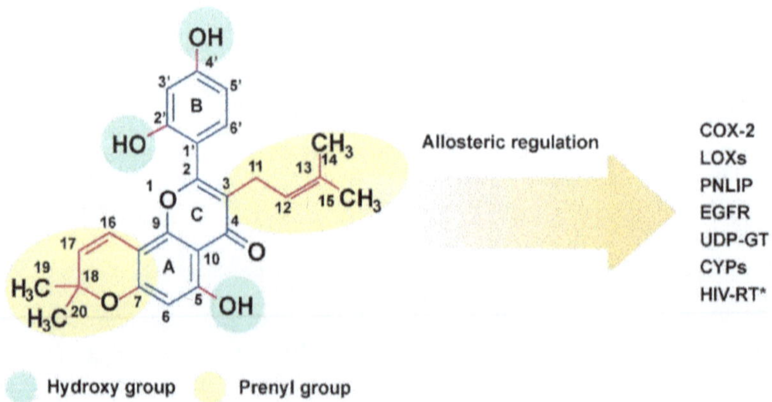

Figure 1. The chemical structure of morusin. Molecular targets of morusin include cyclooxygenase-2 (COX-2), lipoxygenases (LOXs), pancreatic lipase (PNLIP), epidermal growth factor receptor (EGFR), UDP-glucuronosyltransferase, cytochrome P450 (CYP) and HIV reverse transcriptase (* in silico analysis only).

2. Results

2.1. Chemical Properties of Morusin

Various classes of prenylated flavonoids have been isolated from the root bark of *Morus*. Among these, morusin has been highlighted for its versatile effects on human physiology and pathology. Morusin is a prenylated flavone with strong antioxidant capacity, which is structurally characterized by (1) a prenyl unit at position 3, (2) hydroxy groups at 5, 2′′ and 4′ and (3) a 2,2-dimethyl pyran group across positions 7 and 8 (Figure 1) [6]. The structure-bioactivity relationship was analyzed by the comparison of many prenylated flavonoids in the number and position of prenyl moieties [7–10]. In general, attachment of hydroxy group and prenyl group on the flavone backbone affects the bioactivities of a compound depending on the position and number of functional groups. Prenylation results in a more lipophilic compound to provide high affinity with the cell membrane, while leads to decreased bioavailability and plasma absorption [10]. The prenylations at the C-3 and C-7 position in the flavone backbone of morusin greatly contribute to its high cytotoxicity against murine P-388 cells [11,12]. However, the change from a cyclic form of the prenyl group at the C-8 position of morusin to a free prenyl group in kuwanon C slightly reduced the cytotoxic effects, while inhibitory activity against β-secretase and anti-bacterial activity against *E.coli* as well as *S. typhimurium* were markedly increased [8,9]. On the other hand, cyclization of the prenyl group at C-7 of morusin reduced inhibitory activity against tyrosinase and α-glucosidase [7].

Morusin may allosterically regulate several enzymes including cyclooxygenase-2, lipoxygenases, pancreatic lipase, epidermal growth factor receptor, UDP-glucuronosyltransferase, acetylcholine esterase (AChE), matrix metalloproteinases (MMP-9 and MMP-2), cytochrome P450 and HIV reverse transcriptase (further discussed in later sections) [13–17]. Such abilities have been decoupled from its antioxidant and anti-inflammatory capacities, since each carbon residue appears to play differential roles in mediating biological functions. For example, a prenyl substitution on C-3 of morusin caused noncompetitive inhibition characteristics toward AChE, while non-prenylated flavonoids showed mixed inhibition kinetics [18]. In addition, hydroxylation at C-5′ of morusin determined the selective inhibition of different oxygenases. Morusin, which lacks the 5′-hydroxyl group of artonin E, was a less potent 5-lipoxygenase inhibitor. Morusin displayed broad inhibitory activities against several lipoxygenases, while artonin E, produced by hydroxylation at C-5′ of morusin, displayed inhibitory activity against 5-lipoxygenase higher by one order of magnitude [19]. Molecular docking analysis between morusin and the cytochrome isoform CYP3A4 indicated that oxygens in the pyran ring, C-5 and C-2′ are involved in hydrogen bonds with CYP3A4 and the B ring structure of morusin is involved in π-π interaction with Phe108 of CYP3A4 [16,20]. Hydrogen bond formation between the 2′-hydroxy group of morusin and Tyr393 of MMP-9 and cation-pi interactions between the flavone backbone of morusin and Tyr423 of MMP-9 were also associated with morusin binding to MMP-9 [21]. Docking analysis of morusin with 5-lipoxygenase (5-LOX) indicated that the 2′- and 4′-hydroxy groups of the B ring structure of morusin form hydrogen bonds with Val127 of both the B and F chain in 5-LOX, respectively. Moreover, the 3-prenyl group of morusin aligned into the hydrophobic groove generated by Val127, Ala128, Leu124, Leu135 and Phe131 of the 5-LOX active site [22], indicating that hydroxylation and prenylation of morusin are crucial in molecular recognition of the target protein. Molecular modeling of proteins and docking analysis was utilized to screen for ligands of GABA transporter 1. In this study, morusin was identified as the strongest potential ligand for GABA transporter 1, in which Tyr140 and Ser396 of GABA transporter 1 may be involved in hydrogen bonding with morusin [23].

We briefly mentioned the chemical properties of morusin here, since this review mainly focuses on the biological aspects of morusin treatment with various clinical relevance. More detailed physiochemical properties of flavonoids including morusin and their purification processes are well described elsewhere [3,6,7].

2.2. *The Effects of Morusin Treatment on Cellular Processes*

Multiple physiological processes have been proposed as effectors of morusin treatment in various biochemical studies (Figure 2). In this section, we describe the cellular processes specifically implicated in morusin-targeted human pathophysiology discussed later on.

2.2.1. Inflammation

Many studies have confirmed the anti-inflammatory capacity of morusin in different in vivo and in vitro contexts—(1) Morusin treatment inhibits secretion of cytokines such as CCL5 and CCL17 in TNFα- and IFN-γ-stimulated keratinocytes and also inhibits the release of histamine and LTC_4 in A23187-stimulated MC/9 mast cells [24]. (2) Morusin treatment inhibits PMA-induced MUC5AC production in a human pulmonary mucoepidermoid cell line NCI-H292, showing prominent anti-inflammatory effects in vitro [25]. (3) Morusin treatment ameliorates IL-1β-induced chondrocyte inflammation and abrogates osteoarthritis in destabilization of the medial meniscus model in vivo [26]. (4) Morusin treatment inhibits NO production from LPS-induced RAW264.7 cells [27]. (5) Morusin treatment protects against 2,4,6-trinitrobenzensulfonic acid (TNBS)-induced colitis in rats [28]. (6) Morusin treatment alleviates inflammatory signaling thereby controlling the outgrowth of *mycoplasma pneumonia* [29].

This explosion of observations on the anti-inflammatory effects of morusin have raised questions of the molecular and biochemical targets of morusin and the underlying mechanisms. In this context, the antioxidant capacity of morusin has been linked to the suppression of iNOS induction, thereby reducing nitric oxide formation [27,30,31]. This mechanism may also be associated with the neuroprotective function of morusin against NO-induced cell death in SH-SY5Y cells [31]. Although the mechanism underlying iNOS regulation by morusin has not been directly addressed yet, it may be recapitulated by the NF-κB pathway, a key intracellular pathway that governs pro-inflammatory signaling in multiple ways including reactive oxygen species (ROS) production, cytokine production and immune cell activation [32,33]. Furthermore, morusin suppressed STAT1-mediated cytokine secretion in TNF-α- and IFN-γ-stimulated keratinocytes [24]. Indeed, many studies have already emphasized that the versatile effects of morusin may be associated with downregulation of the NF-κB pathway and its crosstalk with STAT1, STAT3 and Wnt/β-catenin signaling [24–26,29,34–40]. Given that activation of the NF-κB and STAT signaling pathways is one of the downstream events of EGF receptor signaling and since morusin directly binds to the catalytic domain of the EGF receptor for inactivation of EGFR [14], it is plausible that morusin-mediated blocking of EGFR signaling may contribute to the down-regulation of NF-κB-mediated iNOS induction and NO synthesis, as well as STAT1-mediated cytokine secretion, to result in anti-inflammatory effects. In addition, several in vitro studies argued that the anti-inflammatory effects of morusin are partly attributed to its role as an allosteric inhibitor of cyclooxygenase-2 (COX-2) and lipoxygenases (LOXs), which are key rate enzymes implicated in arachidonic acid metabolism [41]. Given that arachidonic acid is a key molecule spiking intracellular pro-inflammatory signaling followed by immune cell activation [42], such regulations may be clinically relevant to diseases associated with aberrant inflammation.

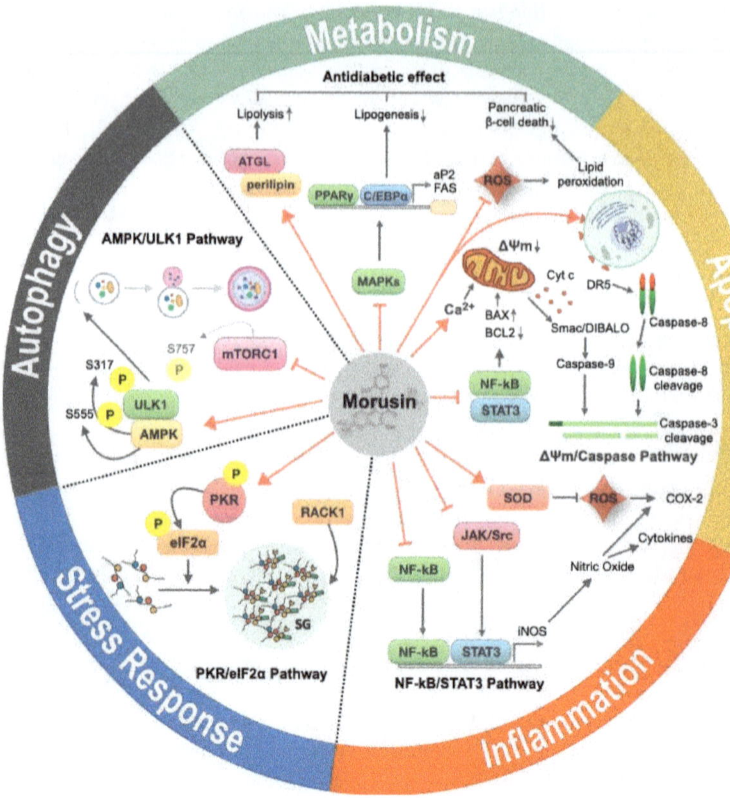

Figure 2. The effects of morusin on cellular processes. Morusin elicits various cellular processes including suppression of inflammation, induction of apoptosis, autophagy and stress granule formation and is involved in the homeostasis of glucose and lipid metabolism. Red solid arrow and solid bar indicates activation and suppression of target molecule, respectively. Upwards and downwards arrow indicates an increase and decrease of target protein or pathway, respectively.

2.2.2. Apoptosis

Although cytotoxicity upon morusin treatment has been observed across multiple cancer cell lines, biochemical characterization of the underlying mechanisms and clinical potential in vivo are still active areas of research. In particular, multiple recent studies have performed a thorough assessment of the biochemical targets of morusin, allowing for the identification of apoptosis as a critical tumor-killing effect of morusin.

Apoptosis is a type of tightly regulated programmed cell death, characterized by cell shrinkage, nuclear fragmentation, chromatin condensation and chromosomal DNA fragmentation [43]. Apoptosis can be subclassified into extrinsic and intrinsic pathways according to the origin of stimuli (intracellular vs extracellular signals). This biological process has been appreciated as an attractive druggable target in numerous pathological contexts especially including cancer [44]. Morusin effects on apoptosis in cancer cells has been mostly linked to its capacity to suppress the NF-κB pathway, since suppressors of the intrinsic apoptotic pathway, such as cIAP and Bcl-xL, are representative downstream targets of p50/p65, the core transcription factor of the NF-κB pathway [43]. The potential involvement of the NF-κB pathway was first suggested by a study in which morusin treatment led to inhibition of the NF-κB pathway and activation of intrinsic apoptosis in human colorectal cancer cells [35]. Such observations were also substantiated by studies across multiple cancer cell lines in which morusin treatment resulted in concomitant NF-κB pathway inhibition and activation of apoptosis [40,45,46].

In addition, several studies have suggested the possibility of STAT3 as a component of the mechanisms underlying the pro-apoptotic effects of morusin [14,36,38,47,48], which may be relevant since (1) ROS formation is both a consequence and a driving force of STAT3 activation [49] and (2) the STAT3 and NF-κB pathways synergistically promote transcriptional activation of suppressors of the intrinsic apoptotic pathway [50]. Consistently, STAT3 target genes, such as anti-apoptotic genes encoding Bcl-xL, Bcl-2, XIAP, survivin and cell cycle regulators (c-Myc and cyclin D), are down-regulated upon morusin treatment, while pro-apoptotic Bax expression was induced [36,38,40,47,51]. In addition, the administration of morusin reduces mitochondrial membrane potential resulting in the release of cytochrome c and Smac/DIABLO and thus, apoptosis is facilitated by the activation of caspase-9 and caspase-3 [35,48,52].

Turning off signaling pathways implicated in cell survival and proliferation has been functionally associated with the activation of apoptosis [53]. In this regard, morusin also exerts its pro-apoptotic effects by inhibiting relevant pathways including the PI3K-AKT and MAPK signaling pathways [35,54,55]. However, it is still unclear if such mechanisms can be dissociated from morusin effects on the NF-κB pathway, which has cross-talk with the PI3K-AKT, MAPK pathways in many physiological and pathological contexts [54,55].

Finally, some evidence has suggested that morusin treatment also results in the activation of extrinsic apoptotic pathways, in which morusin strongly increased expression of the death receptor DR5 at the transcriptional level and conferred sensitization of glioblastoma to TRAIL signaling [35,38]. However, thorough assessment of the underlying mechanisms may be further required to interpret morusin effects on the extrinsic apoptotic pathways, considering the intimate crosstalk between the intrinsic and extrinsic apoptotic pathways [56].

2.2.3. Metabolism

Metabolism plays key roles in energy homeostasis and signal transduction, which are fundamental aspects of organisms. One of the unique aspects of metabolism is the remarkable metabolic flexibility in response to various stimuli [57]; metabolic pathways are dramatically rewired as an adaptive response to specific physiological contexts such as starvation and energy overload. Such adaptations have also been proposed as either the consequence or cause of numerous human pathologies including metabolic disorders, cardiovascular diseases and cancer.

Morusin treatment was shown to exert beneficial effects on systemic and cellular metabolism. Those effects have also been mostly associated with the antioxidant capacity of morusin. Multiple studies have shown that morusin treatment reduces reactive oxygen species in metabolically relevant organs, in which redox homeostasis is tightly associated with general metabolic fitness including increased mitochondrial performance and the amelioration of dysregulated fuel metabolism [58]. For example, treatment of *Morus alba* root bark extract including morusin reduces lipid peroxidation, thereby leading to hypoglycemic effects in vivo in a streptozotocin-induced mouse model of type I diabetes. Such beneficial effects may be associated with decreasing oxidative stress and preservation of pancreatic β-cell integrity by reducing cell death [59,60].

It has recently been shown that morusin also serves as an allosteric inhibitor of metabolic enzymes including pancreatic lipase, UDP-glucuronosyltransferase (UGT) and cytochrome p450 (CYP450) [16,61]. Such findings may be clinically relevant, since these enzymes have been implicated in the dysregulation of lipid homeostasis and endoplasmic reticulum stress, which play key roles in the pathogenesis of various metabolic disorders including diabetes [62,63]. Of note, CYP450 and other ER-resident lipid oxygenases including CYP1A2, CYP2C9, CYP2D6, CYP2E1, CYP3A4 and CYP2C19 have been proposed as metabolizing enzymes of morusin *per se* [17,20], although it is unclear whether the mechanisms contribute to the drug resistance or whether the intermediates mediate the beneficial effects of morusin.

Morusin has also been highlighted as a potent *de novo* lipogenesis inhibitor as well as a lipolysis stimulator [64]. The study was performed in 3T3-L1 and primary adipocytes in which morusin treatment leads to reduced lipid build-up and increased lipid breakdown. Such observations may be associated with changes in multiple processes including down-regulation of the MAPK pathway downstream of insulin receptor signaling, adipogenic transcription factors (PPARγand C/EBPα) and lipogenic factors (aP2 and FAS), while expression of lipolytic factors (HSL, ATGL and perilipin) are enhanced by morusin administration in differentiated 3T3-L1 adipocytes. However, the trans-differentiation of glioblastoma multiforme (GBM) cancer stem cells into adipocyte-like cells was induced by morusin treatment, in which the expression of adipogenic proteins including PPARγ, Adipsin D and aP2 were enhanced in a dose-dependent manner [39]. A similar result was observed with breast cancer cells in which differentiation of breast cancer cells into adipocyte-like cells was induced by morusin treatment and concomitantly expression of adipogenic protein was increased [65]. The detailed clinical potential of morusin treatment for metabolic disorders including diabetes has been discussed in a later section.

2.2.4. Autophagy

Autophagy is a physiological process through which unnecessary and dysfunctional cellular components are removed or recycled [66]. Various extracellular and intracellular signals induce autophagy and thus autophagy plays a key role in the intricate regulation of cellular homeostasis including cell fate determination, fuel metabolism and mitochondrial fitness. In this context, autophagy has been implicated in numerus human pathologies particularly associated with energy stress and mitochondrial dysfunction, such as cancer and neurodegeneration, respectively. This notion, together with previous reports in which drug resistance of cancer cells are tightly associated with the capacity of cells to induce autophagy [67], brought us to the question as to whether cancer cells employ autophagy as a protective mechanism against morusin-induced cell death. Biochemical evidence in our study showed that morusin treatment leads to mTOR1 inhibition and the subsequent activation of AMPK, resulting in ULK1-mediated autophagy activation. Such autophagy activation is associated with the reduced apoptosis induced by morusin (Figure 3), corroborating the idea that activation of autophagy confers drug resistance to cancer cells [68]. This mechanism may be noteworthy in the context of the therapeutic relevance of morusin for neurological disorders, given the anti-cell death effects of morusin on neuronal cells [31] and the intimate connection between autophagy and neurodegenerative diseases [69].

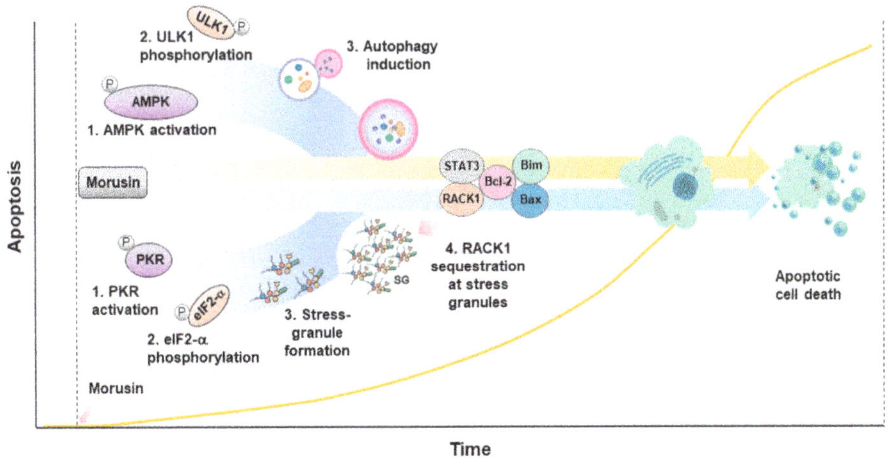

Figure 3. Morusin-induced autophagy and stress granule (SG) formation inhibits induction of apoptosis at early times of the stress response. Morusin induces apoptosis, autophagy and SG formation. However, induction of autophagy and SG formation occurs ahead of apoptosis induction at the early times of morusin treatment, resulting in a delay of the induction of apoptosis.

2.2.5. Stress Response

Cells have evolved multiple physiological strategies to cope with diverse environmental and intracellular stresses. Among these is stress granule (SG) formation, during which macromolecular aggregates of mRNA and proteins are assembled, leading to the attenuation of translation and activation of various physiological pathways for DNA damage repair and cell survival [70]. Since study of the SGs has mainly focused on biochemical or mechanistic perspectives, the physiological and pathological relevance of the SGs is still enigmatic. Recent in vitro and in vivo clinical evidence has proposed the potential implication of SGs in cancers and neurodegenerative diseases, warranting a more detailed examination of the underlying mechanisms [71,72]. In particular, the SGs are induced by various types of cytotoxic drugs, thereby contributing to drug resistance [73]. Our recent study showed that SG formation is induced by morusin treatment [74]. PKR was identified as a target of morusin in an un-biased systemic phospho-antibody array. PKR activation and subsequent eIF2α phosphorylation are key biochemical events for morusin-induced SG formation, which leads to the retention of RACK1 within the SGs. Since RACK1 is a pro-apoptotic molecule that activates intrinsic apoptotic pathways and together with our data in which RACK1 manipulation decouples SG formation and morusin-induced cell death, we suggested that morusin-mediated cell death may at least in part be attributed to RACK1 activation (Figure 3).

Although our study only highlighted the potential connection between SG formation and the intrinsic apoptotic pathway, it may also be of interest to study the biochemical mechanism(s) and the pathological relevance of interplay with other pathways including the extrinsic apoptotic pathways also activated by morusin, as well as inflammatory signaling, given the involvement of PKR in this mechanism.

2.3. Clinical Potential of Morusin Treatment in Pathophysiology

Exploding number of causality studies linking specific cellular processes to human pathologies, together with morusin implication in the processes, warranted assessment of the clinical potential of morusin in various disease settings in vivo and in vitro. Here, we discuss recent advances in the understanding of the potential application of morusin in treating common human diseases.

2.3.1. Inflammatory Diseases

The anti-inflammatory capacity of morusin has been tested in various models of inflammatory diseases, which was initiated with several in vitro studies on morusin-mediated inhibition of iNOS, cyclooxygenases and lipoxygenases [27,41]—(1) Lee et al. demonstrated that morusin reduces musin secretion in NCI-H292 cancer cells with mucoepidermoid characteristics following chronic inflammation [25]. This in vitro observation was concordant with in vivo results in which morusin treatment alleviates the hypersecretion of airway mucin in a rat model of bronchitis induced by sulfur dioxide treatment. (2) Morusin displays potential anti-allergic and anti-inflammatory effects on atopic dermatitis, a common chronic inflammatory skin disease, which was well-demonstrated by Jin et al., in their in vitro study with MC/9 mast cells and HaCat keratinocytes [24]. Mechanistically, they proposed the dual roles of morusin in modulating inflammatory signals in both keratinocytes and immune cells via STAT1/NF-κB and lipoxygenase, respectively. (3) Morusin treatment inhibits the NF-κB signaling pathway and thereby dampens IL-1β-induced chondrocyte inflammation and osteoarthritis using mouse chondrocytes and destabilization of the medial meniscus (DMM) model, a mouse model of osteoarthritis [26]. (4) Morusin treatment has been tested in a chemical-induced rat model of colitis. Histological observations following morusin gavage in the rats showed a significant reduction of tissue damage score and pro-inflammatory markers (TGF-β1 and IL-1β) and an increase in the level of antioxidant enzymes (superoxide dismutase and catalase), indicating that morusin may have therapeutic potential in treating inflammatory bowel disease [28].

Numerous microbes and viruses take advantage of host cellular processes and machineries to maximize their infection capacities. One such strategy is to provoke cellular and systemic inflammatory responses. Morusin has been administered to a few models of infectious diseases, such as an in vivo mouse model of mycoplasma pneumonia infection, in which morusin treatment efficiently suppressed mycoplasma pneumonia via the inhibition of Wnt/β-catenin and NF-κB signaling [29]. Morusin also appears to exert anti-microbial and anti-viral activities on microorganisms *per se*; Pang et al. and We et al., demonstrated that morusin has the remarkable capacity to target *Staphylococcus aureus* and clinical methicillin-resistant *Staphylococcus aureus* (MRSA) in vitro and in vivo [75,76]. The mechanism of action may be associated with the chemical properties of morusin as an isopentenyl leading to an increase in membrane permeability, inhibition of the phospholipid-repair system and dissipation of the proton motive force of the bacteria [77]. Mechanisms underlying the anti-viral effects of morusin are less clear. One possible explanation may be that morusin exerts its effects by allosteric inhibition of essential viral components. Of note, an in silico analysis proposed that morusin could directly inhibit HIV-1 reverse transcriptase activity [78] and SARS-CoV-2 protease [79].

2.3.2. Neurological Disorders

In Chinese medicine, *Morus alba* has been used as a neuroprotective herb. The neuroprotective functions of morusin and the implications for disease have recently been appreciated at the cellular and molecular levels. For example, morusin treatment protects neuroblastoma SH-SY5Y cells from nitric oxide-induced cell death [31]. This in vitro observation was also recapitulated by a follow-up study in which the in vivo relevance of the antioxidant capacities of morusin in neuroprotection was tested [80]. In the study, authors used a rat model of memory disorder induced by aluminum trichloride ($AlCl_3$), showing that morusin ameliorates the impaired memory and learning capacity by decreasing the $AlCl_3$-induced rise in brain acetylcholinesterase (AChE) activity and brain oxidative stress levels. Although in vivo studies of the effects of morusin on neurological disorders are very limited, morusin may exert beneficial functions against neurodegenerative diseases including Alzheimer's disease, given that (1) morusin shows inhibitory activity against AChE, butyrylcholinesterase (BChE) and β-site amyloid precursor protein cleaving enzyme 1 (BACE1), which play important roles in the prevention and treatment of Alzheimer's disease [18,81] and (2) morusin treatment activates autophagy in normal cells and the dysregulation of autophagy significantly contributes to the pathogenesis of neurodegenerative disorders [68,69]. Accordingly, an integrative approach of pharmacokinetics and a structural bioinformatics approach of 210 plant compounds revealed that morusin displays high potential as an anti-Alzheimer drug [82], corroborating the idea of the clinical potential of morusin in targeting neurodegenerative diseases.

2.3.3. Diabetes

Diabetes mellitus is a pathological condition in which the body displays impaired capacity to maintain coordinated systemic fuel metabolism upon various nutrient perturbations, particularly perturbations of glucose, due to (1) pancreatic β-cell dysfunction followed by reduced serum levels of insulin (Type I diabetes), and/or (2) reduced insulin sensitivity of systemic cells and relatively reduced insulin secretion from pancreatic β-cells (Type II diabetes) [83]. Although numerous factors have been implicated in such confounding metabolic disorders, the causalities and underlying mechanisms are not fully understood. Among these, the reduced lipid storage capacity of adipocytes and aberrant lipid metabolism with ROS accumulation in metabolically relevant organs including adipocytes, liver, muscle and pancreatic β-cells have been tightly associated with the pathogenesis of the disease [84,85]. In this context, multiple in vivo and in vitro evidence linked the versatile effects of morusin on the metabolic pathways to its clinical potential in treating diabetes. For example, two independent studies showed that morusin treatment ameliorates hyperglycemia and dysregulated lipid homeostasis in a mice model of type I diabetes induced by streptozotocin treatment that specifically kills pancreatic β-cells [59,60]. Such mechanisms may involve morusin-mediated alleviation of the peroxidation of lipid species spilled over from adipocytes, due to insulin secretion levels insufficient to maintain the fat storage capacity of adipocytes.

Although no studies have yet tested the effects of morusin on the pathogenesis of type II diabetes using relevant in vivo models (e.g., *ob/ob*, *db/db* mouse or HFD-fed mice), morusin may also exert anti-diabetes effects in this context, considering that morusin potentially regulates systemic ROS levels, lipogenesis, lipolysis, C/EBPβ and PPARγ signaling, arachidonic acid metabolism and ER-resident proteins implicated in ER stress [16,59,64], all of which have been frequently implicated in type II diabetes as well as type I diabetes [42,63,86,87].

2.3.4. Cancer

The implication of morusin in multiple biological pathways that cancers take advantage of, described above, led to many studies testing the antitumoral capacities of morusin in vitro, with recent initiation of research on its clinical potential in vivo. Indeed, very early studies on the chemical properties of morusin already appreciated its antitumoral potential [6], connecting the strong antioxidant capacity of morusin to the aberrant ROS production broadly relevant to the pathogenesis of cancers [88].

The effects of morusin on specific subtypes of cancers and the underlying biochemical mechanisms have recently been studied using various in vitro and in vivo model systems (Table 1). For example, Wang et al., demonstrated the clinical potential of morusin in targeting the stemness capacity of human cervical carcinoma in vitro [40]. Such an approach may be relevant, as NF-κB is a well-established upstream transcription factor of genes for stemness and metastatic capacities of various cancers including cervical cancers [33]. In this study, morusin treatment was shown to decrease the expression of Oct4, SOX2, ALDH1, as well as epithelial-to-mesenchymal markers, thereby decreasing the stemness signature of the cancer cells. Morusin-mediated inhibition of the NF-κB pathway also leads to reduced expression of Bcl-2 and upregulation of pro-apoptotic proteins including Bax and caspase-3, which results in the apoptotic death of the cancer cells. Morusin may also display antitumor effects independent of its capacity to activate apoptosis at non-cytotoxic concentrations (lower than 10 μM), since morusin promotes VDAC-mediated Ca^{2+} influx into mitochondria leading to mitochondrial Ca^{2+} overload and mitochondrial dysfunction and subsequent paraptosis-like cell death observed in an in vitro and in vivo model of epithelial ovarian cancer with apoptotic resistance [37].

Table 1. The Effects of Morusin on Subtypes of Cancer.

Organ	Cell Type	Effects on Cellular Processes	System	Reference
Liver	LO2 HepG2 Hep3B	Apoptosis induction & angiogenesis inhibition • increases the expression of caspase-3 and the Bax/Bcl-2 expression ratio • inhibits tube formation of HUVECs in vitro and suppresses constitutive as well as IL-6-induced STAT3 phosphorylation	in vitro in vivo	[47]
Liver	SK-Hep1	Anti-tumor progression through suppressing STAT3 and NF-kB • suppresses cell-matrix adhesion, cell motility and cell invasion at non-cytotoxic concentration • increases the expression of E-cadherin and decreases the expression of vimentin and α2-, α6-, β1-integrin	in vitro in vivo	[34]
Lung	A549	Apoptosis induction & cell migration suppression • induces apoptosis by loss of mitochondrial function and increases the antioxidant activities by up-regulation of SOD • inhibits the invasion and migration by down-regulation of COX2 and VEGF at the transcription level	in vitro	[52]
Lung	H1299 H460 H292	Apoptosis induction by suppression of EGFR/STAT3 activation • inhibits phosphorylation of EGFR and STAT3 and shows potential to treat advanced NSCLC with acquired resistance to EGFR tyrosine kinase inhibitor • The docking analysis: morusin directly binds to the tyrosine kinase domain of EGFR	in vitro	[14]
Breast	MCF-10A 4T1 MCF-7	Suppression of cancer cell growth through C/EBPβ- and PPARγ-mediated lipoapoptosis • inhibits human breast cancer cell proliferation and increases the expression of C/EBPβ, PPARγ, adipsin D and perilipin • induces adipogenic differentiation, apoptosis and lipoapoptosis of cancer cells	in vitro in vivo	[65]
Breast	MCF-10A MCF-7 MDA-MB231	Apoptosis induction • The apoptosis marker proteins, cleaved caspase-3 and caspase-9 were consistently upregulated • suppresses the expression of the anti-apoptotic Survivin and induces pro-apoptotic Bax expression	in vitro	[51]
Brain	U87 GI-1 HCN-1A	Morusin-loaded nanoparticles for targeted glioblastoma therapy • Morusin was loaded in chlorotoxin-modified PLGA nanoparticles which target chloride channels and MMP-2 in glioma tumor cells • Morusin-loaded nanoparticles inhibit growth of U87 and GI-1 glioma cells by ROS generation, enhanced caspase activity, cytoskeletal destabilization and inhibition of MMP activity	in vitro	[89]
Brain	WJ1 WJ2	Inhibition of glioblastoma stem cell growth through stemness attenuation, adipocyte transdifferentiation • increases adipogenic markers, such as PPARγ, adipsin D, aP2 and perilipin and induces apoptosis • reduces stemness of GSCs by inhibition of the expression of stemness markers (CD133, nestin, Sox2 and Oct4)	in vitro in vivo	[39]
Brain	U251MG LN18 U87MG	TRAIL sensitization by regulating EGFR and DR5 in human glioblastoma cells • Combinatorial treatment of TRAIL with morusin synergistically decreased cell viability and increased apoptosis • induces the expression of DR5 and decreases anti-apoptotic survivin and XIAP by reduced expression of EGFR and pSTAT3	in vitro	[38]

Table 1. *Cont.*

Organ	Cell Type	Effects on Cellular Processes	System	Reference
Skin	JB6 P+	Blocking TPA-induced malignant transformation of mouse epidermal cells • reduces the TPA-induced ROS production, AP1 and NF-κB in JB6 P+ cells at non-cytotoxic concentration • decreases TPA-upregulation of COX-2, N-cadherin and Vimentin	in vitro	[45]
Stomach	MKN45 SGC7901	Inhibition of cell proliferation and tumor growth by down-regulating c-Myc • suppresses tumor growth and down-regulates CDKs and cyclins, such as CDK2, CDK4, cyclin D1 and cyclin E1. • reduces the expression of c-Myc and c-Myc protein binding at the E-Box regions	in vitro in vivo	[90]
Pancreas	AsPC-1 BxPC-3 MIAPaCa-2 PANC-1	Apoptosis induction and inhibition of invasion by blockage of STAT3 signaling pathway • inhibits STAT3 activation and suppresses activation of upstream JAK1, JAK2 and c-Src kinases. • arrest cell cycle at G1/G0 or G2/M phase and causes induction of apoptosis and loss of mitochondrial membrane potential	in vitro	[48]
Bone	U2OS HOS	Inhibition of human osteosarcoma via PI3K-AKT signaling pathway • promotes apoptosis and reduces the migration and invasion of osteosarcoma • inhibits the PI3K/AKT signaling pathway and induces the expression of caspase-3 and caspase-8	in vitro	[55]
Ovary	A2780 SKOV-3 HO-8910	Paraptosis-like cell death induction through mitochondrial calcium overload and dysfunction • causes mitochondrial Ca^{2+} influx and induces paraptosis-like cell death via mitochondrial Ca^{2+} overload • increases ROS and decreases mitochondrial membrane potential and inhibits the growth of SKOV-3 xenograft in nude mice	in vitro in vivo	[37]
Prostate	DU145 PC-3 LNCaP RWPE-1	Cell death induction through inactivating STAT3 signaling • reduces STAT3 activity by suppressing kinase activities of JAK2 and Src and increases SHP1 phosphatase activity • down-regulates the expression of STAT3 target genes encoding Bcl-xL, Bcl-2, Survivin, c-Myc and cyclin D1	in vitro	[36]
Cervix	HeLa	Apoptosis induction & inhibition of human cervical cancer stem cell growth and migration through attenuation of NF-κB activity • decreases the proliferation, tumor sphere formation and migration of human cervical CSCs and increases apoptosis • decreases the expression levels of NF-κB/p65 and Bcl-2, while increases expression levels of Bax and caspase-3	in vitro	[40]
Kidney	769-P 786-O OSRC-2	Anti-cancer activity by disturbing MAPK signaling pathways • inhibits cell growth and migration, induces cell apoptosis and induces the cell cycle arrest in the G1 phase • up-regulates P-p38 and P-JNK levels, while the down-regulates P-ERK level	in vitro in vivo	[54]
Liver Spleen	H22	Inhibition of transplanted H22 hepatocarcinoma • inhibits the tumor growth of transplanted H22 hepatocarcinoma in mice by reducing the expression of NF-κB • increases the expression of p53, Survivin, cyclin B1 and caspase-3	in vitro in vivo	[46]

Table 1. *Cont.*

Organ	Cell Type	Effects on Cellular Processes	System	Reference
Colon Liver Breast	HT-29 Hep3B MCF-7	Apoptosis induction & suppression of NF-κB activity • inhibits the phosphorylation of IKK-α, IKK-β and IκB-α and suppresses NF-κB nuclear localization and its DNA binding • causes activation of caspase-8, change of mitochondrial membrane potential, release of Cytochrome c and Smac/DIABLO and activation of caspase-9 and caspase-3	in vitro	[35]
Cervix Breast Bone	HeLa U2OS ZR75B	Attenuation of RACK1-mediated apoptotic cell death by stress granule (SG) formation • induces activation of PKR and subsequent eIF2α phosphorylation for SG formation • sequestration of RACK1 within the SGs contributes to protection of cells from cell death	in vitro	[74]
Cervix Breast Bone Colon	HeLa MCF-7 U2OS HCT116	Autophagy induction inhibits cell death • induces AMPK activation and inhibits mTOR activity, resulting in LC3-II accumulation and ULK1 activation for autophagy • autophagy induction is an impediment for morusin-induced apoptosis	in vitro	[68]

Morusin treatment exerts beneficial effects against liver cancers as well. The intravenous injection of morusin in a xenograft mouse model of hepatocarcinoma significantly reduced tumor growth without any side effects. The histological and biochemical evidence provided in this study indicated that the mechanism may be associated with the activation of p53, survivin, cyclin B1 and caspase-3 and a decrease in the expression of NF-κB [47]. Moreover, morusin treatment also suppresses the metastatic capacity of the human hepatoma SK-Hep1 cell line, which may be biochemically associated with inhibition of STAT3 and NF-κB followed by the reduction of the expression of metastatic markers, such as vimentin, α2- and β1-integrins [34]. Such in vitro observations were also confirmed in an experiment with immunocompromised mice xenografted with SK-Hep1 cells, in which morusin treatment significantly decreased lung colonization of the cancer cells. Following the report that morusin effectively inhibits the growth of tumors in mice transplanted with H22 liver tumor cells [46], similar inhibitory effects of morusin on tumor xenografts were observed for liver, ovary, breast, kidney and gastric cancer cells [37,47,54,65,90]. In addition to inhibitory effects on tumor growth, tube formation of human umbilical vein endothelial cells (HUVEC) and expression of angiogenesis-related genes were also inhibited by morusin treatment in vitro [47,52], suggesting the anti-metastatic potential of morusin.

Morusin also targets the STAT3 pathway and its downstream targets, including survivin, cyclin B1, to exert antitumor activity in prostate and pancreatic cancers [36,48]. In the case of pancreatic cancer, pancreatic lipase may be another relevant target of morusin, considering (1) the strong association between pancreatic cancer and the levels of pancreatic lipase [48] and (2) the potential role of morusin in the allosteric regulation of pancreatic lipase [61].

The antitumor capacity of morusin has also been tested in glioblastoma. For example, the stemness capacity of glioblastoma multiforme (GBM) cancer stem cells (GSCs) were targeted by morusin treatment in vivo and in vitro [39]. This observation may be mechanistically attributable to the expression of genes downstream of NF-κB, which was already proposed in the previously described study with the cervical carcinoma model [40]. Of note, the reduced stemness of the cancers upon morusin treatment results in adipocyte-like trans-differentiation, followed by activation of intrinsic apoptosis [39]. The underlying mechanisms may also involve the activation of extrinsic apoptosis, as morusin treatment promotes expression of an extrinsic receptor, Death receptor 5, thereby sensitizing the glioblastoma to TRAIL, an antitumor chemical that mimics ligands for the extrinsic receptor [38].

Morusin may also employ other various cellular processes independent of NF-κB for targeting cancers—(1) An in vitro study of gastric cancer showed that morusin governs the expression of c-Myc and multiple downstream genes including CDKs and cyclins implicated in cell proliferation, thereby exerting antitumor capacity [90]. (2) The PI3K/AKT pathway has been proposed as a target of morusin in treating osteosarcoma in vitro [55]. (3) Morusin may directly target the MAPK pathway in renal carcinoma in vitro, although this should be further validated considering the intimate crosstalk between the NF-κB and MAPK pathway [54]. (4) C/EBPβ- and PPARγ-mediated lipoapoptosis may be therapeutically relevant targets of morusin in breast cancers, proposed in a study using multiple breast cancer cell lines and a xenograft model of breast cancers [64]. This may be of great importance considering that the capacity of mitochondrial fatty acid oxidation and fatty acid availability associated with host metabolism have recently emerged as key factors contributing to the pathogenesis of breast cancers [91,92]. (5) Studies with a panel of lung cancer cell lines showed that morusin treatment not only modulates conventional targets including NF-κB, STAT3 and downstream genes such as VEGF but may also directly bind and dephosphorylate EGFR on an active site, which would contribute to the further reduction of STAT3/NF-κB activity [14].

A cancer cell continuously undergoes dynamic remodeling of numerous cellular processes, which confer various benefits such as drug resistance. In this context, the two recent studies from our lab have highlighted the biochemical mechanisms by which cancer cells gain resistance against the cytotoxicity of morusin [68,74]. As described in the previous section, cancer cells promote autophagy and SG formation in response to morusin treatment, thereby desensitizing the apoptotic signals exerted by morusin treatment. Indeed, autophagy and SGs have already been proposed as the 'shield pathways' of cancers against various drug treatments. This notion, together with multiple biochemical studies in which morusin synergized with inhibitors of either autophagy or SG formation to boost its antitumor capacity, proposed the cotreatment of morusin and the inhibitors as a therapeutically relevant strategy for targeting cancers with drug resistance. However, whether such resistant mechanisms are still relevant in the context of various subtypes of cancers in vivo should be addressed.

3. Conclusions

The well-acknowledged functions of the root bark of *Morus* in Chinese herbal medicine have been expanded to the potential application of individual phytochemicals extracted from such arcane plants to treat multiple chronic diseases common in modern society. Morusin has been appreciated as a phytochemical with the unique physiochemical properties of isoprene flavonoids, allowing for versatile salutary effects including antioxidant and tumor killing activities. Morusin reduces ROS formation and inflammation at non-cytotoxic concentrations (lower than 10 μM), while it induces ROS formation and cytotoxicity against cancer cells at high concentrations (higher than 20 μM). A growing body of evidence, described here, highlighted the biological properties of morusin in multiple in vitro and in vivo pathophysiological settings and proposed clinical potential. The as yet unraveled clinical potential of morusin is waiting for thorough scientific assessment of the effects of morusin on human pathology with an accompanying understanding of the underlying biochemical mechanisms and effective platforms, which may allow for *bona fide* application of morusin in patients with various pathologies in the near future.

Author Contributions: Conceptualization, D.W.C., S.W.C. and C.Y.C.; writing—original draft preparation, D.W.C., S.W.C., S.-G.L. and C.Y.C.; writing—review and editing, S.-G.L. and C.Y.C.; visualization, S.W.C. and D.W.C.; supervision, C.Y.C.; project administration, C.Y.C.; funding acquisition, C.Y.C. All authors have read and agreed to the published version of the manuscript.

Funding: This work was supported by the National Research Foundation of Korea grant (2020-R1F1A1068120 and SRC-2017R1A5A1014560 to C.Y.C.) funded by the Korea Ministry of Science and ICT (Information and Communications Technology).

Conflicts of Interest: The authors declare no conflict of interest. The funders had no role in the design of the study; in the collection, analyses, or interpretation of data; in the writing of the manuscript, or in the decision to publish the results.

References

1. Yang, S.; Wang, B.L.; Li, Y. Advances in the pharmacological study of *Morus alba* L. *Yao Xue Xue Bao* **2014**, *49*, 824–831. [PubMed]
2. Nomura, T.; Fukai, T.; Matsumoto, J.; Ohmori, T. Constituents of the cultivated mulberry tree. *Planta Med.* **1982**, *46*, 28–32. [CrossRef] [PubMed]
3. Nomura, T.; Fukai, T.; Shimada, T.; Chen, I.S. Components of Root Bark of Morus australis. *Planta Med.* **1983**, *49*, 90–94. [CrossRef] [PubMed]
4. Nomura, T.; Fukai, T.; Hano, Y. Constituents of the Chinese crude drug "sang-bai-pi" (Morus root bark). *Planta Med.* **1983**, *47*, 30–34. [CrossRef] [PubMed]
5. Hano, Y.; Hirakura, K.; Nomura, T.; Terada, S.; Fukushima, K. Components of root bark of morus lhoul 1. Structures of two new natural diels-alder adducts, kuwanons N and o. *Planta Med.* **1984**, *50*, 127–130. [CrossRef]
6. Nomura, T.; Fukai, T.; Hano, Y.; Yoshizawa, S.; Suganuma, M.; Fujiki, H. Chemistry and anti-tumor promoting activity of Morus flavonoids. *Prog. Clin. Biol. Res.* **1988**, *280*, 267–281.
7. Yan, J.; Ruan, J.; Huang, P.; Sun, F.; Zheng, D.; Zhang, Y.; Wang, T. The structure-activity relationship review of the main bioactive constituents of Morus genus plants. *J. Nat. Med.* **2020**, *74*, 331–340. [CrossRef]
8. Cho, J.K.; Ryu, Y.B.; Curtis-Long, M.J.; Kim, J.Y.; Kim, D.; Lee, S.; Lee, W.S.; Park, K.H. Inhibition and structural reliability of prenylated flavones from the stem bark of Morus lhou on β-secretase (BACE-1). *Bioorg. Med. Chem. Lett.* **2011**, *21*, 2945–2948. [CrossRef]
9. Syah, Y.M.; Juliawaty, L.D.; Achmad, S.A.; Hakim, E.H.; Takayama, H.; Said, I.M.; Latip, J. Phenolic constituents from the wood of Morus australis with cytotoxic activity. *Z. Naturforsch. C J. Biosci.* **2008**, *63*, 35–39. [CrossRef]
10. Chen, X.; Mukwaya, E.; Wong, M.S.; Zhang, Y. A systematic review on biological activities of prenylated flavonoids. *Pharm. Biol.* **2014**, *52*, 655–660. [CrossRef]
11. Hakim, E.H.; Achmad, S.A.; Juliawaty, L.D.; Makmur, L.; Syah, Y.M.; Aimi, N.; Kitajima, M.; Takayama, H.; Ghisalberti, E.L. Prenylated flavonoids and related compounds of the Indonesian Artocarpus (Moraceae). *J. Nat. Med.* **2006**, *60*, 161–184. [CrossRef] [PubMed]
12. Dat, N.T.; Binh, P.T.; le Quynh, T.P.; Van Minh, C.; Huong, H.T.; Lee, J.J. Cytotoxic prenylated flavonoids from Morus alba. *Fitoterapia* **2010**, *81*, 1224–1227. [CrossRef]
13. Bahramann, E. The pathology of testicular neoplasms. *Z. Urol. Nephrol.* **1976**, *69*, 91–97. [PubMed]
14. Park, H.J.; Min, T.R.; Chi, G.Y.; Choi, Y.H.; Park, S.H. Induction of apoptosis by morusin in human non-small cell lung cancer cells by suppression of EGFR/STAT3 activation. *Biochem. Biophys. Res. Commun.* **2018**, *505*, 194–200. [CrossRef] [PubMed]
15. Parant, M.R.; Vial, H.J. Rapid and serial determination of protein kinase C activity and of the associated [3H]PDBu binding using a 96-well microtiter plate and a cell harvester. *Anal. Biochem.* **1990**, *184*, 283–290. [CrossRef]
16. Shi, X.; Yang, S.; Zhang, G.; Song, Y.; Su, D.; Liu, Y.; Guo, F.; Shan, L.; Cai, J. The different metabolism of morusin in various species and its potent inhibition against UDP-glucuronosyltransferase (UGT) and cytochrome p450 (CYP450) enzymes. *Xenobiotica* **2016**, *46*, 467–476. [CrossRef] [PubMed]
17. Hou, C.; Liu, W.; Liang, Z.; Han, W.; Li, J.; Ye, L.; Liu, M.; Cai, Z.; Zhao, J.; Chen, Y.; et al. UGT-mediated metabolism plays a dominant role in the pharmacokinetic behavior and the disposition of morusin in vivo and in vitro. *J. Pharm. Biomed. Anal.* **2018**, *154*, 339–353. [CrossRef]
18. Kim, J.Y.; Lee, W.S.; Kim, Y.S.; Curtis-Long, M.J.; Lee, B.W.; Ryu, Y.B.; Park, K.H. Isolation of cholinesterase-inhibiting flavonoids from Morus lhou. *J. Agric. Food Chem.* **2011**, *59*, 4589–4596. [CrossRef]
19. Reddy, G.R.; Ueda, N.; Hada, T.; Sackeyfio, A.C.; Yamamoto, S.; Hano, Y.; Aida, M.; Nomura, T. A prenylflavone, artonin E, as arachidonate 5-lipoxygenase inhibitor. *Biochem. Pharmacol.* **1991**, *41*, 115–118. [CrossRef]
20. Shi, X.; Mackie, B.; Zhang, G.; Yang, S.; Song, Y.; Su, D.; Liu, Y.; Shan, L. Identification of the Metabolic Enzyme Involved Morusin Metabolism and Characterization of Its Metabolites by Ultraperformance Liquid Chromatography Quadrupole Time-of-Flight Mass Spectrometry (UPLC/Q-TOF-MS/MS). *Evid. Based Complement. Altern. Med. eCAM* **2016**, *2016*, 9240103. [CrossRef]

21. Agarwal, S.; Mohamed, M.S.; Raveendran, S.; Rochani, A.K.; Maekawa, T.; Kumar, D.S. Formulation, characterization and evaluation of morusin loaded niosomes for potentiation of anticancer therapy. *RSC Adv.* **2018**, *8*, 32621–32636. [CrossRef]
22. Tseng, T.-H.; Chuang, S.-K.; Hu, C.-C.; Chang, C.-F.; Huang, Y.-C.; Lin, C.-W.; Lee, Y.-J. The synthesis of morusin as a potent antitumor agent. *Tetrahedron* **2010**, *66*, 1335–1340. [CrossRef]
23. Nitya Bankupalli, S.K.G.; Shaik, M. Comparative modeling of sodium- and chloride-dependent GABA transporter 1 and docking studies with natural compounds. *J. Appl. Biol. Biotechnol.* **2020**, *8*, 12–21.
24. Jin, S.E.; Ha, H.; Shin, H.K.; Seo, C.S. Anti-Allergic and Anti-Inflammatory Effects of Kuwanon G and Morusin on MC/9 Mast Cells and HaCaT Keratinocytes. *Molecules* **2019**, *24*, 265. [CrossRef]
25. Lee, H.J.; Ryu, J.; Park, S.H.; Woo, E.R.; Kim, A.R.; Lee, S.K.; Kim, Y.S.; Kim, J.O.; Hong, J.H.; Lee, C.J. Effects of Morus alba L. and Natural Products Including Morusin on In Vivo Secretion and In Vitro Production of Airway MUC5AC Mucin. *Tuberc. Respir. Dis.* **2014**, *77*, 65–72. [CrossRef]
26. Jia, Y.; He, W.; Zhang, H.; He, L.; Wang, Y.; Zhang, T.; Peng, J.; Sun, P.; Qian, Y. Morusin Ameliorates IL-1β-Induced Chondrocyte Inflammation and Osteoarthritis via NF-κB Signal Pathway. *Drug Des. Dev. Ther.* **2020**, *14*, 1227–1240. [CrossRef]
27. Cheon, B.S.; Kim, Y.H.; Son, K.S.; Chang, H.W.; Kang, S.S.; Kim, H.P. Effects of prenylated flavonoids and biflavonoids on lipopolysaccharide-induced nitric oxide production from the mouse macrophage cell line RAW 264.7. *Planta Med.* **2000**, *66*, 596–600. [CrossRef]
28. Vochyanova, Z.; Pokorna, M.; Rotrekl, D.; Smekal, V.; Fictum, P.; Suchy, P.; Gajdziok, J.; Smejkal, K.; Hosek, J. Prenylated flavonoid morusin protects against TNBS-induced colitis in rats. *PLoS ONE* **2017**, *12*, e0182464. [CrossRef]
29. Chen, C.; Wang, J.; Chen, J.; Zhou, L.; Wang, H.; Chen, J.; Xu, Z.; Zhu, S.; Liu, W.; Yu, R.; et al. Morusin alleviates mycoplasma pneumonia via the inhibition of Wnt/beta-catenin and NF-kappaB signaling. *Biosci. Rep.* **2019**, *39*. [CrossRef]
30. Yang, Z.G.; Matsuzaki, K.; Takamatsu, S.; Kitanaka, S. Inhibitory effects of constituents from Morus alba var. multicaulis on differentiation of 3T3-L1 cells and nitric oxide production in RAW264.7 cells. *Molecules* **2011**, *16*, 6010–6022. [CrossRef]
31. Lee, H.J.; da Lyu, H.; Koo, U.; Nam, K.W.; Hong, S.S.; Kim, K.O.; Kim, K.H.; Lee, D.; Mar, W. Protection of prenylated flavonoids from Mori Cortex Radicis (Moraceae) against nitric oxide-induced cell death in neuroblastoma SH-SY5Y cells. *Arch. Pharm. Res.* **2012**, *35*, 163–170. [CrossRef] [PubMed]
32. Arias-Salvatierra, D.; Silbergeld, E.K.; Acosta-Saavedra, L.C.; Calderon-Aranda, E.S. Role of nitric oxide produced by iNOS through NF-κB pathway in migration of cerebellar granule neurons induced by Lipopolysaccharide. *Cell Signal.* **2011**, *23*, 425–435. [CrossRef] [PubMed]
33. Zhang, Q.; Lenardo, M.J.; Baltimore, D. 30 Years of NF-kappaB: A Blossoming of Relevance to Human Pathobiology. *Cell* **2017**, *168*, 37–57. [CrossRef] [PubMed]
34. Lin, W.L.; Lai, D.Y.; Lee, Y.J.; Chen, N.F.; Tseng, T.H. Antitumor progression potential of morusin suppressing STAT3 and NFkappaB in human hepatoma SK-Hep1 cells. *Toxicol. Lett.* **2015**, *232*, 490–498. [CrossRef]
35. Lee, J.C.; Won, S.J.; Chao, C.L.; Wu, F.L.; Liu, H.S.; Ling, P.; Lin, C.N.; Su, C.L. Morusin induces apoptosis and suppresses NF-kappaB activity in human colorectal cancer HT-29 cells. *Biochem. Biophys. Res. Commun.* **2008**, *372*, 236–242. [CrossRef]
36. Lim, S.L.; Park, S.Y.; Kang, S.; Park, D.; Kim, S.H.; Um, J.Y.; Jang, H.J.; Lee, J.H.; Jeong, C.H.; Jang, J.H.; et al. Morusin induces cell death through inactivating STAT3 signaling in prostate cancer cells. *Am. J. Cancer Res.* **2015**, *5*, 289–299.
37. Xue, J.; Li, R.; Zhao, X.; Ma, C.; Lv, X.; Liu, L.; Liu, P. Morusin induces paraptosis-like cell death through mitochondrial calcium overload and dysfunction in epithelial ovarian cancer. *Chem. Biol. Interact.* **2018**, *283*, 59–74. [CrossRef]
38. Park, D.; Ha, I.J.; Park, S.Y.; Choi, M.; Lim, S.L.; Kim, S.H.; Lee, J.H.; Ahn, K.S.; Yun, M.; Lee, S.G. Morusin Induces TRAIL Sensitization by Regulating EGFR and DR5 in Human Glioblastoma Cells. *J. Nat. Prod.* **2016**, *79*, 317–323. [CrossRef]
39. Guo, H.; Liu, C.; Yang, L.; Dong, L.; Wang, L.; Wang, Q.; Li, H.; Zhang, J.; Lin, P.; Wang, X. Morusin inhibits glioblastoma stem cell growth in vitro and in vivo through stemness attenuation, adipocyte transdifferentiation and apoptosis induction. *Mol. Carcinog.* **2016**, *55*, 77–89. [CrossRef]

40. Wang, L.; Guo, H.; Yang, L.; Dong, L.; Lin, C.; Zhang, J.; Lin, P.; Wang, X. Morusin inhibits human cervical cancer stem cell growth and migration through attenuation of NF-κB activity and apoptosis induction. *Mol. Cell. Biochem.* **2013**, *379*, 7–18. [CrossRef]
41. Chi, Y.S.; Jong, H.G.; Son, K.H.; Chang, H.W.; Kang, S.S.; Kim, H.P. Effects of naturally occurring prenylated flavonoids on enzymes metabolizing arachidonic acid: Cyclooxygenases and lipoxygenases. *Biochem. Pharmacol.* **2001**, *62*, 1185–1191. [CrossRef]
42. Hanna, V.S.; Hafez, E.A.A. Synopsis of arachidonic acid metabolism: A review. *J. Adv. Res.* **2018**, *11*, 23–32. [CrossRef] [PubMed]
43. Elmore, S. Apoptosis: A review of programmed cell death. *Toxicol. Pathol.* **2007**, *35*, 495–516. [CrossRef] [PubMed]
44. Pistritto, G.; Trisciuoglio, D.; Ceci, C.; Garufi, A.; D'Orazi, G. Apoptosis as anticancer mechanism: Function and dysfunction of its modulators and targeted therapeutic strategies. *Aging* **2016**, *8*, 603–619. [CrossRef] [PubMed]
45. Cheng, P.S.; Hu, C.C.; Wang, C.J.; Lee, Y.J.; Chung, W.C.; Tseng, T.H. Involvement of the antioxidative property of morusin in blocking phorbol ester-induced malignant transformation of JB6 P(+) mouse epidermal cells. *Chem. Biol. Interact.* **2017**, *264*, 34–42. [CrossRef]
46. Wan, L.Z.; Ma, B.; Zhang, Y.Q. Preparation of morusin from Ramulus mori and its effects on mice with transplanted H22 hepatocarcinoma. *Biofactors* **2014**, *40*, 636–645. [CrossRef]
47. Gao, L.; Wang, L.; Sun, Z.; Li, H.; Wang, Q.; Yi, C.; Wang, X. Morusin shows potent antitumor activity for human hepatocellular carcinoma in vitro and in vivo through apoptosis induction and angiogenesis inhibition. *Drug Des. Dev. Ther.* **2017**, *11*, 1789–1802. [CrossRef]
48. Kim, C.; Kim, J.H.; Oh, E.Y.; Nam, D.; Lee, S.G.; Lee, J.; Kim, S.H.; Shim, B.S.; Ahn, K.S. Blockage of STAT3 Signaling Pathway by Morusin Induces Apoptosis and Inhibits Invasion in Human Pancreatic Tumor Cells. *Pancreas* **2016**, *45*, 409–419. [CrossRef]
49. Simon, A.R.; Rai, U.; Fanburg, B.L.; Cochran, B.H. Activation of the JAK-STAT pathway by reactive oxygen species. *Am. J. Physiol.* **1998**, *275*, C1640–C1652. [CrossRef]
50. Grivennikov, S.I.; Karin, M. Dangerous liaisons: STAT3 and NF-kappaB collaboration and crosstalk in cancer. *Cytokine Growth Factor Rev.* **2010**, *21*, 11–19. [CrossRef]
51. Kang, S.; Kim, E.O.; Kim, S.H.; Lee, J.H.; Ahn, K.S.; Yun, M.; Lee, S.G. Morusin induces apoptosis by regulating expression of Bax and Survivin in human breast cancer cells. *Oncol. Lett.* **2017**, *13*, 4558–4562. [CrossRef] [PubMed]
52. Yin, X.L.; Lv, Y.; Wang, S.; Zhang, Y.Q. Morusin suppresses A549 cell migration and induces cell apoptosis by downregulating the expression of COX-2 and VEGF genes. *Oncol. Rep.* **2018**, *40*, 504–510. [CrossRef] [PubMed]
53. Portt, L.; Norman, G.; Clapp, C.; Greenwood, M.; Greenwood, M.T. Anti-apoptosis and cell survival: A review. *Biochim. Biophys. Acta* **2011**, *1813*, 238–259. [CrossRef] [PubMed]
54. Yang, C.; Luo, J.; Luo, X.; Jia, W.; Fang, Z.; Yi, S.; Li, L. Morusin exerts anti-cancer activity in renal cell carcinoma by disturbing MAPK signaling pathways. *Ann. Transl. Med.* **2020**, *8*, 327. [CrossRef] [PubMed]
55. Zhang, Y.; Weng, Q.; Chen, J.; Han, J. Morusin inhibited human osteosarcoma via PI3K-AKT signaling pathway. *Curr. Pharm. Biotechnol.* **2020**. [CrossRef]
56. Roy, S.; Nicholson, D.W. Cross-talk in cell death signaling. *J. Exp. Med.* **2000**, *192*, F21–F25. [CrossRef]
57. Goodpaster, B.H.; Sparks, L.M. Metabolic Flexibility in Health and Disease. *Cell Metab.* **2017**, *25*, 1027–1036. [CrossRef]
58. Apel, K.; Hirt, H. Reactive oxygen species: Metabolism, oxidative stress and signal transduction. *Annu. Rev. Plant Biol.* **2004**, *55*, 373–399. [CrossRef]
59. Singab, A.N.; El-Beshbishy, H.A.; Yonekawa, M.; Nomura, T.; Fukai, T. Hypoglycemic effect of Egyptian Morus alba root bark extract: Effect on diabetes and lipid peroxidation of streptozotocin-induced diabetic rats. *J. Ethnopharmacol.* **2005**, *100*, 333–338. [CrossRef]
60. Abd El-Mawla, A.M.; Mohamed, K.M.; Mostafa, A.M. Induction of Biologically Active Flavonoids in Cell Cultures of Morus nigra and Testing their Hypoglycemic Efficacy. *Sci. Pharm.* **2011**, *79*, 951–961. [CrossRef]
61. Hou, X.D.; Ge, G.B.; Weng, Z.M.; Dai, Z.R.; Leng, Y.H.; Ding, L.L.; Jin, L.L.; Yu, Y.; Cao, Y.F.; Hou, J. Natural constituents from Cortex Mori Radicis as new pancreatic lipase inhibitors. *Bioorg. Chem.* **2018**, *80*, 577–584. [CrossRef] [PubMed]

62. Foufelle, F.; Fromenty, B. Role of endoplasmic reticulum stress in drug-induced toxicity. *Pharmacol. Res. Perspect.* **2016**, *4*, e00211. [CrossRef] [PubMed]
63. Elfaki, I.; Mir, R.; Almutairi, F.M.; Duhier, F.M.A. Cytochrome P450: Polymorphisms and Roles in Cancer, Diabetes and Atherosclerosis. *Asian Pac. J. Cancer Prev.* **2018**, *19*, 2057–2070. [CrossRef] [PubMed]
64. Lee, M.R.; Kim, J.E.; Choi, J.Y.; Park, J.J.; Kim, H.R.; Song, B.R.; Park, J.W.; Kang, M.J.; Choi, Y.W.; Kim, K.M.; et al. Morusin Functions as a Lipogenesis Inhibitor as Well as a Lipolysis Stimulator in Differentiated 3T3-L1 and Primary Adipocytes. *Molecules* **2018**, *23*, 2004. [CrossRef] [PubMed]
65. Li, H.; Wang, Q.; Dong, L.; Liu, C.; Sun, Z.; Gao, L.; Wang, X. Morusin suppresses breast cancer cell growth in vitro and in vivo through C/EBPβ and PPARγ mediated lipoapoptosis. *J. Exp. Clin. Cancer Res.* **2015**, *34*, 137. [CrossRef] [PubMed]
66. Dikic, I.; Elazar, Z. Mechanism and medical implications of mammalian autophagy. *Nat. Rev. Mol. Cell Biol.* **2018**, *19*, 349–364. [CrossRef]
67. Li, X.; Zhou, Y.; Li, Y.; Yang, L.; Ma, Y.; Peng, X.; Yang, S.; Liu, J.; Li, H. Autophagy: A novel mechanism of chemoresistance in cancers. *Biomed. Pharmacother.* **2019**, *119*, 109415. [CrossRef]
68. Cho, S.W.; Na, W.; Choi, M.; Kang, S.J.; Lee, S.G.; Choi, C.Y. Autophagy inhibits cell death induced by the anti-cancer drug morusin. *Am. J. Cancer Res.* **2017**, *7*, 518–530.
69. Fujikake, N.; Shin, M.; Shimizu, S. Association Between Autophagy and Neurodegenerative Diseases. *Front. Neurosci.* **2018**, *12*, 255. [CrossRef]
70. Buchan, J.R.; Parker, R. Eukaryotic stress granules: The ins and outs of translation. *Mol. Cell* **2009**, *36*, 932–941. [CrossRef]
71. Wolozin, B.; Ivanov, P. Stress granules and neurodegeneration. *Nat. Rev. Neurosci.* **2019**, *20*, 649–666. [CrossRef] [PubMed]
72. Protter, D.S.W.; Parker, R. Principles and Properties of Stress Granules. *Trends Cell Biol.* **2016**, *26*, 668–679. [CrossRef] [PubMed]
73. Yague, E.; Raguz, S. Escape from stress granule sequestration: Another way to drug resistance? *Biochem. Soc. Trans.* **2010**, *38*, 1537–1542. [CrossRef] [PubMed]
74. Park, Y.J.; Choi, D.W.; Cho, S.W.; Han, J.; Yang, S.; Choi, C.Y. Stress Granule Formation Attenuates RACK1-Mediated Apoptotic Cell Death Induced by Morusin. *Int. J. Mol. Sci.* **2020**, *21*, 5360. [CrossRef]
75. Zuo, G.Y.; Yang, C.X.; Han, J.; Li, Y.Q.; Wang, G.C. Synergism of prenylflavonoids from Morus alba root bark against clinical MRSA isolates. *Phytomedicine* **2018**, *39*, 93–99. [CrossRef]
76. Pang, D.; Liao, S.; Wang, W.; Mu, L.; Li, E.; Shen, W.; Liu, F.; Zou, Y. Destruction of the cell membrane and inhibition of cell phosphatidic acid biosynthesis in Staphylococcus aureus: An explanation for the antibacterial mechanism of morusin. *Food Funct.* **2019**, *10*, 6438–6446. [CrossRef]
77. Wu, S.C.; Han, F.; Song, M.R.; Chen, S.; Li, Q.; Zhang, Q.; Zhu, K.; Shen, J.Z. Natural Flavones from Morus alba against Methicillin-Resistant Staphylococcus aureus via Targeting the Proton Motive Force and Membrane Permeability. *J. Agric. Food Chem.* **2019**, *67*, 10222–10234. [CrossRef]
78. Syahdi, R.R.; Mun'im, A.; Suhartanto, H.; Yanuar, A. Virtual screening of Indonesian herbal database as HIV-1 reverse transcriptase inhibitor. *Bioinformation* **2012**, *8*, 1206–1210. [CrossRef]
79. Singh, S.; Florez, H. Coronavirus disease 2019 drug discovery through molecular docking. *F1000Research* **2020**, *9*, 502. [CrossRef]
80. Gupta, G.; Chellappan, D.K.; Agarwal, M.; Ashwathanarayana, M.; Nammi, S.; Pabreja, K.; Dua, K. Pharmacological Evaluation of the Recuperative Effect of Morusin Against Aluminium Trichloride (AlCl3)-Induced Memory Impairment in Rats. *Cent. Nerv. Syst. Agents Med. Chem.* **2017**, *17*, 196–200. [CrossRef]
81. Kuk, E.B.; Jo, A.R.; Oh, S.I.; Sohn, H.S.; Seong, S.H.; Roy, A.; Choi, J.S.; Jung, H.A. Anti-Alzheimer's disease activity of compounds from the root bark of *Morus alba* L. *Arch. Pharm. Res.* **2017**, *40*, 338–349. [CrossRef] [PubMed]
82. Borah, K.; Sharma, S.; Silla, Y. Structural bioinformatics-based identification of putative plant based lead compounds for Alzheimer Disease Therapy. *Comput. Biol. Chem.* **2019**, *78*, 359–366. [CrossRef] [PubMed]
83. Kharroubi, A.T.; Darwish, H.M. Diabetes mellitus: The epidemic of the century. *World J. Diabetes* **2015**, *6*, 850–867. [CrossRef] [PubMed]
84. Guilherme, A.; Virbasius, J.V.; Puri, V.; Czech, M.P. Adipocyte dysfunctions linking obesity to insulin resistance and type 2 diabetes. *Nat. Rev. Mol. Cell Biol.* **2008**, *9*, 367–377. [CrossRef] [PubMed]

85. Athyros, V.G.; Doumas, M.; Imprialos, K.P.; Stavropoulos, K.; Georgianou, E.; Katsimardou, A.; Karagiannis, A. Diabetes and lipid metabolism. *Hormones* **2018**, *17*, 61–67. [CrossRef]
86. Tsalamandris, S.; Antonopoulos, A.S.; Oikonomou, E.; Papamikroulis, G.A.; Vogiatzi, G.; Papaioannou, S.; Deftereos, S.; Tousoulis, D. The Role of Inflammation in Diabetes: Current Concepts and Future Perspectives. *Eur. Cardiol.* **2019**, *14*, 50–59. [CrossRef]
87. Kim, H.I.; Ahn, Y.H. Role of peroxisome proliferator-activated receptor-gamma in the glucose-sensing apparatus of liver and beta-cells. *Diabetes* **2004**, *53* (Suppl. 1), S60–S65. [CrossRef]
88. Kumari, S.; Badana, A.K.; Murali, M.G.; Shailender, G.; Malla, R. Reactive Oxygen Species: A Key Constituent in Cancer Survival. *Biomark. Insights* **2018**, *13*, 1177271918755391. [CrossRef]
89. Agarwal, S.; Mohamed, M.S.; Mizuki, T.; Maekawa, T.; Sakthi Kumar, D. Chlorotoxin modified morusin-PLGA nanoparticles for targeted glioblastoma therapy. *J. Mater. Chem. B* **2019**, *7*, 5896–5919. [CrossRef]
90. Wang, F.; Zhang, D.; Mao, J.; Ke, X.X.; Zhang, R.; Yin, C.; Gao, N.; Cui, H. Morusin inhibits cell proliferation and tumor growth by down-regulating c-Myc in human gastric cancer. *Oncotarget* **2017**, *8*, 57187–57200. [CrossRef]
91. Hao, J.; Zhang, Y.; Yan, X.; Yan, F.; Sun, Y.; Zeng, J.; Waigel, S.; Yin, Y.; Fraig, M.M.; Egilmez, N.K.; et al. Circulating Adipose Fatty Acid Binding Protein Is a New Link Underlying Obesity-Associated Breast/Mammary Tumor Development. *Cell Metab.* **2018**, *28*, 689–705.e685. [CrossRef] [PubMed]
92. Wang, T.; Fahrmann, J.F.; Lee, H.; Li, Y.J.; Tripathi, S.C.; Yue, C.; Zhang, C.; Lifshitz, V.; Song, J.; Yuan, Y.; et al. JAK/STAT3-Regulated Fatty Acid beta-Oxidation Is Critical for Breast Cancer Stem Cell Self-Renewal and Chemoresistance. *Cell Metab.* **2018**, *27*, 1357. [CrossRef] [PubMed]

International Journal of
Molecular Sciences

MDPI

Article

Double Gamers—Can Modified Natural Regulators of Higher Plants Act as Antagonists against Phytopathogens? The Case of Jasmonic Acid Derivatives

Nicolò Orsoni [1], Francesca Degola [1,*], Luca Nerva [2,3], Franco Bisceglie [1], Giorgio Spadola [1], Walter Chitarra [2,3], Valeria Terzi [4], Stefano Delbono [4], Roberta Ghizzoni [4], Caterina Morcia [4], Agnieszka Jamiołkowska [5], Elżbieta Mielniczuk [5], Francesco M. Restivo [1] and Giorgio Pelosi [1]

1 Department of Chemistry, Life Sciences and Environmental Sustainability, University of Parma, Parco Area delle Scienze 11/A, 43124 Parma, Italy; nicolo.orsoni@studenti.unipr.it (N.O.); franco.bisceglie@unipr.it (F.B.); giorgio.spadola1@studenti.unipr.it (G.S.); restivo@unipr.it (F.M.R.); giorgio.pelosi@unipr.it (G.P.)
2 Council for Agricultural Research and Economics—Research Centre for Viticulture and Enology CREA-VE, Via XXVIII Aprile 26, 31015 Conegliano (TV), Italy; luca.nerva@crea.gov.it (L.N.); walter.chitarra@crea.gov.it (W.C.)
3 Institute for Sustainable Plant Protection, CNR, Strada delle Cacce 73, 10135 Torino, Italy
4 Council for Agricultural Research and Economics—Research Centre for Genomics and Bioinformatics CREA-GB, Via San Protaso 302, 29017 Fiorenzuola d'Arda (PC), Italy; valeria.terzi@crea.gov.it (V.T.); stefano.delbono@crea.gov.it (S.D.); roberta.ghizzoni@crea.gov.it (R.G.); caterina.morcia@crea.gov.it (C.M.)
5 Department of Plant Protection, University of Life Sciences in Lublin, Leszczyńskiego 7, 20069 Lublin, Poland; aguto@wp.pl (A.J.); elzbieta.mielniczuk@up.lublin.pl (E.M.)
* Correspondence: francesca.degola@unipr.it

Received: 14 October 2020; Accepted: 13 November 2020; Published: 17 November 2020

Abstract: As key players in biotic stress response of plants, jasmonic acid (JA) and its derivatives cover a specific and prominent role in pathogens-mediated signaling and hence are promising candidates for a sustainable management of phytopathogenic fungi. Recently, JA directed antimicrobial effects on plant pathogens has been suggested, supporting the theory of oxylipins as double gamers in plant-pathogen interaction. Based on these premises, six derivatives (dihydrojasmone and cis-jasmone, two thiosemicarbazonic derivatives and their corresponding complexes with copper) have been evaluated against 13 fungal species affecting various economically important herbaceous and woody crops, such as cereals, grapes and horticultural crops: *Phaeoacremonium minimum, Neofusicoccum parvum, Phaeomoniella chlamydospora, Fomitiporia mediterranea, Fusarium poae, F. culmorum, F. graminearum, F. oxysporum* f. sp. *lactucae, F. sporotrichioides, Aspergillus flavus, Rhizoctonia solani, Sclerotinia* spp. and *Verticillium dahliae*. The biological activity of these compounds was assessed in terms of growth inhibition and, for the two mycotoxigenic species *A. flavus* and *F. sporotrichioides*, also in terms of toxin containment. As expected, the inhibitory effect of molecules greatly varied amongst both genera and species; cis-jasmone thiosemicarbazone in particular has shown the wider range of effectiveness. However, our results show that thiosemicarbazones derivatives are more effective than the parent ketones in limiting fungal growth and mycotoxins production, supporting possible applications for the control of pathogenic fungi.

Keywords: jasmonic acid; jasmone derivatives; phytopathogenic fungi; crop protection; thiosemicarbazones; mycotoxins; mycopesticides

---

1. Introduction

Synthesized from α-linolenic acid originating from chloroplast membranes, jasmonic acid (JA) is regarded as a key player among the molecules that play a role in plant defense, signaling and development; it is known to be involved, together with other phytohormones (e.g., salicylic acid, ethylene, indole-3-acetic acid, abscisic acid, cytokinin and gibberellin), in the host defense response and immune signaling modulation. Above all, it is well known to be involved in the regulation of plant adaptations to biotic (such as pathogen infection and herbivore attack) as well as abiotic stresses [1,2]. Alongside an increasing number of studies showing that JA intervenes in a remarkable number of plant developmental events, including primary root growth, reproductive development and leaf senescence, it has been long-time proven that some plant hormones are also able to affect fungal physiology. In particular, it is worth noting that JA can have an impact on mycelium growth and development, conidiation, spore germination and sexual reproduction, suggesting that these molecules can interfere with specific signal transduction processes in fungi, although the effect strongly depends on the species [3–7]. Some of these molecules were successfully used for the discovery of novel and diverse natural products, being effective in activating silent biosynthetic genes/pathways of fungal secondary metabolism [8]. At the same time, some fungi (and not only plant-interacting ones) have the ability to synthetize plant hormones and their analogs [9–11]. With the idea of exploiting the potential of JA derivatives to interfere with fungal development and metabolism, a panel of six JA-derived molecules (two ketones of natural origin, di-hydrojasmone and cis-jasmone, their thiosemicarbazonic derivatives and the corresponding copper complexes) was evaluated against 13 fungal species affecting various economically important herbaceous and woody crops such as cereals, grape and horticultural plants. All these fungi cause relevant yield and quality losses and are characterized by different life cycle, pathogenicity behavior, transmission mode and effects on both plants and derived products. We grouped the investigated fungal species in three classes according to the prevalent host plant: (1) *Neofusicoccum parvum*, *Phaeoacremonium minimum* and *Phaeomoniella chlamydospora*, belonging to the Ascomycota phylum and previously characterized as associated to esca disease in grapevine, one of the main constraints now emerging in viticulture [12]; *Fomitiporia mediterranea*, belonging to the Basidiomycota phylum and known to be the causal agent of white rot in grape wood trunks. (2) *Fusarium culmorum*, *F. graminearum*, *F. sporotrichioides*, *F. poae* and *Aspergillus flavus*, all Ascomycota species affecting grain cereals. *Fusarium* spp. are responsible for *Fusarium* Head Blight (FHB), a disease of great economic concern for both cereal producers and grain processing industry [13]. This disease is most often caused by *F. graminearum* and *F. culmorum* and in recent years also by *F. poae*. The participation of individual species in causing FHB is highly dependent on weather conditions [14–16]. *Aspergillus flavus*, as well as *Fusarium* spp., can contaminate the colonized grains with several classes of mycotoxins, secondary metabolites dangerous to human and animal health, severely affecting yield production and lowering the quality and safety of the final products [17–19]. (3) *Rhizoctonia solani*, *Sclerotinia* spp., *Fusarium oxysporum* and *Verticillium dahliae*, all soil-borne fungal pathogens, belonging to Basidiomycota and Ascomycota divisions, etiological agents of several diseases in a wide range of plants worldwide. Even though different in their life cycle, they share the responsibility for affecting agricultural yield losses, especially in the horticultural crops cultivated in temperate regions [20].

2. Results

2.1. Chemistry

As stated in the introduction, a panel of six molecules was used to test their effect on the selected fungal species. The compounds are two ketones of natural origin, cis-jasmone and dihydrojasmone, their thiosemicarbazone derivatives and the corresponding copper complexes (Scheme 1). The choice for a derivatization has fallen on thiosemicarbazones because they are a versatile class of compounds that possess biological properties such as antibacterial, antivirus, antiamoebic and antitumor in human pharmacology [21]. It is also well known that inorganic substances, like copper salts, have long been used for their capacity of inhibiting the development of molds and bacteria and can have effect on the growth of fungi and on aflatoxin production [22]. Moreover, thiosemicarbazone metal complexes and those of copper(II) in particular, are known to produce interesting biological effects showing that the coordination compounds own effective properties at lower concentrations with respect to the parent organic molecules and these properties are often related to reactive oxygen species (ROS) production or other pathways involving the metal ion [23]. All the compounds used in this study were characterized by standard methods and for one of them (JTS) the structure by X-ray diffraction (XRD) was also obtained (Figure 1). The structure is particularly interesting because it reveals that the hydrophobic lateral chain of the jasmone moiety molecule possesses a certain degree of freedom. The asymmetric unit of JTS (CCDC 2019086) consists of three independent molecules—two of them (A and B) form a dimeric structure while the third molecule (C) forms a dimer with a symmetry related molecule C. Molecule A in particular differs from molecules B and C by the orientation of the hydrophobic chain. Both dimers (A:B and C:C′) present a double intermolecular hydrogen bond between the N2 of one molecule and the S1 of the other (Table 1). Bond lengths in the thiosemicarbazone moiety of the three molecules were calculated and reported as average in Table 2.

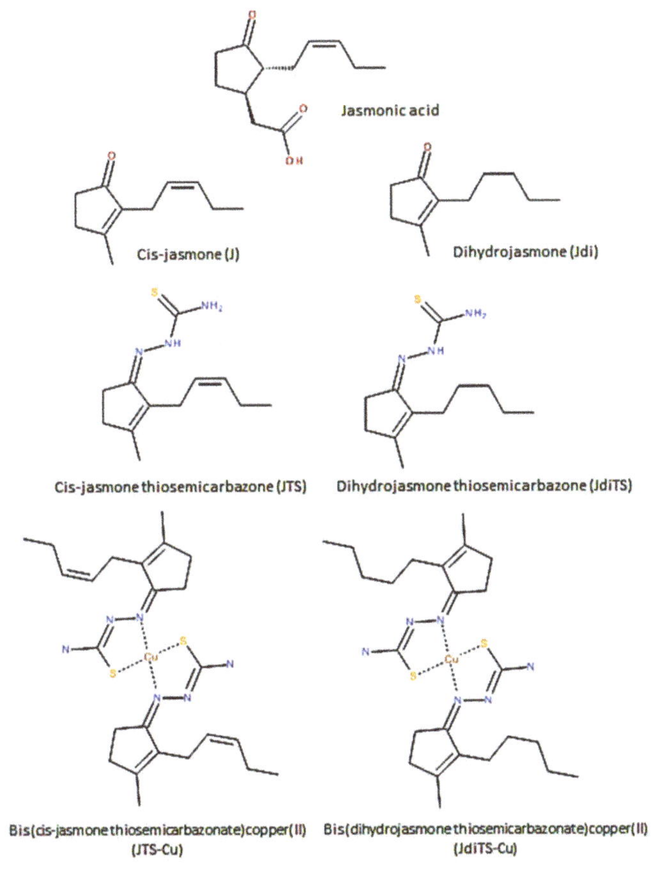

Scheme 1. Schematic representation of the structures of Jasmonic acid (JA) derivatives (J, Jdi), their thiosemicarbazones (JTS, JdiTS), and their copper complexes (JTS-Cu, JdiTS-Cu).

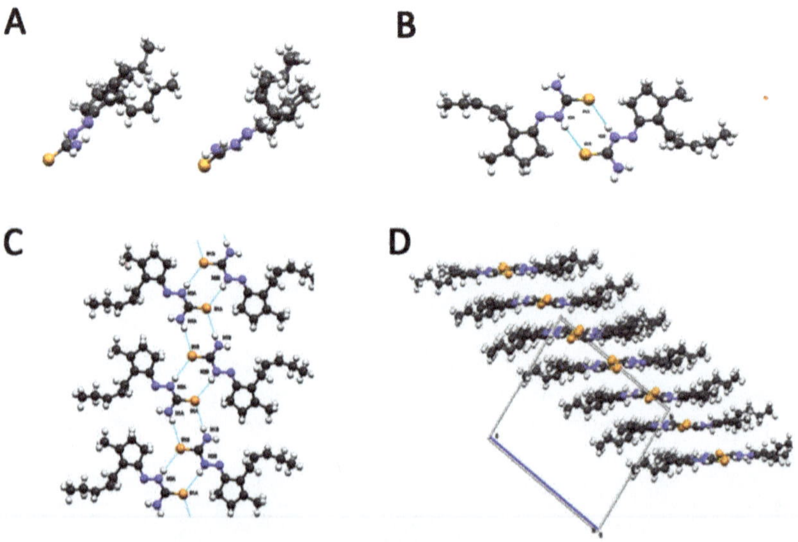

Figure 1. (**A**) Overlapping of molecules A/B (left) and A/C (right). (**B**) Hydrogen bonding scheme between molecule A and B. (**C**) Hydrogen bond scheme along the ribbons formed by molecules A/B. (**D**) packing of the ribbons. Sulfur is represented in yellow, nitrogen in blue and carbon in gray.

Table 1. The different hydrophobic chain dihedral angles for the three independent molecules.

		A	B	C
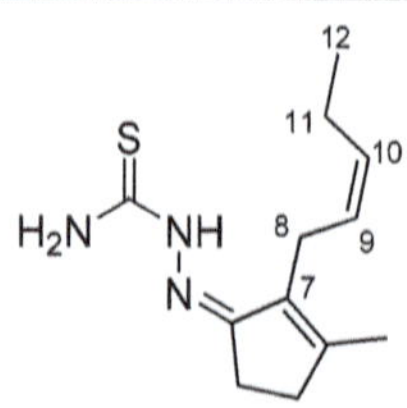	C7–C8–C9–C10	138.9 (8)°	−125.1 (8)°	−130.9 (6)°
	C8–C9–C10–C11	−4.3 (16)°	−2.0 (15)°	−2.0 (15)°
	C9–C10–C11–C12	−153.6 (13)°	134.3 (11)°	123.9 (8)°

Table 2. Average bond lengths in the thiosemicarbazone moiety of the three molecules.

	Atoms	Average Length (Å)
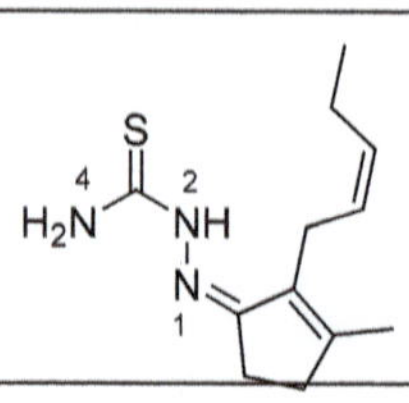	C-N4	1.320
	C-S	1.695
	C-N2	1.346
	N2-N1	1.389

Sulfur S1 of the thiosemicarbazone is involved in a second intermolecular hydrogen bond with N4-H of a third molecule (Figure 1C). This bond is responsible for the formation of a second type of dimer-like structures which form ribbons developing along the *a* axis. As shown in Figure 1D, the packing of the crystal can be described as due to the stacking of planes formed alternatively by ribbons of heterodimers (A and B) and homodimers (C and C′). These layers are kept together by hydrophobic and π-π stacking interactions of the thiosemicarbazone moiety).

2.2. Scavenging Potential Determination of Compounds

The scavenging potential of thiosemicarbazones and of their metal complexes was assessed against the 2,2-diphenyl-1-picrylhydrazyl radical (DPPH), by measuring the bleaching of a purple colored methanol solution of the stable DPPH radical [24]. The DPPH radical scavenging ability of JdiTS-Cu traced those of 0.3 mM ascorbic acid (100%) at each concentration tested, while JTS-Cu showed a dose-dependent potential that reached 45% of DPPH radical inhibition at 40 μM concentration and overcame 70% at 50 μM (Figure 2). On the contrary, both aldehydes (J and Jdi) and their thiosemicarbazones derivatives did not show any appreciable scavenging activity at any tested concentration. These observations suggested that the complexation with copper significantly increases the otherwise slight proton-donating ability of ligands, providing them with more interesting potential as free radical inhibitors - or scavengers, acting possibly as primary antioxidants.

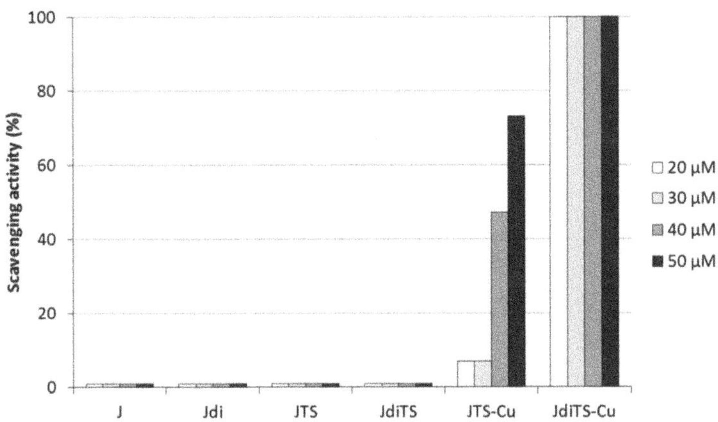

Figure 2. Scavenging assay (2,2-diphenyl-1-picrylhydrazyl radical (DPPH)). The in vitro antioxidant ability of compounds is expressed as percentage of inhibition with respect to ascorbic acid 30 mM scavenging activity (100%).

2.3. Antifungal Activity on Grapevine Esca-Associated Pathogens

The ability of the selected jasmone-derived molecules to impair fungal growth was assessed on four fungal strains associated to esca disease in grapevine, previously characterized [25]. The investigated molecules were added to the Czapek Dox AgarCZA growth media at 25 and 50 μM concentrations. As reported in Figure 3, not all the molecules were effective against the fungal growth at 25 μM concentration: JTS and its Cu-complex (JTS-Cu) resulted in a complete impairment of colony development only in *P. chlamydospora*, almost completely inhibited by both the compounds up to 15 days. On the contrary, some variances were observed among the different species: 100% inhibition was achieved with JTS on *P. minimum* at 5 day post inoculation (dpi), but the effect was completely lost at 10 and 15 dpi. Conversely, JTS-Cu displayed a significant containment on fungal growth (about 30%) only at 10 and 15 dpi. The effect of JTS on *N. parvum* proved to be longer lasting effective, impairing the fungal growth from 65% (2 dpi) to 50% (7 dpi), whereas its complex JTS-Cu, although displaying a higher inhibition ability at 2 dpi (about 90%), completely lost its activity at 7 dpi. Both the cis-jasmone thiosemicarbazones slightly affect growth in *F. mediterranea* at 4 and 7 dpi, while at 2 dpi JTS-Cu was more effective (100% inhibition) than its ligand, showing that for this compound a growth recovery occurred.

Observations conducted on all the relevant species at 50 μM concentration of test compounds showed the achievement, by JTS and JTS-Cu, of a total impairment of colony development at every dpi of evaluation (Figure S1).

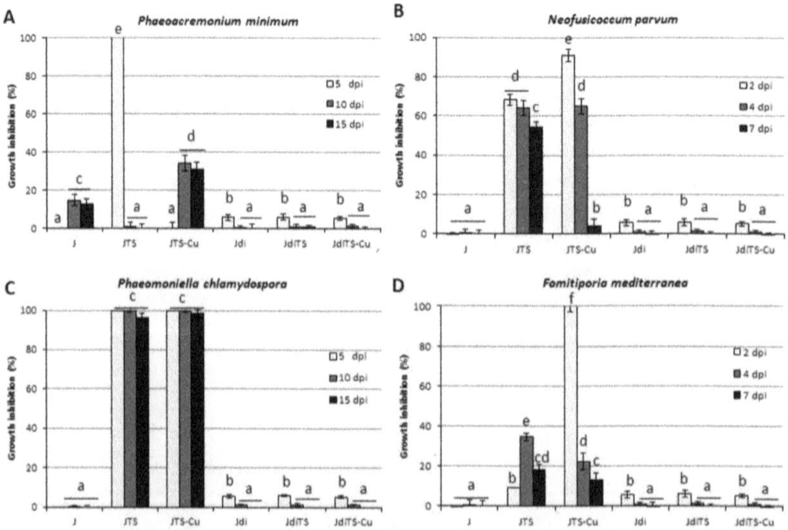

Figure 3. Effect of compounds on the growth of esca disease-linked species (**A**) *P. minimum*, (**B**) *N. parvum*, (**C**) *P. chlamydospora* and (**D**) *F. mediterranea*. Culture medium was amended with molecules at 25 µM concentration; values are reported as inhibition percentage as compared to the control (Dimethyl sulfoxide, DMSO, 0.25% *v/v* amended cultures) ± S.D. Different letters indicate statistically significant differences ($p < 0.05$).

2.4. Antifungal Activity on Cereals Pathogens

The ability of four selected molecules was tested against *Fusarium culmorum*, *F. graminearum*, *F. poae* and *F. sporotrichioides*, studying the effect on their growth in vitro. Four selected molecules (J, JTS, Jdi and JdiTS) were added to growth medium (PDA) at increasing concentrations (5, 25 and 50 µM); control plates contained DMSO at the relevant concentrations. Inhibition of fungal colony growth was monitored at 2, 4 and 6 days post inoculum. As given in Figure 4, compound JTS resulted to be the most effective in reducing fungal development: all strains of *Fusarium* spp. were strongly inhibited but the most affected was *F. poae*, being visibly reduced in growth from 2 dpi also at the lowest concentration (5 µM), while 25 and 50 µM concentrations reached respectively 45.9% and 59.1% inhibition and maintained a similar level until 6 dpi (54.2% and 63.0% respectively). Other compounds did not have any appreciable activity (Figure 4A). In *F. culmorum*, the fungistatic effect of JTS was obtained with 25 and 50 µM concentration and recorded on 2 and 4 dpi (52.6% and 55.6% respectively), while it slightly decreased at 6 dpi (Figure 4B); however, together with *F. graminearum*, this species resulted more susceptible than *F. poae* to other molecules: although below the threshold of 40% inhibition, J, Jdi and JdiTS exerted a time-dependent containment effect, that reached a maximum at 4 dpi (Figure 4C); J molecule inhibited the growth of *F. graminearum* in the initial growth period at 50 µM (28.7%) and of *F. culmorum* 4 dpi at 5 µM (23.4%). Jdi was most effective on *F. culmorum* at 5 µM, at 4 dpi (29.9%) and *F. graminearum* at 50 µM, at 2 dpi (33.1%). JdiTS mainly affected *F. graminearum* and *F. culmorum* growth on day 4 at 50 µM (37.5% and 32.0% respectively). Compound JTS resulted particularly effective in inhibiting *F. sporotrichioides* mycelium growth, reaching a maximum inhibition level of 70%, whereas J and Jdi reduced of 20%–30% the fungal growth only in the initial phase of the exposure. On the contrary, JdiTS showed an intermediate efficacy, reaching a maximum inhibition level of 50% after 4 days of exposure and only at the highest concentration (Figure 4D).

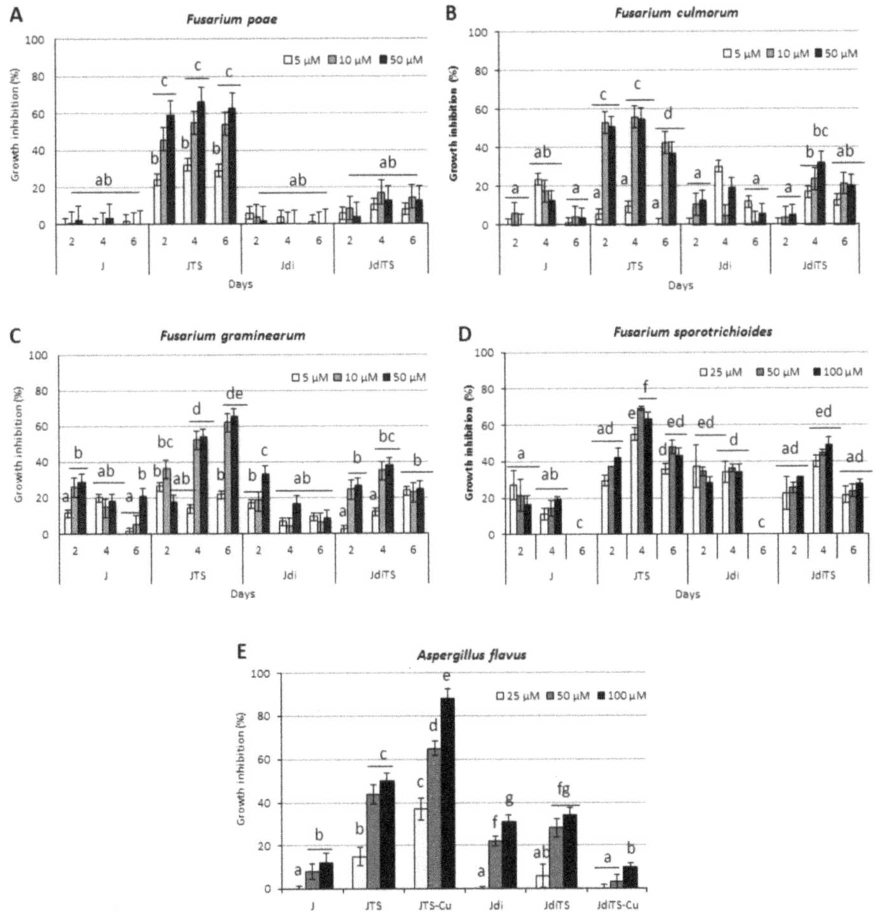

Figure 4. Effects of the compounds on the growth of cereals pathogenic fungi (**A**) *F. poae*, (**B**) *F. culmorum*, (**C**) *F. graminearum*, (**D**) *F. sporotrichioides* and (**E**) *A. flavus*. Culture medium was amended with an increasing concentration of molecules (from 5 to 100 µM) on the basis of the species; values are reported as inhibition percentage as compared to the control (DMSO amended cultures) ± S.D. Different letters indicate statistically significant differences ($p < 0.05$).

The whole panel of compounds was tested against *Aspergillus flavus*: when administrated at increasing concentration (from 25 to 100 µM) and evaluated for the inhibition of mycelium biomass production at 6 dpi, neither J nor Jdi molecule showed any fungistatic effect, while the relevant ligands proved to affect fungal growth at a higher level (Figure 4E). Complexation with copper induced an increase of the fungistatic effect only in the case of JTS-Cu, while almost reducing JdiTS fungistatic activity.

2.5. *Antifungal Activity on Horticultural Crop Pathogens—Sclerotinia spp., V. dahliae, Rhizoctonia solani and F. oxysporum*

The antimicrobial potentialities of J, JTS, Jdi and JdiTS at 25, 50 and 100 µM concentration were evaluated, in terms of colony growth inhibition, on four fungal species affecting a wide range of horticultural crops; Figure 5 reports the percentages of growth inhibition respect to control, obtained at 2, 4 and 6 dpi. As major points, it was found that all the fungal species showed a similar trend of susceptibility to all compounds, even at different extent: *Rhizoctonia solani*, the less sensitive, touched the maximum level of growth inhibition (20–30%) when exposed to JTS, Jdi and JdiTS. On the contrary, *Sclerotinia* proved to be the most affected by the exposure to compounds. Additionally, not all the molecules were effective in containing the fungal growth: in fact, only JTS resulted in a complete impairment of colony development for *Sclerotinia* and *V. dahliae*.

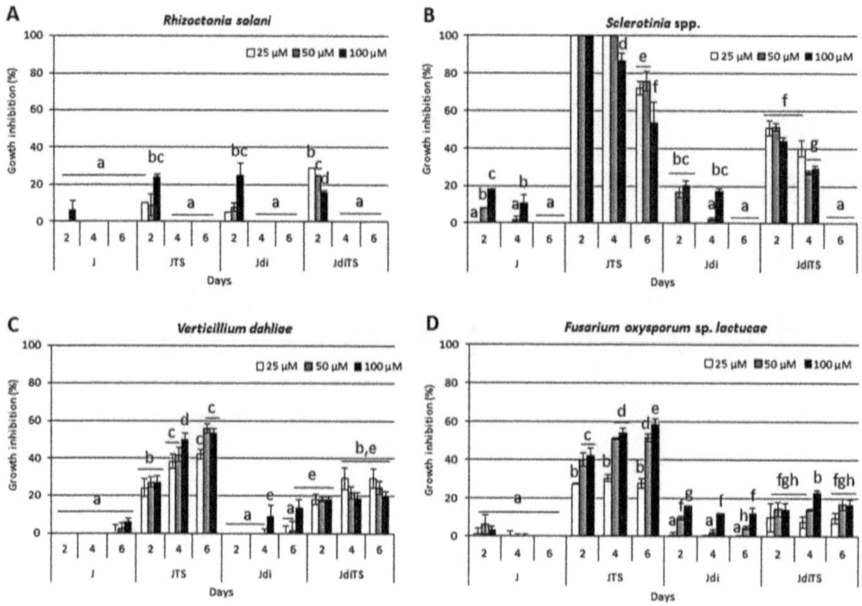

Figure 5. Effect of compounds on the growth of horticultural crop pathogens (**A**) *R. solani*, (**B**) *Sclerotinia* spp., (**C**) *V. dahliae* and (**D**) *F. oxysporum*. Culture medium was amended with increasing concentrations of molecules (from 25 to 100 µM) on the basis of the species; values are reported as inhibition percentage as compared to the control (DMSO amended cultures) ± S.D. Different letters indicate statistically significant differences ($p < 0.05$).

2.6. Interference with Fungal Secondary Metabolism—Effect on Mycotoxin Production and Sclerotia Development

Besides the inhibitory potential against mycelium growth, the possible interference of some compounds with fungal secondary metabolisms was assessed: more specifically, the possible impairment of mycotoxins biosynthesis and sclerotia production in vitro was evaluated for a few relevant species. The production of T-2/HT-2 toxin by *F. sporotrichioides* treated with different concentrations of J, Jdi, JTS and JdiTS showed to be differently affected: the exposure to natural ketones did not induced any reduction, even at the highest concentrations (Figure 6A). On the contrary, both ligands were effective in lowering the mycotoxin biosynthesis: from the 40% inhibition at 5 µM to 75% at 25 µM concentration. *Aspergillus flavus* response to molecules provided interesting findings about ligands and their metal complexes: as observed in *F. sporotrichioides*, ketones reached a maximum of 30% inhibition, at 100 µM, while a stronger containment of aflatoxin (up to 98%) was achieved with JTS and JdiTS. However, the addition of a copper ion to the thiosemicarbazonic molecule did not enhance their anti-mycotoxigenic activity (Figure 6B). Coherently with aflatoxin inhibition, when tested on sclerotia biogenesis J and Jdi resulted completely ineffective and both ligands proved to be able to reduce sclerotia production up to 71.9% and 62.5% (for JTS and JdiTS respectively); surprisingly, contrariwise to what observed for the aflatoxin metabolism, exposure to JTS-Cu determined an enhanced activity with respect to JTS, while the copper complex JdiTS-Cu resulted in a complete abatement of JdiTS effect (Table 3).

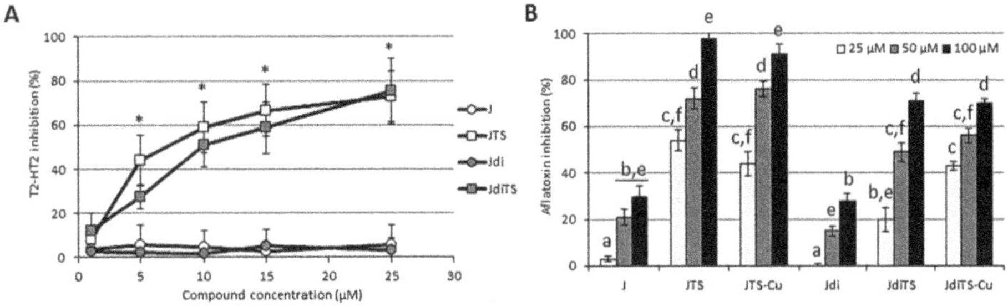

Figure 6. Effect of compounds on the mycotoxin production in two cereals pathogenic fungi. Toxins T-2 and HT-2 were quantified in *F. sporotrichioides* cultures exposed to increasing concentration (from 1 to 25 µM) of J, JTS, Jdi and JdiTS compounds (**A**). Aflatoxins were quantified in *A. flavus* cultures exposed to increasing concentration (from 25 to 100 µM) of all the compounds (**B**). Values are reported as inhibition percentage as compared to the control (DMSO amended cultures) ± S.D. Asterisks or different letters indicate statistically significant differences ($p < 0.05$).

Table 3. Effect of compounds on sclerotia biogenesis. Sclerotia biomass (*A. flavus*) or number (*Sclerotinia* spp.) was determined in cultures exposed to 100 µM concentration of compounds. Values are reported as inhibition percentage as compared to the control (DMSO amended cultures) ± S.D. Asterisks indicate statistically significant differences ($p < 0.05$).

Compound	*Aspergillus flavus* Inhibition (%) mg/Colony Area (cm^2)	*Sclerotinia* spp. Inhibition (%) nr/Colony Area (cm^2)
JTS	71.9 ± 3.0 *	55.3 ± 7.1 *
JTS-Cu	96.9 ± 1.6 *	n.d.
JdiTS	62.5 ± 5.7 *	52.6 ± 3.5 *
JdiTS-Cu	0.0 ± 0.2	n.d.

When evaluated on *Sclerotinia* spp., the two ligands showed a different behavior with respect to *A. flavus*: the inhibitory activity on the sclerotia development was similar for JTS and JdiTS (Table 3) and determined a reduction of the number of sclerotia *per* colony area (cm^2) of 55.3% and 52.6% respectively.

3. Discussion

Jasmonic acid (JA) and its derivatives (such as methyl-jasmonate) belong to the plant oxylipins family, a large group of metabolites derived from polyunsaturated fatty acids commonly associated with the modulation of defense response [26]. Recently, the role of oxylipins in plant defense against biotic stressors has been rediscovered, suggesting, in addition to the induction and regulation of gene expression in the plant host, a direct antimicrobial effect on microbial pathogens. On this path, in vitro investigations on growth inhibitory activity of various natural oxylipins proved that some of them are able to heavily affect mycelial growth and spore germination in a number of eukaryotic phytopathogens, indicating a contribution of such compounds to host resistance through related but different mechanisms [27]. Six molecules structurally related to JA was thus tested against a set of 13 phytopathogenic fungal species, with the aim of compare the potential of modified JA derivatives in controlling fungal diseases of economically important crops. Due to their synthesis low yield, the two metal complexes were tested only on pathogens relevant to maize and grape, since they are considered model systems for herbaceous/monocots and woody/dicots respectively. Relative to the biological activity of compounds on both primary (fungal growth) and secondary metabolisms (mycotoxins accumulation and sclerotia biogenesis), some interesting differences between the JA-derivatives were found—natural ketones did not show any fungistatic effect on fungal pathogens affecting grapevine and linked to grapevine esca disease, as well as pathogens affecting horticulture crops; however, a slight inhibitory activity was recorded on cereals contaminating species (*Fusarium* and *Aspergillus* spp). Thiosemicarbazones were found more effective than their parent compounds, even if to a different, species-specific extent; JTS was undoubtedly the most promising, being able to prejudice the growth, at the highest level, of almost all the pathogens tested, with the only exception of *R. solani*, which, indeed, revealed to be the less susceptible to the exogenous administration of JA derivatives.

From a developmental perspective, the results obtained analyzing the response of *A. flavus* to our compounds were intriguing: in fact, if in *F. sporotrichioides* the inhibitory effect on toxin T2-HT2 accumulation seemed to be coupled with a comparable fungistatic activity, in *A. flavus* the highest fungistatic and anti-aflatoxigenic activity of thiosemicarbazonic ligand—in comparison with its natural ketone—was observed for JTS but not for JdiTS, that, with respect to Jdi, showed an increased effect on aflatoxin accumulation but not on fungal growth suppression. Even more interesting was the behavior of the two copper complexes that, in contrast with their free ligands, showed a different activity trend—the complexation with the metallic nucleus appeared in fact to determine an intensification of the fungistatic potential only for JTS-Cu, whose inhibitory effect on aflatoxin accumulation seems to be justified by the mycelial reduction. On the other hand, JdiTS did not differ from JdiTS for the anti-aflatoxigenic activity but the addition of the copper ion almost deleted its fungistatic effect. This antagonist effect of JA-derivatives against aflatoxins accumulation in *A. flavus* with no apparent effect on mycelial growth was already described [28].

Results obtained for the esca-related pathogens are of great interest: esca syndrome is now becoming the most devastating issue in viticulture, with an impact on both yield and quality [29], since the nature of this syndrome, which derives from the simultaneous colonization of the woody tissues from several fungi, makes it a difficult target for common antimicrobial treatments [30]. For this reason, up to date, a cure for this syndrome is not available and only preventive techniques can be undertaken. In this light, the ability of JAs to impair fungal growth can be implemented and exploited to counteract the esca syndrome.

Interesting results in terms of inhibitory effect were also obtained against *Fusarium* spp., the cause of the economically important disease of cereals (FHB) and mycotoxins contaminant agent: the high fungistatic activity of JTS against *Fusarium* spp. strains, which was maintained even after 6 days of growth, suggests that such compound can be used in the future to formulate new fungicides for the protection of cereal ears against FHB; additionally, the observation of the thiosemicarbazones different effectiveness depending on fungus species confirmed the findings previously described [31,32].

Intriguing observations can also be raised about the antioxidant activity if compared to sclerotia biogenesis impairment, since in many fungal species the redox balance is considered a prominent factor involved in the control of secondary metabolism and several synthetic and natural compounds that modulate secondary pathways—such as AF biosynthesis and sclerotia biogenesis—proved to possess a scavenging activity against ROS. In particular, the sclerotia metamorphosis in filamentous fungi appears to be triggered by oxidative stress [33–35]. In this study, no correlation was found between the predicted antioxidant scavenging ability of compounds, as determined in vitro with the DPPH assay and their effect on the secondary metabolism (intended as both sclerotia biogenesis and mycotoxin biosynthesis), consistently with previously obtained results [23,36]: the Cu ion provided the thiosemicarbazones with a scavenging potential otherwise absent in both ketones and ligands. Copper was thought to be the redox active center of compounds but actually it resulted in a highly variable in vivo effect. In consideration of these findings, the hypothesis of a correspondence between in vitro scavenging potential of Jasmone-derived TSs and their in vivo biological effect on *A. flavus* has to be argued: hereto, the anti-aflatoxigenic effect of JTS and JdiTS is accompanied by the inhibition of sclerotia biogenesis, suggesting that their target(s) could be involved in biochemical processes shared by both the aflatoxin and the sclerotia metabolism. Conversely, the highly significant anti-toxigenic effect of JTS-Cu seems to be unspecific and mainly due to an impairment of fungal primary metabolism, that in turn strongly affects aflatoxin biosynthesis and sclerotia production, as a consequence. In this sense, JdiTS-Cu resulted in the most interesting compound—the outcomes suggest a specificity of action whose target, differently from JTS and JdiTS, might be located in a metabolic knot closely related to the aflatoxin biosynthetic pathway, as supported by the observation that the prominent containment activity of this compound on aflatoxin accumulation is not supported by a significant impact on sclerotia development and hyphae growth.

4. Materials and Methods

4.1. Chemical Synthesis

The desired TSs (ligands) were obtained by mixing an equimolar amount of thiosemicarbazide with the appropriate aldehyde in absolute ethanol. A small amount of hydrochloric acid was added to catalyze the condensation. The mixture was refluxed under stirring for 8 h and left overnight at 0 °C. The precipitate was filtered out, washed with cold ethanol and dried under vacuum.

Cis-jasmonethiosemicarbazone: thiosemicarbazide (0.22 g, 2.4 mmol), cis-jasmone (0.40 mg, 2.4 mmol). White powder. Yield 26%. Mp. 149 °C. Fourier-transform infrared spectra (FT-IR) (cm^{-1}): 3401 (s), 3130 (m), 2964 (m), 1590 (s), 1507 (s) 874 (m). ^{1}H-NMR (δ, ppm; DMSO-d_6): 0.93 (t, J = 7.5 Hz, 3H), 1.89 (s, 3H), 2.13 (m, 2H), 2.44 (m, 2H), 2.59 (m, 2H), 2.98 (d, J = 9.2 Hz, 2H), 5.28 (m, 2H), 734 (s, 1H), 8.02 (s, 2H), 9.87 (s, 1H).

Dihydrojasmonethiosemicarbazone: thiosemicarbazide (0.22 g, 2.41 mmol), dihydrojasmone (0.40 g, 2.41 mmol). Pale yellow powder. Yield 41%. Mp. 177 °C. FT-IR (cm^{-1}): 3403 (s), 3130 (m), 2925 (m), 2852 (s), 1590 (m), 1507 (s), 718 (m). ^{1}H-NMR (δ, ppm; DMSO-d_6): 0.85 (t, J = 7.0 Hz, 3H), 1.25 (m, 4H), 1.37 (m, 2H), 2.22 (t, J = 7.4 Hz, 2H), 2.43 (m, 2H), 2.58 (m, 2H), 7.30 (s, 1H), 8.01 (s, 2H), 9.83 (s, 2H).

Metal complexes were obtained by dripping the solution containing the copper salt into the ligand solution, which rapidly turned to dark brown. The mixture was stirred at room temperature for 2 h and then the solvent was removed under reduced pressure. The solid formed was collected and washed with diethyl ether, then dried under vacuum.

Bis(cis-jasmonethiosemicarbazonate) Cu(II): copper(II) acetate (0.04 g, 0.21 mmol), 38 (0.10 g, 0.42 mmol). Dark brown powder. Yield 95%. FT-IR (cm^{-1}): 3416 (s), 2959 (m), 1570 (m), 1511 (m), 701 (m). ESI-MS (+) m/z calc. 537.28, found 537.39. *Bis(dihydrojasmonethiosemicarbazonate) Cu(II)*: copper(II) acetate (0.04 g, 0.21 mmol), 39 (0.10 mg, 0.42 mmol). Dark brown powder. Yield 74%. FT-IR (cm^{-1}): 2920 (s), 2851 (m), 1550 (m), 618 cm^{-1} (m). ESI-MS (+) m/z calc. 541.31, found 542.42.

All compounds were dissolved in di-methyl sulfoxide (DMSO) to achieve 10 mM stocks.

4.2. X-ray Analysis

X-ray diffraction data collection of compound JTS was carried out with a Bruker-Siemens SMART AXS 1000 diffractometer equipped with a charged-coupled device CCD detector, Mo Kα radiation (λ = 0.71069). The phase problem was solved by direct methods and the structure was refined by full-matrix least-squares on all F2 using SHELXL97 [37] as implemented in the Olex package [38]. Figures have been obtained using the Mercury software [39].

Crystal data: triclinic, $P_{\bar{1}}$, a = 8.164(5) Å, b = 15.645(9) Å, c = 16.434(9) Å, α = 84.723(10), β = 82.036(10), γ = 84.632(10), V = 2063(2) Å^3; Z =2; dcalc = 1.146 mg/cm^3, F(000) = 768, mu = 2.15, Tot. refl. = 19,159, hkl range = $-9 < h < 9$, $-19 < k < 19$, $-20 < l < 20$; Theta range 1.25–25.80, unique reflections = 7881, number of parameters = 433, GooF = 1.033, R = 0.0835, wR2 = 0.2039.

CCDC 2,019,086 contains the supplementary crystallographic data for this compound. These data can be obtained free of charge from The Cambridge Crystallographic Data Centre via www.ccdc.cam.ac.uk/data_request/cif.

4.3. *Scavenging Activity Assay (DPPH)*

A 90 μM methanol solution of 2,2-diphenyl-1-picrylhydrazyl radical (DPPH) was prepared: an aliquot of 2 mL was then incubated at room temperature with 12.5 μL of DMSO-dissolved sock solution of each compound corresponding to 20, 30, 40 and 50 μM concentration. The scavenging activity of the compound against the DPPH was observed with the decrease in absorbance measured at 518 nm. A solution of ascorbic acid at 0.3 mM was used as positive control to determine the maximal decrease in DPPH absorbance. Values were expressed in percentage of inhibition of DPPH absorbance in relation to the control values without the compound (ascorbic acid maximal inhibition was considered 100% of inhibition).

4.4. *Biological Assays*

4.4.1. Microorganisms

Phaeoacremonium minimum, *Neofusicoccum parvum*, *Phaeomoniella chlamydospora* and *Fomitiporia mediterranea* were chosen amongst grapewine diseases-associated fungal species which were previously characterized among the Research Centre for Viticulture and Enology (CREA-VE, Conegliano Veneto (TV), Italy) esca-associated fungal collection [25]. *Fusarium poae* (OZ45), *F. culmorum* (OZ47), *F. graminearum* (OZ161) strains were isolated from barley and oat grains and deposited in the fungal collection of Plant Protection Department at the University of Life Sciences in Lublin (Poland).

Aspergillus flavus strains CR10+ and TOφ strains, previously characterized and described [40,41], belong to the fungal collection of the Laboratory of Mycotoxicology (Dept. of Chemistry, Life Sciences and Environmental Sustainability, University of Parma, Italy). Strains of *Fusarium sporotrichioides*, *Rhizoctonia solani*, *Fusarium oxysporum* f.sp *lactucae*, *Verticillium dahliae* and *Sclerotinia* spp. were provided by the Università Cattolica del Sacro Cuore (Piacenza, Italy). All fungal strains were maintained on Potato Dextrose Agar (Difco, Becton, Dickinson & Co., Le Pont-de-Claix, France).

4.4.2. Grapewine Diseases-Associated Fungal Species Growth Inhibition Assay

Fungal isolates were grown on CYA media as previously reported [42]. Compounds were added to the CYA growth media at 25 and 50 μM concentrations, then fungal growth was compared to a mock amended plate (control). Fungi were grown for at least 3 subcultures prior to be used in the inhibition assay. For *P. minimum* and *P. chlamydospora* conidia were harvested on a 15 days old colony using water with 0.02% Tween 20 and 0.05% agar. The suspension was then then diluted to 10^6 conidia per ml and 2 μL were used to inoculate plates. Since *N. parvum* and *F. mediterranea* do not produce conidia the inoculum was made by mycelia plugs of 25 mm diameter harvested from the edge of 5 days old colonies. Due to different growth rates, *P. minimum* and *P. chlamydospora* were checked at 5, 10 and 15 days after inoculation whereas for *N. parvum* and *F. mediterranea* growth measurements were done at 2, 4 and 7 days post inoculation. Results were expressed as percentage of growth inhibition compared to the control. Experiments were performed in triplicate.

4.4.3. *Fusarium* spp. and Soil-Borne Pathogens Growth Inhibition Assay

The mycelial growth of *Fusarium* strains, *Rhizoctonia solani*, *Verticillium dahliae* and *Sclerotinia* spp. was evaluated on Potato Dextrose Agar (PDA) medium amended with the compounds at increasing concentration (from 5 to 100 μM). The Petri dish method was applied, according to Thanassoulopoulos et al. [43]. Control cultures were prepared amending PDA with equivalent volumes of DMSO (≥99.9%, Sigma Aldrich, St Louis, MO, USA), ranging from 0.25% to 1%. The experiments were conducted twice in triplicate in 60 mm petri dishes inoculated in the center with 8 mm PDA plugs from actively growing cultures. Each molecule was added in appropriate concentration to sterile Petri dishes of 90 mm diameter poured with liquid medium and then inoculated with fungi colonies of 3 mm diameter. Inoculum originated from 10-day-old single-spore colonies of tested species grown on PDA (Difco, Becton, Dickinson & Co., France). Three replications of each experimental combination were made.

Plates were incubated up to 10 days at 25 °C, measuring colonies at 2, 4 and 6 days post-inoculation. The antifungal effect was expressed as percentage inhibition, calculated according to the formula: I = [100 − (T/C)] × 100, where I is the percentage of inhibition, C is the control plate colony diameter in mm and T is the treated plate colony diameter in mm [44].

4.4.4. *F. sporotrichioides* T2-HT2 Toxins Determination

The whole mycelium, including the underlying PDA, was collected from control (PDA amended with DMSO ranging from 0.01% to 1%) and treated plates (PDA amended with J, Jdi, JTS and JdiTS ranging from 1 to 100 μM) six days after inoculation with *F. sporotrichioides*. The collected samples were extracted with 5 volumes of 70% methanol and vigorously shaken for 15 min before filtering. The amount of T-2, HT-2 toxins (as sum of toxins) was determined using the kit Veratox® for T-2, HT-2 (Product code 8230, Neogen Corporation, Lansing, MI, USA) a competitive direct enzyme-linked immunosorbent assay (ELISA). The photometric reading was done at 630 nm, according to the manufacturer's instructions.

4.4.5. Aflatoxin, Biomass and Sclerotia Inhibition Assays in *A. flavus*

For biological assays on *Aspergillus flavus* the aflatoxigenic strain CR10 and the non-aflatoxigenic strain TOφ were used [42,43]. Aflatoxin production was evaluated directly in the culture medium (coconut clarified medium, CCM) by a microplate high throughput procedure [45]. Briefly, increasing concentrations of compounds were added to CCM-filled microplate wells, then CCM cultures were incubated at 25 °C in the dark, under stationary conditions for 6 days. AFs accumulation was determined by fluorescence emission (TECAN SpectraFluor Plus microplate reader, Männedorf, Switzerland; λ_{ex} = 360 nm; λ_{em} = 465 nm; manual gain = 83; lag time = 0 μs; number of flashes = 3; integration time = 200 μs). Compounds were tested at 25, 50 and 100 μM concentration. Results were expressed as percentage inhibition respect to control (0.25, 0.5 and 1% *v/v* DMSO-treated cultures respectively). Cultures were inoculated in quadruplicate and experiments were conducted in triplicate.

Biomass production was assessed by recovering, after six days of incubation, single mycelia from the microplate wells used for aflatoxin accumulation—samples were slightly dried on hands paper and weighted and then values were converted in percentage inhibition respect to control (DMSO-treated cultures). Cultures were inoculated in quadruplicate and experiments were conducted in triplicate.

4.4.6. Sclerotia Inhibition Assays in *A. flavus* and *Sclerotinia* spp.

Sclerotia biogenesis in *A. flavus* was measured by point-inoculating 5 μL of the aflatoxigenic strain CR10 conidial suspension (approximately 10^6 conidia/mL) in Petri dishes (Ø = 5 cm) poured with Czapek Dox Agar (CZA) medium added with 100 μM compounds; control plates were amended with 1% (v/v) DMSO. After two weeks of incubation at 30 °C in darkness, sclerotia were manually scraped from the colonies surface, washed with a 70% ethanol solution, dried for three days at 60 °C then weighted. Inhibition rate on sclerotia production was expressed as percentage respect to the control (mg/colony area). Plates were inoculated in triplicate. For *Sclerotinia* spp., a mycelium plug was cultured in Petri dishes (Ø = 5 cm) poured with Czapek Dox Agar (CZA) medium added with 100 μM compounds and incubated as above. Sclerotia were manually collected from plates and counted. Inhibition rate was expressed as a percentage with respect to the control (nr/colony area).

4.4.7. Statistical Analysis of Biological Data

One-way analysis of variance (Past 3.x software) was used for data analyses. Tukey's test was applied to the data relative to mycelial growth, mycotoxin accumulation and sclerotia production; differences were regarded as significant at $p < 0.05$.

5. Conclusions

Recently, in Europe even more restriction rules have been introduced for Cu containing molecules in the management of defense strategies against fungal pathogens. This because Cu, if massively used, tends to accumulate in soils, so far becoming a pollutant for the environment threatening the human health [46]. For these reasons, an urgent need for the development of alternatives to reduce chemical inputs and environmental Cu accumulation is required, particularly in agricultural soils [47]. In the last decades, research efforts led to a better understanding of molecular and biochemical mechanisms driving plant and fungal response to certain phytohormones, including JA; however, only a few findings suggested a possible direct interaction between JA-related compounds and fungal plant pathogens, also in terms of mycotoxins production. Our results definitely assessed how some thiosemicarbazonic JA-derived molecules (and in particular JTS) can exert a containment effect on both fungal growth/development and secondary metabolism in phytopathogenic species infecting agriculturally important crops, such as grape, cereals and horticultural species, being thus extremely interesting for plant health control strategies and food security purposes.

Supplementary Materials: Supplementary materials can be found at http://www.mdpi.com/1422-0067/21/22/8681/s1, Figure S1: Effect of compounds on the growth of esca disease-linked species. Culture medium was amended with molecules at 50 μM concentration; values are reported as inhibition percentage as compared to the control (DMSO 0.5% *v/v* amended cultures) ± S.D. Different letters indicate statistically significant differences ($p < 0.05$).

Author Contributions: Conceptualization, F.D. and G.P.; Chemistry, N.O., F.B., G.P.; Biological assays, G.S., L.N., C.M. and E.M.; Investigation, F.D.; Resources, G.P., F.M.R., V.T., W.C. and E.M.; Writing—Original Draft Preparation, F.D.; Writing—Review & Editing, G.P., F.B., A.J., W.C., F.M.R., S.D., R.G., V.T.; Supervision, F.D.; Project Administration, G.P.; Funding Acquisition, G.P., V.T., W.C., A.J. and F.M.R. All authors have read and agreed to the published version of the manuscript.

Funding: This research has financially been supported by the Programme "FIL-Quota Incentivante" of the University of Parma and co-sponsored by Fondazione Cariparma (Projec: "A sustainable approach to curb natural food poisoning by aflatoxins"). Part of the project was funded by the Italian Ministry of Agriculture and Forestry in the frame of the DiBio–BIOPRIME project (Prot. 76381, MiPAAF PQAI I).

Acknowledgments: The Authors are grateful to G. Chiusa and G. Bolli (Università Cattolica del Sacro Cuore, Piacenza, Italy) for providing some fungal strains.

Conflicts of Interest: Authors declare no conflict of interest. The funders had no role in the design of the study; in the collection, analyses, or interpretation of data; in the writing of the manuscript, or in the decision to publish the results.

References

1. Huang, H.; Liu, B.; Liu, L.; Song, S. Jasmonate action in plant growth and development. *J. Exp. Bot.* **2017**, *68*, 1349–1359. [CrossRef]
2. Irankhah, S.; Chitarra, W.; Nerva, L.; Antoniou, C.; Lumini, E.; Volpe, V.; Ganjeali, A.; Cheniany, M.; Mashregh, M.; Fotopoulos, V.; et al. Impact of an arbuscular mycorrhizal fungal inoculum and exogenous MeJA on fenugreek secondary metabolite production under water deficit. *Environ. Exp. Bot.* **2020**, *176*, 104096. [CrossRef]
3. Chanclud, E.; Morel, J.B. Plant hormones: A fungal point of view. *Mol. Plant Pathol.* **2016**, *17*, 1289–1297. [CrossRef] [PubMed]
4. Kamisaka, S.; Yanagishima, N.; Masuda, Y. Effect of auxin and gibberellin on sporulation in yeast. *Physiol. Plant.* **1967**, *20*, 90–97. [CrossRef]
5. Nakamura, T.; Mukai, C.; Ozaki, Y.; Saotome, M.; Murayama, T. Effects of auxin and gibberellin on conidial germination and elongation of young hyphae in a cyclic 3′:5′ adenosine monophosphate-dependent protein kinase mutant of *Neurospora crassa*. *Plant Growth Regul.* **1988**, *7*, 201–207.
6. Degani, O.; Drori, R.; Goldblat, Y. Plant growth hormones suppress the development of *Harpophora maydis*, the cause of late wilt in maize. *Physiol. Mol. Biol. Plants* **2015**, *21*, 137–149. [CrossRef]
7. Lin, Y.; Qasim, M.; Hussain, M.; Akutse, K.S.; Avery, P.B.; Dash, C.K.; Wang, L. The herbivore-induced plant volatiles methyl salicylate and menthol positively affect growth and pathogenicity of entomopathogenic fungi. *Sci. Rep.* **2017**, *7*, 40494. [CrossRef] [PubMed]

8. Morishita, Y.; Okazaki, Y.; Luo, Y.Y.; Nunoki, J.; Taniguchi, T.; Oshima, Y.; Asai, T. Use of plant hormones to activate silent polyketide biosynthetic pathways in *Arthrinium sacchari*, a fungus isolated from a spider. *Org. Biomol. Chem.* **2019**, *17*, 780. [CrossRef]
9. Norman, S.M.; Bennett, R.D.; Maier, V.P.; Poling, S.M. Cytokinins inhibit abscisic acid biosynthesis in *Cercospora rosicola*. *Plant Sci. Lett.* **1983**, *28*, 255–263. [CrossRef]
10. Tudzynski, B.; Hölter, K. Gibberellin biosynthetic pathway in *Gibberella fujikuroi*: Evidence for a gene cluster. *Fungal Genet. Biol.* **1998**, *25*, 157–170. [CrossRef]
11. Takino, J.; Kozaki, T.; Sato, Y.; Liu, C.; Ozaki, T.; Minami, A.; Oikawa, H. Unveiling biosynthesis of the phytohormone abscisic acid in fungi: Unprecedented mechanism of core scaffold formation catalyzed by an unusual sesquiterpene synthase. *J. Am. Chem. Soc.* **2018**, *140*, 12392–12395. [CrossRef]
12. Nerva, L.; Zanzotto, A.; Gardiman, M.; Gaiotti, F.; Chitarra, W. Soil microbiome analysis in an ESCA diseased vineyard. *Soil Biol. Biochem.* **2019**, *135*, 60–70. [CrossRef]
13. Covarelli, L.; Beccari, G.; Prodi, A.; Generotti, S.; Etruschi, F.; Juan, C.; Ferrer, E.; Mañes, J. Fusarium species, chemotype characterisation and trichothecene contamination of durum and soft wheat in an area of central Italy. *J. Sci. Food Agric.* **2015**, *95*, 540–551. [CrossRef] [PubMed]
14. Salgado, J.D.; Madden, L.V.; Paul, P.A. Efficacy and economics of integrating in-field and harvesting strategies to manage Fusarium head blight of wheat. *Plant Dis.* **2014**, *98*, 1407–1421. [CrossRef] [PubMed]
15. Ferrigo, D.; Raiola, A.; Causin, R. Fusarium toxins in cereals: Occurrence, legislation, factors promoting the appearance and their management. *Molecules* **2016**, *21*, 627. [CrossRef]
16. Mielniczuk, E.; Skwaryło-Bednarz, B. Fusarium Head Blight, mycotoxins and strategies for their reduction. *Agronomy* **2020**, *10*, 509. [CrossRef]
17. Tournas, V.H. Spoilage of vegetable crops by bacteria and fungi and related health hazards. *Crit. Rev. Microbiol.* **2005**, *31*, 33–44. [CrossRef]
18. Gross-Steinmeyer, K.; Eaton, D.L. Dietary modulation of the biotransformation and genotoxicity of aflatoxin B(1). *Toxicology* **2012**, *299*, 69–79. [CrossRef]
19. Schatzmayr, G.; Streit, E. Global occurrence of mycotoxins in the food and feed chain: Facts and figures. *World Mycotoxin J.* **2013**, *6*, 213–222. [CrossRef]
20. Garibaldi, A.; Gullino, M.L. Emerging soilborne diseases of horticultural crops and new trends in their management. *Acta Hortic.* **2010**, *883*, 37–48. [CrossRef]
21. Pelosi, G. Thiosemicarbazone metal complexes: From structure to activity. *Open Crystall. J.* **2010**, *3*, 16–28. [CrossRef]
22. Gowda, N.K.S.; Malathi, V.; Suganthi, R.U. Effect of some chemical and herbal compounds on growth of *Aspergillus parasiticus* and aflatoxin production. *Anim. Feed Sci. Technol.* **2004**, *116*, 281–291. [CrossRef]
23. Rogolino, D.; Gatti, A.; Carcelli, M.; Pelosi, G.; Bisceglie, F.; Restivo, F.M.; Degola, F.; Buschini, A.; Montalbano, S.; Feretti, D.; et al. Thiosemicarbazone scaffold for the design of antifungal and antiaflatoxigenic agents: Evaluation of ligands and related copper complexes. *Sci. Rep.* **2017**, *7*, 11214. [CrossRef] [PubMed]
24. Choi, C.W.; Kim, S.C.; Hwang, S.S.; Choi, B.K.; Ahn, H.J.; Lee, M.Y.; Park, S.H.; Kim, S.K. Antioxidant activity and free radical scavenging capacity between Korean medicinal plants and flavonoids by assay-guided comparison. *Plant Sci.* **2002**, *163*, 1161–1168. [CrossRef]
25. Nerva, L.; Turina, M.; Zanzotto, A.; Gardiman, M.; Gaiotti, F.; Gambino, G.; Chitarra, W. Isolation, molecular characterization and virome analysis of culturable wood fungal endophytes in esca symptomatic and asymptomatic grapevine plants. *Environ. Microbiol.* **2019**, *21*, 2886–2904. [CrossRef] [PubMed]
26. Carvalhais, L.C.; Schenk, P.M.; Dennis, P.G. Jasmonic acid signalling and the plant holobiont. *Curr. Opin. Microbiol.* **2017**, *37*, 42–47. [CrossRef] [PubMed]
27. Prost, I.; Dhondt, S.; Rothe, G.; Vicente, J.; Rodriguez, M.J.; Kift, N.; Carbonne, F.; Griffiths, G.; Esquerré-Tugayé, M.-T.; Rosahl, S.; et al. Evaluation of the antimicrobial activities of plant oxylipins supports their involvement in defense against pathogens. *Plant Physiol.* **2005**, *139*, 1902–1913. [CrossRef]
28. Goodrich-Tanrikulu, M.; Mahoney, N.E.; Rodriguezt, S.B. The plant growth regulator methyl jasmonate inhibits aflatoxin production by *Aspergillus flavus*. *Microbiology* **1995**, *141*, 2831–2837. [CrossRef]
29. Fischer, M.; Peighami-Ashnaei, S. Grapevine, esca complex and environment: The disease triangle. *Phytopathol. Mediterr.* **2019**, *58*, 17–37.

30. Fischer, J.; Beckers, S.J.; Yiamsawas, D.; Thines, E.; Landfester, K.; Wurm, F.R. Targeted Drug delivery in plants: Enzyme-responsive lignin nanocarriers for the curative treatment of the worldwide grapevine trunk disease Esca. *Adv. Sci.* **2019**, *6*, 1802315. [CrossRef]
31. Paiva, R.; Kneipp, L.F.; Goular, C.M.; Albuquerque, M.A.; Echevarria, A. Antifungal activities of thiosemicarbazones and semicarbazones against mycotoxigenic fungi. *Ciên. Agrotec.* **2014**, *38*, 531–537. [CrossRef]
32. Degola, F.; Morcia, C.; Bisceglie, F.; Mussi, F.; Tumino, G.; Ghizzoni, R.; Pelosi, G.; Terzi, V.; Buschini, A.; Restivo, F.M.; et al. In vitro evaluation of the activity of thiosemicarbazone derivates against mycotoxigenic fungi affecting cereals. *Int. J. Food Microbiol.* **2015**, *200*, 104–111. [CrossRef] [PubMed]
33. Georgiou, C.D.; Patsoukis, N.; Papapostolou, I.; Zervoudakis, G. Sclerotial metamorphosis in filamentous fungi is induced by oxidative stress. *Integr. Comp. Biol.* **2006**, *46*, 691–712. [CrossRef] [PubMed]
34. Patsoukis, N.; Georgiou, C.D. Thiol redox state and oxidative stress affect sclerotial differentiation of the phytopathogenic fungi *Sclerotium rolfsii* and *Sclerotinia sclerotiorum*. *J. Appl. Microbiol.* **2007**, *104*, 42–50. [CrossRef]
35. Papapostolou, I.; Georgiou, C.D. Superoxide radical induces sclerotial differentiation in filamentous phytopathogenic fungi: A superoxide dismutase mimetics study. *Microbiology* **2010**, *156*, 960–966. [CrossRef]
36. Degola, F.; Bisceglie, F.; Pioli, M.; Palmano, S.; Elviri, L.; Pelosi, G.; Lodi, T.; Restivo, F.M. Structural modification of cuminaldehyde thiosemicarbazone increases inhibition specificity toward aflatoxin biosynthesis and sclerotia development in *Aspergillus flavus*. *Appl. Microbiol. Biotechnol.* **2017**, *101*, 6683–6696. [CrossRef] [PubMed]
37. Sheldrick, G.M. A short history of SHELX. *Acta Cryst.* **2008**, *A64*, 112–122. [CrossRef] [PubMed]
38. Dolomanov, L.J.; Bourhis, R.J.; Gildea, J.A.; Howard, K.; Puschmann, H. OLEX2: A complete structure solution refinement and analysis program. *J. Appl. Cryst.* **2009**, *42*, 339–341. [CrossRef]
39. Macrae, C.F.; Edington, P.R.; McCabe, P.; Pidcock, E.; Shields, G.P.; Taylor, R.; Towler, M.; van de Streek, J. Mercury: Visualization and analysis of crystal structures. *J. Appl. Cryst.* **2006**, *39*, 453–457. [CrossRef]
40. Bartoli, J.; Montalbano, S.; Spadola, G.; Rogolino, D.; Pelosi, G.; Bisceglie, F.; Restivo, F.M.; Degola, F.; Serra, O.; Buschini, A.; et al. Antiaflatoxigenic thiosemicarbazones as cropprotective agents: A cytotoxic and genotoxic study. *J. Agric. Food Chem.* **2019**, *67*, 10947–10953. [CrossRef]
41. Bisceglie, F.; Degola, F.; Rogolino, D.; Giannelli, G.; Orsoni, N.; Spadola, G.; Pioli, M.; Restivo, F.M.; Carcelli, M.; Pelosi, G. Sisters in structure but different in character, some benzaldehyde and cinnamaldehyde derivatives differentially tune *Aspergillus flavus* secondary metabolism. *Sci. Rep.* **2020**, *10*, 17686. [CrossRef] [PubMed]
42. Nerva, L.; Forgia, M.; Ciuffo, M.; Chitarra, W.; Chiapello, M.; Vallino, M.; Varese, G.C.; Turina, M. The mycovirome of a fungal collection from the sea cucumber *Holothuria polii*. *Virus Res.* **2019**, *273*, 197737. [CrossRef] [PubMed]
43. Thanassoulopoulos, C.C.; Giannopolitis, C.N.; Vitsos, G.T. Evaluation of sensitiveness and development of resistance of *Fusarium oxysporum* f. sp. *lycopersici* to benomyl. *Phytopathol. Zeitsch.* **1971**, *70*, 114–120.
44. Jamiołkowska, A.; Kowalski, R. Laboratory effect of *Silphium perfoliatum* L. on the growth of tested fungi. *Acta Sci. Pol. Hortorum Cultus* **2012**, *11*, 43–55.
45. Degola, F.; Dall'Asta, C.; Restivo, F.M. Development of a simple and high-throughput method for detecting aflatoxins production in culture media. *Lett. Appl. Microbiol.* **2012**, *55*, 82–89. [CrossRef]
46. Lamichhane, J.R.; Osdaghi, E.; Behlau, F.; Köhl, J.; Jones, J.B.; Aubertot, J.-N. Thirteen decades of antimicrobial copper compounds applied in agriculture. A review. *Agron. Sustain. Dev.* **2018**, *38*, 28. [CrossRef]
47. La Torre, A.; Iovino, V.; Caradonia, F. Copper in plant protection: Current situation and prospects. *Phytopathol. Mediterr.* **2018**, *57*, 201–236.

International Journal of
Molecular Sciences

MDPI

Article

The Biochemical Pathways of Nicotinamide-Derived Pyridones

Faisal Hayat [1,2,†], Manoj Sonavane [1,2,3,†], Mikhail V. Makarov [2], Samuel A. J. Trammell [4], Pamela McPherson [2], Natalie R. Gassman [2,3] and Marie E. Migaud [1,2,*]

1 Department of Pharmacology, College of Medicine, University of South Alabama, Mobile, AL 36688, USA; fhayat@health.southalabama.edu (F.H.); msonavane@health.southalabama.edu (M.S.)
2 Mitchell Cancer Institute, College of Medicine, University of South Alabama, Mobile, AL 36604, USA; mmakarov@health.southalabama.edu (M.V.M.); pvm1821@jagmail.southalabama.edu (P.M.); nrgassman@southalabama.edu (N.R.G.)
3 Department of Physiology & Cell Biology, College of Medicine, University of South Alabama, Mobile, AL 36688, USA
4 Novo Nordisk Foundation, Center for Basic Metabolic Research, University of Copenhagen, 2200 Copenhagen, Denmark; trammell@sund.ku.dk
* Correspondence: mmigaud@southalabama.edu
† Shared first authorship.

Abstract: As catabolites of nicotinamide possess physiological relevance, pyridones are often included in metabolomics measurements and associated with pathological outcomes in acute kidney injury (AKI). Pyridones are oxidation products of nicotinamide, its methylated form, and its ribosylated form. While they are viewed as markers of over-oxidation, they are often wrongly reported or mislabeled. To address this, we provide a comprehensive characterization of these catabolites of vitamin B3, justify their nomenclature, and differentiate between the biochemical pathways that lead to their generation. Furthermore, we identify an enzymatic and a chemical process that accounts for the formation of the ribosylated form of these pyridones, known to be cytotoxic. Finally, we demonstrate that the ribosylated form of one of the pyridones, the 4-pyridone-3-carboxamide riboside (4PYR), causes HepG3 cells to die by autophagy; a process that occurs at concentrations that are comparable to physiological concentrations of this species in the plasma in AKI patients.

Keywords: NAD; redox cofactor; nicotinamide; pyridones

Citation: Hayat, F.; Sonavane, M.; Makarov, M.V.; Trammell, S.A.J.; McPherson, P.; Gassman, N.R.; Migaud, M.E. The Biochemical Pathways of Nicotinamide-Derived Pyridones. *Int. J. Mol. Sci.* **2021**, *22*, 1145. https://doi.org/10.3390/ijms22031145

Academic Editor: Maurizio Battino
Received: 28 December 2020
Accepted: 19 January 2021
Published: 24 January 2021

1. Introduction

Through multiple yet convergent biosynthetic pathways, all components of vitamin B3 are precursors to the nicotinamide-derived redox cofactor, nicotinamide adenine dinucleotide (NAD), its reduced form NADH, its phosphate parent NADP, and the reduced form NADPH, collectively referred to as NAD(P)(H) [1]. The water-soluble components that constitute vitamin B3 are obtained through the diet and/or supplementation. Its regular intake must compensate for the daily losses stemming from the degradation of the cofactors [2]. Some of the NAD(P)(H) degradation products (Figure 1) include *N*-methyl-4-pyridone-3-carboxamide (*N*-Me-4PY) and *N*-methyl-6-pyridone-3-carboxamide, often reported as *N*-methyl-2-pyridone-5-carboxamide (*N*-Me-6PY). Both pyridones are detected in plasma and urine as major vitamin B3 degradation products and are oxidized forms of methyl-nicotinamide (Me-Nam) [3,4]. The ribosylated form of the pyridone carboxamides, 4-pyridone-3-carboxamide riboside (4PYR), and 6-pyridone-3-carboxamide riboside also known as 2-pyridone-5-carboxamide riboside (6PYR), are also detected in urine [5]. In blood, these ribosylated pyridones are mostly present intracellularly in their triphosphorylated (4PYR-TP and 6PYR-TP) [6,7] or adenine dinucleotidic forms [8]. *N*-Methyl-2-pyridone-3-carboxamide (Me-2PY) and 2-pyridone-3-carboxamide riboside (2PYR) are the least prominent pyridones in the literature, as they are much less often detected in biological samples [9].

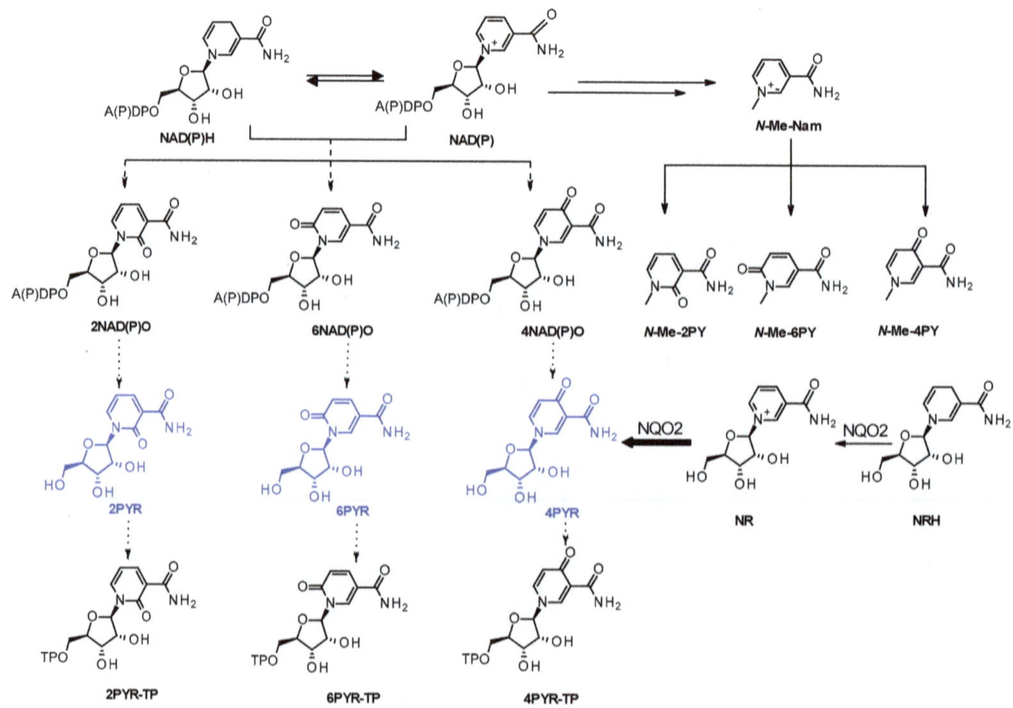

Figure 1. NAD(P)(H)-derived pyridone catabolites. An additional route to the triphosphate from the monophosphate ribosylated pyridone generated from the dinucleotides can also be envisaged. NAD(P)(H): nicotinamide adenine dinucleotide (phosphate)(reduced form); NAD(P)O: hyper- oxidized forms of NAD(P) containing pyridone moieties; PYR-TP: pyridone riboside triphosphate; PYR: pyridone riboside; The 2, 4, and 6 positions of the carbonyl unit are indicated in the abbreviation; *N*-Me-PY: *N*-methyl pyridone.

As waste products, these by-products of NAD(P)(H) metabolism associate with aging and nephrotic dysfunction as their levels associate with advanced kidney injuries (AKI) and kidney failure [4,6,10–12]. Liquid chromatography combined with mass spectrometry is the most common method used to detect nicotinamide-derived pyridones in biological samples [13–15]. However, much confusion arises from their detection, identification, and reporting as additional nicotinamide metabolites that are the non-methylated or ribosylated pyridones are also often mistaken for their methylated species in reports [5,9,11,15–18]. The misperception stems from confusing nomenclature, compounded by poor quantification due to facile fragmentation and adduct formation under mass spectrometry conditions.

Critically, the chemical nature of each form of the pyridone informs on the type of catabolic pathway the nicotinamide-derived species have undertaken. For instance, only the methylated form of nicotinamide (*N*-Me-Nam) can be the precursor of the three methylated pyridones (*N*-Me-2PY, *N*-Me-4PY, and *N*-Me-6PY, Figure 1) [19–22]. Similarly, only the ribosylated form of nicotinamide can be the source of the pyridone ribosides [17]. Each of the catabolic routes indicates dysregulation in NAD-dependent metabolism [23,24]. Pyridone ribosides, unlike the methylated species, can be converted to the adenine dinucleotide or the triphosphate form [17,25]. The 4PYR isomer not only circulates in plasma once generated but is also internalized and metabolized to species that interfere with enzymatic processes [26–28]. Possessing a systematic, reliable account of the pyridones' distribution would greatly help identify whether these species are pathophysiologically significant.

Here, we present a comprehensive comparative characterization of each species and explore the mechanisms by which their under-detection and misidentification can arise.

We also propose a mechanism for the wide-spread production of the ribosylated pyridone 4PYR and rationalize selective cytotoxicity by comparing how exogenous exposure to ribosylated pyridones affects cell survival.

2. Results

The *N*-methylated 6-pyridone were easily generated from their respective chloronicotinic acid precursors as were the ribosylated 6-pyridone according to published syntheses [6,29–32]. However, *N*-Me-2PY, 2PYR, *N*-Me-4PY, and 4PYR converted readily to the methylated ester instead of the amide under aqueous methanolic conditions. Alterations to the published synthetic sequences were deemed necessary to ensure sample authenticity and are described in detail in the experimental and methods section (Schemes 1–8 in Methods). All synthetic compounds were purified by separation by flash column, and their chemical structures were fully characterized by both NMR and mass spectrometry (Supplementary Material Table S1 and Spectra). More details on structural characterization can be found in the supplementary material section. In Tables 1–3, we have compiled the ^{1}H NMR experimental data for the pyridones (Table 1), the methylated pyridones (Table 2), and the ribosylated pyridones (Table 3), while a comprehensive mass spectrometry report can be found in the supplementary material section. The ^{1}H NMR of each one of the pyridone is very characteristic of each species according to the position of the carbonyl moiety as well as the type of substitution present on the heterocycle.

The simple 4PY was synthesized in a three-step process as reported by Slominska et al. [25] First, 4-chloronicotinic acid was hydrolyzed then converted to the acyl chloride with thionyl chloride in MeOH solution. The 4-pyridone-formylchloride was treated with aq.NH_4OH in a sealed tube to generate 4PY. For the 6PY, 5-hydroxy-nicotinic was activated with HOBt in DMF and subsequently treated with aq.NH_4OH in a sealed tube at rt to afford 6PY in 70% yield. Finally, 2PY was generated from 2-hydroxy-nicotinic acid, which following its conversion to the acyl halide using $SOCl_2$ reagent was converted to the amide using aq.NH_4OH in a sealed tube at rt. When stirred with aq. MeOH, the acyl chloride hydrolyzes to its nicotinate salt that exists in both keto and enol forms.

The synthesis of ribosylated forms of 2PY, 4PY, and 6PY (2PYR, 4PYR, and 6PYR, respectively) is based on Vorbrüggen glycosylation reaction [31,32]. Following silylation of the nucleobase using either HMDS (2-PY and 4-PY) or BSTFA (6-PY), the silylated pyridones were mixed with beta-D-riboside tetraacetate in the presence of a Lewis acid. For the ribosylation of 2PY and 4PY, TMSOTf in 1,2-DCE/acetonitrile was applied, while $SnCl_4$ was used for the ribosylation of 6PY. While the deacetylation should have been straightforward, it had to be optimized for 2PYR and 4PYR to prevent the loss of the amine moiety and generation of the corresponding carboxylic acid derivatives, materials that we have also fully characterized for reasons described below. Finally, the *N*-methyl derivatives of 4PY and 6PY (*N*-Me-2PY, *N*-Me-4PY, and *N*-Me-6PY, respectively) were prepared by methylation with methyl iodide. *N*-methyl-2-pyridone, *N*-Me-2PY was obtained by oxidation of *N*-methyl nicotinamide with $K_3Fe(CN)_6$ under basic conditions [33,34].

Table 1. Tables of ^{1}H NMR chemical shift for the non-substituted pyridones.

Compound (Solvent)	Chemical Shift (Multiplicity, J Values in Hz)					
	NH	H2	H3	H4	H5	H6
2PY (DMSO-d_6)	12.3 (s), 9.05 (s), 7.51 (s)			7.66 (d, J = 4.32 Hz)	6.41 (t, J = 6.66 Hz)	8.29 (d, J = 5.28 Hz)
2PY (DMSO-d_6) [25]	11.80 (s), 9.02 (s), 7.57 (s)			7.69 (q, 1H)	6.44 (t)	8.31 (q)
4PY (CD_3OD)		8.54 (s)			6.56 (d, J = 6.53 Hz)	7.79 (d, J = 5.56 Hz)
4PY (CD_3OD) [25]		8.56 (d, J = 1.7 Hz)			6.58 (d, J = 7.2 Hz)	7.80 (dd, J = 7.2, 1.7 Hz)
6PY (DMSO-d_6)	11.92(s), 7.72 (s) 7.18 (s)	8.00 (s)		7.85 (d, J = 9.6 Hz)	6.32 (d, J = 9.56 Hz)	
6PY (DMSO-d_6) [31]	11.84 (s), 7.71 (s), 7.18 (s)	7.99 (d, J = 2.6 Hz)		7.82–7.87 (dd, J = 2.6 & 9.5 Hz)	6.32 (d, J = 9.5 Hz)	

Table 2. Tables of ^{1}H NMR chemical shift for the methylated pyridones.

Compound (Solvent)	Chemical Shift (Multiplicity, J Values in Hz)						
	NH	H2	H3	H4	H5	H6	CH_3
N-Me-2PY (DMSO-d_6)	9.06 (s) 7.54 (s)			8.01 (dd, J = 2.16 & 2.16 Hz)	6.44 (t, J = 6.88 Hz)	8.27 (dd, J = 2.20; 2.16 Hz)	3.08 (s)
N-Me-2PY * (DMSO-d_6) [29]	9.09 (brs, 1H, H-Nb), 7.57 (brs, 1H, H-Na)			8.31 (dd, J = 7.21, 2.21 Hz)	6.47 (dd, J = 7.10, 7.10 Hz)	8.04 (dd, J = 6.55, 2.21 Hz)	3.56 (s)
N-Me-4PY (DMSO-d_6)	9.36 (s), 7.54 (s)	8.54 (s)			6.53 (d, J = 7.4 Hz)	7.87 (d, J = 5.08 Hz)	3.82 (s)
N-Me-4PY * (DMSO-d_6)^2 [29]	9.56 (brs), 7.41 (brs)	7.74 (dd, J = 2.44, 7.49 Hz)	6.38 (d, J = 7.49 Hz)			8.44 (d, J = 2.44 Hz)	3.75 (s)
N- Me-6PY (D_2O)		8.38 (s)		7.95 (d, 1H, J = 6.69 Hz)	6.53 (d, J = 9.44 Hz)		3.62 (s)
N- Me-6PY * (DMSO-d_6) [29]	7.23 (brs), 7.69 (brs)		6.38 (d, J = 9.49 Hz)	7.85 (dd, J = 9.49, 2.54 Hz)		8.36 (d, J = 2.54 Hz)	3.46 (s)

*: non-IUPAC atom numbering.

Table 3. Tables of ^{1}H NMR chemical shift for the ribosylated pyridones.

Compound (Solvent)	Chemical Shift (Multiplicity, J Values in Hz)											
	NH	H2	H3	H4	H5	H6	$H_{1'}$	$H_{2'}$	$H_{3'}$	$H_{4'}$	$H_{5'}$	CH_3
2PYR (D_2O)				8.46 (dd, J = 5.4, 7.04 Hz, 2H, H4 & H6),			6.14 (s),	4.12 (bs, 3H, $C_{2',3',4'}$H)			3.97 (d, J = 12.9 Hz), 3.80 (d, J = 11.56 Hz)	
					6.55 (t, J = 6.96 Hz)							
2PYR (DMSO-d_6) [32]	7.63 and 8.96 (2 brs, $CONH_2$)			8.32 (m)	6.55 (t)	8.42 (m)	6.08 (d, J = 1.17 Hz)	5.55 (d, 1H ($C_{2'}$OH)	5.10 (d, 1H, $C_{3'}$OH)		5.23 (t, 1H, $C_{5'}$OH), 3.64 (m)	
								3.79–3.97 (m, 3H)				
4PYR (D_2O)		8.67 (d, J = 2.24 Hz)			6.65 (d, J = 7.6 Hz)	7.94 (dd, J = 2.24 Hz, 2.24 Hz)	5.55 (d, J = 5.5 Hz)	4.23 (t, J = 5.28 Hz)	4.19–4.13 (m)		3.81 and 3.73 (ABX, 2H, JAB = 12.7 Hz, JAX = 3.04 Hz, JBX = 4.0 Hz)	
4PYR (D_2O) [25]		8.66 (d, J = 2.4 Hz)			6.58 (d, J = 7.8)	7.93 (dd, J = 7.8, 2.4 Hz	5.53 (d, J = 5.9)	4.22 (t, J = 5.4)	4.16 (dd, J = 5.1, 3.5)	4.14 (ddd, J = 4.0, 3.5, 3.0 Hz)	3.72 (dd, J = 12.6, 3.0 Hz), 3.79 (dd, J = 12.6, 3.0 Hz)	
6PYR (D_2O)		8.60 (s)		7.84 (d, J = 9.2 Hz)	6.54 (d, J = 9.44 Hz)		6.01 (s)	4.19–4.13 (m)			3.97 (d, J = 12.92 Hz), 3.80 (d, J = 12.88 Hz)	
6PYR (DMSO-d_6) [31]	7.54(1 H, Amid-NH), 7.30 (s)	8.52 (d, J = 1.9 Hz)		7.81–7.86 (dd, J = 1.9 Hz, 9.5 Hz)	6.40 (d, J = 9.5 Hz)		6.01 (d, J = 3.5 Hz)	5.45 (d, J = 4.6 Hz,1H, $C_{2'}$ OH)	5.07 (d, J = 4.9 Hz, 1 H, $C_{3'}$ OH)		5.17 (t, J = 4.5 Hz, $C_{5'}$OH, 3.57–3.76 (m_b)	
								3.92–4.08 (m, 3H, $C_{2',3',4'}$H)				
4PYR ACID (D_2O)		8.21 (d, J = 2.4 Hz)			6.52 (d, J = 7.6 Hz)	7.89 (dd, J = 7.6, 2.4 Hz)	5.51 (d, J = 5.6 Hz)	4.26 (dd, J = 5.6 and 5.3 Hz)	4.20 (dd, J = 5.3 and 3.6 Hz)	4.15 (dd, J = 8.3, 3.6 Hz)	3.82 (dd, J = 12.8, 4.5 Hz), 3.74 (dd, J = 12.8, 4.5 Hz)	

As described above, *N*-methyl-nicotinamide oxidizes to three possible forms, *N*-Me-2PY, *N*-Me-4PY, and *N*-Me-6PY. By mass spectrometry (ESI-MS, Supplementary Material Table S1), it was found that the three isomers fragment were similar to each other but differ in the relative abundance of the predominant fragmentation products (Supplementary Material Table S1). The fragmentation pattern of each pyridone derivative is a fingerprint of that pyridone and can be used to identify these species unambiguously (Supplementary Material Spectra). Crucially, the functional group exchange between the amide and the carboxylic acid described above is observed under simple electrospray conditions (Supplementary Material Table S1 and Spectra). All three methylated pyridone forms (*N*-Me-2PY, *N*-Me-4PY, and *N*-Me-6PY) are deamidated on the carboxamide to form an ion at *m/z* 136. However, this fragment is the predominant product at low energy when fragmenting *N*-Me-2PY, and *N*-Me-4PY but more minor when fragmenting *N*-Me-6PY. The same fragmentation profile is observed for 2PYR, 4PYR, and 6PYR (Supplementary Material Table S1). Critically, when the mass spectrometry analyses are conducted in the presence of water and/or methanol, the ionized carbonium fragment (Figure 2) can react with in situ solvents to generate adducts that in turn get ionized and thus get detected. We have provided a comparative characterization of these materials to address the possible mischaracterization of these metabolites by MS. While not endogenously generated, they can be major contributors to the pyridone pool they originated from.

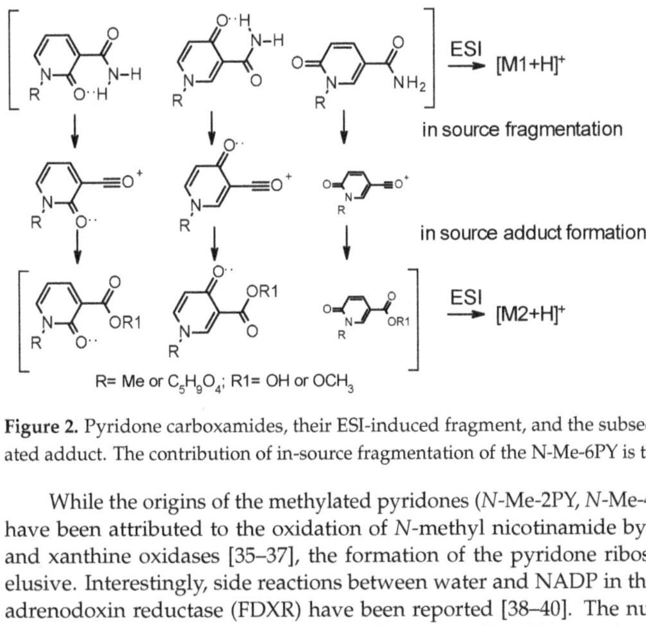

Figure 2. Pyridone carboxamides, their ESI-induced fragment, and the subsequent in source generated adduct. The contribution of in-source fragmentation of the N-Me-6PY is thought to be minimal.

While the origins of the methylated pyridones (*N*-Me-2PY, *N*-Me-4PY, and *N*-Me-6PY) have been attributed to the oxidation of *N*-methyl nicotinamide by aldehyde oxidases and xanthine oxidases [35–37], the formation of the pyridone ribosides remains more elusive. Interestingly, side reactions between water and NADP in the binding pocket of adrenodoxin reductase (FDXR) have been reported [38–40]. The nucleophilic addition of water on the ribosylated nicotinamide moiety of NADP was thought to promote the formation of 4NADPO. We explored the possibility that nicotinamide riboside, NR, and its reduced form, NRH are oxidized by the FAD-dependent reductase, dihydronicotinamide riboside: quinone reductase 2 via a similar mechanism, using LC-MS combined to mass spectrometry. The oxidation of NRH by the FAD-dependent NQO2 was monitored at 340 nM by LC-UV-spectroscopy, while the appearance of NR was confirmed by monitoring at 254 nm. Similarly, when seeking to oxidize NR, the reaction was monitored at 254 nm to follow NR consumption. We observed not only that NRH was oxidized to NR in the presence of FAD and menadione, but that the pyridone riboside, 4PYR, was also generated over time, as confirmed by LC-MS-MS. This chemistry was not observed when NR was used as the starting material. This indicates that the mechanism responsible for the oxidation of

NRH to 4PYR by NQO2 is not the same as the mechanism observed and characterized for the oxidation of NADP by adrenodoxin reductase.

Seeking to establish the mechanism of 4PYR formation by NQO2, we turned to Fenton Chemistry. In the presence of $K_3(Fe(CN)_6)$ in 1 M NaOH, methyl-nicotinamide is readily oxidized, and *N*-Me-2PY and *N*-Me-6PY were detected. A carboxylic acid $[C_7H_7NO_3+H^+]$ was also detected by direct ESI-MS. This outcome could be consistent with the synthetic challenges we experienced in the synthesis of *N*-Me-4PY. When we applied this chemistry to nicotinamide riboside (NR), we detected 2PYR [41] and 6PYR [42,43], the two isomers of 4PYR (Figure 3), along with nicotinamide, the product formed from NR hydrolysis. These outcomes indicate that the NQO2-catalyzed over-oxidation of NRH favors one isomer, 4PYR, and occurs via a mechanism that differs from Fenton chemistry as it is more regioselective. It is also different from adrenodoxin reductase (FDXR) oxidation of NADP, as the over-oxidation of NRH by NQO2 requires the initial release of a hydride for superoxide formation via FAD.

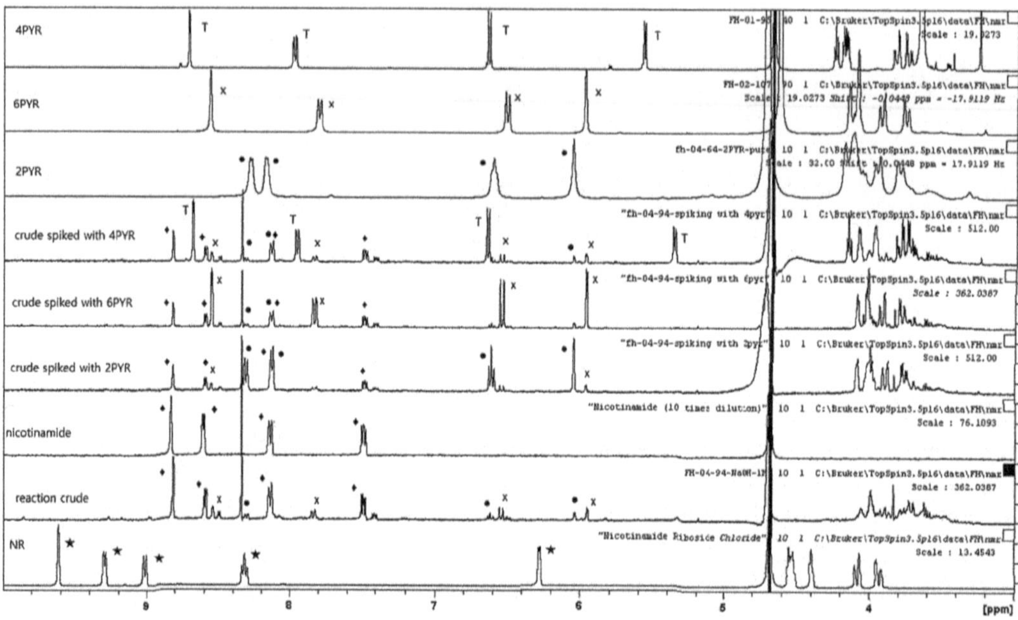

Figure 3. Fenton chemistry facilitates the formation of 2PYR and 6PYR from NR. Overlay ^1H NMR of nicotinamide riboside (bottom spectrum) and crude reaction mixture (above). NR decomposes to nicotinamide under basic conditions (3rd spectrum up). The crude reaction mixture was spiked with each pyridone ribosides (4th–6th spectra). Pure PYR ^{1}H NMR spectra (7th–9th spectra). Only 2PYR and 6PYR are present in the reaction mixture. • 2PYR; x 6PYR; T 4PYR; ★Nicotinamide Riboside; ♦ Nicotinamide.

Acetaminophen, also known as paracetamol is oxidized to a quinone by the FDXR-dependent cytochrome CYP450, CYP2E1, for which acetaminophen administration stimulates the overexpression [44]. acetaminophen exposure also promotes the overexpression of NADPH: quinone oxide-reductase 1 (NQO1) [45], an homolog of NQO2, which catalyzes the reduction of the quinone back to acetaminophen. Catabolites of vitamin B3 found in urine include nicotinuric acid, trigonelline, methyl-nicotinamide, Me-2PY, Me-4PY, and 4PYR. In a human preliminary study, we measured the levels of Me-Nam, *N*-Me-4PY, and 4PYR in urine over a 24 h period and compared these levels with those obtained following the ingestion of 1 gm of acetaminophen (Supplementary Figure S3). We observed that acetaminophen consumption increases the urinary levels of 4PYR by more than 40-fold but does not change the levels of the non-ribosylated species. This observation would indicate that acetaminophen only promotes the oxidation of the ribosylated forms of nicotinamide, i.e., NR, NMN, or NAD(P) and this via a mechanism which is independent of the aldehyde oxidase, the enzyme responsible for the generation of *N*-Me-2PY and *N*-Me-4PY from methyl-nicotinamide in vivo.

The studies evaluating the PYR-family cytotoxicity are somewhat confusing and imprecise, as not all cell-lines appear to be sensitive to exposure to these nucleosides [26,46]. Given the literature, we decided to explore which one if not all the pyridone ribosides were cytotoxic to two very metabolically different cell lines, the human embryonic kidney HEK293T cell line and the human hepatocarcinoma HepG3 cell line. Compared to HEK293T that is more glycolytic, HepG3 is more reliant on mitochondrial respiration for energy production, processes that are NAD(P) dependent and therefore potentially sensitive to the presence of PYR-derived dinucleotides. Using a CellTiter-Fluor^{TM} viability assay as a reporter [47], cytotoxicity assays on HEK293T and HepG3 cell cultures indicated that none of the PYR-derivates were cytotoxic to HEK293T cells but that the 4PYR reduced the cell viability of HepG3 cells by 62% after 24 h of treatment at 100 μM followed by 48 h recovery in cell medium and by 45% after 72 h of continuous exposure (Figure 4).

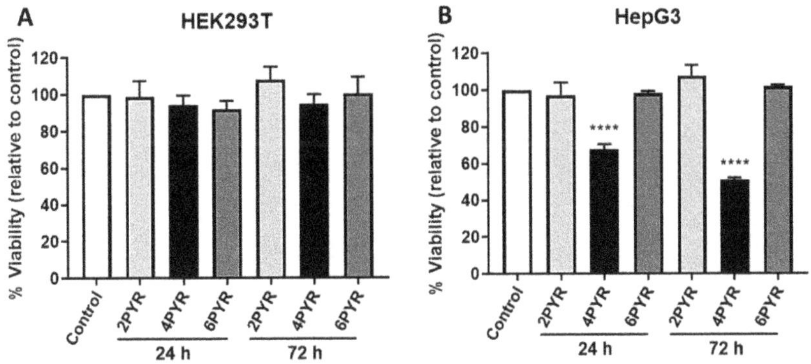

Figure 4. 4PYR is cytotoxic to HepG3 cells but not HEK293T. Cell viability after 72 h was assessed for HEK293T (**A**) and HepG3 (**B**) cells. Cells were exposed to 100 μM of 2PYR, 4PYR, or 6PYR for 24 h with pyridones removed and fresh medium replaced for an additional 48 h or continuously exposed to the pyridone isomers for 72 h. After 72 h, CellTiter-Fluor^{TM} viability assay was performed with results expressed as the mean fluorescence intensity relative to control, vehicle-treated cells (% Viability) ± standard error of the mean (SEM). Statistical significance: **** $p < 0.0001$.

A dose-dependent sensitivity assay of HepG3 cells exposed to continuous 4PYR treatment showed that 4PYR was cytotoxic to HepG3 in a dose-dependent manner with an IC50 at 50 μM (Figure 5). Crucially, even at higher concentrations in this cell line, surviving cells remain, and the loss of cell viability appears to plateau. We also characterized the effects of 4PYR on HepG3 cells in a clonogenic survival assay, with HepG3 cell density less than 50% after 6 days at 25 μM of 4PYR (Figure 6). Together, these data provided some indications of the possible mode of action of 4PYR in HepG3 cells, which likely promotes senescence in the remaining viable cells. Therefore, we sought to establish whether markers of autophagy could be detected in 4PYR-treated HepG3. Immunoblotting showed an increase in protein expression levels of the autophagy marker LC3BII, Beclin-1, and a decrease in phospho-mTOR compared to vehicle-treated control cells (Figure 7). Furthermore, we confirmed that apoptosis was not observed over the same period (data not shown).

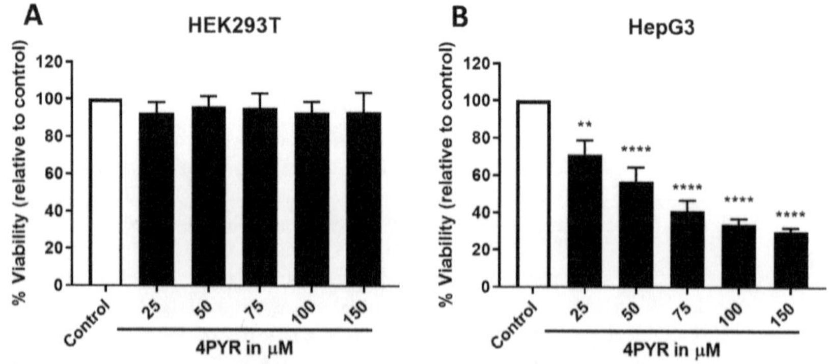

Figure 5. Dose-dependent sensitivity of HepG3 cells to continuous 4PYR treatment. HEK239T (**A**) and HepG3 (**B**) cells were exposed to 25, 50, 75, 100, and 150 μM continuously for 72 h. CellTiter-Fluor^{TM} Viability assay was performed with results expressed as the mean fluorescence intensity relative to control, vehicle-treated cells (% viability) ± SEM. Statistical significance: ** $p < 0.005$, **** $p < 0.0001$.

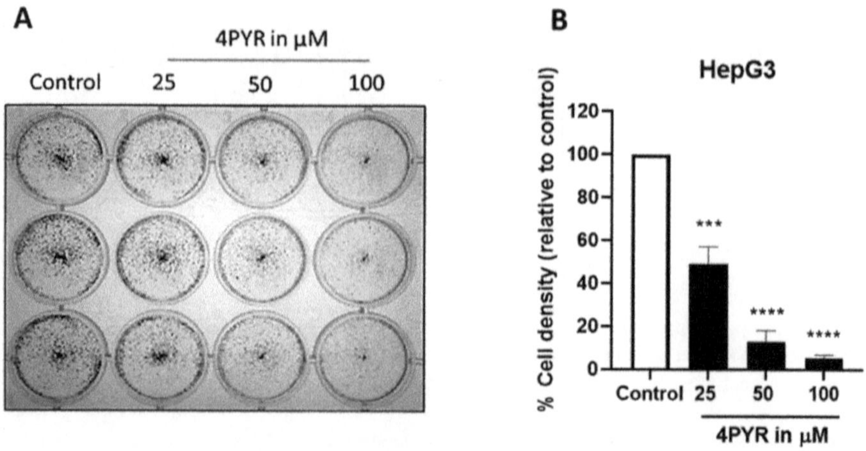

Figure 6. Clonogenic survival of HepG3 cells continuously exposed to 4PYR. HepG3 cells were exposed to 25, 50, and 100 μM of 4PYR for six days then stained with crystal violet (**A**). Cell density was assessed by measuring the area fraction of the crystal violet-stained cells after 4PYR exposure (**B**). Results are presented as percentage cell density to control ± SEM of three independent experiments. Statistical significance: *** $p < 0.001$, **** $p < 0.0001$.

Finally, we exposed HEK293T and HepG3 cells to 100 μM NRH, the reduced form of nicotinamide riboside, for 1, 4, and 24 h and measured the cell extract for the presence of the over-oxidized form of NAD, NADO by LC-MS. While HEK293T did not produce NADO under such conditions, HepG3 cells produced 4NADO intracellularly, and its abundance increased with the length of exposure (Figure 8).

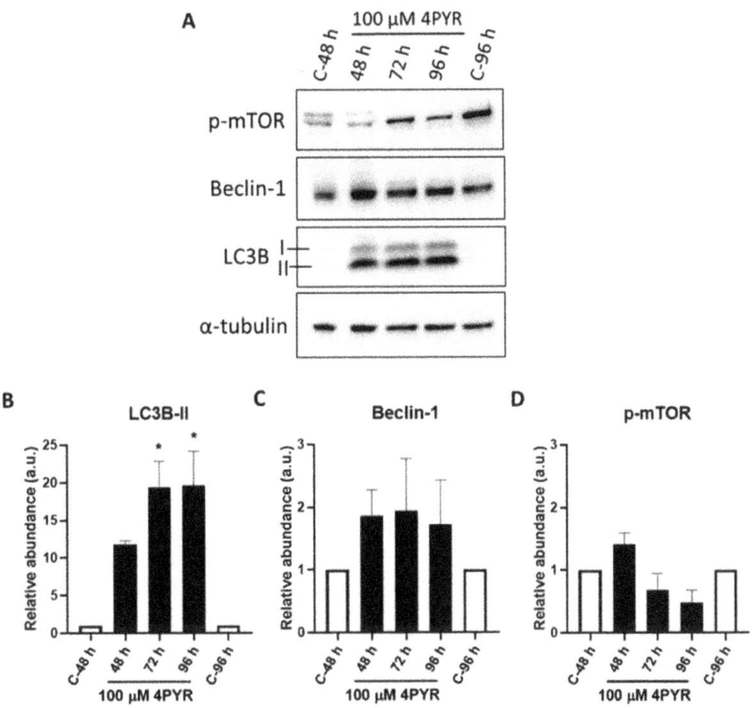

Figure 7. 4PYR induced autophagy in HepG3 cells. HepG3 cells were continuously dosed with 100 μM of 4PYR for 48, 72, and 96 h. (**A**) Increased protein expression levels of the autophagy marker LC3BII, Beclin-1, and decrease in phospho-mTOR were assessed using immunoblot in 4PYR treated HepG3 cells compared to vehicle-treated control cells. α-tubulin was used as a loading control for immunoblotting. The graph shows quantified protein expression levels relative to controls for (**B**) LC3B-II, (**C**) Beclin-1 and (**D**) phosphor-mTOR in HepG3 cells. Results are expressed as relative abundance (a.u.) ± SEM of two biological replicates. Statistical significance: * $p < 0.05$.

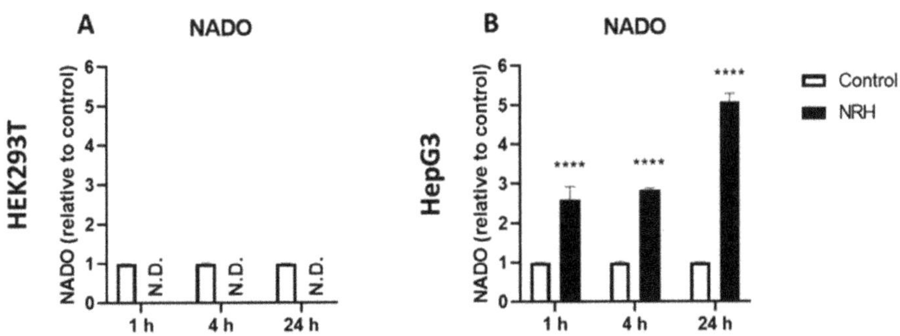

Figure 8. NRH induced 4NADO production in HepG3 cells but not in HEK293T. HEK293T (**A**) and HepG3 (**B**) cells were continuously dosed with 100 μM of 4PYR for 1, 4, and 24 h. The crude cell extract was measured by LC-MS for its NADO content. The NADO detected matches the 4NADO standard. Results are presented as relative abundance to control ± SEM of three independent experiments. Statistical significance: **** $p < 0.0001$. N.D: Not Detected.

3. Discussion

Pyridones and methyl-nicotinamide are measured in clinical settings to identify vitamin B3 deficiencies [48,49] and in metabolomics as they associate with increased NAD consumption [50,51] and cellular dysfunctions in tissues [51–55]. Unfortunately, over the years the nomenclature used to describe the pyridone's nature thus characterized has been somewhat unsystematic and exquisitely confusing (e.g., [5,11,56]). For instance, the term "pyridone" can describe both the methylated or non-methylated forms, abbreviated PY. Similarly, PY and PYR have been used to describe the ribosylated or non-ribosylated species indiscriminately (e.g., [28,57]). Surely, there is a need for consistency, as these species come from very distinct pathways and are sought systematically during physiological and pathophysiological investigations [11,55]. We proposed that by implementing the IUPAC nomenclature, inconsistencies can be readily rectified. We have named, characterized, and compared each of the pyridone species to facilitate this transition to provide clarity to the field.

The abundance of urinary pyridones derived from nicotinamide has been shown to correlate with metabolic syndromes and aging [4,58]. While *N*-Me-2PY, *N*-Me-6PY, 4PYR, and 6PYR are detected in human urine, *N*-Me-2PY and 2PYR could go undetected because of their rapid conversion to the free acid or an ester during acidic or basic extraction protocols [59–61]. When the amide bond can share hydrogen bond interactions with the neighboring carbonyl, as is the case in *N*-Me-4PY, 4PYR, *N*-Me-2PY, and 2PYR, the loss of ammonia occurs readily, as we observed in solution chemistry, a process that could readily occur during sample processing.

Mass spectrometry detects all metabolites in ionic forms. For the NAD metabolome, in some cases, such as for NR, the metabolite exists in an ionic form before reaching the detector; however, in most cases, the metabolite requires protonation/deprotonation or adduction, usually with sodium, potassium, chloride, ammonium. Many metabolites included in the NAD metabolome react to the most common ionization source, electrospray ionization (ESI) [62,63]. ESI employs a steady stream of nitrogen and heat to produce ions. *N*-Me-4PY, 4PYR, *N*-Me-2PY, and 2PYR react within the ESI chamber in a potentially confounding way because of an MS/MS fragmentation and formation of an acylium ion. In the ESI-MS instrument, such fragments, if stabilized or produced at a high enough rate, can react with water or methanol to form the acid or methyl ester analogs of *N*-Me-4PY, 4PYR, *N*-Me-2PY, and 2PYR (Figure 2). Here, the in-source reaction can lead to a loss of signal and potentially the false identification of metabolites that likely do not exist in the biological sample. Furthermore, metabolite derivatives, artifacts of ESI, could go unmeasured. When a low-resolution instrument is used, the carboxylic acids may be non-resolved from the carboxamide form (Supplementary Materials). However, high-resolution mass spectrometers are more than capable of resolving the amide form from the carboxylic acid [64]. In quantitative metabolomics, internal isotopically labeled standards are used to identify and quantify specific metabolites of interest [63]. In a more encompassing but less accurate, semi-quantitative metabolomics, identifying a species often relies upon the m/z ratio and fragmentation. As such, the materials can go undetected or underrepresented unless dedicated internal standards are used for their quantification. It is with this in mind that one can rationalize the often-contradictory reports of pyridones' quantifications.

Crucially, the ribosylated pyridones derived from nicotinamide can only be generated from the ribosylated forms of nicotinamide [17]. Such forms include nicotinamide riboside (NR), its mononucleotide parent (NMN) or NAD(P), and their reduced forms [1]. Circulating endogenous and exogenous 4PYR is readily intracellularized and metabolized by blood and tissues where it is phosphorylated and converted to either 4PYR-triphosphate (4PYR-TP) or the NAD derivative 4NADO, all inhibitors of intracellular enzymes [25,27,46,65].

The cells' ability to convert exogenous pyridone ribosides to NAD-like species might select for 4PYR's capacity to promote cell death in a cell-specific manner, as it has been observed here and by others [26,28]. Crucially, we observed similar cell line selectivity regarding cell survival when the adenine dinucleotide form of 4PYR, 4NADO, is generated endogenously following exposure to the reduced form of NR, NRH (Figure 8) [47]. Still, the causality of cell death observed in HepG3 cells exposed to NRH, and the formation of 4NADO remain to be explored. Finally, exogenous 2PYR and 6PYR do not affect HEK293T and HepG3 cells' viability (Figure 6). Yet, their function, if generated endogenously, remains unknown. Therefore, the physiological properties of endogenously produced toxins derived from the oxidation of NAD, NADP, and NR warrant further investigations.

Finally, aldehyde oxidases are responsible for the formation of the methylated pyridones (*N*-Me-6PY, and *N*-Me-4PY) from methyl-nicotinamide [36]. Yet, the generation of the three ribosylated pyridone derivatives, 2PYR, 4PYR, and 6PYR, remains unexplained. A hint as to the source of the ribosylated pyridones was obtained when the FAD (flavin adenine dinucleotide)-dependent FDXR was shown to generate the oxidized form of NADP, NADPO, in a side-reaction. In the proposed mechanism, NADP, instead of the normal substrate NADPH, binds FAD-bound FDXR [39]. The 1,4-nucleophilic addition of an active site water molecule on the enzyme-bound NADP leads to a 4-hydroxylated dihydropyridine derivative that can then enter in hydride-transfer type catalysis with the bound FAD to generate the 4-pyridone form of NADP (i.e., 4NADPO). At first sight and given the abundance of the urinary ribosylated pyridones, the kinetics of this side-reaction enzymatic process would preclude it to be the major source of 4PYR. Furthermore, this enzymatic process only accounts for the generation of the dinucleotide form of 4PYR and cannot explain the formation of 2PYR, known to also be in circulation [66]. NQO1 and NQO2 are FAD-dependent redox enzymes that bind NAD(P)H and NRH, respectively, and could potentially generate 4NAD(P)O and 4PYR via a mechanism like that of FDXR. Recent work on NQO2 and acetaminophen in the presence of NRH has identified superoxide HOO- as a by-product of the NQO2-catalyzed turn-over, especially in the presence of O_2 [45]. We envisage that the detection of 4PYR in the NQO2-catalyzed oxidation of NRH is promoted by the NQO2-catalyzed production of superoxide upon the reduction of NRH, with O_2 as a co-oxidant and that the production of superoxide within the vicinity of a still FAD-NQO2-bound NR, favors the addition of superoxide onto the electrophilic NR. Superoxide is likely to react within the constraints of the enzyme binding pocket and lead to a favored isomer. As such, this type of side reaction or a variation thereof could apply to other NAD(P)H binding FAD-dependent oxidoreductases and explain the generation of a substantial amount of ribosylated pyridone metabolites in vivo. It is important to note that CYP450 and FAD-dependent aryl hydroxylation has long been recognized as a source of hydroxylated catabolites of aryl amino acids and riboflavin, e.g., 2/3-hydroxylated tyrosine and 7/8-hydroxyriboflavins, respectively [67,68]. In such a process, the transient formation of the ROS species is responsible for the outcomes of the conversion. We, therefore, explored whether Fenton chemistry [69], often used to emulate ROS-chemistry in these systems, could promote PYR formation from NR and found that only 2PYR and 6PYR but not 4PYR were readily detected. We have therefore identified two related oxidative mechanisms that could be wide-spread and responsible for the formation of functionalized ribosylated pyridones.

4. Methods

4.1. Syntheses

4.1.1. Synthesis of *N*-methyl-4-pyridone 3-carboxamide (**1**)

The synthesis of *N*-methyl-4-pyridone 3-carboxamide was conducted according to Scheme 1.

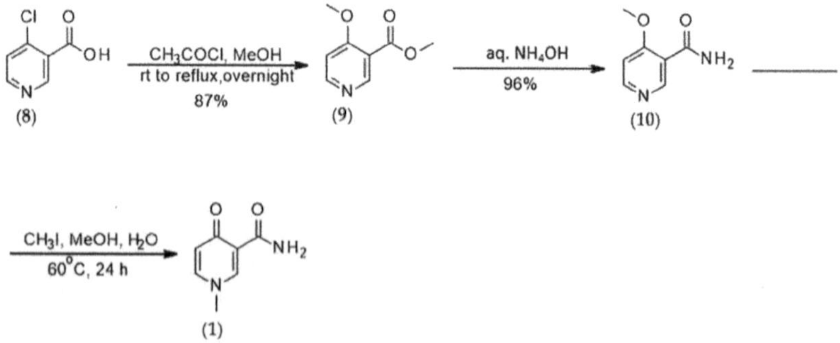

Scheme 1. Synthesis of N-methyl-4-pyridone 3-carboxamide.

Synthesis of O-methyl 4-methoxypyridine-3-carboxylate (**9**) *(Step-1)*: To a solution of 4-chloronicotinic acid (**8**) (1.0 g, 6.34 mmol) in anhydrous methanol (18.5 mL) was added acetyl chloride (1.42 mL) at 0 °C. The reaction mixture was stirred under nitrogen overnight at 65–70 °C. Upon completion of the reaction, the reaction mixture was concentrated under reduced pressure. The crude was dissolved in saturated $NaHCO_3$ (pH = 8–9) and then extracted with $CHCl_3$ (2 × 15 mL). The combined organics were washed with brine (30 mL), dried over Na_2SO_4, filtered, and concentrated under reduced pressure to afford the desired product as a white solid (87%). ^{1}H NMR (MeOD), δ, ppm: 8.74 (s, 1H, H1), 8.51 (d, J = 6.04 Hz, 1H, H6), 7.18 (d, J = 6.04 Hz, 1H, H5), 3.96 (s, 3H, COOMe), 3.87 (s, 3H, –OMe). ^{13}C NMR (MeOD), δ, ppm: 165.58 (C4), 164.92 (C7), 153.52 (C2), 151.37 (C6), 116.45 (C3), 107.93 (C5), 55.33 (C9), 51.31 (C8) [29].

Synthesis of 4-methoxypyridine-3-carboxamide (**10**) *(Step-2)*: O-Methyl-4-methoxypyridine-3-carboxylate (**9**) (0.5 g, 2.99 mmol) in NH_4OH (27%, 20 mL) was stirred at rt for 15 in a 50 mL sealed tube and the reaction was monitored by ^{1}H NMR. After the reaction had gone to completion, the solvent was evaporated under reduced pressure and the obtained white solid was used for the next step without purification (96%). ^{1}H NMR (MeOD), δ, ppm: 8.87 (s, 1H, H1), 8.51 (d, J = 5.96 Hz, 1H, H6), 7.21 (d, J = 6.0 Hz, 1H, H5), 4.04 (s, 3H, –OMe). ^{13}C NMR (MeOD), δ, ppm: 166.50 (C4), 164.35 (C7), 153.02 (C2), 151.37 (C6), 118.16 (C3), 107.51 (C5), 55.55 (C8) [29].

Synthesis of 1-N-methyl-4-oxo-pyridine-3-carboxamide (**1**) *(Step-3)*: To a solution of 4-methoxypyridine-3-carboxamide (**10**) (438 mg, 2.88 mmol) in methanol:water (9:1) mixture (10 mL) was added iodomethane (1.74 mL, 28 mmol). The reaction was stirred under nitrogen for 24 h at 60 °C. After stirring for 24 h at 60 °C with an oil bath, the mixture was cooled in an iced bath for 1 h and the resulting yellow precipitate was filtered off, washed with methanol, and dried under high vacuum. The final product did not require purification. Yield: 65%. ^{1}H NMR (DMSO, d^6), δ, ppm: 9.36 (s, 1H, NH), 8.54 (s, 1H, H2), 7.87 (d, J = 5.08 Hz, 1H, H6), 7.54 (s, 1H, NH), 6.53 (d, J = 7.4 Hz, 1H, H5), 3.82 (s, 3H, Me). ^{13}C NMR (MeOD), δ, ppm: 175.66 (C4), 165.51 (C7), 146.51 (C2), 143.27 (C6), 119.66 (C3), 118.97 (C5), 44.37 (C8), HRMS calcd for $C_7H_9N_2O_2$ $[M + H]^+$ 153.0664 found 153.0652 [29].

4.1.2. Synthesis of O-methyl, 1-N-methyl-4-oxo-pyridine-3-carboxylate (**12**)

The synthesis of O-methyl, 1-N-methyl-4-oxo-pyridine-3-carboxylate was conducted according to Scheme 2.

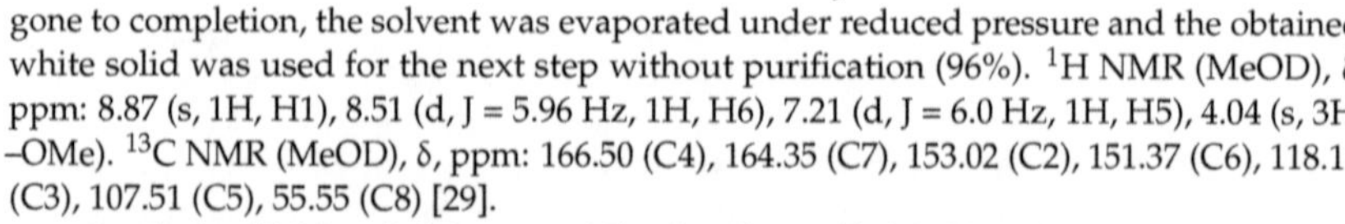

Scheme 2. Synthesis of the O-methyl, 1-N-methyl-4-oxo-pyridine-3-carboxylate (**12**).

Synthesis of O-methyl 4-pyridone-3-carboxylate (**11**) *(Step-1)*: To a solution of the 4-chloronicotinic acid (**8**) (5 g, 31.7 mmol) in H_2O (50 mL) was refluxed for 3 h and then concentrated under reduced pressure. The obtained yellow solid was azeotroped with toluene and then dissolved in anhydrous MeOH (50 mL). To the resulting mixture was added $SOCl_2$ (4.99 mL, 8.19 g, 68.9 mmol) at room temperature, and then the reaction mixture was heated to 75 °C. After stirring at 75 °C for 15 h, the mixture was concentrated under reduced pressure, and the residual product was treated with saturated $NaHCO_3$. As the remaining amount of thionyl chloride got neutralized by $NaHCO_3$, the desired product precipitated. The precipitate was filtered, washed twice with water, and used directly for the next step without further purification. E.g., Scheme Yield 62%. ^{1}H NMR (D_2O), δ, ppm: 8.44 (s, 1H, H2), 7.76 (d, J = 7.2 Hz, 1H, H6), 6.53 (d, J = 6.53 Hz, 1H, H5), 3.77 (s, 3H, OMe). ^{13}C NMR ($CDCl_3$), δ, ppm: 178.34 (C4), 166.77 (COOCH3), 144.41 (C2), 139.15 (C6), 120.11 (C5), 117.13 (C3), 52.26 (Me), HRMS calcd for $C_7H_8NO_3$ $[M + H]^+$ 154.0504 found 154.0490. Spectral data are consistent with published literature [25].

Synthesis of O-methyl, 1-N-methyl 4-pyridone-3-carboxylate (**12**) *(Step-2)*: In a 50-mL round bottom flask, equipped with a magnetic stirring bar, NaH (15.30 mg, 0.65 mmol) was added to 2 mL of anhydrous methanol in a portion-wise manner, and then methyl 4-pyridone-3-carboxylate (**11**) (50 mg, 0.33 mmol) was added. The resulting reaction mixture was stirred at rt for 10 min. Then, CH_3I (81.17 μL, 1.30 mmol) was added, and the resulting mixture was stirred again at the same temperature for 36 h. The reaction progress was monitored by TLC. Upon completion of the reaction, the resulting mixture was concentrated under reduced pressure, and the desired product was precipitated by adding EtOAc (2 mL). (Yield = 70%). ^{1}H NMR (MeOD), δ, ppm: 8.47 (s, 1H, H2), 7.72 (dd, J = 2.28 & 2.28 Hz, 1H, H6), 6.49 (d, J = 7.52 Hz, 1H, H5), 3.83 (s, 3H, OMe), 3.80 (s, 3H, –NMe). ^{13}C NMR ($CDCl_3$), δ, ppm: 176.38 (C4), 164.88 (C7), 147.58 (C2), 142.16 (C6), 120.79 (C3), 117.26 (C5), 50.86 (–OCH_3), 43.23 (–CH_3). HRMS calcd for $C_8H_{10}NO_3$ $[M + H]^+$ 168.0661 found 168.0648. Spectral data are consistent with published $[M^+ + H]$.

4.1.3. Synthesis of *N*-methyl-6-pyridone 3-carboxamide (**2**)

The synthesis of *N*-methyl-6-pyridone 3-carboxamide was conducted according to Scheme 3.

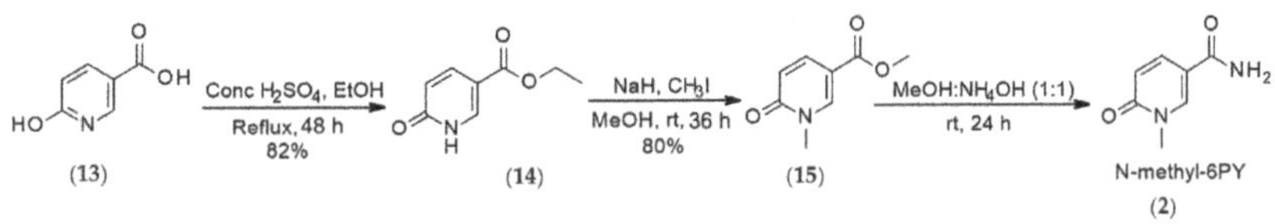

Scheme 3. Synthesis of *N*-methyl-6-pyridone 3-carboxamide.

Synthesis of O-ethyl 6-oxo-1H-pyridine-3-carboxylate (**14**) *(Step-1)*: To a stirred solution of 6-hydroxynicotinic acid (**13**) (5 g, 3.59 mmol) in absolute ethanol (25 mL) was added sulfuric acid (0.2 mL) at room temperature. The mixture was heated to reflux for 48 h. After cooling down to room temperature, water (2.5 mL) was added, and the reaction mixture was neutralized to pH = 6–7 by portion-wise addition of sodium hydrogen carbonate (482 mg). Upon completion of the reaction, the reaction mixture was concentrated under reduced pressure, and the residue was extracted with ethyl acetate (3 × 10 mL). The combined organic layers were washed with brine, dried over Na_2SO_4, and evaporated under reduced pressure leading to the pure ethyl 6-hydroxynicotinate (**14**) (yield = 82%). ^{1}H NMR ($CDCl_3$), δ, ppm: 8.16 (s, 1H, H6), 7.95 (dd, J = 2.44 & 2.44 Hz, 1H, H4), 6.51 (d, J = 9.56 Hz, 1H, H3), 4.25 (q, J = 7.14 Hz, 2H, -CH2), 1.28 (t, J = 7.12 Hz, 3H, Me). ^{13}C NMR ($CDCl_3$), δ, ppm: 165.61 ($COCH_3$), 164.00 (C2), 141.11 (C6), 139.69 (C4), 119.36 (C3), 111.44 (C5), 60.68 (–CH_2), 14.23 (Me) [30].

Synthesis of O-ethyl-1-methyl-6-oxo-pyridine-3-carboxylate (**15**) *(Step-2)*: In a 50-mL round bottom flask, equipped with a magnetic stirring bar, NaH (56.11, 2.39 mmol) was added to

3 mL of anhydrous methanol in a portion-wise manner, and then ethyl 6-hydroxynicotinate (**14**) (0.2 g, 1.19 mmol) was added, the resulting reaction mixture was stirred at rt for 10 min, and then CH_3I (0.297 mL, 4.78 mmol) was added, and the resulting mixture was stirred again at the same temperature for 36 h. The reaction progress was monitored by TLC analysis. Upon completion of the reaction, the resulting mixture was concentrated under vacuum and dissolved in EtOAc (15 mL). The organic layer was washed with water (3 × 10 mL), dried over Na_2SO_4, and concentrated under reduced pressure to afford the product (**15**) as white solid (Yield = 80%). ^{1}H NMR ($CDCl_3$), δ, ppm: 8.12 (s, 1H, H6), 7.77 (dd, J = 2.52 & 2.52 Hz, 1H, H4), 6.46 (d, J = 9.52 Hz, 1H, H3), 3.79 (s, 3H, OMe), 3.52 (s, 3H, Me). ^{13}C NMR ($CDCl_3$), δ, ppm: 165.61 ($COCH_3$), 162.89 (C2), 143.55 (C6), 138.63 (C4), 119.40 (C3), 108.58 (C5), 52.04 (OMe), 38.20 (Me). HRMS calcd for $C_8H_{10}NO_3$ $[M + H]^+$ 168.0661 found 168.0647.

Synthesis of 1-*N-methyl-6-pyridone-3-carboxamide* (**2**) *(Step-3)*: A sealed tube equipped with a Teflon-coated magnetic stirring bar was charged with *O*-methyl, 1-*N*-methyl-6-oxo-pyridine-3-carboxylic acid (**15**) (2 g, 13.0 mmol) and 10 mL mixture of NH_4OH solution (28–30%) and methanol (1:1). The resulting mixture was stirred at room temperature for 24 h. After 24 h, reaction progress was monitored by TLC. Upon completion of the reaction, the solvent was evaporated under reduced pressure, and the resulting crude (**2**) was filtered off and washed with ethyl acetate. Yield 66%. ^{1}H NMR (D_2O), δ, ppm: 8.38 (s, 1H, H2), 7.95 (d, 1H, J = 6.69 Hz, 1H, H4), 6.53 (d, 1H, J = 9.44 Hz, H5), 3.62 (s, 3H, OMe). ^{13}C NMR ($CDCl_3$), δ, ppm: 167.17 ($CONH_2$), 163.69 (C2), 142.03 (C6), 138.61 (C4), 117.89 (C3), 113.52 (C5), 37.27 (Me). HRMS calcd for $C_7H_9N_2O_2$ $[M + H]^+$ 153.0664 found 153.0663.

4.1.4. Synthesis of *N*-methyl-6-pyridone 3-carboxamide (**2**)

The synthesis of 1-*N*-methyl-2-pyridone-3-carboxamide was conducted according to Scheme 4.

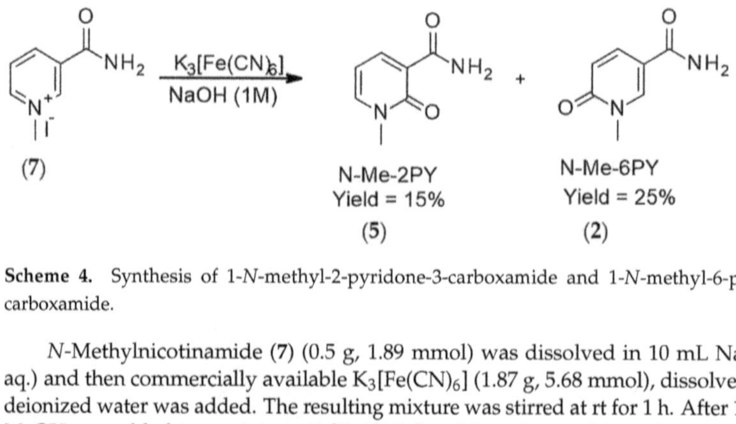

Scheme 4. Synthesis of 1-*N*-methyl-2-pyridone-3-carboxamide and 1-*N*-methyl-6-pyridone-3-carboxamide.

N-Methylnicotinamide (**7**) (0.5 g, 1.89 mmol) was dissolved in 10 mL NaOH (1M, aq.) and then commercially available $K_3[Fe(CN)_6]$ (1.87 g, 5.68 mmol), dissolved in 2 mL deionized water was added. The resulting mixture was stirred at rt for 1 h. After 1 h, 25 mL MeOH was added to precipitate $K_4[Fe(CN)_6]$ and then the resulting mixture was filtered off. The filtrate was concentrated on rotavap and the crude was purified by using hexane, ethyl acetate and methanol as an eluent to afford 1-methyl-2-oxo-pyridine-3-carboxamide (**5**) (15%) and 1-methyl-6-oxo-pyridine-3-carboxamide (**2**) (25%). ^{1}H NMR (DMSO-d_6), δ, ppm: 9.06 (s, 1H, NH), 8.27 (dd, J = 2.2 Hz & 2.16 Hz, 1H, H6), 8.01 (dd, J = 2.16 & 2.16 Hz, 1H, H4), 7.54 (s, 1H, NH), 6.44 (t, J = 6.88 Hz, 1H, H5), 3.08 (s, 3H, Me). ^{13}C NMR (DMSO-d_6), δ, ppm: 165.08 ($CONH_2$), 162.22 (CO), 144.43 (C6), 143.70 (C4), 120.45 (C3), 106.10 (C5), 38.16 (CH_3) HRMS calcd for $C_7H_9N_2O_2$ $[M + H]^+$ 153.0664 found 153.0648 [29].

4.1.5. Synthesis of the Ribosylated Pyridone Carboxamides

Synthesis of 4-pyridone-3-carboxamide riboside (3)

The synthesis of 4-pyridone-3-carboxamide riboside was conducted according to Scheme 5.

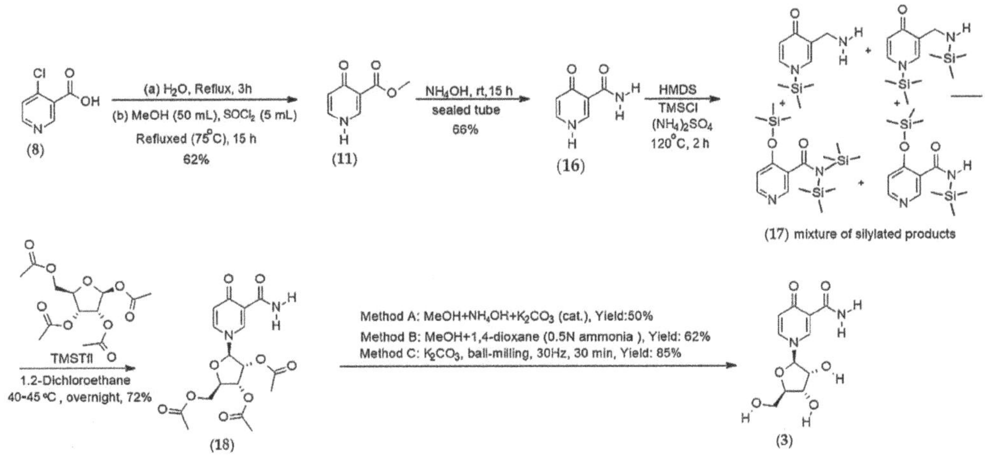

Scheme 5. Synthesis of 4-pyridone-3-carboxamide riboside.

Synthesis of 4-pyridone-3-carboxamide **(16)** *(Step-1)*: A sealed tube equipped with a Teflon-coated magnetic stirring bar was charged with methyl 4-pyridone-3-carboxylate (**11**) (2 g, 13.0 mmol) and 50 mL of NH_4OH methanolic solution (28–30%). The resulting mixture was stirred at room temperature for 15 h. Upon completion of the reaction, the solvent was evaporated at reduced pressure, and the residue was triturated with water. The solid (**16**) was filtered off and washed with water. Yield 66%. ^{1}H NMR (MeOD), δ, ppm: 8.54 (s, 1H, H2), 7.79 (d, J = 5.56 Hz, 1H, H6), 6.56 (d, J = 6.53 Hz, 1H, H5). ^{13}C NMR ($CDCl_3$), δ, ppm: 179.04 (C4), 168.48 (CONH2), 142.95 (C2), 139.32 (C6), 119.45 (C5), 117.69 (C3); HRMS calcd for $C_6H_7N_2O_2$ $[M + H]^+$ 139.0508 found 139.0495 [25].

Synthesis of 1-(2′,3′,5′-tri-O-acetyl-β-D-ribofuranosyl)-4-pyridone-3-carboxamide **(18)** *(Step-2):* Silylation of 4-pyridone-3-carboxamide (**17**): 4-pyridone-3-carboxamide (**16**) (0.5 g, 3.62 mmol): Hexamethyldisilazane (10 mL) and trimethylchlorosilane (1.5 mL, 1.30 g, 12 mmol) were added sequentially in a 50 mL round bottom flask under nitrogen. The resulting mixture was stirred at 120 °C for 2 h. Upon completion of the reaction, the solution was cooled to room temperature and concentrated under reduced pressure. The residue was co-evaporated with anhydrous toluene (2 to 3 times) to afford the mixture of mono and bis-silylated 4-pyridone-3-carboxamide (**17**), which was used directly for the next step.

Vorbrüggen glycosylation *(Step-3)*: The crude mono and bis-silylated 4-pyridone-3-carboxamide (**17**) mixture was dissolved in 6 mL of anhydrous 1,2- dichloroethane. Then, a solution of 1,2,3,5-tetra-O-acetyl-β-D-ribofuranoside (0.6 g, 2 mmol) in 2 mL of 1,2-dichloroethane was added, followed by the addition of trimethylsilyl triflate (0.4 mL, 2.3 mmol). The resulting mixture was stirred at 40–45 °C for overnight. The reaction was monitored by ^{1}H NMR analysis of the crude mixture. After completion of the reaction, the resulting solution was cooled down and poured into a mixture of ice-cooled saturated aq. $NaHCO_3$ solution. The mixture was extracted with DCM (3 × 10 mL). The organic phases were collected, washed with a saturated NaCl solution, and dried over anhydrous Na_2SO_4. Then, the filtrate was concentrated under reduced pressure and the residue was purified by flash column chromatography using DCM: Acetone (7:3) as an eluent to afford the pure product (**18**) (72%). ^{1}H NMR (MeOD), δ, ppm: 8.75 (d, J = 2.4 Hz, 1H, H2), 7.87

(dd, J = 2.4 Hz, 2.4 Hz, 1H, H6), 6.51 (d, J = 7.6 Hz, 1H, H5), 5.80 (d, J = 5.6 Hz, 1H, H1′), 5.38–5.25 (m, 2H, H2′ and H3′), 4.45 (q, J = 3.07 Hz,1H, H4′), 4.36 and 4.27 (AB part of ABX system, 2H, JAB = 12.56 Hz, JAX = 3.12 Hz, JBX = 2.76 Hz, H5′), 2.08 (s, 3H, Ac), 2.04 (s, 3H, Ac), 1.98 (s, 3H, Ac). ^{13}C NMR (MeOD), δ, ppm: 178.69 (C4), 170.73 (CO), 169.88 (CO), 169.69 (CO), 166.32 (CONH2), 141.78 (C2), 138.23 (C6), 119.91 (C5), 119.04 (C3), 94.59 (C1′), 81.48 (C4′), 74.48 (C2′), 70.43 (C3′), 62.83 (C5′), 19.28 (Me), 18.96 (Me), 18.73 (Me); HRMS calcd for $C_{17}H_{21}N_2O_9$ $[M + H]^+$ 397.1249 found 397.1229.

Synthesis of 4-Pyridone-3-carboxamide-1-β-D-ribofuranoside (**3**) *(Step-4)*: Method A: A mixture of compound (**18**) (0.078 g, 0.196 mmol), 1 mL NH_4OH (27–30%) and cat. K_2CO_3 in methanol (10 mL) was stirred at rt overnight, and the reaction progress was monitored by ^{1}H NMR analysis of the crude reaction. Upon completion of the reaction, the reaction mixture was filtered and evaporated in vacuo to afford the desired product 3 as semisolid. Yield 50%.

Method B: A sealed tube was charged with a mixture of compound (**18**) (0.408 g, 1.02 mmol), 2 mL 1,4-dioxane (saturated with 0.5N NH_3), and 15 mL methanol. The resulting mixture was stirred at rt overnight, and the reaction progress was monitored by NMR analysis of the crude mixture. After overnight stirring at rt, the ^{1}H NMR of the crude showed complete removal of the acetates. The resulting mixture was concentrated under reduced pressure to afford the desired product 3. Yield: 62%.

Method C: A PTFE jar was charged with pure compound (**18**) (0.2 g, 0.50 mmol), K_2CO_3 (6.91 mg, 0.1 mmol) and 20 microliter MeOH and was vibrated at a rate of 1800 rpm (30 Hz) at room temperature for 30 min. The reaction progress was monitored by ^{1}H NMR analysis. Yield: 85%. ^{1}H NMR (D_2O), δ, ppm: 8.67 (d, J = 2.24 Hz, 1H, H2), 7.94 (dd, J = 2.24 Hz, 2.24 Hz, 1H, H6), 6.65 (d, J = 7.6 Hz, 1H, H5), 5.55 (d, J = 5.5 Hz, 1H, H1′), 4.23 (t, 1H, J = 5.28 Hz, H2′), 4.19–4.13 (m, 2H, H4′ and H3′), 3.81 and 3.73 (AB part of ABX system, 2H, JAB = 12.7 Hz, JAX = 3.04 Hz, JBX = 4.0 Hz, H5′). ^{13}C NMR (D_2O), δ, ppm: 179.13 (C4), 167.75 ($CONH_2$), 142.72 (C2), 138.73 (C6), 120.35 (C5), 118.25 (C3), 96.96 (C1′), 86.03 (C4′), 75.61 (C2′), 70.02 (C3′), 60.89 (C5′); HRMS calcd for $C_{11}H_{15}N_2O_6$ $[M + H]^+$ 271.0930 found 271.0919.

Synthesis of 6-pyridone 3-carboxamide riboside (**4**)

The synthesis of 6-pyridone 3-carboxamide riboside was conducted according to Scheme 6.

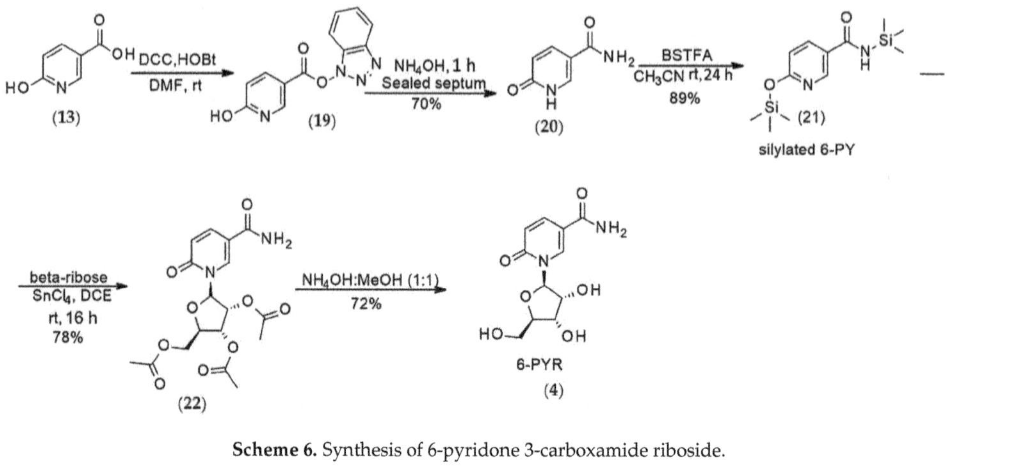

Scheme 6. Synthesis of 6-pyridone 3-carboxamide riboside.

Synthesis of 6-pyridone 3-carboxamide (6PY) (**20**) *(Step-1)*: To an oven-dried 50-mL round bottom flask equipped with a magnetic stir bar, 6-hydroxynicotinic acid (**13**) (5 g, 3.59 mmol), HOBt (0.540 g, 4.0 mmol), DCC (0.928 g, 4.5 mmol), and anhydrous DMF

(10 mL) were added. The RBF was placed under nitrogen sealed with a septum. The reaction mixture was stirred at rt for 2 h under nitrogen. The reaction progress was monitored by ^{1}H NMR of the crude reaction mixture. Upon completion of the reaction, the solid residue (**19**) was filtered off and washed with fresh DMF (10 mL). The organic layer was concentrated under reduced pressure, and the residue was dissolved in 10 mL NH_4OH (28–30%) and stirred at rt for 1 h. The reaction progress was monitored by ^{1}H NMR of the crude reaction mixture. After completion of the reaction, the solvent was evaporated under high vacuum to afford the 6-pyridone-3-carboxamide (**20**) (6PY) as a brown solid. Yield 70%. ^{1}H NMR (DMSO, d^{6}), δ, ppm: 11.92 (s, 1H, NH), 8.00 (s, 1H, H2), 7.85 (d, J = 9.6 Hz, 1H, H4), 7.72 (s, 1H, NH), 7.18 (s, 1H, NH), 6.32 (d, J = 9.56 Hz, 1H, H5). HRMS calcd for $C_{12}H_{16}NO_7$ $[M + H]^+$ 139.0508 found 139.0499 [31].

Silylation of 6-pyridone-3-carboxamide (**21**) *(Step-2)*: 6PY (**20**) (0.3 g, 2.17 mmol) and *N,O*-Bis (trimethylsilyl) trifluoroacetamide (2.3 mL, 2.23 g, 8.69 mmol) were dissolved in anhydrous acetonitrile (5 mL) at rt, and the reaction mixture was stirred at rt for 24 h under nitrogen. After 24 h, the reaction mixture was concentrated under reduced pressure, and the crude product was used directly in the next step.

Synthesis of 1-(2′,3′,5′-tri-O-acetyl-β-D-ribofuranosyl)-6-pyridone-3-carboxamide (**22**) *(Step-3):* To a mixture of pre-silylated 6-pyridone-3-carboxamide (**21**) (0.5 g, 1.6 mmol) and 1, 2, 3, 5-tetra-*O*-acetyl-β-ribofuranoside (0.767 g, 2.4 mmol) in anhydrous 1,2-dichloroethane (20 mL) was added $SnCl_4$ (0.754 mL, 1.67 g, 6.4 mmol) and stirred at rt for 16 h under nitrogen gas environment. Once the reaction was complete, the reaction mixture was poured into water and extracted with $CHCl_3$ (10 mL × 3), washed with saturated $NaHCO_3$ and dried over anhydrous Na_2SO_4. The organic layer was concentrated, and the oily product was purified by flash column chromatography. Yield 78%. ^{1}H NMR ($CDCl_3$), δ, ppm: 8.30 (br.s., 1H, H6), 7.69 (d, J = 9.04 Hz, 1H, H4), 6.45 (d, J = 9.44 Hz, 1H, H3), 6.26 (br.s, 1H, H1′), 5.31 (t, J = 4.58 Hz, 1H, H2′), 5.20 (t, J = 5.20 Hz, 1H, H3′), 4.56–4.51 (m, 1H, H4′), 4.38 (br. s, 1H, H5′ ABX system), 4.26 (d, J = 12.36 Hz, H5′ ABX system), 2.13 (s, 3H, Ac), 2.05 (s, 3H, Ac), 2.04 (s, 3H, Ac). ^{13}C NMR ($CDCl_3$), δ, ppm: 171.43 (C2), 169.60 (CO), 169.36 (CO), 165.73 (CO), 161.80 ($CONH_2$), 141.66 (C6), 135.66 (C4), 121.59 (C3), 119.89 (C5), 87.96 (C1′), 80.10 (C4′), 74.13 (C2′), 69.74 (C3′), 62.74 (C5′), 20.88 (Me), 20.48 (Me), 20.45 (Me). MS (ES): *m/z* 397.63 $[M^+ + H]$ and 419.56 $[M^+ + Na]$.

Synthesis of 6-pyridone-3-carboxamide-1-β-D-ribofuranoside (**4**) *(Step-4)*: The triacetylated precursor (**22**) (0.360 g, 0.9 mmol) was dissolved in 2 mL ordinary methanol and stirred at rt for 15 min. After 15 min, 2 mL NH_4OH (28–30%) was added, and the resulting mixture was stirred at rt for 4 h. The reaction progress was monitored by ^{1}H NMR analysis of the crude reaction mixture. Upon completion of the reaction, the solvent was evaporated under reduced pressure, and the crude was kept at rt overnight before treating with acetone for 24 h. The solid was then filtered off and washed with fresh acetone to afford 6-PYR (**4**) as a white solid. Yield 72%. ^{1}H NMR (D_2O), δ, ppm: 8.60 (s, 1H, H6), 7.84 (d, J = 9.2 Hz, 1H, H4), 6.54 (d, J = 9.44 Hz, 1H, H5), 6.01 (s, 1H, H1′), 4.19–4.13 (m, H2′, H3′ & H4′), 3.97 (d, J = 12.92 Hz, 1H, H5′, ABX system), 3.80 (d, J = 12.88 Hz, 1H, H5′, ABX system). ^{13}C NMR ($CDCl_3$), δ, ppm: 169.12 ($CONH_2$), 163.80 (C2), 139.30 (C6), 139.02 (C4), 118.80 (C3), 114.30 (C5), 90.75 (C1′), 83.56 (C4′), 74.87 (C2′), 68.24 (C3′), 59.83 (C5′), HRMS calcd for $C_{11}H_{15}N_2O_6$ $[M + H]^+$ 271.0930 found 271.0938.

Synthesis of *O*-methyl-4-pyridone-3-carboxylate-1-β-D-ribofuranoside (**26**)

The synthesis of *O*-methyl-4-pyridone-3-carboxylate-1-β-D-ribofuranoside was conducted according to Scheme 7.

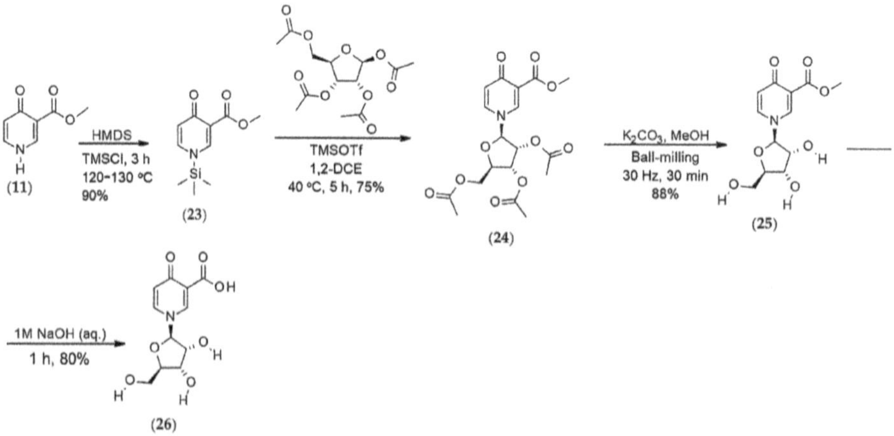

Scheme 7. Synthesis of the methyl-ester and carboxylate ribosyl analogs.

Silylated 4-pyridone-3-carboxylic methyl ester (**23**) *(Step-1):* O-Methyl-4-pyridone-3-carboxylic ester (**11**) (3.0 g, 1.98 mmol) was dissolved in 50 mL of hexamethyl-disilazane (HMDS) in an RBF under nitrogen, 5 mL of trimethylsilyl chloride (TMSCl) (0.494 mL, 0.423 g, 3.9 mmol) was then added. The mixture was refluxed under nitrogen for 3 h at 120–130 °C. The excess reagents were evaporated and co-evaporated with toluene (10 mL). The product was a dark yellow oily residue. Yield 90%. ^{1}H NMR (400 M Hz, $CDCl_3$, δ, ppm): 8.92 (s, 1H, H2), 8.44 (d, J = 5.6 Hz, 1H, H6), 6.71 (d, J = 5.6 Hz, 1H, H5), 3.83 (s, 3H, OMe), 0.27 (s, 9H, $Si(CH_3)_3$).

Synthesis of 1-(2′,3′,5′-Tri-O-acetyl-β-D-ribofuranosyl)-4-pyridone-3-carboxylic methyl ester (**24**) *(Step-2):* Silylated 4-pyridone-3-carboxylic methyl ester (**23**) was dissolved in 25 mL of 1,2-dichloroethane (DCE) under nitrogen. 1,2,3,5-tetra-O-acetyl-δ-D-ribofuranoside (6.30 g, 1.98 mmol) was dissolved in DCE (25 mL) and then added to the ester solution. An additional 50 mL of DCE and 4.1 mL of trimethylsilyl triflate (TMSOTf) (2.28 mmol) were added to the mixture. The resulting mixture was stirred at 40 °C for 5 h under nitrogen. Upon completion of the reaction, the reaction mixture was cooled to 0 °C and poured into iced water. The reaction flask was rinsed with DCM (50 mL) and added to the quenched reaction solution. The mixture was neutralized with 35 mL of a saturated aq $NaHCO_3$ solution. The resulting phases separated, and the organic layer was washed with saturated NaCl solution (50 mL). Once separated, the organic phase was dried over anhydrous sodium sulfate, filtered, and evaporated. The crude product was purified by flash column chromatography using n-hexane and ethyl acetate as an eluent system to afford the desired product (**24**) as a light brown crystalline solid (6.42 g, 75%). ^{1}H NMR (400 M Hz, $CDCl_3$, δ, ppm): 8.40 (d, J = 2.5 Hz, 1H, H-2), 7.45 (dd, J = 7.9 and 2.5 Hz, 1H, H-6), 6.52 (d, J = 7.9 Hz, 1H, H-5), 5.49 (d, J = 5.9 Hz, 1H, H-1′), 5.30 (dd, J = 5.6 and 3.6 Hz, 1H, H-3′), 5.21 (dd, J = 5.9 and 5.6 Hz, 1H, H-2′), 4.33–4.46 (m, 3H, H-5′A, H-5′B, H-4′), 3.88 (s, 3H, O-CH_3), 2.19 (s, 3H, Ac), 2.14 (s, 3H, Ac), 2.10 (s, 3H, Ac); ^{13}C NMR (100 M Hz, $CDCl_3$, δ, ppm): 175.7 (COO), 170.1 (CO), 169.4 (CO), 169.3 (CO), 165.6 (C-4), 142.4 (C-2), 135.1 (C-6), 122.9 (C-5), 119.4 (C-3), 94.22 (C-1′), 81.34 (C-4′), 74.29 (C-2′), 70.21 (C-3′), 62.76 (C-5′), 52.40 (–OCH_3), 20.47 (CH_3), 20.31 (CH_3); HRMS calcd for $C_{18}H_{22}NO_{10}$ $[M + H]^+$ 412.1243 found 412.1243.

Synthesis of O-methyl-4-pyridone-3-carboxylate ribonucleoside (**25**) *(Step-3):* A mixture of 1-(2′,3′,5′-tri-O-acetyl-β-D-ribofuranosyl)-4-pyridone-3-carboxylic methyl ester (**24**) (3.20 g, 7.79 mmol), K_2CO_3 (0.103 g, 0.78 mmol), and MeOH (1.2 mL, 29.6 mmol) was ball-milled for 30 min at 30 Hz. Following removal of methanol, a brown solid with a taffy-like consistency was obtained (1.81 g, 95%). ^{1}H NMR (400 M Hz, MeOD, δ, ppm): 3.76 (dd, J = 12.1 and 2.7 Hz, 1H, H-5′A), 3.83 (s, 3H, -CH_3), 3.85 (dd, J = 2.7 Hz and J = x, 1H, H-5′B), 4.15 (dd, J = 6.3 and 2.6 Hz, 1H, H-4′), 4.17–4.22 (m, 2H, H-2′, H-3′), 5.49 (d, J = 5.1 Hz, 1H, H-1′),

6.53 (d, J = 7.7 Hz, 1H, H-5), 8.09 (dd, J = 7.7 and 2.3 Hz, 1H, H-6), 8.78 (d, J = 2.3 Hz, 1H, H-2); ^{13}C NMR (100 M Hz, MeOD, δ, ppm): 50.89 (–OCH_3), 61.19 (C-5′), 70.94 (C-3′), 76.33 (C-2′), 87.11 (C-4′), 97.57 (C-1′), 117.7 (C-3), 120.6 (C-5), 137.7 (C-6), 143.8 (C-2), 164.8 (C-4), 177.4 (COO); HRMS calcd for $C_{12}H_{16}NO_7$ $[M + H]^+$ 286.0926 found 286.0929.

Synthesis of 4-pyridone-3-carboxylic acid ribonucleoside (**26**) *(Step-4):* A mixture of 4-pyridone-3-carboxylic ester riboside (**25**) (2.43 g, 8.85 mmol) was dissolved in a 1M NaOH aqueous solution (10 mL) and stirred for 1 h. Upon completion of the reaction, the resulting mixture was neutralized with 1 M HCl solution (5 mL) and evaporated on high vacuum (80%). ^{1}H NMR (400 M Hz, D_2O, δ, ppm): 8.21 (d, J = 2.4 Hz, 1H, H-2), 7.89 (dd, J = 7.6 and 2.4 Hz, 1H, H-6), 6.52 (d, J = 7.6 Hz, 1H, H-5), 5.51 (d, J = 5.6 Hz, 1H, H-1′), 4.26 (dd, J = 5.6 and 5.3 Hz, 1H, H-2′), 4.20 (dd, J = 5.3 and 3.6 Hz, 1H, H-3′), 4.15 (dd, J = 8.3 and 3.6 Hz, 1H, H-4′), 3.82 (dd, J = 12.8 and 4.5 Hz, 1H, H-5′B), 3.74 (dd, J = 12.8 and 4.5 Hz, 1H, H-5′A); ^{13}C NMR (100 M Hz, D_2O, δ, ppm): 60.97 (C-5′), 70.03 (C-3′), 5.36 (C-2′), 85.81 (C-4′), 96.67 (C-1′), 118.9 (C-5), 127.3 (C-6), 138.4 (C-2), 138.8 (C-3), 172.0 (C-4), 178.1 (COO). HRMS calcd for $C_{11}H_{14}NO_7$ $[M + H]^+$ 272.0770 found 272.0768.

Synthesis of 2-pyridone-3-carboxamide Riboside (**6**)

The synthesis of 2-pyridone-3-carboxamide riboside was conducted according to Scheme 8.

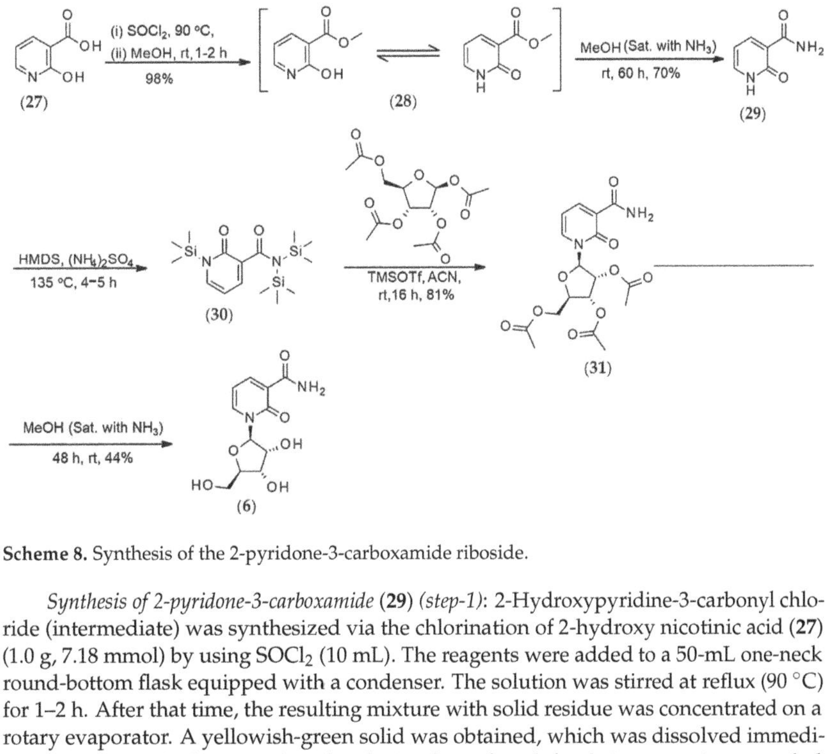

Scheme 8. Synthesis of the 2-pyridone-3-carboxamide riboside.

Synthesis of 2-pyridone-3-carboxamide (**29**) *(step-1)*: 2-Hydroxypyridine-3-carbonyl chloride (intermediate) was synthesized via the chlorination of 2-hydroxy nicotinic acid (**27**) (1.0 g, 7.18 mmol) by using $SOCl_2$ (10 mL). The reagents were added to a 50-mL one-neck round-bottom flask equipped with a condenser. The solution was stirred at reflux (90 °C) for 1–2 h. After that time, the resulting mixture with solid residue was concentrated on a rotary evaporator. A yellowish-green solid was obtained, which was dissolved immediately in 15 mL anhydrous methanol and stirred at rt for 1 h for their conversion to methyl 2-hydroxypyridine-3-carboxylate (**28**). After completion of the reaction, the solvent was removed, and the crude 2-hydroxypyridine-3-carboxylate (**28**) was dissolved in 10 mL NH_3 saturated methanol, and then the round bottom flask was sealed with a septum. The resulting mixture was stirred at rt for the next 60 h. After 60 h, the precipitated white solid was filtered off and washed with cold methanol (2–3 times) and dried to yield 70% of 2-oxo-1H-pyridine-3-carboxamide (**29**). ^{1}H NMR (DMSO-d_6), δ, ppm: 12.3 (s, 1H, NH),

9.05 (s, 1H, NH), 8.29 (d, J = 5.28 Hz, 1H, H6), 7.66 (d, J = 4.32 Hz, 1H, H4), 7.51 (s, 1H, NH), 6.41 (t, J = 6.66 Hz, 1H, H5). ^{13}C NMR (DMSO-d_6), δ, ppm: 165.09 ($CONH_2$), 162.71 (CO), 144.69 (C4), 140.04 (C6), 121.27 (C3), 160.44 (C5); HRMS calcd for $C_6H_6N_2O_2$ [M + H]$^+$ 139.0507 found 139.0493.

Silylation of 2-pyridone-3-carboxamide (**30**) *(Step-2):* 2-pyridone-3-carboxamide (**29**) (426 mg, 3.08 mmol), hexamethyldisilazane (10 mL) and $(NH_4)_2SO_4$ (cat. amount) were added sequentially to a 50-mL round bottom flask under nitrogen. The resulting mixture was stirred at 135 °C for 2 h. Upon completion of the reaction, the resulting solution was cooled to room temperature and then concentrated under vacuum. The residue was co-evaporated with anhydrous toluene (2 to 3 times) to afford the mixture of mono and bis-silylated 2-pyridone-3-carboxamide (**30**), which was used directly for the next step.

Vorbrüggen glycosylation (Step-3): The crude mono and bis-silylated 2-pyridone-3-carboxamide (**30**) was dissolved in 10 mL of anhydrous CH_3CN. Then, a solution of 1,2,3,5-tetra-*O*-acetyl-β-D-ribofuranoside (1.017 g, 3.20 mmol) in 2 mL of anhydrous acetonitrile was added, followed by the addition of a solution trimethylsilyl triflate (1.67 mL, 9.24 mmol). The resulting mixture was stirred at rt for 24 h, and the reaction progress was monitored by NMR analysis of crude mixture. After completion of the reaction, the resulting solution was concentrated under vacuum, and the residue was purified by flash column chromatography using n-hexane: ethyl acetate (1:1.5) as an eluent to afford pure products (**31**) (81%).

Deacetylation of compound **31** *(Step 4)*: Compound **31** (1.01 g, 2.52 mmol) was dissolved in 10 mL methanol (saturated with ammonia) and stirred at rt for 48 h under sealed condition. The reaction progress was monitored by the NMR analysis of crude reaction mixture. Upon completion of the reaction, solvent was evaporated on high vacuum and the crude product was stirred with acetone and filtered off and washed with fresh acetone to afford 2-PYR (**6**) as white solid. The obtained product was enough pure. Yield 44%. ^{1}H NMR (MeOD), δ, ppm: 8.46 (dd, J = 5.4 Hz & 7.04 Hz, 2H, H4 & H6), 6.55 (t, J = 6.96 Hz, 1H, H5), 6.14 (s, 1H, H1′), 4.12 (bs, 3H, H2′, H3′ & H4′), 3.97 (d, J = 12.9 Hz, 1H, H_a5'), 3.80 (d, J = 11.56 Hz, 1H, H_b5'). ^{13}C NMR (MeOD), δ, ppm: 166.62 ($CONH_2$), 161.79 (CO), 144.02 (C6), 138.21 (C4), 119.69 (C3), 105.90 (C5), 91.11 (C1′), 84.38 (C4′), 75.71 (C2′), 68.58 (C3′), 59.88 (C5′), HRMS calcd for $C_{11}H_{15}N_2O_6$ [M + H]$^+$ 271.0930 found 271.0909 [32].

Synthesis of 2PYR and 6PYR via Fenton Chemistry

Nicotinamide riboside chloride (0.20 g, 0.78 mmol) was dissolved in aqueous NaOH (1 M) at 0 °C and rapidly added to a solution of $K_3[Fe(CN)_6]$ (0.773 g, 2.35 mmol) in 10 mL H_2O. The resulting mixture was stirred at 0 °C for 1 h and then warmed to rt. After 1 h stirring at rt, 50 mL of MeOH was added. The mixture was stirred at rt for 2 h. A thick yellowish precipitate formed during stirring and was filtered off and washed with MeOH (2–3 times). NMR of the crude reaction mixture showed the formation of 2PYR and 6PYR but not that of 4PYR.

4.2. Enzymatic Conversions

4.2.1. Synthesis of 4PYR via NQO2 Catalyzed Hyper-Oxidation of the Reduced Form of Nicotinamide Riboside (NR), Abbreviated NRH

To prepare the sample, 1 mM NR in 450 µL buffer (50 mM potassium phosphate buffer, 125 mM NaCl, 5 µM FAD, 1.0 mg/mL BSA) was reacted with 20 µL of the NQO2 enzyme (Sigma-Aldrich, Q0380) and allowed to incubate overnight at room temperature. NRH was also incubated overnight with the same buffer composition for comparison. An Agilent 1200 series HPLC was used for the purification and isolation of compounds. Five microliters of each sample were injected onto an Agilent C18 reversed-phase column (2.1 × 150 mm) equipped with a C18 guard column using an autosampler. Solvent A consisted of H_2O with 2.0% acetonitrile (ACN) and 0.1% trifluoroacetic acid (TFA), and solvent B consisted of ACN with 2.0% H_2O and 0.1% TFA. A flow rate of 100 µL per minute was used for the entirety of the run. The solvent was held at 5.0% B for the first 5 min, from

5 to 20 min a gradient from 5.0 to 80.0% B was used, then solvents were held at 80% B for 5 min and returned to 5.0% B for the remainder of the run. Two injections were performed for each sample, one scanning at 260 nm (for the UV detection of NR and 4PYR) and the other injection scanning at 340 nm (for the UV detection of NRH). Two fractions were collected based on the retention time of previously analyzed standards. The first from 5 to 8 min for the collection of NRH and 4PYR, and the second fraction from 19 to 26 min for the collection of NR. The collected fractions were dried using a speedvac concentrator (without heat applied). Samples were dissolved into 50 µL of 10 mM ammonium bicarbonate (ABC) for analysis by mass spectrometry.

4.2.2. LC-MS/MS Analyses

Liquid chromatography-mass spectrometry analyses were performed on the individual HPLC fractions that were previously prepared. Each fraction was analyzed separately. An Agilent 1200 Series HPLC coupled to a Thermo LTQ Orbitrap XL mass spectrometer was used for sample analysis. An autosampler was used to inject the samples onto an Agilent Zorbax 300SB-C18 reversed-phase column (2.1 × 150 mm, 5-micron) using a 4.0 µL injection volume for each. Solvent A consisted of 10 mM ABC in H_2O, and Solvent B consisted of neat ACN. A flow rate of 50 µL per minute was used for the first 5 min of the run at 5.0% B solvent, followed by a gradient from 5.0–80.0% B from 5 to 15 min along with an increased flow rate of 100 µL per minute. The solvent was held at 80.0% B from 15 to 25 min and returned to 5.0% B for the remainder of the run. A HESI (heated electrospray ionization) source was used with positive polarity, a capillary temperature of 200 °C, the source voltage of 3.0 kV, tube lens voltage of 110, and a sheath gas flow rate of 8.0. One full scan from 80–800 m/z was performed at 15,000 resolution, followed by targeted MS2 scans of NR (255.1 m/z), NRH (257.1 m/z), and 4PYR (271.1 m/z) at 7500 resolution with an isolation width of 1.0 m/z. Normalized CID collision energy was set to 35. All scans were performed in the FTMS. The total run time was 30 min. A blank was run in-between each sample to minimize and monitor for carryover.

4.3. Cell-Based Assays

4.3.1. Cell Culture

Human embryonic kidney containing the SV40 T-antigen (HEK293T) and the human hepatoma cell line (HepG3) were purchased from ATCC (Manassas, VA, USA). The cells were grown at 37 °C in a 5% CO_2 incubator in Dulbecco's modified Eagle's medium (DMEM, Hyclone, Logan, UT, USA; 4.5 g/L glucose and L-glutamine) supplemented with 10% fetal bovine serum (FBS, Atlanta Biologicals, Flowery Branch, GA, USA), and 1% sodium pyruvate (Gibco, Carlsbad, CA, USA). Cells were routinely tested for mycoplasma contamination using the Lonza MycoAlert kit (Walkersville, MD, USA) and found to be free of mycoplasma contamination.

4.3.2. Cytotoxicity Studies

Cell viability was determined by CellTiter-Fluor^{TM} Cell Viability assays (Promega Corporation, Madison, WI, USA). HEK293T and HepG3 cells were seeded at a density of 5000 cells/well and 10,000 cells/well respectively in a 96-well clear bottom black plate and incubated overnight (ON) at 37 °C in a 5% CO_2 incubator. For HEK293T cells, a poly D-lysine-coated black plate was used to prevent washing off the cells during medium replacement. For 24 h exposures, the cells were exposed to 100 µM concentrations alone or increasing doses (25, 50, 75, 100, and 150 µM) of pyridone derivatives (i.e., 2-PYR, 4-PYR, and 6-PYR) for 24 h, then medium containing NRH was replaced with fresh medium, and cells were further incubated for another 48 h (total 72 h) at 37 °C in a 5% CO_2 incubator. For continuous exposures, the cells were exposed to 100 µM concentrations alone or increasing doses (25, 50, 75, 100, and 150 µM) of pyridone derivatives (i.e., 2-PYR, 4-PYR, and 6-PYR) for 72 h. The 2-PYR, 4-PYR, and 6-PYR were dissolved in Dulbecco's phosphate-buffered saline (PBS, Hyclone, Logan, UT, USA) at 10 mM and then diluted to the final working concentrations in the cell growth medium. At the end of the 72-h incubation period, 100 μL of 2X CellTiter-Fluor™ Viability reagent was added in each well, mixed briefly, and then the plates were incubated for another 30 min at 37 °C in a 5% CO_2 incubator. Fluorescence intensity was measured in a microplate reader (Infinite® M1000 PRO, TECAN, Männedorf, Switzerland) with a fluorometer at excitation/emission (Ex/Em) of 380/505 nm. Cells were plated in triplicate for each chemical concentration and exposure durations with at least three biological repeats performed. Fluorescent values are expressed as the number of cells in drug-treated wells relative to cells in vehicle-treated, control wells (%viability) ± standard error of the mean (SEM).

4.3.3. Clonogenic Assays

A clonogenic assay was used to determine the cell growth and colony formation after exposure. HepG3 cells were plated at a density of 10,000 cells/well in each well of a 12-well plate and incubated ON at 37 °C in a 5% CO_2 incubator. The following day, cells were exposed to 25, 50, and 100 μM 4-PYR continuously for six days. At the end of the incubation period, plates were placed on ice, washed twice with ice-cold 1 X PBS before fixing the cells with ice-cold methanol for 10 min. Following the fixation step, cells were incubated with 0.5% crystal violet solution (made in 25% methanol) for 10 min at room temperature (RT, ~23 °C). The plates were rinsed with double-distilled water to remove the excess crystal violet stain and allowed to dry ON at RT. Stained plates were imaged using ChemiDoc™ MP Imaging System (Bio-Rad, Hercules, CA, USA). For analysis, the area fraction of the crystal violet-stained cells was generated from each image using NIS-Elements software. The percentage of the average area fraction of crystal violet staining per well was calculated. Each treatment was performed in technical triplicates and averaged over three biological replicates with final cell density percentage reported relative to control ± SEM.

4.3.4. Immunoblotting

The cell death mechanism of 4-PYR induced cytotoxicity was assessed by immunoblotting. Briefly, HepG3 cells were plated in 65-mm dishes at a density of 2×10^6 cells per dish and incubated ON at 37 °C in a 5% CO_2 incubator. The cells were treated with 100 μM 4-PYR for 48, 72, and 96 h. Cell pellets were collected at the end of the desired exposure period and stored at −80 °C. The cell pellets were thawed on ice and resuspended in a lysis buffer of 25 mM β-glycerolphosphate, 50 mM Tris-HCl, pH 7.5, 150 mM NaCl, 0.2% Triton X-100, and 0.3% NP-40 supplemented with 1X Halt protease and phosphatase inhibitor (Pierce, Waltham, MA, USA). Resuspended cells were incubated for 30 min on ice. Lysates were centrifuged at 12,000 rpm for 15 min at 4 °C, and the supernatant fraction containing protein was retained. Protein concentrations were quantified using Bradford Quick Start protein assay (Bio-Rad, Hercules, CA, USA). About 30 μg of each protein sample was separated on a 4–15% SDS-PAGE gel and transferred onto a nitrocellulose membrane (Bio-Rad). Membranes were blocked in 5% skim milk in Tris-buffered saline (TBS, VWR Life Sciences) containing 0.1% Tween 20 (TBS-T) and incubated with the following antibodies: LC3B (1:1000, PA1-46286) from ThermoFisher (Rockford, IL, USA); Beclin-1 (1:1000) and phosphor-mTOR (1:1000, Ser2448, D92C) from Cell Signaling Technology, Inc. (Danvers, MA, USA); and α-tubulin (1:5000, T9026) from Millipore Sigma (St. Louis, MO, USA).

5. Conclusions

Here, we have synthesized and systematically characterized all pyridones-derived from nicotinamide known to have some physiological relevance and sought the biochemical origin of the PYR series that had remained unknown. Indeed, we report that 4PYR can be generated from NRH by the FAD-dependent NQO2 enzyme and provide evidence that 2PYR and 6PYR can be generated from NR via Fenton chemistry, thus further supporting the hypothesis that reductive stress, as well as oxidative stress, promotes the formation of ribosylated pyridones.

Supplementary Materials: The following are available online at https://www.mdpi.com/1422-0067/22/3/1145/s1. Figure S1: ^{1}H NMR, ^{13}C NMR and HRMS Spectra. Figure S2: NQO2 catalyzed 4PYR formation. Table S1: Isotopic pattern with relative abundance of pyridones and their intermediates.

Author Contributions: Conceptualization, M.E.M.; methodology, N.R.G. and M.E.M.; validation, F.H., M.S. and M.V.M.; formal analysis, F.H. and M.S.; investigation, F.H., M.S., M.V.M. and P.M.; data curation, F.H. and M.S.; writing—original draft preparation, F.H., M.S., S.A.J.T.; writing—review and editing, N.R.G. and M.E.M.; supervision, N.R.G. and M.E.M.; project administration, M.E.M.; funding acquisition, M.E.M. All authors have read and agreed to the published version of the manuscript.

Funding: We thank Elysium Health and the Mitchell Cancer Institute for the financial support of this work.

Institutional Review Board Statement: The preliminary study was conducted according to the guidelines of the Declaration of Helsinki and approved by the Institutional Review Board of the University of Alabama (protocol code 19-465 and 12 August 2020).

Informed Consent Statement: Informed consent was obtained from all subjects involved in the study.

Data Availability Statement: Raw NMR and MS data files can be obtained upon request made directly to M.M.

Acknowledgments: We thank the Mass Spectrometry Core facilities at the MCI and Lindsay Schambeau for performing the mass spec. analyses.

Conflicts of Interest: The authors declare no conflict of interest.

Abbreviations

minute (min), hour (h), room temperature (rt), methanol (MeOH), dimethylsufoxide (DMSO), acetonitrile (CH_3CN, ACN), Hydroxybenzotriazole (HOBt), dimethylformamide (DMF), dicyclohexylcarbodiimide (DCC), dichloromethane (DCM), nicotinamide riboside (NR), nicotinamide riboside reduced form (NRH), nicotinamide adenine dinucleotide, (NAD), nicotinamide adenine dinucleotide phosphate (NADP), nicotinamide adenine dinucleotide reduced form, (NADH), nicotinamide adenine dinucleotide phosphate reduced form (NADPH), pyridone (PY), methyl-pyridones (*N*-Me-PY), pyridone ribosides (PYR), flavin adenine dinucleotide (FAD), Electrospray ionization (ESI), mass spectrometry (MS), liquid chromatography coupled with mass spectrometry (LS-MS), proton nuclear magnetic resonance spectroscopy (^{1}H NMR), calculated (calcd).

References

1. Makarov, M.V.; Trammell, S.A.J.; Migaud, M. The chemistry of the vitamin B3 metabolome. *Biochem. Soc. Trans.* **2018**, *47*, 131–147. [CrossRef] [PubMed]
2. McReynolds, M.R.; Chellappa, K.; Baur, J.A. Age-related NAD(+) decline. *Exp. Gerontol.* **2020**, *134*, 110888. [CrossRef] [PubMed]
3. Lenglet, A.; Liabeuf, S.; Bodeau, S.; Louvet, L.; Mary, A.; Boullier, A.; Lemaire-Hurtel, A.S.; Jonet, A.; Sonnet, P.; Kamel, S.; et al. *N*-methyl-2-pyridone-5-carboxamide (2PY)—Major Metabolite of Nicotinamide: An Update on an Old Uremic Toxin. *Toxins* **2016**, *8*, 339. [CrossRef] [PubMed]
4. Slominska, E.M.; Rutkowski, P.; Smolenski, R.T.; Szutowicz, A.; Rutkowski, B.; Swierczynski, J. The age-related increase in *N*-methyl-2-pyridone-5-carboxamide (NAD catabolite) in human plasma. *Mol. Cell. Biochem.* **2004**, *267*, 25–30. [CrossRef]
5. Rutkowski, B.; Rutkowski, P.; Słomińska, E.; Swierczyński, J. Distribution of Purine Nucleotides in Uremic Fluids and Tissues. *J. Ren. Nutr.* **2010**, *20*, 7–10. [CrossRef]
6. Synesiou, E.; Fairbanks, L.D.; Simmonds, H.A.; Slominska, E.M.; Smolenski, R.T.; Carrey, E.A. 4-Pyridone-3-carboxamide-1-beta-D-ribonucleoside triphosphate (4PyTP), a novel NAD metabolite accumulating in erythrocytes of uremic children: A biomarker for a toxic NAD analogue in other tissues? *Toxins* **2011**, *3*, 520–537. [CrossRef]
7. Carrey, E.A.; Synesiou, E.; Simmonds, H.A.; Fairbanks, L.D. The Novel Nucleotide 4KNTP, in High Concentrations in Erythrocytes of Renal Failure Children: A Comparison with Accumulation of Other Putative Precursors in the Plasma. *Nucleosides Nucleotides Nucleic Acids* **2006**, *25*, 1051–1054. [CrossRef]
8. Rutkowski, B.; Rutkowski, P.; Słomińska, E.; Smolenski, R.T.; Świerczyński, J. Cellular Toxicity of Nicotinamide Metabolites. *J. Ren. Nutr.* **2012**, *22*, 95–97. [CrossRef]

9. Delaney, J.; Hodson, M.P.; Thakkar, H.; Connor, S.C.; Sweatman, B.C.; Kenny, S.P.; McGill, P.J.; Holder, J.C.; Hutton, K.A.; Haselden, J.N.; et al. Tryptophan-NAD$^+$ pathway metabolites as putative biomarkers and predictors of peroxisome proliferation. *Arch. Toxicol.* **2004**, *79*, 208–223. [CrossRef]
10. Deen, C.P.J.; van der Veen, A.; Gomes-Neto, A.W.; Geleijnse, J.M.; Berg, K.J.B.-V.D.; Heiner-Fokkema, M.R.; Kema, I.P.; Bakker, S.J.L. Urinary Excretion of N1-methyl-2-pyridone-5-carboxamide and N1-methylnicotinamide in Renal Transplant Recipients and Donors. *J. Clin. Med.* **2020**, *9*, 437. [CrossRef]
11. Deen, C.P.J.; Veen, A.V.; Gomes-Neto, A.W.; Geleijnse, J.M.; Berg, K.; Heiner-Fokkema, M.R.; Kema, I.P.; Bakker, S.J.L. Urinary Excretion of N(1)-Methylnicotinamide and N(1)-Methyl-2-Pyridone-5-Carboxamide and Mortality in Kidney Transplant Recipients. *Nutrients* **2020**, *12*, 2059. [CrossRef] [PubMed]
12. Gooding, J.; Cao, L.; Ahmed, F.; Mwiza, J.M.; Fernander, M.; Whitaker, C.; Acuff, Z.; McRitchie, S.; Sumner, S.; Ongeri, E.M. LC-MS-based metabolomics analysis to identify meprin-beta-associated changes in kidney tissue from mice with STZ-induced type 1 diabetes and diabetic kidney injury. *Am. J. Physiol. Renal Physiol.* **2019**, *317*, 1034–1046. [CrossRef] [PubMed]
13. Shibata, K.; Kawada, T.; Iwai, K. Microdetermination of N1-methyl-2-pyridone-5-carboxamide, a major metabolite of nicotinic acid and nicotinamide, in urine by high-performance liquid chromatography. *J. Chromatogr. B Biomed. Sci. Appl.* **1987**, *417*, 173–177. [CrossRef]
14. Shibata, K.; Matsuo, H. Levels of NAD, NADP and their related compounds in rat blood. *Teikoku Gakuen Kiyo* **1989**, *15*, 9–12.
15. Slominska, E.M.; Adamski, P.; Lipiński, M.; Swierczynski, J.; Smolenski, R.T. Liquid Chromatographic/Mass Spectrometric Procedure for Measurement of NAD Catabolites in Human and Rat Plasma and Urine. *Nucleosides Nucleotides Nucleic Acids* **2006**, *25*, 1245–1249. [CrossRef]
16. Slominska, E.M.; Yuen, A.; Osman, L.; Gebicki, J.; Yacoub, M.H.; Smolenski, R.T. Cytoprotective Effects of Nicotinamide Derivatives in Endothelial Cells. *Nucleosides Nucleotides Nucleic Acids* **2008**, *27*, 863–866. [CrossRef]
17. Slominska, E.M.; Orlewska, C.; Yuen, A.; Osman, L.; Romaszko, P.; Sokolowska, E.; Foks, H.; Simmonds, H.A.; Yacoub, M.H.; Smolenski, R.T. Metabolism of 4-pyridone-3-carboxamide-1-beta-D-ribonucleoside triphosphate and its nucleoside precursor in the erythrocytes. *Nucleosides Nucleotides Nucleic Acids* **2008**, *27*, 830–834. [CrossRef]
18. Garcia-Perez, I.; Posma, J.M.; Serrano-Contreras, J.I.; Boulangé, C.L.; Chan, Q.; Frost, G.; Stamler, J.; Elliott, P.; Lindon, J.C.; Holmes, E.; et al. Identifying unknown metabolites using NMR-based metabolic profiling techniques. *Nat. Protoc.* **2020**, *15*, 2538–2567. [CrossRef]
19. Horitsu, K. Specific basal bioconversion (biochemical conversion) related to nicotinamide methylation (metabolic process) found in the hepatocytes of rat and mouse regarding Ehrlich ascites tumor host. *Kenkyu Kiyo—Tokyo Kasei Daigaku 2 Shizen Kagaku* **1998**, *38*, 1–6.
20. Ulanovskaya, O.A.; Zuhl, A.M.; Cravatt, B.F. NNMT promotes epigenetic remodeling in cancer by creating a metabolic methylation sink. *Nat. Chem. Biol.* **2013**, *9*, 300–306. [CrossRef]
21. Bockwoldt, M.; Houry, D.; Niere, M.; Gossmann, T.I.; Reinartz, I.; Schug, A.; Ziegler, M.; Heiland, I. Identification of evo-lutionary and kinetic drivers of NAD-dependent signaling. *Proc. Natl. Acad. Sci. USA* **2019**, *116*, 15957–15966. [CrossRef] [PubMed]
22. Loring, H.S.; Thompson, P.R. Kinetic Mechanism of Nicotinamide N-Methyltransferase. *Biochemistry* **2018**, *57*, 5524–5532. [CrossRef] [PubMed]
23. Mierzejewska, P.; Gawlik-Jakubczak, T.; Jablonska, P.; Czajkowski, M.; Kutryb-Zajac, B.; Smolenski, R.T.; Matuszewski, M.; Slominska, E.M. Nicotinamide metabolism alterations in bladder cancer: Preliminary studies. *Nucleosides Nucleotides Nucleic Acids* **2018**, *37*, 687–695. [CrossRef] [PubMed]
24. Jablonska, P.; Mierzejewska, P.; Kutryb-Zajac, B.; Rzyman, W.; Dziadziuszko, R.; Polanska, J.; Sitkiewicz, M.; Smolenski, R.T.; Slominska, E.M. Increased plasma concentration of 4-pyridone-3-carboxamide-1-ss-D-ribonucleoside (4PYR) in lung cancer. Pre-liminary studies. *Nucleosides Nucleotides Nucleic Acids* **2019**, *38*, 781–787. [CrossRef]
25. Slominska, E.M.; Carrey, E.A.; Foks, H.; Orlewska, C.; Wieczerzak, E.; Sowinski, P.; Yacoub, M.H.; Marinaki, A.M.; Simmonds, H.A.; Smolenski, R.T. A novel nucleotide found in human erythrocytes, 4-pyridone-3-carboxamide-1-beta-D-ribonucleoside triphosphate. *J. Biol. Chem.* **2006**, *281*, 32057–32064. [CrossRef]
26. Pelikant-Malecka, I.; Sielicka, A.; Kaniewska, E.; Smolenski, R.T.; Slominska, E.M. 4-Pyridone-3-carboxamide-1beta-D-ribonucleoside metabolism in endothelial cells and its impact on cellular energetic balance. *Nucleosides Nucleotides Nucleic Acids* **2014**, *33*, 338–341. [CrossRef]
27. Pelikant-Malecka, I.; Kaniewska-Bednarczuk, E.; Szrok, S.; Sielicka, A.; Sledzinski, M.; Orlewska, C.; Smolenski, R.T.; Slominska, E.M. Metabolic pathway of 4-pyridone-3-carboxamide-1beta-d-ribonucleoside and its effects on cellular energetics. *Int. J. Biochem. Cell. Biol.* **2017**, *88*, 31–43. [CrossRef]
28. Pelikant-Malecka, I.; Smolenski, R.T.; Slominska, E.M. Metabolism of 4-pyridone-3-carboxamide-1beta-d-ribonucleoside (4PYR) in primary murine brain microvascular endothelial cells (mBMECs). *Nucleosides Nucleotides Nucleic Acids* **2018**, *37*, 639–644. [CrossRef]
29. Lang, R.; Wahl, A.; Skurk, T.; Yagar, E.F.; Schmiech, L.; Eggers, R.; Hauner, H.; Hofmann, T. Development of a Hydrophilic Liquid Interaction Chromatography−High-Performance Liquid Chromatography−Tandem Mass Spectrometry Based Stable Isotope Dilution Analysis and Pharmacokinetic Studies on Bioactive Pyridines in Human Plasma and Urine after Coffee Consumption. *Anal. Chem.* **2010**, *82*, 1486–1497. [CrossRef]

30. Landelle, G.; Schmitt, E.; Panossian, A.; Vors, J.-P.; Pazenok, S.; Jeschke, P.; Gutbrod, O.; Leroux, F.R. Tri- and difluoro-methoxylated N-based heterocycles—Synthesis and insecticidal activity of novel F3CO—And F2HCO—Analogues of Imidacloprid and Thiacloprid. *J. Fluor. Chem.* **2017**, *203*, 155–165. [CrossRef]
31. Frister, H.; Kemper, K.; Boos, K.-S.; Schlimme, E. Darstellung des Coenzymmetaboliten 1,6-Dihydro-6-oxo-1-(β-D-ribofuranosyl)-3-pyridincarbonsäureamid. *Eur. J. Org. Chem.* **1985**, *1985*, 510–516. [CrossRef]
32. Hanna, N.B.; Joshi, R.V.; Larson, S.B.; Robins, R.K.; Revankar, G.R. Synthesis of certain 1-β-D-ribofuranosyl-1,2-dihydro-2-oxopyridines structurally related to nicotinamide ribonucleoside. *J. Heterocycl. Chem.* **1989**, *26*, 1835–1843. [CrossRef]
33. Holman, W.I.M.; Wiegand, C. The chemical conversion of nicotinic acid and nicotinamide to derivatives of *N*-methyl-2-pyridone by methylation and oxidation. *Biochem. J.* **1948**, *43*, 423–426. [CrossRef] [PubMed]
34. Wiegand, C.; Holman, W.I.M. Synthesis of Derivatives of N-Methyl-2-Pyridone from Nicotinic Acid and Nicotinamide. *Nat. Cell. Biol.* **1948**, *162*, 659–660. [CrossRef] [PubMed]
35. Peretz, H.; Watson, D.G.; Blackburn, G.; Zhang, T.; Lagziel, A.; Shtauber-Naamati, M.; Morad, T.; Keren-Tardai, E.; Greenshpun, V.; Usher, S.; et al. Urine metabolomics reveals novel physiologic functions of human aldehyde oxidase and provides biomarkers for typing xanthinuria. *Metabolomics* **2011**, *8*, 951–959. [CrossRef]
36. Kitamura, S.; Nitta, K.; Tayama, Y.; Tanoue, C.; Sugihara, K.; Inoue, T.; Horie, T.; Ohta, S. Aldehyde Oxidase-Catalyzed Metabolism of N1-Methylnicotinamide in Vivo and in Vitro in Chimeric Mice with Humanized Liver. *Drug Metab. Dispos.* **2008**, *36*, 1202–1205. [CrossRef] [PubMed]
37. Sugihara, K.; Kitamura, S.; Tatsumi, K.; Asahara, T.; Dohi, K. Differences in aldehyde oxidase activity in cytosolic prepara-tions of human and monkey liver. *Biochem. Mol. Biol. Int.* **1997**, *41*, 1153–1160.
38. Hanukoglu, I. Conservation of the Enzyme—Coenzyme Interfaces in FAD and NADP Binding Adrenodoxin Reductase—A Ubiquitous Enzyme. *J. Mol. Evol.* **2017**, *85*, 205–218. [CrossRef]
39. de Rosa, M.; Pennati, A.; Pandini, V.; Monzani, E.; Zanetti, G.; Aliverti, A. Enzymatic oxidation of NADP+ to its 4-oxo de-rivative is a side-reaction displayed only by the adrenodoxin reductase type of ferredoxin-NADP+ reductases. *FEBS J.* **2007**, *274*, 3998–4007. [CrossRef]
40. Bossi, R.T.; Aliverti, A.; Raimondi, D.; Fischer, F.; Zanetti, G.; Ferrari, D.; Tahallah, N.; Maier, C.S.; Heck, A.J.R.; Rizzi, M.; et al. A covalent modification of NADP+ revealed by the atomic resolution structure of FprA, a Mycobacterium tuberculosis oxidoreductase. *Biochemistry* **2002**, *41*, 8807–8818. [CrossRef]
41. Huntley, C.M.; Cotterill, A.S.; Maillard, J.-Y.; Balzarini, J.; Simons, C. Synthesis and biological evaluation of pyridine-2-one nucleosides. *Nucleosides Nucleotides Nucleic Acids* **2001**, *20*, 731–733. [CrossRef] [PubMed]
42. Godoy, A.T.; Eberlin, M.N.; Simionato, A.V.C. Targeted metabolomics: Liquid chromatography coupled to mass spectrometry method development and validation for the identification and quantitation of modified nucleosides as putative cancer biomarkers. *Talanta* **2020**, *210*, 120640. [CrossRef] [PubMed]
43. Willmann, L.; Erbes, T.; Krieger, S.; Trafkowski, J.; Rodamer, M.; Kammerer, B. Metabolome analysis via comprehensive two-dimensional liquid chromatography: Identification of modified nucleosides from RNA metabolism. *Anal. Bioanal. Chem.* **2015**, *407*, 3555–3566. [CrossRef] [PubMed]
44. Chen, W.; Koenigs, L.L.; Thompson, S.J.; Peter, R.M.; Rettie, A.E.; Trager, W.F.; Nelson, S.D. Oxidation of Acetaminophen to Its Toxic Quinone Imine and Nontoxic Catechol Metabolites by Baculovirus-Expressed and Purified Human Cytochromes P450 2E1 and 2A6. *Chem. Res. Toxicol.* **1998**, *11*, 295–301. [CrossRef] [PubMed]
45. Braver-Sewradj, S.P.D.; Braver, M.W.D.; Toorneman, R.M.; van Leeuwen, S.; Zhang, Y.; Dekker, S.J.; Vermeulen, N.P.E.; Commandeur, J.N.M.; Vos, J.C. Reduction and Scavenging of Chemically Reactive Drug Metabolites by NAD(P)H:Quinone Oxidoreductase 1 and NRH:Quinone Oxidoreductase 2 and Variability in Hepatic Concentrations. *Chem. Res. Toxicol.* **2018**, *31*, 116–126. [CrossRef]
46. Slominska, E.M.; Borkowski, T.; Rybakowska, I.; Abramowicz-Glinka, M.; Orlewska, C.; Smolenski, R.T. In Vitro and Cellular Effects of 4-pyridone-3-carboxamide riboside on Enzymes of Nucleotide Metabolism. *Nucleosides Nucleotides Nucleic Acids* **2014**, *33*, 353–357. [CrossRef]
47. Sonavane, M.; Hayat, F.; Makarov, M.; Migaud, M.E.; Gassman, N. Dihydronicotinamide riboside promotes cell-specific cytotoxicity by tipping the balance between metabolic regulation and oxidative stress. *PLoS ONE* **2020**, *15*, e0242174. [CrossRef]
48. Irie, J.; Inagaki, E.; Fujita, M.; Nakaya, H.; Mitsuishi, M.; Yamaguchi, S.; Yamashita, K.; Shigaki, S.; Ono, T.; Yukioka, H.; et al. Effect of oral administration of nicotinamide mononucleotide on clinical parameters and nicotinamide metabolite levels in healthy Japanese men. *Endocr. J.* **2020**, *67*, 153–160. [CrossRef]
49. Hiratsuka, C.; Sano, M.; Fukuwatari, T.; Shibata, K. Time-Dependent Effects of L-Tryptophan Administration on Urinary Excretion of L-Tryptophan Metabolites. *J. Nutr. Sci. Vitaminol.* **2014**, *60*, 255–260. [CrossRef]
50. Pelantová, H.; Bugáňová, M.; Holubová, M.; Sediva, B.; Zemenová, J.; Sýkora, D.; Kavalkova, P.; Haluzík, M.; Železná, B.; Maletínská, L.; et al. Urinary metabolomic profiling in mice with diet-induced obesity and type 2 diabetes mellitus after treatment with metformin, vildagliptin and their combination. *Mol. Cell. Endocrinol.* **2016**, *431*, 88–100. [CrossRef]
51. Diguet, N.; Trammell, S.A.; Tannous, C.; Deloux, R.; Piquereau, J.; Mougenot, N.; Gouge, A.; Gressette, M.; Manoury, B.; Blanc, J.; et al. Nicotinamide Riboside Preserves Cardiac Function in a Mouse Model of Dilated Cardiomyopathy. *Circulation* **2018**, *137*, 2256–2273. [CrossRef] [PubMed]
52. Kelly, R.S.; Sordillo, J.; Lasky-Su, J.; Dahlin, A.; Perng, W.; Rifas-Shiman, S.L.; Weiss, S.T.; Gold, D.R.; Litonjua, A.A.; Hivert, M.-F.; et al. Plasma metabolite profiles in children with current asthma. *Clin. Exp. Allergy* **2018**, *48*, 1297–1304. [CrossRef] [PubMed]

53. Kim, K.H.; Joo, J.; Park, B.; Park, S.J.; Lee, W.J.; Han, S.S.; Kim, T.H.; Hong, E.K.; Woo, S.M.; Yoo, B.C. Reduced levels of N′-methyl-2-pyridone-5-carboxamide and lysophosphatidylcholine 16:0 in the serum of patients with intrahepatic cholangiocarci-noma, and the correlation with recurrence-free survival. *Oncotarget* **2017**, *8*, 112598–112609. [CrossRef]
54. Lenglet, A.; Liabeuf, S.; el Esper, N.; Brisset, S.; Mansour, J.; Lemaire-Hurtel, A.-S.; Mary, A.; Brazier, M.; Kamel, S.; Mentaverri, R.; et al. Efficacy and safety of nicotinamide in haemodialysis patients: The NICOREN study. *Nephrol. Dial. Transplant.* **2016**, *32*, 870–879. [CrossRef] [PubMed]
55. Tsalik, E.L.; Willig, L.K.; Rice, B.J.; van Velkinburgh, J.C.; Mohney, R.P.; McDunn, J.E.; Dinwiddie, D.L.; Miller, N.A.; Mayer, E.S.; Glickman, S.W.; et al. Renal systems biology of patients with systemic inflammatory response syndrome. *Kidney Int.* **2015**, *88*, 804–814. [CrossRef]
56. Shibata, K.; Fukuwatari, T.; Suzuki, C. Pharmacological doses of nicotinic acid and nicotinamide are independently metab-olized in rats. *J. Nutr. Sci. Vitaminol.* **2014**, *60*, 86–93. [CrossRef]
57. Gillmor, H.A.; Bolton, C.H.; Hopton, M.; Moore, W.P.T.; Perrett, D.; Bingley, P.J.; Gale, E.A. Measurement of nicotinamide and *N*-methyl-2-pyridone-5-carboxamide in plasma by high performance liquid chromatography. *Biomed. Chromatogr.* **1999**, *13*, 360–362. [CrossRef]
58. Peron, G.; Zengin, G.; Sut, S. Supplementation with resveratrol as Polygonum cuspidatum Sieb. et Zucc. extract induces changes in the excretion of urinary markers associated to aging in rats. *Fitoterapia* **2018**, *129*, 154–161. [CrossRef]
59. Zhou, H.; Li, L.; Wu, C.; Kurtán, T.; Mándi, A.; Liu, Y.; Gu, Q.; Zhu, T.; Guo, P.; Li, D. Penipyridones A–F, Pyridone Alkaloids from Penicillium funiculosum. *J. Nat. Prod.* **2016**, *79*, 1783–1790. [CrossRef]
60. Trammell, S.A.J.; Schmidt, M.S.; Weidemann, B.J.; Redpath, P.; Jaksch, F.; Dellinger, R.W.; Philip, R.; Abel, E.D.; Migaud, M.E.; Brenner, C. Nicotinamide riboside is uniquely and orally bioavailable in mice and humans. *Nat. Commun.* **2016**, *7*, 12948. [CrossRef]
61. Maeta, A.; Sano, M.; Fukuwatari, T.; Shibata, K. Simultaneous measurement of nicotinamide and its catabolites, nicotinamideN-oxide, N1-methyl-2-pyridone-5-carboxamide, andN1-methyl-4-pyridone-3-carboxamide, in mice urine. *Biosci. Biotechnol. Biochem.* **2014**, *78*, 1306–1309. [CrossRef] [PubMed]
62. Steckel, A.; Schlosser, G. An Organic Chemist's Guide to Electrospray Mass Spectrometric Structure Elucidation. *Molecules* **2019**, *24*, 611. [CrossRef] [PubMed]
63. Trammell, S.A.; Brenner, C. Targeted, LCMS-based Metabolomics for Quantitative Measurement of NAD(+) Metabolites. *Comput. Struct. Biotechnol. J.* **2013**, *4*, e201301012. [CrossRef] [PubMed]
64. Kremer, J.I.; Gömpel, K.; Bakuradze, T.; Eisenbrand, G.; Richling, E. Urinary Excretion of Niacin Metabolites in Humans After Coffee Consumption. *Mol. Nutr. Food Res.* **2018**, *62*, e1700735. [CrossRef] [PubMed]
65. Pelikant-Malecka, I.; Sielicka, A.; Kaniewska, E.; Smolenski, R.T.; Slominska, E.M. Influence of 4-pyridone-3-carboxamide-1Beta-D-ribonucleoside (4PYR) on activities of extracellular enzymes in endothelial human cells. *Nucleosides Nucleotides Nucleic Acids* **2016**, *35*, 732–736. [CrossRef]
66. Shibata, K.; Fukuwatari, T. Pyridone compounds, catabolites of NAD are new uremic toxins. *Bitamin* **2007**, *81*, 571–574.
67. Yagi, K.; Ohishi, N. *Hydroxylation of Riboflavin 7- and 8-Methyl Groups in Mammals*; De Gruyter: Berlin, Germany, 1984; pp. 819–832.
68. Houee-Levin, C.; Bobrowski, K.; Horakova, L.; Karademir, B.; Schoeneich, C.; Davies, M.J.; Spickett, C.M. Exploring oxidative modifications of tyrosine: An update on mechanisms of formation, advances in analysis and biological consequences. *Free Radic. Res.* **2015**, *49*, 347–373.
69. Halliwell, B. The Chemistry of Free Radicals. *Toxicol. Ind. Health* **1993**, *9*, 1–21. [CrossRef]

International Journal of
Molecular Sciences

Article

Chrysin Inhibits TNFα-Induced TSLP Expression through Downregulation of EGR1 Expression in Keratinocytes

Hyunjin Yeo [1,†], Young Han Lee [1,†], Sung Shin Ahn [1], Euitaek Jung [1], Yoongho Lim [2] and Soon Young Shin [1,*]

1 Department of Biological Sciences, Sanghuh College of Lifesciences, Konkuk University, Seoul 05029, Korea; jini1606@konkuk.ac.kr (H.Y.); yhlee58@konkuk.ac.kr (Y.H.L.); wendy713@konkuk.ac.kr (S.S.A.); mylife4sci@konkuk.ac.kr (E.J.)
2 Division of Bioscience and Biotechnology, BMIC, Konkuk University, Seoul 05029, Korea; yoongho@konkuk.ac.kr
* Correspondence: shinsy@konkuk.ac.kr; Tel.: +82-2-2030-7946
† H.Y. and Y.H.L. contributed equally.

Abstract: Thymic stromal lymphopoietin (TSLP) is an epithelial cell-derived cytokine that acts as a critical mediator in the pathogenesis of atopic dermatitis (AD). Various therapeutic agents that prevent TSLP function can efficiently relieve the clinical symptoms of AD. However, the downregulation of TSLP expression by therapeutic agents remains poorly understood. In this study, we investigated the mode of action of chrysin in TSLP suppression in an AD-like inflammatory environment. We observed that the transcription factor early growth response (EGR1) contributed to the tumor necrosis factor alpha (TNFα)-induced transcription of *TSLP*. Chrysin attenuated TNFα-induced TSLP expression by downregulating EGR1 expression in HaCaT keratinocytes. We also showed that the oral administration of chrysin improved AD-like skin lesions in the ear and neck of BALB/c mice challenged with 2,4-dinitrochlorobenzene. We also showed that chrysin suppressed the expression of EGR1 and TSLP by inhibiting the extracellular signal-regulated kinase (ERK) 1/2 and c-Jun N-terminal kinase (JNK) 1/2 mitogen-activated protein kinase pathways. Collectively, the findings of this study suggest that chrysin improves AD-like skin lesions, at least in part, through the downregulation of the ERK1/2 or JNK1/2-EGR1-TSLP signaling axis in keratinocytes.

Keywords: atopic dermatitis; chrysin; 2,4-dinitrochlorobenzene; early growth response 1; thymic stromal lymphopoietin

Citation: Yeo, H.; Lee, Y.H.; Ahn, S.S.; Jung, E.; Lim, Y.; Shin, S.Y. Chrysin Inhibits TNFα-Induced TSLP Expression through Downregulation of EGR1 Expression in Keratinocytes. *Int. J. Mol. Sci.* **2021**, *22*, 4350. https://doi.org/10.3390/ijms22094350

Academic Editor: Kenji Izuhara

Received: 25 March 2021
Accepted: 16 April 2021
Published: 21 April 2021

1. Introduction

Atopic dermatitis (AD), also known as atopic eczema, is a chronic inflammatory skin disease characterized by the development of recurrent eczematous lesions and intense pruritus [1]. The prevalence of AD is constantly growing worldwide over the past 30 years, and nowadays, AD affects about 10% of adults and up to 20% of children [2]. AD is associated with multiple comorbid chronic disorders, such as asthma, allergic rhinitis, respiratory infection, mental disorders, metabolic syndrome, gastrointestinal problems, and cardiovascular disease [3]. A recent cohort study has revealed a variety of clinical forms of AD in adult-onset and childhood-onset types, which may be a crucial factor in determining the appropriate therapeutic medications [4]. AD treatment comprises several types of therapies, such as topical versus systemic application and small molecule inhibitors versus biological agents. Topical therapy includes the application of corticosteroids, antihistamine, and immunosuppressants (e.g., calcineurin inhibitors and phosphodiesterase inhibitors) [5]. Systemic administration comprises immunosuppressant-modulators (e.g., cyclosporine), anti-metabolites (e.g., methotrexate and azathioprine), cytokine signaling inhibitor (e.g., JAK kinase inhibitor), antibiotics, and biological agents (e.g., targeted monoclonal antibodies) [6–9]. Topical steroids and immunosuppressants have been used as the primary agents; however, their value is limited by local side effects and insufficient efficacy [10]. Oral steroids or immunosuppressants may be used in adults with severe chronic symptoms, but their use is often unsatisfactory due to considerable long-term side effects [11]. Various herbal medicines have been reported to have beneficial effects in the treatment of AD; however, there is not enough evidence to support the use of herbal medicine [10]. Therefore, new systemic therapies with fewer side effects and more efficacious are needed to treat moderate-to-severe chronic AD.

The cause of AD has not been identified in sufficient detail; however, the onset of AD is known to be influenced by genetic and environmental factors, epidermal barrier abnormalities, and impaired cutaneous immune functions [1,12]. In most cases, AD pathogenesis is primarily driven by a milieu of pro-inflammatory cytokines produced by $CD4^+$ T helper (Th) lymphocytes, including Th2, Th22, and Th17 cells, as well as pro-inflammatory immune cells, including mast cells, neutrophils, and macrophages/monocytes [13–15]. The prominent clinical features of AD include cutaneous inflammation and the chronic itch–scratch–itch cycle, which cause persistent irritation to the skin lesions and impair the skin barrier function. Itching is induced by the stimulation of peripheral sensory neurons by pruritogens. Histamine secreted by mast cells can mediate acute itch in skin inflammation; meanwhile, Th2-associated cytokines, including interleukin (IL)-4, IL-13, and IL-31, directly stimulate sensory neurons [16–18].

Thymic stromal lymphopoietin (TSLP) is an IL-17-like cytokine that has been identified and characterized in murine thymic stromal Z210R.1 cells [19,20]. It is produced by various cell types, including stromal cells, epithelial cells, smooth muscle cells, fibroblasts, dendritic cells, mast cells, and epidermal keratinocytes [21]. TSLP promotes the differentiation and growth of B cells and the activation of CD4+/CD8+ T cells and dendritic cells [19,20,22–25]. In the early stage of AD, keratinocyte-derived TSLP activates dendritic cells to induce the release of various chemokines, which leads to the expansion of the Th2 and Th22 cell populations and induces the release of IL-4, IL-5, IL-13, IL-22, and tumor necrosis factor alpha (TNFα) in large quantities [26]. Ultimately, this results in the persistent activation of Th1 and Th17 cells, impairs epidermal barrier function, accelerates skin inflammation, and promotes the development of AD [27,28]. TSLP also potentially activates mast cells, thus promoting the production of high levels of Th2-like cytokines [29]. TSLP directly stimulates itch sensory neurons independent of Th2 cytokines [30]. Hence, these studies suggest the role of TSLP as a crucial mediator of AD pathogenesis [31] and a potential drug target [32]. Various therapeutic agents that prevent TSLP function can effectively relieve clinical symptoms [33]. TSLP expression is regulated by various cytokines, including pro-inflammatory cytokines, such as TNFα and IL-1, and Th2-related cytokines, such as IL-4, IL-13, and IL-33 [34]. However, the mechanisms underlying TSLP suppression by therapeutic agents remain poorly understood.

Chrysin (5,7-dihydroxyflavone, Figure 1A) is a flavonoid found in large quantities in honey, propolis, mushrooms, and carrot. It exhibits multiple pharmacological and therapeutic properties, including neuroprotective, anti-inflammatory, and anticancer properties [35,36]. Notably, chrysin is known to alleviate AD by inhibiting the production of multiple pro-inflammatory cytokines and chemokines [37–39]. Choi et al. [37] have demonstrated that chrysin significantly inhibits the production of cytokines, Th2 chemokines, CCL17, and CCL22 by the downregulation of p38 MAPK, NF-κB, and STAT1 in TNFα/IFNγ-stimulated HaCaT keratinocytes. However, despite the beneficial effects of chrysin in AD therapy, the mechanism underlying the suppression of TSLP expression by chrysin remains unclear.

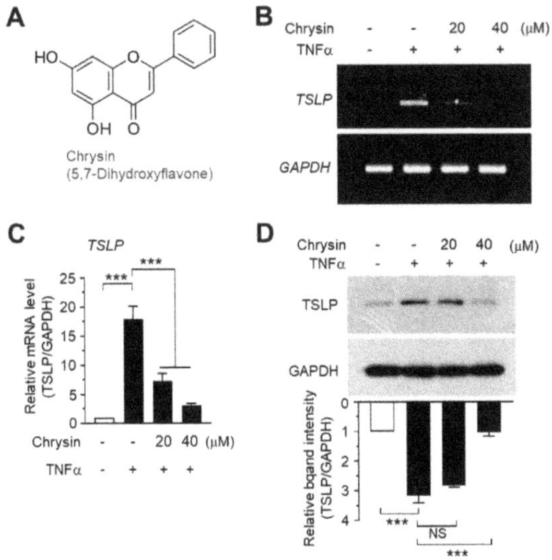

Figure 1. Effect of chrysin on the suppression of TNFα-induced *TSLP* expression. (**A**) Chemical structure of chrysin (5,7-dihydroxyflavone). (**B**) HaCaT cells were pretreated with chrysin (20 and 40 μM) for 30 min before stimulation with 10 ng/mL TNFα. After 12 h, total RNA was isolated, and the levels of *TSLP* mRNA were measured using RT-PCR. *GAPDH* mRNA was used as an internal control. Minus (−), vehicle treatment; Plus (+), TNFα treatment. (**C**) HaCaT cells were treated as in (**B**), and total RNA was isolated. *TSLP* mRNA levels were quantified using quantitative real-time PCR with SYBR Green-based fluorescent probes. The relative expression was normalized to the *GAPDH* mRNA levels. The relative *TSLP* mRNA level in the untreated cells was designated 1. Data are expressed as mean ± SD (n = 3); *** $p < 0.001$ by Dunnett's multiple comparisons test. Minus (−), vehicle treatment; Plus (+), TNFα treatment. (**D**) HaCaT cells were pretreated with chrysin (20 and 40 μM) for 30 min and then stimulated with 10 ng/mL TNFα for 24 h. The quantity of TSLP protein was measured using Western blot analysis. The band intensity corresponding to each TSLP protein was normalized to the GAPDH level using ImageJ v1.52a software. Data are expressed as mean ± SD (n = 3). NS, not significant; *** $p < 0.001$ by Dunnett's multiple comparisons test. Minus (−), vehicle treatment; Plus (+), TNFα treatment.

In this study, we attempted to elucidate the role of chrysin in TSLP suppression in keratinocytes. We found that chrysin inhibited TNFα-induced TSLP expression by downregulating mitogen-activated protein kinase (MAPK)-mediated EGR1 expression in HaCaT keratinocytes. In addition, we demonstrated that the oral administration of chrysin suppressed EGR1 and TSLP expression in AD-like skin lesions in BALB/c mice.

2. Results

2.1. Chrysin Inhibits TNFα-Induced TSLP Expression in HaCaT Keratinocytes

Previous studies have shown that chrysin alleviates AD-like skin lesions in a mouse model [37] and reverses the NF-κB-mediated inhibition of C-C motif chemokine ligand (CCL) 5 [39]. TSLP plays a key role in AD progression, and TSLP upregulation is considered a hallmark of AD pathogenesis [31,32]. TNFα is a pro-inflammatory cytokine that promotes inflammation by inducing the production of various other inflammatory cytokines and chemokines [40]. TNFα production was enhanced in a mouse model of 2.4-dinitrobenzene (DNCB)-induced contact allergy [41], and TNFα induced TSLP expression in skin keratinocytes [42]. To investigate whether chrysin modulates TSLP expression, we used TNFα as a positive signal to induce TSLP expression. As reported in a previous study [42], the *TSLP* mRNA levels were enhanced upon TNFα stimulation, as shown using reverse transcription (RT)-PCR (Figure 1B). However, chrysin pretreatment abrogated the ability of TNFα to induce *TSLP* mRNA expression. The changes in *TSLP* mRNA levels were measured using quantitative real-time PCR (Q-PCR) with *TSLP*-specific SYBR Green-based fluorescent probes. TNFα increased the *TSLP* mRNA level by 17.9 ± 2.52-fold compared to that in the control; however, upon treatment with 20 and 40 μM chrysin, the *TSLP* mRNA levels decreased by 7.40 ± 1.45- and 2.97 ± 0.397-fold, respectively, compared to the levels in the control (Figure 1C). Chrysin consistently suppressed TNFα-induced TSLP accumulation in a dose-dependent manner (Figure 1D). These data suggest that chrysin inhibits TNFα-induced TSLP expression at the mRNA level.

2.2. The Chrysin Response Element Is Located between the −369 and +18 Positions in the TSLP Promoter

To elucidate the effect of chrysin on the inhibition of TNFα-induced *TSLP* expression, we established a series of *TSLP* promoter deletion constructs: −1384/+18, −1338/+18, −1214/+18, −1017/+18, and −369/+18. These constructs harbored the luciferase reporter gene. Each of these promoter-reporters was transiently transfected into HaCaT cells, and the luciferase activity was measured following TNFα stimulation. As shown in Figure 2A, TNFα-induced *TSLP* promoter–reporter activity was persistently repressed in cells transfected with the shortest construct (−369/+18), suggesting that the chrysin response element is located between the −369 and +18 positions.

A

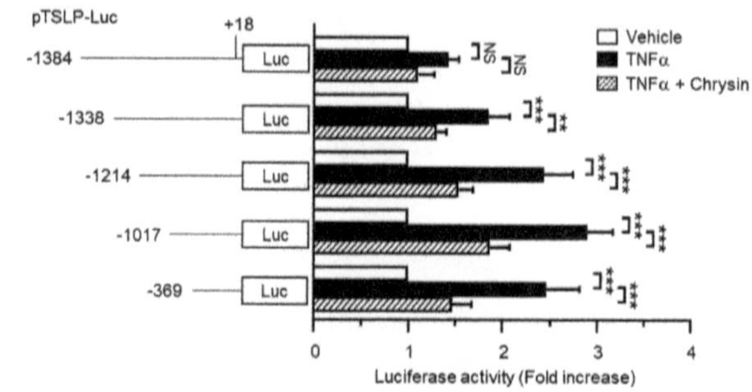

B

```
-369 gggacatatgcaaggactccaaatccttacagtggctgctggtttcagaa
-319 ggtgtttctcccaaatagaaaggattggtcacggaaatgccctgtaggag
-268 aaagacactggtatccgtttcttaaaggacattcacacttatgaagtgtt
-219 tagggcaaagcaaaaaggaggaaggtgagggaaattcctgatgactttac
                     EBS          NF-κB
-169 tgatgttaaaattccatagatttagctactccttttttttttcctttca
-119 catcatcaattcagaattctataaaggaagagaatgacatggtagaaaat
-69  cattggcctaggagaaaagagcccgtaggcgtttaggtgt[tatatagtgc]
-19  agccagaaagctctggagcatcagggagactccaact        TATA
```

Figure 2. Effect of chrysin on the inhibition of TNFα-induced *TSLP* promoter activity. (**A**) HaCaT cells were transfected with 0.2 μg of a set of 5′-deletion constructs of *TSLP* promoter-reporter plasmids. At 48 h post-transfection, the cells were treated with 10 ng/mL TNFα in the absence or presence of 40 μM chrysin. After 8–12 h, the cells were harvested, and the luciferase reporter activities were measured. The schematic diagram shows the set of deletion constructs of the *TSLP* promoter–reporter plasmid. Data are expressed as mean ± SD ($n = 3$). NS, not significant; ** $p < 0.01$; *** $p < 0.001$ by Sidak's multiple comparisons test. (**B**) Nucleotide sequence of the 5′-regulatory region of human *TSLP* spanning between the positions −369 and +18. The EGR1-binding sequence and NF-κB sites are underlined. The TATA box (−29/−22) is indicated using the box.

To identify the chrysin response elements, we analyzed the transcription factor-binding sites between the −369 and +18 positions using MatInspector program (Genomatix Software, Munich, Gertmany). The nuclear factor kappa B (NF-κB)-binding site was found to overlap with a putative early growth response 1 (EGR1)-binding sequence (EBS) located in the region between positions −206 and −187 (Figure 2B).

2.3. *Chrysin Inhibits the DNA-Binding Activity of EGR1*

Previous studies have shown the role of NF-κB in mediating TNFα-induced TSLP expression in human airway smooth muscle cells [43] and IL-1β-induced TSLP expression in intestinal epithelial cells [44]. The transcription factor EGR1 mediates IL33-induced TSLP expression in keratinocytes [45]. However, the role of EBS in the −369/+18 region of the *TSLP* promoter remains elusive. We focused on the role of EGR1 in chrysin-mediated TSLP suppression. To determine whether EGR1 transactivates the EBS in the −369/+18 construct, we co-transfected the −369/+18 construct and an expression plasmid for EGR1 (pcDNA3.1/Egr1) and measured the luciferase reporter activity. Exogenous EGR1 expression increased the promoter–reporter activity of the −369/+18 construct in a plasmid concentration-dependent manner (Figure 3A), suggesting that the putative EBS in the −369/+18 construct could be a functional *cis*-acting element for EGR1 that participates in TNFα-induced *TSLP* transcription.

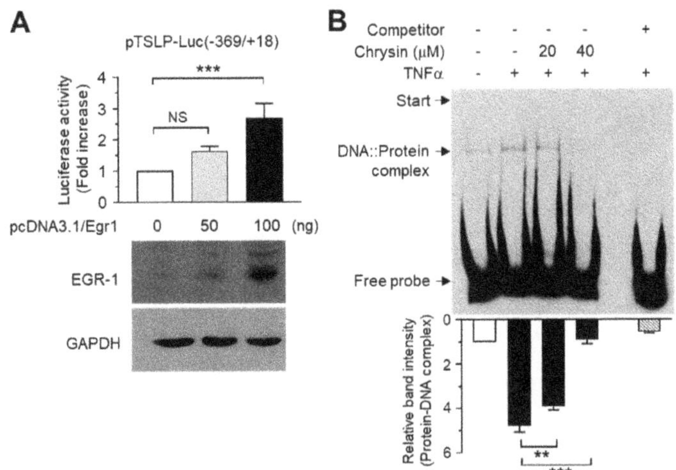

Figure 3. Chrysin inhibits the DNA-binding activity of EGR1. (**A**) HaCaT cells were co-transfected with the pTSLP-Luc(−369/+18) reporter plasmid at increasing concentrations of the EGR1 expression plasmid. After 48 h, the cells were harvested, and the luciferase activities were measured (*top graph*). Bars represent means ± SD ($n = 3$). NS, not significant; *** $p < 0.001$ by Dunnett's multiple comparisons test. Expression of EGR1 post-transfection was confirmed using Western blotting (*bottom panels*). Glyceraldehyde-3-phosphate dehydrogenase (GAPDH) was used as an internal control. (**B**) HaCaT cells were treated with or without 10 ng/mL TNFα for 1 h in the presence or absence of chrysin (20 and 40 μM). Nuclear extracts (3 μg) were prepared and incubated with a biotinylated *EGR1*-binding oligonucleotide probe (50 fmole) in the absence or presence of an unlabeled competitor (2500 fmole). The samples were separated by electrophoresis in non-denaturing 6% polyacrylamide gels and incubated with streptavidin-conjugated horseradish peroxidase. Protein–DNA complexes were visualized using a Western blotting detection kit (*top panel*). The intensity of the protein–DNA complexes was measured using ImageJ v1.52a software (*bottom graph*). ** $p < 0.01$; *** $p < 0.001$ by Dunnett's multiple comparisons test 2.4. Minus (−), vehicle treatment; Plus (+), TNFα or competitor treatment.

To determine whether chrysin affects the binding of EGR1 to the putative EBS in the −369/+18 region, we performed the electrophoretic mobility shift assay (EMSA). Nuclear extracts from HaCaT cells treated with TNFα in the presence or absence of chrysin were incubated with a biotinylated EBS oligonucleotide probe, and the DNA-binding proteins were analyzed using streptavidin-conjugated horseradish peroxidase. Unlabeled EBS competitors were administered at a fifty-fold excess (2.5 pmol) concentration to indicate the specific reaction of the DNA–protein complex formation. Figure 3B shows that TNFα promoted the formation of the DNA–protein complex; however, the concentration of this complex was significantly ($p < 0.01$) reduced upon chrysin pretreatment, suggesting that EGR1 interacts with the putative EBS in the −369/+18 region of the *TSLP* promoter.

2.4. *Chrysin Downregulates EGR1 Expression to Inhibit TNFα-Induced TSLP Expression*

To further confirm whether EGR1 is required for TNFα-induced *TSLP* expression, we silenced EGR1 expression by expressing the control scrambled shRNA (shCT) or *EGR1* shRNA (shEgr1) in HaCaT cells. The knockdown of *EGR1* expression was confirmed using RT-PCR (Figure 4A) and Q-PCR (Figure 4C). The *TSLP* mRNA expression-inducing potential of TNFα was significantly ($p < 0.001$) inhibited in HaCaT/shEgr1 cells compared to HaCaT/shCT cells, as revealed by RT-PCR (Figure 4B). The decrease in *TSLP* mRNA levels by shEgr1 expression was quantitated using Q-PCR analysis. TNFα-induced *TSLP* mRNA expression increased 9.50 ± 0.755-fold in HaCaT/shCT cells but only 1.80 ± 0.300-fold in HaCaT/shEgr1 cells (Figure 4D). These data suggest that EGR1 plays a critical role in TNFα-induced *TSLP* transcription.

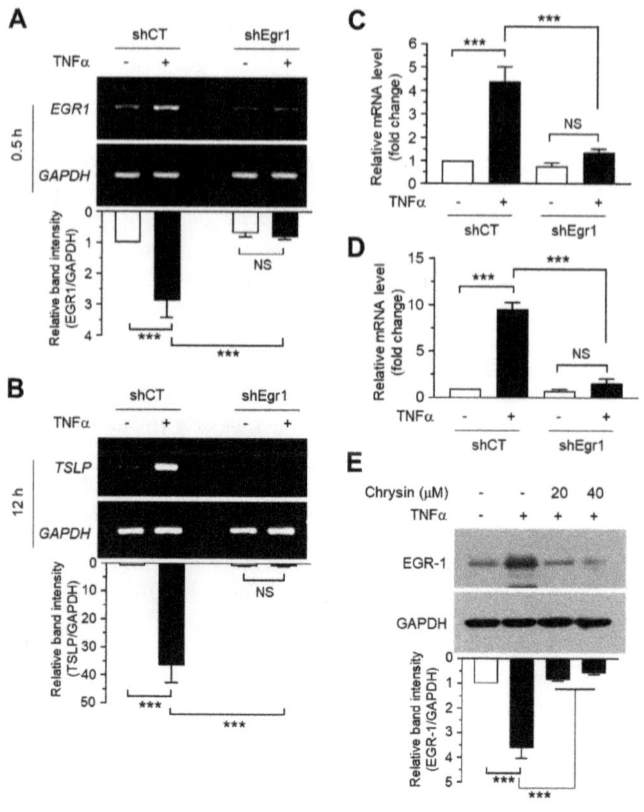

Figure 4. Inhibition of *TSLP* mRNA expression via *EGR1* knockdown and the effect of chrysin on the

Isuppression of TNFα-induced EGR1 expression. (**A**,**B**) HaCaT transfectants expressing scrambled (shCT) or *EGR1* shRNA (shEgr1) were treated with 10 ng/mL TNFα for 30 min (**A**) or 12 h (**B**). Total RNA was isolated, and *EGR1* (**A**,**C**) or *TSLP* mRNA expression (**B**,**D**) was measured using RT-PCR (**A**,**B**) and quantitative real-time PCR (**C**,**D**). The *GAPDH* mRNA level was measured as an internal control. RT-PCR product intensities were measured using the ImageJ v1.52a software. Data are presented as mean ± SD (n = 3). NS, not significant; *** $p < 0.001$ by Sidak's multiple comparisons test. (**E**) HaCaT cells expressing scrambled (shCT) or short-hairpin *EGR1* shRNA (shEgr1) were incubated with 0.5% serum for 24 h, followed by treatment with 10 ng/mL TNFα for 1 h in the presence or absence of chrysin. The cell lysates were immunoblotted using anti-EGR1 antibodies. Glyceraldehyde-3-phosphate dehydrogenase (GAPDH) was used as an internal control. The band intensity corresponding to each EGR1 protein was normalized to the GAPDH level using ImageJ v1.52a software. *** $p < 0.001$ by Dunnett's multiple comparisons test. Minus (−), vehicle treatment; Plus (+), TNFα treatment.

Then, we determined whether chrysin affects EGR1 expression. Serum-starved HaCaT cells were treated with 10 ng/mL TNFα for 1 h in the presence or absence of chrysin, and the EGR1 levels were measured using immunoblotting. TNFα-induced EGR1 accumulation was significantly ($p < 0.001$) abrogated after chrysin pretreatment (Figure 4E). These results suggested that chrysin downregulated EGR1 expression to suppress *TSLP* transcription.

2.5. Oral Administration of Chrysin Attenuates 2,4-Dinitrochlorobenzene (DNCB)-Induced AD-Like Skin Lesions in BALB/c Mice

DNCB has been widely used as an inducer of AD-like skin lesions in mouse models [46]. Chrysin was shown to attenuate DNCB-induced skin lesions [37]. To confirm the effect of chrysin on in vivo TSLP suppression, we induced AD-like skin inflammation by topical sensitization with SDS and DNCB (Figure 5A). The ear skin subjected to repeated DNCB applications exhibited typical signs of AD-like skin lesions, such as superficial erosion; however, the signs of DNCB-induced skin erosion were substantially attenuated by the oral administration of chrysin (25 mg/kg) compared to those in the DNCB-challenged group (Figure 5B). Skin edema is a typical sign of skin inflammation in mouse models. We monitored ear swelling by measuring the ear thickness throughout the experimental period of 21 days. DNCB-challenged mice exhibited ear swelling in a time-dependent manner; however, the oral administration of chrysin significantly reduced the ear thickness on day 21 (Figure 5C). Hematoxylin and eosin (H&E) staining of the tissue sections revealed that oral chrysin administration attenuated DNCB-induced epidermal hyperplasia of the ear and neck skin tissues (Figure 5D). DNCB-induced AD-like skin lesions are also characterized by the massive infiltration of various immune cells, including T cells and mast cells, into the inflammatory regions [47]. We studied the effect of chrysin on the inhibition of immune cell infiltration using toluidine blue (TB) staining [48]. DNCB application increased the infiltration of TB-positive cells, whereas the oral administration of chrysin substantially suppressed the DNCB-induced infiltration of TB-positive cells (Figure 5E). These data confirmed the beneficial effect of chrysin on DNCB-induced AD-like skin lesions in a mouse model.

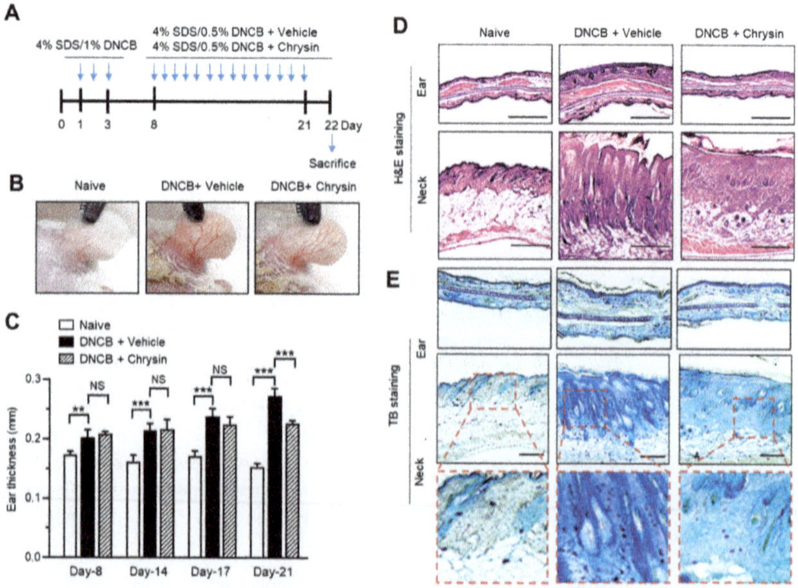

Figure 5. Effect of oral chrysin administration on the attenuation of skin lesions in DNCB-challenged BALB/c mice. (**A**) Illustration of the experimental schedule for the induction of atopic dermatitis-like skin lesions and oral chrysin administration. (**B**) Representative images of the ear and neck skin of BALB/c mice; untreated control (naive), DNCB + vehicle (PBS), and DNCB + chrysin (25 mg/kg). Images were acquired on day 22 immediately after the mice were euthanized. (**C**) Ear thickness was measured using a micro caliper after DNCB challenge. Data are expressed as mean ± SD ($n = 3$). NS, not significant; ** $p < 0.01$; *** $p < 0.001$ by Sidak's multiple comparisons test. (**D**,**E**) Paraffin-embedded ear and neck skin tissues of BALB/c mice were prepared on day 22, and H&E (**D**) and TB staining (**E**) were performed. The enlarged version of each image is provided in the dotted boxes. Scale bars, 400 μm.

2.6. Oral Administration of Chrysin Reduces EGR1 and TSLP Expression in DNCB-Induced Skin Lesions in BALB/c Mice

We next evaluated whether the oral administration of chrysin could suppress EGR1 and TSLP expression in AD-like skin lesions in BALB/c mice. Immunohistochemical analysis of the skin sections showed that DNCB increased EGR1-positive staining in the epidermis of the ear (Figure 6A) and neck (Figure 6C). Notably, EGR1-positive staining induced under DNCB challenge was substantially suppressed in response to the oral administration of chrysin. Similarly, immunofluorescence staining showed that the levels of TSLP-positive staining in the epidermis of the ear (Figure 6B) and neck (Figure 6D) reduced upon the oral administration of chrysin. These results support the notion that chrysin inhibits TSLP expression by downregulating EGR1 in inflammatory microenvironments.

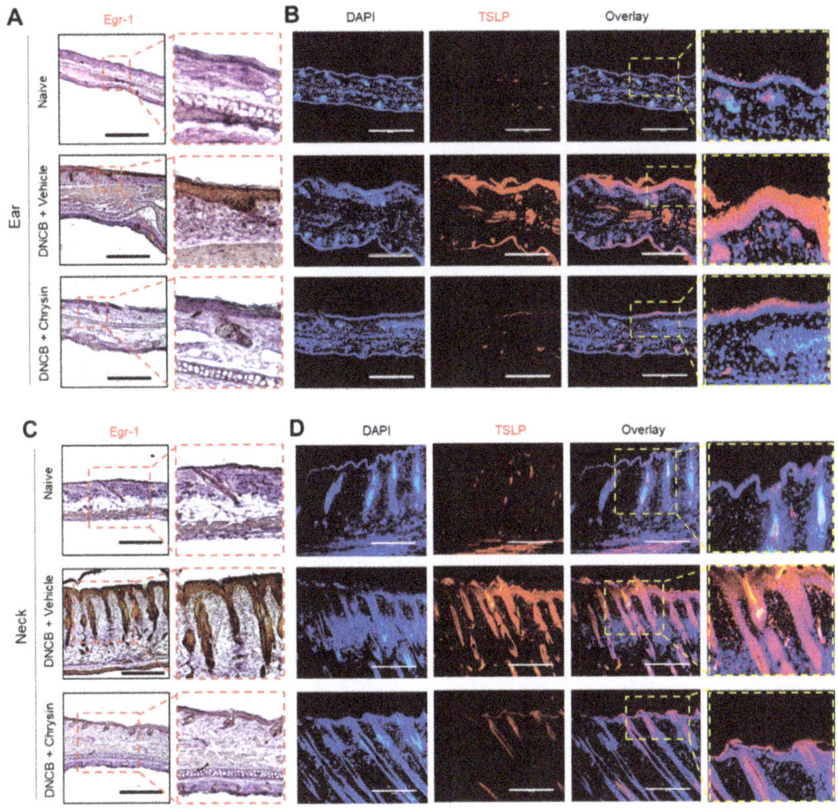

Figure 6. Effect of chrysin on the suppression of EGR1 and TSLP expression in DNCB-challenged BALB/c mice. (**A**,**C**) BALB/c mice either untreated (naive) or treated with DNCB + vehicle (PBS) and DNCB + chrysin. Paraffin-embedded ear (**A**) and neck (**C**) tissue sections were prepared on day 22, and immunohistochemical staining was performed for EGR1. The sections were counterstained with H&E. Scale bars, 400 μm. The areas in the dashed boxes are enlarged in the bottom panels. (**B**,**D**) Paraffin-embedded ear (**B**) and neck (**D**) tissue sections were prepared and subjected to immunofluorescence staining with an anti-TSLP antibody and rhodamine red-X-conjugated secondary antibody (red). The nuclei were counterstained with Hoechst 33258 (blue). Scale bars, 400 μm. The areas in the dashed boxes are enlarged in the panels to the right.

2.7. Chrysin Inhibits the MAPK Pathways

We investigated the mode of action of chrysin, which is considered to inhibit EGR1 expression and downregulate TSLP expression. MAPK pathways regulate EGR1 expression in various cell types [49–51]. The levels of phosphorylated ERK1/2, JNK1/2, and p38 kinase increased rapidly within 10 min following TNFα treatment, whereas the total quantity of each MAPK protein did not change (Figure 7A). Under these experimental conditions, the effect of chrysin on MAPK phosphorylation was examined. We observed that chrysin significantly ($p < 0.001$ in all cases) inhibited the TNFα-induced phosphorylation of ERK1/2 and JNK1/2, but not of p38 kinase (Figure 7B). These data suggest that while the three major MAPKs are activated by TNFα in HaCaT cells, chrysin selectively inhibits only the ERK1/2 and JNK1/2 MAPK pathways.

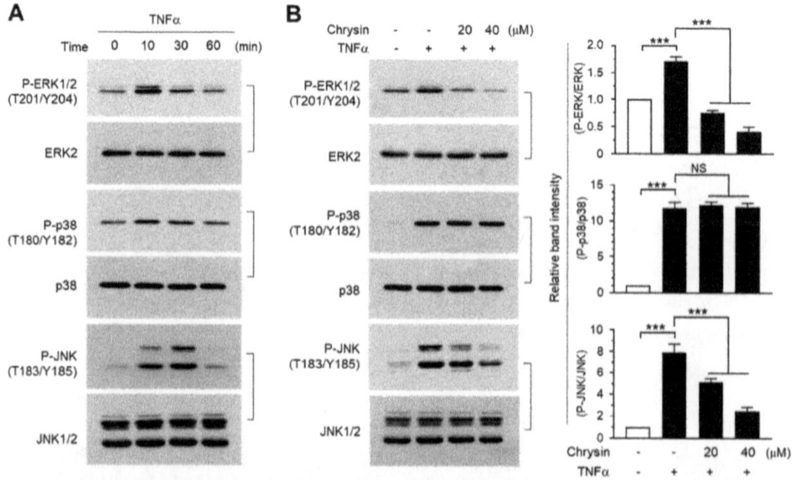

Figure 7. Effect of chrysin on the inhibition of mitogen-activated protein kinases (MAPKs). (**A**) HaCaT cells were treated with 10 ng/mL TNFα for 0–60 min. (**B**) HaCaT cells were treated with 10 ng/mL TNFα for 10 min in the presence or absence of chrysin at different concentrations (20 and 40 μM). Whole-cell lysates were prepared, and Western blotting was performed using phospho-specific and total MAPK protein antibodies. The band intensities of the phosphorylated proteins were normalized relative to those of total proteins, using ImageJ v1.52a software. Data are expressed as mean ± SD ($n = 3$) in the graphs. NS, not significant; *** $p < 0.001$ by Dunnett's multiple comparisons test. Minus (−), vehicle treatment; Plus (+), TNFα treatment.

2.8. MAPK Pathways Are Involved in TNFα-Induced EGR1 and TSLP Expression in HaCaT Keratinocytes

To determine the potential relationship between MAPK activation and TNFα-induced EGR1 expression, we used pharmacological inhibitors of the MAPK pathway. Pretreatment with the MAPK kinase inhibitor U0126, p38 kinase inhibitor SB203580, or JNK inhibitor SP600125 significantly ($p < 0.001$ in all cases) decreased TNFα-induced EGR1 accumulation, as revealed in the Western blot analysis (Figure 8A). Findings from the RT-PCR (Figure 8B) and real-time PCR (Figure 8C) analyses indicated that TNFα-induced *TSLP* mRNA expression was significantly inhibited by all three MAPK inhibitors ($p < 0.001$ in all cases). These data suggest that all three MAPKs mediate TNFα-induced *TSLP* expression via EGR1, but chrysin selectively inhibits only the TNFα-induced ERK1/2 and JNK1/2 pathways to downregulate *TSLP* expression.

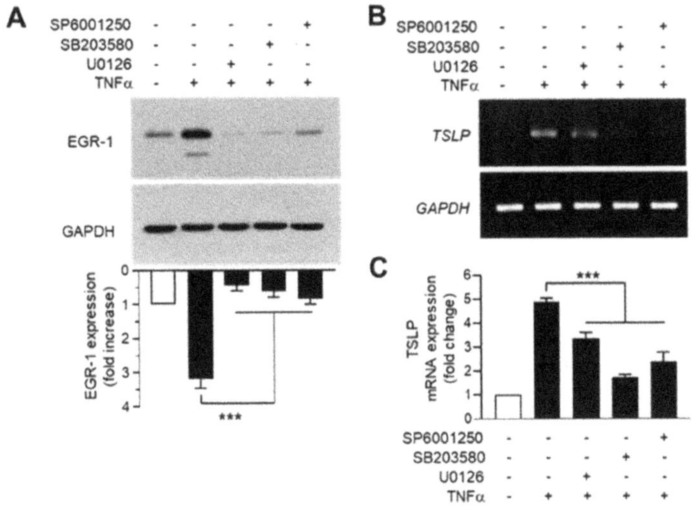

Figure 8. Effect of mitogen-activated protein kinase inhibition on the expression of TNFα-induced EGR1 and *TSLP*. (**A**) HaCaT cells were pretreated with SB203580 (20 μM), U0126 (10 μM), or SP600125 (20 μM) for 30 min, followed by treatment with 10 ng/mL TNFα for 1 h. Whole-cell lysates were prepared, and Western blotting was performed using anti-EGR1 antibodies. Glyceraldehyde-3-phosphate dehydrogenase (GAPDH) was used as an internal control. The band intensity corresponding to EGR1 was normalized to the GAPDH level using the ImageJ v1.52a software. The graphical data show mean ± SD values ($n = 3$). *** $p < 0.001$ using Dunnett's multiple comparisons test. (**B**,**C**) HaCaT cells were pretreated with SB203580 (20 μM), U0126 (10 μM), or SP600125 (20 μM) for 30 min, followed by treatment with 10 ng/mL TNFα for 12 h. Total RNA was isolated, and the levels of *TSLP* mRNA were measured using RT-PCR (**B**) and quantitative real-time PCR (**C**). *GAPDH* mRNA was used as an internal control. Data are expressed as mean ± SD ($n = 3$). *** $p < 0.001$ by Dunnett's multiple comparisons test. Minus (−), vehicle treatment; Plus (+), TNFα or inhibitor treatment.

3. Discussion

Chrysin has a pharmacological property that helps alleviate the clinical symptoms of AD by inhibiting the secretion of pro-inflammatory cytokines and chemokines [37–39]. TNFα is a major pro-inflammatory cytokine that is released from various immune cells and stromal cells. It promotes the production of multiple AD-related inflammatory cytokines. TSLP upregulation is a hallmark of AD pathogenesis. To further elucidate the molecular action of chrysin in AD with respect to therapeutic efficacy, we focused on the effect exerted by chrysin on TSLP suppression in TNFα-stimulated keratinocytes and in a clinically relevant animal model with AD-like skin lesions induced upon DNCB challenge. We showed that chrysin suppresses TSLP expression by inhibiting the MAPKs ERK1/2 and JNK1/2 pathways and downregulating EGR1 expression in the inflammatory environment.

Various transcription factors, such as vitamin D3 receptor, NF-κB, and AP1 [43,44,52], are involved in the transcriptional regulation of *TSLP* based on the stimuli applied. To identify the *cis*-acting element responsible for mediating the effects of chrysin, we established a series of *TSLP* promoter–reporter constructs and evaluated the effect of chrysin on *TSLP* promoter activity in a luciferase activity assay. Cells transfected with the shortest reporter construct (−369/+18) continued to exhibit chrysin activity, suggesting that the chrysin response element is between the positions −369 and +18. We found a putative EGR1-binding motif, EBS, between the positions −206/−187 in the −369/+18 construct. *EGR1*, also named zinc finger protein 225 (*Zif268*), *Drosophila* Kr finger probe 24 (*Krox24*), tetradecanoyl phorbol acetate-induced sequence 8 (TIS8), and nerve growth factor-induced clone A (*NGFI-A*), is an immediate-early response gene that encodes a transcription factor containing three Cys2-His2-

type zinc finger DNA-binding domains [53]. It regulates various cellular pathophysiological responses, including inflammation, synaptic plasticity, female reproduction, cell proliferation, apoptosis, and carcinogenesis [54–59].

EGR1 is expressed at high levels in damaged skin tissues [60]. EGR1 regulates the expression of genes encoding inflammation-related proteins, such as IL-33-induced TSLP [45], IL-17-induced psoriasin [61], and IL-13-induced kallikrein-related peptidase 7 (KLK7) in keratinocytes [62]. Recently, we demonstrated that immune cell infiltration in AD-like skin lesions was substantially attenuated in *Egr1*-knockout mice, and the TNFα-induced expression of cytokines, including TSLP, IL-1β, IL-6, CXCL1, CCL2, and CCL5, was inhibited in response to EGR1 knockdown [42]. Furthermore, the AB1711 compound, a small-molecule inhibitor targeting the EGR1 zinc-finger DNA-binding domains, was shown to abrogate the expression of EGR1-regulated inflammatory cytokines in keratinocytes and improve both skin inflammation and itching in DNCB-challenged NC/Nga mice [42]. These findings suggest that the inhibition of EGR1 transcriptional activity is a promising therapeutic strategy for improving therapeutic efficacy in chronic skin inflammation. In this study, the functional importance of the EBS within the *TSLP* promoter was analyzed via transient transfection experiments. We observed that the transient transfection of EGR1 enhanced the promoter-reporter activity of the −369/+18 construct. We further investigated whether chrysin inhibits EGR1 to downregulate TSLP expression.

We found that chrysin prevented TNFα-induced EGR1 DNA-binding activity, as observed using EMSA. In addition, chrysin inhibited TNFα-induced EGR1 expression in HaCaT keratinocytes. We also confirmed that the oral administration of chrysin attenuated both EGR1 and TSLP expression in vivo in the AD-like skin lesions of DNCB-challenged mice. These findings suggest that chrysin downregulates EGR1 expression to inhibit TSLP expression in the inflammatory microenvironment. One of the best-characterized transcription factors that regulate EGR1 expression is the Ets-like protein-1 (ELK-1), which is phosphorylated and activated by ERK1/2, p38 kinase, and JNK1/2 in response to mitogens and TNFα [63]. Our data showed that chrysin inhibited the TNFα-induced phosphorylation of ERK1/2 and JNK1/2 but not of p38 kinase, suggesting that chrysin downregulates EGR1 expression by differentially inhibiting the MAPK signaling pathways in HaCaT keratinocytes.

NF-κB is a transcription factor expressed ubiquitously in almost all tissues, including skin keratinocytes. TNFα strongly activates the NF-κB signaling pathway in HaCaT keratinocytes [64]. NF-κB mediates TNFα-induced TSLP expression in human airway smooth muscle cells [43] and IL-1β-induced TSLP expression in intestinal epithelial cells [44]. We have previously reported that chrysin inhibits NF-κB activity by targeting the inhibitor of κB kinase, a protein encoded upstream of NF-κB and is involved in the proteolysis of the NF-κB inhibitor IκB [39]. Choi et al. [37] also reported that chrysin inhibits TNFα/IFNγ-induced degradation of IκB, leading to the inhibition of nuclear translocation of NF-κB in HaCaT keratinocytes. Therefore, chrysin may downregulate TSLP expression by inhibiting both EGR1 and NF-κB.

4. Materials and Methods

4.1. Materials

Chrysin, DNCB, TB, and H&E staining kits were purchased from Sigma-Aldrich (St. Louis, MO, USA). TNFα was purchased from ProSpec-Tany TechnoGene, Ltd. (Ness-Ziona, Israel). A firefly luciferase assay system was obtained from Promega (Madison, WI, USA). Anti-TSLP antibody was obtained from Novus Biologicals (Centennial, CO, USA), and phospho-ERK1/2 (Thr202/Tyr204), phospho-p38 (Thr180/Tyr182), and phospho-JNK1/2 (Thr183/Tyr185) antibodies were obtained from Cell Signaling Technology (Danvers, MA, USA). Anti-GAPDH and anti-EGR1 antibodies were purchased from Santa Cruz Biotechnology (Dallas, TX, USA). A secondary antibody conjugated to rhodamine red-X was obtained from Jackson ImmunoResearch Laboratories (West Grove, PA, USA).

4.2. Cells and Cell Culture

Human keratinocyte HaCaT cells were obtained from the Cell Line Service (Eppelheim, Germany). The cells were cultured in Dulbecco's modified Eagle's medium supplemented with 10% fetal bovine serum (HyClone, Logan, UT, USA) and penicillin-streptomycin (Sigma-Aldrich).

4.3. RT-PCR

Total RNA was isolated from the HaCaT cells using a TRIzol RNA Extraction Kit (Invitrogen, Carlsbad, CA, USA), and cDNA was synthesized using an iScript cDNA Synthesis Kit (Bio-Rad, Hercules, CA, USA). RT-PCR was performed using reverse transcriptase (Promega) and gene-specific PCR primers. The PCR primers used in this study were as follows:

- EGR1 forward, 5′-CAG CAG TCC CAT TTA CTC AG-3′;
- EGR1 reverse, 5′-GAC TGG TAG CTG GTA TTG-3;
- TSLP forward, 5′-TAG CAA TCG GCC ACA TTG CCT-3′;
- TSLP reverse, 5′-GAA GCG ACG CCA CAA TCC TTG-3;
- GAPDH forward, 5′-CCA AGG AGT AAG AAA CCC TGG AC-3′;
- GAPDH reverse, 5′-GGG CCG AGT TGG GAT AGG G-3′.

The thermal cycling conditions were as follows: denaturation at 94 °C for 5 min, followed by 30 cycles of denaturation at 94 °C for 30 s, annealing at 58 °C for 30 s, and elongation at 72 °C for 1 min. The amplified PCR products were separated by electrophoresis in a 2% agarose gel containing ethidium bromide and visualized under UV transillumination.

4.4. Quantitative Real-Time PCR (Q-PCR)

The mRNA levels of the genes were quantified using an iCycler iQ system with an iQ SYBR Green Supermix kit (Bio-Rad). Validated Q-PCR primers and SYBR Green-based fluorescent probes specific for *TSLP* (id: qHsaCIP0030468), *EGR1* (qHsaCEP0039196), and *GAPDH* (id: qHsaCEP0041396) were obtained from Bio-Rad. The thermal cycling conditions used for PCR were as follows: denaturation at 95 °C for 2 min, followed by 40 cycles of denaturation at 95 °C for 10 s and 60 °C for 45 s. The relative mRNA levels of *TSLP* or *EGR1* were normalized to those of *GAPDH* using the software provided by the manufacturer.

4.5. Western Blot Analysis

HaCaT cells were lysed in ice-cold cell lysis buffer supplemented with 50 mM Tris-HCl (pH 7.4), 1% NP-40, 0.25% Na-deoxycholate, 500 mM NaCl, 1 mM EDTA, 1 mM Na_3VO_4, 1 mM NaF, 10 µg/mL leupeptin, and 1 mM PMSF. The proteins were separated by electrophoresis in a 10% SDS-polyacrylamide gel and transferred to nitrocellulose membranes. After treatment with the appropriate primary and secondary antibodies, the blots were developed and observed using an enhanced chemiluminescence detection system (GE Healthcare, Piscataway, NJ, USA).

4.6. Construction of Human TSLP Promoter-Reporter Constructs

A *TSLP* promoter fragment spanning nucleotides −1384 to +18 upstream of the transcription start site was synthesized from human genomic DNA (Promega) via PCR using the primers 5′-CGT CCA ACC TCC TTT CTC CG -3′ (forward −1384F) and 5′-TTG GAG TCT CCC TGA TGC TCC AG-3′ (reverse, +18R). The amplified PCR products were ligated to a T&A vector (RBC Bioscience, Taipei County, Taiwan) and digested using *Kpn*I and *Hind*III. The products were ligated at the *Kpn*I and *Hind*III sites of the pGL4-basic vector (Promega), yielding pTSLP-Luc(−1384/+18). Several deletion constructs of the human *TSLP* promoter fragments were synthesized using PCR, for which the pTSLP-Luc(−1384/+18) construct was used as the template. The forward primer sequences were as follows:

- −1338F: 5′-GGA CCA GAG CGA TGC AGG-3′
- −1214F: 5′-CAT GAG CCA AGC CAG GGA G-3′
- −1017F: 5′-AAA TCT GAG CCC GCC ATC TC-3′
- −369F: 5′-GGG ACA TAT GCA AGG ACT CC-3′

One reverse primer, +18R, was used to generate the deletion constructs. The amplified PCR products were ligated to the T&A vector and then to the pGL4-basic vector. The insert sequence of each construct was confirmed using DNA sequencing (Macrogen, Seoul, Korea).

4.7. Luciferase Promoter–Reporter Assay

HaCaT keratinocytes cultured in 12-well plates were transfected with 0.2 μg of each *TSLP* promoter–reporter construct using Lipofectamine™ 2000 (Invitrogen) according to the manufacturer's instruction. After 48 h of transfection, the cells were treated with TNFα in the presence or absence of chrysin (20 or 40 μM). After 8–12 h, the cells were harvested, and the firefly luciferase activity was measured using the Dual-Glo™ Luciferase assay system (Promega) following the manufacturer's instruction. The relative luciferase activity of the untreated cells was assigned the value 1. Luminescence was measured using a dual luminometer (Centro LB960; Berthold Tech, Bad Wildbad, Germany).

4.8. EMSA

EMSA was performed using a LightShift Chemiluminescence EMSA kit (Thermo Fisher Scientific, Waltham, MA, USA), according to the manufacturer's instruction. A biotin-labeled deoxyoligonucleotide probe corresponding to the EBS (5′-CAA AAA GGA GGA AGG TGA GGG AA-biotin-3′) was synthesized by Macrogen. Nuclear extracts (3 μg samples) prepared from the HaCaT keratinocytes were mixed with 50 fmole biotin-labeled EGR1-binding oligonucleotide probes with 1 μg poly(dI-dC) (Amersham Biosciences, Piscataway, NJ, USA). For the competition assay, 2.5 pmol of the unlabeled EGR1-binding oligonucleotide was added. DNA–protein complexes were separated in non-denaturing 6% polyacrylamide gels, and the antibody-reactive bands were visualized using chemiluminescence, according to the manufacturer's instructions.

4.9. Induction of AD-Like Skin Lesions in the Ear and Neck of Mice

BALB/c mice (7-week-old, male) were obtained from Orient Bio, Inc. (Seongnam, Korea). The mice were housed in a specific pathogen-free environment at 20 ± 2 °C and a relative humidity of 50% ± 10%. The mice were randomly divided into three groups (based on the treatment administered): Group I, naive; Group II, DNCB + vehicle; and Group III, DNCB + chrysin (n = 5 in each group). In addiion to those in the naive group, all mice were sensitized with 4% SDS on both the neck and ear skin to disrupt the skin barrier; after 4 h, the SDS-sensitized areas were challenged with 1% DNCB dissolved in an acetone:olive oil mixture (1:3, v/v). The DNCB challenge was repeated once daily for 3 days. After a 4-day break, sensitization with 4% SDS followed by the application of 0.5% DNCB was repeated five times per week for 2 weeks (days 8–21). Chrysin powder was dissolved in dimethyl sulfoxide (250 mg/mL) to prepare a stock solution and then diluted using PBS to a final concentration of 25 mg/mL. The mice in Group III were administered chrysin (25 mg/kg) orally from day 7 (once daily, five times per week for 2 weeks). On day 22, all mice were euthanized, and tissue sections were prepared. The animal experiments were conducted in accordance with the guidelines for animal experiments and procedures approved by the Konkuk University Institutional Animal Care and Use Committee (IACUC). All experimental methods were confirmed to be in accordance with the relevant guidelines and regulations (approval number KU19129).

4.10. Histological Analysis

Skin sections of the neck and ear with AD-like lesions were fixed in 100% acetone and embedded in paraffin. Each section was cut (5 μm) using a microtome (Leica Microsystems,

Wetzlar, Germany). The paraffin-embedded ear and neck skin sections were deparaffinized by treating with xylene (three times for 10 min) and hydrated using a graded ethanol series. After deparaffinization and rehydration, the tissue sections were stained with H&E. The infiltrated mast cells were stained with 0.1% TB. Images of each section were captured using a light microscope (EVOS FL Auto, Bothell, WA, USA).

4.11. Immunohistochemical and Immunofluorescence Analysis

Immunostaining of the skin sections from the ear and neck was performed as previously described [39]. Briefly, after deparaffinization with xylene (three times for 10 min) and hydration with a graded ethanol series, the tissue sections were placed in 1 mM EDTA (pH 8.0) at 70 °C for 20 min. After rinsing with PBS, the tissue sections were incubated with 7% goat serum for 1 h to block non-specific binding of immunoglobulin (Ig). For EGR1 immunostaining, the sections were treated overnight with primary rabbit anti-EGR1 antibodies (1:100 dilution) at 4 °C. After washing three times with PBS, the sections were treated with biotinylated goat anti-rabbit IgG (1:100 dilution) at 25 °C for 1 h. Immunoreactivity was visualized after treatment with a diaminobenzidine substrate for 5 min, followed by counterstaining with H&E.

For TSLP immunofluorescence staining, the tissue sections were treated overnight with an anti-TSLP antibody (1:100 dilution) at 4 °C. After washing, the cells were treated with a rhodamine red-X-conjugated secondary antibody (1:300 dilution) at 25 °C for 1 h. The nuclei were counterstained with Hoechst 33258 solution for 10 min. After extensive washing with PBS, the slides were mounted using the ProLong Gold Antifade reagent (Invitrogen). Fluorescent images were captured using an EVOS FL fluorescence microscope (Advanced Microscopy Group; Bothell, WA, USA).

4.12. Statistical Analysis

Data are expressed as mean ± standard deviation. Statistical analysis was performed using one-way analysis of variance, followed by Dunnett's or Sidak's multiple comparisons test using GraphPad Prism (version 9.0.1; GraphPad Software, Inc., La Jolla, CA, USA). Statistical significance was set at $p < 0.05$.

5. Conclusions

To the best of our knowledge, this is the first study to demonstrate that chrysin suppresses *TSLP* expression by downregulating ERK1/2- and JNK1/2-dependent EGR1 expression in the skin inflammatory microenvironment. We believe that the results of this study will improve our understanding of the mode of action of chrysin and its therapeutic efficacy in AD.

Author Contributions: Y.H.L., Y.L. and S.Y.S. conceived and designed the experiments; H.Y., S.S.A. and E.J. carried out cellular and animal experiments and analyzed the data; Y.H.L. and S.Y.S. wrote the manuscript; S.Y.S. supervised the project; All authors have read and agreed to the published version of the manuscript.

Funding: This research received no external funding.

Institutional Review Board Statement: The study was conducted according to the guidelines of the Declaration of Helsinki, and approved by the Institutional Review Ethics Committee of Konkuk University Institutional Animal Care and Use Committee (IACUC) (protocol code KU19129 at 17 September 2019).

Informed Consent Statement: Not applicable.

Data Availability Statement: The data presented in this study are available on request from the corresponding author.

Acknowledgments: The paper was supported by Konkuk University in 2017.

Conflicts of Interest: The authors declare that there is no conflict of interest.

Abbreviations

AD	atopic dermatitis
CCL	C-C motif chemokine ligand
DNCB	2,4-dinirochlorobenzene
EBS	EGR1-binding sequence
EGR1	early rrowth response 1
shEgr1	EGR1 shRNA
EMSA	electrophoretic mobility shift assay
ERK	Extracellular signal-regulated kinases
GAPDH	glyceraldehyde 3-phosphate dehydrogenase
H&E	hematoxylin and eosin
Ig	immunoglobulin
IKK	inhibitor of κB kinase
IL	interleukin
JNK	c-Jun N-terminal kinase
MAPK	mitogen-activated protein kinases
NF-κB	nuclear factor-κB
Q-PCR	quantitative real-time PCR
RT-PCR	reverse-transcription polymerase chain reaction
shCT	scrambled shRNA
TB	toluidine blue
Th2	T helper cell 2
TNFα	tumor necrosis factor-alpha
TSLP	thymic stromal lymphopoietin

References

1. Weidinger, S.; Novak, N. Atopic dermatitis. *Lancet* **2016**, *387*, 1109–1122. [CrossRef]
2. Maarouf, M.; Vaughn, A.R.; Shi, V.Y. Topical micronutrients in atopic dermatitis—An evidence-based review. *Dermatol. Ther.* **2018**, *31*, e12659. [CrossRef]
3. Bekic, S.; Martinek, V.; Talapko, J.; Majnaric, L.; Vasilj Mihaljevic, M.; Skrlec, I. Atopic dermatitis and comorbidity. *Healthcare* **2020**, *8*, 70. [CrossRef] [PubMed]
4. Nettis, E.; Ortoncelli, M.; Pellacani, G.; Foti, C.; Di Leo, E.; Patruno, C.; Rongioletti, F.; Argenziano, G.; Ferrucci, S.M.; Macchia, L.; et al. A multicenter study on the prevalence of clinical patterns and clinical phenotypes in adult atopic dermatitis. *J. Investig. Allergol. Clin. Immunol.* **2020**, *30*, 448–450. [CrossRef]
5. Smith, C.H. New approaches to topical therapy. *Clin. Exp. Dermatol.* **2000**, *25*, 567–574. [CrossRef] [PubMed]
6. Sidbury, R.; Hanifin, J.M. Systemic therapy of atopic dermatitis. *Clin. Exp. Dermatol.* **2000**, *25*, 559–566. [CrossRef]
7. Cline, A.; Bartos, G.J.; Strowd, L.C.; Feldman, S.R. Biologic Treatment options for pediatric psoriasis and atopic dermatitis. *Children* **2019**, *6*, 103. [CrossRef]
8. Welsch, K.; Holstein, J.; Laurence, A.; Ghoreschi, K. Targeting JAK/STAT signalling in inflammatory skin diseases with small molecule inhibitors. *Eur. J. Immunol.* **2017**, *47*, 1096–1107. [CrossRef]
9. Newsom, M.; Bashyam, A.M.; Balogh, E.A.; Feldman, S.R.; Strowd, L.C. New and emerging systemic treatments for atopic dermatitis. *Drugs* **2020**, *80*, 1041–1052. [CrossRef]
10. Ring, J.; Alomar, A.; Bieber, T.; Deleuran, M.; Fink-Wagner, A.; Gelmetti, C.; Gieler, U.; Lipozencic, J.; Luger, T.; Oranje, A.P.; et al. Guidelines for treatment of atopic eczema (atopic dermatitis) part I. *J. Eur. Acad. Dermatol. Venereol.* **2012**, *26*, 1045–1060. [CrossRef]
11. Dattola, A.; Bennardo, L.; Silvestri, M.; Nistico, S.P. What's new in the treatment of atopic dermatitis? *Dermatol. Ther.* **2019**, *32*, e12787. [CrossRef] [PubMed]
12. Williams, H.C. Epidemiology of atopic dermatitis. *Clin. Exp. Dermatol.* **2000**, *25*, 522–529. [CrossRef]
13. Akdis, M.; Aab, A.; Altunbulakli, C.; Azkur, K.; Costa, R.A.; Crameri, R.; Duan, S.; Eiwegger, T.; Eljaszewicz, A.; Ferstl, R.; et al. Interleukins (from IL-1 to IL-38), interferons, transforming growth factor β, and TNF-α: Receptors, functions, and roles in diseases. *J. Allergy Clin. Immunol.* **2016**, *138*, 984–1010. [CrossRef] [PubMed]
14. Trier, A.M.; Kim, B.S. Cytokine modulation of atopic itch. *Curr. Opin. Immunol.* **2018**, *54*, 7–12. [CrossRef]
15. Girolomoni, G.; Sebastiani, S.; Albanesi, C.; Cavani, A. T-cell subpopulations in the development of atopic and contact allergy. *Curr. Opin. Immunol.* **2001**, *13*, 733–737. [CrossRef]
16. Chan, L.S.; Robinson, N.; Xu, L. Expression of interleukin-4 in the epidermis of transgenic mice results in a pruritic inflammatory skin disease: An experimental animal model to study atopic dermatitis. *J. Investig. Dermatol.* **2001**, *117*, 977–983. [CrossRef]
17. Zheng, T.; Oh, M.H.; Oh, S.Y.; Schroeder, J.T.; Glick, A.B.; Zhu, Z. Transgenic expression of interleukin-13 in the skin induces a pruritic dermatitis and skin remodeling. *J. Investig. Dermatol.* **2009**, *129*, 742–751. [CrossRef]

18. Cevikbas, F.; Wang, X.; Akiyama, T.; Kempkes, C.; Savinko, T.; Antal, A.; Kukova, G.; Buhl, T.; Ikoma, A.; Buddenkotte, J.; et al. A sensory neuron-expressed IL-31 receptor mediates T helper cell-dependent itch: Involvement of TRPV1 and TRPA1. *J. Allergy Clin. Immunol.* **2014**, *133*, 448–460. [CrossRef]
19. Sims, J.E.; Williams, D.E.; Morrissey, P.J.; Garka, K.; Foxworthe, D.; Price, V.; Friend, S.L.; Farr, A.; Bedell, M.A.; Jenkins, N.A.; et al. Molecular cloning and biological characterization of a novel murine lymphoid growth factor. *J. Exp. Med.* **2000**, *192*, 671–680. [CrossRef]
20. Friend, S.L.; Hosier, S.; Nelson, A.; Foxworthe, D.; Williams, D.E.; Farr, A. A thymic stromal cell line supports in vitro development of surface IgM + B cells and produces a novel growth factor affecting B and T lineage cells. *Exp. Hematol.* **1994**, *22*, 321–328.
21. Soumelis, V.; Reche, P.A.; Kanzler, H.; Yuan, W.; Edward, G.; Homey, B.; Gilliet, M.; Ho, S.; Antonenko, S.; Lauerma, A.; et al. Human epithelial cells trigger dendritic cell mediated allergic inflammation by producing TSLP. *Nat. Immunol.* **2002**, *3*, 673–680. [CrossRef]
22. Liu, Y.J.; Soumelis, V.; Watanabe, N.; Ito, T.; Wang, Y.H.; Malefyt Rde, W.; Omori, M.; Zhou, B.; Ziegler, S.F. TSLP: An epithelial cell cytokine that regulates T cell differentiation by conditioning dendritic cell maturation. *Ann. Rev. Immunol.* **2007**, *25*, 193–219. [CrossRef]
23. Rochman, I.; Watanabe, N.; Arima, K.; Liu, Y.J.; Leonard, W.J. Cutting edge: Direct action of thymic stromal lymphopoietin on activated human CD4 + T cells. *J. Immunol.* **2007**, *178*, 6720–6724. [CrossRef]
24. Rochman, Y.; Leonard, W.J. The role of thymic stromal lymphopoietin in CD8 + T cell homeostasis. *J. Immunol.* **2008**, *181*, 7699–7705. [CrossRef]
25. Liu, Y.J. Thymic stromal lymphopoietin: Master switch for allergic inflammation. *J. Exp. Med.* **2006**, *203*, 269–273. [CrossRef]
26. Watanabe, N.; Hanabuchi, S.; Soumelis, V.; Yuan, W.; Ho, S.; de Waal Malefyt, R.; Liu, Y.J. Human thymic stromal lymphopoietin promotes dendritic cell-mediated CD4 + T cell homeostatic expansion. *Nat. Immunol.* **2004**, *5*, 426–434. [CrossRef]
27. Ito, T.; Wang, Y.H.; Duramad, O.; Hori, T.; Delespesse, G.J.; Watanabe, N.; Qin, F.X.; Yao, Z.; Cao, W.; Liu, Y.J. TSLP-activated dendritic cells induce an inflammatory T helper type 2 cell response through OX40 ligand. *J. Exp. Med.* **2005**, *202*, 1213–1223. [CrossRef]
28. Liu, Y.J. Thymic stromal lymphopoietin and OX40 ligand pathway in the initiation of dendritic cell-mediated allergic inflammation. *J. Allergy Clin. Immunol.* **2007**, *120*, 238–244. [CrossRef] [PubMed]
29. Allakhverdi, Z.; Comeau, M.R.; Jessup, H.K.; Yoon, B.R.; Brewer, A.; Chartier, S.; Paquette, N.; Ziegler, S.F.; Sarfati, M.; Delespesse, G. Thymic stromal lymphopoietin is released by human epithelial cells in response to microbes, trauma, or inflammation and potently activates mast cells. *J. Exp. Med.* **2007**, *204*, 253–258. [CrossRef]
30. Wilson, S.R.; The, L.; Batia, L.M.; Beattie, K.; Katibah, G.E.; McClain, S.P.; Pellegrino, M.; Estandian, D.M.; Bautista, D.M. The epithelial cell-derived atopic dermatitis cytokine TSLP activates neurons to induce itch. *Cell* **2013**, *155*, 285–295. [CrossRef]
31. Ziegler, S.F.; Artis, D. Sensing the outside world: TSLP regulates barrier immunity. *Nat. Immunol.* **2010**, *11*, 289–293. [CrossRef]
32. Adhikary, P.P.; Tan, Z.; Page, B.D.G.; Hedtrich, S. TSLP as druggable target—A silver-lining for atopic diseases? *Pharmacol. Ther.* **2021**, *217*, 107648. [CrossRef]
33. Matera, M.G.; Rogliani, P.; Calzetta, L.; Cazzola, M. TSLP inhibitors for asthma: Current status and future prospects. *Drugs* **2020**, *80*, 449–458. [CrossRef] [PubMed]
34. Takai, T. TSLP expression: Cellular sources, triggers, and regulatory mechanisms. *Allergol. Int.* **2012**, *61*, 3–17. [CrossRef] [PubMed]
35. Mani, R.; Natesan, V. Chrysin: Sources, beneficial pharmacological activities, and molecular mechanism of action. *Phytochemistry* **2018**, *145*, 187–196. [CrossRef]
36. Naz, S.; Imran, M.; Rauf, A.; Orhan, I.E.; Shariati, M.A.; Iahtisham, U.H.; Iqra, Y.; Shahbaz, M.; Qaisrani, T.B.; Shah, Z.A.; et al. Chrysin: Pharmacological and therapeutic properties. *Life Sci.* **2019**, *235*, 116797. [CrossRef]
37. Choi, J.K.; Jang, Y.H.; Lee, S.; Lee, S.R.; Choi, Y.A.; Jin, M.; Choi, J.H.; Park, J.H.; Park, P.H.; Choi, H.; et al. Chrysin attenuates atopic dermatitis by suppressing inflammation of keratinocytes. *Food Chem. Toxicol.* **2017**, *110*, 142–150. [CrossRef]
38. Song, H.Y.; Kim, W.S.; Mushtaq, S.; Park, J.M.; Choi, S.H.; Cho, J.W.; Lim, S.T.; Byun, E.B. A novel chrysin derivative produced by γ irradiation attenuates 2,4-dinitrochlorobenzene-induced atopic dermatitis-like skin lesions in Balb/c mice. *Food Chem. Toxicol.* **2019**, *128*, 223–232. [CrossRef] [PubMed]
39. Yeo, H.; Lee, Y.H.; Koh, D.; Lim, Y.; Shin, S.Y. Chrysin inhibits NF-κB-dependent CCL5 Transcription by targeting IκB kinase in the atopic dermatitis-like inflammatory microenvironment. *Int. J. Mol. Sci.* **2020**, *21*, 7348. [CrossRef] [PubMed]
40. Turner, M.D.; Nedjai, B.; Hurst, T.; Pennington, D.J. Cytokines and chemokines: At the crossroads of cell signalling and inflammatory disease. *Biochim. Biophys. Acta* **2014**, *1843*, 2563–2582. [CrossRef] [PubMed]
41. Schottelius, A.J.; Zugel, U.; Docke, W.D.; Zollner, T.M.; Rose, L.; Mengel, A.; Buchmann, B.; Becker, A.; Grutz, G.; Naundorf, S.; et al. The role of mitogen-activated protein kinase-activated protein kinase 2 in the p38/TNF-α pathway of systemic and cutaneous inflammation. *J. Investig. Dermatol.* **2010**, *130*, 481–491. [CrossRef]
42. Yeo, H.; Ahn, S.S.; Lee, J.Y.; Jung, E.; Jeong, M.; Kang, G.S.; Ahn, S.; Lee, Y.; Koh, D.; Lee, Y.H.; et al. Disrupting the DNA binding of EGR-1 with a small-molecule inhibitor ameliorates 2,4-dinitrochlorobenzene-induced skin inflammation. *J. Investig. Dermatol.* **2021**, in press. [CrossRef]

43. Redhu, N.S.; Saleh, A.; Halayko, A.J.; Ali, A.S.; Gounni, A.S. Essential role of NF-κB and AP-1 transcription factors in TNF-α-induced TSLP expression in human airway smooth muscle cells. *Am. J. Physiol. Lung Cell. Mol. Physiol.* **2011**, *300*, L479–L485. [CrossRef] [PubMed]
44. Cultrone, A.; de Wouters, T.; Lakhdari, O.; Kelly, D.; Mulder, I.; Logan, E.; Lapaque, N.; Dore, J.; Blottiere, H.M. The NF-κB binding site located in the proximal region of the TSLP promoter is critical for TSLP modulation in human intestinal epithelial cells. *Eur. J. Immunol.* **2013**, *43*, 1053–1062. [CrossRef] [PubMed]
45. Ryu, W.I.; Lee, H.; Kim, J.H.; Bae, H.C.; Ryu, H.J.; Son, S.W. IL-33 induces Egr-1-dependent TSLP expression via the MAPK pathways in human keratinocytes. *Exp. Dermatol.* **2015**, *24*, 857–863. [CrossRef]
46. Kitagaki, H.; Fujisawa, S.; Watanabe, K.; Hayakawa, K.; Shiohara, T. Immediate-type hypersensitivity response followed by a late reaction is induced by repeated epicutaneous application of contact sensitizing agents in mice. *J. Investig. Dermatol.* **1995**, *105*, 749–755. [CrossRef]
47. Ando, T.; Matsumoto, K.; Namiranian, S.; Yamashita, H.; Glatthorn, H.; Kimura, M.; Dolan, B.R.; Lee, J.J.; Galli, S.J.; Kawakami, Y.; et al. Mast cells are required for full expression of allergen/SEB-induced skin inflammation. *J. Investig. Dermatol.* **2013**, *133*, 2695–2705. [CrossRef]
48. Sridharan, G.; Shankar, A.A. Toluidine blue: A review of its chemistry and clinical utility. *J. Oral Maxillofac. Pathol.* **2012**, *16*, 251–255. [CrossRef]
49. Utreras, E.; Futatsugi, A.; Rudrabhatla, P.; Keller, J.; Iadarola, M.J.; Pant, H.C.; Kulkarni, A.B. Tumor necrosis factor-α regulates cyclin-dependent kinase 5 activity during pain signaling through transcriptional activation of p35. *J. Biol. Chem.* **2009**, *284*, 2275–2284. [CrossRef]
50. Mishra, J.P.; Mishra, S.; Gee, K.; Kumar, A. Differential involvement of calmodulin-dependent protein kinase II-activated AP-1 and c-Jun N-terminal kinase-activated EGR-1 signaling pathways in tumor necrosis factor-α and lipopolysaccharide-induced CD44 expression in human monocytic cells. *J. Biol. Chem.* **2005**, *280*, 26825–26837. [CrossRef]
51. Son, S.W.; Min, B.W.; Lim, Y.; Lee, Y.H.; Shin, S.Y. Regulatory mechanism of TNFα autoregulation in HaCaT cells: The role of the transcription factor EGR-1. *Biochem. Biophys. Res. Commun.* **2008**, *374*, 777–782. [CrossRef]
52. Li, M.; Hener, P.; Zhang, Z.; Kato, S.; Metzger, D.; Chambon, P. Topical vitamin D3 and low-calcemic analogs induce thymic stromal lymphopoietin in mouse keratinocytes and trigger an atopic dermatitis. *Proc. Natl. Acad. Sci. USA* **2006**, *103*, 11736–11741. [CrossRef]
53. Pavletich, N.P.; Pabo, C.O. Zinc finger-DNA recognition: Crystal structure of a Zif268-DNA complex at 2.1 A. *Science* **1991**, *252*, 809–817. [CrossRef] [PubMed]
54. Milbrandt, J. A nerve growth factor-induced gene encodes a possible transcriptional regulatory factor. *Science* **1987**, *238*, 797–799. [CrossRef]
55. Gashler, A.; Sukhatme, V.P. Early growth response protein 1 (Egr-1): Prototype of a zinc-finger family of transcription factors. *Prog. Nucleic Acid Res. Mol. Biol.* **1995**, *50*, 191–224. [PubMed]
56. Duclot, F.; Kabbaj, M. The role of early growth response 1 (EGR1) in brain plasticity and neuropsychiatric disorders. *Front. Behav. Neurosci.* **2017**, *11*, 35. [CrossRef] [PubMed]
57. McMahon, S.B.; Monroe, J.G. The role of early growth response gene 1 (egr-1) in regulation of the immune response. *J. Leukoc. Biol.* **1996**, *60*, 159–166. [CrossRef] [PubMed]
58. Li, T.T.; Liu, M.R.; Pei, D.S. Friend or foe, the role of EGR-1 in cancer. *Med. Oncol.* **2019**, *37*, 7. [CrossRef]
59. Bhattacharyya, S.; Fang, F.; Tourtellotte, W.; Varga, J. Egr-1: New conductor for the tissue repair orchestra directs harmony (regeneration) or cacophony (fibrosis). *J. Pathol.* **2013**, *229*, 286–297. [CrossRef]
60. Bryant, M.; Drew, G.M.; Houston, P.; Hissey, P.; Campbell, C.J.; Braddock, M. Tissue repair with a therapeutic transcription factor. *Hum. Gene Ther.* **2000**, *11*, 2143–2158. [CrossRef]
61. Jeong, S.H.; Kim, H.J.; Jang, Y.; Ryu, W.I.; Lee, H.; Kim, J.H.; Bae, H.C.; Choi, J.E.; Kye, Y.C.; Son, S.W. Egr-1 is a key regulator of IL-17A-induced psoriasin upregulation in psoriasis. *Exp. Dermatol.* **2014**, *23*, 890–895. [CrossRef]
62. Yeo, H.; Ahn, S.S.; Lee, J.Y.; Shin, S.Y. EGR-1 acts as a transcriptional activator of KLK7 under IL-13 stimulation. *Biochem. Biophys. Res. Commun.* **2021**, *534*, 303–309. [CrossRef] [PubMed]
63. Sabio, G.; Davis, R.J. TNF and MAP kinase signalling pathways. *Semin. Immunol.* **2014**, *26*, 237–245. [CrossRef]
64. An, E.J.; Kim, Y.; Lee, S.H.; Choi, S.H.; Chung, W.S.; Jang, H.J. Ophiopogonin D ameliorates DNCB-induced atopic dermatitis-like lesions in BALB/c mice and TNF-α- inflamed HaCaT cell. *Biochem. Biophys. Res. Commun.* **2020**, *522*, 40–46. [CrossRef]

International Journal of *Molecular Sciences*

MDPI

Article

Active Components from *Cassia abbreviata* Prevent HIV-1 Entry by Distinct Mechanisms of Action

Yue Zheng [1], Xian-Wen Yang [1], Dominique Schols [2], Mattia Mori [3], Bruno Botta [4], Andy Chevigné [5], Martin Mulinge [5,6], André Steinmetz [1], Jean-Claude Schmit [5,7] and Carole Seguin-Devaux [5,*]

1 Laboratory of Cellular and Molecular Oncology, Luxembourg Institute of Health, L-1445 Luxembourg, Luxembourg; yue.zheng@path.utah.edu (Y.Z.); yangxianwen@tio.org.cn (X.-W.Y.); andre.steinmetz@lih.lu (A.S.)
2 Laboratory of Virology and Chemotherapy, Department of Microbiology, Immunology and Transplantation, Rega Institute for Medical Research, KU Leuven, 3000 Leuven, Belgium; dominique.schols@kuleuven.be
3 Department of Biotechnology, Chemistry and Pharmacy, University of Siena, 53100 Siena, Italy; mattia.mori@unisi.it
4 Department of Chemistry and Technology of Drugs, Sapienza University of Rome, 00185 Rome, Italy; bruno.botta@uniroma1.it
5 Department of Infection and Immunity, Luxembourg Institute of Health, L-4354 Esch-sur-Alzette, Luxembourg; andy.chevigne@lih.lu (A.C.); mmulinge@uonbi.ac.ke (M.M.); Jean-Claude.Schmit@ms.etat.lu (J.-C.S.)
6 Department of Biochemistry, School of Medicine, University of Nairobi, Nairobi, Kenya
7 Service National of Infectious Diseases, Centre Hospitalier de Luxembourg, L-1210 Luxembourg, Luxembourg
* Correspondence: carole.devaux@lih.lu

Citation: Zheng, Y.; Yang, X.-W.; Schols, D.; Mori, M.; Botta, B.; Chevigné, A.; Mulinge, M.; Steinmetz, A.; Schmit, J.-C.; Seguin-Devaux, C. Active Components from *Cassia abbreviata* Prevent HIV-1 Entry by Distinct Mechanisms of Action. *Int. J. Mol. Sci.* **2021**, *22*, 5052. https://doi.org/10.3390/ijms22095052

Academic Editors: Se-Kwon Kim

Received: 18 February 2021
Accepted: 4 May 2021
Published: 10 May 2021

Abstract: *Cassia abbreviata* is widely used in Sub-Saharan Africa for treating many diseases, including HIV-1 infection. We have recently described the chemical structures of 28 compounds isolated from an alcoholic crude extract of barks and roots of *C. abbreviata*, and showed that six bioactive compounds inhibit HIV-1 infection. In the present study, we demonstrate that the six compounds block HIV-1 entry into cells: oleanolic acid, palmitic acid, taxifolin, piceatannol, guibourtinidol-(4α→8)-epiafzelechin, and a novel compound named as cassiabrevone. We report, for the first time, that guibourtinidol-(4α→8)-epiafzelechin and cassiabrevone inhibit HIV-1 entry (IC_{50} of 42.47 μM and 30.96 μM, respectively), as well as that piceatannol interacts with cellular membranes. Piceatannol inhibits HIV-1 infection in a dual-chamber assay mimicking the female genital tract, as well as HSV infection, emphasizing its potential as a microbicide. Structure-activity relationships (SAR) showed that pharmacophoric groups of piceatannol are strictly required to inhibit HIV-1 entry. By a ligand-based in silico study, we speculated that piceatannol and norartocarpetin may have a very similar mechanism of action and efficacy because of the highly comparable pharmacophoric and 3D space, while guibourtinidol-(4α→8)-epiafzelechin and cassiabrevone may display a different mechanism. We finally show that cassiabrevone plays a major role of the crude extract of *CA* by blocking the binding activity of HIV-1 gp120 and CD4.

Keywords: natural products; *Cassia abbreviata*; HIV-1 entry; piceatannol; structure-activity relationship; pharmacophoric studies; norartocarpetin

1. Introduction

Human Immunodeficiency Virus (HIV) infection was affecting about 38 million people (http://www.unaids.org/ accessed on 18 February 2021) around the world in 2019. The vast majority of HIV-infected patients live in low- and middle-income countries, particularly in Sub-Saharan Africa. Combined antiretroviral therapy (cART) has largely improved the life of HIV-infected patients [1], and is the leading factor in reducing the number of new HIV-infected cases worldwide. Current cART is, nevertheless, facing many challenges for life-long adherence [2]; treatment-experienced patients encountered viral mutants resistant to multiple drugs, and women are more prone to HIV infection than men in Sub-Saharan Africa [3]. The development of new anti-HIV drugs and potent microbicides is, therefore, highly required.

Around 30 antiretroviral drugs have been released for clinical practice, inhibiting HIV-1 at different stages of the viral life cycle [4]. Novel therapeutic strategies could take advantage of the multiple events involved in the entry process [5]. HIV-1 attaches to the cell membrane, engaging its surface envelope glycoprotein gp120 to bind subsequently the CD4 receptor and either to the C-X-C chemokine receptor type 4 (CXCR4) or C-C chemokine receptor type 5 (CCR5) co-receptor, and further triggering conformational changes in HIV-1 envelope glycoproteins leading to the membrane fusion process [6]. New compounds targeting interactions with cellular proteins/membranes or viral membranes might be less prone to selecting for drug resistance, should reduce the formation of viral reservoirs, and could additionally display broad spectrum antiviral effects than compounds targeting viral replication [7].

In the past decades, a high number of drugs have been developed from natural products and showed their critical power in medical therapies, such as the anti-malarial drug artemisinin and the anticancer drug paclitaxel [8]. The 2015 half Nobel Prize award for artemisinin's discovery recently emphasized the enormous value of traditional medicine and ethnopharmacology. In this regard, *Cassia abbreviata* (*C. abbreviata*) is a tropical tree in *Cassia* genus, *Fabaceae* family, which is indigenous to South-East Africa, widespread in African countries and commonly used in the African local medicines [9]. Roots, barks, and leaves are taken as decoction or chewed, for healing abdominal pain, fever, cough, snake bite, malaria, infections, and, in particular, HIV-1 infection [10,11]. We have recently reported the isolation of 28 compounds from a crude extract of bark and roots of *C. abbreviata* [12]. Six bioactive compounds showed anti-HIV activity. Leteane et al. [13] has previously reported that a tanning free crude extract of *C. abbreviata* root inhibited HIV-1c (MJ4) replication in human peripheral blood mononuclear cells (PBMCs). It is known that a variety of compounds, including alkaloids, tannins, anthraquinones, and flavonoids, may contribute to the biological effect of plant medicines [10], while the active components of *C. abbreviata*'s, as well as their mode of action, are still unknown.

To better understand the mode of action of the traditional medicine plant *C. abbreviate*, we elucidated the antiviral activity of *C. abbreviata's* crude extract (pulverized from barks and roots and extracted with 95% ethanol) against several HIV-1 strains, and demonstrated that six compounds isolated from *C. abbreviata were* inhibiting HIV-1 entry, including a novel compound cassiabrevone. We showed here the different mode of action of the active compounds in both structure studies and biological tests. Cassiabrevone plays a major role in the CE of *C. abbreviata* by inhibiting the binding activity of gp120 and CD4. Piceatannol blocks HIV-1 entry using a different mechanism by targeting cell and viral membrane. Optimized synthetic derivatives from cassiabrevone and piceatannol could be used as a microbicide.

2. Results

2.1. The Crude Extract and Purified Compounds of C. abbreviata Inhibit HIV-1 Entry

The anti-HIV-1 activity of *C. abbreviata* was first assessed on MT4 cells and human peripheral blood mononuclear cells (PBMCs) from healthy donors using the HIV-1 reference strain IIIB (X4 tropic virus) and ADA-M (R5 tropic virus), as well as 2 non-B HIV-1 primary clinical isolates carrying several drug resistance mutations to nucleoside/nucleotide reverse transcriptase inhibitor (NRTI), to non-nucleoside reverse transcriptase inhibitor (NNRTI), and to protease inhibitors (PI). The crude extract (CE) inhibited HIV-1 infection in MT4 cells infected with the reference strain HIV-1-IIIB (X4 virus, IC50 = 21.75 ± 1.20 μg/mL) at non-toxic concentrations (CC50 above 1000 μg/mL). As shown in Figure 1A, CE inhibited HIV-1 infection in PBMCs with IC_{50} ranging from 10.47 to 40.77 μg/mL. The NNRTI efavirenz (EFV), the NRTI azidothymidine (AZT), and the fusion inhibitor enfuvirtide (T20) were used as positive controls. As expected, EFV and AZT did not inhibit viral infection of the respective clinical isolates which carried resistance mutations to NRTI and NNRTI, while T20 inhibited both HIV-1 reference strains and clinical isolates. We also examined the cytotoxicity of CE in PBMCs and found that CE was not toxic after 2 days treatment (Figure 1B). To further determine whether CE induces cell apoptosis, we measured the apoptotic cells by Annexin-V/PI staining after incubating CE and PBMCs for 48 h. As shown in Figure 1C, 30 μM H_2O_2, used as a positive control, induced 20% apoptosis, while CE did not induce any apoptosis as in non-treated cells. These data indicate that CE has an anti-HIV-1 activity without inducing any cytotoxicity or apoptosis.

A.

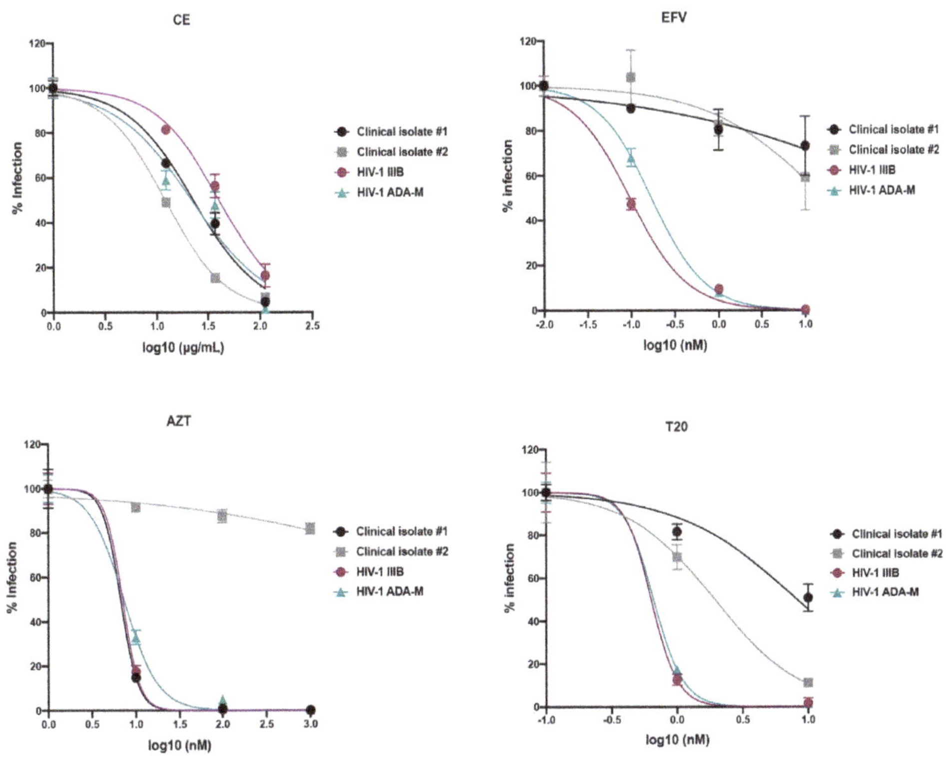

	IC_{50}			
	Clinical isolate #1	Clinical isolate #2	IIIB	ADA-M
CE (μg/mL)	23.06	10.47	40.77	13.53
EFV (nM)	> 10	> 10	0.10	0.17
AZT (nM)	6.70	> 1000	7.07	7.31
T20 (nM)	8.25	1.93	0.64	0.67

Figure 1. *Cont.*

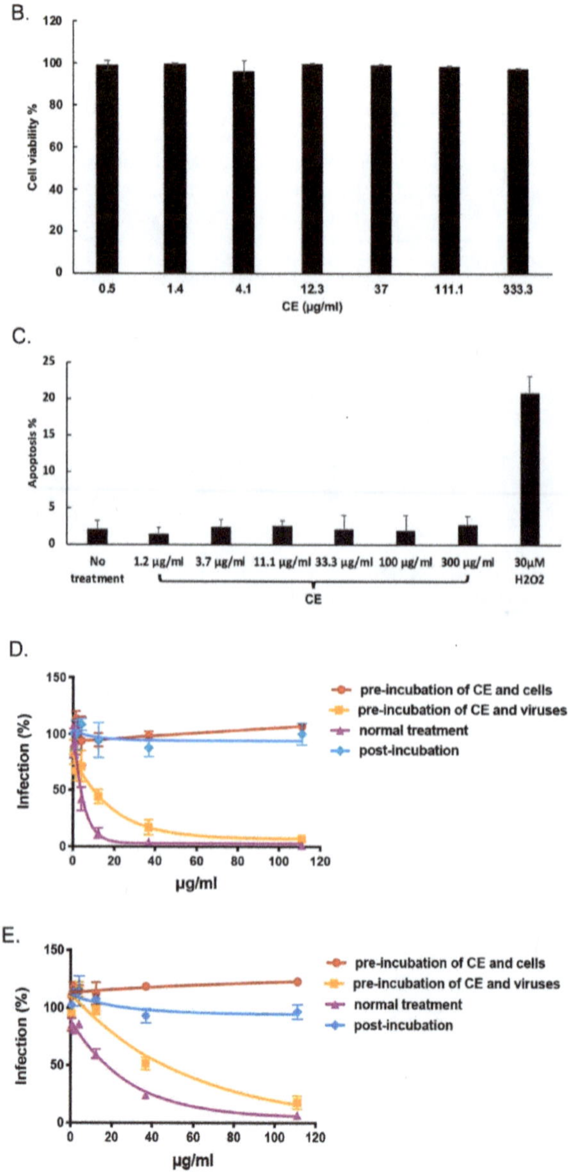

Figure 1. Crude extract (CE) of *C. abbreviata* inhibits HIV-1 entry into cells. (**A**) PBMCs isolated from healthy donor were treated with CE, or the NNRTI inhibitor efavirenz (EFV), the NRTI inhibitor azidothymidine (AZT) or the fusion inhibitor enfuvirtide (T20) for 7 days during infection. HIV-1 infection was assessed by measuring P24 in cell supernatants via ELISA. (**B**) PBMCs were treated with CE for 2 days. Cell viability was measured by flow cytometry. (**C**) PBMCs treated with CE were stained with Annexin-V/PI and measured by flow cytometry. Apoptosis level was calculated by counting both early apoptotic cells (Annexin-V+) and late apoptotic cells (PI+) (**D**,**E**). CE was tested in a multi-dosing time assay (**D**) using U373-CD4-CXCR4 cells against pseudotype virus pNL4.3Δ*Env*Luc-HXB2 and (**E**) using U373-CD4-CCR5 cells against pseudotype virus pNL4.3Δ*Env*Luc-BAL. CE was either pre-incubated with cells or viruses for 2 h, or incubated with cells and viruses during the 2 h spinoculation infection (normal treatment), or added to the cells after infection (post-incubation). HIV-1 infection was assessed by measuring luciferase in the cell supernatant. Three independent experiments were performed in triplicates.

To further characterize at which step of infection CE inhibited HIV-1, multi-dosing time assay experiments were performed with U373-CD4-CXCR4/CCR5 cells against infection of pseudotyped virus pNL4.3Δ*Env*Luc-HXB2 and pNL4.3Δ*Env*Luc-BAL.Pseudotyped viruses pNL4.3Δ*Env*Luc-HXB2/BAL are only able for one cycle of viral infection and allow us to assess the infection level by measuring the luciferase value in the cell supernatant. As shown in Figure 1D,E, the inhibitory effect of CE was apparent when CE was pre-incubated with the virus for 2 h cells (IC_{50} HXB2 = 2.57 μg/mL, IC_{50} BAL = 11.41 μg/mL) but not with the cells, as well as when CE was incubated with both cells and viruses during 2 h spinoculation infection (normal treatment) (IC_{50} HXB2 = 13.37 μg/mL, IC_{50} BAL = 67.40 μg/mL) but not after infection (CE was added during the first 2 h after the spinoculation). These results indicate that CE inhibits HIV-1 infection at an early stage of the HIV-1 infection independently of co-receptor usage

Since we identified 6 compounds with an anti-HIV-1 activity [12], the anti-HIV activities of isolated compounds from *C. abbreviata* were further assessed against pseudotype viruses pNL4.3Δ*Env*Luc-HXB2 when the compounds were added only at the time of infection to identify the compounds targeting HIV-1 entry. Four known anti-HIV compounds were characterized: oleanolic acid, palmitic acid, taxifolin, and piceatannol (Figure 2A). In our experiment, oleanolic acid, palmitic acid, taxifolin and piceatannol inhibited HIV-1 entry at non-cytotoxic concentration showing IC_{50} at 34.87 ± 9.09 μM, 87.48 ± 16.12 μM, 240 ± 3 μM, and 10.28 ± 5.74 μM, respectively (Figure 2B). Moreover, we identified two other flavonoids that prevent HIV-1 entry: guibourtinidol-(4α→8)-epiafzelechin (IC_{50} 42.47 ± 3.88 μM) and a novel compound we named as cassiabrevone (IC_{50} 30.96 ± 5.02 μM). We will elucidate their mode of action in the last part of this work. We first focused our study on piceatannol since it was never described as an HIV-1 entry inhibitor.

2.2. *Piceatannol Interacts with Both Cell and Viral Membranes and has a Synergistic Effect with HIV-1 Entry Inhibitors*

We first confirmed that piceatannol protected PBMCs against HIV-1 IIIB and HIV-1 ADA-M (IC_{50} = 24.22 ± 7.13 μM and IC_{50} = 19.91 ± 0.22 μM, respectively), and against two non-B HIV-1 primary clinical isolates harboring multi-drug resistance to NRTI, NNRTI, and PI (IC_{50} of 37.72 ± 12.54 and 8.04 ± 3.07 μM) (Figure 3A). We also examined the cytotoxicity and apoptosis in the PBMCs treated with piceatannol for 2 days by flow cytometry. As shown in Figure 3B,C, piceatannol was not cytotoxic and did not induce any apoptosis. We next deciphered whether piceatannol targets cells or viruses using the multi-dosing time assay (Figure 3D). In contrast to CE, piceatannol inhibited HIV-1 infection at high concentration when it was pre-incubated with cells but not pre-incubated with the virus, indicating that piceatannol's action is maximal when added together on both surface of cells and viruses. We did not find any effect of piceatannol when added after the step of infection in our pseudotyped assay although the integrase inhibitor raltegravir inhibited HIV-1 infection when added only after the time of infection.

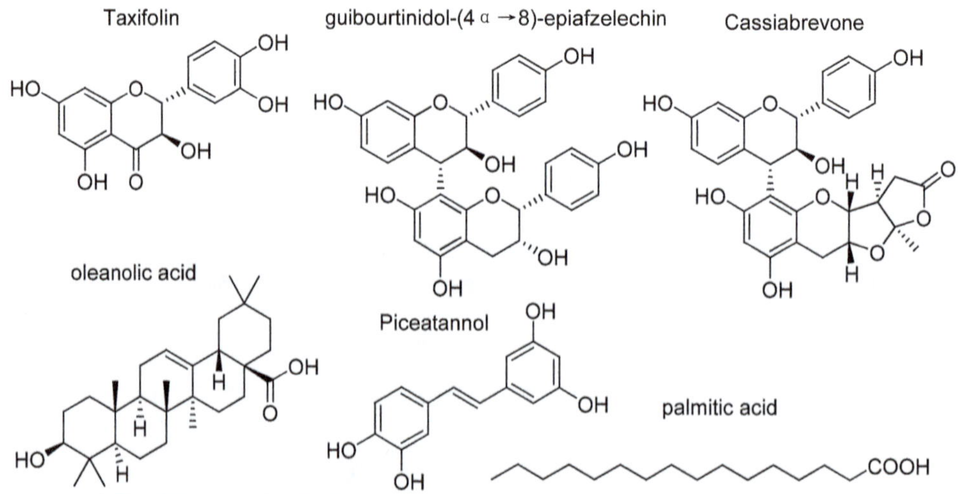

(**A**)

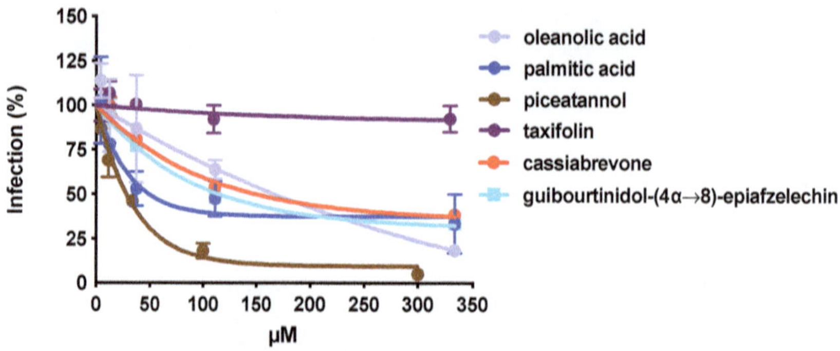

(**B**)

Figure 2. Anti-HIV-1 entry activity of oleanolic acid, palmitic acid, piceatannol, taxifolin, 4′7-dihydroxyflavan-(4,6)-3, 4′, 5,7-tetrahydroxyflavan and cassiabrevone. (**A**) Chemical structures of the compounds. (**B**) Oleanolic acid, palmitic acid, piceatannol, taxifolin, guibourtinidol-(4α→8)-epiafzelechin, and cassiabrevone were tested on U373-CD4-CXCR4 cells infected with pseudotype virus pNL4.3Δ*Env*Luc-HXB2 when added only at the time of infection. Percentage of infection versus the control cells infected without any compounds is represented.

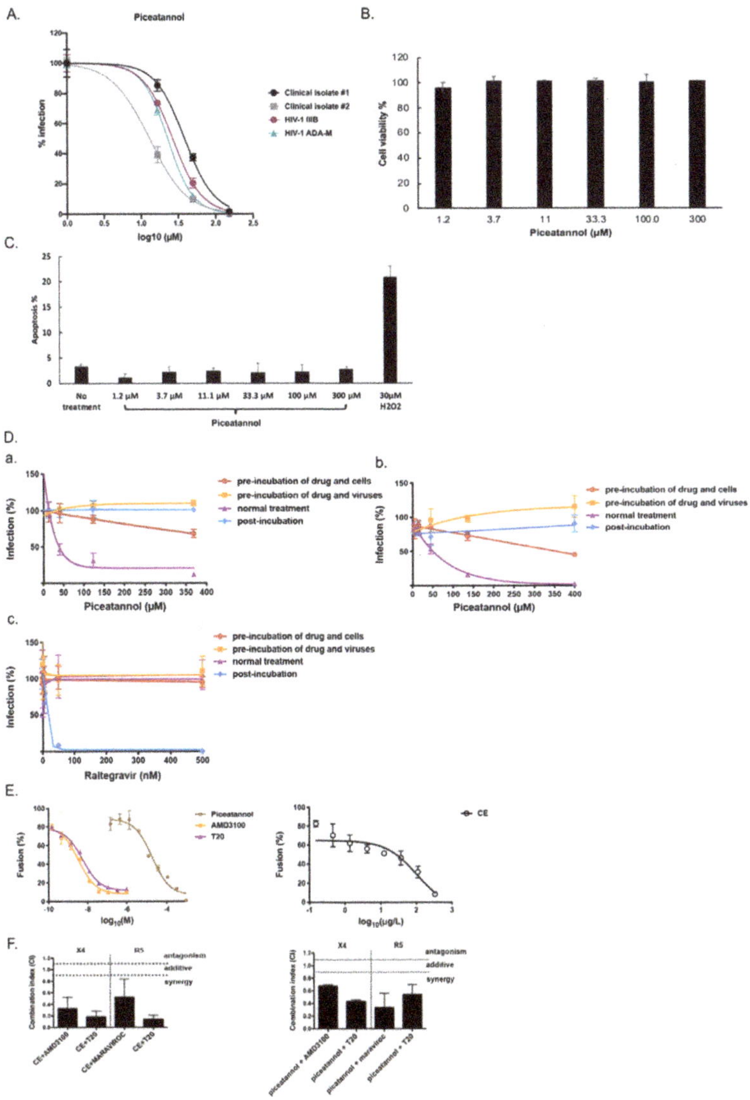

Figure 3. Piceatannol inhibits HIV-1 entry into cells. (**A**) PBMCs isolated from healthy donor were incubated with piceatannol during infection of HIV-1 reference strains and clinical isolates for 7 days. HIV-1 infection level was assessed by measuring P24 in cell supernatants via ELISA. (**B**) PBMCs were treated with piceatannol for 2 days. Cell viability was measured by flow cytometry. (**C**) PBMCs treated with piceatannol were stained with Annexin-V/PI and measured by flow cytometry. Apoptosis level was calculated by counting both early apoptotic cells (Annexin-V+) and late apoptotic cells (PI+). (**D**) Multi-dosing time assay with piceatannol was performed (a) using U373-CD4-CXCR4 cells against pseudotype virus pNL4.3Δ*Env*Luc-HXB2 and (b) using U373-CD4-CCR5 cells against pseudotype virus pNL4.3Δ*Env*Luc-BAL. (c) Raltegravir was tested in the multi-dosing time assay using U373-CD4-CXCR4 cells against pseudotype virus pNL4.3Δ*Env*Luc-HXB2. Percentage of infection relative to no drug treatment is represented. (**E**) Hela-ENV-Lai cells pretreated with piceatannol, CE, T20, or AMD3100 for 2 h were incubated with Hela-P4-CXCR4-LTRLacZ cells. HIV-1 fusion level was assessed by measuring β-Galactosidase activity. (**F**) Drug combination of CE or piceatannol and AMD3100/maraviroc/T20 was tested in U373-CD4-CXCR4/U373-CD4-CCR5 cells against pseudotype particles pNL4.3Δ*Env*Luc-HXB2/BAL. Combination Index (CI) at 95% maximal effective concentration (EC95) level was calculated using CompuSyn (ComboSyn, Paramus, NJ, USA).

To determine if CE and piceatannol block the HIV-1 fusion step, we treated Hela-ENV-Lai cells with CE, piceatannol, the CXCR4 inhibitor AMD100 or the fusion inhibitor T20 for 2 h and then co-cultured with Hela-P4-CXCR4LTRLacZ cells for 2 days. We can assess the effect of CE and piceatannol on the fusion process by measuring the β-galactosidase activity when the fusion step occurs between Hela-P4-CXCR4 cell and Hela-ENV-Lai cell. As shown in Figure 3E, both CE and piceatannol inhibit HIV-1 fusion efficiently with an IC_{50} of 22 μM and 33.35 μg/mL, respectively.

To further investigate if CE and piceatannol have a synergistic effect with known inhibitors of the fusion step, including the fusion inhibitor T20, the CXCR4 inhibitor AMD3100, or the CCR5 inhibitor Maraviroc, we treated U373-CD4-CXCR4/U373-CD4-CCR5 cells with single drug or combined drugs during the infection of pseudotyped viruses pNL4.3Δ*Env*Luc-HXB2/pNL4.3Δ*Env*Luc-BAL for 2 h. By calculating the combination index (CI) using the qualitative Chou and Talalay's method [14], both CE and piceatannol showed synergistic effect with the three known inhibitors T20, AMD100 and Maraviroc (CI < 0.9), which indicates that CE and piceatannol employ a different mechanism to prevent HIV-1 entry than the fusion inhibitor T20 or the CXCR4/CCR5 inhibitors.

2.3. CE of *C. abbreviata* Affects gp120/CD4 Binding Whereas Piceatannol Interacts with Cellular Membranes

To further examine whether CE and piceatannol affects the binding activity of HIV-1 gp120 on the host CD4 receptor, we performed an ELISA assay to analyze if CE or piceatannol could compete with gp120 to bind CD4. As shown in Figure 4A, CE blocked the binding between CD4 and gp120 in a dose-dependent manner, while piceatannol did not. We then further tested if CE and piceatannol inhibited CD4-gp120 binding by targeting CD4. The binding activity of two anti-CD4 monoclonal antibodies targeting the CD4 domain 1 and 3, in the presence or absence of CE and piceatannol was assessed. As shown in Figure 4B, the positive control soluble CD4 (sCD4) inhibited the two anti-CD4 antibodies binding to CD4, while neither CE nor piceatannol did. Thus, CE inhibits CD4-gp120 binding by targeting gp120.

We observed in Figures 1 and 3 that CE and piceatannol inhibited viral infection of both X4 and R5 viruses, suggesting no specific effect of CE or piceatannol on the co-receptors CXCR4 and CCR5. We next measured CXCR4/CCR5 binding with their respective chemokine in the presence or absence of CE and piceatannol. As shown in Figure 4C, the chemokines CXCL12 (C-X-C motif chemokine ligand 12) and CCL5 (C-C motif chemokine ligand 5) inhibited CXCR4/CCR5 binding, respectively, while neither CE nor piceatannol did.

In line with this evidence, piceatannol, but not CE, inhibited the infection of pseudotype particles of vesicular stomatitis virus (VSV) G proteins (IC_{50} = 79.23 ± 17.20 μM, Table 1), indicating that piceatannol inhibits HIV-1 entry independently of any receptors. To determine if CE or piceatannol had a broad-spectrum antiviral activity or was specific to HIV-1, we tested both on diverse viruses. CE and piceatannol demonstrated anti-viral activity on simplex virus infection HSV-1 and HSV-2 (IC_{50} around 45.0 μM) but not on influenza, para-influenza, HCV, coxsackie virus, RSV, reovirus, sindbisvirus, punta toro virus, yellow river virus, feline corona virus, and feline herpes virus (Table 1).

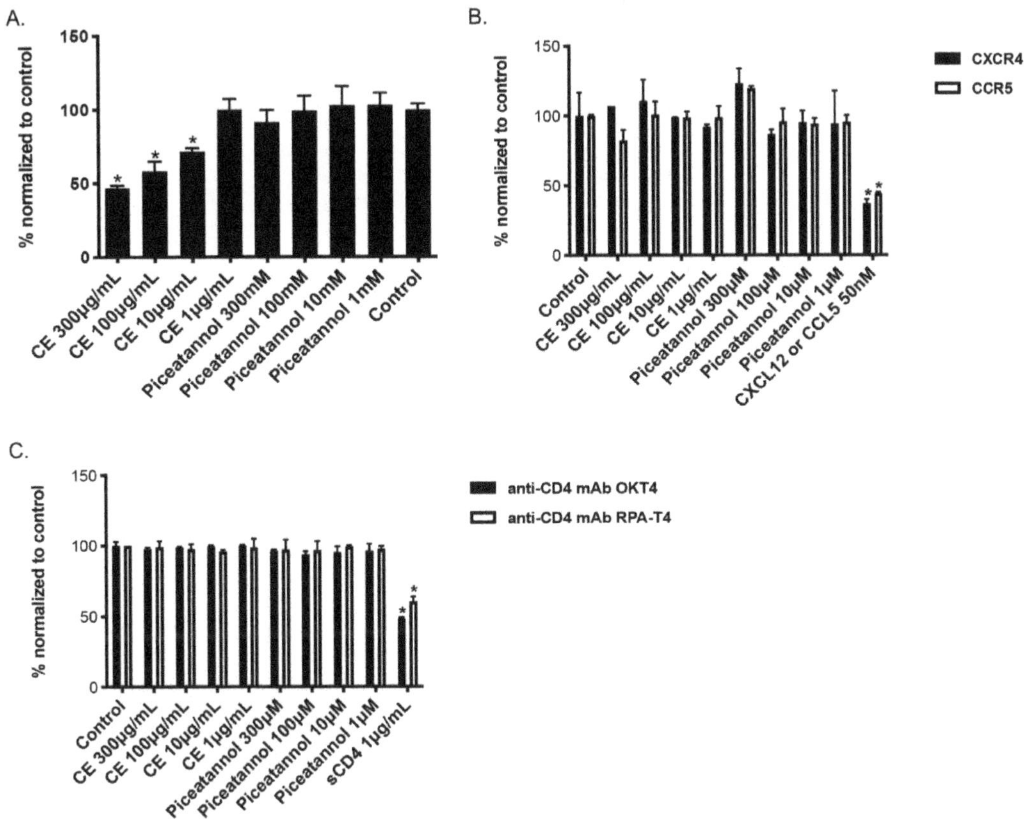

Figure 4. Binding activity of *C. abbreviata*'s crude extract and piceatannol. (**A**). CE and piceatannol were tested in an in-house ELISA assay against gp120 binding activity. Data were normalized to control cells without CE/piceatannol treatment. (**B**) U373-CD4-CXCR4/CCR5 cells were incubated with either anti-CD4 antibody OKT4 or RPA-T4 in the presence of CE and piceatannol. Soluble CD4 (sCD4) was used as a positive control for binding anti-CD4 antibodies. (**C**) U373-CD4-CXCR4/CCR5 cells were incubated with either anti-CXCR4 antibody 12G5 or anti-CCR5 antibody 2D7 in the presence of CE and piceatannol. The chemokines CXCL12 and CCL5 were used as positive controls for binding CXCR4 and CCR5, respectively. Data were normalized to the MFI of the control cells. Three independent assays were performed for each binding assay, and ANOVA analysis was done for statistical significance. * $p < 0.05$.

Table 1. Evaluation of CE and piceatannol's activity against various viruses.

Cells	Viruses	CE (μg/mL)	Piceatannol (μM)
HEL	Herpes simplex virus-1 (KOS)	46.7 ± 2.9	47.5 ± 3.5
HEL	Herpes simplex virus-2 (G)	39.5 ± 5.5	45.0 ± 1.8
HEL	Herpes simplex virus-1 TK KOS ACV	45.0 ± 2.6	45.4 ± 4.0
U87	Vesicular stomatitis virus	>100	79.2 ± 17
MDCK	Influenza A/H1N1 A/Ned/378/05	>100	>100
MDCK	Influenza A/H3N2 A/HK/7/87	>100	>100
MDCK	Influenza B B/Ned/537/05	>100	>100
Huh 7-D	Hepatitis C virus (Jc1)	>100	>100
HeLa	Coxsackie virus B4	>100	>100
vero	Coxsackie virus B4	>100	>100
HeLa	Respiratory syncytial virus	>100	>100
vero	Para-influenza-3 virus	>100	>100
vero	Reovirus-1	>100	>100
vero	sindbisvirus	>100	>100
vero	Punta toro virus	>100	>100
vero	Yellow fever virus	>100	>100
CRFK	Feline corona virus (FIPV)	>100	>100
CRFK	Feline herpes virus	>100	>100
HEL	Human corona virus	>100	>100

IC_{50} (inhibitory concentration of viral replication by 50%), CE, crude extract of *C. abbreviata*.

2.4. Microbicide Activity of Piceatannol

Since both CE and piceatannol were active against HIV-1, HSV-1, and HSV-2, we further checked its potential as a microbicide. We next tested piceatannol in a dual-chamber system mimicking the epithelium female genital tract (Figure 5A). Both CE and piceatannol inhibited HIV-1 infection of TZM-Bl cells without affecting the confluence layer's TEER and epithelial cells viability. At their IC_{50} concentration, CE decreased by more than 4 times HIV-1 infection (from 100 to 18.41%), while piceatannol diminished it by 33% (from 100 to 66.9%). For potential microbicide application, anti-HIV compounds should not stimulate target cells. PHA increased the expression of CD25 (early activation marker) and CD69 (late activation marker), whereas CE and piceatannol did not have any effect (Figure 5B). Finally, CE and piceatannol did not prevent transmission of DC-SIGN (Dendritic Cell-Specific Intercellular adhesion molecule-3-Grabbing Non-integrin)-captured virus to $CD4^{+}$ T cells in vitro (data not shown).

2.5. Structure-Activity Relationship of Piceatannol

Piceatannol emerged from this study as a promising anti-HIV-1 compounds extracted from *C. abbreviata*. Since molecular details of piceatannol interaction with its target(s) are currently not available, we performed a structure-activity relationship (SAR) study with the aim to understand the chemical and pharmacophoric features of piceatannol that are relevant for its anti-HIV activity. Seventeen chemical analogues of piceatannol (molecules **1–17**, Figure 6) were designed and retrieved from commercial vendors and from an in-house library of natural products and their derivatives [15–19] and tested in vitro against HIV-1 entry and HIV-1 infection using the pseudotype and the MTT assay, respectively. Compounds **1–13** are stilbene derivatives endowed with subtle chemical modifications with respect to piceatannol (i.e., molecules **4** and **5**) or bearing larger substituents (i.e., **6–11** and longistilines **12** and **13**) showing no anti-HIV activity. Besides, the two flavones hispidulin **14** and norartocarpetin **15**, and the two styrene derivatives **16** and methyl ferulate **17** were tested. Only compounds that do not bear the stilbene scaffold showed some activity: methyl ferulate **17** at high concentrations (IC_{50} = 600 ± 22 μM) and norartocarpetin **15** (IC_{50} = 39 ±5 μM, CC_{50} > 500 μM), the latest showing antiretroviral activity similar to piceatannol. Norartocarpetin is structurally related to taxifolin, although it is noticeably more potent. This result clearly indicates that pharmacophoric features of piceatannol are highly specific, and that subtle modifications to its chemical structure determine a significant drop of anti-HIV-1 activity. Accordingly, it is not surprising that stilbene derivatives bearing larger substitutions, such as molecules **1**, **2**, **6–11**, and longistilines **12** and **13** proved inactive, as well. Similar to piceatannol, the comparison between structure and activity of **14** and **15** further highlights the key role of polyphenols in providing anti-HIV activity.

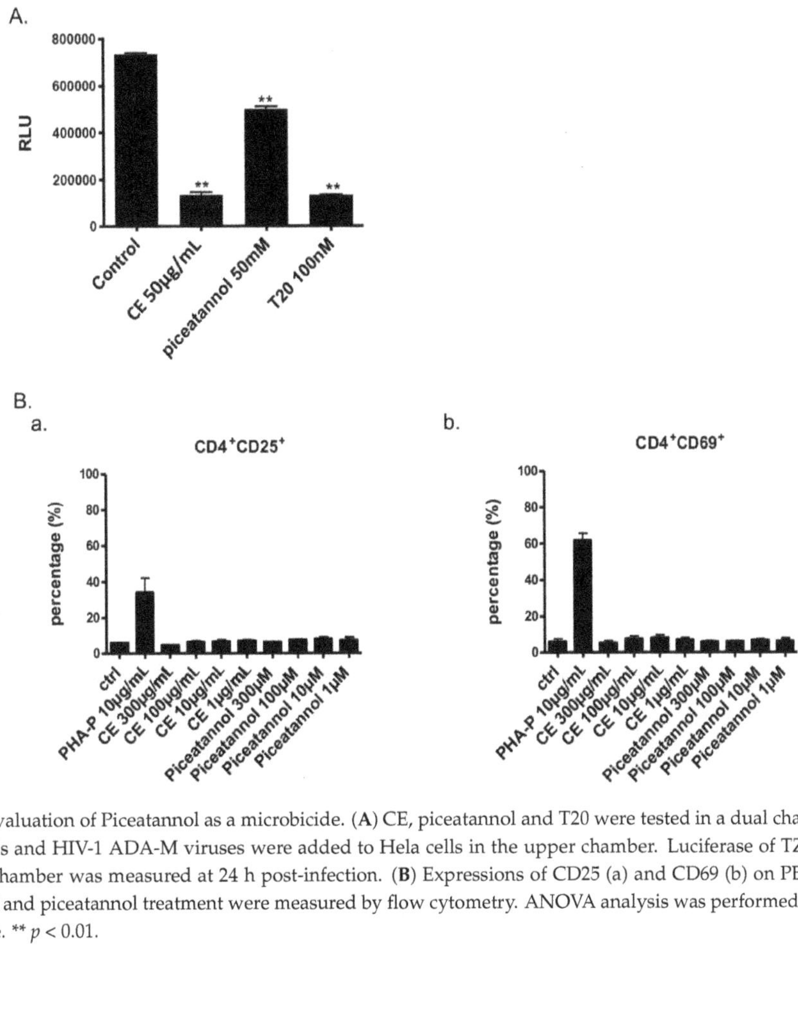

Figure 5. Evaluation of Piceatannol as a microbicide. (**A**) CE, piceatannol and T20 were tested in a dual chamber system. Compounds and HIV-1 ADA-M viruses were added to Hela cells in the upper chamber. Luciferase of TZM-bl cells in the lower chamber was measured at 24 h post-infection. (**B**) Expressions of CD25 (a) and CD69 (b) on PBMCs with or without CE and piceatannol treatment were measured by flow cytometry. ANOVA analysis was performed for statistical significance. ** $p < 0.01$.

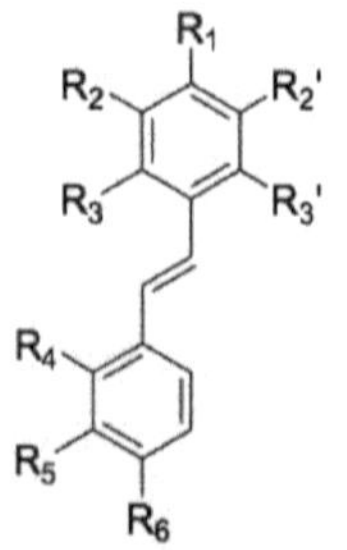

1–13

mol	R_1	R_2	R_2'	R_3	R_3'	R_4	R_5	R_6
1	H	OCH_3	OCH_3	H	H	H	H	OH
2	H	$OCOCH_3$	$OCOCH_3$	H	H	H	H	$OCOCH_3$
3	H	OH	OH	H	H	H	H	OCH_3
4	H	OH	OH	H	H	OH	H	OH
5	H	OH	OH	H	H	H	OCH_3	OH
6	H	OCH_3	OCH_3	H	H	H	OCH_3	OCH_3
7	H	OCH_3	H	H	H	H	H	OH
8	OH	OCH_3	H	H	H	H	H	H
9	H	OH	H	H	H	H	H	H
10	OCH_3	OCH_3	H	H	H	H	OCH_3	OCH_3
11	H	OCH_3	OH	H	H	H	H	OH
12	H	OH	OH	prenyl	prenyl	H	H	H
13	H	OH	OCH_3	H	prenyl	H	H	H

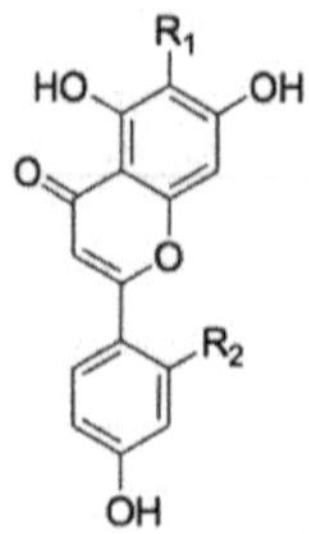

14 (Hispidulin): R_1 = OCH3; R_2 = H
15 (Norartocarpetin): R_1 = H; R_2 = OH
IC50 = 39 ± 5 μM

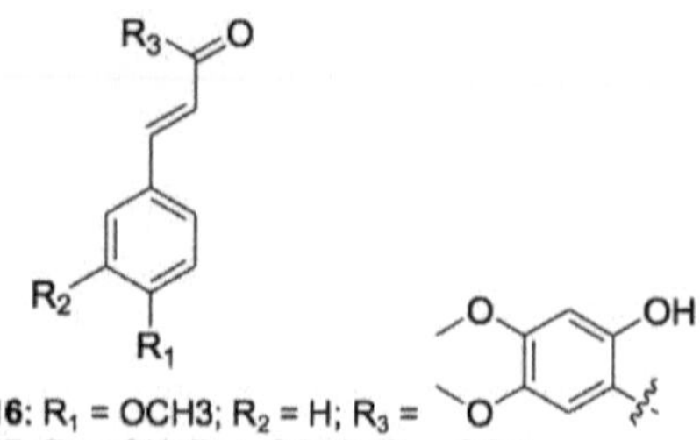

16: R_1 = OCH3; R_2 = H; R_3 =
17: R_1 = OH; R_2 = OCH3; R_3 = OCH_3
IC50 = 600 ± 22 μM

Figure 6. SAR of piceatannol. The chemical structure of compounds **1–17** used to afford SAR of piceatannol is shown. Molecules are grouped based on their chemical structures: (i) stilbene derivatives **1–13**; (ii) flavones **14** and **15**; (iii) styrene derivatives **16** and **17**. IC_{50} values of the compounds are mentioned when the compound displayed anti-HIV-1 activity.

2.6. Mode of Action of the Two Flavonoids Inhibiting HIV-1 Entry

To develop the mode of action, we first performed a ligand-based studies to provide insights into the mechanism of action of piceatannol, norartocarpetin, cassiabrevone, and 4′7-Dihydroxyflavan (4,6)-3, 4′, 5,7- tetrahydroxyflavan using the Rapid Overlay of Chemical Structures (ROCS) program (OpenEye Scientific Software). As shown in Figure 7A, a normalized similarity of 0.63 was obtained by comparing piceatannol to norartocarpetin, suggesting that these compounds cover a highly similar pharmacophoric and 3D space. In contrast, 4′7-dihydroxyflavan (4,6)-3, 4′, 5,7- tetrahydroxyflavan and cassiabrevone share a moderate similarity with piceatannol, (normalized similarity = 0.37 and 0.23, respectively, Figure 7B–C), which suggests a different mechanism of action for these molecules compared to piceatannol and norartocarpetin.

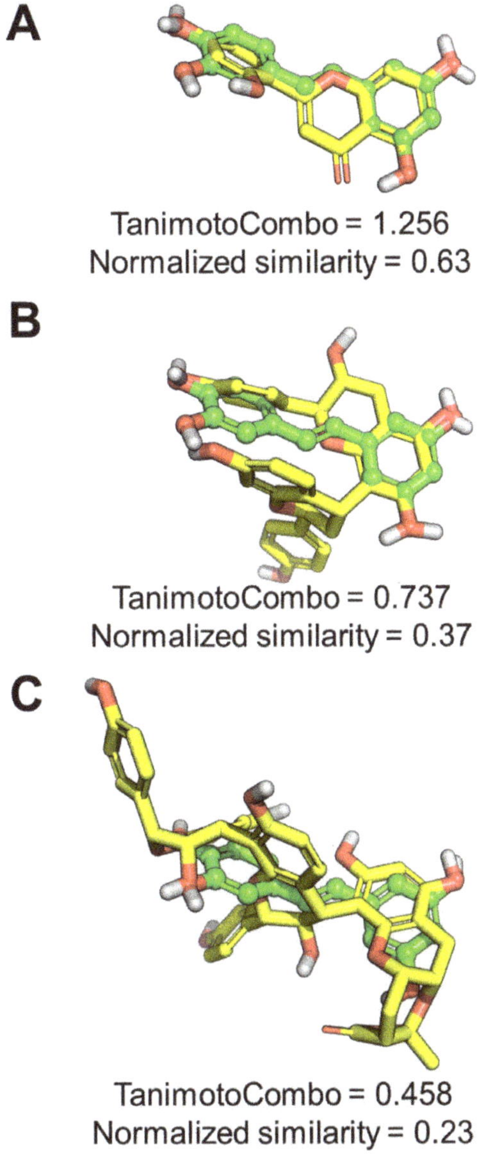

Figure 7. Pharmacophoric studies. (**A**) ROCS alignment between the query built on piceatannol and norartocarpetin, 4′7-Dihydroxyflavan (4,6)-3, 4′, 5,7- tetrahydroxyflavan (**B**), and cassiabrevone (**C**). Piceatannol is shown as green balls and sticks with the same orientation in the three panels. Aligned ligands are shown as yellow sticks. Non-polar H atoms were omitted. The TanimotoCombo and normalized similarity scores are reported for each pair of compounds.

For further validation, we then tested cassiabrevone and guibourtinidol-(4α→8)-epiafzelechin in a fusion assay and showed that both two compounds inhibited viral fusion similarly to CE (Figure 8A). Moreover, cassiabrevone but not guibourtinidol-(4α→8)-epiafzelechin inhibited significantly gp120/CD4 binding activity in a dose-dependent manner as observed for CE (Figure 8B), indicating that cassiabrevone may play the major role of the crude extract of *C. abbreviata* and display different mode of action from piceatannol (Figure 3D).

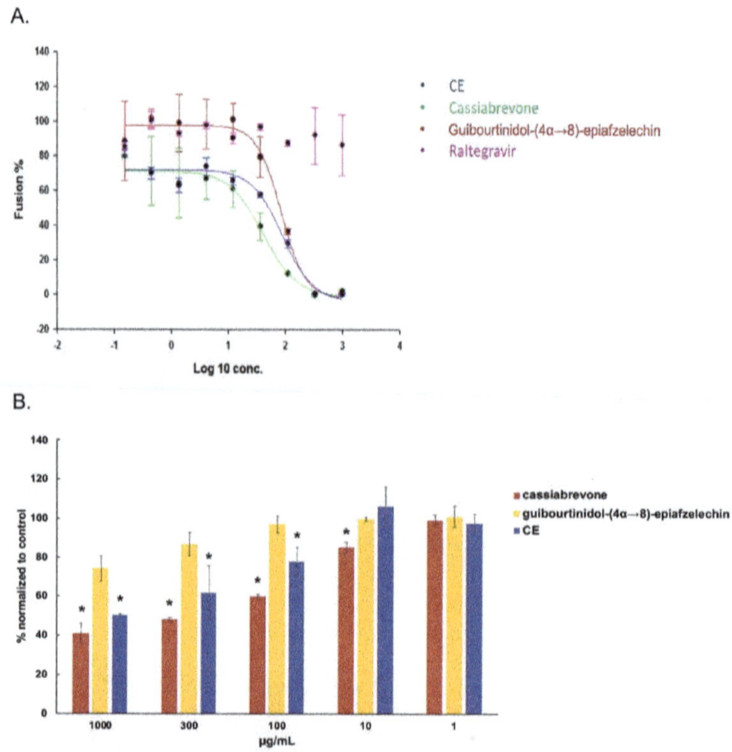

Figure 8. Cassiabrevone inhibits gp120/CD4 interaction. (**A**) Hela-P4-CXCR4 cells pretreated with CE, guibourtinidol-(4α→8)-epiafzelechin), and cassiabrevone were incubated with hela-ENV-Lai cells. β-Galactosidase activity was accessed. Data were normalized to control cells without any drugs to give the percentage of fusion. (**B**) Cassiabrevone and guibourtinidol-(4α→8)- -epiafzelechin were tested in an in-house ELISA assay against gp120 binding activity. Data were normalized to control cells without compound treatment. ANOVA analysis was performed for statistical significance. * $p < 0.05$.

Finally, taken together, our studies confirmed that the traditional medicine plant *C. abbreviata* displayed anti-HIV-1 activity and identified six active components that inhibits HIV-1 entry. Importantly, we revealed a novel compound cassiabrevone and showed that cassiabrevone played the major role of the crude extract of *C. abbreviata*. Moreover, we delineated the anti-HIV mechanism of piceatannol and showed its potential as a microbicide.

3. Discussion

Many antiviral agents have been identified from plant sources but their mechanisms of action are poorly understood. *C. abbreviata* is indigenous to Africa, and commonly used against infectious diseases [9,11]. Nevertheless, *C. abbreviata*'s bioactive compounds were not characterized. The anti-HIV-1 activity of a *C. abbreviata*'s ethanol extract was first described by Leteane's group [13]. In our study, we have performed a secondary ethyl acetate partition after a primary ethanol extraction to obtain a crude extract (CE) with enhanced anti-HIV-1 activity and lower cytotoxicity. We sought here (i) to determine at which step of infection the CE of *C. abbreviata* inhibited HIV-1, (ii) to screen active compounds against HIV-1 entry, and (iii) to unravel the mechanism of action of isolated compounds preventing HIV-1 entry.

Our results indicate that CE inhibited HIV-1 infection at an early stage of the HIV-1 entry process independent of co-receptor usage and interacted with HIV-1 gp120. By testing compounds only at the time of infection against pseudotyped viruses, we found that oleanolic acid, palmitic acid, taxifolin, and piceatannol inhibited HIV-1 entry at the micromolar range for non-cytotoxic concentrations. These results are consistent with previous studies showing that palmitic acid and oleanolic acid blocked gp120-CD4 interaction [20,21] and that oleanolic acid, piceatannol and taxifolin inhibited HIV-1 infection [22–25]. According to our findings, the major active components of CE inhibited the interaction between gp120 and CD4, and, especially, cassiabrevone, a flavonoid with a new structure purified from CE, reproduced this feature. CD4 and co-receptor binding sites, as well as variable loops and glycans of gp120, were already proposed as therapeutic targets. For instance, several small-molecule attachment inhibitors targeting the conserved CD4 binding region within gp120 have been described [26–29]. Fostemsavir, a prodrug of the HIV-1 attachment inhibitor temsavir, was recently approved in clinic for heavily treatment-experienced HIV-1 patients. Using homology models, fostemsavir was proposed to bind to the unliganded conformation of gp120 within the structurally conserved outer domain adjacent to the CD4 binding loop. It is tempting to speculate that other compounds derived from natural products may display similar activity on gp120 conformation or binding with CD4. In our hands, cassiabrevone may display such potential and might explain the main effects of *C. abbreviata* against HIV-1 entry.

In this study, we have shown, for the first time, that piceatannol inhibits HIV-1 entry and focused our work to explain its mode of action, as well as its potency as microbicide and against other viruses. Clouser et al. have reported that piceatannol inhibited HIV-1 replication with an IC_{50} of 21.4 μM in accordance with our results [24]. In our hands, time dose-dependent experiments and fusion assays did not support piceatannol as an HIV-1 integrase inhibitor as previously reported in non-cell-based assays [25,30]. We did not find any effect of piceatannol when added after the step of infection in our pseudotype assay although the integrase inhibitor raltegravir inhibited HIV-1 infection when added only after the time of infection (Figure 3). Taking into consideration that piceatannol did not affect neither CD4 and CCR5/CXCR4 binding, nor the interaction between CD4 and gp120, as well as the fusion process, we assume that piceatannol may interact with virus attachment by adsorbing at either the cell surface or the virus surface. Piceatannol also showed a synergistic effect with both co-receptor and fusion inhibitors, which indicates that piceatannol employs a different mechanism to prevent HIV entry. This hypothesis is reinforced by the observation that piceatannol inhibited VSV infection since VSV enters into target cells by endocytosis and not by interfering with a specific cellular receptor. From a chemical standpoint, piceatannol (trans-3,3′,4,5′-tetrahydroxystilbene) is a natural analogue of resveratrol (trans-3,4′,5′-trihydroxystilbene). Many analogues of resveratrol have been reported, as well, to inhibit HIV-1 infection [31], and trans-3,3′,4,4′,5,5′-hexahydroxystilbene was shown recently to inhibit HIV-1 entry before the fusion step [32]. Both piceatannol and 3,3′,4,4′,5,5′-hexahydroxy-trans-stilbene display better anti-HIV activity than resveratrol, suggesting that the additional hydroxyl groups to the basic stilbene rings may increase the anti-HIV activity by strengthening compound affinity or by facilitating its action on the membrane surface. Furthermore curcumin, having two phenols connected by a carbon chain, has structural similarity to piceatannol and can affect viral membrane fluidity to block viral entry [33] similarly to rigid amphipathic fusion inhibitors (RAFIs) and LJ001, which affects the membrane of HIV [7,34]. Interestingly, resveratrol was also shown to inhibit simplex virus vaginal infection in a mouse model [35].

The SAR and ligand-based studies highlighted the highly specific chemical and pharmacophoric features of piceatannol that are required to exert anti-HIV-1 activity. Moreover, the flavone norartocarpetin was identified as a conformational restrained analogue of piceatannol endowed with comparable efficacy. Piceatannol and norartocarpetin share a number of pharmacophoric and shape similarities, indicating a common mechanism of action. In contrast, 4′7-dihydroxyflavan (4,6)-3, 4´, 5,7- tetrahydroxyflavan and cassiabrevone share only a moderate similarity with piceatannol, which suggests that these molecules might inhibit HIV-1 replication through a mechanism that is different from that observed for piceatannol itself and norartocarpetin. The SAR study also emphasized the key role of polyphenols in providing anti-HIV activity, in agreement with recent reports [32,36,37]. One may note that piceatannol has a catechol moiety, widespread in natural products, which is considered as a protein-reactive species [38,39]. However, the anti-HIV-1 effect of piceatannol was found in this work to be highly specific by targeting only the entry step of the envelope virus HIV, HSV, and VSV without significant cytotoxicity.

Importantly, we have shown that both CE and piceatannol inhibited HIV infection in an in vitro dual-chamber model, mimicking the epithelium of the female genital tract [35], suggesting that the active components can cross the epithelial barrier without any toxicity on epithelial cells and no further activation of PBMCs. These data indicate the potential of piceatannol or cassiabrevone to be used as a lead structure for microbicides, although it did not prevent cell-to-cell and DC-SIGN-mediated viral transmission in vitro. In addition, the dual HIV/HSV activity of piceatannol could be crucial for microbicide applications since genital HSV-2 infection has been shown to potentiate HIV transmission and infection [40].

4. Conclusions

There is still an ongoing need for new potent classes of antiretroviral drugs with improved safety and tolerability profiles to sustain long-life antiretroviral therapy. In the present work, we have shown that 6 active components block HIV-1 entry. Importantly, we have isolated a novel flavonoid named as cassiabrevone and identified cassiabrevone as one of the active components from *C. abbreviata* that prevents HIV-1 entry by targeting gp120. Similar to piceatannol, cassiabrevone might have the potential to be used as a microbicide. Optimized synthetic derivatives from cassiabrevone and/or piceatannol should be resolved to reach a future therapeutic efficacy in humans.

5. Materials and Methods

5.1. Plant Extraction and Compounds Purification

Barks and roots of *C. abbreviata* were collected from mature shrubs in Makueni County, Kenya. Its identity was confirmed by DNA barcoding approach in the lab. All materials were pulverized before extraction. The crude extract (CE) was obtained through a first extraction of barks and roots from *C. abbreviata* that were pulverized with 95% ethanol, and a second extraction with ethyl acetate, and dried. The extracts were combined and concentrated to a small volume to provide a crude extract. The concentrate of the ethanol phase was suspended in deionized water, successively partitioned with $CHCl_3$, EtOAc, and *n*-BuOH and subjected to column chromatography over silica gel. The $CHCl_3$ and EtOAc extracts were combined and subjected to column chromatography over silica gel eluting with a gradient $CHCl_3$-MeOH (0- > 100%), followed by column chromatography over octadecyl silane and sephadex LH-20. Preparative thin layer chromatography was used to purify the compounds, and purity over 95% was verified by HPLC-UV. To characterize the compounds, UV and NMR data were collected from UV-2550 spectrometer and Bruker Avance 500 or 600 NMR spectrometers.

5.2. Cell Cultures

MT-4, U373-CD4-CXCR4, U373-CD4-CCR5, HeLa and TZM-Bl cell lines were obtained through the NIH AIDS Reagent Program. HEK 293, Vero, MDCK, and CRFK cells were purchased from ATCC (Manassas, VA, USA). HeLa-P4-CXCR4-LTRLacZ and Hela-ENV-Lai cells [26] were kindly given by Dr. Marc Alizon, Institute Pasteur, Paris. MT-4 cells were cultured in RPMI 1640 (Lonza, Wijchen, The Netherlands) containing 10% heat-inactivated fetal bovine serum (FBS) (Lonza, Netherlands) and 2mM L-glutamine (Invitrogen, Gosselies, Belgium). U373-CD4-CXCR4 and U373-CD4-CCR5 were cultured in RPMI 1640 containing 10% FBS, 2mM L-glutamine, 200 µg/mL geneticin (Invitrogen, Belgium), and 100 µg/mL hygromicin B (Invitrogen). HEK 293, Vero, Hela, and TZM-Bl were cultured in dulbecco's modified eagle medium (DMEM) (Lonza) containing 10% FBS and 2mM L-glutamine. MDCK and CRFK cells were in eagle's minimum essential medium (Lonza) containing 10% FBS. HeLa-P4-CXCR4-LTRLacZ cells were cultured in DMEM containing 10% FBS, 2mM L-glutamine, and 500 µg/mL geneticin. Hela-ENV-Lai cells were cultured in DMEM containing 10% FBS, 2mM L-glutamine, and 2 µM methotrexate (Sigma-Aldrich, Liège, Belgium).

5.3. Viral Infection with HIV-1 Reference Strains and Clinical Isolates

Peripheral blood mononuclear cells (PBMCs) were isolated from healthy donors' buffy coats (Red Cross of Luxembourg, Luxembourg, Luxembourg) using Ficoll-Hypaque gradient as indicated previously (Sigma-Aldrich, Liège, Belgium). PBMCs were stimulated using 10 µg/mL phytohemagglutinin (PHA-P, Sigma Aldrich) for 48 h and recombinant IL-2 (10 U/mL, Roche, Sigma-Aldrich, Liège, Belgium) for another 24 h. Stimulated PBMCs were infected by the HIV-1 reference strains IIIB/ADA-M or primary clinical isolates expanded in culture from anonymized left-over samples (Centre Hospitalier de Luxembourg, Luxembourg, Luxembourg) in the presence or absence of drugs replaced every other day during 7 days. P24 production was measured in supernatants by ELISA (Perkin Elmer, Brussels, Belgium). Efavirenz (EFV) and azidothymidine (AZT) were obtained from Sigma_Aldrich. Enfuvirtide (T20) was purchased from Eurogentec (Seraing, Belgium).

5.4. Cytotoxicity and Apoptosis Assays

To assess drug cytotoxicity, PBMCs were incubated with or without drugs in a 96-well round bottom plate (Thermo Fisher, Asse, Belgium) at 2×10^5/200 µL/well. After 2 days, cells were washed with PBS and stained with 0.1 µL near-IR fluorescent reactive dye (Life technologies, Ghent, Belgium) in 100 µL PBS for 20 min at room temperature in dark. Stained cells were measured by flow cytometry (FACSCanto, BD biosciences, Belgium).

To measure apoptosis, PBMCs were incubated with or without drugs in a 96-well round bottom plate at 2×10^5/200 µL/well. After 2 days, cells were washed with PBS and stained with 2.5 µL Annexin-V-APC (Biosciences, Aalst, Belgium) in 100 µL staining buffer containing 10 mmol/L HEPES pH7.4, 140 mmol/L NaCl, and 2.5 mmol/L $CaCl_2$ for 20 min at room temperature in dark. Cells were then washed with staining buffer and incubated with 100 µL 0.1 µg/mL propidium iodide (PI) (Thermo Fisher, Asse, Belgium) for 5 min at room temperature in dark. Stained cells were measured by flow cytometry (FACSCanto, BD, Aalst, Belgium).

5.5. Multi-Dosing Time Assay and Drug Combination Assay

For normal treatment, U373-CD4-CXCR4/U373-CD4-CCR5 cells were infected by pseudotyped virus pNL4.3Δ*Env*Luc-HXB2/pNL4.3Δ*Env*Luc-BAL [41], respectively, in the presence or absence of drugs through 2 h spinoculation at 1200× *g*. After infection, cells were then cultured in fresh culture medium for 48 h. Luciferase activity in cell lysates expressed as relative light units was measured via Luciferase System kit (Promega, Leiden, Netherlands).

In multi-dosing time assay, four different treatments were performed: 2 h pre-incubation of drugs on U373-CD4-CXCR4/U373-CD4-CCR5 cells, 2 h pre-incubation of drugs on pseudotyped virus pNL4.3Δ*Env*Luc-HXB2/pNL4.3Δ*Env*Luc-BAL, 2 h co-incubation of drugs, cells and viruses during spinoculation (normal treatment), and 2 h incubation of drugs on infected cells after spinoculation (post-infection).

In drug combination assay, U373-CD4-CXCR4/U373-CD4-CCR5 cells were co-incubated with CE or piceatannol combined with the CXCR4 inhibitor AMD3100 (Sigma, Belgium) or with the CCR5 inhibitor Maraviroc (Sigma, Belgium) or with the fusion inhibitor T20 during infection with the pseudotyped viruses by spinoculation as described above (normal treatment). The combination index (CI) was calculated at the EC_{95}-level using CompuSyn software (ComboSyn, USA). According to Chou and Talalay's method, CI > 1.1 means antagonism, CI < 0.9 means synergy, and 0.9 < CI < 1.1 means additive effect of the drugs.

5.6. Fusion Assay

Fusion inhibition was evaluated as previously described [26]. Briefly, HeLa-ENV-Lai cells were pre-incubated with or without drugs for 2 h and added to HeLa-P4-CXCR4-LTRLacZ cells which were placed in a 96-well plate 1 day before. After 24 h, cells were washed with PBS and then incubated with 50 μL containing 0.5% NP-40 (Sigma, Belgium) for 15 min at room temperature. Fifty microliters chlorophenol red-β-D-galactopyranoside (CPRG) reagent (Roche, Sigma-Aldrich, Liège, Belgium) was then added into the plate. After 30 min incubation at room temperature in dark, β−galactosidase was assessed by measuring OD_{562} of the cells via POLARstar Omega plate reader (BMG Labtech, Ortenberg, Germany).

5.7. Broad Spectrum Antiviral Activity

Antiviral assays were previously described [27] for the following viruses: HSV-1 (KOS), HSV-2 G, HSV-1 TK KOS ACV; VSV, coxsackie virus B4, RSV, Coxsackie virus B4, parainfluenza-3 virus, reovirus-1, sindbisvirus, punta toro virus, yellow fever virus, influenza A/H1N1 A/Ned/378/05, influenza A/H3N2 A/HK/7/87, influenza B B/Ned/537/05; human corona virus, feline corona virus, feline herpes virus, and HCV (Jc1).

5.8. CD4-gp120 Interaction Assay

Human soluble CD4 (sCD4) was immobilized in a 96-well plate by adding 100 ng sCD4 (R&D Systems, Belgium) in 100 μL PBS per well and incubated at 4 °C. After 24 h, the CD4 coated plate was washed with 1% BSA-PBS and blocked by adding 100μL 5% BSA-PBS per well for 1 h at 4 °C. The coated plate was then washed and incubated with or without drugs for 1 h at 4 °C. After washing, 100 ng HIV-1 gp120 protein (Fitzgerald, Acton, MA, USA) was added into the plate and kept for 1 h at 4 °C. The plate was washed, and anti-HIV-1 gp120 (Aalto Bio Reagents, Dublin, Ireland) (100 ng/100 μL PBS) was added. After 1 h, the plate was washed and anti-sheep HRP (Sigma, Belgium) (100 ng/100 μL PBS) was added. After 1 h incubation at 4 °C, 100 μL o-phenylenediamine dihydrochloride (OPD) (Thermo Fisher, Asse, Belgium) was added and incubated for 20 min in dark. One hundred microliters 0.5M H_2SO_4 was finally added to stop the reaction. OD_{492} and OD_{630} were measured by POLARstart Omega Plate Reader. $OD_{492} - OD_{630}$ was calculated.

5.9. Binding Assays and Co-Receptor Internalization

Binding competition between increasing concentrations of compounds and FITC-conjugated mouse anti-human CD4 clone RPA-4 (Biolegend, Amsterdam, Netherlands), PE-conjugated mouse anti-Human CXCR4 clone 12G5, and PE-conjugated mouse anti-human CCR5 clone 2D7 (BD Pharma, Aalst, Belgium) to U373-CD4-CXCR4 and U373-CD4-CCR5 cells was first tested by flow cytometry (FACSCanto). The chemokines CXCL12 and CCL5 (50 nM) (Peprotech, London, United Kingdom) were used as positive controls, and near-IR fluorescent reactive dye was added simultaneously to evaluate cell viability. After 1 h of incubation at 4 °C, cells were washed with FACS buffer (1% bovine serum albumin, 0.1% N3Na in PBS, Sigma), and Mean Fluorescence Intensity (MFI) was measured. To study co-receptor internalization, compounds were incubated alone (agonist mode) or in the presence of 50 nM chemokines (CXCL12/CCL5) (antagonist mode) with MT-4 and U373-CD4-CCR5 cells at 37 °C for 1 h. Internalization was stopped after 30 min by addition of NaN_3 (0.1%) on ice. Cells were stained with anti-human CXCR4 clone 4G10 (BD, Belgium) or anti-human CCR5 (CD195) clone T21/8, (eBioscience, Asse, Belgium), Near-IR fluorescent reactive dye during 1 h at 4 °C, washed with FACS buffer, and a secondary PE-conjugated goat anti-mouse antibody (Jackson ImmunoResearch, West Grove, PA, USA).

5.10. Dual-Chamber and DC-SIGN Transmission Assays

Dual chamber transmission assay was performed as previously described [28]. HeLa cells were seeded into an upper chamber of a transwell plate (Sigma-Aldrich, Liège, Belgium), while TZM-bl cells were seeded in the lower chamber. Trans Epithelial Electric Resistance (TEER) was measured using Millicell-ERS Volt-Ohm Meter. HIV-1 ADA-M (200 pg) and drugs were added to the upper chamber after 4 days when TEER reached 150 Ohm/cm^2. 24 h after infection luciferase value of TZM-bl cells lysate was measured using the Luciferase System Kit and the POLARstar Omega Plate Reader. Data were analyzed using GraphPrism. The HIV-1 DC-SIGN transmission assay was performed as previously described [27]. To investigate the cellular activation induced by CE or piceatannol, CD25 and CD69 expression was measured on PBMCs after incubation with CE/piceatannol or 10 µg/mL PHA-P for 24 h at 37 °C using FITC-conjugated anti-CD4, PE/Cy7-conjugated anti-CD25, PE-conjugated anti-CD69 mAbs (Biolegend, Amsterdam, Netherlands), and the near-IR fluorescent reactive dye.

5.11. In Silico Ligand-Based Study

For SAR purposes, molecules **1–11** were purchased from MolPort (Riga, Latvia), while molecules **12–17** (purity > 95% by HPLC) were retrieved from an in-house library of natural products previously characterized [29,42]. To provide insights into the mechanism of action of bioactive hits, a ligand-based study was conducted using the ROCS program (Openeye Scientific Software) version 3.3.0.3 [43]. A query was built on the chemical structure of piceatannol and used to screen the conformational database of compounds **1-17** that was generated by OMEGA (OpenEye Scientific Software) version 3.1.0.3 [44] using default settings. Ligands similarity was assessed by the TanimotoCombo scores, while normalized similarity scores were calculated by dividing the TanimotoCombo value by two. All the possible stereochemical configurations of cassiabrevone and 4′7-Dihydroxyflavan (4,6)-3, 4′, 5,7- tetrahydroxyflavan were analyzed; the reported normalized similarity and TanimotoCombo scores refer to the isomers endowed with the highest similarity to piceatannol.

Author Contributions: Y.Z. performed all anti-HIV-1 experiments, analyzed the data, and wrote the manuscript. X.Y. and A.S. purified the active compounds from *C. abbreviata*, D.S. performed the screening of the crude extract and piceatannol against various viruses, M.M. (Mattia Mori) and B.B. performed the in silico ligand-based studies, A.C. performed the binding assays and co-receptor internalization, M.M. (Martin Mulinge). provided the barks of *C. abbreviata* and set-up the extraction of the plant with A.S., J.C.S provided the clinical samples and designed the study, and C.S.-D. designed the study, analyzed the data, and wrote the manuscript. All authors contributed to the article and approved the submitted version.

Funding: This work was supported by the "Fonds National de la Recherche" of Luxembourg [PHD AFR grant 1189522], "la Fondation Recherche sur le SIDA", the KU Leuven (GOA 10/014 and PF/10/018), the Foundation of Scientific Research (FWO no. G-0528-12), and the Luxembourg Institute of Health (MESR grant 20150415).

Institutional Review Board Statement: Ethical review and approval were waived for this study, due to the use of anonymized left over clinical primary viral strains or PBMCs from the red Blood cross of Luxembourg that did not require both ethical approval.

Informed Consent Statement: Inform consents were waived for this study, due to the use of anonymized left over clinical primary viral strains or PBMCs from the red Blood cross of Luxembourg that did not require both inform consents of the patients.

Data Availability Statement: All the data are contained within the article.

Acknowledgments: Authors thank Sandra Claes, Evelyne Van Kerckhove, Eric Fonteyn, Leentje Persoons, Lies Van den Heurck (died 3 September 2018), Morgane Lemaire, Jean-Yves Servais, Manuel Counson, and Gilles Iserentant for excellent technical assistance. Authors acknowledge networking contribution by the COST Action CM1407 "Challenging organic syntheses inspired by nature—from natural products chemistry to drug discovery". MM wish to thank the OpenEye Free Academic Licensing Program for providing a free academic license for molecular modeling and chemoinformatics software.

Conflicts of Interest: The authors declare no conflict of interest.

References

1. Este, J.A.; Cihlar, T. Current status and challenges of antiretroviral research and therapy. *Antivir. Res.* **2010**, *85*, 25–33. [CrossRef]
2. Palmisano, L.; Vella, S. A brief history of antiretroviral therapy of HIV infection: Success and challenges. *Ann. Ist. Super Sanita* **2011**, *47*, 44–48. [PubMed]
3. Glynn, J.R.; Carael, M.; Auvert, B.; Kahindo, M.; Chege, J.; Musonda, R.; Kaona, F.; Buve, A. Study Group on the Heterogeneity of, H.I.V.E. i. A. C., Why do young women have a much higher prevalence of HIV than young men? A study in Kisumu, Kenya and Ndola, Zambia. *AIDS* **2001**, *15* (Suppl. 4), S51–S60. [CrossRef] [PubMed]
4. Mehellou, Y.; De Clercq, E. Twenty-six years of anti-HIV drug discovery: Where do we stand and where do we go? *J. Med. Chem.* **2010**, *53*, 521–538. [CrossRef] [PubMed]
5. Vigant, F.; Santos, N.C.; Lee, B. Broad-spectrum antivirals against viral fusion. *Nat. Rev. Microbiol.* **2015**, *13*, 426–437. [CrossRef]
6. Wilen, C.B.; Tilton, J.C.; Doms, R.W. HIV: Cell binding and entry. *Cold Spring Harb. Perspect. Med.* **2012**, *2*, a006866. [CrossRef]
7. St Vincent, M.R.; Colpitts, C.C.; Ustinov, A.V.; Muqadas, M.; Joyce, M.A.; Barsby, N.L.; Epand, R.F.; Epand, R.M.; Khramyshev, S.A.; Valueva, O.A.; et al. Rigid amphipathic fusion inhibitors, small molecule antiviral compounds against enveloped viruses. *Proc. Natl. Acad. Sci. USA* **2010**, *107*, 17339–17344. [CrossRef]
8. Dias, D.A.; Urban, S.; Roessner, U. A historical overview of natural products in drug discovery. *Metabolites* **2012**, *2*, 303–336. [CrossRef] [PubMed]
9. Maurice, M.I. *Handbook of African Medicinal Plants*, 2nd ed.; CRC Press: Boca Raton, FL, USA, 1993.
10. Mongalo, N.I.; Mafoko, B.J. Cassia abbreviata Oliv, a review of its ethnomedicinal uses, toxicology, phytochemistry, possible propagation techniques and pharmacology. *Afr. J. Pharm. Pharmacol.* **2013**, *7*, 6.
11. Ribeiro, A.; Romeiras, M.M.; Tavares, J.; Faria, M.T. Ethnobotanical survey in Canhane village, district of Massingir, Mozambique: Medicinal plants and traditional knowledge. *J. Ethnobiol. Ethnomed.* **2010**, *6*, 33.
12. Yang, X.; He, Z.; Zheng, Y.; Wang, N.; Mulinge, M.; Schmit, J.C.; Steinmetz, A.; Seguin-Devaux, C. Chemical Constituents of Cassia abbreviata and Their Anti-HIV-1 Activity. *Molecules* **2021**, *26*, 2455. [CrossRef]
13. Leteane, M.M.; Ngwenya, B.N.; Muzila, M.; Namushe, A.; Mwinga, J.; Musonda, R.; Moyo, S.; Mengestu, Y.B.; Abegaz, B.M.; Andrae-Marobela, K. Old plants newly discovered: Cassia sieberiana D.C. and Cassia abbreviata Oliv. Oliv. root extracts inhibit in vitro HIV-1c replication in peripheral blood mononuclear cells (PBMCs) by different modes of action. *J. Ethnopharmacol.* **2012**, *141*, 48–56. [CrossRef]
14. Chou, T.C. Drug combination studies and their synergy quantification using the Chou-Talalay method. *Cancer Res.* **2010**, *70*, 440–446. [PubMed]
15. Cirigliano, A.; Stirpe, A.; Menta, S.; Mori, M.; Dell'Edera, D.; Pick, E.; Negri, R.; Botta, B.; Rinaldi, T. Yeast as a tool to select inhibitors of the cullin deneddylating enzyme Csn5. *J. Enzym. Inhib. Med. Chem.* **2016**, *31*, 1632–1637. [CrossRef] [PubMed]
16. Ghirga, F.; Quaglio, D.; Mori, M.; Cammarone, S.; Iazzetti, A.; Goggiamani, A.; Ingallina, C.; Botta, B.; Calcaterra, A. A unique high-diversity natural product collection as a reservoir of new therapeutic leads. *Org. Chem. Front.* **2021**, *8*, 996–1025. [CrossRef]
17. Infante, P.; Alfonsi, R.; Ingallina, C.; Quaglio, D.; Ghirga, F.; D'Acquarica, I.; Bernardi, F.; Di Magno, L.; Canettieri, G.; Screpanti, I.; et al. Inhibition of Hedgehog-dependent tumors and cancer stem cells by a newly identified naturally occurring chemotype. *Cell Death Dis.* **2016**, *7*, e2376.
18. Infante, P.; Mori, M.; Alfonsi, R.; Ghirga, F.; Aiello, F.; Toscano, S.; Ingallina, C.; Siler, M.; Cucchi, D.; Po, A.; et al. Gli1/DNA interaction is a druggable target for Hedgehog-dependent tumors. *EMBO J.* **2015**, *34*, 200–217. [CrossRef] [PubMed]
19. Mascarello, A.; Mori, M.; Chiaradia-Delatorre, L.D.; Menegatti, A.C.; Delle Monache, F.; Ferrari, F.; Yunes, R.A.; Nunes, R.J.; Terenzi, H.; Botta, B.; et al. Discovery of Mycobacterium tuberculosis protein tyrosine phosphatase B (PtpB) inhibitors from natural products. *PLoS ONE* **2013**, *8*, e77081.
20. Lee, D.Y.; Lin, X.; Paskaleva, E.E.; Liu, Y.; Puttamadappa, S.S.; Thornber, C.; Drake, J.R.; Habulin, M.; Shekhtman, A.; Canki, M. Palmitic Acid Is a Novel CD4 Fusion Inhibitor That Blocks HIV Entry and Infection. *Aids Res. Hum. Retrovir.* **2009**, *25*, 1231–1241. [PubMed]
21. Harada, S.; Ogihara, K.; Hikichi, Y.; Matano, T.; Narumi, K.Y. Oleanolic acid derivative OKS3-019 as a novel bifunctional HIV-1 entry inhibitor. In Proceedings of the Conference on Retroviruses and Opportunistic Infections, Seattle, WA, USA, 13–16 February 2017.
22. Clouser, C.L.; Chauhan, J.; Bess, M.A.; van Oploo, J.L.; Zhou, D.; Dimick-Gray, S.; Mansky, L.M.; Patterson, S.E. Anti-HIV-1 activity of resveratrol derivatives and synergistic inhibition of HIV-1 by the combination of resveratrol and decitabine. *Bioorg. Med. Chem. Lett.* **2012**, *22*, 6642–6646.
23. Kashiwada, Y.; Wang, H.K.; Nagao, T.; Kitanaka, S.; Yasuda, I.; Fujioka, T.; Yamagishi, T.; Cosentino, L.M.; Kozuka, M.; Okabe, H.; et al. Anti-AIDS agents. 30. Anti-HIV activity of oleanolic acid, pomolic acid, and structurally related triterpenoids. *J. Nat. Prod.* **1998**, *61*, 1090–1095. [CrossRef] [PubMed]

24. Mengoni, F.; Lichtner, M.; Battinelli, L.; Marzi, M.; Mastroianni, C.M.; Vullo, V.; Mazzanti, G. In vitro anti-HIV activity of oleanolic acid on infected human mononuclear cells. *Planta Med.* **2002**, *68*, 111–114. [CrossRef] [PubMed]
25. Min, B.S.; Lee, H.K.; Lee, S.M.; Kim, Y.H.; Bae, K.H.; Otake, T.; Nakamura, N.; Hattori, M. Anti-human immunodeficiency virus-type 1 activity of constituents from Juglans mandshurica. *Arch Pharm. Res.* **2002**, *25*, 441–445. [CrossRef] [PubMed]
26. Bar, S.; Alizon, M. Role of the ectodomain of the gp41 transmembrane envelope protein of human immunodeficiency virus type 1 in late steps of the membrane fusion process. *J. Virol.* **2004**, *78*, 811–820. [CrossRef]
27. Gordts, S.C.; Ferir, G.; D'Huys, T.; Petrova, M.I.; Lebeer, S.; Snoeck, R.; Andrei, G.; Schols, D. The Low-Cost Compound Lignosulfonic Acid (LA) Exhibits Broad-Spectrum Anti-HIV and Anti-HSV Activity and Has Potential for Microbicidal Applications. *PLoS ONE* **2015**, *10*, e0131219. [CrossRef]
28. Pasetto, S.; Pardi, V.; Murata, R.M. Anti-HIV-1 activity of flavonoid myricetin on HIV-1 infection in a dual-chamber in vitro model. *PLoS ONE* **2014**, *9*, e115323. [CrossRef]
29. Sakamoto, Y.; Ohmoto, N.; Tamotsu, K.; Tomimori, T.; Miyaichi, Y.; Shirataki, Y.; Monache, F.; Botta, B. On the relationship between the chemical structure and the cyclic AMP phosphodiesterase inhibitory activity of flavonoids as studied by carbon13 NMR. *Bull. Chem. Soc. Jpn.* **1989**, *62*, 2450–2452. [CrossRef]
30. Panthong, P.; Bunluepuech, K.; Boonnak, N.; Chaniad, P.; Pianwanit, S.; Wattanapiromsakul, C.; Tewtrakul, S. Anti-HIV-1 integrase activity and molecular docking of compounds from Albizia procera bark. *Pharm. Biol.* **2015**, *53*, 1861–1866. [CrossRef]
31. Piao, Z.S.; Feng, Y.B.; Wang, L.; Zhang, X.Q.; Lin, M. Synthesis and HIV-1 inhibitory activity of natural products isolated from Gnetum parvifolium and their analogues. *Yao Xue Xue Bao* **2010**, *45*, 1509–1515.
32. Han, Y.S.; Quashie, P.K.; Mesplede, T.; Xu, H.; Quan, Y.; Jaeger, W.; Szekeres, T.; Wainberg, M.A. A resveratrol analog termed 3,3′,4,4′,5,5′-hexahydroxy-trans-stilbene is a potent HIV-1 inhibitor. *J. Med. Virol.* **2015**, *87*, 2054–2060. [CrossRef]
33. Anggakusuma Colpitts, C.C.; Schang, L.M.; Rachmawati, H.; Frentzen, A.; Pfaender, S.; Behrendt, P.; Brown, R.J.; Bankwitz, D.; Steinmann, J.; Ott, M.; et al. Turmeric curcumin inhibits entry of all hepatitis C virus genotypes into human liver cells. *Gut* **2014**, *63*, 1137–1149. [CrossRef]
34. Wolf, M.C.; Freiberg, A.N.; Zhang, T.; Akyol-Ataman, Z.; Grock, A.; Hong, P.W.; Li, J.; Watson, N.F.; Fang, A.Q.; Aguilar, H.C.; et al. A broad-spectrum antiviral targeting entry of enveloped viruses. *Proc. Natl. Acad. Sci. USA* **2010**, *107*, 3157–3162. [CrossRef]
35. Docherty, J.J.; Fu, M.M.; Hah, J.M.; Sweet, T.J.; Faith, S.A.; Booth, T. Effect of resveratrol on herpes simplex virus vaginal infection in the mouse. *Antivir. Res.* **2005**, *67*, 155–162. [CrossRef] [PubMed]
36. Malancona, S.; Mori, M.; Fezzardi, P.; Santoriello, M.; Basta, A.; Nibbio, M.; Kovalenko, L.; Speziale, R.; Battista, M.R.; Cellucci, A.; et al. 5,6-Dihydroxypyrimidine Scaffold to Target HIV-1 Nucleocapsid Protein. *ACS Med. Chem. Lett* **2020**, *11*, 766–772. [CrossRef] [PubMed]
37. Mori, M.; Ciaco, S.; Mely, Y.; Karioti, A. Inhibitory Effect of Lithospermic Acid on the HIV-1 Nucleocapsid Protein. *Molecules* **2020**, *25*, 5434. [CrossRef] [PubMed]
38. Huth, J.R.; Mendoza, R.; Olejniczak, E.T.; Johnson, R.W.; Cothron, D.A.; Liu, Y.; Lerner, C.G.; Chen, J.; Hajduk, P.J. ALARM NMR: A rapid and robust experimental method to detect reactive false positives in biochemical screens. *J. Am. Chem. Soc.* **2005**, *127*, 217–224. [CrossRef] [PubMed]
39. Mori, M.; Kovalenko, L.; Malancona, S.; Saladini, F.; De Forni, D.; Pires, M.; Humbert, N.; Real, E.; Botzanowski, T.; Cianferani, S.; et al. Structure-Based Identification of HIV-1 Nucleocapsid Protein Inhibitors Active against Wild-Type and Drug-Resistant HIV-1 Strains. *ACS Chem. Biol.* **2018**, *13*, 253–266. [CrossRef]
40. Baeten, J.M.; Benki, S.; Chohan, V.; Lavreys, L.; McClelland, R.S.; Mandaliya, K.; Ndinya-Achola, J.O.; Jaoko, W.; Overbaugh, J. Hormonal contraceptive use, herpes simplex virus infection, and risk of HIV-1 acquisition among Kenyan women. *AIDS* **2007**, *21*, 1771–1777. [CrossRef] [PubMed]
41. Baatz, F.; Nijhuis, M.; Lemaire, M.; Riedijk, M.; Wensing, A.M.; Servais, J.Y.; van Ham, P.M.; Hoepelman, A.I.; Koopmans, P.P.; Sprenger, H.G.; et al. Impact of the HIV-1 env genetic context outside HR1-HR2 on resistance to the fusion inhibitor enfuvirtide and viral infectivity in clinical isolates. *PLoS ONE* **2011**, *6*, e21535. [CrossRef]
42. Bargellini, G.; Marini-Bettolo, G.B. Flavone, flavanone and flavonol derivatives of hydroxyhydroquinone. *Gazz. Chim. Ital.* **1940**, *70*, 170–178.
43. Hawkins, P.C.; Skillman, A.G.; Nicholls, A. Comparison of shape-matching and docking as virtual screening tools. *J. Med. Chem.* **2007**, *50*, 74–82. [CrossRef] [PubMed]
44. Hawkins, P.C.; Skillman, A.G.; Warren, G.L.; Ellingson, B.A.; Stahl, M.T. Conformer generation with OMEGA: Algorithm and validation using high quality structures from the Protein Databank and Cambridge Structural Database. *J. Chem. Inf. Model* **2010**, *50*, 572–584. [CrossRef] [PubMed]

International Journal of *Molecular Sciences*

MDPI

Review

Immunological Aspects of SARS-CoV-2 Infection and the Putative Beneficial Role of Vitamin-D

Ming-Yieh Peng [1,†], Wen-Chih Liu [2], Jing-Quan Zheng [3,4,5,†], Chien-Lin Lu [6], Yi-Chou Hou [3,7], Cai-Mei Zheng [8,9,10], Jenn-Yeu Song [11,12], Kuo-Cheng Lu [13,*] and You-Chen Chao [12,14]

1 Division of Infectious Disease, Department of Medicine, Taipei Tzu Chi Hospital, Buddhist Tzu Chi Medical Foundation, New Taipei City 231, Taiwan; 16044@s.tmu.edu.tw
2 Division of Nephrology, Department of Medicine, Taipei Hospital, Ministry of Health and Welfare, New Taipei City 242, Taiwan; wayneliu55@gmail.com
3 Graduate Institute of Clinical Medicine, College of Medicine, Taipei Medical University, Taipei 11031, Taiwan; jingquan235@gmail.com (J.-Q.Z.); Athletics910@gmail.com (Y.-C.H.)
4 Division of Pulmonary Medicine, Department of Internal Medicine, School of Medicine, College of Medicine, Taipei Medical University, Taipei 11031, Taiwan
5 Division of Pulmonary Medicine, Department of Internal Medicine, Shuang Ho Hospital, Taipei Medical University, New Taipei City 23561, Taiwan
6 Division of Nephrology, Department of Medicine, Fu Jen Catholic University Hospital, School of Medicine, Fu Jen Catholic University, New Taipei City 242, Taiwan; janlin0123@gmail.com
7 Division of Nephrology, Department of Medicine, Cardinal-Tien Hospital, School of Medicine, Fu-Jen Catholic University, New Taipei City 234, Taiwan
8 Taipei Medical University-Research Center of Urology and Kidney (TMU-RCUK), Taipei Medical University, Taipei 110, Taiwan; 11044@s.tmu.edu.tw
9 Division of Nephrology, Department of Internal Medicine, Taipei Medical University Shuang Ho Hospital, New Taipei City 235, Taiwan
10 Division of Nephrology, Department of Internal Medicine, School of Medicine, College of Medicine, Taipei Medical University, Taipei 110, Taiwan
11 Division of Cardiovascular Surgery, Department of Surgery, Taipei Tzu Chi Hospital, Buddhist Tzu Chi Medical Foundation, New Taipei City 231, Taiwan; ttwyl123@yahoo.com
12 School of Medicine, Tzu Chi University, Hualien 970, Taiwan; thanthanwinnge@gmail.com
13 Division of Nephrology, Department of Medicine, Taipei Tzu Chi Hospital, Buddhist Tzu Chi Medical Foundation, New Taipei City 231, Taiwan
14 Division of Gastroenterology, Department of Internal Medicine, Taipei Tzu Chi Hospital, Buddhist Tzu Chi Medical Foundation, New Taipei City 231, Taiwan
* Correspondence: kuochenglu@gmail.com
† These authors contributed equally to this work.

Citation: Peng, M.-Y.; Liu, W.-C.; Zheng, J.-Q.; Lu, C.-L.; Hou, Y.-C.; Zheng, C.-M.; Song, J.-Y.; Lu, K.-C.; Chao, Y.-C. Immunological Aspects of SARS-CoV-2 Infection and the Putative Beneficial Role of Vitamin-D. *Int. J. Mol. Sci.* **2021**, *22*, 5251. https://doi.org/10.3390/ijms22105251

Academic Editor: Maurizio Battino

Received: 10 April 2021
Accepted: 12 May 2021
Published: 16 May 2021

Abstract: Coronavirus disease 2019 (COVID-19), caused by severe acute respiratory syndrome coronavirus-2 (SARS-CoV-2) is still an ongoing global health crisis. Immediately after the inhalation of SARS-CoV-2 viral particles, alveolar type II epithelial cells harbor and initiate local innate immunity. These particles can infect circulating macrophages, which then present the coronavirus antigens to T cells. Subsequently, the activation and differentiation of various types of T cells, as well as uncontrollable cytokine release (also known as cytokine storms), result in tissue destruction and amplification of the immune response. Vitamin D enhances the innate immunity required for combating COVID-19 by activating toll-like receptor 2. It also enhances antimicrobial peptide synthesis, such as through the promotion of the expression and secretion of cathelicidin and β-defensin; promotes autophagy through autophagosome formation; and increases the synthesis of lysosomal degradation enzymes within macrophages. Regarding adaptive immunity, vitamin D enhances $CD4^+$ T cells, suppresses T helper 17 cells, and promotes the production of virus-specific antibodies by activating T cell-dependent B cells. Moreover, vitamin D attenuates the release of pro-inflammatory cytokines by $CD4^+$ T cells through nuclear factor κB signaling, thereby inhibiting the development of a cytokine storm. SARS-CoV-2 enters cells after its spike proteins are bound to angiotensin-converting enzyme 2 (ACE2) receptors. Vitamin D increases the bioavailability and expression of ACE2, which may be responsible for trapping and inactivating the virus. Activation of the renin–angiotensin–aldosterone system (RAS) is responsible for tissue destruction, inflammation, and organ failure related to SARS-CoV-2. Vitamin D inhibits renin expression and serves as a negative RAS regulator. In conclusion, vitamin D defends the body against SARS-CoV-2 through a novel complex mechanism that operates through interactions between the activation of both innate and adaptive immunity, ACE2 expression, and inhibition of the RAS system. Multiple observation studies have shown that serum concentrations of 25 hydroxyvitamin D are inversely correlated with the incidence or severity of COVID-19. The evidence gathered thus far, generally meets Hill's causality criteria in a biological system, although experimental verification is not sufficient. We speculated that adequate vitamin D supplementation may be essential for mitigating the progression and severity of COVID-19. Future studies are warranted to determine the dosage and effectiveness of vitamin D supplementation among different populations of individuals with COVID-19.

Keywords: coronavirus disease 2019; severe acute respiratory syndrome coronavirus 2; vitamin D; adaptive immunity; innate immunity; angiotensin-converting enzyme 2; renin–angiotensin–aldosterone system

1. Introduction

Coronavirus disease 2019 (COVID-19), a rapidly spreading respiratory illness caused by the severe acute respiratory syndrome (SARS) coronavirus 2 (SARS-CoV-2) virus [1], constitutes a global health emergency. Notably, immune responses activated by the coronavirus, including adaptive immune responses in the earlier and asymptomatic stages, prevent further disease progression. Patients' immune status plays a pivotal role in predicting their prognosis. Patients with immunodeficiency or aberrant immunity may promote viral replication and subsequent tissue damage with multiple organ failure. At the other end of the spectrum, overactive immune responses are correlated with immunopathological conditions and further tissue destruction.

Vitamin D insufficiency or deficiency is a highly prevalent global problem, affecting over a billion people worldwide [2,3]. In the COVID-19 era, it is interestingly found that vitamin D deficiency related with higher risks for SARS-CoV-2 infection [4,5]. It also associated with respiratory immune impairment and increased COVID-19 severity and mortality [6,7]. Moreover, vitamin D deficiency leads to cytokine storms [8] and to excessive tissue damage and mortality in patients with SARS-CoV-2 infection [6,9]. However, the clinical effects of vitamin D supplementation in COVID-19 era is still controversial [10–12]. It is critically important to understand how vitamin D influence the impact of COVID-19 and determine appropriate dosages of vitamin D among different patient population.

COVID-19 world map evidence the variations of COVID-19 occurrence, spread, severity, and mortality varied around the world, and this has been found to overlap with vitamin D deficiency areas. Understanding the immunological differences in individuals and populations is essential. With the continuing spread of infection, the lack of targeted therapy presents a major problem [2]. Adequate supplementation of vitamin D might be helpful for preventing those at higher risk of vitamin D deficiency (e.g., older adults, patients, individuals with comorbid conditions, and immunocompromised individuals) from contracting COVID-19. In this review, we will discuss the details of human immunity defense against SARS-CoV-2, molecular mechanisms underlying vitamin D-related anti-SARS-CoV-2 immunity and different clinical situations.

2. COVID-19 Pathophysiology

During the initial 1–2 days of infection, SARS-CoV-2 either passes through the upper respiratory tract or directly enters the lower respiratory tract, infecting both bronchial and alveolar cells [13,14]. Angiotensin-converting enzyme (ACE) 2 (ACE2), which is expressed in most human cells, acts as a receptor for SARS-CoV-2 [15]. In vitro studies have indicated that SARS coronavirus (SARS-CoV) primarily infects the ciliated cells which is high in ACE2 expression in the conducting airways [16], and that ACE2 expression is lower within other cells in these airways [13,17]. Local propagation of the virus occurs when innate immunity cannot be fully initiated. At this stage, the virus can be detected in nasal swabs. As the infection progresses, the virus moves through the respiratory tract, and a robust innate immune response is triggered. By the time the infection is clinically manifested, viremia has already occurred; that is, the virus has already entered the peripheral bloodstream [18,19]. As the disease progresses, the virus continues to affect other organs that mainly express ACE2 (e.g., the heart and its blood vessels, the kidneys, and the gastrointestinal tract). Thus, further disease progression and systemic organ damage tend to occur in patients with severe pulmonary symptoms [19]. Cytokine storms, characterized by strong inflammatory responses in response to immunological threats, are possibly responsible for systemic organ dysfunction. The clinical symptomatology of COVID-19 varies, from a lack of symptoms to the presentation of symptoms related to local infection (pneumonia), followed by either recovery or disease progression with systemic manifestations [20].

In almost 20% of patients infected with SARS-CoV2, the disease progresses to the point where pulmonary infiltrates are developed [1]—a point at which the inhaled virus reaches the terminal airway and mainly infects alveolar type II cells [21]. Most alveoli in the peripheral and subpleural regions that are infected [22] undergo substantial apoptosis as SARS-CoV2 propagates locally [23]. Pathological data from a case report indicated diffuse alveolar damage with subsequent infiltration of multinucleated giant cells and the formation of fibrin-rich hyaline membranes [19]. Notably, aberrant healing from SARS-CoV-2 results in more severe scarring and fibrosis compared with that from other types of acute respiratory distress syndrome (ARDS).

2.1. Viral Cell Entry

In order to enter the host cell, the SARS-CoV-2 firstly binds to the cell surface receptor for virus attachment, then enters the endosome, and finally fuses the viral membrane and the lysosomal membrane. The virus has several structural proteins, such as S (Spike), M (Membrane), N (Nucleocapsid), and E (Envelope) and HE (Hemagglutinin esterase). Protein S consists of 2 sub-units: S1 and S2. S1 plays a role in viral attachment. SS (signal sequence), NTD (terminal domain N), RBD (receiver link domain), and RBM (receiver link pattern) are parts of S1 proteins. The S2 subunit consists of the fusion peptide (FP), protease cleavage site (S2′), central helix (CH), connector domain (CD), heptad repeat (HR) 1 and 2, transmembrane domain (TM), and cytoplasmic tail (CT). The S2 subunit fuses the virus with the host cells [24].

On mature viruses, the spike protein exists in the form of trimers, and there are three S1 heads that bind to the receptor on the three-trimeric membrane fusion S2 stalk. SRAS-CoV S1 contains a receptor binding domain (RBD) which specifically recognizes the ACE2 as its receptor. RBD continuously switches between the standing position (for the connection of the receiver) and the lying position (to escape immunity) [25]. The furin preactivation of the spike for enhanced entry into cells by keeping RBD in a standing position which enhance the virus binding to host cell membrane ACE2. In order to fuse the membrane, the SRAS-CoV spike needs to be proteolytically activated at the S1/S2 boundary to dissociate S1, and S2. The cleavage of the S1–S2 protein, which is required for the conformational changes of the S2 subunit and processing of viral fusion, is regulated by the host proteases, including TMPRSS2 (independently) and cathepsin L (during endocytosis). The TMPRSS2 may cleave the S protein [26]. The viral spike protein mediates SARS-CoV-2 entry into host cells and harbors a S1/S2 cleavage site containing multiple arginine residues (multibasic) not found in closely related animal coronaviruses [27]. After protease cleavage of the S1 protein, the FP subunit of S2 undergoes membrane fusion with the host cell membrane [28,29]. These entrance characteristics of SARS-CoV contribute to its rapid spread and result in severe symptoms and high mortality in infected patients [30].

After entry into the host cell, SRAS-CoV-2 lowers the ACE2 expression, which in turn regulates angiotensin II (Ang II). ACE2 is a type 1 integral membrane glycoprotein that is constitutively expressed by the epithelial cells of the lungs, kidneys, intestine, and blood vessels. In normal physiology, ACE2 breaks down Ang II and, to a lesser extent, angiotensin-I (Ang I) to smaller peptides, angiotensin 1–7 and angiotensin 1–9, respectively [31]. ACE2/Ang 1–7 system plays an important anti-inflammatory and anti-oxidant role protecting the lung against ARDS; indeed, ACE2 has been shown to be protective against lethal avian influenza A H5N1 infection [32].

The ACE2 molecule, besides being a receptor of SARS-CoV and SARS-CoV-2, reduces the activity of the renin–angiotensin system (RAS) by converting Ang I and Ang II into Ang 1–9 and Ang 1–7 respectively [33]. Thus, the ACE2 protein has been shown to play an important role in protecting against some disorders such as cardiovascular complications, chronic obstructive pulmonary disease (COPD) and diabetes, among other COVID-19 comorbidities [34]. The ACE2/Ang 1–7 axis counterbalances the ACE/Ang II-I axis by decreasing Ang II levels, the activation of angiotensin type 1 receptors (AT1Rs) and, thus, leads to decreased pathophysiological effects on tissues, such as inflammation and fibrosis [35].

Ang II interacts with its receptor, Ang II receptor type 1 (AT1R), and modulates the gene expression of several inflammatory cytokines via nuclear factor κB (NF-κB) signaling. This interaction also promotes macrophage activation and results in the production of inflammatory cytokines that may cause ARDS or macrophage activation syndrome (MAS). Some metalloproteases, such as ADAM17, cleave these pro-inflammatory cytokines and ACE2 receptors, resulting in their release as soluble forms. This contributes to the loss of the protective function of surface ACE2 and potentially exacerbates SARS-CoV-2 pathogenesis [36]. SARS-CoV-2 infects the mononuclear phagocyte system, the cells in which produce different pro-inflammatory cytokines to trigger focal inflammation and systemic inflammatory response, a phenomenon known as a cytokine storm. Altogether, these events play a fundamental role in severe presentations of COVID-19, including ARDS and death [37].

Role of Pulmonary Alveolar Type II Epithelial Cells and Macrophages in SARS-CoV-2 Infection

As mentioned, SARS-CoV-2 infects alveolar type II epithelial cells and downregulates ACE2 expression, leading to the upregulation of and metabolic dysfunction in Ang II. Ang II acts on the AT1R, activating macrophages and releasing various inflammatory cytokines that trigger tissue inflammation and destruction. As type II alveolar cells, these macrophages also express furin and TMPRSS2, which are responsible for SARS virus exposure [38,39], as well as ADAM17, which acts as a sheddase of ACE2 [40]. Theoretically, after the invasion of macrophages and type II alveolar cells, viruses replicate quickly within macrophages and dendritic cells and trigger the aberrant production of pro-inflammatory cytokines [41] (Figure 1). However, recent studies have detected the increased expression of pro-inflammatory chemokines in human macrophages even in the absence of SARS-CoV-2 replication [42] or antiviral cytokine production [43]. Pulmonary dendritic cells also exert anti-inflammatory effects through antigen presentation and regulation of T cell reactions. Notably, a study revealed that human dendritic cells are susceptible to SARS-CoV-2 infection and cannot maintain viral replication [34].

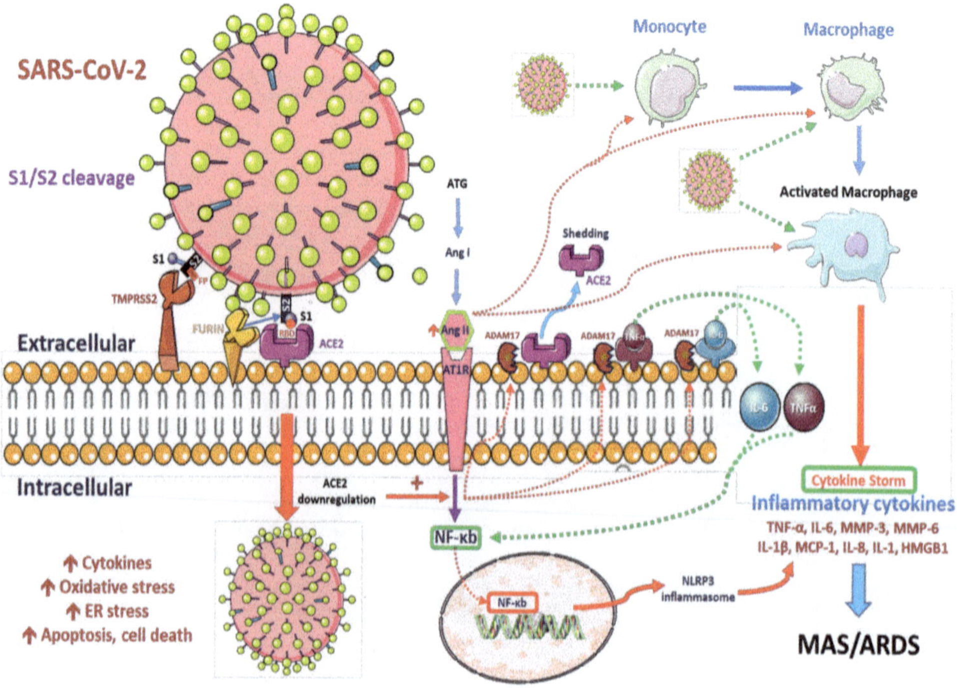

Figure 1. Possible pathophysiological pathways after the entry of severe acute respiratory syndrome coronavirus 2 (SARS-CoV-2). SARS-CoV-2 infects both alveolar macrophages and type II alveolar cells by binding to angiotensin-converting 2 (ACE2) receptors. Before SRAS-CoV-2 enters the host cells, the spike protein 1 (S1) should be pre-activated by the host furin, a convertase proprotein, which will expose the receptor binding domain (RBD) of S1. RBD has a strong binding affinity for the host cell membrane ACE2 for effective entry. After binding of RBD and ACE2, the type 2 transmembrane protease (TMPRSS2) will proteolytically activated the S1/S2 boundary through cleavage the S1–S2 protein which will cause drastic structural changes with further expose the fusion peptide (FP) of S2 which will facilitate the processing of viral-host cell fusion [44]. Immediately after entry, S protein activation is mediated by lysosomal cathepsins and/or furin within the TGN [14,15]. SARS-CoV-2 replication is suppressed by synthetic furin inhibitors [16]. After entry into the host cell, the virus downregulates ACE2 expression, which in turn upregulates angiotensin II (Ang II). Ang II interacts with its receptor, Ang II receptor type 1, and modulates the gene expression of several inflammatory cytokines via nuclear factor κB signaling. This interaction also promotes macrophage activation and results in the production of inflammatory cytokines that may cause acute respiratory distress syndrome or macrophage activation syndrome. Some metalloproteases, such as ADAM metallopeptidase domain 17, cleave these pro-inflammatory cytokines and ACE2 receptors, resulting in their release as soluble forms. This contributes to the loss of the protective function of surface ACE2 and potentially exacerbates SARS-CoV-2 pathogenesis [39]. Monocytes and macrophages in the mononuclear phagocyte system, infected with SARS-CoV-2, produce various pro-inflammatory cytokines and chemokines, a process critical for the induction of local and systemic inflammatory responses known as cytokine storms [18].

2.2. Innate Immune Responses

2.2.1. Local (Pulmonary) Innate Immune Responses to COVID-19 Infection

After harboring SARS-CoV-2, alveolar epithelial cells initiate innate immunity within the lungs [45]. Viral RNA, which constitutes a pathogen-associated molecular pattern (PAMP), is detected by various sensors including toll-like receptors (TLRs) 3, 7, and 8, as well as retinoic acid-inducible gene I-like receptors [46]. TLRs upregulate antiviral and pro-inflammatory mediators and trigger NF-kB signaling-mediated inflammatory pathways within the lungs [47]. In alveolar epithelial cells and macrophages infected with SARS-CoV-2, NF-κB contributes crucially to the production of various inflammatory cytokines and the development of cytokine storms [48]. In a recent study involving meta-transcriptomic sequencing, a robust innate immune response, hypercytokinemia, and expression of interferon (IFN)-stimulated genes (ISGs) in the bronchoalveolar lavage fluid were found in patients with COVID-19 [49]. Considering that SARS-CoV-2 robustly triggers the expression of numerous ISGs with immunopathogenic potential, with overrepresentation of genes involved in inflammation, ISGs can be used to determine disease severity.

2.2.2. Systemic Innate Immune Responses against COVID-19 Infection

T cell activation and differentiation occur after macrophage processing and the presentation of SARS-CoV-2 particles to T cells. This is followed by immune response amplification, as indicated by a massive release of cytokines [50], including interleukins (ILs) 1, 6, 8, and 21, as well as tumor necrosis factor-β and CCL2. Subsequently, activated lymphocytes and leukocytes are recruited to the site of infection [50]. Viral infection further induces the expression of cathelicidin and defensins, the antimicrobial peptides within the infected cells. Human cathelicidin peptide LL-37, a small, cationic, and amphipathic particle, facilitates the effective binding of cathelicidin and prevents viral invasion [51,52]. During antimicrobial attacks, cathelicidin removes the outer membrane of the virus through a single event rather than a gradual process [53]. This causes the leakage of viral components, which in turn leads to the death of the virus [54,55]. A study on Venezuelan equine encephalitis virus infections reported that cathelicidin inhibited virus entry and modulated the expression of IFN-β1 expression in the infected host cells, eliciting an antiviral response through the inhibition of viral replication [56]. Defensin, another antimicrobial peptide associated with the first line of immunological defense [57], also suppresses viral infection, either through the direct blockage of viral particles or indirectly through the indirect disruption of the viral life cycle [58]. Growing evidence suggests that antiviral activity related to defensins occurs not only during viral entry [59] but also modifies innate immune responses to viral infections. The most important mechanisms of these responses include the activation of T cells, recruitment of macrophages and dendritic cells, differentiation and maturation of dendritic cells, and production of pro-inflammatory cytokines by macrophages and mast cells [60] (Figure 2).

2.3. *Adaptive Immune Response: T Cell Differentiation and Inflammatory Cytokines in COVID-19*

Compared with COVID-19, fewer cases of severe Middle East Respiratory Syndrome coronavirus and SARS-CoV cases result in fatal lower respiratory tract infections and systemic extrapulmonary manifestations [61]. Cytotoxic T cells account for approximately 80% of inflammatory cells related to SARS-CoV in the pulmonary interstitium. Thus, they play a vital role in clearing the viral particles and are responsible for most immune-related injuries [50]. Adaptive immunity occurs through activation of $CD4^+$ T cells, which promotes the production of virus-specific antibodies by B cells. Depletion of $CD4^+$ T cells results in the reduced production of neutralizing antibodies and cytokines in patients with interstitial pneumonitis [62]. A 2020 study reported that T cell activation was substantially higher in patients with COVID-19 pneumonia, and that the T cells preferentially differentiate into Th17 cells [63].

The persistent response of T cells to S proteins and other structural proteins (M and N proteins) in SARS coronavirus is well established. This serves as a valuable reference for the development of SARS vaccines, particularly with respect to the induction of long-term memory in T cell and immune responses. However, another report suggested that protective antibodies against SARS-CoV-2 infection may not last long [64]. In another study, serological surveys in early convalescence revealed reduced levels of immunoglobulin G and neutralizing antibodies [65]. Clinical studies have indicated that the level of naïve $CD4^+$ T cells, regulatory T cells (Tregs), and Th2 cells is higher in patients with poor prognosis than that in patients with good prognosis, as reflected by progression in disease severity [66,67]. Immunosuppression and an enhanced inflammatory response were implicated in disease progression among patients with poor prognosis [68].

On the other hands, dendritic cells exhibit anti-inflammatory activity through antigen presentation and the regulation of T cell reactions to SARS-CoV. This will decrease the degree of CD4$^+$ T cells induce pro-inflammatory cytokine production through the activation of the NF-kB pathways [48]. They also produce IL-17, which recruits more inflammatory cells to the infection site, with further activation of downstream cytokine and chemokine cascades [50,69].

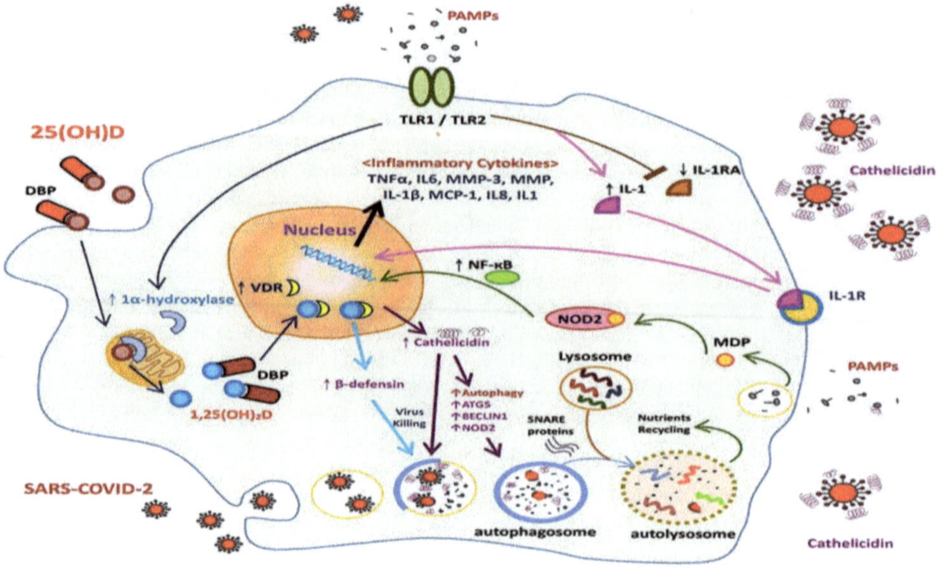

Figure 2. Putative vitamin D-related innate immunity (anti-infection activity) and autophagy responses to coronavirus disease 2019 (COVID-19) infection. The activation of monocyte toll-like receptors (TLR1/TLR2) by pathogen-associated molecular patterns (PAMPs) induces the expression of the cytokine interleukin-1 (IL-1) and suppresses the expression of the IL-1 receptor antagonist, thereby enhancing intracrine signaling by IL-1 and increasing the activity of nuclear factor κB (NF-κB). Pathogen phagocytosis increases the intracellular concentrations of muramyl dipeptide (MDP), which can then bind to the intracellular pathogen recognition receptor NOD2 and increase NF-κB activity. In addition, the activation of TLR1/TLR2 by PAMP results in the transcriptional induction of vitamin D receptor (VDR) and the activation of 1α-hydroxylase expression. Circulating 25-hydroxyvitamin D [25(OH)D] bound to serum vitamin D-binding protein enters monocytes in its free form and is converted to active 1,25-dihydroxyvitamin D [1,25(OH)2D] by mitochondrial 1α-hydroxylase. It then binds to VDR and acts as a transcription factor, induces the expression of cathelicidin and β-defensin 4A, and promotes autophagy through autophagosome formation. NF-κB also enhances the transcriptional induction of cathelicidin and β-defensin 4A. In the presence of increased cathelicidin, immune cells induce the activity of NOD2/CARD15-β-defensin 2, autophagy-related protein 5 (ATG5), and BECLIN1, and they then induce autophagy. Cathelicidin, β-defensin 4A, and mature autophagosomes then work in concert to eliminate bacteria. Cytoplasm SNAP receptor proteins mediate fusion between autophagosomes and lysosomes, and various lysosomal enzymes further hydrolyze proteins, lipids, and nucleic acids. Digestive nutrients may be recycled and utilized by the cells. The net efficacy of such a response is highly dependent on vitamin D status, as well as the availability of circulating 25(OH)D for intracrine conversion to active 1,25(OH)2D by the enzyme 1α-hydroxylase. Activation of TLR1 and TLR2 by PAMP induces the expression of cytokines and inflammatory pathways. Adequate vitamin D supplementation may strengthen the innate immune response against COVID-19 through TLR activation and autophagy, enhance antimicrobial peptide synthesis, and increase the generation of lysosomal degradation enzymes within macrophages.

2.4. *Renin–Angiotensin System and COVID-19*

The activation of the renin–angiotensin–aldosterone system (RAS) and Ang-II-related inflammation and fibrosis play vital roles in COVID-19 infection and mortality.

Renin converts angiotensinogen into angiotensin I, which is again metabolized to Ang II by the dipeptide carboxypeptidase ACE. The pro-inflammatory effects of Ang II [70] are exerted in concert with the AT1R. In a recent study, the ACE2 receptor and the downstream signaling pathway were identified as an essential counter-regulatory mechanism to RAS activation. Aldosterone reduces membrane ACE2 expression. Under favorable conditions, Ang II can be converted to Ang 1–7 via ACE2, the counter-regulatory effects of which are mediated by the Mas receptor [71].

Strains of both SARS-CoV and SARS-CoV-2 have been shown to use ACE2 receptors and enter the affected cells [72]. The identification of ACE2 and its modulation on the RAS constitutes an interesting topic for the development of therapeutic targets [73]. ACE2 exists in both membrane-bound and soluble forms, and SARS-CoV-2 infection involves the binding of the S protein to the membrane form [74]. A study reported that poor prognosis in SARS-CoV infection was accompanied by ACE-2 downregulation [75]. The virus attaches to ACE-2; the complexes then enter the cells by endocytosis. Viral complexes that are not endocytosed are digested by ADAM17 and lead to critical illness [76]. Less entry of viral particles into cells is associated with better clinical outcomes. A viewpoint paper (2020) revealed that reduced ACE-2 levels are correlated with more severe clinical presentations and harmful end organ damage [77].

3. Role of Vitamin D in the COVID-19 Era

3.1. Antiviral Activity of Vitamin D and the Innate Immune Response

The promotion of antiviral immunity by vitamin D, which is of great relevance to the current discussion on COVID-19, involves various mechanisms that overlap with antibacterial responses, such as the induction of cathelicidin and defensins, which can block viral entry into cells as well as suppress viral replication [65,78]. Another property of vitamin D relevant both to antibacterial and antiviral mechanisms acts through the promotion of autophagy [79], a fundamental biological process that maintains cellular homeostasis through the encapsulation of damaged organelles and misfolded proteins by intracellular membranes. Autophagy is also an essential mechanism by which cells respond to viral invasion. Specifically, autophagic encapsulation packages viral particles for lysosomal degradation and subsequent antigen presentation and activation of adaptive antiviral immune responses [80]. Thus, autophagy facilitates the creation of a cellular environment that is hostile to viruses but does not guarantee one.

In addition to its established function in bone homeostasis, vitamin D modulates and regulates multiple processes, including host defense, inflammation, immunity, and epithelial repair. Patients with respiratory disease are frequently presented with deficient in vitamin D; supplementation might provide substantial benefits to this population [81]. After binding to serum vitamin D binding protein, circulating 25-hydroxyvitamin D enters monocytes and increases the intracellular level of active 1,25-dihydroxyvitamin D (1,25D), which after binding to vitamin-D receptor (VDR) induces the expression of antimicrobial peptides cathelicidin and β-defensin 4A and promotes autophagy through autophagosome formation [82]. In humans, cathelicidin [83] and β-defensin [84] are produced through a vitamin D-dependent antimicrobial pathway. Our previous study also demonstrated that vitamin D-treated uremic hyperparathyroidism can efficiently increase serum cathelicidin levels [85]. Taken together, vitamin D promotes innate immunity through the expression of both cathelicidin and β-defensin, enhances autophagy through TLR activation, and affects complement activation. Figure 2 presents the putative immune-related mechanisms of vitamin D linked to COVID-19.

3.2. Vitamin D Regulates Adaptive Immunity

The adaptive immune system is initiated by the activation of antigen-presenting cells, such as dendritic cells and macrophages, which in turn activate the antigen-recognizing cells, T lymphocytes and B lymphocytes, which are major determinants of the immune response [86]. 1α,25-Dihydroxyvitamin D directly modulates inflammatory cytokines that are dependent on NF-κB activity in numerous types of cells, including macrophages, by blocking NF-κB p65 activation via the upregulation of the NF-κB inhibitory protein IκBα [87]. TLRs are transmembrane proteins that recognize conserved molecular motifs of viral and bacterial origin and initiate innate immune responses. TLR3 recognizes viral double-stranded RNA or synthetic double-stranded RNA (polyinosinic: polycytidylic acid) and is primarily involved in viral defense. Vitamin D treatment has been demonstrated to attenuate the expression of IL-8 in respiratory epithelial cells by double-stranded RNA-TLR3 [86,88].

Circulating T cells, B cells, and dendritic cells express the vitamin D-activating enzyme CYP27B1 (1α-hydroxylase) and the VDR, which then utilize the circulating 25D through intracrine conversion to active 1,25D. Increased intracellular 1,25D inhibits the maturation of dendritic cells and thus modulates the function of $CD4^+$ T cells. Systemic active vitamin D (1,25D) also regulates VDR-expressing $CD4^+$ T cells in a similar way. In essence, vitamin D inhibits the activation of type 1 T helper cells and cellular immune responses related to tissue destruction. In addition, vitamin D promotes the association of Th2 cells with humorally mediated immunity. In general, vitamin D regulates immunity by promoting the shift from Th1 to Th2 cells. Vitamin D also mitigates inflammation and tissue damage by inhibiting the development of Th17 cells. Similarly, Tregs suppress inflammation in response to vitamin D [82]. In short, vitamin D was assumed to modulate adaptive immunity against COVID-19 in several ways. For example, it can suppress the maturation of dendritic cells and weaken the antigenic presentation, and then increase cytokine production induced by $CD4^+$ T cells and promote the efficiency of Treg lymphocytes. Recent clinical study revealed that severe immunosuppression but not a prominent cytokine storm characterizes COVID-19 infections [68]. Vitamin D can also suppress the secretion of Th1 and Th17 cytokines and related tissue destruction (Figure 3), these effects are hypothesized to occur even during COVID-19, suggesting that appropriate vitamin D supplementation may attenuate immunosuppression and enhance anti-inflammatory effects against COVID-19.

3.3. *Vitamin D Modulates ACE2 and the RAS*

Vitamin D deficiency is a known pandemic and global public health problem which varies with age, ethnicity, and latitude. Environmental factors, including the lack or reduction of sunlight (UV-B), life with air pollution, and smoking, are responsible for vitamin D deficiency. The presence of comorbid diseases, such as septicemia, diabetes mellitus, chronic respiratory diseases, and cancer, is closely linked to vitamin D deficiency [89]. In this COVID-19 pandemic, a similarity in prevalent areas and the nature of SARS-CoV-2 infection and vitamin D deficiency was observed [9], which might explain the importance of vitamin D supplementation in COVID-19 [90]. Adequate vitamin D supplementation is also required to reduce RAS activity and increase ACE2 concentrations in acute lung injury. Specifically, sufficient vitamin D supplementation induces the ACE2/Ang 1–7 and suppresses the renin axis and the ACE/Ang II/AT1R axis [91].

It is well known that lower levels of 25 hydroxyvitamin D (25[OH]D) are strongly and independently associated with an increased risk of developing high blood pressure [92]. In line with this finding, both animal and human studies suggest that vitamin D deficiency may increase RAS activity within the kidneys both systemically and locally [93]. A study reported that individuals with vitamin D insufficiency and deficiency (15–29.9 ng and <15 ng/mL, respectively) had higher circulating Ang II concentrations than did those whose 25(OH)D concentrations were sufficient (≥ 30 ng/mL) [94]. The patients deficient in vitamin D also had a significantly reduced renal plasma flow response to Ang II infusion [95].

The COVID-19 prognosis of older adults, smokers, and individuals with obesity or other comorbidities such as hypertension and diabetes mellitus is poor [96]. RAS-acting agents that increase ACE2 levels serve as substrates for SARS-CoV-2 infection [97]. Circulating ACE2 is regarded as a biomarker of hypertension and heart failure [98] as well as diabetes [99]. SARS-CoV-2 infection downregulates ACE2 activity and accumulates toxic Ang II and metabolites, which subsequently develop into ARDS or fulminant myocarditis [97]. Vitamin D sufficiency can lower RAS activity through several pathways, including transcriptional suppression of renin, ACE, and Ang II expression [100] and increased ACE2 concentration in lipoprotein (LPS)-induced acute lung injury(ALI) [91]. In other words, vitamin D mitigates LPS-induced ALI by inducing the ACE2/Ang 1–7 axis and by suppressing both renin and the ACE/Ang II/AT1R axis [91]. Vitamin D treatment also increases soluble ACE2 (sACE2) [101] which retains the enzyme activity of ACE2 and can bind to the S-protein of SARS-CoV. So sACE2 may block S protein and prevent cells from being infected. In vivo studies on diabetic rats supplemented with active vitamin D reported that calcitriol decreased the ACE concentration and ACE/ACE2 ratio and increased ACE2 concentration [102]. Expression of ACE2 is reduced in patients with DM possibly due to glycosylation [31,103]; this could explain the increased predisposition to severe pulmonary lesions and ARDS with COVID-19. As a result, we can speculate on the beneficial effect of the vitamin-D supplement on diabetic patients with COVID-19.

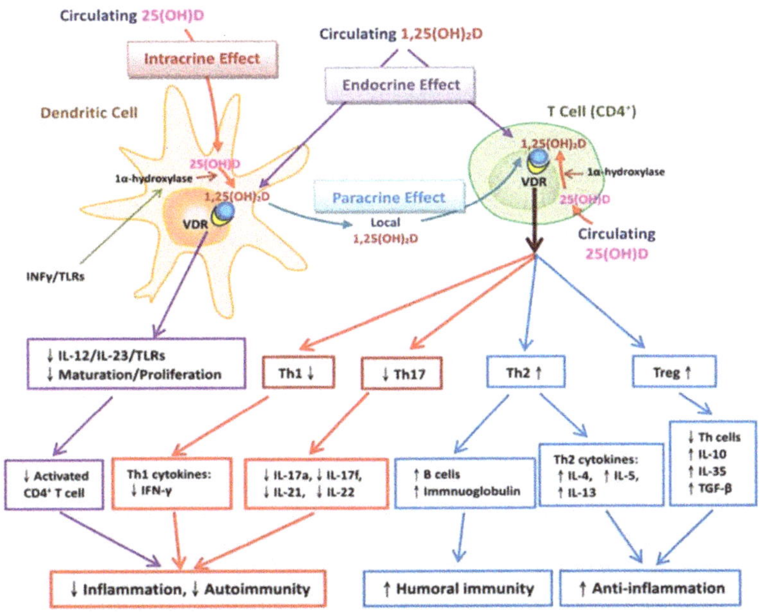

Figure 3. Vitamin D-related adaptive immune responses to COVID-19. Dendritic cells expressing 1α-hydroxylase and the vitamin D receptor (VDR) can utilize circulating 25-hydroxyvitamin D [25(OH)D] for intracrine responses through localized conversion to active vitamin D [1,25(OH)2D]. Intracrine synthesis of 1,25(OH)2D inhibits the maturation of dendritic cells, thereby modulating CD4$^+$ T cell function. CD4$^+$ T cell responses to 25(OH)D may also be mediated in a paracrine manner, with 1,25(OH)2D acting on VDR-expressing CD4$^+$ T cells. VDR-expressing CD4$^+$ T cells are also potential targets for systemic 1,25(OH)2D (endocrine effect). Vitamin D acts on dendritic cells to stimulate effector CD4$^+$ cells to differentiate into one of the four types of CD4$^+$ cells. Activated T cells also express VDR. Under normal circumstances, vitamin D increases T helper (Th) 2 (Th2) cytokines (e.g., IL-10) and the efficiency of regulatory T (Treg) lymphocytes. Vitamin D inhibits the development of Th1 cells, which are associated with the cellular immune response. In addition, vitamin D promotes the association of Th2 cells with humorally mediated immunity. Thus, vitamin D promotes the shift from Th1 to Th2 cells. Vitamin D also inhibits the development of Th17 cells, which play roles in tissue damage and inflammation. The fourth group of CD4$^+$ T cells, Tregs, suppress the function of vitamin D. Circulating and local active vitamin D acts through intracrine, paracrine, and endocrine effects to regulate adaptive immunity in SARS-CoV infection. First, it suppresses the maturation of dendritic cells and weakens the antigenic presentation. Second, it increases cytokine production by CD4$^+$ T cells and promotes the efficiency of Treg lymphocytes. Finally, it suppresses Th1 and Th17 cytokine secretion, as well as related tissue destruction [82].

Vitamin D increases the expression and bioavailability of ACE2, a mechanism that may be responsible for the trapping and inactivation of viruses. This suggests the potential benefits of adequate vitamin D supplementation, which requires further exploration. In sum, vitamin D may be able to combat COVID-19 and the related induction of MAS and ARDS by targeting ACE2 downregulation and the unbalanced RAS (Figure 4).

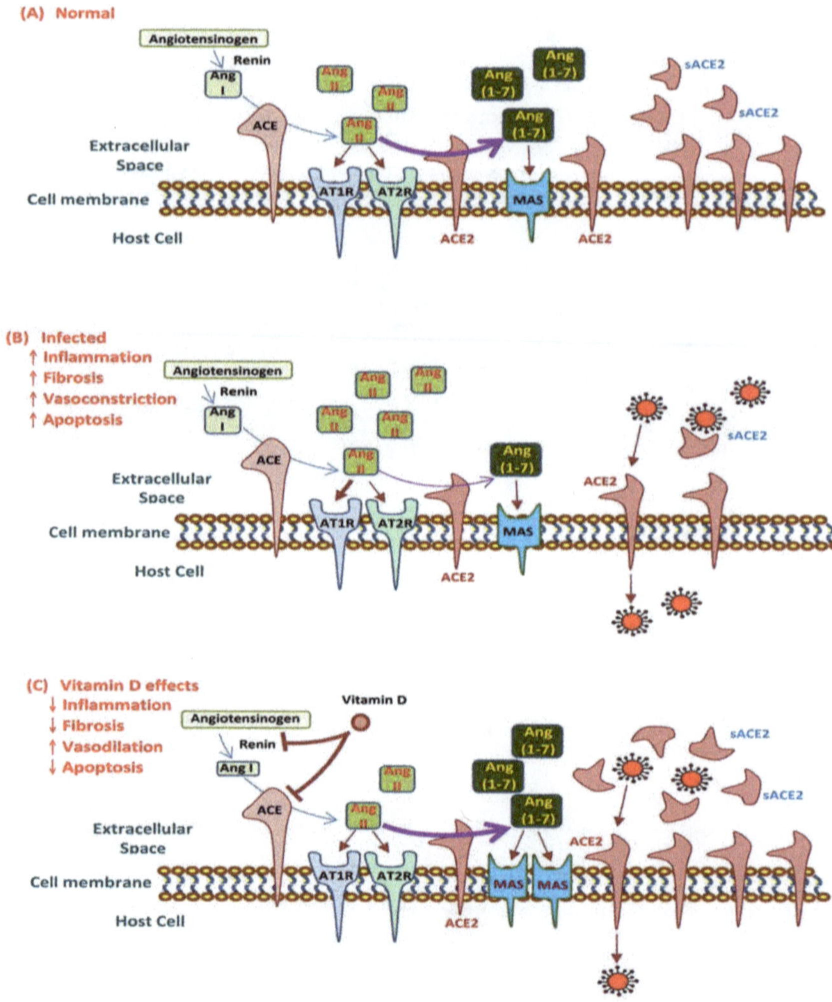

Figure 4. Effects of vitamin D on angiotensin-converting enzyme 2 (ACE2) and the renin–angiotensin–aldosterone system (RAS) in response to the coronavirus disease 2019 (COVID-19). (**A**) Schematic of the RAS under normal circumstances, with physiological steps of the generation of angiotensin (Ang) II and Ang 1–7 shown, as well as their activity on specific receptors. (**B**) Interaction of the severe acute respiratory syndrome coronavirus 2 with the RAS. (**C**) Possible therapeutic effects of vitamin D for COVID-19 and related acute respiratory distress syndrome or macrophage activation syndrome. The ACE2 molecule, besides being a receptor of SARS-CoV-2, reduces the activity of the renin–angiotensin system by converting Ang I and Ang II into Ang 1–9 and Ang 1–7 respectively [33]. Thus, the ACE2 protein has been shown to play an important role in protecting against some disorders such as cardiovascular complications, chronic obstructive pulmonary disease (COPD) and diabetes, among other COVID-19 comorbidities [34]. The ACE2/Ang 1–7 axis counterbalances the ACE/Ang II-I axis by decreasing Ang II levels, the activation of angiotensin type 1 receptors (AT1Rs) and, thus, leads to decreased pathophysiological effects on tissues, such as inflammation and fibrosis [35].

4. Controversial Findings from Clinical Studies

In a recent study on 20 European countries, a close association was found between lower levels of vitamin D and higher numbers of COVID-19 cases and mortality [2]. Vitamin D insufficiency or deficiency is more common in patients with obesity and diabetes, conditions that appear to lead to higher COVID-19 mortality rates [104]. However, recent data from a UK Biobank study did not reveal a relationship between vitamin D levels and the risk of SARS-CoV-2 infection. Moreover, vitamin D insufficiency or deficiency could not explain the ethnic differences in SARS-CoV-2 infection rates [105]. Although it is unclear whether vitamin D can help mitigate the risk of COVID-19 infection or its outcomes, it can be concluded that vitamin D deficiency is not beneficial. A 2020 retrospective cohort study conducted at an urban academic medical center indicated that vitamin D deficiency was associated with increased COVID-19 incidence [5]. Although epidemiological data appear to link vitamin D with COVID-19, conclusive evidence regarding the role of vitamin D in preventing or mitigating the severe respiratory complications of COVID-19 is lacking [106]. Multiple observation studies [107] have shown that serum concentrations of 25 hydroxyvitamin D (25(OH)D) are inversely correlated with the incidence or severity of COVID-19. Evidence to date has generally met Hill's causality criteria [108] in a biological system, although experimental verification is not sufficient. Vitamin D deficiency constitutes only one of numerous determinants of COVID-19 outcomes, but it can be corrected safely and inexpensively [108,109]. Evidence for the contribution of vitamin D to reducing the risk of COVID-19 includes the speculation that the outbreak is more serious in cooler regions and during the winter months, when 25(OH)D levels are the lowest, as well as the fact that case fatality rates from the disease are higher in patients who are potentially deficient in vitamin D (e.g., older adults and those with comorbidities) [6]. The fact that people tend to gather indoors in cooler regions and winter months versus outdoors in summer may account for the higher COVID-19 incidence observed in winter. Thus, ensuring adequate vitamin D intake (in line with national recommendations) probably be advisable.

Specifically, vitamin D supplements at therapeutic doses might not be dangerous to COVID-19 patients; it might even mitigate the exacerbation or severity of the disease. Of course, considering the emergency of the current situation, sufficient evidence to support this claim has yet to be collected. The recommended intake for individuals at higher risk of COVID-19 is 10,000 IU/day for 1 to 2 weeks and subsequently 5000 IU daily, with target 25(OH)D concentrations of over 40 to 60 ng/mL [6]. Well-conducted randomized controlled trials of vitamin D supplementation in patients with COVID-19 remain an urgent need.

5. Conclusions

Vitamin D levels are closely associated with COVID-19 severity and mortality. Adequate vitamin D supplementation may enhance the innate immune response against the disease through the increased synthesis of antimicrobial peptides to kill the virus intracellularly (e.g., intravesicularly). Regarding adaptive immunity, in addition to suppressing cytokine storms, vitamin D enhances $CD4^+$ T cells and promotes the production of virus-specific antibodies by activating T cell-dependent B cells. Vitamin D induces ACE2/Ang 1–7/MasR axis activity and inhibits the renin and ACE/Ang II/AT1R axis, which increases the concentration of ACE2, MasR, and Ang 1–7, and subsequently reduces RAS-related tissue inflammation, destruction, and fibrosis. The increased shedding of soluble ACE2 may provide additional protective effects. Clinical evidence on vitamin D supplementation in patients with COVID-19 remains inconclusive, perhaps due to varied study end-points, variations in epidemiological characteristics (e.g., race and dietary habits) and differed clinical settings. However, it is cost effectiveness to give vitamin D to boost immunity from prevention in COVID-19 era. It is important to conduct more randomized studies among different geographical populations to determine variations in requirements of vitamin D supplementation during COVID-19 era.

Author Contributions: M.-Y.P. drew up the manuscript and obtained revision documents. W.-C.L. and J.-Q.Z. performed a data analysis. C.-M.Z., C.-L.L. and K.-C.L. carried out the research. Y.-C.H. and J.-Y.S. was responsible to collect examination data. K.-C.L. and Y.-C.C. designed this research. All authors have read and agreed to the published version of the manuscript.

Funding: The study was funded by the Taipei Tzu Chi Hospital, Tzu Chi Buddhist Health Foundation. (TCRD-TPE-110-RT-1(1/3)).

Data Availability Statement: This is a narrative review article. The primary collection of documents for analysis and review comes from PubMed.

Acknowledgments: This research was fully supported administratively by the Department of Medical Research at Taipei Tzu Chi Hospital. Not only provided the necessary assistance in terms of workforce, but also in terms of finances, and provided considerable assistance in the publication of this article.

Conflicts of Interest: The authors declare no conflict of interest.

References

1. Mason, R.J. Pathogenesis of COVID-19 from a cell biology perspective. *Eur. Respir. J.* **2020**, *55*. [CrossRef] [PubMed]
2. Ali, N. Role of vitamin D in preventing of COVID-19 infection, progression and severity. *J. Infect. Public Health* **2020**, *10*, 1373–1380. [CrossRef]
3. Holick, M.F. The vitamin D deficiency pandemic: Approaches for diagnosis, treatment and prevention. *Rev. Endocr. Metab. Disord.* **2017**, *18*, 153–165. [CrossRef] [PubMed]
4. D'Avolio, A.; Avataneo, V.; Manca, A.; Cusato, J.; De Nicolo, A.; Lucchini, R.; Keller, F.; Cantu, M. 25-Hydroxyvitamin D Concentrations Are Lower in Patients with Positive PCR for SARS-CoV-2. *Nutrients* **2020**, *12*, 1359. [CrossRef]
5. Meltzer, D.O.; Best, T.J.; Zhang, H.; Vokes, T.; Arora, V.; Solway, J. Association of Vitamin D Status and Other Clinical Characteristics With COVID-19 Test Results. *JAMA Netw. Open* **2020**, *3*, e2019722. [CrossRef]
6. Grant, W.B.; Lahore, H.; McDonnell, S.L.; Baggerly, C.A.; French, C.B.; Aliano, J.L.; Bhattoa, H.P. Evidence that Vitamin D Supplementation Could Reduce Risk of Influenza and COVID-19 Infections and Deaths. *Nutrients* **2020**, *12*, 988. [CrossRef] [PubMed]
7. Zemb, P.; Bergman, P.; Camargo, C.A., Jr.; Cavalier, E.; Cormier, C.; Courbebaisse, M.; Hollis, B.; Joulia, F.; Minisola, S.; Pilz, S.; et al. Vitamin D deficiency and the COVID-19 pandemic. *J. Glob. Antimicrob. Resist.* **2020**, *22*, 133–134. [CrossRef]
8. Meftahi, G.H.; Jangravi, Z.; Sahraei, H.; Bahari, Z. The possible pathophysiology mechanism of cytokine storm in elderly adults with COVID-19 infection: The contribution of "inflame-aging". *Inflamm. Res.* **2020**, *69*, 825–839. [CrossRef] [PubMed]
9. Kara, M.; Ekiz, T.; Ricci, V.; Kara, O.; Chang, K.V.; Ozcakar, L. 'Scientific Strabismus' or two related pandemics: Coronavirus disease and vitamin D deficiency. *Br. J. Nutr.* **2020**, *124*, 736–741. [CrossRef] [PubMed]
10. Cereda, E.; Bogliolo, L.; Lobascio, F.; Barichella, M.; Zecchinelli, A.L.; Pezzoli, G.; Caccialanza, R. Vitamin D supplementation and outcomes in coronavirus disease 2019 (COVID-19) patients from the outbreak area of Lombardy, Italy. *Nutrition* **2021**, *82*, 111055. [CrossRef]
11. Ling, S.F.; Broad, E.; Murphy, R.; Pappachan, J.M.; Pardesi-Newton, S.; Kong, M.F.; Jude, E.B. High-Dose Cholecalciferol Booster Therapy is Associated with a Reduced Risk of Mortality in Patients with COVID-19: A Cross-Sectional Multi-Centre Observational Study. *Nutrients* **2020**, *12*, 3799. [CrossRef] [PubMed]
12. Annweiler, G.; Corvaisier, M.; Gautier, J.; Dubee, V.; Legrand, E.; Sacco, G.; Annweiler, C. Vitamin D Supplementation Associated to Better Survival in Hospitalized Frail Elderly COVID-19 Patients: The GERIA-COVID Quasi-Experimental Study. *Nutrients* **2020**, *12*, 3377. [CrossRef] [PubMed]
13. Bourgonje, A.R.; Abdulle, A.E.; Timens, W.; Hillebrands, J.L.; Navis, G.J.; Gordijn, S.J.; Bolling, M.C.; Dijkstra, G.; Voors, A.A.; Osterhaus, A.D.; et al. Angiotensin-converting enzyme 2 (ACE2), SARS-CoV-2 and the pathophysiology of coronavirus disease 2019 (COVID-19). *J. Pathol.* **2020**, *251*, 228–248. [CrossRef]
14. Sungnak, W.; Huang, N.; Becavin, C.; Berg, M.; Queen, R.; Litvinukova, M.; Talavera-Lopez, C.; Maatz, H.; Reichart, D.; Sampaziotis, F.; et al. SARS-CoV-2 entry factors are highly expressed in nasal epithelial cells together with innate immune genes. *Nat. Med.* **2020**, *26*, 681–687. [CrossRef] [PubMed]
15. Wan, Y.; Shang, J.; Graham, R.; Baric, R.S.; Li, F. Receptor Recognition by the Novel Coronavirus from Wuhan: An Analysis Based on Decade-Long Structural Studies of SARS Coronavirus. *J. Virol.* **2020**, *94*. [CrossRef]
16. Sims, A.C.; Baric, R.S.; Yount, B.; Burkett, S.E.; Collins, P.L.; Pickles, R.J. Severe acute respiratory syndrome coronavirus infection of human ciliated airway epithelia: Role of ciliated cells in viral spread in the conducting airways of the lungs. *J. Virol.* **2005**, *79*, 15511–15524. [CrossRef]
17. Hamming, I.; Timens, W.; Bulthuis, M.L.; Lely, A.T.; Navis, G.; van Goor, H. Tissue distribution of ACE2 protein, the functional receptor for SARS coronavirus. A first step in understanding SARS pathogenesis. *J. Pathol.* **2004**, *203*, 631–637. [CrossRef]
18. Netland, J.; Meyerholz, D.K.; Moore, S.; Cassell, M.; Perlman, S. Severe acute respiratory syndrome coronavirus infection causes neuronal death in the absence of encephalitis in mice transgenic for human ACE2. *J. Virol.* **2008**, *82*, 7264–7275. [CrossRef]

19. Zhou, F.; Yu, T.; Du, R.; Fan, G.; Liu, Y.; Liu, Z.; Xiang, J.; Wang, Y.; Song, B.; Gu, X.; et al. Clinical course and risk factors for mortality of adult inpatients with COVID-19 in Wuhan, China: A retrospective cohort study. *Lancet* **2020**, *395*, 1054–1062. [CrossRef]
20. Arentz, M.; Yim, E.; Klaff, L.; Lokhandwala, S.; Riedo, F.X.; Chong, M.; Lee, M. Characteristics and Outcomes of 21 Critically Ill Patients with COVID-19 in Washington State. *JAMA* **2020**. [CrossRef]
21. Mossel, E.C.; Wang, J.; Jeffers, S.; Edeen, K.E.; Wang, S.; Cosgrove, G.P.; Funk, C.J.; Manzer, R.; Miura, T.A.; Pearson, L.D.; et al. SARS-CoV replicates in primary human alveolar type II cell cultures but not in type I-like cells. *Virology* **2008**, *372*, 127–135. [CrossRef] [PubMed]
22. Wu, J.; Wu, X.; Zeng, W.; Guo, D.; Fang, Z.; Chen, L.; Huang, H.; Li, C. Chest CT Findings in Patients With Coronavirus Disease 2019 and Its Relationship With Clinical Features. *Investig. Radiol.* **2020**, *55*, 257–261. [CrossRef] [PubMed]
23. Qian, Z.; Travanty, E.A.; Oko, L.; Edeen, K.; Berglund, A.; Wang, J.; Ito, Y.; Holmes, K.V.; Mason, R.J. Innate immune response of human alveolar type II cells infected with severe acute respiratory syndrome-coronavirus. *Am. J. Respir. Cell Mol. Biol.* **2013**, *48*, 742–748. [CrossRef] [PubMed]
24. Shang, J.; Wan, Y.; Luo, C.; Ye, G.; Geng, Q.; Auerbach, A.; Li, F. Cell entry mechanisms of SARS-CoV-2. *Proc. Natl. Acad. Sci. USA* **2020**, *117*, 11727–11734. [CrossRef] [PubMed]
25. Walls, A.C.; Park, Y.J.; Tortorici, M.A.; Wall, A.; McGuire, A.T.; Veesler, D. Structure, Function, and Antigenicity of the SARS-CoV-2 Spike Glycoprotein. *Cell* **2020**, *181*, 281–292.e6. [CrossRef] [PubMed]
26. Iwata-Yoshikawa, N.; Okamura, T.; Shimizu, Y.; Hasegawa, H.; Takeda, M.; Nagata, N. TMPRSS2 Contributes to Virus Spread and Immunopathology in the Airways of Murine Models after Coronavirus Infection. *J. Virol.* **2019**, *93*. [CrossRef] [PubMed]
27. Hoffmann, M.; Kleine-Weber, H.; Pohlmann, S. A Multibasic Cleavage Site in the Spike Protein of SARS-CoV-2 Is Essential for Infection of Human Lung Cells. *Mol. Cell* **2020**, *78*, 779–784.e5. [CrossRef] [PubMed]
28. Ju, B.; Zhang, Q.; Ge, J.; Wang, R.; Sun, J.; Ge, X.; Yu, J.; Shan, S.; Zhou, B.; Song, S.; et al. Human neutralizing antibodies elicited by SARS-CoV-2 infection. *Nature* **2020**, *584*, 115–119. [CrossRef] [PubMed]
29. Dong, M.; Zhang, J.; Ma, X.; Tan, J.; Chen, L.; Liu, S.; Xin, Y.; Zhuang, L. ACE2, TMPRSS2 distribution and extrapulmonary organ injury in patients with COVID-19. *Biomed. Pharmacother.* **2020**, *131*, 110678. [CrossRef]
30. Seyedpour, S.; Khodaei, B.; Loghman, A.H.; Seyedpour, N.; Kisomi, M.F.; Balibegloo, M.; Nezamabadi, S.S.; Gholami, B.; Saghazadeh, A.; Rezaei, N. Targeted therapy strategies against SARS-CoV-2 cell entry mechanisms: A systematic review of in vitro and in vivo studies. *J. Cell Physiol.* **2021**, *236*, 2364–2392. [CrossRef]
31. Tikellis, C.; Thomas, M.C. Angiotensin-Converting Enzyme 2 (ACE2) Is a Key Modulator of the Renin Angiotensin System in Health and Disease. *Int. J. Pept.* **2012**, *2012*, 256294. [CrossRef]
32. Zou, Z.; Yan, Y.; Shu, Y.; Gao, R.; Sun, Y.; Li, X.; Ju, X.; Liang, Z.; Liu, Q.; Zhao, Y.; et al. Angiotensin-converting enzyme 2 protects from lethal avian influenza A H5N1 infections. *Nat. Commun.* **2014**, *5*, 3594. [CrossRef] [PubMed]
33. Mascolo, A.; Scavone, C.; Rafaniello, C.; Ferrajolo, C.; Racagni, G.; Berrino, L.; Paolisso, G.; Rossi, F.; Capuano, A. Renin-Angiotensin System and Coronavirus Disease 2019: A Narrative Review. *Front. Cardiovasc. Med.* **2020**, *7*, 143. [CrossRef] [PubMed]
34. Patel, V.B.; Zhong, J.C.; Grant, M.B.; Oudit, G.Y. Role of the ACE2/Angiotensin 1–7 Axis of the Renin-Angiotensin System in Heart Failure. *Circ. Res.* **2016**, *118*, 1313–1326. [CrossRef]
35. Gheblawi, M.; Wang, K.; Viveiros, A.; Nguyen, Q.; Zhong, J.C.; Turner, A.J.; Raizada, M.K.; Grant, M.B.; Oudit, G.Y. Angiotensin-Converting Enzyme 2: SARS-CoV-2 Receptor and Regulator of the Renin-Angiotensin System: Celebrating the 20th Anniversary of the Discovery of ACE2. *Circ. Res.* **2020**, *126*, 1456–1474. [CrossRef] [PubMed]
36. Banu, N.; Panikar, S.S.; Leal, L.R.; Leal, A.R. Protective role of ACE2 and its downregulation in SARS-CoV-2 infection leading to Macrophage Activation Syndrome: Therapeutic implications. *Life Sci.* **2020**, *256*, 117905. [CrossRef]
37. Jafarzadeh, A.; Chauhan, P.; Saha, B.; Jafarzadeh, S.; Nemati, M. Contribution of monocytes and macrophages to the local tissue inflammation and cytokine storm in COVID-19: Lessons from SARS and MERS, and potential therapeutic interventions. *Life Sci.* **2020**, *257*, 118102. [CrossRef]
38. Vijaykrishna, D.; Smith, G.J.; Zhang, J.X.; Peiris, J.S.; Chen, H.; Guan, Y. Evolutionary insights into the ecology of coronaviruses. *J. Virol.* **2007**, *81*, 4012–4020. [CrossRef]
39. Reyfman, P.A.; Walter, J.M.; Joshi, N.; Anekalla, K.R.; McQuattie-Pimentel, A.C.; Chiu, S.; Fernandez, R.; Akbarpour, M.; Chen, C.I.; Ren, Z.; et al. Single-Cell Transcriptomic Analysis of Human Lung Provides Insights into the Pathobiology of Pulmonary Fibrosis. *Am. J. Respir. Crit. Care Med.* **2019**, *199*, 1517–1536. [CrossRef] [PubMed]
40. Ge, X.Y.; Li, J.L.; Yang, X.L.; Chmura, A.A.; Zhu, G.; Epstein, J.H.; Mazet, J.K.; Hu, B.; Zhang, W.; Peng, C.; et al. Isolation and characterization of a bat SARS-like coronavirus that uses the ACE2 receptor. *Nature* **2013**, *503*, 535–538. [CrossRef]
41. Lau, S.K.; Woo, P.C.; Li, K.S.; Huang, Y.; Tsoi, H.W.; Wong, B.H.; Wong, S.S.; Leung, S.Y.; Chan, K.H.; Yuen, K.Y. Severe acute respiratory syndrome coronavirus-like virus in Chinese horseshoe bats. *Proc. Natl. Acad. Sci. USA* **2005**, *102*, 14040–14045. [CrossRef] [PubMed]
42. Lam, T.T.; Jia, N.; Zhang, Y.W.; Shum, M.H.; Jiang, J.F.; Zhu, H.C.; Tong, Y.G.; Shi, Y.X.; Ni, X.B.; Liao, Y.S.; et al. Identifying SARS-CoV-2-related coronaviruses in Malayan pangolins. *Nature* **2020**. [CrossRef]
43. Lan, L.; Xu, D.; Ye, G.; Xia, C.; Wang, S.; Li, Y.; Xu, H. Positive RT-PCR Test Results in Patients Recovered From COVID-19. *JAMA* **2020**, *20*. [CrossRef] [PubMed]

44. Abassi, Z.; Knaney, Y.; Karram, T.; Heyman, S.N. The Lung Macrophage in SARS-CoV-2 Infection: A Friend or a Foe? *Front. Immunol.* **2020**, *11*, 1312. [CrossRef] [PubMed]
45. Whitsett, J.A.; Alenghat, T. Respiratory epithelial cells orchestrate pulmonary innate immunity. *Nat. Immunol.* **2015**, *16*, 27–35. [CrossRef] [PubMed]
46. Chow, K.T.; Gale, M., Jr.; Loo, Y.M. RIG-I and Other RNA Sensors in Antiviral Immunity. *Annu. Rev. Immunol.* **2018**, *36*, 667–694. [CrossRef] [PubMed]
47. Yoshikawa, T.; Hill, T.; Li, K.; Peters, C.J.; Tseng, C.T. Severe acute respiratory syndrome (SARS) coronavirus-induced lung epithelial cytokines exacerbate SARS pathogenesis by modulating intrinsic functions of monocyte-derived macrophages and dendritic cells. *J. Virol.* **2009**, *83*, 3039–3048. [CrossRef]
48. Channappanavar, R.; Perlman, S. Pathogenic human coronavirus infections: Causes and consequences of cytokine storm and immunopathology. *Semin. Immunopathol.* **2017**, *39*, 529–539. [CrossRef] [PubMed]
49. Zhou, Z.; Ren, L.; Zhang, L.; Zhong, J.; Xiao, Y.; Jia, Z.; Guo, L.; Yang, J.; Wang, C.; Jiang, S.; et al. Heightened Innate Immune Responses in the Respiratory Tract of COVID-19 Patients. *Cell Host Microbe* **2020**, *27*, 883–890.e2. [CrossRef]
50. Li, G.; Fan, Y.; Lai, Y.; Han, T.; Li, Z.; Zhou, P.; Pan, P.; Wang, W.; Hu, D.; Liu, X.; et al. Coronavirus infections and immune responses. *J. Med. Virol.* **2020**, *92*, 424–432. [CrossRef]
51. Ganz, T. The role of antimicrobial peptides in innate immunity. *Integr. Comp. Biol.* **2003**, *43*, 300–304. [CrossRef]
52. Ahmed, A.; Siman-Tov, G.; Hall, G.; Bhalla, N.; Narayanan, A. Human Antimicrobial Peptides as Therapeutics for Viral Infections. *Viruses* **2019**, *11*, 704. [CrossRef]
53. Mangoni, M.L.; McDermott, A.M.; Zasloff, M. Antimicrobial peptides and wound healing: Biological and therapeutic considerations. *Exp. Dermatol.* **2016**, *25*, 167–173. [CrossRef]
54. Weidmann, J.; Craik, D.J. Discovery, structure, function, and applications of cyclotides: Circular proteins from plants. *J. Exp. Bot.* **2016**, *67*, 4801–4812. [CrossRef]
55. Agarwal, G.; Gabrani, R. Antiviral Peptides: Identification and Validation. *Int. J. Pept. Res. Ther.* **2020**, 1–20. [CrossRef]
56. Ahmed, A.; Siman-Tov, G.; Keck, F.; Kortchak, S.; Bakovic, A.; Risner, K.; Lu, T.K.; Bhalla, N.; de la Fuente-Nunez, C.; Narayanan, A. Human cathelicidin peptide LL-37 as a therapeutic antiviral targeting Venezuelan equine encephalitis virus infections. *Antiviral. Res.* **2019**, *164*, 61–69. [CrossRef] [PubMed]
57. Klotman, M.E.; Chang, T.L. Defensins in innate antiviral immunity. *Nat. Rev. Immunol.* **2006**, *6*, 447–456. [CrossRef] [PubMed]
58. Pachon-Ibanez, M.E.; Smani, Y.; Pachon, J.; Sanchez-Cespedes, J. Perspectives for clinical use of engineered human host defense antimicrobial peptides. *FEMS Microbiol. Rev.* **2017**, *41*, 323–342. [CrossRef] [PubMed]
59. Castaneda-Sanchez, J.I.; Dominguez-Martinez, D.A.; Olivar-Espinosa, N.; Garcia-Perez, B.E.; Lorono-Pino, M.A.; Luna-Herrera, J.; Salazar, M.I. Expression of Antimicrobial Peptides in Human Monocytic Cells and Neutrophils in Response to Dengue Virus Type 2. *Intervirology* **2016**, *59*, 8–19. [CrossRef]
60. Findlay, F.; Proudfoot, L.; Stevens, C.; Barlow, P.G. Cationic host defense peptides; novel antimicrobial therapeutics against Category A pathogens and emerging infections. *Pathog. Glob. Health* **2016**, *110*, 137–147. [CrossRef]
61. Sato, K.; Misawa, N.; Takeuchi, J.S.; Kobayashi, T.; Izumi, T.; Aso, H.; Nagaoka, S.; Yamamoto, K.; Kimura, I.; Konno, Y.; et al. Experimental Adaptive Evolution of Simian Immunodeficiency Virus SIVcpz to Pandemic Human Immunodeficiency Virus Type 1 by Using a Humanized Mouse Model. *J. Virol.* **2018**, *92*. [CrossRef]
62. Chen, J.; Lau, Y.F.; Lamirande, E.W.; Paddock, C.D.; Bartlett, J.H.; Zaki, S.R.; Subbarao, K. Cellular immune responses to severe acute respiratory syndrome coronavirus (SARS-CoV) infection in senescent BALB/c mice: CD4+ T cells are important in control of SARS-CoV infection. *J. Virol.* **2010**, *84*, 1289–1301. [CrossRef] [PubMed]
63. De Biasi, S.; Meschiari, M.; Gibellini, L.; Bellinazzi, C.; Borella, R.; Fidanza, L.; Gozzi, L.; Iannone, A.; Lo Tartaro, D.; Mattioli, M.; et al. Marked T cell activation, senescence, exhaustion and skewing towards TH17 in patients with COVID-19 pneumonia. *Nat. Commun.* **2020**, *11*, 3434. [CrossRef]
64. Liu, T.; Zeng, G.; Tao, H.; Shi, Y.; Group, C.-i.C.P.R.; Wang, T.; Liu, T.; Guo, F.; Zhou, F.; Wang, X. Low prevalence of IgG antibodies to SARS-CoV-2 in cancer patients with COVID-19. *Int. J. Cancer* **2020**. [CrossRef]
65. Long, Q.X.; Tang, X.J.; Shi, Q.L.; Li, Q.; Deng, H.J.; Yuan, J.; Hu, J.L.; Xu, W.; Zhang, Y.; Lv, F.J.; et al. Clinical and immunological assessment of asymptomatic SARS-CoV-2 infections. *Nat. Med.* **2020**, *26*, 845–848. [CrossRef] [PubMed]
66. Levine, A.G.; Mendoza, A.; Hemmers, S.; Moltedo, B.; Niec, R.E.; Schizas, M.; Hoyos, B.E.; Putintseva, E.V.; Chaudhry, A.; Dikiy, S.; et al. Stability and function of regulatory T cells expressing the transcription factor T-bet. *Nature* **2017**, *546*, 421–425. [CrossRef] [PubMed]
67. Wei, L.L.; Wang, W.J.; Chen, D.X.; Xu, B. Dysregulation of the immune response affects the outcome of critical COVID-19 patients. *J. Med. Virol.* **2020**. [CrossRef] [PubMed]
68. Remy, K.E.; Mazer, M.; Striker, D.A.; Ellebedy, A.H.; Walton, A.H.; Unsinger, J.; Blood, T.M.; Mudd, P.A.; Yi, D.J.; Mannion, D.A.; et al. Severe immunosuppression and not a cytokine storm characterizes COVID-19 infections. *JCI Insight* **2020**, *5*. [CrossRef]
69. Bunte, K.; Beikler, T. Th17 Cells and the IL-23/IL-17 Axis in the Pathogenesis of Periodontitis and Immune-Mediated Inflammatory Diseases. *Int. J. Mol. Sci.* **2019**, *20*, 3394. [CrossRef]
70. Holmberg, J.; Bhattachariya, A.; Alajbegovic, A.; Rippe, C.; Ekman, M.; Dahan, D.; Hien, T.T.; Boettger, T.; Braun, T.; Sward, K.; et al. Loss of Vascular Myogenic Tone in miR-143/145 Knockout Mice Is Associated With Hypertension-Induced Vascular Lesions in Small Mesenteric Arteries. *Arterioscler. Thromb. Vasc. Biol.* **2018**, *38*, 414–424. [CrossRef]

71. Brojakowska, A.; Narula, J.; Shimony, R.; Bander, J. Clinical Implications of SARS-CoV-2 Interaction With Renin Angiotensin System: JACC Review Topic of the Week. *J. Am. Coll. Cardiol.* **2020**, *75*, 3085–3095. [CrossRef]
72. Weiss, S.R.; Navas-Martin, S. Coronavirus pathogenesis and the emerging pathogen severe acute respiratory syndrome coronavirus. *Microbiol. Mol. Biol. Rev.* **2005**, *69*, 635–664. [CrossRef] [PubMed]
73. Thomas, M.C.; Pickering, R.J.; Tsorotes, D.; Koitka, A.; Sheehy, K.; Bernardi, S.; Toffoli, B.; Nguyen-Huu, T.P.; Head, G.A.; Fu, Y.; et al. Genetic Ace2 deficiency accentuates vascular inflammation and atherosclerosis in the ApoE knockout mouse. *Circ. Res.* **2010**, *107*, 888–897. [CrossRef] [PubMed]
74. Hoffmann, M.; Kleine-Weber, H.; Schroeder, S.; Kruger, N.; Herrler, T.; Erichsen, S.; Schiergens, T.S.; Herrler, G.; Wu, N.H.; Nitsche, A.; et al. SARS-CoV-2 Cell Entry Depends on ACE2 and TMPRSS2 and Is Blocked by a Clinically Proven Protease Inhibitor. *Cell* **2020**, *181*, 271–280.e8. [CrossRef] [PubMed]
75. Heurich, A.; Hofmann-Winkler, H.; Gierer, S.; Liepold, T.; Jahn, O.; Pohlmann, S. TMPRSS2 and ADAM17 cleave ACE2 differentially and only proteolysis by TMPRSS2 augments entry driven by the severe acute respiratory syndrome coronavirus spike protein. *J. Virol.* **2014**, *88*, 1293–1307. [CrossRef]
76. Sfera, A.; Osorio, C.; Jafri, N.; Diaz, E.L.; Campo Maldonado, J.E. Intoxication With Endogenous Angiotensin II: A COVID-19 Hypothesis. *Front. Immunol.* **2020**, *11*, 1472. [CrossRef]
77. AlGhatrif, M.; Cingolani, O.; Lakatta, E.G. The Dilemma of Coronavirus Disease 2019, Aging, and Cardiovascular Disease: Insights From Cardiovascular Aging Science. *JAMA Cardiol.* **2020**, *5*, 747–748. [CrossRef]
78. Gwyer Findlay, E.; Currie, S.M.; Davidson, D.J. Cationic host defence peptides: Potential as antiviral therapeutics. *BioDrugs* **2013**, *27*, 479–493. [CrossRef]
79. Campbell, G.R.; Spector, S.A. Vitamin D inhibits human immunodeficiency virus type 1 and Mycobacterium tuberculosis infection in macrophages through the induction of autophagy. *PLoS Pathog.* **2012**, *8*, e1002689. [CrossRef] [PubMed]
80. Tian, Y.; Wang, M.L.; Zhao, J. Crosstalk between Autophagy and Type I Interferon Responses in Innate Antiviral Immunity. *Viruses* **2019**, *11*, 132. [CrossRef]
81. Zdrenghea, M.T.; Makrinioti, H.; Bagacean, C.; Bush, A.; Johnston, S.L.; Stanciu, L.A. Vitamin D modulation of innate immune responses to respiratory viral infections. *Rev. Med. Virol.* **2017**, *27*. [CrossRef] [PubMed]
82. Liu, W.C.; Zheng, C.M.; Lu, C.L.; Lin, Y.F.; Shyu, J.F.; Wu, C.C.; Lu, K.C. Vitamin D and immune function in chronic kidney disease. *Clin. Chim. Acta* **2015**, *450*, 135–144. [CrossRef]
83. Matsumura, T.; Sugiyama, N.; Murayama, A.; Yamada, N.; Shiina, M.; Asabe, S.; Wakita, T.; Imawari, M.; Kato, T. Antimicrobial peptide LL-37 attenuates infection of hepatitis C virus. *Hepatol. Res.* **2016**, *46*, 924–932. [CrossRef]
84. Su, D.; Nie, Y.; Zhu, A.; Chen, Z.; Wu, P.; Zhang, L.; Luo, M.; Sun, Q.; Cai, L.; Lai, Y.; et al. Vitamin D Signaling through Induction of Paneth Cell Defensins Maintains Gut Microbiota and Improves Metabolic Disorders and Hepatic Steatosis in Animal Models. *Front. Physiol.* **2016**, *7*, 498. [CrossRef] [PubMed]
85. Zheng, J.Q.; Hou, Y.C.; Zheng, C.M.; Lu, C.L.; Liu, W.C.; Wu, C.C.; Huang, M.T.; Lin, Y.F.; Lu, K.C. Cholecalciferol Additively Reduces Serum Parathyroid Hormone and Increases Vitamin D and Cathelicidin Levels in Paricalcitol-Treated Secondary Hyperparathyroid Hemodialysis Patients. *Nutrients* **2016**, *8*. [CrossRef] [PubMed]
86. Balla, M.; Merugu, G.P.; Konala, V.M.; Sangani, V.; Kondakindi, H.; Pokal, M.; Gayam, V.; Adapa, S.; Naramala, S.; Malayala, S.V. Back to basics: Review on vitamin D and respiratory viral infections including COVID-19. *J. Community Hosp. Intern. Med. Perspect.* **2020**, *10*, 529–536. [CrossRef] [PubMed]
87. Chen, Y.; Zhang, J.; Ge, X.; Du, J.; Deb, D.K.; Li, Y.C. Vitamin D receptor inhibits nuclear factor kappaB activation by interacting with IkappaB kinase beta protein. *J. Biol. Chem.* **2013**, *288*, 19450–19458. [CrossRef]
88. Hansdottir, S.; Monick, M.M.; Hinde, S.L.; Lovan, N.; Look, D.C.; Hunninghake, G.W. Respiratory epithelial cells convert inactive vitamin D to its active form: Potential effects on host defense. *J. Immunol.* **2008**, *181*, 7090–7099. [CrossRef]
89. Schleicher, R.L.; Sternberg, M.R.; Looker, A.C.; Yetley, E.A.; Lacher, D.A.; Sempos, C.T.; Taylor, C.L.; Durazo-Arvizu, R.A.; Maw, K.L.; Chaudhary-Webb, M.; et al. National Estimates of Serum Total 25-Hydroxyvitamin D and Metabolite Concentrations Measured by Liquid Chromatography-Tandem Mass Spectrometry in the US Population during 2007–2010. *J. Nutr.* **2016**, *146*, 1051–1061. [CrossRef]
90. Ho, P.; Zheng, J.Q.; Wu, C.C.; Hou, Y.C.; Liu, W.C.; Lu, C.L.; Zheng, C.M.; Lu, K.C.; Chao, Y.C. Perspective Adjunctive Therapies for COVID-19: Beyond Antiviral Therapy. *Int. J. Med. Sci.* **2021**, *18*, 314–324. [CrossRef]
91. Xu, J.; Yang, J.; Chen, J.; Luo, Q.; Zhang, Q.; Zhang, H. Vitamin D alleviates lipopolysaccharideinduced acute lung injury via regulation of the reninangiotensin system. *Mol. Med. Rep.* **2017**, *16*, 7432–7438. [CrossRef] [PubMed]
92. Kunutsor, S.K.; Apekey, T.A.; Steur, M. Vitamin D and risk of future hypertension: Meta-analysis of 283,537 participants. *Eur. J. Epidemiol.* **2013**, *28*, 205–221. [CrossRef] [PubMed]
93. McMullan, C.J.; Borgi, L.; Curhan, G.C.; Fisher, N.; Forman, J.P. The effect of vitamin D on renin-angiotensin system activation and blood pressure: A randomized control trial. *J. Hypertens.* **2017**, *35*, 822–829. [CrossRef] [PubMed]
94. Santoro, D.; Caccamo, D.; Lucisano, S.; Buemi, M.; Sebekova, K.; Teta, D.; De Nicola, L. Interplay of vitamin D, erythropoiesis, and the renin-angiotensin system. *BioMed Res. Int.* **2015**, *2015*, 145828. [CrossRef]
95. Tomaschitz, A.; Pilz, S.; Ritz, E.; Grammer, T.; Drechsler, C.; Boehm, B.O.; Marz, W. Independent association between 1,25-dihydroxyvitamin D, 25-hydroxyvitamin D and the renin-angiotensin system: The Ludwigshafen Risk and Cardiovascular Health (LURIC) study. *Clin. Chim. Acta* **2010**, *411*, 1354–1360. [CrossRef]

96. Cure, E.; Cumhur Cure, M. Comment on "Organ-protective effect of angiotensin-converting enzyme 2 and its effect on the prognosis of COVID-19". *J. Med. Virol.* **2020**, *14*. [CrossRef]
97. Hanff, T.C.; Harhay, M.O.; Brown, T.S.; Cohen, J.B.; Mohareb, A.M. Is There an Association Between COVID-19 Mortality and the Renin-Angiotensin System-a Call for Epidemiologic Investigations. *Clin. Infect. Dis.* **2020**. [CrossRef]
98. Uri, K.; Fagyas, M.; Manyine Siket, I.; Kertesz, A.; Csanadi, Z.; Sandorfi, G.; Clemens, M.; Fedor, R.; Papp, Z.; Edes, I.; et al. New perspectives in the renin-angiotensin-aldosterone system (RAAS) IV: Circulating ACE2 as a biomarker of systolic dysfunction in human hypertension and heart failure. *PLoS ONE* **2014**, *9*, e87845. [CrossRef]
99. Soro-Paavonen, A.; Gordin, D.; Forsblom, C.; Rosengard-Barlund, M.; Waden, J.; Thorn, L.; Sandholm, N.; Thomas, M.C.; Groop, P.H.; FinnDiane Study, G. Circulating ACE2 activity is increased in patients with type 1 diabetes and vascular complications. *J. Hypertens.* **2012**, *30*, 375–383. [CrossRef] [PubMed]
100. Yuan, W.; Pan, W.; Kong, J.; Zheng, W.; Szeto, F.L.; Wong, K.E.; Cohen, R.; Klopot, A.; Zhang, Z.; Li, Y.C. 1,25-dihydroxyvitamin D3 suppresses renin gene transcription by blocking the activity of the cyclic AMP response element in the renin gene promoter. *J. Biol. Chem.* **2007**, *282*, 29821–29830. [CrossRef] [PubMed]
101. Jia, H.P.; Look, D.C.; Tan, P.; Shi, L.; Hickey, M.; Gakhar, L.; Chappell, M.C.; Wohlford-Lenane, C.; McCray, P.B., Jr. Ectodomain shedding of angiotensin converting enzyme 2 in human airway epithelia. *Am. J. Physiol. Lung Cell Mol. Physiol.* **2009**, *297*, 84–96. [CrossRef] [PubMed]
102. Lin, M.; Gao, P.; Zhao, T.; He, L.; Li, M.; Li, Y.; Shui, H.; Wu, X. Calcitriol regulates angiotensin-converting enzyme and angiotensin converting-enzyme 2 in diabetic kidney disease. *Mol. Biol. Rep.* **2016**, *43*, 397–406. [CrossRef]
103. Wu, C.; Chen, X.; Cai, Y.; Xia, J.; Zhou, X.; Xu, S.; Huang, H.; Zhang, L.; Zhou, X.; Du, C.; et al. Risk Factors Associated With Acute Respiratory Distress Syndrome and Death in Patients With Coronavirus Disease 2019 Pneumonia in Wuhan, China. *JAMA Intern. Med.* **2020**, *180*, 934–943. [CrossRef] [PubMed]
104. Weir, E.K.; Thenappan, T.; Bhargava, M.; Chen, Y. Does vitamin D deficiency increase the severity of COVID-19? *Clin. Med. (Lond)* **2020**, *20*, 107–108. [CrossRef] [PubMed]
105. Hastie, C.E.; Mackay, D.F.; Ho, F.; Celis-Morales, C.A.; Katikireddi, S.V.; Niedzwiedz, C.L.; Jani, B.D.; Welsh, P.; Mair, F.S.; Gray, S.R.; et al. Vitamin D concentrations and COVID-19 infection in UK Biobank. *Diabetes Metab. Syndr.* **2020**, *14*, 561–565. [CrossRef] [PubMed]
106. Chandran, M.; Chan Maung, A.; Mithal, A.; Parameswaran, R. Vitamin D in COVID-19: Dousing the fire or averting the storm?—A perspective from the Asia-Pacific. *Osteoporos. Sarcopenia* **2020**. [CrossRef]
107. Mercola, J.; Grant, W.B.; Wagner, C.L. Evidence Regarding Vitamin D and Risk of COVID-19 and Its Severity. *Nutrients* **2020**, *12*, 3361. [CrossRef]
108. Hill, A.B. The environment and disease: Association or causation? *J. R. Soc. Med.* **2015**, *108*, 32–37. [CrossRef]
109. Mitchell, F. Vitamin-D and COVID-19: Do deficient risk a poorer outcome? *Lancet Diabetes Endocrinol.* **2020**, *8*, 570. [CrossRef]

International Journal of
Molecular Sciences

MDPI

Article

A Phenylacetamide Resveratrol Derivative Exerts Inhibitory Effects on Breast Cancer Cell Growth

Adele Chimento [1,*,†], Anna Santarsiero [2,†], Domenico Iacopetta [1], Jessica Ceramella [1], Arianna De Luca [1], Vittoria Infantino [2], Ortensia Ilaria Parisi [1], Paola Avena [1], Maria Grazia Bonomo [2,3], Carmela Saturnino [2,3,*], Maria Stefania Sinicropi [1,‡] and Vincenzo Pezzi [1,*,‡]

1 Department of Pharmacy and Health and Nutritional Sciences, University of Calabria, Via Pietro Bucci, Arcavacata di Rende, 87036 Cosenza, Italy; domenico.iacopetta@unical.it (D.I.); jessicaceramella@gmail.com (J.C.); ariannadl@hotmail.it (A.D.L.); ortensiailaria.parisi@unical.it (O.I.P.); paox1982@hotmail.it (P.A.); s.sinicropi@unical.it (M.S.S.)
2 Department of Science, University of Basilicata, Viale dell'Ateneo Lucano 10, 85100 Potenza, Italy; anna.santarsiero@unibas.it (A.S.); vittoria.infantino@unibas.it (V.I.); mariagrazia.bonomo@unibas.it (M.G.B.)
3 Spinoff TNcKILLERS, Viale dell'Ateneo Lucano 10, 85100 Potenza, Italy
* Correspondence: adele.chimento@unical.it (A.C.); carmela.saturnino@unibas.it (C.S.); v.pezzi@unical.it (V.P.); Tel.: +39-0984-493-184 (A.C.); +39-0971-26442 (C.S.); +39-0984-493-148 (V.P.)
† Shared first authors.
‡ Shared last authors.

Abstract: Resveratrol (RSV) is a natural compound that displays several pharmacological properties, including anti-cancer actions. However, its clinical application is limited because of its low solubility and bioavailability. Here, the antiproliferative and anti-inflammatory activity of a series of phenylacetamide RSV derivatives has been evaluated in several cancer cell lines. These derivatives contain a monosubstituted aromatic ring that could mimic the RSV phenolic nucleus and a longer flexible chain that could confer a better stability and bioavailability than RSV. Using MTT assay, we demonstrated that most derivatives exerted antiproliferative effects in almost all of the cancer cell lines tested. Among them, derivative **2**, that showed greater bioavailability than RSV, was the most active, particularly against estrogen receptor positive (ER+) MCF7 and estrogen receptor negative (ER-) MDA-MB231 breast cancer cell lines. Moreover, we demonstrated that these derivatives, particularly derivative **2**, were able to inhibit NO and ROS synthesis and PGE2 secretion in lipopolysaccharide (LPS)-activated U937 human monocytic cells (derived from a histiocytoma). In order to define the molecular mechanisms underlying the antiproliferative effects of derivative **2**, we found that it determined cell cycle arrest at the G1 phase, modified the expression of cell cycle regulatory proteins, and ultimately triggered apoptotic cell death in both breast cancer cell lines. Taken together, these results highlight the studied RSV derivatives, particularly derivative **2**, as promising tools for the development of new and more bioavailable derivatives useful in the treatment of breast cancer.

Keywords: resveratrol; phenylacetamide RSV derivatives; breast cancer cell lines; antiproliferative activity; anti-inflammatory activity; cell cycle arrest; cell death

Citation: Chimento, A.; Santarsiero, A.; Iacopetta, D.; Ceramella, J.; De Luca, A.; Infantino, V.; Parisi, O.I.; Avena, P.; Bonomo, M.G.; Saturnino, C.; et al. A Phenylacetamide Resveratrol Derivative Exerts Inhibitory Effects on Breast Cancer Cell Growth. *Int. J. Mol. Sci.* **2021**, *22*, 5255. https://doi.org/10.3390/ijms22105255

Academic Editor: Maurizio Battino

Received: 7 April 2021
Accepted: 11 May 2021
Published: 17 May 2021

1. Introduction

Many foods and their bioactive components, including polyphenols, are beneficial for various diseases. Several in vitro and in vivo studies have indicated that a diet based on the consumption of cereals, legumes, vegetables, and fruits containing high levels of polyphenols prevents many diseases, including cancer [1,2]. Among them, a promising candidate is resveratrol (3,5,4′-trihydroxystilbene) (RSV), a naturally occurring non-flavonoid polyphenol [3]. RSV is the major phytoalexin produced by plants in response to stress, injury, infection, or UV radiation and is mainly present in peanuts, grapes, red wine, and some berries [4]. Although some of these plants and their extracts have been used for therapeutic purposes by ancient cultures, RSV itself was first described in 1939, when it was isolated from *Veratrum grandiflorum* [5]. Chemically, RSV is a stilbene compound and consists of two aromatic rings hydroxyl substituted with a C=C bond between them. It exists as *cis* and *trans* isomeric forms, with *trans* to *cis* isomerization facilitated by UV exposure, during the fermentation of skins, or under high pH conditions [6]. The *trans* form has greater stability and biological activity than the *cis* form, and is responsible for the induction of cellular responses. The increased interest in the *trans* isoform is also closely related to epidemiological studies showing an inverse relationship between moderate wine consumption and cardiovascular disease (the so-called "French paradox"). Moreover in vitro and in vivo studies have also demonstrated beneficial effects of RSV on human health [7,8], even though the mechanisms have not yet been fully elucidated. RSV has been reported to exhibit numerous activities, including antioxidant [9], anti-inflammatory [10], cardioprotective [11], neuroprotective [12] and anticancer properties [13,14] and so on, by acting through several molecular mechanisms [7]. RSV is able to protect against oxidative stress mainly by: (i) reducing reactive oxygen species (ROS) generation; (ii) directly scavenging free radicals; (iii) improving endogenous antioxidant enzymes (e.g., superoxide dismutase (SOD), catalase (CAT), and glutathione reductase (GSH)); (iv) promoting antioxidant molecules and the expression of related genes involved in mitochondrial energy biogenesis, mainly through AMPK/SIRT1/Nrf2, ERK/p38 MAPK, and PTEN/Akt signaling pathways; and (v) inducing autophagy via the mTOR-dependent or TFEB-dependent pathway [7]. Furthermore, RSV exerts anti-inflammatory effects through the inhibition of cyclooxygenase-1 (COX-1), cyclooxygenase-2 (COX-2), and 5-lipoxygenase catalytic activity, and consequent suppression of prostaglandins, thromboxanes, and leukotriene formation [15]. RSV was found to reduce LPS-induced nitric oxide (NO) and tumor necrosis factor alpha (TNF-α) production in primary microglia [16], prevent LPS-induced microglial BV-2 cell activation [17], and inhibit prostaglandin E2 (PGE2) and free radical production by rat primary microglia [18], modulating inflammatory responses. Since inflammation is a critical component of tumor progression and plays a key role in the tumor microenvironment [19], RSV represents a promising candidate for cancer prevention and/or treatment. There is a large body of in vitro studies that have demonstrated the proapoptotic and antiproliferative actions of RSV in several tumors [13], including lymphoblastic leukemia [20], colon [21], pancreatic [22], melanoma [23], gastric [24], cervical [25], ovarian [26], endometrial [27], liver [28], prostate [29] and breast [30] cancers, which confirmed its antitumor properties. These cancer chemopreventive effects are also corroborated by preclinical in vivo studies and clinical trials [31–33].

Although several studies have confirmed the beneficial effects of RSV on health, its therapeutic application is limited due to its short biological half-life and rapid metabolism and elimination, which limit its systemic bioavailability [34,35]. Indeed, in enterocytes, RSV undergoes phase II of drug metabolism, producing polar metabolites, which are easily excreted from the body. Specifically, it is conjugated with sulfate and glucoronate, and after these reactions, RSV metabolites can: (a) be transported across the apical membrane and reach the intestinal lumen, where they can be processed by the intestinal microbiota generating dihydroresveratrol (DHR), lunularin (L), and 3,4′-dihydroxy-trans-stilbene; (b) pass through the basolateral membrane and enter the bloodstream through which, by binding to blood proteins, they reach other tissues, such as the liver, kidneys and other peripheral tissues [36].

In the last decade, in order to overcome the limitations of low water solubility, absorption, transport across the membranes, and poor bioavailability of RSV, new synthetic derivatives and bio-isosteric analogues have been prepared [14,37–41]. Some of these possess a stronger pharmacological potency and a better pharmacokinetic profile than RSV itself [38,39]. It has been reported that methoxylation increases the molecule's lipophilicity, thus promoting cell permeability with a greater metabolic stability and bioavailability [37]. The beneficial properties of other RSV derivatives have also attracted increased interest in recent years [37]. RSV glycosylation, alkylation, halogenation, hydroxylation, methylation, and prenylation could lead to the development of new derivatives with enhanced bioavailability and pharmacological activity [37].

In this work, the inhibitory effects on cancer cell growth of a series of synthetic RSV phenylacetamide analogues, synthetized by some of us [42], have been evaluated. These derivatives possess a monosubstituted aromatic ring that can mimic the RSV phenolic nucleus and a longer flexible chain that can confer better stability and bioavailability than RSV (Figure 1). We demonstrated that most derivatives exerted antiproliferative effects in almost all of the cancer cell lines used, and inhibited LPS-induced ROS and NO production and PGE2 secretion. We found that derivative **2** displayed a higher bioavailability compared to RSV and was the most active in exerting anti-inflammatory and anti-proliferative effects; this derivative was able to induce cell cycle arrest and apoptosis of estrogen receptor positive (ER+) MCF7 and estrogen receptor negative (ER-) MDA-MB231 breast cancer cells.

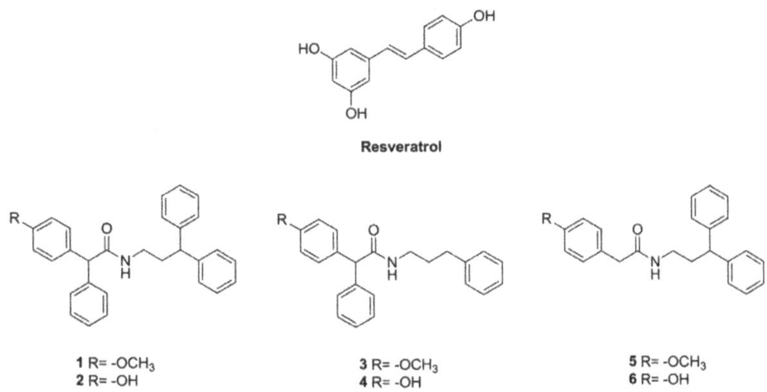

Figure 1. Molecular structure of RSV and the tested RSV derivatives (**1–6**).

2. Results

2.1. *Phenylacetamide RSV Derivatives Display Antiproliferative Effects on Several Cancer Cell Lines*

The antiproliferative effects of phenylacetamide RSV derivatives were evaluated in several cancer cell lines, testing a wide range of concentrations from 2.5 to 40 μM (Figure 2). We observed that derivative **1** causes a significant reduction only in MDA-MB231 and U937 cell viability, starting from the lowest dose of 5 μM (Figure 2A); starting from the same dose, derivatives **2** (Figure 2B) and **6** (Figure 2F) exhibited inhibitory effects in the MCF7 cell line. Moreover, derivative **2** displayed significant cytotoxic action in MDA-MB231 (starting from the 10 μM dose), in U937 (starting from the 2.5 μM dose), and in H295R cells (only at the highest dose of 40 μM) (Figure 2B). The same higher dose (40 μM) of derivative **3** reduced MDA-MB231, H295R, and U937 cell viability (Figure 2C). The 10 μM dose of derivatives **3** (Figure 2C), **4** (Figure 2D), and **5** (Figure 2E), and the 20 and 40 μM doses of derivative **4**, inhibited the MCF7 and MDA-MB231 and U937 cell viability, respectively. At the latter higher concentrations, derivatives **5** (Figure 2E) and **6** (Figure 2F) exerted inhibitory effects on MDA-MB231 (derivative **5**) and R2C and U937 (derivative **6**) cells. Interestingly, only derivatives **2** (Figure 2B) and **1** (Figure 2A) did not exhibit significant cytotoxic effects in the normal mouse fibroblast immortalized 3T3L1 cell line, whereas all other derivatives tested, even at different percentages, showed inhibitory effects. Moreover, we wanted to compare the effects of the same concentration ranges of derivative **2** with those of RSV on the proliferative behavior of tumor cells. The results reported in Supplementary Figure S1 show that RSV exhibited less inhibitory activity than derivative **2**, particularly in breast cancer cells. Moreover, derivative **2** did not show an antiproliferative effect on 3T3L1 cells, as RSV does (Figure S1).

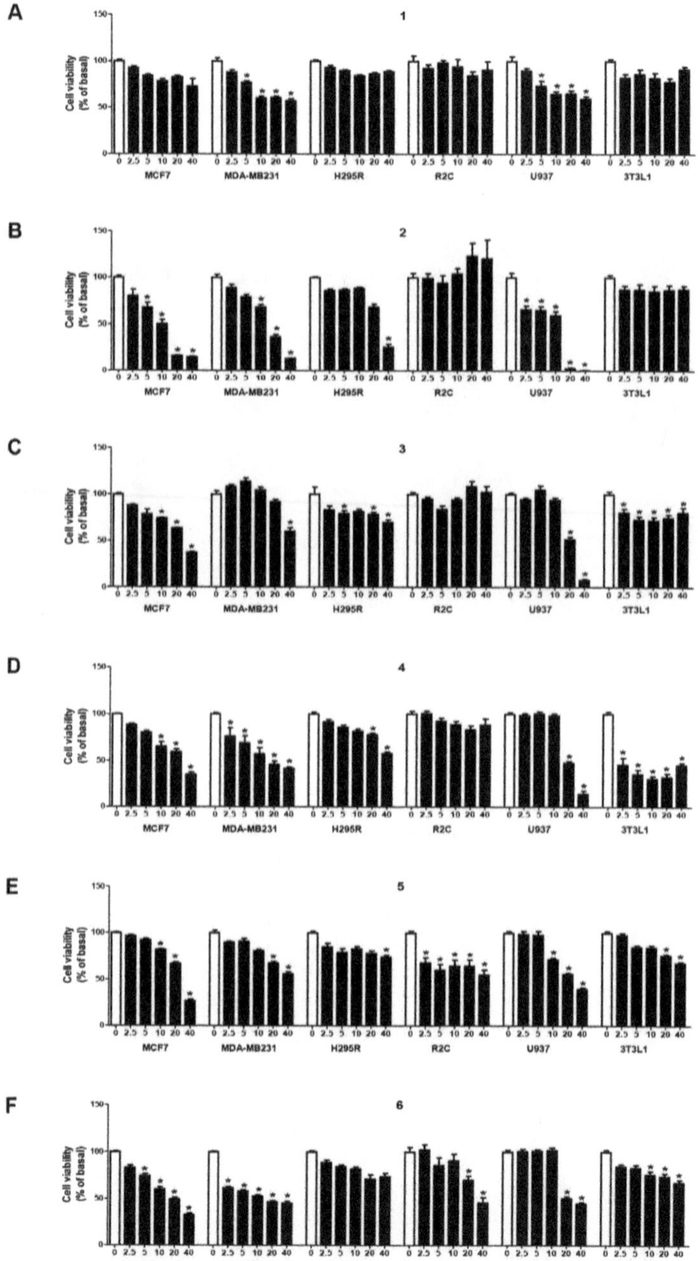

Figure 2. Effects of phenylacetamide RSV derivatives on cancer cell viability. MCF7, MDA-MB231, H295R, R2C, U937, and 3T3L1 cells were treated with vehicle DMSO (0) or the following phenylacetamide RSV derivatives: **1** (**A**), **2** (**B**), **3** (**C**), **4** (**D**), **5** (**E**), or **6** (**F**) at the indicated concentrations (2.5, 5, 10, 20, and 40 μM). Cell viability was assessed by MTT assay after 72 h exposure. Results are expressed as mean ± SE of three separate experiments (* $p < 0.05$ with respect to control (0)).

2.2. Phenylacetamide RSV Derivatives Decrease LPS-Induced NO and ROS Production

It has been reported that inflammation is associated with tumorigenesis. Immune cells and inflammatory cells infiltrate solid tumors, influencing most tumorigenesis stages [19,43]. It is known that LPS induces an inflammatory state characterized by the release of pro-inflammatory mediators such as ROS and NO, two well-characterized chemokines that play a key role in inflammation [44,45]. RSV and its phenylacetamide derivatives were evaluated for their ability to affect NO and ROS levels in LPS-activated U937 cells (Figure 3B,C and Figure S2). Since no effects on cell viability were shown for RSV and all derivatives at doses lower than 0.1 μM in U937 cells (Figure 3A and Figure S2A), we chose this dose as the optimal concentration to evaluate the anti-inflammatory activity of the phenylacetamide RSV derivatives, and we used RSV as a reference compound. U937 cells were differentiated into macrophages with PMA and pre-incubated for 1 h with DMSO or phenylacetamide RSV derivatives at 0.1 μM. Then, macrophages were induced with 400 ng/mL LPS. Following 24 h incubation at 37 °C, NO and ROS levels were detected. LPS induced a marked increase in NO (Figure 3B and Figure S2B) and ROS production (Figure 3C and Figure S2C) in PMA-treated U937 cells. The majority of the tested derivatives inhibited LPS-induced NO and ROS synthesis to various degrees. Derivative **6** was the most active in reducing ROS levels (by about 40%) (Figure 3C). Derivatives **1** (Figure 3C) and **2** (Figure 3C and Figure S2C) were also able to lower ROS production in LPS-activated macrophages. All tested phenylacetamide RSV derivatives significantly decreased NO levels. However, the best results were obtained with derivatives **2**, **4**, and **5**, which abolished the LPS-induced increase (Figure 3B). In accordance with previous findings [46], RSV was not able to significantly reduce NO at the tested concentration (Figure S2B). Moreover, derivative **2** was more effective than RSV in lowering ROS in this LPS-activated macrophage cell line (Figure S2C).

2.3. Phenylacetamide RSV Derivatives Reduce LPS-Triggered PGE2 Secretion

Taking into consideration the putative anti-inflammatory proprieties of the phenylacetamide RSV derivatives, we wondered whether they were able to reduce the secretion of PGE2, a potent inflammatory mediator derived from arachidonic acid by the activity of cyclooxygenase-2. To this end, U937/PMA cells treated with 0.1 μM of RSV or derivatives **1–6** were activated with LPS for 48 h and PGE2 levels were measured in cell culture media. All phenylacetamide RSV derivatives, except derivative **1**, lowered PGE2 secretion (Figure 3D). The most active derivative was **2**, which reduced PGE2 levels by about 30% when compared to the LPS-treated cells (Figure 3D). Notably, derivative **2** was more active than RSV in diminishing PGE2 secretion following the LPS activation of macrophages (Figure S2D).

2.4. Derivative 2 Determines Cell Cycle Arrest at G1 Phase in Both Human MCF7 and MDA-MB231 Breast Cancer Cell Lines

In order to better investigate the molecular mechanism underlying the antiproliferative effects of derivative **2** in ER+ MCF7 and ER- MDA-MB231 breast cancer cells, we performed cell cycle analysis using flow cytometry (Figure 4). Cell cycle changes produced by this RSV derivative were measured and the DNA histograms are reported in Figure 4. We observed a significant increase in the G1 cell population after 24 h for the 20 and 40 μM doses, accompanied by a variable decrease in the S and G2 phase cells, compared to the control, in derivative **2**-treated cells. These results confirmed a cell cycle arrest in the G1 phase in both MCF7 (Figure 4A) and MDA-MB231 (Figure 4C) cells. To explore the molecular alterations that might underpin cell cycle arrest, changes in the expression of specific cell cycle regulatory proteins were determined by Western blot analysis. Our results showed that treatment with derivative **2** caused a dose-dependent decrease in cyclin D1 (CCND1) and cyclin-dependent kinase 4 (CDK4) protein expression in MCF7 (Figure 4B) and MDA-MB231 (Figure 4D) cells. These results confirmed that exposure to this derivative invokes a signaling cascade that culminates in cell cycle arrest.

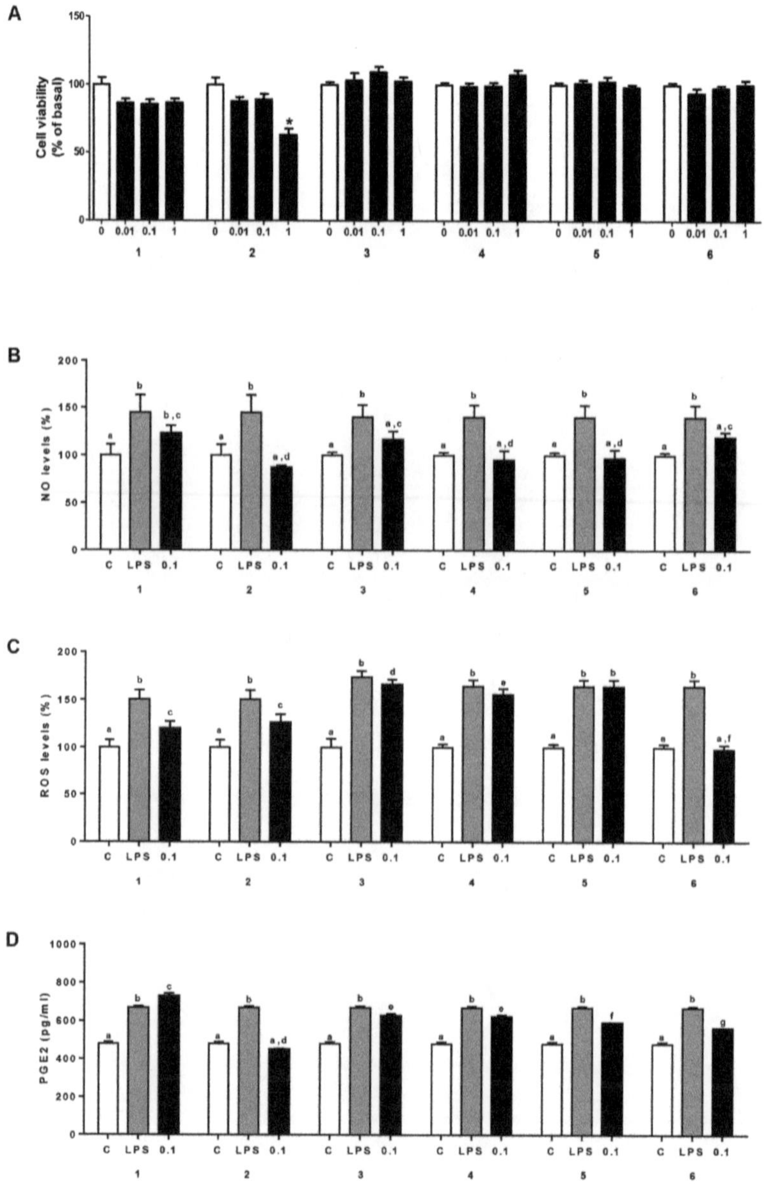

Figure 3. Effects of low doses of phenylacetamide RSV derivatives on U937 cell viability, NO and ROS levels, and PGE2 production. (**A**) U937 cells were treated with vehicle DMSO (0) or the phenylacetamide RSV derivatives (**1–6**) at the indicated concentrations (0.01, 0.1, or 1 μM). Cell viability was assessed by MTT assay after 72 h exposure. Results are expressed as mean ± SE of three separate experiments (* $p < 0.05$ with respect to control (0)). (**B–D**) In PMA-treated U937 cells, unstimulated (C) or activated with LPS alone (LPS), or in the presence of 0.1 μM of the phenylacetamide RSV derivatives (**1–6**), NO (**B**), ROS (**C**) and PGE2 (**D**) levels were quantified. Means ± SE of four independent experiments are shown. In (**B–D**), different letters indicate significant differences between treatments at $p < 0.05$ (Tukey's post hoc test).

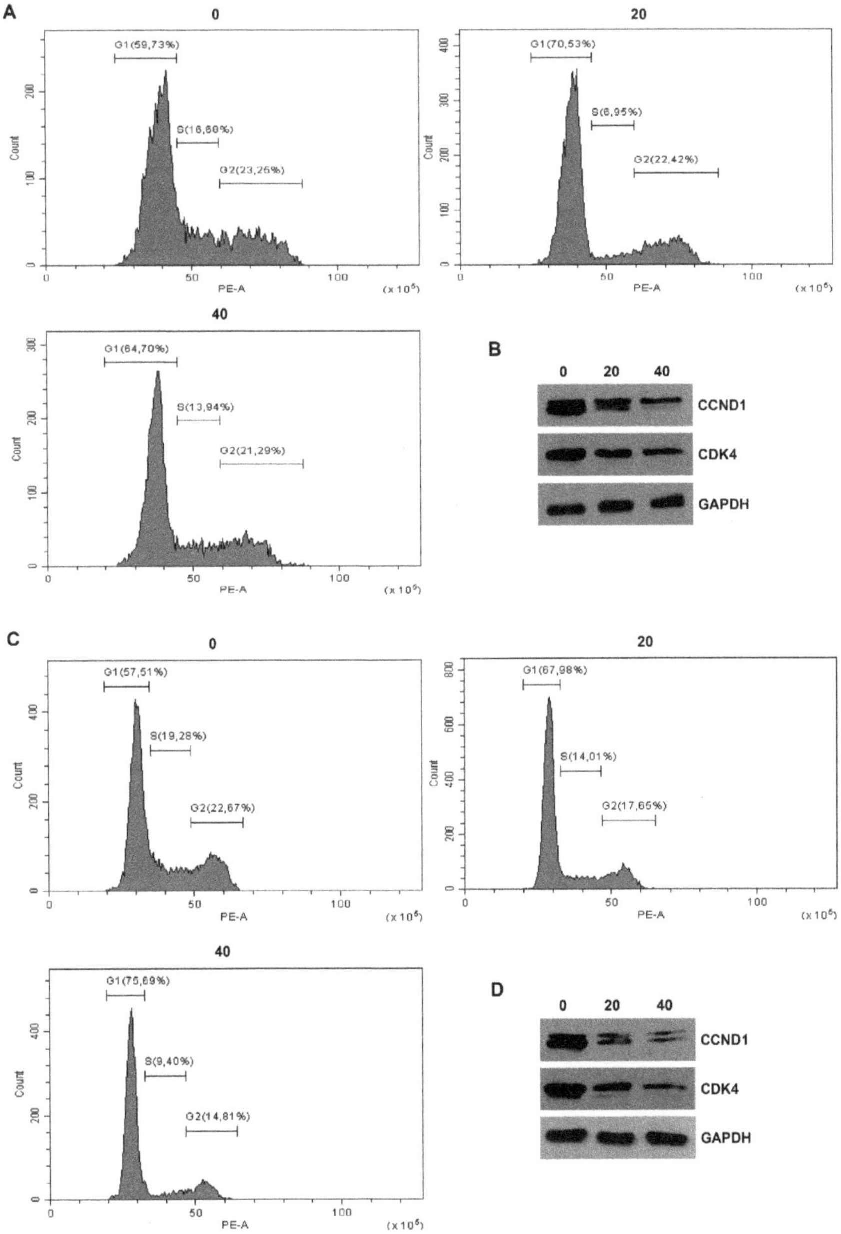

Figure 4. Effects of derivative **2** on the MCF7 and MDA-MB231 cell cycle distribution. (**A**,**C**) MCF7 (**A**) and MDA-MB231 (**C**) cells were synchronized in serum-free media for 12 h and then exposed to vehicle (0) or derivative **2** for 24 h at different concentrations (20 and 40 μM). The distribution of MCF7 and MDA-MB231 in the cell cycle was determined by flow cytometry using propidium iodide stained nuclei. (**B**,**D**) Western blot analysis of CCND1 and CDK4 was performed on equal amounts of total proteins extracted from MCF7 (**B**) and MDA-MB231 (**D**) cells treated with derivative **2** (20 and 40 μM) for 24 h. Blots are representative of three independent experiments with similar results. GAPDH was used as a loading control.

2.5. *Derivative* **2** *Induces Gross Morphological Changes and Apoptosis in Both Human MCF7 and MDA-MB231 Breast Cancer Cell Lines*

To further characterize whether the growth inhibitory activity of derivative **2** on MCF7 and MDA-MB231 cells was related to the induction of cell death, we examined its effects on cell morphology and apoptosis. We found that derivative **2** determines a marked decrease in cell number and gross morphological changes in MCF7 (Figure 5A) and MDA-MB231 cells (Figure 5D); these are dying cells, in which the 20 and 40 μM doses clearly determine an aberrant morphology as compared with the untreated cells (vehicle only) and the presence of many round floating cells.

The ability of derivative **2** to trigger apoptosis in human MCF7 and MDA-MB231 breast cancer cell lines was confirmed by the evaluation of DNA fragmentation. TUNEL staining demonstrated the presence of increased positive cells following derivative **2** treatment (Figure 5B,E). In addition, DAPI staining evidenced that the untreated breast cancer cells had round nuclei with regular contours, while nuclei from cells treated with derivative **2** appeared shrunken and irregularly shaped or degraded with condensed DNA. Apoptosis regulation requires the involvement of proteins that belong to the bcl-2 family; this proteins family includes both proapoptotic and antiapoptotic members, such as bax and bcl2, respectively [47]. Using Western blot analysis, we found that the presence of derivative **2** increased bax expression and decreased bcl-2 in both MCF7 (Figure 5C) and MDA-MB231 (Figure 5F) cells. Particularly, in MDA-MB231 cells, this derivative determined p21 bax cleavage to the p18 form, which may serve to increase the intrinsic cytotoxic properties of bax and enhance its cell death function in the mitochondria [48]. We also detected the cleavage of parp1, which is considered one of the most important biochemical features of cells undergoing apoptosis [49]. Treatment with derivative **2** (20–40 μM) induced parp1 cleavage in both cell lines (Figure 5C,F). These results confirmed that in the studied human breast cancer cell lines, derivative **2** determines the induction of the apoptotic mechanism.

2.6. *In Vitro Bioavailability of Derivative* 2

The in vitro bioavailability of derivative **2** was investigated using simulated gastrointestinal digestion according to the dialysis tubing procedure. The obtained results were expressed as bioavailability percentage, calculated as the percentage of the tested compound recovered in the bioaccessible fraction after in vitro gastric and intestinal digestions compared to the initial undigested sample (Equation (1)):

$$\text{Bioavailability (\%)} = \text{(bioaccessible content)}/\text{(total undigested content)} \times 100 \qquad (1)$$

After the first 2 h of simulated gastric digestion, derivative **2** bioavailability was equal to 23.1 ± 1.1%, while 36.3 ± 0.6% was recovered from the bioaccessible fraction after a further 4 h of simulated intestinal digestion (Table 1). Therefore, the cumulative bioavailability of the tested derivative after the two sequential digestion steps was around 59%. The obtained results showed an increased in vitro bioavailability of derivative **2** compared to RSV, characterized by a bioavailability of 13.4 ± 0.7% and 23.3 ± 0.6% after gastric and intestinal digestions, respectively, reaching a cumulative bioavailability of around 36% (Table 1).

Table 1. Bioaccessibility (%) of derivative **2** and RSV.

Sample	Bioaccessibility (%) after Gastric Digestion	Bioaccessibility (%) after Intestinal Digestion	Cumulative Bioaccessibility (%)
Derivative **2**	23.1 ± 1.1	36.3 ± 0.6	~59
RSV	13.4 ± 0.7	23.3 ± 0.6	~36

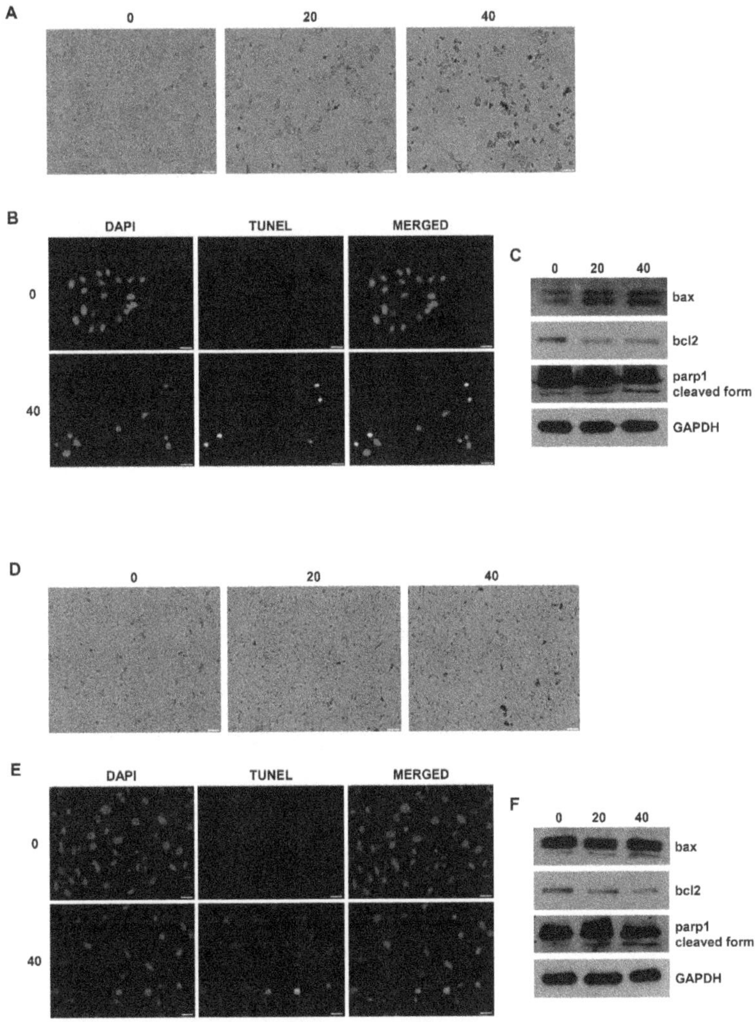

Figure 5. Effects of derivative **2** on MCF7 and MDA-MB231 cell morphology and apoptosis. MCF7 (**A**) and MDA-MB231 (**D**) cells were untreated (0) or treated with derivative **2** (20 and 40 μM) for 24 h; after treatment, cells were examined with a phase-contrast microscope (×10 objective). Images are from a representative experiment. Scale bar: 100 μm. (**B**,**E**) MCF7 (**B**) and MDA-MB231 (**E**) cells were untreated (0) or treated with derivative **2** (10 μM) for 72 h; after treatment, cells were fixed with paraformaldehyde and processed for TUNEL staining. Nuclei counterstaining was performed using DAPI. Fluorescent signals were observed under a fluorescent microscope (×20 objective). Images are from a representative experiment. Scale bar: 50μm. (**C**,**F**) MCF7 (**C**) and MDA-MB231 (**F**) cells were untreated (0) or treated with derivative **2** (20 and 40 μM) for 24 h. Western blot analyses of bax, bcl-2, and parp1 were performed on equal amounts of total proteins. Blots are representative of three independent experiments with similar results. GAPDH was used as a loading control.

3. Discussion

Amongst the phytochemicals present in food that exhibit considerable prospects in the treatment and management of human diseases, RSV, a stilbene-type aromatic phytoalexin, is one of the most widely studied [7]. In the last two decades, further studies have aimed to elucidate the mechanisms underlying RSV activity in reducing cancer initiation, promotion, and progression, as well as the occurrence of inflammation in various cancer models [7,13,14]. By interfering with diverse signal transduction pathways, RSV has been shown to control cell growth and division, apoptosis, metastasis, and angiogenesis [13]. To date, experimental in vitro and in vivo data and several clinical trials have demonstrated the potential role of RSV as a preventive or therapeutic agent for a wide range of cancers [13,20–33]. This is due to its ability to modulate different targets and act on different pathways that are usually altered in cancer, together with a low toxicity [50]. Moreover, the combined treatment of RSV with diverse chemotherapeutics revealed enhanced antineoplastic effects [51]. Even though human clinical trials have produced positive results, some critical concerns, partly due to the dosing protocols and low bioavailability, have arisen [8,34,35]. To overcome these limitations, numerous efforts focused on the formulation of delivery systems and the development of more bioavailable derivatives have been made [37]. From this point of view, we designed and synthesized a series of phenylacetamide RSV analogues that are characterized by a monosubstituted aromatic ring, mimicking the RSV phenolic nucleus, and a longer flexible chain that could confer higher stability and bioavailability. Using a panel of cancer cell lines, we evaluated the anticancer activity of these RSV analogues and found that almost all possessed anticancer activity. Particularly, derivative **2** was the most effective in reducing cancer cell viability, together with a lack of cytotoxicity against the murine fibroblast 3T3L1 cell line, which was used as a normal cell model. Moreover, it displayed enhanced inhibitory activity against estrogen receptor positive (ER+) MCF7 and estrogen receptor negative (ER-) MDA-MB231 breast cancer cell growth and had a higher bioavailability of 23% compared to RSV.

Next, we wondered whether derivative **2** could also act as modulator of inflammatory responses in U937 cells differentiated into macrophages. Indeed, it is noteworthy that inflammation is usually associated with the induction or the exacerbation of about a quarter of cancer cases, very quickly resulting in a chronic situation that leads to higher destruction and lower healing of the affected tissue(s) [43]. Once again, RSV has been shown to modulate pro-inflammatory proteins or their signal transduction pathways, as well as other pathways producing precursors of inflammation [7,10,11]. Our outcomes clearly indicate that derivative **2**, when used at a concentration that is not toxic for the adopted cell model (viz. 0.1 μM), is able to reduce NO synthesis and PGE2 levels (released in the culture medium) in U937/PMA cells exposed to LPS, by about 40% and 30%, respectively. The other analogues shared a similar anti-inflammatory activity, with a certain inter-series variability, but to a lesser extent. The anti-inflammatory activity synergizes with another property of derivative **2**, its ability to significantly decrease ROS levels in the same cellular context previously adopted, even though the best antioxidant performance was obtained with the use of derivative **6**. Taken together, these data proved that the lead derivative **2** possesses the ability to decrease cancer cell viability and diminish, contemporaneously, two important mediators of inflammation and ROS production. Next, we performed flow cytometry and Western blot analysis to further study the mechanisms involved in the observed anticancer activity of derivative **2**. Cell cycle analysis performed on the two breast cancer cell lines revealed a block at the G1 phase after exposure at the indicated doses of derivative **2**. These results have been also confirmed by Western blot analysis. A dose-dependent decrease in the CCND1 and CDK4 protein expression, in both MCF7 and MDA-MB231 cells, has been detected. DNA damage most likely occurs as a result of derivative **2** treatment; indeed, the role of the G1 checkpoint in limiting new cancer cell formation, with damaged DNA, is irrefutable nowadays, because alterations or mutations in several proteins involved in the next phase transition have been linked to specific tumors [52]. If the cells arrive in the G1 phase with damaged DNA, in one way or another, they sense this damage and send signals that cause alterations in the cyclins and CDKs that "put the brake on", producing a block in G1 phase until the DNA damage is repaired. G1-phase arrest of cell cycle progression gives cells the ability to repair damage or die from apoptosis [52,53]. Therefore, we studied the mechanism by which breast cancer cells, blocked at the G1 phase, died following derivative **2** treatment. The morphological cell changes observed are, undoubtedly, a signal of cell death [54]; indeed, derivative **2** is able to induce apoptotic mechanisms, with DNA damage, as determined by the means of TUNEL and parp1 cleavage assays. Moreover, we observed the increased expression of the pro-apoptotic protein bax, accompanied by a decrease in bcl2 levels. These proteins are the major members of the bcl-2 family that play a key role in tumor progression and inhibition of the intrinsic apoptotic pathway [47]. Therefore, the balance between pro- and anti-apoptotic members of this family determines the cellular fate; in this case, the increased bax/bcl2 ratio confirmed the apoptosis of breast cancer cell lines that was previously observed. It should be highlighted that the downregulation and upregulation of bcl2 and bax expression, respectively, triggered by derivative **2** are important markers that have already been clinically associated with better survival in patients affected by some types of tumors [55–57]. Thus, closing the circle, the latter data strengthen our previous cell cycle results, indicating that, following derivative **2** exposure, breast cancer cells are blocked at the G1 phase of the cell cycle, due to their inability to repair DNA damage, and are directed toward cell death by the apoptotic mechanism. In conclusion, the revealed features of this lead derivative make it a valuable molecule that is able to modulate several biological aspects of cancer cell growth, shedding light on the research into new and more active RSV derivatives.

4. Materials and Methods

4.1. Cell Cultures and Treatments

MCF7 and MDA-MB231 breast cancer cells, H295R adrenocortical cancer cells, R2C rat Leydig tumor cells, and 3T3L1 mouse embryonic fibroblast cells were purchased from the American Type Culture Collection (ATCC) (Manassas, VA, USA). MCF7 cells (ER positive breast cancer cells) were maintained as previously described [58,59]. MDA-MB231 cells (ER, PR, and HER2 triple negative breast cancer cells) were maintained in Dulbecco's Modified Eagle Medium/F12 (DMEM/F12) medium supplemented with 10% fetal bovine serum (FBS), 1% L-glutamine (L-Glu), and 1% penicillin/streptomycin (P/S) (Sigma-Aldrich, Milano, Italy) (complete medium). H295R adrenocortical cancer cells were cultured in DMEM/F12 medium supplemented with 1% Insulin-Transferrin-Selenium (ITS) Liquid Media Supplement, 10% FBS), 1% L-Glu, and 1% P/S (complete medium). R2C rat Leydig tumor cells were maintained as previously described [60]. The 3T3L1 mouse embryonic fibroblast cell line, obtained from ATCC, was cultured in DMEM with phenol red supplemented with 10% FBS, 1% L-Glu, and 1% P/S (Sigma-Aldrich) (complete medium). The human pro-monocytic cells derived from a histiocytoma (U937 cells) were purchased from the Interlab Cell Line Collection (ICLC) (IRCCS AOU San Martino IST) and grown in Roswell Park Memorial Institute (RPMI) 1640 medium supplemented with 10% fetal bovine serum, 2 mM L-Glu, 100 U/mL penicillin, and 100 μg/mL streptomycin. All cells were maintained at 37 °C in a humidified atmosphere of 95% air and 5% CO_2 and were screened periodically for Mycoplasma contamination. U937 cells were differentiated to macrophages by adding 10 ng/mL phorbol 12-myristate 13-acetate (PMA, Sigma-Aldrich) for 16 h. Where indicated, U937/PMA cells were stimulated for 24 or 48 h with 400 ng/mL of lipopolysaccharides from *Salmonella enterica* serotype typhimurium (LPS, Sigma-Aldrich) after one hour of treatment with the phenylacetamide RSV derivatives **1–6**. All cell lines were treated in complete medium at the times and concentrations indicated with phenylacetamide RSV derivatives (**1–6**) or RSV.

4.2. Cell Viability Assay

Cell viability was determined as previously described [61] by colorimetric 3-(4,5-dimethylthiazol-2-yl)-2,5-diphenyltetrazolium bromide (MTT) assay, which measures mitochondrial activity in viable cells. Specifically, the cells (2.5×10^4) were plated in a 48-well plate and, after 48 h, were treated with phenylacetamide RSV derivatives (**1–6**) at increasing concentrations (2.5, 5, 10, 20, and 40 μM) for 72 h. Control groups were treated with

Dimethyl Sulfoxide (DMSO) (Sigma-Aldrich) equal to the highest percentage of (<0.1%) solvent used in the experimental conditions for the MTT assay. After treatment, fresh MTT, resuspended in phosphate buffered saline (PBS), was added to each well (final concentration 0.5 mg/ml) and the plate was incubated at 37 °C for 3 h in a humidified 5% CO_2 incubator. After incubation, media were removed, and formazan crystals were dissolved in 200 μL of DMSO (Sigma-Aldrich) for 10 min with gentle agitation. Each experiment was performed with six replicates three times; the optical density was measured at 570 nm with a spectrophotometer (Synergy H1 plate reader, BioTek Instruments, Inc., Winooski, VT, USA).

4.3. Flow Cytometric Analysis of DNA Content

MCF7 and MDA-MB231 breast cancer cells (5×10^5/well) were seeded in six multiwell plates for 24 h. Sub-confluent monolayers were depleted of serum for 12 h and treated for an additional 24 h with derivative **2** (20 and 40 μM). The cells were harvested by trypsinization and resuspended with 0.5 mL of DNA staining solution (0.1 mg/mL propidium iodide (PI), 0.1% sodium citrate, and 0.1% Triton X-100, 0.02 mg/mL RNase; Sigma). The DNA content was measured using a CytoFLEX flow cytometer (Beckman Coulter SRL, Milano, Italy). Nuclei (10,000 events) were analyzed from each sample. The percentage of cells in the G1, S and G2/M phases of the cell cycle were determined by analysis with CytExpert software (Beckman Coulter SRL, Milano, Italy).

4.4. Western Blot Analysis

Total proteins were subjected to Western blot analysis [60]. Blots were incubated overnight at 4 °C with antibodies against cyclin D1 (CCND1), cyclin-dependent kinase 4 (CDK4), bax, bcl2, and parp1 (all from Santa Cruz Biotechnology, Santa Cruz, CA, USA). Membranes were incubated with horseradish peroxidase (HRP)-conjugated secondary antibodies (Amersham Pharmacia Biotech, Piscataway, NJ, USA) and immunoreactive bands were visualized with the ECL Western blotting detection system (Amersham Pharmacia Biotech). To ensure equal loading of proteins, membranes were stripped and incubated overnight with glyceraldehyde 3-phosphate dehydrogenase (GAPDH) antibody (Santa Cruz Biotechnology).

4.5. Phase Contrast Microscopy for Morphological Evaluation

MCF7 and MDA-MB231 cells were seeded in six-well plates at a density of 5×10^5 cells/well for 24 h and then left untreated (0) or treated with derivative **2** (20 and 40 μM) for 24 h. Subsequent to treatment, culture plates were observed using an inverted phase contrast microscope (Olympus CKX53 inverted microscope, Waltham, MA, United States) and images were captured (×10 objective).

4.6. Terminal Deoxynucleotidyl Transferase-Mediated dUTP Nick-End Labelling (TUNEL) Assay

Apoptosis was detected using the TUNEL assay, according to the guidelines of the manufacturer (CF™488A TUNEL Assay Apoptosis Detection Kit, Biotium, Hayward, CA, USA). The cells were grown on glass coverslips and then treated with derivative **2**. After paraformaldehyde fixation, the TUNEL reaction mixture containing the terminal deoxynucleotidyl transferase (TdT) enzyme was added, as reported by Iacopetta et al. [62,63]. Samples were washed, incubated with DAPI (Sigma; 0.2 μg/mL) for nuclei counterstaining, and then observed and imaged under a fluorescence microscope (Leica DM6000, Leica Microsystems GmbH, Wetzlar, Germany; magnification ×20, $\lambda_{ex/em}$ maxima of 490/515 nm for CF™488A or 350/460 nm for DAPI). Images are representative of three independent experiments.

4.7. ROS and NO Detection

For reactive oxygen species and nitric oxide analysis, U937 cells were differentiated in macrophages with PMA and then activated by LPS in the presence (or absence) of

0.1 μM of the phenylacetamide RSV derivatives. After 24 h, ROS and NO levels were measured using 6-carboxy-2′,7′-dichlorodihydrofluorescein diacetate (DCF-DA, Thermo Fisher Scientific, San Jose, CA, USA) and 4-amino-5-methylamino-2′,7′-difluorofluorescein diacetate (DAF-FM Diacetate, Thermo Fisher Scientific), respectively [64]. The fluorescence was analyzed on a microplate reader (Glomax, Promega, Madison, WI, USA).

4.8. PGE2 Quantification

U937/PMA cells were treated with phenylacetamide RSV derivatives (0.1 μM) for one hour, and then stimulated with 400 ng/mL LPS. After 48 h, the cell culture media were collected to measure PGE2 levels using the DetectX® Prostaglandin E2 High Sensitivity Immunoassay Kit (Arbor Assays, Ann Arbor, MI, USA) according to the manufacturer's instructions.

4.9. In Vitro Bioavailability Studies

In vitro bioavailability studies were carried out following the dialysis tubing procedure, which consisted of two sequential digestion steps involving pepsin and pancreatin, respectively [65]. Simulated gastric digestion was performed by preparing a digestion mixture consisting of 100 μL of a derivative **2** solution (10 mM in DMSO), 1.0 mL of HCl (0.85 N), and 3 mL of a sodium cholate solution (2% *w*/*v* in distilled water). Then, the mixture was placed into a dialysis bag (Spectrum Laboratories Inc., MWCO: 12,000–14,000 Dalton, GA, USA) and dialyzed against 10 mL of a 0.85 N HCl solution (pH 1.0) at 37 °C. After 2 h, a 2 mL aliquot was withdrawn and analyzed to evaluate the amount of derivative **2** in the bioaccessible fraction. For sequential intestinal digestion, 11 mg of amylase, 11 mg of esterase, and 1.3 mL of a 0.8 M $NaHCO_3$ solution containing 22.60 mg porcine pancreatin/mL were introduced into the dialysis bag, which was then dialyzed against 10 mL of buffer solution at pH 7.0 for a further 4 h at 37 °C. Then, 2 mL of the external solution was withdrawn and analyzed to assess derivative **2** concentration. UV/Vis spectroscopy (UV/Visible Spectrophotometer Evolution 201, Thermo Fisher Scientific) was used to determine the concentration of derivative **2** in the samples obtained after simulated gastric and intestinal digestions. The in vitro bioavailability studies were carried out using RSV as a reference derivative. Experiments were performed in triplicate.

4.10. Statistical Analysis

All experiments were performed at least three times. Data are expressed as mean values ± standard error (SE). The statistical significance between control (basal, 0) and treated samples was analyzed using GraphPad Prism 5.0 software (GraphPad Software, Inc., San Diego, CA, USA). Control and treated groups were compared using the analysis of variance (ANOVA) with Bonferroni, Dunn's, or Tukey's post hoc tests. Differences were considered as significant when $p < 0.05$.

Supplementary Materials: The following are available online at https://www.mdpi.com/article/10.3390/ijms22105255/s1, Figure S1: Comparative effects of derivative **2** and RSV on cancer cell viability. MCF7, MDA-MB231, H295R, R2C, U937, and 3T3L1 cells were treated in complete medium with vehicle DMSO (0), derivative **2** (A), or RSV (B) at the indicated concentrations (2.5, 5, 10, 20, and 40 μM). Cell viability was assessed by MTT assay after 72 h exposure. Results are expressed as mean ± SE of three separate experiments (* $p < 0.05$ with respect to control (0)), Figure S2: Comparative effects of low doses of derivative **2** and RSV on U937 cell viability, NO and ROS levels, and PGE2 release. (A) U937 cells were treated in complete medium with vehicle DMSO (0) or derivative **2** and RSV at the indicated concentrations (0.01, 0.1, and 1 μM). Cell viability was assessed by MTT assay after 72 h exposure. Results are expressed as mean ± SE of three separate experiments (* $p < 0.05$ with respect to control (0)). (B–D) In U937/PMA cells, unstimulated (C) or activated with LPS alone (LPS) or in presence of 0.1 μM derivative 2 (**2**) or RSV, NO (B) ROS (C), and PGE2 (D) levels were quantified. Means ± SE of three independent experiments are shown. Different letters indicate significant differences between treatments at $p < 0.05$ (Tukey's post hoc test).

Author Contributions: A.C. and A.S.: conceptualization; A.C., D.I., M.S.S., A.S. and V.P.: writing—original draft preparation; A.C., A.S., V.I., D.I., J.C., A.D.L., O.I.P., P.A., M.G.B.: investigation, methodology, validation, and formal analysis; A.C., A.S., C.S., M.S.S. and V.P.: data curation; C.S., V.I., M.S.S. and V.P.: supervision. All authors have read and agreed to the published version of the manuscript.

Funding: This work was supported by a special award (Department of Excellence, Italian Law 232/2016) from the Italian Ministry of Research and University (MIUR) to the Department of Pharmacy, Health and Nutritional Sciences of the University of Calabria (Italy), and by MIUR ex 60% (VP) and by AIRC (Associazione Italiana per la Ricerca sul Cancro), projects n. IG20122. A.D.L. was supported by a fellowship from PAC (Progetto Strategico Regionale Calabria Alta Formazione) Calabria 2014/2020—Asse Prioritario 12, Linea B, Azione 10.5.12; P.A. was supported by a post-doc fellowship 2021 from Fondazione Umberto Veronesi (FUV).

Institutional Review Board Statement: Not applicable.

Informed Consent Statement: Not applicable.

Conflicts of Interest: The authors declare no conflict of interest.

References

1. Cory, H.; Passarelli, S.; Szeto, J.; Tamez, M.; Mattei, J. The Role of Polyphenols in Human Health and Food Systems: A Mini-Review. *Front. Nutr.* **2018**, *5*, 87. [CrossRef]
2. Cappello, A.R.; Dolce, V.; Iacopetta, D.; Martello, M.; Fiorillo, M.; Curcio, R.; Muto, L.; Dhanyalayam, D. Bergamot (Citrus bergamia Risso) Flavonoids and Their Potential Benefits in Human Hyperlipidemia and Atherosclerosis: An Overview. *Mini Rev. Med. Chem.* **2016**, *16*, 619–629. [CrossRef] [PubMed]
3. Koushki, M.; Amiri-Dashatan, N.; Ahmadi, N.; Abbaszadeh, H.A.; Rezaei-Tavirani, M. Resveratrol: A miraculous natural compound for diseases treatment. *Food Sci. Nutr.* **2018**, *6*, 2473–2490. [CrossRef] [PubMed]
4. Pervaiz, S. Resveratrol: From grapevines to mammalian biology. *FASEB J.* **2003**, *17*, 1975–1985. [CrossRef]
5. Takaoka, M. Resveratrol, a new phenolic compound, from Veratrum grandiflorum. *J. Chem. Soc. Jpn.* **1939**, *60*, 1090–1100.
6. Tokusoglu, O.; Unal, M.K.; Yemis, F. Determination of the phytoalexin resveratrol (3,5,4′-trihydroxystilbene) in peanuts and pistachios by high-performance liquid chromatographic diode array (HPLC-DAD) and gas chromatography-mass Spectrometry (GC-MS). *J Agric. Food Chem.* **2005**, *53*, 5003–5009. [CrossRef] [PubMed]
7. Meng, X.; Zhou, J.; Zhao, C.N.; Gan, R.Y.; Li, H.B. Health Benefits and Molecular Mechanisms of Resveratrol: A Narrative Review. *Foods* **2020**, *9*, 340. [CrossRef]
8. Berman, A.Y.; Motechin, R.A.; Wiesenfeld, M.Y.; Holz, M.K. The therapeutic potential of resveratrol: A review of clinical trials. *NPJ Precis. Oncol.* **2017**, *1*, 35. [CrossRef] [PubMed]
9. Meng, Q.W.; Guo, T.; Li, G.Q.; Sun, S.S.; He, S.Q.; Cheng, B.J.; Shi, B.M.; Shan, A.S. Dietary resveratrol improves antioxidant status of sows and piglets and regulates antioxidant gene expression in placenta by Keap1-Nrf2 pathway and Sirt1. *J. Anim. Sci. Biotechnol.* **2018**, *9*, 34. [CrossRef]
10. Nunes, S.; Danesi, F.; Del Rio, D.; Silva, P. Resveratrol and inflammatory bowel disease: The evidence so far. *Nutr. Res. Rev.* **2018**, *31*, 85–97. [CrossRef]
11. Javkhedkar, A.A.; Quiroz, Y.; Rodriguez-Iturbe, B.; Vaziri, N.D.; Lokhandwala, M.F.; Banday, A.A. Resveratrol restored Nrf2 function, reduced renal inflammation, and mitigated hypertension in spontaneously hypertensive rats. *Am. J. Physiol. Regul. I* **2015**, *308*, R840–R846. [CrossRef]
12. Corpas, R.; Grinan-Ferre, C.; Rodriguez-Farre, E.; Pallas, M.; Sanfeliu, C. Resveratrol Induces Brain Resilience against Alzheimer Neurodegeneration Through Proteostasis Enhancement. *Mol. Neurobiol.* **2019**, *56*, 1502–1516. [CrossRef]
13. Rauf, A.; Imran, M.; Butt, M.S.; Nadeem, M.; Peters, D.G.; Mubarak, M.S. Resveratrol as an anti-cancer agent: A review. *Crit. Rev. Food Sci. Nutr.* **2018**, *58*, 1428–1447. [CrossRef] [PubMed]
14. Chimento, A.; Sirianni, R.; Saturnino, C.; Caruso, A.; Sinicropi, M.S.; Pezzi, V. Resveratrol and Its Analogs as Antitumoral Agents for Breast Cancer Treatment. *Mini Rev. Med. Chem.* **2016**, *16*, 699–709. [CrossRef] [PubMed]
15. Kutil, Z.; Temml, V.; Maghradze, D.; Pribylova, M.; Dvorakova, M.; Schuster, D.; Vanek, T.; Landa, P. Impact of Wines and Wine Constituents on Cyclooxygenase-1, Cyclooxygenase-2, and 5-Lipoxygenase Catalytic Activity. *Mediat. Inflamm.* **2014**, *2014*, 178931. [CrossRef] [PubMed]
16. Bi, X.L.; Yang, J.Y.; Dong, Y.X.; Wang, J.M.; Cui, Y.H.; Ikeshima, T.; Zhao, Y.Q.; Wu, C.F. Resveratrol inhibits nitric oxide and TNF-alpha production by lipopolysaccharide-activated microglia. *Int. Immunopharmacol.* **2005**, *5*, 185–193. [CrossRef]
17. Capiralla, H.; Vingtdeux, V.; Zhao, H.; Sankowski, R.; Al-Abed, Y.; Davies, P.; Marambaud, P. Resveratrol mitigates lipopolysaccharide- and Abeta-mediated microglial inflammation by inhibiting the TLR4/NF-kappaB/STAT signaling cascade. *J. Neurochem.* **2012**, *120*, 461–472. [CrossRef]

18. Candelario-Jalil, E.; de Oliveira, A.C.; Graf, S.; Bhatia, H.S.; Hull, M.; Munoz, E.; Fiebich, B.L. Resveratrol potently reduces prostaglandin E2 production and free radical formation in lipopolysaccharide-activated primary rat microglia. *J. Neuroinflammation.* **2007**, *4*, 25. [CrossRef]
19. Grivennikov, S.I.; Greten, F.R.; Karin, M. Immunity, inflammation, and cancer. *Cell* **2010**, *140*, 883–899. [CrossRef]
20. Zhou, W.; Wang, S.Q.; Ying, Y.; Zhou, R.Q.; Mao, P. miR-196b/miR-1290 participate in the antitumor effect of resveratrol via regulation of IGFBP3 expression in acute lymphoblastic leukemia. *Oncol. Rep.* **2017**, *37*, 1075–1083. [CrossRef]
21. Li, D.; Wang, G.C.; Jin, G.G.; Yao, K.; Zhao, Z.J.; Bie, L.Y.; Guo, Y.J.; Li, N.; Deng, W.Y.; Chen, X.B.; et al. Resveratrol suppresses colon cancer growth by targeting the AKT/STAT3 signaling pathway. *Int. J. Mol. Med.* **2019**, *43*, 630–640. [CrossRef]
22. Cheng, L.; Yan, B.; Chen, K.; Jiang, Z.D.; Zhou, C.C.; Cao, J.Y.; Qian, W.K.; Li, J.; Sun, L.K.; Ma, J.G.; et al. Resveratrol-Induced Downregulation of NAF-1 Enhances the Sensitivity of Pancreatic Cancer Cells to Gemcitabine via the ROS/Nrf2 Signaling Pathways. *Oxidative Med. Cell. Longev.* **2018**, *2018*, 9482018. [CrossRef] [PubMed]
23. Heo, J.R.; Kim, S.M.; Hwang, K.A.; Kang, J.H.; Choi, K.C. Resveratrol induced reactive oxygen species and endoplasmic reticulum stress-mediated apoptosis, and cell cycle arrest in the A375SM malignant melanoma cell line. *Int. J. Mol. Med.* **2018**, *42*, 1427–1435. [CrossRef] [PubMed]
24. Wu, X.X.; Xu, Y.D.; Zhu, B.R.; Liu, Q.; Yao, Q.F.; Zhao, G. Resveratrol induces apoptosis in SGC-7901 gastric cancer cells. *Oncol. Lett.* **2018**, *16*, 2949–2956. [CrossRef]
25. Li, L.; Qiu, R.L.; Lin, Y.; Cai, Y.; Bian, Y.; Fan, Y.; Gao, X.J. Resveratrol suppresses human cervical carcinoma cell proliferation and elevates apoptosis via the mitochondrial and p53 signaling pathways. *Oncol. Lett.* **2018**, *15*, 9845–9851. [CrossRef] [PubMed]
26. Liu, Y.; Tong, L.; Luo, Y.; Li, X.; Chen, G.W.; Wang, Y.F. Resveratrol inhibits the proliferation and induces the apoptosis in ovarian cancer cells via inhibiting glycolysis and targeting AMPK/mTOR signaling pathway. *J. Cell. Biochem.* **2018**, *119*, 6162–6172. [CrossRef]
27. Fukuda, T.; Oda, K.; Wada-Hiraike, O.; Sone, K.; Inaba, K.; Ikeda, Y.; Makii, C.; Miyasaka, A.; Kashiyama, T.; Tanikawa, M.; et al. Autophagy inhibition augments resveratrol-induced apoptosis in Ishikawa endometrial cancer cells. *Oncol. Lett.* **2016**, *12*, 2560–2566. [CrossRef] [PubMed]
28. Zhang, B.C.; Yin, X.N.; Sui, S.G. Resveratrol inhibited the progression of human hepatocellular carcinoma by inducing autophagy via regulating p53 and the phosphoinositide 3-kinase/protein kinase B pathway. *Oncol. Rep.* **2018**, *40*, 2758–2765. [CrossRef]
29. Martinez-Martinez, D.; Soto, A.; Gil-Araujo, B.; Gallego, B.; Chiloeches, A.; Lasa, M. Resveratrol promotes apoptosis through the induction of dual specificity phosphatase 1 and sensitizes prostate cancer cells to cisplatin. *Food Chem. Toxicol.* **2019**, *124*, 273–279. [CrossRef] [PubMed]
30. Hu, C.Y.; Liu, Y.; Teng, M.Y.; Jiao, K.L.; Zhen, J.; Wu, M.X.; Li, Z. Resveratrol inhibits the proliferation of estrogen receptor-positive breast cancer cells by suppressing EZH2 through the modulation of ERK1/2 signaling. *Cell Biol. Toxicol.* **2019**, *35*, 445–456. [CrossRef]
31. Ko, J.H.; Sethi, G.; Um, J.Y.; Shanmugam, M.K.; Arfuso, F.; Kumar, A.P.; Bishayee, A.; Ahn, K.S. The Role of Resveratrol in Cancer Therapy. *Int. J. Mol. Sci.* **2017**, *18*, 2589. [CrossRef]
32. Howells, L.M.; Berry, D.P.; Elliott, P.J.; Jacobson, E.W.; Hoffmann, E.; Hegarty, B.; Brown, K.; Steward, W.P.; Gescher, A.J. Phase I Randomized, Double-Blind Pilot Study of Micronized Resveratrol (SRT501) in Patients with Hepatic Metastases-Safety, Pharmacokinetics, and Pharmacodynamics. *Cancer Prev. Res.* **2011**, *4*, 1419–1425. [CrossRef]
33. Ramirez-Garza, S.L.; Laveriano-Santos, E.P.; Marhuenda-Munoz, M.; Storniolo, C.E.; Tresserra-Rimbau, A.; Vallverdu-Queralt, A.; Lamuela-Raventos, R.M. Health Effects of Resveratrol: Results from Human Intervention Trials. *Nutrients* **2018**, *10*, 1892. [CrossRef] [PubMed]
34. Walle, T. Bioavailability of resveratrol. *Ann. N. Y. Acad. Sci.* **2011**, *1215*, 9–15. [CrossRef] [PubMed]
35. Walle, T.; Hsieh, F.; DeLegge, M.H.; Oatis, J.E.; Walle, U.K. High absorption but very low bioavailability of oral resveratrol in humans. *Drug Metab. Dispos.* **2004**, *32*, 1377–1382. [CrossRef] [PubMed]
36. Springer, M.; Moco, S. Resveratrol and Its Human Metabolites-Effects on Metabolic Health and Obesity. *Nutrients* **2019**, *11*, 143. [CrossRef] [PubMed]
37. Chimento, A.; De Amicis, F.; Sirianni, R.; Sinicropi, M.S.; Puoci, F.; Casaburi, I.; Saturnino, C.; Pezzi, V. Progress to Improve Oral Bioavailability and Beneficial Effects of Resveratrol. *Int. J. Mol. Sci.* **2019**, *20*, 1381. [CrossRef]
38. Chimento, A.; Sala, M.; Gomez-Monterrey, I.M.; Musella, S.; Bertamino, A.; Caruso, A.; Sinicropi, M.S.; Sirianni, R.; Puoci, F.; Parisi, O.I.; et al. Biological activity of 3-chloro-azetidin-2-one derivatives having interesting antiproliferative activity on human breast cancer cell lines. *Bioorg. Med. Chem. Lett.* **2013**, *23*, 6401–6405. [CrossRef]
39. Sala, M.; Chimento, A.; Saturnino, C.; Gomez-Monterrey, I.M.; Musella, S.; Bertamino, A.; Milite, C.; Sinicropi, M.S.; Caruso, A.; Sirianni, R.; et al. Synthesis and cytotoxic activity evaluation of 2,3-thiazolidin-4-one derivatives on human breast cancer cell lines. *Bioorg. Med. Chem. Lett.* **2013**, *23*, 4990–4995. [CrossRef]
40. Lizard, G.; Latruffe, N.; Vervandier-Fasseur, D. Aza- and Azo-Stilbenes: Bio-Isosteric Analogs of Resveratrol. *Molecules* **2020**, *25*, 605. [CrossRef] [PubMed]
41. Iacopetta, D.; Lappano, R.; Mariconda, A.; Ceramella, J.; Sinicropi, M.S.; Saturnino, C.; Talia, M.; Cirillo, F.; Martinelli, F.; Puoci, F.; et al. Newly Synthesized Imino-Derivatives Analogues of Resveratrol Exert Inhibitory Effects in Breast Tumor Cells. *Int. J. Mol. Sci.* **2020**, *21*, 7797. [CrossRef]

42. Urbani, P.; Ramunno, A.; Filosa, R.; Pinto, A.; Popolo, A.; Bianchino, E.; Piotto, S.; Saturnino, C.; De Prisco, R.; Nicolaus, B.; et al. Antioxidant activity of diphenylpropionamide derivatives: Synthesis, biological evaluation and computational analysis. *Molecules* **2008**, *13*, 749–761. [CrossRef] [PubMed]
43. Grivennikov, S.I.; Karin, M. Inflammation and oncogenesis: A vicious connection. *Curr. Opin. Genet. Dev.* **2010**, *20*, 65–71. [CrossRef]
44. Infantino, V.; Convertini, P.; Cucci, L.; Panaro, M.A.; Di Noia, M.A.; Calvello, R.; Palmieri, F.; Iacobazzi, V. The mitochondrial citrate carrier: A new player in inflammation. *Biochem. J.* **2011**, *438*, 433–436. [CrossRef] [PubMed]
45. Infantino, V.; Iacobazzi, V.; Menga, A.; Avantaggiati, M.L.; Palmieri, F. A key role of the mitochondrial citrate carrier (SLC25A1) in TNF alpha- and IFN gamma-triggered inflammation. *BBA Gene Regul. Mech.* **2014**, *1839*, 1217–1225.
46. Tsai, S.H.; Lin-Shiau, S.Y.; Lin, J.K. Suppression of nitric oxide synthase and the down-regulation of the activation of NFkappaB in macrophages by resveratrol. *Br. J. Pharm.* **1999**, *126*, 673–680. [CrossRef]
47. Cory, S.; Adams, J.M. The BCL_2 family: Regulators of the cellular life-or-death switch. *Nat. Rev. Cancer* **2002**, *2*, 647–656. [CrossRef]
48. Wood, D.E.; Newcomb, E.W. Cleavage of Bax enhances its cell death function. *Exp. Cell Res.* **2000**, *256*, 375–382. [CrossRef]
49. Soldani, C.; Scovassi, A.I. Poly(ADP-ribose) polymerase-1 cleavage during apoptosis: An update. *Apoptosis* **2002**, *7*, 321–328. [CrossRef]
50. Athar, M.; Back, J.H.; Kopelovich, L.; Bickers, D.R.; Kim, A.L. Multiple molecular targets of resveratrol: Anti-carcinogenic mechanisms. *Arch. Biochem. Biophys.* **2009**, *486*, 95–102. [CrossRef]
51. Xiao, Q.; Zhu, W.; Feng, W.; Lee, S.S.; Leung, A.W.; Shen, J.; Gao, L.; Xu, C. A Review of Resveratrol as a Potent Chemoprotective and Synergistic Agent in Cancer Chemotherapy. *Front. Pharm.* **2018**, *9*, 1534. [CrossRef] [PubMed]
52. Molinari, M. Cell cycle checkpoints and their inactivation in human cancer. *Cell Prolif.* **2000**, *33*, 261–274. [CrossRef] [PubMed]
53. Hall, M.; Peters, G. Genetic alterations of cyclins, cyclin-dependent kinases, and Cdk inhibitors in human cancer. *Adv. Cancer Res.* **1996**, *68*, 67–108. [PubMed]
54. Ziegler, U.; Groscurth, P. Morphological features of cell death. *News Physiol. Sci.* **2004**, *19*, 124–128. [CrossRef]
55. Wang, P.G.; Li, Y.T.; Pan, Y.; Gao, Z.Z.; Guan, X.W.; Jia, L.; Liu, F.T. Lower expression of Bax predicts poor clinical outcome in patients with glioma after curative resection and radiotherapy/chemotherapy. *J. Neuro Oncol.* **2019**, *141*, 71–81. [CrossRef] [PubMed]
56. Ogston, K.N.; Miller, I.D.; Schofield, A.C.; Spyrantis, A.; Pavlidou, E.; Sarkar, T.K.; Hutcheon, A.W.; Payne, S.; Heys, S.D. Can patients' likelihood of benefiting from primary chemotherapy for breast cancer be predicted before commencement of treatment? *Breast Cancer Res. Treat.* **2004**, *86*, 181–189. [CrossRef] [PubMed]
57. Krajewski, S.; Blomqvist, C.; Franssila, K.; Krajewska, M.; Wasenius, V.M.; Niskanen, E.; Nordling, S.; Reed, J.C. Reduced expression of proapoptotic gene BAX is associated with poor response rates to combination chemotherapy and shorter survival in women with metastatic breast adenocarcinoma. *Cancer Res.* **1995**, *55*, 4471–4478.
58. Sirianni, R.; Capparelli, C.; Chimento, A.; Panza, S.; Catalano, S.; Lanzino, M.; Pezzi, V.; Ando, S. Nandrolone and stanozolol upregulate aromatase expression and further increase IGF-I-dependent effects on MCF-7 breast cancer cell proliferation. *Mol. Cell. Endocrinol.* **2012**, *363*, 100–110. [CrossRef]
59. Saturnino, C.; Barone, I.; Iacopetta, D.; Mariconda, A.; Sinicropi, M.S.; Rosano, C.; Campana, A.; Catalano, S.; Longo, P.; Ando, S. N-heterocyclic carbene complexes of silver and gold as novel tools against breast cancer progression. *Future Med. Chem.* **2016**, *8*, 2213–2229. [CrossRef] [PubMed]
60. Sirianni, R.; Chimento, A.; Malivindi, R.; Mazzitelli, I.; Ando, S.; Pezzi, V. Insulin-like growth factor-I, regulating aromatase expression through steroidogenic factor 1, supports estrogen-dependent tumor Leydig cell proliferation. *Cancer Res.* **2007**, *67*, 8368–8377. [CrossRef]
61. Sirianni, R.; Zolea, F.; Chimento, A.; Ruggiero, C.; Cerquetti, L.; Fallo, F.; Pilon, C.; Arnaldi, G.; Carpinelli, G.; Stigliano, A.; et al. Targeting Estrogen Receptor-alpha Reduces Adrenocortical Cancer (ACC) Cell Growth in Vitro and in Vivo: Potential Therapeutic Role of Selective Estrogen Receptor Modulators (SERMs) for ACC Treatment. *J. Clin. Endocrinol. Metab.* **2012**, *97*, E2238–E2250. [CrossRef]
62. Tundis, R.; Iacopetta, D.; Sinicropi, M.S.; Bonesi, M.; Leporini, M.; Passalacqua, N.G.; Ceramella, J.; Menichini, F.; Loizzo, M.R. Assessment of antioxidant, antitumor and pro-apoptotic effects of Salvia fruticosa Mill. subsp thomasii (Lacaita) Brullo, Guglielmo, Pavone & Terrasi (Lamiaceae). *Food Chem. Toxicol.* **2017**, *106*, 155–164. [PubMed]
63. Iacopetta, D.; Mariconda, A.; Saturnino, C.; Caruso, A.; Palma, G.; Ceramella, J.; Muia, N.; Perri, M.; Sinicropi, M.S.; Caroleo, M.C.; et al. Novel Gold and Silver Carbene Complexes Exert Antitumor Effects Triggering the Reactive Oxygen Species Dependent Intrinsic Apoptotic Pathway. *Chemmedchem* **2017**, *12*, 2054–2065. [CrossRef] [PubMed]
64. Infantino, V.; Iacobazzi, V.; Palmieri, F.; Menga, A. ATP-citrate lyase is essential for macrophage inflammatory response. *Biochem. Biophys. Res. Commun.* **2013**, *440*, 105–111. [CrossRef] [PubMed]
65. Grande, F.; Parisi, O.I.; Mordocco, R.A.; Rocca, C.; Puoci, F.; Scrivano, L.; Quintieri, A.M.; Cantafio, P.; Ferla, S.; Brancale, A.; et al. Quercetin derivatives as novel antihypertensive agents: Synthesis and physiological characterization. *Eur. J. Pharm. Sci. Off. J. Eur. Fed. Pharm. Sci.* **2016**, *82*, 161–170. [CrossRef] [PubMed]

International Journal of
Molecular Sciences

MDPI

Article

Prenylated Flavonoids with Potential Antimicrobial Activity: Synthesis, Biological Activity, and In Silico Study

Mauricio Osorio [1,*], Marcela Carvajal [2], Alejandra Vergara [2], Estefania Butassi [3], Susana Zacchino [3], Carolina Mascayano [4], Margarita Montoya [5], Sophia Mejías [5], Marcelo Cortez-San Martín [6] and Yesseny Vásquez-Martínez [7,*]

1 Laboratorio de Productos Naturales, Departamento de Química, Universidad Técnica Federico Santa María, Valparaíso 2390123, Chile
2 Centro de Biotecnología CB-DAL, Universidad Técnica Federico Santa María, Valparaíso 2390136, Chile; marcela.carvajal@usm.cl (M.C.); alejandrajj@gmail.com (A.V.)
3 Farmacognosia, Facultad de Ciencias Bioquímicas y Farmacéuticas, Universidad Nacional de Rosario, Suipacha 531, Rosario 2000, Argentina; fefabutassi@hotmail.com (E.B.); szaabigil@gmail.com (S.Z.)
4 Departamento de Ciencias del Ambiente, Facultad de Química y Biología, Universidad de Santiago de Chile, Santiago 9170022, Chile; carolina.mascayano@usach.cl
5 Laboratorio Bioquímica Celular, Departamento de Biología, Facultad de Química y Biología, Universidad de Santiago de Chile, Santiago 9170022, Chile; margarita.montoya@usach.cl (M.M.); sofia.mejias@usach.cl (S.M.)
6 Laboratorio de Virología Molecular y Control de Patógenos, Departamento de Biología, Facultad de Química y Biología, Universidad de Santiago de Chile, Santiago 9170022, Chile; marcelo.cortez@usach.cl
7 Programa Centro de Investigaciones Biomédicas Aplicadas, Escuela de Medicina, Facultad de Ciencias Médicas, Universidad de Santiago de Chile, Santiago 9170022, Chile
* Correspondence: mauricio.osorio@usm.cl (M.O.); yesseny.vasquez@usach.cl (Y.V.-M.)

Citation: Osorio, M.; Carvajal, M.; Vergara, A.; Butassi, E.; Zacchino, S.; Mascayano, C.; Montoya, M.; Mejías, S.; Cortez-San Martín, M.; Vásquez-Martínez, Y. Prenylated Flavonoids with Potential Antimicrobial Activity: Synthesis, Biological Activity, and In Silico Study. *Int. J. Mol. Sci.* **2021**, *22*, 5472. https://doi.org/10.3390/ijms22115472

Academic Editor: Maurizio Battino

Received: 27 April 2021
Accepted: 18 May 2021
Published: 22 May 2021

Abstract: Prenylated flavonoids are an important class of naturally occurring flavonoids with important biological activity, but their low abundance in nature limits their application in medicines. Here, we showed the hemisynthesis and the determination of various biological activities of seven prenylated flavonoids, named **7–13**, with an emphasis on antimicrobial ones. Compounds **9**, **11**, and **12** showed inhibitory activity against human pathogenic fungi. Compounds **11**, **12** (flavanones) and **13** (isoflavone) were the most active against clinical isolated *Staphylococcus aureus* MRSA, showing that structural requirements as prenylation at position C-6 or C-8 and OH at positions C-5, 7, and 4′ are key to the antibacterial activity. The combination of **11** or **12** with commercial antibiotics synergistically enhanced the antibacterial activity of vancomycin, ciprofloxacin, and methicillin in a factor of 10 to 100 times against drug-resistant bacteria. Compound **11** combined with ciprofloxacin was able to decrease the levels of ROS generated by ciprofloxacin. According to docking results of *S* enantiomer of **11** with ATP-binding cassette transporter showed the most favorable binding energy; however, more studies are needed to support this result.

Keywords: prenylated flavonoids; synthesis; antibacterial; MRSA; synergism; antifungal

1. Introduction

Prenylated flavonoids are a subclass of natural flavonoids characterized by the presence of a prenylated side chain (i.e., prenyl, geranyl and lavandulyl) attached to the flavonoid skeleton. Since prenylated flavonoids are generally more bioactive than their non-prenylated precursors, they are of interest as lead compounds for producing new drugs [1–3]. However, since prenylated flavonoids often exist at trace levels in their natural sources that make further biological studies difficult, it may be useful to synthesize prenylated flavonoids in appreciable quantities and at low costs to allow future biological studies [4]. Compounds of this type exhibit significant effects in the insect–plant interaction [5] as well as antifungal [6], antimicrobial [7], antiviral [8,9], anti-inflammatory [10], and anticancer [11] activities and anti-lipid properties, both in vitro and in vivo [12]. The C-prenylation of flavonoids seems to be crucial to the biological activities of these compounds, which may lead to enhanced cell membrane targeting and thus increased intracellular activity [13]. From a chemistry point of view, the C-prenylation is complicated, inefficient, and time-consuming, so other methods must be explored; i.e., biocatalysis may be the solution to these problems [14,15]. The prenylation of the flavonoid core increases lipophilicity and membrane permeability, which is one of the proposed reasons for the enhanced biological activities of prenylated flavonoids [16]. Efficient chemistry methods to obtain 8-prenylnaringenin with reasonable yields (20–33%) have been reported via Claisen and Claisen–Cope rearrangement [17] in many steps. A good approximation to obtain these kinds of compounds may be the direct prenylation of commercially available flavonoids with 3-methyl-2-buten-1-ol using $ZnCl_2$ as Lewis acid by coupling via an aromatic electrophilic substitution reaction (ArES), whose conditions have been used to obtain prenylated phenols in our laboratory [18].

To determine the structural requirements of the prenylated flavonoids to exert any remarkable biological activity, it is necessary to obtain numerous and varied series of these compounds and to test them against several types of biological tests. These results will provide enough data to obtain a reliable structure–activity relationship [17].

Multi-drug-resistant organisms are one of the major causes for the alarming level of infectious diseases worldwide [19]. The discovery of new drugs with potent antimicrobial activity, particularly against the resistant strains, is therefore highly desirable. Since many of the currently available drugs have undesirable side effects, are ineffective against new or reemerging fungi, or develop a rapid resistance [20], there is an urgent need for the next generation of new antimicrobial agents that overcome these disadvantages. In this sense, prenylated flavonoids have been demonstrated to have potent antimicrobial activity [21]. *Escherichia coli*, *Klebsiella pneumoniae*, and *Staphylococcus aureus* are the three human pathogens of greatest concern, associated with hospital- and community-acquired infections, according to the latest global estimates of antibiotic resistance worldwide [22].

Synergism between natural compounds and traditional antibiotics is a promising strategy that could solve the problem of antibiotic resistance that some bacterial strains have been able to develop [23].

As endogenous free radicals and human 5-lipoxygenase (5-hLOX) have been shown to have major roles in the pathogenesis of various health problems including inflammatory disorders [24,25], the compounds obtained here were tested as 5-hLOX inhibitors and DPPH (2,2-diphenyl-1-picrylhydrazyl) radical scavengers. Since inflammation is related to bacterial-mediated infections, a dual anti-inflammatory, antibacterial agent with an improved safety profile is a goal for improved therapeutic benefits and better patient compliance [26].

In this work, we have synthesized seven prenylated flavonoids and studied their antimicrobial, anti-inflammatory, and antioxidant activities to evaluate the relevance of adding prenyl groups on the structure of non-prenylated flavonoids and to propose the use of naturally occurring organic compounds as possible therapeutic agents.

2. Results and Discussion

2.1. *Synthesis*

The synthesis of prenylated phenols was carried out in one step via the EAS mechanism as shown in Scheme 1, using a prenylation methodology applied to hydroxylated acetophenones and benzaldehydes by us [18], with some modifications. The one-step reaction of different flavonoids (**1–5**) with 3-methyl-2-buten-1-ol (**6**) in the presence of $ZnCl_2$ in ethyl acetate produced prenylated flavonoids (**7–13**) in moderate yields (23–36%) and side products, mainly. Only 5′-prenyl fisetin (**10**) is new. When the flavonoids chrysin (**1**) and naringenin (**4**) were alkylated, we isolated two C-alkylated compounds at positions 8 and 6 of the A-ring. Complex mixtures were found with quercetin (**2**) and fisetin (**3**), probably because their electron-donating substituents are more susceptible to oxidation [27]. Electron-rich aromatic rings can induce polyalkylations and decompositions, reducing the yield.

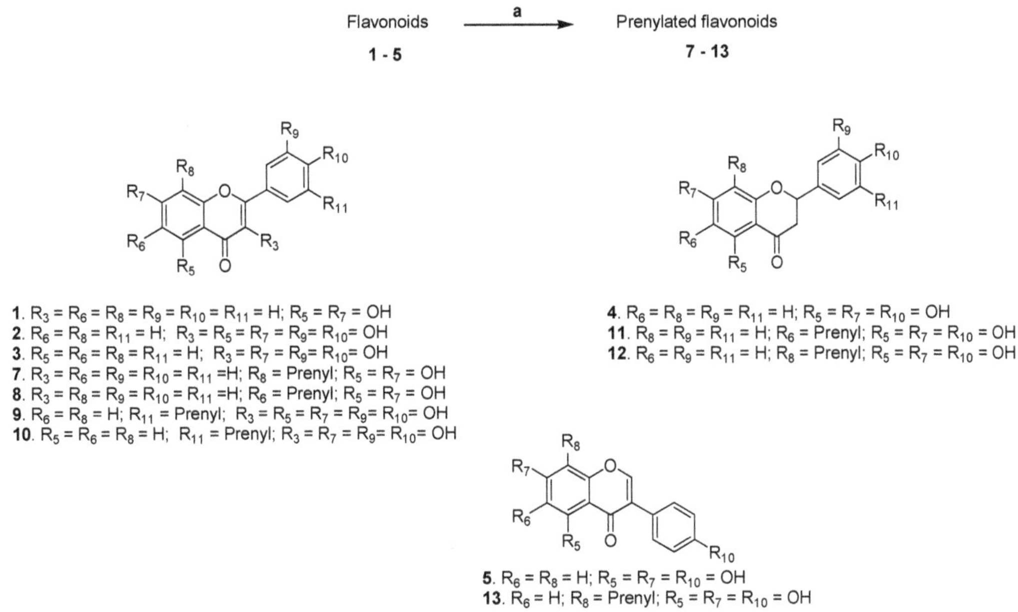

Conditions and reagents: (a) 3-methyl-2-buten-1-ol (*6*), $ZnCl_2$/AcOEt, 40 °C–reflux, 4 h → 7 -13 (23–36%).

Scheme 1. Representative synthesis of prenylated flavonoids in one step.

Structure Determination

The place of C-alkylation was determined by 2D ^{1}H- ^{13}C HMBC NMR spectra. The main J_3 and J_2 interactions between the methylene proton CH_2 of the prenyl chain and aromatic peaks are schematically shown in Figure 1. Only the interaction between the methylene protons with the quaternary 8a carbon in the structure occurs when the C-prenylation is produced in C-8. Correlations J_3 and J_2 between the methylene protons CH_2 and quaternary 8a or 4a carbon are impossible when the prenylation is bounded at Position 6. For the rest of the molecules, the same reasoning was used to determine the place of C-alkylation.

The presence of the carbonyl and hydroxyl groups was confirmed by observing the typical signals in IR spectra. Stretching absorption values between 3550 and 3200 cm^{-1} as a broad signal indicated the presence of phenol groups and stretching absorption values between 1750 and 1680 cm^{-1} as one of the strongest singlet IR absorptions confirmed the presence of carbonyl groups.

The hydrogen bond to carbons was determined by J_1 interactions in the 2D ^{1}H-^{13}C HSQC NMR spectra. 2D NMR spectra (HSQC and HMBC) for all prenylated compounds were included in Supplementary Materials.

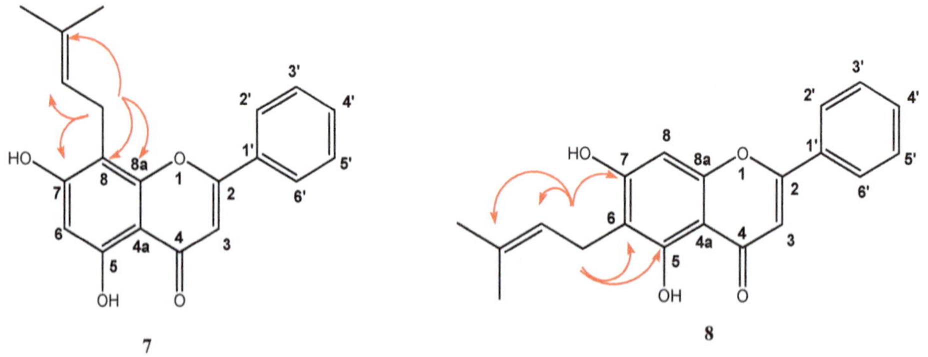

Figure 1. 2D ^{1}H-^{13}C HMBC spectrum obtained for compounds **7** and **8**. Here, the projections of the full spectrum ^{1}H NMR and projections between 100 and 158 ppm for ^{13}C NMR spectrum are represented. Only the main correlations between aromatic peaks and prenyl chain are shown.

2.2. In Vitro Antibacterial Effects on Human Pathogen Bacteria

From the results of bacterial growth inhibitory activity on sensitive and resistant *S. aureus* (Table 1), we can observe that not all prenylated flavonoids showed a good antibacterial effect, being prenylated flavanones **7** to **10**, weak to moderate inhibitors. In Table 2, compounds **11** and **12**, corresponding to prenylated naringenin, together with compound **13** (prenylated genistein), showed strong activity both on the sensitive strain and on the resistant strains, with MIC values ranging from 5 to 50 µg/mL. Compound **11** is the one that shows the best antibacterial action, especially on the resistant strain MRSA 97-7 with an MIC lower than that of the commercial agents tested.

Table 1. In vitro antibacterial assay. The expressed values are the percent inhibition of bacterial growth achieved by prenylated flavonoids at a concentration of 50 µg/mL.

Compound	MRSA 97-7	MRSA 622-4	ATCC 6538
7	21%	0%	6%
8	5%	0%	26%
9	61%	3%	52%
10	13%	18%	17%
11	95%	93%	100%
12	98%	98%	99%
13	91%	96%	99%

Table 2. Effect of prenyl substituent in the flavonoids on the antibacterial activity expressed as minimum inhibitory concentration (MIC) (µg/mL).

Compound	MRSA 97-7	MRSA 622-4	ATCC6538
4	>50	>50	>50
11	5	25	10
12	15	25	15
13	10	50	10
Vancomycin	25	5	3.5
Ciprofloxacin	100	10	2.5
Methicillin	50	50	10

The results show that the saturation of the double bond at the C2–C3 position appears to be important for the antibacterial activity of flavonoids (compounds **11** and **12**). This coincides with those reported in the literature, where prenylated flavanones show better activity than flavones [28,29]. The position of the hydroxyl substituents, as well as that of the prenyl group, appears to be important in the inhibitory activity of bacterial growth, observing that derivatives with OH at positions C-5, 7, and 4′ show better inhibitory activity, as well as the presence of prenyl groups in the C6 and C8 positions of Ring A, as previously reported [30–32]. Araya–Coultier et al. (2018) showed that the position of the prenyl group affects the antimicrobial activity of flavonoids, when comparing the effect of lupiwighteone (5,7,4′-trihydroxy-8-prenylisoflavone) and wighteone (5,7,4′-trihydroxy-6-prenylisoflavone), on the Gram-positive strain *L. monocytogenes*. Lupiwighteone, chain prenylated at C-8, showed no inhibition, whereas wighteone, chain prenylated at position C-6, had high antibacterial activity [33].

When comparing compound **4** with compounds **11** and **12**, where the base structure is maintained and only one prenyl substituent has been added in different positions, we can observe that the introduction of a prenyl group has an important effect on the inhibitory activity, reducing by up to at least one order of magnitude the MIC in MRSA 97-7 (Table 2). It has been reported that prenylated flavonoids are more hydrophobic than common flavonoids, facilitating this characteristic the ability to penetrate the cell membrane, thus improving their action at the active site [29].

Mun et al. (2014) proposed that prenylated flavonoids act through interaction with the bacterial membrane, binding directly with peptidoglycan, observing morphological changes when MRSA strains were treated with the prenylated flavonoid sophora-flavanone B [34].

Tables 3 and 4 show that a synergistic effect is produced between compounds **11**, **12**, and **13** when acting in conjunction with commercial antibiotics, showing that the molecules **11** and **12** reduce the MIC of ciprofloxacin over MRSA 97-7 by 10 times (FICI values 0.6 and 0.16, respectively) and **11**, **12**, and **13** reduce 100 times its MIC over MRSA 622-4 with FICI values from 0.1 to 0.2 (results shown in Table 4). They also acted synergistically with Vancomycin (FICI values from 0.15 to 0.43), decreasing its MIC 10 times against MRSA 97-7. Compound **11** also acted synergistically with Methicillin with an FICI value of 0.01, decreasing its MIC 100 times against MRSA 97-7.

Table 3. Synergistic effects of prenylated flavonoids with commercial antibiotics on isolated MRSA 97-7.

Agents	MIC Alone (μg/mL)	MIC Combined (μg/mL)	FICI	Outcome	Fold Reduction in Commercial Antibiotic MIC
11 Vancomycin	5 25	0.5 2.5	0.2	Synergy	10
12 Vancomycin	15 25	5 2.5	0.43	Synergy	10
13 Vancomycin	10 25	0.5 2.5	0.15	Synergy	10
11 Ciprofloxacin	5 100	2.5 10	0.6	Partial Synergy	10
12 Ciprofloxacin	15 100	1 10	0.16	Synergy	10
11 Methicillin	5 50	0.02 0.5	0.01	Synergy	100

This potentiating effect of the antimicrobial agent action exerted by prenylated molecules is not observed when combining them with each other, as observed in Table 5, where, when combining compounds **11** and **12**, an FICI value between 1.1 and 2 was achieved, given an account of an indifferent effect. This difference in the behavior of prenylated molecules, when compared to the action they exert when combined with antimicrobial agents, may be due to the fact that prenylated flavonoids act by a different mechanism than the commercial antibiotics evaluated, e.g., through the inhibition of pumps of expulsion such as ABC transporters, thus being able to enhance the action of the commercial antibiotic, as proposed by Wagner et al. (2009), who pointed out that the synergistic effect of phytodrugs in conjunction with antimicrobial agents can contribute to effects on the permeability of the cell wall of bacteria, the inhibition of the efflux pump, and the inhibition of bacterial enzymes, and that all these mechanisms would help the antimicrobial agent to reach its site of action [35].

Table 4. Synergistic effects of prenylated flavonoids with commercial antibiotics on isolated MRSA 622-4.

Agents	MIC Alone (μg/mL)	MIC Combined (μg/mL)	FICI	Outcome	Fold Reduction in Commercial Antibiotic MIC
11	25	2.5	0.1	Synergy	100
Ciprofloxacin	10	0.1			
12	25	5	0.2	Synergy	100
Ciprofloxacin	10	0.1			
13	50	5	0.1	Synergy	100
Ciprofloxacin	10	0.1			

Table 5. Synergistic effects between prenylated flavonoids on an isolated MRSA strain.

Strain	Agents	MIC Alone	MIC Combined	FICI	Outcome
MRSA 97-7	**11**	5	5	2	Indifferent
	12	15	15		
MRSA 622-4	**11**	25	2.5	1.1	Indifferent
	12	25	25		

To corroborate the synergistic action of compound **11** with vancomycin, a time kill assay experiment was carried out, the result of which is shown in Figure 2. We can see that compound **11** alone manages to slightly lower the bacterial growth curve, decreasing by less than 2 log the CFU/mL, showing a moderate bacteriostatic action. Vancomycin achieved a greater effect, lowering CFU/mL by more than 3 log10, but this effect is enhanced by acting in conjunction with compound **11**, lowering CFU/mL by more than 5 log10, thus verifying the synergistic effect that compound **11** would exert on ciprofloxacin on MRSA 97-7. It has been described in the literature that prenylated flavonoids isolated from *Morus alba* L. show a broad spectrum of synergistic action towards *Staphylococcus aureus*, acting in conjunction with antibacterial agents such as aminoglycosides, β-lactams, glycopeptides, and fluoroquinolones [28].

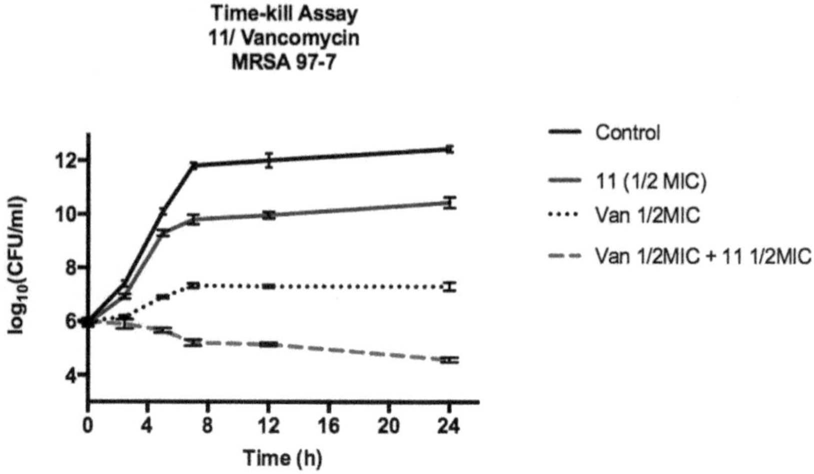

Figure 2. Time kill assay vancomycin with compound **11** on MRSA 97-7.

The results of the bacterial ROS level are shown in Figure 3. We can see that there is a significant decrease in ROS levels inside the bacteria as compounds **11**, **12**, and **13** are present, showing a protective effect by reducing ROS, with an effect as important as that shown by the agent Trolox, a known ROS species trapper. It has been described in the literature that polyhydroxylated flavonoids can act as antioxidant agents, trapping ROS [36], and this antioxidant property can be related to the ROS-reducing effect observed in this assay.

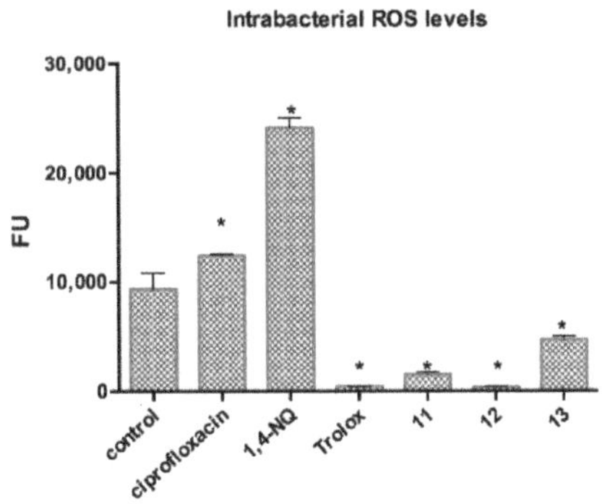

Figure 3. Effect of prenylated compounds in intrabacterial ROS levels in *Staphylococcus aureus* ATCC6538. Control: DMSO. Trolox: ROS scavenger control. 1,4-NQ (1,4-Naphtoquinone): ROS generator control. Ciprofloxacin (0.05 µg/mL): commercial antibiotic control. **11**–**13**: prenylated compounds in evaluation (1/2 MIC concentration). Statistical analysis: Mann–Whitney U test (*) $p < 0.05$, between each treatment with the negative control.

It has been previously reported that ciprofloxacin generates oxidative stress in *S. aureus*, generating an increase in ROS levels within the bacteria [37]. This result was corroborated in this study, where the levels of ROS generated in the bacteria in the presence of ciprofloxacin were statistically higher than those of the control. As a way to explain the synergistic action observed in the previous experiments, the effect on intrabacterial ROS levels was evaluated by combining ciprofloxacin with compound **11**. As we can see in Figure 4, this combination markedly decreases the levels of ROS when compared to those generated by ciprofloxacin alone. From these results, it can be deduced that the prenylated compounds may be acting as trappers for the ROS generated by ciprofloxacin.

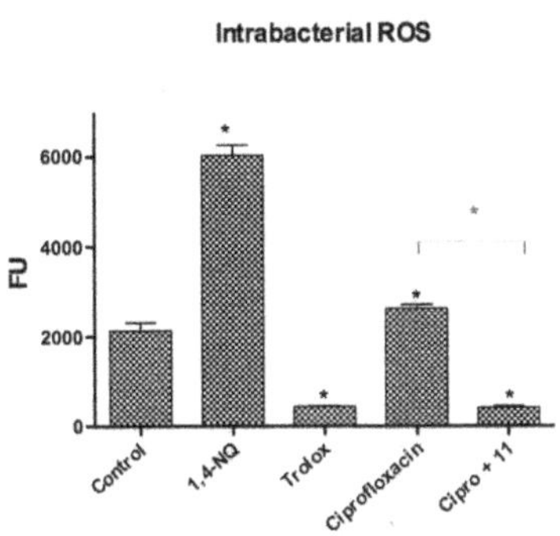

Figure 4. Effect of prenylated Compound **11** on intrabacterial ROS levels produced by ciprofloxacin in *Staphylococcus aureus ATCC6538*. Control: DMSO. Ciprofloxacin (0.05 µg/mL): commercial antibiotic control. Cipro + **11** (ciprofloxacin 0.05 µg/mL + **11** (1/2 MIC concentration). Statistical Analysis: Mann–Whitney U test (*) $p < 0.05$.

Docking of **11** (S and R Enantiomers) and **8** with ATP-Binding Cassette (Multi-Drug ABC Transporter Sav1866 from *S. aureus* in Complex with AMP-PNP)

We used a fast and predictive tool as the molecular docking, which allowed us to obtain by the genetic algorithm the most suitable location of ligand on macromolecule without using molecular dynamic simulations. Thus, we carried out the analysis according to the most favorable binding energy, by grouping the results in a cluster. To analyze the good result of **11** and their correlation with the antibacterial activity, we used the in silico tool with a known and essential pharmacological target linked to this property. For the above, the ABC transporter was selected, which is crucial for balancing nutrients in bacteria and humans. This transporter is also responsible for several processes, such as drug-resistant bacteria [38]. Based on these reasons, we used docking tools to observe which interactions were relevant, especially with the most promising molecule **11**, although it is worth taking into account that the biological tests were carried out with a racemic mixture. In this study, we decided to study both enantiomers in order to be able to discriminate them and determine whether the binding was stereoselective. The results showed a priority for the *S* enantiomer (−5.06 Kcal/mol ± 0.012) versus the *R* enantiomer (−3.37 Kcal/mol ± 0.016) (a difference of approximately 2 Kcal/mol between them) both binding in the similar place of the transporter. These analyses of results effectively indicated a different binding mode: The *S* enantiomer showed two hydrogen bonds with Glu320 and Asp321 of the ABC transporter; on the other hand, the *R* enantiomer showed two hydrogen bonds but with Glu353 and Gly378 (Figure 5), and in both molecules, the hydrophobic interaction is important to the binding. We carried out a similar analysis by docking studies with **8** which has a bad antibacterial activity. The binding energy (-2.39 Kcal/mol) obtained was worse than the **11** enantiomers and the molecule only showed one hydrogen bond with Asp220 for stabilization in the binding area. Finally, we observed that the molecule is located in the same area as the isomers as we observed in Figure 5.

a) b) c)

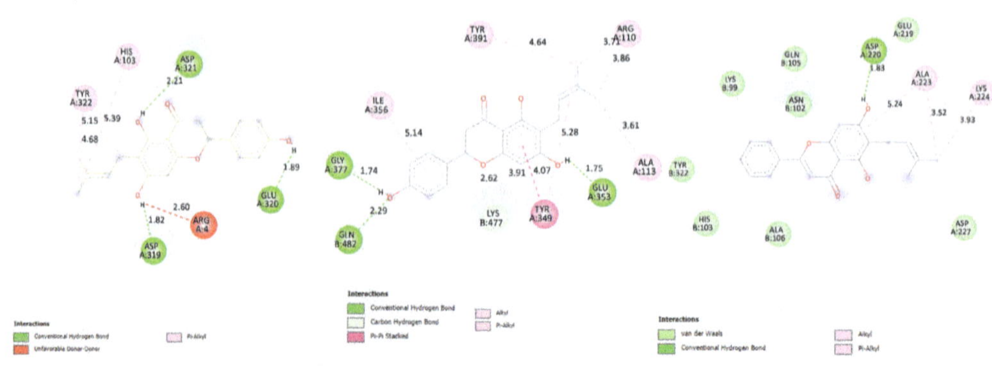

11-R
- 3.37 Kcal/mol

11-S
- 5.06 Kcal/mol

8
- 2.39 Kcal/mol

Figure 5. Docking results: main interactions of the stereoisomers of **11** ((**a**) (*R*)-**11** and (**b**) (*S*)-**11**) and **8** (**c**) with ATP binding cassette.

2.3. *In Vitro Antifungal Effects on Human Pathogenic fungi*

In order to have a look into the potential usefulness of these compounds as candidates for the development of new antifungal drugs, we investigated the antifungal properties of compounds **1–5** and **7–13** against a panel of clinically important fungal species including two yeasts (*Candida albicans* and *Cryptococcus neoformans*), three *Aspergillus* spp. (*A. niger, A. fumigatus,* and *A. flavus*), and three dermatophytes (*Trichophyton rubrum, T. mentagrophytes,* and *Microsporum gypseum*) (voucher numbers are included in Table 6). The minimum inhibitory concentration (MIC) and minimum fungicidal concentration (MFC) of all compounds were determined with the microbroth dilution methods M27-A3 and M38-A2 of the Clinical and Laboratory Standards Institute (CLSI) [39,40]. The evaluation of the minimum fungicidal concentration (MFC) of the prenylated flavonoids showed whether the compounds kill fungi in addition to inhibiting them. This characteristic is highly appreciated in an antifungal drug since the recurrence of fungistatic drugs is mostly avoided because long treatments are no longer necessary.

Table 6. Antifungal activity (MIC/MFC in μg/mL) of prenylated flavonoids **7–13** against pathogenic fungi from American Type Culture Collection (ATCC) and *Centro de Referencia en Micología* Culture Collection (CCC). Amph: Amphotericin B; Terb: Terbinafine.

Comp.	*Candida albicans ATCC10231*	*Cryptococcus neoformans ATCC32264*	*Aspergillus flavus ATCC 9170*	*Aspergillus fumigatus ATCC26934*	*Aspergillus niger ATCC9029*	*Microsporum gypseum CCC115,*	*Trichophyton rubrum CCC113*	*Trichophyton mentagro-phytes ATCC9972*
1	>250	>250	>250	>250	>250	>250	>250	>250
2	>250	>250	>250	>250	>250	>250	>250	>250
3	>250	>250	>250	>250	>250	>250	>250	>250
4	>250	>250	>250	>250	>250	>250	>250	>250
5	>250	>250	>250	>250	>250	>250	>250	>250
7	>250	>250	>250	>250	>250	>250	>250	>250
8	>250	>250	>250	>250	>250	>250	>250	>250
9	62.5/62.5	62.5/62.5	>250	>250	>250	62.5/62.5	62.5/62.5	62.5/62.5
11	>250	125/125	>250	>250	>250	62.5/62.5	62.5/62.5	62.5/62.5
12	125/250	62.5/125	>250	>250	>250	31.25/31.25	15.6/31.25	15.6/31.25
10	>250	250/>250	>250	>250	>250	250/250	250/250	250/250
13	>250	>250	>250	>250	>250	>250	>250	>250
Amph	1.0/1.0	1.0/2.0	2.0/2.0	2.0/2.0	2.0/2.0	-	-	-
Terb	-	-	-	-	-	0.008/0.015	0.004/0.008	0.004/0.015

Compounds with MICs of >250 μg/mL were considered inactive; those between 250 and 125 μg/mL were considered lowly active; those in the range 62.5–31.25 μg/mL were considered moderately active; and MICs of <31.25 μg/mL were considered highly active.

The results showed (Table 6) that the non-prenylated structures (the flavone **1**, the flavonols **2** and **3**, the flavanone **4**, and the isoflavone **5**) were all inactive.

When these flavonoids were prenylated, the following results were obtained: (i) The prenylation of the flavone **1** or the isoflavone **5** did not improve the antifungal activity, since **7**, **8**, and **13** were inactive (MIC $\geq$ 250 μg/mL). (ii) The prenylation of the flavonols **2** and **3** led to **9** (which showed activity against the yeasts *C. albicans* and *C. neoformans* and dermatophytes at 62.5 μg/mL) and **10**, which was inactive. Both compounds possessed the prenyl substituent at position 5′ but differed in that **9** possesses a hydroxyl group in position 5. This structural difference could play a role in the antifungal activity of **9**. (iii) The prenylation of the flavanone **4** led to both **11** (MIC = 125 μg/mL against *C. neoformans* and 62.5 μg/mL against dermatophytes) and the most active compound **12**, which displayed MIC = 125 μg/mL against *C. albicans*, MIC = 62.5 μg/mL against *C. neoformans*, MIC = 15.6 μg/mL against the dermatophytes *T. rubrum* and *T. mentagrophytes*, and 31.2 μg/mL against *M. gypseum*). Probably the prenyl substituent at position 8 of the trihydroxyflavanone naringenin confers high activity to the whole flavanone molecule.

Considering that *C. neoformans* is the main cause of cryptococcal meningoencephalitis among HIV patients, which many times has not responded to any antifungal and has led to death [41], we decided to test compounds **9**, **11**, and **12** (which showed some activity against the ATCC strain) against a panel of five clinical *C. neoformans* strains. The results showed (Table 7) that the behavior of the three compounds against clinical isolates is similar to that against the standard ATCC strain.

Table 7. MIC/MFC (in µg/mL) of selected prenylated compounds against clinical isolates of *C. neoformans* from Malbrán Institute (IM, Buenos Aires Argentina). Amph: Amphotericin B.

Compound	*C. neoformans* Clinical Isolates				
	IM983040	IM972724	IM042074	IM983036	IM00319
9	125/250	125/250	125/125	125/125	125/125
11	250/>250	250/>250	250/>250	250/>250	250/>250
12	125/125	125/125	62.5/62.5	62.5/125	125/125
Amph	0.25	0.25	0.12	0.25	0.5

In turn, since *C. albicans* is the fourth leading cause of nosocomial bloodstream infection (BSI) in intensive care units, causing fatal invasive candidiasis in a high percentage of patients [42], we decided to test **9** and **12** against five clinical isolates of *C. albicans*, in order to determine whether these two compounds have the potential to be further developed. The results showed (Table 8) that both compounds behave against *C. albicans* clinical isolates similarly to how they behave against the standard ATCC strain.

Table 8. MIC/MFC (in µg/mL) of selected prenylated compounds against clinical isolates of *C. albicans* from *Centro de Referencia en Micología*, (CCC, Rosario, Argentina): Amph: Amphotericin B.

Compound	*C. albicans* Clinical Strains				
	C. albicans CCC 126	*C. albicans* CCC 127	*C. albicans* CCC 128	*C. albicans* CCC 129	*C. albicans* CCC 130
9	62.5/62.5	62.5/62.5	62.5/62.5	62.5/62.5	62.5/62.5
12	125/125	125/125	125/125	125/125	125/125
Amph	1.56	0.78	1.56	0.78	0.50

2.4. *In Vitro Antibacterial Effects on Plant Pathogen bacteria*

The results of the antibacterial activity against phytopathogenic bacteria showed that all the prenylated compounds were able to inhibit the in vitro growth of the three bacteria tested. Interestingly, most of the compounds significantly inhibited the growth of the three bacterial species in high percentages, obtaining IC_{50} values of less than 3.9 µM for several of them. Despite the above considerations, the **3**, **4**, and **5** molecules showed the highest IC_{50} values in the series (Table 9). Statistical analysis revealed that the values obtained from inhibition are significant ($p < 0.05$) compared to the negative control.

Table 9. IC_{50} and MIC values of compounds towards phytopathogenic bacteria.

Compound	IC_{50}/MIC (µM)		
	P. syringae	*A. tumefaciens*	*P. carotovorum*
1	>31.3/>250	>250/>250	>250/>250
2	>62.5/>250	>250/>250	>250/>250
3	>250/>250	>250/>250	>250/>250
4	>250/>250	>250/>250	>250/>250
5	>250/>250	>250/>250	>250/>250
7	<3.9/>250	<3.9/>250	<3.9/>250
8	<3.9/>250	<3.9/>250	<3.9/>250
9	<3.9/>250	<3.9/>250	<3.9/>250
11	<3.9/>250	<3.9/>250	<3.9/>250
12	<3.9/>250	<3.9/>250	<7.8/>250
10	<15.6/>250	<3.9/>250	<3.9/>250
13	<3.9/125	<3.9/>250	<3.9/125

Antibacterial activities are shown as IC_{50} and MIC values in µM concentrations; IC_{50} values were calculated from the first kinetic assay. MIC: minimum inhibitory concentration.

IC_{50} values for each compound indicated that the bacteria tested had different levels of sensitivity to the compounds tested. The most sensitive bacteria were *P. syringae*, and the most resistant were *A. tumefaciens*. This behavior could be due to the different mechanisms of action of the compounds on each tested Gram-negative pathogen.

After the first kinetic test, the bactericidal capacity of each compound was evaluated in a second growth kinetics test, using an inoculum of the first kinetic plate [43]. This new culture was incubated for 7 h, and the absorbance was measured. It was observed that the bacterial culture was able to recover its growth rate (with respect to the culture without compound) by 50–85%. This means that the compounds were bacteriostatic but not bactericide at the concentrations tested. Compound **13** showed selectivity against *P. syringae* and *P. carotovorum* and to a lesser extent against *A. tumefaciens* (>250, MIC).

2.5. In Vitro 5-hLOX Enzyme Inhibition Assay

The results showed that, of all the evaluated molecules (Table 10), only two (compounds **9** and **10**) had a relevant inhibitory effect. The structural characteristic that differentiates these compounds from the rest of the structures is the catechol group (1,2-dihydroxybenzene) in Ring B. Many known inhibitors of 5-LOX possess the catechol group that is a relevant characteristic for their inhibitory properties [44].

Table 10. Activity of prenylated phenols as 5-hLOX inhibitors.

Compound	% Inhibition
7	4
8	1
9	51
10	38
11	20
12	12
13	17

2.6. DPPH Radical Scavenging Activity

The results obtained from antioxidant activity assays are shown in Table 11. All compounds were compared with Trolox. All compounds possessing the catechol moiety were active. It is known that the flavonoids quercetin (**2**) and fisetin (**3**) have better antioxidant activity than Trolox [45]; however, their prenylated derivatives (**9** and **10**) show an even greater degree of this activity. This behavior follows the same trend found in some of our previous work [18].

Table 11. Screening results of DPPH radical scavenging activity of flavonoids and their prenylated derivatives.

Compound	IC_{50} μM ± SD	Compound	IC_{50} μM ± SD
1	NA	**7**	NA
2	13.22 ± 0.49	**8**	NA
3	12.82 ± 0.48	**9**	9.49 ± 0.36
4	NA	**10**	10.32 ± 0.03
5	NA	**11**	NA
		12	NA
Trolox	22.54 ± 0.61	**13**	NA

Antioxidant activity is shown as IC_{50} values in μM concentrations; NA = no activity. All compounds were analyzed in triplicate, and the results expressed as average ± standard deviation.

2.7. Cytotoxic Activity of 11 and 12

Anticancer activity for naringenin in diverse tumor cell lines has been previously reported [46–48]. However, in vitro activities were commonly observed using high flavonoid concentrations often exceeding 100 μM, which have limited physiological significance and must be interpreted with caution. Given this anticancer activity and that prenylated flavonoids have enhanced biological activities, we measured cytotoxic activity prenylated naringenin at positions 6 and 8 (**11** and **12**) against MDA-MB-231 (the human breast adenocarcinoma cell line), B16-F10 (mouse melanoma cells), and MEF (primary mouse embryonic fibroblasts). Our results in Table 12 show that the cytotoxic effect of both prenylated naringenins were low for both cancer cell lines and did not show any selectivity compared to the effect on non-cancerous cells (MEFs). When the cytotoxic effect of **11** and **12** against cell lines were compared with commonly used chemotherapeutic drugs, it could be verified that both prenylated flavonoids showed from 100- to 250-fold lower activity than Taxol and from 30- to 50-fold lower activity than etoposide. It is possible that the anticancer activity described for naringenin in vivo could be related to antiangiogenic and immunostimulating effects [49,50].

Table 12. Cytotoxic activities of **11** and **12** against cancer and non-cancer cells.

Compound	IC_{50} (μM) ± SD		
	MDA-MB-231	B16-F10	MEF
11	53.94 ± 9.66	49.14 ± 3.38	48.45 ± 3.44
12	59.02 ± 3.25	45.53 ± 3.82	54.66 ± 3.35
Taxol	0.21 ± 0.05	0.48 ± 0.08	0.40 ± 0.07
Etoposide	1.33 ± 0.28	1.74 ± 0.32	1.03 ± 0.17

SD: standard deviation; MDA-MB-231: human breast adenocarcinoma cell line; B16-F10: mouse melanoma cells; MEF: primary mouse embryonic fibroblast.

On the other hand, compounds **11** and **12** exhibited the most effective antibacterial activity, which had a synergistic effect with commercial antibiotics. A significant synergic activity was observed using approximately 7 μM for **11** and 15 μM for **12** (Table 4), which did not generate a cytotoxic effect in MEF cells. This clearly suggests that **11** and **12** could be useful for in vivo treatment against pathogenic resistant bacteria, although further studies are needed to confirm this possibility.

3. Materials and Methods

3.1. Chemistry

3.1.1. General Data

All chemical reagents obtained were purchased from Merck (Darmstadt, Germany), Sigma–Aldrich (St. Louis, MO, USA), or Alfa Aesar (Kandel, Germany), were of the highest commercially available purity, and were used without previous purification. Melting points (mp: °C) were measured on a melting point apparatus Stuart-Scientific SMP3 and are uncorrected. IR spectra were recorded as a KBr disk in a Thermo Scientific Nicolet 6700 FT-IR spectrometer (Massachusetts, USA), and frequencies are reported in cm^{-1}. High-resolution mass spectra were recorded on an Exactive^{TM} Plus Orbitrap spectrometer (Thermo Scientific, Bremen, Germany) by infusion, applying a voltage of 5 or 8 kV in the positive ionization mode and 4 or 7 kV in the negative ionization mode. The spectra were recorded using full-scan mode, covering a mass range from m/z 214 to 470. The resolution was set to 140,000, and the maximum loading time for the ion cyclotron resonance (ICR) cell was set to 200 ms. ^{1}H-, ^{13}C-(DEPT 135), 2D HSQC, and 2D HMBC spectra were recorded in DMSO-d_6 solutions and referenced to the residual peaks of DMSO at δ 2.50 ppm for ^{1}H and δ 39.5 ppm for ^{13}C, respectively, on a Bruker Avance 400 Digital NMR spectrometer, operating at 400.1 MHz for ^{1}H and 100.6 MHz for ^{13}C. Chemical shifts are reported in δ ppm and coupling constants (*J*) are given in Hz. Silica gel (Merck 200–400 mesh) was used for C.C. and silica gel plates HF-254 for TLC. TLC spots were detected by both under a UV lamp and heating after drenching in 10% H_2SO_4 in H_2O. Antioxidant determinations were performed in a Thermo Scientific Multiskan GO 96-well plate photometer.

3.1.2. General Experimental Procedure for the Prenylation of Flavonoids

A solution of flavonoid (1 mol equiv) and dry $ZnCl_2$ (4 mol equiv) was placed in a round bottom flask and dissolved in ethyl acetate (100 mL). Under vigorous stirring, a solution of 3-methyl-2-buten-1-ol (**6**) (4 mol equiv) in ethyl acetate (10 mL) was added dropwise for 1 h at 40 °C. The reaction mixture was then allowed to warm up to reflux temperature and the stirring continued. After 4 h, water at pH 1 (100 mL) was added to the reaction mixture to decompose the $ZnCl_2$. The organic layer was then separated, and a new extraction with ethyl acetate was made. The organic solutions obtained after extractions were mixed and dried over anhydrous sodium sulphate and filtered, and the solvent was evaporated under reduced pressure. The mixture was then subjected to silica gel flash column chromatography (ethyl acetate/hexane or dichloromethane/methanol or acetone/hexane mixtures were used as mobile phases) to obtain pure products.

All structures were confirmed by IR and NMR spectra as discussed below.

3.1.3. Physical Data of Prenylated Flavonoids

2-phenyl-8-(3-methyl-2-buten-1-yl)-5,7-dihydroxy-4*H*-chromen-4-one (**7**) and 2-phenyl-6-(3-methyl-2-buten-1-yl)-5,7-dihydroxy-4*H*-chromen-4-one (**8**) were obtained from chrysin (**1**) (3.9 mmol), 3-methyl-2-buten-1-ol (**6**) (15.7 mmol), and $ZnCl_2$ (15.7 mmol), as described above. The crude mixture was purified using ethyl acetate-hexane in a gradient system (0 → 70% of ethyl acetate) as the mobile phase to afford **7** as a yellow powder (329 mg, 26%); mp: 222 °C; HRMS *m/z*, observed: 321.1131; $C_{20}H_{17}O_4$ $[M–H]^-$ requires: 321.1127. IR (KBr): ν_{max} cm^{-1}: 3431, 2960, 2927, 2858, 1731, 1647, 1615, 1579, 1507, 1384, 1363, 1275. ^{1}H-NMR (DMSO-d_6) δ ppm: 12.75 (s, 1H, ArO*H*-5); 10.6–11.0 (br. s, 1H, ArO*H*-7); 8.03 (d, 2H, *J* = 6.8 Hz, 2′,6′-Ar*H*); 7.59–7.57 (m, 3H, 3′,4′,5′-Ar*H*); 6.95 (s, 1H, Ar*H*-3); 6.30 (s, 1H, Ar*H*-6); 5.18 (br. t., 1H, C*H* = C(CH_3)$_2$); 3.44 (d, 2H, *J* = 6.7 Hz, C*H*$_2$CH = C(CH_3)$_2$); 1.74 (s, 3H, -CH$_2$CH = CC*H*$_3$CH$_3$); 1.61 (s, 3H, -CH$_2$CH = CCH$_3$C*H*$_3$). ^{13}C-NMR (DMSO-d_6) δ ppm: 17.8 (-CH = CCH$_3$CH$_3$); 21.3 (-CH$_2$CH = CCH$_3$CH$_3$); 25.4 (CH$_2$CH = CCH$_3$CH$_3$); 98.5 (ArC-6); 103.9 (ArC-4a); 105.0 (ArC-3); 106.3 (ArC-8); 122.4 (CH = C(CH_3)$_2$); 126.3 (ArC-2′,6′); 129.2 (ArC-3′,5′); 131.1 (CH = C(CH_3)$_2$); 132.0 (ArC-4′); 154.6 (ArC-8a); 159.1 (ArC-5); 161.9 (ArC-7); 163.1 (ArC-2); 182.2 (4-C = O).

Compound **8** was obtained as a yellow powder (291 mg, 23%) mp: 221–222 °C (lit. [51] 213–214 °C). HRMS *m/z*, observed: 321.1130; $C_{20}H_{17}O_4$ $[M–H]^-$ requires: 321.1127. IR (KBr): ν_{max} cm^{-1}: 3400, 2969, 2918, 2651, 1639, 1580, 1553, 1483, 1449; ^{1}H-NMR (DMSO-d_6) δ ppm: 13.08 (s, 1H, ArO*H*-5); 8.05 (d, 2H, *J* = 7.1 Hz, 2′,6′-Ar*H*); 7.60–7.56 (m, 3H, 3′,4′,5′-Ar*H*); 6.95 (s, 1H, Ar*H*-3); 6.57 (s, 1H, Ar*H*-8); 5.17 (br. t, 1H, C*H* = C(CH_3)$_2$); 3.22 (d, 2H, *J* = 6.8 Hz, -C*H*$_2$CH = C(CH_3)$_2$); 1.72 (s, 3H, -CH$_2$CH = CC*H*$_3$CH$_3$); 1.61 (s, 3H, -CH$_2$CH = CCH$_3$C*H*$_3$). ^{13}C-NMR (DMSO-d_6) δ ppm: 17.7 (-CH = CCH$_3$CH$_3$); 21.0 (-CH$_2$CH = CCH$_3$CH$_3$); 25.5 (CH$_2$CH = CCH$_3$CH$_3$); 93.4 (ArC-8); 103.7 (ArC-4a); 105.0 (ArC-3); 111.2 (ArC-6); 122.1 (CH = C(CH_3)$_2$); 126.4 (ArC-2′,6′); 129.1 (ArC-3′,5′); 130.7 (CH = C(CH_3)$_2$); 130.8 (ArC-1′); 131.9 (ArC-4′); 155.2 (ArC-8a); 158.3 (ArC-5); 162.2 (ArC-7); 162.9 (ArC-2); 181.9 (4-C = O).

5,7-dihydroxy-6-(3-methyl-2-buten-1-yl)-2-(4-hydroxyphenyl)chroman-4-one (**11**) and 5,7-dihydroxy-8-(3-methyl-2-buten-1-yl)-2-(4-hydroxyphenyl)chroman-4-one (**12**) were obtained as racemic mixtures because these were synthesized from (±) naringenin (**4**) (3.7 mmol) commercially available, 3-methyl-2-buten-1-ol (**6**) (7.3 mmol), and $ZnCl_2$ (7.3 mmol) as described above. The crude mixture was purified using ethyl acetate-hexane in a gradient system (0 → 70% of ethyl acetate) as the mobile phase to afford **11** as a yellow powder (378 mg, 30%); mp: 206–207 °C (lit. [52] 212–214 °C); HRMS *m/z*, observed: 341.1384; $C_{20}H_{21}O_6$ $[M + H]^+$ requires: 341.1389. IR (KBr): ν_{max} cm^{-1}: 3420, 2960, 2926, 2857, 1728, 1633, 1586, 1519, 1457, 1384, 1296. ^{1}H-NMR (DMSO-d_6) δ ppm: 12.40 (s, 1H, Ar-O*H*-5); 10.74 (br. s, 1H, ArO*H*-7); 9.55 (br. s, 1H, ArO*H*-4′) 7.29 (d, 2H, *J* = 8.4 Hz, Ar*H*-2′,6′-); 6.78 (d, J =

8.4 Hz, 2H, Ar*H*-3′,5′); 5.94 (s, 1H, Ar*H*-8); 5.39 (dd, 1H, J_1 = 12.5 Hz, J_2 = 2.6 Hz, C*H*-2); 5.11 (br. t, 1H, C*H* = C(CH$_3$)$_2$); 3.22 (dd, 1H, J_1 = 17.1Hz, J_2 = 12.7 Hz, CH*H*-3); 3.10 (d, 2H, *J* = 7.0 Hz, C*H*$_2$CH = C(CH$_3$)$_2$); 2.66 (dd, 1H, J_1 = 14.3 Hz, J_2 = 2.9 Hz, C*H*H-3); 1.68 (s, 3H, CH$_2$CH = CC*H*$_3$CH$_3$); 1.60 (s, 3H, CH$_2$CH = CCH$_3$C*H*$_3$). ^{13}C-NMR (DMSO-d_6) δ ppm: 17.6 (-CH = CCH$_3$CH$_3$); 20.6 (-CH$_2$CH = CCH$_3$CH$_3$); 25.4 (-CH$_2$CH = CCH$_3$CH$_3$); 42.0 (CH$_2$-3); 78.3 (ArC-2); 94.3 (ArC-8); 101.5 (ArC-4a); 107.5 (ArC-6); 115.1 (ArC-3′,5′); 122.6 (CH = C(CH$_3$)$_2$); 128.3 (ArC-2′,6′); 128.98 (ArC-1′); 130.2 (CH = *C*(CH$_3$)$_2$); 157.7 (ArC-4′); 160.5 (ArC-7,8a); 164.3 (ArC-5); 196.4 (4-*C* = O).

Compound **12** was obtained as a yellow powder (453 mg, 36%) mp: 193 °C (lit. [53] 183–184 °C, ±-8-prenylnaringenin); HRMS *m/z*, observed: 339.1236; $C_{20}H_{19}O_5$ [M–H]$^-$ requires: 339.1232. IR (KBr): ν_{max} cm^{-1}: 3169, 2966, 2912, 1635, 1519, 1439, 1383, 1347. ^{1}H-NMR (DMSO-d_6) δ ppm: 12.09 (s, 1H, Ar-O*H*-5); 10.75 (s, 1H, ArO*H*-7); 9.56 (s, 1H, ArO*H*-4′) 7.30 (d, 2H, *J* = 8.4 Hz, Ar*H*-2′,6′); 6.78 (d, *J* = 8.4 Hz, 2H, Ar*H*-3′,5′); 5.95 (s, 1H, Ar*H*-6); 5.40 (dd, 1H, J_1 = 12.4 Hz, J_2 = 2.7 Hz, C*H*-2); 5.07 (br. t, 1H, C*H* = C(CH$_3$)$_2$); 3.19 (dd, 1H, J_1 = 17.1 Hz, J_2 = 12.6 Hz, CH*H*-3); 3.06 (d, 2H, *J* = 7.1 Hz, C*H*$_2$CH = C(CH$_3$)$_2$); 2.70 (dd, 1H, J_1 = 17.2 Hz, J_2 = 2.7 Hz, C*H*H-3); 1.57 (s, 3H, CH$_2$CH = CC*H*$_3$CH$_3$); 1.52 (s, 3H, CH$_2$CH = CCH$_3$C*H*$_3$). ^{13}C-NMR (DMSO-d_6) δ ppm: 17.6 (-CH = CCH$_3$CH$_3$); 21.2 (-CH$_2$CH = CCH$_3$CH$_3$); 25.5 (-CH$_2$CH = CCH$_3$CH$_3$); 41.9 (CH$_2$-3); 78.2 (ArC-2); 95.3 (ArC-6); 101.7 (ArC-4a); 106.9 (ArC-8); 115.1 (ArC-3′,5′); 122.6 (CH = C(CH$_3$)$_2$); 128.0 (ArC-2′,6′); 129.2 (ArC-1′); 130.1 (CH = *C*(CH$_3$)$_2$); 157.5 (ArC-4′); 159.7 (ArC-8a); 161.1 (ArC-5); 164.3 (ArC-7); 196.6 (4-*C* = O).

2-(3,4-dihydroxy-5-(3′-methyl-2′-buten-1′-yl)phenyl)-3,5,7-trihydroxy-4*H*-chromen-4-one (**9**; Uralenol) was obtained from quercetin (**2**) (3.3 mmol), 3-methyl-2-buten-1-ol (**6**) (13.2 mmol), and $ZnCl_2$ (13.2 mmol) as described above. The crude mixture was purified using ethyl acetate-hexane in a gradient system (0 → 80% of ethyl acetate) as the mobile phase to afford the title compound as a pale green powder (417 mg, 34%); mp: 205–206 °C (lit. [54] 176–178 °C); HRMS *m/z*, observed: 369.0974; $C_{20}H_{17}O_7$ [M–H]$^-$ requires: 369.0972. IR (KBr): ν_{max} cm^{-1}: 3599, 3390, 2969, 2914, 1655, 1635, 1601, 1566, 1518, 1462, 1363, 1313, 1283, 1247, 1167. ^{1}H-NMR (DMSO-d_6) δ ppm: 12.52 (s, 1H, ArO*H*-5); 10.75 (br. s, 1H, ArO*H*-7); 9.31 (s, 1H, ArO*H*-3′); 9.03 (s, 1H, ArO*H*-4′); 8.93 (s, 1H, ArO*H*-3); 6.83 (s, 1H, Ar*H*-2′); 6.66 (s, 1H, Ar*H*-6′); 6.29 (d, 1H, *J* = 1.6 Hz, Ar*H*-8); 6.18 (d, 1H, *J* = 1.5 Hz, Ar*H*-6); 5.10 (br. t, 1H, C*H* = C(CH$_3$)$_2$); 3.12 (d, 2H, *J* = 7.0 Hz, C*H*$_2$CH = C(CH$_3$)$_2$); 1.56 (s, 3H, CH$_2$CH = CC*H*$_3$CH$_3$); 1.47 (s, 3H, CH$_2$CH = CCH$_3$C*H*$_3$). ^{13}C-NMR (DMSO-d_6) δ ppm: 17.5 (CH = CCH$_3$CH$_3$); 25.4 (CH$_2$CH = CCH$_3$CH$_3$); 31.1 (CH$_2$CH = CCH$_3$CH$_3$); 93.3 (ArC-8); 98.2 (ArC-6); 103.5 (ArC-4a); 116.5 (ArC-6′); 117.1 (ArC-2′); 120.3 (ArC-5′); 123.5 (CH = C(CH$_3$)$_2$); 131.0 (CH = *C*(CH$_3$)$_2$); 132.2 (ArC-3′); 136.3 (ArC-2); 142.9 (ArC-1′); 147.1 (ArC-4′); 150.0 (ArC-3); 156.7 (ArC-8a); 160.9 (ArC-5); 163.8 (ArC-7); 176.2 (4-*C* = O).

2-(3,4-dihydroxy-5-(3′-methyl-2′-buten-1′-yl)phenyl)-3,7-dihydroxy-4*H*-chromen-4-one (**10**) was obtained from fisetin (**3**) (1.75 mmol), 3-methyl-2-buten-1-ol (**6**) (7 mmol), and $ZnCl_2$ (7 mmol) as described above. The crude mixture was purified using ethyl acetate-hexane in a gradient system (0 → 80% of ethyl acetate) as the mobile phase to afford the title compound as a dark green powder (155 mg, 25%); mp: 229–230 °C; HRMS *m/z*, observed: 355.1176; $C_{18}H_{19}O_6$ [M + H]$^+$ requires: 355.1182. IR (KBr): ν_{max} cm^{-1}: 3350, 2971, 2926, 2855, 1698, 1612, 1596, 1508, 1458, 1417, 1272. ^{1}H-NMR (DMSO- d_6) δ ppm: 10.68 (br. s, 1H, ArO*H*-7); 9.25 (s, 1H, ArO*H*-3′); 9.01 (s, 1H, ArO*H*-4′); 8.62 (s, 1H, ArO*H*-3); 7.93 (d, 1H, *J* = 8.6 Hz, Ar*H*-5); 6.89 (d, 1H, *J* = 8.2 Hz, Ar*H*-6); 6.83 (s, 1H, Ar*H*-2′); 6.78 (s, 1H, Ar*H*-8); 6.66 (s, 1H, Ar*H*-6′); 5.11 (br. t, 1H, C*H* = C(CH$_3$)$_2$); 3.12 (d, 2H, *J* = 6.8 Hz, C*H*$_2$CH = C(CH$_3$)$_2$); 1.54 (s, 3H, CH$_2$CH = CC*H*$_3$CH$_3$); 1.44 (s, 3H, CH$_2$CH = CCH$_3$C*H*$_3$). ^{13}C-NMR (DMSO-d_6) δ ppm: 17.5 (CH = CCH$_3$CH$_3$); 25.4 (CH$_2$CH = CCH$_3$CH$_3$); 31.2 (CH$_2$CH = CCH$_3$CH$_3$); 101.8 (ArC-8);114,6 (ArC-6); 114.8 (ArC-4a); 116.4 (ArC-6′); 117.1 (ArC-2′); 120.9 (ArC-5′); 123.6 (CH = C(CH$_3$)$_2$); 126.5 (ArC-5); 130.8 (CH = *C*(CH$_3$)$_2$); 132.0 (ArC-3′); 137.8 (ArC-1′); 142.8 (ArC-2); 146.8 (ArC-4′); 148.1 (ArC-3); 156.7 (ArC-8a); 162.1 (ArC-7); 172.1 (4-*C* = O).

3-(4-hydroxyphenyl)-8-(3-methyl-2-buten-1-yl)-5,7-dihydroxy-4*H*-chromen-4-one (**13**; lupiwighteone) was obtained from genistein (**5**) (1.9 mmol), 3-methyl-2-buten-1-ol (**6**)

(7.4 mmol), and $ZnCl_2$ (7.4 mmol) as described above. The crude mixture was purified using ethyl acetate-hexane in a gradient system (0 → 80% of ethyl acetate) as the mobile phase to afford the title compound as a white powder (231 mg, 36 %); mp: 140 °C (lit. [55] 133–135 °C); HRMS *m/z*, observed: 339.1228; $C_{20}H_{19}O_5$ $[M + H]^+$ requires: 339.1232. IR (KBr): ν_{max} cm^{-1}: 3350, 3174, 2921, 1705, 1651, 1613, 1570, 1513, 1428, 1364, 1301, 1257, 1198. ^{1}H-NMR (DMSO-d_6): δ ppm: 12.98 (s, 1H, ArO*H*-5); 9.58 (br. s, 1H, ArO*H*-4′); 8.38 (s, 1H, Ar*H*-2); 7.37 (d, 2H, *J* = 8.6 Hz, Ar*H*-2′,6′); 6.80 (d, 2H, *J* = 8.6 Hz, Ar*H*-3′,5′); 6.30 (s, 1H, Ar*H*-6); 5.14 (br. t, 1H, *CH* = $C(CH_3)_2$); 3,34 (overload with H_2O, $CH_2CH = C(CH_3)_2$); 1.74 (s, 3H, -$CH_2CH = CCH_3CH_3$); 1.62 (s, 3H, -$CH_2CH = CCH_3CH_3$). ^{13}C-NMR (DMSO- d_6) δ ppm: 17.7 (-$CH = CCH_3CH_3$); 21.0 (-$CH_2CH = CCH_3CH_3$); 25.4 ($CH_2CH = CCH_3CH_3$); 98.5 (ArC-6); 104.4 (ArC-4a); 105.8 (ArC-8); 115.0 (ArC-3′,5′); 121.3 (ArC-1′); 121.9 (ArC-3); 122.1 ($CH = C(CH_3)_2$); 130.1 (ArC-2′,6′); 131.0 ($CH = C(CH_3)_2$); 154.0 (ArC-2); 154.8 (ArC-8a); 157.3 (ArC-4′); 159.5 (ArC-5); 161.7 (ArC-7); 180.5 (4-C = O).

3.2. Biological Assays

3.2.1. In Vitro Antibacterial Activity Assays: Human Pathogens

Minimum Inhibitory Concentration Assay

Two clinical isolates of methicillin-resistant *S. aureus* (622-4 and 97-7) and one clinical isolate of *E. coli* 33.1 were kindly donated by Dr. Marcela Wilkens from Universidad de Santiago de Chile. *S. aureus* (NCTC8325-4) and *E. coli* (ATCC25922) were used as the control strains. The antimicrobial activities of the isolated compounds against *E. coli* (ATCC25922), multi-resistant *E. coli* (33.1), *S. aureus* (NCTC8325-4), and methicillin-resistant *S. aureus* (MRSA) 97-7 strains were assessed using the Clinical and Laboratory Standards Institute (CLSI) microdilution method [56]. Briefly, stock solutions (5 mg/mL) of compounds in DMSO were two-fold diluted in Mueller–Hinton broth (MHB). The final concentration of DMSO was ≤2.5% and does not affect the microbial growth. The obtained solution was added to MHB and serially two-fold diluted in a 96-well microplate. A sample of 100 mL of inoculum 1.5×10^6 colony-forming units (CFU) ml^{-1} in MHB was added and then incubated at 37 °C for 18 h. The assay was repeated three times. Wells containing MHB, 100 mL of inoculum, and DMSO served as negative controls. The MIC was defined as the lowest concentration of compounds resulting in the complete inhibition of visible growth [57].

Checkerboard Dilution Test

The assessments of synergy of prenylated flavonoids **11** and **12** combined with methicillin were investigated with the checkerboard method [57]. Briefly, the uppermost row (A) of a 96-well microtiter plate contained substance **X** in a concentration of about four times the expected MIC of the microorganism examined. Each following row (B–H) contained half the concentration of the previous one. The same procedure was carried out along the columns (1–12) with substance Y, but not necessarily with the same starting concentration. Therefore, each well contained a unique combination of the two substances (X and Y). Lastly, 100 μL of Mueller–Hinton broth containing about 10^5 (CFU/mL) were added to the wells and incubated at 37 °C for 24 h. The concentrations of the first wells without visible growth along the stepwise boundary between inhibition and growth were used to calculate the FICI values. The fractional inhibitory concentration index (FICI) was calculated as follows: FICI = fractional inhibitory concentration of A (FIC_A) + (FIC_B) = (MIC of drug A in combination/MIC of drug A alone) + (MIC of drug B in combination/MIC of drug B alone). FIC index was interpreted as follows: synergy, <0.5; partial synergy, 0.5–0.75; additive effect, 0.76–1.0; indifference, >1.0–4.0; antagonism, >4.0 [58].

Time Kill Assay

Time kill curves were conducted with a final concentration of antimicrobial agent at four times the MIC [59]. Flasks containing 50 mL of MHB with the appropriate antimicrobial agent were inoculated with 50 mL of the test organism in a logarithmic growth phase adjusted to the appropriate density. Aliquots were removed, diluted, and plated at the 0, 2, 4, 6, and 24 h time points. The plates were incubated for 24 h, and the viable counts were determined. Bactericidal activity was defined as a reduction of 99.9% (≥3 log10) of the total count of CFU/mL in the original inoculum [56]. Bacteriostatic activity was defined as the maintenance of or a reduction of less than 99.9% (3 log10) of the total count of CFU/mL in the original inoculum.

Determination of Reactive Oxygen Species (ROS) Intrabacterial Levels

The production of ROS by *S. aureus* ATCC 6538 after treatment with the prenylated compounds was evaluated using the peroxynitrite indicator 2′,7′-dichlorodihydrofluorescein diacetate (DCFH-DA) (Sigma–Aldrich), which can detect a broad range of ROS including nitric oxide and hydrogen peroxide. The adjusted bacterial culture (0.5 McFarland exponential phase bacteria culture) was treated with a concentration of each compound of 2 μM, using ciprofloxacin as an antibiotic control, Trolox as a ROS scavenger control, 1,4-naphtoquinone as a ROS generation control, and the presence of DCFH-DA at a final concentration of 5 mM in 0.85% saline and incubated at 37 °C aerobically at 200 rpm for 24 h. Untreated bacterial culture was served as the negative control. The fluorescence emission of DCFH-DA was measured at 525 nm using a Tecan microtiter plate reader with an excitation wavelength of 485 nm. The background fluorescence of 0.85% saline and autofluorescence of the bacterial cells incubated without the probe was measured to calculate the net fluorescence emitted from the assay itself. The experiment was conducted in triplicate [60].

3.2.2. In Vitro Antifungal Activity Assays against Human Pathogens

Microorganisms and Media

For the antifungal evaluation, standardized strains from the American Type Culture Collection (ATCC, Manassas, VA, USA), *Centro de Referencia en Micología* (CCC), Facultad de Ciencias Bioquímicas y Farmacéuticas, Suipacha 531-(2000)-Rosario, Argentina and Malbrán Institute (IM), Av. Velez Sársfield 563, Buenos Aires, Argentina were used. The voucher specimens of standardized strains are as follows: *C. albicans* ATCC 10231, *C. neoformans* ATCC 32264, *A. flavus* ATCC 9170, *A. fumigatus* ATTC 26934, A. *niger* ATCC 9029, *T. rubrum* CCC 110, *T. mentagrophytes* ATCC 9972, and *M. gypseum* CCC 115. Clinical isolates of *C. neoformans* (n = 5) were provided by IM. They included five strains whose voucher specimens are presented in Table 7. Clinical isolates of *C. albicans* (n = 5) were provided by CCC; their voucher specimens are presented in Table 8.

Strains were grown on Sabouraud-chloramphenicol agar slants for 48 h at 30 °C, maintained on slopes of Sabouraud-dextrose agar (SDA, Oxoid) and sub-cultured every 15 days to prevent pleomorphic transformations. Inocula were obtained in accordance with reported procedures [39,40] and adjusted to 1–5 × 10^3 cells with colony-forming units (CFUs)/mL.

Antifungal Susceptibility Testing

The minimum inhibitory concentration (MIC) of each compound was determined by using broth microdilution techniques according to the guidelines of the CLSI for yeasts (M27-A3) [39] and for filamentous fungi (including dermatophytes) M38-A2 [40]. MIC values were determined in RPMI-1640 (Sigma–Aldrich) and buffered to pH 7.0 with MOPS. Microtiter trays were incubated at 35 °C for yeasts and *Aspergillus* spp. and at 28–30 °C for dermatophyte strains in a moist, dark chamber. MICs were visually recorded at 48 h for yeasts, and at a time according to the control fungus growth, for the rest of the fungi. For the assay, stock solutions of pure compounds were two-fold diluted with RPMI-1640 from 250 to 0.98 μg/mL (final volume = 100 μL) and a final DMSO concentration of ≤1%. A volume of 100 μL of inoculum suspension was added to each well with the exception of the sterility control, where sterile water was added to the well instead. Terbinafine (Novartis Co, Basel, Switzerland) and amphotericin B (Sigma–Aldrich) were used as positive controls.

Endpoints were defined as the lowest concentration of drug resulting in the total inhibition of visual growth compared to the growth in the control wells containing no antifungal drug.

The MFC of each compound against an isolate was determined as follows: After determining the MIC, an aliquot of 5 μL was withdrawn from each clear well of the microtiter tray and plated onto a 150 mm RPMI-1640 agar plate buffered with MOPS (Remel Inc., Lenexa, KS, USA). Inoculated plates were incubated at 30 °C, and the MFC was recorded after 48 h. The MFC was defined as the lowest concentration of each compound that resulted in the total inhibition of visible growth.

3.2.3. In Vitro Antibacterial Activity against Plant Pathogens

Broth microdilution methods were used to evaluate the effects of the compounds on the growth of *P. carotovorum* (NCPPB 312), *A. tumefaciens* (strain C58C1), and *P. syringae* (NCPPB 281). Bacteria were grown in sterile tubes with 10 mL of Mueller–Hinton (MH) medium and incubated at 27 °C for 12 h with shaking to produce an initial culture. The antimicrobial activity was evaluated by observing the growth response of both microorganisms in samples with different concentrations of the compounds [61–63]. All assays were performed on sterile 96-well microplates with a final volume of 200 μL containing Mueller–Hinton broth (MH) inoculated with 1 μL of bacterial suspension (10^5–10^6 UFC/mL, initial culture) in the presence of different concentrations of test compounds (3.9, 7.8, 15.6, 31.3, 62.5, 125, and 250 μM). MH was used as the negative control [C(−)], and MH with streptomycin [64] was used as the positive control [C(+)]. The plates were incubated for 7 h at 27 °C. Bacterial growth was monitored by measuring the optical density at 595 nm every hour with a microplate reader. All tests were performed in 10 repetitions for each microorganism evaluated. Bacterial growth was shown as the arithmetic mean expressed in terms of the negative control (100% growth). The lowest concentration of the compound preventing the appearance of turbidity was considered to be the minimal inhibitory concentration (MIC).

The first experiment (first kinetic assay) in which the compound was exposed to the bacterial cultures in MH was carried out over a period of 6 h. Subsequently, to determine the minimal bactericidal concentration (MBC), a second experiment (second kinetic assay) was conducted; this experiment involved taking inoculum from the first kinetic assay and adding it to MH in a new 96-well microplate containing culture medium, which was then cultured for 7 h. The aim of this second culture (second kinetic assay) was to determine whether the compounds have bactericidal or bacteriostatic properties [65].

3.2.4. Statistics

The Mann–Whitney U test (*) with $p < 0.05$ was performed to identify significant differences among the treatment and control groups and between the two treatments.

3.3. *In Vitro 5-LOX Enzyme Inhibition Assay*

The commercially available enzyme by Cayman Chemicals Inc., Ann Arbor, MI, USA, was diluted (1:500) in the assay buffer (HEPES 50 mM, EDTA 2 mM, ATP 10 μM, and $CaCl_2$ 10 μM at pH 7.5) and mixed with 10 μM H_2DCFDA dye in the reaction mixture and incubated for 15 min in the assay plate. Subsequently, 280 μL of buffer was added, and 10 μL of inhibitor with a final concentration of 10 μM was placed per well in the reaction mixture and incubated for 30 min. The reaction was started by the addition of a suitable concentration of arachidonic acid (0.5 μM). The fluorescence was read in a multimode detector Synergy™ HT Multi-Mode Microplate Reader (Biotek) at 480 nm excitation/520 nm emission after 1 h of incubation at room temperature. The inhibition percentage was obtained for analysis of oxidation of H_2DCFDA dye to the highly fluorescent 2′,7′-dichloro-fluorescein (DCF) product.

3.4. Docking of **11** (S and R Enantiomers) and with ABC Transporter

The prenylated flavonoids structures were built with the Molecular Operating Environment software [66]. ChelpG charges were obtained at the B3LYP/6-31G** level theory, employing the Gaussian 09 package [67]. Docking was done with the crystal structure of the ABC transporter (PDB code: 2ONJ, 3.40 Å resolution) using the AutoDock4 package [68] and a Lamarckian algorithm, assuming total flexibility of the inhibitors. The grid maps were made up first to 126 × 126 × 126 points and later to 60 × 60 × 60, with a grid-point spacing of 0.375 Å on the grid map. The AutoTors option was used to define the ligand torsions, and the docking results were then analyzed by a ranked cluster analysis, resulting in conformations with the highest overall binding energy (the most negative Gibbs free energy binding value, –ΔG).

3.5. General Procedure to Determine the DPPH Radical Scavenging Activity

The radical scavenging activity of the prenylated compounds and starting materials towards the 2,2-diphenyl-1-picrylhydrazyl (DPPH) radical was measured as described [18], adapted to a screen to 96-well plates. Briefly, stock solutions of each compound were prepared in methanol at a 1-mM concentration (10 mL). Dilutions (1–200 μM) were prepared from the stock solutions. Methanol (90 μL), dilutions (150 μL), and DPPH (60 μL, Sigma–Aldrich) in methanol (0.5 mM), resulting in a final concentration of 0.1 mM of DPPH were added in a 96-well plate. Methanol was used as the blank sample. The mixtures were left for 30 min at room temperature, and the absorbances were then measured at 517 nm. Trolox was used as the standard antioxidant. The radical scavenging activity was calculated as follows: % Inhibition = [(blank absorbance − sample absorbance)/blank absorbance] × 100. The mean of three IC_{50} (concentration causing 50% inhibition) values for each compound was determined graphically.

3.6. In Vitro Anticancer Activity of **11** and **12**

The cell lines used in this work included MDA-MB-231 human breast adenocarcinoma cells, B16-F10 mouse metastatic melanoma cells, and MEF primary mouse embryonic fibroblasts. Cells were maintained in a DMEM high glucose medium (Mediatech, Manassas, VA, USA) supplemented with 10% (MDA-MB-231 and B16-F10) or 15% (MEF) heat-inactivated fetal bovine serum (HyClone Laboratories), 100 IU/mL penicillin, and 100 μg/mL streptomycin and maintained at 37 °C in a 5% CO_2 humidified atmosphere. Cell viability was measured using CyQuant®Direct Cell Proliferation Assay Kit (Life Technologies) following the manufacturer's instruction. Briefly, 5.000 cells/well were seeded onto a flat-bottomed 96-well plate in a 200 μL final volume. Six hours after seeding, the culture medium was replaced with the medium containing the tested compounds at concentrations ranging from 0 to 100 μM dissolved in DMSO (a 0.1% final concentration) for 72 h. The concentrations used to calculate the IC_{50} values were 100, 30, 10, 3, 1, 0.3, 0.1, 0.01, and 0 μM. Untreated cells (medium containing 0.1% DMSO) were used as controls. At the end of the incubation, 100 μL of culture medium was removed from each experimental well and replaced by 2× detection reagent. Cells were incubated for 1 h, and fluorescence emission was measured at 535 nm with excitation at 480 nm in a microplate reader (Infinite 200 PRO, Tecan). At least four independent experiments were performed for each concentration. The results from each experiment were transformed to a percentage of controls, and the IC_{50} values were graphically obtained from the dose–response curves. The IC_{50} value was obtained adjusting the dose–response curve to sigmoidal curves (variable slope), generated using GraphPad Prisma 6.0 software [69].

4. Conclusions

Our results demonstrated that the 6-prenylated-(±)-naringenin (**11**), 8-prenylated-(±)-naringenin (**12**) and the 8-prenylated genistein (**13**) are active molecules against *Staphylococcus aureus*-resistant bacteria with a strong synergistically effect in combination with commercial antibiotics vancomycin, ciprofloxacin, and methicillin, enhancing their effects in a factor of 10–100 times against drug-resistant bacteria. Compound **12** is the most active against the dermatophytes *T. rubrum*, *T. mentagrophytes*, and *M. gypseum fungus*. Prenylation substitution was shown to be fundamental for these biological activities, specifically at positions 6 and 8 of the (±)-naringenin (**11** and **12**, respectively); however, others structural requirements are important as well, such as hydroxylation at positions 5, 7 of ring A, and 4′ of ring B. According to the docking results of **11** with ATP-binding cassette transporter, the *S* isomer showed better interaction energy than the *R* isomer. It is relevant to continue this study with enantiomerically pure isomers of prenylated flavanones **11** and **12**.

Supplementary Materials: Supplementary materials can be found at https://www.mdpi.com/article/10.3390/ijms22115472/s1.

Author Contributions: Synthesis of prenylated compounds and antioxidant assay, M.O.; phytopathogenic bacteria assays, M.C. and A.V.; human-resistant bacteria studies, Y.V.-M. and M.C.-S.M.; human pathogenic fungi studies, E.B. and S.Z.; 5-hLOX and docking studies, C.M.; cytotoxic studies, M.M. and S.M.; writing—original draft preparation, M.O. and Y.V.-M. All authors have read and agreed to the published version of the manuscript.

Funding: This work was supported by CONICYT, grant number 1130924, by the Dirección General de Investigación, Innovación y Emprendimiento (DGIIE) of the Universidad Técnica Federico Santa María, and by DICYT-VRIDEI of the Universidad de Santiago de Chile, grant numbers 021901VM and 021641MC. E.B. acknowledges CONICET.

Institutional Review Board Statement: Not applicable.

Informed Consent Statement: Not applicable.

Acknowledgments: M.O. acknowledges Luis Espinoza for lab facilities (NMR, IR, and other equipment).

Conflicts of Interest: The authors declare no conflict of interest.

References

1. Oh, I.; Yang, W.-Y.; Chung, S.-C.; Kim, T.-Y.; Oh, K.-B.; Shin, J. In vitro sortase A inhibitory and antimicrobial activity of flavonoids isolated from the roots of Sophora flavescens. *Arch. Pharm. Res.* **2011**, *34*, 217–222. [CrossRef] [PubMed]
2. Yazaki, K.; Sasaki, K.; Tsurumaru, Y. Prenylation of aromatic compounds, a key diversification of plant secondary metabolites. *Phytochemistry* **2009**, *70*, 1739–1745. [CrossRef]
3. Kushwaha, P.P.; Prajapati, S.K.; Pothabathula, S.V.; Singh, A.K.; Shuaib, M.; Joshi, K.; Kumar, S. Prenylated flavonoids as a promising drug discovery candidate. In *Phytochemicals as Lead Compounds for New Drug Discovery*; Elsevier: Amsterdam, The Netherlands, 2020; pp. 347–355. ISBN 9780128178911.
4. Kumano, T.; Richard, S.B.; Noel, J.P.; Nishiyama, M.; Kuzuyama, T. Chemoenzymatic syntheses of prenylated aromatic small molecules using Streptomyces prenyltransferases with relaxed substrate specificities. *Bioorg. Med. Chem.* **2008**, *16*, 8117–8126. [CrossRef] [PubMed]
5. Simmonds, M.S.J. Importance of flavonoids in insect–plant interactions: Feeding and oviposition. *Phytochemistry* **2001**, *56*, 245–252. [CrossRef]
6. Quiroga, E.N.; Sampietro, D.A.; Sgariglia, M.A.; Soberón, J.R.; Vattuone, M.A. Antimycotic activity of 5′-prenylisoflavanones of the plant Geoffroea decorticans, against Aspergillus species. *Int. J. Food Microbiol.* **2009**, *132*, 42–46. [CrossRef]
7. Edziri, H.; Mastouri, M.; Mahjoub, M.A.; Mighri, Z.; Mahjoub, A.; Verschaeve, L. Antibacterial, Antifungal and Cytotoxic Activities of Two Flavonoids from Retama raetam Flowers. *Molecules* **2012**, *17*, 7284–7293. [CrossRef]
8. Du, J.; He, Z.-D.; Jiang, R.-W.; Ye, W.-C.; Xu, H.-X.; But, P.P.-H. Antiviral flavonoids from the root bark of Morus alba L. *Phytochemistry* **2003**, *62*, 1235–1238. [CrossRef]
9. Liu, M.; Hansen, P.; Wang, G.; Qiu, L.; Dong, J.; Yin, H.; Qian, Z.; Yang, M.; Miao, J. Pharmacological Profile of Xanthohumol, a Prenylated Flavonoid from Hops (Humulus lupulus). *Molecules* **2015**, *20*, 754–779. [CrossRef]
10. Jin, J.H.; Kim, J.S.; Kang, S.S.; Son, K.H.; Chang, H.W.; Kim, H.P. Anti-inflammatory and anti-arthritic activity of total flavonoids of the roots of Sophora flavescens. *J. Ethnopharmacol.* **2010**, *127*, 589–595. [CrossRef]
11. Molčanová, L.; Janošíková, D.; Dall'Acqua, S.; Šmejkal, K. C-prenylated flavonoids with potential cytotoxic activity against solid tumor cell lines. *Phytochem. Rev.* **2019**, *18*, 1051–1100. [CrossRef]
12. Yang, Z.-G.; Matsuzaki, K.; Takamatsu, S.; Kitanaka, S. Inhibitory Effects of Constituents from Morus alba var. multicaulis on Differentiation of 3T3-L1 Cells and Nitric Oxide Production in RAW264.7 Cells. *Molecules* **2011**, *16*, 6010–6022. [CrossRef]

13. Venturelli, S.; Burkard, M.; Biendl, M.; Lauer, U.M.; Frank, J.; Busch, C. Prenylated chalcones and flavonoids for the prevention and treatment of cancer. *Nutrition* **2016**, *32*, 1171–1178. [CrossRef] [PubMed]
14. Yang, X.; Yang, J.; Jiang, Y.; Yang, H.; Yun, Z.; Rong, W.; Yang, B. Regiospecific synthesis of prenylated flavonoids by a prenyltransferase cloned from Fusarium oxysporum. *Sci. Rep.* **2016**, *6*, 24819. [CrossRef] [PubMed]
15. Levisson, M.; Araya-Cloutier, C.; de Bruijn, W.J.C.; van der Heide, M.; Salvador López, J.M.; Daran, J.-M.; Vincken, J.-P.; Beekwilder, J. Toward Developing a Yeast Cell Factory for the Production of Prenylated Flavonoids. *J. Agric. Food Chem.* **2019**, *67*, 13478–13486. [CrossRef] [PubMed]
16. Sasaki, K.; Tsurumaru, Y.; Yamamoto, H.; Yazaki, K. Molecular Characterization of a Membrane-bound Prenyltransferase Specific for Isoflavone from Sophora flavescens. *J. Biol. Chem.* **2011**, *286*, 24125–24134. [CrossRef]
17. Yang, X.; Jiang, Y.; Yang, J.; He, J.; Sun, J.; Chen, F.; Zhang, M.; Yang, B. Prenylated flavonoids, promising nutraceuticals with impressive biological activities. *Trends Food Sci. Technol.* **2015**, *44*, 93–104. [CrossRef]
18. Osorio, M.E.; Qiuroz, K.A.; Carvajal, M.A.; Vergara, A.P.; Sánchez, E.Y.; González, C.E.; Catalán, K.S. Synthesis, anti-phytopathogenic and DPPH radical Scavenging activities of C-prenylated ace tophenones and benzaldehydes. *J. Chil. Chem. Soc.* **2016**, *61*, 3095–3101. [CrossRef]
19. *Global Antimicrobial Resistance Surveillance System (GLASS)*; Report; WHO: Geneva, Switzerland, 2017; ISBN 9789241513449.
20. Vandeputte, P.; Ferrari, S.; Coste, A.T. Antifungal Resistance and New Strategies to Control Fungal Infections. *Int. J. Microbiol.* **2012**, *2012*, 1–26. [CrossRef]
21. Araya-Cloutier, C.; den Besten, H.M.W.; Aisyah, S.; Gruppen, H.; Vincken, J.-P. The position of prenylation of isoflavonoids and stilbenoids from legumes (Fabaceae) modulates the antimicrobial activity against Gram positive pathogens. *Food Chem.* **2017**, *226*, 193–201. [CrossRef] [PubMed]
22. *State of the World's Antibiotics, 2015*; Center for Disease Dynamics, Economics & Policy: Washington, DC, USA, 2015.
23. Cheesman, M.; Ilanko, A.; Blonk, B.; Cock, I. Developing new antimicrobial therapies: Are synergistic combinations of plant extracts/compounds with conventional antibiotics the solution? *Pharm. Rev.* **2017**, *11*, 57. [CrossRef]
24. Sousa Carvalho, G.F.; Marques, L.K.; Sousa, H.G.; Silva, L.R.; Leão Ferreira, D.C.; Pires de Moura do Amaral, F.; Martins Maia Filho, A.L.; Figueredo-Silva, J.; Alves, W.d.S.; Oliveira, M.d.D.A.d.; et al. Phytochemical study, molecular docking, genotoxicity and therapeutic efficacy of the aqueous extract of the stem bark of Ximenia americana L. in the treatment of experimental COPD in rats. *J. Ethnopharmacol.* **2020**, *247*, 112259. [CrossRef] [PubMed]
25. Zhao, X.; Chen, R.; Shi, Y.; Zhang, X.; Tian, C.; Xia, D. Antioxidant and Anti-Inflammatory Activities of Six Flavonoids from Smilax glabra Roxb. *Molecules* **2020**, *25*, 5295. [CrossRef] [PubMed]
26. Tonk, R.K.; Bawa, S.; Chawla, G.; Deora, G.S.; Kumar, S.; Rathore, V.; Mulakayala, N.; Rajaram, A.; Kalle, A.M.; Afzal, O. Synthesis and pharmacological evaluation of pyrazolo [4, 3-c] cinnoline derivatives as potential anti-inflammatory and antibacterial agents. *Eur. J. Med. Chem.* **2012**, *57*, 176–184. [CrossRef]
27. Atala, E.; Fuentes, J.; Wehrhahn, M.J.; Speisky, H. Quercetin and related flavonoids conserve their antioxidant properties despite undergoing chemical or enzymatic oxidation. *Food Chem.* **2017**, *234*, 479–485. [CrossRef]
28. Zuo, G.-Y.; Yang, C.-X.; Han, J.; Li, Y.-Q.; Wang, G.-C. Synergism of prenylflavonoids from Morus alba root bark against clinical MRSA isolates. *Phytomedicine* **2018**, *39*, 93–99. [CrossRef] [PubMed]
29. Sohn, H.-Y.; Son, K.H.; Kwon, C.-S.; Kwon, G.-S.; Kang, S.S. Antimicrobial and cytotoxic activity of 18 prenylated flavonoids isolated from medicinal plants: Morus alba L., Morus mongolica Schneider, Broussnetia papyrifera (L.) Vent, Sophora flavescens Ait and Echinosophora koreensis Nakai. *Phytomedicine* **2004**, *11*, 666–672. [CrossRef] [PubMed]
30. Tsuchiya, H.; Sato, M.; Miyazaki, T.; Fujiwara, S.; Tanigaki, S.; Ohyama, M.; Tanaka, T.; Iinuma, M. Comparative study on the antibacterial activity of phytochemical flavanones against methicillin-resistant Staphylococcus aureus. *J. Ethnopharmacol.* **1996**, *50*, 27–34. [CrossRef]
31. Sato, M.; Tanaka, H.; Tani, N.; Nagayama, M.; Yamaguchi, R. Different antibacterial actions of isoflavones isolated from Erythrina poeppigiana against methicillin-resistant Staphylococcus aureus. *Lett. Appl. Microbiol.* **2006**, *43*, 243–248. [CrossRef]
32. das Chagas Almeida, A.; Azevedo Rodrigues, L.; dos Santos Paulino, G.; Pereira Aguilar, A.; Andrade Almeida, A.; Olavo Ferreira, S.; Brandão, G.C.; Viana Leite, J.P.; de Oliveira Barros Ribon, A. Prenylated flavonoid-enriched fraction from Maclura tinctoria shows biological activity against Staphylococcus aureus and protects Galleria mellonella larvae from bacterial infection. *BMC Complement. Altern. Med.* **2019**, *19*, 189. [CrossRef]
33. Araya-Cloutier, C.; Vincken, J.-P.; van de Schans, M.G.M.; Hageman, J.; Schaftenaar, G.; den Besten, H.M.W.; Gruppen, H. QSAR-based molecular signatures of prenylated (iso)flavonoids underlying antimicrobial potency against and membrane-disruption in Gram positive and Gram negative bacteria. *Sci. Rep.* **2018**, *8*, 1–14. [CrossRef]
34. Mun, S.-H.; Joung, D.-K.; Kim, S.-B.; Park, S.-J.; Seo, Y.-S.; Gong, R.; Choi, J.-G.; Shin, D.-W.; Rho, J.-R.; Kang, O.-H.; et al. The Mechanism of Antimicrobial Activity of Sophoraflavanone B Against Methicillin-Resistant Staphylococcus aureus. *Foodborne Pathog. Dis.* **2014**, *11*, 234–239. [CrossRef] [PubMed]
35. Wagner, H.; Ulrich-Merzenich, G. Synergy research: Approaching a new generation of phytopharmaceuticals. *Phytomedicine* **2009**, *16*, 97–110. [CrossRef]

36. Jung, H.A.; Jung, M.J.; Kim, J.Y.; Chung, H.Y.; Choi, J.S. Inhibitory activity of flavonoids from Prunus davidiana and other flavonoids on total ROS and hydroxyl radical generation. *Arch. Pharm. Res.* **2003**, *26*, 809–815. [CrossRef] [PubMed]
37. Becerra, M.; Albesa, I. Oxidative stress induced by ciprofloxacin in Staphylococcus aureus. *Biochem. Biophys. Res. Commun.* **2002**, *297*, 1003–1007. [CrossRef]
38. Lage, H. ABC-transporters: Implications on drug resistance from microorganisms to human cancers. *Int. J. Antimicrob. Agents* **2003**, *22*, 188–199. [CrossRef]
39. CLSI. *Reference Method for Broth Dilution Antifungal Susceptibility Testing of Yeast*, Approved Standard-Third Edition; CLSI document M27-A3; Clinical and Laboratory Standards Institute: Wayne, PA, USA, 2008.
40. CLSI. *Reference Method for Broth Dilution Antifungal Susceptibility Testing of Filamentous Fungi*, Approved Standard-Second Edition; CLSI document M38-A2; Clinical and Laboratory Standards Institute: Wayne, PA, USA, 2008.
41. Pfaller, M.A.; Messer, S.A.; Boyken, L.; Rice, C.; Tendolkar, S.; Hollis, R.J.; Doern, G.V.; Diekema, D.J. Global Trends in the Antifungal Susceptibility of Cryptococcus neoformans (1990 to 2004). *J. Clin. Microbiol.* **2005**, *43*, 2163–2167. [CrossRef] [PubMed]
42. Pfaller, M.A.; Diekema, D.J. Epidemiology of Invasive Candidiasis: A Persistent Public Health Problem. *Clin. Microbiol. Rev.* **2007**, *20*, 133–163. [CrossRef] [PubMed]
43. Pankey, G.A.; Sabath, L.D. Clinical Relevance of Bacteriostatic versus Bactericidal Mechanisms of Action in the Treatment of Gram-Positive Bacterial Infections. *Clin. Infect. Dis.* **2004**, *38*, 864–870. [CrossRef]
44. Valdés, E.; González, C.; Díaz, K.; Vásquez-Martínez, Y.; Mascayano, C.; Torrent, C.; Cabezas, F.; Mejias, S.; Montoya, M.; Cortez-San Martín, M.; et al. Biological Properties and Absolute Configuration of Flavanones from Calceolariathyrsiflora Graham. *Front. Pharm.* **2020**, *11*, 1125. [CrossRef] [PubMed]
45. Zhang, Q.; Yang, W.; Liu, J.; Liu, H.; Lv, Z.; Zhang, C.; Chen, D.; Jiao, Z. Identification of Six Flavonoids as Novel Cellular Antioxidants and Their Structure-Activity Relationship. *Oxid. Med. Cell. Longev.* **2020**, *2020*, 1–12. [CrossRef]
46. Choi, J.; Lee, D.-H.; Jang, H.; Park, S.-Y.; Seol, J.-W. Naringenin exerts anticancer effects by inducing tumor cell death and inhibiting angiogenesis in malignant melanoma. *Int. J. Med. Sci.* **2020**, *17*, 3049–3057. [CrossRef] [PubMed]
47. Stompor, M.; Uram, Ł.; Podgórski, R. In Vitro Effect of 8-Prenylnaringenin and Naringenin on Fibroblasts and Glioblastoma Cells-Cellular Accumulation and Cytotoxicity. *Molecules* **2017**, *22*, 1092. [CrossRef] [PubMed]
48. Wang, R.; Wang, J.; Dong, T.; Shen, J.; Gao, X.; Zhou, J. Naringenin has a chemoprotective effect in MDA-MB-231 breast cancer cells via inhibition of caspase-3 and -9 activities. *Oncol. Lett.* **2018**, *17*, 1217–1222. [CrossRef] [PubMed]
49. Li, Q.; Wang, Y.; Zhang, L.; Chen, L.; Du, Y.; Ye, T.; Shi, X. Naringenin exerts anti-angiogenic effects in human endothelial cells: Involvement of ERRα/VEGF/KDR signaling pathway. *Fitoterapia* **2016**, *111*, 78–86. [CrossRef]
50. Maatouk, M.; Elgueder, D.; Mustapha, N.; Chaaban, H.; Bzéouich, I.M.; Loannou, I.; Kilani, S.; Ghoul, M.; Ghedira, K.; Chekir-Ghedira, L. Effect of heated naringenin on immunomodulatory properties and cellular antioxidant activity. *Cell Stress Chaperones* **2016**, *21*, 1101–1109. [CrossRef]
51. Beirne, J.J.; Carroll, N.M.; O'Sullivan, W.I.; Woods, J. Alkylation of the chrysin dianion. *Tetrahedron* **1975**, *31*, 265–267. [CrossRef]
52. Hänsel, R.; Schulz, J. Desmethylxanthohumol: Isolierung aus Hopfen und Cyclisierung zu Flavanonen. *Arch. Pharm.* **1988**, *321*, 37–40. [CrossRef]
53. Dong, X.; Fan, Y.; Yu, L.; Hu, Y. Synthesis of Four Natural Prenylflavonoids and Their Estrogen-like Activities. *Arch. Pharm.* **2007**, *340*, 372–376. [CrossRef]
54. Kawamura, T.; Hayashi, M.; Mukai, R.; Terao, J.; Nemoto, H. The First Synthesis of Uralenol, 5′-Prenylated Quercetin, via Palladium-Catalyzed O-Dimethylallylation Reaction with Concurrent Acetyl Migration. *Synthesis* **2013**, *46*, 170–174. [CrossRef]
55. Al-Maharik, N.; Botting, N.P. Synthesis of lupiwighteone via a para-Claisen–Cope rearrangement. *Tetrahedron* **2003**, *59*, 4177–4181. [CrossRef]
56. CLSI. *Reference Method for Determining Bactericidal Activity of Antibacterial Agents*; Approved Guideline. CLSI document M26-A; Clinical and Laboratory Standards Institute: Wayne, PA, USA, 1999.
57. Iten, F.; Saller, R.; Abel, G.; Reichling, J. Additive Antmicrobial Effects of the Active Components of the Essential Oil of Thymus vulgaris—Chemotype Carvacrol. *Planta Med.* **2009**, *75*, 1231–1236. [CrossRef]
58. Doern, C.D. When Does 2 Plus 2 Equal 5? A Review of Antimicrobial Synergy Testing. *J. Clin. Microbiol.* **2014**, *52*, 4124–4128. [CrossRef] [PubMed]
59. Petersen, P.J.; Jones, C.H.; Bradford, P.A. In vitro antibacterial activities of tigecycline and comparative agents by time-kill kinetic studies in fresh Mueller-Hinton broth. *Diagn. Microbiol. Infect. Dis.* **2007**, *59*, 347–349. [CrossRef]
60. Ong, K.S.; Cheow, Y.L.; Lee, S.M. The role of reactive oxygen species in the antimicrobial activity of pyochelin. *J. Adv. Res.* **2017**, *8*, 393–398. [CrossRef]
61. Cole, M.D. Key antifungal, antibacterial and anti-insect assays-a critical review. *Biochem. Syst. Ecol.* **1994**, *22*, 837–856. [CrossRef]
62. Zampini, I.C.; Vattuone, M.A.; Isla, M.I. Antibacterial activity of Zuccagnia punctata Cav. ethanolic extracts. *J. Ethnopharmacol.* **2005**, *102*, 450–456. [CrossRef]
63. Cos, P.; Vlietinck, A.J.; Berghe, D.V.; Maes, L. Anti-infective potential of natural products: How to develop a stronger in vitro 'proof-of-concept'. *J. Ethnopharmacol.* **2006**, *106*, 290–302. [CrossRef]
64. McManus, P.S.; Stockwell, V.O.; Sundin, G.W.; Jones, A.L. Antibiotic use in plant agriculture. *Annu. Rev. Phytopathol.* **2002**, *40*, 443–465. [CrossRef] [PubMed]

65. Carvajal, M.A.; Vergara, A.P.; Santander, R.; Osorio, M.E. Chemical Composition and Anti-phytopathogenic Activity of the Essential Oil of Beilschmiedia miersii. *Nat. Prod. Commun.* **2016**, *11*, 1367–1372. [CrossRef] [PubMed]
66. *Molecular Operating Environment (MOE), 2013.08*; Chemical Computing Group ULC: Montreal, QC, Canada, 2018.
67. Frisch, M.J.; Trucks, G.W.; Schlegel, H.B.; Scuseria, G.E.; Robb, M.A.; Cheeseman, J.R.; Scalmani, G.; Barone, V.; Petersson, G.A.; Nakatsuji, H.; et al. *Gaussian 09. Revision A.02*; Gaussian Inc.: Wallingford, CT, USA, 2016.
68. Morris, G.M.; Huey, R.; Lindstrom, W.; Sanner, M.F.; Belew, R.K.; Goodsell, D.S.; Olson, A.J. AutoDock4 and AutoDockTools4: Automated docking with selective receptor flexibility. *J. Comput. Chem.* **2009**, *30*, 2785–2791. [CrossRef] [PubMed]
69. *GraphPad Prism, Version 6.00 for Mac OS X*; GraphPad Software: La Jolla, CA, USA, 2015.

International Journal of
Molecular Sciences

MDPI

Article

Carnosic Acid Attenuates an Early Increase in ROS Levels during Adipocyte Differentiation by Suppressing Translation of Nox4 and Inducing Translation of Antioxidant Enzymes

Dae-Kun Lee and Hae-Dong Jang *

Department of Food and Nutrition, Hannam University, Daejeon 34504, Korea; ldk7195@naver.com
* Correspondence: haedong@hnu.kr; Tel.: +82-42-629-8795

Abstract: The objective of this study was to investigate molecular mechanisms underlying the ability of carnosic acid to attenuate an early increase in reactive oxygen species (ROS) levels during MDI-induced adipocyte differentiation. The levels of superoxide anion and ROS were determined using dihydroethidium (DHE) and 2′-7′-dichlorofluorescin diacetate (DCFH-DA), respectively. Both superoxide anion and ROS levels peaked on the second day of differentiation. They were suppressed by carnosic acid. Carnosic acid attenuates the translation of NADPH (nicotinamide adenine dinucleotide phosphate) oxidase 4 (Nox4), $p47^{phox}$, and $p22^{phox}$, and the phosphorylation of nuclear factor-kappa B (NF-κB) and NF-κB inhibitor (IkBa). The translocation of NF-κB into the nucleus was also decreased by carnosic acid. In addition, carnosic acid increased the translation of heme oxygenase-1 (HO-1), γ–glutamylcysteine synthetase (γ-GCSc), and glutathione S-transferase (GST) and both the translation and nuclear translocation of nuclear factor erythroid 2-related factor 2 (Nrf2). Taken together, these results indicate that carnosic acid could down-regulate ROS level in an early stage of MPI-induced adipocyte differentiation by attenuating ROS generation through suppression of NF-κB-mediated translation of Nox4 enzyme and increasing ROS neutralization through induction of Nrf2-mediated translation of phase II antioxidant enzymes such as HO-1, γ-GCS, and GST, leading to its anti-adipogenetic effect.

Keywords: carnosic acid; ROS generation; ROS neutralization; Nox4 enzyme translation; NF-κB translocation; IkBα phosphorylation; HO-1 translation; Nrf2 translocation

Citation: Lee, D.-K.; Jang, H.-D. Carnosic Acid Attenuates an Early Increase in ROS Levels during Adipocyte Differentiation by Suppressing Translation of Nox4 and Inducing Translation of Antioxidant Enzymes. *Int. J. Mol. Sci.* **2021**, *22*, 6096. https://doi.org/10.3390/ijms22116096

Academic Editor: Maurizio Battino

Received: 3 May 2021
Accepted: 1 June 2021
Published: 5 June 2021

1. Introduction

Obesity is a chronic disease directly caused by abnormal increases in both the size and number of adipocytes. In white adipose tissues, several types of cells, including pre-adipocytes, mature adipocytes, fibroblasts, pericytes, macrophages, neutrophils, lymphocytes, endothelial cells, and adipose stem cells, exist [1]. Pre-adipocyte originating from adipose stem cells are known to be differentiated into adipocytes. This process is called adipogenesis [2]. To control obesity, several strategies such as reducing energy/food intake, increasing energy expenditure, inhibiting adipogenesis, inhibiting lipogenesis, and increasing lipolysis and fat oxidation have been proposed [3]. Among them, decreasing adipogenesis might be one of the better strategies for preventing obesity caused by enhanced adipogenesis [4].

For studying adipogenesis, the murine 3T3-L1 cell line is often used as a reliable in vitro model [5–7]. Adipogenesis of 3T3-L1 pre-adipocyte into adipocytes involves the following steps: growth arrest, mitotic clonal expansion, early differentiation, and terminal differentiation [8]. Adipogenesis is a process with tight redox regulation [8–10]. Different levels of reactive oxygen species (ROS) are generated in the four steps of adipogenesis of 3T3-L1 pre-adipocytes into adipocytes [8,11]. In particular, a transient increase in ROS levels can facilitate adipogenesis by accelerating mitotic clonal expansion [12]. Based on this evidence, the transient increase in ROS levels, which is non-toxic, can play a role as a transmitter in signaling pathways during differentiation of 3T3-L1 pre-adipocyte into adipocyte [13,14]. Such an increase in ROS levels may occur at the initiation and termination stages of adipogenesis [8].

Carnosic acid is a bioactive compound present in *Rosmarinus officinalis* L. [15]. It has several biological functions including antioxidant [16,17], anti-inflammatory [18,19], anti-bacterial [20], anti-cancer [21,22], and anti-osteoclastic activities [23,24]. In addition, it possesses an anti-adipogenesis activity both in vitro and in vivo [25–27]. A recent study has reported that carnosic acid can inhibit the adipogenesis of murine 3T3-L1 pre-adipocytes into adipocytes by down-regulating the redox state and increasing the expression of phase II antioxidant enzyme [28]. However, how carnosic acid attenuates ROS level in the early stage of adipogenesis in aspects of ROS generation and neutralization has not been elucidated.

Thus, the aim of this study was to investigate changes in ROS level during methyl-isobuthylxanthine, dexamethasone, and insulin (MDI)-induced differentiation of 3T3-L1 pre-adipocytes into adipocytes and clarify the molecular mechanisms underlying the down-regulation of the early increase in ROS levels by carnosic acid that led to its anti-adipogenetic effect.

2. Results

2.1. Carnosic Acid Suppresses Early Increase in Superoxide Anion and ROS Levels during Adipocyte Differentiation

During the differentiation of 3T3-L1 pre-adipocytes into adipocytes, ROS is known to be involved in promoting both early differentiation and later maturation of adipocytes [9]. Therefore, levels of superoxide anion and ROS were monitored using their specific fluorescent probes of dihydroethidium (DHE) and 2′-7′-dichlorofluorescin diacetate (DCFH-DA), respectively. As shown in Figure 1A,B, the intensity of DHE fluorescence was increased at day two, and the intensity of DCF fluorescence was increased at days two and seven during the differentiation. It has been reported that the anti-adipogenic effect of carnosic acid is through ROS control [22]. To determine whether superoxide anion and ROS levels in the early stage of adipocyte differentiation might be down-regulated by carnosic acid, superoxide anion and ROS levels at day two of adipocyte differentiation were analyzed. Carnosic acid at 10–20 μM attenuated the early increase in both superoxide anion and ROS levels were significantly ($p < 0.001$) induced by an MDI hormone mixture (Figure 1C,D). Notably, carnosic acid treatment at 20 μM reduced DCF fluorescence to 125.7%, which was enhanced (163.0%) by the MDI hormone cocktail as compared to an undifferentiated treatment. These observations suggested that the early increase in superoxide anion and ROS levels at day two during the differentiation of 3T3-L1 pre-adipocytes to adipocytes could be abrogated by carnosic acid. These results showed that the potent anti-adipogenetic effect of carnosic acid on the differentiation of 3T3-L1 pre-adipocytes into adipocytes might be attributable to the attenuation of ROS level induced by the MDI hormone cocktail during the early stage of adipocyte differentiation.

2.2. Carnosic Acid Attenuates the Translation of Nox4, p47 ^{phox}, and p22^{phox} and Nuclear Transport of NF-κB by Inhibiting IκBα Phosphorylation

Nox4 is a multi-component protein that consists of Nox4, p47^{phox}, and p22^{phox}. It is responsible for superoxide anion generation during MDI-induced differentiation of 3T3-L1 pre-adipocytes to adipocytes [10]. Therefore, the inhibitory effect of carnosic acid on the translation of Nox4 components as Nox4, p47^{phox}, and p22^{phox} was analyzed. Results of Western blotting analysis revealed that translations of Nox4, p47^{phox}, and p22^{phox} were all significantly ($p < 0.001$) induced by the MDI hormone mixture and notably attenuated by carnosic acid at 1 to 20 μM (Figure 2A–D).

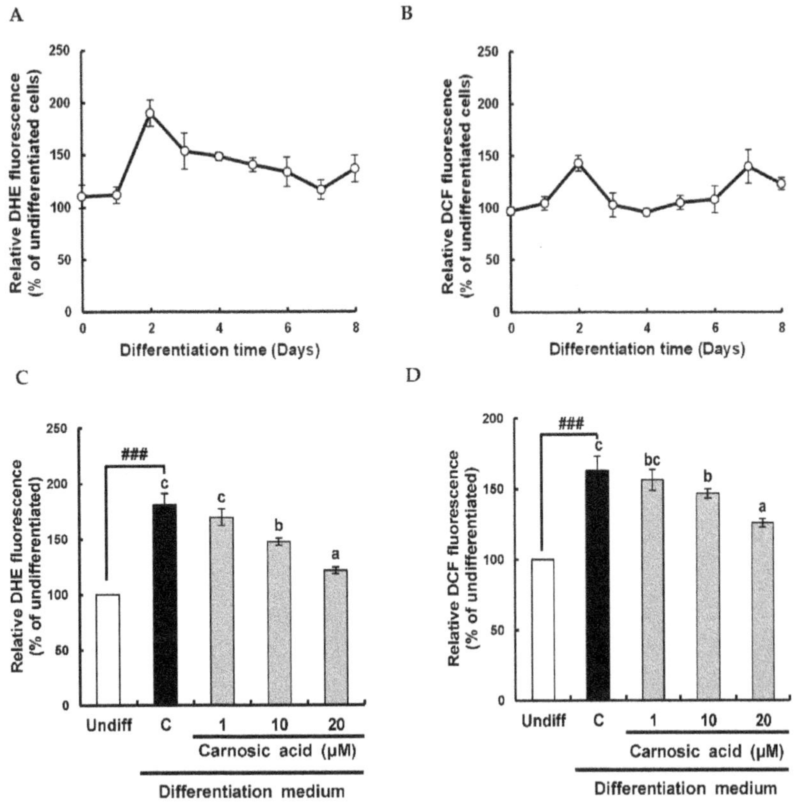

Figure 1. Carnosic acid suppresses superoxide anion and ROS levels in MDI-induced differentiation of 3T3-L1 pre-adipocytes into adipocytes. (**A**) Change of superoxide anion levels during adipogenesis. (**B**) Change of ROS levels during adipogenesis. (**C**) Carnosic acid suppressed superoxide anion levels after differentiation for two days. (**D**) Carnosic acid suppressed ROS levels after differentiation for two days. Data are presented as averages of three independent experiments performed in triplicates and expressed as percentages of the value of the control (means ± standard error mean, $n = 3$). ###: $p < 0.001$ vs. Undiff. Different corresponding letters indicate significant differences at $p < 0.05$ using Duncan's test. (Undiff: undifferentiated, which was not treated with the MDI hormone mixture; C: control, which was treated with the MDI hormone mixture).

To investigate the involvement of NF-κB/IκBα signal pathway in the translation of Nox4 enzyme components, translation and phosphorylation NF-κB and IκBα with nuclear translocation of NF-κB (p65) levels were analyzed by Western blotting after differentiation for two days (Figure 3A). As shown in Figure 3B, p-NF-κB (p65) in whole cells and nuclei were enhanced by the MDI hormone mixture but obviously attenuated by carnosic acid at 1–20 μM. The suppressive effect of carnosic acid on p-NF-κB in the nucleus was dose-dependent (Figure 3B). This attenuation of NF-κB in the nucleus by carnosic acid was also visually observed by fluorescence imaging using a confocal microscope (Figure 3D). To further make clear the underlying mechanism of transport of NF-κB into the nucleus, the translation and phosphorylation of IκBα, an NF-κB inhibitor, were analyzed by Western blotting. Results showed that IκBα translation was not changed by the MDI hormone mixture or carnosic acid treatment (Figure 3A). However, IκBα phosphorylation was significantly ($p < 0.001$) enhanced by the MDI hormone mixture and diminished by carnosic acid treatment at 1–20 μM (Figure 3C). Taken together, these data suggest that carnosic acid could attenuate the translation of Nox4, p47^{phox}, and p22^{phox}, the phosphorylation of NF-κB and IκBα, and the nuclear translocation of NF-κB in the early stage of MDI-induced adipocyte differentiation. Accordingly, these results imply that carnosic acid could interrupt the formation of the Nox4 enzyme complex by interfering with the translocation of NF-κB into the nucleus through inhibition of the NF-κB/IκBα signal pathway.

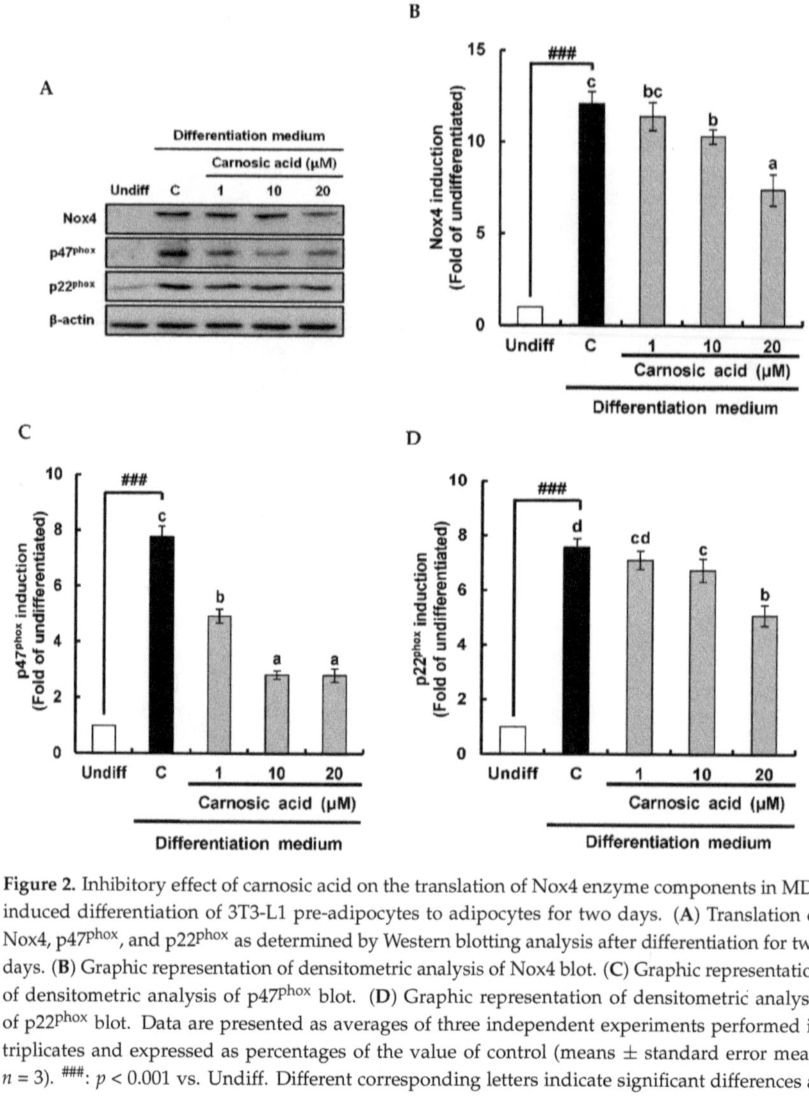

Figure 2. Inhibitory effect of carnosic acid on the translation of Nox4 enzyme components in MDI-induced differentiation of 3T3-L1 pre-adipocytes to adipocytes for two days. (**A**) Translation of Nox4, p47^{phox}, and p22^{phox} as determined by Western blotting analysis after differentiation for two days. (**B**) Graphic representation of densitometric analysis of Nox4 blot. (**C**) Graphic representation of densitometric analysis of p47^{phox} blot. (**D**) Graphic representation of densitometric analysis of p22^{phox} blot. Data are presented as averages of three independent experiments performed in triplicates and expressed as percentages of the value of control (means ± standard error mean, $n = 3$). $^{###}$: $p < 0.001$ vs. Undiff. Different corresponding letters indicate significant differences at $p < 0.05$ using Duncan's test. (Undiff: undifferentiated, which was not treated with the MDI hormone mixture; C: control, which was treated with the MDI hormone mixture).

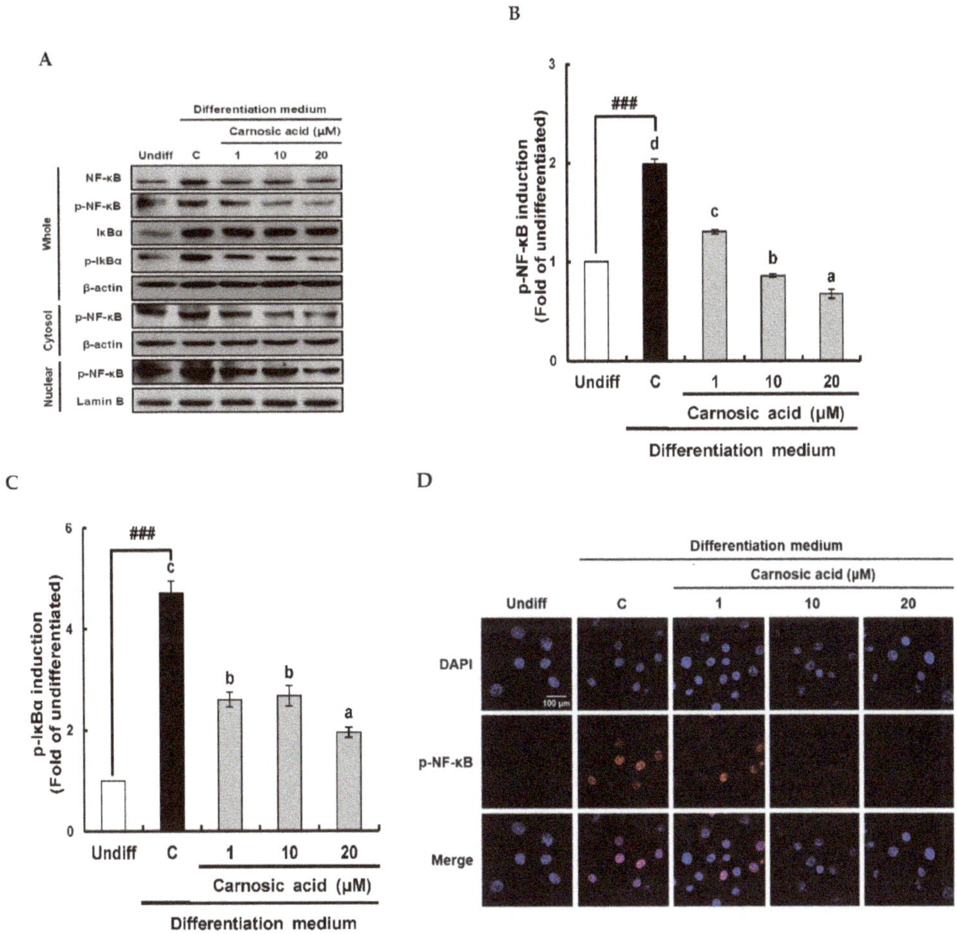

Figure 3. Inhibitory effect of carnosic acid on translation and phosphorylation of NF-κB and IκBα, and translocation of NF-κB into the nucleus during the differentiation of MDI-induced 3T3-L1 pre-adipocytes into adipocytes for two days. (**A**) Translated and phosphorylated NF-κB and IκBα in whole cells by Western blotting analysis after differentiation for two days. (**B**) Graphic representation of densitometric analysis of nuclear p-NF-κB blot. (**C**) Graphic representation of densitometric analysis of p-IκBα blot. (**D**) Immunofluorescence staining of 3T3-L1 pre-adipocyte cells after treatment with 1–20 μM carnosic acid using anti-p-NF-κB (p65) for staining. Data are presented as averages of three independent experiments performed in triplicates and expressed as percentages of the value of the control (means ± standard error mean, $n = 3$). $^{###}$: $p < 0.001$ vs. Undiff. Different corresponding letters indicate significant differences at $p < 0.05$ using Duncan's test. (Undiff: undifferentiated, which was not treated with the MDI hormone mixture; C: control, which was treated with MDI hormone mixture; DAPI: 4′,6-diamidino-2-phenylindole).

2.3. Carnosic Acid Inhibits Adipocyte Differentiation by Attenuating Nox4-Mediated ROS Generation via Interruption of NF-κB/IκBα Signal Pathway

To determine whether the NF-κB/IκBα signal pathway was involved in the down-regulation of Nox4 enzyme expression by carnosic acid, Bay11-7082 (BAY), an inhibitor of IκBα phosphorylation, was used. As shown in Figure 4A, both carnosic acid and BAY diminished the translation of Nox4, $p47^{phox}$, and $p22^{phox}$. As expected, both carnosic acid and BAY attenuated phosphorylation levels of NF-κB and IκBα that were notably enhanced by the MDI hormone mixture (Figure 4B). Furthermore, both carnosic acid and BAY obviously suppressed the translation of CCAAT/enhancer-binding protein α (C/EBPα), C/EBPβ, C/EBPγ, and adiponectin as adipogenesis biomarkers, which were induced by the MDI hormone mixture compared to the undifferentiated treatment (Figure 4C). Taken together, these findings indicate that the translation of Nox4 enzyme components, including Nox4, $p47^{phox}$, and $p22^{phox}$, might be closely associated with the NF-κB/IκBα signal pathway and that the down-regulation of ROS level by carnosic acid in the early stage of adipogenesis might be attributable to the attenuation of Nox4 enzyme translation through interruption of the NF-κB/IκBα signal pathway.

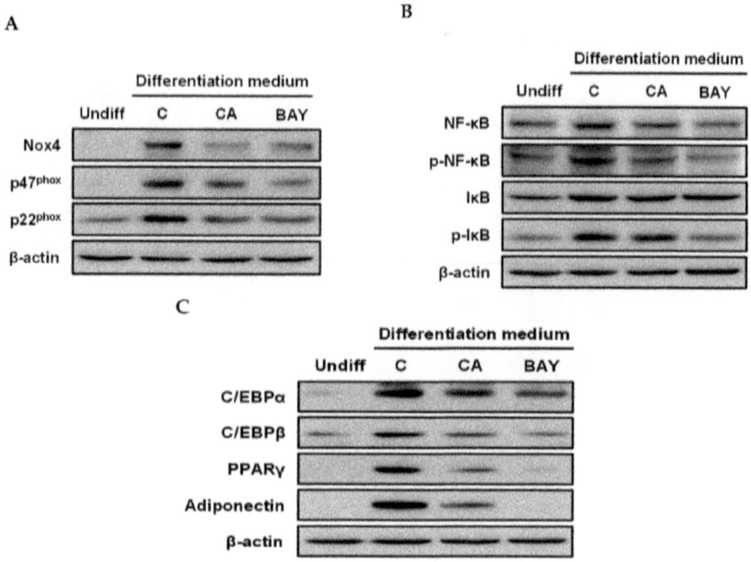

Figure 4. Effects of carnosic acid on the translation of Nox4 enzyme components, phosphorylation of NF-κB and IκBα, and translation of adipogenesis biomarkers in MDI-induced differentiation of 3T3-L1 pre-adipocytes to adipocytes for two days. (**A**) Translation levels of Nox4, $p47^{phox}$, and $p22^{phox}$ by Western blotting analysis after differentiation for two days. (**B**) Translation and phosphorylation levels of NF-κB and IκBα by Western blotting analysis after differentiation for two days. (**C**) Translation levels of C/EBPα, C/EBPβ, C/EBPγ, and adiponectin by Western blotting analysis after differentiation for two days. Data are presented as representative images of three independent experiments performed in triplicates). (Undiff; undifferentiated, which was not treated with MDI hormone mixture; C: control, which was treated with MDI hormone mixture; BAY: Bay11-7082).

2.4. *Carnosic Acid induces Nuclear Factor Erythroid 2-Related Factor 2 (Nrf2)-Mediated Translation of Phase II Antioxidant Enzymes*

To investigate how carnosic acid down-regulated ROS level in term of ROS neutralization in early stage of MDI-induced adipocyte differentiation, Nrf2-mediated translation levels of phase II antioxidant enzyme as heme oxygenase-1 (HO-1), γ–glutamylcysteine synthetase (γ-GCSc), and glutathione S-transferase (GST) were analyzed by Western blotting. As shown in Figure 5A,B, HO-1 was significantly ($p < 0.001$) reduced by MDI hormone mixture compared to that of an undifferentiated treatment. However, it was enhanced by carnosic acid at 1–20 μM dose-dependently in comparison with the control. On the other hand, protein levels of γ-GCSc and GST is known to be required for glutathione synthesis were induced by the MDI hormone mixture and slightly increased by carnosic acid in comparison with the control (Figure 5C,D). The translation of Nrf2, as a key transcription factor of phase II antioxidant enzyme, was also examined by Western blotting. Results showed that Nrf2 translation was significantly enhanced by MDI hormone mixture and dose-dependently induced by carnosic acid at 1–20 μM in comparison with the control (Figure 6A,B). The translocation of Nrf2 into the nucleus was enhanced by carnosic acid was also confirmed in fluorescence imaging using a confocal microscope (Figure 6C). Collectively, these findings showed that Nrf2-mediated translation of phase II antioxidant enzymes such as HO-1, γ-GCSc, and GST might be induced by carnosic acid, thus contributing to the down-regulation of ROS level through neutralization in the early stage of MDI-induced adipocyte differentiation.

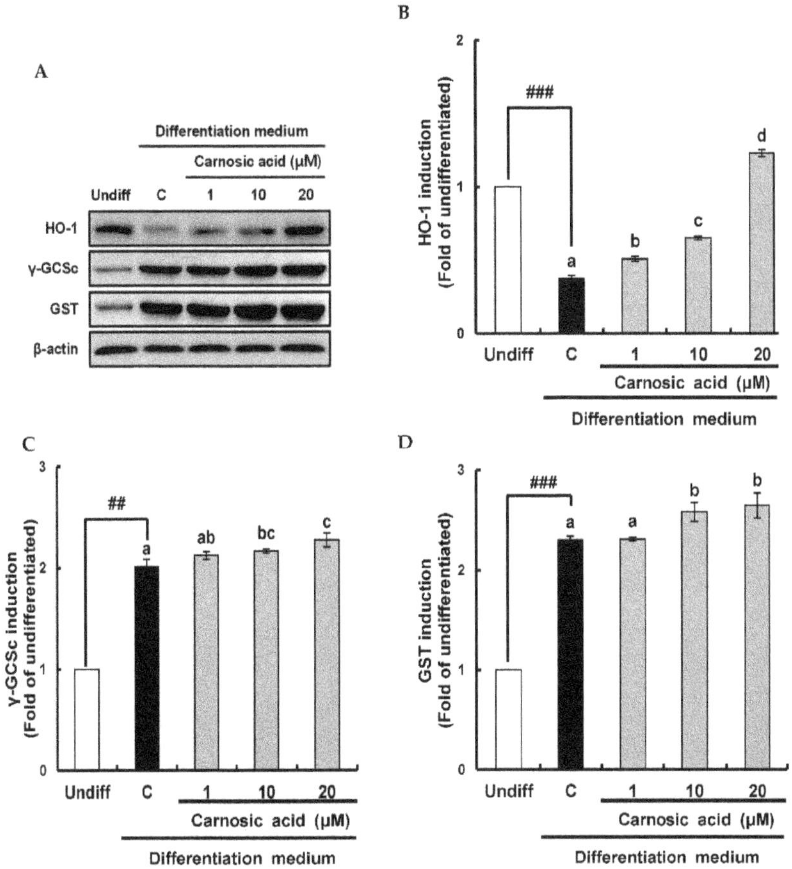

Figure 5. Effects of carnosic acid on the translation of phase II antioxidant enzyme during MDI-induced differentiation of 3T3-L1 pre-adipocytes to adipocytes for two days. (**A**) Translation of phase II antioxidant enzyme by Western blotting analysis after differentiation for two days. (**B**) Graphic representation of densitometric analysis of HO-1 blot. (**C**) Graphic representation of densitometric analysis of phosphorylated γ-GCSc blot. (**D**) Graphic representation of densitometric analysis of GST blot. Data are presented as averages of three independent experiments performed in triplicates and expressed as percentages of the value of the control (means ± standard error mean, $n = 3$). ##: $p < 0.01$ and ###: $p < 0.001$ vs. Undiff. Different corresponding letters indicate significant differences using Student's t-test. (Undiff: undifferentiated, which was not treated with the MDI hormone mixture; C: control, which was treated with the MDI hormone mixture).

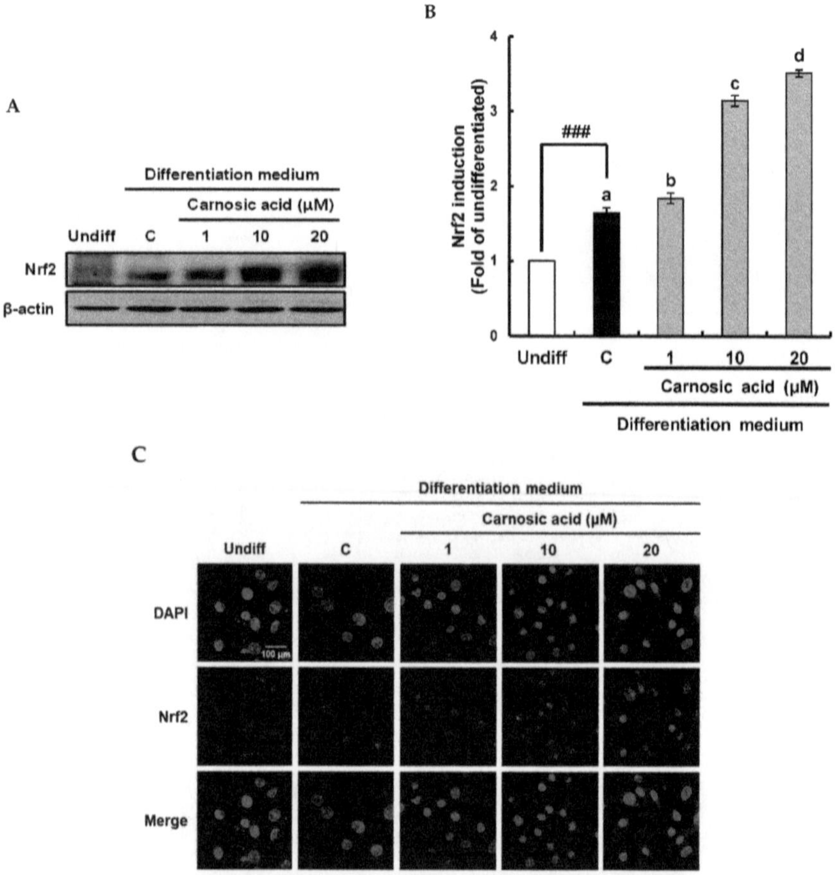

Figure 6. Effects of carnosic acid on translation and nuclear translocation of Nrf2 during MDI-induced differentiation of 3T3-L1 pre-adipocytes to adipocytes for two days. (**A**) Translation of Nrf2 by Western blotting analysis after differentiation for two days. (**B**) Graphic representation of densitometric analysis of Nrf2 blot. (**C**) Immunofluorescence staining of 3T3-L1 pre-adipocyte cells subjected to 1–20 μM carnosic acid using anti-Nrf2. Data are presented as averages of three independent experiments performed in triplicates and expressed as percentages of the value of the control (means ± standard error mean, $n = 3$). ###: $p < 0.001$ vs. Undiff. Different corresponding letters indicate significant differences by Student's t-test. (Undiff: undifferentiated, which was not treated with the MDI hormone mixture; C: control, which was treated with the MDI hormone mixture; DAPI: 4′,6-diamidino-2-phenylindole).

3. Discussion

In this study, changes in ROS levels during MDI-induced adipocyte differentiation were monitored using fluorescent probes such as DHE and DCF-DA specific for detecting superoxide anion and ROS, respectively. An early increase in both superoxide anion and ROS levels was observed on the second day of adipocyte differentiation. In addition, a second increase in ROS levels was found on the seventh day of adipocyte differentiation. This is the first study that monitors ROS changes during MDI-induced adipocyte differentiation. The results were in good agreement with previous reports showing a transient ROS increase was induced by insulin in the stage of mitotic clonal expansion or the early stage of the adipocyte differentiation process [8,13].

Based on the above results and the potent anti-adipogenetic effect of carnosic acid, it was hypothesized that carnosic acid could down-regulate the early increase in ROS levels during MDI-induced adipocyte differentiation, thus attenuating the adipogenesis process. As expected, carnosic acid potently abrogated both superoxide anion and ROS levels on the second day of differentiation, although levels of both superoxide anion and ROS were markedly augmented by the MDI hormone mixture. This is the first study reporting that the anti-adipogenetic effect of carnosic acid is due to down-regulation of the early increase in ROS level during MDI-induced adipocyte differentiation. An anti-adipogenetic effect with down-regulation of ROS has also found for caffeic acid phenethyl ester [29], α–lipoic acid [30], (−)-epigallocatechin-3-gallate [31], and natural extracts of *Granteloupia lanceolata* (Okamura) Kawaguchi [32], buckwheat sprouts [33], and unripe kiwi fruit (*Actinidia deliciosa*) [34], although their working mechanisms seem to be different from that of carnosic acid. It is noteworthy that the potent inhibitory effect of carnosic acid on MDI-induced adipogenesis of 3T3-L1 pre-adipocyte cells primarily depends on the attenuation of the early increase in ROS levels. Accumulating evidence has indicated that the redox state during MDI-induced differentiation of pre-adipocytes into adipocytes is regulated by ROS generation through NADPH oxidase 4 (Nox4) and ROS neutralization through antioxidant systems [8]. Therefore, two aspects of ROS generation and neutralization in down-regulation of ROS level by carnosic acid were taken into consideration in this study. Carnosic acid might attenuate ROS generation and/or induce antioxidant systems to down-regulate ROS levels during the early stage of adipocyte differentiation. During adipogenesis, ROS can be primarily generated as hydrogen peroxide because superoxide anion is generated by Nox4, which is highly expressed and immediately converted into hydrogen peroxide by endogenous superoxide dismutase [35]. It has also been suggested that hydrogen peroxide is likely to play a vital role as a signal mediator for cellular functions, including differentiation, because it can easily permeate cells and stay for a longer time than superoxide anion [10,36].

To investigate how carnosic acid down-regulated ROS levels through suppression of ROS generation in the early stage of adipocyte differentiation, suppressive effects of carnosic acid on the translation of Nox4 enzyme components as Nox4, $p47^{phox}$, and $p22^{phox}$ were investigated. According to the results of Western blot analysis, carnosic acid obviously attenuated the translation of Nox4, $p47^{phox}$, and $p22^{phox}$. This implied that carnosic acid might inhibit the formation of active Nox4 enzyme in the early stage of differentiation of 3T3-L1 pre-adipocytes into adipocytes, leading to the attenuation of the early increase in ROS levels. To further understand the underlying mechanism of the inhibitory effect of carnosic acid on Nox4 enzyme translation, the upper signaling pathway, including transcription factors, needs to be investigated. Previous studies have suggested that the translation of the Nox4 enzyme is closely related to redox-sensitive transcription factors such as NF-κB, which is known to play a vital role as a master key in the expression of various proteins as Nox4 during MDI-induced adipogenesis [13,37]. Therefore, the NF-κB/IκBα signal pathway was examined in the present study. NF-κB exists in the cytosol as a complex with IκBα, an inhibitory protein of NF-κB [38]. For the translation of concerned genes including Nox4, $p47^{phox}$, and $p22^{phox}$, NF-κB has to be translocated into the nucleus after IκBα is phosphorylated to be released from the NF-κB/IκBα complex. The reduction in NF-κB translocation into the nucleus by carnosic acid was confirmed by Western blotting and confocal microscopy. The plausible explanation for this observation was that carnosic acid could inhibit IκBα phosphorylation, leading to the maintenance of NF-κB/IκBα complex and the decrease in NF-κB due to degradation by the proteosome. The inhibition of IκBα phosphorylation by carnosic acid was checked using BAY as an inhibitor of IκBα phosphorylation. As shown in Figure 4, both carnosic acid and Bay treatment markedly suppressed the translation of Nox4, $p47^{phox}$, $p22^{phox}$, adiponectin, and NF-κB as well as the phosphorylation of IκBα and NF-κB, which were enhanced by MDI hormone mixture as compared to the undifferentiated treatment. These results implied that carnosic acid might interrupt IκBα phosphorylation to decrease NF-κB levels in the nucleus, which could cause attenuated translation of Nox4, p47^{phox}, and p22^{phox}. In this study, how carnosic acid inhibited IκBα phosphorylation was not elucidated. However, recent studies have proposed that catechol-type polyphenols could be oxidized into electrophilic quinones by ROS transiently and temporarily increased in cell [39–41]. Therefore, carnosic acid carrying catechol moiety might be oxidized by ROS produced during the early stage of adipocyte differentiation into its quinone and interact with sensible sulfhydryl groups of cysteine residues existing on the surface of IκBα and/or its kinase, leading to their conformational changes, which might be associated with an interruption of IκBα phosphorylation.

With respect to ROS neutralization by carnosic acid in the early stage of adipocyte differentiation, induction of the antioxidant system by carnosic acid was confirmed. HO-1 is known to be one important antioxidant enzyme involved in down-regulating ROS during the early stage of adipogenesis [42,43], and γ-GCS and GST are key enzymes in glutathione synthesis [28], these three antioxidant enzymes were chosen in this study. HO-1 was markedly attenuated in an early stage of MDI-induced adipocyte differentiation while ROS level was increased, consistent with earlier reports showing the import role of HO-1 in ROS neutralization during the early stage of adipogenesis [44,45]. However, HO-1 was notably induced by carnosic acid, which was closely related to the strong suppressive effect of carnosic acid on ROS level. On the other hand, γ-GCSc, and GST were notably induced by the MDI hormone mixture and slightly increased by carnosic acid compared to controls. These results imply that ROS levels in the early stage of adipogenesis might depend on HO-1, and that could be down-regulated through induction of HO-1 translation by carnosic acid. For further understanding of the induction of HO-1, γ-GCSc, and GST translation by carnosic acid, Nrf2/keap1 signaling pathway was investigated by Western blot and confocal microscopy. Both translation and nuclear translocation of Nrf2 were slightly enhanced by the MDI hormone mixture, leading to substantial increases of γ-GCSc, and GST translation, which might be necessary for the expression of some proteins needed in the early stage of adipogenesis [28]. On the other hand, the translation and nuclear translocation of Nrf2 were remarkably induced by carnosic acid, which might lead to the translation of HO-1 rather than γ-GCSc or GST. Consequently, carnosic acid can down-regulate ROS level during the early stage of adipocyte differentiation by neutralizing ROS through Nrf2-mediated induction of phase II antioxidant enzymes such as HO-1, γ-GCSc, and GST.

4. Materials and Methods

4.1. Chemical and Reagents

Carnosic acid was purchased from Cayman Chemical (Ann Arbor, MI, USA). Fetal bovine serum (FBS) was obtained from Biowest (Riverside, MO, USA). Dulbecco's modified Eagle's medium (DMEM), bovine calf serum (BCS), fetal bovine serum (FBS), trypsin-EDTA, phosphate-buffered saline (PBS, pH 7.4), and Hank's balanced salt solution (HBSS) were purchased from Welgene Inc. (Gyeongsan, Korea). Dimethylsulfoxide (DMSO), antibiotic-antimycotic solution 100X, insulin, 3-isobutyl-1-methyl-xanthine (IBMX), dexamethasone, isopropanol, formaldehyde, DCFH-DA, DHE, protease inhibitor, phenylmethanesulfonyl fluoride (PMSF), Triton X-100, 1,4-dithiothreitol (DTT), skim milk, and DPAI (4′,6-diamidino-2-phenylindole) were purchased from Sigma-Aldrich (St. Louis, MO, USA). Antibodies of Nox4, p47^{phox}, p22^{phox}, NF-κB (p65), phospho-NF-κB (p65), IκBα, phospho-IκBα, C/EBPα, C/EBPβ, C/EBPγ, HO-1, γ-GCSc, GST, and Nrf2 were purchased from Santa Cruz Biotechnology (Santa Cruz, CA, USA). Anti-adiponectin was obtained from Cell Signaling Technology (Beverly, MA, USA). 3T3-L1 cells were purchased from the American Type Culture Collection (ATCC, Rockville, MD, USA).

4.2. Cell Culture and Differentiation

The differentiation of pre-adipocytes into adipocytes was performed by a published procedure [46]. Briefly, 3T3-L1 pre-adipocyte cells were purchased from ATCC and grown to confluence in six-well plates containing DMEM supplemented with 10% BCS and antibiotics (100 U/mL penicillin and 100 μg/mL streptomycin) under 5% CO_2 humidified atmosphere at 37 °C. After two days of culturing, cells reached confluence, and the medium was replaced with a differentiation medium containing 10% FBS and MDI (0.5 mM 1-methy-3-isobuthyllxanthine, 1 μM dexamethasone, and 5 μg/mL insulin) hormone mixture [47]. Two days later, the differentiation medium was replaced with DMEM medium supplemented with 10% FBS and 5 μg/mL insulin. After two days, the medium was changed to DMEM supplemented with 10% FBS every two days until day 8 of differentiation. Carnosic acid was added two days before cells reached confluence and kept for four days until the second day.

4.3. Determination of Intracellular Superoxide Anion and ROS

Fluorescent probes, DCFH-DA (Ex/Em = 485 nm/535 nm) and DHE (Ex/Em = 518 nm/605 nm) that could specifically react with ROS and superoxide anion, respectively, were used to check ROS and superoxide anion levels during the differentiation of 3T3-L1 cells into adipocyte for eight days and to investigate the suppressive effect of carnosic acid on ROS and superoxide anion [48]. Briefly, 3T3-L1 cells were seeded into 96-well plates (1×10^5 cells/well) containing DMEM plus 10% BCS and antibiotics and incubated until cells reached confluence. After the confluence of cells was reached, the medium was replaced with DMEM supplemented with 10% BCS and carnosic acid. Two days later, the medium was then replaced with the differentiation medium. After 1 to 8 days of incubation, the medium was removed, and wells were washed twice successively with PBS and HBSS. A fluorescent probe was then added to the culture plates at a final concentration of 50 μM. After incubating in the dark for 30 min at 37 °C, fluorescence intensities of DHE and DCFH-DA were measured using a fluorometric plate reader (SpectraMax i3, Molecular Devices, San Jose, CA, USA).

4.4. Western Blot Analysis

3T3-L1 cells were seeded into 60 mm cell culture dishes (1×10^5 cells/dish) containing DMEM medium plus 10% BCS and antibiotics and incubated until cells reached confluence. After the confluence of cells was reached, the medium was replaced with DMEM containing 10% BCS and carnosic acid. Two days later, the medium was replaced with a differentiation medium. After incubation for another two days, cells were washed twice with PBS. 3T3-L1 cells were then lysed with ice-cold radio-immunoprecipitation assay (RIPA) buffer which was composed of 50 mM Tris–HCl (pH 8.0), 1% NP-40, 0.5% sodium deoxycholate, 150 mM NaCl, 1 mM PMSF, and a protease inhibitor cocktail [49]. The protein from lysed cells was then subjected to sodium dodecylsulfate-polyacrylamide gel electrophoresis (SDS-PAGE) and transferred to nitrocellulose membranes. Membranes were blocked with 5% skimmed milk in Tris-buffered saline containing Tween 20 (TBST) for 1 h at room temperature and then incubated with primary antibodies overnight at 4 °C. Proper horseradish peroxide-conjugated secondary antibodies were then added and incubated with the membranes for 1 h at room temperature. Proteins present in the membranes were visualized with an EZ-Western Lumi pico detection kit (DoGEN, Seoul, Korea) and quantified using a FUSION SOLO S (Vilber Lourmat, Collégien, France).

4.5. Nuclear Protein Extraction

Nuclear protein extraction was performed using a nuclear extraction kit (Abcam, Cambridge, UK). The harvested cell suspension was centrifuged at 300× *g* for 5 min. The supernatant was carefully removed and discarded. Afterward, 0.15 mL of a 1X pre-extraction buffer was added to the cell pellet followed by vigorous vortexing for 10 s, incubation on ice for 10 min, and centrifugation at 15,800× *g* for 1 min to remove cytosolic proteins from the nuclear pellet. The supernatant which contained the cytosolic protein was removed and transferred to a new tube (cytosol protein). Then, 0.06 mL of extraction buffer containing DTT and protease inhibitor cocktail (PIC) was added to the nuclear pellet. The extraction mixture was incubated on ice for 15 min. It was vortexed for 5 s every 3 min.

The suspension was then centrifuged at 18,000× *g* for 10 min at 4 °C. The supernatant was then transferred into a new microcentrifuge vial as a nuclear protein extract.

4.6. Confocal Microscopy

Microscopical observations of cells were performed by confocal imaging as described previously [50]. Briefly, following adipocyte differentiation for 2 days, 3T3-L1 cells were washed twice with PBS and fixed with 3.7% paraformaldehyde in PBS for 15 min at room temperature. After cells were washed with PBS and incubated with PBS containing 0.2% Triton X-100 at room temperature for 10 min, they were blocked with 3% skimmed milk in PBS for 45 min at room temperature. The blocking solution was removed, and the remaining cells were incubated with 1:250 solution of p-NF-κB primary antibody in 3% skimmed milk overnight at 4 °C. Cells were then washed three times with PBS (5 min each wash) and incubated with a 1:2500 solution of AlexaFlour 546 conjugated goat anti-rabbit secondary antibody (Invitrogen, Carlsbad, CA, USA) in a dark condition for 45 min at room temperature. Cells were washed three times with PBS (5 min each wash). A LSM5 live configuration Vario Two VRGB microscope (Zeiss, Jena, Germany) equipped with an oil immersion lens was used for the imaging of cells. Fluorescence imaging was obtained using 405 nm and 535 nm lasers for detecting nuclear staining of DAPI and anti-p-NF-κB (p65), respectively.

4.7. Statistical Analysis

All data are presented as mean ± SEM. All statistical analyses were executed using the statistical package SPSS 11 program (Statistical Package for Social Science 11, SPSS Inc., Chicago, IL, USA). The statistical significance for each group was verified with one-way analysis of variance (ANOVA) followed by Duncan's test at $p < 0.05$ or Student's *t*-test ($^{\#}$: $p < 0.05$; $^{\#\#}$: $p < 0.01$; $^{\#\#\#}$: $p < 0.001$).

5. Conclusions

During MDI-induced differentiation of 3T3-L1 pre-adipocytes into adipocytes, ROS levels peaked on the second day and could be potently attenuated by carnosic acid treatment. Carnosic acid could down-regulate ROS levels during the early stage of MPI-induced adipocyte differentiation by attenuating ROS generation through suppression of NF-B mediated translation of Nox4 enzyme and increasing ROS neutralization through induction of Nrf2-mediated translation of phase II antioxidant enzymes such as HO-1, γ-GCSc, and GST, leading to its anti-adipogenetic effect.

Author Contributions: H.-D.J. designed the experiments and directed the study process, including data analysis. D.-K.L. prepared the Introduction, the Material and Methods, and the Discussion sections of the manuscript. All authors have read and agreed to the published version of the manuscript.

Funding: This work was supported by the National Research Foundation of Korea(NRF) grant funded by the Korea government(MSIT). (2016R1A2B4011837).

Institutional Review Board Statement: Not applicable.

Informed Consent Statement: Not applicable.

Conflicts of Interest: The authors declare no conflict of interest.

Abbreviations

ROS	Reactive oxygen species
DHE	Dihyroethidium
DCFH-DA	2′-7′-dichlorofluorescin diacetate
Nox4	NADPH (nicotinamide adenine dinucleotide phosphate) oxidase 4
NF-κB	Nuclear factor-kappa B
IκBα	NF-κB inhibitor
Nrf2	Nuclear factor erythroid 2-related factor 2
DMEM	Dulbecco's modified Eagle's medium
MDI	Methyl-isobuthylxanthine, dexamethasone, and insulin
HBSS	Hank's balanced salt solution
PMSF	Phenylmethanesulfonyl fluoride
DMSO	Dimethylsulfoxide
RIPA	Radio-immunoprecipitation assay
BAY	Bay11-7082
DPAI	4′,6-diamidine-2-phenylindole
HO-1	Heme oxygenase-1
γ-GCSc	γ–Glutamylcysteine synthetase
GST	Glutathione S-transferase
SPSS	Statistical Package for Social Science

References

1. Wang, Q.A.; Scherer, P.E.; Gupta, R.E. Improved methodologies formthe study of adipose biology: Insights gained and opportunities ahead. *J. Lipid Res.* **2014**, *55*, 605–624. [CrossRef] [PubMed]
2. Hausman, D.B.; DiGirolamo, M.; Bartness, T.J.; Hausman, G.J.; Martin, R.J. The biology of white adipocyte proliferation. *Obes. Rev.* **2001**, *2*, 239–254. [CrossRef]
3. Wang, Y.W.; Jones, P.J. Conjugated linoleic acid and obesity control: Efficacy and mechanisms. *Int. J. Obes. Relat. Metab. Disord.* **2004**, *28*, 941–955. [CrossRef]
4. Ji, E.; Jung, M.Y.; Park, J.H.; Seo, C.R.; Park, K.W.; Lee, E.K.; Yeom, C.H.; Lee, S. Inhibition of adipogenesis in 3T3-L1 cells and suppression of abdominal fat accumulation in high-fat diet-feeding C57BL/6J mice after dowmregulation of hyaluronic acid. *Int. J. Obes.* **2014**, *38*, 1035–1043. [CrossRef] [PubMed]
5. Vishwanath, D.; Srinivasan, H.; Patil, M.S.; Seetarama, S.; Kumar, S.A.; Dixit, M.N. Novel method to differentiate 3T3 L1 cells in vitro to produce highly sensitive adipocytes for a GLUT4 mediated glucose uptake using fluorescent glucose analog. *J. Cell Commun. Signal.* **2013**, *7*, 129–140. [CrossRef] [PubMed]
6. Zebisch, K.; Voigt, V.; Wabitsch, M.; Brandsch, M. Protocol for effective differentiation of 3T3-L1 cells to adipocytes. *Anal. Biochem.* **2012**, *425*, 88–90. [CrossRef]
7. Ruiz-Ojeda, F.J.; Rupérez, A.I.; Gomez-Llorente, C.; Gil, A.; Aguilera, M. Cell models and their application for studying adipogenic differentiation in relation to obesity: A review. *Int. J. Mol. Sci.* **2016**, *17*, 1040. [CrossRef]
8. Wang, X.; Hai, C. Redox modification of adipocyte differentiation: Hypothesis of "Redox Chain" and novel insights into intervention of adipogenesis and obesity. *Free Radic. Biol. Med.* **2015**, *89*, 99–125. [CrossRef]
9. Kanda, Y.; Hinata, T.; Kang, S.W.; Watanabe, Y. Reactive oxygen species mediate adipocyte differentiation in mesenchymal stem cells. *Life Sci.* **2011**, *89*, 250–258. [CrossRef]
10. Liu, G.S.; Chan, E.C.; Higuchi, M.; Dusting, G.J.; Jiang, F. Redox mechanisms in regulation of adipocyte differentiation: Beyond a general stress response. *Cells* **2012**, *1*, 976–993. [CrossRef] [PubMed]
11. Furukawa, S.; Fujita, T.; Shimabukuro, M.; Iwaki, M.; Yamada, Y.; Nakajima, Y.; Nakayama, O.; Makishima, M.; Matsuda, M.; Shimomura, I. Increased oxidative stress in obesity and its impact on metabolic syndrome. *J. Clin. Invest.* **2004**, *114*, 1752–1761. [CrossRef]
12. Lee, H.M.; Lee, Y.J.; Choi, H.J.; Ko, E.H.; Kim, J.W. Reactive oxygen species facilitate adipocyte differentiation by accelerating mitotic clonal expansion. *J. Biol. Chem.* **2009**, *284*, 10601–10609. [CrossRef]
13. Benard, K.; Krause, K.H. The NOX family of ROS-generating NADPH oxidases: Physiology and pathophysiology. *Physiol. Rev.* **2007**, *87*, 245–313.
14. Schöder, K.; Wandzioch, K.; Helmeke, I.; Brandes, R.P. Nox4 acts as a switch between differentiation and proliferation in preadipocytes. *Arterioscler. Thromb. Vasc. Biol.* **2008**, *2*, 239–245.
15. Birtic, S.; Dussort, P.; Pierre, F.X.; Bily, A.C.; Roller, M. Carnosic acid. *Phytochemistry* **2015**, *115*, 9–19. [CrossRef]
16. Erkan, N.; Ayranci, G.; Ayranci, E. Antioxidant activity of rosemary (*Rosmarinus officinalis* L.) extract, blackseed (*Nigella sativa* L.) essential oil, carnosic acid, rosmarinic acid and sesamol. *Food Chem.* **2008**, *110*, 76–82. [CrossRef] [PubMed]

17. Maheswarappa, N.B.; Subbaiah, V.; Muthupalani, M.; Yamagani, P.K.; Mohan, K.; Keshapaga, U.R.; Asokan, S.V.; Kalappurakkal, R.C. Antioxidant activity of carnosic acid and rosmarinic acid in raw and cooked ground chicken patties. *J. Sci. Food Agric.* **2014**, *94*, 273–279. [CrossRef]
18. Hadad, N.; Levy, R. The synergistic anti-inflammatory effects of lycopene, lutein, β-carotene, and carnosic acid combinations via redox-based inhibition of NF-κB signaling. *Free Radical Biol. Med.* **2012**, *53*, 1381–1391. [CrossRef]
19. Yu, Y.M.; Lin, C.H.; Chan, H.C.; Tsai, H.D. Carnosic acid reduce cytokine-induced adhesion molecules expression and monocyte adhesion to endothelial cells. *Eur. J. Nutr.* **2009**, *48*, 101–106. [CrossRef] [PubMed]
20. Moreno, S.; Scheyer, T.; Romano, C.S.; Vojnov, A.A. Antioxidant and antimicrobial activities of rosemary extracts linked their polyphenol composition. *Free Radic. Res.* **2006**, *40*, 223–231. [CrossRef] [PubMed]
21. Kim, D.H.; Park, K.W.; Chae, I.G.; Kundu, J.; Kim, E.H.; Kundu, J.K.; Chen, K.S. Carnosic acid inhibits STAT3 signaling and induces apoptosis through generation of ROS in human colon cancer HCT116 cells. *Mol. Carcinog.* **2016**, *55*, 1096–1100. [CrossRef]
22. Yesil-Celiktas, O.; Sevimli, C.; Bedir, E.; Vardar-Sukan, F. Inhibitory effects of rosemary extracts, carnosic acid and rosmarinic acid on the growth of various human cancer cell lines. *Plant Foods Hum. Nutr.* **2010**, *65*, 158–163. [CrossRef] [PubMed]
23. Rhummuri, D.; Naidu, V.G.M.; Chaudhari, P. Carnosic acid attenuates RANKL-induced oxidative stress and osteoclastogenesis via induction of Nrf2 and suppression of NF-κ**B** and MAPK signaling. *J. Mol. Med.* **2017**, *95*, 1065–1076. [CrossRef] [PubMed]
24. Hagiwara, H.; Basnet, R.; Wiyasihati, S.; Nakata, K.; Hagiwara, K.; Miyazaki, H.; Yoshida, K. Carnosic acid inhibits the formation of osteoclasts through attenuation of expression of RANKL. *PharmaNutrion* **2015**, *3*, 1–6. [CrossRef]
25. Park, M.Y.; Sung, M.K. Carnosic acid inhibits lipid accumulation in 3T3-L1 adipocytes and through attenuation of fatty acid desaturation. *J. Cancer Prev.* **2015**, *20*, 41–49. [CrossRef] [PubMed]
26. Zhao, Y.; Sedighi, R.; Wang, P.; Chen, H.; Zhu, Y.; Sang, S. Carnosic acid as major bioactive component in rosemary extract ameliorates high-fat-induced obesity and metabolic syndrome. *J. Agric. Food Chem.* **2015**, *63*, 4843–4852. [CrossRef]
27. Gaya, M.; Repetto, V.; Toneatto, J.; Anesini, C.; Piwien-Pilipuk, G.; Moreno, S. Antiadipogenetic effect of carnosic acid, a natural compound present in *Rosmarinus officinalis*, is exerted through the C/EBPs and PPARγ pathways at the onset of the differentiation program. *Biochim. Biophys. Acta* **2013**, *1831*, 3796–3806. [CrossRef]
28. Takahashi, T.; Tabuchi, T.; Tamaki, Y.; Kosaka, K.; Takikawa, Y.; Sato, T. Carnosic acid and carnosol inhibit adipocyte differentiation in mouse 3T3-L1 cells through induction of phase2 enzymes and activation of glutathione metabolism. *Biochem. Biophys. Res. Commun.* **2009**, *382*, 549–554. [CrossRef]
29. Kwon, Y.B.; Wang, B.B.; Jang, H.D. Anti-osteoclastic effect of caffeic acid phenethyl ester in murine macrophages depends upon the suppression of superoxide anion production through the prevention of can active-Nox1 complex formation. *J. Nutr. Biochem.* **2018**, *58*, 158–168. [CrossRef]
30. Cho, K.J.; Moon, H.E.; Moini, M.; Pecker, L.; Yoon, D.Y.; Chung, A.S. Alpha-lipoic acid inhibits adipocyte differentiation by regulating pro-adipogenic transcription factors via mitogen-activated protein kinase pathways. *J. Biol. Chem.* **2003**, *278*, 34823–34833. [CrossRef]
31. Wang, C.T.; Chang, H.H.; Hsiao, C.H.; Lee, M.J.; Ku, H.C.; Hu, Y.J.; Kao, Y.H. The effects of green tea (-)-epigallocatechin-3-gallate on reactive oxygen species in 3T3-L1 preadipocytes and adipocytes depends on the glutathione and 67 kDa laminin receptor pathways. *Mol. Nutr. Food Res.* **2009**, *53*, 349–360. [CrossRef]
32. Seo, M.J.; Choi, H.S.; Lee, O.H.; Lee, B.Y. *Granteloupia lanceolata* (Okamura) Kawaguchi, the edible red seaweed, inhibits lipid accumulation and reactive oxygen species production during differentiation in 3T3-L1 cells. *Phytother. Res.* **2013**, *27*, 655–663. [CrossRef] [PubMed]
33. Lee, Y.J.; Kim, K.J.; Park, B.R.; Yoon, J.H.; Lim, O.H.; Lee, O.H. Buckwheat (*Fago-pyrum esculetum* M.) sprout treated with methyl jasmonate (MeJA) improved anti-adipogenic activity associated with the oxidative stress system in 3T3-L1 adipocyte. *Int. J. Mol. Sci.* **2013**, *14*, 1428–1442. [CrossRef]
34. Abe, D.; Saito, Y.; Kubo, Y.; Nakamura, Y.; Sekiya, K. A fraction of unripe kiwi fruit extract regulates adipocyte differentiation and function in 3T3-L1 cells. *Biofactors* **2010**, *36*, 52–59. [CrossRef]
35. Martyn, K.D.; Frederick, L.M.; Loehneysen, K.V.; Dinauer, M.C.; Knaus, U.G. Functional analysis of Nox4 reveals unique characteristics compared to other NADPH oxidases. *Cell Signal* **2006**, *18*, 69–82. [CrossRef]
36. Brown, D.I.; Griendling, K.K. Nox proteins in signal transduction. *Free Radic. Biol. Med.* **2009**, *47*, 1239–1253. [CrossRef] [PubMed]
37. Sauer, H.; Wartenberg, M.; Hescheler, J. Reactive oxygen species as intracellular messengers during cell growth and differentiation. *Cell Physiol. Biochem.* **2001**, *11*, 173–186. [CrossRef] [PubMed]
38. Gloire, G.; Legrand-Poels, S.; Piette, J. NF-κB activation by reactive oxygen species: Fifteen year later. *Biochem. Pharmacol.* **2006**, *72*, 1493–1505. [CrossRef]
39. Ishi, T.; Ishikawa, M.; Miyoshi, N.; Yasunaga, M.; Akagawa, M.; Uchida, K.; Nakamura, Y. Catechol type polyphenol is a potential modifier of protein sulfhydryls: Development and application of a new probe for understanding dietary polyphenol actions. *Chem. Res. Toxicol.* **2009**, *22*, 1689–1698. [CrossRef] [PubMed]
40. Dinkova-Kostova, A.T.; Wang, X.J. Induction of the Keap1/Nrf2/ARE pathway by oxidizable diphenols. *Chem. Biol. Interact.* **2011**, *192*, 101–106. [CrossRef] [PubMed]
41. Satoh, T.; McKercher, S.; Lipton, S. Nrf2/ARE-mediated antioxidant actions of pro-electric drugs. *Free Radic. Biol. Med.* **2013**, *65*, 645–657. [CrossRef] [PubMed]

42. Wagner, G.; Lindroos-Christensen, J.; Einwallner, E.; Husa, J.; Zapf, T.; Lipp, K.; Rauscher, S.; Gröger, M.; Spittler, A.; Loewe, R.; et al. HO-1 inhibits preadipocyte proliferation and differentiation at the onset of obesity via ROS dependent activation of Akt2. *Sci. Rep.* **2017**, *7*, 40881. [CrossRef]
43. Koh, E.J.; Kim, K.J.; Seo, Y.J.; Choi, J.; Lee, B.Y. Modulation of HO-1 by ferulic acid attenuates adipocyte differentiation in 3T3-L1 cells. *Molecules* **2017**, *22*, 745. [CrossRef]
44. Kim, D.H.; Burgess, A.; Li, M.; Tsenovoy, P.L.; Addabo, F.; McClung, J.A.; Puri, N.; Abraham, N.G. Heme oxygenase-mediated increases in adiponectin decrease fat content and inflammatory cytokine tumor necrosis-α and interleuin-6 in Zucker rats and reduce adipogenesis in human mesenchymal stem cells. *J. Phamacol. Exp. Ther.* **2008**, *325*, 833–840. [CrossRef]
45. Vanella, L.; Sodhi, K.; Kim, D.H.; Puri, N.; Maheshwan, M.; Hinds, T.D.; Bellner, L.; Goldstein, D.; Peterson, S.; Shapiro, J.I.; et al. Increased heme-oxygenase-1 expression in mesenchymal stem cell-derived adipocytes decreases differentiation and lipid accumulation via upregulation of the canonical Wnt signaling cascade. *Stem Cell Res. Ther.* **2013**, *4*, 28. [CrossRef] [PubMed]
46. He, Y.; Li, Y.; Zhao, T.; Wang, Y.; Sun, C. Ursolic acid inhibits adipogenesis in 3T3-L1 adipocytes through LKB1/AMPK pathway. *PLoS ONE* **2013**, *8*, e70135. [CrossRef] [PubMed]
47. Kang, S.I.; Ko, H.C.; Shin, H.S.; Kim, H.M.; Hong, Y.S.; Lee, N.H.; Kim, S.J. Fucoxanthin exerts differing effects on 3T3-L1 cells according to differentiation stage and inhibits glucose uptake in mature adipocytes. *Biochem. Biophys. Res. Commun.* **2011**, *409*, 769–774. [CrossRef]
48. Han, Y.H.; Kim, S.H.; Kim, S.Z.; Park, W.H. Caspase inhibitor decreases apoptosis in pyrogallol-treated lung cancer Calu-6 cells via the prevention of GSH depletion. *Int. J. Oncol.* **2008**, *33*, 1099–1105. [PubMed]
49. Daniel, J.; Macphee, D.J. Methodological considerations for improving Western blot analysis. *J. Pharmacol. Toxicol. Methods* **2010**, *2*, 171–177.
50. Grindel, S.J.; Rohe, B.; Safford, S.E.; Bennett, J.J.; Farach-Carson, M.C. Tumor necrosis factor-α treatment of HepG2 cells mobilizes a cytoplasmic pool of Erp57/1.25D3-MARRS to the nucleus. *J. Cell Biochem.* **2011**, *112*, 2606–2615. [CrossRef] [PubMed]

International Journal of *Molecular Sciences*

MDPI

Article

Influence of the Anthocyanin and Cofactor Structure on the Formation Efficiency of Naturally Derived Pyranoanthocyanins

Gonzalo Miyagusuku-Cruzado †, Danielle M. Voss † and M. Monica Giusti *

Department of Food Science and Technology, The Ohio State University, 2015 Fyffe Rd., Columbus, OH 43210 1007, USA; miyagusukucruzado.1@osu.edu (G.M.-C.); voss.129@osu.edu (D.M.V.)
* Correspondence: giusti.6@osu.edu; Tel.: +1-614-247-8016
† These authors contributed equally to this work.

Abstract: Pyranoanthocyanins are anthocyanin-derived pigments with higher stability to pH and storage. However, their slow formation and scarcity in nature hinder their industrial application. Pyranoanthocyanin formation can be accelerated by selecting anthocyanin substitutions, cofactor concentrations, and temperature. Limited information is available on the impacts of the chemical structure of the cofactor and anthocyanin; therefore, we evaluated their impacts on pyranoanthocyanin formation efficiency under conditions reported as favorable for the reaction. Different cofactors were evaluated including pyruvic acid, acetone, and hydroxycinnamic acids (*p*-coumaric, caffeic, ferulic, and sinapic acid) by incubating them with anthocyanins in a molar ratio of 1:30 (anthocyanin:cofactor), pH 3.1, and 45 °C. The impact of the anthocyanin aglycone was evaluated by incubating delphinidin, cyanidin, petunidin, or malvidin derivatives with the most efficient cofactor (caffeic acid) under identical conditions. Pigments were identified using UHPLC-PDA and tandem mass spectrometry, and pyranoanthocyanin formation was monitored for up to 72 h. Pyranoanthocyanin yields were the highest with caffeic acid (~17% at 72 h, $p < 0.05$). When comparing anthocyanins, malvidin-3-*O*-glycosides yielded twice as many pyranoanthocyanins after 24 h (~20%, $p < 0.01$) as cyanidin-3-*O*-glycosides. Petunidin- and delphinidin-3-*O*-glycosides yielded <2% pyranoanthocyanins. This study demonstrated the importance of anthocyanin and cofactor selection in pyranoanthocyanin production.

Keywords: hydroxyphenyl-pyranoanthocyanins; naturally derived pigments; accelerated formation; 10-catechyl-pyranoanthocyanins

Citation: Miyagusuku-Cruzado, G.; Voss, D.M.; Giusti, M.M. Influence of the Anthocyanin and Cofactor Structure on the Formation Efficiency of Naturally Derived Pyranoanthocyanins. *Int. J. Mol. Sci.* **2021**, *22*, 6708. https://doi.org/10.3390/ijms22136708

Academic Editor: Maurizio Battino

Received: 14 May 2021
Accepted: 17 June 2021
Published: 23 June 2021

1. Introduction

Anthocyanins (ACNs) are dietary flavonoids with bright colors that can range from red to blue [1,2]. The use of ACN-rich extracts as food colorants has increased in recent years due to potential behavioral concerns associated with the consumption of artificial dyes [3,4]. Also, interest in ACN consumption has grown due to their potential bioactive and health-promoting properties [5]. However, from an industrial point of view, the application of ACN-rich extracts as food colorants is restricted due to limited long-term stability and color expression [1]. Additionally, common components in the food matrix such as ascorbic acid can bleach ACNs, resulting in their degradation [6]. Several mechanisms have been proposed for the stabilization of ACNs in foods such as copigmentation with phenolic compounds [7], complexation with proteins [8], encapsulation within polysaccharides [9], and chelation with metals [10].

Studies have shown that prolonged interaction of ACNs with hydroxycinnamic acids [7], acetone [11], or pyruvic acid (PA) [12] can result in the formation of ACN-derived pigments called pyranoanthocyanins (PACNs). This process occurs during winemaking as a result of the interaction between ACNs and yeast metabolites [13,14]. These ACN-derived pigments can express color across all pH values [11] and have better long-term storage stability [6,11]. Moreover, PACNs showed enhanced resistance to bleaching by ascorbic acid [6] and sulfur dioxide [11,15]. These pigments are formed through the cycloaddition reaction of a reactive cofactor on position C4 and the 5-OH group of the ACN molecule, resulting in the formation of an additional pyran ring [16] and consequently, in the unavailability of C4 to participate in degradation reactions [17]. Studies on the synthesis pathways of different PACNs from ACN-rich extracts have shown that carboxy-PACNs can result from the reaction with PA, hydroxyphenyl-PACNs from the reaction with hydroxycinnamic acids, and methyl-PACNs from the reaction with acetone [11,18]. Due to this formation process, ACNs with substitutions on position C5 cannot form PACNs, thus excluding many pigment sources commonly used in the food industry from being PACN precursors. ACN sources commonly used in the food industry capable of forming PACNs include black carrot [7,19], elderberries [7], and some grape extracts [20].

Although more stable, PACNs are hard to find in nature with limited quantities reported in onions [21] and strawberries [22]. Additionally, PACNs can be found in a small number of foods such as wine [18], sumac [23], fruit wines [24], and fruit juices after extended periods of time [19]. Most studies have focused on the occurrence and identification of PACNs [18–20,25], their stability [6,11], and their unique color characteristics [26]. However, little has been reported on the optimization of their formation. Previous studies reported that PACN formation efficiency is increased with 1-6 di-glycosylated ACNs [12], the removal of ACN aromatic acylating groups [27], solution pH ~3.0 [28], as well as higher incubation temperatures [29] and molar cofactor ratios [28,29]. In addition, studies comparing different cofactors showed greater PACN yields with caffeic acid (CA) than with PA [27]. However, the impact on PACN formation efficiency of other cofactors, such as acetone, and modifications dependent on cofactor type and ACN structure are still underreported. This information is important if PACNs are to be used by the food industry as it would help in the development of more stable, naturally derived colorants. Therefore, the objective of this study was to determine the effect of the chemical structure of the cofactor and ACN on PACN formation efficiency under accelerated conditions (pH 3.1, ACN to cofactor molar ratio of 1:30, incubation at 45 °C).

2. Results

2.1. Evaluation of Pyranoanthocyanin Formation Efficiency Using Different Types of Cofactors

CA, acetone, and PA were tested as cofactors for the formation of PACNs from cyanidin-glycosides obtained from saponified black carrot ACN extracts (sBC) with results shown in Figure 1. The ACN extract was mainly composed of cyanidin-3-*O*-xylosyl-glucosyl-galactoside (peak 1, C3XyGlGa, ~64% of total area at 475–520 nm) and cyanidin-3-*O*-xylosyl-galactoside (peak 2, C3XyGa, ~34% of total area at 475–520 nm). The 475–520 nm max plot chromatogram was used because it accounted for the $\lambda_{vis\text{-}max}$ of all pigments evaluated in this study. After incubation with CA, two new peaks were detected in the PDA chromatogram. Their hypsochromic shift of $\lambda_{vis\text{-}max}$ compared to the precursor ACNs, later retention times, and MS spectra revealed that these new peaks corresponded to 10-catechyl-PACNs derived from the ACNs previously identified in sBC. Consequently, peak 3 was ascribed to 10-catechyl-pyranocyanidin-3-*O*-xylosyl-glucosyl-galactoside and peak 4 to 10-catechyl-pyranocyanidin-3-*O*-xylosyl-galactoside. No new peaks were detected in the cofactor-free control and samples with acetone and PA after 72 h of incubation.

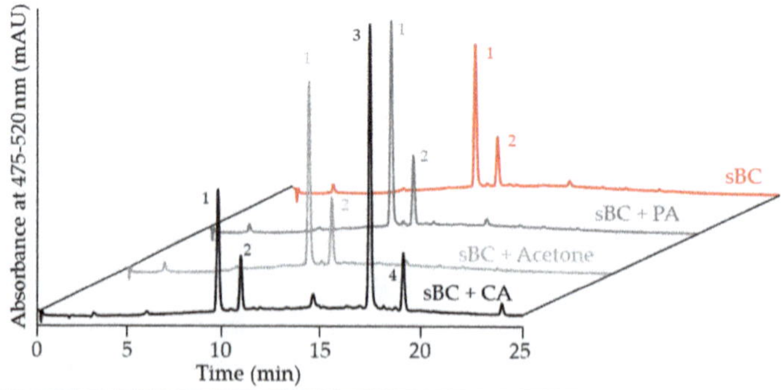

Peak	$\lambda_{vis\text{-}max}$ (nm)	[M]$^+$ (m/z)	Aglycone [M-X]$^+$ (m/z)	ID
1	517	743$^+$	287$^+$	Cyanidin-3-*O*-xylosyl-glucosyl-galactoside (C3XyGlGa)
2	518	581$^+$	287$^+$	Cyanidin-3-*O*-xylosyl-galactoside (C3XyGa)
3	509	875$^+$	419$^+$	10-catechyl-pyranoC3XyGlGa
4	507	713$^+$	419$^+$	10-catechyl-pyranoC3XyGa

Figure 1. Comparing the PACN formation after 72 h of anthocyanin (saponified black carrot, sBC) incubation with caffeic acid (CA), acetone, or pyruvic acid (PA). Chromatograms show the max plot in the 475–520 nm range. Table shows the wavelength of maximum absorption in the visible range ($\lambda_{vis\text{-}max}$), mass per charge ratio (m/z) of the main ion and its aglycone, and tentative identity (ID).

Yields of PACNs, calculated as percentages compared to the initial ACN content using Equation (1), increased from 3.9 ± 0.2% at 24 h to 14.1 ± 0.5% and 17.2 ± 0.6% after 48 and 72 h of incubation with CA at 45 °C, respectively (Figure 2). Based on the peak areas calculated at each time point using Equation (3), PACNs comprised 4.6 ± 0.2, 38.4 ± 3.5, and 63.5 ± 4.2% of the total pigment at 24, 48, and 72 h, respectively. No PACNs were detected in samples incubated with acetone or PA. The pigment remaining after 72 h of incubation was the highest in the samples with CA (27.4 ± 2.6%). This content was higher than the one observed in samples with PA (16.4 ± 1.8%), but not significantly so ($p = 0.06$). Additionally, the pigment remaining in samples with CA was significantly higher than the ones observed in cofactor-free control samples (9.1 ± 1.9%, $p < 0.01$) and samples with acetone (12.1 ± 2.8%, $p < 0.01$).

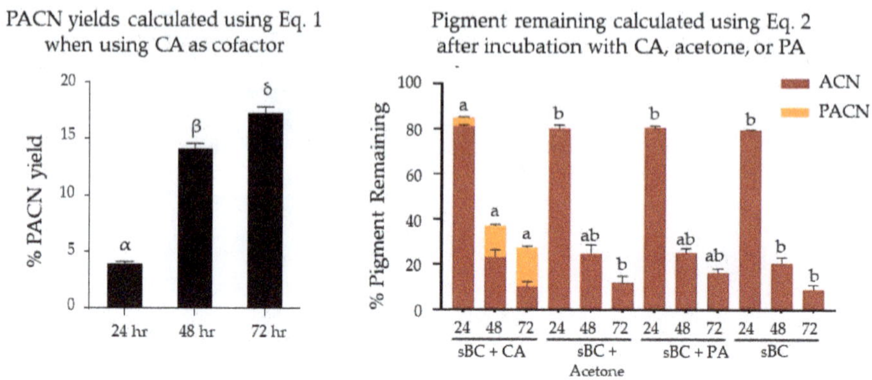

Figure 2. Pyranoanthocyanin yield (%PACN) with caffeic acid (CA) and pigment remaining (%) after incubation with different cofactors for 24, 48, and 72 h at 45 °C with cyanidin-glycosides (from saponified black carrot, sBC). Different Greek letters show significant differences among time points at a 0.05 level. Different letters show statistically significant differences among cofactors at the same time point at a 0.05 level. Results are expressed as means ± standard error ($n = 3$). ACN: anthocyanin, PA: pyruvic acid.

2.2. Evaluation of Pyranoanthocyanin Formation Efficiency Using Different Hydroxycinnamic Acids

After incubation for up to 72 h, CA was the only cofactor forming PACN under our experimental conditions, strongly suggesting that the chemical structure of the cofactor can impact PACN yields. Studies have shown different reactivity rates of hydroxycinnamic acids with ACNs or PACNs resulting in the formation of new pigments [30,31]. In these studies, it was hypothesized that the nucleophilic nature of the hydroxycinnamic acid impacted its reactivity [31], with di- and trisubstituted ones more reactive than monosubstituted ones [30]. Therefore, we hypothesized that minor structural differences among hydroxycinnamic acids (CA, *p*-coumaric acid (pCA), ferulic acid (FA), and sinapic acid (SA)) may also affect PACN yields under accelerated conditions. Analysis with UHPLC-PDA coupled to a tandem mass spectrometer with electrospray ionization (ESI-MS/MS) showed that the identities of peaks 1–4 (deriving from incubation with CA) in Figure 3 were consistent with the ones previously reported in Figure 1. The incubation with the other hydroxycinnamic acids resulted in the formation of new compounds denoted by peaks 5–10 with later retention times and hypsochromic effects on $\lambda_{vis\text{-}max}$ (Figure 3). These characteristics along with the $[M]^+$ and $[M–X]^+$ values were consistent with hydroxyphenyl-PACNs derived from non-acylated ACNs present in sBC. Peaks 5 and 6 were attributed to 10-phenyl-pyranocyanidin-3-*O*-xylosyl-glucosyl-galactoside and 10-phenyl-pyranocyanidin-3-*O*-xylosyl-galactoside, respectively. Similarly, peaks 7 and 8 were recognized as 10-guaiacyl derivatives and peaks 9 and 10 as 10-syringyl derivatives (Figure 3).

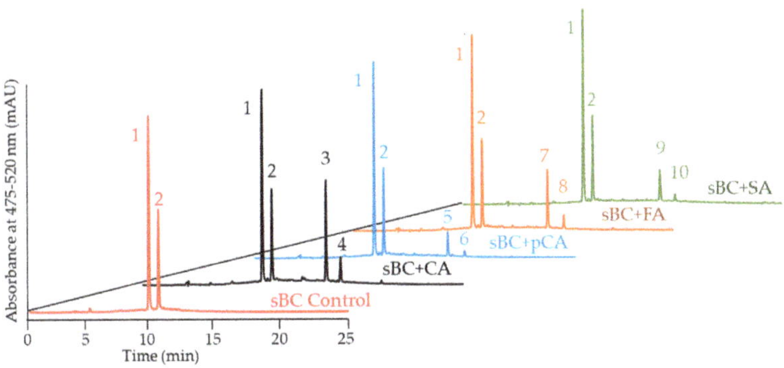

Peak	$\lambda_{vis\text{-}max}$ (nm)	$[M]^+$ (*m/z*)	Aglycone $[M\text{-}X]^+$ (*m/z*)	ID	Peak	$\lambda_{vis\text{-}max}$ (nm)	$[M]^+$ (*m/z*)	Aglycone $[M\text{-}X]^+$ (*m/z*)	ID
1	517	743+	287+	C3XyGlGa	6	508	697+	403+	10-phenyl-pyranoC3XyGa
2	518	581+	287+	C3XyGa	7	503	873+	417+	10-guaiacyl-pyranoC3XyGlGa
3	509	875+	419+	10-catechyl-pyranoC3XyGlGa	8	501	711+	417+	10-guaiacyl-pyranoC3XyGa
4	507	713+	419+	10-catechyl-pyranoC3XyGa	9	513	887+	431+	10-syringyl-pyranoC3XyGlGa
5	509	859+	403+	10-phenyl-pyranoC3XyGlGa	10	512	725+	431+	10-syringyl-pyranoC3XyGa

Figure 3. Comparing the pyranoanthocyanin formation after 72 h of anthocyanin (saponified black carrot, sBC) incubation with caffeic acid (CA), *p*-coumaric acid (pCA), ferulic acid (FA), and sinapic acid (SA). Chromatograms show the max plot in the 475–520 nm range. Table shows the wavelength of maximum absorption in the visible range ($\lambda_{vis\text{-}max}$), mass-per-charge ratio (*m/z*) of the main ion and its aglycone, and tentative identity (ID). C3XyGlGa: cyanidin-3-*O*-xylosyl-glucosyl-galactoside, C3XyGa: cyanidin-3-*O*-xylosyl-galactoside.

Yields calculated using Equation (1) and displayed in Figure 4 showed that after 24 h of incubation, PACN yield with CA (12.5 ± 0.2%) was significantly greater than with FA (8.3 ± 0.2%, $p < 0.01$). Moreover, FA had significantly higher yields than pCA (3.5 ± 0.3%, $p < 0.01$) and SA (4.0 ± 0.2%, $p < 0.01$), with no significant differences among these last two cofactors ($p > 0.05$). After 48 h of incubation, the same efficiency pattern was observed. After 72 h of incubation, CA and FA were significantly more efficient than pCA ($p = 0.0476$ and $p = 0.0246$, respectively). Yields with SA were lower than with CA and FA, but not significantly so ($p > 0.05$ for both CA and FA). When analyzing the percent of pigment remaining, results in Figure 4 showed that after 24 h, incubation with FA (41.2 ± 2.6%) or SA (35.8 ± 3.2%) resulted in a significantly lower pigment remaining ($p = 0.0367$ and $p = 0.3350$, respectively) compared to the cofactor-free control (55.8 ± 1.4%). However, these differences were not significant after 72 h of incubation ($p > 0.05$).

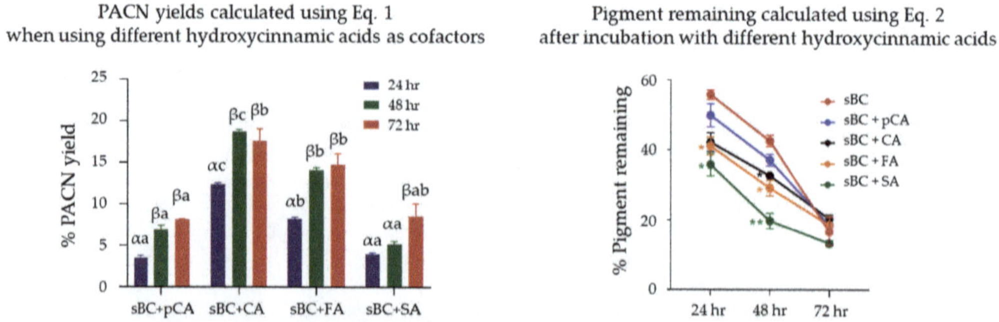

Figure 4. Pyranoanthocyanin yield (%PACN) with different hydroxycinnamic acids and pigment remaining (%) after incubation for 24, 48, and 72 h at 45 °C with cyanidin-glycosides (from saponified black carrot, sBC). Different Greek letters show significant differences among time points at a 0.05 level. Different letters show significant differences among cofactors at the same time point at a 0.05 level. Asterisks (*) and (**) indicate significant differences against the cofactor-free control (sBC) at a 0.05 and 0.01 level, respectively. Results are expressed as means ± standard error ($n = 3$). pCA: *p*-coumaric acid. CA: caffeic acid, FA: ferulic acid, SA: sinapic acid, Eq: equation.

2.3. Evaluation of Pyranoanthocyanin Formation Efficiency Using Different Anthocyanins

Minor structural differences among hydroxycinnamic acids impacted PACN yields. Therefore, we hypothesized that minor structural differences among ACNs could also impact PACN yields. *Berberis boliviana* was chosen as the pigment source for this comparison because it contains simple ACNs deriving from four different aglycones in similar proportions [32]. Results confirmed the presence of delphinidin, cyanidin, petunidin, and malvidin derivatives in the *Berberis boliviana* ACN extract used in this study. In addition, results showed the presence of two glycosylation patterns consistent with glucoside, denoted by a neutral loss of 162 m/z, and rutinoside, denoted by a neutral loss of 308 m/z (Figure 5). After fractionation using semi-preparative HPLC-PDA, four fractions were obtained each with glucoside and rutinoside derivatives of different ACN aglycones. Further UHPLC-PDA-ESI-MS/MS analyses were carried out to identify the individual ACNs in each fraction, and their identities were compared against previous literature. The pigments in the delphinidin-derivatives fraction (Dp) were delphinidin-3-*O*-glucoside (Dp1, ~91%) and delphinidin-3-*O*-rutinoside (Dp2, ~7%), pigments in the cyanidin-derivatives fraction (Cy) were cyanidin-3-*O*-glucoside (Cy1, ~65%) and cyanidin-3-*O*-rutinoside (Cy2, ~27%) with a minor content of petunidin-3-*O*-glucoside (Pt1, ~8%), pigments in the petunidin-derivatives fraction (Pt) were petunidin-3-*O*-glucoside (Pt1, ~73%) and petunidin-3-*O*-rutinoside (Pt2, ~27%), and pigments in the malvidin-derivatives fraction (Mv) were malvidin-3-*O*-glucoside (Mv1, ~77%) and malvidin-3-*O*-rutinoside (Mv2, ~22%).

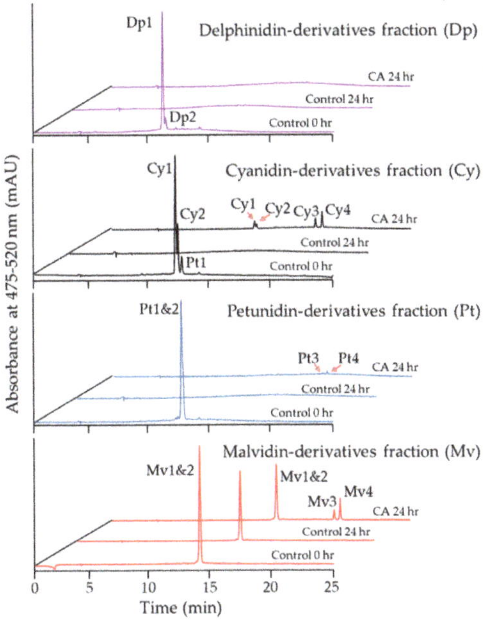

Peak	$\lambda_{vis\text{-}max}$ (nm)	M^+ (m/z)	Aglycone M^+ (m/z)	ID
Dp1	525	465^+	303^+	Delphinidin-3-*O*-glucoside
Dp2	526	611^+	303^+	Delphinidin-3-*O*-rutinoside
Cy1	517	449^+	287^+	Cyanidin-3-*O*-glucoside
Cy2	519	595^+	287^+	Cyanidin-3-*O*-rutinoside
Cy3	514	727^+	419^+	10-catechyl-pyranoCy2
Cy4	511	581^+	419^+	10-catechyl-pyranocCy1
Pt1	526	479^+	317^+	Petunidin-3-*O*-glucoside
Pt2	526	625^+	317^+	Petunidin-3-*O*-rutinoside
Pt3	ND	757^+	449^+	10-catechyl-pyranoPt2
Pt4	ND	611^+	449^+	10-catechyl-pyranoPt1
Mv1	528	493^+	331^+	Malvidin-3-*O*-glucoside
Mv2	528	639^+	331^+	Malvidin-3-*O*-rutinoside
Mv3	511	771^+	463^+	10-catechyl-pyranoMv2
Mv4	514	625^+	463^+	10-catechyl-pyranoMv1

Figure 5. Anthocyanins from fractions with different aglycones and PACN formation after 24 h of 45 °C incubation with caffeic acid (CA). Chromatograms show the max plot in the 475–520 nm range. Table shows the wavelength of maximum absorption in the visible range ($\lambda_{vis\text{-}max}$), mass-per-charge ratio (m/z) of the main ion and its aglycone, and tentative identity (ID).

After 24 h incubation with CA, fractions with cyanidin-, petunidin-, or malvidin-derived ACNs developed peaks with later retention times and hypsochromic $\lambda_{vis\text{-}max}$. UHPLC-PDA-ESI-MS/MS analyses showed that these new peaks were consistent with 10-catechyl derivatives of the ACNs present in each fraction. Indeed, peaks Cy3 and Cy4 were named 10-catechyl-pyranocyanidin derivatives, peaks Pt3 and Pt4 10-catechyl-pyranopetunidin derivatives, and peaks Mv3 and Mv4 10-catechyl-pyranomalvidin derivatives (Figure 5). It is worth noting that incubation of delphinidin-derivatives with CA yielded no PACNs under our experimental conditions.

The highest PACN yield (Figure 6) was observed for malvidin-derivatives followed by cyanidin-derivatives (19.8 ± 1.2% and 9.2 ± 0.8%, respectively, after 24 h of incubation, $p < 0.01$). Furthermore, using malvidin-derivatives not only yielded more PACNs but also resulted in a significantly higher percent of pigment remaining after incubation (64.1 ± 5.2%, $p < 0.01$), followed by cyanidin-derivatives (16.0 ± 2.2%) and then petunidin-derivatives (1.9 ± 0.1%).

When analyzing the relative content of PACNs with a rutinoside substitution as a percentage of total PACNs (Figure 5), this content was 36.1 ± 0.2% in the cyanidin-derivatives fraction, 32.4 ± 1.1% in the petunidin-derivatives fraction, and 31.4 ± 0.4% in the malvidin-derivatives fraction. These values were higher than the ACN-3-*O*-rutinoside proportion in the original ACN extract analyzed in our preliminary analysis (data not shown).

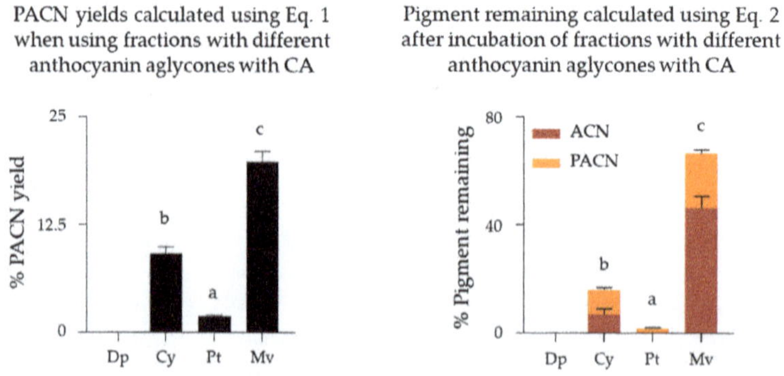

Figure 6. Pyranoanthocyanin yield (%PACN) and pigment remaining (%) after incubation of anthocyanin fractions with different aglycones with caffeic acid (CA) for 24 h at 45 °C. Different letters show significant differences among different fractions at a 0.05 level. Results are expressed as mean ± standard error (n = 3). ACN: anthocyanin, Dp: delphinidin-derivatives fraction, Cy: cyanidin-derivatives fraction, Pt: petunidin-derivatives fractions, Mv: malvidin-derivatives fraction, Eq: equation.

3. Discussion

Under accelerated formation conditions (pH 3.1, 1:30 ACN to cofactor molar ratio, and 45 °C incubation temperature), PACNs were detected in as little as 24 h with yields up to 19.8% using malvidin-derived ACNs and CA as the cofactor. These yields were at least six times higher than those reported for malvidin-3-*O*-glucoside with CA after four months using wine-like systems (pH 3.2, ~1:2 ACN to cofactor molar ratio, and incubation at 15 °C) [31]. Additionally, the PACN yields obtained under our experimental conditions were at least five times higher than those reported for malvidin-3-*O*-glucoside with CA or FA and 20 times higher than those reported with pCA in synthetic grape mediums after 14 days (pH 3.5, ~1:0.8 ACN-to-cofactor molar ratio, and incubation at 30 °C) [33]. The higher yields observed in our study could be attributed to the use of conditions previously reported to accelerate PACN formation [27–29]. However, our results demonstrated that other factors such as the cofactor chemical structure and the ACN type also played a key role in PACN formation efficiency under accelerated formation conditions.

The PACN yields observed in this study using cyanidin-derived ACNs and CA were comparable to previous reports evaluating yields under similar conditions. Straathof and Giusti reported PACN yields of ~20% after 24 h of incubation at 65 °C using elderberry ACNs with CA [29], which was higher than those obtained under our experimental conditions when comparing different hydroxycinnamic acids as cofactors. This higher yield could be attributed to higher incubation temperatures and molar cofactor ratio used (1:50) as well as the different glycosylation patterns of elderberry ACNs. To evaluate potential differences in PACN yields as a result of the use of different cofactors and ACN types, 45 °C instead of 65 °C was selected to reduce pigment degradation as reported by Straathof and Giusti [29] at temperatures of 65 °C or above. In addition, our PACN yields after 72 h of incubation were similar to the ones reported by Zhu and Giusti after seven days of incubation at 25 °C [27]. In that study, incubation for seven additional weeks led to higher PACN yields (~71–85%) that were not obtained under the accelerated conditions used in our experiment.

To compare different cofactors, black carrot was selected as a pigment source because its ACNs do not have glycosylations on position C5 [7], and this allowed for the cycloaddition reaction with a reactive adduct resulting in the formation of a pyran ring between the C4 and 5-OH groups of the ACN molecule [16]. Moreover, the tri- and di-glycosylation patterns at C3 of these ACNs allowed for better solubility of their derived PACNs, preventing their precipitation as observed in our preliminary studies with ACNs from other sources.

As sinapoyl acylating groups have been shown to decrease PACN formation efficiency [27], alkaline hydrolysis was carried out to remove the aromatic acylating moiety.

Among the cofactors tested (CA, acetone, and PA), CA was the only one that produced PACNs under our experimental conditions. Cofactors with an aromatic ring, such as hydroxycinnamic acids, can copigment with ACNs [7,34]. Through intermolecular interactions, the aromatic cofactor is brought in closer proximity to the ACN, which can stabilize the chromophore during the incubation period and increase the likelihood of a reaction [27,28,35]. Copigments must contain a benzene ring for π-π interactions with ACNs to take place [36]. Aliphatic cofactors such as PA and acetone are not expected to copigment with ACNs, which could reduce their interaction with ACNs thus resulting in a lower PACN yield compared to CA [27]. Previous studies have found PACN formation yields 3.5 to 4 times greater with CA as a cofactor than with PA [27,28]. While we did not detect PACNs forming from PA even after 72 h, the longer incubation periods of 42 and 56 days and greater molar ratios of PA (1:100 and 1:200) used in those studies may have compensated for the less efficient cofactor–ACN interaction to facilitate PACN formation [27,28].

CA was the most efficient cofactor even when compared to other hydroxycinnamic acids, yielding the highest amount of PACNs after only 24 h of incubation (CA > FA > SA = pCA). However, after 72 h, no statistical differences were observed in the PACN yield between samples incubated with CA, FA, or SA, although yields with SA were lower and not statistically different than yields with pCA. CA and FA are widely distributed in natural sources. CA is abundant in plants such as aronia berries [37] and black carrots [38], and FA is found in plant cell walls and can be released into a solution by alkaline hydrolysis [39]. Therefore, the saponification process we used to remove the acylating group in the ACN molecule making it more predisposed for PACN formation [27] can, at the same time, release one of the most efficient cofactors into a solution.

The efficiency pattern of hydroxycinnamic acids observed in our results is similar to the one reported in wine-like model solutions [31,40], despite differences in cofactor molar ratios, incubation temperatures, and PACN yields. Additionally, similar reactivity patterns were observed for hydroxycinnamic acids during the formation of portisins [30]. It has been hypothesized that bond formation between C4 of the ACN and the alpha carbon of the hydroxycinnamic acid is affected by the nucleophilic nature of the cofactor [40]. The stronger nucleophilic nature of CA as a result of the two hydroxyl substitutions in the phenolic ring [41] could explain its higher PACN formation efficiency. Additionally, steric hindrance impacts PACN formation efficiency [12,26]; this hindrance may explain the higher PACN yields with FA over SA as cofactors.

When evaluating the impact of the ACN aglycone on PACN formation using CA as a cofactor, malvidin was the most efficient followed by cyanidin (malvidin > cyanidin > petunidin). Malvidin-3-*O*-glucoside has been shown to have stronger copigmentation interactions with CA than other ACN-3-*O*-glucosides [34]. As copigmentation with the cofactor may facilitate PACN formation [27], the increased interactions may contribute to greater PACN formation rates with malvidin-derived ACNs. This stronger copigmentation with CA may also help to explain the significantly higher pigment remaining after incubation for 24 h. Additionally, the higher content of ACNs remaining in the malvidin fraction after 24 h incubation (~45% of initial pigment) may indicate that greater PACN yields may be achieved with longer incubation times. Results also showed very small amounts of PACNs in samples with petunidin-derived pigments (~1.9%) and no PACNs detected in samples with delphinidin-derivatives. Furthermore, after incubation at 45 °C for 24 h, there was little to no pigment remaining in these samples; this negates the possibility of PACN formation with increased incubation times. Delphinidin's and petunidin's propensity to degrade during prolonged heating [42,43] may explain the absence or reduced formation of PACNs in these samples. The significantly higher yields of malvidin- and cyanidin-derived PACNs may explain why most reports have especially identified these derivatives [11,14,16,19,25,44,45].

The ACN fractions used to compare the impact of the aglycone type contained two glycosylation patterns (glucoside and rutinoside). Glycosylation patterns have been reported to impact PACN yields with cyanidin-3-*O*-rutinoside yielding more PACN than cyanidin-3-*O*-glucoside when incubated with PA at 25 °C [12]. This observation was consistent with our results; indeed ACN-3-*O*-rutinosides seemed to form PACNs more efficiently than their glucoside counterparts denoted by an increased content of rutinoside-derivatives in the PACN fraction after incubation with CA (from ~27% to ~37% in the Cy fraction, from ~27% to ~33% in the Pt fraction, and from ~22% to ~32% in the Mv fraction). Despite having similar glucoside to rutinoside proportions, PACN yields with CA were significantly different between the Cy, Pt, and Mv fractions. This difference suggests that the impact of the glycosylation may be secondary to the impact of the aglycone type on PACN yield, possibly due to the aglycone influence on stability and copigmentation ability.

4. Materials and Methods

4.1. Plant Material, Chemical, and Reagents

Berberis boliviana freeze-dried berries were generously donated by Carla del Carpio from Universidad Nacional de San Antonio Abad del Cusco (UNSAAC, Cusco, Peru). A commercial black carrot color powder (*Daucus carota*) was provided by D.D. Williamson (Louisville, KY, USA). CA, FA, SA, and PA were purchased from Sigma-Aldrich (St. Louis, MO, USA). Acetone, chloroform, acetonitrile, UHPLC-grade water, pCA, sodium benzoate, and sodium hydroxide were obtained from Fisher Scientific (Pittsburgh, PA, USA). Potassium sorbate was obtained from Spectrum (New Brunswick, NJ, USA). All other reagents and solvents were of at least analytical or HPLC grade unless otherwise indicated.

4.2. Anthocyanin Preparation

ACN extraction was carried out following the methodology described by Rodriguez-Saona and Wrolstad [46] with minor modifications. Black carrot color powders were prepared for extraction by reconstituting them in water. *Berberis boliviana* berries were deseeded manually, soaked in water, and frozen with liquid nitrogen before extraction. Then, the frozen berries and reconstituted black carrot color powder were each blended with 0.01% HCl acetone (v/v) using a Waring laboratory blender. The slurry was filtered through an N° 4 Whatman filter paper, and the cake was re-extracted using 0.01% HCl aqueous acetone (70% v/v) until a faint pink solution was obtained. The filtrate was then partitioned using chloroform for a final acetone:chloroform proportion of 1:2 (v/v) in a separatory funnel and stored overnight at 4 °C. The upper aqueous layer containing ACNs and other phenolic compounds was collected, and residual solvents were evaporated using a Büchi rotavapor at 45 °C (Büchi, Flawil, Switzerland).

4.3. Saponification of Black Carrot Anthocyanins

Alkaline hydrolysis of acylated ACNs from black carrot was carried out to remove the acylation moiety in their structure following the methodology described by Giusti and Wrolstad [47]. Briefly, solutions rich in black carrot ACNs were mixed with 10% KOH (w/v) in a proportion of 1:10 and left to stand for 8 min in the dark at room temperature, after which the pH was neutralized using 2N HCl. Neutralized, saponified extracts were subjected to solid phase extraction (SPE) following the methodology described in Section 4.4. The obtained semi-purified, sBC was later used for comparison of cofactor efficiency.

4.4. Pigment Semi-Purification

ACN semi-purification was performed using SPE following the methodology described by Rodriguez-Saona and Wrolstad [46] with minor modifications. Briefly, crude ACN extracts were diluted in acidified water prior to SPE using Waters Sep-pak C18 cartridges (Waters, Milford, MA, USA). Cartridges were activated with methanol and then washed with 0.01% HCl (v/v) acidified water, after which the crude extract was applied to the cartridge. Salts, sugars, and organic acids were removed using two volumes of acidified water followed by washing with three to four volumes of ethyl acetate to remove less polar phenolics. Semi-purified pigments were eluted from the column using 0.01% HCl (*v/v*) acidified methanol. The solvent was then evaporated using a rotavapor at 45 °C, and pigments were resolubilized in acidified water and stored under refrigeration until further use.

4.5. Fractionation of Different Anthocyanins

Fractions of different ACNs from semi-purified *Berberis boliviana* extract prepared in Section 4.4 were obtained using semi-preparative HPLC-PDA (Shimadzu, Columbia, MD, USA). Reverse phase chromatographic separation was achieved using a Synergi 4 μm Max-RP 80 Å column of 250 × 21.2 mm dimensions (Phenomenex, Torrance, CA, USA) and a binary solvent system composed of A: 4.5% formic acid in water (*v/v*) and B: acetonitrile at a flow rate of 10 mL/min. Elution gradient started at 10–20% B from 0 to 20 min, 20–40% B from 20 to 21 min, and 40% B from 21 to 24 min. Collected fractions were concentrated using SPE to remove formic acid and acetonitrile, and residual methanol was evaporated using a rotavapor at 45 °C. Isolated fractions were resolubilized with acidified water and stored under refrigeration until further use.

4.6. Anthocyanin and Pyranoanthoycanin Identification

ACN and PACN tentative identification were carried out using a Nexera-i-LC-2040 3D ultra-high performance liquid chromatograph (Shimadzu, Columbia, MD, USA) coupled with an LCMS-8040 triple quadrupole mass spectrometer with an ESI interface (Shimadzu, Columbia, MD, USA). Chromatographic separation was achieved using a Synergi 4 μm Max RP-80 Å 250 × 4.6 mm column (Phenomenex, Torrance, CA, USA) and a binary solvent system consisting of A: 4.5% formic acid and B: acetonitrile with a gradient of 5–45% B from 0 to 20 min and 45% B from 20 to 25 min. After chromatographic separation, a volume of approximately 0.2 mL/min was diverted into the MS/MS for analyses. Tentative identification of pigments was carried out based on their elution time, UV-Vis spectral characteristics, and corresponding m/z using ESI-MS/MS. Mass spectrometry analyses were conducted under positive ion mode with 1.5 L/min nebulizing gas flow, 15 L/min drying gas, a desolvation gas temperature of 230 °C, and collision energy of -35 eV. Spectral data were acquired using total ion scan mode from m/z 100 to 1000 and precursor ion scan mode for the most common six anthocyanidins in nature (271, 287, 301, 303, 317, and 331 m/z). Identification of new colored compounds was carried out using precursor ion scans with the expected PACN aglycone m/z listed in the tables in Figures 1, 3 and 5 for each experimental section. Data analysis and interpretation were performed using Lab Solutions Software Ver. 1 (Shimadzu, Columbia, MD, USA).

4.7. Monomeric Anthocyanin Quantitation

Monomeric ACNs were quantified using the pH differential method [48]. Briefly, semi-purified ACN extracts or isolated ACN fractions were diluted in 0.025 M potassium chloride buffer at pH 1 or 0.4 M sodium acetate buffer at pH 4.5. The difference of absorbance at their respective $\lambda_{vis\text{-}max}$ and 700 nm was measured using a SpectraMax M2 plate reader (Molecular Devices, Sunnyvale, CA, USA). The monomeric ACN content in sBC was expressed as cyanidin-3-*O*-glucoside equivalents while the content in each ACN fraction from *Berberis boliviana* was expressed in equivalents of its most abundant ACN (delphinidin-3-*O*-glucoside, cyanidin-3-*O*-glucoside, petunidin-3-*O*-glucoside, or malvidin-3-*O*-glucoside) using the corresponding molecular weight and molar absorptivities reported in the literature [48,49].

4.8. Pyranoanthocyanin Formation—Comparing Different Types of Cofactors

PACN formation was carried out following the methodology described by Straathof and Giusti [29] with minor modifications. Briefly, stock solutions of sBC and three different types of cofactors (PA, acetone, and CA) were prepared using pH 3.1 acidified water containing 0.1% potassium sorbate (w/v) and 0.1% sodium benzoate (w/v). Stock solutions of pigment and cofactor were mixed and diluted to a pigment concentration of 80 μM and a cofactor concentration of 2.4 mM (1:30 pigment to cofactor molar ratio). The final solution was adjusted to pH 3.1 ± 0.05 with 1 M NaOH and 2N HCl when needed. Samples were placed into HPLC vials and stored in the dark in a Roto-Therm™ H2020 benchtop incubator (Benchmark Scientific, Edison, NJ, USA) set at 45 °C. Every 24 h, the vial was removed, run on the HPLC, and returned to the incubator for additional incubation, up to 72 h total.

4.9. Pyranoanthocyanin Formation—Comparing Different Hydroxycinnamic Acids

The effects of minor structural differences among different hydroxycinnamic acids on PACN formation were evaluated following the methodology described in Section 4.8. The sBC extract was used as the pigment source, and four different hydroxycinnamic acids (CA, pCA, FA, and SA) were tested as cofactors at equivalent 1:30 ACN to cofactor molar ratios and incubated at 45 °C in pH 3.1 solution in the dark.

4.10. Pyranoanthocyanin Formation—Comparing Different Aglycones

The effect of the aglycone structure on the formation efficiency of PACNs was evaluated using ACN-rich fractions containing a different aglycone and the cofactor identified as producing the highest PACN yield (in this case, CA) following the methodology described in Section 4.8. *Berberis boliviana* was selected as the pigment source because it contains four of the six most common ACNs found in nature in relatively similar proportions with glycosylations (either glucose or rutinose) only on position C3 [32]. CA was used as a cofactor at a 1:30 ACN to cofactor molar ratio, and samples were incubated at 45 °C in pH 3.1 solution in the dark.

4.11. Monitoring Pyranoanthocyanin Formation and Anthocyanin Changes Over Time

Total pigment content and PACN formation were monitored during incubation using HPLC-PDA (Shimadzu, Columbia, MD, USA) and calculated using the areas under the curve (AUC) of the compounds of interest. The system consisted of two LC-20AD pumps, a CBM-20A controller, a SIL-20AC refrigerated autosampler, an SPD-M20A PDA detector, and a CTO-20A column oven. Chromatographic separation was achieved using the same column, solvents, and gradient parameters as in Section 4.6.

PACN formation yield was calculated using Equation (1):

$$PACN\ formation\ yield\ (\%) = \frac{(AUC_{475-520\ nm}\ PACNs\ at\ t_n)}{(AUC_{475-520\ nm}\ ACNs\ at\ t_0)} * 100 \quad (1)$$

The percent of pigment remaining was calculated using Equation (2):

$$Percent\ of\ Pigment\ Remaining = \frac{(AUC_{475-520\ nm}\ ACNs + PACNs\ at\ t_n)}{(AUC_{475-520\ nm}\ ACNs\ at\ t_0)} * 100 \quad (2)$$

The content of PACNs, as the percentage of total pigment at a given time point, was calculated using Equation (3):

$$Percent\ of\ PACN\ to\ total\ pigment\ (\%) = \frac{(AUC_{475-520\ \mathrm{nm}}\ PACNs\ at\ t_n)}{(AUC_{475-520\ \mathrm{nm}}\ ACNs + PACNs\ at\ t_n)} * 100 \quad (3)$$

4.12. Statistical Evaluation of Data

Data from each replication were collected from duplicate samples and experiments were conducted in triplicate. Data were expressed as mean ± standard error of means. Differences among treatments were analyzed using a one-way analysis of variance (ANOVA) with Bonferroni post-hoc tests. Analyses were conducted using GraphPad Prism (GraphPad Software Inc., La Jolla, CA, USA). A p-value lower than 0.05 was considered significant.

5. Conclusions

The chemical structure of the cofactor and ACN type impacted PACN yields with significant differences detected after only 24 h of incubation. CA and FA were the most efficient cofactors with similar PACN yields when using cyanidin-glycosides from sBC as the ACN source (~19% and ~14%, respectively). Among the different types of aglycones tested, malvidin-glycosides were the most efficient for PACN formation with CA (~20% after 24 h). Overall, this research demonstrated that cyanidin- or malvidin-derived ACNs in combination with CA or FA can produce high amounts of PACNs under accelerated formation conditions. These results highlight the importance of the ACN source and cofactor selection for the efficient production of PACNs. Given their scarcity in nature, this efficient production could facilitate their use by the industry as naturally derived colorants with increased stability.

Author Contributions: Conceptualization, G.M.-C. and M.M.G.; methodology, G.M.-C. and D.M.V.; validation, G.M.-C. and D.M.V.; formal analysis, G.M.-C. and D.M.V.; investigation, G.M.-C. and D.M.V.; resources, M.M.G.; data curation, G.M.-C. and D.M.V.; writing—original draft preparation, G.M.-C. and D.M.V.; writing—review and editing, G.M.-C., D.M.V. and M.M.G.; visualization, G.M.-C., D.M.V. and M.M.G.; supervision, M.M.G.; project administration, G.M.-C. and M.M.G.; funding acquisition, M.M.G. All authors have read and agreed to the published version of the manuscript.

Funding: This research was supported in part by FONDECYT-CONCYTEC, grant number 225-2015-FONDECYT and the USDA National Institute of Food and Agriculture, Hatch Project OHO01423, Accession Number 1014136.

Institutional Review Board Statement: Not applicable.

Informed Consent Statement: Not applicable.

Data Availability Statement: The data that supports the findings of this study are available from the corresponding author upon reasonable request.

Acknowledgments: The authors would like to thank Yesen Cheng for his technical assistance in this project. The authors would also like to thank Molly J. Davis for editing the manuscript.

Conflicts of Interest: The authors declare no conflict of interest.

References

1. Mazza, G.; Brouillard, R. Recent Developments in the Stabilization of Anthocyanins in Food Products. *Food Chem.* **1987**, *25*, 207–225. [CrossRef]
2. Sigurdson, G.T.; Tang, P.; Giusti, M.M. Natural Colorants: Food Colorants from Natural Sources. *Annu. Rev. Food Sci. Technol.* **2017**, *8*, 261–280. [CrossRef] [PubMed]
3. Bateman, B.; Warner, J.O.; Hutchinson, E.; Dean, T.; Rowlandson, P.; Gant, C.; Grundy, J.; Fitzgerald, C.; Stevenson, J. The Effects of a Double Blind, Placebo Controlled, Artificial Food Colourings and Benzoate Preservative Challenge on Hyperactivity in a General Population Sample of Preschool Children. *Arch. Dis. Child.* **2004**, *89*, 506–511. [CrossRef] [PubMed]
4. McCann, D.; Barrett, A.; Cooper, A.; Crumpler, D.; Dalen, L.; Grimshaw, K.; Kitchin, E.; Lok, K.; Porteous, L.; Prince, E.; et al. Food Additives and Hyperactive Behaviour in 3-Year-Old and 8/9-Year-Old Children in the Community: A Randomised, Double-Blinded, Placebo-Controlled Trial. *Lancet* **2007**, *370*, 1560–1567. [CrossRef]
5. He, J.; Monica Giusti, M. Anthocyanins: Natural Colorants with Health-Promoting Properties. *Annu. Rev. Food Sci. Technol.* **2010**, *1*, 163–187. [CrossRef]
6. Farr, J.E.; Giusti, M.M. Investigating the Interaction of Ascorbic Acid with Anthocyanins and Pyranoanthocyanins. *Molecules* **2018**, *23*, 744. [CrossRef] [PubMed]
7. Pangestu, N.P.; Miyagusuku-Cruzado, G.; Giusti, M.M. Copigmentation with Chlorogenic and Ferulic Acid Affected Color and Anthocyanin Stability in Model Beverages Colored with Sambucus Peruviana, Sambucus Nigra and Daucus Carota during Storage. *Foods* **2020**, *9*, 1476. [CrossRef] [PubMed]
8. Miyagusuku-Cruzado, G.; Jiménez-Flores, R.; Giusti, M.M. Whey Protein Addition and Its Increased Light Absorption and Tinctorial Strength of Model Solutions Colored with Anthocyanins. *J. Dairy Sci.* **2020**, *104*, 6449–6462. [CrossRef]
9. Weber, F.; Boch, K.; Schieber, A. Influence of Copigmentation on the Stability of Spray Dried Anthocyanins from Blackberry. *LWT Food Sci. Technol.* **2017**, *75*, 72–77. [CrossRef]
10. Tang, P.; Giusti, M.M. Metal Chelates of Petunidin Derivatives Exhibit Enhanced Color and Stability. *Foods* **2020**, *9*, 1426. [CrossRef]

11. Sun, J.; Li, X.; Luo, H.; Ding, L.; Jiang, X.; Li, X.; Jiao, R.; Bai, W. Comparative Study on the Stability and Antioxidant Activity of Six Pyranoanthocyanins Based on Malvidin-3-Glucoside. *J. Agric. Food Chem.* **2020**, *68*, 2783–2794. [CrossRef]
12. Farr, J.E.; Sigurdson, G.T.; Giusti, M.M. Influence of Cyanidin Glycosylation Patterns on Carboxypyranoanthocyanin Formation. *Food Chem.* **2018**, *259*, 261–269. [CrossRef] [PubMed]
13. Velenosi, M.; Crupi, P.; Perniola, R.; Marsico, A.D.; Salerno, A.; Alexandre, H.; Archidiacono, N.; Ventura, M.; Cardone, M.F. Color Stabilization of Apulian Red Wines through the Sequential Inoculation of Starmerella Bacillaris and Saccharomyces Cerevisiae. *Molecules* **2021**, *26*, 907. [CrossRef] [PubMed]
14. Marquez, A.; Serratosa, M.P.; Merida, J. Pyranoanthocyanin Derived Pigments in Wine: Structure and Formation during Winemaking. *J. Chem.* **2013**, *2013*, 713028. [CrossRef]
15. Bakker, J.; Timberlake, C.F. Isolation, Identification, and Characterization of New Color-Stable Anthocyanins Occurring in Some Red Wines. *J. Agric. Food Chem.* **1997**, *45*, 35–43. [CrossRef]
16. Mateus, N.; Silva, A.M.S.; Vercauteren, J.; de Freitas, V. Occurrence of Anthocyanin-Derived Pigments in Red Wines. *J. Agric. Food Chem.* **2001**, *49*, 4836–4840. [CrossRef]
17. He, J.; Carvalho, A.R.F.; Mateus, N.; de Freitas, V. Spectral Features and Stability of Oligomeric Pyranoanthocyanin-Flavanol Pigments Isolated from Red Wines. *J. Agric. Food Chem.* **2010**, *58*, 9249–9258. [CrossRef] [PubMed]
18. de Freitas, V.; Mateus, N. Formation of Pyranoanthocyanins in Red Wines: A New and Diverse Class of Anthocyanin Derivatives. *Anal. Bioanal. Chem.* **2011**, *401*, 1467–1477. [CrossRef]
19. Schwarz, M.; Wray, V.; Winterhalter, P. Isolation and Identification of Novel Pyranoanthocyanins from Black Carrot (*Daucus Carota* L.) Juice. *J. Agric. Food Chem.* **2004**, *52*, 5095–5101. [CrossRef]
20. Blanco-Vega, D.; López-Bellido, F.J.; Alía-Robledo, J.M.; Hermosín-Gutiérrez, I. HPLC-DAD-ESI-MS/MS Characterization of Pyranoanthocyanins Pigments Formed in Model Wine. *J. Agric. Food Chem.* **2011**, *59*, 9523–9531. [CrossRef]
21. Fossen, T.; Andersen, Ø.M. Anthocyanins from Red Onion, Allium Cepa, with Novel Aglycone. *Phytochemistry* **2003**, *62*, 1217–1220. [CrossRef]
22. Andersen, Ø.M.; Fossen, T.; Torskangerpoll, K.; Fossen, A.; Hauge, U. Anthocyanin from Strawberry (*Fragaria ananassa*) with the Novel Aglycone, 5-Carboxypyranopelargonidin. *Phytochemistry* **2004**, *65*, 405–410. [CrossRef]
23. Peng, Y.; Zhang, H.; Liu, R.; Mine, Y.; McCallum, J.; Kirby, C.; Tsao, R. Antioxidant and Anti-Inflammatory Activities of Pyranoanthocyanins and Other Polyphenols from Staghorn Sumac (*Rhus Hirta* L.) in Caco-2 Cell Models. *J. Funct. Foods* **2016**, *20*, 139–147. [CrossRef]
24. Liu, S.; Laaksonen, O.; Yang, W.; Zhang, B.; Yang, B. Pyranoanthocyanins in Bilberry (*Vaccinium Myrtillus* L.) Wines Fermented with Schizosaccharomyces Pombe and Their Evolution during Aging. *Food Chem.* **2020**, *305*, 125438. [CrossRef]
25. Rein, M.J.; Ollilainen, V.; Vahermo, M.; Yli-Kauhaluoma, J.; Heinonen, M. Identification of Novel Pyranoanthocyanins in Berry Juices. *Eur. Food Res. Technol.* **2005**, *220*, 239–244. [CrossRef]
26. Mateus, N.; Oliveira, J.; Haettich-Motta, M.; de Freitas, V. New Family of Bluish Pyranoanthocyanins. *J. Biomed. Biotechnol.* **2004**, *2004*, 299–305. [CrossRef] [PubMed]
27. Zhu, X.; Giusti, M.M. Pyranoanthocyanin Formation Rates and Yields as Affected by Cyanidin-3-Substitutions and Pyruvic or Caffeic Acids. *Food Chem.* **2021**, *345*, 128776. [CrossRef] [PubMed]
28. Hoehn, M.E. Altering pH, Temperature and Cofactors to Increase the Formation of the More Stable Anthocyanin Derived Pyranoanthocyanin. Master's Thesis, The Ohio State University, Columbus, OH, USA, 2019.
29. Straathof, N.; Giusti, M.M. Improvement of Naturally Derived Food Colorant Performance with Efficient Pyranoanthocyanin Formation from Sambucus Nigra Anthocyanins Using Caffeic Acid and Heat. *Molecules* **2020**, *25*, 5998. [CrossRef]
30. Oliveira, J.; de Freitas, V.; Silva, A.M.S.; Mateus, N. Reaction between Hydroxycinnamic Acids and Anthocyanin-Pyruvic Acid Adducts Yielding New Portisins. *J. Agric. Food Chem.* **2007**, *55*, 6349–6356. [CrossRef] [PubMed]
31. Schwarz, M.; Wabnitz, T.C.; Winterhalter, P. Pathway Leading to the Formation of Anthocyanin-Vinylphenol Adducts and Related Pigments in Red Wines. *J. Agric. Food Chem.* **2003**, *51*, 3682–3687. [CrossRef] [PubMed]
32. del Carpio Jiménez, C.; Serrano Flores, C.; He, J.; Tian, Q.; Schwartz, S.J.; Giusti, M.M. Characterisation and Preliminary Bioactivity Determination of Berberis Boliviana Lechler Fruit Anthocyanins. *Food Chem.* **2011**, *128*, 717–724. [CrossRef]
33. Topić Božič, J.; Ćurko, N.; Kovačević Ganić, K.; Butinar, L.; Albreht, A.; Vovk, I.; Korte, D.; Mozetič Vodopivec, B. Synthesis of Pyranoanthocyanins from Pinot Noir Grape Skin Extract Using Fermentation with High Pyranoanthocyanin Producing Yeasts and Model Wine Storage as Potential Approaches in the Production of Stable Natural Food Colorants. *Eur. Food Res. Technol.* **2020**, *246*, 1141–1152. [CrossRef]
34. Eiro, M.J.; Heinonen, M. Anthocyanin Color Behavior and Stability during Storage: Effect of Intermolecular Copigmentation. *J. Agric. Food Chem.* **2002**, *50*, 7461–7466. [CrossRef] [PubMed]
35. Fan, L.; Wang, Y.; Xie, P.; Zhang, L.; Li, Y.; Zhou, J. Copigmentation Effects of Phenolics on Color Enhancement and Stability of Blackberry Wine Residue Anthocyanins: Chromaticity, Kinetics and Structural Simulation. *Food Chem.* **2019**, *275*, 299–308. [CrossRef] [PubMed]
36. Boulton, R. The Copigmentation of Anthocyanins and Its Role in the Color of Red Wine: A Critical Review. *Am. J. Enol. Vitic.* **2001**, *52*, 67–87.
37. Zheng, W.; Wang, S.Y. Oxygen Radical Absorbing Capacity of Phenolics in Blueberries, Cranberries, Chokeberries, and Lingonberries. *J. Agric. Food Chem.* **2003**, *51*, 502–509. [CrossRef]

38. Frond, A.D.; Iuhas, C.I.; Stirbu, I.; Leopold, L.; Socaci, S.; Andreea, S.; Ayvaz, H.; Andreea, S.; Mihai, S.; Diaconeasa, Z.; et al. Phytochemical Characterization of Five Edible Purple-Reddish Vegetables: Anthocyanins, Flavonoids, and Phenolic Acid Derivatives. *Molecules* **2019**, *24*, 1536. [CrossRef] [PubMed]
39. Torre, P.; Aliakbarian, B.; Rivas, B.; Domínguez, J.M.; Converti, A. Release of Ferulic Acid from Corn Cobs by Alkaline Hydrolysis. *Biochem. Eng. J.* **2008**, *40*, 500–506. [CrossRef]
40. Schwarz, M.; Winterhalter, P. A Novel Synthetic Route to Substituted Pyranoanthocyanins with Unique Colour Properties. *Tetrahedron Lett.* **2003**, *44*, 7583–7587. [CrossRef]
41. Gaspar, A.; Garrido, E.M.; Esteves, M.; Quezada, E.; Milhazes, N.; Garrido, J.; Borges, F. New Insights into the Antioxidant Activity of Hydroxycinnamic Acids: Synthesis and Physicochemical Characterization of Novel Halogenated Derivatives. *Eur. J. Med. Chem.* **2009**, *44*, 2092–2099. [CrossRef]
42. Song, B.J.; Sapper, T.N.; Burtch, C.E.; Brimmer, K.; Goldschmidt, M.; Ferruzzi, M.G. Photo- and Thermodegradation of Anthocyanins from Grape and Purple Sweet Potato in Model Beverage Systems. *J. Agric. Food Chem.* **2013**, *61*, 1364–1372. [CrossRef]
43. Wang, F.; Li, H.; Qin, Y.; Mao, Y.; Zhang, B.; Deng, Z. Effects of Heat, Ultrasound, and Microwave Processing on the Stability and Antioxidant Activity of Delphinidin and Petunidin. *J. Food Biochem.* **2019**, *43*, e12818. [CrossRef] [PubMed]
44. Benito, S.; Morata, A.; Palomero, F.; González, M.C.; Suárez-Lepe, J.A. Formation of Vinylphenolic Pyranoanthocyanins by Saccharomyces Cerevisiae and Pichia Guillermondii in Red Wines Produced Following Different Fermentation Strategies. *Food Chem.* **2011**, *124*, 15–23. [CrossRef]
45. Hillebrand, S.; Schwarz, M.; Winterhalter, P. Characterization of Anthocyanins and Pyranoanthocyanins from Blood Orange [*Citrus Sinensis* (L.) Osbeck] Juice. *J. Agric. Food Chem.* **2004**, *52*, 7331–7338. [CrossRef] [PubMed]
46. Rodríguez-Saona, L.E.; Wrolstad, R.E. Extraction, Isolation, and Purifification of Anthocyanins. *Curr. Protoc. Food Anal. Chem.* **2001**, F1.1.1–F1.1.11. [CrossRef]
47. Giusti, M.M.; Wrolstad, R.E. Characterization of Red Radish Anthocyanins. *J. Food Sci.* **1996**, *61*, 322–326. [CrossRef]
48. Giusti, M.M.; Wrolstad, R.E. Characterization and Measurement of Anthocyanins by UV-Visible Spectroscopy. *Curr. Protoc. Food Anal. Chem.* **2001**, 1–13. [CrossRef]
49. Cabrita, L.; Fossen, T.; Andersen, Ø.M. Colour and Stability of the Six Common Anthocyanidin 3-Glucosides in Aqueous Solutions. *Food Chem.* **2000**, *68*, 101–107. [CrossRef]

International Journal of
Molecular Sciences

MDPI

Article

Glycyrrhizin Attenuates Portal Hypertension and Collateral Shunting via Inhibition of Extrahepatic Angiogenesis in Cirrhotic Rats

Chon Kit Pun [1,2,3,†], Hui-Chun Huang [1,2,3,4,†], Ching-Chih Chang [1,2,4], Chiao-Lin Chuang [1,2,4], Chun-Hsien Yen [3], Shao-Jung Hsu [1,2,3,*], Fa-Yauh Lee [1,2,3], Ming-Chih Hou [1,2,3] and Yi-Hsiang Huang [1,2,3]

1 Faculty of Medicine, National Yang-Ming University School of Medicine, Taipei 11217, Taiwan; ckpan2@vghtpe.gov.tw (C.K.P.); hchuang2@vghtpe.gov.tw (H.-C.H.); ccchang7@vghtpe.gov.tw (C.-C.C.); clchuang@vghtpe.gov.tw (C.-L.C.); fylee@vghtpe.gov.tw (F.-Y.L.); mchou@vghtpe.gov.tw (M.-C.H.); yhhuang@vghtpe.gov.tw (Y.-H.H.)
2 Faculty of Medicine, National Yang Ming Chiao Tung University, Taipei 11217, Taiwan
3 Division of Gastroenterology and Hepatology, Department of Medicine, Taipei Veterans General Hospital, Taipei 11217, Taiwan; sam1022yen86@gmail.com
4 Division of General Medicine, Department of Medicine, Taipei Veterans General Hospital, Taipei 11217, Taiwan
* Correspondence: sjhsu@vghtpe.gov.tw; Tel.: +886-2-28712121 (ext. 2065)
† Chon Kit Pun and Hui-Chun Huang contributed equally to this study.

Abstract: Portal hypertension develops along with liver cirrhosis then induces the formation of portal-systemic collaterals and lethal complications. Extrahepatic angiogenesis plays an important role. Glycyrrhizin has been found to exhibit anti-angiogenic features, which leads to its extensive use. However, the relevant effects of glycyrrhizin on liver cirrhosis and portal hypertension have not been evaluated. This study thus aimed to investigate the impact of glycyrrhizin on portal hypertension-related derangements in cirrhotic rats. Male Sprague-Dawley rats received bile duct ligation (BDL) to induce cirrhosis or sham operation as control. The rats were subdivided to receive glycyrrhizin (150 mg/kg/day, oral gavage) or vehicle beginning on the 15th day post operation, when BDL-induced liver fibrosis developed. The effects of glycyrrhizin were determined on the 28th day, the typical timing of BDL-induced cirrhosis. Glycyrrhizin significantly reduced portal pressure ($p = 0.004$). The splanchnic inflow as measured by superior mesenteric arterial flow decreased by 22% ($p = 0.029$). The portal-systemic collateral shunting degree reduced by 30% ($p = 0.024$). The mesenteric angiogenesis and phospho-VEGFR2 protein expression were also downregulated ($p = 0.038$ and 0.031, respectively). Glycyrrhizin did not significantly influence the liver biochemistry data. Although glycyrrhizin tended to reverse liver fibrosis, statistical significance was not reached ($p = 0.069$). Consistently, hepatic inflow from portal side, hepatic vascular resistance, and liver fibrosis-related protein expressions were not affected. Glycyrrhizin treatment at the stage of hepatic fibrosis still effectively attenuated portal hypertension and portosystemic collateral shunting. These beneficial effects were attributed to, at least in part, the suppression of mesenteric angiogenesis by VEGF signaling pathway downregulation.

Keywords: liver cirrhosis; portal hypertension; angiogenesis; portosystemic collateral shunting; glycyrrhizin

Citation: Pun, C.K.; Huang, H.-C.; Chang, C.-C.; Chuang, C.-L.; Yen, C.-H.; Hsu, S.-J.; Lee, F.-Y.; Hou, M.-C.; Huang, Y.-H. Glycyrrhizin Attenuates Portal Hypertension and Collateral Shunting via Inhibition of Extrahepatic Angiogenesis in Cirrhotic Rats. *Int. J. Mol. Sci.* **2021**, *22*, 7662. https://doi.org/10.3390/ijms22147662

Academic Editor: Maurizio Battino

Received: 15 April 2021
Accepted: 15 July 2021
Published: 17 July 2021

1. Introduction

Portal hypertension, a phenomenon attributed to both systematic and portal hemodynamic derangements, develops along with the progression of liver cirrhosis. These hemodynamic changes include increased splanchnic blood flow, portal inflow, and enhanced hepatic vascular resistance, which lead to excess blood retention in the portal system [1]. To drain the stagnant blood out from the portal system to systemic circulation, portal-systemic collateral vasculature gradually forms. Nevertheless, portal-systemic collateral in itself poses threats such as gastroesophageal variceal hemorrhage and hepatic encephalopathy. Traditionally, the excess portal inflow was mainly due to overt splanchnic vasodilatation. Recently, evidence suggested that angiogenesis, the generation of new blood vessels, participates in the formation of portal-systemic collaterals, aggravation of splanchnic hyperemia, and increase of portal inflow [2]. During the process, the actions of vascular endothelium growth factor (VEGF) and activation of its receptor, VEGF receptor 2 (VEGFR-2) are considered the major factors [3]. In brief, attenuation of angiogenesis is a reasonable strategy to ameliorate portal hypertension and its complications.

Glycyrrhizin, the major bioactive compound of licorice roots extraction, has been widely used as an herbal medicine for anti-tumor, anti-inflammation, and anti-virus therapy in Asia [4]. A clinical trial further demonstrated that licorice exposure increased large arterial stiffness and systemic vascular resistance [5]. Glycyrrhizin has also been proved to have a hepatic protection effect. Glycyrrhizin and its metabolite, glycyrrhetinic acid, ameliorated bile-induced hepatotoxicity in rats through inhibiting apoptosis and necrosis of hepatocytes [6]. In rat models with hepatic injury, glycyrrhizin reduced plasma levels of aspartate aminotransferase (AST) and alanine aminotransferase (ALT), indicators of liver injury, as compared with those of the vehicle control group [7,8]. Moreover, a randomized control trial showed that glycyrrhizin plus tenofovir significantly reduced serum AST and ALT levels and decreased MELD score compared with tenofovir alone in patients with chronic hepatitis B with severe acute exacerbation [9].

Interestingly, a recent study found that glycyrrhizin effectively suppresses angiogenesis. Glycyrrhizin inhibited tumor growth and angiogenesis in vivo, and attenuated migration, invasion, and tube formation of endothelial cells [10]. Another study showed that glycyrrhizin suppressed angiogenesis activity of endothelial cells. Furthermore, it inhibited tumor growth and neovascularization in mice [11]. Angiogenesis plays an important role in portal hypertension. Furthermore, glycyrrhizin also induced hepatic vascular relaxation in rats with CCL_4-induced liver cirrhosis [12]. However, the effects of glycyrrhizin on chronic liver cirrhosis and portal hypertension-related derangements have not been surveyed.

We therefore hypothesized that glycyrrhizin may exert beneficial effects on portal hypertension. Via a rat model with liver cirrhosis and portal hypertension, glycyrrhizin was administered at the stage of liver fibrosis, which is more relevant to the clinical condition.

2. Results

2.1. Effects of Glycyrrhizin on Body Weight and Systemic Circulation

The rats received common bile duct ligation (BDL) to induce cirrhosis or sham operation as surgical control. The rats were then subdivided to receive glycyrrhizin or vehicle treatment beginning on the 15th day post operation, when BDL-induced liver fibrosis developed. Experiments were performed on the 28th day, when cirrhosis developed in the BDL group (Figure 1).

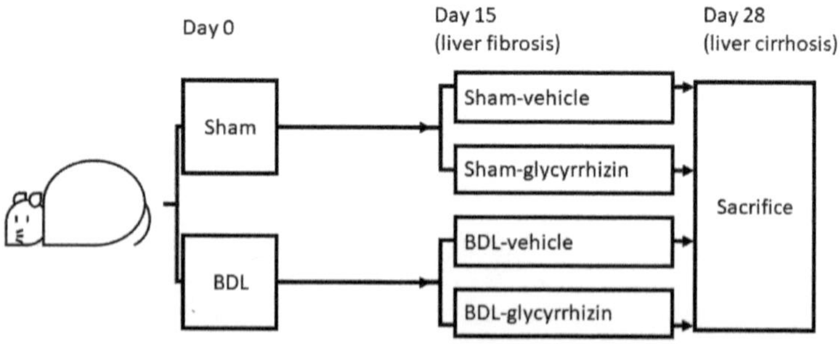

Figure 1. Experimental design. The rats received bile duct ligation (BDL) or sham operation. Treatments were started on the 15th day after operations, when liver fibrosis developed in BDL groups. After 2 weeks of treatments, experiments were performed on the 28th day after operations.

Table 1 depicts the results of BW and parameters of splanchnic and systemic circulation of experimental groups. The cirrhotic rats had significantly lower BW compared with sham-operated rats ($p < 0.001$). The MAP and SVR decreased while the CI increased significantly in cirrhotic groups, reflecting the feature of hyperdynamic circulation in portal hypertension (MAP: $p = 0.012$; SVR: $p < 0.001$; CI: $p < 0.001$). Glycyrrhizin treatment did not affect the BW, MAP, HR, SVR, and CI in sham-operated groups and BDL groups ($p > 0.05$).

Table 1. Hemodynamic parameters in sham or BDL rats treated with vehicle or glycyrrhizin.

	Sham Vehicle	Sham Glycyrrhizin	BDL Vehicle	BDL Glycyrrhizin	*p* Value *
	$n = 6$	$n = 6$	$n = 6$	$n = 6$	
BW (g)	452 ± 9	428 ± 15	375 ± 10 †	386 ± 17	0.605
MAP (mmHg)	141 ± 8	142 ± 7	114 ± 4 †	102 ± 4	0.075
HR (beats/min)	337 ± 19	304 ± 24	322 ± 13	302 ± 18	0.386
PP (mmHg)	9.5 ± 0.8	8.7 ± 0.4	16.8 ± 1.7 †	12.3 ± 1.0 *	0.004
Systemic circulation					
CI (mL/min/100 g)	31.5 ± 1.4	33.2 ± 2.4	43.2 ± 1.4 †	40.8 ± 3.7	0.560
SVR (mmHg/mL/min/100 g)	4.5 ± 0.3	4.3 ± 0.3	2.6 ± 0.1 †	2.6 ± 0.3	0.976

BDL: common bile duct ligation; BW: body weight; MAP: mean arterial pressure; HR: heart rate; PP: portal pressure; CI: cardiac index; SVR: systemic vascular resistance. * Glycyrrhizin-treated groups compared with paralleled vehicle groups; † $p < 0.05$, BDL-vehicle group compared with sham-vehicle group.

2.2. *Effects of Glycyrrhizin on Portal Hypertension*

Compared with sham-operated groups, cirrhotic rats had significantly higher PP ($p = 0.004$). Glycyrrhizin treatment from the stage of liver fibrosis still significantly reduced PP in cirrhotic rats. In sham-operated groups, glycyrrhizin did not affect PP ($p = 0.356$).

2.3. *Effects of Glycyrrhizin on Extrahepatic Systems and Mesenteric Angiogenesis*

Portal hypertension is driven by abnormal splanchnic inflow and hepatic outflow. In cirrhotic (BDL)-vehicle rats, the SMA flow significantly increased compared with the sham-vehicle group, representing the abnormally high splanchnic blood flow (Figure 2, sham-vehicle vs. BDL-vehicle (mL/min/100 g): 5.48 ± 0.53 vs. 7.05 ± 0.42, $p = 0.042$). Consistently, the SMA resistance decreased in cirrhotic rats (sham-vehicle vs. BDL-vehicle (mmHg/mL/min/100 g): 26.0 ± 4.1 vs. 13.9 ± 0.8, $p = 0.032$). Glycyrrhizin significantly reduced SMA flow in cirrhotic rats (BDL-vehicle vs. BDL-glycyrrhizin: (mL/min/100 g): 7.05 ± 0.42 vs. 5.51 ± 0.44, $p = 0.029$). Interestingly, the SMA resistance was not affected by glycyrrhizin ((mmHg/mL/min/100 g): 13.9 ± 0.8 vs. 16.7 ± 1.3, $p = 0.100$).

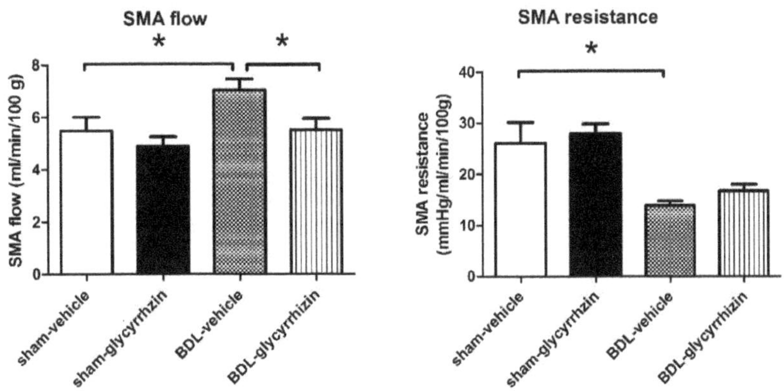

Figure 2. Effects of glycyrrhizin on splanchnic system. In rats with BDL-induced cirrhosis, the superior mesenteric artery (SMA) flow significantly increased compared with the sham-vehicle group. Glycyrrhizin significantly attenuated SMA flow in cirrhotic rats. The SMA resistance decreased in cirrhotic rats and was not affected by glycyrrhizin (n = 6, 6, 6, 6; BDL, bile duct ligation). * $p < 0.05$.

Mesenteric angiogenesis plays an important role in increasing splanchnic blood inflow and formation of portal-systemic collateral vascular system (Figure 3A). The mesenteric vascular density was evaluated by CD31 immunofluorescent staining (Figure 3B). In cirrhotic rats, the mesenteric vascular density was markedly higher than that of the sham-operated rats (sham-vehicle vs. BDL-vehicle (%): 6.2 ± 0.7 vs. 12.8 ± 1.2, $p = 0.002$). Glycyrrhizin treatment significantly reduced mesenteric vascular density (BDL-vehicle vs. BDL-glycyrrhizin (%): 12.8 ± 1.2 vs. 8.6 ± 1.3, $p = 0.038$). The collateral vascular system was evaluated by color microsphere method (Figure 3C). The results show that the shunting degree was significantly decreased in the glycyrrhizin-treated group ((%): 68.1 ± 2.5 vs. 47.8 ± 6.8, $p = 0.024$). This suggests that glycyrrhizin attenuates splanchnic blood inflow and portal-systemic collateral shunting though inhibition of splanchnic angiogenesis. This also explains why the SMA flow decreased in the glycyrrhizin-treated group with unaltered SMA resistance.

The angiogenic protein expressions of mesentery were assessed. A parallel series of experiments were performed to compare the protein expressions between sham-vehicle and BDL-vehicle groups (supplementary Figure S1). The results reveal that the phospho-eNOS, COX1, COX2, phospho-VEGFR2, and VEGF protein expressions significantly increased in the BDL group (sham-vehicle vs. BDL-vehicle, phospho-eNOS: 0.46 ± 0.15 vs. 1.00 ± 0.16, $p = 0.032$; COX1: 0.54 ± 0.10 vs. 0.80 ± 0.06, $p = 0.047$; COX2: 0.28 ± 0.02 vs. 0.58 ± 0.06, $p = 0.001$; phospho-VEGFR2: 0.49 ± 0.09 vs. 0.79 ± 0.04, $p = 0.022$; VEGF: 0.19 ± 0.03 vs. 0.78 ± 0.06, $p < 0.001$).

Figure 4 discloses the protein expression of BDL groups that received vehicle or glycyrrhizin. The results show that phospho-VEGFR2 was significantly downregulated by glycyrrhizin ((/β-actin): BDL-vehicle vs. BDL-glycyrrhizin: 0.95 ± 0.05 vs. 0.71 ± 0.08, $p = 0.031$). On the other hand, VEGF, phospho-eNOS, iNOS, COX1, and COX2 expressions were unaffected (VEGF: 0.56 ± 0.08 vs. 0.56 ± 0.04, $p = 0.996$; phospho-eNOS: 1.00 ± 0.04 v.s. 1.00 ± 0.01, $p = 0.954$; iNOS: 1.04 ± 0.13 v.s. 0.85 ± 0.07, $p = 0.211$; COX1: 0.74 ± 0.10 vs. 0.82 ± 0.06, $p = 0.534$; COX2: 0.70 ± 0.05 v.s. 0.68 ± 0.08, $p = 0.882$). The uncropped membranes are shown in supplementary Figure S2.

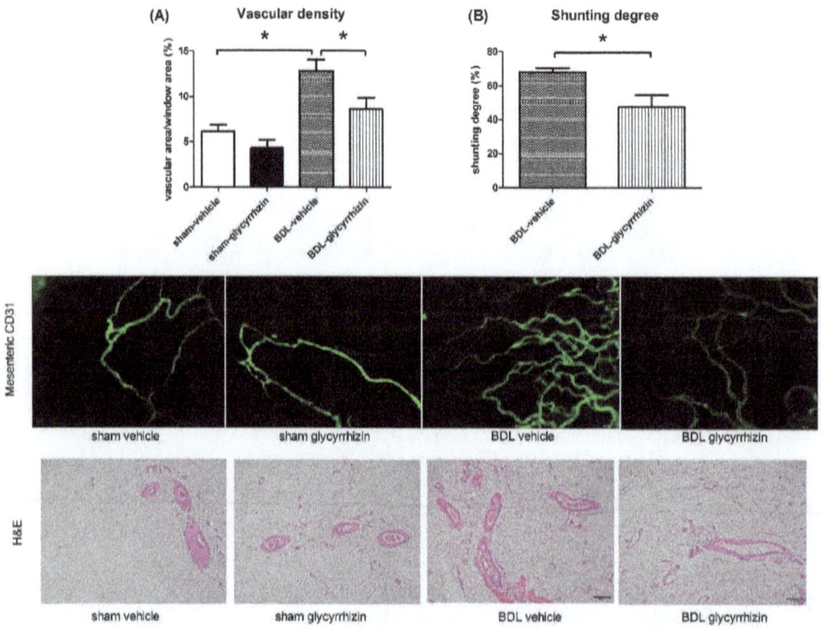

Figure 3. Effects of glycyrrhizin on mesenteric angiogenesis and portal-systemic collateral shunting. (**A**) Mesenteric angiogenesis was evaluated by vascular density of mesenteric window. In cirrhotic rats, the mesenteric vascular density was markedly higher than sham rats. Glycyrrhizin significantly reduced mesenteric vascular density (n = 5, 6, 6, 6). (**B**) The shunting degree was evaluated by color microsphere method. The shunting degree was significantly decreased in glycyrrhizin-treated group (n = 7, 7). Scale bar = 200 μm. * $p < 0.05$.

2.4. Effects of Glycyrrhizin on Hepatic System

The plasma liver injury markers ALT, AST, and total bilirubin were determined as well and shown in Figure 5A. The ALT, AST, and total bilirubin levels significantly increased in BDL groups (sham-vehicle vs. BDL-vehicle, ALT (IU/L): 50 ± 5 vs. 146 ± 20, $p = 0.001$; AST (IU/L): 97 ± 5 vs. 714 ± 105, $p < 0.001$; total bilirubin (mg/dl): 0.04 ± 0.01 vs. 8.30 ± 0.54, $p < 0.001$). However, glycyrrhizin did not influence ALT, AST, and total bilirubin levels in cirrhotic rats (BDL-vehicle vs. BDL-glycyrrhizin: ALT (IU/L): 146 ± 20 vs. 197 ± 67, $p = 0.479$; AST (IU/L): 714 ± 105 vs. 906 ± 372, $p = 0.630$; total bilirubin (mg/dl): 8.30 ± 0.54 vs. 7.37 ± 0.52, $p = 0.244$).

Figure 5B shows the hemodynamic parameters of the hepatic system. The HVR tended to increase in cirrhotic rats compared with sham-operated control rats (sham-vehicle vs. BDL-vehicle (mmHg/mL/min/100 g): 1.7 ± 0.1 vs. 2.4 ± 0.3, $p = 0.065$). There was no significant difference between the BDL-control and BDL-glycyrrhizin group in HVR ((mmHg/mL/min/100 g): 2.4 ± 0.3 vs. 2.3 ± 0.3, $p = 0.785$).

The severity of liver fibrosis was evaluated by fibrosis area ratio of the whole liver section, stained by Sirius red (Figure 6A). The area ratio was significantly increased in BDL-control group compared with the sham-control group ((%) 4.8 ± 0.4 vs. 22.4 ± 1.3, $p < 0.001$). Glycyrrhizin did not affect liver fibrosis (22.4 ± 1.3 vs. 24.2 ± 2.0, $p = 0.476$). To further validate the result, the severity of fibrosis of the liver sections were classified blindly by the Metavir scoring system by an independent investigator (Figure 6B). Consistently, BDL rats had significantly more severe liver fibrosis compared with sham rats ($p = 0.002$). Glycyrrhizin did not affect liver fibrosis in sham or BDL groups.

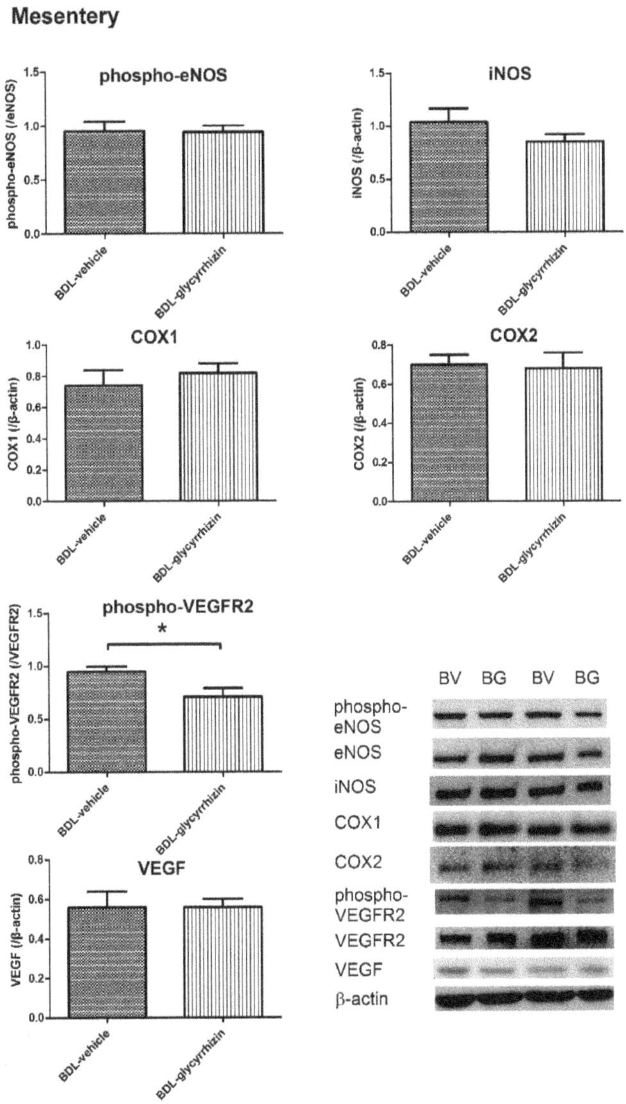

Figure 4. The angiogenic protein expressions in the mesentery. Glycyrrhizin downregulated phospho-VEGFR2 expression. VEGF, Phospho-eNOS, iNOS, COX1, and COX2 expressions were unaffected (n = 6, 6; BC, BDL-control; BG, BDL-glycyrrhizin). * $p < 0.05$.

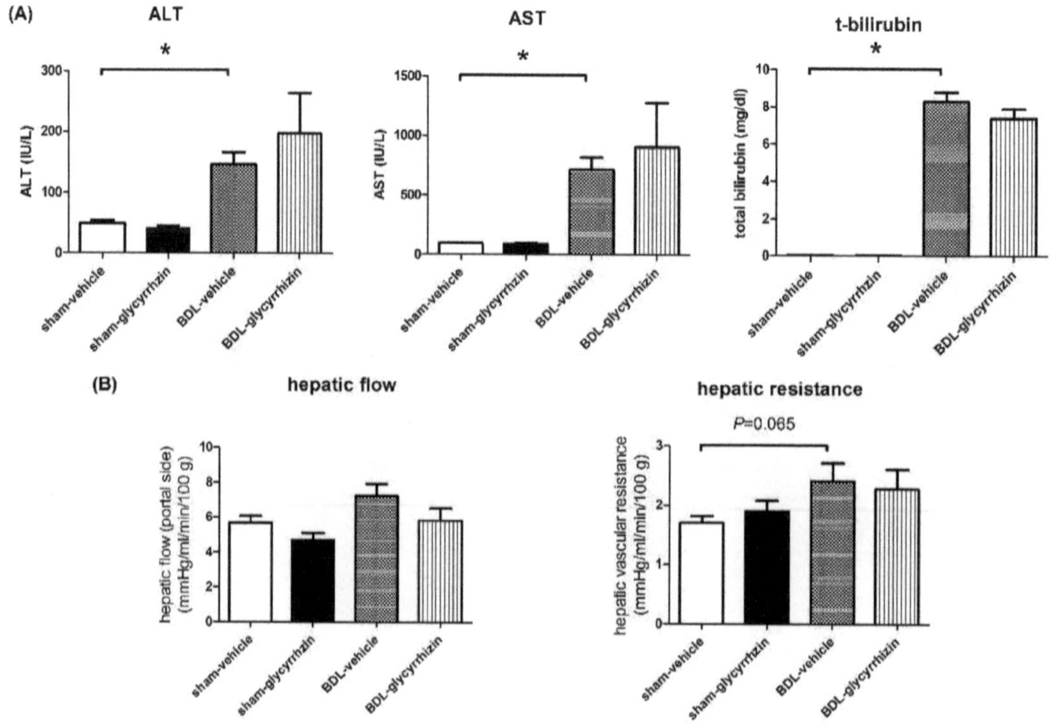

Figure 5. Effects of glycyrrhizin on hepatic system. (**A**) The levels of plasma liver injury markers alanine transaminase (ALT), aspartate transaminase (AST), and total bilirubin were determined. The levels of liver injury markers increased markedly in BDL groups. However, glycyrrhizin did not affect ALT, AST, and total bilirubin levels in cirrhotic rats (n = 6, 6, 6, 6). (**B**) There was a trend toward increased hepatic vascular resistance (HVR) in cirrhotic rats compared with sham-operated control rats. The HVR of the BDL-vehicle and BDL-glycyrrhizin groups was not significantly different (n = 6, 6, 6, 6). * $p < 0.05$.

Intrahepatic angiogenesis was analyzed by comparing the vascular numbers in portal or sinusoidal area. In sinusoidal area, liver sinusoidal endothelial cells were stained with CD31 (Figure 7A). In the portal area, vessels were stained with α-SMA, a smooth muscle cell marker (Figure 7B). The two evaluation methods showed the consistent findings that there was significant intrahepatic angiogenesis in cirrhotic rats (sham-vehicle vs. BDL-vehicle, sinusoidal area (counts/field): 323 ± 19 vs. 420 ± 24, $p = 0.010$; portal area (counts/field): 14 ± 2 vs. 46 ± 6, $p = 0.001$). Glycyrrhizin did not affect intrahepatic angiogenesis (BDL-vehicle vs. BDL-glycyrrhizin, sinusoidal area (counts/field): 420 ± 24 vs. 360 ± 26, $p = 0.121$; portal area (counts/field): 46 ± 6 vs. 45 ± 8, $p = 0.922$). Intrahepatic angiogenesis promotes liver fibrosis and vice versa. This result further supports the non-significant effects of glycyrrhizin on intrahepatic circulation and liver fibrosis in this experimental setting.

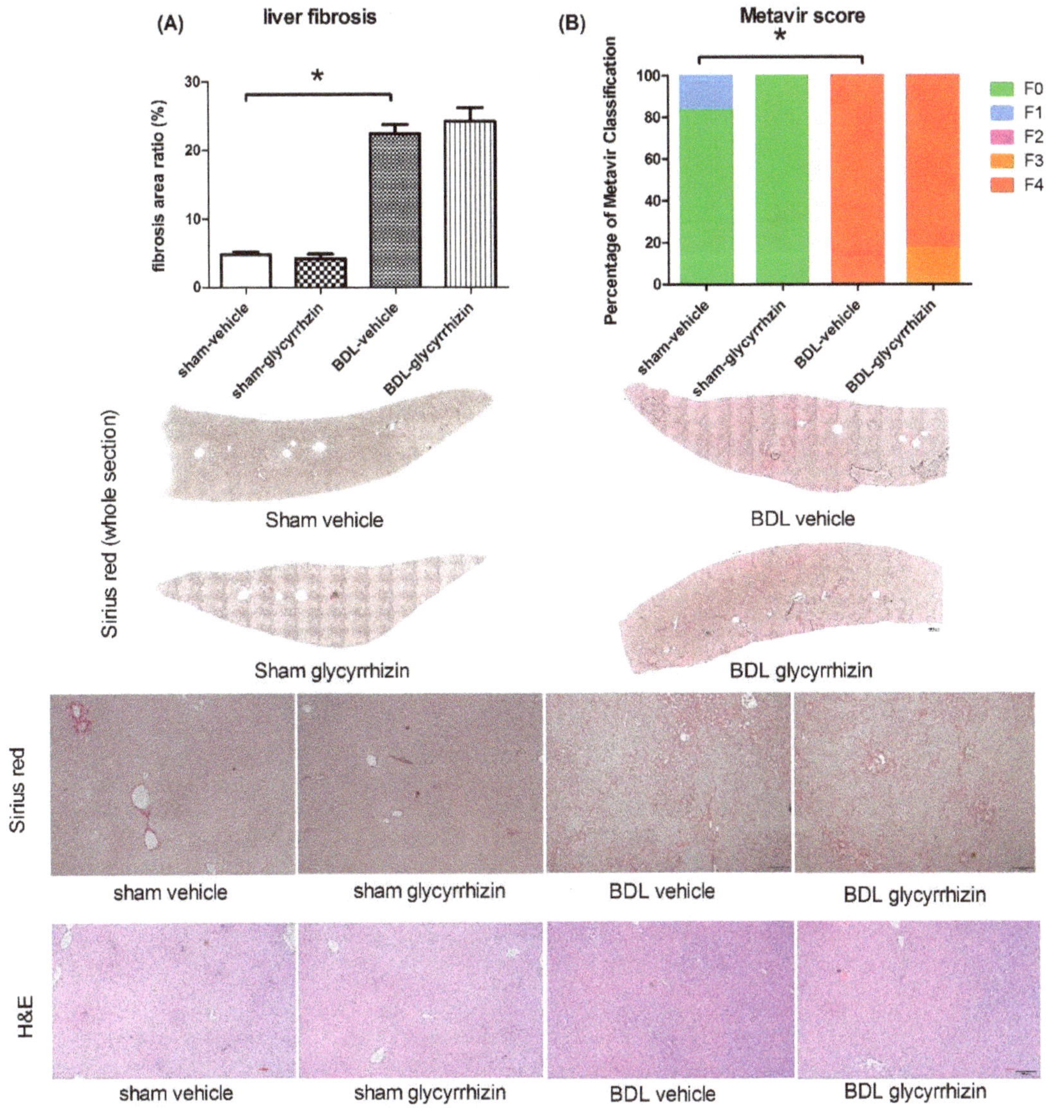

Figure 6. Effects of glycyrrhizin on liver fibrosis. (**A**) The severity of liver fibrosis was evaluated by fibrosis area ratio. The area ratio was significantly increased in BDL-vehicle group compared with sham-vehicle group. Glycyrrhizin treatment tended to attenuate the fibrosis severity ($n = 6, 6, 6, 6$). (**B**) The severity of liver fibrosis was further classified blindly by Metavir scoring system by an independent investigator. The results were compatible with the findings of the area ratio analyses that BDL significantly increased the fibrosis severity. Glycyrrhizin did not affect liver fibrosis in sham or BDL groups ($n = 6, 6, 6, 6$). Scale bar of upper panel = 1000 μm, scale bar of lower panel = 200 μm. * $p < 0.05$.

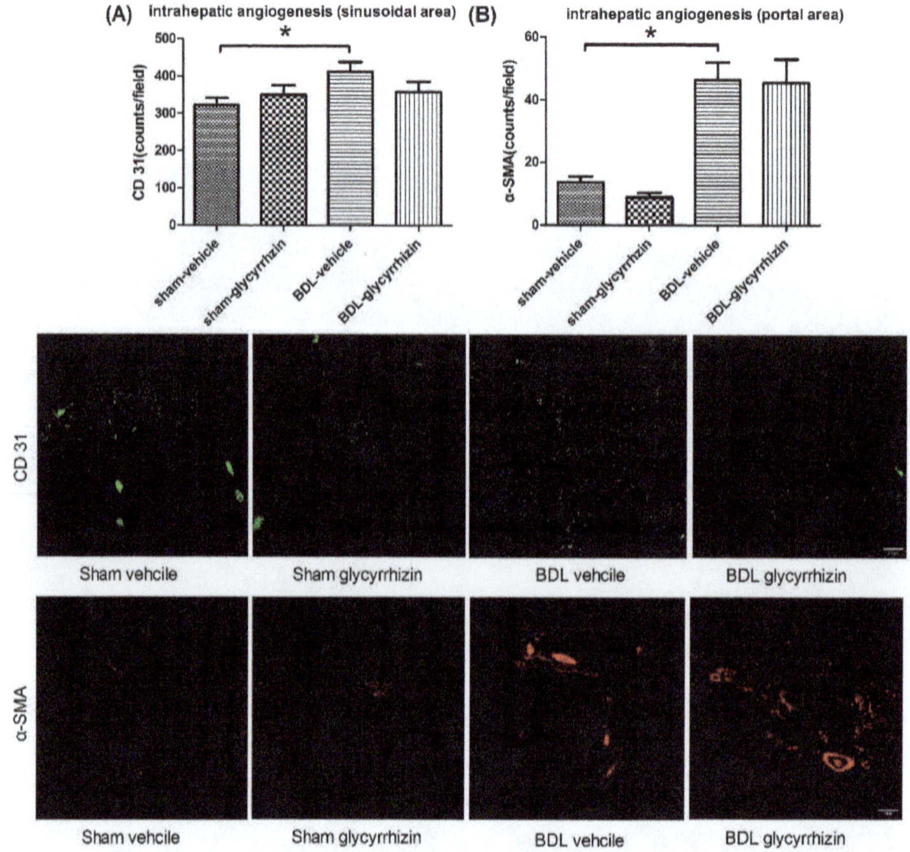

Figure 7. Effects of glycyrrhizin on intrahepatic angiogenesis. (**A**) Intrahepatic angiogenesis over sinusoidal area was evaluated by CD31 staining targeting endothelial cells. (**B**) Intrahepatic angiogenesis over portal area was investigated by α-SMA staining for vascular smooth muscle cells. The results were consistent. Intrahepatic vessels increased significantly in BDL-vehicle group compared with sham-vehicle group. Glycyrrhizin did not significantly influence the vascular numbers in the liver ($n = 6, 6, 6, 6$). Scale bar = 200 μm. * $p < 0.05$.

The protein expressions of fibrogenesis factors in the liver of sham-vehicle and BDL-vehicle groups were investigated in another parallel series of experiments (supplementary Figure S3). In cirrhotic rats, α–SMA, procollagen Iα1, TIMP1, and TGF-β expressions increased and MMP13 decreased significantly (α–SMA: 0.21 ± 0.03 vs. 0.77 ± 0.10, $p = 0.002$; procollagen Iα1: 0.25 ± 0.03 vs. 0.57 ± 0.10, $p = 0.020$; MMP13: 0.72 ± 0.07 vs. 0.26 ± 0.06, $p = 0.001$; TIMP1: 0.34 ± 0.05 vs. 0.75 ± 0.10, $p = 0.004$; TGF-β: 0.34 ± 0.03 vs. 0.92 ± 0.05, $p < 0.001$).

The protein expressions were then determined in BDL-vehicle and BDL-glycyrrhizin groups (Figure 8). The protein expressions of α–SMA, procollagen Iα1, MMP13, TIMP1, and TGF-β in cirrhotic rats were not significantly influenced by glycyrrhizin (BDL-vehicle vs. BDL-glycyrrhizin (/β-actin): α–SMA: 0.96 ± 0.03 vs. 0.93 ± 0.03, $p = 0.614$; procollagen Iα1: 0.77 ± 0.09 vs. 0.80 ± 0.03, $p = 0.747$, MMP13: 1.08 ± 0.06 vs. 1.18 ± 0.06, $p = 0.227$; TIMP1: 0.92 ± 0.05 vs. 0.88 ± 0.07, $p = 0.645$; TGF-β: 0.88 ± 0.03 vs. 0.85 ± 0.05, $p = 0.588$). The results suggest that glycyrrhizin administration since the stage of liver fibrosis did not ameliorate the severity of fibrosis. The uncropped membranes are listed in supplementary Figure S4.

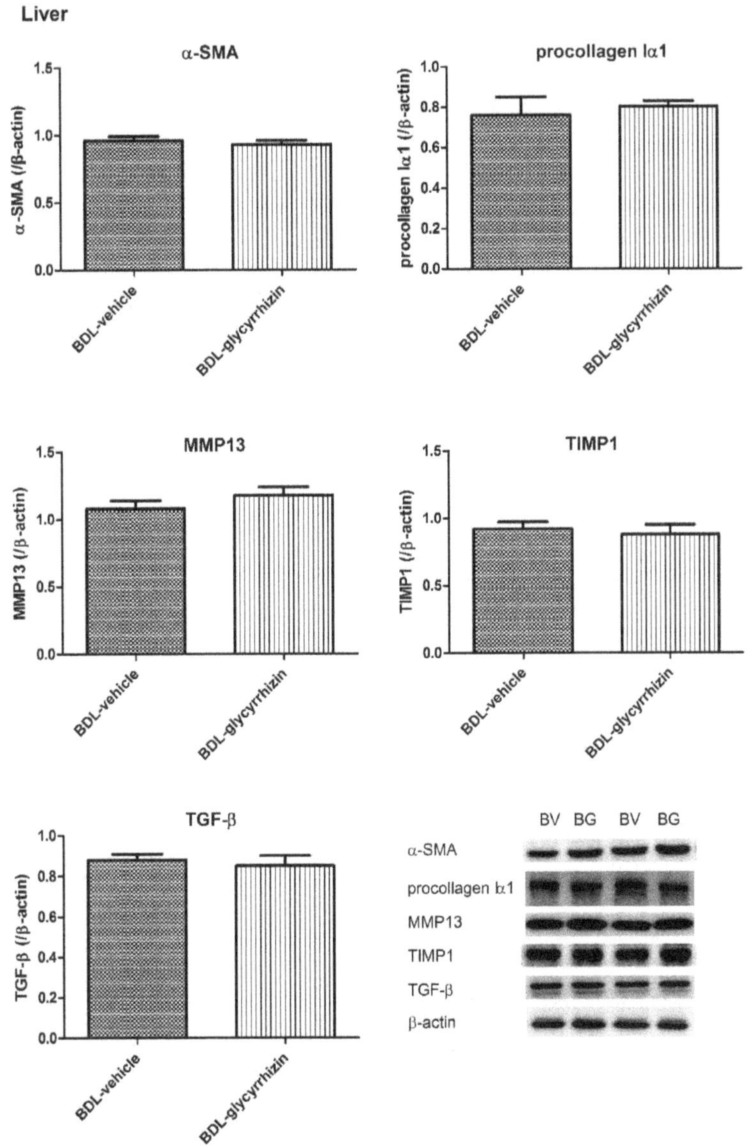

Figure 8. The fibrogenic protein expressions in the liver. The protein expressions of α–SMA, procollagen Iα1, MMP13, TIMP1, and TGF-β were not significantly different between BDL-vehicle (BV) and BDL-glycirrhizin (BG) groups (n = 6, 6).

3. Discussion

In this study, glycyrrhizin effectively attenuated portal hypertension and portal-systemic collateral shunting degree. The actions and mechanism are shown in Figure 9. Glycyrrhizin was given starting at the 15th day after BDL, when liver fibrosis had developed. As a result, the findings could be more relevant to a clinical condition, implying that the administration of glycyrrhizin at the stage of liver fibrosis is still effective in ameliorating portal hypertension.

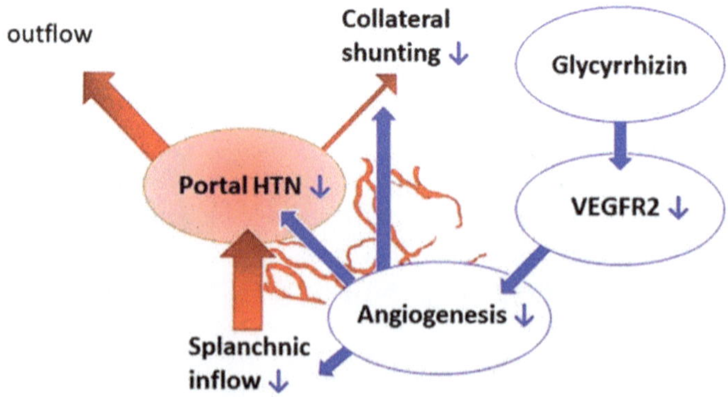

Figure 9. Effects of glycyrrhizin on portal hypertension-related derangements. Glycyrrhizin suppressed mesenteric angiogenesis through VEGFR2 downregulation. The splanchnic blood inflow and collateral shunting thus decreased. The hepatic system was unaffected by glycyrrhizin. The net effect of glycyrrhizin is the amelioration of portal hypertension.

Portal hypertension develops along with the progression of liver cirrhosis. The collagen fiber in the liver interferes with the hepatic outflow. On the other hand, the splanchnic inflow increases pathologically. The abnormal blood flow becomes stagnant in the portal system and results in portal hypertension. To cope with the problem, the portal-systemic collateral vascular system develops to shunt the abnormal flow to the systemic circulation with a relatively "lower pressure". Unfortunately, severe and fatal complications such as gastroesophageal variceal hemorrhage and hepatic encephalopathy ensue. Regarding the previous reports on the influences of glycyrrhizin in liver fibrosis, several studies have disclosed the anti-fibrotic effect of glycyrrhizin: Intraperitoneal injection of 3 mL of 0.2% glycyrrhizin solution three times a week beginning the first day of carbon tetrachloride administration markedly attenuated liver injury in a rat liver fibrosis model [13]. In another study, rats received carbon tetrachloride for 8 weeks to induce liver fibrosis. Treatment with glycyrrhizin beginning the first day of liver injury significantly ameliorated liver fibrosis [14]. Furthermore, in a mice liver fibrosis model induced by concanavalin A, glycyrrhizin treatment beginning the first day effectively ameliorated liver fibrosis [15]. In this study, glycyrrhizin treatment beginning at the stage of liver fibrosis did not significantly attenuate liver cirrhosis. Although the liver fibrosis ratio tended to decrease in the glycyrrhizin-treated group, further study showed that glycyrrhizin did not affect liver injury-related enzyme levels and liver fibrosis-related protein expressions. The contrary results may be ascribed to the timing to initiate treatment. Furthermore, the BDL-cirrhosis model adopted in this study leads to a relatively more severe liver damage as compared with that induced by carbon tetrachloride and concanavalin A [16]. Taken together, glycyrrhizin exerts a neutral effect on liver fibrosis under the current experimental setting. Indeed, there are several animal models of liver cirrhosis with various features. We chose the BDL model because it is a reproducible and nontoxic model. Since BDL-induced cholestatic liver injury may not be representative of all clinical conditions, other animal models simulating various liver injuries may be worth investigating in the future.

Abnormal angiogenic activity in splanchnic circulation further deteriorates portal hypertension. The portal blood flow is mainly supplied by the splanchnic system. During cirrhosis progression, abnormal angiogenesis in the splanchnic system further aggravates portal inflow [17]. Indeed, it has been demonstrated that inhibition of mesenteric angiogenesis ameliorated portal hypertension [3]. Several studies support that glycyrrhizin decreased angiogenesis activity. Glycyrrhizin also suppressed tumor growth and angiogenesis in mice [11]. In a rat colon precancerous model, glycyrrhizin suppressed the growth of lesions by the inhibition of angiogenesis [18]. In this study, glycyrrhizin decreased mesenteric vascular density, which suggests that glycyrrhizin effectively attenuated extrahepatic angiogenesis.

Activation of VEGFR2 is considered the main trigger of the most important pathway of angiogenesis in liver cirrhosis [3]. Fernandez et al. demonstrated that pharmaceutical blockade, either by antagonist or monoclonal antibody of the VEGF signaling molecules or receptors, effectively impeded neovascularization of portal-systemic collaterals and decreased portal blood inflow in rats with portal hypertension [17]. Our group has also identified that caffeine, through perturbing the VEGF signaling, exhibits beneficial effects toward hemodynamic derangements [19]. Taken together, the current and previous studies support the idea that extrahepatic angiogenesis due to VEGF pathway upregulation plays an important role in portal hypertension and manifests VEGF signaling antagonism as a potent therapeutic strategy in portal hypertension.

Several studies revealed the anti-VEGF properties of glycyrrhizin. Oral gavage of glycyrrhizin diminished VEGF in DMH-induced precancerous lesions in the colon of rats [18]. Glycyrrhizin significantly suppressed advanced glycation end product-induced VEGF production in rat retinal ganglion cell line [20]. Furthermore, glycyrrhizin attenuated VEGF and its receptor expression in a mice model with mammary cancer [21]. In this study, the protein expression of phospho-VEGFR2 in mesentery decreased significantly after glycyrrhizin administration. This finding supports prior research showing that glycyrrhizin inhibits angiogenesis via inhibition of the VEGF pathway.

The role of the spleen in portal hypertension should also be taken into consideration: Splenomegaly and increased splenic blood flow are important features of cirrhotic patients with portal hypertension. Increased splenic blood flow contributes to increased portal venous blood inflow, splanchnic hyperemia and angiogenesis, and aggravation of portal-systemic collateral vasculature [22]. A link of splenic blood flow in the pathogenesis of portal hypertension has also been suggested: The portal hypotensive effect of terlipressin, a drug used to control gastroesophageal variceal hemorrhage, is correlated with a decrease in splenic blood flow [23]. Therefore, glycyrrhizin may reduce the splenic size and/or splenic venous blood flow along with the reduction of portal pressure, which is something worth investigating.

The portosystemic collateral system diverts abnormal blood flow from the portal system. However, it results in severe complications such as esophageal variceal bleeding and hepatic encephalopathy. In this study, glycyrrhizin markedly attenuated collateral shunting. Since we have found that glycyrrhizin decreased splanchnic inflow but that the hepatic outflow (portal side) was not influenced, the current findings support the idea that the blood flow drained by the collateral system is reduced by glycyrrhizin, subsequently reducing the degree of collateral shunting.

4. Materials and Methods

4.1. Animal Model: Common Bile Duct Ligation (BDL)

Male Sprague-Dawley rats weighing 240–270 g at the time of surgery were used for experiments. The rats were allowed free access to food and water. Rats with secondary biliary cirrhosis were induced with common bile duct ligation [24]. Under anesthesia (Zoletil 50 mg/kg BW, intramuscularly), the common bile duct was doubly ligated with 3-0 silk. The first ligature was made below the junction of the hepatic ducts and the second ligature above the entrance of the pancreatic duct, followed by section of the common bile duct between the ligatures. The rats were allowed to recover. Liver cirrhosis developed two weeks after BDL, and a high yield of secondary biliary cirrhosis could be observed four weeks after BDL [25]. To avoid the coagulation defects, BDL rats received weekly vitamin K injection (50 μg/kg intramuscularly) [26].

This study was approved by Taipei Veterans General Hospital Animal Committee (IACUC 2017-081). All experimental procedures were performed at Taipei Veterans General Hospital Animal Laboratory and were conducted in accordance with the standard procedures indicated in the principles of laboratory animal care (Guide for the Care and

Use of Laboratory Animals, DHEW publication No. (NIH) 85-23, rev. 985, Office of Science and Health Reports, DRR/NIH, Bethesda, MD, USA).

4.2. Experiment Design

Liver cirrhosis and portal hypertension were induced in male Sprague-Dawley rats through BDL. Sham-operated rats were controls. Sham and BDL rats receive glycyrrhizin (150 mg/kg/day, oral gavage) [27] or vehicle (distilled water) beginning at the 15th day after operations, when BDL-induced liver fibrosis had developed [26]. The effects of glycyrrhizin were evaluated on the 28th day.

4.3. Measurement of Systemic and Portal Hemodynamics

The right carotid artery was cannulated with a PE-50 catheter that was connected to a pressure transducer. Continuous recordings of mean arterial pressure (MAP), heart rate (HR), and PP were performed on a multi-channel recorder (MP45, Biopac Systems Inc., Goleta, CA, USA). The external zero reference was placed at the level of the mid-portion of the rat. The abdomen was then opened with a mid-line incision, and the mesenteric vein was cannulated with a PE-50 catheter connected to the transducer [28].

Superior mesenteric artery (SMA) was identified at its aortic origin and a 5 mm segment was gently dissected free from surrounding tissues. Then a pulsed-Doppler flow transducer (TS420, Transonic system Inc., Ithaca, NY, USA) was placed to measure the SMA flow [29]. Hepatic inflow via the portal vein (portal part) was also measured by placing a flow probe around the portal vein as proximal to the liver as possible.

Cardiac output (CO) was measured by thermodilution, as previously described [30]. Briefly, a thermistor was placed in the aortic arch just distal to the aortic valve, and the thermal indicator (100 μL of normal saline) was injected into the right atrium through a PE-50 catheter. The aortic thermistor was connected to a cardiac output computer Cardiomax III (Columbus Instruments International Co., Columbus, OH, USA). Five thermodilution curves were obtained for each cardiac output measurement. The final value was obtained from the arithmetic mean of the data. Cardiac index (CI, mL/min/100 g BW) was calculated as CO per 100 g BW. Systemic vascular resistance (SVR, mmHg/mL/min/100 g BW) was calculated by dividing MAP by CI. SMA resistance (mmHg/mL/min/100 g BW) was calculated by (MAP-PP)/SMA flow per 100 g BW. Hepatic vascular resistance (HVR, mmHg/mL/min/100 g BW) was calculated by PP/hepatic inflow (portal part) per 100 g BW.

4.4. Immunofluorescent Study for the Mesenteric Vascular Density

Mesenteric angiogenesis was quantified by CD31-labelled microvascular networks in rat mesenteric connective tissue windows according to the previous study [29]. From each rat, at least four mesenteric windows (wedge-shaped regions of connective tissue surrounded by the intestinal wall and the ileal blood vessel pairs) were dissected free, washed in PBS, dried on gelatin slides, and fixed in 100% MeOH (−20 °C for 30 min). Slides were then incubated overnight at 4 °C with the primary antibody mouse anti-rat CD31-biotin (AbD Serotec, Oxford, UK). Then, a secondary antibody (CY2-conjugated streptavidin; Jackson ImmunoResearch, West Grove, PA, USA) was applied for 1 h at room temperature. At least four sets of data were obtained for each mesenteric window. Immunofluorescent images at magnification ×100 were assessed using an upright fluorescent microscope (AX80, Olympus, Tokyo, Japan) and thresholded by ImageJ software (ImageJ, Available online: https://imagej.nih.gov/ij/ (accessed on 6 June 2018). The vascular area was measured with the histogram function.

4.5. Color Microsphere Method for Portosystemic Shunting Degree Analysis

Portosystemic shunting degree was determined using the technique described by Chojkier and Groszmann [31], substituting color for radioactive microspheres; 30,000 of 15 μm yellow microspheres (Dye Track; Triton Technology, San Diego, CA, USA) were slowly injected into the spleen. The rats were euthanized, and the livers and lungs were dissected and placed into new polypropylene centrifuge tubes. The number of microspheres in each tissue was determined following the protocol provided by the manufacturer. In brief, 3000 blue microspheres (Dye Track) were added to each tube as an internal control. Tissue was digested overnight with 1 M KOH at 60 °C and thoroughly sonicated. After centrifugation, the supernatant was removed, and the pellet was washed once with 10% Triton X-100 and twice with acidified ethanol. At the end of the process, a minimum pellet containing the microspheres was allowed to dry overnight. The color of the microspheres was diluted with 200 μL of acidified Cellosolve acetate (Spectrum Chemicals, Gardens, CA, USA). The absorbance of the solution was read at 448 nm wavelength (yellow) and 670 nm wavelength (blue) in a spectrophotometer (Shimadzu, Columbia, MD, USA), and the number of microspheres was calculated by comparison with standards. Spillover between wavelengths was corrected with the matrix inversion technique. Portosystemic shunting was calculated as lung microspheres/(liver microspheres plus lung microspheres). Assuming a worst-case scenario in which two-thirds of the microspheres remain trapped in the spleen, this technique detects a minimum shunt of 3.5%. Studies using color microspheres have been shown to provide results similar to those using radioactive microspheres [32].

4.6. Western Blot

Tissue was immediately frozen in liquid nitrogen and stored at −80 °C until required. The protein extracts were made by pulverization in a grinder with liquid nitrogen, using a ratio of 1 mL of lysis buffer (phosphate-buffered solution containing 1% Nonidet P-40, 0.5% sodium deoxycholate, 0.1% sodium dodecyl sulfate (SDS), and 0.05% protease inhibitor cocktail solution (Roche Diagnostics GmbH, Penzberg, Germany)) for each 100 mg powdered sample. Protein concentration was determined for each sample by the Bradford method [33]. An aliquot of 20–40 μg protein from each sample that dissolved in sample buffer (63 mmol/L of Tris-HCL, pH 6.8, containing 2% SDS, 10% glycerol, 5% 2-mercaptoethanol, and 0.005% bomphenol blue) and 10 μg positive control was separated on denaturing SDS-10% polyacrylamide gels by electrophoresis (Mini-PROTEAN® 3 Cell, Bio-Rad Laboratories, Hercules, CA, USA). Prestained proteins markers (SDS-PAGE Standards, Bio-Rad Laboratories, Hercules, CA, USA) were used for molecular weight determinations. Proteins were then transferred to a polyvinylidene difluoride membrane (Immum-BlotTM PVDF Membrane, Bio-Rad Laboratories, Hercules, CA, USA) by a semi-dry electroblotting system (Trans-Blot® SD Semi-dry Electrophoretic Transfer Cell, Bio-Rad Laboratories, Hercules, CA, USA) for 1.5 h at 4 °C. To block non-specific binding, membranes were blocked for 30 min with 3% non-fat dry milk in TBS-T, pH 7.4 (25 mmol/L Tris base-137 mmol/L NaCl-2.7 mmol/L KCL-1% Tween 20). Blots were incubated with the primary antibody, diluted with 3% non-fat dry milk in TBS-T for 90 min at room temperature, and washed. Then the blots were incubated for 90 min with the secondary antibody and washed. The specific proteins were detected by enhanced chemiluminescence (Immobilon Western Chemiluminescent HRP Substrate, Merk Millipore Co., Billerica, MA, USA) and scanned with a computer-assisted video densitometer and digitalized system (BioSpectrum® 600 Imaging System, Ultra-Violet Products Ltd., Upland, CA, USA). The blots were scanned, photographed, then the signal intensity (integral volume) of the appropriate bands was analyzed.

4.7. Hepatic Fibrosis Determination with Sirius Red Staining

Liver paraffin sections were stained with Sirius red staining kit (Polysciences Inc., Warrington, PA, USA). To avoid selection bias, whole liver sections were analyzed. ImageJ was used to measure the percentage of Sirius red-stained area. Briefly, a grayscale image was used, then the red-stained collagen was isolated using the thresholding function. After that, the thresholded area was measured and shown as the percentage of thresholded area per image [29].

4.8. Drugs

Glycyrrhizin was purchased from Merck (Merck KGaA, Darmstadt, Germany). All solutions were freshly prepared on the days of the experiment.

4.9. Statistical Analysis

All results were analyzed using SPSS version 21.0 software [34] and data are expressed as mean ± S.E.M. The Shapiro–Wilk normality test showed that almost all of the data were in the pattern of "normal distribution". This study was composed of 2 variables in 4 groups (sham vs. BDL and vehicle vs. glycyrrhizin). Therefore, we used unpaired Student's *t*-test to check the differences between the following group pairs: 1. sham-vehicle vs. BDL-vehicle; 2. sham-vehicle vs. sham-glycyrrhizin; 3. BDL-vehicle vs. BDL-glycyrrhizin. The results of liver fibrosis severity classified by the Metavir scoring system were analyzed by Fisher's exact test. Results were considered statistically significant at a two-tailed *p*-value less than 0.05.

5. Conclusions

In conclusion, glycyrrhizin administered even at the stage of fibrosis still effectively attenuated portal hypertension and portal-systemic collateral shunting. The beneficial effects were exerted, at least partly, through suppression of extrahepatic angiogenesis via VEGF pathway downregulation. The application of glycyrrhizin in liver cirrhosis in clinical settings deserves further investigation.

Supplementary Materials: The following are available online at https://www.mdpi.com/article/10.3390/ijms22147662/s1.

Author Contributions: Conceptualization, H.-C.H. and S.-J.H.; methodology, S.-J.H.; formal analysis, S.-J.H.; investigation, C.K.P. and C.-H.Y.; resources, M.-C.H.; writing—original draft preparation, C.K.P.; writing—review and editing, H.-C.H. and S.-J.H.; supervision, C.-C.C., C.-L.C., F.-Y.L., and Y.-H.H.; funding acquisition, H.-C.H. All authors have read and agreed to the published version of the manuscript.

Funding: This study was supported by the grant from Taipei Veterans General Hospital (Grant number: V107B-005), Taipei, Taiwan.

Institutional Review Board Statement: This study was approved by Taipei Veterans General Hospital Animal Committee (IACUC 2017-081).

Data Availability Statement: The datasets used and analyzed during the current study are available from the corresponding author on reasonable request.

Conflicts of Interest: All the authors have no conflict of interest to declare.

References

1. Bosch, J.; Pizcueta, P.; Feu, F.; Fernández, M.; Garcia-Pagan, J.C. Pathophysiology of portal hypertension. *Gastroenterol. Clin. N. Am.* **1992**, *21*, 1–14. [CrossRef]
2. Fernandez, M. Molecular pathophysiology of portal hypertension. *Hepatology* **2015**, *61*, 1406–1415. [CrossRef]
3. Mejias, M.; García-Pras, E.; Tiani, C.; Miquel, R.; Bosch, J.; Fernandez, M. Beneficial effects of sorafenib on splanchnic, intrahepatic, and portocollateral circulations in portal hypertensive and cirrhotic rats. *Hepatology* **2008**, *49*, 1245–1256. [CrossRef]
4. Asl, M.N.; Hosseinzadeh, H. Review of pharmacological effects of *Glycyrrhiza* sp. and its bioactive compounds. *Phytother. Res.* **2008**, *22*, 709–724. [CrossRef] [PubMed]
5. Hautaniemi, E.J.; Tahvanainen, A.M.; Koskela, J.K.; Tikkakoski, A.J.; Kähönen, M.; Uitto, M.; Sipilä, K.; Niemelä, O.; Mustonen, J.; Pörsti, I.H. Voluntary liquorice ingestion increases blood pressure via increased volume load, elevated peripheral arterial resistance, and decreased aortic compliance. *Sci. Rep.* **2017**, *7*, 10947. [CrossRef] [PubMed]
6. Gumpricht, E.; Dahl, R.; Devereaux, M.W.; Sokol, R.J. Licorice compounds glycyrrhizin and 18beta-glycyrrhetinic acid are potent modulators of bile acid-induced cytotoxicity in rat hepatocytes. *J. Biol. Chem.* **2005**, *280*, 10556–10563. [CrossRef] [PubMed]
7. Rasool, M.; Iqbal, J.; Malik, A.; Ramzan, H.S.; Qureshi, M.S.; Asif, M.; Qazi, M.H.; Kamal, M.A.; Chaudhary, A.G.A.; Al-Qahtani, M.H.; et al. Hepatoprotective Effects of *Silybum marianum* (Silymarin) and *Glycyrrhiza glabra* (Glycyrrhizin) in Combination: A Possible Synergy. *Evid. Based Complement. Altern. Med.* **2014**, *2014*, 641597. [CrossRef] [PubMed]

8. Lin, G.; Nnane, I.P.; Cheng, T.Y. The effects of pretreatment with glycyrrhizin and glycyrrhetinic acid on the retrorsine-induced hepatotoxicity in rats. *Toxicon* **1999**, *37*, 1259–1270. [CrossRef]
9. Hung, C.-H.; Kee, K.-M.; Chen, C.-H.; Tseng, P.-L.; Tsai, M.-C.; Chen, C.-H.; Wang, J.-H.; Chang, K.-C.; Kuo-Chin, C.; Yen, Y.-H.; et al. A Randomized Controlled Trial of Glycyrrhizin Plus Tenofovir vs. Tenofovir in Chronic Hepatitis B with Severe Acute Exacerbation. *Clin. Transl. Gastroenterol.* **2017**, *8*, e104. [CrossRef]
10. Smolarczyk, R.; Cichoń, T.; Matuszczak, S.; Mitrus, I.; Lesiak, M.; Kobusińska, M.; Kamysz, W.; Jarosz, M.; Sieroń, A.; Szala, S. The Role of Glycyrrhizin, an Inhibitor of HMGB1 Protein, in Anticancer Therapy. *Arch. Immunol. Ther. Exp.* **2012**, *60*, 391–399. [CrossRef]
11. Kim, K.J.; Choi, J.S.; Kim, K.W.; Jeong, J.W. The anti-angiogenic activities of glycyrrhizic acid in tumor progression. *Phytother. Res.* **2013**, *27*, 841–846. [CrossRef]
12. Zhao, X.; Deng, B.; Xu, X.-Y.; Yang, S.-J.; Zhang, T.; Song, Y.-J.; Liu, X.-T.; Wang, Y.-Q.; Cai, D.-Y. Glycyrrhizinate reduces portal hypertension in isolated perfused rat livers with chronic hepatitis. *World J. Gastroenterol.* **2013**, *19*, 6069–6076. [CrossRef] [PubMed]
13. Guo, X.L.; Liang, B.; Wang, X.W.; Fan, F.G.; Jin, J.; Lan, R.; Yang, J.H.; Wang, X.C.; Jin, L.; Cao, Q. Glycyrrhizic acid attenuates CCl_4-induced hepatocyte apoptosis in rats via a p53-mediated pathway. *World J. Gastroenterol.* **2013**, *19*, 3781–3791. [CrossRef] [PubMed]
14. Liang, B.; Guo, X.-L.; Jin, J.; Ma, Y.-C.; Feng, Z.-Q. Glycyrrhizic acid inhibits apoptosis and fibrosis in carbon-tetrachloride-induced rat liver injury. *World J. Gastroenterol.* **2015**, *21*, 5271–5280. [CrossRef] [PubMed]
15. Tu, C.-T.; Li, J.; Wang, F.-P.; Li, L.; Wang, J.-Y.; Jiang, W. Glycyrrhizin regulates CD4+T cell response during liver fibrogenesis via JNK, ERK and PI3K/AKT pathway. *Int. Immunopharmacol.* **2012**, *14*, 410–421. [CrossRef]
16. Wallace, K.; Burt, A.; Wright, M.C. Liver fibrosis. *Biochem. J.* **2008**, *411*, 1–18. [CrossRef]
17. Fernandez, M.; Vizzutti, F.; Garcia-Pagan, J.C.; Rodes, J.; Bosch, J. Anti-VEGF receptor-2 monoclonal antibody prevents por-tal-systemic collateral vessel formation in portal hypertensive mice. *Gastroenterology* **2004**, *126*, 886–894. [CrossRef]
18. Khan, R.; Khan, A.Q.; Lateef, A.; Rehman, M.U.; Tahir, M.; Ali, F.; Hamiza, O.O.; Sultana, S. Glycyrrhizic acid suppresses the develop-ment of precancerous lesions via regulating the hyperproliferation, inflammation, angiogenesis and apoptosis in the colon of Wistar rats. *PLoS ONE* **2013**, *8*, e56020.
19. Hsu, S.-J.; Lee, F.-Y.; Wang, S.-S.; Hsin, I.-F.; Lin, T.-Y.; Huang, H.-C.; Chang, C.-C.; Chuang, C.-L.; Ho, H.-L.; Lin, H.-C.; et al. Caffeine ameliorates hemodynamic derangements and portosystemic collaterals in cirrhotic rats. *Hepatology* **2014**, *61*, 1672–1684. [CrossRef]
20. Lee, J.-J.; Hsiao, C.-C.; Yang, I.-H.; Chou, M.-H.; Wu, C.-L.; Wei, Y.-C.; Chen, C.-H.; Chuang, J.-H. High-mobility group box 1 protein is implicated in advanced glycation end products–induced vascular endothelial growth factor A production in the rat retinal ganglion cell line RGC-5. *Mol. Vis.* **2012**, *18*, 838–850.
21. Park, S.Y.; Kwon, S.J.; Lim, S.S.; Kim, J.-K.; Lee, K.W.; Park, J.H.Y. Licoricidin, an Active Compound in the Hexane/Ethanol Extract of *Glycyrrhiza uralensis*, Inhibits Lung Metastasis of 4T1 Murine Mammary Carcinoma Cells. *Int. J. Mol. Sci.* **2016**, *17*, 934. [CrossRef]
22. Bolognesi, M.; Merkel, C.; Sacerdoti, D.; Nava, V.; Gatta, A. Role of spleen enlargement in cirrhosis with portal hypertension. *Dig. Liver Dis.* **2002**, *34*, 144–150. [CrossRef]
23. Merkel, C.; Gatta, A.; Bolognesi, M.; Finucci, G.; Battaglia, G.; Angeli, P.; Zuin, R. Hemodynamic changes of systemic, hepatic, and splenic circulation following triglycyl-lysin-vasopressin administration in alcoholic cirrhosis. *Dig. Dis. Sci.* **1988**, *33*, 1103–1109. [CrossRef] [PubMed]
24. Franco, D.; Gigou, M.; Szekely, A.M.; Bismuth, H. Portal hypertension after bile duct obstruction. Effect of the bile diversion on portal pressure in the rat. *Arch. Surg.* **1979**, *114*, 1064–1067. [CrossRef] [PubMed]
25. Cameron, G.R.; Hasan, S.M. Disturbances of structure and function in the liver as the result of biliary obstruction. *J. Pathol. Bacteriol.* **1958**, *75*, 333–349. [CrossRef]
26. Kountouras, J.; Billing, B.H.; Scheuer, P.J. Prolonged bile duct ligation obstruction: A new experimental model of cirrhosis in the rat. *Br. J. Exp. Pathol.* **1984**, *65*, 305–311. [PubMed]
27. Dhingra, D.; Parle, M.; Kulkarni, S.K. Memory enhancing activity of *Glycyrrhiza glabra* in mice. *J. Ethnopharmacol.* **2004**, *91*, 361–365. [CrossRef] [PubMed]
28. Lee, F.-Y.; Colombato, L.A.; Albillos, A.; Groszmann, R.J. Administration of Nω-nitro-l-arginine ameliorates portal-systemic shunting in portal-hypertensive rats. *Gastroenterology* **1993**, *105*, 1464–1470. [CrossRef]
29. Huang, H.-C.; Wang, S.-S.; Hsin, I.-F.; Chang, C.-C.; Lee, F.-Y.; Lin, H.-C.; Chuang, C.-L.; Lee, J.-Y.; Hsieh, H.-G.; Lee, S.-D. Cannabinoid receptor 2 agonist ameliorates mesenteric angiogenesis and portosystemic collaterals in cirrhotic rats. *Hepatology* **2012**, *56*, 248–258. [CrossRef]
30. Albillos, A.; Colombato, L.A.; Groszmann, R.J. Vasodilatation and sodium retention in prehepatic portal hypertension. *Gastroenterology* **1992**, *102*, 931–935. [CrossRef]
31. Chojkier, M.; Groszmann, R.J. Measurement of portal-systemic shunting in the rat by using γ-labeled microspheres. *Am. J. Physiol.* **1981**, *240*, G371–G375. [CrossRef] [PubMed]
32. Hodeige, D.; De Pauw, M.; Eechaute, W.; Weyne, J.; Heyndrickx, G.R. On the validity of blood flow measurement using colored microspheres. *Am. J. Physiol. Content* **1999**, *276*, H1150–H1158. [CrossRef] [PubMed]

33. Bradford, M.M. A rapid and sensitive method for the quantitation of microgram quantities of protein utilizing the principle of protein-dye binding. *Anal. Biochem.* **1976**, *72*, 248–254. [CrossRef]
34. *SPSS Statistics, Version 21.0, Software for Statistics Analysis*; IBM Corp: Armonk, NY, USA, 2012.

International Journal of *Molecular Sciences*

MDPI

Review

Targeting Mammalian 5-Lipoxygenase by Dietary Phenolics as an Anti-Inflammatory Mechanism: A Systematic Review

Juan Antonio Giménez-Bastida [1,*], Antonio González-Sarrías [1,*], José Moisés Laparra-Llopis [2], Claus Schneider [3] and Juan Carlos Espín [1]

1 Laboratory of Food and Health, Research Group on Quality, Safety and Bioactivity of Plant Foods, Department Food Science and Technology, CEBAS-CSIC, P.O. Box 164, Campus de Espinardo, 30100 Murcia, Spain; jcespin@cebas.csic.es

2 Group of Molecular Immunonutrition in Cancer, Madrid Institute for Advanced Studies in Food (IMDEA-Food), 28049 Madrid, Spain; moises.laparra@imdea.org

3 Division of Clinical Pharmacology, Department of Pharmacology, Vanderbilt Institute of Chemical Biology, Vanderbilt University Medical School, Nashville, TN 37232, USA; claus.schneider@vanderbilt.edu

* Correspondence: jgbastida@cebas.csic.es (J.A.G.-B.); agsarrias@cebas.csic.es (A.G.-S.)

Abstract: 5-Lipoxygenase (5-LOX) plays a key role in inflammation through the biosynthesis of leukotrienes and other lipid mediators. Current evidence suggests that dietary (poly)phenols exert a beneficial impact on human health through anti-inflammatory activities. Their mechanisms of action have mostly been associated with the modulation of pro-inflammatory cytokines (TNF-α, IL-1β), prostaglandins (PGE_2), and the interaction with NF-κB and cyclooxygenase 2 (COX-2) pathways. Much less is known about the 5-lipoxygenase (5-LOX) pathway as a target of dietary (poly)phenols. This systematic review aimed to summarize how dietary (poly)phenols target the 5-LOX pathway in preclinical and human studies. The number of studies identified is low (5, 24, and 127 human, animal, and cellular studies, respectively) compared to the thousands of studies focusing on the COX-2 pathway. Some (poly)phenolics such as caffeic acid, hydroxytyrosol, resveratrol, curcumin, nordihydroguaiaretic acid (NDGA), and quercetin have been reported to reduce the formation of 5-LOX eicosanoids in vitro. However, the in vivo evidence is inconclusive because of the low number of studies and the difficulty of attributing effects to (poly)phenols. Therefore, increasing the number of studies targeting the 5-LOX pathway would largely expand our knowledge on the anti-inflammatory mechanisms of (poly)phenols.

Keywords: 5-LOX; polyphenols; inflammation; leukotrienes; eicosanoids; hemiketals; arachidonic acid

Citation: Giménez-Bastida, J.A.; González-Sarrías, A.; Laparra-Llopis, J.M.; Schneider, C.; Espín, J.C. Targeting Mammalian 5-Lipoxygenase by Dietary Phenolics as an Anti-Inflammatory Mechanism: A Systematic Review. *Int. J. Mol. Sci.* **2021**, 22, 7937. https://doi.org/10.3390/ijms22157937

Academic Editor: Maurizio Battino

Received: 25 June 2021
Accepted: 21 July 2021
Published: 25 July 2021

1. Introduction

1.1. Lipoxygenases

Lipoxygenases (LOXs) are found widely in nature and are abundant in plants and animals. Polyunsaturated fatty acids (PUFA) containing *cis* double bonds are the substrates of these enzymes. Linoleic and linolenic acids (18-carbon fatty acids) and arachidonic acid (AA; 20-carbon fatty acid) are the most common substrates for LOXs in plants and animals, respectively. The nomenclature of these enzymes is based on the specific position of the carbon oxygenated. Some examples are 9-LOX and 13-LOX, which are important LOXs described in plants, whereas 5-LOX, 12-LOX, and 15-LOX are present in animals [1,2].

LOXs are involved in the modulation of essential biological functions by synthesizing specific hydroperoxides, which are further metabolized into signaling molecules/biological mediators. Among these molecules, divinyl ethers, aldehydes, and jasmonates exert protective effects in plants from abiotic stress and(or) pathogens [3,4], whereas lipoxins or leukotrienes modulate the inflammatory response in humans [5]. LOX-catalyzed reactions are also associated with undesirable effects. Legume spoilage, generation of hay-like flavor, loss of pigments (e.g., carotenoids and chlorophylls), enzymatic browning and/or rancidity are effects linked to LOX oxidation (together with other oxidases) in plants [2]. In humans, an exacerbated activation of 5-LOX produces elevated levels of leukotrienes (LTs) promoting inflammation and related diseases (e.g., bronchoconstriction) [6,7].

The inhibition of LOX-mediated oxidation is an interesting strategy to minimize/avoid the loss of quality of plant-derived foodstuff. Current techniques for inhibition of LOX oxidation include the assay of phenolic compounds, which exert their protective effects through their antioxidant activity [2]. The structural similarities and mechanisms of action between plant and animals LOXs [8] suggest that the phenolic compounds might interfere with animal LOXs, including 5-LOX. However, the mechanisms by which phenolic compounds modulate 5-LOX (and the inflammatory response) go beyond their antioxidant activity, as described in this review.

1.2. 5-Lipoxygenase and Inflammation

Inflammation is a complex physiological process that functions as a network of interconnected elements regulated by many signaling molecules, including cytokines, chemokines, and lipid mediators. The disruption of the equilibrium between these molecules results in chronic inflammation and the development of related diseases [9,10]. AA is a substrate for the biosynthesis of several groups of lipid mediators collectively termed eicosanoids. The formation of prostaglandins (PGs) and LTs are two major pathways of eicosanoid biosynthesis catalyzed by cyclooxygenases (COX)-1/COX-2 and 5-lipoxygenase (5-LOX), respectively. The established role of the latter enzyme is its contribution to (patho)-physiological inflammation by the formation of LTs [11,12]. The enzyme 5-LOX is also central for the biosynthesis of the more recently discovered (and less investigated) 5-LOX-derived metabolites termed hemiketal (HK) eicosanoids [13] that appear to be novel lipid mediators in inflammation. LTs, at low nanomolar concentrations, can modulate the immune response and promote chronic inflammation, implying a role of these eicosanoids in a range of inflammatory diseases, including atherosclerosis, inflammatory bowel diseases, rheumatoid arthritis, and asthma [14,15]. The understanding of the biology of 5-LOX and its LT products has culminated in the development of anti-LT drugs (receptor antagonist and enzyme inhibitors) that are used clinically in the treatment of asthma and that may also provide a clinical benefit in atherosclerosis [11,16,17]. However, although these drugs show therapeutic effects (e.g., against asthma and atherosclerosis), the side-effects associated with their use and/or the poor in vivo efficacy highlight the need for better therapeutic options, including the search for possible alternatives such as natural products that may include dietary (poly)phenolic compounds.

In this regard, dietary (poly)phenols have been tested in numerous preclinical (animal and cellular) models and in a limited number of human studies, showing anti-inflammatory effects by diverse mechanisms of action, including cytokine modulation, inhibition of inducible nitric oxide synthase (iNOS) and nuclear factor kappa B (NF-κB) activation, as well as decreasing PG production by down-regulation of COX-2 [18,19]. Much less is known about the anti-inflammatory effect of (poly)phenols targeting the 5-LOX pathway. It is noteworthy that the number of studies that have investigated the anti-inflammatory effects of dietary (poly)phenols on 5-LOX (around 120 hits for a PubMed search) is much lower compared to COX-2 (around 2500 studies). The preponderance of a focus on the COX-2 pathway is difficult to rationalize given the importance of targeting both COX-2 and 5-LOX pathways to ameliorate undesirable effects of chronic inflammation.

Accordingly, our main objective was to perform a systematic and critical review of the current evidence concerning the anti-inflammatory effect of dietary (poly)phenolics via modulation of the 5-LOX pathway to identify knowledge gaps and future research needs, allowing an increase in the understanding of the anti-inflammatory effects of (poly)phenols.

1.3. 5-Lipoxygenase Pathway

1.3.1. Biosynthesis of 5-LOX-Derived Eicosanoids

Early studies on LT biosynthesis date back to the 1970s when Samuelsson and Borgeat described the formation of 5S-hydroxyeicosatetraenoic acid (5S-HETE) together with new arachidonate metabolites that were later termed LTA_4 and LTB_4 in peripheral leukocytes [20,21]. These investigations established a novel pathway of oxidative transformation of AA catalyzed by 5-LOX.

5-LOX is a key enzyme in the biosynthesis of LTs from AA (Figure 1). The formation of LTs requires activation of phospholipase A_2 (PLA_2) by Ca^{+2}-dependent (such as purinergic stimulation by ATP) or independent (i.e., innate immune "toll-like" receptor (TLR) stimulation by LPS) mechanisms, resulting in the hydrolysis of AA esterified in the membrane phospholipids [22,23]. In intact cells, 5-LOX is activated in response to Ca^{+2} influx and associates with 5-lipoxygenase activating protein (FLAP) to form a 5-LOX/FLAP complex at the nuclear membrane. In this complex, the essential function of FLAP is to present AA as a substrate to 5-LOX. The 5-LOX catalytic reaction involves an initial hydrogen abstraction from carbon 7 of AA and the addition of molecular oxygen to produce 5*S*-hydroperoxyeicosatetraenoic acid (5*S*-HPETE) followed by a second hydrogen abstraction from position 10 to form LTA_4. LTA_4 is unstable and undergoes enzymatic transformation by LTA_4 hydrolase to form LTB_4 or conjugation with glutathione by LTC_4 synthase to produce LTC_4, which is further metabolized by γ-glutamyltransferase and dipeptidase yielding LTD_4 and LTE_4, respectively [24,25]. Alternatively, 5*S*-HPETE can undergo reduction to 5S-HETE, which is in turn oxidized by 5-hydroxyeicosanoid dehydrogenase (5-HEDH), yielding 5-oxo-eicosatetraenoic acid (5-oxo-ETE) [26] (Figure 1).

1.3.2. Transcellular Biosynthesis of Leukotrienes and Lipoxins

LT biosynthesis goes beyond a string of enzymatic transformations in single cells (i.e., granulocytes and mast cells). Namely, LT biosynthesis involves transcellular biosynthesis, a term that describes eicosanoid formation by cell–cell interactions [27]. Early evidence on transcellular biosynthesis of LTs came from in vitro studies describing LTA_4 exchange between neutrophils and erythrocytes or endothelial cells to produce LTB_4 or LTC_4, respectively [28,29]. Subsequent in vivo studies provided evidence that transcellular biosynthesis of LTB_4 and LTC_4 does indeed occur in animal models [30,31].

The formation of lipoxins (lipoxygenase interaction products; LXs) is another paradigmatic example of transcellular biosynthesis (Figure 2). Synthesis of LXs requires the combination of different cell types (i.e., neutrophils, endothelial cells, and(or) platelets) expressing different lipoxygenases (i.e., 5-LOX, 12-LOX and(or) 15-LOX) [32]. For example, the tandem interaction of 5-LOX- and 12-LOX-expressing cells (i.e., neutrophils and platelets, respectively) leads to the formation of LXA_4 from LTA_4 [33,34]. Another transcellular mechanism of LX biosynthesis involves the interaction of 15-LOX (monocytes/macrophages and epithelial cells) and 5-LOX, resulting in the sequential transformation of AA to 15*S*-HPETE followed by the conversion by 5-LOX to yield LXA_4 and LXB_4 [35]. The synthesis of LXs is not limited to the interaction of only lipoxygenases. Acetylated COX-2 retains catalytic activity, forming 15*R*-HETE as the primary product [36,37], which can serve as a substrate for 5-LOX producing the aspirin-triggered LXs (LXA_4 and LXB_4 epimers) known as 15-epi-LXA_4 and 15-epi-LXB_4 [22,38,39]. LXs and aspirin-triggered LXs have anti-inflammatory properties promoting the resolution of inflammation [40–42], which contrasts with the largely pro-inflammatory effects of other products of the 5-LOX pathway.

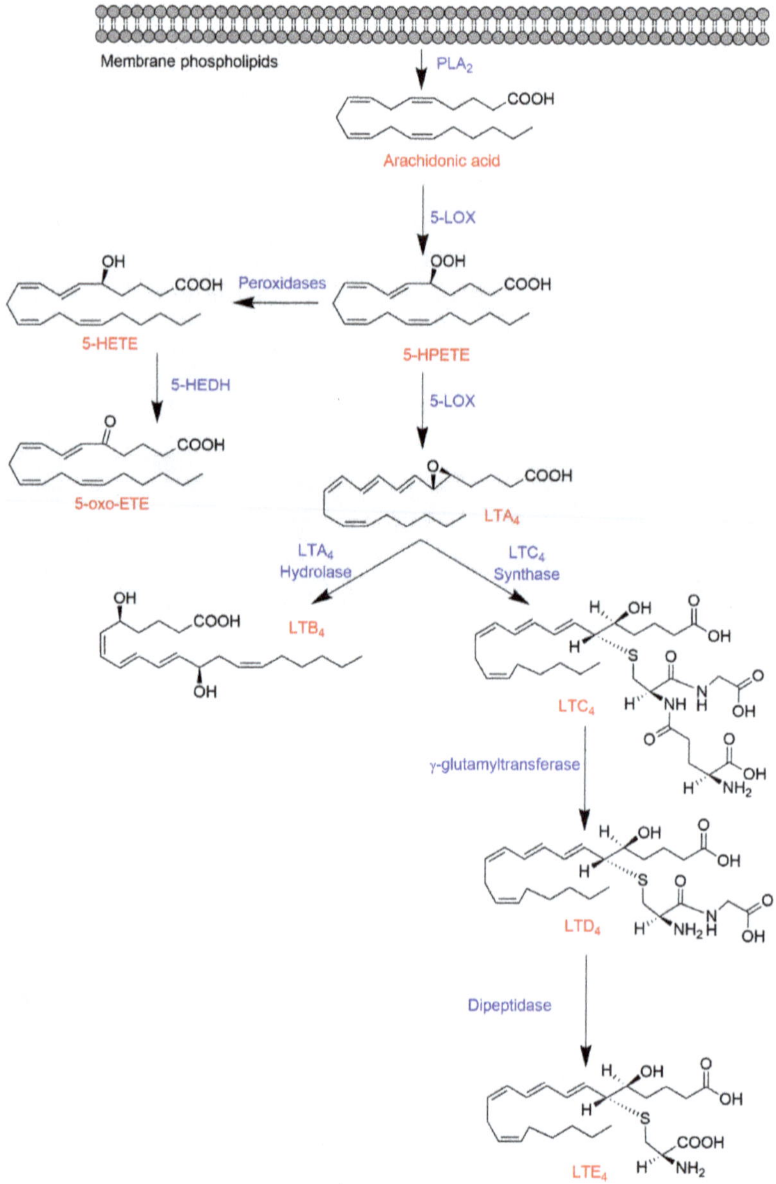

Figure 1. Biosynthesis of 5-HETE, 5-oxo-HETE, and LTs from arachidonic acid.

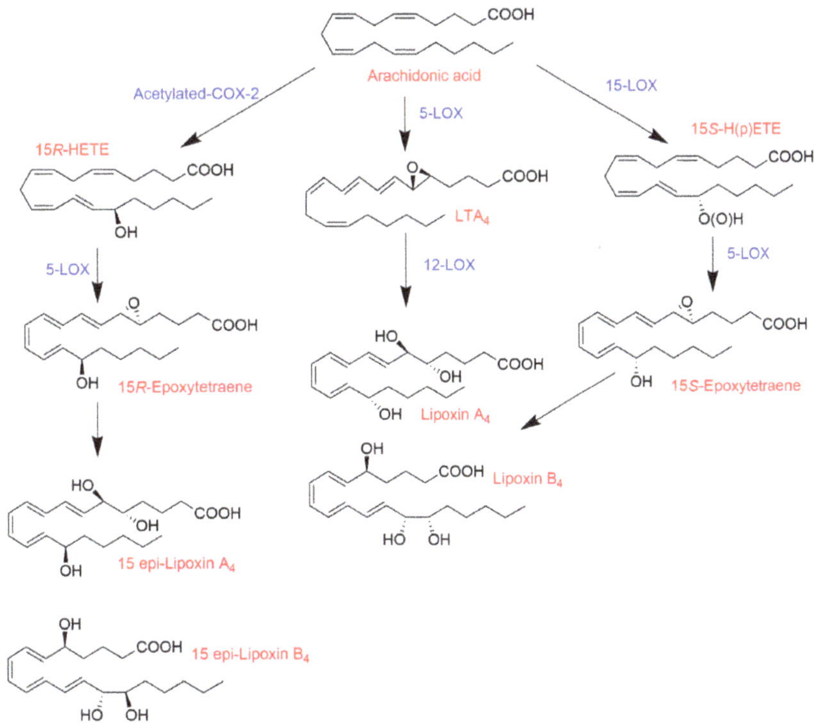

Figure 2. Biosynthesis of LXs and aspirin-triggered LXs.

1.3.3. The 5-LOX/COX-2 Crossover Biosynthetic Pathway

Besides a role in the biosynthesis of LTs and LXs, 5-LOX is also a key enzyme in the biosynthesis of a novel type of eicosanoids recently described. In vitro biochemical studies showed that COX-2 catalyzes the oxidation of the 5-LOX product 5S-HETE (resulting from the reduction of 5S-HPETE) to form a di-endoperoxide [43] and 5-OH-PGH_2 [44], which are equivalent to the prostaglandin endoperoxide PGH_2 of the COX-2 pathway. The reaction also yields two minor compounds identified as 5,11- and 5,15-di-HETE, the 5-hydroxy-analogs of the known COX by-products, 11- and 15-HETE [45,46]. Additional studies of the 5-LOX/COX-2 crossover pathways described the in vitro transformation (enzymatic and non-enzymatic) of the di-endoperoxide to two hemiketals (HKs) named HKE_2 and HKD_2 [13] and of 5-OH-PGH_2 to 5-OH-PGE_2 and 5-OH-PGD_2 [44] as shown in Figure 3.

Biosynthesis of HKE_2 and HKD_2 was established using an ex vivo model of human isolated leukocytes, stimulated with calcium ionophore A23187 and LPS for 5-LOX activation and COX-2 up-regulation, respectively [13]. Analysis of the time-course of the formation of HKs in human leukocyte mixtures revealed that their biosynthesis mainly depends on the availability of the 5S-HETE substrate and, to a lesser extent, on the activity of COX-2 [47]. HK formation in the mixture of human leukocytes may be another example of transcellular biosynthesis, given that a single type of leukocyte is unlikely to exhibit a significant activity of both 5-LOX and COX-2. The dependence of HK biosynthesis on the activation of leukocyte mixtures by both A23187 (inducing 5-LOX) and LPS (inducing COX-2) suggests a role for neutrophils and activated monocytes/macrophages, respectively, although this was not directly established (Figure 4).

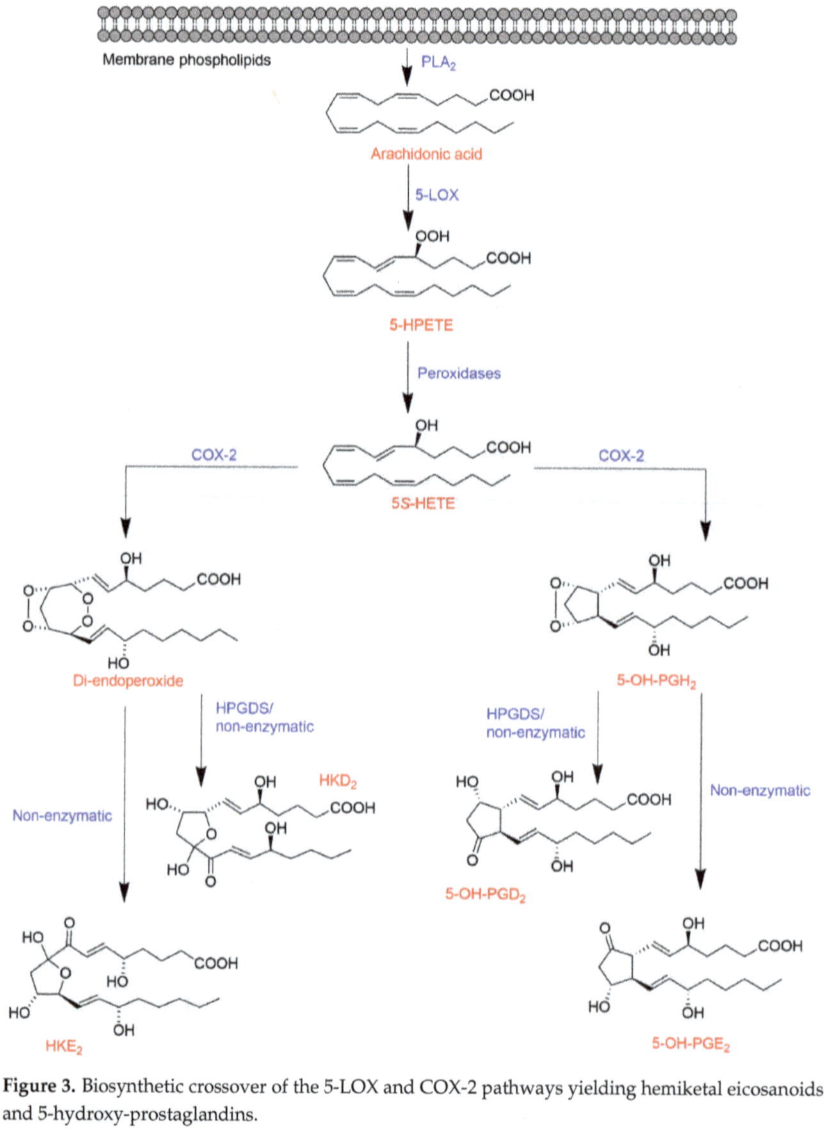

Figure 3. Biosynthetic crossover of the 5-LOX and COX-2 pathways yielding hemiketal eicosanoids and 5-hydroxy-prostaglandins.

1.3.4. Role of 5-Lipoxygenase-Derived Eicosanoids in Inflammation

Decades of intense investigation have identified LTs as potent inducers of inflammation through the interaction with distinct G protein-coupled receptors. LTB_4 was one of the first chemotactic molecules identified [48] and is a well-known pro-inflammatory molecule that exerts its effects through interaction with its high-affinity receptor BLT1 [49,50]. LTB_4 also binds (and activates) a second receptor, BLT2, albeit with much less affinity than that reported for BLT1, and its function via interaction with BLT2 remains elusive [51]. Thus, LTB_4 promotes pro-inflammatory responses such as leukocyte chemoattraction, leukocyte–endothelial cell interaction, and the release of inflammatory mediators at inflammation sites [52]. Here, aberrant inflammation results in tissue damage and impairs adequate function of host innate immune effectors such as neutrophils and activated monocytes/macrophages to recognize, respond, and resolve inflammatory processes properly.

This dysregulated response is implicated in the pathogenesis of chronic diseases such as atherosclerosis, cardiovascular, and inflammatory bowel diseases [15,53–55]. On the other hand, the cysteinyl-LTs (cysLTs), LTC_4, LTD_4, and LTE_4 activate their cognate receptors, CysLT1 and CysLT2, and exert profound effects on airway inflammation leading to bronchoconstriction, vascular permeability, and neutrophil extravasation [6,15,56–59].

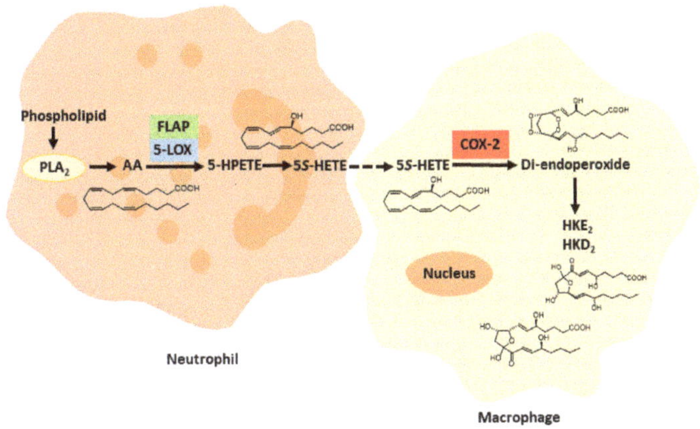

Figure 4. Proposed transcellular biosynthesis of hemiketal eicosanoids.

5-HETE is a particular eicosanoid that shows limited biological activity by itself but serves as a precursor to form biologically active molecules, including 5-oxo-ETE, HKs, and 5-OH-PGs. 5-oxo-ETE is an oxidized metabolite of 5*S*-HETE that binds to the 5-oxo-ETE receptor (OXE), exerting a powerful granulocyte chemoattractant effect [60–62].

The recently described biochemical and chemical synthesis of HKs [63,64] enables investigating the biological role(s) of these newly discovered eicosanoids. An established activity of these molecules is the stimulation of endothelial cell migration and tubulogenesis [13] and the modulation of platelet aggregation [63], implying a possible role in atherosclerosis and CVD. However, unlike LTs, it remains unexplored thus far whether HKE_2 and HKD_2 exert their effects via interaction with a specific receptor (Figure 5). Even less is known about the biological activity of 5-OH-PGs that appear not to activate traditional prostanoid receptors [44].

1.4. *Inflammation as a Target of Dietary (Poly)phenols: Role of Their Bioavailability*

Evidence from epidemiological and observational studies highlight diet as one of the cornerstones in preventing inflammatory diseases such as intestinal inflammation and cardiovascular diseases. Dietary patterns that include a high intake of fruits and vegetables, such as the Mediterranean diet as a source of high levels of phytochemicals, including dietary (poly)phenols, have been shown to significantly ameliorate inflammation [65–67]. In this regard, some clinical trials have provided evidence supporting the beneficial role of dietary (poly)phenols against chronic inflammatory diseases [68,69]. However, while numerous preclinical studies describe the anti-inflammatory effects of many (poly)phenolic compounds through the modulation of a plethora of cellular processes related to inflammation, the evidence of activity in humans remains unclear overall from a nutritional point of view, partly owing to the limited bioavailability of (poly)phenols [18,70].

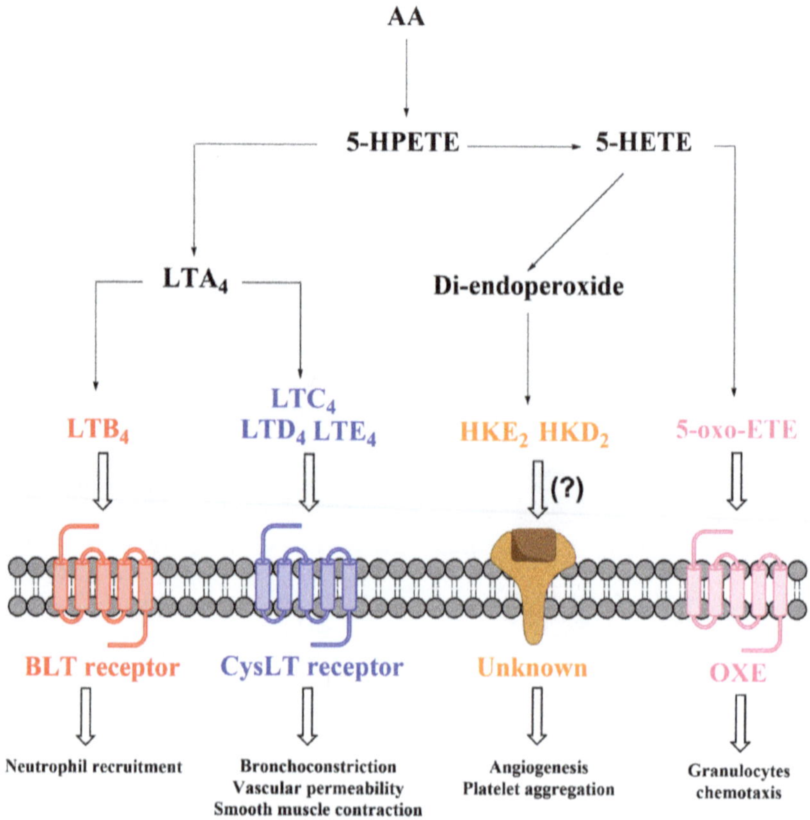

Figure 5. Biological effects of 5-LOX-derived eicosanoids.

The bioavailability of dietary (poly)phenols is essential for a better understanding of the anti-inflammatory effects of (poly)phenolic compounds and to design physiologically relevant studies to corroborate their potential effects. Plant-derived foods (e.g., *Citrus* fruits, walnuts, pomegranates, green tea, soy, grapes, and others) contain phenolic compounds in free form or conjugated with sugar moieties, which are not well absorbed in the small intestine, thus limiting their distribution in systemic tissues in their native form. Upon consumption, (poly)phenols reach the gastrointestinal tract in their original molecular form, mainly as glycosides and complex oligomeric structures, and are hydrolyzed and further metabolized by either intestinal enzymes or by the gut microbiota forming new metabolites [71,72]. For instance, ellagitannins (ETs) (such as punicalagin from pomegranate) and their hydrolysis product ellagic acid (EA) undergo gut microbiota metabolism to yield metabolites collectively termed urolithins (Uro), with the most relevant ones identified as Uro-C, Uro-A, IsoUro-A, and Uro-B. Similarly, isoflavones (IsoFlv) and their aglycones (e.g., daidzin and daidzein, respectively) also undergo microbial metabolism producing equol and(or) *O*-desmethyl-angolensin (ODMA) [71,73]. The flavonoid glycoside rutin (quercetin-3-rutinoside) acts as a precursor (via deglycosylation) of quercetin in the colon [74], whereas curcumin and flavanones (such as the glycoside hesperidin and its aglycone hesperetin) can be detected in the colon for hours in their original form [75,76]. Upon absorption, the (poly)phenolic compounds and the microbial metabolites undergo phase-II metabolism to form conjugated molecules (glucuronides, sulfates, and methyl esters), which are the main molecules detected in the bloodstream, intestinal tissues, bile, feces, urine, and different systemic tissues [77,78]. Animal and human metabolism and bioavailability studies have reported that, at the intestinal level, the parental phenolic compounds (which serve as substrates of the gut microbiota) and their microbial-derived metabolites can achieve concentrations from μM to mM. In contrast, the concentrations reached in the bloodstream by the phase-II metabolites can range from nM to low μM, show anti-inflammatory effects and persist in the circulation for a few days after intake [79]. Among the phase-II metabolites detected in vivo, glucuronides are recognized as the major conjugated molecules, including Uro-A glucuronide (Uro-A glur), Quercetin-3 glur, and curcumin-glur [80–82]. Interestingly, the increasing knowledge about the metabolism of these compounds indicates that, under inflammatory conditions, the circulating glucuronides might also play a role as precursors of their aglycones, including luteolin [83,84], quercetin [85–89], resveratrol [90], Uro-A [91], and curcumin [82]. Besides, a recent trial showed that the intake of a (poly)phenolic cocktail by breast cancer patients allowed detecting relevant concentrations of free curcumin, most likely due to a conjugation-saturation process, in malignant mammary tumors [77].

2. Methods

Search Strategy and Study Selection

This review was conducted and reported following the Preferred Reporting Items for Systematic Review and Meta-Analyses (PRISMA) [92]. A comprehensive literature search was performed using PubMed and Scopus databases. The search strategy included the combination of the following search terms in abstracts and titles and was adapted for each database: (5-lipoxygenase or 5-LOX) and (phenol* or flavonoid* or polyphenol* or curcumin or resveratrol or EGCG or urolithin or procyanidin or proanthocyanidin or flavan* or flavone* or catechin or epicatechin or quercetin or curcumin or valerolactone or punicalagin or ellagic or tannin or lignan or isoflavone or equol or silymarin or thistle). Independent literature searches and article selection were completed from January to April 2021 by the authors. The authors also hand-searched the bibliography of identified articles.

Regarding eligibility criteria and study selection, all human, animal, and cellular model studies that investigated the role of 5-LOX in the anti-inflammatory effect of (poly)phenolic compounds were included in this review. Otherwise, enzymatic, in silico and in vitro studies that used physiologically unrealistic conditions, such as high concentrations (compared to those reported in vivo), irrelevant metabolic forms of (poly)phenols, inappropriate cellular models, and those conducted with plant extracts were excluded.

A total of 1250 articles were found after the literature search in the two electronic databases. Removing duplicates and full screening yielded 872 articles, of which 5 human, 24 animal, and 127 in vitro cell model studies met the inclusion criteria for this systematic review. A summary of the selection of articles included in this study is outlined in Figure 6.

The selected studies are reviewed in detail in the following sections and are summarized in independent tables (Tables 1 and 2 and Table S1) to show the current evidence of the role of 5-LOX as a target of the anti-inflammatory effects of (poly)phenols.

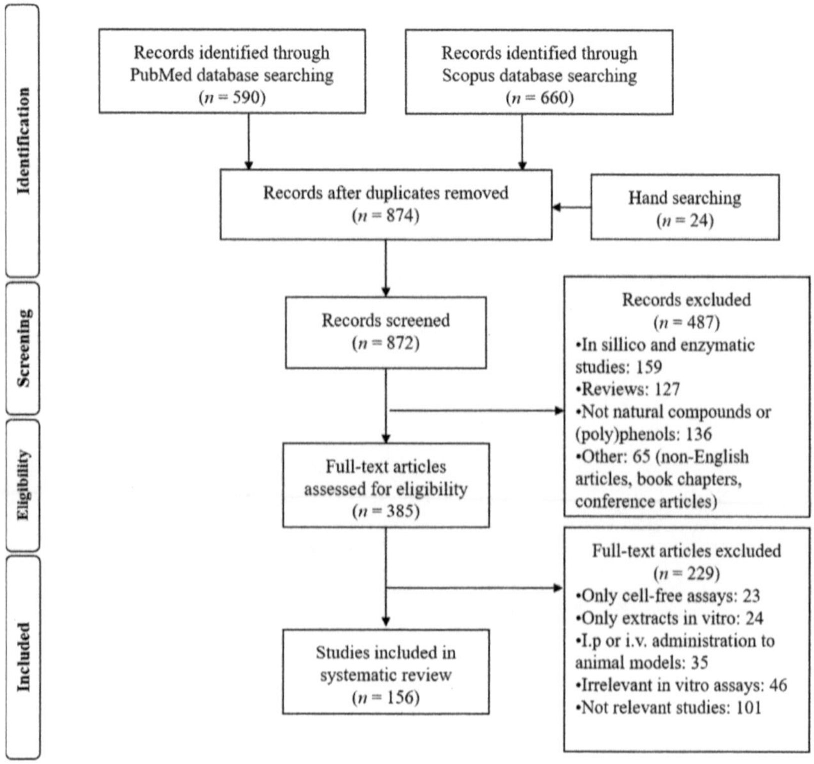

Figure 6. PRISMA flow diagram. Description of the search strategy and exclusion/inclusion criteria.

3. Results and Discussion

3.1. Human Studies

Targeting 5-LOX to address inflammation using a nutritional approach has been tested in only a few human studies. Only five studies determined the level of 5-LOX-derived metabolites in subjects who consumed plant-derived foodstuff (Table 1). These studies are characterized by a small number of subjects (n = 10–18) and short-term duration (up to four weeks). Four investigations using healthy volunteers described the modulation of 5-LOX expression and its metabolites (LTs and lipoxins) after the consumption of soy milk, procyanidin-enriched chocolate bars, enriched beverages, or olive oil (Table 1). One human study evaluated the effect of a diet supplemented with soy isoflavone tablets in patients with asthma. Lung function parameters remained unaltered, while isolated eosinophils showed an attenuated capacity to synthesize LTB_4 ex vivo after stimulation [93]. The absence of a noticeable effect of (possible) modulation of LTs by polyphenolics in healthy subjects can be somewhat predicted and is in accord with 5-LOX-deficient animals, which are largely normal in the absence of an inflammatory stimulus [49]. Therefore, an insufficient number of human studies, mostly conducted in healthy subjects, makes it difficult to draw conclusions or even speculate regarding the beneficial effects of dietary phenolics via modulation of the 5-LOX pathway in humans.

Table 1. Human studies describing the effects of the consumption of (poly)phenols on 5-LOX and its products.

Population of Study	Design of the Study	Foodstuff; Intake and Duration	Main Outcomes	Ref.
Healthy volunteers (n = 10; 20–55 years)	Randomized, crossover, double-blind, placebo-controlled.	Low- (0.09 mg/g) and high-procyanidin (4.0 mg/g) chocolate bars; 37 g (single dose); Duration: blood collection at 2 and 6 h; wash-out period of 1 week between treatments.	↑Epicatechin plasma level (especially the high procyanidins consumer group). ↓CysLTs/PGI_2 ratio (relative to the effects observed in the low procyanidins consumer group).	[94]
Patients suffering mild/moderate persistent asthma (n = 13; 18–65 years)	Quasi-experimental intervention study. No control group.	[1] Soy isoflavones tablets (NOVASOY, ArcherDanielsMidland, Decatur, IL, USA); two 50 mg tablets (once per day); duration: 4 weeks.	↓LTB_4 and FE_{NO} in A23187-stimulated eosinophils (ex vivo); no significant changes in pre-bronchodilator FEV_1 and Juniper Asthma Control Score.	[95]
Healthy volunteers (n = 18; 22–44 years)	Acute consumption, no control group.	Raw virgin olive oil; 50 mL (single dose consumed with 200 g bread); Duration: blood collection at 1 and 6 h; 1 week wash-out period before olive oil consumption.	↑Hydroxytyrosol in plasma. ↓ALOX5AP gene expression at 1 h (basal values at 6 h). The downregulation was inversely correlated with glucose and insulin levels.	[96]
Healthy volunteers of Asian ethnicities (n = 18)	Randomized, crossover, double-blind, placebo-controlled.	[2] Soy milk; 2 daily treatments (20 g); Duration: 3 h after acute consumption followed by 1-week wash-out period and 4-weeks daily exposure.	↓LTB_4 and LXA_4 level in plasma after 3 h; ↓LTB_4 and F_2-isoprostanes in plasma and urine after 4-weeks daily exposure; ↑LXA_4 in plasma after 4-weeks daily exposure; ↓MPO activity, serum lipid hydroperoxides and hsCRP in plasma after 4-weeks daily exposure.	[93]
Senior and young taekwondo athletes (n = 10; 18–57 years old)	Quasi-experimental intervention study, no placebo.	Isotonic beverage enriched with almond (0.3%), sucrose (0.8%), oils (0.2% [3] DHA-S and 0.6% olive oil), and α-tocopherol acetate (vitamin E); duration: 5 weeks (consumed 5 days a week); blood samples taken at the beginning and at the final of the (1 h before and after each stress test).	Beverage supplementation exerted ↓sL-Selectin, sICAM3 and ↑IL-6 in young athletes (after exercise) as well as ↑TNF-α level in plasma in the young group (in absence of exercise) and in the senior group (after exercise). The beverage consumption exerted no significant effects on lipoxin, PGE_2, PGE_1, and NF-κB. Modulation of TNF-α, 15-LOX2, COX-2, IL-1β and IL-8 mRNA expression in PBMC; No Effect on TLR_4, NF-κB, 5-LOX, IL-10, IL-15, HSP72 expression (mRNA) in PBMC.	[97]

Abbreviations: CysLTs, cysteinyl leukotrienes; DHA, docosahexaenoic acid; FE_{NO}, fraction of exhaled nitric oxide; FEV_1, forced expiratory volume; LTB_4, leukotriene B_4; LXA_4, lipoxin A_4; HSP72, heat shock protein 72; hsCRP, high sensitivity C-reactive protein; MPO, myeloperoxidase; sICAM3, soluble intracellular adhesion molecule 3; NF-κB, nuclear factor kappa-light-chain-enhancer of activated B cells; PBMC, peripheral blood mononuclear cells; PGI_2, prostacyclin; TLR-4, toll-like receptor-4; TNF-α, tumor necrosis factor- α, Composition: [1] Soy isoflavones tables: 8 mg glycitein, 28 mg daidzein, 29 mg genistein. [2] Soy milk powder with the BASF Vegapure®: stigmasterol, campesterol, and palmitates of β-sitosterol (maximun 23% w/w, 29% w/w, 55% w/w, respectively); [3] DHA-S; nutritional oil obtained from *Schizochitrium* sp. (marine alga).

3.2. Animal Studies

Table 2 summarizes animal studies that considered 5-LOX (together with other markers) a potential target of plant extracts and phenolic compounds administered through the diet.

Overall, a common anti-inflammatory effect observed in animals fed diets enriched with distinct plant extracts was the reduction of the carrageenan-induced paw edema volume as a prototypical assay to determine anti-inflammatory activity in vivo. Further exploration of the underlying molecular mechanisms showed inhibition of 5-LOX and COX-2 activities, lower LTB_4 and PGE_2 levels, and modulation of related inflammatory markers. Studies using animal models of arthritis described positive effects by ethanolic extracts of *Jasminum laceolarium, Vitex negundo, Dendropanax dentiger*, and *Pterospermun heterophyllum* [98–101] linked to the modulation of 5-LOX and COX-2 expression in PBMCs. However, it is unclear what compound(s) were responsible for the benefits described since crude extracts were used and the composition of the extracts was unexplored.

The consumption of diets supplemented with phenolic compounds (single or mixtures) showed beneficial anti-inflammatory effects via modulation of the 5-LOX pathway. Among the phenolic compounds, curcumin inhibited 5-LOX activity in PMNLs and reduced LTC_4 biosynthesis in paw edema (alone or combined with capsaicin) and in animal models of anaphylaxis [102,103]. Caffeic acid ameliorated brain and liver damage via downregulation of 5-LOX mRNA expression and protein level [104,105]. Flavocoxid, a dual inhibitor of 5-LOX and COX-2, targeted 5-LOX at the intestinal level, reducing LTB_4 levels and MPO activity (exerted by neutrophils) in colitis animal models [106]. As shown in Table 2, other phenolics such as catechins, quercetin, salidroside, nordihydroguaiaretic acid (NDGA), sesamol, and sesamin alleviated induced inflammation through similar molecular mechanisms, including 5-LOX-mediated inflammation.

3.3. In Vitro Studies

In vitro cell models are an essential tool to investigate the underlying molecular mechanisms by which phenolic compounds exert their effects. Numerous in vitro studies indicate that a wide range of phenolic compounds might exert anti-inflammatory effects by targeting the 5-LOX pathway (Table S1). Thus, caffeic acid, hydroxytyrosol, resveratrol, curcumin, NDGA and quercetin are compounds with the capacity to reduce the formation of 5-HETE, LTB_4 (and its ω-oxidized metabolites), and Cys-LTs [107,122–140]. Although desirable, these effects should be considered with caution since inhibition of the 5-LOX pathway could result in higher (pro-inflammatory) COX-2 metabolite levels by shunting the substrate, AA. This phenomenon has been described in stimulated leukocytes and mast cells treated with caffeic acid and NDGA, respectively [141,142].

Table 2. In vivo studies carried out with (poly)phenols-rich extracts or individual compounds in relation to inflammation and 5-LOX modulation.

Animal Model	Extract/Compound Assayed	Dose/Duration	Main Outcomes	Ref.
♂Wistar rats; carrageenin-soaked sponges implanted subcutaneously.	Quercetin and NDGA.	100 mg/kg; administered (16 and) 1 h prior sponge implantation.	No effect on leukocyte infiltration, oedema formation or PGE_2 and LTB_4 formation in A23187-stimulated leukocytes ex vivo.	[107]
♂Wistar rats; ethanol-induced gastric mucosal damage.	NDGA.	100 mg/kg (prepared in 0.25% carboxymethylcellulose); administration of a single dose for 30 min.	↓Gastric lesions and LTC_4 biosynthesis; no changes on PGs and TxB_2 production.	[108]
♂Sprague Dawley rats; cadmium-poisoned rats.	Crude catechin powder. [1]	0.25 and 0.5 g powder/100 g diet; 4 and 20 weeks.	Assays in platelets (ex vivo): ↓PLA_2 and COX-1 activity, ↓TxB_2; assays in aortic slices: ↓6-keto-$PGF_{1\alpha}$ and ↓LTB_4 in A23187-stimulated leukocytes; ↓5-LOX activity (enzyme fraction level).	[109, 110]
♂Wistar rats; carrageenan-induced paw inflammation.	Curcumin, capsaicin, and a mix of curcumin/capsaicin.	Curcumin: 0.2%, capsaicin: 0.015%, curcumin/capsaicin: 0.2/0.015%; 10 weeks.	↓Volume of paw edema; ↓5-LOX activity in the enzyme obtained from PMNL isolated from blood of the rats; ↓histamine release.	[103]
♂KM strain mice; aluminum-induced brain damage.	Caffeic acid.	10 or 30 mg/kg; days.	↓5-LOX mRNA expression in the cortical brain (at 10 and 30 mg/kg) and protein expression in hippocampi (only at 30 mg/kg); improvement of memory and learning functions together with ↓MDA, ↓ChAT, and ↓amyloid β and amyloid precursor protein.	[105]
♂Wistar rats; carrageenan-induced paw edema.	*Bacopa monniera* extracts.	20–200 mg/kg; 3 and 5 h.	↓Volume of paw edema; ↓LPS-induced TNF-α release in whole blood (ex vivo); ↓5-LOX and 15-LOX (IC_{50} = 100 µg/mL) as well as COX-1 (IC_{50} = 15.66 µg/mL) and COX-2 (IC_{50} = 1.22 µg) in A23187-induced rat mononuclear cells (ex vivo).	[111]
♀New Zealand white rabbits; hypercholesterolemic diet.	Quercetin.	25 mg/kg; 90 days.	↓5-LOX, 12-LOX, COX, activity in rabbit mononuclear cells; ↓CRP in plasma, ↓MPO activity in the aorta, and improvement of lipid profile and histopathological aortic features.	[112]

Table 2. *Cont.*

Animal Model	Extract/Compound Assayed	Dose/Duration	Main Outcomes	Ref.
♂Wistar albino rats; carrageenan-induced paw edema.	*Atropa acuminata* ethanolic extract.	62.5–500 mg/kg b.w.; up to 4 h.	↓LTB_4 and PGE_2 in carrageenan-treated paws; ↓leukocyte and neutrophil recruitment (no effect on mononuclear cells); ↓vascular permeability; ↓paw edema and exudate volume; modulation of the antioxidant status.	[113]
ICR mice; IgE/Ag-mediated passive systemic anaphylaxis.	Curcumin.	20 and 50 mg/kg; 1 h.	↓LTC_4, PGD_2 and histamine.	[102]
♂Wistar rats; CFA-induced rheumatoid arthritis.	*Xanthium strumarium* extract.	75 and 300 mg/kg; administered twice a day after the adjuvant arthritis induction for 28 days.	↓5-LOX and COX-2 expression in PBMCs; ↓paw swelling and arthritic score; ↓TNF-α and IL-1β together with ↑IL-10 in serum; improvement of histopathological features.	[114]
♂Wistar albino rats; CFA-induced arthritis.	*Vitex negundo* seed extract.	85 and 340 mg/kg/day; 28 days.	↓Paw swelling (from day 14th) and clinical arthritis score; attenuation of CFA-induced weight loss and index of spleen; ↓synovial lining hyperplasia and massive infiltration of mononuclear cells; ↓TNF-α, IL-1β (at both concentrations) and IL-6 (at 340 mg/kg); ↑IL-10; ↓COX-2 and 5-LOX expression in isolated PBMC.	[101]
♂Sprague Dawley albino rats; Isoproterenol-induced myocardial infarction.	*Ocimum sanctum* methanolic extract.	50–250 mg/kg b.w.; 30 days.	↓TBARS and NF-κB expression in the heart; ↓FLAP and BLT1 (mRNA) expression in the heart; ↓PLA, PLC and PLD activity, whereas ↑SOD activity and phospholipids in the heart; ↓CK-MB, LDH, hsCRP, LTB_4, TxB_2 (in serum); ↓COX-2 and 5-LOX activity in monocytes; attenuation of the effects of isoproterenol on cardiomyocytes.	[115]
♂BALB/c mice; ethanol-induced gastric ulcer.	Salidroside.	20 and 40 mg/kg; 6 days.	↓5-LOX and COX-2 protein expression; ↓LTB_4 level; modulation of the MAPK and NF-κB pathways; ↓IL-6, IL-1β and TNF-α; improvement of gastric histopathological features.	[116]
♂Wistar rats; carrageenan-induced rat paw edema.	*Jasminum laceolarium.*	100–400 mg/kg; 7 days.	↓5-LOX (only at 400 mg/kg) and COX-2 expression in serum; ↓paw edema.	[98]

Table 2. *Cont.*

Animal Model	Extract/Compound Assayed	Dose/Duration	Main Outcomes	Ref.
♀Sprague Dawley albino rats; HCD-fed atherosclerotic rats.	Quercetin.	25 mg/kg b.w.; 60 days.	↓5-LOX and COX activity as well as IL-6 expression (mRNA) in mononuclear cells; ↓NOS activity and CRP in serum; ↓MDA in serum and aorta.	[117]
♂Sprague Dawley rats; DNBS- and DSS-induced colitis.	Flavocoxid.	20 mg/kg/day (twice a day); 4–5 days.	↓LTB_4, PGE_2, 6-keto $PGF_{1\alpha}$, TxB_2, and TNF-α serum level; ↓MPO activity and MDA level in colon tissue; ↓histological damage and apoptosis; ↓CD3 in colon tissue.	[106]
♂Wistar rats; LPS-induced inflammation.	Sesamol, sesamin, and a mix of sesamol/sesamin.	10 mg/kg b.w.; 15 days.	↓LTB_4, LTC_4, MCP-1, IL-1β, CRP, and TNF-α serum level; ↓5-LOX, $cPLA_2$, and BLT-1 protein expression; ↓LTC_4 synthase protein expression (only sesamol and sesamin); ↓MDA (liver tissue and serum); ↓NO serum level (only sesamol); modulation of the antioxidant enzymes.	[118]
♂Wistar rats and New Zealand rabbits; MSU crystal-induced inflammation.	Salidroside.	40–80 mg/kg for rats and 20–40 mg/kg for rabbits; 6 days.	↓LTB_4, PGE_2, and 20-HETE level in synovial fluid macrophages; ↓COX-2, 5-LOX, and CYP4A1 mRNA expression (only at 80 mg/kg) in synovial fluid macrophages; ↓number of leukocytes and neutrophils; binding to the catalytic side of 5-LOX, COX-2, and CYP4A1 (in silico); ↑macrophages polarization; improvement of ankle swelling and histopathological features.	[119]
♂Wistar rats and Swiss albino mice; carrageenan-inflammation model and acetic acid-induced writhes.	*Salix tetrasperma* methanolic extract.	200–600 mg/kg; single dose.	↓COX-2, 5-LOX, PGE_2, TNF-α, iNOS level, and NF-κB activation in sciatic nerve and brain stem; ↓oxidative stress; ↓p53 positive cells in brain stem tissue; ↓paw edema in rats and leukocyte migration in mice; ↓acetic acid-induced writhes; ↑response latency to heat hyperalgesic stimulus; improvement of histopathological features; antipyretic effect.	[120]
♂Sprague Dawley rats; aluminum gluconate-induced liver injury.	Caffeic acid.	30 mg/kg.	↓5-LOX protein expression in the liver (no effect on COX-2); ↓TNF-α, IL-1β, IL-6, MDA and ↑SOD in the liver; improvement of histopathological features; modulation of the alteration of hepatic enzymes;	[104]

Table 2. *Cont.*

Animal Model	Extract/Compound Assayed	Dose/Duration	Main Outcomes	Ref.
♂Sprague Dawley rats; adjuvant-induced arthritis.	*Pterospermun heterophyllum* ethanolic extract.	160–640 mg/kg/day; 22 days.	↓5-LOX, COX-2, and MMP-2 expression in rat-isolated PBMCs; ↓TNF-α, IL-1β, IL-6, IL-17, RF, and CRP serum level; ↑IL-4 and IL-10 serum level; improvement of histopathological features of the knee joint and arthritis markers.	[100]
♂Sprague Dawley rats; adjuvant-induced arthritis.	*Dendropanax dentiger* ethanolic extract.	127.5–510 mg/kg/day; 22 days.	↓5-LOX, COX-2, and MMP-2 expression in rat-isolated PBMCs; ↓TNF-α, IL-1β, IL-6, IL-17, RF, and CRP serum level; ↑IL-4 and IL-10 serum level; improvement of histopathological features of the knee joint and arthritis markers.	[99]
♂Sprague Dawley rats; pharmaco-kinetic study. ♀Wistar rats; MIA-induced knee OA.	A curcumin formulation (NGUC) or turmeric extract.	NGUC or 95% turmeric extract as 100% (w/v) aqueous solution to deliver 200 mg/kg b.w. equivalent of curcuminoids.	Enhanced total curcuminoids bioavailability in NGUC-treated animals; reduced swelling; improvement of joint architecture; ↓IL-6, IL-1β, TNF-α, CRP, COMP, NF-κB, COX-2, MMP-3, 5-LOX, COX-2 in synovial fluid; ↓MDA, SOD, CAT, and GPx level.	[121]

Abbreviations: AA, arachidonic acid; BLT-1, leukotriene B_4 receptor-1; CAT, catalase; ChAT, choline acetyltransferase; CFA, complete Freund's adjuvant; CK-MB, creatinine kinase-MB; COMP, cartilage oligomeric matrix protein; cPLA2, cytoplasmic phospholipase A_2, CRP, C-reactive protein; DNBS, dinitrobenzenesulfonic acid; DSS, dextran sulphate sodium; EpiCat, epicatechin; EpiGal, epigallocatechin; FLAP, 5-lipoxygenase-activating protein; GPx, glutathione peroxidase; HCD, hypercholesterolemic diet; HETE, hydroxyeicosatetraenoic; hsCRP, high sensitive C-reactive protein; iNOS, inducible nitric oxide synthase; LDH, lactate dehydrogenase; LPS, lipopolysaccharide; MAPK, mitogen activated protein kinase; MCP-1, monocyte chemoattractant protein-1; MDA, malondialdehyde; MIA, monosodium iodoacetate; MMP, matrix metalloproteinase; MPO, myeloperoxidase; MSU, monosodium urate; NDGA, nordihydroguiaretic acid; NF-κB, nuclear factor kappa-light-chain-enhancer of activated B cells; NGUC, next generation ultrasol curcumin; NO, nitric oxide; OA, osteoarthritis; PBMCs, peripheral blood mononuclear cells; PLA, phospholipase A; PLA_2, phospholipase-A_2; PLC, phospholipase C; PLD, phospholipase D; RF, rheumatoid factor; SOD, superoxide dismutase; STZ, streptozocin; TBARS, thiobarbituric acid reactive substance; TNF-α, tumor necrosis factor-α; TxB_2, thromboxane B_2. Composition: [1] Crude catechin powder from green tea: 4.56% EpiGal, 4.52% EpiCat, 38.56% EpiGal gallate, 20.76% EpiCat gallate.

Blood- and peritoneal-isolated leukocytes are a widely employed cellular model, allowing to determine whether the phenolic compounds exert dual inhibition on 5-LOX and COX-2. In general, some phenolic compounds such as curcumin, resveratrol, and caffeic acid show the capacity to attenuate the biosynthesis of 5-LOX (5-HETE, LTB_4, Cys-LTs) and COX-2 (PGE_2) metabolites in stimulated leukocytes (Table S1). As shown in Table S1, different molecular mechanisms account for how phenolic compounds modulate the biosynthesis of eicosanoids. One of these mechanisms involves the interaction with the upstream regulator $cPLA_2$. This interaction can include lower levels of $cPLA_2$ (i.e., through down-regulation of mRNA expression), reduced activation (phosphorylation) or translocation, and inhibition of its enzymatic activity. Further mechanisms of action are related to the inhibition of 5-LOX translocation to the nuclear membrane where it becomes active, down-regulation of 5-LOX (and COX-2) mRNA and protein expression, as well as inhibition of 5-LOX (and COX-2) enzymatic activity (Table S1). Despite valuable results gained from these studies, other aspects of how phenolic compounds modulate the 5-LOX pathway are still unclear. For example, information on the interaction of phenolic compounds with (other) essential components of the 5-LOX pathway is scarce. None of the studies included in Table S1 tested whether the phenolic compounds target LT receptors (i.e., BLT1 and BLT2), FLAP, and/or LTC_4 hydrolase. Regarding LTA_4 hydrolase, only one study showed that isoflavones failed to inhibit (or promote) its enzymatic activity [143]. Thus, the characterization of the interaction of phenolic compounds with these elements will be essential in future studies.

Studies using single-cell lines provide essential information about how the phenolic compounds modulate 5-LOX (and COX-2) activity in different leukocytes. Curcumin reduced the formation of LTB_4 and PGE_2 in neutrophils [144] and macrophages [145]. Curcumin also targeted LTC_4 biosynthesis in mast cells by blocking 5-LOX translocation and $cPLA_2$ activation [102]. Quercetin reduced biosynthesis of LTB_4 in RBL-1 basophils, neutrophils, and murine PB-3c mast cells [146–148], while genistein acted on eosinophils decreasing LTC_4 synthesis via inhibition of 5-LOX translocation [95]. Likewise, the effect of silibinin on macrophages (Kupffer cells) was associated with lower LTB_4 levels while sparing the COX-2 pathway [149,150].

A common facet to the majority of in vitro studies (Table S1) is the treatment of leukocytes with glycosides and/or aglycones using conceivable concentrations at the intestinal level. Such conditions indicate that the effects observed on eicosanoid biosynthesis might be relevant in the context of intestinal inflammation [151]. These studies focused on describing the effects on leukocytes, thus overlooking the crucial role of other cells (i.e., intestinal cells). To date, a limited number of studies have investigated the interaction between dietary phenolics and the 5-LOX and COX-2 pathways in intestinal cells. In this regard, NDGA and geraniin (at 10 μM) decreased the synthesis of 5-HETE in stimulated AGS cells [152], whereas methoxy flavonoids isolated from *Chiliadenus montanus* failed to modulate the expression of 5-LOX in Caco-2 cells [153]. The effects of curcumin and tetrahydrocurcumin (THC) on HT-29 colon cancer cells were related to lower AA levels associated with cPLA2 inhibition; yet, how this affected the 5-LOX (or COX-2) pathway was not determined [145].

Leukocytes are abundant in the bloodstream, making them an excellent model to test in vitro the anti-inflammatory effects of phase-II metabolites. However, according to our analysis, only two studies have approached the effect of conjugated metabolites on 5-LOX and COX-2 using leukocytes (Table S1). Among the conjugated metabolites tested, 3′-*O*-methyl-quercetin reduced the biosynthesis of LTB_4 (at 2 μM) and PGE_2 (IC_{50} = 2 μM) [154], whereas the phase II conjugated urolithins (ellagic acid-derived metabolites) were inactive against 5-LOX, COX-2, and the 5-LOX/COX-2 crossover pathways [155]. Under the same conditions, their free forms (quercetin, Uro-A, IsoUro-A, and Uro-C) effectively decreased 5-HETE, LTB_4, PGE_2, HKE_2, and HKD_2 levels in a dose-dependent manner. Although the free forms are hardly found in the bloodstream, their presence in inflammatory environments is conceivable via deconjugation, as described for luteolin, quercetin, resveratrol, Uro-A, and curcumin [77,82–91].

To date, no studies describe the in vitro effect of phenolic compounds or derived metabolites on lipoxin biosynthesis. Only two studies have investigated the effect of phenolics on the 5-LOX/COX-2 crossover pathway and found that curcumin and urolithins (Uro-A, IsoUro-A, and Uro-C) showed the capacity to inhibit HKE_2 and HKD_2 formation in a mixture of stimulated leukocytes [47,155]. This limited evidence underscores the need for future studies on the biosynthesis of complex 5-LOX dependent eicosanoids using isolated cell preparation or co-incubations (i.e., platelets with neutrophils) in the presence of phenolic compounds.

4. Conclusions

This systematic review focuses on the effect of phenolic compounds on the 5-LOX pathway. The current evidence linking modulation of the 5-LOX pathway and the anti-inflammatory effects of phenolic compounds is still weak. One of the main reasons comes from the low number of human studies and clinical trials, which are essential to test the preventive and/or therapeutic effects of phenolic compounds. Thus, well-designed and robust clinical trials in patients suffering from 5-LOX-related inflammatory diseases (e.g., asthma) would be desirable. The number of animal studies is also small and more research is needed using equivalent conditions to those reported in humans, including adequate exposure times and doses of phenolic-rich foodstuff.

As expected, a higher number of in vitro studies describe the anti-inflammatory effects of phenolic compounds targeting the 5-LOX pathway. The in vitro studies focus on using mixtures of leukocytes or individual cells such as neutrophils, eosinophils, basophils, and/or mast cells. New studies should consider enlarging the range of 5-LOX-expressing cells such as dendritic cells and should consider the interaction with other cell lines such as endothelial (i.e., atherosclerosis) or intestinal cells and microbiota (i.e., intestinal inflammation). The specific role of immune-related receptors regarding the modulation of the 5-LOX pathway by phenolic compounds is an additional mechanism (not considered hitherto) that deserves attention, such as the direct effect of phenolic compounds on LT receptors or via modulation of TLR-4, which regulates CysLT1 expression in dendritic cells [156]. These studies should avoid the use of unreasonable concentrations and metabolic forms of phenolic compounds (considering metabolism and bioavailability) since this inadequate design limits the physiological significance (from an in vivo point of view) of these investigations. Hence, the design of future in vitro studies should follow the roadmap set elsewhere [157] to provide physiologically relevant results.

Another key point to contemplate is 5-LOX/COX-2 dual inhibition. Targeting the 5-LOX pathway might lead to undesirable side effects due to AA shunting towards pro-inflammatory COX-2-derived metabolites. Therefore, identifying and studying the biological activity of phenolic compounds that act as dual inhibitors is critical to avoid single inhibition and resulting side effects. Otherwise, a possible drawback of this approach is the reduced biosynthesis of anti-inflammatory eicosanoids due to blocking cyclooxygenases and lipoxygenases (including 5-LOX). Whether the phenolic compounds can exert anti-inflammatory effects by increasing the synthesis of lipoxins is a question not addressed by the available preclinical studies. The expanded analysis of pro- and anti-inflammatory eicosanoids in future studies will improve the understanding of how phenolic compounds modulate the inflammatory response through the 5-LOX pathway.

Supplementary Materials: The following are available online at https://www.mdpi.com/article/10.3390/ijms22157937/s1, Table S1: Modulation of 5-LOX activity and comparison with COX-2 activity by (poly)phenols in cellular models.

Author Contributions: Conceptualization, J.A.G.-B., A.G.-S., and J.C.E.; funding: J.A.G.-B., J.C.E., and C.S.; methodology, J.A.G.-B. and A.G.-S.; investigation, J.A.G.-B. and A.G.-S.; writing—original draft preparation, J.A.G.-B. and A.G.-S.; writing—review and editing, all authors. All authors have read and agreed to the published version of the manuscript.

Funding: This research was supported by the project PID2019-103914RB-I00 from the Ministry of Science and Innovation (MICINN, Spain). J.A.G.-B. was supported by a Standard European Marie Curie Fellowship from the European Commission. This project has received funding from the European Union's Horizon 2020 research and innovation program under the Marie Sklodowska-Curie grant agreement no. 838991. CS is supported by NIH awards GM076592 and GM118412.

Conflicts of Interest: The authors declare no conflict of interest.

References

1. Brash, A.R. Lipoxygenases: Occurrence, Functions, Catalysis, and Acquisition of Substrate. *J. Biol. Chem.* **1999**, *274*, 23679–23682. [CrossRef] [PubMed]
2. Shi, Y.; Mandal, R.; Singh, A.; Singh, A.P. Legume Lipoxygenase: Strategies for Application in Food Industry. *Legum. Sci.* **2020**, *2*, e44. [CrossRef]
3. Baysal, T.; Demirdöven, A. Lipoxygenase in Fruits and Vegetables: A Review. *Enzyme Microb. Technol.* **2007**, *40*, 491–496. [CrossRef]
4. Ogorodnikova, A.V.; Mukhitova, F.K.; Grechkin, A.N. Oxylipins in the Spikemoss Selaginella Martensii: Detection of Divinyl Ethers, 12-Oxophytodienoic Acid and Related Cyclopentenones. *Phytochemistry* **2015**, *118*, 42–50. [CrossRef]
5. Bennett, M.; Gilroy, D.W. Lipid Mediators in Inflammation. *Microbiol. Spectr.* **2016**, *4*. [CrossRef] [PubMed]
6. Lynch, K.R.; O'Neill, G.P.; Liu, Q.; Im, D.S.; Sawyer, N.; Metters, K.M.; Coulombe, N.; Abramovitz, M.; Figueroa, D.J.; Zeng, Z.; et al. Characterization of the Human Cysteinyl Leukotriene CysLT1 Receptor. *Nature* **1999**, *399*, 789–793. [CrossRef] [PubMed]
7. Mashima, R.; Okuyama, T. The Role of Lipoxygenases in Pathophysiology; New Insights and Future Perspectives. *Redox Biol.* **2015**, *6*, 297–310. [CrossRef]
8. Muñoz-Ramírez, A.; Mascayano-Collado, C.; Barriga, A.; Echeverría, J.; Urzúa, A. Inhibition of Soybean 15-Lipoxygenase and Human 5-Lipoxygenase by Extracts of Leaves, Stem Bark, Phenols and Catechols Isolated From Lithraea Caustica (Anacardiaceae). *Front. Pharmacol.* **2020**, *11*, 594257. [CrossRef]
9. Kotas, M.E.; Medzhitov, R. Homeostasis, Inflammation, and Disease Susceptibility. *Cell* **2015**, *160*, 816–827. [CrossRef]
10. Medzhitov, R. Origin and Physiological Roles of Inflammation. *Nature* **2008**, *454*, 428–435. [CrossRef]
11. Haeggström, J.Z. Leukotriene Biosynthetic Enzymes as Therapeutic Targets. *J. Clin. Investig.* **2018**, *128*, 2680–2690. [CrossRef] [PubMed]
12. Haeggström, J.Z.; Funk, C.D. Lipoxygenase and Leukotriene Pathways: Biochemistry, Biology, and Roles in Disease. *Chem. Rev.* **2011**, *111*, 5866–5898. [CrossRef] [PubMed]
13. Griesser, M.; Suzuki, T.; Tejera, N.; Mont, S.; Boeglin, W.E.; Pozzi, A.; Schneider, C. Biosynthesis of Hemiketal Eicosanoids by Cross-over of the 5-Lipoxygenase and Cyclooxygenase-2 Pathways. *Proc. Natl. Acad. Sci. USA* **2011**, *108*, 6945–6950. [CrossRef] [PubMed]
14. Calder, P.C. Eicosanoids. *Essays Biochem.* **2020**, *64*, 423–441. [CrossRef] [PubMed]
15. Peters-Golden, M.; Henderson, W.R. Leukotrienes. *N. Engl. J. Med.* **2007**, *357*, 1841–1854. [CrossRef] [PubMed]
16. Bruno, F.; Spaziano, G.; Liparulo, A.; Roviezzo, F.; Nabavi, S.M.; Sureda, A.; Filosa, R.; D'Agostino, B. Recent Advances in the Search for Novel 5-Lipoxygenase Inhibitors for the Treatment of Asthma. *Eur. J. Med. Chem.* **2018**, *153*, 65–72. [CrossRef] [PubMed]
17. Poff, C.D.; Balazy, M. Drugs That Target Lipoxygenases and Leukotrienes as Emerging Therapies for Asthma and Cancer. *Curr. Drug Targets Inflamm. Allergy* **2004**, *3*, 19–33. [CrossRef]
18. Chen, L.; Teng, H.; Jia, Z.; Battino, M.; Miron, A.; Yu, Z.; Cao, H.; Xiao, J. Intracellular Signaling Pathways of Inflammation Modulated by Dietary Flavonoids: The Most Recent Evidence. *Crit. Rev. Food Sci. Nutr.* **2018**, *58*, 2908–2924. [CrossRef]
19. Yahfoufi, N.; Alsadi, N.; Jambi, M.; Matar, C. The Immunomodulatory and Anti-Inflammatory Role of Polyphenols. *Nutrients* **2018**, *10*, 1618. [CrossRef]
20. Borgeat, P.; Samuelsson, B. Transformation of Arachidonic Acid by Rabbit Polymorphonuclear Leukocytes. Formation of a Novel Dihydroxyeicosatetraenoic Acid. *J. Biol. Chem.* **1979**, *254*, 2643–2646. [CrossRef]
21. Borgeat, P.; Samuelsson, B. Arachidonic Acid Metabolism in Polymorphonuclear Leukocytes: Unstable Intermediate in Formation of Dihydroxy Acids. *Proc. Natl. Acad. Sci. USA* **1979**, *76*, 3213–3217. [CrossRef]
22. Norris, P.C.; Gosselin, D.; Reichart, D.; Glass, C.K.; Dennis, E.A. Phospholipase A2 Regulates Eicosanoid Class Switching during Inflammasome Activation. *Proc. Natl. Acad. Sci. USA* **2014**, *111*, 12746–12751. [CrossRef]
23. Qi, H.-Y.; Shelhamer, J.H. Toll-like Receptor 4 Signaling Regulates Cytosolic Phospholipase A2 Activation and Lipid Generation in Lipopolysaccharide-Stimulated Macrophages. *J. Biol. Chem.* **2005**, *280*, 38969–38975. [CrossRef]
24. Martinez Molina, D.; Wetterholm, A.; Kohl, A.; McCarthy, A.A.; Niegowski, D.; Ohlson, E.; Hammarberg, T.; Eshaghi, S.; Haeggström, J.Z.; Nordlund, P. Structural Basis for Synthesis of Inflammatory Mediators by Human Leukotriene C4 Synthase. *Nature* **2007**, *448*, 613–616. [CrossRef]

25. Murphy, R.C.; Gijón, M.A. Biosynthesis and Metabolism of Leukotrienes. *Biochem. J.* **2007**, *405*, 379–395. [CrossRef] [PubMed]
26. Powell, W.S.; Gravelle, F.; Gravel, S. Metabolism of 5(S)-Hydroxy-6,8,11,14-Eicosatetraenoic Acid and Other 5(S)-Hydroxyeicosanoids by a Specific Dehydrogenase in Human Polymorphonuclear Leukocytes. *J. Biol. Chem.* **1992**, *267*, 19233–19241. [CrossRef]
27. Sala, A.; Folco, G.; Murphy, R.C. Transcellular Biosynthesis of Eicosanoids. *Pharmacol. Rep. PR* **2010**, *62*, 503–510. [CrossRef]
28. Feinmark, S.J.; Cannon, P.J. Endothelial Cell Leukotriene C4 Synthesis Results from Intercellular Transfer of Leukotriene A4 Synthesized by Polymorphonuclear Leukocytes. *J. Biol. Chem.* **1986**, *261*, 16466–16472. [CrossRef]
29. McGee, J.E.; Fitzpatrick, F.A. Erythrocyte-Neutrophil Interactions: Formation of Leukotriene B4 by Transcellular Biosynthesis. *Proc. Natl. Acad. Sci. USA* **1986**, *83*, 1349–1353. [CrossRef] [PubMed]
30. Fabre, J.-E.; Goulet, J.L.; Riche, E.; Nguyen, M.; Coggins, K.; Offenbacher, S.; Koller, B.H. Transcellular Biosynthesis Contributes to the Production of Leukotrienes during Inflammatory Responses in Vivo. *J. Clin. Investig.* **2002**, *109*, 1373–1380. [CrossRef] [PubMed]
31. Zarini, S.; Gijón, M.A.; Ransome, A.E.; Murphy, R.C.; Sala, A. Transcellular Biosynthesis of Cysteinyl Leukotrienes in Vivo during Mouse Peritoneal Inflammation. *Proc. Natl. Acad. Sci. USA* **2009**, *106*, 8296–8301. [CrossRef]
32. Folco, G.; Murphy, R.C. Eicosanoid Transcellular Biosynthesis: From Cell-Cell Interactions to in Vivo Tissue Responses. *Pharmacol. Rev.* **2006**, *58*, 375–388. [CrossRef]
33. Edenius, C.; Heidvall, K.; Lindgren, J.A. Novel Transcellular Interaction: Conversion of Granulocyte-Derived Leukotriene A4 to Cysteinyl-Containing Leukotrienes by Human Platelets. *Eur. J. Biochem.* **1988**, *178*, 81–86. [CrossRef]
34. Romano, M.; Serhan, C.N. Lipoxin Generation by Permeabilized Human Platelets. *Biochemistry* **1992**, *31*, 8269–8277. [CrossRef] [PubMed]
35. Serhan, C.N.; Hamberg, M.; Samuelsson, B. Lipoxins: Novel Series of Biologically Active Compounds Formed from Arachidonic Acid in Human Leukocytes. *Proc. Natl. Acad. Sci. USA* **1984**, *81*, 5335–5339. [CrossRef]
36. Giménez-Bastida, J.A.; Boeglin, W.E.; Boutaud, O.; Malkowski, M.G.; Schneider, C. Residual Cyclooxygenase Activity of Aspirin-Acetylated COX-2 Forms 15 R-Prostaglandins That Inhibit Platelet Aggregation. *FASEB J. Off. Publ. Fed. Am. Soc. Exp. Biol.* **2019**, *33*, 1033–1041. [CrossRef]
37. Lecomte, M.; Laneuville, O.; Ji, C.; DeWitt, D.L.; Smith, W.L. Acetylation of Human Prostaglandin Endoperoxide Synthase-2 (Cyclooxygenase-2) by Aspirin. *J. Biol. Chem.* **1994**, *269*, 13207–13215. [CrossRef]
38. Clària, J.; Serhan, C.N. Aspirin Triggers Previously Undescribed Bioactive Eicosanoids by Human Endothelial Cell-Leukocyte Interactions. *Proc. Natl. Acad. Sci. USA* **1995**, *92*, 9475–9479. [CrossRef] [PubMed]
39. Serhan, C.N.; Oliw, E. Unorthodox Routes to Prostanoid Formation: New Twists in Cyclooxygenase-Initiated Pathways. *J. Clin. Investig.* **2001**, *107*, 1481–1489. [CrossRef] [PubMed]
40. Chiang, N.; Arita, M.; Serhan, C.N. Anti-Inflammatory Circuitry: Lipoxin, Aspirin-Triggered Lipoxins and Their Receptor ALX. *Prostaglandins Leukot. Essent. Fat. Acids* **2005**, *73*, 163–177. [CrossRef] [PubMed]
41. Ringholz, F.C.; Buchanan, P.J.; Clarke, D.T.; Millar, R.G.; McDermott, M.; Linnane, B.; Harvey, B.J.; McNally, P.; Urbach, V. Reduced 15-Lipoxygenase 2 and Lipoxin A4/Leukotriene B4 Ratio in Children with Cystic Fibrosis. *Eur. Respir. J.* **2014**, *44*, 394–404. [CrossRef] [PubMed]
42. Snodgrass, R.G.; Brüne, B. Regulation and Functions of 15-Lipoxygenases in Human Macrophages. *Front. Pharmacol.* **2019**, *10*, 719. [CrossRef] [PubMed]
43. Schneider, C.; Boeglin, W.E.; Yin, H.; Stec, D.F.; Voehler, M. Convergent Oxygenation of Arachidonic Acid by 5-Lipoxygenase and Cyclooxygenase-2. *J. Am. Chem. Soc.* **2006**, *128*, 720–721. [CrossRef] [PubMed]
44. Nakashima, F.; Suzuki, T.; Gordon, O.N.; Golding, D.; Okuno, T.; Giménez-Bastida, J.A.; Yokomizo, T.; Schneider, C. Biosynthetic Crossover of 5-Lipoxygenase and Cyclooxygenase-2 Yields 5-Hydroxy-PGE2 and 5-Hydroxy-PGD2. *JACS Au* **2021**. [CrossRef]
45. Mulugeta, S.; Suzuki, T.; Hernandez, N.T.; Griesser, M.; Boeglin, W.E.; Schneider, C. Identification and Absolute Configuration of Dihydroxy-Arachidonic Acids Formed by Oxygenation of 5S-HETE by Native and Aspirin-Acetylated COX-2. *J. Lipid Res.* **2010**, *51*, 575–585. [CrossRef] [PubMed]
46. Tejera, N.; Boeglin, W.E.; Suzuki, T.; Schneider, C. COX-2-Dependent and -Independent Biosynthesis of Dihydroxy-Arachidonic Acids in Activated Human Leukocytes. *J. Lipid Res.* **2012**, *53*, 87–94. [CrossRef]
47. Giménez-Bastida, J.A.; Shibata, T.; Uchida, K.; Schneider, C. Roles of 5-Lipoxygenase and Cyclooxygenase-2 in the Biosynthesis of Hemiketals E2 and D2 by Activated Human Leukocytes. *FASEB J. Off. Publ. Fed. Am. Soc. Exp. Biol.* **2017**, *31*, 1867–1878. [CrossRef]
48. Ford-Hutchinson, A.W.; Bray, M.A.; Doig, M.V.; Shipley, M.E.; Smith, M.J. Leukotriene B, a Potent Chemokinetic and Aggregating Substance Released from Polymorphonuclear Leukocytes. *Nature* **1980**, *286*, 264–265. [CrossRef]
49. Chen, X.S.; Sheller, J.R.; Johnson, E.N.; Funk, C.D. Role of Leukotrienes Revealed by Targeted Disruption of the 5-Lipoxygenase Gene. *Nature* **1994**, *372*, 179–182. [CrossRef]
50. Dennis, E.A.; Norris, P.C. Eicosanoid Storm in Infection and Inflammation. *Nat. Rev. Immunol.* **2015**, *15*, 511–523. [CrossRef]
51. Sasaki, F.; Yokomizo, T. The Leukotriene Receptors as Therapeutic Targets of Inflammatory Diseases. *Int. Immunol.* **2019**, *31*, 607–615. [CrossRef]
52. Lämmermann, T.; Afonso, P.V.; Angermann, B.R.; Wang, J.M.; Kastenmüller, W.; Parent, C.A.; Germain, R.N. Neutrophil Swarms Require LTB4 and Integrins at Sites of Cell Death in Vivo. *Nature* **2013**, *498*, 371–375. [CrossRef] [PubMed]

53. Pérez-Jeldres, T.; Tyler, C.J.; Boyer, J.D.; Karuppuchamy, T.; Bamias, G.; Dulai, P.S.; Boland, B.S.; Sandborn, W.J.; Patel, D.R.; Rivera-Nieves, J. Cell Trafficking Interference in Inflammatory Bowel Disease: Therapeutic Interventions Based on Basic Pathogenesis Concepts. *Inflamm. Bowel Dis.* **2019**, *25*, 270–282. [CrossRef] [PubMed]
54. Qiu, H.; Gabrielsen, A.; Agardh, H.E.; Wan, M.; Wetterholm, A.; Wong, C.-H.; Hedin, U.; Swedenborg, J.; Hansson, G.K.; Samuelsson, B.; et al. Expression of 5-Lipoxygenase and Leukotriene A4 Hydrolase in Human Atherosclerotic Lesions Correlates with Symptoms of Plaque Instability. *Proc. Natl. Acad. Sci. USA* **2006**, *103*, 8161–8166. [CrossRef]
55. Spanbroek, R.; Grabner, R.; Lotzer, K.; Hildner, M.; Urbach, A.; Ruhling, K.; Moos, M.P.W.; Kaiser, B.; Cohnert, T.U.; Wahlers, T.; et al. Expanding Expression of the 5-Lipoxygenase Pathway within the Arterial Wall during Human Atherogenesis. *Proc. Natl. Acad. Sci. USA* **2003**, *100*, 1238–1243. [CrossRef]
56. Busse, null Leukotrienes and Inflammation. *Am. J. Respir. Crit. Care Med.* **1998**, *157*, S210–S213. [CrossRef] [PubMed]
57. Hui, Y.; Cheng, Y.; Smalera, I.; Jian, W.; Goldhahn, L.; Fitzgerald, G.A.; Funk, C.D. Directed Vascular Expression of Human Cysteinyl Leukotriene 2 Receptor Modulates Endothelial Permeability and Systemic Blood Pressure. *Circulation* **2004**, *110*, 3360–3366. [CrossRef] [PubMed]
58. Moos, M.P.W.; Mewburn, J.D.; Kan, F.W.K.; Ishii, S.; Abe, M.; Sakimura, K.; Noguchi, K.; Shimizu, T.; Funk, C.D. Cysteinyl Leukotriene 2 Receptor-Mediated Vascular Permeability via Transendothelial Vesicle Transport. *FASEB J. Off. Publ. Fed. Am. Soc. Exp. Biol.* **2008**, *22*, 4352–4362. [CrossRef]
59. Samuelsson, B. Leukotrienes: Mediators of Immediate Hypersensitivity Reactions and Inflammation. *Science* **1983**, *220*, 568–575. [CrossRef]
60. Erlemann, K.-R.; Rokach, J.; Powell, W.S. Oxidative Stress Stimulates the Synthesis of the Eosinophil Chemoattractant 5-Oxo-6,8,11,14-Eicosatetraenoic Acid by Inflammatory Cells. *J. Biol. Chem.* **2004**, *279*, 40376–40384. [CrossRef]
61. Grant, G.E.; Rokach, J.; Powell, W.S. 5-Oxo-ETE and the OXE Receptor. *Prostaglandins Other Lipid Mediat.* **2009**, *89*, 98–104. [CrossRef]
62. Powell, W.S.; Gravel, S.; MacLeod, R.J.; Mills, E.; Hashefi, M. Stimulation of Human Neutrophils by 5-Oxo-6,8,11,14-Eicosatetraenoic Acid by a Mechanism Independent of the Leukotriene B4 Receptor. *J. Biol. Chem.* **1993**, *268*, 9280–9286. [CrossRef]
63. Boer, R.E.; Giménez-Bastida, J.A.; Boutaud, O.; Jana, S.; Schneider, C.; Sulikowski, G.A. Total Synthesis and Biological Activity of the Arachidonic Acid Metabolite Hemiketal E2. *Org. Lett.* **2018**, *20*, 4020–4022. [CrossRef]
64. Giménez-Bastida, J.A.; Suzuki, T.; Sprinkel, K.C.; Boeglin, W.E.; Schneider, C. Biomimetic Synthesis of Hemiketal Eicosanoids for Biological Testing. *Prostaglandins Other Lipid Mediat.* **2017**, *132*, 41–46. [CrossRef]
65. Chicco, F.; Magrì, S.; Cingolani, A.; Paduano, D.; Pesenti, M.; Zara, F.; Tumbarello, F.; Urru, E.; Melis, A.; Casula, L.; et al. Multidimensional Impact of Mediterranean Diet on IBD Patients. *Inflamm. Bowel Dis.* **2021**, *27*, 1–9. [CrossRef] [PubMed]
66. Garcia-Marcos, L. Mediterranean Diet as a Protection against Asthma: Still Another Brick in Building a Causative Association. *Allergol. Immunopathol. (Madr.)* **2016**, *44*, 97–98. [CrossRef] [PubMed]
67. Temple, N.J.; Guercio, V.; Tavani, A. The Mediterranean Diet and Cardiovascular Disease: Gaps in the Evidence and Research Challenges. *Cardiol. Rev.* **2019**, *27*, 127–130. [CrossRef] [PubMed]
68. Oz, H.S. Chronic Inflammatory Diseases and Green Tea Polyphenols. *Nutrients* **2017**, *9*, 660. [CrossRef]
69. Widmer, R.J.; Freund, M.A.; Flammer, A.J.; Sexton, J.; Lennon, R.; Romani, A.; Mulinacci, N.; Vinceri, F.F.; Lerman, L.O.; Lerman, A. Beneficial Effects of Polyphenol-Rich Olive Oil in Patients with Early Atherosclerosis. *Eur. J. Nutr.* **2013**, *52*, 1223–1231. [CrossRef] [PubMed]
70. Rahman, I.; Biswas, S.K.; Kirkham, P.A. Regulation of Inflammation and Redox Signaling by Dietary Polyphenols. *Biochem. Pharmacol.* **2006**, *72*, 1439–1452. [CrossRef] [PubMed]
71. Cortés-Martín, A.; Selma, M.V.; Tomás-Barberán, F.A.; González-Sarrías, A.; Espín, J.C. Where to Look into the Puzzle of Polyphenols and Health? The Postbiotics and Gut Microbiota Associated with Human Metabotypes. *Mol. Nutr. Food Res.* **2020**, *64*, e1900952. [CrossRef]
72. Espín, J.C.; González-Sarrías, A.; Tomás-Barberán, F.A. The Gut Microbiota: A Key Factor in the Therapeutic Effects of (Poly)Phenols. *Biochem. Pharmacol.* **2017**, *139*, 82–93. [CrossRef] [PubMed]
73. Frankenfeld, C.L. Cardiometabolic Risk and Gut Microbial Phytoestrogen Metabolite Phenotypes. *Mol. Nutr. Food Res.* **2017**, *61*. [CrossRef] [PubMed]
74. Olthof, M.R.; Hollman, P.C.H.; Buijsman, M.N.C.P.; van Amelsvoort, J.M.M.; Katan, M.B. Chlorogenic Acid, Quercetin-3-Rutinoside and Black Tea Phenols Are Extensively Metabolized in Humans. *J. Nutr.* **2003**, *133*, 1806–1814. [CrossRef] [PubMed]
75. Dei Cas, M.; Ghidoni, R. Dietary Curcumin: Correlation between Bioavailability and Health Potential. *Nutrients* **2019**, *11*, 2147. [CrossRef] [PubMed]
76. Giménez-Bastida, J.A.; Martínez-Florensa, M.; Espín, J.-C.; Tomás-Barberán, F.A.; García-Conesa, M.-T. A Citrus Extract Containing Flavanones Represses Plasminogen Activator Inhibitor-1 (PAI-1) Expression and Regulates Multiple Inflammatory, Tissue Repair, and Fibrosis Genes in Human Colon Fibroblasts. *J. Agric. Food Chem.* **2009**, *57*, 9305–9315. [CrossRef]
77. Ávila-Gálvez, M.Á.; González-Sarrías, A.; Martínez-Díaz, F.; Abellán, B.; Martínez-Torrano, A.J.; Fernández-López, A.J.; Giménez-Bastida, J.A.; Espín, J.C. Disposition of Dietary Polyphenols in Breast Cancer Patients' Tumors, and Their Associated Anticancer Activity: The Particular Case of Curcumin. *Mol. Nutr. Food Res.* **2021**, e2100163. [CrossRef]

78. Ávila-Gálvez, M.Á.; Giménez-Bastida, J.A.; González-Sarrías, A.; Espín, J.C. New Insights into the Metabolism of the Flavanones Eriocitrin and Hesperidin: A Comparative Human Pharmacokinetic Study. *Antioxid. Basel Switz.* **2021**, *10*, 435. [CrossRef]
79. González-Sarrías, A.; Espín, J.C.; Tomás-Barberán, F.A. Non-Extractable Polyphenols Produce Gut Microbiota Metabolites That Persist in Circulation and Show Anti-Inflammatory and Free Radical-Scavenging Effects. *Trends Food Sci. Technol.* **2017**, *69*, 281–288. [CrossRef]
80. Erlund, I.; Kosonen, T.; Alfthan, G.; Mäenpää, J.; Perttunen, K.; Kenraali, J.; Parantainen, J.; Aro, A. Pharmacokinetics of Quercetin from Quercetin Aglycone and Rutin in Healthy Volunteers. *Eur. J. Clin. Pharmacol.* **2000**, *56*, 545–553. [CrossRef]
81. González-Sarrías, A.; Giménez-Bastida, J.A.; García-Conesa, M.T.; Gómez-Sánchez, M.B.; García-Talavera, N.V.; Gil-Izquierdo, A.; Sánchez-Alvarez, C.; Fontana-Compiano, L.O.; Morga-Egea, J.P.; Pastor-Quirante, F.A.; et al. Occurrence of Urolithins, Gut Microbiota Ellagic Acid Metabolites and Proliferation Markers Expression Response in the Human Prostate Gland upon Consumption of Walnuts and Pomegranate Juice. *Mol. Nutr. Food Res.* **2010**, *54*, 311–322. [CrossRef]
82. Kunihiro, A.G.; Brickey, J.A.; Frye, J.B.; Luis, P.B.; Schneider, C.; Funk, J.L. Curcumin, but Not Curcumin-Glucuronide, Inhibits Smad Signaling in TGFβ-Dependent Bone Metastatic Breast Cancer Cells and Is Enriched in Bone Compared to Other Tissues. *J. Nutr. Biochem.* **2019**, *63*, 150–156. [CrossRef] [PubMed]
83. Shimoi, K.; Saka, N.; Nozawa, R.; Sato, M.; Amano, I.; Nakayama, T.; Kinae, N. Deglucuronidation of a Flavonoid, Luteolin Monoglucuronide, during Inflammation. *Drug Metab. Dispos. Biol. Fate Chem.* **2001**, *29*, 1521–1524. [PubMed]
84. Shimoi, K.; Saka, N.; Kaji, K.; Nozawa, R.; Kinae, N. Metabolic Fate of Luteolin and Its Functional Activity at Focal Site. *BioFactors Oxf. Engl.* **2000**, *12*, 181–186. [CrossRef] [PubMed]
85. Galindo, P.; Rodriguez-Gómez, I.; González-Manzano, S.; Dueñas, M.; Jiménez, R.; Menéndez, C.; Vargas, F.; Tamargo, J.; Santos-Buelga, C.; Pérez-Vizcaíno, F.; et al. Glucuronidated Quercetin Lowers Blood Pressure in Spontaneously Hypertensive Rats via Deconjugation. *PLoS ONE* **2012**, *7*, e32673. [CrossRef] [PubMed]
86. Ishisaka, A.; Kawabata, K.; Miki, S.; Shiba, Y.; Minekawa, S.; Nishikawa, T.; Mukai, R.; Terao, J.; Kawai, Y. Mitochondrial Dysfunction Leads to Deconjugation of Quercetin Glucuronides in Inflammatory Macrophages. *PLoS ONE* **2013**, *8*, e80843. [CrossRef]
87. Kawai, Y.; Nishikawa, T.; Shiba, Y.; Saito, S.; Murota, K.; Shibata, N.; Kobayashi, M.; Kanayama, M.; Uchida, K.; Terao, J. Macrophage as a Target of Quercetin Glucuronides in Human Atherosclerotic Arteries: Implication in the Anti-Atherosclerotic Mechanism of Dietary Flavonoids. *J. Biol. Chem.* **2008**, *283*, 9424–9434. [CrossRef] [PubMed]
88. Menendez, C.; Dueñas, M.; Galindo, P.; González-Manzano, S.; Jimenez, R.; Moreno, L.; Zarzuelo, M.J.; Rodríguez-Gómez, I.; Duarte, J.; Santos-Buelga, C.; et al. Vascular Deconjugation of Quercetin Glucuronide: The Flavonoid Paradox Revealed? *Mol. Nutr. Food Res.* **2011**, *55*, 1780–1790. [CrossRef]
89. Terao, J.; Murota, K.; Kawai, Y. Conjugated Quercetin Glucuronides as Bioactive Metabolites and Precursors of Aglycone in Vivo. *Food Funct.* **2011**, *2*, 11–17. [CrossRef]
90. Fernández-Castillejo, S.; Macià, A.; Motilva, M.-J.; Catalán, Ú.; Solà, R. Endothelial Cells Deconjugate Resveratrol Metabolites to Free Resveratrol: A Possible Role in Tissue Factor Modulation. *Mol. Nutr. Food Res.* **2019**, *63*, e1800715. [CrossRef]
91. Ávila-Gálvez, M.A.; Giménez-Bastida, J.A.; González-Sarrías, A.; Espín, J.C. Tissue Deconjugation of Urolithin A Glucuronide to Free Urolithin A in Systemic Inflammation. *Food Funct.* **2019**, *10*, 3135–3141. [CrossRef]
92. Page, M.J.; McKenzie, J.E.; Bossuyt, P.M.; Boutron, I.; Hoffmann, T.C.; Mulrow, C.D.; Shamseer, L.; Tetzlaff, J.M.; Akl, E.A.; Brennan, S.E.; et al. The PRISMA 2020 Statement: An Updated Guideline for Reporting Systematic Reviews. *PLoS Med.* **2021**, *18*, e1003583. [CrossRef] [PubMed]
93. Ho, X.L.; Liu, J.J.H.; Loke, W.M. Plant Sterol-Enriched Soy Milk Consumption Modulates 5-Lipoxygenase, 12-Lipoxygenase, and Myeloperoxidase Activities in Healthy Adults – a Randomized-Controlled Trial. *Free Radic. Res.* **2016**, *50*, 1396–1407. [CrossRef] [PubMed]
94. Schramm, D.D.; Wang, J.F.; Holt, R.R.; Ensunsa, J.L.; Gonsalves, J.L.; Lazarus, S.A.; Schmitz, H.H.; German, J.B.; Keen, C.L. Chocolate Procyanidins Decrease the Leukotriene-Prostacyclin Ratio in Humans and Human Aortic Endothelial Cells. *Am. J. Clin. Nutr.* **2001**, *73*, 36–40. [CrossRef] [PubMed]
95. Kalhan, R.; Smith, L.J.; Nlend, M.C.; Nair, A.; Hixon, J.L.; Sporn, P.H.S. A Mechanism of Benefit of Soy Genistein in Asthma: Inhibition of Eosinophil P38-Dependent Leukotriene Synthesis. *Clin. Exp. Allergy* **2007**, *38*, 103–112. [CrossRef] [PubMed]
96. Konstantinidou, V.; Khymenets, O.; Covas, M.-I.; de la Torre, R.; Muñoz-Aguayo, D.; Anglada, R.; Farré, M.; Fito, M. Time Course of Changes in the Expression of Insulin Sensitivity-Related Genes after an Acute Load of Virgin Olive Oil. *OMICS J. Integr. Biol.* **2009**, *13*, 431–438. [CrossRef] [PubMed]
97. Capó, X.; Martorell, M.; Sureda, A.; Riera, J.; Drobnic, F.; Tur, J.A.; Pons, A. Effects of Almond- and Olive Oil-Based Docosahexaenoic- and Vitamin E-Enriched Beverage Dietary Supplementation on Inflammation Associated to Exercise and Age. *Nutrients* **2016**, *8*, 619. [CrossRef] [PubMed]
98. Yan, W.; Zhang, J.; Zhang, Y.; Meng, D.; Yan, D. Anti-Inflammatory Activity Studies on the Stems and Roots of Jasminum Lanceolarium Roxb. *J. Ethnopharmacol.* **2015**, *171*, 335–341. [CrossRef] [PubMed]
99. Yang, L.; Liu, R.; Fan, A.; Zhong, G.; He, J. Dendropanax Dentiger (Harms) Merr. Root and Its Major Constituents Exert Therapeutic Effect on Adjuvant-Induced Arthritis in Rats. *J. Ethnopharmacol.* **2021**, *267*, 113631. [CrossRef] [PubMed]

100. Yang, L.; Liu, R.; Fan, A.; Zhao, J.; Zhang, Y.; He, J. Chemical Composition of Pterospermum Heterophyllum Root and Its Anti-Arthritis Effect on Adjuvant-Induced Arthritis in Rats via Modulation of Inflammatory Responses. *Front. Pharmacol.* **2020**, *11*, 584849. [CrossRef]
101. Zheng, C.-J.; Zhao, X.-X.; Ai, H.-W.; Lin, B.; Han, T.; Jiang, Y.-P.; Xing, X.; Qin, L.-P. Therapeutic Effects of Standardized Vitex Negundo Seeds Extract on Complete Freund's Adjuvant Induced Arthritis in Rats. *Phytomed. Int. J. Phytother. Phytopharm.* **2014**, *21*, 838–846. [CrossRef] [PubMed]
102. Li, X.; Lu, Y.; Jin, Y.; Son, J.-K.; Lee, S.H.; Chang, H.W. Curcumin Inhibits the Activation of Immunoglobulin E-Mediated Mast Cells and Passive Systemic Anaphylaxis in Mice by Reducing Serum Eicosanoid and Histamine Levels. *Biomol. Ther.* **2014**, *22*, 27–34. [CrossRef] [PubMed]
103. Manjunatha, H.; Srinivasan, K. Protective Effect of Dietary Curcumin and Capsaicin on Induced Oxidation of Low-Density Lipoprotein, Iron-Induced Hepatotoxicity and Carrageenan-Induced Inflammation in Experimental Rats. *FEBS J.* **2006**, *273*, 4528–4537. [CrossRef]
104. Xia, H.; He, Q.; Wang, H.; Wang, Y.; Yang, Y.; Li, Y.; Zhang, J.; Chen, Z.; Yang, J. Treatment with Either COX-2 Inhibitor or 5-LOX Inhibitor Causes No Compensation between COX-2 Pathway and 5-LOX Pathway in Chronic Aluminum Overload-induced Liver Injury in Rats. *Fundam. Clin. Pharmacol.* **2019**, *33*, 535–543. [CrossRef]
105. Yang, J.-Q.; Zhou, Q.-X.; Liu, B.-Z.; He, B.-C. Protection of Mouse Brain from Aluminum-Induced Damage by Caffeic Acid. *CNS Drug Rev.* **2008**, *14*, 10–16. [CrossRef] [PubMed]
106. Pallio, G.; Bitto, A.; Pizzino, G.; Galfo, F.; Irrera, N.; Minutoli, L.; Arcoraci, V.; Squadrito, G.; Macrì, A.; Squadrito, F.; et al. Use of a Balanced Dual Cyclooxygenase-1/2 and 5-Lypoxygenase Inhibitor in Experimental Colitis. *Eur. J. Pharmacol.* **2016**, *789*, 152–162. [CrossRef] [PubMed]
107. Foster, S.J.; McCormick, M.E.; Howarth, A.; Aked, D. Leukocyte Recruitment in the Subcutaneous Sponge Implant Model of Acute Inflammation in the Rat Is Not Mediated by Leukotriene B1. *Biochem. Pharmacol.* **1986**, *35*, 1709–1717. [CrossRef]
108. Peskar, B.M.; Lange, K.; Hoppe, U.; Peskar, B.A. Ethanol Stimulates Formation of Leukotriene C4 in Rat Gastric Mucosa. *Prostaglandins* **1986**, *31*, 283–293. [CrossRef]
109. Choi, J.-H.; Chai, Y.-M.; Joo, G.-J.; Rhee, I.-K.; Lee, I.-S.; Kim, K.-R.; Choi, M.-S.; Rhee, S.-J. Effects of Green Tea Catechin on Polymorphonuclear Leukocyte 5′-Lipoxygenase Activity, Leukotriene B_4 Synthesis, and Renal Damage in Diabetic Rats. *Ann. Nutr. Metab.* **2004**, *48*, 151–155. [CrossRef] [PubMed]
110. Choi, J.-H.; Chang, H.-W.; Rhee, S.-J. Effect of Green Tea Catechin on Arachidonic Acid Cascade in Chronic Cadmium-Poisoned Rats. *Asia Pac. J. Clin. Nutr.* **2002**, *11*, 292–297. [CrossRef]
111. Viji, V.; Helen, A. Inhibition of Lipoxygenases and Cyclooxygenase-2 Enzymes by Extracts Isolated from Bacopa Monniera (L.) Wettst. *J. Ethnopharmacol.* **2008**, *118*, 305–311. [CrossRef]
112. Bhaskar, S.; Kumar, K.S.; Krishnan, K.; Antony, H. Quercetin Alleviates Hypercholesterolemic Diet Induced Inflammation during Progression and Regression of Atherosclerosis in Rabbits. *Nutrition* **2013**, *29*, 219–229. [CrossRef] [PubMed]
113. Nisar, A.; Malik, A.H.; Zargar, M.A. Atropa Acuminata Royle Ex Lindl. Blunts Production of pro-Inflammatory Mediators Eicosanoids., Leukotrienes, Cytokines in Vitro and in Vivo Models of Acute Inflammatory Responses. *J. Ethnopharmacol.* **2013**, *147*, 584–594. [CrossRef]
114. Lin, B.; Zhao, Y.; Han, P.; Yue, W.; Ma, X.-Q.; Rahman, K.; Zheng, C.-J.; Qin, L.-P.; Han, T. Anti-Arthritic Activity of Xanthium Strumarium L. Extract on Complete Freund's Adjuvant Induced Arthritis in Rats. *J. Ethnopharmacol.* **2014**, *155*, 248–255. [CrossRef]
115. Kavitha, S.; John, F.; Indira, M. Amelioration of Inflammation by Phenolic Rich Methanolic Extract of Ocimum Sanctum Linn. Leaves in Isoproterenol Induced Myocardial Infarction. *Indian J. Exp. Biol.* **2015**, *53*, 632–640.
116. Chang, X.; Luo, F.; Jiang, W.; Zhu, L.; Gao, J.; He, H.; Wei, T.; Gong, S.; Yan, T. Protective Activity of Salidroside against Ethanol-Induced Gastric Ulcer via the MAPK/NF-KB Pathway in Vivo and in Vitro. *Int. Immunopharmacol.* **2015**, *28*, 604–615. [CrossRef] [PubMed]
117. Bhaskar, S.; Sudhakaran, P.R.; Helen, A. Quercetin Attenuates Atherosclerotic Inflammation and Adhesion Molecule Expression by Modulating TLR-NF-KB Signaling Pathway. *Cell. Immunol.* **2016**, *310*, 131–140. [CrossRef] [PubMed]
118. Yashaswini, P.S.; Sadashivaiah, B.; Ramaprasad, T.R.; Singh, S.A. In Vivo Modulation of LPS Induced Leukotrienes Generation and Oxidative Stress by Sesame Lignans. *J. Nutr. Biochem.* **2017**, *41*, 151–157. [CrossRef] [PubMed]
119. Liu, Y.; Tang, H.; Liu, X.; Chen, H.; Feng, N.; Zhang, J.; Wang, C.; Qiu, M.; Yang, J.; Zhou, X. Frontline Science: Reprogramming COX-2, 5-LOX, and CYP4A-mediated Arachidonic Acid Metabolism in Macrophages by Salidroside Alleviates Gouty Arthritis. *J. Leukoc. Biol.* **2019**, *105*, 11–24. [CrossRef]
120. Sobeh, M.; Mahmoud, M.F.; Rezq, S.; Alsemeh, A.E.; Sabry, O.M.; Mostafa, I.; Abdelfattah, M.A.; Ait El-Allem, K.; El-Shazly, A.M.; Yasri, A.; et al. Salix Tetrasperma Roxb. Extract Alleviates Neuropathic Pain in Rats via Modulation of the NF-KB/TNF-α/NOX/INOS Pathway. *Antioxidants* **2019**, *8*, 482. [CrossRef]
121. Yabas, M.; Orhan, C.; Er, B.; Tuzcu, M.; Durmus, A.S.; Ozercan, I.H.; Sahin, N.; Bhanuse, P.; Morde, A.A.; Padigaru, M.; et al. A Next Generation Formulation of Curcumin Ameliorates Experimentally Induced Osteoarthritis in Rats via Regulation of Inflammatory Mediators. *Front. Immunol.* **2021**, *12*, 609629. [CrossRef]
122. Beetens, J.R.; Loots, W.; Somers, Y.; Coene, M.C.; De Clerck, F. Ketoconazole Inhibits the Biosynthesis of Leukotrienes in Vitro and in Vivo. *Biochem. Pharmacol.* **1986**, *35*, 883–891. [CrossRef]

123. Billah, M.M.; Bryant, R.W.; Siegel, M.I. Lipoxygenase Products of Arachidonic Acid Modulate Biosynthesis of Platelet-Activating Factor (1-O-Alkyl-2-Acetyl-Sn-Glycero-3-Phosphocholine) by Human Neutrophils via Phospholipase A2. *J. Biol. Chem.* **1985**, *260*, 6899–6906. [CrossRef]
124. Bossù, E.; Aglianò, A.M.; Desideri, N.; Sestili, I.; Porrà, R.; Grandilone, M.; Quaglia, M.G. LTB4 as Marker of 5-LO Inhibitory Activity of Two New N-Omega-Ethoxycarbonyl-4-Quinolones. *J. Pharm. Biomed. Anal.* **1999**, *19*, 539–548. [CrossRef]
125. Bremm, K.D.; Konig, W.; Alouf, J.E. Effect of Thiol-Activated Toxins (Streptolysin 0, Alveolysin, and Theta Toxin) on the Generation of Leukotrienes and Leukotriene- Inducing and -Metabolizing Enzymes from Human Polymorphonuclear Granulocytes. *Infect. Immun.* **1985**, *50*, 8. [CrossRef]
126. Conti, P.; Panara, M.R.; Barbacane, R.C.; Bongrazio, M.; Dempsey, R.A.; Reale, M. Human Recombinant IL-1 Receptor Antagonist (IL-1Ra) Inhibits Leukotriene B4 Generation from Human Monocyte Suspensions Stimulated by Lipopolysaccharide (LPS). *Clin. Exp. Immunol.* **1993**, *91*, 526–531. [CrossRef] [PubMed]
127. Fujimoto, K.; Kubo, K.; Shinozaki, S.; Okada, K.; Matsuzawa, Y.; Kobayashi, T.; Sugane, K. Neutrophil Elastase Inhibitor Reduces Asthmatic Responses in Allergic Sheep. *Respir. Physiol.* **1995**, *100*, 91–100. [CrossRef]
128. Hulkower, K.I.; Pollock, J.S.; Walsh, R.E.; Huang, R.; Otis, E.R.; Brooks, C.D.; Bell, R.L. Leukotrienes Do Not Regulate Nitric Oxide Production in RAW 264.7 Macrophages. *Prostaglandins Leukot. Essent. Fat. Acids* **1996**, *55*, 145–149. [CrossRef]
129. Igarashi, Y.; Lundgren, J.D.; Shelhamer, J.H.; Kaliner, M.A.; White, M.V. Effects of Inhibitors of Arachidonic Acid Metabolism on Serotonin Release from Rat Basophilic Leukemia Cells. *Immunopharmacology* **1993**, *25*, 131–144. [CrossRef]
130. Ishii, K.; Yakuo, I.; Motoyoshi, S.; Nakagawa, H.; Nakamura, H. Inhibition of Leukotriene Production by N-[4-[4-(Diphenylmethyl)-1- Piperazinyl]Butyl]-3-(6-Methyl-3-Pyridyl) Acrylamide (AL-3264), a New Antiallergic Agent. *Jpn. J. Pharmacol.* **1994**, *65*, 19–25. [CrossRef]
131. Johnson-Henry, K.; Wallace, J.L.; Basappa, N.S.; Soni, R.; Wu, G.K.; Sherman, P.M. Inhibition of Attaching and Effacing Lesion Formation Following Enteropathogenic Escherichia Coli and Shiga Toxin-Producing *E. Coli* Infection. *Infect. Immun.* **2001**, *69*, 7152–7158. [CrossRef]
132. Malleron, J.L.; Roussel, G.; Gueremy, G.; Ponsinet, G.; Robin, J.L.; Terlain, B.; Tissieres, J.M. Penta- and Hexadienoic Acid Derivatives: A Novel Series of 5-Lipoxygenase Inhibitors. *J. Med. Chem.* **1990**, *33*, 2744–2749. [CrossRef] [PubMed]
133. Müller, K.; Ziereis, K.; Paper, D.H. Ilex Aquifolium: Protection against Enzymatic and Non-Enzymatic Lipid Peroxidation. *Planta Med.* **1998**, *64*, 536–540. [CrossRef]
134. Ohemeng, K.A.; Nguyen, V.N.; Schwender, C.F.; Singer, M.; Steber, M.; Ansell, J.; Hageman, W. Novel Bishydroxamic Acids as 5-Lipoxygenase Inhibitors. *Bioorg. Med. Chem.* **1994**, *2*, 187–193. [CrossRef]
135. Petroni, A.; Blasevich, M.; Papini, N.; Salami, M.; Sala, A.; Galli, C. Inhibition of Leukocyte Leukotriene B4 Production by an Olive Oil-Derived Phenol Identified by Mass-Spectrometry. *Thromb. Res.* **1997**, *87*, 315–322. [CrossRef]
136. Prasad, N.S.; Raghavendra, R.; Lokesh, B.R.; Naidu, K.A. Spice Phenolics Inhibit Human PMNL 5-Lipoxygenase. *Prostaglandins Leukot. Essent. Fat. Acids* **2004**, *70*, 521–528. [CrossRef]
137. Rotondo, S.; Rajtar, G.; Manarini, S.; Celardo, A.; Rotilio, D.; De Gaetano, G.; Evangelista, V.; Cerletti, C. Effect of *Trans* -Resveratrol, a Natural Polyphenolic Compound, on Human Polymorphonuclear Leukocyte Function: *Trans* -Resveratrol and Human PMN Leukocyte Function. *Br. J. Pharmacol.* **1998**, *123*, 1691–1699. [CrossRef] [PubMed]
138. Tommasini, I.; Guidarelli, A.; Palomba, L.; Cerioni, L.; Cantoni, O. 5-Hydroxyeicosatetraenoic Acid Is a Key Intermediate of the Arachidonate-Dependent Protective Signaling in Monocytes/Macrophages Exposed to Peroxynitrite. *J. Leukoc. Biol.* **2006**, *80*, 929–938. [CrossRef]
139. Van Hilten, J.A.; Elliott, G.R.; Bonta, I.L. Endogenous Lipoxygenase Metabolites Mediate A23187 Induced Macrophage Cytostasis towards P815 Tumor Cells in Vitro. *Agents Actions* **1989**, *26*, 170–172. [CrossRef]
140. Van Hilten, J.A.; Elliott, G.R.; Bonta, I.L. Specific Lipoxygenase Inhibition Reverses Macrophage Cytotasis towards P815 Tumor Cells in Vitro Induced by the Calcium Ionophore A23187. *Prostaglandins Leukot. Essent. Fat. Acids* **1988**, *34*, 187–192. [CrossRef]
141. Kimura, Y.; Okuda, H.; Okuda, T.; Hatano, T.; Arichi, S. Studies on the Activities of Tannins and Related Compounds, X. Effects of Caffeetannins and Related Compounds on Arachidonate Metabolism in Human Polymorphonuclear Leukocytes. *J. Nat. Prod.* **1987**, *50*, 392–399. [CrossRef]
142. Lin, T.-Y.; London, C.A. Characterization and Modulation of Canine Mast Cell Derived Eicosanoids. *Vet. Immunol. Immunopathol.* **2010**, *135*, 118–127. [CrossRef] [PubMed]
143. Tsen, S.Y.; Tan, X.Y.; Tan, Y.M.; Yan, B.Y.; Loke, W.M. Relative Inhibitions of 5-Lipoxygenase and Myeloperoxidase and Free-Radical Scavenging Activities of Daidzein, Dihydrodaidzein, and Equol. *J. Med. Food* **2016**, *19*, 543–548. [CrossRef] [PubMed]
144. Ammon, H.P.T.; Safayhi, H.; Mack, T.; Sabieraj, J. Mechanism of Antiinflammatory Actions of Curcumine and Boswellic Acids. *J. Ethnopharmacol.* **1993**, *38*, 105–112. [CrossRef]
145. Hong, J. Modulation of Arachidonic Acid Metabolism by Curcumin and Related -Diketone Derivatives: Effects on Cytosolic Phospholipase A2, Cyclooxygenases and 5-Lipoxygenase. *Carcinogenesis* **2004**, *25*, 1671–1679. [CrossRef] [PubMed]
146. Gomes, A.; Couto, D.; Alves, A.; Dias, I.; Freitas, M.; Porto, G.; Duarte, J.A.; Fernandes, E. Trihydroxyflavones with Antioxidant and Anti-Inflammatory Efficacy. *BioFactors* **2012**, *38*, 378–386. [CrossRef]
147. Kwon, O.S.; Choi, J.S.; Islam, M.N.; Kim, Y.S.; Kim, H.P. Inhibition of 5-Lipoxygenase and Skin Inflammation by the Aerial Parts of Artemisia Capillaris and Its Constituents. *Arch. Pharm. Res.* **2011**, *34*, 1561–1569. [CrossRef] [PubMed]

148. Takasugi, M.; Muta, E.; Yamada, K.; Arai, H. A New Method to Evaluate Anti-Allergic Effect of Food Component by Measuring Leukotriene B4 from a Mouse Mast Cell Line. *Cytotechnology* **2018**, *70*, 177–184. [CrossRef]
149. Dehmlow, C.; Murawski, N.; de Groot, H. Scavenging of Reactive Oxygen Species and Inhibition of Arachidonic Acid Metabolism by Silibinin in Human Cells. *Life Sci.* **1996**, *58*, 1591–1600. [CrossRef]
150. Dehmlow, C.; Erhard, J.; de Groot, H. Inhibition of Kupffer Cell Functions as an Explanation for the Hepatoprotective Properties of Silibinin. *Hepatology* **1996**, *23*, 749–754. [CrossRef]
151. Jupp, J.; Hillier, K.; Elliott, D.H.; Fine, D.R.; Bateman, A.C.; Johnson, P.A.; Cazaly, A.M.; Penrose, J.F.; Sampson, A.P. Colonic Expression of Leukotriene-Pathway Enzymes in Inflammatory Bowel Diseases. *Inflamm. Bowel Dis.* **2007**, *13*, 537–546. [CrossRef] [PubMed]
152. Park, S.; Han, S.-U.; Lee, K.-M.; Park, K.H.; Cho, S.W.; Hahm, K.-B. 5-LOX Inhibitor Modulates the Inflammatory Responses Provoked by Helicobacter Pylori Infection. *Helicobacter* **2007**, *12*. [CrossRef]
153. Habib, E.S.; El-Bsoumy, E.; Ibrahim, A.K.; Helal, M.A.; El-Magd, M.A.; Ahmed, S.A. Anti-Inflammatory Effect of Methoxyflavonoids from *Chiliadenus Montanus (Jasonia Montana)* Growing in Egypt. *Nat. Prod. Res.* **2020**, 1–5. [CrossRef] [PubMed]
154. Loke, W.M.; Proudfoot, J.M.; Stewart, S.; McKinley, A.J.; Needs, P.W.; Kroon, P.A.; Hodgson, J.M.; Croft, K.D. Metabolic Transformation Has a Profound Effect on Anti-Inflammatory Activity of Flavonoids Such as Quercetin: Lack of Association between Antioxidant and Lipoxygenase Inhibitory Activity. *Biochem. Pharmacol.* **2008**, *75*, 1045–1053. [CrossRef] [PubMed]
155. Giménez-Bastida, J.A.; González-Sarrías, A.; Espín, J.C.; Schneider, C. Inhibition of 5-Lipoxygenase-Derived Leukotrienes and Hemiketals as a Novel Anti-Inflammatory Mechanism of Urolithins. *Mol. Nutr. Food Res.* **2020**, e2000129. [CrossRef]
156. Thivierge, M.; Stankova, J.; Rola-Pleszczynski, M. Toll-like Receptor Agonists Differentially Regulate Cysteinyl-Leukotriene Receptor 1 Expression and Function in Human Dendritic Cells. *J. Allergy Clin. Immunol.* **2006**, *117*, 1155–1162. [CrossRef] [PubMed]
157. Ávila-Gálvez, M.Á.; González-Sarrías, A.; Espín, J.C. In Vitro Research on Dietary Polyphenols and Health: A Call of Caution and a Guide on How To Proceed. *J. Agric. Food Chem.* **2018**, *66*, 7857–7858. [CrossRef]

International Journal of
Molecular Sciences

Article

Oleanolic Acid Alleviates Atopic Dermatitis-like Responses In Vivo and In Vitro

Yun-Mi Kang [1], Hye-Min Kim [1], Minho Lee [2,*] and Hyo-Jin An [1,*]

[1] Department of Pharmacology, College of Korean Medicine, Sangji University, Wonju 26339, Gangwon-do, Korea; yunmi6115@naver.com (Y.-M.K.); mins7576@daum.net (H.-M.K.)
[2] Department of Life Science, Dongguk University-Seoul, Ilsandong-gu, Goyang-si 10326, Gyeonggi-do, Korea
* Correspondence: MinhoLee@dgu.edu (M.L.); sangjipharm@gmail.com (H.-J.A.); Tel.: +82-31-961-5138 (M.L.); +82-33-738-7503 (H.-J.A.); Fax: +82-2-2277-1274 (M.L.); +82-33-730-0679 (H.-J.A.)

Abstract: Oleanolic acid (OA) is a pentacyclic triterpenoid, abundantly found in plants of the *Oleaceae* family, and is well known for its beneficial pharmacological activities. Previously, we reported the inhibitory effect of OA on mast cell-mediated allergic inflammation. In this study, we investigated the effects of OA on atopic dermatitis (AD)-like skin lesions and its underlying mechanism of action. We evaluated the inhibitory effect of OA on AD-like responses and the possible mechanisms using a 1-chloro-2,4-dinitrochlorobenzene (DNCB)-induced AD animal model and tumor necrosis factor (TNF)-α/interferon (IFN)-γ-stimulated HaCaT keratinocytes. We found that OA has anti-atopic effects, including histological alterations, on DNCB-induced AD-like lesions in mice. Moreover, it suppressed the expression of Th2 type cytokines and chemokines in the AD mouse model and TNF-α/IFN-γ-induced HaCaT keratinocytes by blocking the activation of serine-threonine kinase Akt, nuclear factor-κB, and the signal transducer and activator of transcription 1. The results demonstrate that OA inhibits AD-like symptoms and regulates the inflammatory mediators; therefore, it may be used as an effective and attractive therapeutic agent for allergic disorders, such as AD. Moreover, the findings of this study provide novel insights into the potential pharmacological targets of OA for treating AD.

Keywords: oleanolic acid; atopic dermatitis; 2,4-dinitrochlorobenzene; keratinocyte; NF-κB; STAT1

Citation: Kang, Y.-M.; Kim, H.-M.; Lee, M.; An, H.-J. Oleanolic Acid Alleviates Atopic Dermatitis-like Responses In Vivo and In Vitro. *Int. J. Mol. Sci.* **2021**, *22*, 12000. https://doi.org/10.3390/ijms222112000

Academic Editor: Michal Zmijewski

Received: 6 October 2021
Accepted: 2 November 2021
Published: 5 November 2021

1. Introduction

Allergic inflammation is characterized by pathophysiological or hypersensitivity disorders, including allergic asthma, allergic rhinitis, anaphylaxis, and atopic dermatitis (AD), after exposure to allergens [1]. AD is a chronic inflammatory skin disease that arises from the complicated interaction of innate and adaptive immune responses based on genetics, environmental factors, immune abnormalities, and skin barrier functions [2]. The characteristic features of AD include itchy, swollen, red, and cracked skin with inflammatory cell accumulation in AD skin lesions. Although the pathogenesis of AD is not clear, it is known that several cells and factors are associated with its development. The pathological processes of AD are thought to be mediated by Th1/ Th2 balance, which is skewed toward Th2 in AD. Th2 cells are mainly activated in the acute phase of AD, while Th1 cells mediate the alteration of expression in chronic AD [3]. The standard treatment for AD involves the application of topical corticosteroids or the administration of immunosuppressive agents; however, protracted use of these agents can cause various side effects, such as skin atrophy, bleeding, vasodilation, and organ toxicity. For this reason, medicines originating from herbal sources may be preferred to steroids, and may be used in combination with other methods, such as enhancing immunity, reducing house mite dust, and dietary restrictions [4,5].

Numerous intracellular signal transduction triggered by ligand–cell surface receptor binding is mediated by transcription factors. Nuclear factor (NF)-κB and the signal transducer and activator of transcription (STAT)−1 are pivotal transcription factors associated with the allergic inflammatory response [6]. Upon stimulation, the inhibitor κB (IκB)-α protein is phosphorylated, leading to the ubiquitination and proteasomal degradation of IκB. Sequentially activated NF-κB and interferon (IFN)-γ-activated STAT 1 in the cytoplasm translocate into the nucleus, where they engage in the expression of numerous pro-inflammatory mediators. Thus, these transcription factors are important pharmacological targets for the discovery of novel therapeutics to treat allergic disorders [7,8].

Oleanolic acid (OA) is a pentacyclic triterpenoid that is abundant in plants of the *Oleaceae* family, such as *Olea europaea*. OA is ubiquitously found in food and plants, where it exists as a free acid or as an aglycone of triterpenoid saponins, such as ursolic acid, moronic acid, and betulinic acid [9]. To date, various reports have described the pharmacological activities of OA, including its antioxidant [10], anti-inflammatory [11,12], anti-asthmatic [13], anti-diabetic [14,15], anti-tumor [16], hepatoprotective [17], immunomodulatory [18], anti-parasitic [19], and anti-hypertensive [20] properties. Despite the fact that OA is a well-known active component contained in various plants, studies on its effect on AD are insufficient. As it is important to study natural materials that are effective against allergic diseases, we focused on OA that exhibits a wide range of biological activities, such as anti-inflammatory and anti-asthmatic effects, as a feasible active compound for allergic diseases. Previously, we reported the anti-allergic effect of OA, demonstrating that OA exerted an inhibitory effect on mast cell-mediated allergic inflammation in vivo and in vitro [21]. Allergic response and inflammation can trigger AD and worsen the condition, thus, controlling allergic and inflammatory reaction could be important strategy in the manage of AD. These results prompted us to investigate its potential effect on other allergic diseases, such as AD. As AD is mainly the beginning of a series of allergic disorders, we hypothesized that OA would attenuate AD-like symptoms. Thus, in the present study, we aimed to elucidate the effects of OA on AD-like lesions and define the underlying mechanisms of action using DNCB-induced AD mouse models and human keratinocytes.

2. Results

2.1. OA Attenuated AD Lesions in DNCB-Induced AD Mice

The repeated topical application of DNCB on the dorsal skin of mice induces AD skin symptoms. DNCB is a "contact sensitizer" that induces contact hypersensitivity of the skin in mice, which is considered to be a cell-mediated response [22]. To investigate the remedial effects of OA on AD mice, we administered OA following the induction of AD mouse skin. The experimental procedure is summarized in Figure 1A. On the day of sacrifice, severe AD-like lesions, such as erythema, edema, hemorrhage, scarring, dryness, excoriation, and erosion were observed on the dorsal skin of DNCB-induced AD mice. However, topical application of dexamethasone (10 μM), a well-known therapeutic agent for AD, and OA (10 and 50 μM) for 3 weeks significantly alleviated these AD skin symptoms compared to the DNCB group ($p < 0.001$) (Figure 1B,C).

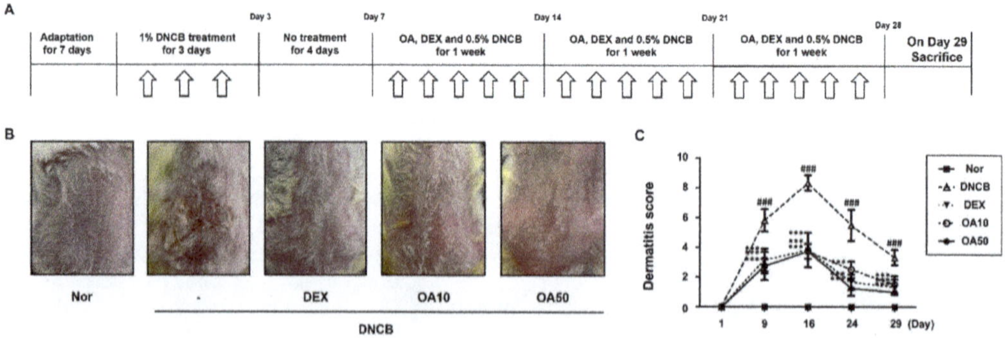

Figure 1. Effects of OA on DNCB-induced AD skin lesions in ICR mice. (**A**) Experimental schedule for the induction of AD. (**B**) Effect of OA on clinical features of DNCB-induced AD skin lesions. White arrows indicated DNCB treatment. (**C**) Effects of OA on dermatitis score. Densitometric analysis was performed using Bio-Rad Quantity One® Software. The data shown represent mean ± S.D. ($n = 6$) of three independent experiments. ### $p < 0.001$ vs. the control group; *** $p < 0.001$ vs. DNCB-treated group.

2.2. *OA Improved the Histological Observations and Histamine Release in DNCB-Induced AD Mice*

Histological alterations, such as epidermal hyperplasia and infiltration of lymphocytes and mast cells in the skin, are the main hallmarks of AD [23]. Improvements in clinical skin conditions following OA treatment were confirmed by histological examination. Histological analysis was performed on atopic skin tissues. The excised skin from each group was stained with hematoxylin and eosin (H&E) or toluidine blue, and histological alterations were observed microscopically. H&E-stained tissue sections revealed that the thickness of epidermal and dermal tissues was greater in the DNCB-treated group (91.84 ± 7.60, 474.66 ± 43.65 μm, respectively, $p < 0.001$) than the control group due to edema, hyperkeratosis, and hyperplasia (Figure 2A). However, treatment with 10 and 50 μM OA markedly attenuated the epidermal (47.16 ± 5.98 and 52.65 ± 9.56 μm, $p < 0.001$) and dermal thickening (258.65 ± 17.56 and 292.65 ± 25.61 μm, $p < 0.001$) (Figure 2B,C). In the toluidine blue-stained tissue sections, mast cell infiltration, an indicator of inflammation, was noticeably increased in the DNCB-treated group compared to the control group (73.67 ± 12.06 cells, $p < 0.001$). Treatment with 10 and 50 μM OA attenuated the infiltration of inflammatory cells, particularly mast cells, as evidenced by toluidine blue staining (28.5 ± 6.98, 25.5 ± 3.73 cells, respectively, $p < 0.001$) (Figure 3A,B). As mast cells are sources of histamine, which is the most potent mediator involved in AD symptoms [24], histamine levels in the serum were also examined. The results showed that treatment with 50 μM OA remarkably inhibited histamine release (271.91 ± 35.75 ng/mL, $p < 0.001$) compared to the DNCB-treated group (411.81 ± 60.12 ng/mL, $p < 0.001$) (Figure 3C).

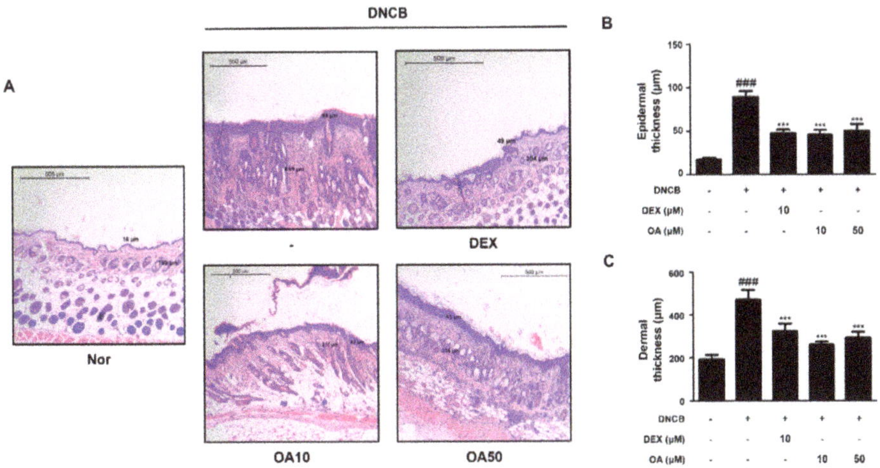

Figure 2. Effect of OA on epidermal and dermal thickness in DNCB-induced AD skin lesions. (**A**) H&E stained AD mouse skin lesions (scale bar = 500 μm). (**B**) Determination of epidermal thickness and (**C**) dermal thickness. Epidermal and dermal thickness in H&E stained sections were measured under a microscope. The data shown represent mean ± S.D. (n = 6) of three independent experiments. ### $p < 0.001$ vs. the control group; *** $p < 0.001$ vs. DNCB-treated group.

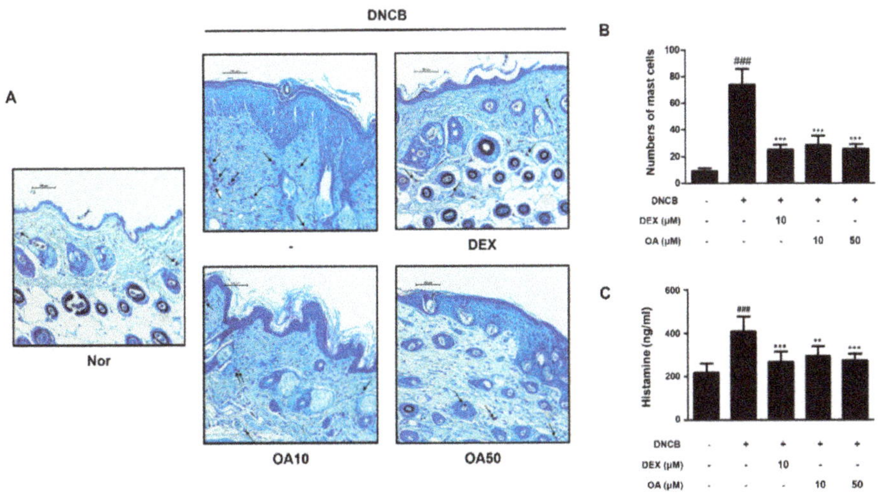

Figure 3. Effect of OA on mast cell infiltration and serum histamine level in DNCB-induced AD skin lesions. (**A**) Toluidine blue stained AD mouse skin lesions (scale bar = 100 μm). Black arrows indicated stained mast cells. (**B**) Number of mast cells per mm section. Mast cell infiltration in toluidine blue stained sections is expressed as the average total count in five fields. (**C**) Histamine release in mouse serum was measured using an ELISA kit. The data shown represent mean ± S.D. (n = 6) of three independent experiments. ### $p < 0.001$ vs. the control group; ** $p < 0.01$ and *** $p < 0.001$ vs. DNCB-treated group.

2.3. OA Suppressed the mRNA Expression of AD-Related Cytokines and Activation of IκB and STAT1 in DNCB-Induced AD Mice

Next, we investigated whether OA inhibited the signature cytokines of AD in the dorsal tissues of DNCB-induced AD mice. Above all, thymus and activation-regulated chemokine (TARC)/CCL17, are members of the CC chemokine subfamily and are involved in the recruitment of Th2 lymphocytes and the continuation of Th2 immune responses [25]. In addition, thymic stromal lymphopoietin (TSLP) is known to provoke dendritic cell-mediated Th2 responses and is highly expressed in activated mast cells and skin of AD, which triggers allergic inflammation. Therefore, these cytokines are considered mediators of inflammatory skin diseases, such as AD [5–26]. As shown in Figure 4A, the mRNA expression levels of TSLP and TARC were markedly ($p < 0.001$) increased by repetitive treatment with DNCB, while OA reduced the expression levels of TSLP and TARC by approximately basal levels ($p < 0.001$). In line with these results, Th2-type cytokines, including IL-4, IL-5, and IL-13, were downregulated by OA treatment compared to DNCB-induced AD mice ($p < 0.001$) (Figure 4B).

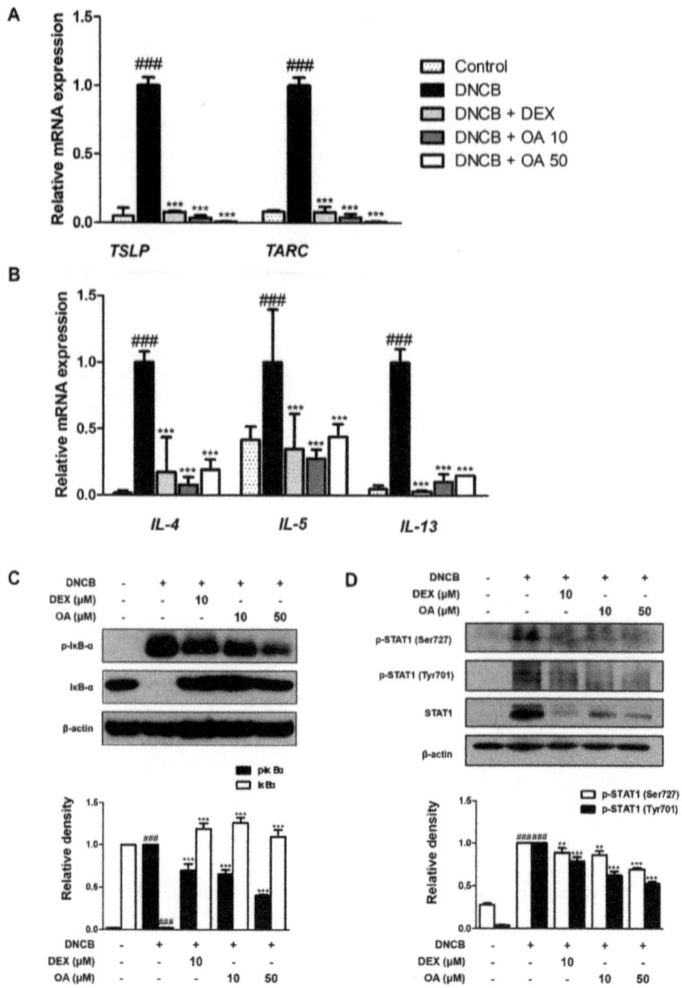

Figure 4. Effect of OA on AD cytokines and IκB, STAT1 activation in DNCB-induced AD skin lesions. Total RNA prepared from the dorsal tissue, and the level of (**A**) TARC, TSLP, (**B**) IL-4, IL-5, and IL-13 were determined by quantitative qRT-PCR. Expression of IκB (**C**) and STAT1 (**D**) was determined by Western blot analysis using specific antibodies. Densitometric analysis was performed using Bio-Rad Quantity One® Software. The data shown represent mean ± S.D. (n = 6) of three independent experiments. ### $p < 0.001$ vs. the control group; ** $p < 0.01$ and *** $p < 0.001$ vs. DNCB-treated group.

To investigate the signaling pathways involved in the inhibitory effect of OA on cytokine production, we examined the phosphorylation and degradation of IκB and activation of STAT1 in DNCB-induced AD mice. The results demonstrated that the phosphorylation and degradation of IκB induced by DNCB were significantly ($p < 0.001$) inhibited by treatment with OA (Figure 4C). In addition, OA inhibited the DNCB-induced phosphorylation of STAT1 at residues Ser727 and Tyr701 with significance (Figure 4D). Considering our results, it can be presumed that the NF-κB and STAT1 signaling pathways are involved in the inhibitory effect of OA on the cytokine profiles of DNCB-induced AD-like skin.

2.4. OA Suppressed the Expression of AD Cytokines and Activation of NF-κB and Akt in TNF-α/IFN-γ-Stimulated HaCaT Keratinocytes

To test in vivo findings in vitro, we investigated the effects of OA on pro-inflammatory cytokine expression in tumor necrosis factor (TNF)-α/IFN-γ-stimulated HaCaT keratinocytes. HaCaT human keratinocytes are the most commonly used cell line in the AD model, which produces various AD-related pro-inflammatory mediators in response to a variety of stimuli [27]. OA had no signs of toxicity up to 100 μM in HaCaT keratinocytes (Supplementary Figure S1). Many studies have shown that TSLP, TARC, and RANTES (Regulated on Activation, Normal T Cell Expressed and Secreted) play an active role in the pathogenesis of AD, and TNF-α/IFN-γ synergistically induces production of these mediators by human keratinocytes and HaCaT cells [28], and activation of NF-κB and STAT1 activation is involved in the production of chemokines and cytokines in keratinocytes [29]. Therefore, we investigated whether OA suppressed the production of AD cytokines and chemokines in TNF-α/IFN-γ-stimulated HaCaT cells by suppressing the NF-κB and STAT1 signaling pathways. First, we evaluated the mRNA levels of TSLP, TARC, and RANTES in TNF-α/IFN-γ-stimulated HaCaT cells. OA noticeably suppressed TNF-α/IFN-γ-stimulated mRNA expression of TSLP, TARC, and RANRES ($p < 0.001$) (Figure 5A), Th2 cytokines ($p < 0.001$) (Figure 5B), and pro-inflammatory cytokines, TNF-α and IL−6 ($p < 0.001$) (Figure 5C). These data suggest that OA can control the expression of Th2-related mediators and pro-inflammatory factors in TNF-α/IFN-γ-stimulated HaCaT cells. Next, we examined the effects of OA on the phosphorylation of NF-κB and Akt, which are signaling pathways that regulate inflammatory factors in human keratinocytes [30]. The results showed that OA reversed the nuclear translocation of NF-κB by treatment with 25 μM OA ($p < 0.001$), phosphorylation ($p < 0.001$), and degradation ($p < 0.05$) of IκB. Moreover, treatment with 25 μM OA prevented phosphorylation of Akt ($p < 0.001$), which promotes the activation of NF-κB by directly phosphorylating IκB kinase in response to stimuli, but it did not affect the total amount of Akt in TNF-α/IFN-γ-stimulated HaCaT cells (Figure 5D,E). These results suggest that OA might suppress TNF-α/IFN-γ-induced cytokines and chemokines by suppressing NF-κB and Akt in HaCaT keratinocytes.

2.5. OA Inhibited STAT1 Activation in TNF-α/IFN-γ-Stimulated HaCaT Keratinocytes

To investigate the mechanism responsible for the inhibitory effect of OA in parallel with the inhibition of NF-κB, we investigated its effect on TNF-α/IFN-γ-induced STAT1 activation. As shown in Figure 6A,B, treatment with TNF-α/IFN-γ resulted in the activation of STAT1 ($p < 0.001$). However, OA decreased the phosphorylation at Tyr701 with significance and alleviated the nuclear translocation of STAT1 in TNF-α/IFN-γ-stimulated HaCaT cells (Figure 6B). In addition, the expression of cytokines, such as IL-2 and IL-12, which are involved in the phosphorylation of STAT1, was downregulated by OA treatment compared to TNF-α/IFN-γ-treated HaCaT cells ($p < 0.001$) (Figure 6C). The results demonstrated that the anti-inflammatory effects of OA in HaCaT and DNCB-induced AD mice are associated with the STAT1 signaling pathway.

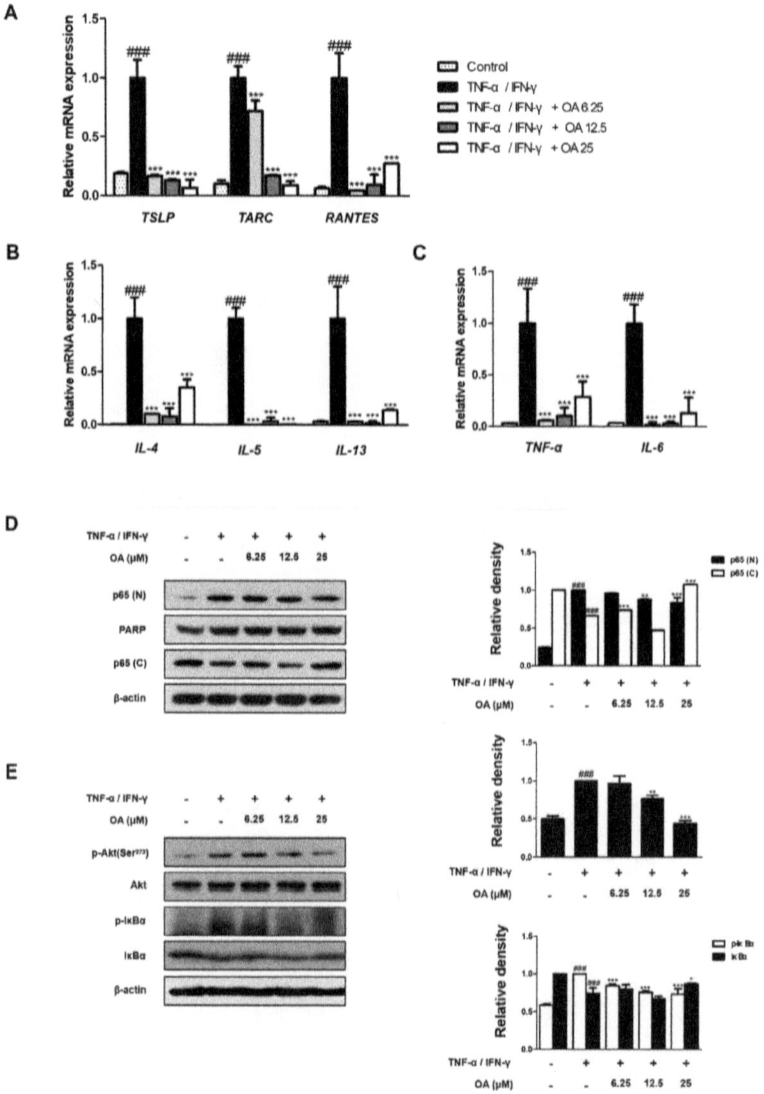

Figure 5. Effect of OA on chemokine and NF-κB and Akt activation in TNF-α/IFNγ-stimulated HaCaT cells (**A–C**) The mRNA level of cytokines and chemokines in HaCaT cells were determined by qRT-PCR. (**D**,**E**) Total proteins were prepared, and Western blot analysis was performed using specific antibodies. β-actin was used as internal control. Densitometric analysis was performed using Bio-Rad Quantity One® Software. The data shown represent mean ± S.D. of three independent experiments. ### $p < 0.001$ vs. the control group; * $p < 0.05$, ** $p < 0.01$, and *** $p < 0.001$ vs. TNF-α/IFN-γ-treated group.

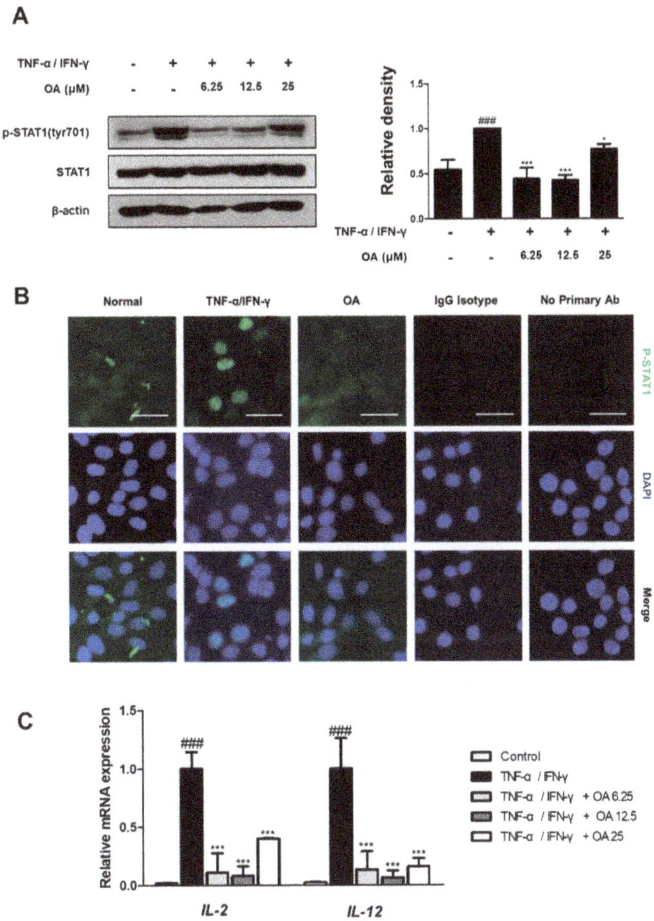

Figure 6. Effect of OA on STAT1 activation in TNF-α/IFNγ-stimulated HaCaT cells. (**A**) Total proteins were prepared, and Western blot analysis was performed using specific antibodies. β-actin was used as internal control. Densitometric analysis was performed using Bio-Rad Quantity One® Software. (**B**) Immunofluorescence staining of p-STAT1 (green) in TNF-α/IFN-γ-stimulated HaCaT keratinocytes. Nuclei were counterstained using DAPI. IgG isotype control and secondary antibody only control were used as negative control (scale bar = 100 μm). (**C**) The mRNA level of IL-2, IL-12, and IL-8 in HaCaT cells were determined by qRT-PCR. The data shown represent mean ± S.D. of three independent experiments. $^{###}$ $p < 0.001$ vs. the control group; * $p < 0.05$, and *** $p < 0.001$ vs. TNF-α/IFN-γ-treated group.

3. Discussion

3.1. AD Phenotypes Reflected in This Study

AD is a complex disease, whose relationship with allergy remains controversial and can be viewed from different perspectives. AD is recognized as a type I and type IV complex according to the Coombs and Gell classification system [31]. Epidermal keratinocytes provide a functional skin barrier on the frontline of the defense system. In the skin of patients with AD, allergic sensitization, due to pathogen invasion, irritants, and allergens, occurs through a damaged skin barrier that leads to the stimulation of immune responses in keratinocytes, with increased skin inflammation [23–32]. The expression of various inflammatory cytokines and chemokines derived from keratinocytes plays important roles in the pathogenesis of AD [33,34]. Although a biphasic immune response is observed, a Th2-biased immune response (IL-4, IL-13, TSLP, and eosinophils) is predominant in the initial and acute phases of AD [35].

Based on the well-known anti-allergic and anti-inflammatory properties of OA, we adopted the DNCB-induced AD model to evaluate the effect of OA on AD mice in vivo. DNCB induces contact dermatitis accompanying skin and immunological alterations in mice, which is similar to aspects observed in human patients with AD, including clinical features, such as skin dryness, erosion, edema, hemorrhage, erythema, and increased serum IgE levels [2]. In this study, we confirmed that OA alleviated the symptoms in patients with AD, such as hypertrophy of the epidermal layer (Figures 1 and 2), recruitment of inflammatory cells, and histamine release (Figure 3), using histological methods. Since AD is thought to be a Th2-dominant inflammatory skin disease in the acute phase, TARC is thought to be implicated in the pathogenesis of AD [36], correlating with the severity of disease in some chronic allergic pathologies, such as asthma and AD [37]. TSLP is an epithelial cell-derived cytokine [38] and is strongly implicated in the pathogenesis of Th2 cell-mediated allergic disorders. Eventually, TSLP specifically increased the frequency of IL-4-, IL-5-, and IL-13-expressing effector cells, which enhanced cytokine production [39]. In this regard, we investigated the effect of OA on the mRNA expression of TSLP and TARC. The results showed that OA significantly decreased the mRNA levels of TSLP and TARC in the DNCB-induced AD animal model (Figure 4A). Furthermore, OA inhibited the DNCB-induced phosphorylation and degradation of IκB and the phosphorylation of STAT1 at both residues. These results indicate the possibility that the promoter of TSLP or TARC includes NF-κB- and STAT-binding sequences, so that these transcription factors may regulate the transcription of TSLP and TARC, as reported previously [40,41].

3.2. The Mechanisms of OA in This Study

These results prompted us to confirm how OA regulates AD development. Therefore, we investigated the mechanism of action of OA in AD using human-derived keratinocytes. OA suppressed the mRNA levels of TSLP, TARC, RANTES, and Th1/Th2 type cytokines in TNF-α/IFN-γ-stimulated HaCaT keratinocytes. Moreover, the TNF-α/IFN-γ-induced phosphorylation and degradation of IκB, translocation of NF-κB, and phosphorylation of Akt and STAT1 (Tyr701) were inhibited by pretreatment with OA. These results indicate that the regulation of cytokine and chemokine expression in HaCaT keratinocytes is related to the Akt, NF-κB, and STAT1 signaling pathways. Previous studies have shown that the mitogen-activated protein kinase (MAPK) signaling pathway is involved in the regulation of NF-κB and STAT1 in response to cytokines, such as TNF-α and IFN-γ [42–44]. However, the results indicated that OA did not suppress TNF-α/IFN-γ-induced phosphorylation of MAPKs including ERK, JNK/SAPK, and p38 MAPK (data not shown), suggesting that the expression of cytokines and chemokines in TNF-α/IFN-γ-stimulated HaCaT keratinocytes via suppression of other pathways, such as other STAT family members under the action of OA, judging from the multiple roles of cytokines. However, it is necessary to examine how OA exerts its action via a certain receptor at the upstream signaling pathway. Previous reports have described that anti-inflammatory activities of OA were related to inhibition of toll-like receptors (TLRs) signaling pathway and consequent inflammatory responses [45,46]. TLRs induce the MyD88-dependent pathway to mediate downstream inflammatory pathways through activation of NF-κB [47], supporting our results regarding downregulation of the NF-κB pathway. Moreover, engagement of IFN-γ receptor signaling, which leads to STAT1 activation and the receptors of various cytokines also should be considered to clarify the roles of OA in AD pathology. Despite these limitations, in this study, we provided potential mechanisms through which OA exerts its role in allergic inflammation. The activated Akt, NF-κB, and STAT1 pathways mediated the signal transduction triggered by allergen, and the mediators induced the development AD symptoms, including skin inflammatory response, immune cell infiltration, and Th2 cytokines production. However, OA alleviates AD symptoms through the downregulation of Akt, NF-κB, STAT1 signaling pathways and the expression of AD-associated cytokines, suggesting the potential therapeutic targets.

3.3. For Improving the Activity of OA in Skin

We conducted the present study to evaluate the anti-AD effects of OA in in vivo and in vitro models. Consistent with a previous study that found that oral administration of oleanolic acid acetate, the derivative of oleanolic acid, suppressed DNCB-induced atopic skin symptoms in BALB/c animal models [48], we confirmed that OA alleviated atopic symptoms, despite the difference in mouse strains. Our study showed that OA decreased the serum histamine levels which is induced by DNCB (Figure 3C), indicating that the topical application of OA can affect the systemic immune system as well as local skin barrier. Due to its low water solubility, there have been several attempts to improve the water solubility and bioavailability of OA such as polymeric micelles [49], nanoemulsions [50], nanoliposomes [51] containing oleanolic acid. Studies on routes of administration of OA including topical medication are required for preclinical efficacy. It is expected that developed dosage forms of OA can be widely used in skin disorders. Furthermore, it is necessary to clarify the upstream event of the identified molecular mechanisms of OA, along with its effect on skin barrier-related markers. Further studies based on immune responses, such as the effect of OA on toll-like receptors, innate lymphoid cells, and T cell responses, are needed to understand the anti-allergic inflammatory effect of OA and develop a therapeutic approach to treat allergic disorders.

In conclusion, OA inhibited AD-like responses, suppressing the pathway to make up the atopic environments dominated by Th2 cells via the inhibition of cytokines derived from skin keratinocytes via the blockade of Akt, NF-κB, and STAT1 signaling pathways. Considering this, we suggest that OA can be used as a potential therapeutic agent for the prevention or treatment of allergic inflammatory diseases, such as AD. We believe that these findings can be further elucidated in future studies to determine the specific roles of OA.

4. Materials and Methods

4.1. Chemicals and Reagents

For the present study, OA (O5504, ≥97%), 3-(4,5-Dimethylthiazol-2-yl)-2,5-diphenyl tetrazolium bromide (MTT, >98%), dimethyl sulfoxide (DMSO, ≥99.9%), and all other chemicals were purchased from Sigma; EMD Millipore (Billerica, MA, USA). Recombinant human TNF-α and recombinant human IFN-γ were purchased from Bio-Techne Ltd. (Abingdon, OX, UK). Dulbecco's modified Eagle medium (DMEM), fetal bovine serum (FBS), penicillin, and streptomycin were obtained from Life Technologies Inc. (Grand Island, NY, USA). Primary antibodies against NF-κB p65 (cat no. 8242), p-Akt (cat no. 9271), p-STAT1 (Tyr701; cat no. 9167), and p-STAT1 (Ser727; cat no. 8826), STAT1 (cat no. sc-14994), and IgG XP® Isotype (cat no. 3900) were obtained from Cell Signaling Technology, Inc. (Danvers, MA, USA). Primary antibodies for p-IκB-α (cat no. sc-8404), IκB-α (cat no. sc-203), Akt1/2/3 (cat no. sc-8312), PARP (cat no. sc-9542), α-tubulin (cat no. sc-8035), and β-actin (cat no. sc-81178) were purchased from Santa Cruz Biotechnology, Inc. (Dallas, TX, USA). Horseradish peroxidase-conjugated secondary antibodies were purchased from Jackson ImmunoResearch laboratories, Inc. (West Grove, PA, USA). The histamine enzyme-linked immunosorbent assay (ELISA) kit was obtained from Enzo life Sciences, Inc. (Farmingdale, NY, USA). SYBR Premix Ex Taq was purchased from Takara Bio, Inc. (Kusatsu, Japan). Oligonucleotide primers were purchased from Bioneer Corporation (Daejeon, Korea).

4.2. DNCB-Induced AD Model

A total of 30 ICR female mice (6 weeks old; 20–25 g body weight) were obtained from Charles River Laboratories (Harlan laboratories, inc., Wilmington, MA, USA) and maintained under constant conditions at a humidity of 40–60%, temperature of 20–25 °C, and a

12 h light/dark cycle. The mice were randomly assigned to one of five groups ($n = 6$ per group). To induce AD-like symptoms and skin lesions, 1-Chloro-2,4-dinitrochlorobenzene (DNCB, 97%) was used. Briefly, the mice were sensitized topically with 100 µL of 1% DNCB dissolved in 4:1 v/v mixture of acetone and corn oil and topically applied on the shaved area of the dorsal surface of mice for three days, followed by no treatment for four days. The same volume of vehicle (acetone/corn oil) was applied to the Normal group. After the first challenge inducing the AD-like symptoms, the treatment was repeated with 100 µL of 0.5% DNCB for 21 days. The mice were topically applied vehicle, dexamethasone (≥97%, 10 µM), or OA (10 and 50 µM) 4 h after DNCB treatment once a day. Dexamethasone is dissolved in PBS:100% EtOH:Cremophor (6:1:3). OA is dissolved in 100% EtOH:Cremophor (7:3). Mice were sacrificed on day 29 of the experiment. The experimental scheme is summarized in Figure 1A. Skin tissues from the back of the mice were obtained and subjected to histological analysis, qRT-PCR examination. All procedures were performed in accordance with university guidelines and approved by the Instructional Animal Care and Use Committee (IACUC) of Korean Medicine, Sangji University (Wonju, Korea; approval no. 2015-06).

4.3. Evaluation of Dermatitis Severity

Clinical dermatitis severity was tested using the method described by Yamamoto and colleagues (35). The severity of dermatitis was evaluated at the experiment start day and end day. The development of erythema/hemorrhage, scarring/dryness, edema, and excoriation/erosion was scored as follows: 0, none; 1, mild (<20%); 2, moderate (20–60%); 3, severe (>60%). The scores were determined in agreement between three observers, and the sum of the individual scores was used as the dermatitis score.

4.4. Histopathological Analysis

At the end of the study period, the dorsal skin of mice was collected. The samples were fixed in 10% buffered formalin, embedded in paraffin, sectioned into 4 µm thick, and stained with hematoxylin and eosin (H&E) and toluidine blue to detect epidermal thickness and inflammatory cells. Pathological changes of all stained skin sections were observed using a DM IL LED microscope (Leica, Wetzlar, Germany) and photographed using a DFC295 (Leica, Wetzlar, Germany). Digital images were taken from each slide (2 per group), and measured using Leica Application Suite (Leica, Wetzlar, Germany).

4.5. Histamine Assay

Blood from the mice was collected from each mouse at the end of the experiment. Serum was obtained by centrifugation at 1700× g for 30 min and stored at −80 °C until analysis. The release of histamine was measured using an ELISA kit in accordance with the manufacturer's protocol.

4.6. Cell Culture and Sample Treatment

HaCaT keratinocytes (passage 19) were provided by Professor Jae-Young Um (Kyung Hee University, Seoul, Korea), and were grown at 37 °C in DMEM supplemented with 10% FBS, penicillin (100 U/mL) and streptomycin (100 µg/mL) in a humidified atmosphere of 5% CO_2. HaCaT keratinocytes were seeded at a density of 1×10^5 cell per well, starved with 0.1% FBS media for 24 h, and treated with OA at 6.25, 12.5, and 25 µM for 1 h at 37 °C in humidified air with 5% CO_2, and then stimulated with 10 ng/mL of TNF-α/IFN-γ at 37 °C for indicated time. The cells were either treated with DMSO as a control. The OA was dissolved in DMSO.

4.7. Western Blot Analysis

Segments of cells, or dorsal tissue were suspended in PRO-PREP™ protein extraction solution (Intron Biotechnology, Inc., Seoul, Korea) and incubated for 20 min at 4 °C. Cell debris was removed via micro-centrifugation 11,000× g for 30 min at 4 °C, followed by rapid freezing of the supernatant. The protein concentration was determined using Bio-Rad protein assay reagent (Bio-Rad Laboratories, Inc., Hercules, CA, USA) according to the manufacturer's protocol. Cellular proteins from the treated and untreated cell extracts (10–30 μL) were electroblotted onto a polyvinylidene fluoride membrane following separation via 8–12% SDS-PAGE. The membrane was incubated for 1 h with blocking solution (5% skim milk) at room temperature, followed by overnight incubation with the primary antibodies (1:1000) at 4 °C. The blots were washed three times with Tween 20/Tris-buffered saline (T/TBS) and incubated with horseradish peroxidase-conjugated secondary antibody (1:2000) for 2 h at room temperature. The blots were washed three times with T/TBS, and then developed via enhanced chemiluminescence (GE Healthcare Life Sciences, Chalfont Saint Giles, UK). Densitometric analysis was performed using Bio-Rad Quantity One software version 4.3.0 (Bio-Rad Laboratories, Inc., Hercules, CA, USA).

4.8. Reverse Transcription-Quantitative Polymerase Chain Reaction (RT-qPCR) Analysis

Total RNA was isolated from the cells, or dorsal tissue using an Easy Blue kit (Intron Biotechnology, Inc., Seoul, Korea) according to the manufacturer's protocol. Total RNA was quantified using an Epoch micro-volume spectrophotometer system (BioTek Instruments, Inc., Winooski, VT, USA). cDNA was obtained using isolated total RNA (2 μg), d (T)16 primer, and Avian Myeloblastosis Virus reverse transcriptase with genomic DNA remover. The relative gene expression was quantified using RT-qPCR analysis (Real Time PCR System 7500; Applied Biosystems; Thermo Fisher Scientific, Inc., Waltham, MA, USA) with SYBR Premix Ex Taq. Fold changes of gene expression were calculated using the comparative quantification cycle (Cq) method. The Cq values of target genes were normalized to that of GAPDH using the ABI gene express 2.0 program (Applied Biosystems; Thermo Fisher Scientific, Inc., Waltham, MA, USA).

4.9. Immunofluorescence Staining

Cells were seeded in a chamber at 1×10^5 cells/mL, fixed with 100% methanol for 30 min at 20 °C, and blocked in 10% NGS in 0.3% Triton-X100 (Sigma). The samples were then incubated with p-STAT1 primary antibody or IgG isotype control overnight at 4 °C. After washing, the samples were incubated with the secondary antibody with Alexa-Fluor 488-conjugated goat anti-rabbit IgG (Invitrogen, Waltham, MA, USA). Coverslips were mounted on to the glass slides and the images were captured on a confocal laser-scanning fluorescence microscope Leica TCS SP5 (LAS AF) suite (Leica Microsystems, Wetzlar, Germany).

4.10. Statistical Analysis

The data are expressed as the mean ± standard deviation of triplicate experiments. Statistically significant differences were compared using one-way analysis of variance (ANOVA) with Dunnett's post hoc test. $p < 0.05$ was considered to indicate a statistically significant difference. Statistical analysis was performed using GraphPad Prism (version 5).

Supplementary Materials: The following are available online at https://www.mdpi.com/article/10.3390/ijms222112000/s1.

Author Contributions: Conceptualization; H.-J.A., data curation; M.L., formal analysis; Y.-M.K., funding acquisition; M.L., investigation; Y.-M.K. and H.-M.K., methodology; Y.-M.K., supervision; H.-J.A., validation; M.L., visualization; Y.-M.K., roles/writing—original draft; Y.-M.K., writing—review and editing; Y.-M.K. and H.-M.K. All authors have read and agreed to the published version of the manuscript.

Funding: This research was funded by the National Research Foundation of Korea (NRF), grant numbers NRF-2017R1C1B2008617, NRF-2017M3A9B6061511; and the Research fund from Sangji University Graduate School.

Institutional Review Board Statement: All procedures were performed in accordance with university guidelines and approved by the Instructional Animal Care and Use Committee (IACUC) of Korean Medicine, Sangji University (Wonju, Korea; approval no. 2015-06).

Informed Consent Statement: Not applicable.

Data Availability Statement: The data presented in this study are available on request from the corresponding author.

Conflicts of Interest: The authors declare no conflict of interest.

References

1. Jeong, H.J.; Kim, H.Y.; Kim, H.M. Molecular mechanisms of anti-inflammatory effect of chrysophanol, an active component of AST2017-01 on atopic dermatitis in vitro models. *Int. Immunopharmacol.* **2018**, *54*, 238–244.
2. Hou, D.D.; Di, Z.H.; Qi, R.Q.; Wang, H.X.; Zheng, S.; Hong, Y.X.; Guo, H.; Chen, H.D.; Gao, X.H. Sea Buckthorn (*Hippophae rhamnoides* L.) Oil Improves Atopic Dermatitis-Like Skin Lesions via Inhibition of NF-kappaB and STAT1 Activation. *Skin Pharmacol. Physiol.* **2017**, *30*, 268–276.
3. Yang, G.; Cheon, S.-Y.; Chung, K.-S.; Lee, S.-J.; Hong, C.-H.; Lee, K.-T.; Jang, D.-S.; Jeong, J.-C.; Kwon, O.-K.; Nam, J.-H.; et al. *Solanum tuberosum* L. cv. Jayoung Epidermis Extract Inhibits Mite Antigen-Induced Atopic Dermatitis in NC/Nga Mice by Regulating the Th1/Th2 Balance and Expression of Filaggrin. *J. Med. Food* **2015**, *18*, 1013–1021. [CrossRef]
4. Jung, M.; Lee, T.H.; Oh, H.J.; Kim, H.; Son, Y.; Lee, E.H.; Kim, J. Inhibitory effect of 5,6-dihydroergosteol-glucoside on atopic dermatitis-like skin lesions via suppression of NF-kappaB and STAT activation. *J. Dermatol. Sci.* **2015**, *79*, 252–261. [CrossRef]
5. Lee, H.; Ha, H.; Lee, J.K.; Park, S.J.; Jeong, S.I.; Shin, H.K. The Leaves of Broussonetia kazinoki Siebold Inhibit Atopic Dermatitis-Like Response on Mite Allergen-Treated Nc/Nga Mice. *Biomol. Ther.* **2014**, *22*, 438–444.
6. Choi, J.K.; Jang, Y.H.; Lee, S.; Lee, S.R.; Choi, Y.A.; Jin, M.; Choi, J.H.; Park, J.H.; Park, P.H.; Choi, H.; et al. Chrysin attenuates atopic dermatitis by suppressing inflammation of keratinocytes. *Food Chem. Toxicol.* **2017**, *110*, 142–150.
7. Jung, M.R.; Lee, T.H.; Bang, M.H.; Kim, H.; Son, Y.; Chung, D.K.; Kim, J. Suppression of thymus- and activation-regulated chemokine (TARC/CCL17) production by 3-O-beta-D-glucopyanosylspinasterol via blocking NF-kappaB and STAT1 signaling pathways in TNF-alpha and IFN-gamma-induced HaCaT keratinocytes. *Biochem. Biophys. Res. Commun.* **2012**, *427*, 236–241. [CrossRef] [PubMed]
8. Han, E.H.; Hwang, Y.P.; Choi, J.H.; Yang, J.H.; Seo, J.K.; Chung, Y.C.; Jeong, H.G. Psidium guajava extract inhibits thymus and activation-regulated chemokine (TARC/CCL17) production in human keratinocytes by inducing heme oxygenase-1 and blocking NF-kappaB and STAT1 activation. *Environ. Toxicol. Pharmacol.* **2011**, *32*, 136–145. [CrossRef]
9. Ayeleso, T.B.; Matumba, M.G.; Mukwevho, E. Oleanolic Acid and Its Derivatives: Biological Activities and Therapeutic Potential in Chronic Diseases. *Molecules* **2017**, *22*, 1915. [CrossRef]
10. Sultana, N.; Ata, A. Oleanolic acid and related derivatives as medicinally important compounds. *J. Enzyme Inhib. Med. Chem.* **2008**, *23*, 739–756. [CrossRef] [PubMed]
11. Giner-Larza, E.M.; Manez, S.; Recio, M.C.; Giner, R.M.; Prieto, J.M.; Cerda-Nicolas, M.; Rios, J.L. Oleanonic acid, a 3-oxotriterpene from Pistacia, inhibits leukotriene synthesis and has anti-inflammatory activity. *Eur. J. Pharmacol.* **2001**, *428*, 137–143.
12. Dzubak, P.; Hajduch, M.; Vydra, D.; Hustova, A.; Kvasnica, M.; Biedermann, D.; Markova, L.; Urban, M.; Sarek, J. Pharmacological activities of natural triterpenoids and their therapeutic implications. *Nat. Prod. Rep.* **2006**, *23*, 394–411. [PubMed]
13. Lee, J.Y.; Moon, H.; Kim, C.J. Effects of hydroxy pentacyclic triterpene acids from Forsythia viridissima on asthmatic responses to ovalbumin challenge in conscious guinea pigs. *Biol. Pharm. Bull.* **2010**, *33*, 230–237. [CrossRef]
14. Li, M.; Han, Z.; Bei, W.; Rong, X.; Guo, J.; Hu, X. Oleanolic Acid Attenuates Insulin Resistance via NF-kappaB to Regulate the IRS1-GLUT4 Pathway in HepG2 Cells. *Evid. Based Complement. Altern. Med.* **2015**, *2015*, 643102. [CrossRef] [PubMed]
15. Li, Y.; Wang, J.; Gu, T.; Yamahara, J.; Li, Y. Oleanolic acid supplement attenuates liquid fructose-induced adipose tissue insulin resistance through the insulin receptor substrate-1/phosphatidylinositol 3-kinase/Akt signaling pathway in rats. *Toxicol. Appl. Pharmacol.* **2014**, *277*, 155–163. [CrossRef]
16. Petronelli, A.; Pannitteri, G.; Testa, U. Triterpenoids as new promising anticancer drugs. *Anticancer Drugs* **2009**, *20*, 880–892. [CrossRef]
17. Wang, X.; Ye, X.-L.; Liu, R.; Chen, H.-L.; Bai, H.; Liang, X.; Zhang, X.-D.; Wang, Z.; Li, W.-L.; Hai, C.-X. Antioxidant activities of oleanolic acid in vitro: Possible role of Nrf2 and MAP kinases. *Chem. Biol. Interact.* **2010**, *184*, 328–337. [CrossRef]
18. Martin, R.; Hernandez, M.; Cordova, C.; Nieto, M.L. Natural triterpenes modulate immune-inflammatory markers of experimental autoimmune encephalomyelitis: Therapeutic implications for multiple sclerosis. *Br. Pharmacol. J.* **2012**, *166*, 1708–1723.
19. Sifaoui, I.; Lopez-Arencibia, A.; Martin-Navarro, C.M.; Reyes-Batlle, M.; Mejri, M.; Valladares, B.; Lorenzo-Morales, J.; Abderabba, M.; Pinero, J.E. Selective activity of Oleanolic and Maslinic Acids on the Amastigote form of *Leishmania* spp. *Iran J. Pharm. Res.* **2017**, *16*, 1190–1193.
20. Somova, L.I.; Shode, F.O.; Ramnanan, P.; Nadar, A. Antihypertensive, antiatherosclerotic and antioxidant activity of triterpenoids isolated from *Olea europaea*, subspecies africana leaves. *J. Ethnopharmacol.* **2003**, *84*, 299–305. [PubMed]

21. Kang, Y.M.; Lee, M.; An, H.J. Oleanolic acid protects against mast cell-mediated allergic responses by suppressing Akt/NF-kappaB and STAT1 activation. *Phytomedicine* **2021**, *80*, 153340.
22. Zhang, E.Y.; Chen, A.Y.; Zhu, B.T. Mechanism of dinitrochlorobenzene-induced dermatitis in mice: Role of specific antibodies in pathogenesis. *PLoS ONE* **2009**, *4*, e7703. [CrossRef]
23. Yang, I.J.; Lee, D.U.; Shin, H.M. Inhibitory Effect of Valencene on the Development of Atopic Dermatitis-Like Skin Lesions in NC/Nga Mice. *Evid. Based Complement. Altern. Med.* **2016**, *2016*, 9370893. [CrossRef] [PubMed]
24. Cho, M.S.; Park, W.S.; Jung, W.K.; Qian, Z.J.; Lee, D.S.; Choi, J.S.; Lee, D.Y.; Park, S.G.; Seo, S.K.; Kim, H.J.; et al. Caffeic acid phenethyl ester promotes anti-inflammatory effects by inhibiting MAPK and NF-kappaB signaling in activated HMC-1 human mast cells. *Pharm. Biol.* **2014**, *52*, 926–932. [PubMed]
25. Park, E.J.; Park, K.C.; Eo, H.; Seo, J.; Son, M.; Kim, K.H.; Chang, Y.-S.; Cho, S.-H.; Min, K.-U.; Jin, M.; et al. Suppression of spontaneous dermatitis in NC/Nga murine model by PG102 isolated from *Actinidia arguta*. *J. Investig. Dermatol.* **2007**, *127*, 1154–1160. [PubMed]
26. Liu, Y.J. Thymic stromal lymphopoietin: Master switch for allergic inflammation. *J. Exp. Med.* **2006**, *203*, 269–273. [CrossRef] [PubMed]
27. Sung, Y.Y.; Kim, Y.S.; Kim, H.K. Illicium verum extract inhibits TNF-alpha- and IFN-gamma-induced expression of chemokines and cytokines in human keratinocytes. *J. Ethnopharmacol.* **2012**, *144*, 182–189. [CrossRef]
28. Choi, H.J.; Lee, J.H.; Jung, Y.S. (+)-Nootkatone inhibits tumor necrosis factor alpha/interferon gamma-induced production of chemokines in HaCaT cells. *Biochem. Biophys. Res. Commun.* **2014**, *447*, 278–284. [CrossRef]
29. Park, J.H.; Kim, M.S.; Jeong, G.S.; Yoon, J. Xanthii fructus extract inhibits TNF-alpha/IFN-gamma-induced Th2-chemokines production via blockade of NF-kappaB, STAT1 and p38-MAPK activation in human epidermal keratinocytes. *J. Ethnopharmacol.* **2015**, *171*, 85–93. [CrossRef] [PubMed]
30. Lee, B.S.; Shim, S.M.; Heo, J.; Pae, H.O.; Seo, B.Y.; Han, S.Y.; Sohn, D.H.; Jang, S.I.; Chung, H.T. Wogonin suppresses TARC expression induced by mite antigen via heme oxygenase 1 in human keratinocytes. Suppressive effect of wogonin on mite antigen-induced TARC expression. *J. Dermatol. Sci.* **2007**, *46*, 31–40. [CrossRef]
31. Toncic, R.J.; Lipozencic, J. Atopy patch test—When is it useful? *Acta Med. Croat.* **2011**, *65*, 97–106.
32. Hajar, T.; Gontijo, J.R.V.; Hanifin, J.M. New and developing therapies for atopic dermatitis. *An. Bras. Dermatol.* **2018**, *93*, 104–107. [CrossRef]
33. Kawahara, T.; Tsutsui, K.; Nakanishi, E.; Inoue, T.; Hamauzu, Y. Effect of the topical application of an ethanol extract of quince seeds on the development of atopic dermatitis-like symptoms in NC/Nga mice. *BMC Complement. Altern. Med.* **2017**, *17*, 80. [CrossRef] [PubMed]
34. Zhu, T.H.; Zhu, T.R.; Tran, K.A.; Sivamani, R.K.; Shi, V.Y. Epithelial Barrier Dysfunctions in Atopic Dermatitis: A Skin-Gut-Lung Model Linking Microbiome Alteration and Immune Dysregulation. *Br. J. Dermatol.* **2018**, *179*, 570–581. [CrossRef]
35. Nutten, S. Atopic dermatitis: Global epidemiology and risk factors. *Ann. Nutr. Metab.* **2015**, *66* (Suppl. 1), 8–16. [CrossRef]
36. Saeki, H.; Tamaki, K. Thymus and activation regulated chemokine (TARC)/CCL17 and skin diseases. *J. Dermatol. Sci.* **2006**, *43*, 75–84. [CrossRef]
37. Liddiard, K.; Welch, J.S.; Lozach, J.; Heinz, S.; Glass, C.K.; Greaves, D.R. Interleukin-4 induction of the CC chemokine TARC (CCL17) in murine macrophages is mediated by multiple STAT6 sites in the TARC gene promoter. *BMC Mol. Biol.* **2006**, *7*, 45. [CrossRef]
38. Licona-Limon, P.; Kim, L.K.; Palm, N.W.; Flavell, R.A. TH2, allergy and group 2 innate lymphoid cells. *Nat. Immunol.* **2013**, *14*, 536–542.
39. Rochman, Y.; Dienger-Stambaugh, K.; Richgels, P.K.; Lewkowich, I.P.; Kartashov, A.V.; Barski, A.; Khurana Hershey, G.K.; Leonard, W.J.; Singh, H. TSLP signaling in CD4(+) T cells programs a pathogenic T helper 2 cell state. *Sci. Signal.* **2018**, *11*, eaam8858. [PubMed]
40. Choi, J.H.; Jin, S.W.; Han, E.H.; Park, B.H.; Kim, H.G.; Khanal, T.; Hwang, Y.P.; Do, M.T.; Lee, H.S.; Chung, Y.C.; et al. Platycodon grandiflorum root-derived saponins attenuate atopic dermatitis-like skin lesions via suppression of NF-kappaB and STAT1 and activation of Nrf2/ARE-mediated heme oxygenase-1. *Phytomedicine* **2014**, *21*, 1053–1061. [CrossRef] [PubMed]
41. Cultrone, A.; de Wouters, T.; Lakhdari, O.; Kelly, D.; Mulder, I.; Logan, E.; Lapaque, N.; Dore, J.; Blottiere, H.M. The NF-kappaB binding site located in the proximal region of the TSLP promoter is critical for TSLP modulation in human intestinal epithelial cells. *Eur. J. Immunol.* **2013**, *43*, 1053–1062. [CrossRef] [PubMed]
42. Darnell, J.E., Jr.; Kerr, I.M.; Stark, G.R. Jak-STAT pathways and transcriptional activation in response to IFNs and other extracellular signaling proteins. *Science* **1994**, *264*, 1415–1421. [CrossRef] [PubMed]
43. Ahn, S.; Siddiqi, M.H.; Aceituno, V.C.; Simu, S.Y.; Zhang, J.; Perez, Z.E.; Kim, Y.J.; Yang, D.C. Ginsenoside Rg5:Rk1 attenuates TNF-alpha/IFN-gamma-induced production of thymus- and activation-regulated chemokine (TARC/CCL17) and LPS-induced NO production via downregulation of NF-kappaB/p38 MAPK/STAT1 signaling in human keratinocytes and macrophages. *In Vitro Cell. Dev. Biol. Anim.* **2016**, *52*, 287–295. [PubMed]
44. Park, J.H.; Lee, K.Y.; Park, B.; Yoon, J. Suppression of Th2 chemokines by crocin via blocking of ERK-MAPK/NF-kappaB/STAT1 signalling pathways in TNF-alpha/IFN-gamma-stimulated human epidermal keratinocytes. *Exp. Dermatol.* **2015**, *24*, 634–636. [CrossRef] [PubMed]

45. Kim, M.; Lee, S.; Lim, H.; Lee, J.; Park, J.-Y.; Kwon, H.-J.; Lee, I.-C.; Ryu, Y.-B.; Kim, J.; Shin, T.; et al. Oleanolic Acid Acetate Alleviates Symptoms of Experimental Autoimmune Encephalomyelitis in Mice by Regulating Toll-Like Receptor 2 Signaling. *Front. Pharmacol.* **2020**, *11*, 556391. [CrossRef]
46. Dong, N.; Xue, C.; Zhang, L.; Zhang, T.; Wang, C.; Bi, C.; Shan, A. Oleanolic acid enhances tight junctions and ameliorates inflammation in Salmonella typhimurium-induced diarrhea in mice via the TLR4/NF-kappaB and MAPK pathway. *Food Funct.* **2020**, *11*, 1122–1132. [CrossRef]
47. Kawai, T.; Akira, S. Signaling to NF-kappaB by Toll-like receptors. *Trends Mol. Med.* **2007**, *13*, 460–469. [CrossRef]
48. Choi, J.K.; Oh, H.M.; Lee, S.; Park, J.W.; Khang, D.; Lee, S.W.; Lee, W.S.; Rho, M.C.; Kim, S.H. Oleanolic acid acetate inhibits atopic dermatitis and allergic contact dermatitis in a murine model. *Toxicol. Appl. Pharmacol.* **2013**, *269*, 72–80.
49. An, J.Y.; Yang, H.S.; Park, N.R.; Koo, T.S.; Shin, B.; Lee, E.H.; Cho, S.H. Development of Polymeric Micelles of Oleanolic Acid and Evaluation of Their Clinical Efficacy. *Nanoscale Res. Lett.* **2020**, *15*, 133. [CrossRef]
50. Alvarado, H.L.; Abrego, G.; Souto, E.B.; Garduno-Ramirez, M.L.; Clares, B.; Garcia, M.L.; Calpena, A.C. Nanoemulsions for dermal controlled release of oleanolic and ursolic acids: In vitro, ex vivo and in vivo characterization. *Colloids Surf. B Biointerfaces* **2015**, *130*, 40–47. [CrossRef] [PubMed]
51. Liu, Y.; Luo, X.; Xu, X.; Gao, N.; Liu, X. Preparation, characterization and in vivo pharmacokinetic study of PVP-modified oleanolic acid liposomes. *Int. J. Pharm.* **2017**, *517*, 1–7. [PubMed]

International Journal of
Molecular Sciences

MDPI

Article

Transcriptomic Characterization of Cow, Donkey and Goat Milk Extracellular Vesicles Reveals Their Anti-inflammatory and Immunomodulatory Potential

Samanta Mecocci [1,2], Daniele Pietrucci [3,4], Marco Milanesi [3], Luisa Pascucci [1], Silvia Filippi [5], Vittorio Rosato [6], Giovanni Chillemi [3,4,*], Stefano Capomaccio [1,2,†] and Katia Cappelli [1,2,†]

1 Department of Veterinary Medicine, University of Perugia, 06123 Perugia, Italy; samanta.mecocci@studenti.unipg.it (S.M.); luisa.pascucci@unipg.it (L.P.); stefano.capomaccio@unipg.it (S.C.); katia.cappelli@unipg.it (K.C.)
2 Sports Horse Research Center (CRCS), University of Perugia, 06123 Perugia, Italy
3 Department for Innovation in Biological, Agro-Food and Forest Systems (DIBAF), University of Tuscia, 01100 Viterbo, Italy; daniele.pietrucci.89@gmail.com (D.P.); marco.milanesi@unitus.it (M.M.)
4 Institute of Biomembranes, Bioenergetics and Molecular Biotechnologies, IBIOM, CNR, 70126 Bari, Italy
5 Department of Ecological and Biological Sciences, University of Tuscia, 01100 Viterbo, Italy; silvia.filippi@unitus.it
6 Genechron Srl, Via Giunio Antonio Resti 63, 00143 Roma, Italy; vittorio.rosato@genechron.com
* Correspondence: gchillemi@unitus.it; Tel.:+39-0761-357429
† Stefano Capomaccio and Katia Cappelli contributed equally to this paper.

Abstract: Milk extracellular vesicles (mEVs) seem to be one of the main maternal messages delivery systems. Extracellular vesicles (EVs) are micro/nano-sized membrane-bound structures enclosing signaling molecules and thus acting as signal mediators between distant cells and/or tissues, exerting biological effects such as immune modulation and pro-regenerative activity. Milk is also a unique, scalable, and reliable source of EVs. Our aim was to characterize the RNA content of cow, donkey, and goat mEVs through transcriptomic analysis of mRNA and small RNA libraries. Over 10,000 transcripts and 2000 small RNAs were expressed in mEVs of each species. Among the most represented transcripts, 110 mRNAs were common between the species with cow acting as the most divergent. The most represented small RNA class was miRNA in all the species, with 10 shared miRNAs having high impact on the immune regulatory function. Functional analysis for the most abundant mRNAs shows epigenetic functions such as histone modification, telomere maintenance, and chromatin remodeling for cow; lipid catabolism, oxidative stress, and vitamin metabolism for donkey; and terms related to chemokine receptor interaction, leukocytes migration, and transcriptional regulation in response to stress for goat. For miRNA targets, shared terms emerged as the main functions for all the species: immunity modulation, protein synthesis, cellular cycle regulation, transmembrane exchanges, and ion channels. Moreover, donkey and goat showed additional terms related to epigenetic modification and DNA maintenance. Our results showed a potential mEVs immune regulatory purpose through their RNA cargo, although in vivo validation studies are necessary.

Keywords: milk; extracellular vesicles; mEVs; RNAseq; miRNA; anti-inflammatory; immunomodulatory; cow; donkey; goat

Citation: Mecocci, S.; Pietrucci, D.; Milanesi, M.; Pascucci, L.; Filippi, S.; Rosato, V.; Chillemi, G.; Capomaccio, S.; Cappelli, K. Transcriptomic Characterization of Cow, Donkey and Goat Milk Extracellular Vesicles Reveals Their Anti-inflammatory and Immunomodulatory Potential. *Int. J. Mol. Sci.* **2021**, *22*, 12759. https://doi.org/10.3390/ijms222312759

Academic Editor: Jesús Osada

Received: 29 October 2021
Accepted: 23 November 2021
Published: 25 November 2021

Publisher's Note: MDPI stays neutral with regard to jurisdictional claims in published maps and institutional affiliations.

1. Introduction

Other than being a valuable nutrition source, milk represents a sophisticated signaling system that delivers maternal messages to the offspring [1]. This property seems to be mostly mediated by molecules enclosed in micro/nano-sized membrane-bound structures called extracellular vesicles (EVs) [2]. EVs are involved in cell-cell communication by transporting DNAs, mRNAs, microRNAs, lipids, and proteins to close and distant cells or tissues, protecting their labile content against degradation and providing a vehicle for cargo uptake by recipient cells. They can exert a plethora of biological effects with immunomodulating, anti-inflammatory, anti-cancer and pro-regenerative activities [3–15]. Signals can be as simple as an antigenic stimuli by EVs proteins or more complex as epigenetic modifications mediated by their miRNA cargo [16].

Food can be considered a valuable source for the mass production of EVs for their multiple uses; since milk is a widely available and inexpensive raw material for their production [17], milk-derived EVs (mEVs) have emerged as a reliable and scalable EVs source [18–23]. EVs from colostrum and breast milk were first isolated in 2007 in the human species [23], then subsequent studies reported the isolation from other mammals (cow, camel, buffalo, pig, and sheep) [24–28]. Milk immunomodulatory and anti-inflammatory activity, mainly due to the miRNA content [29,30], has been proven for human, cow, and donkey [31,32].

It is interesting that EVs and their RNA content are unaffected by acidification of the gastric environment [24,33] and different types of cells can uptake EVs from human and bovine milk [20,29,34]. For example, porcine milk EVs and their miRNAs are internalized by IPEC-J2 intestinal epithelial cells, modifying target gene expression and promoting intestinal cell proliferation [25]. Recent evidences demonstrate that milk EVs and miRNAs, taken up by human intestinal cells, can also reach the systemic circulation modifying gene expression of distant cells [35,36]. Moreover, besides the determination of high conservation level in miRNA coding genes through different species, these studies suggest a broad gene regulation in recipient cells, since each one of them targets several protein-coding genes [18,25,37–39].

MiRNAs adsorption through EV endocytosis after milk ingestion may play a role in the regulation of innate and adaptive immunity even through epigenetic changes [30,40,41]. Milk miRNA-mediated epigenetic regulation is a physiological phenomenon representing a natural channel for genetic material transfer to offspring and milk consumers [40]. Milk miRNAs also seem to be positively involved in resistance to allergy development and to autoimmune diseases [29,30].

Several studies have shown that miRNAs are associated with inflammatory processes in chronic diseases, regulating several mechanisms involved in inflammation initiation and resolution [42,43]. Within this path, the role of EV-mediated immune responses in the pathogenesis of chronic inflammation, such as inflammatory bowel disease (IBD), was recently investigated [44]. Non-coding RNAs play a pivotal role [45,46], taking part in the regulation of inflammatory processes, immune system modulation, and in the mucosal and microbiota homeostasis [44]. IBD, including Crohn's disease (CD) and ulcerative colitis (UC), is a chronic relapsing inflammatory disorder of the intestine of unclear etiology that has become a global disease with an increasing incidence in newly industrialized countries [47].

In addition, there are evidences that miRNAs themselves can be regulated by feeding, suggesting that diet manipulation could be a promising therapeutic approach for modulating the risk of chronic diseases [31,48–50].

Moreover, the comparative study of mEVs in domestic species, while highlighting the particular cellular messages in different milks, might help to understand how the selection of mutational changes in miRNAs and/or miRNA binding sites could have provided a mechanism to generate some of the traits that differentiate domesticated species [51].

With this work, we aimed to complete our journey [52] in characterizing the molecular content of cow, donkey, and goat mEVs through a comprehensive RNA analysis to search for future applications as a nutraceutical in inflammatory conditions such as IBD.

2. Results

A complete overview of the different experiments and analyses carried out in the study is shown in Figure 1, wherein the upper part of the figure represents the milk sampling, the EV isolation method, and the morphological analysis with the relative synthetic representation of results. Below, the two bioinformatic pipelines used for RNA sequencing data analysis are detailed in the two arrows: the orange one for mRNAs and the green one for small RNAs. The references to figures and tables generated from the different analytical steps and representing the results obtained from this work are reported between the two arrows.

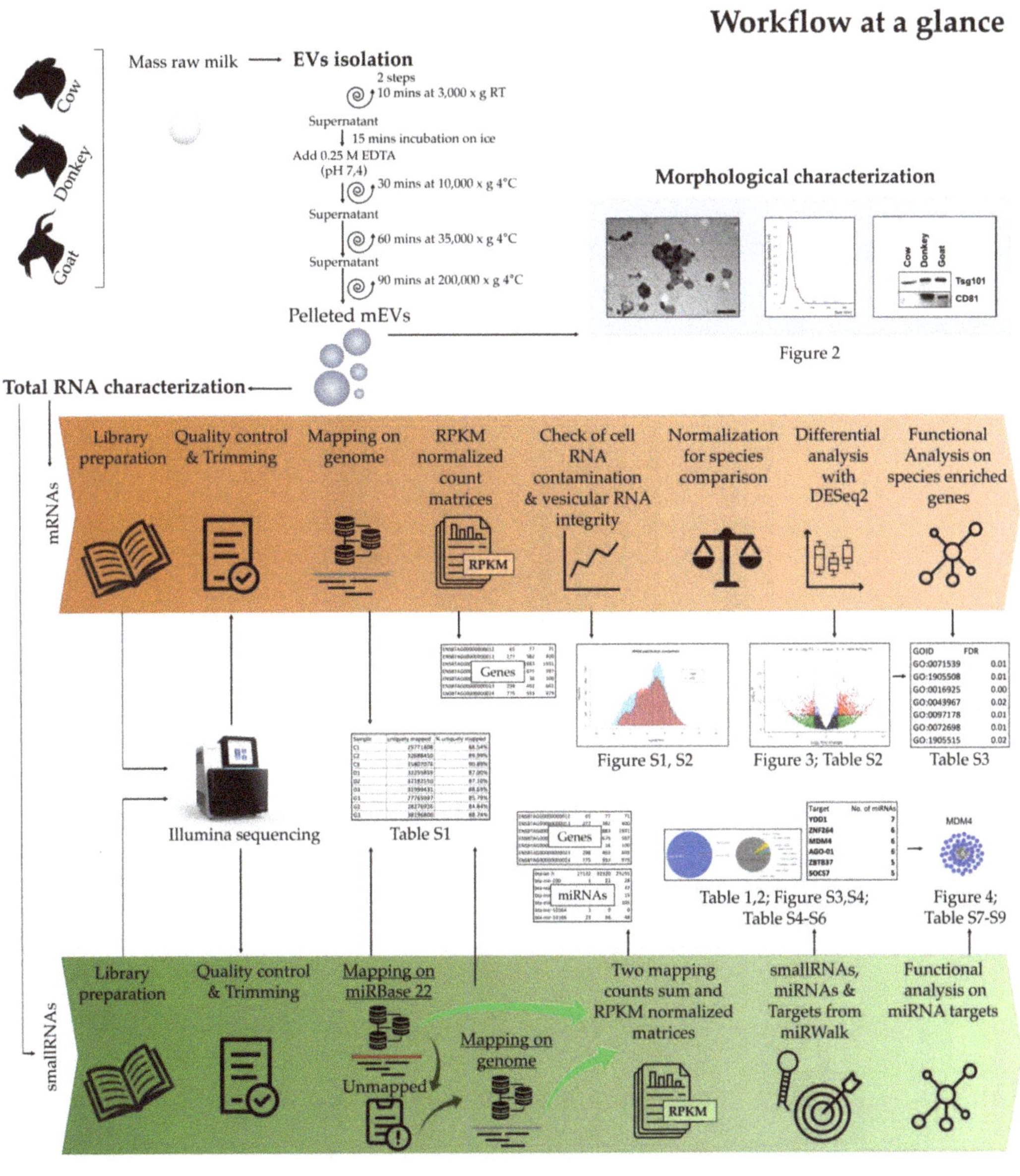

Figure 1. Workflow overview cartoon.

2.1. Morphologic Characterization of mEVs

The presence and purity of mEVs in the pellet generated from the milk of the three species along with size range and shape assessment were determined, respectively, by

Western blotting, Transmission Electron Microscopy (TEM), and Nanoparticle Tracking Analysis (NTA). Figure 2 shows the electron micrographs, the size distributions observed via NTA, and the antibody reaction against mEV antigens.

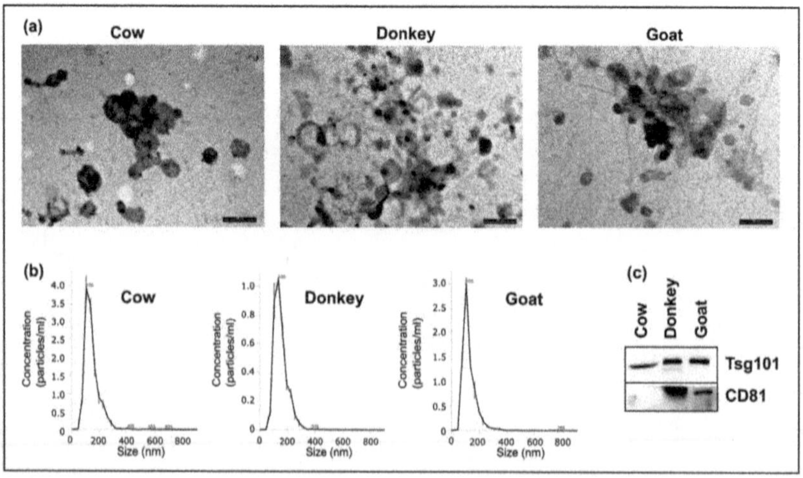

Figure 2. Morphological characterization of mEVs: (**a**) Transmission electron microscopy (TEM) revealed the presence of single and aggregated vesicles mainly in the range of 30–150 nm. Scale bar: 200 nm; (**b**) Size distribution of vesicles measured using the Nanosight NS300 nanoparticle tracking analysis system. The peak of the mEV size distribution was 105 nm for cow and goat and 135 for donkey; (**c**) Western blot results of Tsg101 (Tumor Susceptibility gene 101 protein) and CD81 (Cluster of Differentiation 81) tested mEV antigens.

At TEM, mEVs appeared rather homogeneous in shape and showed mainly intact limiting membrane. They ranged from 30 to 150 nm and were sometimes arranged in aggregates. No cell debris and minimal background were detected, revealing the efficiency of the purification procedure (Figure 2a). Based on the NTA data, the mEV mean (±standard error) diameter was 142.7 ± 2.9 nm for cow, 150.5 ± 3.2 for donkey, and 124.1 ± 2.3 for goat (Figure 2b). The distribution of mEV populations was "gaussian-like" for all the species with a unique peak and a narrow range (cow: D10 = 92.6 ± 1.7 nm and D90 = 215.2 ± 4.1 nm; donkey: D10 = 98.3 ± 1.7 nm and D90 = 219.7 ± 4.7 nm; goat: D10 = 75.9 ± 1.6 nm and D90 = 184.5 ± 3.5 nm), demonstrating homogeneity both within and between species. NTA also determines nanoparticle densities, reporting the mean concentration (particles/mL) (±standard deviation) of five measurements after a pellet resuspension in 400 µL of PBS that resulted as quite similar for the three specie: 1.22×10^{12} ($\pm 3.63 \times 10^{10}$) for cow; 3.51×10^{11} ($\pm 1.22 \times 10^{10}$) for donkey; and 7.39×10^{11} ($\pm 1.57 \times 10^{10}$) for goat. Western blot assay proved the mEV presence, showing a positive reaction for Tumor Susceptibility gene 101 protein (Tsg101) in all samples. Cluster of Differentiation 81 (CD81) expression was positive in donkey and goat, but not in cow (Figure 2c).

2.2. *Molecular Characterization of mEVs RNAs*

2.2.1. Messenger RNAs

A total of 326 million raw reads were generated from single-end sequencing of three replicates per species on Illumina platform with an average of more than 36 million reads per sample. Over 99% of the produced sequences were retained after quality control and trimming procedures; of these, 84.8–90.9% were uniquely mapped on the reference sequences. Mapping statistics details are reported in Table S1.

To further confirm mEV isolation purity and exclude cell contamination, a gross comparison between the number of expressed genes (reads per kilobase per million mapped reads-RPKM > 0) in mEV and mammary glands, from publicly available studies, was conducted. Only bovine mRNAs were used since udder mRNA sequencing data for the other two species are not yet available. The mean number of expressed genes in mEVs (14,407.67 ± 339.37) was significantly lower (p-value = 0.0286) than mammary glands (19,117.25 ± 1424.25). At the same time, the mean expression profile was significantly different (p-value < 2.2×10^{-16}) between mEVs (1.16 ± 4.09) and mammary glands (0.48 ± 4.56). A graphical representation of the two gene sets is available in Figure S1.

Gene body coverage analysis carried out through the RSeQC package revealed a good representation in almost the entire transcript set length both in cow and donkey while goat showed a general reduction at the 3′-end (Figure S2).

A total of 10,656, 9464 and 11,143 expressed genes (RPKM > 1) were found for cow, donkey, and goat, respectively. For the differential expression analysis between mEV mRNAs of the three species through DESeq2 software, a core of 5907 orthologous genes (with a one-to-one type of orthology) was taken into account. Pairwise comparisons were applied, showing cow as divergent species compared to the others (Figure 3) with twice as many up-regulated genes (DEGs are reported in Table S2).

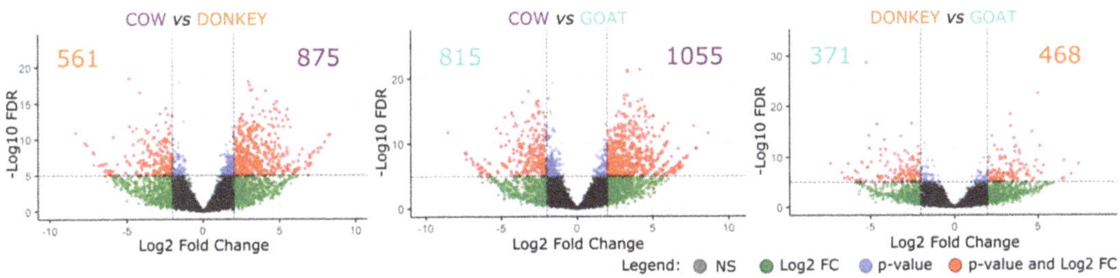

Figure 3. Volcano plots of differentially expressed genes through mEV RNA pairwise comparisons of the three species using DESeq2 for data analysis. The X-axis represents $\log_2$ fold change ($\log_2$ FC) of the pairwise comparison, while the Y-axis indicates significance of differential expression (False Discovery Rate - FDR). The gray dots indicate no significant (NS) changes in the gene expression ($|\log_2 FC| < 2$, FDR > 0.05), the green dots denote genes with $|\log_2 FC| > 2$ but not significant (FDR > 0.05), light blue dots indicate very small changes in gene expression ($|\log_2 FC| < 2$), while red dots represent significantly up- and down-regulated genes ($|\log_2 FC| > 2$, FDR < 0.05). Purple numbers indicate the significantly up-regulated genes in cow compared to the other reported species, the orange numbers indicate those up-regulated in donkey, and the teal numbers refer to goat over-expressed genes.

For each species, transcripts found to be up-regulated in both the pairwise comparisons were considered enriched (Table S2) and selected for functional analysis of the three gene ontology (GO) vocabularies (Table S3). All of these enriched transcripts showed an RPKM value equal to or greater than six in at least one species, assessing a relevant expression level. The 78 GO terms significantly enriched in cow are representative of 19 functional groups, including the regulation of telomere maintenance and histone modifications, ncRNAs transport, and regulation of cytokines stimulus response. Terms of lipid catabolism, oxidative stress, and vitamin metabolism were enriched for donkey among the 63 GO terms enclosed in 13 functional groups. Goat showed 28 GO terms representative of 11 functional groups including terms related to chemokine receptor interaction, leukocytes migration, histone modification, and transcriptional regulation in response to stress. All species show significant enrichment in GO terms related to microtubule function. All GO details are reported in Table S3.

2.2.2. Small RNAs

Concerning small RNA features in cow and goat, almost all the reads were aligned to the microRNA class (miRNAs), comprising more than over 99% of total RPKM. Although this was the most representative class also in donkey (57% of total RPKM), the remaining portion of sequences fell into miscellaneous RNA, mostly Y-RNA and Vault classes (Figure S3).

In Table 1 are reported the higher expressed miRNAs (chosen from the features covered by 95% of total RPKM, Table S4) resulting in 41, 28, and 40 miRNA for cow, donkey, and goat, respectively. As shown in Figure S4, 10 of these are shared among the three species. Details on sharing and uniqueness are provided in Table 1.

Table 1. Most expressed miRNAs (covered by 95% RPKM) in mEV species cargo.

Enclosing mEVs	No. of miRNAs	miRNAs		
Cow–Donkey–Goat	10	miR-151a miR-191 miR-30b let-7i	miR-27b miR-30d miR-186	miR-200b miR-148b miR-21
Cow–Goat	15	miR-141 miR-23b miR-93 miR-25 miR-361	let-7b miR-660 let-7g miR-30a miR-200a	miR-125a miR-20a miR-29a miR-140 miR-200c
Donkey–Goat	6	miR-148a miR-143	miR-26b miR-30e	miR-223 miR-155
Cow–Donkey	1	miR-375		
Cow	10	miR-340 mir181a miR-362 miR-99a	miR-532 miR-26b miR-182	miR-429 miR-425 miR-34a
Donkey	10	miR-374a miR-152 miR-101 miR-98	miR-590 miR-320a let-7d	miR-23a miR-145 let-7c
Goat	8	miR-26a let-7f miR-27a	miR-22 miR-224 miR-16	miR-125b miR-190a

MiRNAs listed in Table 1 were used as input for MiRWalk analysis, obtaining their validated targets differentiated for target binding sites (3′-UTR, 5′-UTR, and CDS) (Table 2). However, for chi-miR-3431, eca-miR-3548, bta-miR-3600, bta-miR-669, bta-miR-148c, and bta-miR-148d it was not possible to identify the human homolog. Therefore, they were excluded from the target analysis. For each species, a unique target list was generated and filtered according to the number of miRNA hits (five for cow and goat and four for donkey) (Table S5).

Table 2. Number of total and selected targets retrieved for each species.

Species	3′-UTR Targets	5′-UTR Targets	CDS Targets	Total Targets	Filtered Targets
Cow	1351	221	1196	2296	3
Donkey	814	109	603	1313	15
Goat	1537	256	1283	2568	14

Filtered targets were used to generate a Protein-Protein Interaction Network (PPI) using the IMEx database: a total of 123 nodes with 253 edges were produced for cow genes, donkey had 1025 nodes and 2272 edges, while goat showed 988 nodes and 2899 edges. Among the numerous protein clusters identified by the analysis, we selected those with an interaction threshold greater than 25, resulting in three clusters for cow and and eight for both donkey and goat (Figure 4). Briefly, a cluster is characterized by a central node, corresponding to the protein with the highest number of interactions with other proteins correlated to similar biological functions.

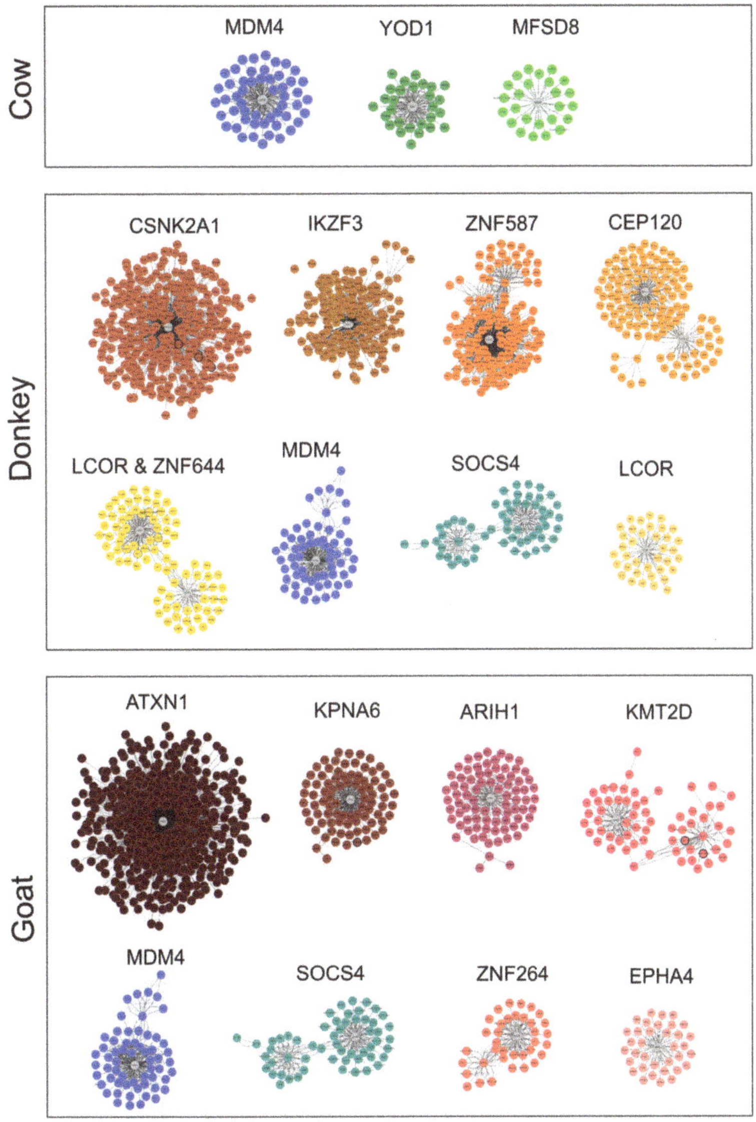

Figure 4. Clusters of proteins generated by clusterMaker 2.0 Cytoscape application starting from the PPI network. Central nodes (grey) indicate proteins with the higher number of interconnections. A consistent colour scheme was applied for clusters with the same central node in the different species.

MDM4 was a central node protein for all the species, while donkey and goat shared an additional central node, SOCS4 (Figure 4; Table S6). A GO enrichment analysis was then carried out on these clusters through the ClueGO app and results are reported in Tables S7–S9 for cow, donkey, and goat clusters, respectively.

3. Discussion

In the last two decades, extracellular vesicles (EVs) gained global attention, tickling researchers minds for new ideas and potential applications thanks to their intrinsic messenger carrier role [53]. In this study, the total RNA content of EVs isolated from cow, donkey, and goat milk (mEVs) was characterized, and their potential functional role highlighted, both for the mRNA and miRNA cargo. A focus was placed on their anti-inflammatory and immunomodulant potential as a conditional prerequisite to use mEVs for attenuating symptoms in inflammatory bowel diseases (IBD).

Three different species were chosen to evaluate the mEV RNA cargo since domestication and selection may have influenced metabolic phenotypes in modern livestock and during the domestication process, as pointed out in an in silico analysis [51].

One of the major flaws that one can encounter in the characterization of EVs content is based on the quality of the isolated material. Indeed, as a prerequisite for the RNAs analysis, we verified that our samples were essentially enriched in mEVs, ruling out protein and cellular contamination as reported by the TEM analysis (Figure 2a). All considered parameters such as shape and the dimensional range as well as the positive reaction to known markers (CD81 and Tsg101) corroborated a correct mEV isolation (Figure 2). Moreover, a check on sequenced reads was then carried out, comparing mEV mRNA RPKM with sequencing data of mammary glands deposited on SRA database. As expected, a reduction in the number of expressed genes was observed in mEV RNA (Figure S1).

Then, analyzing the mEV RNA cargo, the total number of genes with an RPKM greater than one was quite similar for the three species, with approximately 10,000 transcripts. This number is in line with what previously observed for cow [54], porcine [55], and rat [56] mEVs; however, differences in RNA integrity were observed.

Although precise functions of EV mRNAs in target cells have not been clarified, it is known that complete mRNAs or fragmented mRNAs that keep the 5′ starting codon can be transferred in the receiving cell and directly translated into protein, modifying the global protein balance and consequently influencing metabolic pathways and biological processes [57,58]. On the other hand, previous experiences show widespread mRNA fragmentation mainly affecting the 5′-end of the typical transcript from mEVs [54] and from EV mRNA in general [59,60], hypothesizing a regulatory role (competitor, regulator of stability, localization, and translational activity of mRNAs) in target cells [60,61].

In detail, we detected high-quality mRNA in our mEVs, especially for cow and donkey with a peak at the 5′-end, maintaining good coverage for the rest of the sequence and a slight reduction for the very 3′-end (Figure S2). Goat mEVs were characterized by a general higher degradation while maintaining a good representativity at the 5′-end.

Then we evaluated if genes enclosed in mEVs were differentially expressed among the three species, although a direct comparison of expression levels is not trivial. For this reason, we implemented a procedure based on orthologous genes and proper normalization. The resulting pairwise comparisons show similarity between donkey and goat (Figure 3), while cow have a higher up-regulated number of genes (DEGs, reported in Table S2). The greatest part of the selected orthologous genes (5907) was stable in terms of expression, but species signatures emerged. Cow, for example, showed 875 and 1055 up-regulated genes compared to donkey and goat, respectively. The donkey had 561 over-expressed genes with respect to cow and 468 with respect to goat. Finally, 615 and 371 genes in goat were found over-expressed with respect to cow and donkey. These data may suggest that the three cargos deliver different messages to the target cells. The GO results, where different terms for each species emerged (Table S3), seem to corroborate this hypothesis. In cow, one of the major functionalities for mEV mRNAs is related to epigenetic regulation and DNA preservation, with enrichment of terms related to histones acetylation, regulation of mRNA metabolic process, ncRNA export from the nucleus, tRNA modification, and regulation of telomere maintenance. Another enriched GO term in cow, "Positive regulation of stress-activated MAPK cascade", points towards regulatory properties. Ideed, this is a crucial signaling pathway for cell homeostasis, which activates the transcription of various regulatory proteins [62]. For donkey, an activity focused on the vitamin metabolism and the fatty acid catabolism, with terms linked to peroxisome and oxidation, appeared to be enriched. In line, human and camel milk-derived EVs have been proven to improve oxidative stress conditions in pathological environments [63]. "CCR chemokine receptor binding" was the main functional group of GO terms (i.e., with the highest number of associated GO terms and genes) in goat, indicating immune system cells as a possible source of these mEVs. Milk EVs, indeed, are highly heterogeneous and originate from a variety of cell populations residing in the source organism's mammary gland, such as immune cells [64].

Concering our data for small RNAs, only donkey showed an important fraction of Y-RNAs. Although this finding could be biased by a reduced assembly quality of the ass genome and a suboptimal annotation, many read counts were attributed to some Y-RNAs. These small ncRNAs have recently gained interest due to the discovery of their abundance in some EVs types, especially released in biofluids, showing some kind of involvement in the immune system [65].

Concerning miRNAs, the majority of molecules highly expressed and shared between the three species (let-7i, miR-148b, miR-151a, miR-186, miR-191, miR-200b, miR-21, miR-27b, miR-30b, and miR-30d) are known to be enriched in milk [20,66–68] and involved in immunomodulation [22,63,67–73]. Although several studies have investigated the miRNA content in milk EVs from individual species [67], very few are about EV-associated miRNAs among different species [68]. Interestingly, in all studies, several abundant miRNAs were shared between species and are coherent with our results. These miRNAs were implicated in immune-related functions, regulation of cell growth, and signal transduction [68].

The most abundant miRNA in breast, cow and goat milk is miR-148, which is highly conserved in mammals [66,68]. Moreover, miR-151 and miR-186 were proposed as potential breast milk biomarkers [73]. MiR-148 is consistently reported in virtually all milk small RNA sequencing, supporting its evolutionary importance in lactation and for the newborn healthy development [74,75].

Regarding the most expressed common miRNAs in mEVs from all species and focusing on the potential immunomodulatory and anti-inflammatory activity, Liu and coauthors found that miR-151 interacts with *STAT3* mRNA, inducing down-regulation in mouse macrophages in response to LPS stimulation, and amplifying the initial innate immune response [71]. In addition, it is also known that miR-151 suppresses the expression of Th1 cytokines such as IL-2, IL-12, and IFN-γ [76].

MiR-148 regulates DNA methyl-transferase 1 (DNMT1), suggesting that milk consumption might affect the epigenome signatures [40,67] through mEVs uptake. Interestingly, milk miR-148a-3p and miR-29a-3p can also downregulate DNMT3B [77], a protein responsible for Forkhead box P3 (FOXP3) epigenetic inhibition [78]. Therefore, by preventing its epigenetic inhibition by DNMT3B, milk miR-148a and miR-29a could enhance *FOXP3* expression, driving the differentiation of T lymphocytes towards the anti-inflammatory regulatory T cell phenotype. This effect could be an explanation for mEVs' beneficial role in rheumatoid arthritis [79]. Recent research linked milk-derived miRNA-148a to pancreatic beta-cell differentiation, promoting a potential protective role against type 2 diabetes mellitus development [80]. The biological effects of these miRNAs corroborate the mRNA enrichment findings already described with processes involved in the epigenetic modification and DNA maintenance.

Regarding other common mEV miRNAs, let-7 together with miR-148, regulate transcription factor NF-κB in vivo, with the immune response suppression as an outcome [67,68]. Moreover, miR-30b and miR-200, along with miR-148, have been designated as major immune-related miRNAs [81] and are highly represented in milk [82] and also present in our core molecules in mEVs (Table 1).

MiR-200a-3p [83] also promotes the proliferation of intestinal epithelial cells through the epithelial-mesenchymal transition inhibition via TGF-β [84] that, as recently reported, exerts a pleiotropic effect towards the immune network balance via SMAD activities in the nucleus [85]. SMAD transcription factors act through the transforming growth factor-beta/SMAD (TGF-β/SMAD) signaling pathway that sets the balance between immune activation for pathogen clearing and the immune suppression to limit damage to self [85]. Since SMAD members are increased in Crohn's disease patients [85] and are validated targets in our analyses (Table S5), the mEV message (miRNA content and their targets) could acquire particular importance in the IBD pathogenesis.

MiR-21, another most abundant small RNA in our mEVs, beyond its presence in milk, is one of the most expressed members of the small non-coding microRNA family in many mammalian cell types; its expression is enhanced in many diseases and in inflamed tissues. MiR-21 is part of a complex regulatory feedback cascade where it can negatively regulate the pro-inflammatory response induced by different stimuli; in particular, miR-21 has emerged as a key mediator of the anti-inflammatory response in macrophages and a potential marker of immune cell activation [69].

In addition to the shared miRNAs, we found many peculiar miRNAs in each species (Table 1, Figure S4). Concerning cow, many miRNAs were found to have a pivotal role in inflammation, such as miR-340 which targets JAK, cyclin D1, and MMP2, that in turn down-regulate cytokine production [86]. This miRNA is often downregulated during the development of autoimmune diseases. Moreover, miR-340 directly targets the pro-inflammatory cytokines IL-4 [87] and IL-17A [88] in psoriasis mice model, where Th17 cells and the consequent IL-17A production are implicated in the pathogenesis of this autoimmune disease as in others such as IBD [89]. Similar effects seem to be addressed to miR-181a with anti-inflammatory effects mediated by the down-regulation of IL-1α [90] and the inhibition of NF-κB activation [91]. Furthermore, miR-99a can reduce macrophage M1 phenotype and increase M2 activation by targeting TNFα [92]; miR-182-5p can modulate the TLR4/NF-κB signaling pathway, inhibiting the inflammatory response induced by LPS treatment [93], while miR-429 modulates the IL-8 production through the NF-κB inhibition, reducing inflammation in epithelial cells [94]. Therefore, it was proposed as a candidate for anti-colitis therapy thanks to its capability of modulating mucin secretion in human colorectal cells and mouse colitis tissues [95].

An ability to modulate many pro-inflammatory cytokines such as IL-6, IL-1β, and TNFα in different tissues was shown by miR-26a, miR-27a, and miR-125b that we found most abundant in goat mEVs [86]. Concerning abundant miRNAs in donkey, miR-320a through its targets modulates the NF-κB signaling in IBD, while miR-23a and miR-145 inhibit TLR4 and IL-17 signaling, decreasing IL-6, MCP-1, and metalloproteinases production [86].

Target genes and protein network analyses (Figure 4, Table S6) on highly expressed miRNAs in each species underline the EV message finalized to protein synthesis and cellular cycle regulation (Tables S7–S9). Moreover, we retrieved donkey and goat enriched GO terms involved in epigenetic modification and DNA maintenance (donkey cluster 1 "chromatin maintenance", "chromatin organization"—donkey cluster 5 "histone methyltransferase binding", "regulation of gene expression, epigenetic "—goat cluster 4 "positive regulation of histone H3-K4 methylation "—goat cluster 2 "ATP-dependent chromatin remodeling") (Tables S8 and S9). GO terms that refer to transmembrane exchanges and ion channels were found for all species (cow cluster 1 "calcium-dependent protein binding", "positive regulation of intracellular transport"—goat cluster 5 "calcium-dependent protein binding", "positive regulation of intracellular protein transport"—donkey cluster 6 "calcium-dependent protein binding", "positive regulation of intracellular transport") (Tables S7–S9). These categories are congruent with those of mEV mRNA GO analyses.

In addition, peculiar mEV miRNA targets appeared to be linked to immunity (cow cluster 1 "CD40 receptor complex"—goat and donkey cluster 1 "I-kappaB kinase/NF-kappaB signaling"—donkey cluster 1 "positive regulation of NF-kappaB transcription factor activity", "positive regulation of leukocyte apoptotic process" "interleukin-12-mediated signaling pathway"—donkey cluster 2 "T cell costimulation", "regulation of B cell proliferation"—donkey cluster 5 "antimicrobial humoral immune response mediated by antimicrobial peptide"—goat cluster 1 "response to cytokine", "positive regulation of tumor necrosis factor-mediated signaling pathway"—goat cluster 2 "interleukin-2 production") and signal transduction (cow cluster 3 "adenylate cyclase-modulating G protein-coupled receptor signaling pathway", "phospholipase C-activating G protein-coupled receptor signaling pathway"—donkey cluster 2 "AMP-activated protein kinase activity"—goat cluster 3 "cAMP response element binding"—goat cluster 9 "signaling receptor complex adaptor activity") (Tables S7–S9).

The abundance of the molecules both present and targeted reinforce the hypothesis that mEVs may attenuate intestinal inflammation in IBD conditions [93], although in vivo validation studies are necessary.

Indeed, while not directly assessed here as this is a characterization study, mEV RNA cargo has promising potential biological effects as recent literature comfirmed that transported miRNAs in EVs are partially stable during gastrointestinal digestion, and bioavailable since they are capable of reaching target tissues [96]. While this is intriguing, it will be important to improve research in the minimum amount of transcripts for impacting gene expression, the possible degradation within the gastrointestinal tract, the amount that might reach the circulation and the cell to exert a biological effect, which are all aspects that have been poorly investigated in in vitro/in vivo experiments [68,97].

4. Materials and Methods

4.1. Milk Collection

Three samples for each species (cow, goat, and donkey) were collected from mass milk, to cope with the individual variability. Milk was sampled from farms surveilled by the Veterinary Medicine Department, University of Perugia: the Didactic Zootechnical Farm of the University for bovine and two distinct owned farms for donkey and goat. The goat milk was sampled from animals derived from a crossbreed of Camosciata delle Alpi and Saanen; donkey milk was collected from crossbreed animals. In both the cases, samples were taken during the mid-lactation period and animals were fed with grass, cereals, and hay in extensive pastoralism fashion. Cow milk was collected from Holstein Friesian cattle breed farm, where the greatest part of the animals was in mid-lactation. Rearing conditions are referable to standard intensive farms, with unifeed and mechanical milking. Milk was immediately processed or stored at 4 °C for less than 24 h before, avoiding cryo-preservation to minimize artifacts.

4.2. EVs Isolation

Milk EVs were isolated by serial differential centrifugations (DC) and a step in ethylenediaminetetraacetic acid tetrasodium salt dihydrate (EDTA), following the protocol of Vaswani and collaborators [98], with little modifications hereafter reported. In brief, preliminary centrifugation steps were used to eliminate fat globules on the surface and cellular debris and protein complexes in the pellet, recuperating the intermediate phase. Three hundred (300) mL of each milk were subjected to two consecutive 3000× g centrifugations for 10 min at room temperature (Eppendorf® Centrifuge 5810R with a F34-6-38 rotor); then, an equal volume of 0.25 M EDTA (pH 7.4) was added to the supernatant, incubated for 15 min on ice, and centrifuged at 10,000× g for 1 h at 4 °C. Then, centrifugation at 35,000× g for 1 h at 4 °C was carried out using polyallomer tubes in a Beckman Coulter Optima L-100 XP with an SW41 Ti rotor. A final ultracentrifugation at 200,000× g for 90 mins at 4 °C was carried out collecting mEVs in the 24 separate pellets used for morphological analysis and RNA characterization.

4.3. Milk EVs Characterization

4.3.1. Western Blotting

RIPA Buffer (20–40 µL) was added to a pellet for each species to extract proteins, incubating the solution for 15 min in ice. The cell lysates were centrifuged at 13,000 rpm for 5 min and the supernatant containing the proteins was recovered, according to Botta et al.

2019 [99]. Proteins were quantified at Qubit® 3.0 fluorometer (Thermo Fisher Scientific, Waltham, MA, USA) and 25–30 μg protein samples were analyzed by immunoblotting on 10% SDS-PAGE as previously described [100]. CD81 (bs-6934R, Bioss Antibodies, Woburn, MA, USA) diluted 1:500 and TSG-101 (sc-7964, Santa Cruz Biotechnology, Santa Cruz, CA, USA) diluted 1:400 were used as primary antibodies. A HRP-conjugated IgG was used as secondary antibody (Vector Labs, San Francisco, CA, USA) diluted 1:10000. The signal was detected using the enhanced chemiluminescence method following the manufacturer's instructions (Amersham) using Chemi Doc XRS system (Bio-Rad Laboratories Ltd., Hemel Hempstead, UK) and images processed with ImageLab (BioRad Laboratories Ltd., Hemel Hempstead, UK).

4.3.2. Transmission Electron Microscopy (TEM) and Nanoparticle Tracking Assay (NTA)

A drop of mEV suspension (one pellet) was placed on Parafilm. mEVs were allowed to adere to the surphase of a Formvar-coated copper grid (Electron Microscopy Sciences) placed on the top of each drop for about 20 min. Grids were then washed in PBS and distilled water and then contrasted with 2% uranyl acetate for 5 min. The observation was performed using a Philips EM208 transmission electron microscope equipped with a digital camera (University Centre of Electron Microscopy—CUME).

A Malvern Panalytical NanoSight NS300 nanoparticle tracking analysis (NTA) system (Malvern, Worcestershire, UK) was used to assess the concentration and size distribution of isolated mEVs. One mEV pellet of the three species was resuspended and diluted in filtered (0.22 μm pore size) phosphate buffered saline (PBS) (Sigma, St. Louis, MI, USA) to be suitable for the NTA system's working concentration range and five measurements were performed for each one. Concentration and diameter results are reported as mean ± 1 standard error of the mean.

4.4. RNA Extraction and Library Preparation

To obtain a sufficient RNA amount suitable for sequencing procedure, 21 mEV pellets, obtained as described in 4.2, were immediately treated with 100 μL (each) of TRIzol™ (Thermo Fisher Scientific, Waltham, MA, USA) and total RNA was extracted through the miRNeasy Mini Kit (QIAGEN, Germantown, MD, USA) following the manufacturer's instructions. After RNA extraction, samples were subjected to a DNase digestion using the TURBO DNA-freeTM Kit (Thermo Fisher Scientific, Waltham, MA, USA), then quantified through the NanoDrop 2000 spectrophotometer (Thermo Fisher Scientific, Waltham, MA, USA) and quality tested by the Agilent 2100 Bioanalyzer RNA assay (Agilent technologies, Santa Clara, CA, USA). The RNA extracted from mEVs was subjected to Next Generation Sequencing for both mRNA and small RNA; for each species, three libraries were prepared for both. Universal Plus mRNA-Seq kit (library type: fr-secondstrand) (Tecan Genomics, Redwood City, CA, USA) and QIAseq miRNA library kit (QIAGEN, Germantown, MD, USA) were used for the two libraries' preparation following the manufacturer's instructions. Final libraries were checked with both Qubit 2.0 Fluorometer (Invitrogen, Carlsbad, CA, USA) and Agilent Bioanalyzer DNA assay and sequenced on single-end 75 bp mode on NextSeq 500 (Illumina, San Diego, CA, USA).

4.5. Bioinformatic Analysis

4.5.1. From Sequencing to Datasets

Raw reads (deposited in the Sequence Read Archive (SRA), see reference submission number at the Data Availability Statement section of the article) were checked for quality through the FastQC tool (https://www.bioinformatics.babraham.ac.uk/projects/fastqc/ accessed on 1 October 2021) and adaptors trimmed using TrimGalore (https://github.com/FelixKrueger/TrimGalore accessed on 1 October 2021). STAR aligner [101] was used for mRNA reads mapping to the relative reference genomes (ARS-UCD1.2 for cow, ARS1 for goat) using Ensembl annotation (release 102) [102], while ASM303372v1 downloaded from the National Center for Biotechnology Information (NCBI) website was used for donkey.

FeatureCounts [103] was used to generate the count matrix on which the differential gene expression analysis was carried out.

For small RNA reads, BowTie2 [104] was used, setting parameters for short sequences and performing a two-step alignment procedure: a first passage on miRBase (v.22) [105] database and a second to reference the genome to retrieve accurate information on miRNAs and other small RNAs with the unmapped sequences from the first passage. Concerning miRBase database, the hairpin version of the species annotated sequences were used for cow and goat, while, for donkey, the horse annotation was adopted. A single count matrix was produced, merging the information generated from the two alignments through the count sum of the same miRNAs if covered by reads for both the miRBase and the genomic mapping. Uniquely mapped reads were selected for normalization of mRNA and smallRNA count matrices through the reads per kilobase per million mapped reads (RPKM) method and only features with an RPKM greater than 1 were considered for downstream analysis. This threshold was chosen to be as inclusive as possible in this step, preferring further filtering in downstream analyses.

Cow mRNA mEV profile was compared to mRNA data of mammary glands downloaded from the SRA database [106] (BioSamples: SAMN12831050, SAMN12831049, SAMN14600526 and SAMN14600525). The number of genes expressed (i.e., RPKM > 0) in the two samples was compared (Wilcoxon rank sum one-tail test). A frequency histogram comparing the average expression profile, expressed in log_2(RPKM) at intervals of 0.5 was generated. A statistic test (Wilcoxon rank sum test) was conducted. RNA quality in terms of transcript integrity was evaluated through RSeQC package [107].

4.5.2. Differential Gene Expression Analysis between the Three Species

For cargo mRNAs of mEVs, a differential expression analysis was carried out, basing on a unique list of orthologous genes, selecting those with the one-to-one type of orthology available for all the three species. The RPKM matrix was log-2 transformed and it underwent a normalization process that allows the comparison of gene expression levels across different species. The normalization process consists of the calculation, for each sample, of median-scaling factors across the 1000 most conserved genes. Then, the scaling factors were used to normalize all the genes in all samples [108,109]. The scaling factors were computed by using the normalization function available at the GitHub repository [108]. After the normalization process, data were squared to handle normalization data with RPKM values < 0. Finally, statistical analyses to identify all differential express genes were performed in R 3.5.3 using the DESeq2 package [110]. Using the contrast parameter in the results function, we evaluated all the pairwise comparisons (cow vs. donkey, cow vs. goat, and donkey vs. goat). For each comparison, only genes with a False Discovery Rate (FDR) less than 0.05 (FDR < 0.05) and a log_2 Fold Change (FC) less than −2 or greater than 2 ($|log_2 FC| > 2$) were considered as differential expressed genes. The R package EnhancedVolcano was used to produce the volcano plots for all the comparisons [111]. For each species, a list composed of the up-regulated genes found to be over-expressed in both comparisons with the other two species was generated. Since the list of genes used for differential analysis was created with a relaxed threshold, the risk of taking into account scarcely expressed genes was real. To overcome this, the initial RPKM relative to "one-to-one" transcripts was verified before proceeding with the functional analysis: all the RPKM values were equal to or greater than 6 in at least one species, ensuring relevant and reliable expression level for the functional analysis. These transcripts were used for gene ontology (GO) enrichment analysis through the ClueGO [112] application of Cytoscape (v. 7.1) suite [113].

4.5.3. miRNA Targets Retrieving and Functional Analysis

For the most expressed miRNAs (covered by 95% of total RPKM) of the three species, human orthologues were retrieved through miRBase (v.22) and used to identify all the validated target genes. In order to select unique molecules for each miRNA, we traced back the portion sequenced in our experiment (5p or 3p) and the human homolog was used as input in the MiRWalk 3.0 (http://mirwalk.umm.uni-heidelberg.de/ accessed on 1 October 2021) database. A unique list of all targets for each miRNA was produced, specifying also the miRNA site of action on the targeted mRNA: 3′-UTR, 5′-UTR, or CDS (coding region). The targets were filtered for the number of miRNA hits, taking into account those genes targeted by five miRNAs for cow and goat and four miRNAs for donkey. The different miRNA hits threshold between donkey and the other two species is explained by the lower number of most expressed miRNAs in this species compared to the others. A Protein-Protein Interaction Network (PPI) was build starting from targets through the Cytoscape 3.7.1 suite [113], using the IMEx database [114], which contains non-redundant information deriving from the major public protein databases. Then, the clusterMaker 2.0 app [115] with the "gLay" option was used to highlight different clusters within the network based on the number and type of connections between the nodes. Gene Ontology (GO) enrichment analysis was carried out on clusters with a number of interactions greater than 25 through the ClueGO application [112], and results were filtered for a FDR < 0.05 (Benjamini Hockberg correction).

5. Conclusions

This study allowed the in-depth characterization of mRNA and small RNA content of cow, goat, and donkey mEVs. Key molecules relevant for immune and inflammatory signals regulation and biological processes involved in transcription regulation and cell homeostasis were found, denoting similar functions for the mEVs of the three species. However, some peculiar features emerged. In particular, goat and donkey showed as the main functions those related to transmembrane ion channels and gene transcription, especially epigenetic regulation. A particular role in lipid metabolism and response to oxidative stress appeared for donkey mEVs.

The anti-inflammatory and immunomodulatory potential of cow, donkey, and goat mEVs highlighted in this study due to the characteristic RNA cargo is prodromal for further investigations through in vitro and in vivo models of inflammatory-based diseases such as IBD.

Supplementary Materials: The following are available online at https://www.mdpi.com/article/10.3390/ijms222312759/s1.

Author Contributions: Conceptualization, K.C., S.M. and G.C.; methodology, K.C., S.C., D.P., M.M., S.M., L.P., S.F.; software, S.C., D.P., M.M., V.R.; validation, S.C., D.P., M.M., V.R. and G.C.; formal analysis, S.M., S.C., D.P., M.M. and G.C.; investigation, K.C., S.M., L.P., S.F.; resources, K.C., S.C., G.C.; data curation, S.M., S.C., D.P., M.M.; writing—original draft preparation, S.M., K.C.; writing—review and editing, S.C., D.P., M.M., L.P. and G.C.; visualization, S.M., S.C.; supervision, G.C., K.C.; project administration, K.C.; funding acquisition, S.C. and G.C. All authors have read and agreed to the published version of the manuscript.

Funding: This research was funded by Dipartimento di Medicina Veterinaria di Perugia—ricerca di base dipartimentale, grant number RB20172019CAPOM, by the 'Departments of Excellence-2018' Program (Dipartimenti di Eccellenza) of the Italian Ministry of Education, University and Research, DIBAF-Department of University of Tuscia, Project 'Landscape 4.0–food, wellbeing and environment', and by the CEF Highlander project, co-financed by the Connecting European Facility Programme of the European Union, grant agreement no. INEA/CEF/ICT/A2018/1815462.

Institutional Review Board Statement: Not applicable.

Informed Consent Statement: Not applicable.

Data Availability Statement: The raw sequencing data generated during this study are openly available in the Sequence Read Archive (SRA) at the National Center for Biotechnology Information (NCBI) servers under the Bio-Project PRJNA771627, with the following accession numbers: SAMN22322873, SAMN22322874, SAMN22322875, SAMN22322876, SAMN22322877, SAMN22322878, SAMN22322879, SAMN22322880, SAMN22322881, SAMN22322882, SAMN22322883, SAMN22322884, SAMN22322885, SAMN22322886, SAMN22322887, SAMN22322888, SAMN22322889, SAMN22322890.

Acknowledgments: This work was partly supported by the 'Departments of Excellence-2018' Program (Dipartimenti di Eccellenza) of the Italian Ministry of Education, University and Research, DIBAF-Department of University of Tuscia, Project 'Landscape 4.0–food, wellbeing and environment' and by the CEF Highlander project, co-financed by the Connecting European Facility Programme of the European Union, grant agreement no. INEA/CEF/ICT/A2018/1815462.

Conflicts of Interest: The authors declare no conflict of interest.

References

1. Verduci, E.; Giannì, M.L.; Vizzari, G.; Vizzuso, S.; Cerasani, J.; Mosca, F.; Zuccotti, G.V. The Triad Mother-Breast Milk-Infant as Predictor of Future Health: A Narrative Review. *Nutrients* **2021**, *13*, 486. [CrossRef]
2. Raposo, G.; Stoorvogel, W. Extracellular Vesicles: Exosomes, Microvesicles, and Friends. *J. Cell Biol.* **2013**, *200*, 373–383. [CrossRef]
3. Robbins, P.D.; Morelli, A.E. Regulation of Immune Responses by Extracellular Vesicles. *Nat. Rev. Immunol.* **2014**, *14*, 195–208. [CrossRef]
4. Mentkowski, K.I.; Snitzer, J.D.; Rusnak, S.; Lang, J.K. Therapeutic Potential of Engineered Extracellular Vesicles. *AAPS J.* **2018**, *20*, 50. [CrossRef]
5. Ma, Z.; Wang, Y.; Li, H. Applications of Extracellular Vesicles in Tissue Regeneration. *Biomicrofluidics* **2020**, *14*, 011501. [CrossRef]
6. Vella, L.J.; Hill, A.F.; Cheng, L. Focus on Extracellular Vesicles: Exosomes and Their Role in Protein Trafficking and Biomarker Potential in Alzheimer's and Parkinson's Disease. *Int. J. Mol. Sci.* **2016**, *17*, 173. [CrossRef]
7. Lucchetti, D.; Ricciardi Tenore, C.; Colella, F.; Sgambato, A. Extracellular Vesicles and Cancer: A Focus on Metabolism, Cytokines, and Immunity. *Cancers* **2020**, *12*, 171. [CrossRef] [PubMed]
8. Huang, X.; Yuan, T.; Tschannen, M.; Sun, Z.; Jacob, H.; Du, M.; Liang, M.; Dittmar, R.L.; Liu, Y.; Liang, M.; et al. Characterization of Human Plasma-Derived Exosomal RNAs by Deep Sequencing. *BMC Genom.* **2013**, *14*, 319. [CrossRef] [PubMed]
9. Greening, D.W.; Gopal, S.K.; Xu, R.; Simpson, R.J.; Chen, W. Exosomes and Their Roles in Immune Regulation and Cancer. *Semin. Cell Dev. Biol.* **2015**, *40*, 72–81. [CrossRef]
10. Dickhout, A.; Koenen, R.R. Extracellular Vesicles as Biomarkers in Cardiovascular Disease; Chances and Risks. *Front. Cardiovasc. Med.* **2018**, *5*, 113. [CrossRef]
11. Burrello, J.; Monticone, S.; Gai, C.; Gomez, Y.; Kholia, S.; Camussi, G. Stem Cell-Derived Extracellular Vesicles and Immune-Modulation. *Front. Cell Dev. Biol.* **2016**, *4*, 83. [CrossRef]
12. Aryani, A.; Denecke, B. Exosomes as a Nanodelivery System: A Key to the Future of Neuromedicine? *Mol. Neurobiol.* **2016**, *53*, 818–834. [CrossRef]
13. Capomaccio, S.; Cappelli, K.; Bazzucchi, C.; Coletti, M.; Gialletti, R.; Moriconi, F.; Passamonti, F.; Pepe, M.; Petrini, S.; Mecocci, S.; et al. Equine Adipose-Derived Mesenchymal Stromal Cells Release Extracellular Vesicles Enclosing Different Subsets of Small RNAs. *Stem Cells Int.* **2019**, *2019*, 1–12. [CrossRef] [PubMed]
14. Van Niel, G.; D'Angelo, G.; Raposo, G. Shedding Light on the Cell Biology of Extracellular Vesicles. *Nat. Rev. Mol. Cell Biol.* **2018**, *19*, 213–228. [CrossRef]
15. Kalluri, R.; LeBleu, V.S. The Biology, Function, and Biomedical Applications of Exosomes. *Science* **2020**, *367*, eaau6977. [CrossRef]
16. Melnik, B.C.; Stremmel, W.; Weiskirchen, R.; John, S.M.; Schmitz, G. Exosome-Derived MicroRNAs of Human Milk and Their Effects on Infant Health and Development. *Biomolecules* **2021**, *11*, 851. [CrossRef]
17. Wang, Q.; Zhuang, X.; Mu, J.; Deng, Z.-B.; Jiang, H.; Zhang, L.; Xiang, X.; Wang, B.; Yan, J.; Miller, D.; et al. Delivery of Therapeutic Agents by Nanoparticles Made of Grapefruit-Derived Lipids. *Nat. Commun.* **2013**, *4*, 1867. [CrossRef] [PubMed]
18. Munagala, R.; Aqil, F.; Jeyabalan, J.; Gupta, R.C. Bovine Milk-Derived Exosomes for Drug Delivery. *Cancer Lett.* **2016**, *371*, 48–61. [CrossRef]
19. Lin, D.; Chen, T.; Xie, M.; Li, M.; Zeng, B.; Sun, R.; Zhu, Y.; Ye, D.; Wu, J.; Sun, J.; et al. Oral Administration of Bovine and Porcine Milk Exosome Alter MiRNAs Profiles in Piglet Serum. *Sci. Rep.* **2020**, *10*, 6983. [CrossRef] [PubMed]
20. Izumi, H.; Tsuda, M.; Sato, Y.; Kosaka, N.; Ochiya, T.; Iwamoto, H.; Namba, K.; Takeda, Y. Bovine Milk Exosomes Contain MicroRNA and MRNA and Are Taken up by Human Macrophages. *J. Dairy Sci.* **2015**, *98*, 2920–2933. [CrossRef] [PubMed]
21. El-Andaloussi, S.; Lee, Y.; Lakhal-Littleton, S.; Li, J.; Seow, Y.; Gardiner, C.; Alvarez-Erviti, L.; Sargent, I.L.; Wood, M.J.A. Exosome-Mediated Delivery of SiRNA in Vitro and in Vivo. *Nat. Protoc.* **2012**, *7*, 2112–2126. [CrossRef] [PubMed]
22. Benmoussa, A.; Laugier, J.; Beauparlant, C.J.; Lambert, M.; Droit, A.; Provost, P. Complexity of the MicroRNA Transcriptome of Cow Milk and Milk-Derived Extracellular Vesicles Isolated via Differential Ultracentrifugation. *J. Dairy Sci.* **2020**, *103*, 16–29. [CrossRef]
23. Admyre, C.; Johansson, S.M.; Qazi, K.R.; Filen, J.-J.; Lahesmaa, R.; Norman, M.; Neve, E.P.A.; Scheynius, A.; Gabrielsson, S. Exosomes with Immune Modulatory Features Are Present in Human Breast Milk. *J. Immunol.* **2007**, *179*, 1969–1978. [CrossRef]
24. Hata, T.; Murakami, K.; Nakatani, H.; Yamamoto, Y.; Matsuda, T.; Aoki, N. Isolation of Bovine Milk-Derived Microvesicles Carrying MRNAs and MicroRNAs. *Biochem. Biophys. Res. Commun.* **2010**, *396*, 528–533. [CrossRef]
25. Chen, T.; Xie, M.-Y.; Sun, J.-J.; Ye, R.-S.; Cheng, X.; Sun, R.-P.; Wei, L.-M.; Li, M.; Lin, D.-L.; Jiang, Q.-Y.; et al. Porcine Milk-Derived Exosomes Promote Proliferation of Intestinal Epithelial Cells. *Sci. Rep.* **2016**, *6*. [CrossRef]
26. Badawy, A.A.; El-Magd, M.A.; AlSadrah, S.A. Therapeutic Effect of Camel Milk and Its Exosomes on MCF7 Cells In Vitro and In Vivo. *Integr. Cancer Ther.* **2018**, *17*, 1235–1246. [CrossRef]

27. Baddela, V.S.; Nayan, V.; Rani, P.; Onteru, S.K.; Singh, D. Physicochemical Biomolecular Insights into Buffalo Milk-Derived Nanovesicles. *Appl. Biochem. Biotechnol.* **2016**, *178*, 544–557. [CrossRef]
28. Quan, S.; Nan, X.; Wang, K.; Jiang, L.; Yao, J.; Xiong, B. Characterization of Sheep Milk Extracellular Vesicle-MiRNA by Sequencing and Comparison with Cow Milk. *Animals* **2020**, *10*, 331. [CrossRef] [PubMed]
29. Golan-Gerstl, R.; Elbaum Shiff, Y.; Moshayoff, V.; Schecter, D.; Leshkowitz, D.; Reif, S. Characterization and Biological Function of Milk-Derived MiRNAs. *Mol. Nutr. Food Res.* **2017**, *61*, 1700009. [CrossRef]
30. Zhao, S.; Wang, Y.; Liang, Y.; Zhao, M.; Long, H.; Ding, S.; Yin, H.; Lu, Q. MicroRNA-126 Regulates DNA Methylation in CD4+ T Cells and Contributes to Systemic Lupus Erythematosus by Targeting DNA Methyltransferase 1. *Arthritis Rheum.* **2011**, *63*, 1376–1386. [CrossRef]
31. Quintanilha, B.; Reis, B.; Duarte, G.; Cozzolino, S.; Rogero, M. Nutrimiromics: Role of MicroRNAs and Nutrition in Modulating Inflammation and Chronic Diseases. *Nutrients* **2017**, *9*, 1168. [CrossRef] [PubMed]
32. Trinchese, G.; Cavaliere, G.; De Filippo, C.; Aceto, S.; Prisco, M.; Chun, J.T.; Penna, E.; Negri, R.; Muredda, L.; Demurtas, A.; et al. Human Milk and Donkey Milk, Compared to Cow Milk, Reduce Inflammatory Mediators and Modulate Glucose and Lipid Metabolism, Acting on Mitochondrial Function and Oleylethanolamide Levels in Rat Skeletal Muscle. *Front. Physiol.* **2018**, *9*, 32. [CrossRef]
33. Benmoussa, A.; Lee, C.H.C.; Laffont, B.; Savard, P.; Laugier, J.; Boilard, E.; Gilbert, C.; Fliss, I.; Provost, P. Commercial Dairy Cow Milk MicroRNAs Resist Digestion under Simulated Gastrointestinal Tract Conditions. *J. Nutr.* **2016**, *146*, 2206–2215. [CrossRef] [PubMed]
34. Kusuma, R.J.; Manca, S.; Friemel, T.; Sukreet, S.; Nguyen, C.; Zempleni, J. Human Vascular Endothelial Cells Transport Foreign Exosomes from Cow's Milk by Endocytosis. *Am. J. Physiol. Cell Physiol.* **2016**, *310*, C800–C807. [CrossRef]
35. Alsaweed, M.; Hartmann, P.; Geddes, D.; Kakulas, F. MicroRNAs in Breastmilk and the Lactating Breast: Potential Immunoprotectors and Developmental Regulators for the Infant and the Mother. *Int. J. Environ. Res. Public Health* **2015**, *12*, 13981–14020. [CrossRef]
36. Liao, Y.; Du, X.; Li, J.; Lönnerdal, B. Human Milk Exosomes and Their MicroRNAs Survive Digestion in Vitro and Are Taken up by Human Intestinal Cells. *Mol. Nutr. Food Res.* **2017**, *61*, 1700082. [CrossRef]
37. Hou, J.; An, X.; Song, Y.; Cao, B.; Yang, H.; Zhang, Z.; Shen, W.; Li, Y. Detection and Comparison of MicroRNAs in the Caprine Mammary Gland Tissues of Colostrum and Common Milk Stages. *BMC Genet.* **2017**, *18*, 1–8. [CrossRef] [PubMed]
38. Li, R.; Dudemaine, P.-L.; Zhao, X.; Lei, C.; Ibeagha-Awemu, E.M. Comparative Analysis of the MiRNome of Bovine Milk Fat, Whey and Cells. *PloS ONE* **2016**, *11*, e0154129. [CrossRef]
39. Sedykh, S.E.; Purvinish, L.V.; Monogarov, A.S.; Burkova, E.E.; Grigor'eva, A.E.; Bulgakov, D.V.; Dmitrenok, P.S.; Vlassov, V.V.; Ryabchikova, E.I.; Nevinsky, G.A. Purified Horse Milk Exosomes Contain an Unpredictable Small Number of Major Proteins. *Biochim. Open* **2017**, *4*, 61–72. [CrossRef] [PubMed]
40. Melnik, B.C.; Schmitz, G. MicroRNAs: Milk's Epigenetic Regulators. *Best Pract. Res. Clin. Endocrinol. Metab.* **2017**, *31*, 427–442. [CrossRef]
41. Wang, J.; Bian, Y.; Wang, Z.; Li, D.; Wang, C.; Li, Q.; Gao, X. MicroRNA-152 Regulates DNA Methyltransferase 1 and Is Involved in the Development and Lactation of Mammary Glands in Dairy Cows. *PLoS ONE* **2014**, *9*, e101358. [CrossRef] [PubMed]
42. Contreras, J.; Rao, D.S. MicroRNAs in Inflammation and Immune Responses. *Leukemia* **2012**, *26*, 404–413. [CrossRef] [PubMed]
43. Hulsmans, M.; De Keyzer, D.; Holvoet, P. MicroRNAs Regulating Oxidative Stress and Inflammation in Relation to Obesity and Atherosclerosis. *FASEB J.* **2011**, *25*, 2515–2527. [CrossRef] [PubMed]
44. Ocansey, D.K.W.; Zhang, L.; Wang, Y.; Yan, Y.; Qian, H.; Zhang, X.; Xu, W.; Mao, F. Exosome-mediated Effects and Applications in Inflammatory Bowel Disease. *Biol. Rev.* **2020**, *95*, 1287–1307. [CrossRef]
45. Ghafouri-Fard, S.; Eghtedarian, R.; Taheri, M. The Crucial Role of Non-Coding RNAs in the Pathophysiology of Inflammatory Bowel Disease. *Biomed. Pharmacother.* **2020**, *129*, 110507. [CrossRef]
46. Lin, L.; Zhou, G.; Chen, P.; Wang, Y.; Han, J.; Chen, M.; He, Y.; Zhang, S. Which Long Noncoding RNAs and Circular RNAs Contribute to Inflammatory Bowel Disease? *Cell Death Dis.* **2020**, *11*, 456. [CrossRef]
47. Harrison's Principles of Internal Medicine, 18e | AccessMedicine | McGraw-Hill Medical. Available online: http://accessmedicine.mhmedical.com/book.aspx?bookid=331 (accessed on 26 March 2018).
48. Ortega, F.J.; Cardona-Alvarado, M.I.; Mercader, J.M.; Moreno-Navarrete, J.M.; Moreno, M.; Sabater, M.; Fuentes-Batllevell, N.; Ramírez-Chávez, E.; Ricart, W.; Molina-Torres, J.; et al. Circulating Profiling Reveals the Effect of a Polyunsaturated Fatty Acid-Enriched Diet on Common MicroRNAs. *J. Nutr. Biochem.* **2015**, *26*, 1095–1101. [CrossRef]
49. Tian, L.; Song, Z.; Shao, W.; Du, W.W.; Zhao, L.R.; Zeng, K.; Yang, B.B.; Jin, T. Curcumin Represses Mouse 3T3-L1 Cell Adipogenic Differentiation via Inhibiting MiR-17-5p and Stimulating the Wnt Signalling Pathway Effector Tcf7l2. *Cell Death Dis.* **2017**, *8*, e2559. [CrossRef]
50. Tomé-Carneiro, J.; Larrosa, M.; Yáñez-Gascón, M.J.; Dávalos, A.; Gil-Zamorano, J.; Gonzálvez, M.; García-Almagro, F.J.; Ruiz Ros, J.A.; Tomás-Barberán, F.A.; Espín, J.C.; et al. One-Year Supplementation with a Grape Extract Containing Resveratrol Modulates Inflammatory-Related MicroRNAs and Cytokines Expression in Peripheral Blood Mononuclear Cells of Type 2 Diabetes and Hypertensive Patients with Coronary Artery Disease. *Pharmacol. Res.* **2013**, *72*, 69–82. [CrossRef]

51. Braud, M.; Magee, D.A.; Park, S.D.E.; Sonstegard, T.S.; Waters, S.M.; MacHugh, D.E.; Spillane, C. Genome-Wide MicroRNA Binding Site Variation between Extinct Wild Aurochs and Modern Cattle Identifies Candidate MicroRNA-Regulated Domestication Genes. *Front. Genet.* **2017**, *8*, 3. [CrossRef]
52. Mecocci, S.; Gevi, F.; Pietrucci, D.; Cavinato, L.; Luly, F.R.; Pascucci, L.; Petrini, S.; Ascenzioni, F.; Zolla, L.; Chillemi, G.; et al. Anti-Inflammatory Potential of Cow, Donkey and Goat Milk Extracellular Vesicles as Revealed by Metabolomic Profile. *Nutrients* **2020**, *12*, 2908. [CrossRef] [PubMed]
53. Teng, F.; Fussenegger, M. Shedding Light on Extracellular Vesicle Biogenesis and Bioengineering. *Adv. Sci.* **2021**, *8*, 2003505. [CrossRef]
54. Zempleni, J.; Aguilar-Lozano, A.; Sadri, M.; Sukreet, S.; Manca, S.; Wu, D.; Zhou, F.; Mutai, E. Biological Activities of Extracellular Vesicles and Their Cargos from Bovine and Human Milk in Humans and Implications for Infants. *J. Nutr.* **2017**, *147*, 3–10. [CrossRef] [PubMed]
55. Chen, T.; Xi, Q.-Y.; Sun, J.-J.; Ye, R.-S.; Cheng, X.; Sun, R.-P.; Wang, S.-B.; Shu, G.; Wang, L.-N.; Zhu, X.-T.; et al. Revelation of MRNAs and Proteins in Porcine Milk Exosomes by Transcriptomic and Proteomic Analysis. *BMC Vet. Res.* **2017**, *13*, 101. [CrossRef]
56. Izumi, H.; Kosaka, N.; Shimizu, T.; Sekine, K.; Ochiya, T.; Takase, M. Time-Dependent Expression Profiles of MicroRNAs and MRNAs in Rat Milk Whey. *PLoS ONE* **2014**, *9*, e88843. [CrossRef]
57. Herrera, M.B.; Fonsato, V.; Gatti, S.; Deregibus, M.C.; Sordi, A.; Cantarella, D.; Calogero, R.; Bussolati, B.; Tetta, C.; Camussi, G. Human Liver Stem Cell-Derived Microvesicles Accelerate Hepatic Regeneration in Hepatectomized Rats. *J. Cell Mol. Med.* **2010**, *14*, 1605–1618. [CrossRef]
58. Valadi, H.; Ekström, K.; Bossios, A.; Sjöstrand, M.; Lee, J.J.; Lötvall, J.O. Exosome-Mediated Transfer of MRNAs and MicroRNAs Is a Novel Mechanism of Genetic Exchange between Cells. *Nat. Cell Biol.* **2007**, *9*, 654–659. [CrossRef] [PubMed]
59. Wei, Z.; Batagov, A.O.; Schinelli, S.; Wang, J.; Wang, Y.; El Fatimy, R.; Rabinovsky, R.; Balaj, L.; Chen, C.C.; Hochberg, F.; et al. Coding and Noncoding Landscape of Extracellular RNA Released by Human Glioma Stem Cells. *Nat. Commun.* **2017**, *8*, 1145. [CrossRef]
60. Batagov, A.O.; Kurochkin, I.V. Exosomes Secreted by Human Cells Transport Largely MRNA Fragments That Are Enriched in the 3′-Untranslated Regions. *Biol. Direct.* **2013**, *8*, 12. [CrossRef] [PubMed]
61. Di Liegro, C.M.; Schiera, G.; Di Liegro, I. Extracellular Vesicle-Associated RNA as a Carrier of Epigenetic Information. *Genes* **2017**, *8*, 240. [CrossRef] [PubMed]
62. Plotnikov, A.; Zehorai, E.; Procaccia, S.; Seger, R. The MAPK Cascades: Signaling Components, Nuclear Roles and Mechanisms of Nuclear Translocation. *Biochim. Biophys. Acta (BBA) Mol. Cell Res.* **2011**, *1813*, 1619–1633. [CrossRef]
63. Qi, H.; Wang, Y.; Fa, S.; Yuan, C.; Yang, L. Extracellular Vesicles as Natural Delivery Carriers Regulate Oxidative Stress Under Pathological Conditions. *Front. Bioeng. Biotechnol.* **2021**, *9*, 752019. [CrossRef]
64. Sanwlani, R.; Fonseka, P.; Chitti, S.V.; Mathivanan, S. Milk-Derived Extracellular Vesicles in Inter-Organism, Cross-Species Communication and Drug Delivery. *Proteomes* **2020**, *8*, 11. [CrossRef] [PubMed]
65. Driedonks, T.A.P.; Nolte-'t Hoen, E.N.M. Circulating Y-RNAs in Extracellular Vesicles and Ribonucleoprotein Complexes; Implications for the Immune System. *Front. Immunol.* **2019**, *9*, 3164. [CrossRef]
66. Benmoussa, A.; Provost, P. Milk MicroRNAs in Health and Disease. *Compr. Rev. Food Sci. Food Saf.* **2019**, *18*, 703–722. [CrossRef] [PubMed]
67. Tingö, L.; Ahlberg, E.; Johansson, L.; Pedersen, S.A.; Chawla, K.; Sætrom, P.; Cione, E.; Simpson, M.R. Non-Coding RNAs in Human Breast Milk: A Systematic Review. *Front. Immunol.* **2021**, *12*, 725323. [CrossRef]
68. Van Herwijnen, M.J.C.; Driedonks, T.A.P.; Snoek, B.L.; Kroon, A.M.T.; Kleinjan, M.; Jorritsma, R.; Pieterse, C.M.J.; Nolte-'t Hoen, E.N.M.; Wauben, M.H.M. Abundantly Present MiRNAs in Milk-Derived Extracellular Vesicles Are Conserved Between Mammals. *Front. Nutr.* **2018**, *5*, 81. [CrossRef]
69. Sheedy, F.J. Turning 21: Induction of MiR-21 as a Key Switch in the Inflammatory Response. *Front. Immunol.* **2015**, *6*, 19. [CrossRef] [PubMed]
70. Manca, S.; Upadhyaya, B.; Mutai, E.; Desaulniers, A.T.; Cederberg, R.A.; White, B.R.; Zempleni, J. Milk Exosomes Are Bioavailable and Distinct MicroRNA Cargos Have Unique Tissue Distribution Patterns. *Sci. Rep.* **2018**, *8*, 11321. [CrossRef] [PubMed]
71. Liu, X.; Su, X.; Xu, S.; Wang, H.; Han, D.; Li, J.; Huang, M.; Cao, X. MicroRNA in Vivo Precipitation Identifies MiR-151-3p as a Computational Unpredictable MiRNA to Target Stat3 and Inhibits Innate IL-6 Production. *Cell Mol. Immunol.* **2018**, *15*, 99–110. [CrossRef]
72. Kakimoto, Y.; Matsushima, Y.; Tanaka, M.; Hayashi, H.; Wang, T.; Yokoyama, K.; Ochiai, E.; Osawa, M. MicroRNA Profiling of Gastric Content from Breast-Fed and Formula-Fed Infants to Estimate Last Feeding: A Pilot Study. *Int. J. Legal Med.* **2020**, *134*, 903–909. [CrossRef]
73. Carrillo-Lozano, E.; Sebastián-Valles, F.; Knott-Torcal, C. Circulating MicroRNAs in Breast Milk and Their Potential Impact on the Infant. *Nutrients* **2020**, *12*, 3066. [CrossRef]
74. Zempleni, J. Milk Exosomes: Beyond Dietary MicroRNAs. *Genes Nutr.* **2017**, *12*, 1–4. [CrossRef]
75. Friedrich, M.; Pracht, K.; Mashreghi, M.-F.; Jäck, H.-M.; Radbruch, A.; Seliger, B. The Role of the MiR-148/-152 Family in Physiology and Disease. *Eur. J. Immunol.* **2017**, *47*, 2026–2038. [CrossRef] [PubMed]

76. Szczepankiewicz, D.; Langwiński, W.; Kołodziejski, P.; Pruszyńska-Oszmałek, E.; Sassek, M.; Nowakowska, J.; Chmurzyńska, A.; Nowak, K.W.; Szczepankiewicz, A. Allergic Inflammation Alters MicroRNA Expression Profile in Adipose Tissue in the Rat. *Genes* **2020**, *11*, 1034. [CrossRef] [PubMed]
77. Chou, C.-H.; Shrestha, S.; Yang, C.-D.; Chang, N.-W.; Lin, Y.-L.; Liao, K.-W.; Huang, W.-C.; Sun, T.-H.; Tu, S.-J.; Lee, W.-H.; et al. MiRTarBase Update 2018: A Resource for Experimentally Validated MicroRNA-Target Interactions. *Nucleic Acids Res.* **2018**, *46*, D296–D302. [CrossRef] [PubMed]
78. Pion, M.; Jaramillo-Ruiz, D.; Martínez, A.; Muñoz-Fernández, M.A.; Correa-Rocha, R. HIV Infection of Human Regulatory T Cells Downregulates Foxp3 Expression by Increasing DNMT3b Levels and DNA Methylation in the FOXP3 Gene. *AIDS* **2013**, *27*, 2019–2029. [CrossRef]
79. Arntz, O.J.; Pieters, B.C.H.; Oliveira, M.C.; Broeren, M.G.A.; Bennink, M.B.; de Vries, M.; van Lent, P.L.E.M.; Koenders, M.I.; van den Berg, W.B.; van der Kraan, P.M.; et al. Oral Administration of Bovine Milk Derived Extracellular Vesicles Attenuates Arthritis in Two Mouse Models. *Mol. Nutr. Food Res.* **2015**, *59*, 1701–1712. [CrossRef]
80. Melnik, B.C. Milk Exosomal MiRNAs: Potential Drivers of AMPK-to-MTORC1 Switching in β-Cell de-Differentiation of Type 2 Diabetes Mellitus. *Nutr. Metab.* **2019**, *16*, 85. [CrossRef] [PubMed]
81. Stephen, B.J.; Pareek, N.; Saeed, M.; Kausar, M.A.; Rahman, S.; Datta, M. Xeno-MiRNA in Maternal-Infant Immune Crosstalk: An Aid to Disease Alleviation. *Front. Immunol.* **2020**, *11*, 404. [CrossRef] [PubMed]
82. Zhao, Z.; Yu, S.; Xu, M.; Li, P. Effects of Microwave on Extracellular Vesicles and MicroRNA in Milk. *J. Dairy Sci.* **2018**, *101*, 2932–2940. [CrossRef]
83. Sun, Q.; Chen, X.; Yu, J.; Zen, K.; Zhang, C.-Y.; Li, L. Immune Modulatory Function of Abundant Immune-Related MicroRNAs in Microvesicles from Bovine Colostrum. *Protein Cell* **2013**, *4*, 197–210. [CrossRef] [PubMed]
84. Chen, Y.; Xiao, Y.; Ge, W.; Zhou, K.; Wen, J.; Yan, W.; Wang, Y.; Wang, B.; Qu, C.; Wu, J.; et al. MiR-200b Inhibits TGF-B1-Induced Epithelial-Mesenchymal Transition and Promotes Growth of Intestinal Epithelial Cells. *Cell Death Dis.* **2013**, *4*, e541. [CrossRef]
85. Malhotra, N.; Kang, J. SMAD Regulatory Networks Construct a Balanced Immune System. *Immunology* **2013**, *139*, 1–10. [CrossRef]
86. Nejad, C.; Stunden, H.J.; Gantier, M.P. A Guide to MiRNAs in Inflammation and Innate Immune Responses. *FEBS J.* **2018**, *285*, 3695–3716. [CrossRef] [PubMed]
87. Kim, E.S.; Choi, Y.E.; Hwang, S.J.; Han, Y.-H.; Park, M.-J.; Bae, I.H. IL-4, a Direct Target of MiR-340/429, Is Involved in Radiation-Induced Aggressive Tumor Behavior in Human Carcinoma Cells. *Oncotarget* **2016**, *7*, 86836–86856. [CrossRef]
88. Bian, J.; Liu, R.; Fan, T.; Liao, L.; Wang, S.; Geng, W.; Wang, T.; Shi, W.; Ruan, Q. MiR-340 Alleviates Psoriasis in Mice through Direct Targeting of IL-17A. *J. Immunol.* **2018**, *201*, 1412–1420. [CrossRef] [PubMed]
89. Schmitt, H.; Neurath, M.F.; Atreya, R. Role of the IL23/IL17 Pathway in Crohn's Disease. *Front. Immunol.* **2021**, *12*, 622934. [CrossRef] [PubMed]
90. Xie, W.; Li, M.; Xu, N.; Lv, Q.; Huang, N.; He, J.; Zhang, Y. MiR-181a Regulates Inflammation Responses in Monocytes and Macrophages. *PLoS ONE* **2013**, *8*, e58639. [CrossRef] [PubMed]
91. Su, Y.; Yuan, J.; Zhang, F.; Lei, Q.; Zhang, T.; Li, K.; Guo, J.; Hong, Y.; Bu, G.; Lv, X.; et al. MicroRNA-181a-5p and MicroRNA-181a-3p Cooperatively Restrict Vascular Inflammation and Atherosclerosis. *Cell Death Dis.* **2019**, *10*, 365. [CrossRef]
92. Jaiswal, A.; Reddy, S.S.; Maurya, M.; Maurya, P.; Barthwal, M.K. MicroRNA-99a Mimics Inhibit M1 Macrophage Phenotype and Adipose Tissue Inflammation by Targeting TNFα. *Cell Mol. Immunol.* **2019**, *16*, 495–507. [CrossRef]
93. Zhu, M.; Li, Y.; Sun, K. MicroRNA-182-5p Inhibits Inflammation in LPS-Treated RAW264.7 Cells by Mediating the TLR4/NF-KB Signaling Pathway. *Int. J. Clin. Exp. Pathol.* **2018**, *11*, 5725–5734.
94. Kawasaki, H.; Amano, H. Anti-Inflammatory Role of MicroRNA-429 in Human Gingival Epithelial Cells-Inhibition of IL-8 Production through Direct Binding to IKKβ MRNA. *Mol. Med. Rep.* **2021**, *24*, 581. [CrossRef] [PubMed]
95. Mo, J.-S.; Alam, K.J.; Kim, H.-S.; Lee, Y.-M.; Yun, K.-J.; Chae, S.-C. MicroRNA 429 Regulates Mucin Gene Expression and Secretion in Murine Model of Colitis. *J. Crohn's Colitis* **2016**, *10*, 837–849. [CrossRef]
96. López de las Hazas, M.-C.; del Pozo-Acebo, L.; Hansen, M.S.; Gil-Zamorano, J.; Mantilla-Escalante, D.C.; Gómez-Coronado, D.; Marín, F.; Garcia-Ruiz, A.; Rasmussen, J.T.; Dávalos, A. Dietary Bovine Milk MiRNAs Transported in Extracellular Vesicles Are Partially Stable during GI Digestion, Are Bioavailable and Reach Target Tissues but Need a Minimum Dose to Impact on Gene Expression. *Eur. J. Nutr.* **2021**, online ahead of print. [CrossRef]
97. Del Pozo-Acebo, L.; López de las Hazas, M.-C.; Tomé-Carneiro, J.; Gil-Cabrerizo, P.; San-Cristobal, R.; Busto, R.; García-Ruiz, A.; Dávalos, A. Bovine Milk-Derived Exosomes as a Drug Delivery Vehicle for MiRNA-Based Therapy. *IJMS Int. J. Mol. Sci.* **2021**, *22*, 1105. [CrossRef]
98. Vaswani, K.; Mitchell, M.D.; Holland, O.J.; Qin Koh, Y.; Hill, R.J.; Harb, T.; Davies, P.S.W.; Peiris, H. A Method for the Isolation of Exosomes from Human and Bovine Milk. *J. Nutr. Metab.* **2019**, *2019*, 5764740. [CrossRef]
99. Botta, L.; Filippi, S.; Bizzarri, B.M.; Meschini, R.; Caputo, M.; Proietti-De-Santis, L.; Iside, C.; Nebbioso, A.; Gualandi, G.; Saladino, R. Oxidative Nucleophilic Substitution Selectively Produces Cambinol Derivatives with Antiproliferative Activity on Bladder Cancer Cell Lines. *Bioorg. Med. Chem. Lett.* **2019**, *29*, 78–82. [CrossRef]
100. Cavinato, L.; Genise, E.; Luly, F.R.; Di Domenico, E.G.; Del Porto, P.; Ascenzioni, F. Escaping the Phagocytic Oxidative Burst: The Role of SODB in the Survival of Pseudomonas Aeruginosa Within Macrophages. *Front. Microbiol.* **2020**, *11*, 326. [CrossRef] [PubMed]

101. Dobin, A.; Davis, C.A.; Schlesinger, F.; Drenkow, J.; Zaleski, C.; Jha, S.; Batut, P.; Chaisson, M.; Gingeras, T.R. STAR: Ultrafast Universal RNA-Seq Aligner. *Bioinformatics* **2013**, *29*, 15–21. [CrossRef] [PubMed]
102. Howe, K.L.; Achuthan, P.; Allen, J.; Allen, J.; Alvarez-Jarreta, J.; Amode, M.R.; Armean, I.M.; Azov, A.G.; Bennett, R.; Bhai, J.; et al. Ensembl 2021. *Nucleic Acids Res.* **2021**, *49*, D884–D891. [CrossRef] [PubMed]
103. Liao, Y.; Smyth, G.K.; Shi, W. FeatureCounts: An Efficient General Purpose Program for Assigning Sequence Reads to Genomic Features. *Bioinformatics* **2014**, *30*, 923–930. [CrossRef]
104. Langmead, B.; Salzberg, S.L. Fast Gapped-Read Alignment with Bowtie 2. *Nat. Methods* **2012**, *9*, 357–359. [CrossRef] [PubMed]
105. Kozomara, A.; Birgaoanu, M.; Griffiths-Jones, S. MiRBase: From MicroRNA Sequences to Function. *Nucleic Acids Res.* **2019**, *47*, D155–D162. [CrossRef] [PubMed]
106. Leinonen, R.; Sugawara, H.; Shumway, M. The Sequence Read Archive. *Nucleic Acids Res.* **2011**, *39*, D19–D21. [CrossRef]
107. Wang, L.; Wang, S.; Li, W. RSeQC: Quality Control of RNA-Seq Experiments. *Bioinformatics* **2012**, *28*, 2184–2185. [CrossRef]
108. Darbellay, F.; Necsulea, A. Comparative Transcriptomics Analyses across Species, Organs, and Developmental Stages Reveal Functionally Constrained LncRNAs. *Mol. Biol. Evol.* **2020**, *37*, 240–259. [CrossRef]
109. Brawand, D.; Soumillon, M.; Necsulea, A.; Julien, P.; Csárdi, G.; Harrigan, P.; Weier, M.; Liechti, A.; Aximu-Petri, A.; Kircher, M.; et al. The Evolution of Gene Expression Levels in Mammalian Organs. *Nature* **2011**, *478*, 343–348. [CrossRef] [PubMed]
110. Love, M.I.; Huber, W.; Anders, S. Moderated Estimation of Fold Change and Dispersion for RNA-Seq Data with DESeq2. *Genome Biol.* **2014**, *15*. [CrossRef]
111. Blighe, K.; Rana, S.; Turkes, E.; Ostendorf, B.; Grioni, A.; Lewis, M. *EnhancedVolcano: Publication-Ready Volcano Plots with Enhanced Colouring and Labeling Bioconductor Version Release (3.13)*. 2021. Available online: https://www.bioconductor.org/packages/release/bioc/vignettes/EnhancedVolcano/inst/doc/EnhancedVolcano.htmL (accessed on 1 October 2021).
112. Bindea, G.; mLecnik, B.; Hackl, H.; Charoentong, P.; Tosolini, M.; Kirilovsky, A.; Fridman, W.-H.; Pagès, F.; Trajanoski, Z.; Galon, J. ClueGO: A Cytoscape Plug-in to Decipher Functionally Grouped Gene Ontology and Pathway Annotation Networks. *Bioinformatics* **2009**, *25*, 1091–1093. [CrossRef]
113. Shannon, P.; Markiel, A.; Ozier, O.; Baliga, N.S.; Wang, J.T.; Ramage, D.; Amin, N.; Schwikowski, B.; Ideker, T. Cytoscape: A Software Environment for Integrated Models of Biomolecular Interaction Networks. *Genome Res.* **2003**, *13*, 2498–2504. [CrossRef] [PubMed]
114. The IMEx Consortium Curators; del-Toro, N.; Duesbury, M.; Koch, M.; Perfetto, L.; Shrivastava, A.; Ochoa, D.; Wagih, O.; Piñero, J.; Kotlyar, M.; et al. Capturing Variation Impact on Molecular Interactions in the IMEx Consortium Mutations Data Set. *Nat. Commun.* **2019**, *10*, 1–14. [CrossRef]
115. Morris, J.H.; Apeltsin, L.; Newman, A.M.; Baumbach, J.; Wittkop, T.; Su, G.; Bader, G.D.; Ferrin, T.E. ClusterMaker: A Multi-Algorithm Clustering Plugin for Cytoscape. *BMC Bioinform.* **2011**, *12*, 436. [CrossRef] [PubMed]

International Journal of
Molecular Sciences

MDPI

Article

Dityrosine Crosslinking of Collagen and Amyloid-β Peptides Is Formed by Vitamin B_{12} Deficiency-Generated Oxidative Stress in *Caenorhabditis elegans*

Kyohei Koseki [1], Aoi Yamamoto [2], Keisuke Tanimoto [3], Naho Okamoto [1], Fei Teng [4], Tomohiro Bito [1,2,3,*], Yukinori Yabuta [1,2,3], Tsuyoshi Kawano [1,2,3] and Fumio Watanabe [1,2,3]

1 The United Graduate School of Agricultural Sciences, Tottori University, Tottori 680-8553, Japan; kyouhei.ganbalu@gmail.com (K.K.); mizonbon26@gmail.com (N.O.); yabuta@tottori-u.ac.jp (Y.Y.); kawano@tottori-u.ac.jp (T.K.); watanabe@tottori-u.ac.jp (F.W.)
2 Department of Agricultural Science, Graduate School of Sustainability Science, Tottori University, Tottori 680-8553, Japan; m21j7037b@edu.tottori-u.ac.jp
3 Department of Agricultural, Life and Environmental Sciences, Faculty of Agriculture, Tottori University, Tottori 680-8553, Japan; b18a5103b@edu.tottori-u.ac.jp
4 Department of Food Quality and Safety, College of Food Science, Northeast Agricultural University, Harbin 150030, China; tengfei@neau.edu.cn
* Correspondence: bito@tottori-u.ac.jp; Tel.: +81-857-31-5443

Abstract: (1) Background: Vitamin B_{12} deficiency in *Caenorhabditis elegans* results in severe oxidative stress and induces morphological abnormality in mutants due to disordered cuticle collagen biosynthesis. We clarified the underlying mechanism leading to such mutant worms due to vitamin B_{12} deficiency. (2) Results: The deficient worms exhibited decreased collagen levels of up to approximately 59% compared with the control. Although vitamin B_{12} deficiency did not affect the mRNA expression of prolyl 4-hydroxylase, which catalyzes the formation of 4-hydroxyproline involved in intercellular collagen biosynthesis, the level of ascorbic acid, a prolyl 4-hydroxylase coenzyme, was markedly decreased. Dityrosine crosslinking is involved in the extracellular maturation of worm collagen. The dityrosine level of collagen significantly increased in the deficient worms compared with the control. However, vitamin B_{12} deficiency hardly affected the mRNA expression levels of *bli-3* and *mlt-7*, which are encoding crosslinking-related enzymes, suggesting that deficiency-induced oxidative stress leads to dityrosine crosslinking. Moreover, using GMC101 mutant worms that express the full-length human amyloid β, we found that vitamin B_{12} deficiency did not affect the gene and protein expressions of amyloid β but increased the formation of dityrosine crosslinking in the amyloid β protein. (3) Conclusions: Vitamin B_{12}-deficient wild-type worms showed motility dysfunction due to decreased collagen levels and the formation of highly tyrosine-crosslinked collagen, potentially reducing their flexibility. In GMC101 mutant worms, vitamin B_{12} deficiency-induced oxidative stress triggers dityrosine-crosslinked amyloid β formation, which might promote its stabilization and toxic oligomerization.

Keywords: Alzheimer's disease; ascorbic acid; *Caenorhabditis elegans*; collagen; dityrosine crosslinking; oxidative stress; vitamin B_{12} deficiency

Citation: Koseki, K.; Yamamoto, A.; Tanimoto, K.; Okamoto, N.; Teng, F.; Bito, T.; Yabuta, Y.; Kawano, T.; Watanabe, F. Dityrosine Crosslinking of Collagen and Amyloid-β Peptides Is Formed by Vitamin B_{12} Deficiency-Generated Oxidative Stress in *Caenorhabditis elegans*. *Int. J. Mol. Sci.* **2021**, *22*, 12959. https://doi.org/10.3390/ijms222312959

Academic Editor: Maurizio Battino

Received: 25 October 2021
Accepted: 26 November 2021
Published: 30 November 2021

1. Introduction

Vitamin B_{12} (B_{12}) functions as the two coenzymes 5′-deoxyadenosylcobalamin and methylcobalamin of methylmalonyl-CoA mutase (EC 5.4.99.2) [1] and methionine synthase (MS; EC 2.1.1.13) [2], respectively, in mammals. Individuals deficient in B_{12} reportedly showed a significant increase of intracellular homocysteine (Hcy), a potent prooxidant [3], due to the reduced activity of MS, catalyzing methionine synthesis from Hcy and N5′-methyltetrahydrofolate [4]. Severe B_{12} deficiency leads to various symptoms, such as megaloblastic anemia, infertility, and neuropathy [3]. However, the underlying disease mechanisms are not fully understood [5,6].

As its molecular and cellular processes are similar to those of humans, *Caenorhabditis elegans* has been widely used as a model organism for genetic and biochemical studies. Our preceding study of B_{12} deficiency using *C. elegans* showed the occurrence of B_{12}-deficient worms with specific morphological abnormalities, like the short and plump "dumpy" mutant phenotype induced by the disordered biosynthesis of cuticular collagen, the main extracellular matrix component of the worm cuticle [7]. Mammalian collagen biosynthesis is well known to involve various posttranslational modifications [8], such as proline hydroxylation of collagen polypeptide chains in the rough endoplasmic reticulum and lysine oxidation of collagen triple helices in the extracellular space, catalyzed by prolyl 4-hydroxylase (EC 1.14.11.2) [9] and lysyl oxidase (EC 1.14.11.4) [10], respectively. Lysyl oxidase-derived linkages are predominantly formed in mammals during the extracellular maturation of the collagen molecule. However, lysyl crosslinking is absent from the *C. elegans* cuticle collagen [11]: it is replaced by dityrosine crosslinking [12]. No evidence has indicated whether B_{12} deficiency could affect the intracellular biosynthesis and subsequent extracellular maturation of the worm cuticle collagen.

In this study, we demonstrated that B_{12} deficiency results in significantly decreased collagen levels due to decreased ascorbic acid, a prolyl 4-hydroxylase coenzyme, and increased dityrosine crosslinking formation, leading to motility dysfunction. Furthermore, to the best of our knowledge, this study is the first to report that the dityrosine crosslinking of collagen was induced by oxidative stress generated by B_{12} deficiency. The reactive oxygen species that induce dityrosine crosslinking are reportedly formed in the amyloid-β (Aβ) oligomers involved in Alzheimer's disease (AD) pathogenesis [13]. Therefore, our finding could be applied to evaluate whether B_{12} deficiency could promote AD development. Using GMC101 mutant worms producing Aβ peptides in their muscle cells, we also discussed how B_{12} deficiency could affect the dityrosine crosslinking level of the Aβ peptide in inducing oligomerization and toxicity.

2. Results

2.1. Effect of B_{12} Deficiency on Collagen Biosynthesis in C. elegans

Figure 1A shows how B_{12} deficiency affected worm body collagen levels, calculated from the hydroxyproline content determined by the amino acid analysis of hydrolyzed worm body proteins. The worm collagen level significantly decreased during B_{12} deficiency. The decreased collagen level (approximately 8.1 ± 0.7 mg/g wet weight) was approximately 59% of that of control worms. B_{12} supplementation of B_{12}-deficient worms completely recovered the reduced collagen level to that of the control (approximately 14.8 ± 0.4 mg/g wet weight).

To clarify the underlying mechanism of significantly reduced collagen levels during B_{12} deficiency, the prolyl 4-hydroxylase α (PHY-1 and PHY-2) and β (PDI) subunits' mRNA expressions were determined. As shown in Figure 1B, B_{12} deficiency did not affect the prolyl 4-hydroxylase α (*dpy-18* and *phy-2*) and β (*pdi-2*) subunits' mRNA expressions.

Prolyl 4-hydroxylase requires *L*-ascorbic acid (AsA) as a coenzyme. Therefore, the AsA level was determined in the homogenates of the control and B_{12}-deficient worms (Figure 1C). B_{12} deficiency significantly reduced the AsA level (approximately 22.3 ± 5.2 μg/g wet weight), reaching approximately 52% of that of the control worms. B_{12} supplementation of B_{12}-deficient worms showed that the reduced AsA levels were completely recovered to that of the control (approximately 45.3 ± 2.6 μg/g wet weight). These results indicated that the significantly decreased collagen biosynthesis in B_{12}-deficient worms was mainly due to the reduced level of AsA as a coenzyme of prolyl 4-hydroxylase, as the mRNA expression levels of the enzyme subunits were completely unaffected by the B_{12} deficiency.

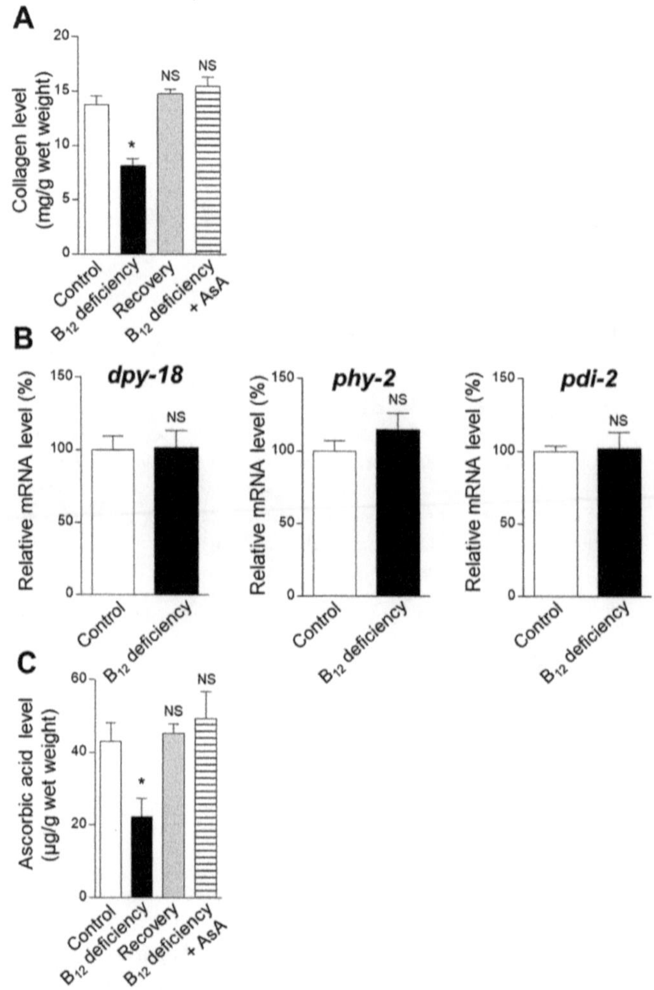

Figure 1. Effects of collagen biosynthesis during B_{12} deficiency in *C. elegans*. (**A**) Collagen levels calculated from the hydroxyproline content, (**B**) the mRNA expression levels of genes encoding complex prolyl 4-hydroxylase proteins, and (**C**) vitamin C (AsA) levels determined in B_{12}-supplemented worms (Control), B_{12}-deficient worms (B_{12} deficiency), B_{12}-deficient worms grown for three generations under B_{12}-supplemented conditions (Recovery), and B_{12}-deficient worms grown for three generations under an AsA-supplemented condition (B_{12} deficiency + AsA). Shown are the mRNA expression levels of human prolyl 4-hydroxylase subunit α1 (P4HA1) and prolyl 4-hydroxylase subunit α2 (P4HA2), and of prolyl 4-hydroxylase subunit β (P4HB) and their genetic orthologs *dpy-18*, *phy-2*, and *pdi-2*, respectively. The data represent the mean ± SEM of three independent experiments (n = 3). * $p < 0.05$ versus the control group. NS represents no significant differences.

2.2. Effect of B_{12} Deficiency on the Dityrosine Crosslinking Level of Worm Collagen

To clarify whether B_{12} deficiency affects dityrosine crosslinking in the extracellular maturation of cuticular collage, dityrosine levels were assayed in B_{12}-deficient worms. Dityrosine levels (per mg collagen) significantly increased in the B_{12}-deficient worms compared with control worms (Figure 2A). These results indicate that, although B_{12} deficiency significantly reduces collagen levels, the dityrosine crosslinking level of collagen was high. *bli-3* [14] and *mlt-7* [11] mRNA expression levels, encoding enzymes involved in the dityrosine crosslinking of worm collagen, were also tested. As shown in Figure 2B, B_{12} deficiency hardly affected *bli-3* and *mlt-7* mRNA expression levels compared with those of the control. These results suggest that the dityrosine crosslinking of collagen is triggered by B_{12} deficiency-induced oxidative stress.

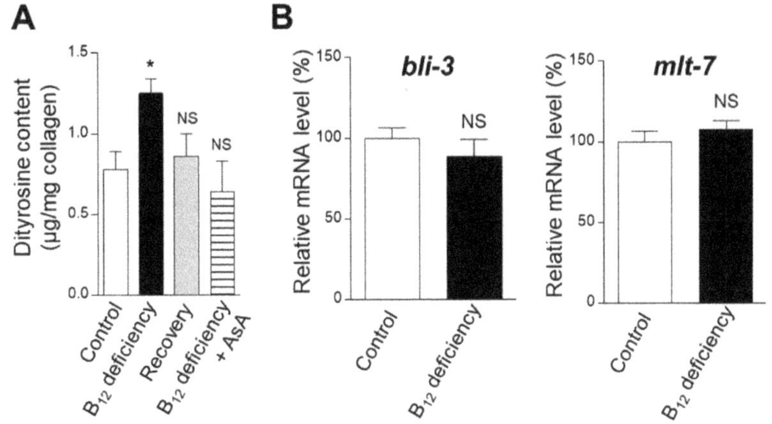

Figure 2. Effects of dityrosine and mRNA expression levels of enzymes involved in dityrosine crosslinking during B_{12} deficiency in *C. elegans*. (**A**) Dityrosine (per 1 mg collagen) and (**B**) mRNA expression levels (*bli-3* and *mlt-7*) determined in B_{12}-supplemented worms (Control), B_{12}-deficient worms (B_{12} deficiency), B_{12}-deficient worms grown for three generations under B_{12}-supplemented conditions (Recovery), and B_{12}-deficient worms grown for three generations under AsA-supplemented conditions (B_{12} deficiency + AsA). The data represent the mean ± SEM of three independent experiments (n = 3). * $p < 0.05$ versus the control group. NS represents no significant differences.

2.3. Effect of Collagenase Treatment on the Cuticular Extracellular Matrix Epidermal Collagen Layer in Control and B_{12}-Deficient Worms

To evaluate the effect of the highly formed dityrosine crosslinking of collagen, the epidermal collagen layer was treated with collagenase. When control and B_{12}-deficient worms were treated with collagenase solution for 10 min, the epidermal collagen layer was mostly digested in the control worms but remained unaffected in the B_{12}-deficient worms (Figure 3). When B_{12}-deficient worms were grown for three generations under B_{12}-supplemented conditions (recovery), the epidermal collagen layer of the recovery worms was readily digested by the collagenase treatment. These results indicate that the epidermal collagen layer of B_{12}-deficient worms became collagenase-resistant due to the high-level dityrosine crosslinking, implying that cuticular collagen would become structurally stronger and more rigid in the B_{12}-deficient worms compared with the control worms.

2.4. Effect of B_{12} Deficiency on C. elegans Motility Function

The B_{12} deficiency-induced high-level dityrosine crosslinking might potentially affect the physiological functions of the cuticular extracellular matrix. When the whiplash movement in the M9 buffer was evaluated as a motility function, the movement for 30 s in the B_{12}-deficient worms (approximately 99.3 thrashes/30 s) was decreased up to approximately 80% of that of the control worms (approximately 125.9 thrashes/30 s) (Figure 4). The decreased motility function of B_{12}-deficient worms was completely recovered to the control level when grown for three generations under B_{12}-supplemented conditions (recovery). In addition, the AsA supplementation of B_{12}-deficient worms showed that the reduced motility function was almost recovered to the control level. These results show that B_{12} deficiency leads to worm motility dysfunction due to decreased collagen level and increased dityrosine crosslinking formation in collagen, potentially reducing in flexibility of cuticular extracellular matrix.

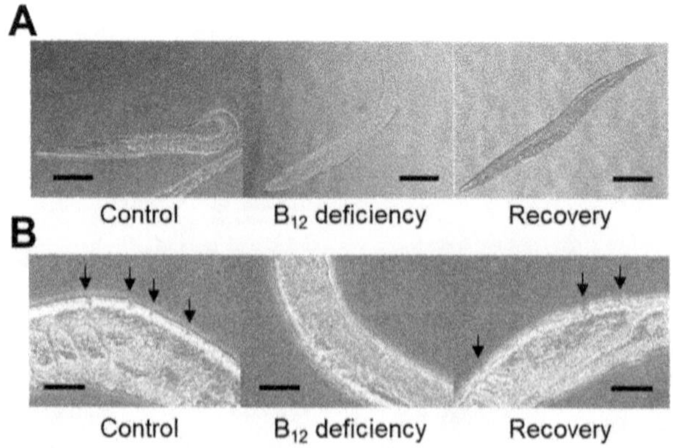

Figure 3. The morphological changes during B_{12} deficiency and collagen layer states after collagenase treatment. (**A**) Each worm (approximately 100 individuals) was washed three times with M9 buffer (3 g KH_2PO_4, 6 g Na_2HPO_4, 0.5 g NaCl, and 1 g NH_4Cl/L) and then imaged using a microscope system. Scale bars = 250 µm. (**B**) Each worm (approximately 100 individuals) was washed three times with the same M9 buffer and then fixed using 4% (*w*/*v*) paraformaldehyde for 10 min at 4 °C. The fixed worms were washed three times with PBS buffer (pH 7.2). Subsequently, the worms were soaked with 1.5 mL of β-mercaptoethanol buffer for 30 min using a rotator; then, the worms were washed three times with PBS buffer (pH 7.2). The washed worms were treated with collagenase solution for 10 min at room temperature and then kept on ice for 30 min to stop the reaction. Immediately after that, the worms were washed three times with PBS buffer (pH 7.2); then, each worm was mounted on glass slides. A microscope was used to observe the collagen layer. The arrow (↓) indicates the collagen layer decomposed by the collagenase treatment. Scale bars = 25 µm. Control, B_{12} deficiency, and Recovery represent B_{12}-supplemented worms, B_{12}-deficient worms, and B_{12}-deficient worms grown for three generations under B_{12}-supplemented conditions, respectively.

2.5. Effect of B_{12} Deficiency on the Dityrosince Crosslinking Level of Aβ Peptides in GMC101 Worms

Dityrosine crosslinks generated by reactive oxygen species are reportedly formed in Aβ oligomers, and these dityrosine links further stabilize the fibrils in developing AD [13]. To elucidate the relationship between B_{12} deficiency and AD, we investigated whether reactive oxygen species induced by the B_{12} deficiency significantly increased the dityrosine crosslinking levels of Aβ peptides using *C. elegans* GMC101 mutant worms producing $A\beta_{1-42}$ peptides in their muscle cells [15]. GMC101 mutant worms were grown until the young adult stage, under control and B_{12}-deficient conditions; then, they were shifted from 20 °C to 25 °C to induce Aβ [15]. Half of the GMC101 worms grown under B_{12}-deficient and control conditions showed signs of paralysis 48 and 96 h after the induction of Aβ, respectively (Figure 5). These results indicate that B_{12} deficiency significantly promoted Aβ-induced paralysis in GMC101 mutant worms.

To evaluate how B_{12} deficiency affects the Aβ mRNA and protein levels in GMC101 mutant worms, quantitative PCR and Western blot analyses were conducted. No significant changes could be observed between the control and B_{12}-deficient mutant worm Aβ mRNA levels (Figure 6A). Figure 6B shows the protein expression levels in N2 wild-type and GMC101 mutant worms grown under B_{12}-deficient and B_{12}-supplemented conditions. Human Aβ antibody-immunoreactive components (A–F) were detected as oligomeric Aβ forms (approximately 17–31 kDa) only in GMC101 mutant worms, and monomeric Aβ (5 kDa) was not detected. An immunoreactive component with the molecular mass of 34 kDa was a non-specific component as the component was found in N2 wild-type worms. As shown in Figure 6C, no significant difference in the immunoreactive component A–F could be detected between the B_{12}-supplemented and B_{12}-deficient GMC101 mutant worms.

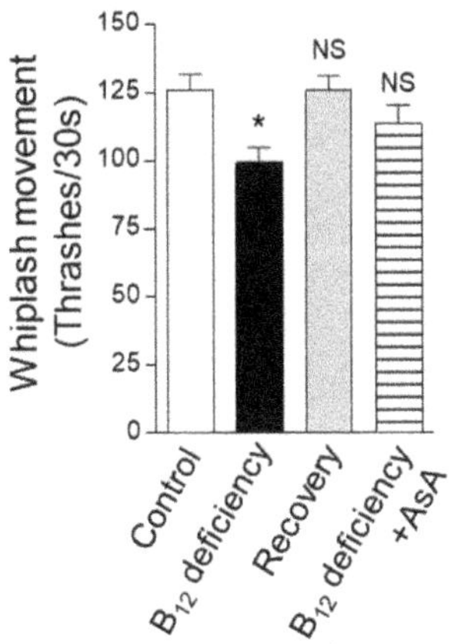

Figure 4. Effects of B_{12} deficiency on *C. elegans* motility. Whiplash movement with underwater condition was determined in B_{12}-supplemented (Control), B_{12}-deficient (B_{12} deficiency), B_{12}-deficient worms grown for three generations under B_{12}-supplemented conditions (Recovery), and B_{12}-deficient worms grown for three generations under AsA-supplemented conditions (B_{12} deficiency + AsA). The data represent the mean ± SEM of whiplash movement of 50 individual animals. * $p < 0.05$ versus the control group. NS represents no significant differences.

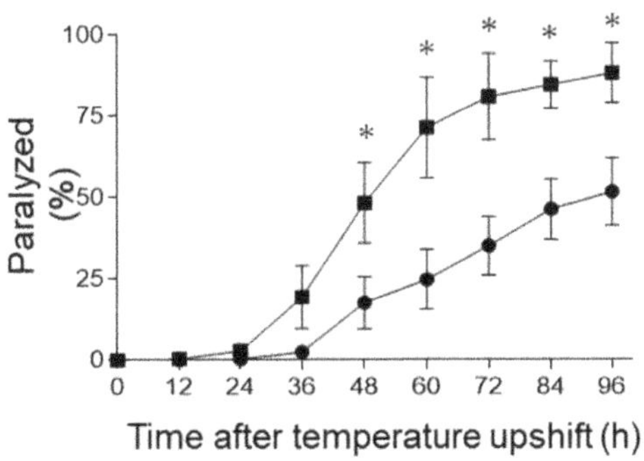

Figure 5. Paralysis rate of B_{12}-supplemented and B_{12}-deficienct GMC101 mutants. When GMC101 mutants grown under B_{12}-supplemented (•) and B_{12}-deficient conditions (■) developed into young adults, each GMC101 strain was shifted from 20 °C to 25 °C for the Aβ induction. The mean percentage of paralyzed worms is plotted against the time post temperature shift (h). All values represent the mean ± SD of five independent experiments (N = 5). Approximately 250 worms were screened for each condition. Asterisks indicate significant differences compared with the B_{12}-supplemented worms at the same time point (* $p < 0.05$).

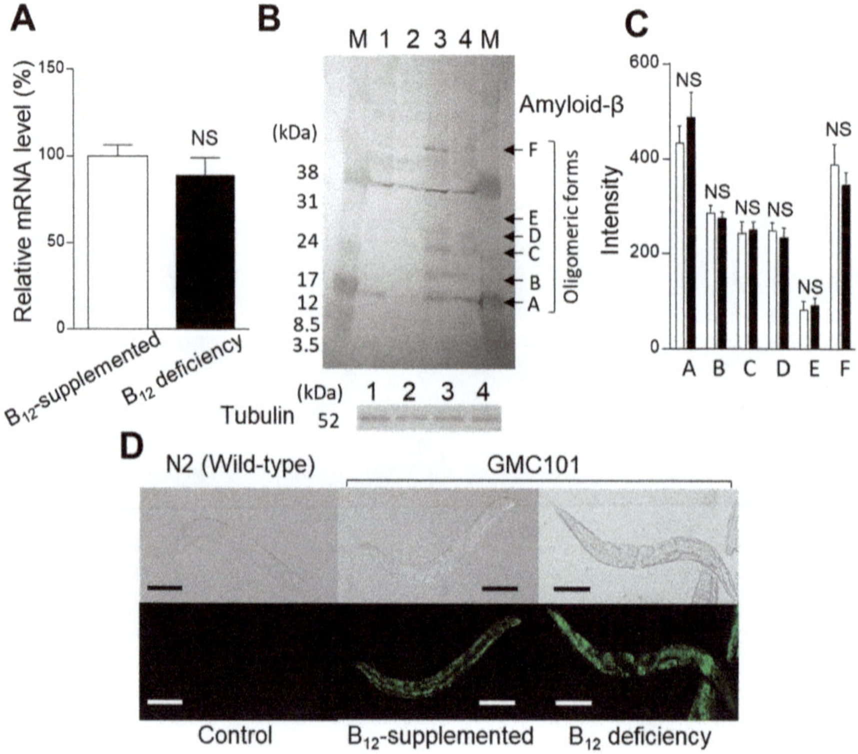

Figure 6. Effects of B_{12} deficiency on the Aβ mRNA and protein expression levels in GMC101 mutants. (**A**) mRNA expression levels in B_{12}-supplemented GMC101 mutant (B_{12}-supplemented) and B_{12}-deficient GMC101 mutant worms (B_{12} deficiency). The data represent the mean ± SEM of three independent experiments. NS: no significant differences. (**B**) protein expression levels in B_{12}-supplemented N2 wild-type worms (1), B_{12}-deficient N2 wild-type worms (2), B_{12}-supplemented GMC101 mutants (3), and B_{12}-deficient GMC101 mutants (4). M: molecular mass marker proteins. A, B, C, D, E, and F: human Aβanti-body-immunoreactive components (oligomeric Aβ forms). (**C**) Relative amounts of oligomeric Aβ forms in B_{12}-supplemented (white bar) and B_{12}-deficient (black bar) GMC101 mutants. Immunoreactive components A-F, detected in panel B, were quantified using the ImageJ software. The data represent the mean ± SEM of three independent experiments. NS: no significant differences. (**D**) Fluorescent images of Aβ in N2 wild-type and GMC101 mutant animals. B_{12}-supplemented N2 (Control), B_{12}-supplemented GMC101 mutants (B_{12}-supplemented), and B_{12}-deficient GMC101 mutants (B_{12} deficiency). Scale bars = 200 μm.

However, B_{12}-deficient GMC101 mutant worms appear to accumulate slightly more Aβ in the pharynx and tail compared with control mutant worms (Figure 6D), although quantitative differences cannot be shown. These results indicate that B_{12} deficiency itself does not increase Aβ mRNA and protein expression levels.

To elucidate the effect of B_{12} deficiency on the dityrosine crosslinking level of Aβ in GMC101 mutant worms, dityrosine crosslinking was investigated using a fluorescent microscope. The dityrosine crosslinking of Aβ was detected in the whole body of B_{12}-deficient mutant, but not in control mutant worms (Figure 7). The significant dityrosine crosslinking found in B_{12}-deficient mutant worms considerably decreased when grown for three generations under B_{12}-supplemented conditions (Recovery). Moreover, the dityrosine crosslinking of Aβ could be hardly found in AsA-supplemented B_{12}-deficient mutant worms.

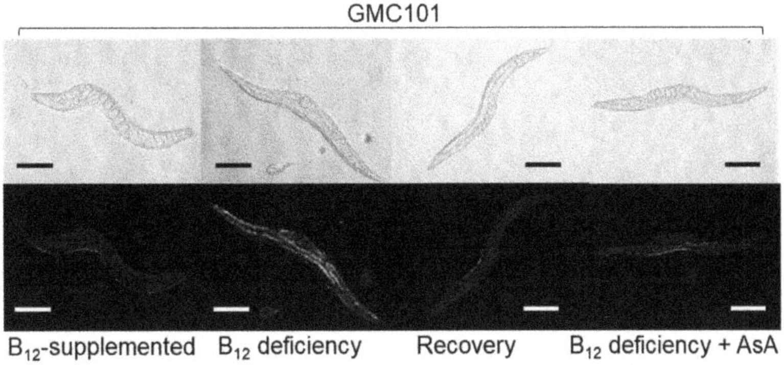

Figure 7. Dityrosine crosslinking accumulation in GMC101 mutants during B_{12} deficiency. Fluorescent images of dityrosine in GMC101 mutants in B_{12}-supplemented mutants (B_{12}-supplemented), B_{12}-deficient mutants (B_{12} deficiency), B_{12}-deficient mutants grown for three generations under B_{12}-supplemented conditions (Recovery), and B_{12}-deficient mutants grown in AsA-supplemented medium for three generations (B_{12} deficiency + AsA). Scale bars = 200 μm.

To clarify whether B_{12} deficiency-induced oxidative stress could promote the dityrosine crosslinking of Aβ, MDA and H_2O_2 levels were determined in GMC101 mutant worms grown under control, B_{12}-deficient, recovery, and AsA-supplemented B_{12}-deficient conditions (Figure 8A,B). The MDA and H_2O_2 levels significantly increased during B_{12} deficiency. The increased MDA and H_2O_2 levels of B_{12}-deficient mutant worms were completely recovered to the control level when grown for three generations under B_{12}-supplemented conditions (Recovery). Moreover, AsA supplementation did not show any increase in MDA and H_2O_2 levels even in B_{12}-deficient mutant worms. To evaluate the effects of B_{12} deficiency on mRNA levels, encoding enzymes involved in cellular antioxidant systems, the superoxide dismutase (*sod-1*), and catalase (*ctl-1*) mRNA levels were measured by qPCR (Figure 8C). No significant changes in the *sod-1* and *ctl-1* mRNA levels were observed during B_{12} deficiency. Although B_{12} deficiency did not affect the mRNA levels of these antioxidant enzymes, it significantly increased the MDA and H_2O_2 levels in GMC101 mutant worms, suggesting that superoxide dismutase and catalase activities decreased due to the oxidative inactivation of the enzymes, as described in B_{12}-deficient N2 wild-type worms [16]. These observations indicate that B_{12} deficiency-induced oxidative stress promoted the dityrosine crosslinking of Aβ.

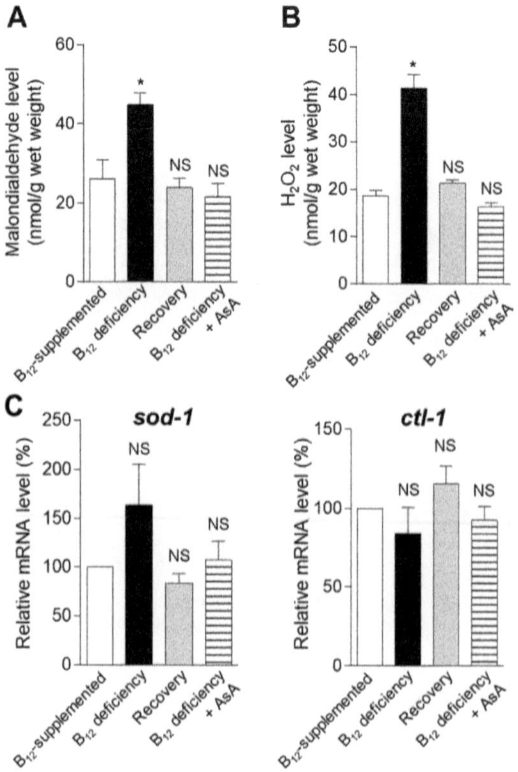

Figure 8. Effect of B_{12} deficiency on oxidative stress marker and oxidative defense enzyme levels in GMC101 mutants. (**A**) MDA and (**B**) H_2O_2 levels in GMC101 mutants are shown in B_{12}-supplemented mutants (B_{12}-supplemented), B_{12}-deficient mutants (B_{12} deficiency), B_{12}-deficient mutants grown for three generations under B_{12}-supplemented conditions (Recovery), and B_{12}-deficient mutants grown in AsA-supplemented medium for three generations (B_{12} deficiency + AsA). (**C**) Levels of mRNAs encoding enzymes involved in oxidant defense enzymes in GMC101 mutant animals. Using qPCR, we determined the levels of mRNAs encoding superoxide dismutase (*sod-1*) and catalase (*ctl-1*). The data represent the mean ± SEM of three independent experiments. * $p < 0.05$ versus the control group. NS represents no significant differences.

3. Discussion

We previously reported that certain B_{12}-deficient worms showed a short and plump phenotype, like "dumpy" mutants [7] formed due to the disordered cuticle collagen biosynthesis [17]. *C. elegans* possesses an external structure known as the cuticle, containing collagen and collagen-like protein major components of approximately 80% of the total cuticular protein content [18]. The cuticle is required to maintain body shape [19–21] and is synthesized five times from late embryogenesis throughout the *C. elegans* lifecycle, since the worm requires a new cuticle after molting in each growth stage [18]. Therefore, these observations indicate that the cuticle is crucial for the development and survival of worms. Worm cuticle collagen is synthesized and maturated by the following steps. Synthesized collagen single polypeptides are modified with proline hydroxylation and disulfide bond formation by the proline hydroxylase complex [DPY-18 (*dpy-18*), PHY-2 (*phy-2*), and PDI (*pdi-2*)] in the rough endoplasmic reticulum. After the secretion of the modified collagen triple helices into the extracellular space, they undergo a final modification of an enzymatic intermolecular tyrosine crosslinking by a dual oxidase BLI-3 (*bli-3*) [11].

Although a *C. elegans* gene ortholog (*duox-2*) of the human dual oxidase 2, involved in collagen intermolecular crosslinking, has been identified [22], *duox-2* was not expressed in the worms [22,23], as this gene might be a pseudo-gene. Furthermore, a heme peroxidase MLT-7 (*mlt-7*), along with BLI-3, reportedly played an essential role in cuticle–collagen crosslinking [11]. In contrast, the occurrence of tyrosine crosslinking is rare in vertebrates, and lysyl oxidase-derived linkages are predominant, while lysyl crosslinking is absent in the *C. elegans* cuticle [11].

As shown in Figures 1B and 2B, B_{12} deficiency did not affect the mRNA expression levels of the proline hydroxylase complex (DPY-18, PHY-2, and PDI), dual oxidase (BLI-3), and heme peroxidase (MLT-7) involved in the posttranslational modification of worm collagen. However, worm collagen level was significantly decreased to approximately 59% of that in the control during B_{12} deficiency (Figure 1A). The decreased collagen level of B_{12}-deficient worms was due to the significant decrease in AsA (approximately 52% of that of the control worms) (Figure 1C) as a coenzyme of prolyl 4-hydroxylase.

During B_{12} deficiency, homocysteine (Hcy) was significantly accumulated in the worm body [7]; leading to the disruption of redox regulation [16] due to severe Hcy-related oxidative stress [24]. Although B_{12} deficiency did not affect the mRNA expression levels of superoxide dismutase (*sod-1*) and catalase (*ctl-1*) involved in the cellular oxidant defense systems, the activities of superoxide dismutase and catalase were significantly reduced due to the oxidative inactivation of these enzymes [16]. Therefore, cellular antioxidant compound levels, such as those of glutathione and AsA were significantly reduced [16]. These observations indicate that AsA was mainly used as an antioxidant to scavenge oxidative stress induced during B_{12} deficiency and, consequently, was significantly decreased. Therefore, the decreased AsA induced decreased collagen biosynthesis in *C. elegans*. AsA deficiency has reportedly interfered with collagen synthesis in guinea pigs [25]. Moreover, it has been reported that Hcy, itself, can disrupt the collagen posttranslational modification in mammalian bones [10,26,27].

As shown in Figure 2A, dityrosine levels (per mg collagen) were significantly increased in the B_{12}-deficient worms compared with the control worms, indicating that B_{12} deficiency significantly increases the crosslinking level of tyrosine residues in collagen. Tyrosine crosslinking was formed by reactive oxygen species, induced by significantly increased Hcy during B_{12} deficiency, as B_{12} deficiency hardly stimulated the mRNA expression levels of dual oxidase BLI-3 (*bli-3*) and the heme peroxidase MLT-7 (*mlt-7*) involved in the tyrosine crosslinking of the cuticular collagen. Lévigne et al. [28] have reported that a deficiency of certain NADPH oxidase involved in the production of reactive oxygen species reduced dityrosine crosslinking, coinciding with the results of this study. Severe oxidative stress caused by aging has reportedly increased the dityrosine crosslinking of worm collagen [11].

The above-presented results suggest that the "dumpy" mutants formed by B_{12} deficiency were due to reduced intracellular collagen biosynthesis. Furthermore, B_{12} deficiency significantly increased the formation of the dityrosine crosslinking of collagen as an extracellular maturation step. As shown in Figure 4, B_{12} deficiency leads to worm motility dysfunction, probably due to such disordered collagen biosynthesis and maturation, themselves being probably due to the reduction of cuticular extracellular matrix flexibility. Although thrashing rates generally reflect worm body wall muscles, we have no information available on whether B_{12} deficiency-induced structural or functional muscular disorders.

Although little information is available on the mechanism by which B_{12} deficiency. contributes to AD pathogenesis, numerous studies have reported that the serum Hcy, elevated by B_{12} deficiency, is associated with AD pathogenesis [29–31]. In this study, dityrosine crosslinking of the cuticular collagen was formed by B_{12} deficiency-induced oxidative stress. Similarly, oxidative stress due to B_{12} deficiency can induce the dityrosine crosslinking of Aβ peptides in developing AD, as Aβ oligomers are formed by the dityrosine crosslinking generated by reactive oxygen species [13]. In GMC101 mutant worms, the MDA and H_2O_2 levels significantly increased during B_{12} deficiency, although no significant changes could be observed in the mRNA levels of the superoxide dismutase (*sod-1*) and catalase (*ctl-1*) involved in cellular oxidant defense systems. Similar results were reported in B_{12}-deficient N2 wild-type worms exhibiting oxidative inactivation of superoxide dismutase and catalase, leading to the disruption of redox regulation [16]. Therefore, we evaluated whether oxidative stress generated by B_{12} deficiency can form toxic Aβ oligomers using a GMC101 mutant worm that expresses full-length human Aβ peptide in the muscle cells. When Aβ oligomerization is formed and its toxicity is induced, mutant worms have shown paralysis [15]. Therefore, the GMC101 mutant worm has been used as a model animal of AD [32]. Our previous study [33] indicated that B_{12} deficiency did not stimulate the production of Aβ peptides in GMC101 mutant worms, although B_{12}-deficient worms exhibited paralysis faster and more severely than B_{12}-sufficient worms (control) did. As shown in Figure 5, similar results were obtained in this study. Furthermore, AsA-supplemented B_{12}-deficient worms rescued the paralysis phenotype. However, AsA supplementation did not affect Aβ peptide aggregations, suggesting that oxidative stress caused by elevated Hcy levels is an important factor in toxicity.

As shown in Figure 7, the dityrosine crosslinking of Aβ was present in the entire bodies of B_{12}-deficient mutant worms, but not in those of control mutant worms. The dityrosine crosslinking of Aβ could hardly be found in AsA-supplemented B_{12}-deficient mutant worms. As shown in Figure 6, no significant changes could be observed in Aβ oligomerization between the control and B_{12}-deficient mutant worms. However, the B_{12}-deficient worms exhibited paralysis due to toxicity faster than control worms did (Figure 5). Sitkiewicz et al. [34] demonstrated that tyrosine crosslinking shifts the equilibrium toward more compact oligomer types, leading to highly toxic fibrils. Maina et al. [35] suggest that dityrosine crosslinking, specifically, promotes the stabilization, but not the induction or facilitation, of Aβ assembly, and Aβ exerts high-level toxicity at a stage when self-assembly is high. These observations and the results presented in worms suggest that the dityrosine crosslinking formed during B_{12} deficiency promotes the formation and stabilization of the compact Aβ oligomers that facilitate self-assembly to induce high toxicity, probably in neuronal cells specifically located in the pharynx and tail (Figure 6D).

4. Materials and Methods

4.1. Organisms

The N2 Bristol wild-type *C. elegans* strain was maintained at 20 °C on nematode growth medium (NGM) plates using the *Escherichia coli* OP50 strain as a food source [36]. B_{12}-supplemented (control) and B_{12}-deficient worms were prepared as previously described [7]. B_{12}-deficient worms were transferred to a B_{12}-supplemented medium for three generations and used as the recovery worms. In the case of *L*-ascorbic acid (AsA)-supplemented experiments, B_{12}-deficient worms were grown in a B_{12}-deficient medium containing AsA 2-glucoside (final concentration of 1 mM) for three generations [16]. The transgenic GMC101 mutant worm strain was obtained from the Caenorhabditis Genetics Center (University of Minnesota, Minneapolis, MN, USA). The mutant worms were backcrossed five times before experimental use. When the mutant worms developed into young adults, their cultivation temperature was shifted from 20 °C to 25 °C to induce Aβ. B_{12} deficiency was also induced in the case of the GMC101 mutant worms through the above-described method. All nematodes were synchronized to obtain an identical developmental stage for experimental use.

4.2. Worm Body Collagen Determination

The acid hydrolysis of worm-body proteins was conducted according to the modified method of Roach and Gehrke [37]. Briefly, the worms (approximately 0.05 g wet weight of each background) were homogenized in 500 μL of 6-M HCl using a hand homogenizer (AS ONE Corp., Osaka, Japan). The homogenates were transferred into glass reaction tubes and supplemented with 6-M HCl (500 μL), and the pressure of the reaction tubes was reduced. After the homogenates were hydrolyzed under reduced pressure at 110 °C for 24 h, the resulting hydrolysates were centrifuged at 15,000× *g* for 10 min at 4 °C. Each supernatant

(250 μL) was diluted with an equal volume of 0.25 mol/L lithium citrate buffer (pH 2.2) (Fujifilm Wako Pure Chemical, Osaka, Japan) and filtered using a Millex®-LH membrane filter (Merck Millipore, Darmstadt, Germany). Hydroxyproline was analyzed using a fully automated amino acid analyzer (JEOL JLC-500/V, Nihon Denshi Datem Corp. Ltd., Tokyo, Japan). The worm collagen content was calculated from the determined hydroxyproline values using a conversion factor of 8.33 [19].

4.3. AsA Determination

The worms (approximately 0.05 g wet weight of each background) were homogenized in 300 μL of 5% (*w*/*v*) metaphosphoric acid solution on ice using a hand homogenizer (AS ONE). After the homogenates were centrifuged at 15,000× *g* for 10 min at 4 °C, the supernatants were used as samples. AsA was assayed according to the 2,4-dinitrophenyl hydrazine derivatization method [38]. Briefly, after AsA was completely oxidized to dehydro-form using indophenol solution, the formed dehydroAsA was derivatized with 2,4-dinitrophenyl hydrazine to form its osazone. The formed derivative compound was determined using a Shimadzu High-Performance Liquid Chromatography (HPLC) system (SPD-6AV UV-VIS Spectrophotometric Detector, LP-6A Liquid Delivery Pump, and CTO-6V Column Oven) with a CDS ver. 5 chromato-data processing system (LAsoft, Ltd., Chiba, Japan). Each sample (20 μL) was applied onto a Normal Phase HPLC Column (Senshu Pak Silica-2150-N, φ 6.0 × 150 mm, Senshu Scientific Corp. Ltd., Tokyo, Japan) and eluted with acetic acid/hexane/ethyl acetate (1:4:5, *v*/*v*/*v*) as a mobile phase at 40 °C. The flow rate was 1.5 mL/min. The derivative compound was monitored by measuring the absorbance at 495 nm.

4.4. Dityrosine Determination

Dityrosine was determined using the Shimadzu HPLC system (PU-2080 Plus Intelligent HPLC Pump, DG-2080-53 Degasser, RF-530 Fluorescence HPLC Monitor, and CTO-20A Column Oven) according to the method of Thein et al. [11]. Briefly, each hydrolyzed worm protein sample (20 μL), prepared as described above, was loaded onto a reversed-phase HPLC column (Luna C18 (2) 5 μm, 250 mm × 4.6 mm 100 Å; Phenomenex., Torrance, CA, USA) and isocratically eluted with 0.1-M KH_2PO_4-phosphoric acid (pH 3.8) as a mobile phase, at 40 °C. The flow rate was 1 mL/min. Dityrosine was monitored by measuring the fluorescence with an excitation and emission at 285 and 410 nm, respectively.

4.5. Collagenase Treatment

Worms (approximately 100 individuals) grown under various conditions were washed three times with M9 buffer (3 g KH_2PO_4, 6 g Na_2HPO_4, 0.5 g NaCl, and 1 g NH_4Cl/L) and then fixed with 4% (*w*/*v*) paraformaldehyde for 10 min at 4 °C. The fixed worms were washed three times with phosphate-buffered saline (PBS) buffer (pH 7.2). After the worms were soaked with 1.5 mL of β-mercaptoethanol buffer containing β-mercaptoethanol (75 μL), distilled water (1222.5 μL), 1 M Tris-HCl (pH 6.9) (187.5 μL), and Triton X-100 (15 μL) for 30 min using a rotator, they were washed three times with PBS buffer (pH 7.2). The washed worms were treated with 450 μL of collagenase solution (1 unit/μL) for 10 min at room temperature (25 °C), and the reaction vessel was placed on ice for 30 min to stop the enzyme reaction. After the treated worms were immediately washed three times with PBS buffer (pH 7.2), each worm was mounted on glass slides. The C. elegans cuticular collagen layer was observed under an ECLIPSE Ts2 microscope (Nikon Corp., Tokyo, Japan).

4.6. Assays of Malondialdehyde and H_2O_2 as Oxidative Stress Markers

Worms (approximately 0.05 g wet weight of each background) grown under various conditions were homogenized in 200 μL of 100 mM potassium-phosphate buffer (pH 7.0) on ice using a hand homogenizer (AS ONE). After the homogenates were centrifuged at 15,000× *g* for 10 min at 4 °C, the supernatants were used as samples. Malondialdehyde (MDA) and H_2O_2 were determined using a TBARS assay kit (ZeptoMetrix Corp., Buffalo,

NY, USA) and a H_2O_2 assay kit (BioVision, Inc., Milpitas, CA, USA), respectively. The MDA–thiobarbituric acid adducts or the reaction product of the OxiRed probe and H_2O_2 in the presence of horseradish peroxidase formed in the samples were determined by measuring the absorbance at 540 or 570 nm, respectively, using a Sunrise Rainbow RC-R microplate reader (Tecan Austria GmbH, Salzburg, Austria).

4.7. Immunofluorescent Staining of Aβ and Dityrosine in GMC 101 Mutant Worms

For visualizing Aβ peptides and their dityrosine crosslinking in the worm body, worms (approximately 100 individuals), grown under veracious conditions, were washed three times with M9 buffer and then fixed using 4% (*w*/*v*) paraformaldehyde for 10 min at 4 °C. The fixed worms were washed three times with PBS buffer (pH 7.2). After the worms were soaked with 1.5 mL of β-mercaptoethanol buffer, as described above, for 30 min using a rotator, they were washed three times with PBS buffer (pH 7.2). The washed worms were treated with 450 μL of collagenase solution (1 unit/μL) for 13 or 15 min at room temperature (25 °C), and the reaction vessel was placed on ice for 30 min to stop the enzyme reaction. The treated worms were immediately washed three times with PBS buffer (pH 7.2) and then were treated with 500 μL of blocking solution [5 mg bovine serum albumin (BSA) and 5 μL Triton X-100 per 1 mL of PBS (pH 7.2)] for 30 min at room temperature (25 °C) to block nonspecific antibody binding. The treated worms were washed three times with a washing buffer [3 mg BSA and 5 μL Triton X-100 per 1 mL of PBS (pH 7.2)]. After the worms were treated with an anti-β amyloid 1-42 rabbit monoclonal antibody (ab180956, Abcam, Cambridge, MA, USA) or an anti-dityrosine mouse monoclonal antibody (Nikken Seil Co., Ltd., Shizuoka, Japan) for 24 h at room temperature (25 °C), they were washed with the abovementioned washing buffer. The worms were treated with an anti-rabbit IgG secondary antibody (20-fold dilution) (ab6717, Abcam) or an anti-mouse IgG secondary antibody (20-fold dilution) (ab6785, Abcam) coupled to fluorescein isothiocyanate under dark conditions for 1 h. Next, the worms were washed with the above-described washing buffer and mounted on glass slides. Aβ and dityrosine visualization in the worms was performed using an ECLIPSE Ts2 fluorescent microscope (Nikon Corp.).

Immunoblot analysis was performed as previously described [33]. Sodium dodecyl sulfate polyacrylamide gel electrophoresis was performed using p-PAGEL slab gels (P-T16.5S; ATTO Corp., Tokyo, Japan). Aβ peptide was detected using a monoclonal anti-$A\beta_{1-42}$ primary antibody (EPR9296, Abcam) and anti-rabbit IgG-horseradish peroxidase conjugate (ab6721, Abcam). Signals were detected using EzWestBlue (ATTO Corp.) according to the manufacturer's instructions. The chemical coloring intensity was quantified using ImageJ (ImageJ Software, Bethesda, MD, USA) for three independent experiments.

4.8. Quantitative Polymerase Chain Reaction (qPCR) Analysis

Worm total RNA was prepared using Sephasol®-RNA1 (Nacalai Tesque Inc., Kyoto, Japan). Poly(A)+ mRNA was prepared from the total RNA using the Poly (A)+ Isolation Kit from Total RNA (Nippon Gene, Tokyo, Japan) and then was used to synthesize cDNA using a PrimeScript™ II 1st Strand cDNA Synthesis Kit (Takara Bio, Otsu, Japan). The primer pairs used for the qPCR analysis were designed using the GENETYX software (GENETYX Corp., Tokyo, Japan) to yield 20–24-nucleotide sequences with approximately 100–150-bp amplification products. A CFX Connect™ Real-Time System (Bio-Rad) with SYBR Premix Ex Taq (Takara Bio) was used to perform qPCR. β-Actin (*act-1*) was used as an internal standard. The qPCR experiments were repeated at least three times for each cDNA prepared from three preparations of worms. Table 1 shows the primers used for qPCR.

Table 1. Primer pairs used for the qPCR analysis.

Genes	Sequence (5′–3′)
dpy-18 (Sense)	CTACCACACTGTGATGTGGATG
dpy-18 (Antisense)	GCGTGCTTCAAGTTGTTCTG
phy-2 (Sense)	GCTTGATGTGTGGATGCAGGTT
phy-2 (Antisense)	TTGCGAGTCGTTTGGTGAGA
pdi-2 (Sense)	CGGAATCGATGATGTTCCATTCGG
pdi-2 (Antisense)	TTGGGTGAGCTTCTCGTCGAAAG
bli-3 (Sense)	GCGCTCAAAACATGTGCTGT
bli-3 (Antisense)	GCCAGATTGTTGTACCATCCGT
mlt-7 (Sense)	TTGCGATCATCACGAGTGGTGT
mlt-7 (Antisense)	AGCAGTTGTCGTGACTGGCAAA
Aβ (Sense)	GCGGATGCAGAATTCCGACATGAC
Aβ (Antisense)	TATGACAACACCGCCCACCATGAG
sod-1 (Sense)	TCTTCTCACTCAGGTCTCCAAC
sod-1 (Antisense)	TCGGACTTCTGTGTGATCCA
ctl-1 (Sense)	ATTATGCTCGTGGTGGAAACCC
ctl-1 (Antisense)	ACAATGTTTGGCGCCCTCAA
act-1 (Sense)	TCCAAGAGAGAGGTATCCTTACCC
act-1 (Antisense)	CTCCATATCATCCCAGTTGGTG

The qPCR primer pairs were designed using the GENETYX software. For normalization, β-actin (*act-1*) served as the internal standard.

4.9. *Swim Locomotion Analysis*

Worm motility function was evaluated using the swim locomotion method as follows. Individual three-day-old worms were placed on NGM agar plates (diameter of 3 cm) filled with 1.5 mL M9 buffer; then, their swimming motion was video-recorded for 1 min using a microscope (ECLIPSE Ts2, Nikon Corp.) equipped with a video system (DS-Fi3 camera unit and DS-L4 DS camera controller unit, Nikon Corp.). One round trip of worm head in the M9 buffer was determined as one whiplash movement.

4.10. *Paralysis Assay*

GMC101 mutant worms were grown at 25 °C after the L4 stage for Aβ induction. Worms that could move their heads but failed to move their bodies were scored as paralyzed [39]. The paralysis assay was performed every 12 h using approximately 50 individual worms.

4.11. *Statistical Analysis*

The results shown in Figure 1A,C, Figures 2A, 4 and 8 were analyzed using one-way ANOVA with Bonferroni's post hoc test using GraphPad Prism 4 (GraphPad Software, La Jolla, CA, USA). The results shown in Figures 1B, 2B, 5 and 6A,C were analyzed using Student's *t*-test for pairwise comparison. All data, except for those presented in Figure 5, are presented as the mean ± SEM. Differences were considered statistically significant at $p < 0.05$.

Author Contributions: Conceptualization, K.K., T.B. and F.W.; methodology, K.K., A.Y., K.T., N.O., F.T. and T.B.; software, K.K., N.O. and T.B.; validation, K.K., A.Y., K.T., N.O., F.T. and T.B.; formal analysis, K.K., A.Y., K.T., N.O., F.T. and T.B.; investigation, K.K., A.Y., K.T., N.O., F.T. and T.B.; resources, Y.Y. and T.K.; data curation, K.K., N.O. and T.B.; writing—original draft preparation, K.K., N.O., T.B. and F.W.; writing—review and editing, K.K., A.Y., K.T., N.O., T.B., Y.Y., T.K. and F.W.; visualization, K.K., A.Y., K.T. and T.B.; supervision, T.B. and F.W.; project administration, T.B. and F.W.; funding acquisition, T.B. and F.W. All authors have read and agreed to the published version of the manuscript.

Funding: This work was supported in part by JSPS KAKENHI Grant Numbers 19K15767 (T.B.) and 20580132 (F.W.).

Institutional Review Board Statement: Not applicable.

Informed Consent Statement: Not applicable.

Data Availability Statement: Data sharing not applicable.

Conflicts of Interest: The authors declare no conflict of interest.

References

1. Sokolovskaya, O.M.; Plessl, T.; Bailey, H.; Mackinnon, S.; Baumgartner, M.R.; Yue, W.W.; Froese, D.S.; Taga, M.E. Naturally occurring cobalamin (B_{12}) analogs can function as cofactor for human methylmalonyl-CoA mutase. *Biochimie* **2021**, *183*, 35–43. [CrossRef]
2. Froese, D.S.; Fowler, B.; Baumgartner, M.R. Vitamin B_{12}, folate, and the methionine remethylation cycle—Biochemistry, pathways, and regulation. *J. Inherit. Metab. Dis.* **2018**, *42*, 673–685. [CrossRef]
3. Institute of Medicine (US) Standing Committee on the Scientific Evaluation of Dietary Reference Intakes and Its Panel on Folate, Other B Vitamins, and Choline. *Dietary Reference Intakes for Thiamin, Riboflavin, Niacin, Vitamin B6, Folate, Vitamin B12, Pantothenic Acid, Biotin, and Choline*; National Academies Press: Washington, DC, USA, 1998.
4. Lu, S.C. S-Adenosylmethionine. *Int. J. Biochem. Cell Biol.* **2000**, *32*, 391–395. [CrossRef]
5. Pepper, M.R.; Black, M.M. B_{12} in fetal development. *Semin. Cell Dev. Biol.* **2011**, *22*, 619–623. [CrossRef]
6. Ebara, S.; Toyoshima, S.; Matsumura, T.; Adachi, S.; Takenaka, S.; Yamaji, R.; Watanabe, F.; Miyatake, K.; Inui, H.; Nakano, Y. Cobalamin deficiency results in severe metabolic disorder of serine and threonine in rats. *Biochim. Biophys. Acta* **2001**, *1568*, 111–117. [CrossRef]
7. Bito, T.; Matsunaga, Y.; Yabuta, Y.; Kawano, T.; Watanabe, F. Vitamin B_{12} deficiency in *Caenorhabditis elegans* results in loss of fertility, extended life cycle, and reduced lifespan. *FEBS Open Bio* **2013**, *3*, 112–117. [CrossRef]
8. Kavitha, O.; Thampan, R.V. Factors influencing collagen biosynthesis. *J. Cell. Biochem.* **2008**, *104*, 1150–1160. [CrossRef]
9. Gorres, K.L.; Raines, R.T. Prolyl 4-hydroxylase. *Crit. Rev. Biochem. Mol. Biol.* **2010**, *45*, 106–124. [CrossRef]
10. Thaler, R.; Agsten, M.; Spitzer, S.; Paschalis, E.P.; Karlic, H.; Klaushofer, K.; Varga, F. Homocysteine suppresses the expression of the collagen cross-linker lysyl oxidase involving IL-6, Fli1, and Epigenetic DNA methylation. *J. Biol. Chem.* **2011**, *286*, 5578–5588. [CrossRef]
11. Thein, M.C.; Winter, A.D.; Stepek, G.; McCormack, G.; Stapleton, G.; Johnstone, I.L.; Page, A.P. Combined extracellular matrix cross-linking activity of the peroxidase MLT-7 and the dual oxidase BLI-3 is critical for post-embryonic viability in *Caenorhabditis elegans*. *J. Biol. Chem.* **2009**, *284*, 17549–17563. [CrossRef]
12. Ewald, C.Y. Redox signaling of NADPH oxidases regulates oxidative stress responses, immunity and aging. *Antioxidants* **2018**, *7*, 130. [CrossRef] [PubMed]
13. Al-Hilaly, Y.K.; Williams, T.L.; Stewart-Parker, M.; Ford, L.; Skaria, E.; Cole, M.; Bucher, W.G.; Morris, K.L.; Sada, A.A.; Thorpe, J.R.; et al. A central role for dityrosine crosslinking of amyloid-β in Alzheimer's disease. *Acta Neuropathol. Commun.* **2013**, *1*, 83. [CrossRef] [PubMed]
14. Edens, W.A.; Sharling, L.; Cheng, G.; Shapira, R.; Kinkade, J.M.; Lee, T.; Edens, H.A.; Tang, X.; Sullards, C.; Flaherty, D.B.; et al. Tyrosine cross-linking of extracellular matrix is catalyzed by Duox, a multidomain oxidase/peroxidase with homology to the phagocyte oxidase subunit gp91phox. *J. Cell Biol.* **2001**, *154*, 879–892. [CrossRef] [PubMed]
15. Huang, J.; Chen, S.; Hu, L.; Niu, H.; Sun, Q.; Li, W.; Tan, G.; Li, J.; Jin, L.; Lyu, J.; et al. Mitoferrin-1 is involved in the progression of Alzheimer's disease through targeting mitochomdrial iron metablosm in a *Caenorhabditis elegans* model of Alzheimer's disease. *Neuroscience* **2018**, *385*, 90–101. [CrossRef]
16. Bito, T.; Misaki, T.; Yabuta, Y.; Ishikawa, T.; Kawano, T.; Watanabe, F. Vitamin B_{12} deficiency results in severe oxidative stress, leading to memory retention impairment in *Caenorhabditis elegans*. *Redox Biol.* **2017**, *11*, 21–29. [CrossRef]
17. Cai, L.; Phong, B.L.; Fisher, A.L.; Wang, Z. Regulation of fertility, survival, and cuticle collagen function by the *Caenorhabditis elegans eaf-1* and *ell-1* genes. *J. Biol. Chem.* **2011**, *286*, 35915–35921. [CrossRef]
18. Page, A.P.; Stepek, G.; Winter, A.D.; Pertab, D. Enzymology of the nematode cuticle: A potential drug target? *Int. J. Parasitol. Drugs Drug Resist.* **2014**, *4*, 133–141. [CrossRef]
19. Kramer, J.M. Structures and functions of collagens in *Caenorhabditis elegans*. *FASEB J.* **1994**, *8*, 329–336. [CrossRef]
20. Page, A.P.; Winter, A.D. Enzymes involved in the biogenesis of the nematode cuticle. *Adv. Parasitol.* **2003**, *53*, 85–148.
21. Wormbook. The Cuticle. Available online: http://www.wormbook.org/index.html (accessed on 2 October 2021).
22. Hill, A.A.; Hunter, C.P.; Tsung, B.T.; Tucker-Kellogg, G.; Brown, E.L. Genomic analysis of expression in *C. elegans*. *Science* **2000**, *290*, 809–812. [CrossRef]
23. Chávez, V.; Mohri-Shiomi, A.; Garsin, D.A. Ce-Duox1/BLI-3 generated reactive oxygen species as a protective innate immune mechanism in *Caenorhabditis elegans*. *Infect. Immun.* **2009**, *77*, 4983–4989. [CrossRef]
24. Škovierová, H.; Vidomanová, E.; Mahmood, S.; Sopková, J.; Drgová, A.; Červeňová, T.; Halašová, E.; Lehotský, J. The molecular and cellular effect of homocysteine metabolism imbalance on human health. *Int. J. Mol. Sci.* **2016**, *17*, 1733. [CrossRef]
25. Mahmoodian, F.; Peterkofsky, B. Vitamin C deficiency in guinea pigs differentially affects the expression of type 4 collagen, laminin, and elastin in blood vessels. *J. Nutr.* **1999**, *129*, 83–91. [CrossRef]
26. Clarke, M.; Ward, M.; Strain, J.J.; Hoey, L.; Dickey, W.; McNulty, H. B-vitamins and bone in health and disease: The current evidence. *Proc. Nutr. Soc.* **2014**, *73*, 330–339. [CrossRef] [PubMed]

27. Saito, M.; Marumo, K. The effects of homocysteine on the skeleton. *Curr. Osteoporos. Rep.* **2018**, *16*, 554–560. [CrossRef]
28. Lévigne, D.; Modarressi, A.; Krause, K.H.; Pittet-Cuénod, B. NADPH oxidase 4 deficiency leads to impaired wound repair and reduced dityrosine-crosslinking, but does not affect myofibroblast formation. *Free Radic. Biol. Med.* **2016**, *96*, 374–384. [CrossRef] [PubMed]
29. Chen, H.; Liu, S.; Ji, L.; Wu, T.; Ma, F.; Ji, Y.; Zhou, Y.; Zheng, M.; Zhang, M.; Huang, G. Associations between Alzheimer's disease and blood homocysteine, vitamin B_{12}, and folate: A case-control study. *Curr. Alzheimer Res.* **2015**, *12*, 88–94. [CrossRef]
30. Liang, S.; Hong-Fang, J. Associations between homocysteine, folic acid, vitamin B_{12} and Alzheimer's disease: Insights from meta-analyses. *J. Alzheimer's Dis.* **2015**, *46*, 777–790.
31. McLimans, K.E.; Martinez, A.D.C.; Mochel, J.P.; Allenspach, K. Serum vitamin B_{12} and related 5-methyltetrahydrofolate-homocysteine methyltransferase reductase and cubilin genotypes predict neural outcomes across the Alzheimer's disease spectrum. *Br. J. Nutr.* **2020**, *124*, 135–145. [CrossRef]
32. Sorrentino, V.; Romani, M.; Mouchiroud, L.; Beck, J.S.; Zhang, H.; D'Amico, D.; Moullan, N.; Potenza, F.; Schmid, A.W.; Rietsch, S.; et al. Enhancing mitochondrial proteostasis reduces amyloid-β proteotoxicity. *Nature* **2017**, *552*, 187–193. [CrossRef]
33. Andra, A.; Tanigawa, S.; Bito, T.; Ishihara, A.; Watanabe, F.; Yabuta, Y. Effects of vitamin B_{12} deficiency on amyloid- toxicity in *Caenorhabditis elegans*. *Antioxidants* **2021**, *10*, 962. [CrossRef] [PubMed]
34. Sitkiewicz, E.; Oledzki, J.; Poznański, J.; Dadlez, M. Di-tyrosine cross-link decreases the collisional cross-section of Aβ peptide dimers and trimers in the gas phase: An ion mobility study. *PLoS ONE* **2014**, *9*, e100200. [CrossRef] [PubMed]
35. Maina, M.B.; Mengham, K.; Burra, G.K.; Al-Hilaly, Y.A.; Serpell, L.C. Dityrosine cross-link trapping of amyloid-β intermediates reveals that self-assembly is required for Aβ-induced cytotoxicity. *bioRxiv* **2020**. [CrossRef]
36. Brenner, S. The genetics of *Caenorhabditis elegans*. *Genetics* **1974**, *77*, 71–94. [CrossRef] [PubMed]
37. Roach, D.; Gehrke, C.W. The hydrolysis of proteins. *J. Chromatogr. A* **1970**, *52*, 393–404. [CrossRef]
38. Standard Tables of Food Composition in Japan-2010. *The Council for Science and Technology, Ministry of Education, Culture, Sports, Science and Technolgy*; Official Gazette Co-Operation of Japan: Tokyo, Japan, 2010.
39. Gutierrez-Zepeda, A.; Santell, R.; Wu, Z.; Brown, M.; Wu, Y.; Khan, I.; Link, C.D.; Zhao, B.; Luo, Y. Soy isoflavone glycitein protects against beta amyloid-induced toxicity and oxidative stress in transgenic *Caenorhabditis elegans*. *BMC Neurosci.* **2005**, *6*, 54. [CrossRef]

International Journal of
Molecular Sciences

MDPI

Review

Natural Dibenzo-α-Pyrones: Friends or Foes?

Georg Aichinger

Laboratory of Toxicology, Department of Health Sciences and Technology, ETH Zurich, 8092 Zurich, Switzerland; georg.aichinger@hest.ethz.ch

Abstract: Natural dibenzo-α-pyrones (DAPs) can be viewed from two opposite angles. From one angle, the gastrointestinal metabolites urolithins are regarded as beneficial, while from the other, the emerging mycotoxin alternariol and related fungal metabolites are evaluated critically with regards to potential hazardous effects. Thus, the important question is: can the structural characteristics of DAP subgroups be held responsible for distinct bioactivity patterns? If not, certain toxicological and/or pharmacological aspects of natural DAPs might yet await elucidation. Thus, this review focuses on comparing published data on the two groups of natural DAPs regarding both adverse and beneficial effects on human health. Literature on genotoxic, estrogenic, endocrine-disruptive effects, as well as on the induction of the cellular anti-oxidative defense system, anti-inflammatory properties, the inhibition of kinases, the activation of mitophagy and the induction of autophagy, is gathered and critically reviewed. Indeed, comparing published data suggests similar bioactivity profiles of alternariol and urolithin A. Thus, the current stratification into hazardous *Alternaria* toxins and healthy urolithins seems debatable. An extrapolation of bioactivities to the other DAP sub-class could serve as a promising base for further research. Conclusively, urolithins should be further evaluated toward high-dose toxicity, while alternariol derivatives could be promising chemicals for the development of therapeutics.

Keywords: functional nutrition; emerging contaminants; urolithins; mycotoxins; chemoprevention; neuroprotection; DNA damage; estrogenicity; microbiome; bioactives

Citation: Aichinger, G. Natural Dibenzo-α-Pyrones: Friends or Foes?. *Int. J. Mol. Sci.* **2021**, 22, 13063. https://doi.org/10.3390/ijms222313063

Academic Editor: Cristobal Miranda

Received: 5 November 2021
Accepted: 30 November 2021
Published: 2 December 2021

1. Introduction

Dibenzo-α-pyrone (DAP, Figure 1A) is the basic scaffold of a group of naturally occurring chemicals, which are mainly formed by microbial species, such as bacteria or filamentous fungi. When substituted with multiple hydroxy groups, those metabolites belong to the chemical class of polyphenols, from which many representatives are regarded as beneficial for human health, mostly due to anti-oxidative and chemopreventive effects [1]. This also applies to some polyphenolic DAP derivatives. For example, urolithin A (UA, Figure 1B), a metabolite formed by ellagitannin-degrading gut bacteria, is extensively researched and marketed as a health-promoting agent in the scope of functional foods [2] or even as a therapeutic agent to improve muscle health [3].

However, there seems to be another side to the story. DAPs biosynthesized by food-contaminating molds, such as mycotoxin alternariol (AOH, Figure 1C), are regarded as potential carcinogens due to their ability to damage the DNA and to potentially induce endocrine-disruptive effects [4,5]. Given the striking structural similarity of these natural DAPs, some rather uncomfortable questions present themselves. Is the division into healthy urolithins and toxic AOH derivatives justified due to distinct bioactivity, or is current research on natural DAPs incomplete due to the different viewpoints formed depending on the source of origin? Could a chemical hazard be hidden in beneficial antioxidants? Can positive health effects be attributed to certain mycotoxins? Do the different substitution patterns of hydroxy and methyl groups result in distinct biological activities, or do we merely observe beneficial effects at moderate doses of exposure and toxic effects at high doses (which seems to be a steady companion in research on polyphenols [6,7])?

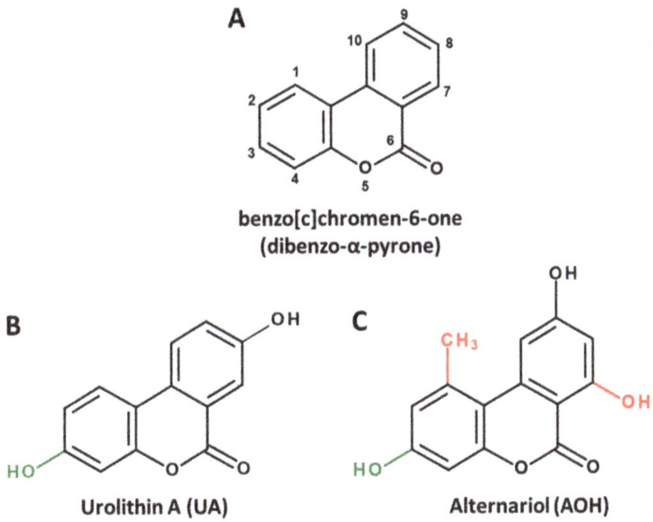

Figure 1. Chemical structures of the basic DAP scaffold (**A**), as well as two signature representatives of natural DAP derivatives: the bacterial polyphenol metabolite urolithin A (**B**) and the *Alternaria* mycotoxin alternariol (**C**). The hydroxy group at C1 (marked green) is a common feature of most natural DAPs. Methylation of C5, as well as hydroxylation of C11 (marked red), are common in biosynthesized DAPs but are not featured in ellagitannin biodegradation products.

This brief review aims at comparing the available literature on urolithins and mycogenic DAPs and focuses on the overlapping area of toxicology and functional nutrition, in order to provide a solid base for a more holistic research approach on DAP bioactivity.

2. Microbial Sources and Associated Structural Peculiarities

There are isolated reports of DAP derivatives being formed by plants [8], but the two main ways that they are produced in or from food commodities are (a) the complete biosynthesis as secondary metabolites of molds or (b) the biodegradation of ellagitannins by intestinal bacteria. The biosynthesis pathway is described mainly for filamentous fungi, particularly the genus *Alternaria*. A polyketide synthase encoded by the *pksJ* gene was found to be critical for the production of the two most prevalent DAPs, AOH and alternariol 9-methyl ether (AME) by *Alternaria alternata* [9]. Furthermore, the production of these and similar DAPs was also reported in other *Alternaria* [10], *Acremonium* [11], *Cephalosporum* [12] and *Hyalodendriella* spp. [13], all described as endophytic molds. Mycogenic DAPs are frequently reported in mold-contaminated grains, fruits, vegetables, etc., and are regarded as food contaminants [14]. As data suggest that they could be responsible for potentially toxic effects, but there are no regulations for maximum contamination levels yet around the globe, AOH and AME are considered to belong to the class of emerging mycotoxins [15].

On the other hand, DAPs that derive from the biodegradation of ellagitannins are uniformly referred to as urolithins. After ingestion, ellagitannins are hydrolyzed by bacteria carrying tannase enzymes to yield ellagic acid [16], which is further catalyzed by a currently unidentified lactonase/decarboxylase enzyme to the 3,4,8,9,10-pentahydroxy-DAP, urolithin M-5 (UM5). From the latter, all other urolithins are formed by subsequent dehydroxylation reactions that are catalyzed by currently unidentified enzymes [2]. However, a few bacterial species that are able to carry out at least a part of these reactions were already discovered. *Gordonibacter pamelae* and *Gordonibacter urolithinfaciens*, two species belonging to the strictly anaerobic family of *Eggerthellaceae*, were reported to decompose ellagic acid and perform dehydroxylations to sequentially yield UM5, urolithin M-6 and urolithin C (UC), the latter being the final metabolite [17]. Recently, another *Eggerthella* species, *Ellagibacter isourolithinfaciens*, was isolated from a human gut microbiome and observed to be capable of further dehydroxylating UC to isourolithin A [18,19]. Another study found *Bifidobacterium pseudocatenulatum* INIA P815 to produce UA and UB under certain growth conditions [20]. In complex microbiomes obtained from human feces, high interindividual differences were observed in the activity of the human microbiome, which allows for its stratification into three main groups [2]. Metabotype 0 (accounting for approximately 10–15% of the population) does not produce urolithins from ellagic acid. In urolithin producers, the final metabolites are either UA (metabotype A) or isourolithin A and urolithin B (UB) (metabotype B) [21].

Notably, the common precursor molecule UM5 predetermines that DAPs deriving from ellagic acid breakdown are only substituted with hydroxyl groups and are not functionalized at C1, C2 and C7 (Figure 1B). This is in stark contrast to biosynthesized DAPs, where substitutions at those positions, particularly the methylation of C1 and the hydroxylation of C7, are the norm (Figure 1C). In addition, based on current knowledge, UA and UB are not further metabolized by microbes, while for biosynthesized DAPs the methylation of functional hydroxy groups is common. For example, AOH is naturally produced as a mixture with AME, probably increasing its bioavailability and potentially its adverse effects [5].

3. Pharmacokinetics

Animal data on pharmacokinetics of urolithins and *Alternaria* toxins are only comparable to a limited extent due to differences in used species and experimental setups. However, according to a quick survey using the SwissADME quantitative structure–activity relationship (QSAR) tool [22], the bioavailability of major urolithins and fungal DAPs is predicted to be very similar (Table 1). UA and AOH, as well as UB and AME, share a comparable lipophilicity, and all four compounds have a 0.55 probability to be at least 10% bioavailable from oral uptake in rats, referred to as "bioavailability score" [23]. One exception is the blood brain barrier (BBB) permeation that is predicted only for UA/UB, not for AOH/AME (Table 1), which might be of high interest in the scope of neuroprotective effects that are proposed for UA. In line with this prediction, the presence of UA in mammalian brains was recently confirmed [24], while AOH was not reported to reach the brains of exposed mice in another study [25].

Table 1. Pharmacokinetic parameters of major natural DAPs. Shown are: octanol-water partition coefficients (PO/W), gastrointestinal (GI) adsorption, blood brain barrier (BBB) permeability and bioavailability score, as predicted by the SwissADME QSAR [21].

	log $P_{O/W}$	GI Absorption	BBB Permeant	Bioavailability Score
UA	2.06	high	yes	0.55
UB	2.48	high	yes	0.55
AOH	2.17	high	no	0.55
AME	2.55	high	no	0.55

Corresponding with the predicted bioavailability score, a recent study on Sprague-Dawley rats found approximately 90% of orally consumed AOH and AME to be excreted via the feces [26]. Based on physico-chemical similarities, it seems reasonable to expect corresponding total uptake ratios for UA and UB.

Hepatic metabolism of natural DAPs is generally assessed qualitatively and lacks the application of state-of-the-art quantitative tools as of yet. Nevertheless, their biotransformation seems to be comparable, with rapid glucuronidation to less bioactive phase II metabolites as the main pathway of hepatic clearance [27,28].

Taken together, it seems very unlikely that differing pharmacokinetics cause a significantly distinct in vivo bioactivity of methylated and non-methylated DAPs.

4. Bioactivity Profiles

To address the question of whether significant differences exist between those compounds in the impact on human health, one must ask if the additional functionalization of AOH/AME, particularly the methylation of the DAP scaffold, might serve as a driving force of toxicity. Thus, the upcoming section will focus on comparing respective bioactivity data of AOH/AME and UA/UB as representative metabolites of each natural DAP class.

4.1. *Topoisomerase Poisoning and Genotoxicity*

In two-digit micromolar concentrations, AOH and AME are well described to act genotoxic in human cancer cell models by poisoning topoisomerase (topo) II, an enzyme critical to the maintenance of DNA integrity during replication and transcription [29,30]. To a lesser extent, the induction of oxidative stress might play a role in genotoxicity as well [31]. To act as topo poison, a molecule has to stabilize the so-called "cleavable complex" between the enzyme and DNA, preventing the ligation of a previously induced gap in the phosphate backbone of the DNA, which may then persist as a strand break [32].

Published data do not suggest that UA or UB might poison topoisomerases and exert corresponding in vivo genotoxicity. UA was found to increase micronuclei formation in cells that were exposed to concentrations of approximately 5 μM for 20 h, but did not cause genotoxicity in vivo or mutagenicity in the Ames test, which is why UA passed the general safety assessment for application as a food supplement [33]. Furthermore, ellagic acid and UM5 were reported to catalytically inhibit topo II by competing with ATP at sub-micromolar concentrations [34], a biological activity that might lead to an increased number of DNA strand breaks at high concentrations due to an impaired management of torsional stress [35]. UA and UB were found to be inactive toward topo II up to 5 μM in decatenation assays [34]. However, it should be noted that with the same method, AOH was found to inhibit topo II only at concentrations above 10 μM [30]. Thus, the testing of urolithins for interactions with topo II seems incomplete, and it should be encouraged to apply higher concentrations and methods to assess topo poisoning, e.g., the in vivo complex of enzyme assay [36].

With regards to genotoxicity, AOH and UA might exert similar biological effects, even if the published data are not fully comparable. Both were shown to induce the formation of micronuclei in cultured cells of different origin, starting from comparable concentrations [33,37], and both were reported not to cause systemic genotoxicity in rodents [25,33].

Natural DAPs in general have not yet been studied regarding potential genotoxic effects in the colon, which is the site where the highest doses of aglycons can occur—in up to millimolar concentrations in the case of UA [38]. For AOH, the gastrointestinal toxicity is considered one of the major knowledge gaps [5,25]. Thus, the respective testing of not only *Alternaria* mycotoxins, but also urolithins, should be encouraged to complement risk assessment.

4.2. *Endocrine Activity*

AOH and AME are reported as estrogen receptor (ER) agonists, resulting in related gene transcription and a growth stimulation of ER-positive cells [37]. Moreover, several metabolites of those compounds were predicted to act estrogenic in a mixed in silico/in vitro approach [39]. AOH was also found to interact synergistically with other xenoestrogens, such as the mycotoxin zearalenone or the soy isoflavone genistein [40,41], and to exert cumulative estrogenic effects with the plasticizer bisphenol A [42] towards estrogenicity. Furthermore, the two biosynthesized DAPs were also reported to activate other steroid receptors, such as the androgen (AR) [43] and progesterone receptor [44]. Together, these findings have sparked concerns about the endocrine-disruptive potential of *Alternaria* toxins [5].

For urolithins, similar concerns have been reported to some degree. UA in particular was found to have a high affinity for estrogen receptors, with IC_{50} values in ERα receptor binding assays being even lower as compared to the well-known dietary phytoestrogen genistein [45]. However, the stimulation of MCF-7 cell growth only took place at comparably high concentrations of about 40 μM, pointing toward a possible antagonistic mechanism. Furthermore, another study reported that 10 μM of UA induced ER-dependent gene transcription in human endometrial cancer cells, but, contradictorily, also suppressed cell proliferation, which might be linked to a differentiated activity toward ERα (antagonism) and ERβ (agonism) [46]. UA is also suspected to interact with AR activation, although the exact mechanism has not yet been elucidated. While a direct agonism or antagonism was not reported in luciferase reporter gene assays in MDA-kb2 cells [47], other studies found UA to decrease AR expression [48] and to increase the proportion of the receptor residing in the cytosol, leading to the hypothesis that this DAP might be a valuable bioactive toward the prevention of prostate cancer [49].

Furthermore, UA and UB were demonstrated to inhibit 17β-hydroxysteroid dehydrogenase (17β-HSD), an enzyme critical for the biosynthesis of the endogenous estrogen 17β-estradiol (E2) in a hybrid in silico/in vitro approach [50]. The resulting decrease in intracellular E2 levels might serve as an alternative explanation for the compounds' anti-proliferative effects in breast cancer cells [50]. AOH and AME were not yet tested for effects toward an interaction with 17β-HSD. However, AOH lead to a reduced proliferation of human endometrial cancer cells, even as ER-mediated gene expression is activated [37]. While this effect can also be attributed to the onset of genotoxicity, the inhibition of 17β-HSD might be reconsidered as an alternative mechanism. Both fungal DAPs were reported to impair progesterone synthesis in porcine cells [51], potentially by interfering with 3-beta-hydroxysteroid dehydrogenase (3β-HSD).

Overall, it seems that natural DAPs generally exert a certain potential to act as steroid receptor agonists and a high potential to impair steroid biosynthesis, regardless of substitution patterns that are characteristic for their biological origin.

4.3. Inhibition of Casein Kinase 2

Casein kinase 2 (CK2) is a highly pleiotropic protein kinase whose overexpression is linked to pro-oncogenic processes [52] and anti-apoptotic effects in cancer treatment [53]. Thus, inhibition of CK2 has emerged as a therapeutic mode of action for overcoming drug resistance in cancers [54]. The DAP backbone seems to be a promising scaffold for this activity, as several representatives were predicted or reported to inhibit CK2. UA inhibited the enzyme with a IC_{50} of 0.39 μM and served as a precursor for the development of a much more potent inhibitor, its 4-bromo-derivative, that reached an IC_{50} of 0.015 nM [55]. For AOH, a similar IC_{50} (0.71 μM) regarding CK2 inhibition was observed in a cell-free assay, and the idea to base CK2-inhibiting drugs on its scaffold was ventilated [56]. In addition to the possible application in chemotherapy, a general chemopreventive effect of dietary CK2 inhibitors is currently discussed [57].

4.4. Mitophagy and Mitochondrial Health

Mitophagy is the cellular process of recycling damaged mitochondria that is central to mitochondrial health and of particular importance to highly stressed tissues, such as muscles [58]. It includes several pathways that can be influenced by extrinsic factors. It is well established that UA promotes mitophagy by stabilizing PTEN-induced kinase 1 (PINK1), responsible for recruiting and activating the protein Parkin, which in turn triggers the ubiquitination and thus degradation of mitochondrial proteins [59,60]. Exploiting this mechanism, the compound has even passed clinical trials as a promotor of mitochondrial and cellular health [61] and is marketed as a supplement to improve muscle health, particularly for elderly people [62].

AOH and AME have not yet been tested within the scope of inducing mitophagy, but given that the exact connection of chemical structure and PINK1 stabilization seems not very well elucidated, it might be a target for DAPs in general. Moreover, there is some evidence that an activation of the nuclear factor erythroid 2–related factor 2 (Nrf2) pathway, a process that not only UA [63] but also AOH and AME are capable of triggering [29], might play an additional role in the promotion of mitophagy [64].

4.5. Nrf2, Anti-Oxidative and Anti-Inflammatory Effects

In addition to the therapeutical application in the context of mitophagy, the propagation of urolithins as healthy dietary metabolites is based on their characterization as antioxidant and anti-inflammatory agents that have been extensively reviewed in recent literature [2,3]. The main mechanism behind their counteracting of oxidative stress is the activation of the Nrf2 pathway. The protein is bound to "Kelch-like ECH-associated protein 1" (Keap1) in the cytosol, which undergoes conformational changes to release Nrf2 in the presence of reactive oxygen species (ROS) or other electrophilic agents [65,66]. It then relocates into the nucleus, where it serves as a transcription factor inducing the expression of endogenous antioxidant agents and enzymes. Additionally, anti-inflammatory effects are mediated by Nrf2 via a crosstalk with NF-κB signaling [67]. Consequently, the beneficial compounds that trigger this pathway usually act slightly pro-oxidatively themselves and/or might even cause cellular oxidative stress at significantly higher concentrations.

Unfortunately, UA and UB have not yet been thoroughly tested for a pro-oxidative potential, but extensive data exist on their protective role against stressors, most commonly H_2O_2, which occurs in vitro at low micromolar concentrations [64,68,69]. On the contrary, the mycotoxins AOH and AME have not been tested for potential protective effects, but only for their direct pro-oxidative potential. In human cancer cells, these mycogenic DAPs lead to an increase in intracellular ROS levels at concentrations ≥ 5 μM, but they also cause an activation of Nrf2-related gene transcription at nanomolar doses [29]. Assessing a potentially corresponding ability to counteract oxidative stressors might be promising and should be encouraged, particularly in light of the Nrf2 activators being extensively tested as therapeutic agents for the treatment of type 2 diabetes mellitus [70].

Regarding a possible anti-inflammatory activity, UA/UB [71–73], as well as AOH/AME [74–77], have been reported to counteract pro-inflammatory stimuli in different cell models. Mechanisms that might play a role are the cholesterol-like intercalation into the cell membranes of macrophages that interferes with immunomodulatory receptors [78] or an interplay with NF-κB signaling to produce and release pro-inflammatory cytokines, potentially mediated again via Nrf2 activation [79].

4.6. Autophagy and Senescence

Autophagy is the process of degradation and recycling of cytosolic proteins of damaged cells, which is mostly associated with beneficial health effects, such as the prevention of cellular stress and tumor progression [79]. Several studies have reported the induction of this process after the exposure of different cells to DAPs of distinct sources. UA was found to induce this autophagy and to thereby inhibit metastasis-related biomarkers in colorectal cancer cells [80] and to protect neural cells from injury by decreasing endoplasmic reticulum (ER) stress [81]. Furthermore, there is some evidence suggesting that a part of the anti-inflammatory properties of UA might be related to increased autophagy in macrophages [82].

Interestingly, AOH was also reported to induce autophagy in macrophages, presumably via the mediation of ER stress that triggers the mTOR pathway [83]. However, in that study, a prolonged exposure of cells to AOH resulted in cellular senescence, a less desirable condition. From the two studies conducted on macrophages, it seems likely that UA and AOH exert similar effects toward the induction of autophagy in human cells.

4.7. Interactions with the Gut Microbiome

Large parts of ingested DAPs, such as AOH and AME, are excreted via the feces [26], and thus the gastrointestinal tract is probably the primarily exposed organ, which has sparked interest on potential interactions with the gut microbiome as an additional mode of action. A recent study addressed these questions and reported inhibitory effects of a complex mixture of *Alternaria* toxins on a multitude of bacterial strains and their ability to form biofilms [84]. However, as the extract that microbes were exposed to contained large amounts of mycotoxins with other chemical structures, a causal relationship with exposure to AOH/AME cannot be established from the published research. Likewise, experiments simulating the gastrointestinal tract revealed pomegranate extract to modulate the composition of the microbiome by increasing the prevalence of *Akkermansia* and *Gordonibacter*, particularly in the distal colon [85]. Furthermore, these changes seemed to have a direct enhancing impact on the formation of urolithins. Again, it seems difficult to draw a direct conclusion on the effects of DAPs on microbial communities due to the chemical complexity of the applied extract.

5. Conclusions

Taken together, the differentiation between "healthy" urolithins and "toxic" AOH derivatives seems to be a direct consequence of the reputation of their respective origins (superfoods vs. molds) and thus should be viewed critically. The two signature compounds of the respective groups, UA and AOH, are predicted to exert similar pharmacokinetic characteristics and share many biological activities, such as in vitro genotoxicity at high doses, the interaction with steroid receptor activation and steroid biosynthesis, the activation of the Nrf2 pathway, related anti-inflammatory effects and the induction of autophagy. However, blind spots on both sides prevent a full comparability of existing data.

On the one hand, the risk assessment of urolithins might not be fully completed yet, particularly regarding potential endocrine effects of higher doses that could hypothetically be reached via the application of pure urolithins as supplements. On the other hand—and somewhat counterintuitively—DAPs produced by *Alternaria* spp. and similar fungi might have a hidden potential in chemoprevention or as scaffolds for the design of therapeutic bioactives. A particularly relevant open question is whether AOH and AME would be able to promote mitophagy in a comparable way as their siblings of bacterial origin. Additionally, sub-genotoxic concentrations of mycogenic DAPs could hypothetically play a role in reducing cellular oxidative stress by activating the Nrf2 pathway, which could be utilized for the design of novel Nrf2-activating agents, e.g., for the treatment of diabetes mellitus.

It should be said that the discrepancies around research on DAPs might serve as an example of how research focus and chosen methodology can shape scientific understanding. A direct comparison of results for the two DAP classes is very challenging due to the different scientific fields (pharmacology vs. toxicology) and the respectively chosen experimentation. Nevertheless, such a comparative approach could allow for a prediction of possible biological targets for compounds of the other DAP class and thus could prove highly valuable in fueling corresponding research. In this light, it seems obvious that studies that test DAPs of different origin with a harmonized methodology would be of great value in evaluating the remaining toxicological or pharmacological questions.

Funding: This research was supported as a fellowship by the Future Food Initiative, a program run by the World Food System Center of ETH Zurich, the Integrative Food and Nutrition Center of EPFL and their industry partners.

Institutional Review Board Statement: Not applicable.

Informed Consent Statement: Not applicable.

Data Availability Statement: Not applicable.

Acknowledgments: I thank Ian Mitchell for kindly proofreading the manuscript.

Conflicts of Interest: The authors declare no conflict of interest.

References

1. Fraga, C.G.; Croft, K.D.; Kennedy, D.O.; Tomás-Barberán, F.A. The effects of polyphenols and other bioactives on human health. *Food Funct.* **2019**, *10*, 514–528. [CrossRef] [PubMed]
2. Tomás-Barberán, F.A.; González-Sarrías, A.; García-Villalba, R.; Núñez-Sánchez, M.A.; Selma, M.V.; García-Conesa, M.T.; Espín, J.C. Urolithins, the rescue of "old" metabolites to understand a "new" concept: Metabotypes as a nexus among phenolic metabolism, microbiota dysbiosis, and host health status. *Mol. Nutr. Food Res.* **2017**, *61*, 1500901. [CrossRef] [PubMed]
3. D'Amico, D.; Andreux, P.A.; Valdés, P.; Singh, A.; Rinsch, C.; Auwerx, J. Impact of the Natural Compound Urolithin A on Health, Disease, and Aging. *Trends Mol. Med.* **2021**, *27*, 687–699. [CrossRef] [PubMed]
4. Arcella, D.; Eskola, M.; Gómez Ruiz, J.A.; European Food Safety Authority. Dietary exposure assessment to Alternaria toxins in the European population. *EFSA J.* **2016**, *14*, e04654. [CrossRef]
5. Aichinger, G.; Del Favero, G.; Warth, B.; Marko, D. *Alternaria* toxins—Still emerging? *Compr. Rev. Food Sci. Food Saf.* **2021**, *20*, 4390–4406. [CrossRef]
6. Hayat, K.; Iqbal, H.; Malik, U.; Bilal, U.; Mushtaq, S. Tea and its consumption: Benefits and risks. *Crit. Rev. Food Sci. Nutr.* **2015**, *55*, 939–954. [CrossRef]
7. Shaito, A.; Posadino, A.M.; Younes, N.; Hasan, H.; Halabi, S.; Alhababi, D.; Al-Mohannadi, A.; Abdel-Rahman, W.M.; Eid, A.H.; Nasrallah, G.K.; et al. Potential adverse effects of resveratrol: A literature review. *Int. J. Mol. Sci.* **2020**, *21*, 2084. [CrossRef] [PubMed]
8. Liang, D.; Luo, H.; Liu, Y.-F.; Hao, Z.-Y.; Wang, Y.; Zhang, C.-L.; Zhang, Q.-J.; Chen, R.-Y.; Yu, D.-Q. Lysilactones A–C, three 6H-dibenzo[b,d]pyran-6-one glycosides from *Lysimachia clethroides*, total synthesis of Lysilactone A. *Tetrahedron* **2013**, *69*, 2093–2097. [CrossRef]
9. Saha, D.; Fetzner, R.; Burkhardt, B.; Podlech, J.; Metzler, M.; Dang, H.; Lawrence, C.; Fischer, R. Identification of a Polyketide Synthase Required for Alternariol (AOH) and Alternariol-9-Methyl Ether (AME) Formation in *Alternaria alternata*. *PLoS ONE* **2012**, *7*, e40564. [CrossRef]
10. Zwickel, T.; Kahl, S.M.; Rychlik, M.; Müller, M.E.H. Chemotaxonomy of mycotoxigenic small-spored *Alternaria* fungi–Do multitoxin mixtures act as an indicator for species differentiation? *Front. Microbiol.* **2018**, *9*, 1368. [CrossRef]
11. Hussain, H.; Krohn, K.; Ullah, Z.; Draeger, S.; Schulz, B. Bioactive chemical constituents of two endophytic fungi. *Biochem. Syst. Ecol.* **2007**, *35*, 898–900. [CrossRef]
12. Zhang, H.-W.; Huang, W.-Y.; Song, Y.-C.; Chen, J.-R.; Tan, R.-X. Four 6H-dibenzo[b,d]pyran-6-one derivatives produced by the endophyte *Cephalosporium acremonium* IFB-E007. *Helv. Chim. Acta* **2005**, *88*, 2861–2864. [CrossRef]
13. Meng, X.; Mao, Z.; Lou, J.; Xu, L.; Zhong, L.; Peng, Y.; Zhou, L.; Wang, M. Benzopyranones from the endophytic fungus *Hyalodendriella* sp. Ponipodef12 and their bioactivities. *Molecules* **2012**, *17*, 11303–11314. [CrossRef]
14. EFSA Panel on Contaminants in the Food Chain (CONTAM). Scientific opinion on the risks for animal and public health related to the presence of *Alternaria* toxins in feed and food. *EFSA J.* **2011**, *9*, 2407. [CrossRef]
15. Gruber-Dorninger, C.; Novak, B.; Nagl, V.; Berthiller, F. Emerging mycotoxins: Beyond traditionally determined food contaminants. *J. Agric. Food Chem.* **2017**, *65*, 7052–7070. [CrossRef]
16. de Las Rivas, B.; Rodríguez, H.; Anguita, J.; Muñoz, R. Bacterial tannases: Classification and biochemical properties. *Appl. Microbiol. Biotechnol.* **2019**, *103*, 603–623. [CrossRef]
17. Selma, M.V.; Beltrán, D.; García-Villalba, R.; Espín, J.C.; Tomás-Barberán, F.A. Description of urolithin production capacity from ellagic acid of two human intestinal *Gordonibacter* species. *Food Funct.* **2014**, *5*, 1779–1784. [CrossRef]
18. Selma, M.V.; Beltrán, D.; Luna, M.C.; Romo-Vaquero, M.; García-Villalba, R.; Mira, A.; Espín, J.C.; Tomás-Barberán, F.A. Isolation of human intestinal bacteria capable of producing the bioactive metabolite isourolithin A from ellagic acid. *Front. Microbiol.* **2017**, *8*, 1521. [CrossRef]
19. Beltrán, D.; Romo-Vaquero, M.; Espín, J.C.; Tomás-Barberán, F.A.; Selma, M.V. *Ellagibacter isourolithinifaciens* gen. nov., sp. nov., a new member of the family *Eggerthellaceae*, isolated from human gut. *Int. J. Syst. Evol. Microbiol.* **2018**, *68*, 1707–1712. [CrossRef]
20. Gaya, P.; Peirotén, Á.; Medina, M.; Álvarez, I.; Landete, J.M. *Bifidobacterium pseudocatenulatum* INIA P815: The first bacterium able to produce urolithins A and B from ellagic acid. *J. Funct. Foods* **2018**, *45*, 95–99. [CrossRef]
21. Cortés-Martín, A.; García-Villalba, R.; González-Sarrías, A.; Romo-Vaquero, M.; Loria-Kohen, V.; Ramírez-de-Molina, A.; Tomás-Barberán, F.A.; Selma, M.V.; Espín, J.C. The gut microbiota urolithin metabotypes revisited: The human metabolism of ellagic acid is mainly determined by aging. *Food Funct.* **2018**, *9*, 4100–4106. [CrossRef] [PubMed]
22. Daina, A.; Michielin, O.; Zoete, V. SwissADME: A free web tool to evaluate pharmacokinetics, drug-likeness and medicinal chemistry friendliness of small molecules. *Sci. Rep.* **2017**, *7*, 42717. [CrossRef] [PubMed]
23. Martin, Y.C. A bioavailability score. *J. Med. Chem.* **2005**, *48*, 3164–3170. [CrossRef] [PubMed]
24. Kujawska, M.; Jourdes, M.; Kurpik, M.; Szulc, M.; Szaefer, H.; Chmielarz, P.; Kreiner, G.; Krajka-Kuźniak, V.; Mikołajczak, P.Ł.; Teissedre, P.-L.; et al. Neuroprotective effects of pomegranate juice against Parkinson's Disease and presence of ellagitannins-derived metabolite—Urolithin A—In the brain. *Int. J. Mol. Sci.* **2020**, *21*, 202. [CrossRef] [PubMed]
25. Schuchardt, S.; Ziemann, C.; Hansen, T. Combined toxicokinetic and in vivo genotoxicity study on *Alternaria* toxins. *EFSA Support. Publ.* **2014**, *11*, 679E. [CrossRef]
26. Puntscher, H.; Aichinger, G.; Grabher, S.; Attakpah, E.; Krüger, F.; Tillmann, K.; Motschnig, T.; Hohenbichler, J.; Braun, D.; Plasenzotti, R.; et al. Bioavailability, metabolism, and excretion of a complex Alternaria culture extract versus altertoxin II: A comparative study in rats. *Arch. Toxicol.* **2019**, *93*, 3153–3167. [CrossRef]

27. Pfeiffer, E.; Schmit, C.; Burkhardt, B.; Altemöller, M.; Podlech, J.; Metzler, M. Glucuronidation of the mycotoxins alternariol and alternariol-9-methyl ether in vitro: Chemical structures of glucuronides and activities of human UDP-glucuronosyltransferase isoforms. *Mycotox Res.* **2009**, *25*, 3–10. [CrossRef]
28. González-Sarrías, A.; Giménez-Bastida, J.A.; Núñez-Sánchez, M.Á.; Larrosa, M.; García-Conesa, M.T.; Tomás-Barberán, F.A.; Espín, J.C. Phase-II metabolism limits the antiproliferative activity of urolithins in human colon cancer cells. *Eur. J. Nutr.* **2014**, *53*, 853–864. [CrossRef]
29. Tiessen, C.; Fehr, M.; Schwarz, C.; Baechler, S.; Domnanich, K.; Böttler, U.; Pahlke, G.; Marko, D. Modulation of the cellular redox status by the *Alternaria* toxins alternariol and alternariol monomethyl ether. *Toxicol. Lett.* **2013**, *216*, 23–30. [CrossRef]
30. Fehr, M.; Pahlke, G.; Fritz, J.; Christensen, M.O.; Boege, F.; Altemöller, M.; Podlech, J.; Marko, D. Alternariol acts as a topoisomerase poison, preferentially affecting the IIalpha isoform. *Mol. Nutr. Food Res.* **2009**, *53*, 441–451. [CrossRef]
31. Aichinger, G.; Beisl, J.; Marko, D. Genistein and delphinidin antagonize the genotoxic effects of the mycotoxin alternariol in human colon carcinoma cells. *Mol. Nutr. Food Res.* **2017**, *61*, 1600462. [CrossRef]
32. Pommier, Y.; Leo, E.; Zhang, H.; Marchand, C. DNA topoisomerases and their poisoning by anticancer and antibacterial drugs. *Chem. Biol.* **2010**, *17*, 421–433. [CrossRef]
33. Heilman, J.; Andreux, P.; Tran, N.; Rinsch, C.; Blanco-Bose, W. Safety assessment of Urolithin A, a metabolite produced by the human gut microbiota upon dietary intake of plant derived ellagitannins and ellagic acid. *Food Chem. Toxicol.* **2017**, *108*, 289–297. [CrossRef]
34. Furlanetto, V.; Zagotto, G.; Pasquale, R.; Moro, S.; Gatto, B. Ellagic acid and polyhydroxylated urolithins are potent catalytic inhibitors of human topoisomerase II: An in vitro study. *J. Agric. Food Chem.* **2012**, *60*, 9162–9170. [CrossRef]
35. Roca, J. Transcriptional inhibition by DNA torsional stress. *Transcription* **2011**, *2*, 82–85. [CrossRef]
36. Nitiss, J.L.; Soans, E.; Rogojina, A.; Seth, A.; Mishina, M. Topoisomerase Assays. *Curr. Protoc. Pharmacol.* **2012**, *57*, 3.3.1–3.3.27. [CrossRef]
37. Lehmann, L.; Wagner, J.; Metzler, M. Estrogenic and clastogenic potential of the mycotoxin alternariol in cultured mammalian cells. *Food Chem. Toxicol.* **2006**, *44*, 398–408. [CrossRef]
38. Mosele, J.I.; Gosalbes, M.-J.; Macià, A.; Rubió, L.; Vázquez-Castellanos, J.F.; Hernández, N.J.; Moya, A.; Latorre, A.; Motilva, M.-J. Effect of daily intake of pomegranate juice on fecal microbiota and feces metabolites from healthy volunteers. *Mol. Nutr. Food Res.* **2015**, *59*, 1942–1953. [CrossRef]
39. Dellafiora, L.; Warth, B.; Schmidt, V.; Del Favero, G.; Mikula, H.; Fröhlich, J.; Marko, D. An integrated *in silico/in vitro* approach to assess the xenoestrogenic potential of *Alternaria* mycotoxins and metabolites. *Food Chem.* **2018**, *248*, 253–261. [CrossRef]
40. Vejdovszky, K.; Hahn, K.; Braun, D.; Warth, B.; Marko, D. Synergistic estrogenic effects of *Fusarium* and *Alternaria* mycotoxins in vitro. *Arch. Toxicol.* **2017**, *91*, 1447–1460. [CrossRef]
41. Vejdovszky, K.; Schmidt, V.; Warth, B.; Marko, D. Combinatory estrogenic effects between the isoflavone genistein and the mycotoxins zearalenone and alternariol in vitro. *Mol. Nutr. Food Res.* **2017**, *61*, 1600526. [CrossRef] [PubMed]
42. Aichinger, G.; Pantazi, F.; Marko, D. Combinatory estrogenic effects of bisphenol A in mixtures with alternariol and zearalenone in human endometrial cells. *Toxicol. Lett.* **2020**, *319*, 242–249. [CrossRef]
43. Stypuła-Trębas, S.; Minta, M.; Radko, L.; Jedziniak, P.; Posyniak, A. Nonsteroidal mycotoxin alternariol is a full androgen agonist in the yeast reporter androgen bioassay. *Environ. Toxicol. Pharmacol.* **2017**, *55*, 208–211. [CrossRef]
44. Frizzell, C.; Ndossi, D.; Kalayou, S.; Eriksen, G.S.; Verhaegen, S.; Sørlie, M.; Elliott, C.T.; Ropstad, E.; Connolly, L. An in vitro investigation of endocrine disrupting effects of the mycotoxin alternariol. *Toxicol. Appl. Pharmacol.* **2013**, *271*, 64–71. [CrossRef] [PubMed]
45. Larrosa, M.; González-Sarrías, A.; García-Conesa, M.T.; Tomás-Barberán, F.A.; Espín, J.C. Urolithins, ellagic acid-derived metabolites produced by human colonic microflora, exhibit estrogenic and antiestrogenic activities. *J. Agric. Food Chem.* **2006**, *54*, 1611–1620. [CrossRef] [PubMed]
46. Zhang, W.; Chen, J.-H.; Aguilera-Barrantes, I.; Shiau, C.-W.; Sheng, X.; Wang, L.-S.; Stoner, G.D.; Huang, Y.-W. Urolithin A suppresses the proliferation of endometrial cancer cells by mediating estrogen receptor-α-dependent gene expression. *Mol. Nutr. Food Res.* **2016**, *60*, 2387–2395. [CrossRef] [PubMed]
47. Skledar, D.G.; Tomašič, T.; Dolenc, M.S.; Mašič, L.P.; Zega, A. Evaluation of endocrine activities of ellagic acid and urolithins using reporter gene assays. *Chemosphere* **2019**, *220*, 706–713. [CrossRef] [PubMed]
48. Dahiya, N.R.; Chandrasekaran, B.; Kolluru, V.; Ankem, M.; Damodaran, C.; Vadhanam, M.V. A natural molecule, urolithin A, downregulates androgen receptor activation and suppresses growth of prostate cancer. *Mol. Carcinog.* **2018**, *57*, 1332–1341. [CrossRef]
49. Stanisławska, I.J.; Piwowarski, J.; Granica, S.; Kiss, A. The effects of urolithins on the response of prostate cancer cells to non-steroidal antiandrogen bicalutamide. *Phytomedicine* **2018**, *46*, 176–183. [CrossRef]
50. Dellafiora, L.; Milioli, M.; Falco, A.; Interlandi, M.; Mohamed, A.; Frotscher, M.; Riccardi, B.; Puccini, P.; Del Rio, D.; Galaverna, G.; et al. A hybrid in silico/in vitro target fishing study to mine novel targets of urolithin A and B: A step towards a better comprehension of their estrogenicity. *Mol. Nutr. Food Res.* **2020**, *64*, 2000289. [CrossRef]
51. Tiemann, U.; Tomek, W.; Schneider, F.; Müller, M.; Pöhland, R.; Vanselow, J. The mycotoxins alternariol and alternariol methyl ether negatively affect progesterone synthesis in porcine granulosa cells in vitro. *Toxicol. Lett.* **2009**, *186*, 139–145. [CrossRef]
52. Seldin, D.C.; Leder, P. Casein kinase II alpha transgene-induced murine lymphoma: Relation to theileriosis in cattle. *Science* **1995**, *267*, 894–897. [CrossRef]
53. Guo, C.; Yu, S.; Davis, A.T.; Wang, H.; Green, J.E.; Ahmed, K. A potential role of nuclear matrix-associated protein kinase CK2 in protection against drug-induced apoptosis in cancer cells. *J. Biol. Chem.* **2001**, *276*, 5992–5999. [CrossRef]

54. Borgo, C.; Ruzzene, M. Role of protein kinase CK2 in antitumor drug resistance. *J. Exp. Clin. Cancer Res.* **2019**, *38*, 287. [CrossRef]
55. Cozza, G.; Gianoncelli, A.; Bonvini, P.; Zorzi, E.; Pasquale, R.; Rosolen, A.; Pinna, L.A.; Meggio, F.; Zagotto, G.; Moro, S. Urolithin as a converging scaffold linking ellagic acid and coumarin analogues: Design of potent protein kinase CK2 inhibitors. *ChemMedChem.* **2011**, *6*, 2273–2286. [CrossRef]
56. Aichinger, G.; Dellafiora, L.; Pantazi, F.; Del Favero, G.; Galaverna, G.; Dall'Asta, C.; Marko, D. *Alternaria* toxins as casein kinase 2 inhibitors and possible consequences for estrogenicity: A hybrid in silico/in vitro study. *Arch. Toxicol.* **2020**, *94*, 2225–2237. [CrossRef]
57. Husain, K.; Williamson, T.T.; Nelson, N.; Ghansah, T. Protein kinase 2 (CK2): A potential regulator of immune cell development and function in cancer. *Immunol. Med.* **2021**, *44*, 159–174. [CrossRef]
58. Youle, R.J.; Narendra, D.P. Mechanisms of mitophagy. *Nat. Rev. Mol. Cell Biol.* **2011**, *12*, 9–14. [CrossRef]
59. Ryu, D.; Mouchiroud, L.; Andreux, P.A.; Katsyuba, E.; Moullan, N.; Nicolet-Dit-Félix, A.A.; Williams, E.G.; Jha, P.; Lo Sasso, G.; Huzard, D.; et al. Urolithin A induces mitophagy and prolongs lifespan in *C. elegans* and increases muscle function in rodents. *Nat. Med.* **2016**, *22*, 879–888. [CrossRef]
60. Fang, E.F.; Hou, Y.; Palikaras, K.; Adriaanse, B.A.; Kerr, J.S.; Yang, B.; Lautrup, S.; Hasan-Olive, M.M.; Caponio, D.; Dan, X.; et al. Mitophagy inhibits amyloid-β and tau pathology and reverses cognitive deficits in models of Alzheimer's disease. *Nat. NeuroSci.* **2019**, *22*, 401–412. [CrossRef]
61. Andreux, P.A.; Blanco-Bose, W.; Ryu, D.; Burdet, F.; Ibberson, M.; Aebischer, P.; Auwerx, J.; Singh, A.; Rinsch, C. The mitophagy activator urolithin A is safe and induces a molecular signature of improved mitochondrial and cellular health in humans. *Nat. Metab.* **2019**, *1*, 595–603. [CrossRef] [PubMed]
62. Mitopure. Available online: https://www.mitopure.com/ (accessed on 21 October 2021).
63. Singh, R.; Chandrashekharappa, S.; Bodduluri, S.R.; Baby, B.V.; Hegde, B.; Kotla, N.G.; Hiwale, A.A.; Saiyed, T.; Patel, P.; Vijay-Kumar, M.; et al. Enhancement of the gut barrier integrity by a microbial metabolite through the Nrf2 pathway. *Nat. Commun.* **2019**, *10*, 1–18. [CrossRef] [PubMed]
64. Gumeni, S.; Papanagnou, E.-D.; Manola, M.S.; Trougakos, I.P. Nrf2 activation induces mitophagy and reverses Parkin/Pink1 knock down-mediated neuronal and muscle degeneration phenotypes. *Cell Death Dis.* **2021**, *12*, 1–12. [CrossRef] [PubMed]
65. Kansanen, E.; Kuosmanen, S.M.; Leinonen, H.; Levonen, A.-L. The Keap1-Nrf2 pathway: Mechanisms of activation and dysregulation in cancer. *Redox Biol.* **2013**, *1*, 45–49. [CrossRef]
66. Robledinos-Antón, N.; Fernández-Ginés, R.; Manda, G.; Cuadrado, A. Activators and inhibitors of NRF2: A review of their potential for clinical development. *Oxidative Med. Cell. Longev.* **2019**, *2019*, 1–20. [CrossRef]
67. Cuadrado, A.; Rojo, A.I.; Wells, G.; Hayes, J.D.; Cousin, S.P.; Rumsey, W.L.; Attucks, O.C.; Franklin, S.; Levonen, A.-L.; Kensler, T.W.; et al. Therapeutic targeting of the NRF2 and KEAP1 partnership in chronic diseases. *Nat. Rev. Drug Discov.* **2019**, *18*, 295–317. [CrossRef]
68. Kim, K.B.; Lee, S.; Kim, J.H. Neuroprotective effects of urolithin A on H_2O_2-induced oxidative stress-mediated apoptosis in SK-N-MC cells. *Nutr. Res. Pract.* **2020**, *14*, 3–11. [CrossRef]
69. Cásedas, G.; Les, F.; Choya-Foces, C.; Hugo, M.; López, V. The metabolite urolithin-A ameliorates oxidative stress in Neuro-2a cells, becoming a potential neuroprotective agent. *Antioxidants* **2020**, *9*, 177. [CrossRef]
70. Matzinger, M.; Fischhuber, K.; Heiss, E.H. Activation of Nrf2 signaling by natural products-can it alleviate diabetes? *Biotechnol. Adv.* **2018**, *36*, 1738–1767. [CrossRef]
71. Lee, G.; Park, J.-S.; Lee, E.-J.; Ahn, J.-H.; Kim, H.-S. Anti-inflammatory and antioxidant mechanisms of urolithin B in activated microglia. *Phytomedicine* **2019**, *55*, 50–57. [CrossRef]
72. Rønning, S.B.; Voldvik, V.; Bergum, S.K.; Aaby, K.; Borge, G.I.A. Ellagic acid and urolithin A modulate the immune response in LPS-stimulated U937 monocytic cells and THP-1 differentiated macrophages. *Food Funct.* **2020**, *11*, 7946–7959. [CrossRef]
73. Komatsu, W.; Kishi, H.; Yagasaki, K.; Ohhira, S. Urolithin A attenuates pro-inflammatory mediator production by suppressing PI3-K/Akt/NF-κB and JNK/AP-1 signaling pathways in lipopolysaccharide-stimulated RAW264 macrophages: Possible involvement of NADPH oxidase-derived reactive oxygen species. *Eur. J. Pharmacol.* **2018**, *833*, 411–424. [CrossRef]
74. Schmutz, C.; Cenk, E.; Marko, D. The *Alternaria* mycotoxin alternariol triggers the immune response of IL-1beta-stimulated, differentiated Caco-2 cells. *Mol. Nutr. Food Res.* **2019**, *63*, 1900341. [CrossRef]
75. Kollarova, J.; Cenk, E.; Schmutz, C.; Marko, D. The mycotoxin alternariol suppresses lipopolysaccharide-induced inflammation in THP-1 derived macrophages targeting the NF-kappaB signalling pathway. *Arch. Toxicol.* **2018**, *92*, 3347–3358. [CrossRef]
76. Grover, S.; Lawrence, C.B. The *Alternaria* alternata mycotoxin alternariol suppresses lipopolysaccharide-induced inflammation. *Int. J. Mol. Sci.* **2017**, *18*, 1577. [CrossRef]
77. Solhaug, A.; Wisbech, C.; Christoffersen, T.E.; Hult, L.O.; Lea, T.; Eriksen, G.S.; Holme, J.A. The mycotoxin alternariol induces DNA damage and modify macrophage phenotype and inflammatory responses. *Toxicol. Lett.* **2015**, *239*, 9–21. [CrossRef]
78. Del Favero, G.; Mayer, R.M.; Dellafiora, L.; Janker, L.; Niederstaetter, L.; Dall'Asta, C.; Gerner, C.; Marko, D. Structural similarity with cholesterol reveals crucial insights into mechanisms sustaining the immunomodulatory activity of the mycotoxin alternariol. *Cells* **2020**, *9*, 847. [CrossRef]
79. Khandia, R.; Dadar, M.; Munjal, A.; Dhama, K.; Karthik, K.; Tiwari, R.; Yatoo, M.I.; Iqbal, H.M.N.; Singh, K.P.; Joshi, S.K.; et al. A comprehensive review of autophagy and its various roles in infectious, non-infectious, and lifestyle diseases: Current knowledge and prospects for disease prevention, novel drug design, and therapy. *Cells* **2019**, *8*, 674. [CrossRef]

80. Zhao, W.; Shi, F.; Guo, Z.; Zhao, J.; Song, X.; Yang, H. Metabolite of ellagitannins, urolithin A induces autophagy and inhibits metastasis in human sw620 colorectal cancer cells. *Mol. Carcinog.* **2018**, *57*, 193–200. [CrossRef]
81. Ahsan, A.; Zheng, Y.R.; Wu, X.L.; Tang, W.D.; Liu, M.R.; Ma, S.J.; Jiang, L.; Hu, W.W.; Zhang, X.N.; Chen, Z. Urolithin A-activated autophagy but not mitophagy protects against ischemic neuronal injury by inhibiting ER stress in vitro and in vivo. *CNS NeuroSci. Ther.* **2019**, *25*, 976–986. [CrossRef]
82. Boakye, Y.D.; Groyer, L.; Heiss, E.H. An increased autophagic flux contributes to the anti-inflammatory potential of urolithin A in macrophages. *Biochim. Biophys Acta Gen. Subj.* **2018**, *1862*, 61–70. [CrossRef] [PubMed]
83. Solhaug, A.; Torgersen, M.L.; Holme, J.A.; Lagadic-Gossmann, D.; Eriksen, G.S. Autophagy and senescence, stress responses induced by the DNA-damaging mycotoxin alternariol. *Toxicology* **2014**, *326*, 119–129. [CrossRef] [PubMed]
84. Crudo, F.; Aichinger, G.; Mihajlovic, J.; Varga, E.; Dellafiora, L.; Warth, B.; Dall'Asta, C.; Berry, D.; Marko, D. In vitro interactions of *Alternaria* mycotoxins, an emerging class of food contaminants, with the gut microbiota: A bidirectional relationship. *Arch. Toxicol.* **2021**, *95*, 2533–2549. [CrossRef] [PubMed]
85. García-Villalba, R.; Vissenaekens, H.; Pitart, J.; Romo-Vaquero, M.; Espín, J.C.; Grootaert, C.; Selma, M.V.; Raes, K.; Smagghe, G.; Possemiers, S.; et al. Gastrointestinal Simulation Model TWIN-SHIME Shows Differences between Human Urolithin-Metabotypes in Gut Microbiota Composition, Pomegranate Polyphenol Metabolism, and Transport along the Intestinal Tract. *J. Agric. Food Chem.* **2017**, *65*, 5480–5493. [CrossRef]

International Journal of *Molecular Sciences*

Review

The Versatility in the Applications of Dithiocarbamates

Timothy O. Ajiboye [1,2], Titilope T. Ajiboye [3], Riadh Marzouki [4,5] and Damian C. Onwudiwe [1,2,*]

1 Material Science Innovation and Modelling (MaSIM) Research Focus Area, Faculty of Natural and Agricultural Sciences, Mafikeng Campus, North-West University, Private Bag X2046, Mmabatho 2735, South Africa; 32480342@student.g.nwu.ac.za
2 Department of Chemistry, Faculty of Natural and Agricultural Sciences, Mafikeng Campus, North-West University, Private Bag X2046, Mmabatho 2735, South Africa
3 Food Security and Safety Niche Area, Faculty of Natural and Agricultural Sciences, Mafikeng Campus, North-West University, Private Bag X2046, Mmabatho 2735, South Africa; 37781294@student.g.nwu.ac.za
4 Chemistry Department, College of Science, King Khalid University, Abha 61413, Saudi Arabia; riadh.marzouki@hotmail.fr
5 Chemistry Department, Faculty of Sciences of Sfax, University of Sfax, Sfax 3029, Tunisia
* Correspondence: Damian.Onwudiwe@nwu.ac.za; Tel.: +27-18-389-2545; Fax: +27-18-389-2420

Abstract: Dithiocarbamate ligands have the ability to form stable complexes with transition metals, and this chelating ability has been utilized in numerous applications. The complexes have also been used to synthesize other useful compounds. Here, the up-to-date applications of dithiocarbamate ligands and complexes are extensively discussed. Some of these are their use as enzyme inhibitor and treatment of HIV and other diseases. The application as anticancer, antimicrobial, medical imaging and anti-inflammatory agents is examined. Moreover, the application in the industry as vulcanization accelerator, froth flotation collector, antifouling, coatings, lubricant additives and sensors is discussed. The various ways in which they have been employed in synthesis of other compounds are highlighted. Finally, the agricultural uses and remediation of heavy metals via dithiocarbamate compounds are comprehensively discussed.

Keywords: dithiocarbamate; metal complexes; medical use; industrial applications; agricultural applications

Citation: Ajiboye, T.O.; Ajiboye, T.T.; Marzouki, R.; Onwudiwe, D.C. The Versatility in the Applications of Dithiocarbamates. *Int. J. Mol. Sci.* **2022**, *23*, 1317. https://doi.org/10.3390/ijms23031317

Academic Editor: Maurizio Battino

Received: 1 December 2021
Accepted: 18 January 2022
Published: 24 January 2022

1. Introduction

Dithiocarbamates are amides formed from dithiocarbamic acid and they have the ability to form stable metal complexes as a result of their exceptional coordination properties [1]. They could generally be classified as heterocyclic dithiocarbamates, symmetric dithiocarbamates, unsymmetric dithiocarbamate, dialkyldithiocarbamates and monoalkyldithiocarbamates [2]. Several methods have been used to synthesize dithiocarbamate compounds. However, the synthesis is commonly achieved by the reaction of carbon disulphide and amine (primary or secondary). The reaction is usually carried out in the presence of electrophiles such as imines, transition metals, epoxides and alkyl halides [3]. The synthesis could be effected without a catalyst or in the presence of an appropriate alkali as shown in Figure 1 through (equation a–n). Their ligands can form complexes with octahedral, square planar or tetrahedral geometry depending on the type of metal ion and also the ratio of the metal-to-ligand [2]. Dimers of dithiocarbamates are also formed by using dilauroyl peroxide as the oxidizing agent [4] (equation o). Other polyfunctional ligands of dithiocarbamate exist but they are rare compared to other forms of dithiocarbamate compounds [5]. Both the dithiocarbamate ligands and complexes are useful in several applications. However, when both ligands and complexes found relevance in similar applications, the complexes appear to be more potent than the ligands. For instance, dithiocarbamate complexes are more active against microbes than the ligands from which the complexes are formed [6]. The choice of dithiocarbamates compared to other related compounds is attributed to its poor solubility in water, ease of preparation under laboratory conditions, and formation of more stable compounds than several complexes made from other common analytical ligands [7].

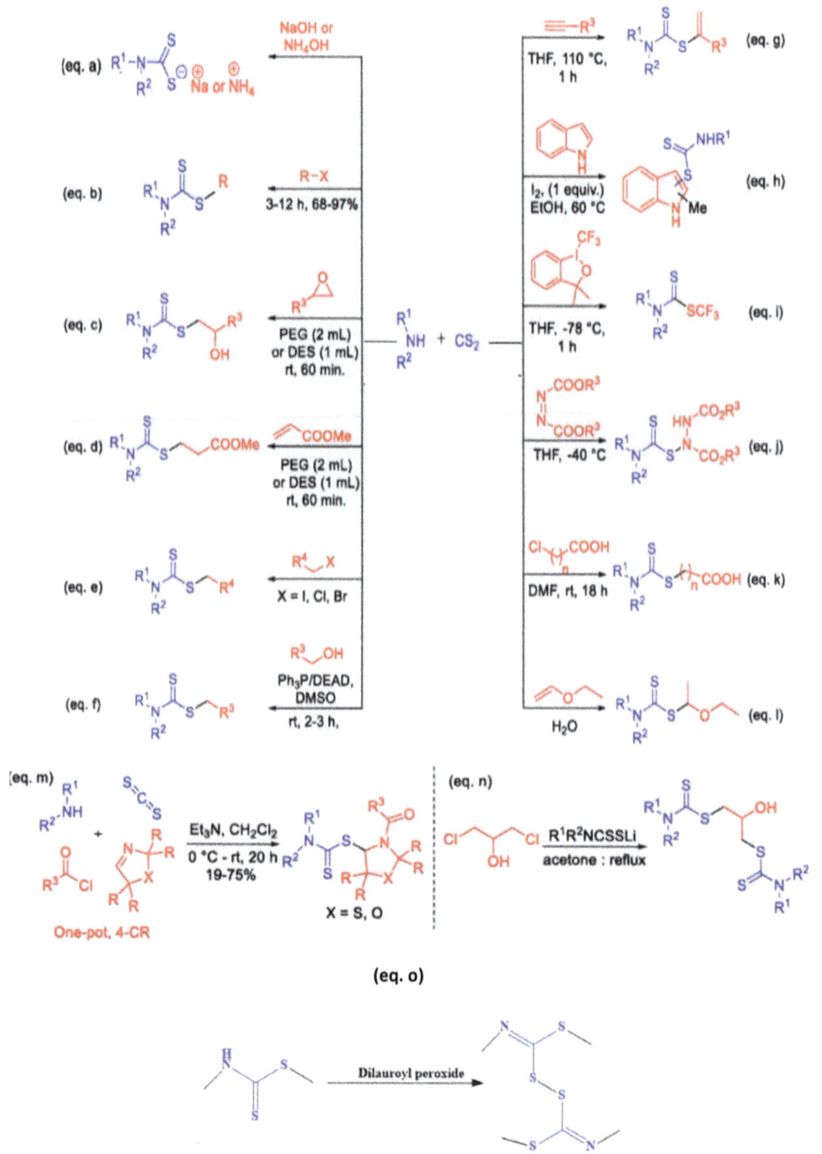

Figure 1. Various routes for the synthesis of dithiocarbamates. Adapted from [3]. Copyright (2020), with permission from Elsevier.

The study and discovery of different novel dithiocarbamate derivatives continues to increase as the different areas of their application are being investigated. As shown in the statistical data obtained from Scopus database (insert date), there are 2406 publications on dithiocarbamate from 2015, out of which 2264 (94.1%) are research articles (Figure 2). The publications on dithiocarbamate reached a peak in 2016, and within this range, chemistry researchers are at the forefront of the dithiocarbamate studies. Szolar reviewed the different ways of identifying and analyzing dithiocarbamates [8], while other reports focused only on some fragments of the applications. However, the need to review the comprehensive applications of the dithiocarbamate compounds is rife [9]. Consequently, this review gives an up-to-date and detailed account of various areas of applications of dithiocarbamate compounds including agriculture, medicine, industries, catalysis and in synthesis. These different areas of applications will be discussed in more detail in the following sections.

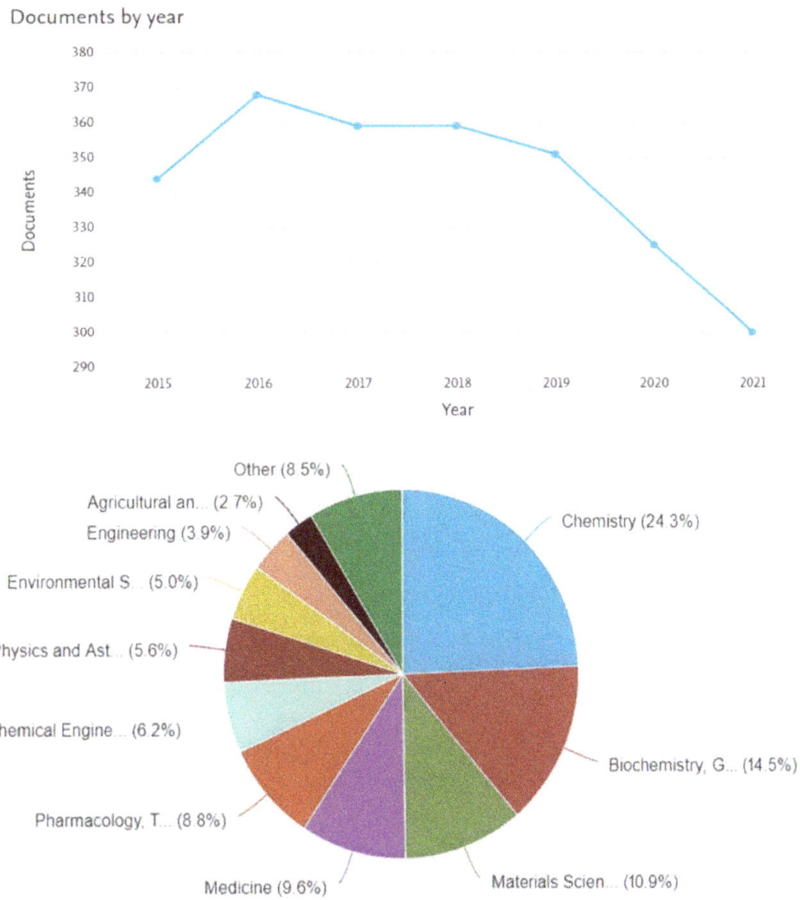

Figure 2. Statistics of publications on dithiocarbamate from 2015 to 2021 from Scopus database, accessed on 21 November 2021.

2. Heavy Metals Concentration and Remediation

Polluted samples usually consist of a mixture of organic (such as parabens, organochlorine pesticides and dyes) [10,11] and inorganic pollutants (such as heavy metals and nitrates) [12,13]. In some cases, there may be a need to remove one pollutant in the presence of other pollutants in the environmental samples. Several strategies have been used for concentrating heavy metals before their removal from the environmental samples. Both concentrating and removal of heavy metals from different media have been achieved through the use of dithiocarbamate compounds. Some of these dithiocarbamates as well as the heavy metals that were concentrated and removed are discussed in this section.

2.1. Heavy Metals' Removal from the Environment through Dithiocarbamate Compounds

The ability of dithiocarbamate to selectively and strongly bind to most metal ions to form organometallic complex makes them a useful candidate for removing heavy metals from the environment [14,15]. The presence of two sulphurs with lone pairs of electrons makes it possible for dithiocarbamates to form chelate with these metals as well. However, it is possible for dithiocarbamate to use one of the sulphur donor atoms to form a bond with the metals. In short, it can act as bidentate or monodentate ligand [14]. Another factor that makes them particularly useful for metals with a variable oxidation state is their ability to stabilize these metals irrespective of their oxidation states and this can be explained by the oxygen bonding ability of the conjugates formed by dithiocarbamates [16,17]. The possibility of sharing electrons between the metal ions, sulphur atoms and nitrogen atoms coupled with the formation of metal complexes that cannot dissolve in water also makes them a better heavy metal chelator from the environmental samples [18].

As a result of these features, the use of dithiocarbamates to solve the problem of heavy metal pollution has been investigated and it was found to be a good metal chelator. In the studies conducted by Ayalew et al. [19], different amine-modified dithiocarbamates were used to successfully remove zinc, nickel and copper from wastewater at a low pH. The modified dithiocarbamate used for the investigation were tetraethylenepentamine-dithiocarbamate, triethylenetetramine-dithiocarbamate, diethylenetriamine-dithiocarbamate and ethylenediamine-dithiocarbamate. However, the three heavy metals were removed, but the amount of copper removed was more than the amount of zinc and nickel removed. Dithiocarbamates were also used to modify carbon compounds and then to remove heavy metals from wastewater. Trivalent arsenic has been removed in this way and dithiocarbamate was used to modify the cellulose that was used for trivalent arsenic removal [20]. Hydrochar is another carbon-based compound that was modified with dithiocarbamate and this also was found to be effective in removing divalent lead from the environment [21]. More than 90% of the heavy metals (lead(II), copper(II), and cadmium(II)) were removed from wastewater within 40 min when dithiocarbamate was grafted into crosslinked polymer made from glutaraldehyde and polyethyleneimine [22]. Other investigations involving the removal of heavy metals via dithiocarbamates are shown in Table 1.

Table 1. Heavy metals remediation via dithiocarbamate.

Dithiocarbamate Compound Used	Heavy Metals Removed	Media/Samples Remediated	Amount Removed/Performance	Ref.
Iron-containing reduced graphene oxide modified with dithiocarbamate	Hg(II), Pb(II), Cd(II) and Cu(II)	wastewater	181.82, 147.06, 116.28 and 113.64 mg/g respectively	[23]
Dithiocarbamate-modified coal	Ni(II)	Aqueous solution	82.37 mg/g	[24]
Al(OH) -poly(acrylamide-dimethyldiallylammonium chloride)-graft-dithiocarbamate	Pb(II) and Cu(II)	Wastewater	17.777 mg/g for Cu and 586.699 mg/g for Pb	[25]
poly-sodium dithiocarbamate and poly-ammonium dithiocarbamate	Zn(II), Ni(II) and Cu(II)	Electroplating wastewater	226.76, 234.47 and 245.53 mg/g, for Zn, Ni and Cu respectively at pH 6 in 20 min	[26]
Heavy metal-dithiocarbamates (using sodium diethyldithiocarbamate)	Zn(II), Pb(II), Ni(II), Mn(II), Fe(II), Cu(II) and Cd(II) ions	Water sample	More than 90% removal	[14]
Sodium polyamidoamine-multi dithiocarbamate (using sodium diethyldithiocarbamate)	Divalent Zn, Cu, Cd and Pb	Soil sediments	Complete precipitation	[27]
sodium tetraethylenepentamine-multi dithiocarbamate	Divalent Cu, Cd and Pb	Soil samples	Near complete precipitation	[28]

The efficiency of heavy metals removal depends on the type of dithiocarbamate used for metal chelating. For instance, the metal chelating ability of diphenyldithiocarbamate ligands was found to be better than the chelating ability of diethyldithiocarbamate that did not contain a phenyl group [18]. Apart from the use of dithiocarbamates in the removal of heavy metals, they have also been used to determine and concentrate heavy metals instead of using surfactants [29–31].

2.2. Trace Elements Concentration and Determination through Dithiocarbamate Compounds

The determination of trace metals usually involves separation and pre-concentration stages. Dithiocarbamate compounds have been used for these purposes and this could be attributed to their selective and chelating properties. Activated carbon coated with phenylpiperazine dithiocarbamate was successfully used to concentrate Pb, Cd, Cu and Mn before they were determined by the flame atomic absorption spectrophotometry (FAAS) method [32]. Ammonium pyrrolidine dithiocarbamate, on glass fibre base, was also used to form a chelate complex with metal ions, which was followed by methyl isobutyl ketone elution and atomization of the metal ions. The quantification of the atomized sample was then carried out through high performance liquid chromatography (HPLC) [33]. When the multi-element determination of heavy metal ions was carried out through HPLC, dithiocarbamate was included in the column to improve the performance of the method [34]. Dithiocarbamate-modified silica gel was also employed for pre-concentration and separation of ions of several precious metals prior to their determination via inductively coupled plasma [35]. Table 2 shows other specific examples of investigations where dithiocarbamates were used to quantify metals.

Table 2. Determination of trace elements using dithiocarbamates.

Dithiocarbamate Compound Used	Metal(s) Determined	Method Used for the Determination	Limit of Detection	Ref.
pyrrolidine dithiocarbamate	Ni(II), Cr(VI), Co(II), and Hg(II)	liquid liquid micro-extraction	0.011–2.0 μg L^{-1}	[36]
Ammonium 1-pyrrolidine dithiocarbamate and Diethylammonium diethyldithiocarbamate	Pb(II), Cu(II) and Cd(II)	Inductively coupled plasma-mass spectroscopy (ICP-MS)	0.13–1.18 pmol L^{-1}	[37]
Ammonium pyrrolidine dithiocarbamate	As(III)	solid phase extraction(SPE)	0.01 μg L^{-1}	[38]
Sodium diethyl dithiocarbamate	Cd(II) and Pb(II)	SPE/ FAAS	0.30 μg L^{-1}	[39]
Dithiocarbamate-functionalized magnetite composite	Hg(II)	Atomic absorption spectrometry with gold amalgamation	1.8 ng L^{-1}	[40]
Pyrrolidine dithicarbamate	Pb(II), Bi(III), Pb(II), Hg(II), Au(III), Se(IV), As(III),Ni(II) and Co(II)	Thin-film microextraction	0.2–0.6 μg/L	[41]

3. Application of Dithiocarbamate Compounds as Stationary Phase in Chromatography

Dithiocarbamate compounds were also used as a component of the stationary phase during ligand exchange chromatography. They were useful for this application due to their strong chelating ability. Yeh and co-workers [42] utilized dithiocarbamate coated on silica as the stationary phase in the separation of heavy metals. It was observed that the amount of mercury taken up by this stationary phase was high, which could be attributed to the presence of extra complexing-nitrogen atoms from dithiocarbamate present in the stationary phase. In the chromatographic determination of multiple heavy metals, diethyldithiocarbamate and pyrrolidinedithiocarbamate were deposited on the Sep-Pak cartridge, which was used as the stationary phase. The method was able to determine these heavy metals even at μg l−1 level [43].

4. Application of Dithiocarbamate Compounds as Catalysts

Catalytic application of dithiocarbamate is another aspect that has attracted lots of research attention. It has been used for the synthesis of catalyst during organic synthesis as well as catalysts in polymerization. Some of these applications are explained in this section.

4.1. Application of Dithiocarbamate Compounds as Catalyst in Organic Transformation

Core/shell nanostructures have been functionalized with magnetic dithiocarbamate deposited on gold and utilized as the catalysts for synthesizing propargyl amines through A^3 coupling reaction [44]. The catalyst displayed good performance for the synthesis of propargyl amines when phenylacetylene, benzaldehyde and morpholine were used as the starting material. Further probe into the mechanism of the reaction showed that the reaction proceeded through a process involving the formation of iminium ion intermediate and C-H activation as shown in Figure 3. The choice of metal dithiocarbamate was as a result of its good solubility in organic solvents, chemical stability and the fact that it can be easily used in the anhydrous form [45,46].

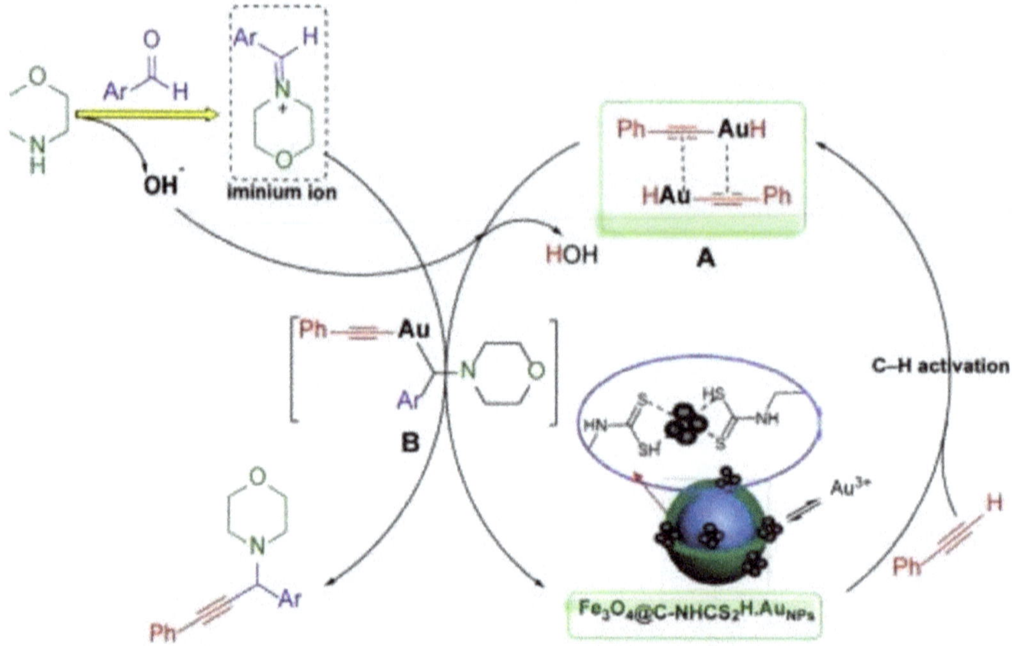

Figure 3. Mechanism for the dithiocarbamate-containing Au-catalyzed A^3 coupling. Reprinted from [44]. Copyright (2021), with permission from Elsevier.

The need to obtain carbon fibres with improved surface energy, roughness and chemical inertness led to the use of dithiocarbamate in its synthesis. Two of the methods that have been used with the incorporation of dithiocarbamates are Markovnikov addition and alkaline synthesis method. Guan et al. [47] utilized nickel dithiocarbamate compound as catalyst for enhancing the properties of carbon fibres by using both Markovnikov addition and alkaline synthesis. These carbon fibre are used as photopolymerization catalysts.

4.2. Application of Dithiocarbamate Compounds as RAFT Agent in Polymerization

Simultaneous control of stereoregularity and molecular weight of polymers is beneficial in polymer synthesis but it is difficult to achieve [48]. The use of RAFT (reversible addition–fragmentation chain transfer) agents has made simultaneous control feasible and different dithiocarbamate compounds have been investigated as RAFT agent [49]. Nitrogen-containing dithiocarbamates are now being used as the most effective RAFT agent with reduced bulky attachment when compared to other RAFT agents [50]. The presence of nitrogen in the dithiocarbamate compound stabilizes the cationic intermediate due to the fact that nitrogen is an electron-donating atom [48,50]. Dithiocarbamate was also used as both emulsifier and RAFT agent in the polymerization of stable latex of vinyl acetate polymer [51]. They are often used along with other RAFT agents for better control of tacticity and molecular weight. For instance, thiocarbonylthiol compound was included in the RAFT agent used for polymerization of vinyl ethers in the presence of Lewis acid catalysts [48].

5. Application of Dithiocarbamate in Synthesis

Dithiocarbamate compounds have been useful in the synthesis of organic intermediate as well as chalcogenides of metals. This section presents some of these synthesized compounds.

5.1. Application of Dithiocarbamate Compounds as Precursors in Material Synthesis

Different synthetic methods have been used to produce metal sulphide nanoparticles and one of these methods involves the use of metal complexes as single source precursors (SSP). Among the metal complexes used as SSP, dithiocarbamate complexes have being the most explored complexes. In our laboratory, we have synthesized some dithiocarbamate complexes, which were thermolyzed to generate metal sulphides [52]. Some of these nanoparticles (especially the bismuth based) have been reviewed by Ajiboye et al. [53]. The use of dithiocarbamate complexes for the synthesis of these nanoparticles is preferred since dithiocarbamate is rich in sulphur; hence, the use of a separate sulphur source will not be required [54]. Generally, the synthesis from the dithiocarbamate complex using the solvothermal method requires the use of capping agents such as oleylamine, octadecene, dodecane thiol, ethylene glycol and hexadecylamine. Their presence in the system controls the growth of the nanoparticles [55], while some of these capping agents (such as oleylamine) can also function as reducing agent, solvent or surfactant in the material synthesis [56]. Table 3 highlights other examples of nanoparticles made from dithiocarbamates.

Table 3. Application of dithiocarbamates in nanoparticle synthesis.

Dithiocarbamate Precursor Used	Nanoparticle(s) Obtained	Temp. Used	Particle Size and (Band Gap)	Ref.
Bis(N-ethylphenyldithiocarbamato) palladium(II)	Palladium sulphide	160, 200 and 240 °C resp.	2.01–2.50 nm, 4.00–4.86 nm and 2.53–4.12 nm (4.90–5.02 eV)	[57]
Bis(N,N-di(4-fluorobenzyl)dithiocarbamato-S,S′)M(II). (M = Cd)	Cadmium sulphide (CdS)	-	–(3.29 eV)	[58]
Cu (II) bis N-methyl-N-phenyl Dithiocarbamate	Copper sulphide (CuS and Cu_5S_9)	≥240 °C	34.7 ± 13.3 nm width size (1.85 eV)	[52]
Dithiocarbamate complexes with varied Ag/In/Ga/Zn ratios	Quinary Ag-In-Ga-Zn-S quantum dots	220 °C	2.0 ± 0.4 nm	[59]
Molybdenum dithiocarbamates	Molybdenium sulphide (MoS_2)	-	40 nm	[60]
N-alkyldithiocarbamate copper(II) complexes with NaBH4	Copper sulphide (Cu_9S_5 and Cu_2S)	180 °C	–(3.0 eV)	[61]
copper(ii) bis-(2,2′-(dithiocarboxyazanediyl)diacetic acid)	Copper sulphide (CuS)	90 °C	8 ± 1 nm	[62]
bis(diethyldithiocarbamato)disulfidothioxo tungsten(VI)	chromium-doped tungsten disulphide (WS_2)	450 °C	-	[63]
tetrakis(N,N-diethyldithiocarbamato)molybdenum(IV)	Molybdenum sulphide (MoS_2)	450 °C	flake thickness of ∼10 nm	[64]
$[V_2S_4(nBu_2dtc)_4]$(dtc=dithiocarbamate)	Vanadium sulphide (VS_2)	150 °C		[65]
Manganese diethyldithiocarbamate trihydrate	Manganese sulphide (MnS)	290 °C	(3.3 eV)	[66]
Tris-(piperidinedithiocarbamato) iron(III) and tris-(tetrahydroquinolinedithiocarbamato)iron(III)	Iron sulphide ($Fe_{0.975}S$ and Fe_3S_4 phases)	350–450 °C	(0.95–2.0 eV)	[67]
lead(II) complexes of morpholine dithiocarbamate	Lead sulphide (PbS)	160 °C	(13.86–36.06 nm)	[68]

5.2. Application of Dithiocarbamate Compounds in the Synthesis of Organic Intermediates

The light-catalyzed reaction of dithiocarbamates in cyclohexane or chlorobenzene solvent leads to the formation of dithiocarbamate-containing lactam. The fact that the product contains dithiocarbamate makes it suitable for other dithiocarbamate-based applications [69]. Examples of lactam produced from dithiocarbamate are shown in Figure 4. Diethyldithiocarbamate has been used for the synthesis of ferrugine through a reaction that involves refluxing in the presence of cyclohexane and light [70].

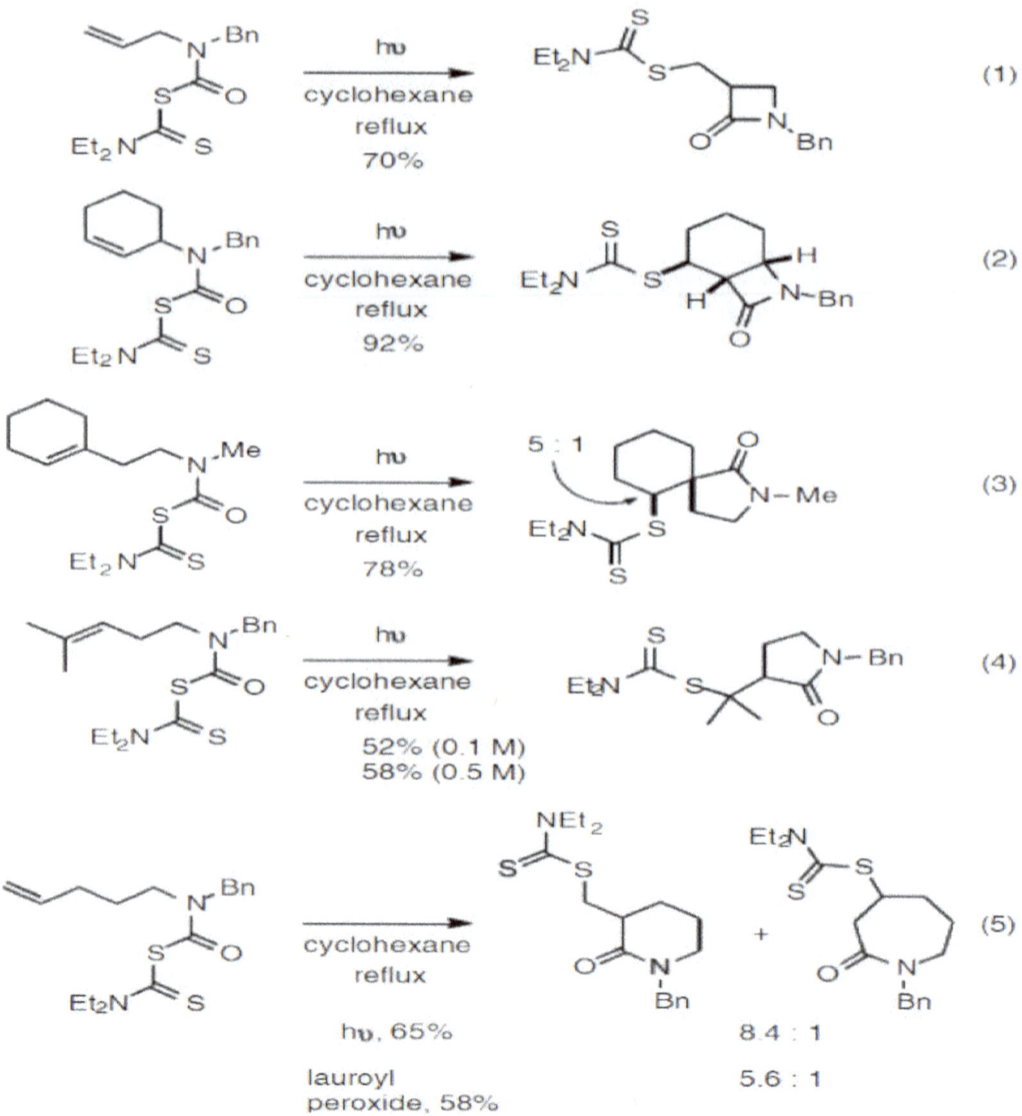

Figure 4. Synthesis of lactams (four-eight membered ring). Reprinted from [69]. Copyright (2007), with permission from Wiley and Sons.

The synthesis of cyanamide, which is an important intermediate for synthesizing pharmaceutical compounds, has been a serious challenge to researchers because its synthesis involves the use of highly toxic cyanogen halide. The synthesis is now carried out in a 'greener' way by using dithiocarbamate for its synthesis. Other reactants used for the synthesis are sodium bicarbonate, molecular iodine and hydrogen peroxide. The hydrogen peroxide functions as the oxidizing and desulphurizing agent. Other intermediates such as 1-phenylthiourea and isothiocyanates were formed during the synthesis [71]. The mechanism of the whole process is shown in Figure 5a. Another intermediate that was synthesized by using dithiocarbamate is thiourea, and synthesis via this procedure was preferred because toxic reagents such as hydrogen sulphide and thiophosgene were not needed [72]. Moreover, harsh reaction conditions such as the use of strong base or acid, elongated time of reaction and high temperature of the reaction are not required, unlike the other known synthetic routes [73]. In short, the synthesis is carried out by reacting dithiocarbamate with either ammonia, primary aliphatic or aromatic amine and a secondary aliphatic amine at 60 degrees Celsius. It could be carried out without using solvent or catalyst [73]. As shown in Figure 5b, thiazolidine-2-thiones synthesis has also been achieved from dithiocarbamate through a three-step method involving iodocyclization, dehydrohalogenation and nucleophilic substitution reactions [74]. The synthesis of novel amide was also feasible when dithiocarbamate compound was used as the starting material [75] as shown in Figure 5c.

Aryanasab and co-workers [76] reacted acid hydrazides with S-alkyl dithiocarbamates for synthesizing 1,3,4-thiadiazoles. The procedure was applauded because its cyclization step does not involve toxic catalysts or dangerous organic solvents. Apart from this specific reaction, it has general applicability. For instance, the reaction was used to prepare 2-amino-1,3,4-thiadiazoles by reacting acid hydrazides with dithiocarbamate.

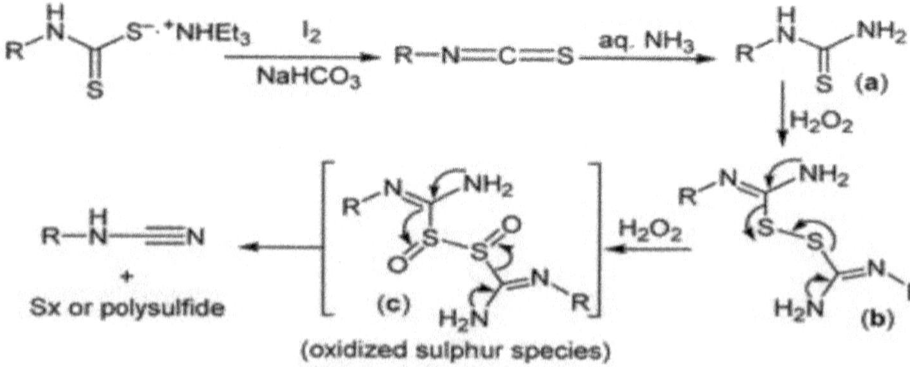

(a)

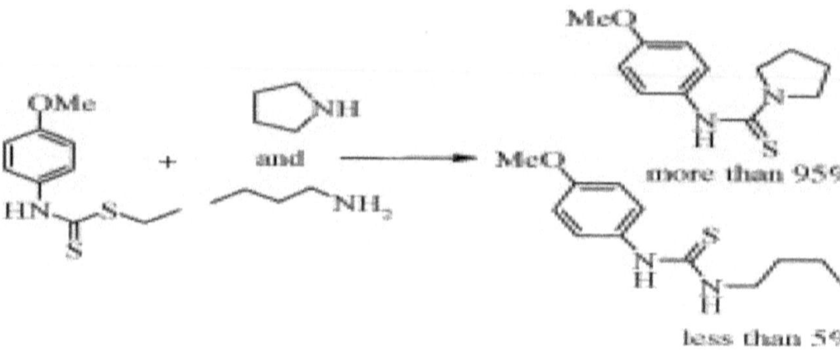

(b)

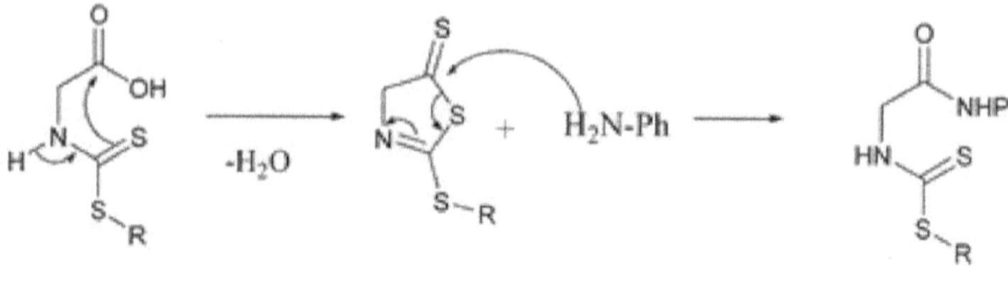

(c)

Figure 5. (**a**) Synthesis of cyanamide from dithiocarbamate. Reproduced from [71]. Copyright (2012), with permission from Taylor and Francis. (**b**) Synthesis of thiourea from dithiocarbamate and amines. Reproduced from [73]. Copyright (2009), with permission from Elsevier. (**c**) Synthesis of amide from dithiocarbamate. Reproduced from [75]. Copyright (2011), with permission from Royal Society of Chemistry.

6. Application of Dithiocarbamate Compounds in Agriculture

One of the uses of dithiocarbamate is in the eradication of diseases of plants and livestock. They have been used as pesticides to either prevent or eliminate plants' diseases. The growth of unwanted plants has also been prevented or eliminated through the use of dithiocarbamate compounds. Some of the dithiocarbamate compounds that have been used for these applications are discussed in this section.

6.1. Application of Dithiocarbamate Compounds as Herbicides

Dithiocarbamate-based herbicides contain groups such as dimethyldithiocarbamate, ethylenebis(dithiocarbamate) and propylenebis(dithiocarbamates). Examples of dithiocarbamate-containing herbicides are Metiram, Dazomet, Thiram, Disulfiram, Propineb, Maneb, Ziram and Zineb [77], although some of them are also used as pesticides. These herbicides are majorly used to prevent the growth of some broadleaf weeds as well as plants such as crabgrass, cheatgrass, bromegrass and foxtail [78]. Even plant that generates oxidants (active oxygen species) was successfully eliminated through dithiocarbamate herbicides [79]. Adjustment of the lipophilic and hydrophilic properties of dithiocarbamate by introducing groups such as sodium salts of dibutyldithiocarbamic acids, hexyl (2-(2- ethoxyethoxy) ethyl) dithiocarbamic acid, butyl (2-(2-ethoxyethoxy) ethyl) dithiocarbamic acid and ethyl (2-(2-ethoxyethoxy) ethyl)-dithiocarbamic acid was found to aid the action of dithiocarbamate as the pesticide. This is because of better penetration of plant cuticles compared to when ordinary sodium diethyldithiocarbamate was used as the herbicide [79]. Diallate, Sulfallate, Dazomet and Triallate are other common dithiocarbamate-based herbicides (Figure 6). Diallate [S-(2,3 dichloroallyl-)diisopropylthiocarbamate] is used to control monocotyledon weeds and it acts by attacking their fatty acids [80].

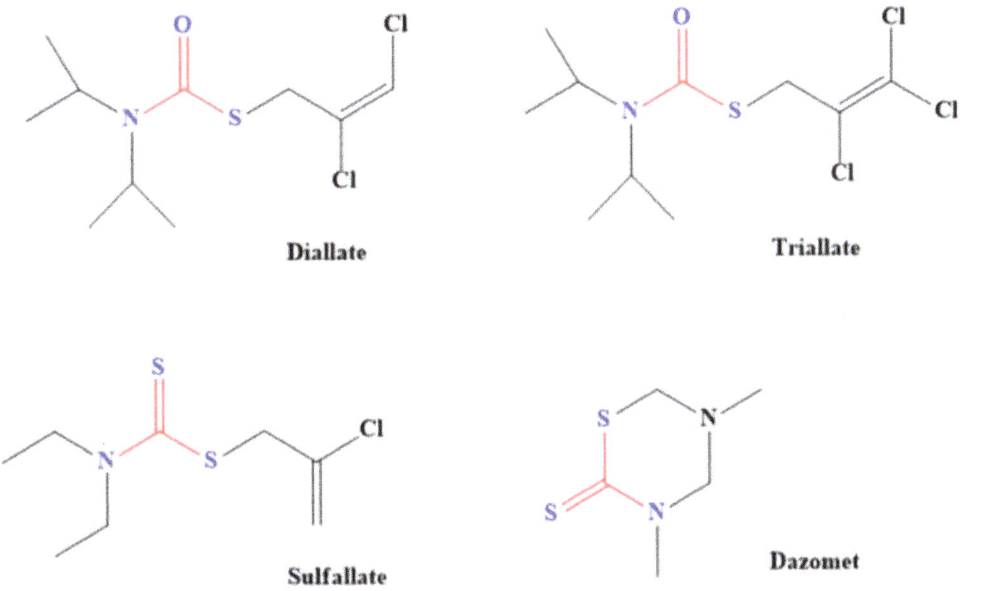

Figure 6. Examples of common dithiocarbamate-based herbicides. (One of the sulphur in dithiocarbamate has been replaced in diallate and triallate).

6.2. Application of Dithiocarbamate Compounds as Pesticides

Pesticides made from dithiocarbamates are used as fungicides for various crops during processes such as shipment, storage and growth [81]. The structures of some of these dithiocarbamate-based pesticides are shown in Figure 7. These pesticides also kill the larva of some pests that cause plants' and farm animals' diseases, thereby boosting food security. For instance, both tricyclohexyltin and triphenyltin N-n-butyldithiocarbamate have been used as larvicide against the larva of *Aedes aegypti* and *Anopheles stephensi* mosquitoes [82]. These dithiocarbamates were found to be effective against the larva of these mosquito species. Moreover, Meloidogyne incognita, which is a disease caused by nematode, was eradicated by using dithiocarbamate derived from chitin oligosaccharide [83]. The derived dithiocarbamate pesticide has high activities for eliminating the nematode. In addition, it inhibits the hatching of eggs, thereby decreasing the population of the nematodes [83]. Specific examples of how these pesticides are being used are shown in Table 4.

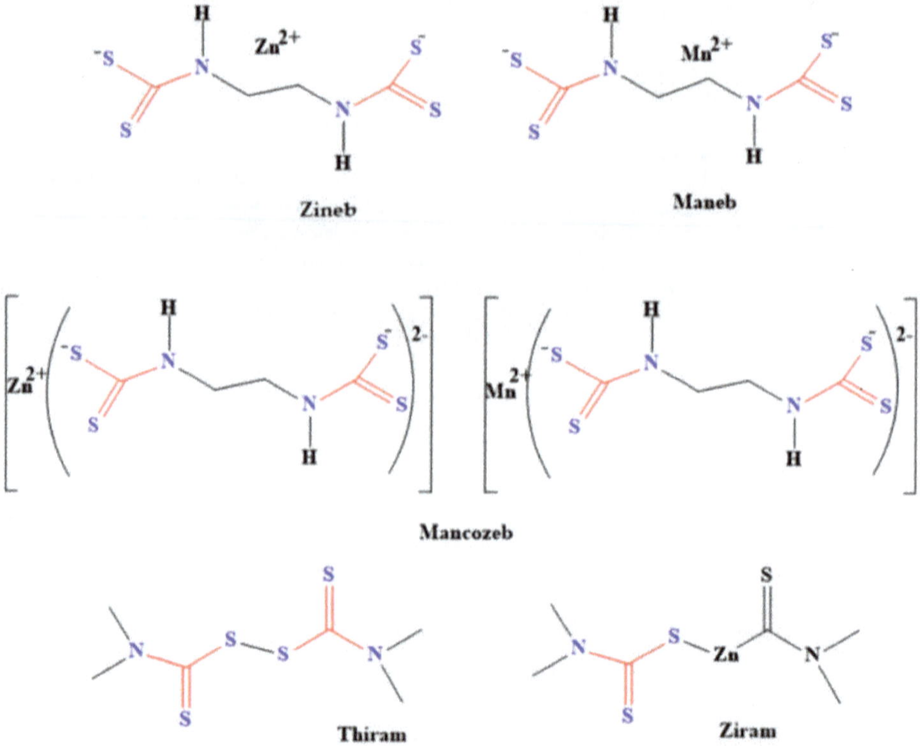

Figure 7. Examples of dithiocarbamate pesticides.

Table 4. Scientific name of common dithiocarbamate pesticides and the organisms affected.

Dithiocarbamate Pesticides (Common Names)	Dithiocarbamate Pesticides (Scientific Names)	Classification	Organism(s) Affected	Ref.
Ferbam	Ferric dimethyldithiocarbamate	Fungicide	Drugs against gastrointestinal flukes, tapeworms, lungworms and roundworms in farm animals	[84]
mancozeb	Zinc;manganese(2+); N-[2-(sulfidocarbothioylamino)ethyl]carbamodithioate	Fungicide	Acts against over 400 micro-organisms that damage agricultural produce such as citrus, grapevine, tomato and potato	[85]
Carbaryl	1-naphthyl methyl carbamate	Insecticide	Acts against 100 species of destructive insects affecting pets, livestock, poultry, shade trees, ornamentals, nuts, lawns, forests, fruit and citrus	[86]
Maneb	Manganese-containing ethylene bis-dithiocarbamate	fungicide	To control the diseases of plants	[87,88]
metam-sodium	Methylisothiocyanate	Fungicide, nematocides and (herbicides)	To fumigate soil prior to planting so as to prevent soilborne diseases	[89]
Metiram	Zinc ammoniate ethlenebis(dithiocarbamate)-poly (ethylene disulphide)	Fungicide	Prevent plants(ornamentals, field, nuts, vegetables and fruits) by inhibiting the spores of the pathogens from germinating	[90,91]
Nabam	Ethylenebis[dithiocarbamic acid] disodium salt	Algaecide, bacteriacide and Fungicide	To prevent fungal diseases in tomato, apple and cotton and to eliminate algae from plant field	[92]
Thiram	Tetramethyl thiuram disulphide	Fungicide	It affects the mucous membrane and skin of microbes	[93]
Propineb	Polymeric zinc 1, 2-propylene bis(dithiocarbamate)	Fungicide	To treat fungal infections such as leaf blotch in apple and other crops.	[94]
Zineb	Zincethylenebis(dithiocarbamate)	Fungicide	To control the diseases of plants	[88]
Ziram	Zinc-dimethyl dithiocarbamate	Fungicide	To repel birds from flowers	[95]
Methiocarb	N-methylcarbamate	Insecticide	To repel birds from plants	[95]

7. Medical Applications of Dithiocarbamate Compounds

The use of dithiocarbamate compounds in medicine has been investigated for more than 40 years [96]. One such application is their use as anti-angiogenic agent and they are usually evaluated for this application by studying their potential to heal wounds. For example, thalidomide dithiocarbamate was evaluated for wound healing to confirm its usage as the anti-angiogenic agent [97]. Dithiocarbamate ligands and complexes have also been studied for magnetic resonance imaging and other radiopharmaceutical imaging [96]. Gold nanoparticles functionalized with biomimetic amino acid dithiocarbamate were used as nanoprobe for cell imaging as a result of their negligible toxicity to human cells. This dithiocarbamate compound showed an enhancement factor of 9.8×10^5 when used for surface-enhanced Raman scattering imaging [98]. Generally, the medical applications could be ascribed to their ability to form metal chelate and the high reactivity of dithiocarbamate anions to other moieties (such as thiol) [98,99]. Other medical applications of dithiocarbamate, which are discussed in this review, are summarized in Figure 8.

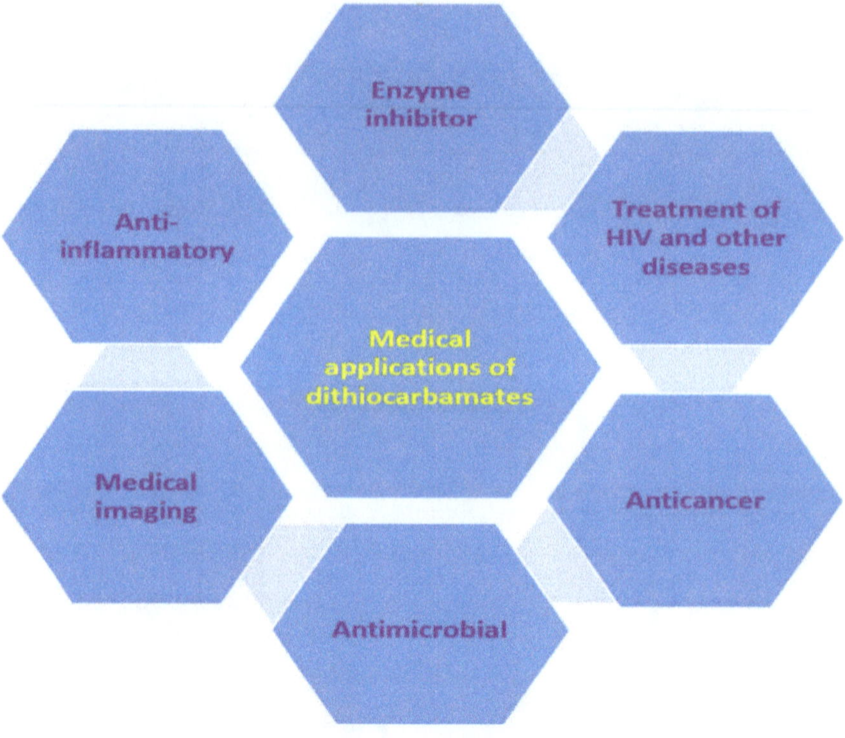

Figure 8. Medical applications of dithiocarbamate compounds.

7.1. Application of Dithiocarbamate Compounds as Enzyme Inhibitor

A hydrolyzing enzyme (α-Glucosidase), which is important in the breaking down of starch and carbohydrate to glucose, is usually a target enzyme in the treatment of diabetes mellitus [100,101]. Among the compounds that has been used for the inhibition of this enzyme, coumarin-dithiocarbamate scaffold has proven to be very effective and this has made it a useful compound in the treatment of type 2 diabetes. Coumarin-dithiocarbamate is a competitive inhibitor of α-glucosidase since it binds to its active site as evidenced by results obtained from molecular docking [101]. Specifically, there is formation of a hydrogen bond between the amino acid (His279) and coumarin moiety [100]. Pyrrolidine dithiocarbamate has also been used as an effective inhibitor of enzymes, specifically the nuclear factor kappa B [102]. A metalloenzyme, carbonic anhydrase, which is involved in the reversible reaction of forming bicarbonate from carbon dioxide in the body has also been inhibited by the derivative of dithiocarbamate-sulfonamide [103]. Its inhibition is usually required when it starts to display abnormal activities in the body of animals, which may lead to physiological disorder such as altitude sickness, epilepsy, glaucoma, cerebral and retina oedema [104]. The treatment of 'superbug' infection has been made possible through the inhibition of metallo-β-lactamases which are responsible for the infection. Dithiocarbamates play significant roles in the inhibition of this enzyme because the carbonyls and hydroxyl group in some dithiocarbamate compounds effectively bind to the zinc in the active site of this enzyme, leading to their inhibition [105]. Dithiocarbamate coupled with phthalimide is a competitive inhibitor of butyrylcholinesterase and acetylcholinesterase. This inhibitive property makes it suitable for the treatment of Alzheimer's Disease [106].

7.2. Application of Dithiocarbamate Compounds in HIV Treatment

Elimination of HIV is very challenging with the current retroviral treatment due to numerous latently infected CD4T cells. This is because the available treatment requires placing the patients on drugs for a long period of time and some of these drugs are associated with known side effects. However, the treatment is active in prolonging the survival of the patient, thereby reducing the mortality associated with HIV infections and minimizing the transmission of the disease [107]. The quest for the improvement of these existing medications has resulted in a continuous search for novel HIV inhibitors. Dithiocarbamate has been investigated as a possible HIV inhibitor. For instance, zinc-dithiocarbamate-*S*,*S*′-dioxidcyclic zinc-dithiocarbamate-*S*,*S*′-dioxide was used to effectively inhibit HIV. Specifically, HIV-1 was inhibited by mediating a cell-to-cell fusion between anti-CXCR4 and CXCR4 that is present on the cell's surface [108]. Dithiocarbamate compounds have also been used to delay the progression of HIV into AIDS. Diethyldithiocarbamate has been reported for this purpose and it was found to be active in delaying the proliferation of the virus [109]. Pyrrolidine dithiocarbamate has also been found to be useful in the treatment of HIV because it inhibits the nuclear factor-κB [110]. It acts as an antioxidant to block the activation of HIV-1 and nuclear factor KB (NF-KB) since oxygen radicals play significant roles in the activation of HIV-1 and NF-KB [111]. Other studies have also established the link between nuclear factor-κB, immune systems and HIV [2,112]. Lang et al. [113] observed that the HIV symptoms were relieved, the immune function was enhanced overall and the progression of HIV was retarded when sodium dithiocarbamate was used as an oral drug for HIV patients. One of the reasons why sodium dithiocarbamate was found to be useful for this purpose was because of its relatively low toxicity when compared to other dithiocarbamates and this was further shown by its lethal dose (LD_{50}), which was measured to be 1.5 g/Kg of body weight from the investigation conducted with rats and mice [114]. Furthermore, sodium dithiocarbamate drugs did not in any way initiate any major biological or clinical side effects [115].

7.3. Application of Dithiocarbamate Compounds in the Treatment of Other Diseases

Dithiocarbamates are also useful antiglaucoma agents, even better than sulfonamide dorzolamide which is a clinically-recognized drug for treating glaucoma [116]. That dithiocarbamate compounds are easy to prepare, coupled with their ability to lower the intraocular pressure, made them the preferred compounds compared to the sulfonamides [116]. The fact that dithiocarbamate compounds could inhibit carbonic anhydrase make them more suitable for treating glaucoma [117]. They can also inhibit carbonic anhydrase, which leads to the treatment of several diseases/disease conditions. Some of these diseases are edema, epilepsy, obesity, hypoxic tumor, inflammatory diseases, neuropathic pain, Alzheimer diseases and cerebral ischemia [117]. Pyrrolidine dithiocarbamate was reported for the repair of damaged lungs (lung edema) instead of lung transplant. Pyrrolidine dithiocarbamate acted by inhibiting NF-Kb, thereby suppressing the activation of immunity during lung reconditioning via ex vivo lung perfusion [118]. In addition to lung treatment, this dithiocarbamate compound was used for the treatment of epilepsy by protecting the piriform cortex of the cerebrium without causing loss of hilar neuronal [119]. There was an increase in the generation of reactive oxygen species in the renal cortical and a decrease in the lipoprotein level of the tested animals that were fed with water containing pyrrolidine dithiocarbamate [120]. In fact, some dithiocarbamate compounds are under clinical trial for the treatment of corona virus (SARS-CoV-2) [9]. Examples of the common dithiocarbamates used for the treatment of various diseases are shown in Figure 9.

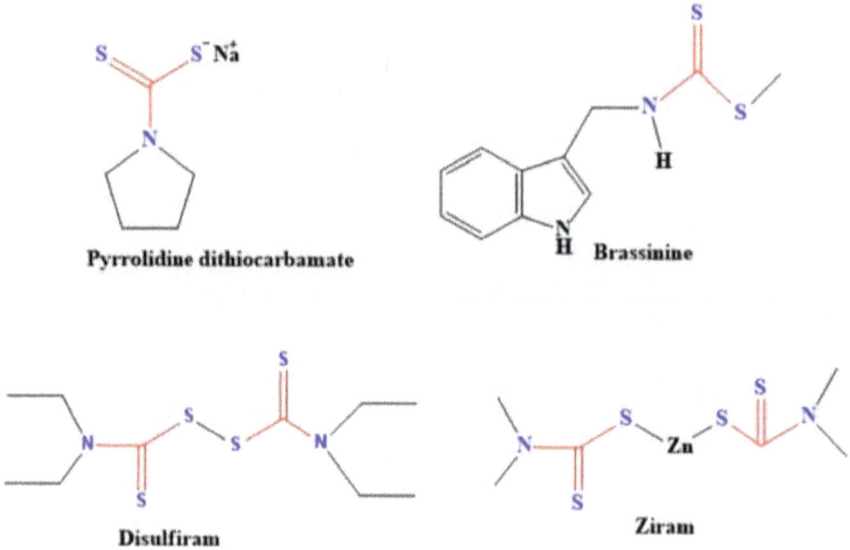

Figure 9. Representative of dithiocarbamate compounds used for the treatment of diseases.

Dithiocarbamate drugs (Propineb, Zineb and Maneb) were also found to be useful in the treatment of leishmaniasis, a protozoan disease [121]. The disease is common in the subtropical and tropical countries and has claimed several lives [122]. Before the discovery of dithiocarbamate drugs, miltefosine, paromomycin and amphotericin, which are expensive, were used but it was discovered that the disease had developed resistance against these drugs and some side effects were also reported [123]. The dithiocarbamate-based drugs were found to be particularly useful because they have no significant effect on the mammalian cells as they lead to the death of *Leishmania* cell with a lethal dose of 50% [121]. Bromine-containing ethylsarcosinedithiocarbamate of gold complex has been used to treat trypanosomiasis caused by *Trypanosoma brucei rhodesiense* and other parasites [124]. Apart from the fact that gold itself has inhibitory properties against these parasites, the amine-end of the dithiocarbamate compound also initiates the generation of reactive oxygen species leading to the death of the parasites [125]. Brassinin, which is a dithiocarbamate compound, and its derivatives have been found to be active against *Trypanosoma cruzi* (trypanosome that cause Chagas diseases). It has displayed a good antiproliferative effects that is similar to benznidazole and nifurtimox, which are well known antichagasic agents [126]. Apart from brassinin, Ochoa et al. [127] synthesized 34 dithiocarbamate compounds (3,5-disubstituted-tetrahydro-2H-1,3,5-thiadiazine-2-thione derivatives). Some of these compounds were reported for the treatment of Chagas diseases. They have the ability to generate reactive oxygen species, leading to oxidative damage of *Trypanosoma cruzi*. One of the psychological implications of diabetes 1 and 2 is anxiety [128]. Studies conducted by using mice showed that pyrrolidine dithiocarbamate showed anxiolytic-like effects [129]. Table 5 presents some of the diseases in which the use of dithiocarbamate compounds have found relevance.

Table 5. Different diseases treated with dithiocarbamate compounds.

Diseases/ Abnormality Treated	Brief Description of the Disease/ Abnormality	Dithiocarbamate Compound Used	Roles of Dithiocarbamate	Ref.
Influenza	Viral disease that affect the respiratory organs	Pyrrolidine dithiocarbamate	It acts against overproduction of reactive oxygen species and inhibit DNA fragmentation	[130]
Hyperglycemia	Too much of glucose in the bloodstream that may be as a result of diabetes mellitus	Allyldithiocar-bamates	Dithiocarbamates improved the sensitivity of insulin instead of the concentration of insulin leading to 18.2% glucose AUC (glucose area under the curve) in 15 days.	[131]
Tuberculosis	Bacterial infection that affect the lung	Several N,N-disubstituted and N-mono-dithiocarbamates	Treatment through the inhibition of carbonic anhydrase enzyme. These dithiocarbamate compounds were more effective as inhibitor than the clinically-approved sulfonamide.	[132]
Alzheimer disease	age-related neurodegenerative disorder	Several coumarin-dithiocarbamate	Treatment through the inhibition of acetylcholinesterase. They were able to reverse the cognative dysfunction	[133]
Dandruff	Fungal disease that affect the scalp leading to the shedding of dead skin cells.	Series of dithiocarbamates	Inhibition of β-class carbonic anhydrase of *Malassezia globosa*	[134]
Myasthenia gravis	An auto-immune disease causing the weakness of muscle	N,N-disubstituted dithiocarbamic acid	Treatment via inhibition of cholinesterase. They possessed better anticholinesterase properties more than Donepezil which is used for treating the disease.	[135]
SARS-CoV-2. (Still on clinical trial) NCT 04485130	Viral respiratory disease also known as coronavirus (COVID-19)	Disulfiram	Inhibition of viral replication and the anti-inflammatory activities leading to the treatment of the disease.	[9]
Alcoholism	Excessive and uncontrollable alcohol intake	Disulfiram	It inhibits acetaldehyde metabolism which is a product obtained from the breakdown of alcohol	[136,137]
Parkinson's disease	Genetic disease associated with the loss of neuron	Pyrrolidine dithiocarbamate	It suppresses the level of glutamate	[138]
Male infertility	Inability to conceive children	Ziram	Reduction of the level of proteineous kinase by damaging the mitochondria ultrastructure thereby inhibiting human sperm motility.	[139]
Scorpionism	Painful condition as a result of scorpion sting	pyrrolidine dithiocarbamate	Inhibition of venom-induced thermal and mechanical hyperalgesia of *Tityus bahiensis*.	[140]

7.4. Anti-Inflammatory Application of Dithiocarbamate Compounds

Aspirin and the non-steroidal anti-inflammatory drugs have side effects [141]. As a result of the side effects, alternative anti-inflammatory drugs that have minimal toxicity and side effects without compromizing the efficiency are required [142]. The dithiocarbamate-based compounds have also been discovered to possess anti-inflammatory properties. Song et al. reported the replacement of nitrogen position of indoles with dithiocarbamate groups at room temperature, which were found to inhibit the release of interleukin-6 and tumor necrosis factor alpha, thereby displaying anti-inflammatory properties [143]. This anti-inflammatory property was found to be useful in the treatment of acute lung injury because it perpetuates, amplifies and mediates anti-inflammatory injury, thereby leading to inflammatory response blockage [144,145]. Another dithiocarbamate compound that has been utilized for anti-inflammatory properties is pyrrolidine dithiocarbamate and one of the reasons why it is considered for this application is its stability at physiological pH in solution, in addition to its ability to traverse the cell membrane [146]. Pyrrolidine dithiocarbamate was effective against chronic and acute inflammation [147].

7.5. Anticancer Application of Dithiocarbamate Compounds

There are more than 10 million cases of cancer every year around the world [148], and it is one of the leading causes of death [149]. Hence, there is need for the synthesis of novel anticancer agents to complement the existing anticancer drugs. Several compounds containing dithiocarbamate have been investigated as anticancer agents and they act by inhibiting enzymes responsible for cancer growth (such as catalase), alter the production of reactive oxygen species or trigger the induction of apoptosis at the mitochondria [150]. For example, the ability of diethyldithiocarbamate to chelate copper was utilized in treating both breast and prostate cancer. This chemotherapeutic cancer treatment worked through the accumulation of copper in the cancerous tissues and cells [109]. The copper complexes also have the ability to initiate the inhibition of proteasome and cause apoptosis in the cancer cells of humans. Similar to copper dithiocarbamate, zinc dithiocarbamate was also found to have a similar effect on cancer cells but it occurs through a different mechanism. Despite the difference in mechanism, caplain is involved in the apoptotic cell death process of dithiocarbamate of both zinc and copper [151]. Dithiocarbamate complexes of trivalent gold have also been found to be effective against cancer cells [152]. Similarly, derivatives of benzoxazole with dithiocarbamate moieties were found to be active in the treatment of breast cancer [153]. Gamma glutamyl transferase was used as trigger for copper diethyldithiocarbamate prodrug and it was used for the treatment of prostate cancer, which is the second most common cancer among men. The drug showed high antiproliferative efficiency within 24 h in prostate cancer cells [154]. Thiocarbonylthiol compounds have been found to be a good anticancer agent with reduced toxicity when compared to cisplatin, a very known anticancer agent. The anticancer activity of thiocarbonylthiol occurs by inducing apoptosis and induction of DNA damage [155]. The anticancer activities of dithiocarbamate often occur via unrestrained cell death as a result of inflammation, hypoxia or other external damage leading to the release of the content of the cytoplasm into the surroundings. This cell damage through these means is termed necrosis [156].

7.6. Antimicrobial Applications of Dithiocarbamate

The presence of donor atom (sulphur) in dithiocarbamate compounds makes them possess good antimicrobial properties. So, they are able to form a chelate with positively charged metal ions. The sulphur atom reduces the polarity of the binding metal through the delocalization of electrons over the entire chelate ring. This process makes the permeability of the microbes feasible [157,158]. In some cases, there is formation of a hydrogen bond between the active center of the microbe and the –N_ (S)SH group of the dithiocarbamate, leading to an interference of the physiological processes of the cells [157,158]. The common micro-organisms that dithiocarbamate compounds have been used against are bacteria, fungi and virus. Different dithiocarbamate compounds that have acted against these microbes will be discussed in this section.

7.6.1. Antibacterial Application of Dithiocarbamate Compounds

There is increased interest in the development of novel antibacterial substances as a result of the increase in the number of drug-resistant bacteria. Silver(I) dithiocarbamate triphenylphosphine complexes have showed better antibacterial properties, greater than ciprofloxacin against Gram (−) and Gram (+) bacteria. The bacteria used for the investigations are *Staphylococcus aureus, Salmonella typhimurium, Escherichia coli, Klebsiella pneumoniae* and *Pseudomonas aeruginosa.* It was observed that this dithiocarbamate complex displayed better antibacterial activity against Gram (+) positive bacteria than the Gram (−) negative bacteria with the exemption of *K. pneumonia.* The reduced activity of the dithiocarbamate complex could be linked to the fact that the cell wall of Gram (−) is made of several layers unlike the cell wall of Gram (+) bacteria, which is made of a single layer. Hence, the penetration of dithiocarbamate is hampered by multiple cell walls [159]. Another novel compound, 1,2,3-triazole-dithiocarbamate-naphthalimides, showed good antibacterial activity against *Staphylococcus aureus, Bacillus subtilis, Escherichia coli* and *Candida albicans.* Notably, this novel compound showed a better antibacterial performance than a common antibacterial drug (Cefuroxim) when it was tested against *B. subtilis* [160]. Derivatives of isatin dithiocarbamate have also been investigated as an antibacterial agent against both Gram (+) bacteria (*Strep. Pneumonia* and Staph. *aureas*) and Gram (−) bacteria (*Pseud aeruginosa* and *Escherichia coli*). It also showed satisfactory antibacterial activities against these bacteria when compared to antibacterial activities of some common drugs [161]. In all the antimicrobial investigations, the methods used were broth dilution, disc diffusion, zebrafish model, well diffusion, tube diffusion, agar dilution, broth micro-dilution methods or the combination of the methods. Other dithiocarbamate investigated for antimicrobial activities are shown in Table 6.

Table 6. Specific examples of antibacterial properties of dithiocarbamates against some bacterial strains.

Dithiocarbamate Compounds	Bacteria	Conc. of Isolation	Min. Inhibitory Conc. Range	Ref.
Phenyl dithiocarbamate mixed ligand metal complexes	*Escherichia coli, Proteus valgaris, Salmonella typhii, Shigella flexneri, Staphylococcus aureus, Bacillus subtilis, Streptococcus pneumonia, Psendomonas aeruginosa, Vibro chlolerae* and *Klebseilla pneumonia*	10 mg/mL	6–8 nm	[6]
sodium cyclohexyldithioc-arbamate and sodium phenyldithiocarbamate	*Salmonella typhi, Proteus mirabilis, Pseudomonas aeruginosa, Bacillus cereus* and *Bacillus subtilis*	15–30 mg/mL	(7.7–16.3 mm) and (8.5–19 mm) respectively	[162]
tris(ephedrinedithiocarbamate) complexes	*Pseudomona aeruginosa, Staphylococcus sciuri, Enterococcus caseofluvialis, Staphylococcus aureus, Enterobacter cloacae, Salmonella dublin, Klebsiella pneumonia* and *Escherichia coli*	25–100 µg/mL	14.6–126.5 µM	[163]
N-ethyl-*N*-phenyldithiocarbamate complexes	*Staphylococcus aureus, Salmonella typhi, Pseudomonas aureginosa* and *Escherichia coli*	100 µg/mL	-	[164]
Dibenzyldithiocarbamate	*Mycobacterium smegmatis Staphylococcus aureus, Pseudomonas aeruginosa and Escherichia coli*	0.5 mg/mL	64–1000 µg/mL	[165]

Table 6. *Cont.*

Dithiocarbamate Compounds	Bacteria	Conc. of Isolation	Min. Inhibitory Conc. Range	Ref.
Rh(III)-morpholine-4-dithiocarbamate	*Salmonella typhai, Pseudomunas aeroginosa, Proteus mirabilis, Yersinia enterocolitica, Enterococcus faecalis Staphylococcus aureus*	50 ppm	5–28 mm	[166]
silver(I) dithiocarbamate triphenylphosphine	*Escherichia coli, Salmonella. typhimurium, Pseudomunas aeruginosa, Klebsiella pneumonia, Staphylococcus aureus*	1000 μg/mL	0.19–75.45 μM/mL	[159]
N-methyl-N-phenyl dithiocarbamate complexes of Cu(II), In(III) and Sb(III)	*Bacillus cereus, Enterococcus faecalis, Enterococcus gallinurium, Listeria monocytogenes, Listeria monocytogenes, Staphylococcus aureus, Escherichia coli, Klebsiella pneumonia, Salmonella enterica* and *Salmonella Typhimurium*	0.022–2.522 μg/mL	7.00–19.33 mm	[167]

7.6.2. Antifungal Application of Dithiocarbamate Compounds

The reduction in the plant yield as a result of fungal infections coupled with the negative impacts of fungi on the health of plants and animals makes the synthesis of efficient antifungal compounds of utmost priority. Dithiocarbamates are one of the numerous compounds that have been investigated as antifungal drugs. For example, two fungi (*Candida albicans* and *Candida tropicalis*) extracted from HIV patients that are also suffering from oral candidiasis were rendered passive in the presence of organotin(IV) dithiocarbamates. The organotin dithiocarbamate was able to achieve this by suppressing the ergosterol synthesis without cytochrome deactivation [168]. Plant pathogenic fungi have also been eradicated by using dithiocarbamate compounds as antifungal agents [169]. *Alternaria porri* and *Fusarium oxysporum,* which are plant pathogens were inhibited by using ammonium dithiocarbamate coupled with chitosan [170]. The inhibitory effect of this dithiocarbamate compound was clearly better than when chitosan alone was used as the antifungal agent. From the investigation conducted by Ferreira et al. [171] dithiocarbamate complexes containing nickel, platinum and palladium were found to be effective against several fungi (*Penicillium citrinum, Aspergillus niger, Aspergillus flavus* and *Aspergillus parasiticus*). The antifungal activities of these dithiocarbamates were found to be better than some known antifungal drugs (nystatin and miconazole nitrate). When the antifungal activities of dithiocarbamate complexes of nickel and palladium were compared by this same group [171], the nickel complexes were more effective against *Aspergillus parasiticus,* whereas palladium complexes were more effective against *Aspergillus flavus.* Three organotin dithiocarbamate compounds (tributyltin dithiocarbamate propionates, tributyltin dithiocarbamates and dibutyltin dithiocarbamates) have been reported to possess antifungal activities against fungi that destroy woods (*Coriolus versicolor, Coniophora puteana* and *Serpula lacrymans*). The antifungal activities of these dithiocarbamate compounds is comparable with that of tris-(benzyltriazolylmethyl)amine, a common antifungal compound [172].

7.6.3. Antiviral Application of Dithiocarbamate Compounds

The treatment of several viral infections have been carried out via dithiocarbamate-containing ligands and complexes. One of the common dithiocarbamates that has been utilized for this purpose is pyrrolidine dithiocarbamate. It was used to alter the pathogenesis of cells infected with dengue virus and its high replication ability was inhibited. In fact, this dithiocarbamate was observed to be more active against dengue virus than gefitinib, which is a receptor inhibitor [173]. Enterovirus 71, which is a viral disease that affects the mouth, foot and hand of animals have been treated with pyrrolidine dithiocarbamate.

There was significant reduction in the yield of the virus after cell culture was treated with this dithiocarbamate [174]. Antiviral properties of pyrrolidine dithiocarbamate have also been investigated against herpes simplex virus, influenza virus, rhinovirus and cox sackie virus B3 [174].

7.7. Application of Dithiocarbamate in Medical Imaging

Two dithiocarbamate ligands, (methoxyisobutyl dithiocarbamate) and tert-butyl dithiocarbamate, were radiolabeled with ^{99m}Tc-nitrido core and used for myocardial imaging. These dithiocarbamate ligands performed better than $^{99m}TcN(NOEt)_2$, which was already on phase III clinical trial for the same imaging application [175]. The synergistic application of magnetic resonance imaging (MRI) and positron emission tomography (PET) was achieved with radio-labelled copper dithiocarbamate bonded to iron trioxide. This dual modality imaging (Figure 10) was possible due to the accumulation of this dithiocarbamate compound in the lymph nodes without translocation of radioactivity to other parts of the tissues. The results were also obtained faster with less dose of radiation required compared to other common dual MRI-PET agents [176].

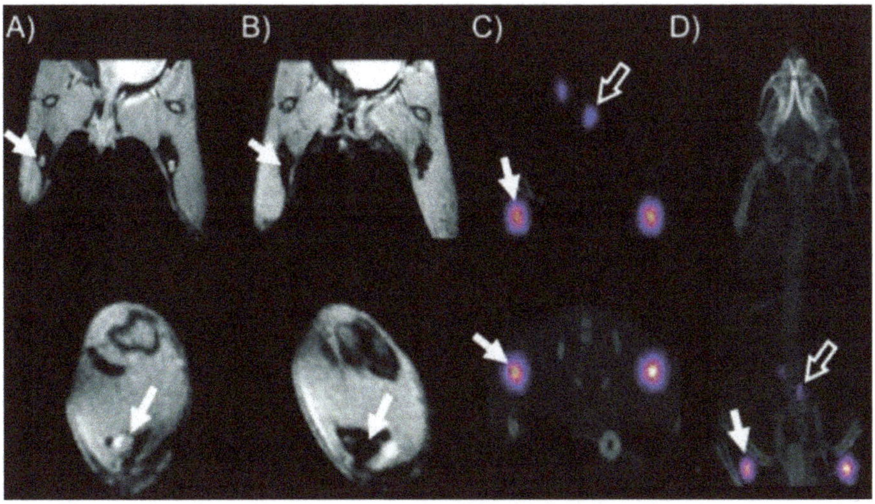

Figure 10. In-vivo dual MRI-PET images obtained from mouse using isotopic-labelled copper dithiocarbamate complex. (**A**,**B**) Popliteal nodes of coronal (**top**) and short axis (**bottom**) MR images of the lower abdominal area and upper hind legs before (**A**) and after (**B**) injecting dithiocarbamate imaging agents. (**C**) Coronal (**top**) and short-axis (**bottom**) images showing the uptake of the dithiocarbamate (**D**) image of the whole body of the mouse. Reprinted from [176]. Copyright (2011), with permission from Wiley and Sons.

Ciprofloxacin dithiocarbamate has been radiolabeled with ^{99m}TcN complex and used for imaging infections in mice. The binding affinity of the complex significantly improved compared with similar compounds without dithiocarbamate. Moreover, the complex was stable for more than 6 h at room temperature [177]. Dithiocarbamate compounds have also been useful in imaging tumor tissues and this is as a result of the good tumour/muscle ratios of these compounds. In addition, their high tumour uptake leading to their accumulation in the site containing tumors makes them suitable for imaging applications. An example of such a compound is ^{99m}Tc(V)-glucoheptonate radiolabeled deoxyglucose dithiocarbamate [178].

8. Application of Dithiocarbamate Compounds in the Industries

Several industries are using dithiocarbamate as the starting materials in different industrial processes and this has spiked the consumption of dithiocarbamate compounds. Some of the industrial uses of dithiocarbamate compounds that will be discussed in this section are shown in Figure 11.

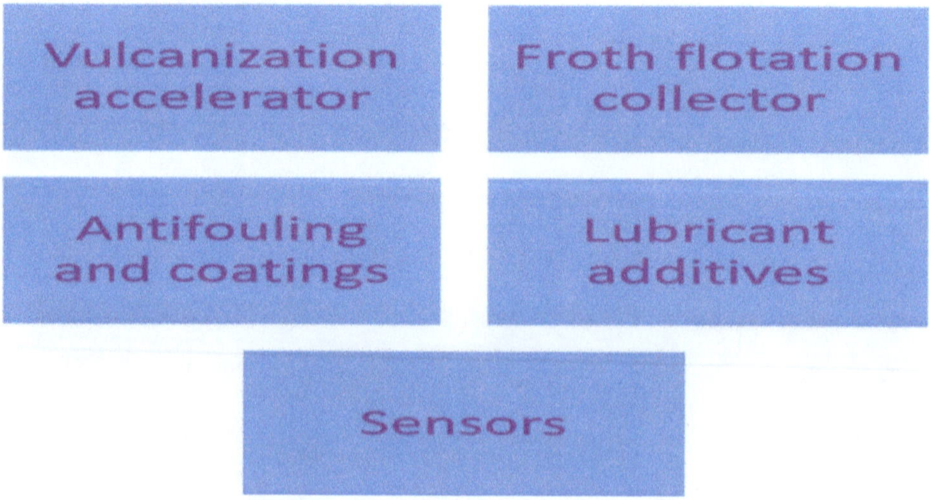

Figure 11. Industrial applications of dithiocarbamates.

8.1. Application of Dithiocarbamate Compounds as Vulcanization Accelerator

Vulcanization accelerator is required for large scale production of rubber becaused it improves the state and rate of crosslinking of rubber during the process [179]. Thiocarbanilide, guanidine and aniline have been used as accelerators, and the vulcanization process (in their presence) was found to be faster than sulphur vulcanization. However, these accelerators showed different levels of toxicity [179,180]. Vulcanization of nitrile butadiene and other types of rubber are now speeded up by dithiocarbamate. This is due to its ability to simultaneously enhance the state and rate of vulcanization [179]. Wang et al. [181] investigated the effect of using sodium, zinc and lanthanium dithiocarbamate as a vulcanization accelerator. The vulcanization carried out with these dithiocarbamates was fast compared to the investigation without dithiocarbamate. Among the metal dithiocarbamates used for the investigation, lanthanium diethyldithiocarbamate was observed to perform better in accelerating vulcanization process. The rate of rubber acceleration further increased when rubber black was also added as an additive to assist dithiocarbamates. The carbon black has functional groups such as lactones and phenolic, which allows it to react with sulphur to form a network during vulcanization [182]. In a similar investigation, samarium lysine dithiocarbamate was reported to accelerate the vulcanization process and also boosted the crosslink of the network. The introduction of stearic acid and zinc oxide as the activators further enhanced the properties of the rubber produced [183].

Amine-containing zinc dithiocarbamates were also found to be effective as a vulcanization accelerator. Some of these dithiocarbamates are zinc (*N*-ethyl piperazino) dithiocarbamate and zinc (*N*-benzyl piperazino) dithiocarbamate. They were found to be safer and were able to improve the ability of rubber to withstand aging unlike zinc dimethyl dithiocarbamate [184]. Apart from using dithiocarbamate directly for accelerating vulcanization, the dithiocarbamates have also been found to be good precursors for preparing other materials that were used as the vulcanization accelerator. For instance, molybdenum dialkyl dithiocarbamate was used as a precursor for preparing molybdenium sulphide nanoparticles, which were then used as catalysts for speeding up the rate of vulcanization [185]. Sometimes, dithiocarbamates could be useful as a bridge for other structures, thereby resulting in a composite with improved vulcanization kinetics. This was demonstrated by using lanthanum glutamic dithiocarbamate to bridge silica with styrene butadiene rubber and the resulting composite was used as a vulcanization accelerator. Other examples of dithiocarbamates that were used as vulcanization accelerator are zinc diisononyldithiocarbamate, zinc isobutyldithiocarbamate, zinc dibenzyldithiocarbamate, zinc dibutyldithiocarbamate and zinc diethyldithiocarbamate [186]. Several modifications have been carried out to improve the performance of these dithiocarbamates as vulcanization accelerator. One such attempt is the use of zinc salts of butyl, isopropyl and ethyl xanthates along with these dithiocarbamates and it has yielded a positive outcome [182]. The introduction of phosphorus into dithiocarbamate to form phosphorylated dithiocarbamates has also been reported as a vulcanization accelerator and this was also discovered to give more positive results compared with ordinary dithiocarbamates [187].

The studies carried out by Nieuwenhuizen et al. [188] showed the use of zinc dithiocarbamate as a vulcanization accelerator. The complex acts as a mediator between the rubber and sulphur. It brings the sulphur atom in the ring of zinc dithiocarbamate and introduces it into the carbon–hydrogen bond through a reaction involving a double bond. The resultant product of this reaction is polythiothiol and some of them further undergo methathesis reaction leading to the formation of polysulphide. Desulfhydration of polythiothiols may also occur, leading to the formation of hydrogen peroxide and sulphides. These reactions and products lead to the increase in the speed of vulcanization.

8.2. Application of Dithiocarbamate Compounds as Froth Flotation Collector

In froth flotation, a collector is needed to capture the mineral that is needed. The principle upon which the collector acts is that the active sites of the mineral interact with the polar region of the collector, while the non-polar region of the collector binds to the bubbles. The combination of adsorbed mineral particles and the collector binds to the surface of the slurry, leading to efficient separation [189]. Several mineral (such as sulphides of lead, zinc and tin) ores have been obtained via the use of collectors as the flotation agent. Xanthates are common collectors that are used for this purpose, but it has been discovered that oxidized mineral ores showed insufficient response to xanthate collector [190]. This slow response necessitated the sulphidation of the oxidized minerals prior to conditioning with the collector so as to improve the performance of the process [191]. The sulphidations are carried out by using ammonium sulphide, sodium hydrosulphide or sodium sulphide [192]. To carry out flotation without sulphidation, hydroxamic acids were used as froth flotation collectors but their performance depends on the nature of the ore [192]. So, there is a need for a more efficient flotation collector.

Dithiocarbamate compounds have been investigated as a possible replacement for these known collectors. For example, 2-hydroxyethyl dibutyldithiocarbamate has been used as surfactant collector for the removal of galena from sphalerite. This was achieved by using 4×10^{-4} mol·L^{-1} of the dithiocarbamate compound. Its effectiveness was proven through the adsorption mechanism, which revealed that the presence of this dithiocarbamate improved the hydrophobicity of the surface of galena via the process of chemisorption. S-benzoyl-*N*,*N*-diethyldithiocarbamate is another flotation surfactant collector and its performance was better than that of isobutyl xanthate and diethyldithiocarbamate. Similar to 2-hydroxyethyl dibutyldithiocarbamate, S-benzoyl-*N*,*N*-diethyldithiocarbamate also displayed enhanced selectivity for galena in the presence of aphalerite [193]. In some cases, dithiocarbamates are used as co-collector along with other known collectors. Ngobeni et al. used both xanthates and sodium di-methyl-dithiocarbamate to separate nickel ores from pentlandite in a South African mine. Their study showed an enhanced nickel recovery when these co-collectors were used together. This indicated that selectivity of the collector improved in the presence of dithiocarbamate [194]. In another investigation, varied ratios of di-n-propyl dithiocarbamates and cyclo-hexyl dithiocarbamates were used as collectors along with other sulphur-containing collectors. The presence of dithiocarbamate resulted in the recovery of more than 80% of the ore. Finally, the recovery of galena from the ore containing several metallic sulphides was enhanced when S-benzyl-*N*-ethoxycarbonyl thiocarbamate was used as the collector. Its performance was even better than that of ammonium dibutyl dithiophosphate and sodium diethyl dithiocarbamate which are conventional collectors [68]. The same performance was observed when N-[(3-hydroxyamino)-propoxy]-*N*-octyl dithiocarbamate was used as the collector for extracting cassiterite [195].

Dithiocarbamate compounds also found application in the extraction of precious metals from their ores. This is connected to their usefulness as froth flotation collectors. S-cyanoethyl *N,N*-diethyl dithiocarbamate and S-cyanoethyl *N,N*-diethyl dithiocarbamate are two dithiocarbamate compounds that have the ability to form an undegraded compound with gold when it is in aqueous form. This ability qualifies them as collector for recovering gold from their ores. In addition, S-cyanoethyl *N,N*-diethyl dithiocarbamate also enhances the floatability of chalcopyrite, which makes it useful in the extraction of high quality copper with minimal arsenic contaminant [196]. Modified dibutyldithiocarbamate and diethyldithiocarbamate performed the same function in the extraction of gold from its ore with a better gold recovery [197].

8.3. Application of Dithiocarbamate Compounds as Antifouling/Electroplating Agents

The control of organisms responsible for fouling in the marine environment has been a subject of research, which has led to the use of dichlorodiphenyltrichloroethane/tributyltin, 8-methyl-N-vanillyl-6-nonenamide and triphenylborane pyridine as antifouling agents. Further research has shown that zinc ethylene(bis) dithiocarbamate can also perform a similar function [198]. Zinc dithiocarbamate was added to some known antifouling agents and the overall effect was discovered to be synergistic, which implies that the dithiocarbamate could be used alone or in a mixed form as antifouling agent [199]. Zwitterionic phenyl phosphorylcholine dithiocarbamate was able to lower the adsorption of protein into the surface of the gold electrode, thereby reducing fouling in these electrodes. The dithiocarbamate-containing zwitterionic phenyl phosphorycholine performed better than when diazonium salt was used to replace dithiocarbamate in the same compound [200].

8.4. Application of Dithiocarbamate Compounds in Coatings

The formation of coatings that is rich in phosphophyllite is possible when phosphate is being used for coating with the addition of long-chain dithiocarbamates. The effect is the rise in the soluble iron, wet adhesion and alkaline stability of the phosphate coatings. In short, the presence of dithiocarbamate as the additive makes electrophoretic deposition feasible [201]. The corrosion resistance and porosity of zinc-phosphated steel was also enhanced when dithiocarbamate compounds such as octadecyldithiocarbamate, hexadecyldithiocarbamate and dodecyldithiocarbamate were used as additives during the coating process [202]. The need to minimize acid mine drainage or acid rock drainage, which cause problems in the environment, led to the coating of the pyrite [203]. Some of the chemical species that have been used for this purposes are oxalic acid, natural lignin, fatty acid, humic acid and acetyl acetone and they all act by slowing down the oxidation of pyrite. However, their usage requires the use of hydrogen peroxide which also has a negative impact on the environment. Besides, coating involving phosphate and silicate has little stability when the pH is too low [204]. To overcome these challenges, sodium triethylenetetramine-bisdithiocarbamate has been used to coat pyrite. It acts by forming a passivating cross-link on the surface of the pyrite and the formed crosslink is not only hydrophobic but it also prevents the release of metals even at a low pH [204].

8.5. Application of Dithiocarbamate Compounds as Lubricant Additives

Improvement of lubricants is vital for the durability and efficiency of energy generated in the machines [205]. One of the strategies adopted to enhance the quality of lubricants is to introduce additives. Other reasons for introducing additives to lubricants are to cut down the gas environmental pollutants and to minimize the consumption of fuels [206]. Dithiocarbamates are also a known sulphur-containing lubricant additive and antiwear agent. Depending on their chemical properties and structures, dithiocarbamate additive promotes the economy of the fuel, boosts its load-carrying potential and reduces the possible wear and tear [207]. Tribological applications of several metal dithiocarbamate complexes have been investigated. Among the tested lubricant additives, molybdenum dialkyl dithiocarbamate was reported to be the most effective among the dithiocarbamate complexes based on the fuel economy, lubricant viscosity and driving cycle results [208]. Introduction of zinc dialkyldithio-phosphate to molybdenium dialky dithiocarbamate further improved the tribological properties of molybdenium dialky dithiocarbamate [209]. Shah et al. [210] investigated the comparative efficiencies of S-hydroxyethyl-*N*,*N*′-dibenzyldithiocarbamate (HE-BzDTC), *S*-(Di-*n*-butyl-borate)-ethyl-*N*,*N*′-dibenzyldithiocarbamate (DBB-EBzDTC), *S*-(Di-*n*-octyl-borate)-ethyl-*N*,*N*′-dibenzyldithiocarbamate (DOB-EBzDTC) and *S*-(Di-*n*-octyl-borate)-ethyl-*N*,*N*′-di-*n*-ethyldithiocarbamate (DOB-EEDTC) as lubricant additives. The performances of these dithiocarbamate compounds were compared with those without dithiocarbamate, and remarkable performance was observed compared to the additives without dithiocarbamates (as shown in Figure 12).

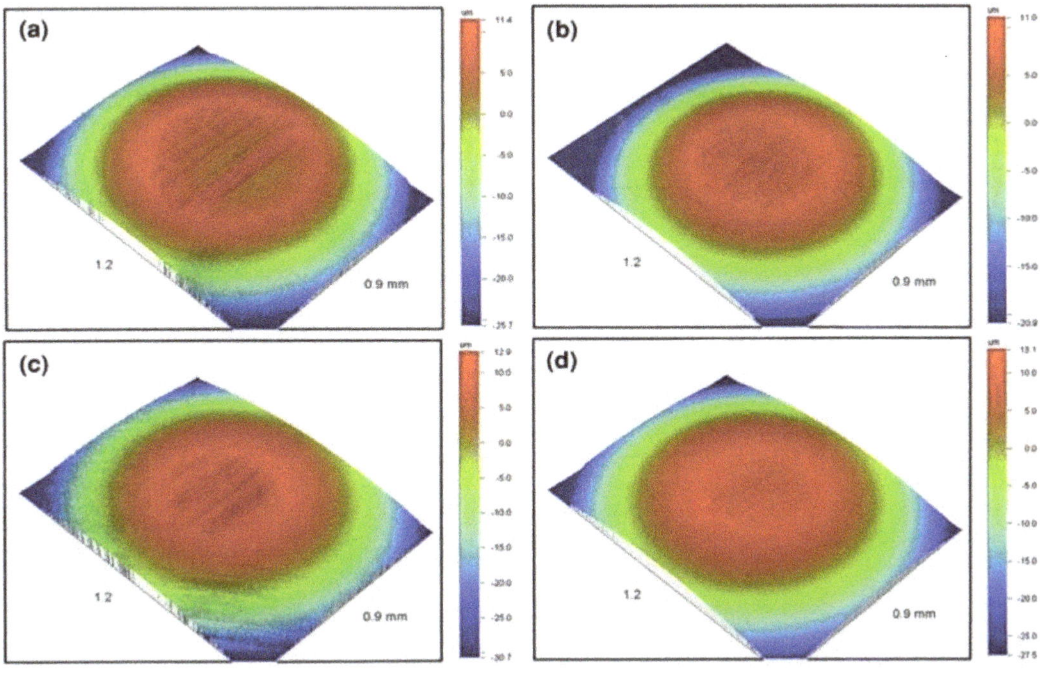

Figure 12. (**a**) The lubricating performance of oil without dithiocarbamate additives compared with the oil incorporated with dithiocarbamates (**b**) DBB-EBzDTC (**c**); DOB-EBzDTC and (**d**) DOB-EEDTC. Reprinted with permission from Springer Nature, Tribology letters [210]. Copyright (2011).

8.6. Application of Dithiocarbamate Compounds as Sensor

Chromogenic properties of dithiocarbamate anions are utilized in detecting both organic and inorganic pollutants. In some cases, dithiocarbamates are attached to other fluorescent moieties to sense pollutants even at a very low pollutant concentration [211]. Dithiocarbamate modified with gold was reported as sensor for divalent zinc through trimodal techniques. This sensor is significant because it can perfectly distinguish divalent cadmium from divalent zinc on the spot [212]. The ability of nickel dithiocarbamate-containing ortho isomer of sulforhodamine B to show a fluorescence increase when it reacts with nitrogen dioxide makes it a good sensor for nitrogen dioxide [213]. Apart from using dithiocarbamates alone as sensor, they have also been used to functionalize other materials used as sensor. For instance, the sensing of series of polyaromatic hydrocarbons has been made possible through the use of silver nanoparticles functionalized with dithiocarbamates [214]. Moreover, sensing of dithiocarbamate fungicide that is present in water and fruit juice was achieved through the use of silver nanoparticles functionalized with dopamine dithiocarbamate [215]. In addition to sensing metals, polyaromatic hydrocarbon, fungicides and gas, and dithiocarbamates have also found use in the sensing of anions. Bromide anion was detected when homoleptic cobalt(III) dithiocarbamate was used as sensor [216]. However, the lability of some of the dithiocarbamate complexes when they are in solution limited their use as sensor, but they become more applicable when they are attached to fluorescent moieties. This was adopted when organotin(IV) dithiocarbamate was added to antracene, which is a good fluorescent moiety to detect *O*-donor anions even when the concentrations of these anions were very low [211]. Other investigations involving the use of dithiocarbamate compounds as sensors are shown in Table 7.

Table 7. Application of dithiocarbamates in sensing.

Dithiocarbamate Compound	Substance Sensed	Detection Limits	Ref.
Chitosan dithiocarbamate	Divalent cadmium	63 nM.	[217]
Dithiocarbamate functionalized silver nanoparticles	Divalent cobalt	14 μM	[218]
ZnS quantum dots doped with glycine dithiocarbamate -functionalized Mn.	Trivalent cerium	2.29×10^{-7} mol.L^{-1}	[219]
Nickel(II) dithiocarbamate complexes	Halide ions	-	[220]
Gold nanoparticles functionalized with Malonamide dithiocarbamate	Divalent mercury and copper	45 nM and 41 nM for Hg^{2+} and Cu^{2+}ions respectively.	[221]
Silver nanoparticles functioalized with Cyclen dithiocarbamate	Paraquat and thiram pesticides	7.21×10^{-6} M and 2.81×10^{-6} M for paraquat and thiram respectively	[222]
Gold nanoparticles functionalized with *p*-amino salicylic acid dithiocarbamate	Trivalent iron	14.82 nM	[223]
Gold nanoparticles decorated with Ractopamine-dithiocarbamate	Pendimethalin herbicide	0.22 μM	[224]
Gold nanoparticles decorated with dithiocarbamate-p-tertbutylcalix[4]arene	Metsulfuron-methyl herbicide	1.9×10^{-7} M	[225]

9. Challenges Associated with the Utilization of Dithiocarbamates

Dithiocarbamates that possess aliphatic chains are vulnerable to acid hydrolysis and liberate CS_2 under acidic or neutral conditions. In a very strong alkaline condition, the aliphatic dithiocarbamates degrade to give mixtures of sulphur-containing compounds such as sulfonates and disulphides [8]. Catalytic oxidation of thiols by dithiocarbamate compounds, leading to the inhibition of pro-apoptotic enzymes, has been reported [226]. Dithiocarbamate compounds also play significant roles in the disruption of the developmental stage of aquatic animals [227]. The product of metabolic degradation of dithiocarbamate (carbon disulphide) also causes notochord distortions in zebra fish [228]. Dithiocarbamate compounds have been found to possess biocidal and cytotoxic properties. Their cytotoxity was discovered to be related to their structures [229]. Disulfiram, thiram and mancozeb cause changes in the cell membrane and block glutamate from binding to the receptor, which results in toxic effects on the brain [230].

10. Conclusions and Future Perspectives

Dithiocarbamate may simply be the solution to the many environmental, medical, agricultural and industrial challenges based on their applications that have been highlighted in this review. The discussion presented herein is believed to inspire more studies and investigations into new applications of dithiocarbamate compounds. For future research, the use of dithiocarbamate complexes in medical imaging is still at the infant stage and it needs to be further explored. For instance, the possibility of using dithiocarbamate compounds to solve the problem of scattering, sensitivity and absorption in medical imaging should be investigated. Toxicity of the metal dithiocarbamate complexes should be thoroughly investigated prior to their use for these applications, so that any possible cytotoxic effect that could emanate from the introduction of dithiocarbamate into the ecosystem could be mitigated for the protection of aquatic and terrestrial lives. Moreover, the fate of the unused dithiocarbamate in the environment and their degradation mechanism through the use of photocatalysis and other removal methods should be studied. Furthermore, the effect of dithiocarbamate on the root exudates of common food crops such as maize and soy bean should be investigated so as to enhance food safety and productivity. While iron dithiocarbamate has been investigated for the removal of nitrogen oxides from air samples [231], this investigation needs to be carried out on other air pollutants. Finally, renewed efforts should be geared towards the synthesis of novel dithiocarbamate ligands and complexes.

Author Contributions: Conceptualization, T.O.A. and D.C.O.; Writing—original draft preparation, T.O.A. and T.T.A.; Writing—review and editing, D.C.O. and R.M.; supervision and funding acquisition, D.C.O. All authors have read and agreed to the published version of the manuscript.

Funding: This research was funded by Deanship of Scientific Research at King Khalid University (RA.KKU/148/43), North-West University, National Research Foundation, South Africa (Grants Ref: UID109333 and UID 116338) and the APC was funded by North-West University, South Africa.

Institutional Review Board Statement: Not applicable.

Informed Consent Statement: Not applicable.

Acknowledgments: The authors acknowledge the Deanship of Scientific Research at King Khalid University (RA.KKU/148/43) and the North-West University South Africa.

Conflicts of Interest: The authors declare no conflict of interest.

References

1. Adeyemi, J.O.; Onwudiwe, D.C. Chemistry and Some Biological Potential of Bismuth and Antimony Dithiocarbamate Complexes. *Molecules* **2020**, *25*, 305. [CrossRef] [PubMed]
2. Cvek, B.; Dvorak, Z. Targeting of nuclear factor-κB and proteasome by dithiocarbamate complexes with metals. *Curr. Pharm. Des.* **2007**, *13*, 3155–3167. [CrossRef] [PubMed]
3. Shinde, S.D.; Sakla, A.P.; Shankaraiah, N. An insight into medicinal attributes of dithiocarbamates: Bird's eye view. *Bioorganic Chem.* **2020**, *105*, 104346. [CrossRef] [PubMed]
4. Chen, N.; Zhong, X.; Li, P.; Xu, J. A Mild Radical Method for the Dimerzation of Dithiocarbamates. *Eur. J. Org. Chem.* **2015**, *2015*, 802–809. [CrossRef]
5. Tan, Y.S.; Yeo, C.I.; Tiekink, E.R.T.; Heard, P.J. Dithiocarbamate Complexes of Platinum Group Metals: Structural Aspects and Applications. *Inorganics* **2021**, *9*, 60. [CrossRef]
6. Ejelonu, B.C.; Olagboye, S.A.; Oyeneyin, O.E.; Ebiesuwa, O.A.; Bada, O.E. Synthesis, characterization and antimicrobial activities of sulfadiazine Schiff base and phenyl dithiocarbamate mixed ligand metal complexes. *Open J. Appl. Sci.* **2018**, *8*, 346. [CrossRef]
7. Kanchi, S.; Singh, P.; Bisetty, K. Dithiocarbamates as hazardous remediation agent: A critical review on progress in environmental chemistry for inorganic species studies of 20th century. *Arab. J. Chem.* **2014**, *7*, 11–25. [CrossRef]
8. Szolar, O.H.J. Environmental and pharmaceutical analysis of dithiocarbamates. *Anal. Chim. Acta* **2007**, *582*, 191–200. [CrossRef]

9. Kaul, L.; Süss, R.; Zannettino, A.; Richter, K. The revival of dithiocarbamates: From pesticides to innovative medical treatments. *iScience* **2021**, *24*, 102092. [CrossRef]
10. Ajiboye, T.O.; Oyewo, O.A.; Onwudiwe, D.C. Photocatalytic removal of parabens and halogenated products in wastewater: A review. *Environ. Chem. Lett.* **2021**, *19*, 3789–3819. [CrossRef]
11. Ajiboye, T.O.; Oyewo, O.A.; Onwudiwe, D.C. Adsorption and photocatalytic removal of Rhodamine B from wastewater using carbon-based materials. *FlatChem* **2021**, *29*, 100277. [CrossRef]
12. Ajiboye, T.O.; Oyewo, O.A.; Onwudiwe, D.C. Simultaneous removal of organics and heavy metals from industrial wastewater: A review. *Chemosphere* **2021**, *262*, 128379. [CrossRef] [PubMed]
13. Ajiboye, T.O.; Kuvarega, A.T.; Onwudiwe, D.C. Recent strategies for environmental remediation of organochlorine pesticides. *Appl. Sci.* **2020**, *10*, 6286. [CrossRef]
14. Kane, S.; Lazo, P.; Ylli, F.; Stafilov, T.; Qarri, F.; Marku, E. Separation of heavy metal from water samples—The study of the synthesis of complex compounds of heavy metal with dithiocarbamates. *J. Environ. Sci. Health Part A* **2016**, *51*, 335–340. [CrossRef] [PubMed]
15. Nabipour, H.; Ghammamy, S.; Ashuri, S.; Aghbolaghc, Z.S. Synthesis of a new dithiocarbamate compound and Study of Its biological properties. *J. Org. Chem.* **2010**, *2*, 75–80.
16. Hogarth, G.; Rainford-Brent, E.-J.C.R.C.R.; Kabir, S.E.; Richards, I.; Wilton-Ely, J.D.E.T.; Zhang, Q. Functionalised dithiocarbamate complexes: Synthesis and molecular structures of 2-diethylaminoethyl and 3-dimethylaminopropyl dithiocarbamate complexes [M{S2CN(CH2CH2NEt2)2}n] and [M{S2CN(CH2CH2CH2NMe2)2}n] (n = 2, M = Ni, Cu, Zn, Pd; n = 3, M = Co). *Inorg. Chim. Acta* **2009**, *362*, 2020–2026. [CrossRef]
17. Tarique, M.; Aslam, M. Bi and Trivalent transition metal complexes of dithiocarbamates derived from 2, 6-diacetyl pyridine. *Orient. J. Chem.* **2008**, *24*, 267.
18. Abu-El-Halawa, R.; Zabin, S.A. Removal efficiency of Pb, Cd, Cu and Zn from polluted water using dithiocarbamate ligands. *J. Taibah Univ. Sci.* **2017**, *11*, 57–65. [CrossRef]
19. Ayalew, Z.M.; Zhang, X.; Guo, X.; Ullah, S.; Leng, S.; Luo, X.; Ma, N. Removal of Cu, Ni and Zn directly from acidic electroplating wastewater by Oligo-Ethyleneamine dithiocarbamate (OEDTC). *Sep. Purif. Technol.* **2020**, *248*, 117114. [CrossRef]
20. Morita, F.; Nakakubo, K.; Yunoshita, K.; Endo, M.; Biswas, F.B.; Nishimura, T.; Mashio, A.S.; Hasegawa, H.; Taniguchi, T.; Maeda, K. Dithiocarbamate-modified cellulose-based sorbents with high storage stability for selective removal of arsenite and hazardous heavy metals. *RSC Adv.* **2020**, *10*, 30238–30244. [CrossRef]
21. Li, B.; Guo, J.Z.; Liu, J.L.; Fang, L.; Lv, J.Q.; Lv, K. Removal of aqueous-phase lead ions by dithiocarbamate-modified hydrochar. *Sci. Total Environ.* **2020**, *714*, 136897. [CrossRef] [PubMed]
22. Zeng, Q.; Hu, S.; Zheng, W.; He, Z.; Zhou, L.; Huang, Y. Spongy Crosslinked Branched Polyethylenimine-Grafted Dithiocarbamate: Highly Efficient Heavy Metal Ion-Adsorbing Material. *J. Environ. Eng.* **2020**, *146*, 04019105. [CrossRef]
23. Fu, W.; Huang, Z. Magnetic dithiocarbamate functionalized reduced graphene oxide for the removal of Cu(II), Cd(II), Pb(II), and Hg(II) ions from aqueous solution: Synthesis, adsorption, and regeneration. *Chemosphere* **2018**, *209*, 449–456. [CrossRef] [PubMed]
24. Liu, Z.; Han, X.; Ho, C.H.; Fan, A. Adsorption of Ni^{2+} from aqueous solution by functionalized coal particles with dithiocarbamate. *J. Hazard. Toxic Radioact. Waste* **2018**, *22*, 04018027. [CrossRef]
25. Liu, Y.; Qian, P.; Yu, Y.; Yu, B.; Wang, Y.; Ye, S.; Chen, Y. Preparation and characterization of a novel hybrid chelating material for effective adsorption of Cu(II) and Pb(II). *J. Environ. Sci.* **2018**, *67*, 224–236. [CrossRef]
26. Chen, H.; Zhao, Y.; Yang, Q.; Yan, Q. Preparation of poly-ammonium/sodium dithiocarbamate for the efficient removal of chelated heavy metal ions from aqueous environments. *J. Environ. Chem. Eng.* **2018**, *6*, 2344–2354. [CrossRef]
27. Deng, T.; Zhang, B.; Li, F.; Jin, L. Sediment washing by EDTA and its reclamation by sodium polyamidoamine-multi dithiocarbamate. *Chemosphere* **2017**, *168*, 450–456. [CrossRef]
28. Wang, Y.; Zhang, B.; Deng, T.; Li, F. Reclamation of EDTA by sodium tetraethylenepentamine-multi dithiocarbamate after soil washing process with EDTA. *Environ. Earth Sci.* **2017**, *76*, 311. [CrossRef]
29. Srinivasan, V.; Subbaiyan, M. Electroflotation Studies on Cu, Ni, Zn, and Cd with Ammonium Dodecyl Dithiocarbamate. *Sep. Sci. Technol.* **1989**, *24*, 145–150. [CrossRef]
30. Soylak, M.; Elci, L. Preconcentration and separation of trace metal ions from sea water samples by sorption on amberlite XAD-16 after complexation with sodium diethyl dithiocarbamate. *Int. J. Environ. Anal. Chem.* **1997**, *66*, 51–59. [CrossRef]
31. Imyim, A.; Daorattanachai, P.; Unob, F. Determination of Cadmium, Nickel, Lead, and Zinc in Fish Tissue by Flame and Graphite Furnace Atomic Absorption after Extraction with Pyrrolidine Dithiocarbamate and Activated Carbon. *Anal. Lett.* **2013**, *46*, 2101–2110. [CrossRef]
32. Cesur, H. Determination of manganese, copper, cadmium and lead by FAAS after solid-phase extraction of their phenylpiperazine dithiocarbamate complexes on activated carbon. *Turk. J. Chem.* **2003**, *27*, 307–314.
33. Lazaridou, E.; Kabir, A.; Furton, K.G.; Anthemidis, A. A Novel Glass Fiber Coated with Sol-Gel Poly-Diphenylsiloxane Sorbent for the On-Line Determination of Toxic Metals Using Flow Injection Column Preconcentration Platform Coupled with Flame Atomic Absorption Spectrometry. *Molecules* **2020**, *26*, 9. [CrossRef] [PubMed]
34. Smith, R.M.; Butt, A.M.; Thakur, A. Determination of lead, mercury and cadmium by liquid chromatography using on-column derivatisation with dithiocarbamates. *Analyst* **1985**, *110*, 35–37. [CrossRef]

35. Losev, V.N.; Parfenova, V.V.; Elsuf'ev, E.V.; Borodina, E.V.; Metelitsa, S.I.; Trofimchuk, A.K. Separation and preconcentration followed by ICP-OES and ICP-MS determination of precious metals using silica gel chemically modified with dithiocarbamate groups. *Sep. Sci. Technol.* **2020**, *55*, 2659–2669. [CrossRef]
36. Laosuwan, M.; Mukdasai, S.; Srijaranai, S. A simple in syringe low density solvent-dispersive liquid liquid microextraction for enrichment of some metal ions prior to their determination by high performance liquid chromatography in food samples. *Molecules* **2020**, *25*, 552. [CrossRef]
37. Lu, Y.; Gao, X.; Chen, C.T.A. Separation and determination of colloidal trace metals in seawater by cross-flow ultrafiltration, liquid-liquid extraction and ICP-MS. *Mar. Chem.* **2019**, *215*, 103685. [CrossRef]
38. Santos, L.B.; de Oliveira, D.M.; de Souza, A.O.; Lemos, V.A. A new method for the speciation of arsenic species in water, seafood and cigarette samples using an eggshell membrane. *J. Iran. Chem. Soc.* **2019**, *16*, 1879–1889. [CrossRef]
39. Kazantzi, V.; Drosaki, E.; Skok, A.; Vishnikin, A.B.; Anthemidis, A. Evaluation of polypropylene and polyethylene as sorbent packing materials in on-line preconcentration columns for trace Pb(II)and Cd(II)determination by FAAS. *Microchem. J.* **2019**, *148*, 514–520. [CrossRef]
40. Tavares, D.S.; Vale, C.; Lopes, C.B.; Trindade, T.; Pereira, E. Reliable quantification of mercury in natural waters using surface modified magnetite nanoparticles. *Chemosphere* **2019**, *220*, 565–573. [CrossRef]
41. de la Calle, I.; Ruibal, T.; Lavilla, I.; Bendicho, C. Direct immersion thin-film microextraction method based on the sorption of pyrrolidine dithiocarbamate metal chelates onto graphene membranes followed by total reflection X-ray fluorescence analysis. *Spectrochim. Acta-Part B At. Spectrosc.* **2019**, *152*, 14–24. [CrossRef]
42. Yeh, C.F.; Chyueh, S.-D.; Chen, W.-S.; Fang, J.-D.; Liu, C.-Y. Application of dithiocarbamate resin-metal complexes as stationary phases in gas chromatography. *J. Chromatogr. A* **1993**, *630*, 275–285.
43. Bond, A.M.; Wallace, G.G. Preparation of metal dithiocarbamate complexes for chromatographic separation and multi-element determinations. *Anal. Chim. Acta* **1984**, *164*, 223–232. [CrossRef]
44. Aghbash, K.O.; Alamgholiloo, H.; Pesyan, N.N.; Khaksar, S.; Rostamnia, S. Gold nanoparticle stabilized dithiocarbamate functionalized magnetite carbon as promise clean nanocatalyst for A3-coupling organic transformation. *Mol. Catal.* **2021**, *499*, 111252. [CrossRef]
45. Pitchaimani, P.; Lo, K.M.; Elango, K.P. Synthesis, crystal structures, luminescence properties and catalytic application of lanthanide(III) piperidine dithiocarbamate complexes. *Polyhedron* **2015**, *93*, 8–16. [CrossRef]
46. Vale, J.A.; Faustino, W.M.; Menezes, P.H.; de Sá, G.F. Eu(iii) dithiocarbamate complex and *N*-p-tolylsulfonylphenylalanine as a novel chiral catalyst for the asymmetric synthesis of cyanohydrins. *Chem. Commun.* **2006**, *31*, 3340–3342. [CrossRef]
47. Guan, S.; Zhong, Z.; Li, J.; Xu, Y.; Ding, L.; Huang, Y.; Liu, L. Preparation of in-situ grown carbon nanotubes via dithiocarbamate in composites with excellent microstructure and mechanical performance. *Compos. Sci. Technol.* **2021**, *203*, 108569. [CrossRef]
48. Uchiyama, M.; Satoh, K.; Kamigaito, M. Stereospecific cationic RAFT polymerization of bulky vinyl ethers and stereoblock poly(vinyl alcohol) via mechanistic transformation to radical RAFT polymerization of vinyl acetate. *Giant* **2021**, *5*, 100047. [CrossRef]
49. Uchiyama, M.; Satoh, K.; Kamigaito, M. Thioether-Mediated Degenerative Chain-Transfer Cationic Polymerization: A Simple Metal-Free System for Living Cationic Polymerization. *Macromolecules* **2015**, *48*, 5533–5542. [CrossRef]
50. Uchiyama, M.; Satoh, K.; Kamigaito, M. Cationic RAFT Polymerization Using ppm Concentrations of Organic Acid. *Angew. Chem. Int. Ed.* **2015**, *54*, 1924–1928. [CrossRef]
51. Huang, Q.; Liao, H.; Hu, X.; Cheng, C. A cardanol-based surface-active dithiocarbamate and its application in emulsion polymerization. *IOP Conf. Ser. Mater. Sci. Eng.* **2019**, *490*, 022009. [CrossRef]
52. Olatunde, O.C.; Onwudiwe, D.C. Temperature Controlled Evolution of Pure Phase Cu9S5 Nanoparticles by Solvothermal Process. *Front. Mater.* **2021**, *8*, 211. [CrossRef]
53. Ajiboye, T.O.; Oyewo, O.A.; Onwudiwe, D.C. The performance of bismuth-based compounds in photocatalytic applications. *Surf. Interfaces* **2021**, *23*, 100927. [CrossRef]
54. Srinivasan, N. Fabrication and photocatalytic properties of Multi–Morphological CdS NSs prepared by the thermolysis of heterocyclic dithiocarbamate Cadmium(II) complexes as precursors. *Dyes Pigments* **2019**, *162*, 786–796. [CrossRef]
55. Sarker, J.C.; Hogarth, G. Dithiocarbamate Complexes as Single Source Precursors to Nanoscale Binary, Ternary and Quaternary Metal Sulfides. *Chem. Rev.* **2021**, *121*, 6057–6123. [CrossRef]
56. Mourdikoudis, S.; Liz-Marzán, L.M. Oleylamine in Nanoparticle Synthesis. *Chem. Mater.* **2013**, *25*, 1465–1476. [CrossRef]
57. Paca, A.M.; Ajibade, P.A. Bis-(*N*-ethylphenyldithiocarbamato)palladium(II) as molecular precursor for palladium sulfide nanoparticles. *J. Mol. Struct.* **2021**, *1243*, 130777. [CrossRef]
58. Eswari, S.; Selvaganapathi, P.; Thirumaran, S.; Ciattini, S. Effect of solvent used for crystallization on structure: Synthesis and characterization of bis(*N*,*N*-di(4-fluorobenzyl)dithiocarbamato-S,S′)M(II) (M = Cd, Hg) and usage as precursor for CdS nanophotocatalyst. *Polyhedron* **2021**, *206*, 115330. [CrossRef]
59. Galiyeva, P.; Rinnert, H.; Balan, L.; Alem, H.; Medjahdi, G.; Uralbekov, B.; Schneider, R. Single-source precursor synthesis of quinary AgInGaZnS QDs with tunable photoluminescence emission. *Appl. Surf. Sci.* **2021**, *562*, 150143. [CrossRef]
60. Tanabe, T.; Osaki, J.; Miyajima, M.; Kitamura, K.; Oyama, Y. Raman and TEM characterization of 2D layered MoS2 crystals grown on non-metal surfaces by friction-induced synthesis. *Appl. Surf. Sci.* **2021**, *561*, 150016. [CrossRef]

61. Duran-García, E.I.; Martínez-Santana, J.; Torres-Gómez, N.; Vilchis-Nestor, A.R.; García-Orozco, I. Copper sulfide nanoparticles produced by the reaction of *N*-alkyldithiocarbamatecopper(II) complexes with sodium borohydride. *Mater. Chem. Phys.* **2021**, *269*, 124743. [CrossRef]
62. Mann, P.B.; McGregor, I.J.; Bourke, S.; Burkitt-Gray, M.; Fairclough, S.; Ma, M.T.; Hogarth, G.; Thanou, M.; Long, N.; Green, M. An atom efficient, single-source precursor route to plasmonic CuS nanocrystals. *Nanoscale Adv.* **2019**, *1*, 522–526. [CrossRef]
63. Murtaza, G.; Venkateswaran, S.P.; Thomas, A.G.; O'Brien, P.; Lewis, D.J. Chemical vapour deposition of chromium-doped tungsten disulphide thin films on glass and steel substrates from molecular precursors. *J. Mater. Chem. C* **2018**, *6*, 9537–9544. [CrossRef]
64. Zeng, N.; Hopkinson, D.G.; Spencer, B.F.; McAdams, S.G.; Tedstone, A.A.; Haigh, S.J.; Lewis, D.J. Direct synthesis of MoS_2 or MoO_3 via thermolysis of a dialkyl dithiocarbamato molybdenum(iv) complex. *Chem. Commun.* **2019**, *55*, 99–102. [CrossRef]
65. Fomenko, I.S.; Gushchin, A.L.; Nadolinny, V.A.; Efimov, N.N.; Laricheva, Y.A.; Sokolov, M.N. Dinuclear Vanadium Sulfide Clusters: Synthesis, Redox Behavior, and Magnetic Properties. *Eur. J. Inorg. Chem.* **2018**, *2018*, 2965–2971. [CrossRef]
66. Peng, L.; Shen, S.; Zhang, Y.; Xu, H.; Wang, Q. Controllable synthesis of MnS nanocrystals from a single-source precursor. *J. Colloid Interface Sci.* **2012**, *377*, 13–17. [CrossRef] [PubMed]
67. Mlowe, S.; Lewis, D.J.; Malik, M.A.; Raftery, J.; Mubofu, E.B.; O'Brien, P.; Revaprasadu, N. Heterocyclic dithiocarbamato-iron(iii) complexes: Single-source precursors for aerosol-assisted chemical vapour deposition (AACVD) of iron sulfide thin films. *Dalton Trans.* **2016**, *45*, 2647–2655. [CrossRef] [PubMed]
68. Dong, Z.; Jiang, T.; Xu, B.; Li, Q.; Zhong, H.; Yang, Y. Selective flotation of galena using a novel collector S-benzyl-*N*-ethoxycarbonyl thiocarbamate: An experimental and theoretical investigation. *J. Mol. Liq.* **2021**, *330*, 115643. [CrossRef]
69. Grainger, R.S.; Innocenti, P. New applications of dithiocarbamates in organic synthesis. *Heteroat. Chem.* **2007**, *18*, 568–571. [CrossRef]
70. Ahmed, S.; Baker, L.A.; Grainger, R.S.; Innocenti, P.; Quevedo, C.E. Thermal Elimination of Diethyldithiocarbamates and Application in the Synthesis of (±)-Ferrugine. *J. Org. Chem.* **2008**, *73*, 8116–8119. [CrossRef]
71. Jamir, L.; Sinha, U.B.; Nath, J.; Patel, B.K. Environmentally Benign One-Pot Synthesis of Cyanamides from Dithiocarbamates Using I2 and H2O2. *Synth. Commun.* **2012**, *42*, 951–958. [CrossRef]
72. Yin, B.; Liu, Z.; Yi, M.; Zhang, J. An efficient method for the synthesis of disubstituted thioureas via the reaction of *N,N′*-di-Boc-substituted thiourea with alkyl and aryl amines under mild conditions. *Tetrahedron Lett.* **2008**, *49*, 3687–3690. [CrossRef]
73. Halimehjani, A.Z.; Pourshojaei, Y.; Saidi, M.R. Highly efficient and catalyst-free synthesis of unsymmetrical thioureas under solvent-free conditions. *Tetrahedron Lett.* **2009**, *50*, 32–34. [CrossRef]
74. Ziyaei-Halimehjani, A.; Marjani, K.; Ashouri, A. A one-pot, three-component synthesis of thiazolidine-2-thiones. *Tetrahedron Lett.* **2012**, *53*, 3490–3492. [CrossRef]
75. Ziyaei Halimehjani, A.; Ranjbari, M.A.; Pasha Zanussi, H. Synthesis of a new series of dithiocarbamate-linked peptidomimetics and their application in Ugi reactions. *RSC Adv.* **2013**, *3*, 22904–22908. [CrossRef]
76. Aryanasab, F.; Halimehjani, A.Z.; Saidi, M.R. Dithiocarbamate as an efficient intermediate for the synthesis of 2-amino-1,3,4-thiadiazoles in water. *Tetrahedron Lett.* **2010**, *51*, 790–792. [CrossRef]
77. Raina-Fulton, R. A Review of Methods for the Analysis of Orphan and Difficult Pesticides: Glyphosate, Glufosinate, Quaternary Ammonium and Phenoxy Acid Herbicides, and Dithiocarbamate and Phthalimide Fungicides. *J. AOAC Int.* **2019**, *97*, 965–977. [CrossRef]
78. Kaufman, D.D. Degradation of carbamate herbicides in soil. *J. Agric. Food Chem.* **1967**, *15*, 582–591. [CrossRef]
79. Rogachev, I.; Kampel, V.; Gusis, V.; Cohen, N.; Gressel, J.; Warshawsky, A. Synthesis, Properties, and Use of Copper-Chelating Amphiphilic Dithiocarbamates as Synergists of Oxidant-Generating Herbicides. *Pestic. Biochem. Physiol.* **1998**, *60*, 133–145. [CrossRef]
80. Abulnaja, K.O.; Harwood, J.L. Thiocarbamate herbicides inhibit fatty acid elongation in a variety of monocotyledons. *Phytochemistry* **1991**, *30*, 1445–1447. [CrossRef]
81. Wang, Z.; Yang, L.; Ye, X.; Huang, C.; Yang, W.; Zhang, L.; Wu, Z.; Fu, F. Multicolor visual screening of total dithiocarbamate pesticides in foods based on sulfydryl-mediated growth of gold nanobipyramids. *Anal. Chim. Acta* **2020**, *1139*, 59–67. [CrossRef] [PubMed]
82. Eng, G.; Song, X.; Duong, Q.; Strickman, D.; Glass, J.; May, L. Synthesis, structure characterization and insecticidal activity of some triorganotin dithiocarbamates. *Appl. Organomet. Chem.* **2003**, *17*, 218–225. [CrossRef]
83. Fan, Z.; Qin, Y.; Liu, S.; Xing, R.; Yu, H.; Chen, X.; Li, K.; Li, R.; Wang, X.; Li, P. The bioactivity of new chitin oligosaccharide dithiocarbamate derivatives evaluated against nematode disease (*Meloidogyne incognita*). *Carbohydr. Polym.* **2019**, *224*, 115155. [CrossRef]
84. Hussain, A.; Pu, H.; Hu, B.; Sun, D.-W. Au@Ag-TGANPs based SERS for facile screening of thiabendazole and ferbam in liquid milk. *Spectrochim. Acta Part A Mol. Biomol. Spectrosc.* **2021**, *245*, 118908. [CrossRef] [PubMed]
85. Runkle, J.; Flocks, J.; Economos, J.; Dunlop, A.L. A systematic review of Mancozeb as a reproductive and developmental hazard. *Environ. Int.* **2017**, *99*, 29–42. [CrossRef] [PubMed]
86. Boran, H.; Altinok, I.; Capkin, E. Histopathological changes induced by maneb and carbaryl on some tissues of rainbow trout, Oncorhynchus mykiss. *Tissue Cell* **2010**, *42*, 158–164. [CrossRef]
87. Roede, J.R.; Jones, D.P. Thiol-reactivity of the fungicide maneb. *Redox Biol.* **2014**, *2*, 651–655. [CrossRef] [PubMed]

88. Nash, R.G.; Beall, M.L., Jr. Fate of maneb and zineb fungicides in microagroecosystem chambers. *J. Agric. Food Chem.* **1980**, *28*, 322–330. [CrossRef]
89. Triky-Dotan, S.; Ofek, M.; Austerweil, M.; Steiner, B.; Minz, D.; Katan, J.; Gamliel, A. Microbial aspects of accelerated degradation of metam sodium in soil. *Phytopathology* **2010**, *100*, 367–375. [CrossRef]
90. Chen, M.; Zhao, Z.; Lan, X.; Chen, Y.; Zhang, L.; Ji, R.; Wang, L. Determination of carbendazim and metiram pesticides residues in reapeseed and peanut oils by fluorescence spectrophotometry. *Measurement* **2015**, *73*, 313–317. [CrossRef]
91. Charles, J.M.; Tobia, A.; van Ravenzwaay, B. Subchronic and Chronic Toxicological Investigations on Metiram: The Lack of a Carcinogenic Response in Rodents. *Toxicol. Sci.* **2000**, *54*, 481–492. [CrossRef] [PubMed]
92. Lin, M.S.; Wang, J.S. Determination of an ethylene bisdithiocarbamate based pesticide (Nabam) by cobalt phthalocyanine modified carbon ink electrode. *Electroanal. Int. J. Devoted Fundam. Pract. Asp. Electroanal.* **2004**, *16*, 904–909. [CrossRef]
93. Zhang, H.; Mehmood, K.; Jiang, X.; Yao, W.; Iqbal, M.; Waqas, M.; Rehman, M.U.; Li, A.; Shen, Y.; Li, J. Effect of tetramethyl thiuram disulfide (thiram) in relation to tibial dyschondroplasia in chickens. *Environ. Sci. Pollut. Res.* **2018**, *25*, 28264–28274. [CrossRef] [PubMed]
94. Zhou, T.; Zhao, H.; Huang, L.; Xi, H.; Zhou, D.; Cheng, J. Efficacy of propineb for controlling leaf blotch caused by Marssonina coronaria and its effect on zinc content in apple leaves. *Acta Phytophylacica Sin.* **2008**, *35*, 519–524.
95. Cummings, J.L.; Mason, J.R.; Otis, D.L.; Davis, J.E., Jr.; Ohashi, T.J. Evaluation of methiocarb, ziram, and methyl anthranilate as bird repellents applied to dendrobium orchids. *Wildl. Soc. Bull.* **1994**, *22*, 633–638.
96. Berry, D.J.; Torres Martin de Rosales, R.; Charoenphun, P.; Blower, P.J. Dithiocarbamate complexes as radiopharmaceuticals for medical imaging. *Mini Rev. Med. Chem.* **2012**, *12*, 1174–1183. [CrossRef]
97. El-Aarag, B.Y.A.; Kasai, T.; Zahran, M.A.H.; Zakhary, N.I.; Shigehiro, T.; Sekhar, S.C.; Agwa, H.S.; Mizutani, A.; Murakami, H.; Kakuta, H.; et al. In vitro anti-proliferative and anti-angiogenic activities of thalidomide dithiocarbamate analogs. *Int. Immunopharmacol.* **2014**, *21*, 283–292. [CrossRef]
98. Adokoh, C.K. Therapeutic potential of dithiocarbamate supported gold compounds. *RSC Adv.* **2020**, *10*, 2975–2988. [CrossRef]
99. Morrison, B.W.; Doudican, N.A.; Patel, K.R.; Orlow, S.J. Disulfiram induces copper-dependent stimulation of reactive oxygen species and activation of the extrinsic apoptotic pathway in melanoma. *Melanoma Res.* **2010**, *20*, 11–20. [CrossRef]
100. Elahabaadi, E.; Salarian, A.A.; Nassireslami, E. Design, Synthesis, and Molecular Docking of Novel Hybrids of Coumarin-Dithiocarbamate Alpha-Glucosidase Inhibitors Targeting Type 2 Diabetes Mellitus. *Polycycl. Aromat. Compd.* **2021**, 1–11. [CrossRef]
101. Mollazadeh, M.; Mohammadi-Khanaposhtani, M.; Valizadeh, Y.; Zonouzi, A.; Faramarzi, M.A.; Kiani, M.; Biglar, M.; Larijani, B.; Hamedifar, H.; Mahdavi, M.; et al. Novel Coumarin Containing Dithiocarbamate Derivatives as Potent alpha-Glucosidase Inhibitors for Management of Type 2 Diabetes. *Med. Chem.* **2021**, *17*, 264–272. [CrossRef] [PubMed]
102. Gao, Y.; Li, L.; Liu, Y.; Li, W.; Wang, Z.; Shou, S.; Chai, Y. Effect of semaphorin-3A on the cellular stability of $CD4^{+}CD25^{+}$ regulatory T cells induced by lipopolysaccharide. *Zhonghua Wei Zhong Bing Ji Jiu Yi Xue* **2020**, *32*, 1454–1460. [PubMed]
103. Sağlık, B.N.; Osmaniye, D.; Çevik, U.A.; Levent, S.; Çavuşoğlu, B.K.; Büyükemir, O.; Nezir, D.; Karaduman, A.B.; Özkay, Y.; Koparal, A.S.; et al. Synthesis, characterization and carbonic anhydrase I and II inhibitory evaluation of new sulfonamide derivatives bearing dithiocarbamate. *Eur. J. Med. Chem.* **2020**, *198*, 112392. [CrossRef] [PubMed]
104. Aspatwar, A.; Parvathaneni, N.K.; Barker, H.; Anduran, E.; Supuran, C.T.; Dubois, L.; Lambin, P.; Parkkila, S.; Winum, J.-Y. Design, synthesis, in vitro inhibition and toxicological evaluation of human carbonic anhydrases I, II and IX inhibitors in 5-nitroimidazole series. *J. Enzym. Inhib. Med. Chem.* **2020**, *35*, 109–117. [CrossRef] [PubMed]
105. Ge, Y.; Xu, L.W.; Liu, Y.; Sun, L.Y.; Gao, H.; Li, J.Q.; Yang, K. Dithiocarbamate as a valuable scaffold for the inhibition of metallo-β-lactmases. *Biomolecules* **2019**, *9*, 699. [CrossRef] [PubMed]
106. Asadi, M.; Ebrahimi, M.; Mohammadi-Khanaposhtani, M.; Azizian, H.; Sepehri, S.; Nadri, H.; Biglar, M.; Amanlou, M.; Larijani, B.; Mirzazadeh, R.; et al. Design, Synthesis, Molecular Docking, and Cholinesterase Inhibitory Potential of Phthalimide-Dithiocarbamate Hybrids as New Agents for Treatment of Alzheimer's Disease. *Chem. Biodivers.* **2019**, *16*, e1900370. [CrossRef] [PubMed]
107. Cihlar, T.; Fordyce, M. Current status and prospects of HIV treatment. *Curr. Opin. Virol.* **2016**, *18*, 50–56. [CrossRef]
108. Takamune, N.; Misumi, S.; Shoji, S. Cyclic Zinc-Dithiocarbamate-S,S′-Dioxide Blocks CXCR4-Mediated HIV-1 Infection1. *Biochem. Biophys. Res. Commun.* **2000**, *272*, 351–356. [CrossRef]
109. Pang, H.; Chen, D.; Cui, Q.C.; Ping Dou, Q. Sodium diethyldithiocarbamate, an AIDS progression inhibitor and a copper-binding compound, has proteasome-inhibitory and apoptosis-inducing activities in cancer cells. *Int. J. Mol. Med.* **2007**, *19*, 809–816. [CrossRef]
110. Watanabe, K.; Kazakova, I.; Furniss, M.; Miller, S.C. Dual activity of pyrrolidine dithiocarbamate on κB-dependent gene expression in U937 cells: I. Regulation by the phorbol ester TPA. *Cell. Signal.* **1999**, *11*, 479–489. [CrossRef]
111. Schreck, R.; Meier, B.; Männel, D.N.; Dröge, W.; Baeuerle, P.A. Dithiocarbamates as potent inhibitors of nuclear factor kappa B activation in intact cells. *J. Exp. Med.* **1992**, *175*, 1181–1194. [CrossRef]
112. Ahlenstiel, C.L.; Suzuki, K.; Marks, K.; Symonds, G.P.; Kelleher, A.D. Controlling HIV-1: Non-Coding RNA Gene Therapy Approaches to a Functional Cure. *Front. Immunol.* **2015**, *6*, 474. [CrossRef] [PubMed]

113. Lang, J.-M.; Trepo, C.; Kirstetter, M.; Herviou, L.; Retornaz, G.; Renoux, G.; Musset, M.; Touraine, J.-L.; Choutet, P.; Falkenrodt, A.; et al. The Aids-Imuthiol French Study, G. Randomised, double-blind, placebo-controlled trial of ditiocarb sodium ('Imuthiol') in human immunodeficiency virus infection. *Lancet* **1988**, *332*, 702–706. [CrossRef]
114. Sunderman, F.W.S. Therapeutic properties of sodium diethyldithiocarbamate: Its role as an inhibitor in the progression of AIDS. *Ann. Clin. Lab. Sci.* **1991**, *21*, 70–81. [PubMed]
115. Hersh, E.M.; Brewton, G.; Abrams, D.; Bartlett, J.; Galpin, J.; Gill, P.; Gorter, R.; Gottlieb, M.; Jonikas, J.J.; Landesman, S.; et al. Ditiocarb Sodium (Diethyldithiocarbamate) Therapy in Patients With Symptomatic HIV Infection and AIDS: A Randomized, Double-blind, Placebo-Controlled, Multicenter Study. *JAMA* **1991**, *265*, 1538–1544. [CrossRef] [PubMed]
116. Bozdag, M.; Carta, F.; Vullo, D.; Akdemir, A.; Isik, S.; Lanzi, C.; Scozzafava, A.; Masini, E.; Supuran, C.T. Synthesis of a new series of dithiocarbamates with effective human carbonic anhydrase inhibitory activity and antiglaucoma action. *Bioorganic Med. Chem.* **2015**, *23*, 2368–2376. [CrossRef] [PubMed]
117. Supuran, C.T. Emerging role of carbonic anhydrase inhibitors. *Clin. Sci.* **2021**, *135*, 1233–1249. [CrossRef] [PubMed]
118. Francioli, C.; Wang, X.; Parapanov, R.; Abdelnour, E.; Lugrin, J.; Gronchi, F.; Perentes, J.; Eckert, P.; Ris, H.-B.; Piquilloud, L. Pyrrolidine dithiocarbamate administered during ex-vivo lung perfusion promotes rehabilitation of injured donor rat lungs obtained after prolonged warm ischemia. *PLoS ONE* **2017**, *12*, e0173916. [CrossRef]
119. Soerensen, J.; Pekcec, A.; Fuest, C.; Nickel, A.; Potschka, H. Pyrrolidine dithiocarbamate protects the piriform cortex in the pilocarpine status epilepticus model. *Epilepsy Res.* **2009**, *87*, 177–183. [CrossRef]
120. Ebenezer, P.J.; Mariappan, N.; Elks, C.M.; Haque, M.; Soltani, Z.; Reisin, E.; Francis, J. Effects of pyrrolidine dithiocarbamate on high-fat diet-induced metabolic and renal alterations in rats. *Life Sci.* **2009**, *85*, 357–364. [CrossRef]
121. Pal, D.S.; Mondal, D.K.; Datta, R. Identification of Metal Dithiocarbamates as a Novel Class of Antileishmanial Agents. *Antimicrob. Agents Chemother.* **2015**, *59*, 2144–2152. [CrossRef] [PubMed]
122. Alvar, J.; Vélez, I.D.; Bern, C.; Herrero, M.; Desjeux, P.; Cano, J.; Jannin, J.; den Boer, M.; Team, W.L.C. Leishmaniasis worldwide and global estimates of its incidence. *PLoS ONE* **2012**, *7*, e35671. [CrossRef]
123. Pandey, K.; Pun, S.B.; Pandey, B.D. Relapse of kala-azar after use of multiple drugs: A case report and brief review of literature. *Indian J. Med. Microbiol.* **2012**, *30*, 227–229. [CrossRef]
124. Massai, L.; Messori, L.; Micale, N.; Schirmeister, T.; Maes, L.; Fregona, D.; Cinellu, M.A.; Gabbiani, C. Gold compounds as cysteine protease inhibitors: Perspectives for pharmaceutical application as antiparasitic agents. *BioMetals* **2017**, *30*, 313–320. [CrossRef]
125. Oliveira, J.W.d.F.; Rocha, H.A.O.; de Medeiros, W.M.T.Q.; Silva, M.S. Application of Dithiocarbamates as Potential New Antitrypanosomatids-Drugs: Approach Chemistry, Functional and Biological. *Molecules* **2019**, *24*, 2806. [CrossRef]
126. Mezencev, R.; Galizzi, M.; Kutschy, P.; Docampo, R. Trypanosoma cruzi: Antiproliferative effect of indole phytoalexins on intracellular amastigotes in vitro. *Exp. Parasitol.* **2009**, *122*, 66–69. [CrossRef]
127. Ochoa, C.; Perez, E.; Roland, P.; Suarez, M.; Ochoab, E.; Rodriguez, H.; Barrio, A.G.; Susana, M.; Nogal, J.J.; Martinez, R.A. Synthesis and antiprotozoan properties of new 3,5-disubstituted-tetrahydro-2H-1,3,5-thiadiazine-2-thione derivatives. *Arzneimittelforschung* **1999**, *49*, 764–769.
128. Alam, U.; Asghar, O.; Azmi, S.; Malik, R.A. General aspects of diabetes mellitus. *Handb. Clin. Neurol.* **2014**, *126*, 211–222.
129. Chu, G.; Lei1, C.; Qiu, P.; Hu, Y.; Meng, X. Pyrrolidine dithiocarbamate alleviated anxiety in diabetic mice. *Indian J. Pharm. Sci.* **2017**, *79*, 149–154.
130. Uchide, N.; Ohyama, K.; Bessho, T.; Yuan, B.; Yamakawa, T. Effect of antioxidants on apoptosis induced by influenza virus infection: Inhibition of viral gene replication and transcription with pyrrolidine dithiocarbamate. *Antivir. Res.* **2002**, *56*, 207–217. [CrossRef]
131. Dighe, S.U.; Yadav, V.D.; Srivastava, R.; Mishra, A.; Gautam, S.; Srivastava, A.K.; Balaramnavar, V.M.; Saxena, A.K.; Batra, S. Reinvestigations into synthesis of allyldithiocarbamates and their intramolecular cyclization: Synthesis and antihyperglycemic activity of 2-thioxothiazolidine-4-alkanoates. *Tetrahedron* **2014**, *70*, 6841–6850. [CrossRef]
132. Maresca, A.; Carta, F.; Vullo, D.; Supuran, C.T. Dithiocarbamates strongly inhibit the β-class carbonic anhydrases from Mycobacterium tuberculosis. *J. Enzym. Inhib. Med. Chem.* **2013**, *28*, 407–411. [CrossRef] [PubMed]
133. Jiang, N.; Huang, Q.; Liu, J.; Liang, N.; Li, Q.; Li, Q.; Xie, S.-S. Design, synthesis and biological evaluation of new coumarin-dithiocarbamate hybrids as multifunctional agents for the treatment of Alzheimer's disease. *Eur. J. Med. Chem.* **2018**, *146*, 287–298. [CrossRef] [PubMed]
134. Vullo, D.; Del Prete, S.; Nocentini, A.; Osman, S.M.; AlOthman, Z.; Capasso, C.; Bozdag, M.; Carta, F.; Gratteri, P.; Supuran, C.T. Dithiocarbamates effectively inhibit the β-carbonic anhydrase from the dandruff-producing fungus Malassezia globosa. *Bioorganic Med. Chem.* **2017**, *25*, 1260–1265. [CrossRef]
135. Mohsen, U.; Kaplancikli, Z.; Özkay, Y.; Yurttaş, L. Synthesis and evaluation of anti-acetylcholinesterase activity of some benzothiazole based new piperazine-dithiocarbamate derivatives. *Drug Res.* **2015**, *65*, 176–183. [CrossRef]
136. Brewer, C. Recent developments in disulfiram treatment. *Alcohol Alcohol.* **1993**, *28*, 383–395. [PubMed]
137. O'Farrell, T.J.; Allen, J.P.; Litten, R.Z. Disulfiram (antabuse) contracts in treatment of alcoholism. *NIDA Res. Monogr.* **1995**, *150*, 65–91.
138. Abdelkader, N.F.; Arafa, N.M.; Attia, A.S.; Ain-Shoka, A.A.; Abdallah, D.M. Pyrrolidine dithiocarbamate ameliorates rotenone-induced Parkinson's disease in rats. *Bull. Fac. Pharm. Cairo Univ.* **2017**, *55*, 107–113. [CrossRef]
139. Wen, Z.; Lei, Z.; Tian, E.; Wang, Y.; Zhong, Y.; Ge, R.S. Inhibition of human sperm motility and capacitation by ziram is mediated by decreasing tyrosine protein kinase. *Ecotoxicol. Environ. Saf.* **2021**, *218*, 112281. [CrossRef]

140. Ferraz, C.R.; Manchope, M.F.; Andrade, K.C.; Saraiva-Santos, T.; Franciosi, A.; Zaninelli, T.H.; Bagatim-Souza, J.; Borghi, S.M.; Cândido, D.M.; Knysak, I.; et al. Peripheral mechanisms involved in Tityus bahiensis venom-induced pain. *Toxicon* **2021**, *200*, 3–12. [CrossRef]
141. Vane, J.R.; Botting, R.M. The mechanism of action of aspirin. *Thromb. Res.* **2003**, *110*, 255–258. [CrossRef]
142. Dinarello, C.A. Anti-inflammatory Agents: Present and Future. *Cell* **2010**, *140*, 935–950. [CrossRef] [PubMed]
143. Song, Z.; Zhou, Y.; Zhang, W.; Zhan, L.; Yu, Y.; Chen, Y.; Jia, W.; Liu, Z.; Qian, J.; Zhang, Y.; et al. Base promoted synthesis of novel indole-dithiocarbamate compounds as potential anti-inflammatory therapeutic agents for treatment of acute lung injury. *Eur. J. Med. Chem.* **2019**, *171*, 54–65. [CrossRef] [PubMed]
144. Lamkanfi, M. Emerging inflammasome effector mechanisms. *Nat. Rev. Immunol.* **2011**, *11*, 213–220. [CrossRef]
145. Li, C.-W.; Chen, Z.-W.; Wu, X.-L.; Ning, Z.-X.; Su, Z.-Q.; Li, Y.-C.; Su, Z.-R.; Lai, X.-P. A Standardized Traditional Chinese Medicine Preparation Named Yejuhua Capsule Ameliorates Lipopolysaccharide-Induced Acute Lung Injury in Mice via Downregulating Toll-Like Receptor 4/Nuclear Factor-kB. *Evid.-Based Complementary Altern. Med.* **2015**, *2015*, 264612. [CrossRef]
146. Topping, R.J.; Jones, M.M. Optimal dithiocarbamate structure for immunomodulator action. *Med. Hypotheses* **1988**, *27*, 55–57. [CrossRef]
147. Cuzzocrea, S.; Chatterjee, P.K.; Mazzon, E.; Dugo, L.; Serraino, I.; Britti, D.; Mazzullo, G.; Caputi, A.P.; Thiemermann, C. Pyrrolidine dithiocarbamate attenuates the development of acute and chronic inflammation. *Br. J. Pharmacol.* **2020**, *135*, 496–510. [CrossRef]
148. Bhalla, Y.; Gupta, V.K.; Jaitak, V. Anticancer activity of essential oils: A review. *J. Sci. Food Agric.* **2013**, *93*, 3643–3653. [CrossRef]
149. Reddy, L.; Odhav, B.; Bhoola, K.D. Natural products for cancer prevention: A global perspective. *Pharmacol. Ther.* **2003**, *99*, 1–13. [CrossRef]
150. Fu, D.J.; Li, J.H.; Yang, J.J.; Li, P.; Zhang, Y.B.; Liu, S.; Li, Z.R.; Zhang, S.Y. Discovery of novel chalcone-dithiocarbamates as ROS-mediated apoptosis inducers by inhibiting catalase. *Bioorganic Chem.* **2019**, *86*, 375–385. [CrossRef]
151. Milacic, V.; Chen, D.; Giovagnini, L.; Diez, A.; Fregona, D.; Dou, Q.P. Pyrrolidine dithiocarbamate-zinc(II) and -copper(II) complexes induce apoptosis in tumor cells by inhibiting the proteasomal activity. *Toxicol. Appl. Pharmacol.* **2008**, *231*, 24–33. [CrossRef] [PubMed]
152. Lawal, M.M.; Lawal, I.A.; Klink, M.J.; Tolufashe, G.F.; Ndagi, U.; Kumalo, H.M. Density functional theory study of gold(III)-dithiocarbamate complexes with characteristic anticancer potentials. *J. Inorg. Biochem.* **2020**, *206*, 111044. [CrossRef] [PubMed]
153. Omar, A.M.M.E.; AboulWafa, O.M.; El-Shoukrofy, M.S.; Amr, M.E. Benzoxazole derivatives as new generation of anti-breast cancer agents. *Bioorganic Chem.* **2020**, *96*, 103593. [CrossRef] [PubMed]
154. Bakthavatsalam, S.; Wiangnak, P.; George, D.J.; Zhang, T.; Franz, K.J. Dithiocarbamate prodrugs activated by prostate specific antigen to target prostate cancer. *Bioorganic Med. Chem. Lett.* **2020**, *30*, 127148. [CrossRef]
155. Wang, H.; Wei, J.; Jiang, H.; Zhang, Y.; Jiang, C.; Ma, X. Design, synthesis and pharmacological evaluation of three novel dehydroabietyl piperazine dithiocarbamate ruthenium (II) polypyridyl complexes as potential antitumor agents: DNA damage, cell cycle arrest and apoptosis induction. *Molecules* **2021**, *26*, 1453. [CrossRef]
156. Syed Annuar, S.N.; Kamaludin, N.F.; Awang, N.; Chan, K.M. Cellular Basis of Organotin(IV) Derivatives as Anticancer Metallodrugs: A Review. *Front. Chem.* **2021**, *9*, 657599. [CrossRef]
157. Adeyemi, J.O.; Onwudiwe, D.C. The mechanisms of action involving dithiocarbamate complexes in biological systems. *Inorg. Chim. Acta* **2020**, *511*, 119809. [CrossRef]
158. Manoussakis, G.; Bolos, C.; Ecateriniadou, L.; Sarris, C. Synthesis, characterization and anti-bacterial studies of mixed-ligand complexes of dithiocarbamato—thiocyanato and iron(III), nickel(II), copper(II) and zinc(II). *Eur. J. Med. Chem.* **1987**, *22*, 421–425. [CrossRef]
159. Oladipo, S.D.; Tolufashe, G.F.; Mocktar, C.; Omondi, B. Ag(I) symmetrical *N,N′*-diarylformamidine dithiocarbamate PPh3 complexes: Synthesis, structural characterization, quantum chemical calculations and in vitro biological studies. *Inorg. Chim. Acta* **2021**, *520*, 120316. [CrossRef]
160. Chen, Q.M.; Li, Z.; Tian, G.X.; Chen, Y.; Wu, X.H. 1,2,3-triazole-dithiocarbamate-naphthalimides: Synthesis, characterization, and biological evaluation. *J. Chem. Res.* **2021**, *45*, 258–264. [CrossRef]
161. Mohammed, M.H.; Leelon, A.A. Synthesis, characterization of isatin dithiocarbamate derivatives with expected biological activities. *Int. J. Drug Deliv. Technol.* **2021**, *11*, 209–212.
162. Ndukwe, G.I.; Nzeneri, J.U.; Abayeh, O.J. Antibacterial assay of two synthesized dithiocarbamate ligands. *Am. J. Chem. Appl.* **2018**, *5*, 51–57.
163. Ariza-Roldán, A.O.; López-Cardoso, E.M.; Rosas-Valdez, M.E.; Roman-Bravo, P.P.; Vargas-Pineda, D.G.; Cea-Olivares, R.; Acevedo-Quiroz, M.; Razo-Hernández, R.S.; Alvarez-Fitz, P.; Jancik, V. Synthesis, characterization, antimicrobial and theoretical studies of the first main group tris(ephedrinedithiocarbamate) complexes of As(III), Sb(III), Bi(III), Ga(III) and In(III). *Polyhedron* **2017**, *134*, 221–229. [CrossRef]
164. Onwudiwe, D.C.; Ekennia, A.C. Synthesis, characterization, thermal, antimicrobial and antioxidant studies of some transition metal dithiocarbamates. *Res. Chem. Intermed.* **2017**, *43*, 1465–1485. [CrossRef]
165. Pastrana-Dávila, A.; Amaya-Flórez, A.; Aranaga, C.; Ellena, J.; Macías, M.; Flórez-López, E.; D'Vries, R.F. Synthesis, characterization, and antibacterial activity of dibenzildithiocarbamate derivates and Ni(II)–Cu(II) coordination compounds. *J. Mol. Struct.* **2021**, *1245*, 131109. [CrossRef]

166. Mansouri, G.; Ghobadi, M.; Notash, B. Synthesis, spectroscopic, structural, DFT and antibacterial studies of cyclometalated rhodium(III) complex based on morpholinedithiocarbamate ligand. *Inorg. Chem. Commun.* **2021**, *130*, 108707. [CrossRef]
167. Ajiboye, T.O.; Oluwarinde, B.O.; Montso, P.K.; Ateba, C.N.; Onwudiwe, D.C. Antimicrobial activities of Cu(II), In(III), and Sb(III) complexes of *N*-methyl-*N*–phenyl dithiocarbamate complexes. *Results Chem.* **2021**, *3*, 100241. [CrossRef]
168. Menezes, D.C.; Vieira, F.T.; de Lima, G.M.; Wardell, J.L.; Cortés, M.E.; Ferreira, M.P.; Soares, M.A.; Vilas Boas, A. The in vitro antifungal activity of some dithiocarbamate organotin(IV) compounds on Candida albicans—A model for biological interaction of organotin complexes. *Appl. Organomet. Chem.* **2008**, *22*, 221–226. [CrossRef]
169. Qin, Y.; Liu, S.; Xing, R.; Yu, H.; Li, K.; Meng, X.; Li, R.; Li, P. Synthesis and characterization of dithiocarbamate chitosan derivatives with enhanced antifungal activity. *Carbohydr. Polym.* **2012**, *89*, 388–393. [CrossRef]
170. Badawy, M.E.I.; Rabea, E.I. Chapter 7-Chitosan and Its Derivatives as Active Ingredients Against Plant Pests and Diseases. In *Chitosan in the Preservation of Agricultural Commodities*; Bautista-Baños, S., Romanazzi, G., Jiménez-Aparicio, A., Eds.; Academic Press: Cambridge, MA, USA, 2016; pp. 179–219.
171. Ferreira, I.P.; de Lima, G.M.; Paniago, E.B.; Takahashi, J.A.; Pinheiro, C.B. Synthesis, characterization and antifungal activity of new dithiocarbamate-based complexes of Ni(II), Pd(II) and Pt(II). *Inorg. Chim. Acta* **2014**, *423*, 443–449. [CrossRef]
172. Fargasova, A.; Reinprecht, L.; Kizlink, J. Efficiency of organotin dithiocarbamate derivatives against wood destroying fungi. *Biologia* **1997**, *52*, 451–455.
173. Duran, A.; Valero, N.; Mosquera, J.; Fuenmayor, E.; Alvarez-Mon, M. Gefitinib and pyrrolidine dithiocarbamate decrease viral replication and cytokine production in dengue virus infected human monocyte cultures. *Life Sci.* **2017**, *191*, 180–185. [CrossRef] [PubMed]
174. Lin, L.; Qin, Y.; Wu, H.; Chen, Y.; Wu, S.; Si, X.; Wang, H.; Wang, T.; Zhong, X.; Zhai, X.; et al. Pyrrolidine dithiocarbamate inhibits enterovirus 71 replication by down-regulating ubiquitin–proteasome system. *Virus Res.* **2015**, *195*, 207–216. [CrossRef] [PubMed]
175. Mathur, A.; Mallia, M.B.; Subramanian, S.; Banerjee, S.; Kothari, K.; Dhotare, B.; Sarma, H.D.; Venkatesh, M. 99mTcN complexes of tert-butyl dithiocarbamate and methoxyisobutyl dithiocarbamate as myocardial and brain imaging agents. *Nucl. Med. Commun.* **2005**, *26*, 1013–1019. [CrossRef]
176. Torres Martin de Rosales, R.; Tavaré, R.; Paul, R.L.; Jauregui-Osoro, M.; Protti, A.; Glaria, A.; Varma, G.; Szanda, I.; Blower, P.J. Synthesis of 64CuII–bis (dithiocarbamatebisphosphonate) and its conjugation with superparamagnetic iron oxide nanoparticles: In vivo evaluation as dual-modality PET–MRI agent. *Angew. Chem.* **2011**, *123*, 5623–5627. [CrossRef]
177. Zhang, J.; Guo, H.; Zhang, S.; Lin, Y.; Wang, X. Synthesis and biodistribution of a novel 99mTcN complex of ciprofloxacin dithiocarbamate as a potential agent for infection imaging. *Bioorganic Med. Chem. Lett.* **2008**, *18*, 5168–5170. [CrossRef]
178. Lin, X.; Jin, Z.; Ren, J.; Pang, Y.; Zhang, W.; Huo, J.; Wang, X.; Zhang, J.; Zhang, Y. Synthesis and Biodistribution of a New 99mTc-oxo Complex with Deoxyglucose Dithiocarbamate for Tumor Imaging. *Chem. Biol. Drug Des.* **2012**, *79*, 239–245. [CrossRef]
179. Hait, S.; Valentín, J.L.; Jiménez, A.G.; Ortega, P.B.; Ghosh, A.K.; Stöckelhuber, K.W.; Wießner, S.; Heinrich, G.; Das, A. Poly(acrylonitrile-co-butadiene) as polymeric crosslinking accelerator for sulphur network formation. *Heliyon* **2020**, *6*, e04659. [CrossRef]
180. Oenslager, G. Organic Accelerators. *Ind. Eng. Chem.* **1933**, *25*, 232–237. [CrossRef]
181. Wang, Y.; Lü, Y.; Hu, S.; Hu, T.; Wen, S.; Liu, L. Application of Lanthanum Diethyldithiocarbamate as Rubber Accelerator Used in Nitrile Butadiene Rubber. *J. Chin. Rare Earth Soc.* **2019**, *37*, 609–616.
182. Palaty, S.; Joseph, R. Synergism of Xanthate/Dithiocarbamate Accelerator in Carbon Black Filled NR Compounds. *Iran. Polym. J.* **2004**, *13*, 85–91.
183. Zou, Y.; He, J.; Tang, Z.; Zhu, L.; Luo, Y.; Liu, F. Effect of multifunctional samarium lysine dithiocarbamate on curing properties, static and dynamic mechanical properties of SBR/silica composites. *RSC Adv.* **2015**, *6*, 269–280. [CrossRef]
184. Alam, M.N.; Mandal, S.K.; Roy, K.; Debnath, S.C. Safe amine based zinc dithiocarbamates for the vulcanization of carbon black reinforced natural rubber. *J. Appl. Polym. Sci.* **2014**, *131*. [CrossRef]
185. Guo, A.J.; Pan, H.H.; Zheng, W.L.; Jiao, S.H.; Wang, F.; Jin, Z.Z.; Liu, H.; Chen, K.; Wang, Z.X. Synthesis of dispersed molybdenum disulfide nano-catalysts and their performance in the hydrogenation of simulated oil slurry. *J. Fuel Chem. Technol.* **2019**, *47*, 629–640.
186. Yang, S.; Liu, L.; Jia, Z.; Jia, D.; Luo, Y. Study on the curing properties of SBR/La-GDTC/SiO_2 composites. *J. Rare Earths* **2011**, *29*, 444–453. [CrossRef]
187. Pudovik, A.N.; Khairullin, V.K.; Il'yasov, A.V.; Vasyanina, M.A.; Aleksandrova, I.A.; Ismayev, I.E.; Ovcharov, V.I. Mechanism of action of phosphorylated dithiocarbamates on the vulcanization of rubbers. *Polym. Sci. USSR* **1988**, *30*, 475–480. [CrossRef]
188. Nieuwenhuizen, P.J.; Ehlers, A.W.; Haasnoot, J.G.; Janse, S.R.; Reedijk, J.; Baerends, E.J. The Mechanism of Zinc(II)-Dithiocarbamate-Accelerated Vulcanization Uncovered; Theoretical and Experimental Evidence. *J. Am. Chem. Soc.* **1999**, *121*, 163–168. [CrossRef]
189. Liu, S.; Dong, Y.; Xie, L.; Liu, G.; Zhong, H.; Zeng, H. Uncovering the hydrophobic mechanism of a novel dithiocarbamate-hydroxamate surfactant towards galena. *Chem. Eng. Sci.* **2021**, *245*, 116765. [CrossRef]
190. Liu, B.; Wang, X.; Du, H.; Liu, J.; Zheng, S.; Zhang, Y.; Miller, J.D. The surface features of lead activation in amyl xanthate flotation of quartz. *Int. J. Miner. Processing* **2016**, *151*, 33–39. [CrossRef]
191. Feng, Q.-C.; Wen, S.-M.; Zhao, W.-J.; Cao, Q.-B.; Lü, C. A novel method for improving cerussite sulfidization. *Int. J. Miner. Metall. Mater.* **2016**, *23*, 609–617. [CrossRef]

192. Elizondo-Álvarez, M.A.; Uribe-Salas, A.; Nava-Alonso, F. Flotation studies of galena (PbS), cerussite (PbCO3) and anglesite (PbSO4) with hydroxamic acids as collectors. *Miner. Eng.* **2020**, *155*, 106456. [CrossRef]
193. Huang, X.; Jia, Y.; Cao, Z.; Wang, S.; Ma, X.; Zhong, H. Investigation of the interfacial adsorption mechanisms of 2-hydroxyethyl dibutyldithiocarbamate surfactant on galena and sphalerite. *Colloids Surf. A Physicochem. Eng. Asp.* **2019**, *583*, 123908. [CrossRef]
194. Ngobeni, W.A.; Hangone, G. The effect of using sodium di-methyl-dithiocarbamate as a co-collector with xanthates in the froth flotation of pentlandite containing ore from Nkomati mine in South Africa. *Miner. Eng.* **2013**, *54*, 94–99. [CrossRef]
195. Qi, J.; Dong, Y.; Liu, S.; Liu, G. A selective flotation of cassiterite with a dithiocarbamate-hydroxamate molecule and its adsorption mechanism. *Appl. Surf. Sci.* **2021**, *538*, 147996. [CrossRef]
196. Matveeva, T.N.; Gromova, N.K.; Lantsova, L.B. Analysis of Complexing and Adsorption Properties of Dithiocarbamates Based on Cyclic and Aliphatic Amines for Gold Ore Flotation. *J. Min. Sci.* **2020**, *56*, 268–274. [CrossRef]
197. Matveeva, T.N.; Chanturia, V.A.; Gromova, N.K.; Lantsova, L.B. New compositions of agents for fine gold recovery from tailings. *Gorn. Zhurnal* **2019**, *2019*, 48–51. [CrossRef]
198. Wang, J.B.; Zhao, F.; Yang, X.L.; Han, W.Y.; Long, K.; Zhou, Y.R. Marine Environmental Risk Assessment Method for Active Substances Used in Antifouling Systems on Ships in China. In *Advanced Materials Research*; Trans Tech Publications Ltd.: Zurich, Switzerland, 2014; Volume 864, pp. 962–972.
199. Nagata, S.; Zhou, X.; Okamura, H. Antagonistic and Synergistic Effects of Antifouling Chemicals in Mixture. In *Encyclopedia of Ecology, Five-Volume Set*; Elsevier: Amsterdam, The Netherlands, 2008; pp. 194–203.
200. Parviz, M.; Darwish, N.; Alam, M.T.; Parker, S.G.; Ciampi, S.; Gooding, J.J. Investigation of the Antifouling Properties of Phenyl Phosphorylcholine-Based Modified Gold Surfaces. *Electroanalysis* **2014**, *26*, 1471–1480. [CrossRef]
201. Narayanan, T.S.N.; Subbaiyan, M. Effect of dithiocarbamates on the phase constituents, alkaline stability, and wet adhesion of phosphate coatings. *Met. Finish.* **1994**, *92*, 33–34.
202. Narayanan, T.S.N.S.; Subbaiyan, M. Effect of surfactants on the porosity and corrosion resistance of zinc-phosphated steel. *Met. Finish.* **1993**, *91*, 43–45.
203. Belzile, N.; Chen, Y.-W.; Cai, M.-F.; Li, Y. A review on pyrrhotite oxidation. *J. Geochem. Explor.* **2004**, *84*, 65–76. [CrossRef]
204. Shu, X.; Dang, Z.; Zhang, Q.; Yi, X.; Lu, G.; Guo, C.; Yang, C. Passivation of metal-sulfide tailings by covalent coating. *Miner. Eng.* **2013**, *42*, 36–42. [CrossRef]
205. Zhou, Y.; Qu, J. Ionic Liquids as Lubricant Additives: A Review. *ACS Appl. Mater. Interfaces* **2017**, *9*, 3209–3222. [CrossRef] [PubMed]
206. Kenbeek, D.; Buenemann, T.; Rieffe, H. *Review of Organic Friction Modifiers-Contribution to Fuel Efficiency*; SAE International: Warrendale, PA, USA, 2000.
207. Rastogi, R.B.; Maurya, J.L.; Jaiswal, V.; Tiwary, D. Lanthanum dithiocarbamates as potential extreme pressure lubrication additives. *Int. J. Ind. Chem.* **2012**, *3*, 32. [CrossRef]
208. Yamamoto, K.; Hiramatsu, T.; Hanamura, R.; Moriizumi, Y.; Heiden, S. *The Study of Friction Modifiers to Improve Fuel Economy for WLTP with Low and Ultra-Low Viscosity Engine Oil*; SAE International: Warrendale, PA, USA, 2019.
209. Wang, Y.; Yue, W.; Kang, J.; Zhu, L.; Fu, Z.; Wang, C. Effect of Surface Nanocrystallization Pretreatment on the Tribological Properties of Plasma Nitrided AISI 316 L Stainless Steel Under Boundary Lubrication. *J. Tribol.* **2019**, *141*, 042102. [CrossRef]
210. Shah, F.U.; Glavatskih, S.; Antzutkin, O.N. Novel Alkylborate–Dithiocarbamate Lubricant Additives: Synthesis and Tribophysical Characterization. *Tribol. Lett.* **2012**, *45*, 67–78. [CrossRef]
211. Fuentes-Martínez, J.P.; Toledo-Martínez, I.; Román-Bravo, P.; Garcia y García, P.; Godoy-Alcántar, C.; López-Cardoso, M.; Morales-Rojas, H. Diorganotin(IV) dithiocarbamate complexes as chromogenic sensors of anion binding. *Polyhedron* **2009**, *28*, 3953–3966. [CrossRef]
212. Gao, R.; Li, D.; Zheng, S.; Gu, H.; Deng, W. Colorimetric/fluorescent/Raman trimodal sensing of zinc ions with complexation-mediated Au nanorod. *Talanta* **2021**, *225*, 121975. [CrossRef]
213. Yan, Y.; Krishnakumar, S.; Yu, H.; Ramishetti, S.; Deng, L.-W.; Wang, S.; Huang, L.; Huang, D. Nickel(II) Dithiocarbamate Complexes Containing Sulforhodamine B as Fluorescent Probes for Selective Detection of Nitrogen Dioxide. *J. Am. Chem. Soc.* **2013**, *135*, 5312–5315. [CrossRef]
214. Guerrini, L.; Garcia-Ramos, J.V.; Domingo, C.; Sanchez-Cortes, S. Sensing Polycyclic Aromatic Hydrocarbons with Dithiocarbamate-Functionalized Ag Nanoparticles by Surface-Enhanced Raman Scattering. *Anal. Chem.* **2009**, *81*, 953–960. [CrossRef]
215. Rohit, J.V.; Solanki, J.N.; Kailasa, S.K. Surface modification of silver nanoparticles with dopamine dithiocarbamate for selective colorimetric sensing of mancozeb in environmental samples. *Sens. Actuators B Chem.* **2014**, *200*, 219–226. [CrossRef]
216. Gurumoorthy, G.; Rani, P.J.; Thirumaran, S.; Ciattini, S. Cobalt(III) dithiocarbamates for anion sensing and preparation of cobalt sulfide and cobalt-iron sulfide nanoparticles: Photocatalytic degradation of dyes with as-prepared nanoparticles. *Inorg. Chim. Acta* **2017**, *455*, 132–139. [CrossRef]
217. Mehta, V.N.; Basu, H.; Singhal, R.K.; Kailasa, S.K. Simple and sensitive colorimetric sensing of Cd^{2+} ion using chitosan dithiocarbamate functionalized gold nanoparticles as a probe. *Sens. Actuators B Chem.* **2015**, *220*, 850–858. [CrossRef]
218. Mehta, V.N.; Mungara, A.K.; Kailasa, S.K. Dopamine dithiocarbamate functionalized silver nanoparticles as colorimetric sensors for the detection of cobalt ion. *Anal. Methods* **2013**, *5*, 1818–1822. [CrossRef]

219. Rofouei, M.K.; Tajarrod, N.; Masteri-Farahani, M.; Zadmard, R. A New Fluorescence Sensor for Cerium (III) Ion Using Glycine Dithiocarbamate Capped Manganese Doped ZnS Quantum Dots. *J. Fluoresc.* **2015**, *25*, 1855–1866. [CrossRef]
220. Sathiyaraj, E.; Gurumoorthy, G.; Thirumaran, S. Nickel(ii) dithiocarbamate complexes containing the pyrrole moiety for sensing anions and synthesis of nickel sulfide and nickel oxide nanoparticles. *New J. Chem.* **2015**, *39*, 5336–5349. [CrossRef]
221. Mehta, V.N.; Kailasa, S.K. Malonamide dithiocarbamate functionalized gold nanoparticles for colorimetric sensing of Cu^{2+} and Hg^{2+} ions. *RSC Adv.* **2015**, *5*, 4245–4255. [CrossRef]
222. Rohit, J.V.; Kailasa, S.K. Cyclen dithiocarbamate-functionalized silver nanoparticles as a probe for colorimetric sensing of thiram and paraquat pesticides via host–guest chemistry. *J. Nanoparticle Res.* **2014**, *16*, 2585. [CrossRef]
223. Mehta, V.N.; Kailasa, S.K.; Wu, H.-F. Sensitive and selective colorimetric sensing of Fe^{3+} ion by using p-amino salicylic acid dithiocarbamate functionalized gold nanoparticles. *New J. Chem.* **2014**, *38*, 1503–1511. [CrossRef]
224. Rohit, J.V.; Kailasa, S.K. Simple and selective detection of pendimethalin herbicide in water and food samples based on the aggregation of ractopamine-dithiocarbamate functionalized gold nanoparticles. *Sens. Actuators B Chem.* **2017**, *245*, 541–550. [CrossRef]
225. Rohit, J.V.; Singhal, R.K.; Kailasa, S.K. Dithiocarbamate-calix[4]arene functionalized gold nanoparticles as a selective and sensitive colorimetric probe for assay of metsulfuron-methyl herbicide via non-covalent interactions. *Sens. Actuators B Chem.* **2016**, *237*, 1044–1055. [CrossRef]
226. Tonkin, E.G.; Valentine, H.L.; Zimmerman, L.J.; Valentine, W.M. Parenteral *N,N*-diethyldithiocarbamate produces segmental demyelination in the rat that is not dependent on cysteine carbamylation. *Toxicol. Appl. Pharmacol.* **2003**, *189*, 139–150. [CrossRef]
227. Tilton, F.; La Du, J.K.; Tanguay, R.L. Sulfhydryl systems are a critical factor in the zebrafish developmental toxicity of the dithiocarbamate sodium metam (NaM). *Aquat. Toxicol.* **2008**, *90*, 121–127. [CrossRef] [PubMed]
228. Tilton, F.; La Du, J.K.; Vue, M.; Alzarban, N.; Tanguay, R.L. Dithiocarbamates have a common toxic effect on zebrafish body axis formation. *Toxicol. Appl. Pharmacol.* **2006**, *216*, 55–68. [CrossRef] [PubMed]
229. Van Leeuwen, C.J.; Maas-Diepeveen, J.L.; Niebeek, G.; Vergouw, W.H.A.; Griffioen, P.S.; Luijken, M.W. Aquatic toxicological aspects of dithiocarbamates and related compounds. I. Short-term toxicity tests. *Aquat. Toxicol.* **1985**, *7*, 145–164. [CrossRef]
230. Lushchak, V.I.; Matviishyn, T.M.; Husak, V.V.; Storey, J.M.; Storey, K.B. Pesticide toxicity: A mechanistic approach. *EXCLI J.* **2018**, *17*, 1101–1136.
231. Fujii, S.; Yoshimura, T. A new trend in iron–dithiocarbamate complexes: As an endogenous NO trapping agent. *Coord. Chem. Rev.* **2000**, *198*, 89–99. [CrossRef]

International Journal of
Molecular Sciences

Review

Exogenous Bioactive Peptides Have a Potential Therapeutic Role in Delaying Aging in Rodent Models

Jianqiang Wang, Yixin Wu, Zhongxu Chen, Yajuan Chen, Qinlu Lin and Ying Liang *

Molecular Nutrition Branch, National Engineering Research Center of Rice and By-Product Deep Processing, College of Food Science and Engineering, Central South University of Forestry and Technology, Changsha 410004, China; 20201100385@csuft.edu.cn (J.W.); 20191200410@csuft.edu.cn (Y.W.); 20191100352@csuft.edu.cn (Z.C.); 20201100406@csuft.edu.cn (Y.C.); linqinlu@hotmail.com (Q.L.)
* Correspondence: liangy@csuft.edu.cn

Abstract: In recent years, some exogenous bioactive peptides have been shown to have promising anti-aging effects. These exogenous peptides may have a mechanism similar to endogenous peptides, and some can even regulate the release of endogenous active peptides and play a synergistic role with endogenous active peptides. Most aging studies use rodents that are easy to maintain in the laboratory and have relatively homogenous genotypes. Moreover, many of the anti-aging studies using bioactive peptides in rodent models only focus on the activity of single endogenous or exogenous active peptides, while the regulatory effects of exogenous active peptides on endogenous active peptides remain largely under-investigated. Furthermore, the anti-aging activity studies only focus on the effects of these bioactive peptides in individual organs or systems. However, the pathological changes of one organ can usually lead to multi-organ complications. Some anti-aging bioactive peptides could be used for rescuing the multi-organ damage associated with aging. In this paper, we review recent reports on the anti-aging effects of bioactive peptides in rodents and summarize the mechanism of action for these peptides, as well as discuss the regulation of exogenous active peptides on endogenous active peptides.

Keywords: bioactive peptide; anti-aging; rodents

Citation: Wang, J.; Wu, Y.; Chen, Z.; Chen, Y.; Lin, Q.; Liang, Y. Exogenous Bioactive Peptides Have a Potential Therapeutic Role in Delaying Aging in Rodent Models. *Int. J. Mol. Sci.* **2022**, *23*, 1421. https://doi.org/10.3390/ijms23031421

Academic Editor: Fidel Toldrá

Received: 21 December 2021
Accepted: 20 January 2022
Published: 26 January 2022

1. Introduction

In modern society, the extension of average life expectancy and the decreased birth rate have led to aging-related burdens across many regions [1,2]. Aging is a dynamic process associated with accumulated cell damage, a decline in biological function, and susceptibility to disease occurring over time [3]. A common and widely recognized mechanism for aging is oxidative damage caused by the accumulation of reactive oxygen species (ROS) [4], resulting from decreased antioxidant capacity, mitochondrial dysfunction, inflammation, etc. [5]. Aging can lead to multiple age-related diseases (ARDs) [6], such as cancer, Alzheimer's disease (AD), cardiovascular disease (CVD), metabolic syndrome, obesity, fatty liver, and many other chronic diseases. The aging process inevitably involves the aging of cells, which is usually caused by damage at the molecular and cellular level by long-term exposure to endogenous and exogenous stressors. These damaged cells eventually lose their proliferative capacity and promote aging at an organism level [7]. These senescent cells can release a variety of pro-inflammatory factors and chemokines to promote cellular dysfunction, causing senescence-related diseases. In the process of skin aging, oxidative stress and inflammation can increase the activity of matrix metalloproteinases (MMPs) and increase the degradation of collagen, resulting in skin sagging and wrinkle formation. In some neurodegenerative diseases, such as AD, oxidative stress and inflammation can increase the accumulation of amyloid plaques (Aβ) and promote lesions in the brain. Oxidative stress and inflammation also play an important role in the aging of several other organs, such as the heart, liver, and kidneys. Collectively, these pathological changes can cause a variety of complications that affect multiple systems in the body. Thus, ARDs seriously impact the quality of life, shorten the lifespan, and bring a heavy burden to families and society. Therefore, in-depth studies of aging are particularly important.

Bioactive peptides are short peptides consisting of 2–20 amino acid residues. They have positive effects on body functions and generally have antibacterial, antihypertensive, antioxidant, and anti-inflammatory effects [8]. Natural bioactive peptides can be generally divided into two categories: endogenous peptides, which are naturally released from precursor proteins and secreted from cells, and exogenous peptides, which are produced by enzymatic hydrolysis of proteins or by biosynthesis or organic synthesis [9,10]. Bioactive peptide resources have been found in plants (soybeans, walnuts, rice bran, etc.), animals (some fish, dairy products, etc.), and some fungi and bacteria (yeast, lactic acid bacteria, etc.). The bioactive peptides used in early research were mainly derived from milk, cheese, and other dairy products. As research has progressed, active peptides have also been derived from other foods, including animal products as well as plant products [11]. They have been widely used in animal research, especially in rodents, but with limited research in humans. This is because rodents are easy to breed in the laboratory setting, have a short life cycle, and can be rapidly bred. Rodents also share similar genes and physiological functions with humans, making them ideal experimental animal models [12,13]. In this paper, we review the recent progress in anti-aging research involving the use of bioactive peptides in animal models, especially in rodents. In addition, we also highlight that aromatic residue, such as Trp, in some of the reported active peptides, can confer their anti-inflammatory and antioxidant activity. Moreover, many studies on active peptides mainly focus on the direct effects of these exogenous active peptides but ignore their indirect effects through regulating endogenous antioxidants in vivo. For example, exogenous active peptides can enhance endogenous antioxidative activity by increasing the levels of glutathione (GSH), superoxide dismutase (SOD), and bone-derived neurotrophic factor (BDNF) [14–16].

We divide this review into several sections based on the anti-aging effects on different organs. In each section, we review the mechanism of aging and the mechanism of action for the anti-aging effect of these bioactive peptides in each organ. We also summarize their common mechanism of action in different organs and the synergistic regulatory effects between endogenous and exogenous active peptides.

2. Bioactive Peptides Delay Skin Aging

2.1. Skin Aging

Skin is the largest organ and the body's first barrier of defense against external pathogens. The skin protects the body from environmental damage and invasion of pathogens, and it is responsible for managing body temperature, sensation, and secretion function. Aging can cause different degrees of skin damage and interfere with the normal physiological function of other organs in the body [17]. The etiology of skin aging includes many factors. Among them, internal aging and photoaging are most common. Aging can alter the structure, function, and appearance of the skin, eventually leading to the increase of wrinkles, loss of elasticity, sagging, and pigment precipitation [18]. The main mechanisms of skin aging are the decrease of antioxidants in the skin, inflammation, and the degradation of collagen by increased MMPs [19]. Anti-aging bioactive peptides often act on these aging mechanisms. For example, oral collagen hydrolysates (CHs) can inhibit the activity of MMPs to reduce the degradation of collagen fibers [20]. Active peptides can reduce skin photoaging by scavenging free radicals [21]. Some bioactive peptides can reduce inflammation. In general, both endogenous and exogenous active peptides can down-regulate the factors causing skin aging. We discuss these in detail below.

2.2. Antioxidant Peptides in Delaying Skin Aging

Bioactive peptides derived from some animal proteins have antioxidant activity. These bioactive peptides can delay skin aging by regulating oxidative stress (Figure 1). For example, the collagen peptide extracted from the swim bladder of Sturgeon can increase the activities of catalase (CAT), SOD, and GSH peroxidase (GSH-PX) and decrease the activity of MMPs in skin tissue from Sprague-Dawley rats, as well as reduce the degradation of collagen by MMPs [22]. In recent years, some insect proteins with biological activity have also been found. For example, Eupolyphaga sinensis walker polypeptides (EPs) is a polypeptide mixture with a molecular weight of less than 3.3 kDa obtained from enzymatic digestion that can significantly improve the activity of antioxidant enzymes and reduce the generation of harmful free radicals. Thus, EPs can reduce the UV-irradiation-induced increase in epidermal thickness and elastic fiber breakage and restore the content of collagen [23]. In both cases, the mechanism of action of these exogenous active peptides is mainly to improve the activity of antioxidant enzymes in the skin and reduce the activity of MMPs and the degradation of collagen.

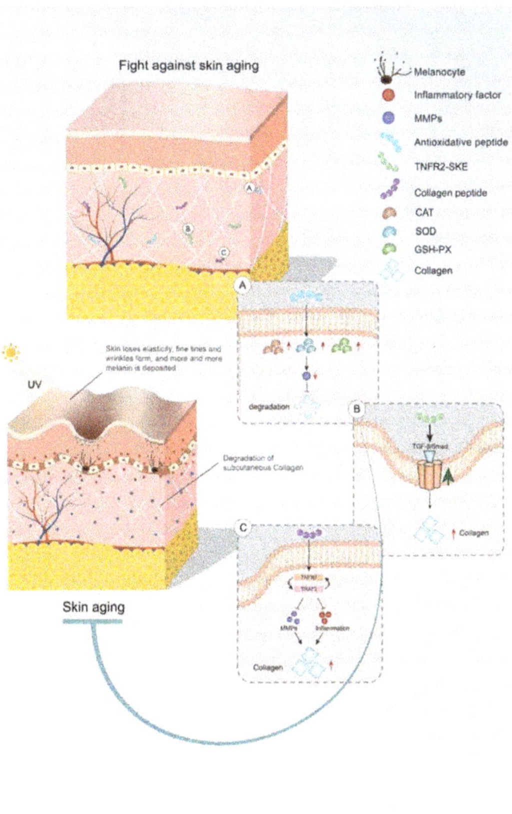

Figure 1. Mechanism of bioactive peptides in delaying skin aging. (**A**) Antioxidant peptides can increase the activity of antioxidant enzymes. (**B**) Bioactive peptides retard skin aging through the TGF-β/Smad pathway. (**C**) Active peptides inhibit inflammation and MMP activity. This figure cannot be reproduced without author permission.

2.3. *Anti-Inflammatory Peptides in Delaying Skin Aging*

Some endogenous peptides with anti-inflammatory effects have been used to delay skin aging. The tripeptide TNFR2-SKE (362.4 Da) derived from the tetrapeptide of TNF receptor-associated factor 2 (TNFR2) showed a good protective effect against skin photoaging. TNFR2-SKE can block the interaction between TNFR1 and TRAF2 and inhibit the inflammation induced by TNF-2 (Figure 1). Intraperitoneal administration of TNFR2-SKE to UVB-irradiated six-week-old male DBA/2 mice was shown to significantly improve epidermal thickness and pigment cell proliferation [24]. MOTS-C is a 16-peptide from the MDP family derived from mitochondria with a molecular weight of 2174.61 Da. This bioactive peptide can regulate cell metabolism and inflammation [25,26]. In a D galactose-induced aging mouse model, treatment with MOTS-c was shown to increase collagen fiber content in the dermis by increasing NRF2 and MFN2 and decreasing interleukin-6 (IL-6). The anti-aging activity of MOTS-c is likely achieved by reducing inflammation [27]. Thus, both TNFR2-SKE and MOTS-C active peptides showed good performance in significantly alleviating skin inflammation and increasing collagen fiber content in mice. These endogenous active peptides can delay skin aging through their anti-inflammatory effects. However, there are many endogenous anti-inflammatory polypeptides in the body, and their anti-aging effects on the skin remain to be explored.

2.4. Peptides in Reducing Collagen Hydrolysis

Collagen is the main component of the dermis, and its content decreases with age. Skin sagging and wrinkles are caused by a decrease in collagen content. It is noteworthy that oral CHs can reduce skin laxity and wrinkles [28] and delay skin aging. Fish skin and fish scales are generally rich in collagen. Two collagen hydrolysates (ACH and CCH) prepared from fish skin can up-regulate the transforming growth factor β (TGF-β)/Smad signaling pathway related to collagen synthesis and increase the amount of collagen. CHs have a good protective effect on skin laxity, as shown in 13-month-old female KM mice [29]. Collagen hydrolysate CPNS (Gly-Pro and Pro-Hyp) [30] and CP [31] prepared from fish scales can significantly attenuate the increase in epidermal thickness and water loss and the decrease in dermal hyaluronic acid (HA) induced by UVB irradiation, as well as recover HA loss by regulating hyaluronan synthases 1 (HAS1), hyaluronan synthases 2 (HAS2), and hyaluronidase 2 (HYAL2). Another elastin hydrolysate (EH) prepared from the bovine artery is composed of four polypeptides: Gly-Leu-Pro-Tyr (GLPY), Pro-Tyr (PY), Gly-Leu-Gly-Pro-Gly-Val-Gly (GLGPGVG), and Gly-Pro-Gly-Gly-Val-Gly-Ala- Leu (GPGGVGAL). EH can inhibit UV-induced skin thickening and sebaceous gland hyperplasia in mice and promote moisturizing of the skin. GLPY and GPGGVGAL have better inhibitory effects on elastase and thus can reduce extracellular matrix (ECM) degradation and improve the activity of UV damaged fibroblasts [32]. Collagen hydrolysis is the main cause of skin sagging, and the supplement of some collagen hydrolytic peptides can reduce the hydrolysis of collagen by MMPs. However, the detailed underlying mechanism is still unclear and needs to be further explored.

3. Bioactive Peptides and Brain Aging

3.1. Brain Aging

In the process of aging, brain function will gradually decline, which is manifested by a decline of learning ability and memory, as well as attention, decision-making ability, sensory perception, and motor ability. The prevalence of some neurodegenerative diseases, such as AD, Parkinson's disease (PD), and stroke, also increases with age. The development of these diseases is related to mitochondrial dysfunction, accumulation of oxidative damage, and increased inflammation [33]. AD is the most common neurodegenerative disease. Currently, abnormal folding of Aβ1-42 produced by the metabolism of amyloid precursor protein (APP) is considered to be the main cause of AD pathology [34]. Iron is involved in many biological processes in the brain and plays an important role in maintaining normal brain function. However, an iron imbalance can cause toxic effects on the brain. When the iron concentration is too high, it can increase the misfolding of Aβ and promote the development of AD [35]. The role of oxidative stress and inflammation in the development of AD is well known, and some new therapeutic targets have become research hotspots. Serotonin receptors (5-HT4R) have been found to reduce Aβ production. Many 5-HT4R agonists have been studied, but their potential therapeutic effect on AD has rarely been studied in vivo [36]. Glycosylation of proteins produces advanced glycation end products

(AGEs) that can cause neurodegeneration. When glyoxalase activity is reduced, the ability of these toxic glycosylated proteins to be eliminated is significantly reduced, leading to neurological disease [37]. The relationship between the gut microbiome and aging and the development of AD has been confirmed, but no clear mechanism has been elucidated. In a recent report, we found that intestinal dysregulation of Firmicutes and Bacteroidetes promotes T helper 1 (Th1) cell infiltration and promotes microglia differentiation in a pro-inflammatory direction. This may be related to the development of AD [38]. Bioactive peptides can exert their anti-aging effect on the brain through various mechanisms. They can increase antioxidant enzyme activity, reduce inflammation, increase the removal ability of iron and AGEs, increase expression of 5-HT receptors, and regulate the gut microbiota.

3.2. Antioxidant Peptides in Delaying Brain Aging

Carnosine (CAR) is an endogenous dipeptide (β-Ala-L-His) existing in muscle, blood, and the brain. CAR has good antioxidant activity and can attenuate neurological diseases caused by aging; CAR supplementation reduces the accumulation of Aβ in the hypothalamus and prefrontal cortex of aging rats and has potential therapeutic effects on AD [39]. After CAR treatment, GSH levels and SOD and GSH-Px activity were increased, whereas acetylcholinesterase (AChE) activity was significantly decreased (Figure 2), and there was a significant reduction in neuronal apoptosis, brain edema, and inflammation in D-galactose treated rats [40]. With aging, iron gradually accumulates and induces the generation of free radicals, promoting the formation of Tau and Aβ oligomers, which are neurotoxic and the main cause of AD [41]. The amount of iron found in the brains of AD patients is much higher than that of normal brains, suggesting that excess iron may be one of the causes of AD [35,42]. To better understand the effects of iron, researchers have synthesized the peptides with the ability to remove iron ions. Pentapeptide YHEDA (Tyr-His-Glu-Asp-Ala) and polypeptide mixture HAYED (5) Five (His-Ala-Tyr-Glu-Asp) repeat sequences are two synthetic active peptides with good iron ion scavenging ability (Figure 2). They can prevent the decrease of blood oxygen metabolism, inhibit the generation of free radicals, and reduce the damage in brain tissue, effectively improving cognitive impairment in senescent (SN) mice (25 months old) [43,44]. However, many high-quality natural antioxidant peptides have yet to be discovered and utilized in anti-aging studies. For example, many plant-derived bioactive peptides have antioxidant activities, and the research and development of these active peptides in aging studies will be of great significance in delaying brain aging [45].

3.3. Anti-Inflammatory Peptide in Delaying Brain Aging

Synthetic bioactive peptides are being increasingly produced for the treatment of different diseases. Liraglutide, a synthetic long-acting glucagon-like peptide 1 (GLP-1) analog, is widely used in the treatment of diabetes mellitus and CVDs. Recently, it has been speculated that liraglutide may have neuroprotective effects [46,47]. In senescence accelerated mouse P8 (SAMP8) mice (model of AD-like dementia), liraglutide treatment can improve spatial long-term memory and increase the number of hippocampal neurons [48,49]. The active peptides in dairy products have been long known, and there have been some reports that these active peptides can delay brain aging, mainly with the improvement of AD symptoms. The Whey protein hydrolysate tryptophan-methionine and tryptophan-tyrosine, extracted from fermented dairy products, can improve the cognitive impairment of AD mice. Inflammation and $A\beta_{1-42}$ deposition in the cerebral cortex and hippocampus are also significantly reduced in 5 × FAD transgenic mice fed tryptophan-tyrosine [50]. Notably, Tryptophan-Tyrosine dipeptide and whey protein hydrolysate GTWY (Gly-Thr-Trp-Tyr) can increase dopamine (DA) content in the hippocampus and frontal cortex of AD mice by inhibiting the activity of monoamine oxidase B (MAO-B) [51–53]. β-lactolin, an active polypeptide extracted from whey protein hydrolysate, has been shown to improve cognitive impairment. Specifically, β-lactolin can reduce amyloid plaque deposition and phosphorylated Tau protein content in the cerebral cortex of 5 × FAD transgenic mice

(AD mice), as well as increase DA and BDNF levels, thereby improving the cognitive impairment [54]. BDNF is one of the most widely distributed neurotrophic factors in the brain, and it plays an important role in regulating synaptic growth, neuroprotection, and affecting memory and cognition in vivo [55]. β-lactolin can increase the expression of BDNF in vivo (Figure 2). This is an example of how exogenous active peptides have a regulatory effect on endogenous active substances. Thus, exogenous active peptides not only play a therapeutic role in some antioxidant and anti-inflammatory pathways but also enhance the expression of endogenous active peptides to treat some diseases. The mechanism of action of exogenous active peptides may differ from endogenous ones, but they can supplement the body's defense system.

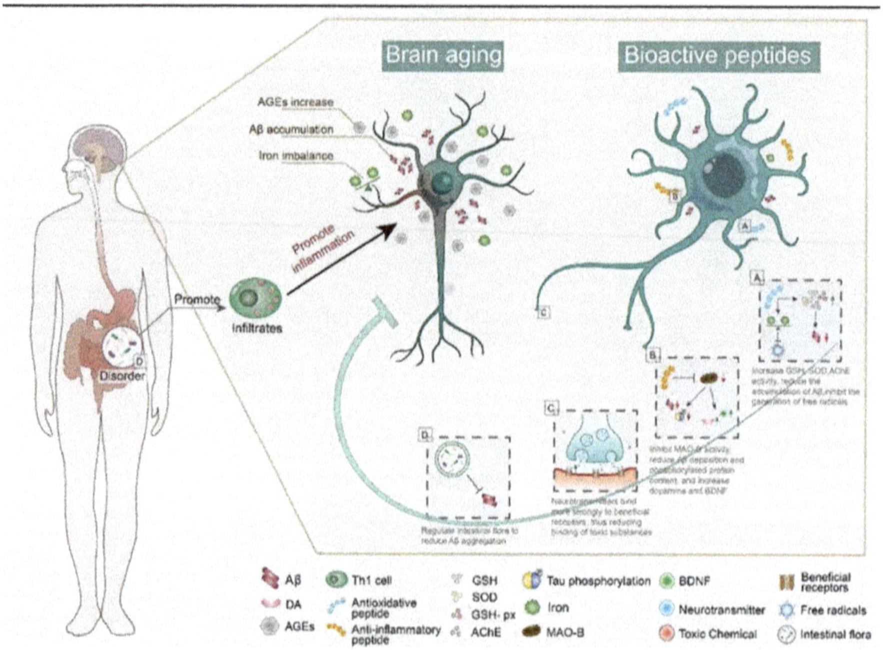

Figure 2. The main mechanism of bioactive peptides in delaying brain aging. (**A**) Bioactive peptides reduce Aβ accumulation by regulating oxidative stress. (**B**) The bioactive peptides inhibit the activity of MAO-B, up-regulate BDNF, and reduce the aggregation of Aβ. (**C**) Bioactive peptides reduce brain damage caused by toxic substances in the brain. (**D**) Bioactive peptides reduce Aβ aggregation by regulating intestinal microbiota. This figure cannot be reproduced without author permission.

3.4. Regulation of Peptide Receptors in Delaying Brain Aging

Serotonin (5-HT) is an important neurotransmitter that is involved in a variety of brain activities and functions. 5-HT receptors decrease gradually in the aging process. Serotonergic neurons are widely distributed in the brain. Reduction of 5-HT receptors can cause functional impairment of these neurons and lead to cognitive impairment. CAR is a dipeptide extracted from the meat. It can enhance 5-HT binding to its receptor and restore the regional senage-induced decrease in serotonin to normal levels [56,57]. Pituitary adenylate cyclase activated polypeptide (PACAP) is an endogenous active polypeptide with 38 amino acid residues and has a neuroprotective effect. It is widely distributed in the brain, pancreas, gonad, and respiratory tract. PACAP38 can be cleaved to form a 27 amino acid polypeptide, PACAP27 [58]. The level of PACAP gradually decreases in the normal aging process, and decreased PACAP levels have been found in the brain tissues of AD patients [59]. PACAP27 and PACAP38 can reduce the accumulation of Aβ in the brain by activating pituitary adenylate cyclase-activating polypeptide (PAC1), which causes the shedding of the receptor for advanced glycation end products (RAGE) of late glycation end products on the cell surface [60]. In summary, these peptides act on receptors, promoting the binding of beneficial receptors in neurons but reducing the binding of toxic substances.

3.5. Intestinal Microbiota Regulation by Peptides in Delaying Brain Aging

The link between the gut microbiota and AD is widely recognized, and many substances, including bioactive peptides, have been reported to regulate the gut microbiota. Some active peptides can regulate the intestinal microbiota in a beneficial direction by reducing Aβ aggregation, which has a potential role in the treatment of AD by regulating the intestinal microbiota [61]. The walnut protein hydrolysate PW5 (Pro-Pro-Lys-Asn-Trp) identified from walnut protein can reduce Aβ aggregation and improve cognitive impairment in mice by regulating intestinal microbiota (Figure 2). PW5 fed to APP/PS mice (AD mice) can increase firmicutes in the intestinal microbiota, which may be associated with reduced Aβ aggregation in mice [62]. The association between intestinal microbiota and AD has been widely recognized, and many bioactive peptides have been used to regulate intestinal microbiota to improve AD symptoms, but the mechanism is still not deeply studied, and further exploration is needed.

4. Bioactive Peptides and Aging in Other Organs

Aging is an irreversible biological process. Organs in the body cannot avoid aging. This leads to a variety of chronic diseases, including CVD, chronic obstructive pulmonary disease (COPD), intermittent lung disease, and asthma [63,64]. The aging processes of these important organs are correlated, and complications of one organ often lead to multi-organ disease. For example, lung aging causes COPD, which causes systemic inflammation and increases the risk of non-alcoholic liver disease. Moreover, people with non-alcoholic liver disease are more likely to have chronic kidney disease (CKD) and CVD. Oxidative stress and inflammation play an important role in the pathogenesis of these diseases. Many bioactive peptides with antioxidant and anti-inflammatory activities have been used in the prevention and treatment of these diseases. However, the role of a peptide in a disease is often limited, and there is still a lack of research on the complications of these diseases.

4.1. Lung Aging

COPD is a major form of lung disease characterized by chronic inflammation of the windpipe. Aging and smoking are the main causes of COPD. People over the age of 65 are five times more likely to develop the disease than younger people [65,66]. COPD is often associated with metabolic abnormalities, CVD, skeletal muscle atrophy, and other chronic diseases. In the later stages of COPD, arteriosclerosis, oxidative stress, and inflammation are the main mechanisms of its progression. Persistent inflammation disrupts the normal function of the lungs and is one of the causes of other complications [67]. Other researchers point to systemic inflammation from COPD as a major cause of non-alcoholic fatty liver disease (NAFLD) [68].

4.1.1. Antioxidant Peptides in Delaying Lung Aging

The human body is rich in peptides that play various biological activities in the body to adapt to different needs. The tripeptide GHK (glycyl-L-histidyl-L-lysine) is an active peptide existing in the human body, which has a high affinity for copper and can form a GHK-Cu complex. GHK-Cu has anti-inflammatory and antioxidant functions and can promote blood vessel growth and increase neural nutrition [69]. GHK-Cu has been shown to improve the symptoms of acute lung injury (ALI), which is usually accompanied by severe oxidative stress and inflammation. In ALI mice treated with GHK-Cu, SOD activity and GSH levels were significantly increased, and the NF-κB signaling pathway was blocked to reduce the release of inflammatory factors [70]. GHK-Cu is also a potential drug candidate for treating some chronic lung diseases such as COPD, asthma, and lung cancer [71].

4.1.2. Anti-Inflammatory Peptide in Delaying Lung Aging

Since many plants are rich in active substances, they are widely studied for use in drug development. The cyclic peptide CPE extracted from hydrolysates of Pseudostellariae can effectively relieve the symptoms of COPD. CPE treatment can significantly reduce the degree of alveolar destruction and lung inflammation, increase alveolar space, and regulate various cytokines. CPE treatment also reduces several mRNAs for TLR4, the adaptor protein MyD88 and activator protein-1 (AP-1), and active phosphorylated forms of proteins (P-JNK, P-P38, and P-TAK1) in alveolar macrophages in a COPD rat model [72]. These results suggest that CPE can act on the TLR4-MyD88-JNK/P38 signaling pathway and inhibit the release of important inflammatory factors to reduce lung inflammation. Thus, CPE has therapeutic potential for treating COPD. As an exogenous active peptide, CPE has a similar mechanism of action as GHK-Cu; both can block inflammatory pathways and reduce the release of inflammatory factors to attenuate lung inflammation. However, it would be interesting to explore whether CPE plays a synergistic role with GHK-Cu in vivo.

4.2. *Liver Aging*

A high-fat diet can cause NAFLD and non-alcoholic hepatitis (NASH), which is one of the major causes of cirrhosis and hepatocellular carcinoma (HCC). According to research, older people are more likely to develop NAFLD [73]. The liver is an important organ in the body. Dysfunction of antioxidant enzymes can reduce the ability of liver cells to remove peroxides, leading to the damage of mitochondrial DNA and mitochondrial dysfunction, resulting in liver aging [74,75]. Changes in the gut microbiome can also cause liver disease. For example, chronic inflammation, known as "metabolic inflammation", caused by changes in the microbial metabolites of the gut microbiome, can lead to NAFLD. Analysis of these altered gut microbes has revealed a significant increase in Proteobacteria, a group of microbes that may be responsible for NAFLD [76]. NAFLD, in turn, can increase the risk of atherosclerosis and accelerate the development of atherosclerosis symptoms. This is supported by a correlation in lesions of several organs [77]. Internal organs also interfere with each other as the body ages. This is exemplified by the fact that the severity of NAFLD increases the risk and severity of CKD [78].

4.2.1. Antioxidant and Anti-Inflammatory Peptides in Delaying Liver Aging

The body secretes some active polypeptides when maintaining normal physiological functions. Adropin is a peptide hormone that is expressed in the liver and can regulate blood glucose and lipid homeostasis. Studies have shown that adropin knockout can increase liver inflammation, liver steatosis, and fibrosis in mice, promoting the development of NASH. The underlying changes due to the knockout, such as reduction of Nrf2 transcriptional activity, GSH level, and mitochondrial membrane potential, can be normalized by adropin supplementation. These results indicate that adropin can delay the development of NASH by maintaining mitochondrial homeostasis and increasing antioxidant enzyme activity [79]. This is an example of how the liver can secrete beneficial active peptides to maintain normal physiological function. However, in the aging process, when the liver has reduced ability to produce such peptides, there is subsequent liver damage. Thus, the supplementation of these active peptides in an aging individual can maintain the normal function of the liver and prevent liver damage. It is worth mentioning that after long-term research on the anti-aging of active peptides, our research group has found that the mice treated with the active peptide derived from rice bran have reduced aging characteristics caused by galactose. This peptide is extracted from rice bran protein hydrolysate, named KF-8 (Lys-His-Asn-Arg-Gly-Asp-Glu-Phe), can reduce oxidative stress in D-gal-treated mouse livers by inhibiting the NF-κB/p38 signal transduction pathway and delaying liver aging [80]. This suggests that some exogenous active peptides with antioxidant activity could be added to the diet as anti-aging supplements.

4.2.2. Intestinal Microbiota Regulation by Peptides in Delaying Liver Aging

Liraglutide (discussed above) can improve the symptoms of NAFLD by regulating the gut microbiome. Specifically, liraglutide treatment can reduce Proteobacteria, a common factor in many diseases, and increase Verucommicrobia, which contributes to intestinal health and glucose homeostasis in the intestines of obese mice. Changes in the abundance of gut microbiota by liraglutide are associated with improvement of NAFLD symptoms and a reduction in inflammatory cell infiltration in the cecum and liver [81]. Mechanistically, liraglutide regulates the gut microbiome to reduce liver inflammation. There are many other bioactive peptides that can regulate the intestinal microbiome; however, the detailed mechanisms of their effects have not yet been elucidated.

4.3. *Kidney Aging*

The physiological function of the kidneys gradually deteriorates with aging, causing some kidney diseases [82]. With the increase of age, kidneys are also more vulnerable to oxidative damage, especially in the mitochondria of the kidney cells. Impaired mitochondrial function and cellular metabolism eventually lead to chronic renal failure [83]. CKD is an important cause of CVD because it can lead to high blood pressure and a decrease in the capillary density of the cardiac tissue. In addition, CKD reduces nitric oxide synthase expression in the vascular endothelium and increases renin–angiotensin system activity, resulting in increased release of superoxide and inflammatory cytokines and subsequent CVD [84].

4.3.1. Antioxidant Peptides in Delaying Renal Aging

In the aging process, increased oxidative stress and chronic inflammation can lead to some kidney diseases. Most of the endogenous active peptides in the endocrine system can reduce oxidative stress and inflammation, and thus, these peptides have potential therapeutic for alleviating kidney diseases. Mitochondrial targeted peptide SBT-20 (also known as SS-20) is a synthetic tetrapeptide that can reduce ROS and maintain the normal production of the electron transport chain and ATP. SBT-20 can reduce the expression of mitochondrial mitotic protein Drp1 and increase the expression of mitochondrial fusion protein 2 (Mfn2) to maintain the normal structure and function of mitochondria. SBT-20 treatment can decrease the expression of inflammatory cytokines IL-1β, IL-6, NF-κB1, and NF-κB2 in the kidney and alleviate the symptoms of chronic renal failure (CRF) in CRF mice [85]. Compared with endogenous active peptides, SBT-20 also has regulatory effects on inflammation and oxidative stress. With the technological development of peptide synthesis in vitro, the synthesis of these active peptides to meet the therapeutic needs of different diseases would be a powerful and desirable approach in the future.

4.3.2. Anti-Inflammatory Peptides in Delaying Renal Aging

Compared with endogenous active peptides produced in animals, exogenous active peptides from plants can also reduce the symptoms of CRF through similar mechanisms. Soy is rich in proteins that can be hydrolyzed into some bioactive peptides. The soybean protein hydrolysate (SPH) can lower blood pressure and maintain normal renal function. We showed that feeding rats (5/6 nephrectomized model) SPH can reduce ACE activity and TNF-α levels. These results suggest that SPH can reduce renal inflammation in CRF by down-regulating TNF-α activity [86]. Thus, exogenous active peptides are similar to endogenous active peptides in slowing CRF progression. Since exogenous active peptides are more widely derived, they hold great promise for providing therapeutic effects for many diseases.

4.4. Aging of the Heart and Blood Vessels

Complications of many diseases can increase the risk of CVD, and CVD can also promote lesions in other organs. The impaired endothelial cell function that is associated with aging can lead to vascular dilation and decreased anti-thrombotic ability, and eventually CVD. The main causes of CVD are activation of inflammatory signals induced by NF-κB, increases in MMP-9, and changes in TGF-β [87,88]. In addition, increased inflammation with aging can induce the expression of vascular endothelial growth factor (VEGF) family proteins. Although VEGF plays a beneficial role in some CVDs, over-expression of VEGF promotes the formation of new blood vessels, which in turn contributes to the development of atherosclerotic pathology [89].

4.4.1. Antioxidant Peptides in Delaying Cardiovascular Aging

Some active peptides extracted from animals and plants have potential therapeutic effects on CVD. For example, some exogenous active peptides extracted from grains may improve CVD [90]. Rice α-globulin hydrolysate Try-Try-Gly-Gly-Glu-Ser-Ser-Ser-Glu-Gln-Gly (YYGGESSSEQG) and Ser-Glu-Ser-Glu-Met (SESEM) extracted from rice can improve the symptoms of atherosclerosis in mice. They can down-regulate the TNF-α pathway and NF-κB in the aorta and aortic root tissues and reduce oxidative stress levels and inflammatory factors in apolipoprotein E-deficient mice [91]. Another exogenous active peptide, zebra blenny protein hydrolysates (ZBPHs), extracted from zebrafish protein, has a good antioxidant effect and can reduce the lipid deposition and apoptosis of cardiac cells induced by a high cholesterol diet in hypercholesterolemic rats. This is a potential therapeutic agent for CVD [92]. Since these two exogenous active peptides are isolated from rice and zebrafish proteins, respectively, they are widely available and can easily be incorporated into the diet.

4.4.2. Anti-Inflammatory Peptide in Delaying Cardiovascular Aging

In normal conditions, the heart can secrete some active peptides to maintain the physiological functions of the cardiovascular system. Natriuretic peptide (NP) is a kind of polypeptide that can maintain normal function of the heart, blood vessels, and kidney, and is associated with some CVDs [93]. NPs mainly exist in two forms: atrial na-triuretic peptide (ANP) and cerebral natriuretic peptide (BNP). BNP has been reported to treat myocardial infarction in mice. BNP can promote endothelial cell proliferation and myocardial vascularization in the infarcted and non-infarcted areas in the hearts of mice following myocardial infarction. The action of BNP is mediated by P38 MAP kinase [94]. Adropin, a peptide used to treat the non-alcoholic liver disease, can also be used to treat CVDs. It can reduce inflammation and migration of vascular smooth muscle cells, improve symptoms, and reduce intravascular plaque significantly [95]. Since adropin can be used to treat liver aging and atherosclerosis, it would have an added therapeutic potential to treat the cardiovascular complications of NASH.

5. Conclusions

Continuous improvement of biomedical research and healthcare has resulted in a significant increase in life span and the aging population. However, this has created a subsequent problem because aging is associated with many diseases. Therefore, great efforts have been made in anti-aging research, and many bioactive peptides have been discovered to have anti-aging activity. Bioactive peptides can be endogenously produced in the body, but more and more bioactive peptides are being exogenously produced from natural products or biosynthesis. The mechanisms of action of these bioactive peptides mainly involve their antioxidant and anti-inflammatory activities. Interestingly, some exogenous and endogenous active peptides have synergistic effects. Several organs in the body share similar aging mechanisms, and one organ disease can affect multiple organs. Thus, bioactive peptides can have anti-aging effects on multiple organs.

Although bioactive peptides have been widely used in anti-aging studies of rodents, it is not clear whether these active peptides exert their anti-aging effects in humans through similar mechanisms. Therefore, anti-aging research using genetically similar animal models, such as primates, is needed before many bioactive peptides can be tested in human clinical trials.

Here, we have reviewed the anti-aging activity of bioactive peptides, most of them from the hydrolysate of some food, as well as some synthetic active peptides and endogenous active peptides. These peptides were ingested by mouth, gavage, and injection in rodent models, and behavioral and physiological changes in these animals demonstrated the protective benefits of these peptides (reduced disease symptoms). It is worth mentioning that these active peptides have no adverse side effects and toxicity to experimental animals within the range of experimental concentrations, which also indicates that bioactive peptides are non-toxic and hypoallergenic active substances.

In this paper, we have introduced some of the anti-aging activities of active peptides, but there is additional research in the literature on other peptides. In view of this, Table 1 provides more information on other active peptides that are not discussed in detail in the text, as well as summarizing the animal models used in the studies of these active peptides and methods and dosages of peptide administration for reference of interested scholars. In addition, the main abbreviations that appear in this article are given in Table A1 of the Appendix A.

Table 1. Bioactive peptides with anti-aging activity.

Classification	Name and Delivery Way	Source	Rodent Model	Target Organ	Mechanism
Food-derived active peptide	Walnut protein hydrolysates(WPH) Oral gavage for 21 days Low: 333 mg/kg High: 666 mg/kg	Walnut	Alzheimer's disease model mice aged 6–8 weeks scopolamine solution (1.0 mg/kg)	Brain	SOD↑ GSH-Px↑ CAT↑ Nrf2↑ BDNF↑ CREB↑ MDA↓ TNFα↓ AchE↓ Trp-, Tyr-, or Phe-containing peptide has high affinity to Keap1 and Ache, so it can increase the activity of NRF2 and reduce the activity of Ache, which ultimately increases antioxidant capacity and anti-inflammatory ability and leads to increased BDNF, CREB transcription [15]
	Walnut protein hydrolysate and its low-molecular-weight fraction (WPH/WPHL) Oral gavage for 21 days WPH: 666 mg/kg WPHL: 666 mg/kg	Walnut	Alzheimer's disease model mice aged 6–8 weeks LPS (300 μg/kg bw)	Brain	SOD↑ GSH-Px↑ CAT↑ MDA↓ TNFα↓ TNFα↓ IL-6↓ IL-1β↓ Trp, Gly, Leu residues, hydrophobic amino acids, and aromatic amino acids in polypeptides can inhibit the expression of pro-inflammatory factors TNF-α, IL-1β, and IL-6 and reduce inflammation [16]
	Tyr-Val-Leu-Leu-Pro-Ser-Pro-Ly (walnut protein hydrolysates) Continuous injection for 4 weeks 60 mg/kg bw	Walnut	Alzheimer's disease model mice (C57BL/6) 5–6 week oldscopolamine solution (1 mg/kg bw)	Brain	ATP↑ PINK1↑ Parkin↑ NRF2↑ LC3 II/LC3 I↑ Beclin↑ KEAP1↓ p62↓ It can increase antioxidant capacity through Nrf2 signaling pathway and increase the expression of Beclin-1, Parkin, and PINK1 to enhance mitochondrial autophagy capacity [96]
	Alcalase potato-protein hydrolysates (IF) Oral administration 3 weeks 1 mg/kg bw	Potato	Senescence-Accelerated mice (SAMP8) 6 months high-fat diet	Liver/heart	pAKT↑ Sirt1↑ pAMPK↑ PGC1α↑ pFOXO3a↑ Bax↓ GOT↓ GPT↓ LDL↓ ANP↓ BNP↓ pGATA4↓ It can down-regulate cardiac hypertrophy markers ANP and BNP, reduce inflammation in the heart and liver, and reduce apoptosis by stimulating the activity of Sirt1 [97]

Table 1. *Cont.*

Classification	Name and Delivery Way	Source	Rodent Model	Target Organ	Mechanism
	Alcalase potato protein hydrolysate (APPH) Oral administration 4 weeks Low: 15 mg/kg/day Middle: 45 mg/kg/day High: 75 mg/kg/day	Potato	Sprague-Dawley (SD) rat 23 months old high-fat diet	Heart	p-p38/p38↓ GSN↓ p-Gata4↓ TGFβ↓ APPH has good lipid solubility and can reduce myocardial hypertrophy and fibrosis in aging rats through TGF-β/GSN pathway [98]
	Casein hydrolysates Continuous injection 10 weeks 200 mg/kg	Casein	Diabetic rat high-fat diet	Liver	NRF2↑ HO-1↑ SOD↑ GSH↑ MDA↓ By enhancing Nrf2 translation, the activity of antioxidant enzymes was enhanced, and the activities of DPP-IV and ACE were inhibited, among which dipeptide WM could inhibit Keap1/Nrf2 interaction [99]
	Wheat germ albumin hydrolysates ((Ala-Asp-Trp-Gly-Gly-Pro-Leu-Pro-His)) Continuous injection 1 week 4 mg/kg	Wheat	Diabetic mice 6 weeks old	Vascular	pAMPK/AMP↑ pPKCζ/PKCζ↓ NOX4↓ ROS↓ pAKT/AKT↓ Inhibition of NOX4 expression through the PKCζ/AMPK signaling pathway reduced oxidative stress levels and the release of inflammatory factors [100]
	Collagen hydrolysate Pro-Hyp Oral administration 4 weeks 210 mg/kg	Porcine skin	Chronic kidney disease mice 6 weeks old	Kidney	Liver iron content↑ EPO↑ HIF-2α↑ Hepcidin↓ TNF-α↓ IL-1β↓ IL-6↓ NF-κB↓ COX2↓ It reduces inflammation by regulating inflammatory pathways and plays a protective role in regulating HIF-2α, EPO, and Hepcidin [101]
	Anchovy hydrolysates Pro-Ala-Tyr-Cys-Ser (PAYCS) 20 days 0.2 mM/kg/day	Anchovy	Alzheimer's disease model mice 6 weeks old Scopolamine solution (1 mg/kg bw)	Brain	Ach↑ AchR↑ Nrf2↑ BDNF↑ SOD↑ The antioxidative effects of PAYCS and PAY may be related to the Try active phenolic structure in the sequence and the hydrogen donor of the sulfhydryl group in Cys. Both active peptides have the ability to promote the binding of Ach and AchR [102]
	Soy protein isolate (SPI) Oral administration 8 weeks	Soy	Obese rat 6 weeks old	Liver	NPTX2↑ GPT↑ INMT↑ HAL↑ The increased expression of NPTX2 reduced the inflammation of the rat liver, the increased expression of GPT may be related to mitochondrial energy metabolism, the increased expression of INMT may be related to the relief of NAFLD symptoms, and the increased expression of HLT can consume excess protein in the liver [103]
	Walnut protein hydrolysate Oral administration Low: 0.32 g/L Middle: 0.96 g/L High: 2.88 g/L	Walnut	Skin-aging model rat Exposed to UV-R	Skin	Elastin↑ Fibrillin-1↑ MMP-1↓ Increasing the expression of Col I, Col III, HYP, and HA and significantly attenuated the activity of MMP-1 [104]
	Eucheuma hydrolysate (EZY-1) 28 day 0.25 mg/kg 0.5 mg/kg 1 mg/kg 50 mg/kg	Eucheuma	Pulmonary fibrosis mice (C57BL/6J) 8 weeks old injected with 3.5 mg/kg of bleomycin	Lung	T-SOD↑ GSH-Px↑ HYP↓ MDA↓ pSmad3↓ EZY- 1 is easily absorbed in the intestinal tract, and its hydrophobic point facilitates the entry of EZY-1 into cells, while EZY-1 can reduce pulmonary fibrosis through TGF-β/Samd signaling pathway. [105]

Table 1. *Cont.*

Classification	Name and Delivery Way	Source	Rodent Model	Target Organ	Mechanism
	Egg white protein hydrolysate (EWPs) Gavage 14 days Low: 50 mg/kg Middle: 100 mg/kg High: 200 mg/kg	Egg	Colitis model mice (BALB/c) administered 3% (*w*/*v*) DSS	Gut	Candidatus_Sacchar-imonas↑ norank_f_Ruminococcaceae↓ Ruminiclostridium↓ TNF-α↓ IL-6↓ IL-8↓ EWPs contain Trp, Try, His, and Met, which make it have good antioxidant activity and can reduce the release of inflammatory factors by increasing the content of Lactobacillus and Candidatus-Saccharimonas in the gut [106]
	Whey protein hydrolysate (WHP) Gavage 30 days Low: 0.3 g/kg Middle: 1.5 g/kg High: 3.0 g/kg	Egg	D-galactose-treated mice (C57BL/6N) 6 months 100 mg/kg	Brain	SOD↑ GSH-Px↑ AChE↑ p-CaMKII↑ MDA↓ TNF-α↓ IL-1β↓ TNF-α↓ WHP can reduce the release of inflammatory factors, increase the activity of antioxidant enzymes, and enhance the activities of AchE and P-CamKII, which play an important role in maintaining synaptic plasticity [107]
A peptide en-crypted from the venom of Tityus ser-rulatus scorpion	Lys-Pro-Pro (KPP)	Scorpion	Mice 10 weeks old	Heart	pPLN/PLN↓ pERK/ERK↓ KPP regulates cellular stress-related proteins and exerts cardioprotective effects through PLN dephosphorylation [108]
Secretory bioactive peptide	Humanin (HNG) Injections 14 months 4 mg/kg	Mitochondria	Aging mice (C57BL/6N) 18 months	Heart	pAKT↑ pGSK3β↓ 4-HNE↓ TGF-β1↓ FGF-2↓ MMP-2↓ HNG down-regulated the expression of GSK-3β through Akt pathway, reduced myocardial apoptosis, down-regulated FGF-2 and MMP-2 expression, and inhibited cardiac fibrosis [109]
Peptide hormone	Melatonin Injections 30 days 10 mg/kg	Pineal gland	Aging mice 8 weeks old D-galactose 100 mg/kg	Brain	SNAP-25↑ PSD95↑ GluR1↑ p-CREB↑ ROS↓ GFAP↓ p-IKKβ↓ NF-κB↓ COX-2↓ NOS2↓ IL-1β↓ TNFα↓ p-JNK↓ Melatonin can reduce synaptic damage caused by oxidative stress and neuroinflammation through RAGE/NFκB/JNK pathway and has a good therapeutic effect on neurodegeneration [110]

Annotation: "↑" Up-regulation, "↓" Down-regulation, the abbreviations in Table 1 are all listed in Table A1.

Author Contributions: J.W. and Y.W.: writing—original draft preparation and data curation; Z.C. and Y.C.: software and investigation; Q.L.: supervision and funding acquisition; Y.L.: conceptualization, writing—reviewing and editing, and visualization. All authors have read and agreed to the published version of the manuscript.

Funding: This work was supported by funding from the Natural Science Foundation for Distinguished Young Scholars of Hunan Province (No. 2021JJ10078), the Natural Science Foundation of Hunan Province (No. 2020JJ4138), Hunan Furong Scholars Program, Huxiang Youth Talents Supporting Program (No. 2016RS3033), Scientific Research Foundation of Hunan Provincial Education Department (No. 18A160), and Grain-Oil Process and Quality Control 2011 Collaborative and Innovative Grant from Hunan Province.

Institutional Review Board Statement: Not applicable.

Informed Consent Statement: Not applicable.

Data Availability Statement: Data available in a publicly accessible repository (https://pubmed.ncbi.nlm.nih.gov, accessed on 21 January 2022).

Conflicts of Interest: The authors declare no conflict of interest.

Appendix A

Table A1. The full name of abbreviations.

Abbreviations	Full Name	Abbreviations	Full Name
SOD	Superoxide dismutase	Samd	Drosophila mothers against decapentaplegic protein
GSH-Px	Glutathione peroxidase	pAMPK/AMP	AMP-activated protein kinase/AMP
GSH	Glutathione	pPKCζ/PKCζ	Anti-phospho-protein kinase ζ
CAT	Catalase	NOX4	Antibodies against NADPH oxidase4
Nrf2	Transcription factor nuclear factor erythroid 2-related factor 2	EPO	Erythropoietin
BDNF	Brain-derived neurotrophic factor	HIF-2α	Hypoxia-inducible factor
CREB	cAMP-response element-binding protein	NF-κB	Nuclear factor-kappa beta
AchE	Acetylcholinesterase	COX2	Cyclooxygenase
MDA	Malondialdehyde	Ach	Acetylcholine
TNFα	Tumour necrosis factor-α	AchR	Cetylcholine receptor
IL-8	Interleukin-8	NPTX2	Neuronal pentraxin 2
IL-6	Interleukin-6	INMT	Indolethylamine N-methyltransferas
IL-1β	Interleukin-1β	HAL	Histamine ammonia-lyase
ATP	Adenosine triphosphate	MMP	Atrix metalloproteinase
PINK1	Mutations in the PTEN-induced kinase 1	ERK	Extracellular signal-regulated kinase
Parkin	Parkin RBR E3 ubiquitin protein ligase	HYP	Hydroxyproline
LC3II/LC3I	Microtubule-associated protein light chain 3	p-CaMKII	Phosphorylated Ca2+/calmodulin-dependent protein kinase II
KEAP1	Kelch-like ECH-associated protein 1	PLN	Dephosphorylation of phospholamban
p62	Protein sequestosome 1/p62	ERK	Extracellular regulated protein kinases
pAKT	Phosphorylated protein kinase B	pGSK3β	Phosphorylated glycogen synthase kinase-3beta
Sirt1	Silencing information regulator 2 related enzyme	4-HNE	4-hydroxynonenal
pAMPK	Phosphorylated AMP-activated protein kinase	FGF-2	Fibroblast growth factor 2
PGC1α	Peroxisome proliferator-activated receptor-γ co-activator-1α	SNAP-25	Synaptosomal associated protein 25
pFOXO3a	Phospho forkhead box O3a	PSD95	Postsynaptic density proteins
GOT	Glutamic oxaloacetic transaminase	GluR1	Anti-phospho-AMPARs
GPT	Glutamic-pyruvic transaminase	p-CREB	Phosphorylated cAMP-response element-binding protein
LDL	Low-density lipoprotein	GFAP	Astrocytosis
ANP	Atrial natriuretic peptide	p-IKKβ	Phosphorylated IKKbeta
BNP	Cerebral natriuretic peptide	NOS2	Nitric oxide synthase-2
pGATA4	Phosphorylated GATA binding protein 4	p-JNK	Hospho-c-JunN-terminal Kinase
p-p38/p38	Phosphorylated p38 kinase/p38 kinase	TGFβ	Transforming growth factor-beta
GSN	Gelsolin		

References

1. Partridge, L.; Fuentealba, M.; Kennedy, B.K. The quest to slow ageing through drug discovery. *Nat. Rev. Drug Discov.* **2020**, *19*, 513–532. [CrossRef] [PubMed]
2. Kane, A.E.; Sinclair, D.A. Sirtuins and NAD+in the Development and Treatment of Metabolic and Cardiovascular Diseases. *Circ. Res.* **2018**, *123*, 868–885. [CrossRef] [PubMed]
3. Luo, J.; Mills, K.; Le Cessie, S.; Noordam, R.; Van Heemst, D. Ageing, age-related diseases and oxidative stress: What to do next? *Ageing Res. Rev.* **2020**, *57*, 100982. [CrossRef] [PubMed]
4. Da Costa, J.P.; Vitorino, R.; Silva, G.M.; Vogel, C.; Duarte, A.C.; Rocha-Santos, T. A synopsis on aging-Theories, mechanisms and future prospects. *Ageing Res. Rev.* **2016**, *29*, 90–112. [CrossRef]
5. Wagner, K.H.; Cameron-Smith, D.; Wessner, B.; Franzke, B. Biomarkers of Aging: From Function to Molecular Biology. *Nutrients* **2016**, *8*, 338. [CrossRef]
6. Gurau, F.; Baldoni, S.; Prattichizzo, F.; Espinosa, E.; Amenta, F.; Procopio, A.D.; Albertini, M.C.; Bonafe, M.; Olivieri, F. Anti-senescence compounds: A potential nutraceutical approach to healthy aging. *Ageing Res. Rev.* **2018**, *46*, 14–31. [CrossRef]
7. Kim, D.H.; Bang, E.; Jung, H.J.; Noh, S.G.; Yu, B.P.; Choi, Y.J.; Chung, H.Y. Anti-aging Effects of Calorie Restriction (CR) and CR Mimetics based on the Senoinflammation Concept. *Nutrients* **2020**, *12*, 422. [CrossRef]
8. Sánchez, A.; Vázquez, A. Bioactive peptides: A review. *Food Qual. Saf.* **2017**, *1*, 29–46. [CrossRef]
9. Gorguc, A.; Gencdag, E.; Yilmaz, F.M. Bioactive peptides derived from plant origin by-products: Biological activities and techno-functional utilizations in food developments—A review. *Food Res. Int.* **2020**, *136*, 109504. [CrossRef]
10. Bechaux, J.; Gatellier, P.; Le Page, J.F.; Drillet, Y.; Sante-Lhoutellier, V. A comprehensive review of bioactive peptides obtained from animal byproducts and their applications. *Food Funct.* **2019**, *10*, 6244–6266. [CrossRef]
11. Rutherfurd-Markwick, K.J. Food proteins as a source of bioactive peptides with diverse functions. *Br. J. Nutr.* **2012**, *108* (Suppl. 2), S149–S157. [CrossRef]
12. Azzu, V.; Valencak, T.G. Energy Metabolism and Ageing in the Mouse: A Mini-Review. *Gerontology* **2017**, *63*, 327–336. [CrossRef]
13. Brunet, A. Old and new models for the study of human ageing. *Nat. Rev. Mol. Cell. Biol.* **2020**, *21*, 491–493. [CrossRef]
14. Gomes, M.J.C.; Lima, S.L.S.; Alves, N.E.G.; Assis, A.; Moreira, M.E.C.; Toledo, R.C.L.; Rosa, C.O.B.; Teixeira, O.R.; Bassinello, P.Z.; De Mejia, E.G.; et al. Common bean protein hydrolysate modulates lipid metabolism and prevents endothelial dysfunction in BALB/c mice fed an atherogenic diet. *Nutr. Metab. Cardiovasc. Dis.* **2020**, *30*, 141–150. [CrossRef]
15. Wang, S.; Su, G.; Zhang, X.; Song, G.; Zhang, L.; Zheng, L.; Zhao, M. Characterization and Exploration of Potential Neuroprotective Peptides in Walnut (Juglans regia) Protein Hydrolysate against Cholinergic System Damage and Oxidative Stress in Scopolamine-Induced Cognitive and Memory Impairment Mice and Zebrafish. *J. Agric. Food Chem.* **2021**, *69*, 2773–2783. [CrossRef]
16. Wang, S.; Zheng, L.; Zhao, T.; Zhang, Q.; Liu, Y.; Sun, B.; Su, G.; Zhao, M. Inhibitory Effects of Walnut (*Juglans regia*) Peptides on Neuroinflammation and Oxidative Stress in Lipopolysaccharide-Induced Cognitive Impairment Mice. *J. Agric. Food Chem.* **2020**, *68*, 2381–2392. [CrossRef]
17. Chambers, E.S.; Vukmanovic-Stejic, M. Skin barrier immunity and ageing. *Immunology* **2020**, *160*, 116–125. [CrossRef]
18. Gu, Y.; Han, J.; Jiang, C.; Zhang, Y. Biomarkers, oxidative stress and autophagy in skin aging. *Ageing Res. Rev.* **2020**, *59*, 101036. [CrossRef]
19. Lephart, E.D. Skin aging and oxidative stress: Equol's anti-aging effects via biochemical and molecular mechanisms. *Ageing Res. Rev.* **2016**, *31*, 36–54. [CrossRef]
20. Wang, Z.; Wang, Q.; Wang, L.; Xu, W.; He, Y.; Li, Y.; He, S.; Ma, H. Improvement of skin condition by oral administration of collagen hydrolysates in chronologically aged mice. *J. Sci. Food Agric.* **2017**, *97*, 2721–2726. [CrossRef]
21. Aguirre-Cruz, G.; León-López, A.; Cruz-Gómez, V.; Jiménez-Alvarado, R.; Aguirre-Álvarez, G. Collagen Hydrolysates for Skin Protection: Oral Administration and Topical Formulation. *Antioxidants* **2020**, *9*, 181. [CrossRef]
22. Wang, L.; Wang, X.; Bai, F.; Fang, Y.; Wang, J.; Gao, R. The anti-skin-aging effect of oral administration of gelatin from the swim bladder of Amur sturgeon (*Acipenser schrenckii*). *Food Funct.* **2019**, *10*, 3890–3897. [CrossRef]
23. Zhang, N.; Zhao, Y.; Shi, Y.; Chen, R.; Fu, X.; Zhao, Y. Polypeptides extracted from Eupolyphaga sinensis walker via enzymic digestion alleviate UV radiation-induced skin photoaging. *Biomed. Pharmacother.* **2019**, *112*, 108636. [CrossRef]
24. Lee, K.J.; Park, K.H.; Hahn, J.H. Alleviation of Ultraviolet-B Radiation-Induced Photoaging by a TNFR Antagonistic Peptide, TNFR2-SKE. *Mol. Cells* **2019**, *42*, 151–160. [CrossRef]
25. Kim, K.H.; Son, J.M.; Benayoun, B.A.; Lee, C. The Mitochondrial-Encoded Peptide MOTS-c Translocates to the Nucleus to Regulate Nuclear Gene Expression in Response to Metabolic Stress. *Cell Metab.* **2018**, *28*, 516–524.e7. [CrossRef]
26. Lee, C.; Zeng, J.; Drew, B.G.; Sallam, T.; Martin-Montalvo, A.; Wan, J.; Kim, S.J.; Mehta, H.; Hevener, A.L.; de Cabo, R.; et al. The mitochondrial-derived peptide MOTS-c promotes metabolic homeostasis and reduces obesity and insulin resistance. *Cell Metab.* **2015**, *21*, 443–454. [CrossRef]
27. Li, Q.; Lu, H.; Hu, G.; Ye, Z.; Zhai, D.; Yan, Z.; Wang, L.; Xiang, A.; Lu, Z. Earlier changes in mice after D-galactose treatment were improved by mitochondria derived small peptide MOTS-c. *Biochem. Biophys. Res. Commun.* **2019**, *513*, 439–445. [CrossRef]
28. Song, H.; Zhang, S.; Zhang, L.; Li, B. Effect of Orally Administered Collagen Peptides from Bovine Bone on Skin Aging in Chronologically Aged Mice. *Nutrients* **2017**, *9*, 1209. [CrossRef]
29. Zhang, L.; Zhang, S.; Song, H.; Li, B. Ingestion of collagen hydrolysates alleviates skin chronological aging in an aged mouse model by increasing collagen synthesis. *Food Funct.* **2020**, *11*, 5573–5580. [CrossRef]

30. Lee, H.J.; Jang, H.L.; Ahn, D.K.; Kim, H.J.; Jeon, H.Y.; Seo, D.B.; Lee, J.H.; Choi, J.K.; Kang, S.S. Orally administered collagen peptide protects against UVB-induced skin aging through the absorption of dipeptide forms, Gly-Pro and Pro-Hyp. *Biosci. Biotechnol. Biochem.* **2019**, *83*, 1146–1156. [CrossRef]
31. Kang, M.C.; Yumnam, S.; Kim, S.Y. Oral Intake of Collagen Peptide Attenuates Ultraviolet B Irradiation-Induced Skin Dehydration In Vivo by Regulating Hyaluronic Acid Synthesis. *Int. J. Mol. Sci.* **2018**, *19*, 3551. [CrossRef] [PubMed]
32. Liu, Y.; Su, G.; Zhou, F.; Zhang, J.; Zheng, L.; Zhao, M. Protective Effect of Bovine Elastin Peptides against Photoaging in Mice and Identification of Novel Antiphotoaging Peptides. *J. Agric. Food Chem.* **2018**, *66*, 10760–10768. [CrossRef] [PubMed]
33. Mattson, M.P.; Arumugam, T.V. Hallmarks of Brain Aging: Adaptive and Pathological Modification by Metabolic States. *Cell Metab.* **2018**, *27*, 1176–1199. [CrossRef] [PubMed]
34. Lane, C.A.; Hardy, J.; Schott, J.M. Alzheimer's disease. *Eur. J. Neurol.* **2018**, *25*, 59–70. [CrossRef]
35. Ward, R.J.; Zucca, F.A.; Duyn, J.H.; Crichton, R.R.; Zecca, L. The role of iron in brain ageing and neurodegenerative disorders. *Lancet Neurol.* **2014**, *13*, 1045–1060. [CrossRef]
36. Lalut, J.; Karila, D.; Dallemagne, P.; Rochais, C. Modulating 5-HT and 5-HT receptors in Alzheimer's disease treatment. *Future Med. Chem.* **2017**, *9*, 781–795. [CrossRef]
37. Hipkiss, A.R. Glycotoxins: Dietary and Metabolic Origins; Possible Amelioration of Neurotoxicity by Carnosine, with Special Reference to Parkinson's Disease. *Neurotox. Res.* **2018**, *34*, 164–172. [CrossRef]
38. Wang, X.; Sun, G.; Feng, T.; Zhang, J.; Huang, X.; Wang, T.; Xie, Z.; Chu, X.; Yang, J.; Wang, H.; et al. Sodium oligomannate therapeutically remodels gut microbiota and suppresses gut bacterial amino acids-shaped neuroinflammation to inhibit Alzheimer's disease progression. *Cell Res.* **2019**, *29*, 787–803. [CrossRef]
39. Banerjee, S.; Mukherjee, B.; Poddar, M.K.; Dunbar, G.L. Carnosine improves aging-induced cognitive impairment and brain regional neurodegeneration in relation to the neuropathological alterations in the secondary structure of amyloid beta (Abeta). *J. Neurochem.* **2021**, *158*, 710–723. [CrossRef]
40. Aydin, A.F.; Coban, J.; Dogan-Ekici, I.; Betul-Kalaz, E.; Dogru-Abbasoglu, S.; Uysal, M. Carnosine and taurine treatments diminished brain oxidative stress and apoptosis in D-galactose aging model. *Metab. Brain Dis.* **2016**, *31*, 337–345. [CrossRef]
41. Derry, P.J.; Hegde, M.L.; Jackson, G.R.; Kayed, R.; Tour, J.M.; Tsai, A.L.; Kent, T.A. Revisiting the intersection of amyloid, pathologically modified tau and iron in Alzheimer's disease from a ferroptosis perspective. *Prog. Neurobiol.* **2020**, *184*, 101716. [CrossRef]
42. Roberts, B.R.; Ryan, T.M.; Bush, A.I.; Masters, C.L.; Duce, J.A. The role of metallobiology and amyloid-beta peptides in Alzheimer's disease. *J. Neurochem.* **2012**, *120* (Suppl. 1), 149–166. [CrossRef]
43. Zou, Z.; Cai, J.; Zhong, A.; Zhou, Y.; Wang, Z.; Wu, Z.; Yang, Y.; Li, X.; Cheng, X.; Tan, J.; et al. Using the synthesized peptide HAYED (5) to protect the brain against iron catalyzed radical attack in a naturally senescence Kunming mouse model. *Free Radic. Biol. Med.* **2019**, *130*, 458–470. [CrossRef]
44. Zou, Z.; Shao, S.; Zou, R.; Qi, J.; Chen, L.; Zhang, H.; Shen, Q.; Yang, Y.; Ma, L.; Guo, R.; et al. Linking the low-density lipoprotein receptor-binding segment enables the therapeutic 5-YHEDA peptide to cross the blood-brain barrier and scavenge excess iron and radicals in the brain of senescent mice. *Alzheimers Dement.* **2019**, *5*, 717–731. [CrossRef]
45. Cui, X.; Lin, Q.; Liang, Y. Plant-Derived Antioxidants Protect the Nervous System From Aging by Inhibiting Oxidative Stress. *Front. Aging Neurosci.* **2020**, *12*, 209. [CrossRef]
46. Kelly, A.S.; Auerbach, P.; Barrientos-Perez, M.; Gies, I.; Hale, P.M.; Marcus, C.; Mastrandrea, L.D.; Prabhu, N.; Arslanian, S.; Investigators, N.N.T. A Randomized, Controlled Trial of Liraglutide for Adolescents with Obesity. *N. Engl. J. Med.* **2020**, *382*, 2117–2128. [CrossRef]
47. Wicinski, M.; Socha, M.; Malinowski, B.; Wodkiewicz, E.; Walczak, M.; Gorski, K.; Slupski, M.; Pawlak-Osinska, K. Liraglutide and its Neuroprotective Properties-Focus on Possible Biochemical Mechanisms in Alzheimer's Disease and Cerebral Ischemic Events. *Int. J. Mol. Sci.* **2019**, *20*, 1050. [CrossRef]
48. Hansen, H.H.; Fabricius, K.; Barkholt, P.; Niehoff, M.L.; Morley, J.E.; Jelsing, J.; Pyke, C.; Knudsen, L.B.; Farr, S.A.; Vrang, N. The GLP-1 Receptor Agonist Liraglutide Improves Memory Function and Increases Hippocampal CA1 Neuronal Numbers in a Senescence-Accelerated Mouse Model of Alzheimer's Disease. *J. Alzheimers Dis.* **2015**, *46*, 877–888. [CrossRef]
49. Yan, W.; Pang, M.; Yu, Y.; Gou, X.; Si, P.; Zhawatibai, A.; Zhang, Y.; Zhang, M.; Guo, T.; Yi, X.; et al. The neuroprotection of liraglutide on diabetic cognitive deficits is associated with improved hippocampal synapses and inhibited neuronal apoptosis. *Life Sci.* **2019**, *231*, 116566. [CrossRef]
50. Ano, Y.; Yoshino, Y.; Uchida, K.; Nakayama, H. Preventive Effects of Tryptophan-Methionine Dipeptide on Neural Inflammation and Alzheimer's Pathology. *Int. J. Mol. Sci.* **2019**, *20*, 3206. [CrossRef]
51. Ano, Y.; Ayabe, T.; Ohya, R.; Kondo, K.; Kitaoka, S.; Furuyashiki, T. Tryptophan-Tyrosine Dipeptide, the Core Sequence of beta-Lactolin, Improves Memory by Modulating the Dopamine System. *Nutrients* **2019**, *11*, 348. [CrossRef]
52. Ano, Y.; Yoshino, Y.; Kutsukake, T.; Ohya, R.; Fukuda, T.; Uchida, K.; Takashima, A.; Nakayama, H. Tryptophan-related dipeptides in fermented dairy products suppress microglial activation and prevent cognitive decline. *Aging* **2019**, *11*, 2949–2967. [CrossRef]
53. Ano, Y.; Ayabe, T.; Kutsukake, T.; Ohya, R.; Takaichi, Y.; Uchida, S.; Yamada, K.; Uchida, K.; Takashima, A.; Nakayama, H. Novel lactopeptides in fermented dairy products improve memory function and cognitive decline. *Neurobiol. Aging* **2018**, *72*, 23–31. [CrossRef]

54. Ano, Y.; Ohya, R.; Takaichi, Y.; Washinuma, T.; Uchida, K.; Takashima, A.; Nakayama, H. beta-Lactolin, a Whey-Derived Lacto-Tetrapeptide, Prevents Alzheimer's Disease Pathologies and Cognitive Decline. *J. Alzheimers Dis.* **2020**, *73*, 1331–1342. [CrossRef]
55. Kowianski, P.; Lietzau, G.; Czuba, E.; Waskow, M.; Steliga, A.; Morys, J. BDNF: A Key Factor with Multipotent Impact on Brain Signaling and Synaptic Plasticity. *Cell Mol. Neurobiol.* **2018**, *38*, 579–593. [CrossRef]
56. Banerjee, S.; Ghosh, T.K.; Poddar, M.K. Carnosine reverses the aging-induced down regulation of brain regional serotonergic system. *Mech. Ageing Dev.* **2015**, *152*, 5–14. [CrossRef]
57. Banerjee, S.; Poddar, M.K. Aging-induced changes in brain regional serotonin receptor binding: Effect of Carnosine. *Neuroscience* **2016**, *319*, 79–91. [CrossRef]
58. Vaudry, D.; Falluel-Morel, A.; Bourgault, S.; Basille, M.; Burel, D.; Wurtz, O.; Fournier, A.; Chow, B.K.; Hashimoto, H.; Galas, L.; et al. Pituitary adenylate cyclase-activating polypeptide and its receptors: 20 years after the discovery. *Pharmacol. Rev.* **2009**, *61*, 283–357. [CrossRef]
59. Reglodi, D.; Atlasz, T.; Szabo, E.; Jungling, A.; Tamas, A.; Juhasz, T.; Fulop, B.D.; Bardosi, A. PACAP deficiency as a model of aging. *Geroscience* **2018**, *40*, 437–452. [CrossRef]
60. Metz, V.V.; Kojro, E.; Rat, D.; Postina, R. Induction of RAGE shedding by activation of G protein-coupled receptors. *PLoS ONE* **2012**, *7*, e41823. [CrossRef]
61. Wu, S.; Bekhit, A.E.-D.A.; Wu, Q.; Chen, M.; Liao, X.; Wang, J.; Ding, Y. Bioactive peptides and gut microbiota: Candidates for a novel strategy for reduction and control of neurodegenerative diseases. *Trends Food Sci. Technol.* **2021**, *108*, 164–176. [CrossRef]
62. Wang, M.; Amakye, W.K.; Guo, L.; Gong, C.; Zhao, Y.; Yao, M.; Ren, J. Walnut-Derived Peptide PW5 Ameliorates Cognitive Impairments and Alters Gut Microbiota in APP/PS1 Transgenic Mice. *Mol. Nutr. Food Res.* **2019**, *63*, e1900326. [CrossRef] [PubMed]
63. Cho, S.J.; Stout-Delgado, H.W. Aging and Lung Disease. *Annu. Rev. Physiol.* **2020**, *82*, 433–459. [CrossRef] [PubMed]
64. Ren, J.; Zhang, Y. Targeting Autophagy in Aging and Aging-Related Cardiovascular Diseases. *Trends Pharmacol. Sci.* **2018**, *39*, 1064–1076. [CrossRef]
65. Easter, M.; Bollenbecker, S.; Barnes, J.W.; Krick, S. Targeting Aging Pathways in Chronic Obstructive Pulmonary Disease. *Int. J. Mol. Sci.* **2020**, *21*, 6924. [CrossRef]
66. Barnes, P.J.; Baker, J.; Donnelly, L.E. Cellular Senescence as a Mechanism and Target in Chronic Lung Diseases. *Am. J. Respir. Crit. Care Med.* **2019**, *200*, 556–564. [CrossRef]
67. Chan, S.M.H.; Selemidis, S.; Bozinovski, S.; Vlahos, R. Pathobiological mechanisms underlying metabolic syndrome (MetS) in chronic obstructive pulmonary disease (COPD): Clinical significance and therapeutic strategies. *Pharmacol. Ther.* **2019**, *198*, 160–188. [CrossRef]
68. Lonardo, A.; Nascimbeni, F.; Ponz de Leon, M. Nonalcoholic fatty liver disease and COPD: Is it time to cross the diaphragm? *Eur. Respir J.* **2017**, *49*, 1700546. [CrossRef]
69. Pickart, L.; Vasquez-Soltero, J.M.; Margolina, A. The human tripeptide GHK-Cu in prevention of oxidative stress and degenerative conditions of aging: Implications for cognitive health. *Oxid. Med. Cell Longev.* **2012**, *2012*, 324832. [CrossRef]
70. Park, J.-R.; Lee, H.; Kim, S.-I.; Yang, S.-R. The tri-peptide GHK-Cu complex ameliorates lipopolysaccharide-induced acute lung injury in mice. *Oncotarget* **2016**, *7*, 58405–58417. [CrossRef]
71. Meiners, S.; Eickelberg, O. Next-generation personalized drug discovery: The tripeptide GHK hits center stage in chronic obstructive pulmonary disease. *Genome Med.* **2012**, *4*, 70. [CrossRef]
72. Lu, F.; Yang, H.; Lin, S.D.; Zhao, L.; Jiang, C.; Chen, Z.B.; Liu, Y.Y.; Kan, Y.J.; Hu, J.; Pang, W.S. Cyclic Peptide Extracts Derived From Pseudostellaria heterophylla Ameliorates COPD via Regulation of the TLR4/MyD88 Pathway Proteins. *Front. Pharmacol.* **2020**, *11*, 850. [CrossRef]
73. Estes, C.; Razavi, H.; Loomba, R.; Younossi, Z.; Sanyal, A.J. Modeling the epidemic of nonalcoholic fatty liver disease demonstrates an exponential increase in burden of disease. *Hepatology* **2018**, *67*, 123–133. [CrossRef]
74. Niemann, J.; Johne, C.; Schroder, S.; Koch, F.; Ibrahim, S.M.; Schultz, J.; Tiedge, M.; Baltrusch, S. An mtDNA mutation accelerates liver aging by interfering with the ROS response and mitochondrial life cycle. *Free Radic. Biol. Med.* **2017**, *102*, 174–187. [CrossRef]
75. Chen, Z.; Tian, R.; She, Z.; Cai, J.; Li, H. Role of oxidative stress in the pathogenesis of nonalcoholic fatty liver disease. *Free Radic. Biol. Med.* **2020**, *152*, 116–141. [CrossRef]
76. Tilg, H.; Zmora, N.; Adolph, T.E.; Elinav, E. The intestinal microbiota fuelling metabolic inflammation. *Nat. Rev. Immunol.* **2020**, *20*, 40–54. [CrossRef]
77. Lonardo, A.; Nascimbeni, F.; Mantovani, A.; Targher, G. Hypertension, diabetes, atherosclerosis and NASH: Cause or consequence? *J. Hepatol.* **2018**, *68*, 335–352. [CrossRef]
78. Musso, G.; Gambino, R.; Tabibian, J.H.; Ekstedt, M.; Kechagias, S.; Hamaguchi, M.; Hultcrantz, R.; Hagstrom, H.; Yoon, S.K.; Charatcharoenwitthaya, P.; et al. Association of non-alcoholic fatty liver disease with chronic kidney disease: A systematic review and meta-analysis. *PLoS Med.* **2014**, *11*, e1001680. [CrossRef]
79. Chen, X.; Xue, H.; Fang, W.; Chen, K.; Chen, S.; Yang, W.; Shen, T.; Chen, X.; Zhang, P.; Ling, W. Adropin protects against liver injury in nonalcoholic steatohepatitis via the Nrf2 mediated antioxidant capacity. *Redox. Biol.* **2019**, *21*, 101068. [CrossRef]
80. Wang, Y.; Cui, X.; Lin, Q.; Cai, J.; Tang, L.; Liang, Y. Active Peptide KF-8 from Rice Bran Attenuates Oxidative Stress in a Mouse Model of Aging Induced by d-Galactose. *J. Agric. Food Chem.* **2020**, *68*, 12271–12283. [CrossRef]

81. Moreira, G.V.; Azevedo, F.F.; Ribeiro, L.M.; Santos, A.; Guadagnini, D.; Gama, P.; Liberti, E.A.; Saad, M.; Carvalho, C. Liraglutide modulates gut microbiota and reduces NAFLD in obese mice. *J. Nutr. Biochem.* **2018**, *62*, 143–154. [CrossRef]
82. Choudhury, D.; Levi, M. Kidney aging—Inevitable or preventable? *Nat. Rev. Nephrol.* **2011**, *7*, 706–717. [CrossRef]
83. Kimura, T.; Isaka, Y.; Yoshimori, T. Autophagy and kidney inflammation. *Autophagy* **2017**, *13*, 997–1003. [CrossRef]
84. Gansevoort, R.T.; Correa-Rotter, R.; Hemmelgarn, B.R.; Jafar, T.H.; Heerspink, H.J.L.; Mann, J.F.; Matsushita, K.; Wen, C.P. Chronic kidney disease and cardiovascular risk: Epidemiology, mechanisms, and prevention. *Lancet* **2013**, *382*, 339–352. [CrossRef]
85. Sun, L.; Xu, H.; Wang, Y.; Ma, X.; Xu, Y.; Sun, F. The mitochondrial-targeted peptide SBT-20 ameliorates inflammation and oxidative stress in chronic renal failure. *Aging* **2020**, *12*, 18238–18250. [CrossRef]
86. Yang, H.-Y.; Chen, J.-R.; Chang, L.-S. Effects of soy protein hydrolysate on blood pressure and angiotensin-converting enzyme activity in rats with chronic renal failure. *Hypertens. Res.* **2008**, *31*, 957–963. [CrossRef]
87. Paneni, F.; Diaz Canestro, C.; Libby, P.; Luscher, T.F.; Camici, G.G. The Aging Cardiovascular System: Understanding It at the Cellular and Clinical Levels. *J. Am. Coll. Cardiol.* **2017**, *69*, 1952–1967. [CrossRef] [PubMed]
88. Donato, A.J.; Machin, D.R.; Lesniewski, L.A. Mechanisms of Dysfunction in the Aging Vasculature and Role in Age-Related Disease. *Circ. Res.* **2018**, *123*, 825–848. [CrossRef]
89. Camare, C.; Pucelle, M.; Negre-Salvayre, A.; Salvayre, R. Angiogenesis in the atherosclerotic plaque. *Redox. Biol.* **2017**, *12*, 18–34. [CrossRef]
90. Gong, X.; An, Q.; Le, L.; Geng, F.; Jiang, L.; Yan, J.; Xiang, D.; Peng, L.; Zou, L.; Zhao, G.; et al. Prospects of cereal protein-derived bioactive peptides: Sources, bioactivities diversity, and production. *Crit. Rev. Food Sci. Nutr.* **2020**, 1–17. [CrossRef]
91. Tong, L.-T.; Ju, Z.; Wang, L.; Qiu, J.; Liu, L.; Zhou, X.; Liang, T.; Geng, D.; Zhou, S. Peptides derived from rice α-globulin reduce atherosclerosis in apolipoprotein E-deficient mice by inhibiting TNF-α-induced vascular endothelial cells injury. *J. Funct. Foods* **2019**, *63*, 103582. [CrossRef]
92. Ktari, N.; Bkhairia, I.; Nasri, R.; Ben Abdallah Kolsi, R.; Ben Slama-Ben Salem, R.; Ben Amara, I.; Zeghal, N.; Ben Salah, B.; Ben Salah, R.; Nasri, M. Zebra blenny protein hydrolysates as a source of bioactive peptides with prevention effect against oxidative dysfunctions and DNA damage in heart tissues of rats fed a cholesterol-rich diet. *Food Res. Int.* **2017**, *100*, 423–432. [CrossRef] [PubMed]
93. Rubattu, S.; Volpe, M. Natriuretic Peptides in the Cardiovascular System: Multifaceted Roles in Physiology, Pathology and Therapeutics. *Int. J. Mol. Sci.* **2019**, *20*, 3991. [CrossRef] [PubMed]
94. Li, N.; Rignault-Clerc, S.; Bielmann, C.; Bon-Mathier, A.C.; Deglise, T.; Carboni, A.; Ducrest, M.; Rosenblatt-Velin, N. Increasing heart vascularisation after myocardial infarction using brain natriuretic peptide stimulation of endothelial and WT1(+) epicardial cells. *Elife* **2020**, *9*, e61050. [CrossRef] [PubMed]
95. Sato, K.; Yamashita, T.; Shirai, R.; Shibata, K.; Okano, T.; Yamaguchi, M.; Mori, Y.; Hirano, T.; Watanabe, T. Adropin Contributes to Anti-Atherosclerosis by Suppressing Monocyte-Endothelial Cell Adhesion and Smooth Muscle Cell Proliferation. *Int. J. Mol. Sci.* **2018**, *19*, 1293. [CrossRef]
96. Zhao, F.; Liu, C.; Fang, L.; Lu, H.; Wang, J.; Gao, Y.; Gabbianelli, R.; Min, W. Walnut-Derived Peptide Activates PINK1 via the NRF2/KEAP1/HO-1 Pathway, Promotes Mitophagy, and Alleviates Learning and Memory Impairments in a Mice Model. *J. Agric. Food Chem.* **2021**, *69*, 2758–2772. [CrossRef]
97. Asokan, S.M.; Wang, T.; Wang, M.F.; Lin, W.T. A novel dipeptide from potato protein hydrolysate augments the effects of exercise training against high-fat diet-induced damages in senescence-accelerated mouse-prone 8 by boosting pAMPK/SIRT1/PGC-1α/pFOXO3 pathway. *Aging* **2020**, *12*, 7334–7349. [CrossRef]
98. Hu, W.S.; Ting, W.J.; Tamilselvi, S.; Day, C.H.; Wang, T.; Chiang, W.D.; Viswanadha, V.P.; Yeh, Y.L.; Lin, W.T.; Huang, C.Y. Oral administration of alcalase potato protein hydrolysate-APPH attenuates high fat diet-induced cardiac complications via TGF-beta/GSN axis in aging rats. *Environ. Toxicol.* **2019**, *34*, 5–12. [CrossRef]
99. Wang, C.; Zheng, L.; Su, G.; Zeng, X.A.; Sun, B.; Zhao, M. Evaluation and Exploration of Potentially Bioactive Peptides in Casein Hydrolysates against Liver Oxidative Damage in STZ/HFD-Induced Diabetic Rats. *J. Agric. Food Chem.* **2020**, *68*, 2393–2405. [CrossRef]
100. Wang, F.; Weng, Z.; Lyu, Y.; Bao, Y.; Liu, J.; Zhang, Y.; Sui, X.; Fang, Y.; Tang, X.; Shen, X. Wheat germ-derived peptide ADWGGPLPH abolishes high glucose-induced oxidative stress via modulation of the PKCzeta/AMPK/NOX4 pathway. *Food Funct.* **2020**, *11*, 6843–6854. [CrossRef]
101. Zhu, S.; Wu, L.; Zhang, J.; Miao, Y.; Zhao, Y.; Zeng, M.; Li, D.; Wu, H. Collagen Hydrolysate Corrects Anemia in Chronic Kidney Disease via Anti-Inflammatory Renoprotection and HIF-2alpha-Dependent Erythropoietin and Hepcidin Regulation. *J. Agric. Food Chem.* **2020**, *68*, 11726–11734. [CrossRef]
102. Zhao, T.; Zheng, L.; Zhang, Q.; Wang, S.; Zhao, Q.; Su, G.; Zhao, M. Stability towards the gastrointestinal simulated digestion and bioactivity of PAYCS and its digestive product PAY with cognitive improving properties. *Food Funct.* **2019**, *10*, 2439–2449. [CrossRef]
103. Kozaczek, M.; Bottje, W.; Greene, E.; Lassiter, K.; Kong, B.; Dridi, S.; Korourian, S.; Hakkak, R. Comparison of liver gene expression by RNAseq and PCR analysis after 8 weeks of feeding soy protein isolate- or casein-based diets in an obese liver steatosis rat model. *Food Funct.* **2019**, *10*, 8218–8229. [CrossRef]
104. Xu, D.; Li, D.; Zhao, Z.; Wu, J.; Zhao, M. Regulation by walnut protein hydrolysate on the components and structural degradation of photoaged skin in SD rats. *Food Funct.* **2019**, *2019*, 6792–6802. [CrossRef]

105. Yu, H.; Zhang, Z.; Huang, H.; Wang, Y.; Lin, B.; Wu, S.; Ma, J.; Chen, B.; He, Z.; Wu, J.; et al. Inhibition of bleomycin-induced pulmonary fibrosis in mice by the novel peptide EZY-1 purified from Eucheuma. *Food Funct.* **2019**, *10*, 3198–3208. [CrossRef]
106. Ge, H.; Cai, Z.; Chai, J.; Liu, J.; Liu, B.; Yu, Y.; Liu, J.; Zhang, T. Egg white peptides ameliorate dextran sulfate sodium-induced acute colitis symptoms by inhibiting the production of pro-inflammatory cytokines and modulation of gut microbiota composition. *Food Chem.* **2021**, *360*, 129981. [CrossRef]
107. Yu, X.C.; Li, Z.; Liu, X.R.; Hu, J.N.; Liu, R.; Zhu, N.; Li, Y. The Antioxidant Effects of Whey Protein Peptide on Learning and Memory Improvement in Aging Mice Models. *Nutrients* **2021**, *13*, 2100. [CrossRef]
108. Gomez-Mendoza, D.P.; Lemos, R.P.; Jesus, I.C.G.; Gorshkov, V.; McKinnie, S.M.K.; Vederas, J.C.; Kjeldsen, F.; Guatimosim, S.; Santos, R.A.; Pimenta, A.M.C.; et al. Moving Pieces in a Cellular Puzzle: A Cryptic Peptide from the Scorpion Toxin Ts14 Activates AKT and ERK Signaling and Decreases Cardiac Myocyte Contractility via Dephosphorylation of Phospholamban. *J. Proteome Res.* **2020**, *19*, 3467–3477. [CrossRef]
109. Qin, Q.; Mehta, H.; Yen, K.; Navarrete, G.; Brandhorst, S.; Wan, J.; Delrio, S.; Zhang, X.; Lerman, L.O.; Cohen, P.; et al. Chronic treatment with the mitochondrial peptide humanin prevents age-related myocardial fibrosis in mice. *Am. J. Physiol. Heart Circ. Physiol.* **2018**, *315*, H1127–H1136. [CrossRef]
110. Ali, T.; Badshah, H.; Kim, T.H.; Kim, M.O. Melatonin attenuates D-galactose-induced memory impairment, neuroinflammation and neurodegeneration via RAGE/NF-K B/JNK signaling pathway in aging mouse model. *J. Pineal Res.* **2015**, *58*, 71–85. [CrossRef]

International Journal of
Molecular Sciences

MDPI

Review

The Role of Nutrients in Maintaining Hematopoietic Stem Cells and Healthy Hematopoiesis for Life

Yuko Tadokoro [1] and Atsushi Hirao [1,2,*]

[1] Division of Molecular Genetics, Cancer Research Institute, Kanazawa University, Kakuma-machi, Kanazawa 920-1192, Japan; tadokoro@staff.kanazawa-u.ac.jp
[2] WPI Nano Life Science Institute (WPI-Nano LSI), Kanazawa University, Kakuma-machi, Kanazawa 920-1192, Japan
* Correspondence: ahirao@staff.kakanazawa-u.ac.jp

Abstract: Nutrients are converted by the body to smaller molecules, which are utilized for both anabolic and catabolic metabolic reactions. Cooperative regulation of these processes is critical for life-sustaining activities. In this review, we focus on how the regulation of nutrient-driven metabolism maintains healthy hematopoietic stem cells (HSCs). For this purpose, we have examined the metabolic regulation of HSCs from two perspectives: (1) the control of intracellular metabolism by the balance of anabolic and catabolic reactions; and (2) the control of organismal metabolic status and hematopoiesis by dietary intake of nutrients. Critical roles of catabolic regulators in stem cell homeostasis are conserved in several types of tissues, including hematopoiesis. These catabolic signals are also major regulators of organismal lifespan in multiple species. In parallel, changes to nutrients via alterations to dietary intake affect not only an organism's metabolic state but also the behavior of its stem cells. While the molecular mechanisms involved in these two aspects of nutrient function may not necessarily overlap, a deeper understanding of these phenomena will point to new avenues of medical research and may furnish new agents for improving human health care.

Keywords: hematopoietic stem cell; nutrient; catabolism; anabolism; dietary intervention

Citation: Tadokoro, Y.; Hirao, A. The Role of Nutrients in Maintaining Hematopoietic Stem Cells and Healthy Hematopoiesis for Life. *Int. J. Mol. Sci.* **2022**, *23*, 1574. https://doi.org/10.3390/ijms23031574

Academic Editor: Maurizio Battino

Received: 28 December 2021
Accepted: 26 January 2022
Published: 29 January 2022

1. Introduction

Nutrients are substances in food that are essential for biological activity in organisms and include carbohydrates, lipids, proteins, vitamins, and minerals. Nutrients are converted by the body into smaller molecules that are utilized for the set of life-sustaining chemical reactions called metabolism. Collectively, reactions that break down food or fuel to obtain energy are called catabolism, whereas the reactions by which larger molecules are synthesized from smaller ones are called anabolism. Anabolic reactions consume the energy produced by catabolic reactions, such that cooperative regulation of both processes is critical for maintaining life.

Hematopoietic stem cells (HSCs) have unique metabolic characteristics that contribute to the maintenance of their homeostasis in blood [1]. In an animal at steady-state, the majority of HSCs within the bone marrow (BM) niche maintain low levels of reactive oxygen species (ROS), ATP production, and protein synthesis (translation); these properties are associated with a state of cell cycle arrest called quiescence or dormancy. Debate has raged for a long time over whether metabolic status controls HSC fate, i.e., whether this relationship is causative or not. It is now evident that nutrient-sensing molecules that govern the balance between anabolic and catabolic processes play critical roles in HSC maintenance by controlling their intracellular metabolism (Figure 1). Knowledge gained over the past 15 years has generated three fundamental conclusions in this field. First, HSCs depend on catabolic regulators to remain healthy. Accumulating evidence has demonstrated that the acquisition of catabolic status (as supported by autophagic and lysosomal activity), coupled with reduced anabolic signaling, maintains HSC function and prevents the development of phenotypes of aging. Second, the essential roles of catabolic regulators in HSC homeostasis are conserved in other types of tissue stem cells that remain dormant in animals at steady-state [2]. Third, these catabolic signals are major regulators of organismal lifespan in multiple species [3].

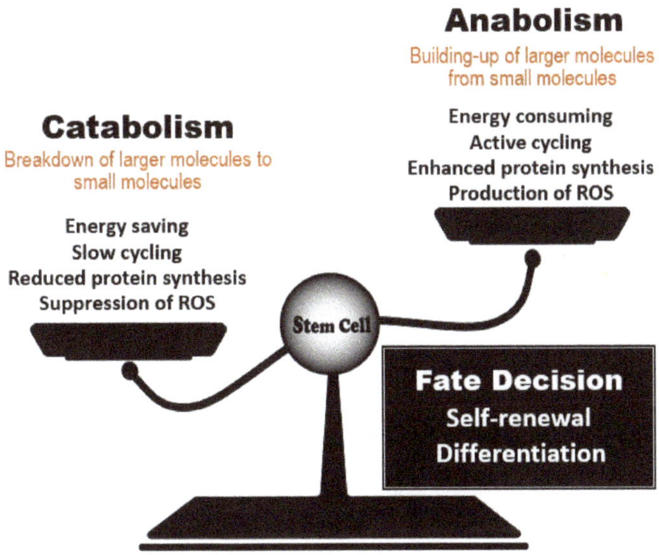

Figure 1. Regulation of stem cell fate decisions by the anabolic/catabolic balance. Stem cell behavior is influenced by a balance between regulators of catabolic and anabolic metabolic reactions. Dominance by catabolic regulators, which act to reduce energy use, protein synthesis and ROS accumulation, favors stem cell quiescence and the steady-state. When anabolic regulators gain dominance, active cycling by a stem cell increases its energy consumption and protein synthesis, enabling the cells to undertake the decision to either self-renew or differentiate. The anabolic/catabolic balance thus has a major influence on stem cell homeostasis.

Another important issue in the relationship between nutrients and healthy HSCs is an organism's dietary intake. Dietary interventions such as dietary restriction (DR), or the feeding of a ketogenic or high-fat diet (HFD), markedly affect an organism's metabolic state and the behavior of its stem cells [4]. There is evidence showing that DR (such as prolonged fasting) contributes positively to HSC functions, whereas the prolonged consumption of a HFD induces aberrant HSC self-renewal. Compared to our knowledge of the effects of nutrients on other stem cell types such as intestinal stem cells (ISCs), our understanding of the influence of nutrients on HSCs is much more limited. However, it is clear that signaling molecules involved in dietary interventions have a significant impact on both HSCs and the cells that surround them in the BM niche.

In this review, we focus on the above two aspects of nutrient-driven metabolic regulation required to maintain healthy HSCs. In the first section, we discuss the control of intracellular metabolism by the balance between anabolism and catabolism and how it helps to maintain quiescent HSCs. In the second section, we present an overview of recent findings on how changes to organismal metabolic status caused by altered dietary intake of nutrients affect HSC homeostasis. While the molecular mechanisms involved in these two aspects of nutrient function may not necessarily overlap, a better understanding of both sets of phenomena may point to ways to improve human health by ensuring healthy HSC homeostasis for life.

2. Effect of the Anabolic/Catabolic Balance on Stem Cell Homeostasis

2.1. Commonality of Nutrient-Sensing Regulators

Several types of tissue stem cells, including HSCs, experience long periods of quiescence or dormancy throughout the life of the host organism [2]. In HSCs, quiescence blocks the cell cycle and maintains the cells in an undifferentiated state that protects them against various insults, including DNA breaks and oxidative stress [1]. Similarly, resident satellite cells in muscle, known as muscle stem cells (MuSCs), are quiescent in an organism at steady-state [5]. In response to injury, MuSCs enter the cell cycle to regenerate the skeletal muscle tissue and replenish the MuSC pool. The same applies to adult neural stem cells (NSCs) in the subventricular zone (SVZ) of the brain, which are preserved in a dormant state in healthy animals [6]. Most adult NSCs cells are derived from a quiescent subpopulation of embryonic NSCs and differentiate into interneurons that migrate into the olfactory bulb.

The catabolic regulators required to maintain the dormancy of tissue stem cells are shared in animals. The regulators can also prolong lifespan or delay aging in multiple species, including yeast, worms, flies, and mice [3]. Conversely, stem cells in the intestinal epithelium and epidermis are actively cycling, and this cycling is vital to their homeostasis [2]. These differences indicate that the regulation of stem cell biology is not simple. In the sub-sections that follow, we review several evolutionarily conserved signaling pathways that are linked to catabolic status and modulate organismal longevity, but which also are important common regulators of the functions of HSCs, NSCs and MuSCs.

2.2. Catabolic Regulators That Maintain Stem Cell Quiescence

2.2.1. Autophagy Regulators

Autophagy is a conserved catabolic system that degrades cellular components which have become engulfed in autophagosomes [7]. Autophagy contributes to maintenance of routine cellular functions, but it is induced to a high level in response to insults such as oxidative stress, infection, and starvation. Autophagy is thus a survival mechanism that maintains cellular homeostasis by eliminating unwanted components such as damaged organelles, protein aggregates, and intracellular bacteria. Autophagy also supplies an energy source under stress conditions.

Autophagy occurs in a series of steps that are controlled by various autophagy-related genes (*Atgs*). These steps include initiation, formation of a phagophore structure, maturation of an autophagosome, fusion of the autophagosome with a lysosome, degradation of the engulfed entity, and recycling of nutrients. The *Atg5* and *Atg7* genes are involved in elongation of the autophagosome membrane. When the Vav-Cre-loxP system was used to delete these genes specifically in murine hematopoietic cells during the fetal period, the animals suffered from weight loss, severe anemia, and a reduction in HSCs that resulted in death [8–12]. However, mice in which the gene was deleted at either the juvenile or adult stage appeared largely healthy. Detailed analysis of HSC functions after birth showed that, although Atg5 deficiency induced no obvious abnormalities at P0, significant defects arose at P7 [12]. In addition, induction of *Atg5* deletion at P5 did not result in remarkable abnormality in hematopoiesis. These data suggest a critical role for autophagy in protecting HSCs against insults during the early neonatal stage, which is essential for healthy long-term hematopoiesis. In adult, HSCs with loss of *Atg12*, which is also involved in elongation of the autophagosome membrane, show impaired functions, increased mitochondrial content, and a bias toward myeloid differentiation [13,14]. At the signaling level, autophagy is induced rapidly upon starvation via a gene expression program that is driven by forkhead transcription factor FoxO3 (FoxO3). Thus, autophagy contributes to the HSC maintenance of their low metabolic state and quiescence by degrading and clearing active healthy mitochondria at steady-state.

Autophagy is also involved in the "aging phenotypes" exhibited by HSCs isolated from aged animals [13,14]. Mice unable to carry out autophagy as adults develop hematopoietic phenotypes resembling those of aged normal mice, such as increased cellularity and a skewed ratio of myeloid versus lymphoid cells in peripheral blood. Thus, regulation of metabolic status via autophagy is crucial for HSC homeostasis throughout the entire life of an animal (Figure 2A).

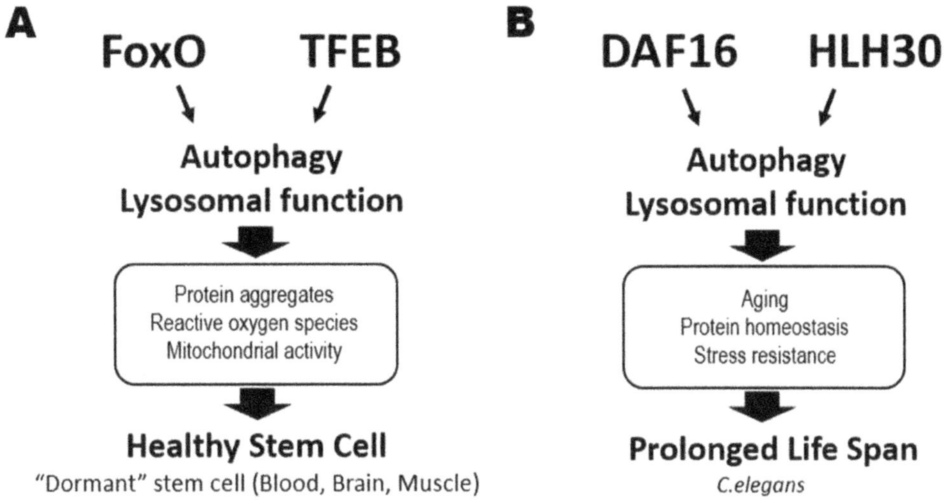

Figure 2. Critical roles of catabolic regulators in organismal life extension and the maintenance of healthy stem cells. (**A**) In mammalian stem cells, activation of the FoxO and TFEB transcription factors triggers autophagy and lysosomal functions that help to maintain tissue stem cell quiescence or dormancy in an animal at steady-state. This process reduces the accumulation of protein aggregates and ROS and suppresses mitochondrial activity in stem cells in blood (HSCs), brains (NSCs) and muscle (MuSCs), sustaining their good health. (**B**) DAF16 and HLH30 are the homologs of FoxO and TFEB, respectively, in *C. elegans*. These transcription factors also cooperate to regulate genes involved in autophagy and lysosomal functions, which prolong a worm's lifespan by maintaining protein homeostasis and increasing stress resistance.

In NSCs, autophagy plays a critical role in maintaining quiescence by reducing protein aggregates in a FoxO3-mediated manner and by decreasing ROS in mitochondria [15]. During aging, NSCs show a reduction in lysosome content and an increase in protein aggregates, suggesting that autophagy-mediated lysosomal activity is critical for the control of the latter [16]. Among MuSCs, autophagic flux is significantly elevated in cells from younger mice but reduced in MuSCs from aged animals [17]. Loss of Atg7 impairs mitophagy and elevates ROS, leading to a loss of the MuSC pool. Pertinently, restoring autophagy improves the function of MuSCs from aged mice.

Collectively, these data indicate that autophagy makes a major contribution to stem cell quiescence, maintenance, and prevention of premature aging.

2.2.2. Transcription Factor EB (TFEB)

Gene transcription associated with catabolic processes is induced by the nuclear translocation and consequent activation of the MiT/TFE family members, such as transcription factor EB (TFEB), a master modulator of autophagy and lysosomal biogenesis (Figure 2A, right). In *C. elegans*, activation of HLH30, its TFEB homolog, prolongs worm lifespan by promoting autophagy and lysosome functions [18] (Figure 2B). Accordingly, TFEB agonists also extend *C. elegans* lifespan [19]. In humans, lysosome functions are regulated dichotomously by TFEB and MYC to balance catabolic and anabolic processes required for activating HSCs and guiding their lineage fate [20]. The TFEB-mediated endolysosomal pathway promotes HSC quiescence and self-renewal, reinforcing the theory that catabolic processes control HSC fate determination. Similarly, TFEB deficiency or chemical inhibition of lysosomal degradation impairs NSC quiescence and function [21]. Quiescent NSCs exhibit large lysosomes containing insoluble protein aggregates, have higher lysosomal activity, and degrade activated EGF receptors by endolysosomal degradation [16,22]. Although the role of lysosomal activity in MuSC homeostasis has yet to be elucidated, the commonalities among HSCs and NSCs suggest that it would not be surprising to find a crucial role for TFEB in this stem cell type as well.

2.2.3. FoxO Family

As mentioned above, FoxO family members are vital for catabolic regulation in tissue stem cells. In *C. elegans*, reduced insulin/IGF-1 signaling extends lifespan in a process mediated by DAF16 (FoxO homolog) [23]. DAF16 and HLH30 (TFEB homolog) cooperatively regulate genes related to aging, protein homeostasis, and stress resistance, helping to prolong lifespan in a harsh environment [24] (Figure 2B). In mammals, FoxO1, FoxO3, FoxO4 and FoxO6 are all downstream targets of PI3K-AKT signaling [25]. In the absence of cellular stimulation by growth factors or insulin, FoxOs localize in the nucleus and activate their transcriptional targets. When a growth factor or insulin binds to the appropriate cell surface receptor, AKT is activated and directly phosphorylates FoxOs, resulting in their nuclear exclusion and degradation in the cytoplasm. Metabolic or oxidative stress can also induce nuclear localization of FoxOs and their transcriptional activity. In HSCs, it is FoxO1 and FoxO3 that are mainly expressed and localized in the nucleus under conditions of AKT inactivation [26,27]. Deficiency of FoxO3 or FoxO1/3/4 in HSCs results in defective quiescence and decreased long-term repopulating capacity [28–30]. This impaired HSC quiescence is linked to elevated ROS and modified expression of genes involved in either cell cycle arrest, such as *Cdkn1b*, *Ccng2*, *Cdkn1c* and *Cdkn1a*, or redox regulation, such as *SOD* [28–30]. However, the roles of FoxOs in ROS-dependent regulation of HSC functions is complex. One study reported that abnormalities of FoxO1/3/4-deficient HSCs could be rescued by the anti-oxidative reagent N-acetyl cysteine (NAC) [29], whereas another group found that NAC did not restore the functions of FoxO3-deficient HSCs [31]. In HSCs from obese mice, FoxO proteins become insensitive to their normal upstream regulators such as AKT, suggesting that hyperglycemia can directly alter the AKT-FoxO axis in these cells [32]. Thus, FoxOs are important for the maintenance of healthy HSCs in animals both at steady-state and in those under environmental stress.

FoxOs are also critical regulators in pluripotent stem cells. In embryonic stem (ES) cells, FoxO1 activates the expression of the pluripotency genes *Oct4* and *Sox2*, promoting ES cell self-renewal capacity [33]. FoxO1 also controls ES pluripotency by directly regulating the expression of core autophagic machinery genes [34]. FoxO3 protects ES cells from hyperglycemic stress by inducing upregulation of SOD2, catalase, p21 and p27, thereby promoting ROS detoxification and cell cycle arrest [35]. In NSCs, loss of FoxO1/3/4 or FoxO3 leads to defects in quiescence and undifferentiated status. Mechanistically, FoxOs control genes regulating the cell cycling and redox state of NSCs. In MuSCs, FoxOs are again critical for quiescence and prevention of premature aging. Niche-derived IGF1-dependent AKT activation negatively controls MuSC functions mediated by FoxOs [36]. While nuclear localization of FoxO3 is observed in $Pax7^{+}$ MuSCs, loss of FoxO3 impairs quiescence, leading to MuSC exhaustion [37]. An intriguing study has revealed how FoxOs act to prevent of premature aging of MuSCs. There are two types of quiescent MuSCs: those of a "genuine state", which exhibit stemness, and those of a "primed state", which commit to myogenic differentiation [36]. Extremely old mice lose "genuine" MuSCs and gain "primed" MuSCs. FoxO1/3/4 deficiency causes a similar loss of "genuine" MuSCs, causing regenerative failure. These data indicate that stem cell aging is controlled by FoxOs through their regulation of catabolism and stress responses.

Collectively, these observations link stem cell maintenance and function to the coordinated cooperation of numerous factors governing intracellular nutrients, including catabolic pathways, autophagy, and TFEB and/or FoxOs activation.

2.2.4. Anabolic Regulators That Control HSC Self-Renewal

One characteristic that is unique to quiescent HSCs is downregulation of protein synthesis [38,39]. This observation suggests that suppression of anabolism may be a critical hallmark of healthy HSCs. In general, protein synthesis in cells is strictly controlled by mTOR complex 1 (mTORC1), which is one of two functionally different protein complexes containing the vital kinase mTOR [40]. mTORC1 has three core components: mTOR, "mammalian lethal with SEC13 protein 8" (mLST8, also known as GβL), and the scaffold protein "regulatory-associated protein of mTOR" (RAPTOR). mTORC1 phosphorylates numerous protein substrates, including S6K1 and eIF4EBPs, to regulate protein translation. mTORC1 also upregulates the selective protein translation of mRNAs with 5′TOP structure. Once activated by mTORC1-mediated phosphorylation, S6K1 phosphorylates ribosomal protein S6, a component of the 40S ribosomal subunit, to upregulate rRNA transcription and promote the translation elongation of spliced transcripts. Quiescent HSCs show profound inhibition of S6 and 4EBP phosphorylation [41], confirming the downregulation of protein synthesis in these cells [42].

The regulation of mTORC1 is critical for HSC functions. Induction of mTORC1 hyperactivation by Tsc1 deletion or Rheb overexpression causes HSC depletion that is associated with cell cycle progression and mitochondrial activation [43–45]. The suppression of mTORC1 signaling in quiescent HSCs depends on Pten, which is a negative regulator of the PI3K-AKT pathway. Deletion of Pten in mice induces hyperactivation of mTORC1 in HSCs, aberrant cell cycle progression, and upregulation of p16Ink4A and p53, resulting in HSC depletion [46]. These abnormalities of HSCs can be reversed by the mTOR inhibitor rapamycin, confirming that mTORC1 hyperactivation leads to HSC failure in vivo. Raptor deficiency shows mTORC1 inactivation, and these cells gradually lose their capacity for hematopoietic reconstitution [47]. Therefore, mTORC1 function appears to be crucial for HSC maintenance regardless of Pten deficiency.

While hyperactivation of mTORC1 induces severe defects in HSC function, lower levels of mTORC1 activity are necessary for HSC self-renewal. Similar to Raptor, Rheb1 is an activator of mTORC1. Loss of Rheb1 suppresses mTORC1 activity, but the level of mTORC1 activity that remains in Rheb1-deficient cells is significantly greater than that in Raptor-deficient cells [48]. While the total number of HSCs as defined by marker phenotype is increased by deficiency of either Raptor or Rheb1, competitive reconstitution assays in vivo have demonstrated a clear difference in the functionality of Raptor-deficient vs. Rheb1-deficient HSCs [48]. Loss of Raptor, but not Rheb1, reduces HSC competitiveness in vivo, suggesting that HSCs that expand due to Raptor deficiency lose their stemness, whereas Rheb1 deficiency induces the expansion of HSCs that retain their stemness. These data suggest that the fine-tuning of moderate levels of mTORC1 activity is needed to preserve HSC self-renewal activity in vivo. Consistent with this finding, rapamycin-mediated inhibition of mTORC1 has made it possible to perform ex vivo HSC expansion for both mice and humans [49,50]. Combining rapamycin with a GSK-3 inhibitor further increases the number of functional HSCs, indicating a role for nutrient-sensing in the maintenance of HSCs in culture [51]. Rapamycin treatment also reverses the natural aging-related decline in HSC function in normal mice [52]. Thus, strict control of mTORC1 activity is the key to maintaining or expanding healthy HSC populations in vitro and in vivo.

Taken together, the findings presented in the first part of this review establish the importance of both catabolic and anabolic regulatory elements in HSC biology.

3. The Effects of Dietary Interventions on Stem Cell Homeostasis

3.1. Beneficial Effects of Dietary Restriction and Prolonged Fasting on HSC Functions

Dietary restrictions (DRs) have been shown to prolong lifespan by delaying aging in multiple species [53,54]. This observation has led to the hypothesis that such dietary interventions can also prevent the aging of HSCs and thus promote their preservation in a healthy state. In general, DR means reduced intake of either total dietary calories or one or more specific components of the diet. Intermittent and periodic fasting are also forms of DR. In the laboratory, differences in experimental DR protocols can generate different results. For example, Lazare et al. investigated the effects of caloric restriction (CR) on HSC function by feeding C57BL/6 (B6) mice on a diet containing 30% fewer calories than a control diet. These researchers showed that lifelong CR in mice did not have any impact on aging-associated HSC phenotypes [55]. However, Tang et al. reported that a 30% DR regimen, in which mice were fed only 70% of the amount of food consumed by ad libitum-fed mice, prevented HSCs from displaying aging phenotypes. The DR-treated mice in this study possessed fewer of the CD150^{high} myeloid-biased HSCs that increased in number with age compared to mice on the control diet. In this case, DR also conferred on HSCs increased dormancy and enhanced reconstitution ability compared with HSCs isolated from ad libitum-fed mice [56]. In addition to effects on HSCs, this form of DR suppressed lymphopoiesis while enhancing erythropoiesis and myelopoiesis in early progenitor cells. Mechanistically, DR suppressed expression levels of IGF-1, IL-6 and IL-7, which could be rescued by feeding the ad libitum diet to DR-treated mice. All three of these molecules proved to be involved in one or more DR-induced hematopoietic changes. For example, enhancement of quiescence of HSCs by DR-treatment was normalized by administration of IGF-1, but impaired lymphopoiesis was not rescued. In contrast, injection of IL-6 or IL-7 was able to rescue lymphopoiesis in DR-treated animals. These results indicate that DR affects hematopoiesis in a cell type-specific manner, and that multiple factors are involved in exerting DR-related effects.

When normal mice are treated with prolonged cycles of fasting and feeding, beneficial effects are conferred on HSCs. For example, Cheng et al. subjected mice to repeated cycles of fasting for 48 h followed by feeding for 12–14 days and examined the effects on hematopoiesis [57]. This group reported that prolonged fasting led to HSC expansion, protection of HSCs against chemotoxicity, and promotion of HSC regeneration capacity coupled to normal differentiation ability. Thus, limitation of nutrient intake by DR or prolonged fasting has a positive effect on HSC functions.

IGF-1 is well known as an aging-associated molecule in several tissues [54], but its physiological role in HSC behavior appears complicated. Cheng et al. found that the beneficial effects of prolonged fasting cycles on HSC function were mediated by a reduction in IGF-1/PKA signaling, which induced upregulation of FoxO1 expression and downregulation of G9a [57] (Figure 3, bottom). A recent report has shown that BM levels of IGF-1, which is expressed mainly by mesenchymal cells, decreased with age [58]. This reduction of IGF-1 in the BM microenvironment then induced myeloid-biased hematopoiesis that could be reversed by IGF-1-mediated stimulation of HSCs. Thus, these data appear to conflict with those from the study mentioned above. It is possible that feeding after fasting may induce sudden and significant changes to IGF-1 levels that alter the metabolic dynamics affecting HSC behavior. Fine-tuning of the IGF-1 signaling pathway within an optimum range may be important for maintaining healthy HSCs. A more detailed dissection of the effects of IGF-1 signaling on HSC capacities for self-renewal and balanced differentiation are needed.

3.2. The Effects of Obesity on HSC Homeostasis

Obesity caused by a sedentary lifestyle and unhealthy eating habits leads to inflammatory myelopoiesis in mice and humans. Because chronic, low-grade inflammation due to obesity exacerbates age-related diseases, it has been assumed that obesity may also accelerate aging-related defects in stem cells. However, the effects of obesity on HSCs remain unclear.

Previous studies using several different models of obesity, including leptin-deficient *ob* mice, leptin receptor-deficient *db* mice, and mice fed a HFD, have indicated that obesity increases numbers of myeloid-committed progenitor cells, including multipotent progenitors (MPPs), common myeloid progenitors (CMPs), and granulocyte and macrophage progenitors (GMPs) [59–63]. In addition, obesity induces an aberrant BM microenvironment [64,65]. The increase in myeloid-committed progenitor cells in obese mice is associated with adipocyte accumulation in the BM, and it is now clear that BM adipocytes can regulate HSC functions. For example, after BM injury by 5-fluorouracil (5-FU) or cyclophosphamide treatment, adiponectin deficiency inhibited the activation of HSCs by attenuating mTORC1 signaling pathway and delayed hematopoietic recovery [66]. The data suggested that adiponectin promotes hematopoietic regeneration by accelerating the entry of HSCs into the cell cycle. Furthermore, BM adipocytes are an important source of stem cell factor (SCF) after irradiation or 5-FU treatment, and BM adipocyte-derived SCF promotes the maintenance of HSCs and hematopoietic regeneration [67]. These observations show a positive role for at least some effects of adipocytes on HSC function.

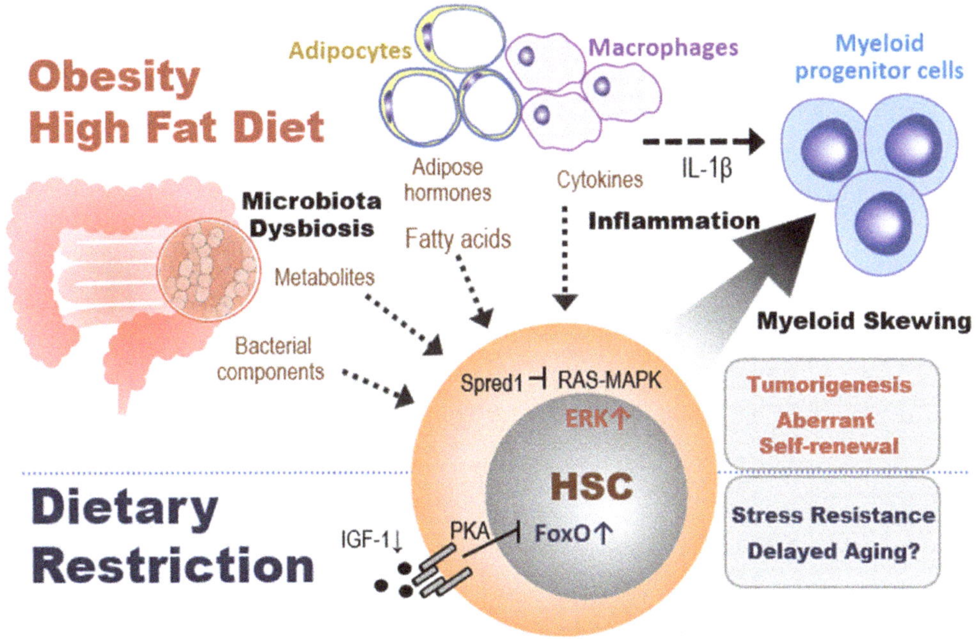

Figure 3. Impacts of dietary interventions on HSCs and myeloid progenitors. Top section: Obesity or HFD alters the spectrum of bacterial and adipocyte products and macrophage-produced cytokines that act on HSCs in the BM. In the presence of inflammation, these changes promote macrophage secretion of cytokines that cause HSCs to undergo abnormal self-renewal and a skewing to myeloid progenitor differentiation. In tumor-prone mice such as Spred1-deficient animals, in which RAS-MAPK-mediated tumor suppression is lost, HFD triggers aberrant myelopoiesis leading to the development of myeloproliferative neoplasm-like diseases. Bottom section. In contrast to obesity, a DR such as fasting reduces IGF-1 and PKA signaling, which activates FoxO1. This FoxO1 activation promotes the quiescence and stress resistance of HSCs and thereby prolongs host lifespan.

In contrast to the above, other studies have suggested that the accumulation of adipocytes within the BM during obesity negatively regulates HSC functions. For example, Ambrosi et al. showed in mice that aging and obesity reprogram the mesenchymal lineage to give rise to adipogenic lineage cells, resulting in expansion and accumulation of BM adipocytes. They also found that BM adipocytes reduced HSC reconstitution ability in a transplant model with isolated BM adipocytes [68]. Others have demonstrated that, in addition to BM adipocytes, other adipose tissue cells are involved in promoting myeloid-biased hematopoiesis during obesity. Nagareddy et al. reported that monocytosis in obesity was associated with infiltration of macrophages into adipose tissue [60]. Transplantation assays of adipose tissue revealed that obesity due to *ob* mutation or HFD promoted NLRP3 inflammasome-dependent IL-1β production by adipose tissue macrophages via the TLR4-

MyD88 signaling pathway (Figure 3, top). IL-1β derived from adipose tissue macrophages stimulates the proliferation of CMPs and GMPs, resulting in enhanced myelopoiesis. These observations show a negative role for at least some effects of obesity-induced adipocyte expansion on hematopoiesis.

While obesity appears to affect mainly myeloid-committed progenitor cells, the capacity of HSCs for total long-term reconstitution is also influenced by obesity-related factors [59]. Whereas a primary transplantation of HSCs from *db/db* or HFD-induced obese mice resulted in significantly increased chimerism in the recipients, a secondary transplantation of donor-derived HSCs revealed a great reduction in their reconstitution ability. This obesity-induced defect in HSCs depended on upregulated expression of the transcriptional repressor *Gfi1*, which was induced by the increased ROS levels in HSCs. *Gfi1* expression in HSCs can also be induced by aging or acute/chronic inflammation but these effects are reversible. In contrast, the obesity-induced defects in HSCs were irreversible. This obesity-specific regulation of HSC functions remains under investigation.

3.3. The Effects of HFD on HSC Homeostasis

In general, a HFD includes animal fats that are rich in undesirable saturated fatty acids. Consumption of a HFD reduces an organism's insulin sensitivity and the efficiency with which glucose enters the tissues. These effects of HFD alter organismal metabolic status by modulating multiple factors, affecting hematopoiesis (Figure 3, top left).

Hermetet et al. reported that a short term consumption of HFD induces exhaustion of HSCs mediated by modulation of lipid raft organization [69]. Lipid rafts are cholesterol-enriched patches located in the plasma membrane, which play a critical role in dormancy of HSCs mediated by TGF-β signaling [27,70]. While lipid rafts were distributed across the surface of HSC isolated from mice fed a normal diet (ND), they were clustered on HSC from mice fed a HFD, associated with down-regulation of TGF-β signaling. Since TGF-β injection restored impaired HSC functions induced by a HFD, they concluded that consuming a HFD induces alteration of the TGF-β signaling-mediated HSC quiescence, leading to exhaustion of HSCs. Although it is unclear how lipid rafts in HSCs are affected by a HFD, systemic metabolic change may cause altered lipid metabolism of HSC's plasma membrane, because it has been implied that dietary fatty acids, such as polyunsaturated fatty acids, affect membrane lipid raft organization [71]. Further investigation for molecular mechanisms of linkage between diet and HSC signaling is needed.

Another systemic effects of a HFD is abnormal intestinal permeability, setting the stage for plasma endotoxemia. Liu et al. reported that HFD-fed mice developed low-grade endotoxemia due to increased levels of serum LPS [63]. This report showed that feeding a HFD or treating with low-dose LPS induced functional changes in myeloid progenitors and CLPs, and that these changes depended on TLR4 signaling. However, the involvement of TLR4 signaling in HSC regulation under conditions of obesity is unclear. Since TLR4 activation by high-dose LPS is known to cause HSC dysfunction [72], it has been assumed that the endotoxemia resulting from HFD consumption may also impair HSC function. Detailed studies of the signaling molecules activated by low-dose bacterial pathogens in HFD-fed and obese animals are needed to clarify their effects on HSC regulation. These results indicate that the impact of HFD-derived factors on the regulation of HSCs and niche cells is in need of further exploration.

HFD consumption also changes the composition of the gut microbiota, often favoring the expansion of Gram-positive bacterial strains [73]. Recent studies have revealed that HFD-induced alterations to the composition of the gut microbiota affect hematopoietic homeostasis. For example, Luo et al. showed that the transfer of stool from HFD-fed mice to mice fed a ND caused the ND animals to exhibit HFD-associated hematopoietic and BM niche abnormalities. In the recipient mice, MPPs, myeloid progenitors, mesenchymal stem cells and adipocytes were all increased, whereas LT-HSCs, CLPs and osteoblasts were decreased. Cytokine expression levels were also altered in BM cells [74].

We and others have recently identified a critical effect of HFD on the self-renewal of HSCs in tumor-prone mice. We demonstrated that the altered gut microbiota composition caused by HFD affected the regulation of HSC self-renewal in Spred1-deficient mice, which develop myeloproliferative neoplasm-like disease and anemia. Spred1 is a negative regulator of RAS-MAPK signaling activated by SCF/c-Kit signaling [72]. Loss of Spred1 in ND mice enhanced HSC self-renewal capacity and success in cell competitions, suggesting that Spred1 negatively regulates HSC self-renewal. The consumption of HFD, but not LPS treatment or aging, induced ERK hyperactivation and aberrant self-renewal in Spred1-deficient HSCs, leading to the appearance of anemia and myeloproliferative neoplasm-like disease (Figure 3, right). Depletion of gut microbiota by antibiotic treatment mitigated these HFD-induced hematopoietic abnormalities, suggesting that the microbiota play a crucial role in regulating HSC homeostasis. Since downregulation of SPRED1 is observed in myeloid leukemia patients with a poor prognosis, [75], HFD-driven metabolic aberration or microbiota dysbiosis may significantly promote pathogenesis of leukemia patients.

3.4. The Effects of Metabolites Induced by Dietary Interventions on Stem Cell Fate Determination

Although our knowledge of how HSC regulation is affected by dietary interventions is currently limited, evidence of significant effects of various metabolites on other tissue stem cells has been accumulating. The next sub-section examines metabolites that control ISC self-renewal. The findings suggest possible roles of these metabolites in HSC homeostasis.

3.4.1. Fatty Acid Composition

Previous reports showed that the treatment with either HFD or fatty acid constituents enhanced ISC self-renewal in a manner dependent on PPAR-fatty acid oxidation (FAO) in the absence of inflammatory signaling [76,77]. Conversely, intermittent fasting enhances self-renewal of ISCs by activation of PPARδ signaling through enhancing FAO [78]. These contradictory data suggest that fatty acids may be critical but complex regulators of HSC homeostasis.

3.4.2. Ketone Bodies and Related Metabolites

During fasting, glucose levels decline and ketone biogenesis increases, releasing ketone bodies such as acetone and β-hydroxybutyrate (β-OHB) into the circulation. β-OHB enhances ISC self-renewal [79] by inhibiting histone deacetylases (HDACs) [80]. A ketogenic diet, which is a high-fat, low-carbohydrate diet, also induces the production of ketone bodies, which are used as alternate carbon sources. β-oxidation of fatty acids induced by a ketogenic diet promotes the activity of the tricarboxylic acid (TCA) cycle mediated by production of acetyl-CoA [81]. While there have been no reports of significant effects of a fasting or ketogenic diet on HSC behavior, it is plausible that such dietary interventions might modulate hematopoiesis and/or HSC aging.

3.4.3. Microbiota Producing Metabolites

Alterations of the gut microbiota composition can induce systemic or local changes in gut microbiota-derived metabolites, which have recently been shown to affect organismal aging, tissue homeostasis and development of several diseases in humans as well as in mice [82–84]. For example, mice fed HFD show expansion of Gram-positive bacterial strains in the intestine. Increased levels of deoxycholic acid (DCA), which is a secondary bile acid produced by the microbiota, appear in the blood plasma of these animals. Bile acids, including secondary bile acids, are known to affect the regulation of HSC functions through their suppression of the ER stress response [85–87]. In addition, difference in intake of dietary fiber can change the spectrum of gut microbiota-derived metabolites in mice. Levels of short-chain fatty acids (SCFAs), including acetate, propionate and butyrate, are modulated by the feeding of dietary fiber, and these molecules can affect differentiation and the functions of immune cells [88,89]. Butyrate can also suppress the proliferation of murine ISCs in a FoxO3-dependent manner [90], implying an important role for this metabolite in regulating HSC biology. The same may be true for many other such metabolites in a broad range of stem cells.

4. Future Perspective

Over the past 15 years, there has been intensive study of the intracellular metabolic regulation in HSCs mediated by the anabolic/catabolic balance. The underlying rationale for these studies may be the hypothesis that the longevity or lifespan of an organism relies on the continued functionality of its stem cells. This body of work has now established that the activation of catabolic regulators is required for the maintenance of HSCs, as well as that of other tissue stem cells such as NSCs and MuSCs. Suppression of aberrant protein aggregation, mitochondrial activation, and oxidative stress has proven vital for the proper regulation of these tissue stem cells. Although mechanisms of control of HSC self-renewal by anabolic regulators have remained unclear, a deeper understanding of how stem cell behavior is regulated by the anabolic/catabolic balance will no doubt yield new strategies for maintaining healthy HSCs in vitro and in vivo.

The effects of catabolism on stem cell homeostasis in the context of cancer cell dormancy have not fully elucidated. Several catabolic regulators have been reported to play critical roles in the maintenance of stem cells in malignant tissues [91,92], but the functions of these regulators were found to be highly variable among cancers. Compared to tissue stem cells, cancer cells are much more complicated because they undergo numerous gene mutations and epigenetic alterations. It is thus difficult to find biological commonalities among such cells. However, cancer cell dormancy might be one area where a shared biological mechanism exists. It has recently been proposed that non-mutational drug resistance mechanisms underlie the survival of residual "drug tolerant persister" (DTP) cancer cells, and that these DTPs display transcriptional and functional similarities to cells in diapause, a reversible state of suspended embryonic development triggered by unfavorable environmental conditions [93,94]. Embryonic diapause is induced by inhibition of Myc [95] or mTOR [96], which are anabolic regulators. This finding suggests an important role for nutrient-sensing signaling in the survival of DTPs. That being said, this hypothesis appears paradoxical because the same signals have been thought to be important factors driving malignant progression. Further delving into how cancer behavior is influenced by nutrients may lead to the development of novel therapeutics useful for treating a variety of cancers.

Unlike the intensive analyses of intracellular metabolic regulators within HSCs, there has been little work performed to understand the molecular mechanisms by which changes to dietary nutrients affect HSC behavior. Although it is evident that the inflammatory machinery is at the center of this phenomenon, crucial players remain hidden. Specifically, studies of the effects of nutrients on cells in the BM niche supporting hematopoiesis and HSC homeostasis should be expanded, and the molecular mechanisms underlying the linkages between intracellular metabolism and systemic metabolic changes, and how they affect the microenvironments controlling stem cell behavior, should be explored. Many different cell types within the BM niche contribute to HSC homeostasis, including vascular cells, mesenchymal stem cells, mesenchymal lineage cells (osteoblasts and adipocytes), and neural cells. However, it remains unclear how these multiple cell types are organized by nutrient-related factors to preserve an environment suitable for healthy HSCs. To properly understand the spatiotemporal dynamics of HSC-niche interactions under various nutrient conditions, studies combining genomics, bioinformatics, metabolomics and imaging should be performed via interdisciplinary collaboration. Only by this multi-pronged approach will the mechanisms underlying the physiological roles of nutrients in the BM niche be clarified.

The knowledge gained from the above investigations may also spur the identification of compounds that can be used to influence HSC survival and behavior. It has been demonstrated that HSCs bearing genetic alterations to the *TET2*, *DNMT3A* or *ASXL1* genes exhibit abnormal differentiation with a bias toward myelopoiesis, promoting chronic inflammation. The myeloid-biased HSCs accelerate atherosclerosis and increase the risk of cardiovascular disease because of myeloid cells secretion of inflammatory cytokines or leakage of gut microbiota, both of which sustain systemic chronic inflammation. If compounds could be found to prevent the generation of aberrant HSCs despite their mutations, such products could have important medical applications in the future. Since these compounds will be natural molecules already present in the human body, it is likely that some may be useable without modification for the prevention of hematopoietic abnormalities, such as aging-related myelodysplasia. In addition, the early detection of upregulation of a metabolite known to be harmful to HSCs could aid in the diagnosis of hematopoietic diseases. While genetic alterations are critical indicators, additional information with biomarkers that reflect inflammatory status would contribute to accurate prediction of prognosis and successful medical intervention.

In conclusion, continued research in the field of the effects of intra- and extra-cellular metabolism on HSCs is both necessary and potentially highly rewarding. These efforts will no doubt make numerous major contributions to society in the form of new tools for medical research and novel therapeutic agents for improving human health care.

Author Contributions: Conceptualization, Y.T. and A.H; writing and editing, Y.T. and A.H. All authors have read and agreed to the published version of the manuscript.

Funding: This work was supported by a Grant-in-Aid for Scientific Research (A) (19H01033) (A.H.); a Grant-in-Aid for Scientific Research (C) (19K08860) (Y.T.) from the Ministry of Education, Culture, Sports, Science, and Technology (MEXT) of Japan; and a Grant-in-Aid for Project for Cancer Research and Therapeutic Evolution (P-CREATE) (21cm0106104h0006) from the Japan Agency for Medical Research and Development (AMED) (A.H.).

Acknowledgments: We thank all members of the Hirao laboratory and D.C. Voon for contributing to invaluable scientific discussions and comments on the manuscript.

Conflicts of Interest: The authors declare no conflict of interest.

References

1. Nakamura-Ishizu, A.; Ito, K.; Suda, T. Hematopoietic Stem Cell Metabolism during Development and Aging. *Dev. Cell* **2020**, *54*, 239–255. [CrossRef]
2. van Velthoven, C.T.J.; Rando, T.A. Stem Cell Quiescence: Dynamism, Restraint, and Cellular Idling. *Cell Stem Cell* **2019**, *24*, 213–225. [CrossRef]
3. Nakamura, S.; Yoshimori, T. Autophagy and Longevity. *Mol. Cells* **2018**, *41*, 65–72. [CrossRef]
4. Novak, J.S.; Baksh, S.C.; Fuchs, E. Dietary interventions as regulators of stem cell behavior in homeostasis and disease. *Genes Dev.* **2021**, *35*, 199–211. [CrossRef]
5. Tedesco, F.S.; Dellavalle, A.; Diaz-Manera, J.; Messina, G.; Cossu, G. Repairing skeletal muscle: Regenerative potential of skeletal muscle stem cells. *J. Clin. Investig.* **2010**, *120*, 11–19. [CrossRef] [PubMed]
6. Zhao, C.; Deng, W.; Gage, F.H. Mechanisms and Functional Implications of Adult Neurogenesis. *Cell* **2008**, *132*, 645–660. [CrossRef] [PubMed]
7. Dikic, I.; Elazar, Z. Mechanism and medical implications of mammalian autophagy. *Nat. Rev. Mol. Cell Biol.* **2018**, *19*, 349–364. [CrossRef]
8. Mortensen, M.; Ferguson, D.; Edelmann, M.; Kessler, B.; Morten, K.J.; Komatsu, M.; Simon, A.K. Loss of autophagy in erythroid cells leads to defective removal of mitochondria and severe anemia in vivo. *Proc. Natl. Acad. Sci. USA* **2010**, *107*, 832–837. [CrossRef]
9. Watson, A.S.; Riffelmacher, T.; Stranks, A.J.; Williams, O.; de Boer, J.; Cain, K.C.; Macfarlane, M.; McGouran, J.; Kessler, B.M.; Khandwala, S.; et al. Autophagy limits proliferation and glycolytic metabolism in acute myeloid leukemia. *Cell Death Discov.* **2015**, *1*, 15008. [CrossRef]
10. Jung, H.E.; Shim, Y.R.; Oh, J.E.; Oh, D.S.; Lee, H.K. The autophagy Protein Atg5 Plays a Crucial Role in the Maintenance and Reconstitution Ability of Hematopoietic Stem Cells. *Immune Netw.* **2019**, *19*, e12. [CrossRef]
11. Mortensen, M.; Soilleux, E.J.; Djordjevic, G.; Tripp, R.; Lutteropp, M.; Sadighi-Akha, E.; Stranks, A.J.; Glanville, J.; Knight, S.; Jacobsen, S.E.; et al. The autophagy protein Atg7 is essential for hematopoietic stem cell maintenance. *J. Exp. Med.* **2011**, *208*, 455–467. [CrossRef] [PubMed]
12. Nomura, N.; Ito, C.; Ooshio, T.; Tadokoro, Y.; Kohno, S.; Ueno, M.; Kobayashi, M.; Kasahara, A.; Takase, Y.; Kurayoshi, K.; et al. Essential role of autophagy in protecting neonatal haematopoietic stem cells from oxidative stress in a p62-independent manner. *Sci. Rep.* **2021**, *11*, 1666. [CrossRef] [PubMed]
13. Warr, M.R.; Binnewies, M.; Flach, J.; Reynaud, D.; Garg, T.; Malhotra, R.; Debnath, J.; Passegué, E. FOXO3A directs a protective autophagy program in haematopoietic stem cells. *Nature* **2013**, *494*, 323–327. [CrossRef] [PubMed]

14. Ho, T.T.; Warr, M.R.; Adelman, E.R.; Lansinger, O.M.; Flach, J.; Verovskaya, E.V.; Figueroa, M.E.; Passegué, E. Autophagy maintains the metabolism and function of young and old stem cells. *Nature* **2017**, *543*, 205–210. [CrossRef]
15. Audesse, A.J.; Dhakal, S.; Hassell, L.-A.; Gardell, Z.; Nemtsova, Y.; Webb, A.E. FOXO3 directly regulates an autophagy network to functionally regulate proteostasis in adult neural stem cells. *PLoS Genet.* **2019**, *15*, e1008097. [CrossRef]
16. Leeman, D.S.; Hebestreit, K.; Ruetz, T.; Webb, A.E.; McKay, A.; Pollina, E.A.; Dulken, B.W.; Zhao, X.; Yeo, R.W.; Ho, T.T.; et al. Lysosome activation clears aggregates and enhances quiescent neural stem cell activation during aging. *Science* **2018**, *359*, 1277–1283. [CrossRef]
17. Garcia-Prat, L.; Martinez-Vicente, M.; Perdiguero, E.; Ortet, L.; Rodriguez-Ubreva, J.; Rebollo, E.; Ruiz-Bonilla, V.; Gutarra, S.; Ballestar, E.; Serrano, A.L.; et al. Autophagy maintains stemness by preventing senescence. *Nature* **2016**, *529*, 37–42. [CrossRef]
18. Lapierre, L.R.; Daniel De Magalhaes Filho, C.; McQuary, P.R.; Chu, C.-C.; Visvikis, O.; Chang, J.T.; Gelino, S.; Ong, B.; Davis, A.E.; Irazoqui, J.E.; et al. The TFEB orthologue HLH-30 regulates autophagy and modulates longevity in Caenorhabditis elegans. *Nat. Commun.* **2013**, *4*, 2267. [CrossRef]
19. Wang, C.; Niederstrasser, H.; Douglas, P.M.; Lin, R.; Jaramillo, J.; Li, Y.; Oswald, N.W.; Zhou, A.; McMillan, E.A.; Mendiratta, S.; et al. Small-molecule TFEB pathway agonists that ameliorate metabolic syndrome in mice and extend *C. elegans* lifespan. *Nat. Commun.* **2017**, *8*, 2270. [CrossRef]
20. García-Prat, L.; Kaufmann, K.B.; Schneiter, F.; Voisin, V.; Murison, A.; Chen, J.; Chan-Seng-Yue, M.; Gan, O.I.; McLeod, J.L.; Smith, S.A.; et al. TFEB-mediated endolysosomal activity controls human hematopoietic stem cell fate. *Cell Stem Cell* **2021**, *28*, 1838–1850 e10. [CrossRef]
21. Yuizumi, N.; Harada, Y.; Kuniya, T.; Sunabori, T.; Koike, M.; Wakabayashi, M.; Ishihama, Y.; Suzuki, Y.; Kawaguchi, D.; Gotoh, Y. Maintenance of neural stem-progenitor cells by the lysosomal biosynthesis regulators TFEB and TFE3 in the embryonic mouse telencephalon. *Stem Cells* **2021**, *39*, 929–944. [CrossRef]
22. Kobayashi, T.; Piao, W.; Takamura, T.; Kori, H.; Miyachi, H.; Kitano, S.; Iwamoto, Y.; Yamada, M.; Imayoshi, I.; Shioda, S.; et al. Enhanced lysosomal degradation maintains the quiescent state of neural stem cells. *Nat. Commun.* **2019**, *10*, 5446. [CrossRef]
23. Kenyon, C.J. The genetics of ageing. *Nature* **2010**, *464*, 504–512. [CrossRef]
24. Lin, X.X.; Sen, I.; Janssens, G.E.; Zhou, X.; Fonslow, B.R.; Edgar, D.; Stroustrup, N.; Swoboda, P.; Yates, J.R., 3rd; Ruvkun, G.; et al. DAF-16/FOXO and HLH-30/TFEB function as combinatorial transcription factors to promote stress resistance and longevity. *Nat. Commun.* **2018**, *9*, 4400. [CrossRef]
25. Eijkelenboom, A.; Burgering, B.M.T. FOXOs: Signalling integrators for homeostasis maintenance. *Nat. Rev. Mol. Cell Biol.* **2013**, *14*, 83–97. [CrossRef]
26. Liang, R.; Rimmelé, P.; Bigarella, C.L.; Yalcin, S.; Ghaffari, S. Evidence for AKT-independent regulation of FOXO1 and FOXO3 in haematopoietic stem and progenitor cells. *Cell Cycle* **2016**, *15*, 861–867. [CrossRef]
27. Yamazaki, S.; Iwama, A.; Takayanagi, S.-I.; Morita, Y.; Eto, K.; Ema, H.; Nakauchi, H. Cytokine signals modulated via lipid rafts mimic niche signals and induce hibernation in hematopoietic stem cells. *EMBO J.* **2006**, *25*, 3515–3523. [CrossRef]
28. Miyamoto, K.; Araki, K.Y.; Naka, K.; Arai, F.; Takubo, K.; Yamazaki, S.; Matsuoka, S.; Miyamoto, T.; Ito, K.; Ohmura, M.; et al. Foxo3a Is Essential for Maintenance of the Hematopoietic Stem Cell Pool. *Cell Stem Cell* **2007**, *1*, 101–112. [CrossRef]
29. Tothova, Z.; Kollipara, R.; Huntly, B.J.; Lee, B.H.; Castrillon, D.H.; Cullen, D.E.; McDowell, E.P.; Lazo-Kallanian, S.; Williams, I.R.; Sears, C.; et al. FoxOs are critical mediators of hematopoietic stem cell resistance to physiologic oxidative stress. *Cell* **2007**, *128*, 325–339. [CrossRef]
30. Yalcin, S.; Zhang, X.; Luciano, J.P.; Mungamuisk, S.K.; Marinkovic, D.; Vercherat, C.; Sarkar, A.; Grisotto, M.; Taneja, R.; Ghaffari, S. Foxo3 Is Essential for the Regulation of Ataxia Telangiectasia Mutated and Oxidative Stress-mediated Homeostasis of Hematopoietic Stem Cells. *J. Biol. Chem.* **2008**, *283*, 25692–25705. [CrossRef]
31. Rimmelé, P.; Liang, R.; Bigarella, C.L.; Kocabas, F.; Xie, J.; Serasinghe, M.N.; Chipuk, J.E.; Sadek, H.; Zhang, C.C.; Ghaffari, S. Mitochondrial metabolism in hematopoietic stem cells requires functional FOXO 3. *EMBO Rep.* **2015**, *16*, 1164–1176. [CrossRef] [PubMed]
32. Govindarajah, V.; Lee, J.-M.; Solomon, M.; Goddard, B.; Nayak, R.; Nattamai, K.; Geiger, H.; Salomonis, N.; Cancelas, J.A.; Reynaud, D. FOXO activity adaptation safeguards the hematopoietic stem cell compartment in hyperglycemia. *Blood Adv.* **2020**, *4*, 5512–5526. [CrossRef] [PubMed]
33. Zhang, X.; Yalcin, S.; Lee, D.-F.; Yeh, T.-Y.J.; Lee, S.-M.; Su, J.; Mungamuisk, S.K.; Rimmelé, P.; Kennedy, M.; Sellers, R.; et al. FOXO1 is an essential regulator of pluripotency in human embryonic stem cells. *Nat. Cell Biol.* **2011**, *13*, 1092–1099. [CrossRef]
34. Liu, P.; Liu, K.; Gu, H.; Wang, W.; Gong, J.; Zhu, Y.; Zhao, Q.; Cao, J.; Han, C.; Gao, F.; et al. High autophagic flux guards ESC identity through coordinating autophagy machinery gene program by FOXO1. *Cell Death Differ.* **2017**, *24*, 1672–1680. [CrossRef]
35. McClelland Descalzo, D.L.; Satoorian, T.S.; Walker, L.M.; Sparks, N.R.; Pulyanina, P.Y.; Zur Nieden, N.I. Glucose-Induced Oxidative Stress Reduces Proliferation in Embryonic Stem Cells via FOXO3A/beta-Catenin-Dependent Transcription of p21(cip1). *Stem Cell Rep.* **2016**, *7*, 55–68. [CrossRef] [PubMed]
36. García-Prat, L.; Perdiguero, E.; Alonso-Martín, S.; Dell'Orso, S.; Ravichandran, S.; Brooks, S.R.; Juan, A.H.; Campanario, S.; Jiang, K.; Hong, X.; et al. FoxO maintains a genuine muscle stem-cell quiescent state until geriatric age. *Nat. Cell Biol.* **2020**, *22*, 1307–1318. [CrossRef]
37. Gopinath, S.D.; Webb, A.E.; Brunet, A.; Rando, T.A. FOXO3 Promotes Quiescence in Adult Muscle Stem Cells during the Process of Self-Renewal. *Stem Cell Rep.* **2014**, *2*, 414–426. [CrossRef]

38. Signer, R.A.; Magee, J.A.; Salic, A.; Morrison, S.J. Haematopoietic stem cells require a highly regulated protein synthesis rate. *Nature* **2014**, *509*, 49–54. [CrossRef]
39. Signer, R.A.; Qi, L.; Zhao, Z.; Thompson, D.; Sigova, A.A.; Fan, Z.P.; DeMartino, G.N.; Young, R.A.; Sonenberg, N.; Morrison, S.J. The rate of protein synthesis in hematopoietic stem cells is limited partly by 4E-BPs. *Genes Dev.* **2016**, *30*, 1698–1703. [CrossRef]
40. Liu, G.Y.; Sabatini, D.M. mTOR at the nexus of nutrition, growth, ageing and disease. *Nat. Rev. Mol. Cell Biol.* **2020**, *21*, 183–203. [CrossRef]
41. Hoshii, T.; Tadokoro, Y.; Naka, K.; Ooshio, T.; Muraguchi, T.; Sugiyama, N.; Soga, T.; Araki, K.; Yamamura, K.-I.; Hirao, A. mTORC1 is essential for leukemia propagation but not stem cell self-renewal. *J. Clin. Investig.* **2012**, *122*, 2114–2129. [CrossRef] [PubMed]
42. Hoshii, T.; Matsuda, S.; Hirao, A. Pleiotropic roles of mTOR complexes in haemato-lymphopoiesis and leukemogenesis. *J. Biochem.* **2014**, *156*, 73–83. [CrossRef] [PubMed]
43. Chen, C.; Liu, Y.; Liu, R.; Ikenoue, T.; Guan, K.L.; Liu, Y.; Zheng, P. TSC-mTOR maintains quiescence and function of hematopoietic stem cells by repressing mitochondrial biogenesis and reactive oxygen species. *J. Exp. Med.* **2008**, *205*, 2397–2408. [CrossRef] [PubMed]
44. Gan, B.; Sahin, E.; Jiang, S.; Sanchez-Aguilera, A.; Scott, K.L.; Chin, L.; Williams, D.A.; Kwiatkowski, D.J.; DePinho, R.A. mTORC1-dependent and -independent regulation of stem cell renewal, differentiation, and mobilization. *Proc. Natl. Acad. Sci. USA* **2008**, *105*, 19384–19389. [CrossRef]
45. Campbell, T.B.; Basu, S.; Hangoc, G.; Tao, W.; Broxmeyer, H.E. Overexpression of Rheb2 enhances mouse hematopoietic progenitor cell growth while impairing stem cell repopulation. *Blood* **2009**, *114*, 3392–3401. [CrossRef]
46. Lee, J.; Nakada, D.; Yilmaz, O.H.; Tothova, Z.; Joseph, N.M.; Lim, M.; Gilliland, D.G.; Morrison, S.J. mTOR Activation Induces Tumor Suppressors that Inhibit Leukemogenesis and Deplete Hematopoietic Stem Cells after Pten Deletion. *Cell Stem Cell* **2010**, *7*, 593–605. [CrossRef]
47. Kalaitzidis, D.; Sykes, S.M.; Wang, Z.; Punt, N.; Tang, Y.; Ragu, C.; Sinha, A.U.; Lane, S.W.; Souza, A.L.; Clish, C.B.; et al. mTOR complex 1 plays critical roles in hematopoiesis and Pten-loss-evoked leukemogenesis. *Cell Stem Cell* **2012**, *11*, 429–439. [CrossRef]
48. Peng, H.; Kasada, A.; Ueno, M.; Hoshii, T.; Tadokoro, Y.; Nomura, N.; Ito, C.; Takase, Y.; Vu, H.T.; Kobayashi, M.; et al. Distinct roles of Rheb and Raptor in activating mTOR complex 1 for the self-renewal of hematopoietic stem cells. *Biochem. Biophys. Res. Commun.* **2017**, *495*, 1129–1135. [CrossRef]
49. Luo, Y.; Li, L.; Zou, P.; Wang, J.; Shao, L.; Zhou, D.; Liu, L. Rapamycin Enhances Long-Term Hematopoietic Reconstitution of Ex Vivo Expanded Mouse Hematopoietic Stem Cells by Inhibiting Senescence. *Transplantation* **2014**, *97*, 20–29. [CrossRef]
50. Rohrabaugh, S.L.; Campbell, T.B.; Hangoc, G.; Broxmeyer, H.E. Ex vivo rapamycin treatment of human cord blood CD34+ cells enhances their engraftment of NSG mice. *Blood Cells, Mol. Dis.* **2011**, *46*, 318–320. [CrossRef]
51. Huang, J.; Nguyen-McCarty, M.; O Hexner, E.; Danet-Desnoyers, G.; Klein, P.S. Maintenance of hematopoietic stem cells through regulation of Wnt and mTOR pathways. *Nat. Med.* **2012**, *18*, 1778–1785. [CrossRef] [PubMed]
52. Chen, C.; Liu, Y.; Liu, Y.; Zheng, P. mTOR regulation and therapeutic rejuvenation of aging hematopoietic stem cells. *Sci. Signal.* **2009**, *2*, ra75. [CrossRef] [PubMed]
53. Green, C.L.; Lamming, D.W.; Fontana, L. Molecular mechanisms of dietary restriction promoting health and longevity. *Nat. Rev. Mol. Cell Biol.* **2021**, *23*, 56–73. [CrossRef] [PubMed]
54. Longo, D.V.; Tano, M.D.; Mattson, M.P.; Guidi, N. Intermittent and periodic fasting, longevity and disease. *Nat. Aging* **2021**, *1*, 47–59. [CrossRef]
55. Lazare, S.; Ausema, A.; Reijne, A.C.; van Dijk, G.; van Os, R.; de Haan, G. Lifelong dietary intervention does not affect hematopoietic stem cell function. *Exp. Hematol.* **2017**, *53*, 26–30. [CrossRef] [PubMed]
56. Tang, D.; Tao, S.; Chen, Z.; Koliesnik, I.O.; Calmes, P.G.; Hoerr, V.; Han, B.; Gebert, N.; Zornig, M.; Loffler, B.; et al. Dietary restriction improves repopulation but impairs lymphoid differentiation capacity of hematopoietic stem cells in early aging. *J. Exp. Med.* **2016**, *213*, 535–553. [CrossRef]
57. Cheng, C.W.; Adams, G.B.; Perin, L.; Wei, M.; Zhou, X.; Lam, B.S.; Da Sacco, S.; Mirisola, M.; Quinn, D.I.; Dorff, T.B.; et al. Prolonged fasting reduces IGF-1/PKA to promote hematopoietic-stem-cell-based regeneration and reverse immunosuppression. *Cell Stem Cell* **2014**, *14*, 810–823. [CrossRef]
58. Young, K.; Eudy, E.; Bell, R.; A Loberg, M.; Stearns, T.; Sharma, D.; Velten, L.; Haas, S.; Filippi, M.-D.; Trowbridge, J.J. Decline in IGF1 in the bone marrow microenvironment initiates hematopoietic stem cell aging. *Cell Stem Cell* **2021**, *28*, 1473–1482. [CrossRef]
59. Lee, J.-M.; Govindarajah, V.; Goddard, B.; Hinge, A.; Muench, D.E.; Filippi, M.-D.; Aronow, B.; Cancelas, J.A.; Salomonis, N.; Grimes, H.L.; et al. Obesity alters the long-term fitness of the hematopoietic stem cell compartment through modulation of Gfi1 expression. *J. Exp. Med.* **2018**, *215*, 627–644. [CrossRef]
60. Nagareddy, P.R.; Kraakman, M.; Masters, S.L.; Stirzaker, R.A.; Gorman, D.J.; Grant, R.W.; Dragoljevic, D.; Hong, E.S.; Abdel-Latif, A.; Smyth, S.S.; et al. Adipose tissue macrophages promote myelopoiesis and monocytosis in obesity. *Cell Metab.* **2014**, *19*, 821–835. [CrossRef]
61. Berg, S.M.V.D.; Seijkens, T.T.P.; Kusters, P.J.H.; Beckers, L.; Toom, M.D.; Smeets, E.; Levels, J.; de Winther, M.P.J.; Lutgens, E. Diet-induced obesity in mice diminishes hematopoietic stem and progenitor cells in the bone marrow. *FASEB J.* **2016**, *30*, 1779–1788. [CrossRef] [PubMed]

62. Singer, K.; DelProposto, J.; Morris, D.L.; Zamarron, B.; Mergian, T.; Maley, N.; Cho, K.W.; Geletka, L.; Subbaiah, P.; Muir, L.; et al. Diet-induced obesity promotes myelopoiesis in hematopoietic stem cells. *Mol. Metab.* **2014**, *3*, 664–675. [CrossRef] [PubMed]
63. Liu, A.; Chen, M.; Kumar, R.; Stefanovic-Racic, M.; O'Doherty, R.M.; Ding, Y.; Jahnen-Dechent, W.; Borghesi, L. Bone marrow lympho-myeloid malfunction in obesity requires precursor cell-autonomous TLR4. *Nat. Commun.* **2018**, *9*, 708. [CrossRef]
64. Chen, H.; Charlat, O.; Tartaglia, L.A.; Woolf, E.A.; Weng, X.; Ellis, S.J.; Lakey, N.D.; Culpepper, J.; More, K.J.; Breitbart, R.E.; et al. Evidence That the Diabetes Gene Encodes the Leptin Receptor: Identification of a Mutation in the Leptin Receptor Gene in db/db Mice. *Cell* **1996**, *84*, 491–495. [CrossRef]
65. Zhang, Y.; Proenca, R.; Maffei, M.; Barone, M.; Leopold, L.; Friedman, J.M. Positional cloning of the mouse obese gene and its human homologue. *Nature* **1994**, *372*, 425–432. [CrossRef]
66. Masamoto, Y.; Arai, S.; Sato, T.; Kubota, N.; Takamoto, I.; Kadowaki, T.; Kurokawa, M. Adiponectin Enhances Quiescence Exit of Murine Hematopoietic Stem Cells and Hematopoietic Recovery Through mTORC1 Potentiation. *Stem Cells* **2017**, *35*, 1835–1848. [CrossRef]
67. Zhou, B.O.; Yu, H.; Yue, R.; Zhao, Z.; Rios, J.J.; Naveiras, O.; Morrison, S.J. Bone marrow adipocytes promote the regeneration of stem cells and haematopoiesis by secreting SCF. *Nat. Cell Biol.* **2017**, *19*, 891–903. [CrossRef]
68. Ambrosi, T.H.; Scialdone, A.; Graja, A.; Gohlke, S.; Jank, A.-M.; Bocian, C.; Woelk, L.; Fan, H.; Logan, D.W.; Schurmann, A.; et al. Adipocyte Accumulation in the Bone Marrow during Obesity and Aging Impairs Stem Cell-Based Hematopoietic and Bone Regeneration. *Cell Stem Cell* **2017**, *20*, 771–784.e6. [CrossRef]
69. Hermetet, F.; Buffiere, A.; Aznague, A.; Pais de Barros, J.P.; Bastie, J.N.; Delva, L.; Quere, R. High-fat diet disturbs lipid raft/TGF-beta signaling-mediated maintenance of hematopoietic stem cells in mouse bone marrow. *Nat. Commun.* **2019**, *10*, 523. [CrossRef]
70. Yamazaki, S.; Iwama, A.; Takayanagi, S.I.; Eto, K.; Ema, H.; Nakauchi, H. TGF-{beta} as a candidate bone marrow niche signal to induce hematopoietic stem cell hibernation. *Blood* **2009**, *113*, 1250–1256. [CrossRef]
71. Shaikh, S.R. Biophysical and biochemical mechanisms by which dietary N-3 polyunsaturated fatty acids from fish oil disrupt membrane lipid rafts. *J. Nutr. Biochem.* **2012**, *23*, 101–105. [CrossRef] [PubMed]
72. Tadokoro, Y.; Hoshii, T.; Yamazaki, S.; Eto, K.; Ema, H.; Kobayashi, M.; Ueno, M.; Ohta, K.; Arai, Y.; Hara, E.; et al. Spred1 Safeguards Hematopoietic Homeostasis against Diet-Induced Systemic Stress. *Cell Stem Cell* **2018**, *22*, 713–725.e8. [CrossRef] [PubMed]
73. Yoshimoto, S.; Loo, T.M.; Atarashi, K.; Kanda, H.; Sato, S.; Oyadomari, S.; Iwakura, Y.; Oshima, K.; Morita, H.; Hattori, M.; et al. Obesity-induced gut microbial metabolite promotes liver cancer through senescence secretome. *Nature* **2013**, *499*, 97–101. [CrossRef] [PubMed]
74. Luo, Y.; Chen, G.-L.; Hannemann, N.; Ipseiz, N.; Krönke, G.; Bäuerle, T.; Munos, L.; Wirtz, S.; Schett, G.; Bozec, A. Microbiota from Obese Mice Regulate Hematopoietic Stem Cell Differentiation by Altering the Bone Niche. *Cell Metab.* **2015**, *22*, 886–894. [CrossRef] [PubMed]
75. Pasmant, E.; Gilbert-Dussardier, B.; Petit, A.; De Laval, B.; Luscan, A.; Gruber, A.; Lapillonne, H.; Deswarte, C.; Goussard, P.; Laurendeau, I.; et al. SPRED1, a RAS MAPK pathway inhibitor that causes Legius syndrome, is a tumour suppressor downregulated in paediatric acute myeloblastic leukaemia. *Oncogene* **2015**, *34*, 631–638. [CrossRef]
76. Beyaz, S.; Mana, M.D.; Roper, J.; Kedrin, D.; Saadatpour, A.; Hong, S.J.; Bauer-Rowe, K.E.; Xifaras, M.E.; Akkad, A.; Arias, E.; et al. High-fat diet enhances stemness and tumorigenicity of intestinal progenitors. *Nature* **2016**, *531*, 53–58. [CrossRef]
77. Mana, M.D.; Hussey, A.M.; Tzouanas, C.N.; Imada, S.; Millan, Y.B.; Bahceci, D.; Saiz, D.R.; Webb, A.T.; Lewis, C.A.; Carmeliet, P.; et al. High-fat diet-activated fatty acid oxidation mediates intestinal stemness and tumorigenicity. *Cell Rep.* **2021**, *35*, 109212. [CrossRef]
78. Mihaylova, M.M.; Cheng, C.-W.; Cao, A.; Tripathi, S.; Mana, M.D.; Bauer-Rowe, K.E.; Abu-Remaileh, M.; Clavain, L.; Erdemir, A.; Lewis, C.A.; et al. Fasting Activates Fatty Acid Oxidation to Enhance Intestinal Stem Cell Function during Homeostasis and Aging. *Cell Stem Cell* **2018**, *22*, 769–778.e4. [CrossRef]
79. Cheng, C.-W.; Biton, M.; Haber, A.L.; Gunduz, N.; Eng, G.; Gaynor, L.T.; Tripathi, S.; Calibasi-Kocal, G.; Rickelt, S.; Butty, V.; et al. Ketone Body Signaling Mediates Intestinal Stem Cell Homeostasis and Adaptation to Diet. *Cell* **2019**, *178*, 1115–1131.e15. [CrossRef]
80. Shimazu, T.; Hirschey, M.D.; Newman, J.; He, W.; Shirakawa, K.; Le Moan, N.; Grueter, C.A.; Lim, H.; Saunders, L.R.; Stevens, R.D.; et al. Suppression of oxidative stress by beta-hydroxybutyrate, an endogenous histone deacetylase inhibitor. *Science* **2013**, *339*, 211–214. [CrossRef]
81. Shi, L.; Tu, B.P. Acetyl-CoA and the regulation of metabolism: Mechanisms and consequences. *Curr. Opin. Cell. Biol.* **2015**, *33*, 125–131. [CrossRef] [PubMed]
82. Agus, A.; Clément, K.; Sokol, H. Gut microbiota-derived metabolites as central regulators in metabolic disorders. *Gut* **2021**, *70*, 1174–1182. [CrossRef]
83. Fan, Y.; Pedersen, O. Gut microbiota in human metabolic health and disease. *Nat. Rev. Microbiol.* **2021**, *19*, 55–71. [CrossRef] [PubMed]
84. Nagpal, R.; Mainali, R.; Ahmadi, S.; Wang, S.; Singh, R.; Kavanagh, K.; Kitzman, D.W.; Kushugulova, A.; Marotta, F.; Yadav, H. Gut microbiome and aging: Physiological and mechanistic insights. *Nutr. Health Aging* **2018**, *4*, 267–285. [CrossRef] [PubMed]

85. Sigurdsson, V.; Takei, H.; Soboleva, S.; Radulovic, V.; Galeev, R.; Siva, K.; Leeb-Lundberg, L.M.; Iida, T.; Nittono, H.; Miharada, K. Bile Acids Protect Expanding Hematopoietic Stem Cells from Unfolded Protein Stress in Fetal Liver. *Cell Stem Cell* **2016**, *18*, 522–532. [CrossRef] [PubMed]
86. Sigurdsson, V.; Haga, Y.; Takei, H.; Mansell, E.; Okamatsu-Haga, C.; Suzuki, M.; Radulovic, V.; Van Der Garde, M.; Koide, S.; Soboleva, S.; et al. Induction of blood-circulating bile acids supports recovery from myelosuppressive chemotherapy. *Blood Adv.* **2020**, *4*, 1833–1843. [CrossRef] [PubMed]
87. Persaud, A.K.; Nair, S.; Rahman, F.; Raj, R.; Weadick, B.; Nayak, D.; McElroy, C.; Shanmugam, M.; Knoblaugh, S.; Cheng, X.; et al. Facilitative lysosomal transport of bile acids alleviates ER stress in mouse hematopoietic precursors. *Nat. Commun.* **2021**, *12*, 1248. [CrossRef] [PubMed]
88. Trompette, A.; Gollwitzer, E.S.; Yadava, K.; Sichelstiel, A.K.; Sprenger, N.; Ngom-Bru, C.; Blanchard, C.; Junt, T.; Nicod, L.P.; Harris, N.L.; et al. Gut microbiota metabolism of dietary fiber influences allergic airway disease and hematopoiesis. *Nat. Med.* **2014**, *20*, 159–166. [CrossRef]
89. Furusawa, Y.; Obata, Y.; Fukuda, S.; Endo, T.A.; Nakato, G.; Takahashi, D.; Nakanishi, Y.; Uetake, C.; Kato, K.; Kato, T.; et al. Commensal microbe-derived butyrate induces the differentiation of colonic regulatory T cells. *Nature* **2013**, *504*, 446–450. [CrossRef]
90. Kaiko, G.E.; Ryu, S.H.; Koues, O.I.; Collins, P.L.; Solnica-Krezel, L.; Pearce, E.J.; Pearce, E.L.; Oltz, E.M.; Stappenbeck, T.S. The Colonic Crypt Protects Stem Cells from Microbiota-Derived Metabolites. *Cell* **2016**, *167*, 1137. [CrossRef]
91. Naka, K.; Hoshii, T.; Muraguchi, T.; Tadokoro, Y.; Ooshio, T.; Kondo, Y.; Nakao, S.; Motoyama, N.; Hirao, A. TGF-β–FOXO signalling maintains leukaemia-initiating cells in chronic myeloid leukaemia. *Nature* **2010**, *463*, 676–680. [CrossRef] [PubMed]
92. Saito, Y.; Chapple, R.H.; Lin, A.; Kitano, A.; Nakada, D. AMPK Protects Leukemia-Initiating Cells in Myeloid Leukemias from Metabolic Stress in the Bone Marrow. *Cell Stem Cell* **2015**, *17*, 585–596. [CrossRef] [PubMed]
93. Rehman, S.K.; Haynes, J.; Collignon, E.; Brown, K.R.; Wang, Y.; Nixon, A.M.; Bruce, J.P.; Wintersinger, J.A.; Mer, A.S.; Lo, E.B.; et al. Colorectal Cancer Cells Enter a Diapause-like DTP State to Survive Chemotherapy. *Cell* **2021**, *184*, 226–242.e21. [CrossRef] [PubMed]
94. Dhimolea, E.; Simoes, R.D.M.; Kansara, D.; Al'Khafaji, A.; Bouyssou, J.; Weng, X.; Sharma, S.; Raja, J.; Awate, P.; Shirasaki, R.; et al. An Embryonic Diapause-like Adaptation with Suppressed Myc Activity Enables Tumor Treatment Persistence. *Cancer Cell* **2021**, *39*, 240–256.e11. [CrossRef]
95. Scognamiglio, R.; Cabezas-Wallscheid, N.; Thier, M.; Altamura, S.; Reyes, A.; Prendergast, Á.; Baumgärtner, D.; Carnevalli, L.; Atzberger, A.; Haas, S.; et al. Myc depletion induces a pluripotent dormant state mimicking diapause. *Cell* **2018**, *164*, 668–680. [CrossRef]
96. Bulut-Karslioglu, A.; Biechele, S.; Jin, H.; Macrae, T.A.; Hejna, M.; Gertsenstein, M.; Song, J.S.; Ramalho-Santos, M. Inhibition of mTOR induces a paused pluripotent state. *Nature* **2016**, *540*, 119–123. [CrossRef]

International Journal of
Molecular Sciences

MDPI

Article

Theophylline Extracted from Fu Brick Tea Affects the Metabolism of Preadipocytes and Body Fat in Mice as a Pancreatic Lipase Inhibitor

Tian-Tian Liu [1,2], Xiao-Tian Liu [3], Gui-Li Huang [4], Long Liu [1,2], Qing-Xi Chen [3] and Qin Wang [3,*]

1 Science Center for Future Foods, Jiangnan University, Wuxi 214122, China; liutiantian@jiangnan.edu.cn (T.-T.L.); longliu@jiangnan.edu.cn (L.L.)
2 Key Laboratory of Carbohydrate Chemistry and Biotechnology, Ministry of Education, Jiangnan University, Wuxi 214122, China
3 School of Life Sciences, Xiamen University, Xiamen 361005, China; 21620181153704@stu.xmu.edu.cn (X.-T.L.); chenqx@xmu.edu.cn (Q.-X.C.)
4 Agricultural Product Storage and Processing Laboratory, Suzhou Academy of Agricultural Sciences, Suzhou 215105, China; huanggl2015@163.com
* Correspondence: qwang@xmu.edu.cn; Tel./Fax: +86-0592-2185695

Abstract: The dramatic increase in obesity is putting people under increasing pressure. Lipase inhibitors, as a kind of effective anti-obesity drug, have attracted more and more researchers' attention in recent years because of their advantages of acting on the intestinal tract and having no side effects on the central nervous system. In this study, lipase inhibitor Fu Brick Theophylline (FBT) was screened based on enzyme molecular dynamics, and the inhibition mechanism of lipase inhibitors on obesity was analyzed and discussed at the cellular level and animal model level. We found that FBT had high inhibition effects of lipase with an IC_{50} of 1.02~0.03 μg/mL. Firstly, the laboratory used 3T3-L1 proadipocytes as models, flow cytometry was used to detect the effects of FBT on the cycle, apoptosis and intracellular ROS activity of proadipocytes. To study the contents of triglyceride, total cholesterol, related metabolites and related gene and protein expression in adipocytes. The results showed that FBT could reduce ROS production and inflammatory factor mRNA expression during cell differentiation. Secondly, by establishing the animal model of high-fat feed ob nutritional obese mice, the morphological observation and gene expression analysis of body weight, fat rate, adipocyte and hepatocyte metabolism of FBT obese mice were further discussed. It was proven that FBT can effectively reduce the degree of fatty liver, prevent liver fibrosis and fat accumulation, and improve the damage of mitochondrial membrane structure. This study provides a theoretical basis for the screening and clinical treatment of lipase inhibitors.

Keywords: obesity; lipase inhibitors; theophylline; Fu Brick tea; preadipocytes

Citation: Liu, T.-T.; Liu, X.-T.; Huang, G.-L.; Liu, L.; Chen, Q.-X.; Wang, Q. Theophylline Extracted from Fu Brick Tea Affects the Metabolism of Preadipocytes and Body Fat in Mice as a Pancreatic Lipase Inhibitor. *Int. J. Mol. Sci.* **2022**, *23*, 2525. https://doi.org/10.3390/ijms23052525

Academic Editor: Maurizio Battino

Received: 28 December 2021
Accepted: 6 February 2022
Published: 25 February 2022

1. Introduction

Adipose tissue is the central organ that maintains homeostasis, and white adipose tissue (WAT) is an important part of the body that stores fat and regulates energy metabolism, but the excessive accumulation of WAT leads to the development of obesity [1]. In WAT, lipid synthesis is directly positively correlated with decomposition efficiency [2]. Many studies have shown that the secretion and expression of various inflammatory factors increase in the white fat of type 2 diabetes patients. These inflammatory factors and inflammatory markers can cause insulin resistance by interfering with normal insulin signal transduction pathways. Hepatic insulin resistance is an important pathophysiological mechanism of glucose and lipid metabolism disorder, which is of great significance for the development of non-alcoholic fatty liver disease (NAFLD) [3].

Lipase plays a key role in human fat metabolism. It breaks down the oil in food into small molecules of glycerol and fatty acids that the body can absorb and metabolize [4]. Screening of pancreatic lipase inhibitory active ingredients has become a research hotspot [5]. In previous studies of pancreatic lipase inhibitors, most inhibitors have shown an effect of improving fatty liver damage and reducing blood lipid levels in the body [6]. Clinical trials have also shown that pancreatic lipase inhibitors can accelerate the emptying of food in the stomach and improve fat metabolism in the body [7]. Chinese herbal medicines are of interest due to their diverse structure, nontoxicity and wide range of sources, and their flavonoids, alkaloids, polyphenols, terpenes and other components show lipase inhibitory effects. At present, studies on the regulation of the metabolism of Fu Brick tea focus only on the effects of polysaccharide extraction on intestinal microorganisms.

As one of the three health drinks, tea enjoys a good reputation in the world. Fu Brick tea is a unique black tea and the most typical tea naturally enriched with selenium which has effects, such as lowering blood sugar and blood fat, anti-oxidation, bacteriostasis and increasing resistance to healthcare effects. In the study of Liu et al., the extraction of polysaccharides from Fu Brick tea improved metabolic disorders by regulating the intestinal microbial population [8]. Fu Brick tea is mainly digested by intestinal flora and has a significant healthcare effect [9]. However, there has been no report on theophylline extracted from Fu Brick tea. In this study, the inhibitory effects and mechanisms of FBT on lipase and the proliferation and differentiation of preadipocytes were studied. The results showed that FBT inhibited the generation of lipid droplets by affecting the differentiation of preadipocytes. Furthermore, FBT can reduce the release of preadipocyte ROS and inflammatory cytokines. At the same time, mice treated with FBT had significantly improved fatty liver and lipid metabolism due to pancreatic lipase inhibition and reduced lipid intake.

2. Result

2.1. The Molecular Structure of FBT

The 98.78% pure of FBT was obtained with a series of specific separation and extraction steps, which are shown in the supplemental materials (Figures S1 and S2). The structural formula of FBT is shown in Figure 1. In order to characterize the group information of FBT, nuclear magnetic resonance analysis (1H-NMR and 13C-NMR) was carried out on FBT. The absorption peak frequency on the NMR spectrum, namely chemical shift δ, is one of the important parameters of the NMR spectrum. Figure 1A shows the 1H-NMR analysis of Fu Brick tea pigment, Δ δ3.17 (3H, s, N1-CH3), Δ δ3.34 (3H, s, N3-CH3), Δ δ3.81 (3H, s, N7-CH3), Δ δ7.92 (1H, s, C8-H) and the 1H-NMR peaks Δ 2.67 (3H, s, N1-CH3), Δ 2.67 (3H, s, N3-CH3), δ3.93 (3H, S, N7-CH3), δ8.05 (1H, S, C8-H) belong to similar positions. In the analysis results, a relatively wide peak at Δ 4.2 position is water peak, which may be due to the deviation of peak shifts of other groups. Figure 1B shows 13C-NMR analysis and judgment. The positions of the absorption peaks of Δ 27.94 (n1-CH3), Δ 29.81 (n3-CH3), Δ 33.57 (n7-CH3), Δ 107.35 (C5), Δ 143.27 (C8), Δ 148.31 (C4), Δ 151.55 (C2), Δ 154.79 (C6) are basically consistent with the positions of the carbon spectrum peaks of theanine in the document. From this, it is judged that this compound is tea alkali in Fu Brick tea, and its chemical structure is shown in Figure 1B. In this paper, it is named Fu Brick Theophylline, abbreviated FBT.

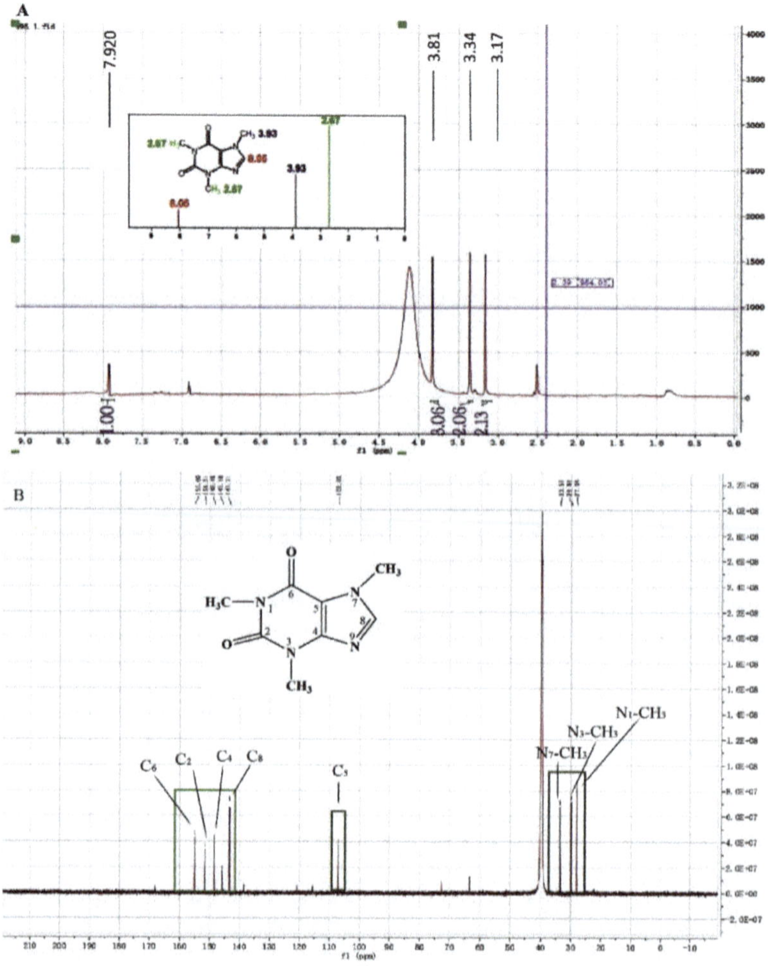

Figure 1. Nuclear magnetic resonance (NMR) analysis of Fu Brick tea element (**A**), carbon spectrum analysis; (**B**), hydrogen spectrum analysis.

2.2. *Inhibitory Effect and Mechanism of FBT on Lipase*

2.2.1. Effect of FBT on Lipase Activity

As shown in Figure 2A, the relative activity of lipase showed a significant downward trend with increasing FBT concentration, and its IC_{50} was 1.02 ± 0.03 μg/mL. The relationship between the remaining enzyme activity and enzyme concentration in the presence of different concentrations of FBT was a family of straight lines that all passed through the origin, indicating that it is the reversible property of the inhibition. According to the Lineweaver-Burk method of double reciprocal mapping, a set of straight lines intersecting in the second quadrant was obtained (Figure 2B). FBT increases the K_m value of the enzyme and decreases the V_m value, indicating that the type of inhibition on lipase is mixed. The calculated data are summarized in Table 1.

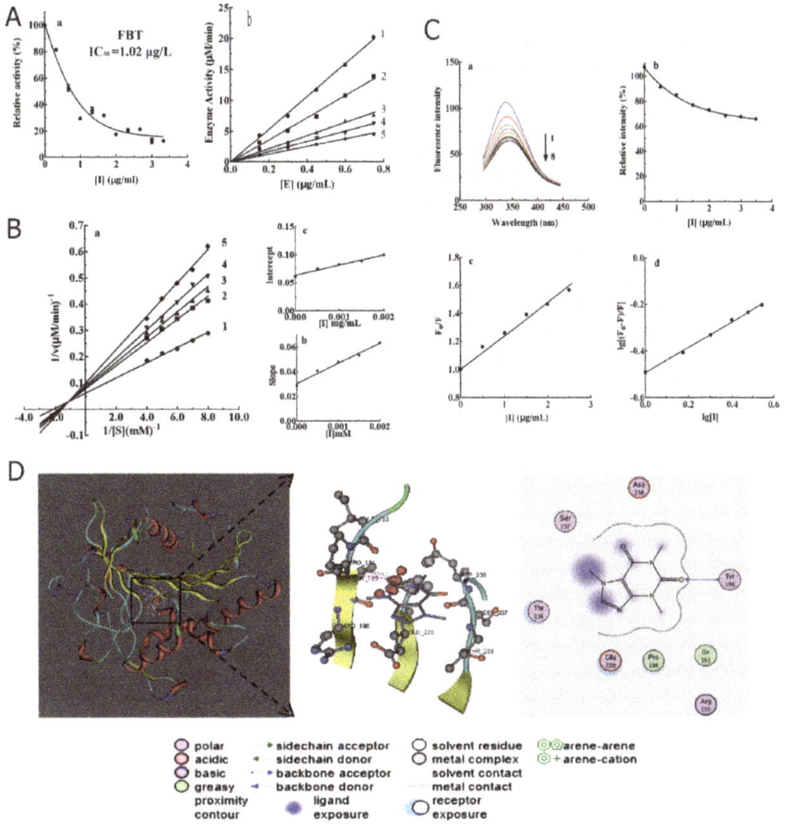

Figure 2. Inhibition of lipase by FBT and its mechanism. (**A**) Effect of FBT on lipase activity. (**A-a**) IC_{50} value of FBT; (**A-b**) Slope diagram of enzyme activity at different FBT concentrations. (**B**) Type of lipase inhibition by FBT. (**B-a**) Lineweaver–Burk double reciprocal curve; (**B-b**) double reciprocal curve slope plotted against effector concentration to determine inhibition constant K_I; (**B-c**) double reciprocal curve intercept versus effector concentration plot to determine the inhibition constant K_{IS}. (**C**) Effect of FBT on lipase emission spectrum. (**C-a**) Stratification of lipase fluorescence emission spectra; (**C-b**) Absorption peak height pair [I]; (**C-c**) Stern–Volmer curve; (**C-d**) Lg [(F_0-F)/F] stands for LG [I]. (**D**) Molecular docking mode of FBT and lipase residues.

Table 1. Inhibition constants of FBT on lipase.

Sample	IC_{50} (mg/mL)		Inhibition Effect		Inhibition Constant (mM)	
			Mechanism	Types	K_I	K_{IS}
FBT	1.02	0.98	Reversible	mixed	0.93	1.78

2.2.2. Fluorescence Quenching of FBT on Lipase

As shown in Figure 2C-a, with the continuous addition of FBT, the fluorescence emission peak of lipase gradually decreased but did not affect the peak shape and peak position of the enzyme fluorescence emission spectrum. This shows that FBT and lipase form an enzyme-FBT complex, which prevents the enzyme from binding to the substrate and achieves the effect of inhibiting the catalytic activity of the enzyme. Figure 2C-b is the peak height value corresponding to Figure 2C-a. The peak of the lipase fluorescence spectrum gradually decreases with the increase in the amount of FBT.

According to the Stern–Volmer equation: $F_0/F = 1 + K_{SV}$ [I], with F_0/F as Y and [I] as X (Figure 2C-c), we can calculate $K_{SV} = 2.39 \times 102\ M^{-1} > 100\ M^{-1}$. Therefore, the quenching process of lipase protein molecules by FBT is static. In this case, the binding constant K_A and the binding site n can be calculated according to the Scatchard equation lg $[(F_0 - F)/F] = \lg K_A + n\lg$ [I]. In Figure 2C-d, using lg $[(F_0 - F)/F]$ as the *Y*-axis and lg [I] as the *X*-axis, the values of n and K_A were obtained according to the slope and intercept of the obtained linear equation. The calculated data are listed in Table 2.

Table 2. The fluorescence parameters between FBT and lipase.

Types	Quenching Type	K_{SV} (M^{-1})	K_q ($M^{-1}s$)	K_A (M^{-1})	n
FBT	Static	2.39×10^2	2.39×10^{10}	3.12×10^3	1.745

2.2.3. Molecular Simulation Docking

The molecular mechanism of lipase inhibition by FBT was studied by the molecular simulation docking software MOE, as shown in Figure 2D. FBT directly interacts with the lipase in the carbonyl oxygen of the carbon 2 bond and has a strong docking force with the Tyr195 residue of the lipase, and the Ser237 Glu220, Asp238, Pro194, Ile193, Arg196, and Thr236 residues also interact with the lipase. FBT has a relatively large molecular size, a part of which is bound to the cavity and a part of which is bound to the outside of the cavity, and the mixed type inhibits the steric hindrance of the substrate and the enzyme active center, which is consistent with the enzyme kinetic analysis [10].

2.3. *Effects of FBT on 3T3-L1 Preadipocytes*

2.3.1. Effects of FBT on 3T3-L1 Preadipocyte Toxicology and Proliferation

The lipid-lowering effects of lipase inhibitors were verified at the cellular level. Changes in the number and volume of fat were determined by the proliferation, differentiation and apoptosis of proadipocytes. Therefore, it is necessary to study the effects of FBT on the proliferation, differentiation and apoptosis-related regulatory genes and proteins of proadipocytes. The A4 quadrant represents the number of apoptotic cells, and the A2 quadrant represents necrotic cells in the flow cytometer. In Figure 3A-(a–e), the results show that FBT caused only a small amount of 3T3-L1 preadipocytes to enter early apoptosis when the concentration was greater than 50 μg/mL, indicating that it has inhibitory effects on preadipocytes but does not cause cell apoptosis and necrosis. In addition, FBT had no significant difference in lactic dehydrogenase (LDH) release from 3T3-L1 preadipocytes ($F(1, 4) = 2.000$, $p = 0.2302$), indicating that FBT is not toxic to 3T3-L1 preadipocytes (Figure 3A-f). The results show that FBT can inhibit the proliferation of 3T3-L1 preadipocytes, as shown by the morphological observation in Figure 3B-(a–f) and MTT assay in Figure 2B-g. According to the two-way ANOVA, the treatment time ($F(5, 48) = 60.22$, $p < 0.0001$) and the concentration of FBT ($F(1, 48) = 6.119$, $p = 0.0170$) had a significant effect on the proliferation of 3T3-L1 preadipocytes, and the two factors had a significant correlation ($F(5, 48) = 2.757$, $p = 0.0287$). The average inhibition rates were 35.2% and 51.3% after treatment for 24 h and 48 h, respectively. The decrease in the number of 3T3-L1 preadipocytes is due to FBT inhibiting proliferation.

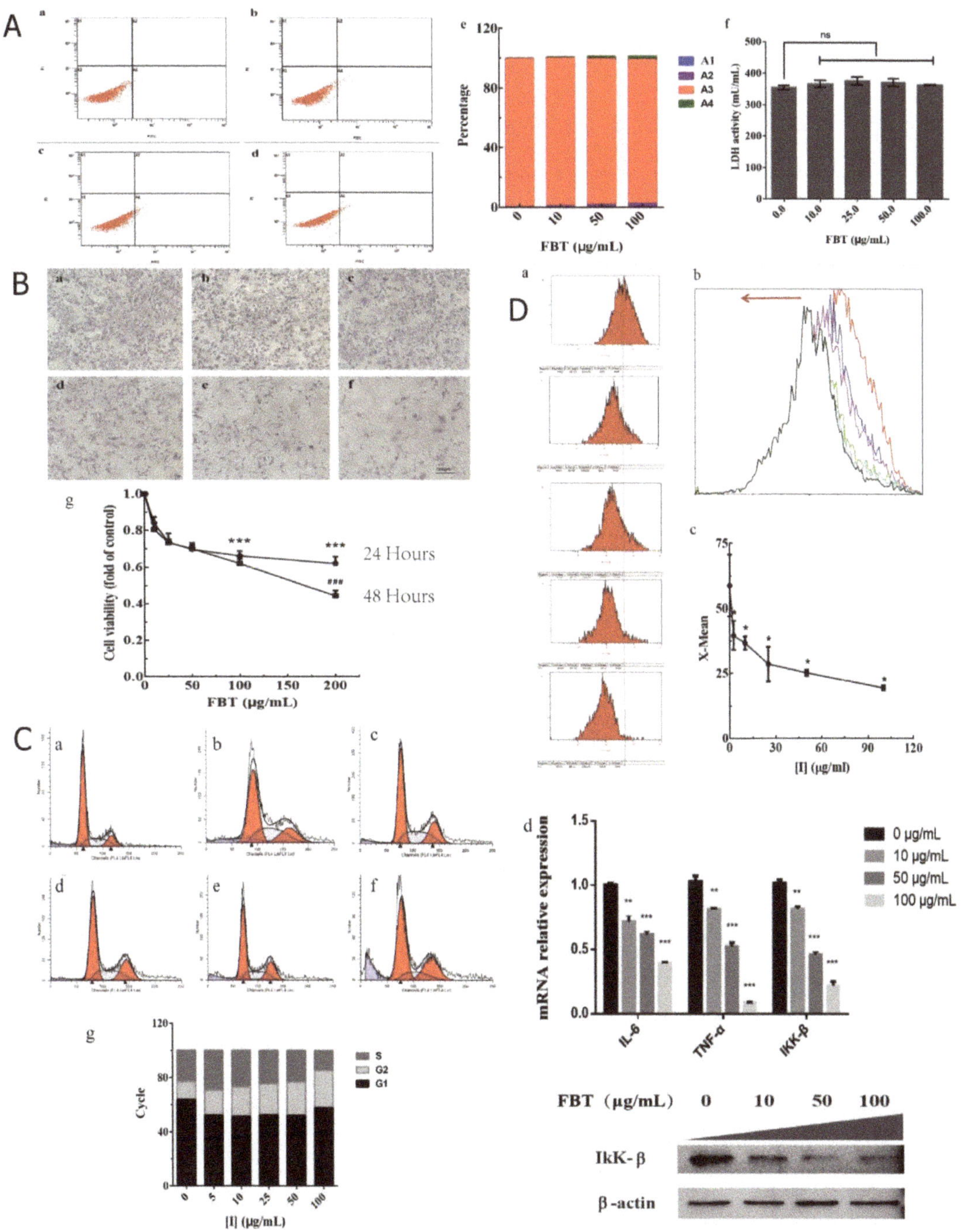

Figure 3. Effects of FBT on 3 T3-L1 preadipocytes. A: Effect of FBT on the apoptosis rate of 3T3-L1 preadipocytes. (**A-a**) Blank control; (**A-(b–d)**) FBT dosage was 10, 50, and 100 μg/mL, respectively;

(**A-e**) histogram quantization; (**A-f**) Effects of FBT on the proliferation and activity of LDH in the supernatant of 3T3-L1 cells. (**B**) Cell 3T3-L1 morphological variation after FBT treatment for 24 h. (**B-(a–f)**) FBT dosage was 0, 10, 25, 50, 100, and 200 μg/mL, respectively; (**B-g**) MTT assay was used to detect the effect of FBT on the proliferation of 3T3-L1 cells.*** $p < 0.0001$ versus the control group; and ### $p < 0.0001$ versus different times (24 h and 48 h) at the same concentration. (**C**) Cell cycle results. (**C-(a–f)**) Cells were treated with FBT at 0, 5, 10, 25, 50, and 100 μg/mL FBT for 48 h. Treated and control cells were stained with PI, and changes in the cell cycle were examined by flow cytometry. (**C-g**) Histogram showing the percentage of cell cycle distribution in each phase of the cell cycle (G0/G1, S, and G2/M). (**D**) FBT decreases intracellular ROS levels in 3T3-L1 cells; (**D-a**) flow peak diagram; (**D-b**) Cumulative displacement of ROS peaks in flow cytometry under different concentrations of FBT, where the FBT concentration gradient is 0, 5, 10, 25, 50, and 100 μg/mL; (**D-c**) X-mean quantization diagram. (**D-d**) Histogram of the influences of FBT on ROS-related genes in 3T3-L1 cells, $p < 0.01$, and *** $p < 0.0001$ versus the control group.

2.3.2. Effects of FBT on the Cell Cycle of 3T3-L1 Preadipocytes

We used different concentrations of FBT (0, 5, 10, 25, 50, and 100 μg/mL) to simultaneously treat logarithmic phase preadipocytes for 48 h, and the results are shown in Figure 3C. With increasing FBT concentration, the proportion of 3T3-L1 preadipocytes entering the S phase decreased, and the proportion of cells in the G2/M phase increased. The G1 phase stabilized after decreasing. Further cell cycle analysis revealed that the cells were blocked in the G2 phase, which was the reason why FBT affected preadipocyte proliferation.

2.3.3. Effects of FBT on ROS Levels in 3T3-L1 Preadipocytes

The DCFH-DA was used as a fluorescent probe of the active oxygen detection kit to combine the active oxygen in the cell with the fluorescent DCFH to generate fluorescent DCF. The level of ROS was measured on a flow cytometer based on the amount of DCF fluorescence. Figure 3D-a shows the ROS peak of adipose cells 24 h after FBT treatment. With the increase of FBT concentration, the mean value of ROS X gradually decreases compared with the control group (3D-c), and the peak value gradually shifts to the left (3D-b). These results suggest that FBT can reduce the ROS level of 3T3-L1 preadipocytes ($F(5, 12) = 15.48$, $p < 0.0001$). Furthermore, the mRNA expression levels of ROS-related genes, including the key kinase interleukin 6 (IL-6) ($F(3, 4) = 126.5$, $p = 0.0002$), tumor necrosis factor (TNF-α) ($F(3, 4) = 457.5$, $p < 0.0001$), and inhibitor kappa B kinase β (IκKβ) ($F(3, 4) = 360.6$, $p < 0.0001$), were downregulated with increasing concentrations of FBT, as shown in Figure 3D-d. TNF-α and IL-6 have a variety of effects on the physiological functions of adipocytes, and IKK-β is a key hub for inflammatory signals to interfere with insulin signals [11], which can mediate chronic inflammation induced by obesity. Their decrease indicates that FBT has a certain restriction on the inflammatory response.

2.3.4. Effects of FBT on 3T3-L1 Preadipocyte Lipid Production

Lipid droplets store various substances required for energy metabolism and are embedded with various proteins on their surfaces. Triglycerides (TGs) are stored in the cytoplasm of fat cells in the form of lipid droplets, while total cholesterol (TC) exists in the plasma membrane of cells and is the main component of lipid droplets in fat cells. In Figure 4A-a,b, by testing TG and TC in the 100 μg/mL FBT and control groups on different days, it was found that the accumulation of TG and TC in cells treated with FBT began to decrease significantly from the second day of cell differentiation. On the eighth day, TG accumulation decreased by 37.5–37.7% and TC accumulation decreased by 53.7–54.7% compared with the control group. After that, we tested the contents of TG and TC at different concentrations of FBT on the sixth day (Figure 4A-(c,d)). The results showed that FBT reduced the contents of TG ($F(5, 6) = 153.6$, $p < 0.0001$) and TC ($F(5, 6) = 107.2$, $p < 0.0001$) in a gradient manner.

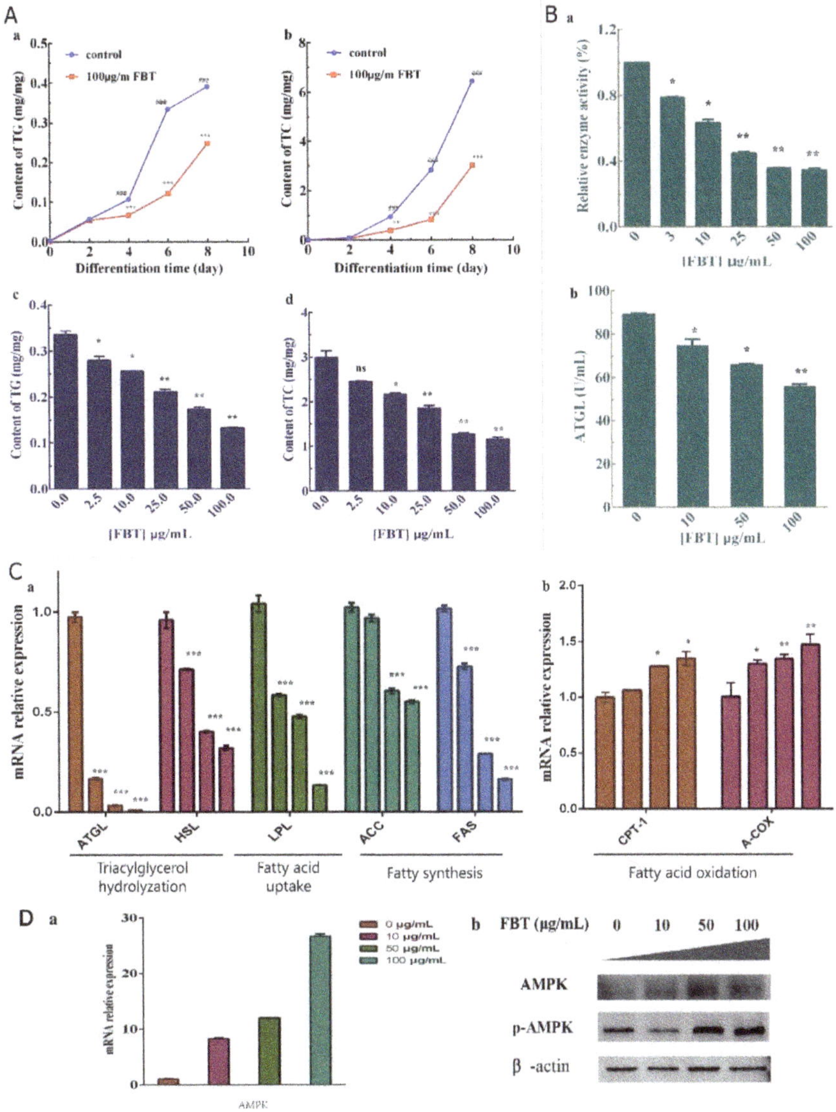

Figure 4. (**A**) The effect of FBT on the changes in TG and TC content during the differentiation of 3T3-L1 preadipocytes. (**A-(a,b)**) Changes in TG and TC contents within eight days of the cell differentiation process under the 100 μg/mL FBT treatment; (**A-(c,d)**) The contents of TG and TC under different concentrations of FBT on the sixth day of differentiation. * $p < 0.05$, ** $p < 0.01$, and *** $p < 0.0001$ versus the control group; ### $p < 0.0001$ versus different times at the same concentration. (**B**) Effects of FBT on the related metabolism of 3T3-L1 cells. (**B-a**) 4-NPP was used as a substrate to detect lipase activity by UV spectroscopy. (**B-b**) ATGL activity detection results. (**C**) Lipid metabolism of 3T3-L1-related gene expression with FBT; (**C-a**) Gene expression related to triacylglycerol hydrolyzation, fatty acid uptake and fatty synthesis under FBT (0, 10, 50 and 100 μg/mL) treatment; (**C-b**) Effects of FBT on the mRNA levels of genes related to fatty acid oxidation. (**D-a**), AMPK gene expression with RNA; (**D-b**), AMPK and phosphorylated AMPK protein expression with Western Blot.

The substrate method was used to verify the lipase in cells directly, demonstrating that FBT could inhibit the decomposition of lipase, and FBT also inhibited the hydrolysis of triglycerides in tissues. The results showed that lipase activity decreased in a concentration-dependent manner with drugs (Figure 4B-a), ($F(5, 18) = 703.6$, $p < 0.0001$). An ATGL kit was used to detect the decreased triglyceride lipase content in fat cells (Figure 4B-b), ($F(3, 12) = 83.50$, $p < 0.0001$), and lipase activity was collected and detected after cell lysis. The substrate method was used to verify the lipase in cells directly, which proved that FBT could inhibit the decomposition of lipase, and FBT also inhibited the hydrolysis of triglyceride in tissues.

2.3.5. Effects of FBT on the Expression of mRNAs Related to Differentiation and Lipid Production in Preadipocyte 3T3-L1 Cells

After treatment with FBT, some mRNAs encoding fat absorption-related and synthesis-related enzyme genes of preadipocytes were downregulated, as shown in Figure 4C-a, the TG-degradation rate-limiting enzyme lipoglycerol lipase (ATGL) ($F (3, 4) = 1206$, $p < 0.0001$) and hormone-sensitive lipase (HSL) ($F (3, 4) = 196.9$, $p < 0.0001$), lipoprotein esterase (LPL) ($F (3, 4) = 328.8$, $p < 0.0001$) were related to fatty acid absorption, and acetyl-CoA carboxylase (ACC) ($F (3, 4) = 221.8$, $p < 0.0001$) and fatty acid synthase (FAS) ($F (3, 4) = 1375$, $p < 0.0001$) were related to fat synthesis. These results indicated that FBT had a certain inhibitory effect on the synthesis of fat. In Figure 4C-b, the mRNA expression levels of fatty acid oxidation rate-limiting enzymes, including carnitine palmitoyl transferase (CPT-1) and acetyl-CoA oxidase (A-COX) genes, were significantly upregulated. These results indicate that FBT may play a certain role in promoting fatty acid oxidation.

Adenosine-activated protein kinase (AMPK) is a crucial energy regulator of cellular anabolism and catabolism. When AMP/ATP levels rise due to cell stress, AMPK will be phosphorylated and then activate multiple downstream target molecules to reduce ATP consumption and inhibit lipid and cholesterol synthesis, etc., [12]. Conversely, the decrease of AMP/ATP will increase the synthesis of ATP and promote the oxidation of fatty acids [13]. Figure 4D-a showed that FBT increased AMPK mRNA expression in adipose tissue. Meanwhile, AMPK and phosphorylated AMPK protein expression were detected (Figure 4D-b), and the difference in AMPK protein expression was not significant, while p-AMPK expression was significantly increased. Our results suggest that FBT is involved in adipose tissue regulation, most likely by activating the AMPK pathway to reduce lipid decomposition and lipid production. At the same time, the expression of P-AMPK was up-regulated, suggesting that the drug might protect mitochondria from ROS oxidative damage by activating AMPK

2.4. FBT Reducing Fat Based on an OB Mouse Obesity Model

2.4.1. Effect of FBT on the Accumulation of Lipid Droplets in Non-Alcoholic Fatty Liver

To investigate the therapeutic effects of FBT on diet-induced obesity, we orally administered vehicle (saline) FBT or orlistat to OB/OB mice for 40 days. Although food intake was similar to that in the control group, FBT supplementation significantly alleviated the perirenal fat and hepatomegaly detected by PET-CT in mice (Figure 5A). Figure 5A-a showed that the fat accumulation and fatty liver volume in the control group increased significantly. Compared to the perirenal fat of mice treated with fuzhuancha extract (Figure 5A-b) to that at the beginning of gavage, we found that the perirenal fat decreased, and there was little difference in the size of the liver at the end of gavage and at the beginning of gavage. At the same time, compared with the control group, the volume of perirenal fat and liver fat decreased significantly in the FBT treatment group after gavage, indicating that the fat accumulation in the abdominal cavity and the hypertrophy of fatty liver can be significantly reduced after Fu Brick tea extract treatment. The results in Figure 5B show that the weight of mice in the control group exhibited an upward trend. The average weight of mice in the control group was 57.2 g after 48 days of modeling and 40 days of continued growth. A significant inhibitory effect was observed in all groups administered by gavage. The average weight of mice in the orlistat group was controlled at 64.3 g, and the weight of mice in the FBT group was reduced to 51.8 g, indicating that FBT had a certain inhibitory effect on the weight of obese mice and even had a better effect on fat reduction than the positive control. Strikingly, liver sections stained by Oil Red O staining (Figure 5C) and hematoxylin & eosin (Figure 5D) showed fewer and smaller lipid vacuoles in the liver of the FBT-treated mice than in the untreated mice. The reduced lipid accumulation in the livers of the treated mice was supported by Oil Red O staining and biochemical analysis, which showed an approximate 30% reduction in liver triglyceride (TC) content and a 50% reduction in liver cholesterol (TG) ($p < 0.05$; Figure 5E).

2.4.2. Effect of FBT Remission on Liver Fibrosis and Degree of Injury in Mice

The livers of OB/OB mice in the control group exhibited hepatocytes with lipid droplets spread throughout the cytoplasm and cells with characteristics of myofibroblasts stained blue near the hepatocytes and lipid drops, thereby indicating the development of fibrosis in Masson's trichrome staining ($p < 0.01$) (Figure 6A). Next, to determine the effects of FBT on glucose homeostasis and insulin sensitivity, GTT and ITT were performed after 40 days of FBT treatment. As shown in Figure 6B, FBT-treated OB/OB mice exhibited lower glucose levels at all time points up to 90 min after intraperitoneal glucose (Figure 6B-a) or insulin injection (Figure 6B-b) compared with control OB/OB mice. The fluctuation of blood glucose levels is small, and the mice in the Fu Brick tea extract group have a higher tolerance to glucose.

However, we found that blood glucose did not decrease but increased rapidly after insulin injection in the control group and orlistat group, which is consistent with the characteristics of type 2 diabetes caused by obesity. The ultrastructural analysis of the liver cells of the OB/OB mice revealed many of the different organelles, such as mitochondria, rough-surfaced endoplasmic reticulum (rER), lipid droplets, lysosomes and nuclei, between the control and treatment groups and the presence of lipid droplets spread throughout the cytoplasm in the control group (Figure 6C-i). In addition, we found that the mitochondrial membrane structure of the control and orlistat groups was damaged, the coloring was deep, and the ridge structure was unevenly distributed (Figure 6C-ii). We found that the apoptotic cells in FBT-treated mouse liver tissues were reduced, and they could be specifically labeled with fluorescence area quantification by TUNEL staining, as shown in Figure 6D. After FBT treatment, the fluctuation in blood glucose levels (4.5 mM/L–6.5 mM/L) in the mice remained relatively small compared with that of the control group both in GTT and ITT. Furthermore, FBT also increased the total antioxidant capacity in tissues and serum detected by the Solarbio kit, as shown in Figure 6E.

2.4.3. Effects of FBT on Lipid Metabolism in Non-Alcoholic Fatty Liver Tissue

Finally, a lipid metabolomics analysis was performed to further verify the lipid inhibition effect of FBT. The metabolomic data is shown in Figure 7. Compared with the FBT-treated group, the contents of 30 triglyceride TAGs in the liver of the control group were significantly higher, and the contents of 10 TAGs (all unsaturated fatty acids) were significantly lower (Figure 7A). At the same time, as shown in Figure 7A,B, the accumulation of a large amount of diglyceride (DAG), fatty acid (FA) and acetylated fatty acid (FAHFA) in the liver tissue of mice without FBT treatment may be due to the inhibition of lipase in the intestine by FBT such that a large number of TAGs cannot be broken down into DAG, FA, and FAHFA and directly excreted. Furthermore, the contents of phosphatidylcholine (PC) and phosphatidylethanolamine (PE) were also clearly different. FBT-treated mice had 20 types of PC and 17 types of PE accumulation, while mice without FBT treatment had only two types of PC and eight types of PE accumulation. PC and PE are often used in the treatment of non-alcoholic fatty liver, which can promote the formation of HDL and promote lipid transport [13].

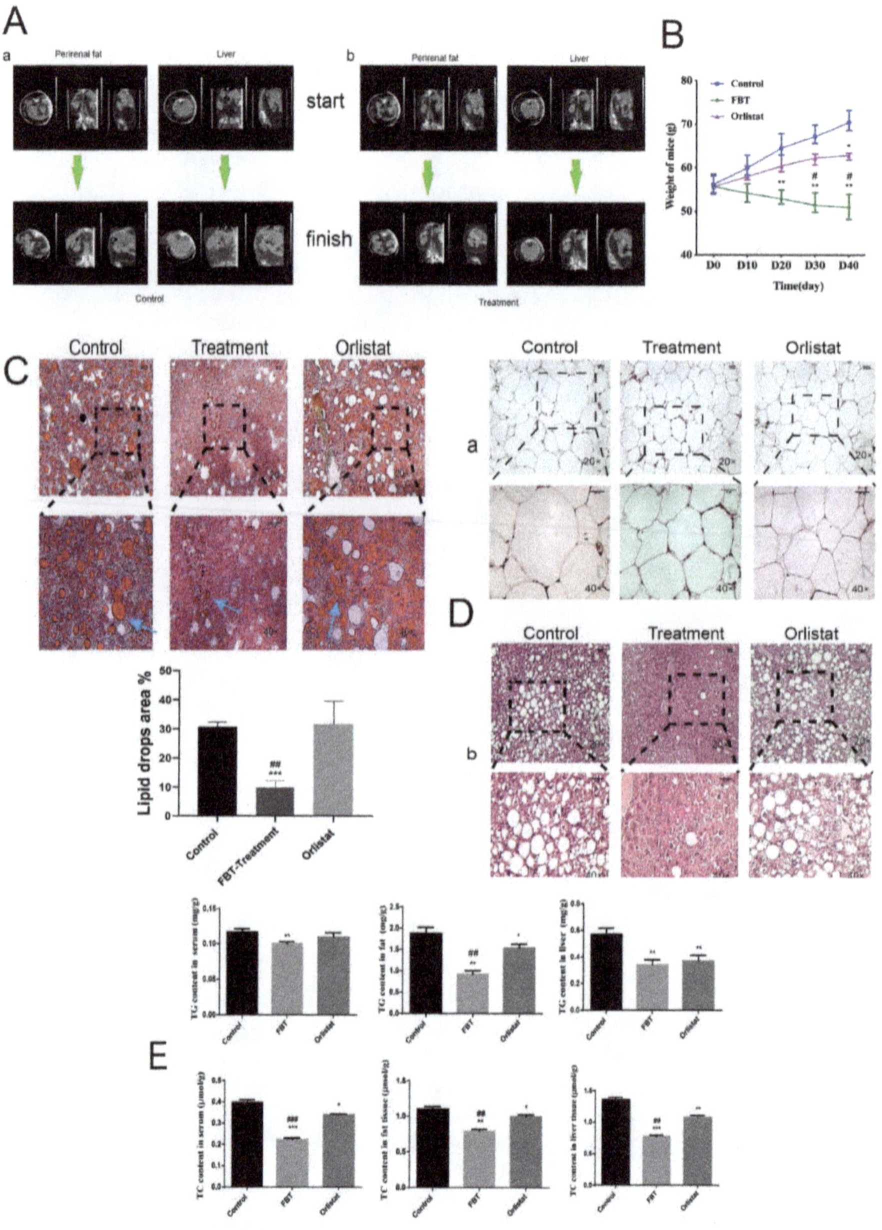

Figure 5. (**A**) Comparison of PET-CT scan results of mice before and after gavage. (**A-a**) PET-CT scan of perirenal fat and liver in the control group; (**A-b**) PET-CT scan of perirenal fat and liver in the FBT treatment group. (**B**) Weight changes in different groups over 40 days. (**C**) Frozen sections of liver were stained with Oil Red O, and the lipid droplet area was quantified using ImageJ. (**D**) HE staining of paraffin sections of mouse white adipose tissue and liver tissue ($n = 6$). (**E**) The contents of TG and TC in mouse liver, white fat and serum. * $p < 0.05$, ** $p < 0.01$, and *** $p < 0.0001$ versus the control group; # $p < 0.05$, ## $p < 0.01$, and ### $p < 0.0001$ versus the orlistat group.

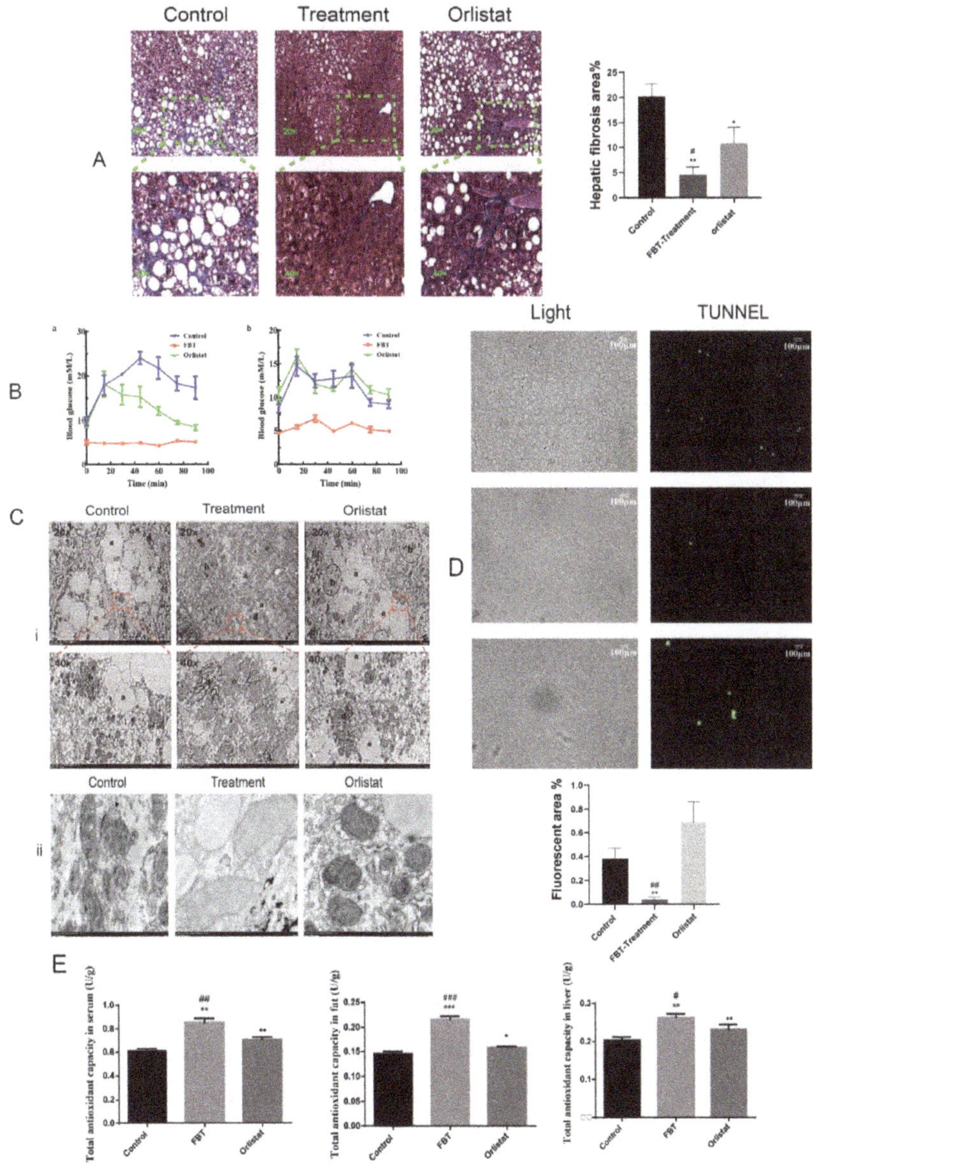

Figure 6. (**A**) Stained by Masson's trichrome represents the fibrotic area, and the fibrotic area was quantified using ImageJ. The blue area is the liver tissue; The red area is the cytoplasm; The purple area is the nucleus. (**B**) Changes in blood glucose concentration within 90 min in insulin resistance and glucose tolerance tests in different groups of mice. (**C**) i; Effects of FBT on the ultrastructure of liver tissue in obese mice under a transmission electron microscope. a; lipid particles, b; nucleus, c; mitochondria. d; endoplasmic reticulum. ii; Ultrastructure of mitochondria in different treatment groups. (**D**) Mouse liver apoptosis and necrotic cells were labeled by TUNEL fluorescence staining, and the fluorescence area was quantified by ImageJ. (**E**) Detection of total antioxidant activity in the liver, fat and serum of mice. * $p < 0.05$, ** $p < 0.01$, and *** $p < 0.0001$ versus the control group; # $p < 0.05$, ## $p < 0.01$, and ### $p < 0.0001$ versus the orlistat group.

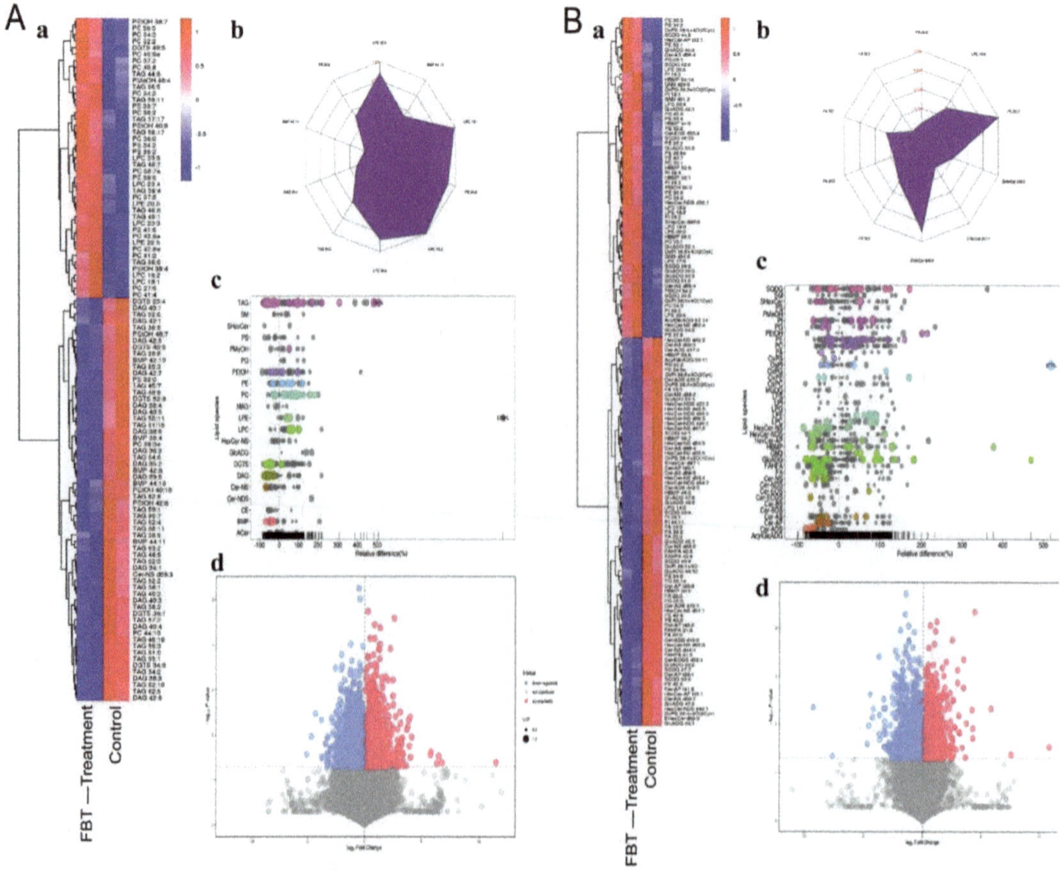

Figure 7. The ionization source of the LC-QTOFMS platform is electrospray ionization, and there are two ionization modes: (**A**) positive ion mode (POS), (**B**) negative ion mode (NEG). The combination of the two modes in the detection of the metabolome can increase the metabolite coverage rate. a: Heatmap of hierarchical clustering analysis for the group (FBT treatment vs. the control). b: Radar chart analysis for group. (FBT treatment vs. te control). c: Lipid species of bubble plot for the group (FBT treatment vs. the control). Each point in the lipid bubble diagram represents a metabolite. The size of the point represents the *p*-value of the Student's *t*-test (taking the negative number of the logarithm base 10), and a larger point represents a smaller *p*-value. Grey dots represent nonsignificant differences with a *p*-value not less than 0.05, while colored dots represent significant differences with a *p*-value less than 0.05 (different colors are marked according to lipid classification). d: Volcano plot for the group (FBT treatment vs. the control).

3. Discussion

In this study, a relatively safe and effective lipase inhibitor, FBT was screened from Fu Brick tea, and the total yield was 1.34%. The compound with a molecular weight of 195 was collected by an automatic purification system, and the possible molecular formula was determined by Q-Exactive high-resolution liquid chromatography-mass spectrometry analysis. 1H-NMR and 13C-NMR scanning spectrograms determine that the compound is theophylline. Its IC_{50} value on lipase is 1.02 μg/mL, and the inhibitory type is reversible mixed inhibition. The endogenous UV fluorescence results show that it has a static quenching effect on the emission spectrum of lipase, and the binding site n is 1.754, which is close to two sites. The molecular simulation results show that FBT may interact with several amino acids of lipase and directly act on the Tyr195 residue. According to the preliminary judgment of lipase kinetics, this active substance has potential in the study of anti-obesity drugs.

Our results demonstrated that the direct effects of FBT on the differentiation of 3T3-L1 cells were significant. Differentiation of preadipocytes into mature adipocytes is considered an important goal for the development of anti-obesity drugs [14]. The results of flow cytometry were analyzed for the changes of intracellular ROS content after FBT treatment. The reduction of ROS can protect the mitochondria of liver cells from damage, which is consistent with the results of in vivo experiments (Figure 6C). Obesity has also been related to antioxidant defense enzymes. Some findings have suggested that the activity of SOD increases at the onset of obesity development in an attempt to combat the increased generation of free radicals [15]. Therefore, understanding the molecular mechanisms that tightly control adipocyte development and adipogenesis will provide valuable information for controlling obesity [16]. We found that FBT can significantly inhibit the proliferation of preadipocytes, but LDH treatment experiments confirmed that FBT is not toxic to cells. The flow cytometry detection of PI single staining and analysis showed that FBT blocked 3T3-L1 preadipocytes in the G2 phase and affected proliferation. However, as FBT had no apoptotic effect on 3T3-L1 preadipocytes in V-FITC/PI double staining, FBT inhibited the proliferation of cells by affecting cell differentiation rather than inducing cell apoptosis and necrosis. Macrophages that infiltrate fat tissues release cytokines, such as TNF-a, IL-6 and IκKβ. These factors are responsible for obesity-related disorders, including hypertension, diabetes, atherosclerosis, insulin resistance, and non-alcoholic fatty liver disease [17]. The upregulation of cytokines, such as IL-6 and TNF-a in adipose tissue serves as a marker of obese adipocytes [18]. Rodents show an increase in oxidative stress and proinflammatory cytokines (TNF-a, IL-6 and IκKβ) in white adipose tissue after a high-fat diet (HFD) [19]. According to our study, anti-inflammatory effects of FBT have been detected at the cellular level and have not been detected in animals. In the flow cytometer test using the DCFH-DA method, the production of ROS during differentiation in FBT-treated cells was reduced in vitro, while the total antioxidant capacity of white fat, liver tissues and serum was improved in FBT-treated mice induced by a high-fat diet in vivo. Acetyl-CoA carboxylase (ACC) produces malonyl-CoA, which inhibits CPT1 and has been shown to regulate the accumulation of body fat [20]. While inhibiting the synthesis of fat, it promotes the oxidation of fatty acids and the absorption of glucose [21]. We observed a decrease in ACC expression in FBT-treated fat progenitor cells, suggesting that it lessened CPT-1 inhibition and promoted its expression. At the same time, inhibiting lipolysis is more conducive to maintaining a healthy energy metabolism balance. Some data show that inhibition of ATGL and HSL expression and activity can effectively inhibit the lipid reaction and improve metabolic disorders [22,23]. The specific knockout of the ATGL gene in mouse white fat cells not only inhibited lipolysis and decreased blood lipids but also inhibited the expression of genes related to fat absorption and synthesis, enhanced the hepatic insulin signaling pathway, and improved glucose tolerance [24]. LPL is closely related to the degradation of TG and the absorption of FA in cells. Adipose tissue-specific LPL deficiency reduced fat storage in OB/OB mice [25]. Other studies have shown that downregulating the expression of LPL and FAS in white fat can effectively inhibit the hypertrophy of fat cells [26]. The results of this study showed that by downregulating the expression of ACC, FAS, LPL, HSL and ATGL, FBT can inhibit the TG synthesis pathway in fat cells, reduce the synthesis of TG and other lipids, and reduce the release of glycerol and FFAs, thus maintaining the stable state of lipid metabolism.

The build-up of lipids in hepatocytes suggests a possible interference with mitochondrial and microsomal function, leading to an interruption in the transport of lipoproteins and a build-up of fatty acids [27]. Fluorescence staining with Tunnel and electron microscope observation of mouse liver showed that the liver status of mice treated with FBT showed decreased fluorescence spots and the mitochondrial status of mice tended to be normal. This confirms that FBT can protect liver cells from apoptosis and inflammation. Non-alcoholic fatty liver disease often leads to apoptosis and inflammation of liver cells and fibrosis of tissues [28,29]. Initially, characterized by hepatic steatosis and defined as liver fat levels in excess of 5% of the liver's weight, NAFLD can progress to fatty hepatitis, fibrosis, and cirrhosis [30]. The reduction in liver triglycerides in the FBT-treated mice may be explained by metabolomic analysis, including enhanced hepatic triglyceride secretion, decreased hepatic lipogenesis, enhanced intracellular lipolysis, and increased hepatic fatty acid oxidation. Excessive fat storage in the liver can be caused by a range of metabolic disorders, including defective fatty acid oxidation, enhanced fat production, impaired triglyceride secretion, and increased intake of fatty acids from circulation [31]. We found that FBT treatment lowered hepatic fat accumulation and improved glucose homeostasis in HFD-fed mice. Through Oil Red O staining, we found that the size and number of lipid particles in the liver of FBT-treated mice decreased. Meanwhile, according to observations under transmission electron microscopy, the mitochondria of the control group were seriously damaged, resulting in damage to fatty acid oxidation and accumulation of TAG, DAG FA and FAHFA. A large amount of evidence shows that insulin resistance leads to fatty liver, and fatty liver may also lead to obesity and insulin resistance to the liver, which further leads to a vicious cycle [32]. In obesity and insulin resistance, increased intrahepatic fatty acid inflow from lipolysis of diet or adipose tissue is generally believed to be a major driver of the development of NAFLD [33,34]. We found that FBT could maintain the stability of blood glucose and insulin in mice, which is closely related to the reduction in FA and FAHFA in the liver. In addition, FBT can inhibit lipase in the intestine, and its mechanism is to reduce the accumulation of triglycerides in fat and liver tissues by affecting the absorption of triglycerides in the intestine. In a study by Melha Benlebna et al., excessive accumulation of acetylated fatty acid FAHFA in non-fat tissues was found to often lead to liver damage and fibrosis [35]. Giovanni Solinas et al. found that the accumulation of acetylated fatty acids FAHFA in non-adipose tissues would lead to excessive accumulation of lipids and metabolic disorders [36]. However, we found in the metabolomics analysis that regardless of whether the accumulation of PC and PE in liver tissues increased after FBT treatment, the accumulation of PC and PE was related to the transport of HDL and VLDL, which could increase the transport of cholesterol and triglycerides [37]. Further reducing the accumulation of lipids in the liver, the specific mechanism of increasing the accumulation of PC and PE in liver tissue after taking FBT remains to be studied. According to Kocelak et al. and Tonstad et al., lipase inhibitors can reduce the level of LDL in the blood [38].

In general, the effect of FBT on obese mice verified the results of adipocyte level research analysis. Figure 8 summarized the possible mechanism of FBT for lipid reduction and weight loss. Combined with the results of cell experiments, FBT can reduce the decomposition and synthesis of fat and the accumulation of triglyceride and cholesterol through AMPK activation. At the same time promote the oxidative decomposition of fatty acids so as to achieve the purpose of reducing fat and weight loss. The liver tissue staining and lipid metabonomics analysis showed that FBT could repair nonalcoholic fatty liver injury in obese mice. This provides a potential possibility for natural product inhibitors as anti-obesity drugs and provides new ideas for further research.

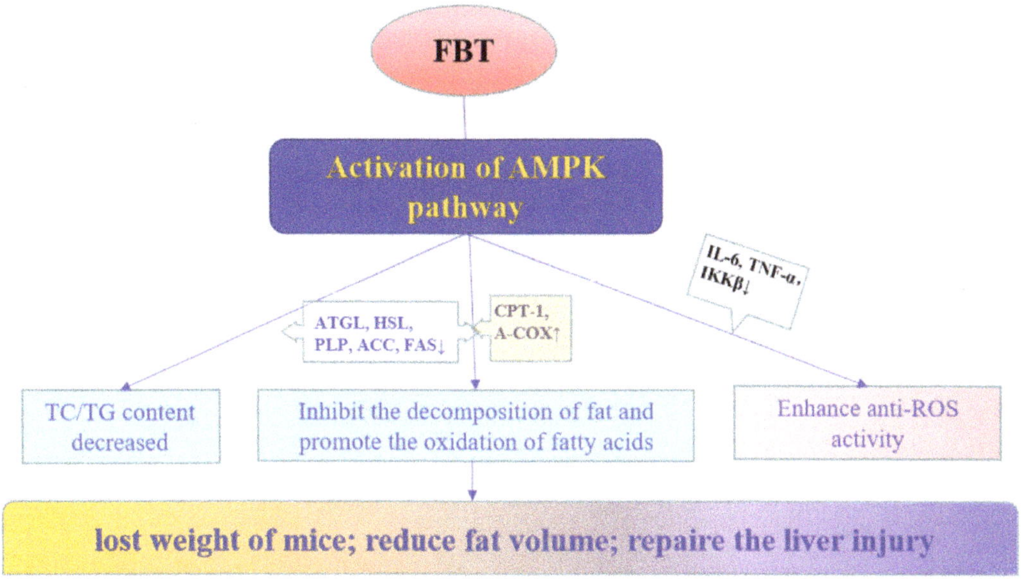

Figure 8. The lipid-lowering mechanism of FBT on fat metabolism in mice.

4. Materials and Methods

4.1. Materials and Sample Preparation

Lipase (EC3.1.1.3) from *Mucor miehei* (lyophilized powder) was the product of Sigma-Aldrich. The specific activity of the enzyme was over 4000 U/mg (using olive oil). Please refer to the explanation in the literature for the reason for choosing this lipase [39]. DMEM, fetal bovine serum (FBS), and antibiotics were purchased from Genetimes Technology Co., Ltd. (Shanghai, China). Lipase, 4-nitrophenyl palmitate (4-NPP), DMSO, 3-isobutyl-1-methylxanthine (IBMX), dexamethasone (DEX), insulin, and Oil Red O powder were obtained from Sigma–Aldrich (Shanghai, China).

4.2. Extraction of Active Ingredients from Fu Brick Tea

In this study, Fu Brick tea, a traditional Chinese fermented black tea, was used as the material (Purchased from China Shannxi Jingyang YiChangMing Fu Brick tea Co., LTD, Shannxi, China). The tea powder was obtained in a pulverizer, weighed to 500 g with 3 L of 90% ethanol three times for extraction, ultrasonically treated for 30 min each time, concentrated by rotary evaporation under reduced pressure at 45 °C, freeze-dried and stored at −20 °C for later use. After removing impurities in the crude extract with petroleum ether and isovolume extraction with ethyl acetate and n-butanol, most of the active substances with inhibitory effects of lipase existed in the ethyl acetate phase. After further passing through an LH20 gel column and eluting and separating with 25%, 50%, 75% and 100% ethanol, respectively. Pure concentrations of Fu Brick tea extract were obtained, and the single product FBT was analyzed through a high-performance liquid phase. The molecular weight and structure of the single substance were determined through Q-Exactive high-resolution LC/MS and nuclear magnetic resonance, which are methods described in the previous literature [40].

4.3. Analysis of Inhibition Kinetics of FBT on Lipase and Molecular Simulation Docking

The inhibition rate, type and mechanism were measured using the method we previously described [41]. The type of lipase inhibition was further explored by FBT. Under the conditions of the most suitable live system, the lipase concentration was fixed (0.2 μg/mL), the concentration of substrate 4-NPP was changed, and the initial reaction rate at different FBT concentrations was determined. Endogenous fluorescence spectroscopy was used to analyze the interaction between FBT and lipase. First, a fluorescence scan was performed on 2 mL of 0.1 mg/mL lipase, followed by the addition of an interval of 2 μL of FBT inhibitor for each data point. The docking simulation of the enzyme molecule and the effector was conducted using the bioinformatics analysis software MOE (Molecular Operation Environment). The energy of the enzyme molecule and the effector was minimized, and the docking site with the highest score was found. The docking parameter setting was according to Chen [42].

4.4. Cell Culture

3T3-L1 preadipocytes were purchased from Shanghai Fuheng Biotechnology Co., Ltd. (Shanghai, China). Cells were maintained in DMEM supplemented with 10% FBS and 1% antibiotics in an atmosphere of 5% CO_2 at 37 °C. Adipocyte differentiation was induced by treating cells for 48 h in media containing 10% FBS, 0.5 mM IBMX, 1 μM DEX, and 10 μg/mL insulin. The medium was replaced with medium supplemented with only 5 μg/mL of insulin every other day. The cells were treated with or without FBT for 8 days during adipogenesis.

4.4.1. Effects of FBT on the Proliferation of 3T3-L1 Preadipocytes

3T3-L1 preadipocytes in the logarithmic growth phase were inoculated into 96-well plates at 2×10^3/well and cultured for 48 h after adherence to media containing different concentrations of FBT. Four parallels were used to determine the absorbance A at 570 nm of each treatment group by MTT assay. Value-added rate% = Sample Group A_{570}/Control Group $A_{570} \times 100\%$. A lactate dehydrogenase kit was used to detect cytotoxicity.

4.4.2. Determination of Intracellular TC/TG

During the differentiation process on Days 2, 4, 6, and 8, the effects of different concentrations of FBT on preadipocyte differentiation and lipid production were detected. After digestion with trypsin, a certain amount of isopropanol was added for ultrasonication for 1 h to help dissolve intracellular lipids and then centrifuged at 3000 rpm for 10 min. The total triglyceride (TC) and total cholesterol (TG) of the supernatant were measured in accordance with the assay kits (Solarbio).

4.4.3. Determination of Glucose Consumption and Adiponectin Production in 3T3-L1 Preadipocytes

Preadipocytes were divided into a control group, model group and dose group. The control group was differentiated into a normal medium; 1 μM DEX and 5 μg/mL insulin were added to the model group; and the dose group was divided into four concentrations of FBT: 10, 20, 100 and 200 μg/mL. The medium was changed every 2 days. After differentiation, the cells were collected, and the supernatant was collected after digestion and lysis. The glucose oxidase assay kit (Pulilai Gene Technology Co., Ltd.) was used to measure the glucose consumption of each concentration group and determine whether the modeling was successful. A mouse adiponectin enzyme-linked immunosorbent assay (ELISA) detection kit (QinCheng Biological) was used to determine the adiponectin content in the cell supernatant.

4.4.4. Cell Cycle Analysis and Determination of ROS Production by Flow Cytometry

The cell cycle was detected with PI single-staining Annexin, apoptosis and necrosis were detected with V-FITC/PI staining, and intracellular reactive oxygen species (ROS) production was detected with the DCFH-DA method; all were detected using a flow cytometer and performed as previously described [43]. All of the reagents above were purchased from Beyotime Biotechnology (Shanghai, China).

4.5. RNA Analysis

Total RNA from tissues and cells was isolated using TRIzol (Sigma-Aldrich, Shanghai, China), and cDNA was synthesized using HiScript II Q Select RT SuperMix (Vazyme, Nanking, China) for qPCR. Quantitative PCR was performed with ChamQ Universal SYBR qPCR Master Mix (Vazyme, Nanking, China) using a ROCHE Light Cycler 96 (ROCHE, Switzerland). The primer sequences are shown in Table S1.

4.6. Animals

The animal experiment was conducted in accordance with the "Laboratory Animal Guideline for Ethical Review of Animal Welfare" (GB/T 35892-1272018) and the Experimental Animal Management and Ethics Committee of Xiamen University (Certificate no. SYXK2013-0006). Five-week-old male OB/OB mice were obtained from GemPharmatech Co., Ltd. (Nanjing, China) and maintained on a chow diet for 1 week with a 12 h light/dark cycle for acclimatization. After adaptation to the experimental environment for 3 days, the mice were fed a HFD (ResearchDiets, #D12492, New Brunswick, NJ, USA) for 10 weeks. The HFD provided 5.21 kcalg^{-1} of energy (20% calories from protein, 60% calories from fat, and 20% calories from carbohydrate). The mice were randomly divided into three groups (n = 6 for each of the treatment groups): the control group (gastric gavage normal saline), FBT treatment group (gastric gavage 60 mg/kg FBT), and orlistat treatment group (gastric gavage 800 mg/kg orlistat). Bodyweight was measured weekly, and food intake was recorded every other day. We used PET-CT to evaluate the effect of FBT on perirenal fat and fatty liver hypertrophy in mice. Glucose and insulin tolerance tests were performed according to a previous paper [44].

4.6.1. Preparation of Liver, Serum and White Fat Tissue Samples

All mice were sacrificed on the 42nd day of intragastric administration. The liver and white fat tissue from the entire brain tissue were immediately separated, and the samples were stored in AllProtectTM (Beyotime Biotechnology, Shanghai, China). The hippocampal tissues were used for other experiments.

4.6.2. Oil Red-O Staining

Mouse adipocytes were fixed with 10% formalin solution for 10 min and stained with freshly prepared Oil Red O solution for 30 min at room temperature. Intracellular lipid contents were measured by extracting Oil Red O with isopropanol, and the absorbance was measured at 500 nm with Thermo Multiskan FC (Shanghai, China). Liver tissue was embedded in OCT after dehydration with 30% sucrose. The frozen sections were operated according to the Oil Red O staining kit purchased from Solarbio. Images were acquired using a Leica DM 4B microscope (Wetzlar, Germany).

4.6.3. Staining of Liver and White Adipose Tissue

Liver and white adipose tissue were dehydrated and sectioned in a routine procedure. Liver and adipose tissue were stained with an HE staining kit. In addition, a Masson staining kit (Beijing Solarbio Science & Technology Co., Ltd., Beijing, China) and a TUNEL fluorescence staining kit (Sangon Biotechnology Co., Ltd. Shanghai, China) was used to evaluate the degree of liver injury. Images were acquired using a Leica DM 4B microscope (Wetzlar, Germany).

4.6.4. Transmission Electron Microscopy

For each group, three animals were analyzed. The liver fragments were fixed overnight in a solution containing 2.5% glutaraldehyde and 4% formaldehyde in 0.1 M cacodylate buffer. The method of transmission electron microscopy was as described in the previous literature [27].

4.6.5. Metabolomics Analysis

The liver lipid LC-QTOFMS-based metabolomic experiments and data analysis were commissioned by Shanghai BIOTREE Biotechnology Co., Ltd.

4.7. Statistical Analysis

The data are presented as the mean ± SD. Differences between the means of individual groups were assessed by one-way analysis of variance followed by a multiple comparisons test; differences were considered significant at $p < 0.05$. The statistical software package Prism6.0 (GraphPad Software, LaJolla, CA, USA) was used for these analyses.

5. Conclusions

In summary, FBT, a highly effective lipase inhibitor, was extracted from Fu Brick tea. The extraction process is convenient and feasible. The inhibition mechanism of FBT on lipase was explored from the perspective of enzyme molecular dynamics. The IC_{50} value of FBT on lipase was 1.02 ± 0.03 μg/mL, which was a reversible mixed inhibition. At the same time, at the cellular and animal experiment levels, FBT downregulated the mRNA expression of ACC, FAS, LPL, HSL, ATGL, upregulated the mRNA expression of A-COX and CPT-1, inhibited the synthesis and breakdown of TG and TC in 3T3-L1 adipocytes, promoted fatty acid peroxidation, and maintained a stable state of lipid metabolism. Surprisingly, FBT also increased sugar consumption and promoted adiponectin activity in the IR model, inhibiting differentiation, eliminating inflammatory factor proinflammatory cytokines (TNF-a, IL-6 and IκKβ) and reducing ROS production. In the animal experiments, we confirmed that FBT can effectively alleviate fatty liver levels, prevent liver fibrosis and accumulation of lipid droplets, reduce liver cell apoptosis, improve damage to mitochondrial membrane structure, and ensure normal β-oxidation. Furthermore, FBT reduces the accumulation of TG and TC in the liver and increases the total antioxidant capacity of the tissues as well as the metabolism of DAG, FA and FAHFA. In this study, FBT was first applied to fat cells and obese mice. The inhibition effect and safety were evaluated, and the lipid-lowering mechanism of FBT influencing fat metabolism was preliminarily discussed. This study provides a new idea for natural product inhibitors as anti-obesity drugs and provides a theoretical basis for clinical treatment.

Supplementary Materials: The following are available online at https://www.mdpi.com/article/10.3390/ijms23052525/s1.

Author Contributions: T.-T.L. performed enzymology and cell experiments, analyzed related data. X.-T.L. mainly responsible for mouse experiments, making tables and graphs. G.-L.H. mainly performed FBT molecular structure determination. L.L., Q.-X.C. and Q.W. mainly performed Experimental design and related technical guidance. All authors have read and agreed to the published version of the manuscript.

Funding: This research was funded by Science and Technology Bureau of Xiamen Municipal (3502Z20193006) and Postdoctoral Science Foundation of China (2021M691283).

Institutional Review Board Statement: The experimental procedures were performed in accordance with the guidelines for the humane treatment of animals set by the Experimental Animal Management and Ethics Committee of Xiamen University (Certificate No. SYXK2013-0006). Xiamen, China.

Informed Consent Statement: Not applicable.

Data Availability Statement: The data presented in this study are available on request from the authors.

Acknowledgments: We gratefully acknowledge the kind cooperation of BIOTREE Biotechnology Co., Ltd.

Conflicts of Interest: The authors declare no conflict of interest.

References

1. Rosen, E.D.; Spiegelman, B.M. Adipocytes as regulators of energy balance and glucose homeostasis. *Nature* **2006**, *444*, 847–853. [CrossRef] [PubMed]
2. Pistor, K.E.; Sepa-Kishi, D.M.; Hung, S.; Ceddia, R.B. Lipolysis, lipogenesis, and adiposity are reduced while fatty acid oxidation is increased in visceral and subcutaneous adipocytes of endurance-trained rats. *Adipocyte* **2015**, *4*, 22–31. [CrossRef]
3. Samuel, V.T.; Liu, Z.X.; Qu, X.; Elder, B.D.; Bilz, S.; Befroy, D.; Romanelli, A.J.; Shulman, G.I. Mechanism of hepatic insulin resistance in non-alcoholic fatty liver disease. *J. Biol. Chem.* **2004**, *279*, 32345–32353. [CrossRef] [PubMed]
4. Park, H.; Bae, S.H.; Park, Y.; Choi, H.S.; Suh, H.J. Lipase-mediated lipid removal from propolis extract and its antiradical and antimicrobial activity. *J. Sci. Food Agric.* **2015**, *95*, 1697–1705. [CrossRef] [PubMed]
5. Buchholz, T.; Melzig, M.F. Polyphenolic Compounds as Pancreatic Lipase Inhibitors. *Planta Med.* **2015**, *81*, 771–783. [CrossRef] [PubMed]
6. Chen, G.; Xie, M.; Dai, Z.; Wan, P.; Ye, H.; Zeng, X.; Sun, Y. Kudingcha and Fuzhuan Brick tea Prevent Obesity and Modulate Gut Microbiota in High-Fat Diet Fed Mice. *Mol. Nutr. Food Res.* **2018**, *62*, e1700485. [CrossRef] [PubMed]
7. Foster, M.T.; Gentile, C.L.; Cox-York, K.; Wei, Y.; Wang, D.; Estrada, A.L.; Reese, L.; Miller, T.; Pagliassotti, M.J.; Weir, T.L. Fuzhuan tea consumption imparts hepatoprotective effects and alters intestinal microbiota in high saturated fat diet-fed rats. *Mol. Nutr. Food Res.* **2016**, *60*, 1213–1220. [CrossRef]
8. Liu, D.; Huang, J.; Luo, Y.; Wen, B.; Wu, W.; Zeng, H.; Zhonghua, L. Fuzhuan Brick tea Attenuates High-Fat Diet-Induced Obesity and Associated Metabolic Disorders by Shaping Gut Microbiota. *J. Agric. Food Chem.* **2019**, *67*, 13589–13604. [CrossRef]
9. Sang, S.; Lambert, J.D.; Ho, C.T.; Yang, C.S. The chemistry and biotransformation of tea constituents. *Pharmacol. Res.* **2011**, *64*, 87–99. [CrossRef]
10. Balcao, V.M.; Paiva, A.L.; Malcata, F.X. Bioreactors with immobilized lipases: State of the art. *Enzym. Microb. Technol.* **1996**, *18*, 392–416. [CrossRef]
11. Shoelson, S.E.; Lee, J.; Yuan, M. Inflammation and the IKK beta/I kappa B/NF-kappa B axis in obesity- and diet-induced insulin resistance. *Int. J. Obes. Relat. Metab. Disord.* **2003**, *27* (Suppl. 3), S49–S52. [CrossRef]
12. Wang, G.; Wu, B.; Zhang, L.; Cui, Y.; Zhang, B.; Wang, H. Laquinimod Prevents Adipogenesis and Obesity by Down-Regulating PPAR-gamma and C/EBPalpha through Activating AMPK. *ACS Omega* **2020**, *5*, 22958–22965. [CrossRef]
13. Corbin, K.D.; Zeisel, S.H. Choline metabolism provides novel insights into nonalcoholic fatty liver disease and its progression. *Curr. Opin. Gastroenterol.* **2012**, *28*, 159–165. [CrossRef]
14. Boss, O.; Bergenhem, N. Adipose targets for obesity drug development. *Expert Opin. Ther. Targets* **2006**, *10*, 119–134. [CrossRef]
15. Bentley, A.R.; Kritchevsky, S.B.; Harris, T.B.; Holvoet, P.; Jensen, R.L.; Newman, A.B.; Lee, J.S.; Yende, S.; Bauer, D.; Cassano, P.A. Dietary antioxidants and forced expiratory volume in 1 s decline: The Health, Aging and Body Composition study. *Eur. Respir. J.* **2012**, *39*, 979–984. [CrossRef]
16. Farmer, S.R. Transcriptional control of adipocyte formation. *Cell Metab.* **2006**, *4*, 263–273. [CrossRef]
17. Diaz-Arjonilla, M.; Schwarcz, M.; Swerdloff, R.S.; Wang, C. Obesity, low testosterone levels and erectile dysfunction. *Int. J. Impot. Res.* **2009**, *21*, 89–98. [CrossRef]
18. Yadav, A.; Kataria, M.A.; Saini, V.; Yadav, A. Role of leptin and adiponectin in insulin resistance. *Clin. Chim. Acta* **2013**, *417*, 80–84. [CrossRef]
19. Mahmoud, A.M.; Ashour, M.B.; Abdel-Moneim, A.; Ahmed, O.M. Hesperidin and naringin attenuate hyperglycemia-mediated oxidative stress and proinflammatory cytokine production in high fat fed/streptozotocin-induced type 2 diabetic rats. *J. Diabetes Complicat.* **2012**, *26*, 483–490. [CrossRef]
20. Choi, C.S.; Savage, D.B.; Abu-Elheiga, L.; Liu, Z.; Kim, S.; Kulkarni, A.; Distefano, A.; Hwang, Y.; Reznick, R.M.; Codella, R.; et al. Continuous fat oxidation in acetyl-CoA carboxylase 2 knockout mice increases total energy expenditure, reduces fat mass, and improves insulin sensitivity. *Proc. Natl. Acad. Sci. USA* **2007**, *104*, 16480–16485. [CrossRef]
21. Sullivan, J.E.; Brocklehurst, K.J.; Marley, A.E.; Carey, F.; Carling, D.; Beri, R.K. Inhibition of lipolysis and lipogenesis in isolated rat adipocytes with aicar, a cell-permeable activator of AMP-activated protein-kinase. *Febs Lett.* **1994**, *353*, 33–36. [CrossRef]
22. Heimann, E.; Nyman, M.; Degerman, E. Propionic acid and butyric acid inhibit lipolysis and de novo lipogenesis and increase insulin-stimulated glucose uptake in primary rat adipocytes. *Adipocyte* **2015**, *4*, 81–88. [CrossRef] [PubMed]
23. van Hees, A.M.J.; Jocken, J.W.E.; Essers, Y.; Roche, H.M.; Saris, W.H.M.; Blaak, E.E. Adipose triglyceride lipase and hormone-sensitive lipase protein expression in subcutaneous adipose tissue is decreased after an isoenergetic low-fat high-complex carbohydrate diet in the metabolic syndrome. *Metabolism* **2012**, *61*, 1404–1412. [CrossRef] [PubMed]
24. Schoiswohl, G.; Stefanovic-Racic, M.; Menke, M.N.; Wills, R.C.; Surlow, B.A.; Basantani, M.K.; Sitnick, M.T.; Cai, L.; Yazbeck, C.F.; Stolz, D.B.; et al. Impact of Reduced ATGL-Mediated Adipocyte Lipolysis on Obesity-Associated Insulin Resistance and Inflammation in Male Mice. *Endocrinology* **2015**, *156*, 3610–3624. [CrossRef]
25. Weinstock, P.H.; LevakFrank, S.; Hudgins, L.C.; Radner, H.; Friedman, J.M.; Zechner, R.; Breslow, J.L. Lipoprotein lipase controls fatty acid entry into adipose tissue, but fat mass is preserved by endogenous synthesis in mice deficient in adipose tissue lipoprotein lipase. *Proc. Natl. Acad. Sci. USA* **1997**, *94*, 10261–10266. [CrossRef]
26. Shiomi, Y.; Yamauchi, T.; Iwabu, M.; Okada-Iwabu, M.; Nakayama, R.; Orikawa, Y.; Yoshioka, Y.; Tanaka, K.; Ueki, K.; Kadowaki, T. A Novel Peroxisome Proliferator-activated Receptor (PPAR)alpha Agonist and PPARgamma Antagonist, Z-551, Ameliorates High-fat Diet-induced Obesity and Metabolic Disorders in Mice. *J. Biol. Chem.* **2015**, *290*, 14567–14581. [CrossRef]

27. Cipriano Torres, D.D.O.; Oliveira Dos Santos, A.C.; Silva, A.K.S.E.; Alvarez Leite, J.I.; Botelho De Souza, J.R.; Carneiro Beltrao, E.I.; Peixoto, C.A. Effect of Maternal Diet Rich in Omega-6 and Omega-9 Fatty Acids on the Liver of LDL Receptor-Deficient Mouse Offspring. *Birth Defects Res. B* **2010**, *89*, 164–170. [CrossRef]
28. Targher, G.; Day, C.P.; Bonora, E. Risk of cardiovascular disease in patients with nonalcoholic fatty liver disease. *N. Engl. J. Med.* **2010**, *14*, 1341–1350. [CrossRef]
29. Nobili, V.; Alisi, A.; Vania, A.; Tiribelli, C.; Pietrobattista, A.; Bedogni, G. The pediatric NAFLD fibrosis index: A predictor of liver fibrosis in children with non-alcoholic fatty liver disease. *BMC Med.* **2009**, *7*, 21. [CrossRef]
30. Schuppan, D.; Schattenberg, J.M. Non-alcoholic steatohepatitis: Pathogenesis and novel therapeutic approaches. *J. Gastroenterol. Hepatol.* **2013**, *28* (Suppl. 1), 68–76. [CrossRef]
31. Koo, S.H. Nonalcoholic fatty liver disease: Molecular mechanisms for the hepatic steatosis. *Clin. Mol. Hepatol.* **2013**, *19*, 210–215. [CrossRef]
32. Loria, P.; Lonardo, A.; Anania, F. Liver and diabetes. A vicious circle. *Hepatol. Res.* **2013**, *43*, 51–64. [CrossRef]
33. Donnelly, K.L.; Smith, C.I.; Schwarzenberg, S.J.; Jessurun, J.; Boldt, M.D.; Parks, E.J. Sources of fatty acids stored in liver and secreted via lipoproteins in patients with nonalcoholic fatty liver disease. *J. Clin. Investig.* **2005**, *115*, 1343–1351. [CrossRef]
34. Jacome-Sosa, M.M.; Parks, E.J. Fatty acid sources and their fluxes as they contribute to plasma triglyceride concentrations and fatty liver in humans. *Curr. Opin. Lipidol.* **2014**, *25*, 213–220. [CrossRef] [PubMed]
35. Benlebna, M.; Balas, L.; Bonafos, B.; Pessemesse, L.; Fouret, G.; Vigor, C.; Gaillet, S.; Grober, J.; Bernex, F.; Landrier, J.F.; et al. Long-term intake of 9-PAHPA or 9-OAHPA modulates favorably the basal metabolism and exerts an insulin sensitizing effect in obesogenic diet-fed mice. *Eur. J. Nutr.* **2021**, *60*, 2013–2027. [CrossRef]
36. Solinas, G.; Boren, J.; Dulloo, A.G. De novo lipogenesis in metabolic homeostasis: More friend than foe? *Mol. Metab.* **2015**, *4*, 367–377. [CrossRef]
37. Li, Z.; Agellon, L.B.; Allen, T.M.; Umeda, M.; Jewell, L.; Mason, A.; Vance, D.E. The ratio of phosphatidylcholine to phosphatidylethanolamine influences membrane integrity and steatohepatitis. *Cell Metab.* **2006**, *3*, 321–331. [CrossRef]
38. Tonstad, S.; Pometta, D.; Erkelens, D.W.; Ose, L.; Moccetti, T.; Schouten, J.A.; Golay, A.; Reitsma, J.; Del, B.A.; Pasotti, E.; et al. The effect of the gastrointestinal lipase inhibitor, orlistat, on serum lipids and lipoproteins in patients with primary hyperlipidaemia. *Eur. J. Clin. Pharmacol.* **1994**, *46*, 405–410. [CrossRef]
39. Tian-Tian, L.; Xiao-Rong, H.; Run-Xue, X.; Xiao-Bing, W.; Ya-Xin, Q.; Jian-Zhong, H.; Qiong-Huan, C.; Qing-Xi, C. Inhibitory mechanism and molecular analysis of furoic acid and oxalic acid on lipase. *Int. J. Biol. Macromol.* **2018**, *120*, 1925–1934.
40. Kang, N.H.; Lee, W.K.; Yi, B.R.; Park, M.A.; Lee, H.R.; Park, S.K.; Hwang, K.A.; Park, H.K.; Choi, K.C. Modulation of lipid metabolism by mixtures of protamine and chitooligosaccharide through pancreatic lipase inhibitory activity in a rat model. *Lab. Anim. Res.* **2012**, *28*, 31–38. [CrossRef]
41. Liu, T.; Su, W.; Chen, Q.; Shen, D.; Zhuang, J. The inhibitory kinetics and mechanism of glycolic acid on lipase. *J. Biomol. Struct. Dyn.* **2020**, *38*, 2021–2028. [CrossRef] [PubMed]
42. Chen, X.X.; Zhang, J.; Chai, W.M.; Feng, H.L.; Xiang, Z.H.; Shen, D.Y.; Chen, Q.X. Reversible and competitive inhibitory kinetics of amoxicillin on mushroom tyrosinase. *Int. J. Biol. Macromol.* **2013**, *62*, 726–733. [CrossRef]
43. Su, W.C.; Lin, Y.F.; Yu, X.P.; Wang, Y.X.; Lin, X.D.; Su, Q.Z.; Shen, D.Y.; Chen, Q.X. Mitochondria-Associated Apoptosis in Human Melanoma Cells Induced by Cardanol Monoene from Cashew Nut Shell Liquid. *J. Agric. Food Chem.* **2017**, *65*, 5620–5631. [CrossRef] [PubMed]
44. Truong, X.T.; Nguyen, T.; Kang, M.J.; Jung, C.H.; Lee, S.; Moon, C.; Moon, J.H.; Jeon, T.I. Pear Extract and Malaxinic Acid Reverse Obesity, Adipose Tissue Inflammation, and Hepatosteatosis in Mice. *Mol. Nutr. Food Res.* **2019**, *63*, e1801347. [CrossRef]

International Journal of
Molecular Sciences

MDPI

Review

Diet and Male Fertility: The Impact of Nutrients and Antioxidants on Sperm Energetic Metabolism

Alessandra Ferramosca * and Vincenzo Zara

Department of Biological and Environmental Sciences and Technologies, University of Salento, I-73100 Lecce, Italy; vincenzo.zara@unisalento.it
* Correspondence: alessandra.ferramosca@unisalento.it

Abstract: Diet might affect male reproductive potential, but the biochemical mechanisms involved in the modulation of sperm quality remain poorly understood. While a Western diet is considered a risk factor for male infertility, the Mediterranean diet seems to protect against male infertility; moreover, the role of a vegetarian habitus in the preservation of sperm quality is controversial. The aim of this review is to analyze the molecular effects of single nutrients on sperm quality, focusing on their involvement in biochemical mechanisms related to sperm bioenergetics. It appears that diets rich in saturated fatty acids (SFA) and low in polyunsaturated fatty acids (PUFA) negatively affect sperm quality, whereas unsaturated fatty acids supplementation ameliorates sperm quality. In fact, the administration of PUFA, especially omega-3 PUFA, determined an increase in mitochondrial energetic metabolism and a reduction in oxidative damage. Carbohydrates and proteins are also nutritional modulators of oxidative stress and testosterone levels, which are strictly linked to sperm mitochondrial function, a key element for sperm quality. Moreover, many dietary natural polyphenols differentially affect (positively or negatively) the mitochondrial function, depending on their concentration. We believe that an understanding of the biochemical mechanisms responsible for sperm quality will lead to more targeted and effective therapeutics for male infertility.

Keywords: spermatozoa; infertility; obesity; fatty acids; sugar; bioactive molecules; mitochondria; metabolism

Citation: Ferramosca, A.; Zara, V. Diet and Male Fertility: The Impact of Nutrients and Antioxidants on Sperm Energetic Metabolism. *Int. J. Mol. Sci.* **2022**, 23, 2542. https://doi.org/10.3390/ijms23052542

Academic Editor: Maurizio Battino

Received: 27 January 2022
Accepted: 24 February 2022
Published: 25 February 2022

1. Introduction

Nutrition can affect, negatively or positively, sperm quality [1–3], and this effect depends on both quantitative and qualitative aspects of a diet, such as the calorie content of each macronutrient (carbohydrates, proteins, and fats), as well as on the specific fatty acid, carbohydrate, and protein profiles.

In this context, unhealthy hypercaloric diets and excessive intake of saturated and trans fatty acids have a negative impact on sperm quality and, therefore, on the fertilization process [4–11]. On the other hand, healthy dietary models are clearly associated with a better sperm quality, suggesting that nutritional interventions could have a key role in the preservation of male fertility [3,12–14]. Moreover, an adequate intake of antioxidant molecules has been quite effective in the prevention and/or in the treatment of male infertility [15–17].

Despite a solid body of evidence showing the effects of nutrients and antioxidant molecules on male reproductive potential, there is little knowledge concerning the potential mechanisms involved in the modulation of sperm quality.

The aim of this review is to summarize the most recent evidence regarding the impact of nutrients and antioxidant molecules on sperm quality, with a particular focus on their involvement in biochemical mechanisms related to sperm bioenergetics.

We believe that an understanding of the biochemical mechanisms responsible for sperm quality will lead to more targeted and effective therapeutics for male infertility.

2. Diets and Male Fertility

There is increasing evidence that dietary behavior is associated with semen quality parameters [18]. In recent decades, the main dietary pattern has become the so-called "Western diet", which is the consequence of the "westernization" of human lifestyle. This diet is characterized by a high intake of industrially processed foods, rich in animal proteins, simple carbohydrates, trans and saturated fats, and poor in dietary fiber and essential unsaturated fatty acids. Recent studies have linked a Western-pattern diet to an increased risk of metabolic diseases, atherosclerosis, neurodegeneration, cancer, as well as infertility [6].

Differently from the Western diet, the Mediterranean diet, which is one of the healthiest dietary patterns, has evident health benefits [19], including benefits in terms of semen quality parameters [20–22]. This diet is characterized by a high consumption of legumes, cereals, fruits, vegetables, a moderate consumption of fish and wine, and a low consumption of dairy products and meat; olive oil is the main source of added fat. Thus, the Mediterranean diet is rich in monounsaturated fatty acids (MUFA), fiber, and antioxidants and low in saturated fat (SFA).

Another dietary model, the vegetarian diet, is similar in dietary composition to the Mediterranean diet, but it does not include meat and meat products, poultry, seafood, and flesh from any other animal. Recently, there has been considerable interest in the impact of this diet on male fertility, because it has been proposed that this dietary model decreased semen quality [23–25].

Western, Mediterranean, and vegetarian diets are the most investigated dietary models in the field of nutrition and male reproduction. Therefore, in the following paragraphs, we will try to discuss the effects of various combinations of nutrients that are the main components of these dietary patterns on sperm quality and metabolism.

2.1. Western Diet as a Risk Factor for Male Infertility

The current western dietary habits generally imply high sugar and high fat consumption, with the consequent intake of unbalanced diets and/or of an excess of calories [26]. Over the last several decades, the Western diet has been therefore a significant contributor to the growing rate of obesity, which has had a significant effect on fertility [3] through changes in hormonal levels, sperm function, and gamete molecular composition. However, the molecular mechanisms responsible for the causal links between obesity and male infertility are not totally clear.

Obesity has been shown to disrupt various components of the hypothalamic–pituitary–gonadal axis, causing hypogonadism [27], which is mainly associated with reduced levels of testosterone and fewer spermatozoa [28].

Increased amounts of adipose tissue cause insulin resistance and have an important role in the development of oxidative stress, thus altering reproductive pathways and sperm function [29,30]. Hyperinsulinemia and hyperglycemia, which are the hallmarks of insulin resistance, seem to be responsible for the reduction in sperm glucose uptake and metabolism [31], thus having a possible role in the impairment of glycolysis in sperm cells. It is important to underline that glycolysis, along with oxidative phosphorylation (OXPHOS), is a metabolic pathway producing adenosine triphosphate (ATP), which is the primary source of energy for spermatozoa [32,33]. According to this hypothesis, diabetic male rats showed a decrease in sperm motility [34,35], which was restored after insulin administration [35].

Glucose uptake and homeostasis may be also modulated by leptin released from fat cells in adipose tissue depots. It has been found that hyperinsulinemia and hyperleptinemia were associated with an increase of insulin and leptin concentrations in seminal plasma, which may negatively impact male reproductive function and sperm quality [36].

Leptin has also an important role in the chronic pro-inflammatory state in the testicular microenvironment and/or excurrent ductal system, thus increasing the level of reactive oxygen species (ROS) which are responsible for the decrease in sperm quality [4,29,37].

Excess fat tissue results in the increased activity of aromatase, which is an enzyme responsible for converting testosterone to estradiol [38]. The consequent decrease in testosterone levels results in low sperm production, because this hormone is the major androgen in the testis involved in the regulation of spermatogenesis. Low testosterone levels seem to be related not only to oxidative stress but also to mitochondrial dysfunction in Leydig cells [39,40], located in the connective tissue surrounding the seminiferous tubules, where the first step of the synthesis of testosterone occurs.

Defects in Leydig cell mitochondria are responsible for oxidative damages in lipids, proteins, and mitochondrial DNA (mtDNA) and cause a decrease in ATP levels and an increase in ROS production [40]. Germ cells and mature spermatozoa are susceptible to oxidative stress, which leads to a decrease in sperm quality (decrease in sperm number and motility and increase in abnormalities in sperm morphology) [41–43].

Different findings suggest that sperm mitochondria play a pivotal role in the decrease of sperm quality caused by ROS. In germ cells, mitochondrial proteins and membrane lipids are damaged, and mtDNA is fragmented. Therefore, ATP synthesis is severely affected, and the decrease in energy production results in meiotic arrest, causing the presence of abnormalities in sperm morphology. High ROS levels also disrupt mitochondrial membranes and induce apoptosis, thus leading to a decrease in sperm number [40]. At the same time, in mature spermatozoa, the mitochondria are the target of ROS, and the decrease in mitochondrial functionality might be one of the causes responsible for the reduction of sperm motility [41].

It has been suggested that also dyslipidemia, which is known to be associated with increased amounts of adipose tissue, may have an impact on semen quality. In this regard, lipid profile alterations have been correlated with male infertility [4,5,42–44]. Dyslipidemia is a term referring to a group of different blood lipid imbalances such as hypercholesterolemia, hypertriglyceridemia, decrease of HDL-cholesterol, or combined hyperlipidemia.

The possible effects of the Western diet on sperm quality are schematized in Figure 1.

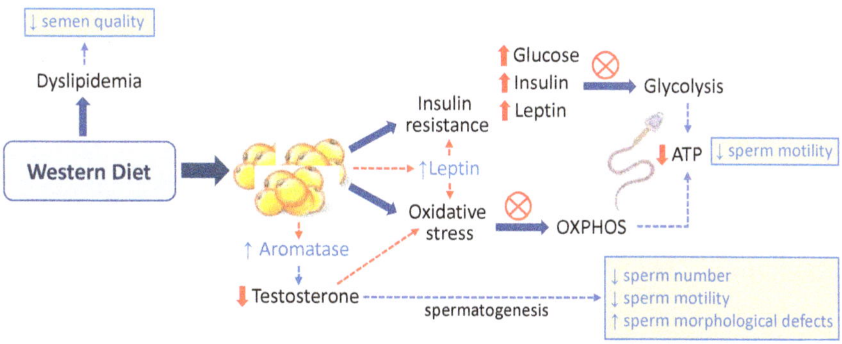

Figure 1. Effects of the Western diet on sperm quality.

2.2. *Mediterranean Diet as a Protection Factor against Male Infertility*

The Mediterranean diet incorporates the traditional healthy living habits of people from countries surrounding the Mediterranean Sea, although differentiated by some food choices and cooking practices. This dietary pattern has been shown to confer multiple health benefits. Interestingly, it seems able to promote good male reproductive health, being associated with an increase in sperm number and quality and with improved chances of conceiving [13,20,45].

One reason why the Mediterranean dietary pattern is so positive for male fertility is because it provides a low level of SFA and trans fatty acids and adequate levels of certain nutrients such as omega-3 fatty acids, antioxidant molecules, and vitamins. In fact, it has been shown that the intake of antioxidant vitamins and carotenoids was related to higher sperm counts [46]. Moreover, higher intakes of fruit, cereals, and vegetables were positively related to sperm motility and concentration [47,48].

Many dietary natural compounds isolated from fruits, vegetables, and edible plants can target the mitochondria, modulating their metabolism, biogenesis, and redox status [49–51]. The protection of mitochondrial function by these compounds may be important in explaining their beneficial effects on male reproductive performance [16].

Another characteristic of the Mediterranean diet is the consumption of olive oil as the main source of fat. It has been demonstrated in animal models that olive oil supplementation significantly increased sperm quality [5,42]. This is because olive oil, which is the main source of MUFA, may modify the sperm membrane lipid composition, reducing oxidative stress damages and restoring mitochondrial function [5].

Moreover, the Mediterranean pattern induces a reduction in omega-6 fatty acids in favor of omega-3 fatty acids, which have been associated with an improvement of sperm energetic metabolism [5].

2.3. *Vegetarian Diet as a Controversial Factor for Male Infertility*

A substantial proportion of the world's population is vegetarian for cultural or ethical values, religious beliefs, environmental concerns, and health considerations.

Existing studies have reported mostly protective associations between a vegetarian pattern and risk factors for chronic diseases [52]. However, the role of a vegetarian diet in the preservation of sperm quality is controversial.

Vegetables and fruits are rich in antioxidant molecules, which can act as sperm ROS regulators by reducing sperm DNA damage and by increasing sperm motility and vitality. At the same time, it has been described that the vegetarian diet reduced sperm concentration and motility, but its effect on infertility was not thoroughly assessed [25]. This effect may be attributed to estrogenic compounds or chemical residues in the diet which had a negative effect on sperm parameters [16,25,53,54].

3. Nutrients Impacts on Molecular Aspects Related to Sperm Quality

The differential impacts of Western, Mediterranean, and Vegetarian diets on male fertility depend on the amount and quality of the nutrients introduced. In the following paragraphs, we will analyze the molecular effects of single nutrients on sperm quality, focusing on their involvement in biochemical mechanisms related to sperm bioenergetics.

3.1. *Dietary Fats*

The negative impact of lipid metabolism disorders on male fertility is now well known, but the underlying molecular mechanisms involved are not totally clear. Dietary fats can influence the lipid composition of sperm cells, having harmful or beneficial consequences on male reproductive potential [55].

3.1.1. Fatty Acids

In sperm cells, fatty acids are constituents of the gamete membrane as well as energy suppliers [56] and can derive from de novo synthesis or from dietary sources.

It is known that 18-carbon chain omega-6 and omega-3 polyunsaturated fatty acids (PUFA) cannot be endogenously synthesized by humans and therefore must be obtained from food. Long-chain PUFA, such as arachidonic acid (ARA; C20:4 omega-6) and eicosapentanoic acid (EPA; C20:5 omega-3), can derive from exogenous and endogenous sources. Vegetable oils, seeds, and nuts are a source of the omega-6 PUFA linoleic acid (LA; C18:2 omega-6) and α-linolenic acid (ALA; C18:3 omega-3). Seafood is a source of EPA and docosahexaenoic acid (DHA; C22:6 omega-3), which are long-chain omega-3 PUFA. Meat and dairy are a source of ARA.

Diets rich in SFA and low in PUFA or with an unbalanced omega-6/omega-3 PUFA ratio negatively affected sperm quality, whereas dietary unsaturated fatty acid supplementation ameliorated sperm quality [57].

In humans, the dietary intake of saturated fat negatively correlated with total sperm count and concentration; conversely, the intake of omega-3 fatty acids showed a positive correlation with sperm quality. However, the role of omega-3 fatty acids in the improvement of sperm quality is not always accompanied by changes in the fatty acid profile of sperm [55]. Therefore, a role of fatty acids in the regulation of sperm metabolism has been proposed [4,5,7].

To investigate the role of dietary fatty acids on sperm dysfunction, researchers used the model of diet-induced obesity in rodents and rabbits. They found that PUFA can influence reproductive processes, directly or indirectly, through several mechanisms. For example, PUFA may promote the loss of body fat, hence preventing obesity and then the development of infertility, but they are also able to act on specific aspects linked to male fertility. In fact, these fatty acids are also components of sperm membrane, whose fluidity and dynamics are necessary to promote fertilization. They are also precursors for eicosanoids synthesis, which can modulate many key enzymes involved in steroid metabolism.

The molecular mechanisms responsible for the effects exerted by PUFA on male fertility may be due to a parallel modulation of lipid metabolism [58] and of sperm mitochondrial function [5] (Figure 2). Regarding this last aspect, a nutritional modulation of specific enzymes involved in sperm bioenergetics pathways has been proposed [5]. Among these enzymes, the sperm lactate dehydrogenase isoenzymatic form (LDH-C4 or LDH-X) is an important site of nutritional modulation by omega-3 PUFA [5]. This enzyme, which is present both in the mitochondrial matrix and in the cytosol of spermatozoa, with a net prevalence in the cytosol [59], catalyzes the conversion of pyruvate to lactate with the concomitant oxidation of NADH (reduced form of nicotinamide adenine dinucleotide) to NAD^+ (oxidized form of nicotinamide adenine dinucleotide), playing a key role in the energy metabolism of spermatozoa [32,33]. In fact, LDH-C4 allows the parallel progress of glycolysis, by regenerating NAD^+, and of OXPHOS, by the transport of reducing equivalents from the cytosol into the mitochondria. Increased activity of LDH-C4 after omega-3 PUFA administration was accompanied by a parallel increase in the activity of respiratory complexes. Moreover, dietary administration of omega-3 PUFA reduced oxidative damage on sperm cells, as also suggested by the increase in the aconitase/fumarase activity ratio [5]. Aconitase and fumarase are two Krebs cycle enzymes, and their activity ratio is used as an indicator of mitochondrial ROS production [60].

Interestingly, the activity of LDH-C4, pyruvate dehydrogenase, and respiratory enzymes was decreased after the administration of a diet rich in SFA and low in PUFA [4], thus suggesting that a specific fatty acid composition of the diet can counteract the negative effects of a high-fat diet on sperm cells.

Recent studies demonstrated that also the administration of MUFA counteracted the negative effects of a hyperlipidic diet on sperm quality. The effects of MUFA were evaluated by adding olive oil or avocado extracts to experimental diets [5,42,61,62]. In fact, olive oil consists mainly of oleic acid (up to 83%), whereas oleic acid contributes to about 60% of the total fatty acid content of avocados.

The Authors found that MUFA supplementation modified sperm membrane composition, reduced oxidative stress damages, and modulated enzymatic activities involved in energetic metabolism [5,7,61] (Figure 2).

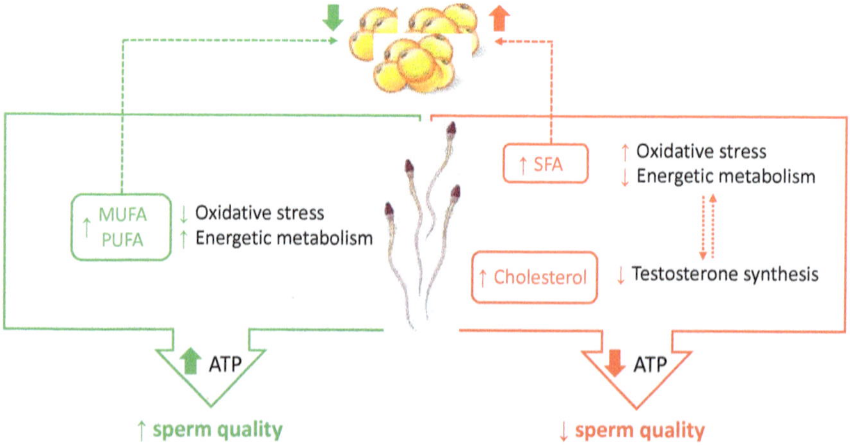

Figure 2. Effects of dietary fat on sperm quality. Dietary fatty acids exert a parallel modulation of lipid metabolism and sperm mitochondrial function. On the one hand, they may promote loss or increase of body fat, hence modulating molecular aspects related to obesity; on the other hand, they modulate oxidative stress and energetic metabolism. A dietary cholesterol excess could induce cholesterol accumulation in testicular Leydig cells, causing a decrease in testosterone production, which is related to oxidative stress mitochondrial dysfunction.

3.1.2. Dietary Cholesterol

The study of the relationships between dietary cholesterol intake and male fertility is difficult to carry out in humans, and most of the information has been generated using rabbits as animal models. In fact, rabbits are sensitive to a cholesterol-enriched diet, showing a lipid metabolism which is closer to that of humans than to that of rodents [42,63–67].

It has been demonstrated that hypercholesterolemia modifies plasma membrane composition and dynamics, thus modifying sperm morphology and function [55]. Although rabbits are the gold standard for studying the relationships between hypercholesterolemia and male fertility, also data obtained in mice and rats suggested that an overload of dietary cholesterol has a very negative impact on male fertility [68–70].

The dietary cholesterol impact on male fertility is still poorly understood at the molecular level. What is known is that cholesterol is an essential lipid for membrane structure and dynamics, as well as for testosterone synthesis. In fact, the cholesterol levels in the sperm membrane affect membrane fluidity, which plays a key role in sperm motility, capacitation, and acrosome reaction [33].

At the same time, dietary cholesterol could induce cholesterol accumulation in testicular Leydig cells. Although cholesterol is the substrate for testosterone biosynthesis, an excess of cholesterol can be dangerous, because high cholesterol levels are responsible for the activation of the endoplasmic reticulum stress, causing the downregulation of steroidogenic enzymes and then a decreased testosterone production [71] (Figure 2). As reported before, low testosterone levels are strictly related to oxidative stress and to mitochondrial dysfunction (Figure 2).

Therefore, cholesterol concentration is critical to assure sperm quality. Cholesterol homeostasis is under the control of the transcription factors known as sterol regulatory element-binding proteins (SREBPs). It has recently been found that, in the short term, high circulating cholesterol levels due to diet decreased the expression of molecules involved in the cholesterol regulatory pathway, such as SREBP2 and its targets. In the long term, this short-term protective effect governed by SREBP2 became deregulated. The consequent increase in the membrane levels of cholesterol may be responsible for sperm abnormalities [72].

3.2. Dietary Carbohydrates

The role of dietary carbohydrates on sperm quality is an aspect still largely unexplored. Sugar is present in almost all fruits and vegetables in the form of glucose and fructose, and a higher intake of fruits and vegetables is associated with improved semen parameters [2,20]. On the other hand, Chiu et al. found that sugar-sweetened beverage consumption was correlated with lower sperm motility in healthy young men [73]. On the other hand, reproductive hormone levels, as well as other sperm quality parameters, remained unaffected.

A link between sugar intake and lower sperm motility may be found in the increase of insulin resistance, which corresponds to a scarce utilization of glucose by sperm [31]. In sperm cells, glucose is the main substrate for glycolysis, where it is metabolized to pyruvate and/or lactate to obtain cellular energy in the form of ATP. Thus, s reduction in sperm glucose uptake and metabolism may correspond to a decrease in ATP concentration, which is necessary to sustain sperm motility.

Increased blood glucose levels were also accompanied by a decrease in testosterone levels and an increase in oxidative stress [74]. Sperm mitochondria are a common target of oxidative stress and testosterone levels, which decreased their functionality. The reduced mitochondrial respiratory efficiency might be responsible for the decrease in sperm motility [41].

Very recently, the reduction in sperm motility observed after the administration of a high-sugar diet was also linked to an alteration of human sperm small RNA profiles [75]. In fact, mature sperm cells have a rich and diverse profile of small RNA, which is determined during the testicular phases of their development and displays a considerable plasticity in response to environmental insults [76,77]. Therefore, small RNA remodeling during post-testicular maturation of mammalian sperm has an essential role in the production of functionally mature spermatozoa [78].

Experiments carried out in rats suggested that a high fructose intake starting at juvenile age can impair the reproductive function. A high fructose intake lowered serum testosterone and sperm count. A possible effect on sperm motility was also proposed [79].

Although complex carbohydrates have been shown to offer multiple advantages in terms of overall health, there are no studies about their role in the preservation of male reproductive potential.

Instead, some studies evaluated the effects on male fertility of artificial sweeteners, that have become increasingly popular as an alternative to sugar. In this context, some animal studies suggest that they may not be safer for male fertility than real sugar. For example, a recent study carried out in mice showed that high doses of aspartame correlated with sperm DNA fragmentation and morphologic defects. These effects were due to the increased production of reactive species, weakening the antioxidant defense system, and the consequent induction of oxidative stresses [80]. Stevia, the natural sweetener native to South America, was also associated with decreased sperm count and lowered testosterone level [81]. On the other hand, it was reported that the most common sucralose does not impair sperm quality and has no effect on sperm glycolysis [82].

3.3. Dietary Proteins

Proteins are not energetic substrates for sperm cells [32,33]; moreover, a high-protein diet had no significant effect on glycemic control [83]. However, a low-protein diet has been considered a potential risk factor for male-factor infertility, causing a significant reduction in testis, epididymis, and seminal vesicle weights, as well as a decrease in serum testosterone [84]. Conversely, the reports on the effect of a high-protein diet present in the literature are contradictory [85].

Therefore, in addition to the level of dietary protein intake, also the type of protein is of significance, as amino acid profiles vary depending on the protein source. For example, plant-based proteins have lower sulfur-containing amino acids (methionine and cysteine) compared to animal proteins. Methionine, cysteine, and phenylalanine can affect sperm quality by decreasing their progressive motility in vitro [86].

A study conducted in monkeys evaluated the influence of animal and plant protein diets on sperm quality. When compared to monkeys receiving a plant-based diet, monkeys fed an animal protein diet showed lower sperm counts and motility and increased sperm abnormalities [87].

3.4. General Aspects concerning Caloric Nutrients and Sperm Metabolism

A wide spectrum of exogenous factors, including caloric nutrients, affect sperm quality and function by acting on sperm energetic metabolism. In fact, since sperm rely on energetic substrates from their microenvironment to fuel their metabolism, it is easy to envision that all temporal changes in nutrient flux are directly reflected in sperm metabolism.

As reported in Table 1, oxidative stress and testosterone levels are the main players of this nutritional modulation and are strictly linked. In fact, low testosterone levels could be the result of mitochondrial defects caused by an excess of ROS in Leydig cells, where testosterone is synthesized. At the same time, a decrease in testosterone level could be a physiological response to reduced oxidative stress, since ROS are produced during steroidogenesis itself.

Table 1. Effects of nutrients on sperm quality.

	Sperm Quality	Molecular Mechanism	References
Cholesterol	↓	↓ membrane fluidity ↓ testosterone synthesis ↑ oxidative stress	[55,71]
SFA	↓	↑ insulin resistance ↓ sperm mitochondrial function ↑ oxidative stress	[4]
MUFA	↑	↑ membrane fluidity ↑ sperm mitochondrial function ↓ oxidative stress	[5,7,61]
PUFA	↑	↓ insulin resistance ↓ lipogenesis ↑ membrane fluidity ↑ sperm mitochondrial function ↓ oxidative stress	[5,57]
Carbohydrates	↓ (high sugar intake)	↑ insulin resistance ↓ testosterone synthesis ↑ oxidative stress small RNA profiles	[74,75,79]
Proteins	↓ (low protein intake)	↓ testosterone synthesis	[84]

Gamete mitochondria are a common target of oxidative stress and testosterone levels and an important source of ROS. These organelles play a key role in sperm functionality [32,88,89] and can be considered a hub of cellular events related to energy production, ROS homeostasis, and steroid hormone biosynthesis. Therefore, all molecules that can influence this crosstalk may affect male fertility by targeting gamete mitochondria [16,54,90].

4. Antioxidants Impacts on Molecular Aspects Related to Sperm Quality

It has been shown that many dietary natural polyphenols (mainly flavonoids) isolated from fruits, vegetables, and edible plants modulate mitochondrial metabolism and biogenesis, as well as ROS homeostasis [49–51]. The modulation of mitochondrial function by these plant bioactive molecules may be important for the improvement of male reproductive performance.

In this context, it is important to underline that the mean polyphenol intake in the European population of 0.5–0.8 mg/day corresponds to polyphenol plasma concentrations of about 10 nM. For vegetarians and vegans, the mean polyphenol intake is 22.4 mg/day, and polyphenol plasma concentrations were estimated to be over 200 nM [91].

Quercetin is a dietary-derived bioflavonoid widely distributed in plants and vegetables, which has attracted considerable attention in the field of the study of male fertility owing to its potent antioxidant properties. However, controversial reports exist in the literature highlighting the antioxidant as well as the prooxidant properties of this flavonoid, leading to the "quercetin paradox in male reproductive dysfunction" [92]. The conflicting biological effects may be explained by a biphasic concentration-dependent response of sperm cells to quercetin. It has been recently demonstrated that quercetin stimulated the active state of mitochondrial respiration at concentrations of 0.1–1000 nM, also causing the uncoupling between electron transport and ATP synthesis in a dose-dependent manner starting from concentrations of 10 nM [16]. At the molecular level, quercetin interacts directly with mitochondrial membranes at the coenzyme Q-binding site, suppressing superoxide generation and stimulating the production of ATP [93–95]. At higher concentrations, quercetin can interact with lipid bilayers and membrane proteins, influencing the electric properties of mitochondrial membranes and uncoupling mitochondrial respiration from ATP synthesis [96].

Resveratrol is one of the most investigated natural polyphenolic compounds contained in several (more than 70) types of plants and in red wine. Some studies reported that resveratrol improves semen quality in humans, acting as a regulator of male reproductive function [97,98]. The effects of resveratrol on mitochondrial function have been investigated in different experimental models, and it has been demonstrated that it possesses antioxidant properties at low concentrations, while at high concentrations, its pro-oxidant properties could be responsible for detrimental effects on sperm mitochondria [16,99]. According to this hypothesis, it has been recently demonstrated that, starting from a concentration of 10 nM, resveratrol significantly uncoupled mitochondrial oxidative phosphorylation [16]. The investigation of the molecular mechanisms showed that resveratrol acts on mitochondrial metabolism via a sirtuin-dependent mechanism [50,75]. In this scenario, the positive effects of resveratrol on the control of mitochondrial metabolism have also been shown in metabolic disorders such as diabetes [100].

An hormetic effect on sperm quality has also been observed for naringenin, a flavanone commonly available in tomatoes, bergamot, and citrus fruits, that received some attention in the field of male reproduction for its antioxidant properties. Other molecules of plant origin that have been studied for their effects on human spermatozoa are apigenin, luteolin, and genistein. All these molecules, as well as quercetin and resveratrol, have also been recognized to display estrogenic activity and are also commonly known as phytoestrogens. Therefore, as multi-functional endocrine disruptors, they interfere with the enzymes needed for steroid biosynthesis and/or degradation [101]. This is an interesting aspect, since soy and soy-derived products, which have become more widely adopted by some vegan and vegetarian diets, contain isoflavones that mimic the actions of estrogens and may exert adverse effects on male fertility [102,103].

Among plant antioxidant molecules, lycopene is a lipophilic reddish carotenoid frequently found in tomatoes and several red fruits, with known antioxidant and free-radical scavenging activities [90]. It has been demonstrated that lycopene has positive effects on testicular mitochondrial function since it is a modulator of lipid peroxidation, antioxidant enzyme activities, and activity of the Krebs cycle [104]. This evidence suggests that lycopene may ameliorate mitochondrial respiration efficiency, and then sperm quality, in situations in which increased mitochondrial membrane lipid peroxidation might also contribute to impair mitochondrial functionality.

Astaxanthin is a xanthophyll carotenoid present in various microorganisms and marine organisms, with strong antioxidant properties, since it is 100-fold to 500-fold more effective than vitamin E in preventing lipid peroxidation [50,105]. A positive effect of astaxanthin on sperm parameters and fertility has been proposed [106–108], whose molecular basis can be explained by the improvement of mitochondrial function. In fact, astaxanthin appears to be able to increase mitochondrial membrane potential and respiratory control [109], which are important measures of mitochondrial functionality.

The main antioxidant vitamins that help regulate free radicals in male reproduction are Vitamin E and Vitamin C [90,110]. Vitamin E is not synthesized by mammalians and includes a group of lipid-soluble compounds—tocopherols and tocotrienols—with alpha-tocopherol being the most active form. Vitamin E is found in plant-based oils, nuts, seeds, fruits, and vegetables and acts as an antioxidant, defending the organism against oxidative stress, thus having an important role in the protection of sperm membranes against ROS and lipid peroxidation. Therefore, it has been suggested that Vitamin E improves mitochondria activity, decreasing the damage to sperm and mitochondria [90,111].

Differently from Vitamin E, Vitamin C (ascorbic acid) is a very potent water-soluble antioxidant molecule, which has also a role in vitamin E recycle. It is found in many fresh fruits like oranges, lemons, limes, grapefruit, cantaloupes, mangoes, papayas, and their juices. Various studies have been carried out on the effects of ascorbic acid supplementation on sperm function, showing that the administration of vitamin C improves seminal quality [112–114]. The mechanism responsible for this effect involves a reduction of oxidative stress and an improvement of the antioxidant status, which change the microenvironment of the testes and enhance the production of energy needed for sperm motility. Molecular targets of Vitamin C are two mitochondrial enzymes, succinate dehydrogenase and ATPase [115].

5. Conclusions

Diet may be an important modifiable determinant of male reproductive potential. Therefore, the role of daily nutrient exposure needs to be highlighted to preserve male fertility or to prevent male infertility. A strong adherence to a healthy dietary pattern based mainly on plant foods and fish is positively correlated with indicators of sperm quality.

Although the picture of the complex relationship between nutrients and sperm quality is far from complete, some indications can be drawn. First, the amount and quality of the nutrients introduced can affect sperm quality by acting on sperm energetic metabolism. Then, diets rich in SFA and low in PUFA or with an unbalanced omega-6/omega-3 PUFA ratio negatively affect sperm quality, whereas dietary unsaturated fatty acid supplementation ameliorates sperm quality. While an excess of simple carbohydrates negatively affects sperm function, there are no studies about the role of complex carbohydrates on male reproductive potential. Lastly, a low-protein diet, as well as the deficiency of some specific amino acids have been considered a potential risk factor for male-factor infertility.

Fats, carbohydrates, and proteins affect sperm quality by acting on oxidative stress and testosterone levels, whose common target are the mitochondria. The mitochondria are key organelle supporting several sperm functions. Since they are involved in energy production, ROS homeostasis, and steroid hormone biosynthesis, all molecules that can influence their functions may affect male fertility. Among these molecules, dietary fatty acids and natural polyphenols act as modulators of sperm mitochondrial function.

In fact, the administration of PUFA, especially omega-3 PUFA, determined an increase in the activities of mitochondrial enzymes involved in gamete energetic metabolism and a reduction in oxidative damage. Moreover, many dietary natural polyphenols (mainly flavonoids) found in fruits and vegetables differentially affect (positively or negatively) the mitochondrial function, depending on their concentration. Therefore, the modulation of sperm mitochondrial function could play a key role in the improvement of sperm quality (Figure 3).

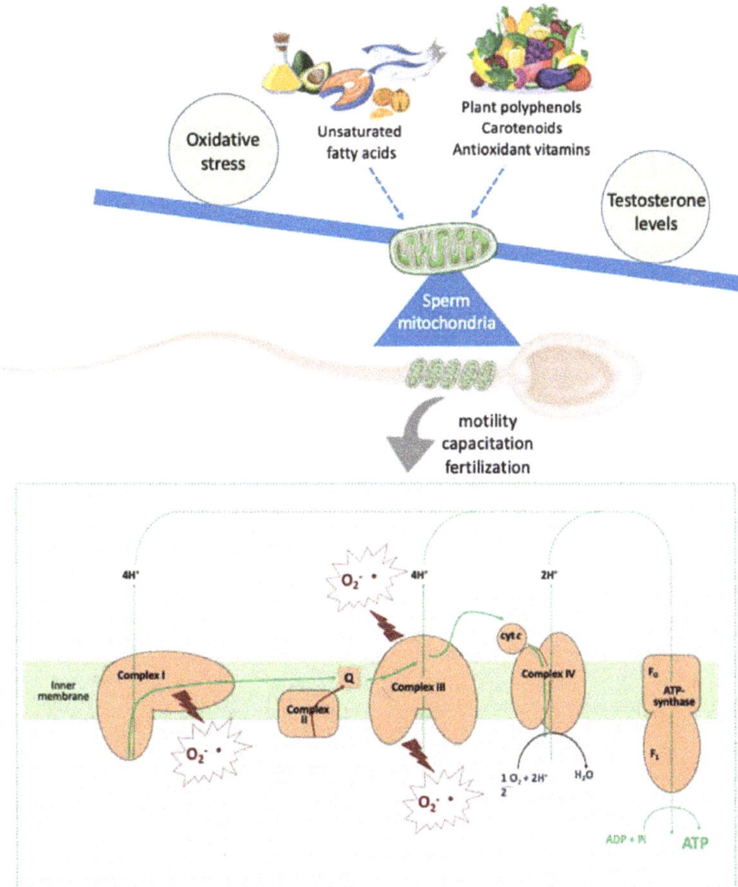

Figure 3. Dietary modulators of sperm mitochondrial function. Nutrients are modulators of oxidative stress and testosterone levels, which are strictly linked to sperm mitochondrial function, a key element related to sperm quality. In fact, in addition to their basic role in ATP synthesis, mitochondria are a major source of ROS, which are key mediators of cellular physiology and pathology. Fatty acids and several plant antioxidant molecules can target the mitochondria, improving and/or restoring their function, by acting on the ATP and ROS levels. These molecules can regulate sperm motility, capacitation, and fertilization, thus affecting sperm quality.

We are aware that our analysis provides only a small contribution to the field of nutrition and male reproduction. However, due to the importance of the role of diet on male infertility, whose frequency is increasing exponentially today, we believe that further investigation of the molecular mechanisms underlying the action of nutrients and natural compounds is necessary to develop new dietary approaches to preserve male reproductive potential.

Author Contributions: A.F. and V.Z. collected the data, drafted, reviewed, and edited the manuscript. All authors have read and agreed to the published version of the manuscript.

Funding: This research received no external funding.

Institutional Review Board Statement: Not applicable.

Informed Consent Statement: Not applicable.

Conflicts of Interest: The authors declare no conflict of interest.

References

1. Arab, A.; Rafie, N.; Mansourian, M.; Miraghajani, M.; Hajianfar, H. Dietary patterns and semen quality: A systematic review and meta-analysis of observational studies. *Andrology* **2018**, *6*, 20–28. [CrossRef] [PubMed]
2. Nassan, F.L.; Chavarro, J.E.; Tanrikut, C. Diet and men's fertility: Does diet affect sperm quality? *Fertil. Steril.* **2018**, *110*, 570–577. [CrossRef] [PubMed]
3. Skoracka, K.; Eder, P.; Łykowska-Szuber, L.; Dobrowolska, A.; Krela-Kaźmierczak, I. Diet and nutritional factors in male (in)fertility—Underestimated factors. *J. Clin. Med.* **2020**, *9*, 1400. [CrossRef] [PubMed]
4. Ferramosca, A.; Conte, A.; Moscatelli, N.; Zara, V. A high-fat diet negatively affects rat sperm mitochondrial respiration. *Andrology* **2016**, *4*, 520–525. [CrossRef]
5. Ferramosca, A.; Moscatelli, N.; Di Giacomo, M.; Zara, V. Dietary fatty acids influence sperm quality and function. *Andrology* **2017**, *5*, 423–430. [CrossRef]
6. Varani, J. Healthful eating, the western style diet and chronic disease. *Approaches Poult. Dairy Vet. Sci.* **2017**, *1*, 3. [CrossRef]
7. Ferramosca, A.; Di Giacomo, M.; Moscatelli, N.; Zara, V. Obesity and male infertility: Role of fatty acids in the modulation of sperm energetic metabolism. *Eur. J. Lipid Sci. Technol.* **2018**, *120*, 1700451. [CrossRef]
8. Danielewicz, A.; Przybyłowicz, K.E.; Przybyłowicz, M. Dietary patterns and poor semen quality risk in men: A cross-sectional study. *Nutrients* **2018**, *10*, 1162. [CrossRef]
9. Molaie, S.; Shahverdi, A.; Sharafi, M.; Shahhoseini, M.; Ghaleno, L.R.; Esmaeili, V.; Abed-Heydari, E.; Bucak, M.N.; Alizadeh, A. Dietary trans and saturated fatty acids effects on semen quality, hormonal levels and expression of genes related to steroid metabolism in mouse adipose tissue. *Andrologia* **2019**, *51*, e13259. [CrossRef]
10. Merino, O.; Sánchez, R.; Gregorio, M.B.; Sampaio, F.; Risopatrón, J. Effect of high-fat and vitamin D deficient diet on rat sperm quality and fertility. *Theriogenology* **2019**, *125*, 6–11. [CrossRef]
11. Suliga, E.; Gluszek, S. The relationship between diet, energy balance and fertility in men. *Int. J. Vitam. Nutr. Res.* **2020**, *90*, 514–526. [CrossRef] [PubMed]
12. Giahi, L.; Mohammadmoradi, S.; Javidan, A.; Sadeghi, M.R. Nutritional modifications in male infertility: A systematic review covering 2 decades. *Nutr. Rev.* **2015**, *74*, 118–130. [CrossRef] [PubMed]
13. Salas-Huetosk, A.; Bullo, M.; Salas-Salvado, J. Dietary patterns, foods and nutrients in male fertility parameters and fecunda-bility: A systematic review of observational studies. *Hum. Reprod. Update* **2017**, *23*, 371–389. [CrossRef] [PubMed]
14. Ricci, E.; Al-Beitawi, S.; Cipriani, S.; Alteri, A.; Chiaffarino, F.; Candiani, M.; Gerli, S.; Viganó, P.; Parazzini, F. Dietary habits and semen parameters: A systematic narrative review. *Andrology* **2018**, *6*, 104–116. [CrossRef]
15. De Cosmi, V.; Parazzini, F.; Agostoni, C.; Noli, S.; Cipriani, S.; La Vecchia, I.; Ferrari, S.; Esposito, G.; Bravi, F.; Ricci, E. Antioxidant vitamins and carotenoids intake and the association with poor semen quality: A cross-sectional analysis of men referring to an Italian fertility clinic. *Front Nutr.* **2021**, *8*, 737077. [CrossRef]
16. Ferramosca, A.; Lorenzetti, S.; Di Giacomo, M.; Lunetti, P.; Murrieri, F.; Capobianco, L.; Dolce, V.; Coppola, L.; Zara, V. Modulation of human sperm mitochondrial respiration efficiency by plant polyphenols. *Antioxidants* **2021**, *10*, 217. [CrossRef]
17. Talebi, S.; Arab, A.; Sorraya, N. The association between dietary antioxidants and semen parameters: A cross-sectional study among Iranian infertile men. *Biol. Trace Element Res.* **2021**, 1–8. [CrossRef]
18. Jurewicz, J.; Radwan, M.; Sobala, W.; Radwan, P.; Bochenek, M.; Hanke, W. Dietary patterns and their relationship with semen quality. *Am. J. Men Health* **2018**, *12*, 575–583. [CrossRef]
19. Guasch-Ferré, M.; Willett, W.C. The Mediterranean diet and health: A comprehensive overview. *J. Intern. Med.* **2021**, *290*, 549–566. [CrossRef]
20. Karayiannis, D.; Kontogianni, M.D.; Mendorou, C.; Douka, L.; Mastrominas, M.; Yiannakouris, N. Association between ad-herence to the Mediterranean diet and semen quality parameters in male partners of couples attempting fertility. *Hum. Reprod.* **2017**, *32*, 215–222.
21. Ricci, E.; Bravi, F.; Noli, S.; Ferrari, S.; De Cosmi, V.; La Vecchia, I.; Cavadini, M.; La Vecchia, C.; Parazzini, F. Mediterranean diet and the risk of poor semen quality: Cross-sectional analysis of men referring to an Italian fertility clinic. *Andrology* **2019**, *7*, 156–162. [CrossRef] [PubMed]

22. Caruso, P.; Caputo, M.; Cirillo, P.; Scappaticcio, L.; Longo, M.; Maiorino, M.I.; Bellastella, G.; Esposito, K. Effects of Mediter-ranean diet on semen parameters in healthy young adults: A randomized controlled trial. *Minerva Endocrinol.* **2020**, *45*, 280–287. [PubMed]
23. West, M.C.L.; Anderson, L.; Mcclure, N.; Lewis, S.E.M. Dietary oestrogens and male fertility potential. *Hum. Fertil.* **2005**, *8*, 197–207. [CrossRef] [PubMed]
24. Liu, Z.H.; Kanjo, Y.; Mizutani, S. A review of phytoestrogens: Their occurrence and fate in the environment. *Water Res.* **2010**, *44*, 567–577. [CrossRef]
25. Orzylowska, E.M.; Jacobson, J.D.; Bareh, G.M.; Ko, E.Y.; Corselli, J.U.; Chan, P.J. Food intake diet and sperm characteristics in a blue zone: A Loma Linda study. *Eur. J. Obstet. Gynecol. Reprod. Biol.* **2016**, *203*, 112–115. [CrossRef]
26. Cordain, L.; Eaton, S.B.; Sebastian, A.; Mann, N.; Lindeberg, S.; Watkins, B.A.; O'Keefe, J.H.; Brand-Miller, J. Origins and evolution of the Western diet: Health implications for the 21st century. *Am. J. Clin. Nutr.* **2005**, *81*, 341–354. [CrossRef]
27. Mihalca, R.; Fica, S. The impact of obesity on the male reproductive axis. *J. Med. Life* **2014**, *7*, 296–300.
28. Bhasin, S.; Cunningham, G.R.; Hayes, F.J.; Matsumoto, A.M.; Snyder, P.J.; Swerdloff, R.S.; Montori, V.M. Testosterone therapy in men with androgen deficiency syndromes: An Endocrine Society clinical practice guideline. *J. Clin. Endocrinol. Metab.* **2010**, *95*, 2536–2559. [CrossRef]
29. Bachir, B.G.; Jarvi, K. Infectious, inflammatory, and immunologic conditions resulting in male infertility. *Urol. Clin. N. Am.* **2014**, *41*, 67–81. [CrossRef]
30. Morielli, T.; O'Flaherty, C. Oxidative stress impairs function and increases redox protein modifications in human spermatozoa. *Reproduction* **2015**, *149*, 113–123. [CrossRef]
31. Dias, T.R.; Alves, M.G.; Silva, B.M.; Oliveira, P.F. Sperm glucose transport and metabolism in diabetic individuals. *Mol. Cell. Endocrinol.* **2014**, *396*, 37–45. [CrossRef] [PubMed]
32. Piomboni, P.; Focarelli, R.; Stendardi, A.; Ferramosca, A.; Zara, V. The role of mitochondria in energy production for human sperm motility. *Int. J. Androl.* **2012**, *35*, 109–124. [CrossRef] [PubMed]
33. Ferramosca, A.; Zara, V. Bioenergetics of mammalian sperm capacitation. *BioMed Res. Int.* **2014**, *2014*, 902953. [CrossRef] [PubMed]
34. Seethalakshmi, L.; Menon, M.; Diamond, D. The effect of streptozotocin-induced diabetes on the neuroendocrine-male reproductive tract axis of the adult rat. *J. Urol.* **1987**, *138*, 190–194. [CrossRef]
35. Hassan, A.A.; Hassouna, M.M.; Taketo, T.; Gagnon, C.; Elhilali, M.M. The effect of diabetes on sexual behavior and repro-ductive tract function in male rats. *J. Urol.* **1993**, *149*, 148–154. [CrossRef]
36. Leisegang, K.; Bouic, P.J.; Menkveld, R.; Henkel, R.R. Obesity is associated with increased seminal insulin and leptin alongside reduced fertility parameters in a controlled male cohort. *Reprod. Biol. Endocrinol.* **2014**, *12*, 34. [CrossRef]
37. Chianese, R.; Pierantoni, R. Mitochondrial Reactive Oxygen Species (ROS) production alters sperm quality. *Antioxidants* **2021**, *10*, 92. [CrossRef]
38. Zhao, J.; Zhai, L.; Liu, Z.; Wu, S.; Xu, L. Leptin level and oxidative stress contribute to obesity-induced low testosterone in murine testicular tissue. *Oxid. Med. Cell. Longev.* **2014**, *2014*, 190945. [CrossRef]
39. Rovira-Llopis, S.; Bañuls, C.; de Marañon, A.M.; Diaz-Morales, N.; Jover, A.; Garzon, S.; Rocha, M.; Victor, V.M.; Hernandez-Mijares, A. Low testosterone levels are related to oxidative stress, mitochondrial dysfunction and altered subclinical atherosclerotic markers in type 2 diabetic male patients. *Free Radic. Biol. Med.* **2017**, *108*, 155–162. [CrossRef]
40. Lunetti, P.; Capobianco, L.; Zara, V.; Ferramosca, A. Physical activity and male reproductive function: A new role for gamete mitochondria. *Exerc. Sport Sci. Rev.* **2021**, *49*, 99–106. [CrossRef]
41. Ferramosca, A.; Provenzano, S.P.; Montagna, D.D.; Coppola, L.; Zara, V. Oxidative stress negatively affects human sperm mitochondrial respiration. *Urology* **2013**, *82*, 78–83. [CrossRef] [PubMed]
42. Saez Lancellotti, T.E.; Boarelli, P.V.; Monclus, M.A.; Cabrillana, M.E.; Clementi, M.A.; Espínola, L.S.; Cid Barría, J.L.; Vincenti, A.E.; Santi, A.G.; Fornés, M.W. Hypercholesterolemia impaired sperm functionality in rabbits. *PLoS ONE* **2010**, *5*, e13457. [CrossRef] [PubMed]
43. Bobjer, J.; Naumovska, M.; Giwercman, Y.L.; Giwercman, A. High prevalence of androgen deficiency and abnormal lipid profile in infertile men with non-obstructive azoospermia. *Int. J. Androl.* **2012**, *35*, 688–694. [CrossRef] [PubMed]
44. Hagiuda, J.; Ishikawa, H.; Furuuchi, T.; Hanawa, Y.; Marumo, K. Relationship between dyslipidaemia and semen quality and serum sex hormone levels: An infertility study of 167 Japanese patients. *Andrologia* **2014**, *46*, 131–135. [CrossRef]
45. Cutillas-Tolín, A.; Mínguez-Alarcón, L.; Mendiola, J.; López-Espín, J.J.; Jørgensen, N.; Navarrete-Muñoz, E.M.; Torres-Cantero, A.M.; Chavarro, J.E. Mediterranean and western dietary patterns are related to markers of testicular function among healthy men. *Hum. Reprod.* **2015**, *30*, 2945–2955. [CrossRef]
46. Mínguez-Alarcón, L.; Mendiola, J.; López-Espín, J.J.; Sarabia-Cos, L.; Vivero-Salmerón, G.; Vioque, J.; Navarrete-Muñoz, E.M.; Torres-Cantero, A.M. Dietary intake of antioxidant nutrients is associated with semen quality in young university students. *Hum. Reprod.* **2012**, *27*, 2807–2814. [CrossRef]
47. Braga, D.P.; Halpern, G.; Rita de Cássia, S.F.; Setti, A.S.; Iaconelli, A., Jr.; Borges, E., Jr. Food intake and social habits in male patients and its relationship to intracytoplasmic sperm injection outcomes. *Fertil. Steril.* **2012**, *97*, 53–59. [CrossRef]
48. Eslamian, G.; Amirjannati, N.; Rashidkhani, B.; Sadeghi, M.R.; Hekmatdoost, A. Intake of food groups and idiopathic asthenozoospermia: A case-control study. *Hum. Reprod.* **2012**, *27*, 3328–3336. [CrossRef]

49. Forbes-Hernández, T.Y.; Giampieri, F.; Gasparrini, M.; Mazzoni, L.; Quiles, J.L.; Alvarez-Suarez, J.M.; Battino, M. The effects of bioactive compounds from plant foods on mitochondrial function: A focus on apoptotic mechanisms. *Food Chem. Toxicol.* **2014**, *68*, 154–182. [CrossRef]
50. Ferramosca, A.; Di Giacomo, M.; Zara, V. Antioxidant dietary approach in treatment of fatty liver: New insights and updates. *World J. Gastroenterol.* **2017**, *23*, 4146–4157. [CrossRef]
51. Di Giacomo, M.; Zara, V.; Bergamo, P.; Ferramosca, A. Crosstalk between mitochondrial metabolism and oxidoreductive homeostasis: A new perspective for understanding the effects of bioactive dietary compounds. *Nutr. Res. Rev.* **2020**, *33*, 90–101. [CrossRef] [PubMed]
52. McEvoy, C.T.; Temple, N.; Woodside, J.V. Vegetarian diets, low-meat diets and health: A review. *Public Health Nutr.* **2012**, *15*, 2287–2294. [CrossRef] [PubMed]
53. Rozati, R.; Reddy, P.P.; Reddanna, P.; Mujtaba, R. Role of environmental estrogens in the deterioration of male factor fertility. *Fertil. Steril.* **2002**, *78*, 1187–1194. [CrossRef]
54. Ferramosca, A.; Lorenzetti, S.; Di Giacomo, M.; Murrieri, F.; Coppola, L.; Zara, V. Herbicides glyphosate and glufosinate ammonium negatively affect human sperm mitochondria respiration efficiency. *Reprod. Toxicol.* **2020**, *99*, 48–55. [CrossRef] [PubMed]
55. Saez, F.; Drevet, J.R. Dietary cholesterol and lipid overload: Impact on male fertility. *Oxid. Med. Cell. Longev.* **2019**, *2019*, 4521786. [CrossRef]
56. Amaral, A.; Castillo, J.; Estanyol, J.M.; Ballescà, J.L.; Ramalho-Santos, J.; Oliva, R. Human sperm tail proteome suggests new endogenous metabolic pathways. *Mol. Cell. Proteom.* **2013**, *12*, 330–342. [CrossRef] [PubMed]
57. Collodel, G.; Castellini, C.; Lee, J.C.-Y.; Signorini, C. Relevance of fatty acids to sperm maturation and quality. *Oxid. Med. Cell. Longev.* **2020**, *2020*, 7038124. [CrossRef]
58. Albracht-Schulte, K.; Kalupahana, N.S.; Ramalingam, L.; Wang, S.; Rahman, S.M.; Robert-McComb, J.; Moustaid-Moussa, N. Omega-3 fatty acids in obesity and metabolic syndrome: A mechanistic update. *J. Nutr. Biochem.* **2018**, *58*, 1–16. [CrossRef]
59. Burgos, C.; Maldonado, C.; Gerez de Burgos, N.M.; Aoki, A.; Blanco, A.; Si, Y.; Okuno, M. Intracellular localization of the testicular and sperm-specific lactate dehydrogenase isozyme C_4 in mice. *Biol. Reprod.* **1995**, *53*, 84–92. [CrossRef]
60. Ferramosca, A.; Conte, A.; Guerra, F.; Felline, S.; Rimoli, M.G.; Mollo, E.; Zara, V.; Terlizzi, A. Metabolites from invasive pests inhibit mitochondrial complex II: A potential strategy for the treatment of human ovarian carcinoma? *Biochem. Biophys. Res. Commun.* **2016**, *473*, 1133–1138. [CrossRef]
61. Mansour, S.W.; Sangi, S.; Harsha, S.; Khaleel, M.A.; Ibrahim, A.R. Sensibility of male rats fertility against olive oil, *Nigella sativa* oil and pomegranate extract. *Asian Pac. J. Trop. Biomed.* **2013**, *3*, 563–568. [CrossRef]
62. Al-Okbi, S.Y.; Mohamed, D.A.; Hamed, T.E.; Esmail, R.S.; Donya, S.M. Prevention of renal dysfunction by nutraceuticals prepared from oil rich plant foods. *Asian Pac. J. Trop. Biomed.* **2014**, *4*, 618–626. [CrossRef] [PubMed]
63. Diaz-Fontdevila, M.; Bustos-Obregon, E. Cholesterol and polyunsaturated acid enriched diet: Effect on kinetics of the acrosome reaction in rabbit spermatozoa. *Mol. Reprod. Dev.* **1993**, *35*, 176–180. [CrossRef] [PubMed]
64. Kolodgie, F.D.; Katocs, A.S., Jr.; Largis, E.E.; Wrenn, S.M.; Cornhill, J.F.; Herderick, E.E.; Lee, S.J.; Virmani, R. Hypercholesterolemia in the rabbit induced by feeding graded amounts of low-level cholesterol. Methodological considerations regarding individual variability in response to dietary cholesterol and development of lesion type. *Arterioscler. Thromb. Vasc. Biol.* **1996**, *16*, 1454–1464. [CrossRef] [PubMed]
65. Diaz-Fontdevila, M.; Peña, W.; Bustos-Obregón, E. Experimental hypercholesterolaemia in rabbits. Effect on lipid domains in homologous spermatozoa. *Andrologia* **1998**, *30*, 15–22. [CrossRef] [PubMed]
66. Fan, J.; Kitajima, S.; Watanabe, T.; Xu, J.; Zhang, J.; Liu, E.; Chen, Y.E. Rabbit models for the study of human atherosclerosis: From pathophysiological mechanisms to translational medicine. *Pharmacol. Ther.* **2015**, *146*, 104–119. [CrossRef]
67. Simón, L.; Funes, A.K.; Yapur, M.A.; Cabrillana, M.E.; Monclus, M.A.; Boarelli, P.V.; Vincenti, A.E.; Saez Lancellotti, T.E.; Fornés, M.W. Manchette-acrosome disorders during spermiogenesis and low efficiency of seminiferous tubules in hypercholesterolemic rabbit model. *PLoS ONE.* **2017**, *12*, e0172994. [CrossRef]
68. Fan, Y.; Liu, Y.; Xue, K.; Gu, G.; Fan, W.; Xu, Y.; Ding, Z. Diet-induced obesity in male C57BL/6 mice decreases fertility as a consequence of disrupted blood-testis barrier. *PLoS ONE* **2015**, *10*, e0120775. [CrossRef]
69. Borges, B.C.; Garcia-Galiano, D.; da Silveira Cruz-Machado, S.; Han, X.; Gavrilina, G.B.; Saunders, T.L.; Auchus, R.J.; Hammoud, S.S.; Smith, G.D.; Elias, C.F. Obesity-induced infertility in male mice is associated with disruption of *Crisp4* expression and sperm fertilization capacity. *Endocrinology* **2017**, *158*, 2930–2943. [CrossRef]
70. Whitfield, M.; Guiton, R.; Rispal, J.; Acar, N.; Kocer, A.; Drevet, J.R.; Saez, F. Dyslipidemia alters sperm maturation and ca-pacitation in LXR-null mice. *Reproduction* **2017**, *154*, 827–842. [CrossRef]
71. Yu, C.; Jiang, F.; Zhang, M.; Luo, D.; Shao, S.; Zhao, J.; Gao, L.; Zuo, C.; Guan, Q. HC diet inhibited testosterone synthesis by activating endoplasmic reticulum stress in testicular Leydig cells. *J. Cell. Mol. Med.* **2019**, *23*, 3140–3150. [CrossRef] [PubMed]
72. Funes, A.K.; Simón, L.; Colombo, R.; Avena, M.V.; Monclús, M.; Crescitelli, J.; Cabrillana, M.E.; Conte, M.I.; Cayado, N.; Boarelli, P.; et al. Impact of high fat diet on the sterol regulatory element-binding protein 2 cholesterol pathway in the testicle. *Mol. Hum. Reprod.* **2021**, *27*, gaab023. [CrossRef]

73. Chiu, Y.H.; Afeiche, M.C.; Gaskins, A.J.; Williams, P.L.; Mendiola, J.; Jørgensen, N.; Swan, S.H.; Chavarro, J.E. Sugar-sweetened beverage intake in relation to semen quality and reproductive hormone levels in young men. *Hum. Reprod.* **2014**, *29*, 1575–1584. [CrossRef] [PubMed]
74. La Vignera, S.; Condorelli, R.A.; Vicari, E.S.D.; D'Agata, R.; Calogero, A.E. Diabetes mellitus and sperm parameters. *J. Androl.* **2012**, *33*, 145–153. [CrossRef]
75. Zhang, Y.; Chen, Q. Human sperm RNA code senses dietary sugar. *Nat. Rev. Endocrinol.* **2020**, *16*, 200–201. [CrossRef] [PubMed]
76. Trigg, N.A.; Eamens, A.L.; Nixon, B. The contribution of epididymosomes to the sperm small RNA profile. *Reproduction* **2019**, *157*, R209–R223. [CrossRef] [PubMed]
77. Zhu, Q.; Kirby, J.A.; Chu, C.; Gou, L.-T. Small noncoding RNAs in reproduction and infertility. *Biomedicines* **2021**, *9*, 1884. [CrossRef]
78. Conine, C.C.; Sun, F.; Song, L.; Rivera-Pérez, J.A.; Rando, O.J. Small RNAs gained during epididymal transit of sperm are essential for embryonic development in mice. *Dev. Cell* **2018**, *46*, 470–480.e3. [CrossRef]
79. Medaglia, D.S.A.; Vieira, H.R.; da Silva Silveira, S.; Siervo, G.E.M.D.L.; da Silva Marcon, M.S.; de Freitas Mathias, P.C.; Fernandes, G.S. High-fructose diet during puberty alters the sperm parameters, testosterone concentration, and histopathology of testes and epididymis in adult Wistar rats. *J. Dev. Orig. Health Dis.* **2021**, *13*, 20–27. [CrossRef]
80. Anbara, H.; Sheibani, M.T.; Razi, M. Long-term effect of aspartame on male reproductive system: Evidence for testicular histomorphometrics, Hsp70-2 protein expression and biochemical status. *Int. J. Fertil. Steril.* **2020**, *14*, 91–101. [CrossRef]
81. Melis, M.S. Effects of chronic administration of *Stevia rebaudiana* on fertility in rats. *J. Ethnopharmacol.* **1999**, *67*, 157–161. [CrossRef]
82. Kille, J.W.; Ford, W.C.; McAnulty, P.; Tesh, J.M.; Ross, F.W.; Willoughby, C.R. Sucralose: Lack of effects on sperm glycolysis and reproduction in the rat. *Food Chem. Toxicol.* **2000**, *38*, S19–S29. [CrossRef]
83. Yu, Z.; Nan, F.; Wang, L.Y.; Jiang, H.; Chen, W.; Jiang, Y. Effects of high-protein diet on glycemic control, insulin resistance and blood pressure in type 2 diabetes: A systematic review and meta-analysis of randomized controlled trials. *Clin. Nutr.* **2020**, *39*, 1724–1734. [CrossRef] [PubMed]
84. Ajuogu, P.K.; Al-Aqbi, M.A.; Hart, R.A.; Wolden, M.; Smart, N.A.; McFarlane, J.R. The effect of dietary protein intake on factors associated with male infertility: A systematic literature review and meta-analysis of animal clinical trials in rats. *Nutr. Health* **2020**, *26*, 53–64. [CrossRef] [PubMed]
85. Kemp, B.; Grooten, H.J.G.; Den Hartog, L.A.; Luiting, P.; Verstegen, M.W.A. The effect of a high protein intake on sperm production in boars at two semen collection frequencies. *Anim. Reprod. Sci.* **1988**, *17*, 103–113. [CrossRef]
86. Lapointe, S.; Sirard, M.A. Catalase and oviductal fluid reverse the decreased motility of bovine sperm in culture medium containing specific amino acids. *J. Androl.* **1998**, *19*, 31–36.
87. Johnson, Q.; Veith, W. Effect of dietary plant and animal protein intake on sperm quality in monkeys. *Arch. Androl.* **2001**, *46*, 145–151. [CrossRef]
88. Amaral, A.; Lourenço, B.; Marques, M.; Ramalho-Santos, J. Mitochondria functionality and sperm quality. *Reproduction* **2013**, *146*, R163–R174. [CrossRef]
89. Barbagallo, F.; La Vignera, S.; Cannarella, R.; Aversa, A.; Calogero, A.E.; Condorelli, R.A. Evaluation of sperm mitochondrial function: A key organelle for sperm motility. *J. Clin. Med.* **2020**, *9*, 363. [CrossRef]
90. Amaral, S.; Tavares, R.S.; Baptista, M.; Sousa, M.I.; Silva, A.; Escada-Rebelo, S.; Paiva, C.P.; Ramalho-Santos, J. Mitochondrial functionality and chemical compound action on sperm function. *Curr. Med. Chem.* **2016**, *23*, 3575–3606. [CrossRef]
91. Zamora-Ros, R.; Knaze, V.; Lujan-Barroso, L.; Kuhnle, G.G.; Mulligan, A.A.; Touillaud, M.; Slimani, N.; Romieu, I.; Powell, N.; Tumino, R.; et al. Dietary intakes and food sources of phytoestrogens in the European Prospective Investigation into Cancer and Nutrition (EPIC) 24-hour dietary recall cohort. *Eur. J. Clin. Nutr.* **2012**, *66*, 932–941. [CrossRef] [PubMed]
92. Ranawat, P.; Pathak, C.M.; Khanduja, K.L. A new perspective on the quercetin paradox in male reproductive dysfunction. *Phytother. Res.* **2013**, *27*, 802–810. [CrossRef] [PubMed]
93. Sandoval-Acuña, C.; Lopez-Alarcón, C.; Aliaga, M.E.; Speisky, H. Inhibition of mitochondrial complex I by various non-steroidal anti-inflammatory drugs and its protection by quercetin via a coenzyme Q-like action. *Chem. Biol. Interact.* **2012**, *199*, 18–28. [CrossRef]
94. de Oliveira, M.R.; Nabavi, S.M.; Braidy, N.; Setzer, W.N.; Ahmed, T.; Nabavi, S.F. Quercetin and the mitochondria: A mechanistic view. *Biotechnol. Adv.* **2016**, *34*, 532–549. [CrossRef] [PubMed]
95. Houghton, M.J.; Kerimi, A.; Tumova, S.; Boyle, J.P.; Williamson, G. Quercetin preserves redox status and stimulates mitochondrial function in metabolically-stressed HepG2 cells. *Free Radic. Biol. Med.* **2018**, *129*, 296–309. [CrossRef] [PubMed]
96. Hendrich, A.B. Flavonoid-membrane interactions: Possible consequences for biological effects of some polyphenolic compounds. *Acta Pharmacol. Sin.* **2006**, *27*, 27–40. [CrossRef]
97. Juan, M.E.; González-Pons, E.; Munuera, T.; Ballester, J.; Rodríguez-Gil, J.E.; Planas, J.M. Trans-resveratrol, a natural antioxidant from grapes, increases sperm output in healthy rats. *J. Nutr.* **2005**, *135*, 757–760. [CrossRef]
98. Pasquariello, R.; Verdile, N.; Brevini, T.A.L.; Gandolfi, F.; Boiti, C.; Zerani, M.; Maranesi, M. The role of resveratrol in mammalian reproduction. *Molecules* **2020**, *25*, 4554. [CrossRef]
99. de Oliveira, M.R.; Nabavi, S.F.; Manayi, A.; Daglia, M.; Hajheidari, Z.; Nabavi, S.M. Resveratrol and the mitochondria: From triggering the intrinsic apoptotic pathway to inducing mitochondrial biogenesis, a mechanistic view. *Biochim. Biophys. Acta (BBA) Gen. Subj.* **2016**, *1860*, 727–745. [CrossRef]

100. Abdelali, A.; Al-Bader, M.; Kilarkaje, N. Effects of trans-resveratrol on hyperglycemia-induced abnormal spermatogenesis, DNA damage and alterations in poly (ADP-ribose) polymerase signaling in rat testis. *Toxicol. Appl. Pharmacol.* **2016**, *311*, 61–73. [CrossRef]
101. Nordeen, S.K.; Bona, B.J.; Jones, D.N.; Lambert, J.R.; Jackson, T.A. Endocrine disrupting activities of the flavonoid nutraceuticals luteolin and quercetin. *Horm. Cancer* **2013**, *4*, 293–300. [CrossRef] [PubMed]
102. Cederroth, C.R.; Auger, J.; Zimmermann, C.; Eustache, F.; Nef, S. Soy, phyto-oestrogens and male reproductive function: A review. *Int. J. Androl.* **2010**, *33*, 304–316. [CrossRef] [PubMed]
103. Rizzo, G.; Baroni, L. Soy, soy foods and their role in vegetarian diets. *Nutrients* **2018**, *10*, 43. [CrossRef] [PubMed]
104. Aly, H.A.; El-Beshbishy, H.A.; Banjar, Z.M. Mitochondrial dysfunction induced impairment of spermatogenesis in LPS-treated rats: Modulatory role of lycopene. *Eur. J. Pharmacol.* **2012**, *677*, 31–38. [CrossRef]
105. Guerin, M.; Huntley, M.E.; Olaizola, M. *Haematococcus* astaxanthin: Applications for human health and nutrition. *Trends Biotechnol.* **2003**, *21*, 210–216. [CrossRef]
106. Comhaire, F.H.; Mahmoud, A. The role of food supplements in the treatment of the infertile man. *Reprod. Biomed. Online* **2003**, *7*, 385–391. [CrossRef]
107. Comhaire, F.H.; El Garem, Y.; Mahmoud, A.; Eertmans, F.; Schoonjans, F. Combined conventional/antioxidant "Astaxanthin" treatment for male infertility: A double blind, randomized trial. *Asian J. Androl.* **2005**, *7*, 257–262. [CrossRef]
108. Dona, G.; Kožuh, I.; Brunati, A.M.; Andrisani, A.; Ambrosini, G.; Bonanni, G.; Ragazzi, E.; Armanini, D.; Clari, G.; Bordin, L. Effect of astaxanthin on human sperm capacitation. *Mar. Drugs* **2013**, *11*, 1909–1919. [CrossRef]
109. Wolf, A.M.; Asoh, S.; Hiranuma, H.; Ohsawa, I.; Iio, K.; Satou, A.; Ishikura, M.; Ohta, S. Astaxanthin protects mitochondrial redox state and functional integrity against oxidative stress. *J. Nutr. Biochem.* **2010**, *21*, 381–389. [CrossRef]
110. Sheweita, S.A.; Tilmisany, A.M.; Al-Sawaf, H. Mechanisms of male infertility: Role of antioxidants. *Curr. Drug Metab.* **2005**, *6*, 495–501. [CrossRef]
111. Yue, D.; Yan, L.; Luo, H.; Xu, X.; Jin, X. Effect of vitamin E supplementation on semen quality and the testicular cell membranal and mitochondrial antioxidant abilities in Aohan fine-wool sheep. *Anim. Reprod. Sci.* **2010**, *118*, 217–222. [CrossRef] [PubMed]
112. Yousef, M.I.; Abdallah, G.A.; Kamel, K.I. Effect of ascorbic acid and vitamin E supplementation on semen quality and bio-chemical parameters of male rabbits. *Anim. Reprod. Sci.* **2003**, *76*, 99–111. [CrossRef]
113. Cyrus, A.; Kabir, A.; Goodarzi, D.; Moghimi, M. The effect of adjuvant vitamin C after varicocele surgery on sperm quality and quantity in infertile men: A double blind placebo controlled clinical trial. *Int. Braz. J. Urol.* **2015**, *41*, 230–238. [CrossRef] [PubMed]
114. Shabanian, S.; Farahbod, F.; Rafieian, M.; Ganji, F.; Adib, A. The effects of vitamin C on sperm quality parameters in laboratory rats following long-term exposure to cyclophosphamide. *J. Adv. Pharm. Technol. Res.* **2017**, *8*, 73–79. [CrossRef] [PubMed]
115. Harikrishnan, R.; Abhilash, P.A.; Das, S.S.; Prathibha, P.; Rejitha, S.; John, F.; Kavitha, S.; Indira, M. Protective effect of ascorbic acid against ethanol-induced reproductive toxicity in male guinea pigs. *Br. J. Nutr.* **2013**, *110*, 689–698. [CrossRef] [PubMed]

International Journal of
Molecular Sciences

MDPI

Review

Gases in Sepsis: Novel Mediators and Therapeutic Targets

Zhixing Zhu [1,2], Stephen Chambers [1], Yiming Zeng [2] and Madhav Bhatia [1,*]

1 Department of Pathology and Biomedical Science, University of Otago, Christchurch 8140, New Zealand; zhixing.zhu@postgrad.otago.ac.nz (Z.Z.); steve.chambers@otago.ac.nz (S.C.)
2 Department of Internal Medicine (Pulmonary and Critical Care Medicine), The Second Clinical Medical School of Fujian Medical University, Quanzhou 362002, China; zeng_yi_ming@126.com
* Correspondence: madhav.bhatia@otago.ac.nz

Abstract: Sepsis, a potentially lethal condition resulting from failure to control the initial infection, is associated with a dysregulated host defense response to pathogens and their toxins. Sepsis remains a leading cause of morbidity, mortality and disability worldwide. The pathophysiology of sepsis is very complicated and is not yet fully understood. Worse still, the development of effective therapeutic agents is still an unmet need and a great challenge. Gases, including nitric oxide (NO), carbon monoxide (CO) and hydrogen sulfide (H_2S), are small-molecule biological mediators that are endogenously produced, mainly by enzyme-catalyzed reactions. Accumulating evidence suggests that these gaseous mediators are widely involved in the pathophysiology of sepsis. Many sepsis-associated alterations, such as the elimination of invasive pathogens, the resolution of disorganized inflammation and the preservation of the function of multiple organs and systems, are shaped by them. Increasing attention has been paid to developing therapeutic approaches targeting these molecules for sepsis/septic shock, taking advantage of the multiple actions played by NO, CO and H_2S. Several preliminary studies have identified promising therapeutic strategies for gaseous-mediator-based treatments for sepsis. In this review article, we summarize the state-of-the-art knowledge on the pathophysiology of sepsis; the metabolism and physiological function of NO, CO and H_2S; the crosstalk among these gaseous mediators; and their crucial effects on the development and progression of sepsis. In addition, we also briefly discuss the prospect of developing therapeutic interventions targeting these gaseous mediators for sepsis.

Keywords: sepsis; pathophysiology; gaseous mediators; NO; CO; H_2S; therapeutic targets

Citation: Zhu, Z.; Chambers, S.; Zeng, Y.; Bhatia, M. Gases in Sepsis: Novel Mediators and Therapeutic Targets. *Int. J. Mol. Sci.* **2022**, 23, 3669. https://doi.org/10.3390/ijms23073669

Academic Editor: Andreas von Knethen

Received: 28 February 2022
Accepted: 25 March 2022
Published: 27 March 2022

1. Introduction

Sepsis is a life-threatening organ dysfunction that arises as a consequence of the host response failure to control invading pathogens, including bacteria, viruses, fungi and protozoa, as well as their toxins and subsequent dysregulation of the immune response. Septic shock refers to a subgroup of sepsis with a higher risk of mortality due to severe circulatory failure, in addition to cellular and metabolic abnormalities [1,2]. Sepsis/septic shock develops in approximately 30 million people every year, and the reported incidence of sepsis/septic shock keeps rising, making sepsis/septic shock a major health crisis worldwide [3,4]. Notably, sepsis/septic shock is not only responsible for about 20% of deaths worldwide, but it also takes a heavy toll on the quality of life of survivors. Patients who survive sepsis/septic shock usually suffer from post-sepsis sequelae, including neurocognitive impairment and psychological deficits [2]. As a result, sepsis/septic shock has been identified as a leading cause of morbidity, mortality and disability [4–6]. As such, sepsis/septic shock has become a global health priority since 2017 [5]. However, the development of effective therapeutic agents for sepsis/septic shock is still an unmet need and a great medical and scientific challenge. In order to address this health burden, gaining an in-depth understanding of the mediators involved in sepsis/septic shock pathogenesis, together with the biomarkers and therapeutic targets of sepsis/septic shock, is urgently needed.

Gaseous mediators refer to several gaseous signaling messengers that are endogenously and enzymatically produced [7,8]. Most of their functions do not occur when they are in a gaseous state, although they are collectively termed gaseous mediators [9–13]. In recent years, gases, such as sulfur dioxide, hydrogen cyanide and methane, have emerged as potential novel gaseous mediators [8,14–16]; however, to date, only three small, chemically reactive and freely permeable molecules comprise the novel signaling gaseous mediator family, which includes nitric oxide (NO), carbon monoxide (CO) and hydrogen sulfide (H_2S) [8]. Importantly, NO, CO and H_2S harbor several features in common. Firstly, these gaseous mediators are widely distributed in various tissues and organs. In addition, the biological half-life of each gas is short, with a range from seconds to minutes. Moreover, although they are hazardous at moderate to high concentrations, they play multiple, indispensable roles in a wide range of critical cellular functions and biological processes in low amounts. Consequently, any disturbance with respect to the homeostasis of these gaseous mediators, including their metabolism and bioavailability, can profoundly affect various physiological and pathological functions in the body, thereby leading to the onset and progression of various pathological conditions [17,18]. Recently, several studies have appeared that indicate NO, CO and H_2S may be involved in the development of different stages of sepsis/septic shock [19–21]. However, the ways in which these gaseous signaling molecules affect the pathophysiology of sepsis/septic shock are not yet fully clear.

A comprehensive review focusing on summarizing and comparing the roles of these three gaseous messengers in sepsis/septic together would contribute to bridging the gaps in our knowledge. In this review, we aim to summarize the state-of-the-art knowledge on the pathophysiology of sepsis/septic shock and provide an update of the biometabolism, bioavailability and biofunction of NO, CO and H_2S, together with the potential interaction among these gaseous mediators. Moreover, we attempt to emphasize the present-day evidence that points to the potential effects of these gaseous mediators on the development and progression of sepsis/septic shock. Additionally, we seek to briefly discuss the prospect of developing therapeutic interventions for sepsis by targeting these gaseous signaling molecules.

2. Pathophysiology of Sepsis

Sepsis affects nearly every organ system in living organisms; however, the lungs, liver, kidneys, heart, central nervous system and hematologic system are found to be most vulnerable and are frequently affected in sepsis [22]. Of note, as a well-characterized hallmark of sepsis, multiple organ dysfunction syndrome serves as a major determinant of the course and outcome of sepsis [23,24]. However, the underlying mechanism by which multiple organ dysfunction syndrome occurs in sepsis is far from fully understood [25].

Many researchers have centered their focus on the involvement of a dysregulated host defense response in sepsis. A normal host defense response to infection can localize and control pathogen invasion and simultaneously initiate the repair process at the site of inflammation. However, the host immune system is profoundly perturbed in sepsis; thus, the host defense response in sepsis appears to be inappropriate and detrimental [26,27]. Inflammation plays a critical role in many other pathological conditions, such as cancer [28]. Specifically, the devastating infection caused by these invasive pathogens can lead to a multifaceted imbalance of proinflammatory response and anti-inflammatory response in the body [29,30]. To make matters worse, this destructive and uncontrolled host defense response can further result in systemic inflammatory response syndrome and cytokine storm, thereby ultimately progressing to multiple organ failure or even death [29,31]. As a result, sepsis has been treated as an exaggerated inflammation-led clinical trajectory from local microorganism infection to multiple organ dysfunction or death [22]. On the other hand, upon infection, the hypoinflammatory response/immune paralysis termed "endotoxin tolerance" is triggered simultaneously with the activation of proinflammatory response [32,33]. Due to the concomitant paralyzed immune system, the host's ability to eradicate the invading microbes is severely compromised, which opens the way for late infections to develop and leads to unresolved organ failure. Hence, a prolonged and disorganized host hypoinflammatory response/immune paralysis is also detrimental and contributes to the increased risk of death in sepsis [34,35]. On top of the extensive involvement of an overactive host defense response, the importance of microcirculation dysfunction, coagulation impairment, mitochondrial damage, endoplasmic reticulum stress, cellular alterations and organ–organ crosstalk is also becoming increasingly evident and recognized [24,25] (Figure 1).

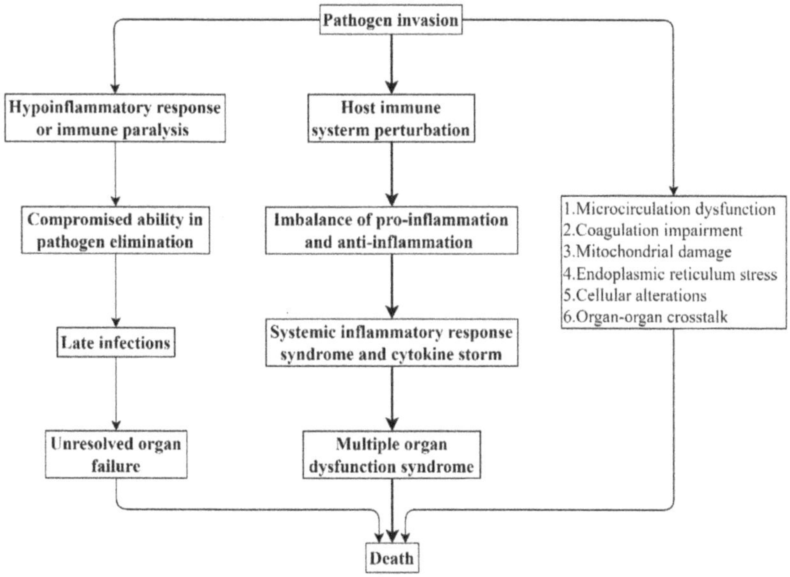

Figure 1. Pathophysiology of sepsis.

3. Gaseous Mediators

NO, CO and H_2S are known for their toxic activity, but they are believed to have contributed to the origin of life and the appearance of eukaryotic animals, which should not be ignored [13]. More importantly, NO, CO and H_2S are known to play crucial roles in diverse physiological and pathological conditions. To that end, increasing attention has been paid to these gaseous signaling mediators.

3.1. Nitric Oxide (NO)

NO, a colorless and toxic gas with a slight odor, was the first identified gaseous mediator. The endogenous synthesis of NO in eukaryotic cells is predominantly catalyzed by nitric oxide synthase (NOS) isozymes [36]. To date, three distinctive NOS isoforms comprise this enzyme family in mammals, which are neuronal NOS (nNOS), endothelial NOS (eNOS) and inducible NOS (iNOS) [37]. While the isoforms of nNOS and eNOS are calcium dependent and are constitutively expressed in cells, the iNOS isoform is calcium independent, and its expression is upregulated in response to many proinflammatory stimuli, including cytokines and lipopolysaccharide (LPS) [37,38]. More recently, the mitochondrial NOS (mtNOS) isozyme has been identified. This is also calcium dependent and constitutively expressed. Even though it is still somewhat controversial, this enzyme is emerging as the fourth member of the NOS family, since it is capable of catalyzing the synthesis of NO [39]. It is noteworthy that all these enzymes catalyze the same stepwise biological process. With the catalysis of these enzymes, L-arginine is hydroxylated and eventually converts to L-citrulline in the presence of molecular oxygen, nicotinamide adenine dinucleotide phosphate (NADP) and several cofactors, such as tetrahydrobiopterin

(BH4). From this two-step enzymatic reaction, NO is liberated [40,41]. NO can be further oxidized to stable end products, namely nitrite and nitrate. Recently, it turned out that nitrate and nitrite are not only the oxidation products of NO but also the major biological reserves of NO [42]. On top of the oxidation of L-arginine, NO can be generated by the reduction of nitrite and nitrate. This noncanonical pathway is thought to be the compensatory mechanism for NO synthesis under hypoxic and acidic circumstances [42,43]. As a gaseous free radical, NO is highly reactive [44].

NO is best known as an endothelium-derived relaxing factor due to its role in mediating the dilation of blood vessels [8,45,46]. NO is also responsible for the regulation of vascular tone, platelet–vessel interactions, blood flow, angiogenesis and heart function [45–48]. Therefore, NO is very important, as the cardiovascular system is involved in various pathological conditions, including cancer [49]. Apart from this, as a versatile cellular signaling messenger, NO exerts a wide range of functions in many other organ systems, such as the nervous system, endocrine system, respiratory system, immune system and digestive system. Specifically, NO plays a significant role in the regulation of nerve development and neurotransmission, insulin secretion, airway tone, immune response reaction, wound healing and intestinal peristalsis [36,50]. In addition, it is becoming apparent that NO also functions as an endogenous regulator in epigenetics, including DNA methylation, DNA demethylation and histone post-translational modifications [51]. Moreover, NO has been shown to counteract oxidative stress, inhibit cell death and facilitate pathogen scavenging [44]. Dysregulated metabolism of NO is central to the pathogenesis of many diseases, including, but not limited to: sepsis, cancers, hypertension, stroke, inflammation, diabetes and retinopathy [19,36,52,53].

3.2. Carbon Monoxide (CO)

CO, a colorless, poisonous and odorless gas, was recognized as a gaseous mediator following NO. The majority of endogenous CO production is catalyzed by heme oxygenase (HO) isozymes [54,55]. Similar to NOS isozymes, this enzyme family also encompasses three different isoforms, namely HO-1, HO-2 and HO-3 [55]. HO-1 and HO-2 are two common and functionally active isoforms of HO. Specifically, HO-1, also known as heat shock protein 32, is inducible and highly dynamic; however, HO-2 is constitutively expressed, and its expression is much less regulated [44,56]. Controversially, a new isoform of HO was discovered in rat tissues and was referred to as HO-3 [56]. Despite HO-3 having significant homology to HO-2 in the amino acid sequence (as high as 90%), its enzyme activity is much inferior to that of HO-2 [55,56]. CO is produced during the degradation of heme. Specifically, HO-catalyzed CO synthesis starts with the catabolism of heme. In this process, heme is firstly oxidized to α-metahydroxyheme in the presence of oxygen and NADP; afterward, α-metahydroxyheme further reacts with oxygen, from which CO is produced [57]. This HO-catalyzed heme catabolism is biologically crucial, as this reaction is responsible for around 85% of CO production in living organisms. In addition to HO, many other enzymes, including cytochrome P450 reductase, human acireductone dioxygenase and tyrosinase, also contribute to the endogenous production of CO. Compared with heme-independent CO synthesis, heme-independent CO production, namely the oxidation of phenols, the peroxidation of lipids and the photo-oxidation of organic compounds, is minor; however, these reactions are also contributory [54,55]. Although CO can be further oxidized to CO_2, the oxygenation of CO in mammals under normal conditions has not been observed. Unlike NO, the predominant catabolism of endogenous CO is either by exhalation (approximately 80%) or by binding to hemoglobin and other heme proteins.

CO is bioactive and plays a pivotal role in a variety of biological systems [58,59]. In the cardiovascular system, although it is not as potent as NO, CO also serves as a vasodilator in the body because of its role in inducing the relaxation of vascular smooth muscle. In addition, CO has been found to promote angiogenesis and inhibit the aggregation and adhesion of platelets and the activation of monocytes [44,60]. In the respiratory system, CO acts as a crucial bronchodilator, as it is essential for reversing methacholine-induced bronchoconstriction and relaxing tracheal smooth muscle [61]. In the digestive system, CO has been found to be gastroprotective due to its involvement in maintaining gastric mucosal integrity and promoting gastric ulcer healing [62]. CO also acts as a neurotransmitter due to its versatile roles in regulating the functions of the nervous system, such as neurodevelopment, long-term memory and sensory discharge [44]. CO is also an endogenous modulator of inflammation, cell death, oxidative stress and immune responses [10,63]. Alternations in CO metabolism have been observed in various pathological disturbances, such as sepsis, lung injury, anemias and liver cirrhosis [64–66].

3.3. Hydrogen Sulfide (H_2S)

H_2S, a colorless, flammable and notorious gas with a pungent odor of rotten eggs, has emerged as the third gaseous signaling mediator. The endogenous generation of H_2S in mammalians mainly arises from the desulfydration of L-cysteine or homocysteine. To date, three enzymes with ascribed roles in this conventional source of H_2S biosynthesis have been recognized. These are cystathionine β-synthase (CBS), cystathionine γ-lyase (CSE) and 3-mercaptopyruvate sulfurtransferase (3-MST) [67–69]. Among these enzymes, CBS and CSE, two pyridoxal 50-phosphate (vitamin B6)-dependent enzymes, catalyze transsulfuration reactions, whereas 3-MST is responsible for L-cysteine catabolism [70]. Specifically, CBS, the first enzyme in the transsulfuration pathway, generates H_2S, mainly by the condensation reaction of L-cysteine and homocysteine, while CSE, the second enzyme in the pathway, produces H_2S, predominantly by the α, β-elimination of L-cysteine [67,71]. While CBS is believed to be the main source of H_2S biosynthesis in the central nervous system, CSE is thought to be the major contributor to endogenous production of H_2S in the peripheral tissues [67,72]. As the third enzyme involved in H_2S biosynthesis, 3-MST, along with cysteine aminotransferase (CAT), also contributes to the endogenous production of H_2S [72]. Particularly, 3-MST-catalyzed H_2S generation requires the presence of reducing substrates, and most of the yielding H_2S is bound in the form of sulfate sulfur. While CBS and CSE are primarily cytosolic enzymes, 3-MST is mainly localized in the mitochondrion and with some in the cytoplasm. Consequently, CBS- or CSE-catalyzed H_2S biosynthesis mainly occurs in the cytoplasm, whereas 3-MST-mediated H_2S generation takes place in both places [70]. More recently, the production of H_2S by 3-MST and the D-amino acid oxidase (DAO) pathway has been identified [73]. Apart from these enzyme-catalyzed reactions, it has become increasingly apparent that many other sources, including the natural liberation of H_2S from persulfides and polysulfide species, also contribute to H_2S biosynthesis [74]. The predominant pathways by which H_2S is metabolized in vivo include oxidation to thiosulfate and sulfate in the mitochondrion and methylation to methanethiol in the cytoplasm. In addition, H_2S can be eliminated by methemoglobin or metallo- or disulfide-containing molecules. Moreover, thiosulfate and sulfate excretion by urinating also leads to the clearance of H_2S [69,75]. Unlike its two counterparts, NO and CO, which employ soluble guanylyl cyclase to transduce their signals, it is yet not clear whether H_2S also has its second messenger. Nonetheless, as a gaseous free radical similar to NO, H_2S is also biologically reactive [76].

H_2S has been found to be a versatile modulator of various organs and systems [77]. In the context of a pivotal role of an endothelium-derived relaxing factor for NO, H_2S is regarded as an endothelium-derived hyperpolarizing factor [46,78]. However, H_2S also possesses a potent vasoconstrictive effect [79]. More significantly, it is reported that endogenous H_2S can regulate the bioavailability of NO in the cardiovascular system [80]. Apart from the regulatory effect on vascular tone, H_2S also participates in regulating the proliferation and death of vascular smooth muscle cells, inhibiting oxidative stress suppression inflammation inhibition and modulating vascular permeability and angiogenesis [81]. In the respiratory system, H_2S profoundly affects various respiratory functions, such as regulating the respiratory rhythm and maintaining the development and homeostasis of the pulmonary vessel [82,83]. In the immune system, H_2S regulates the viability and functions of various immune cells, including neutrophils, macrophages, dendritic cells,

T lymphocytes and B lymphocytes; thus, it greatly shapes the landscape of innate and adaptive immunity [84]. In the central nervous system, H_2S serves as an antioxidant and antineuroinflammatory mediator; thus, H_2S plays a significant role in neuroprotection. Moreover, H_2S is closely associated with neurotransmission [85]. H_2S also affects the functions of many other systems, such as the reproductive system, digestive system and endocrine system [77]. Notably, similar to NO, H_2S is discovered to be an important endogenous epigenetic modulator [51]. Aberrant H_2S metabolism occurs in many pathological states, such as sepsis, inflammation, coronavirus disease 2019, hypertension, atherosclerosis, obstructive respiratory disease, lung injury, macrophage activation, retinal diseases and neurodegenerative disease [77,81,86–89].

3.4. Interplay among NO, CO and H_2S

Accumulating evidence has pointed to the crosstalk among NO, CO and H_2S [86,90]. To date, four interaction mechanisms have been identified, which are competition/synergy for heme in heme-containing proteins, modulation of the generation of other gases and competition for post-translational modification sites in proteins and formation of hybrid molecules [91–93]. Among these four mechanisms, the regulation of the biosynthesis of other gaseous mediators has been widely explored. Firstly, NO has been shown to up-regulate the endogenous production of CO by increasing the level of HO-1, whereas NO can suppress the activity of HO-2 [94,95]. The regulatory effects of NO on H_2S generation are also complex. While NO has been shown to inhibit the activities of CBS and CSE, it can also promote CSE-mediated H_2S production [96–98]. Secondly, CO has been shown to reduce the production of NO and H_2S by suppressing the activities of iNOS and CBS [99]. Thirdly, H_2S also greatly affects the biosynthesis of NO and CO. H_2S can promote or stabilize eNOS activity; however, it can also inhibit eNOS function [97,100,101]. Additionally, in the presence of NO, the activities of nNOS and iNOS were suppressed by H_2S [102]. Moreover, H_2S can inhibit iNOS activity by promoting HO-1 activity [103]. However, NO production can be enhanced by H_2S as it can promote iNOS expression [104]. Importantly, the crosstalk among NO, CO and H_2S may differ in different organs or under different conditions, leading to the complicated or even opposite regulatory effects of one gaseous mediator on the biosynthesis of others.

Table 1 briefly summarizes the characteristics of NO, CO and H_2S.

Table 1. Characteristics of currently recognized gaseous mediators.

	Nitric Oxide	Carbon Monoxide	Hydrogen Sulfide
Formula and molecular weight	NO (30.01 g/mol)	CO (28.01 g/mol)	H_2S (34.08 g/mol)
Biological half-life	Seconds	Several minutes	Seconds–minutes
Chemical reactivity	Very high	Moderate	Very high
Properties of free radicals	Yes	No	Yes
Endogenous production enzymes	nNOS; iNOS; eNOS; mtNOS [a]	HO-1; HO-2; HO-3 [a]	CBS; CSE; 3-MST/CAT; 3-MST/DAO [b]
Main substrates for biosynthesis	L-arginine	Heme	L-cysteine; 3-mercaptopyruvate; D-cysteine [b]
Clearance sources	Oxidization	Being exhaled from the airway; Binding to heme proteins; Oxidization [c]	Oxidization (mitochondrion); Methylation (cytoplasm); Being excreted from urine
End products	Nitrite and nitrate	Carboxyhemoglobin; Carbon dioxide [c]	Thiosulfate and sulfate; methanethiol
Second messenger	sGC	sGC	NA
Involvement in sepsis	Yes (Mainly detrimental)	Yes (Mainly beneficial)	Yes (Mainly detrimental)

Abbreviations: neuronal nitric oxide synthase (nNOS); inducible nitric oxide synthase (iNOS); endothelial nitric oxide synthase (eNOS); mitochondrial nitric oxide synthase (mtNOS); heme oxygenase-1 (HO-1); heme oxygenase-2 (HO-2); heme oxygenase-3 (HO-3); cystathionine β-synthase (CBS); cystathionine γ-lyase (CSE); 3-mercaptopyruvate sulfurtransferase (3-MST); cysteine aminotransferase (CAT); D-amino acid oxidase (DAO); soluble guanylyl cyclase (sGC); Not applicable (NA). [a] Controversial; [b] newly discovered; [c] has not been observed under physiological conditions.

Figure 2 briefly summarizes the ways in which NO, CO and H_2S are enzymatically produced and that in which the biosynthesis of one gaseous mediator is affected by the remaining two.

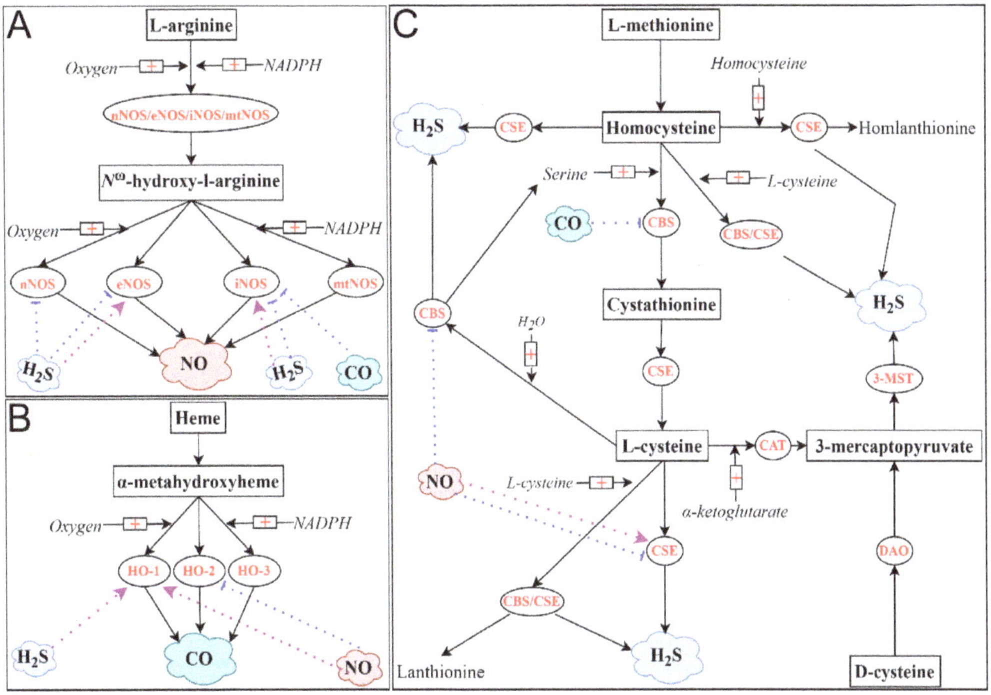

Figure 2. Enzyme-derived endogenous production of NO (**A**), CO (**B**) and H_2S (**C**). Endogenous NO is produced from a two-step biological process catalyzed by NOS. L-arginine is hydroxylated to N^{ω}-hydroxy-l-arginine; the latter is further oxidized to L-citrulline and NO. Endogenous CO is generated from the degradation of heme. Firstly, heme is oxidized to α-metahydroxyheme in the presence of oxygen and NADPH. Secondly, α-metahydroxyheme reacts with oxygen, resulting in the biosynthesis of CO. NADHP is also required for this reaction. Relatively, the biosynthesis of H_2S is more complex as H_2S can be generated from four enzymatic pathways. Briefly, H_2S is naturally produced from the desulfydration of L-cysteine or homocysteine catalyzed by CBS, CSE and 3-MST/CAT. Specifically, CBS and CSE are involved in transsulfuration reactions, whereas 3-MST is responsible for L-cysteine catabolism. Recently, the 3-MST/DAO pathway has gained acceptance as the fourth pathway for H_2S biosynthesis, using D-cysteine as the substrate. The biosynthesis of every gaseous mediator might be affected by the remaining two gases. Purple arrow: activation; blue arrow: suppression.

4. Gaseous Mediators in Sepsis/Septic Shock

As three significant endogenous regulators that play indispensable roles in maintaining the homeostasis of organ systems in living organisms [18], NO, CO and H_2S have been shown to play vital roles in the intricate pathophysiology of sepsis/septic shock.

4.1. NO in Sepsis/Septic Shock

Overproduction of NO throughout the organism resulting from the excessive activation of iNOS is one hallmark of sepsis. NO has emerged as a significant modulator in sepsis, as it has been shown to extensively impact the pathophysiology and outcome of sepsis [105,106]. The level of NO was correlated with the increased severity of sepsis-

and endotoxemia-associated systemic inflammation and organ injury, while the inhibition of NO production mitigated these alterations [107–109]. Specifically, in rats with cecal ligation and puncture (CLP)-induced sepsis or LPS-induced endotoxemia, the level of iNOS in the diaphragm was upregulated. The elevated iNOS and attendant increased production of NO were involved in endotoxemia- and sepsis-induced diaphragm injury, as alterations that occurred in the diaphragm (sarcolemmal injury and myofiber damage) were obviously mitigated by the administration of a nonselective NOS inhibitor named L-NMMA [107]. In addition, it was reported that the level of NO was elevated in mice with *Escherichia coli* infection-induced sepsis, and the increased NO production was associated with sepsis-associated alterations, such as dysregulated systemic inflammation, as indicated by elevated proinflammatory mediators, oxidative stress (increased malondialdehyde) and organ dysfunction (liver failure and kidney failure). However, the pretreatment of L-NAME, a nonselective inhibitor of NOS, significantly reduced sepsis-induced overproduction of NO and consequently mitigated sepsis-associated abnormalities [108]. Moreover, as shown by Luo et al., once the activity of the toll-like receptor 4 (TLR4)/myeloid differentiation primary response 88 (MyD88)/nuclear factor kappa-light-chain-enhancer of activated B cells (NF-κB) pathway was inhibited, iNOS became inactive, leading to an enhancement of vascular responsiveness and an increase in the survival of mice with CLP-induced sepsis (mean survival time increased from 1.7 days to 4.5 days) [109]. Similarly, compared with patients without sepsis, the concentration of serum NO in sepsis patients was significantly upregulated [110]. More importantly, monitoring of the level of NO in serum may contribute to precise of evaluation the severity of sepsis [110]. Among the organs and systems affected by elevated NO in sepsis and endotoxemia, the cardiovascular system has attracted the most attention. While a physiological level of NO is essential to the maintenance of the cardiovascular system, sepsis-associated hemodynamic instability, including vasorelaxation, hypotension and shock, has been attributed to an aberrant NO-induced macrovascular compromise, myocardial dysfunction, vascular hyporesponsiveness, direct cellular toxicity and bioenergetic failure [47]. Furthermore, when the amount of endogenous NO reaches a certain threshold, the production of NO is inhibited, since the high concentration of NO itself can suppress the activity of NOS [37]. Moreover, compared with iNOS, nNOS and eNOS are more sensitive to the autoinhibitory effect of NO. Therefore, once the biosynthesis of NO catalyzed by iNOS is activated in response to the proinflammatory stimuli in sepsis, the essential basal effects of eNOS may be insufficient to support the function of the fragile cardiovascular system in sepsis [46,105]. Thus, the aberrant production of NO catalyzed by active iNOS is thought to be a primary cause of hemodynamic instability in sepsis [47,111]. Worse still, these NO-induced hemodynamic alterations can further result in hypoxia in multiple organ systems, leading to progressive organ dysfunction [112]. In addition, many studies have shown that there is extensive involvement of NO in sepsis- and endotoxemia-induced abnormalities and derangements in the respiratory system, renal system, immune system, central nervous system and digestive system and identified the underlying mechanisms by which these organs and systems are disturbed by dysregulated NO in sepsis and endotoxemia [105].

Not surprisingly, several studies have also concluded that the increased generation of NO may have potential benefits in sepsis, since NO has been shown to facilitate bacterial destruction [112]. For example, in a controlled trial of inhibition of nNOS either by pharmacological blockage or gene deletion, there was an increase in mortality (hazard ratio of death was 1.71) and blood bacterial counts (1.4-fold greater) in mice with sepsis induced by CLP. This was accompanied by an upregulation of proinflammatory mediators, including tumor necrosis factor (TNF)-α and interleukin (IL)-6, and peritoneal lavage cell counts were increased. These results indicate that the nNOS/NO pathway improves survival from sepsis plays an important role in modulating the inflammatory response [113]. In another study, sepsis was induced in wild-type mice and genetically deficient iNOS-knockout mice by infection with *Salmonella typhimurium* (a Gram-negative pathogen). The deletion of the iNOS gene attenuated sepsis-induced systemic inflammation, as evidenced by lower levels of proinflammatory mediators and neutrophil accumulation in the peritoneal cavity. In addition, the deficiency of the iNOS gene also increased the bacterial load, decreased the thymic atrophy, aggregated the hepatic and cardiovascular dysfunction and increased the risk of mortality of mice. By contrast, the pretreatment of iNOS-deficient mice with DETA-NO (a NO donor) significantly attenuated these sepsis-associated abnormalities. These findings revealed the protective roles played by the iNOS/NO pathway in sepsis [114].

4.2. CO in Sepsis/Septic Shock

Increased generation of CO is also commonly observed in sepsis and endotoxemia, and many studies have implicated the beneficial effects of CO in sepsis [115]. Kyokane et al. reported that endogenous production of CO catalyzed by HO-1 in the liver was upregulated in rats with LPS-induced endotoxemia and concluded that the increased production of CO played an important role in protecting the liver from dysfunction [116]. Similarly, HO-1-derived CO resulted in an enhancement of phagocytosis and host defense response directed at microorganism invasion, leading to an enhancement of pathogen clearance without suppressing the host inflammatory response in CLP-induced sepsis in mice. In addition, the administration of the tricarbonyldichlororuthenium (II) dimer, a CO-releasing molecule (CORM), significantly improved the survival probability of mice in sepsis [117]. The protective role of CO has been confirmed in CLP-induced sepsis, as well as LPS-induced endotoxemia, both in vivo and in vitro [118]. Specifically, in vivo, the administration of HO-1 inducers or CORM-2 suppressed the activity of high-mobility group box 1, improving the survival of sepsis. In addition, sepsis-associated systemic inflammation was alleviated by the treatment of HO-1 inducers and CORM-2, as indicated by a drop of proinflammatory cytokines, including TNF-α, IL-1β and interferon-β. In vitro, the induction of LPS-induced endotoxemia activated the activity of high-mobility group box 1, thereby promoting the proinflammatory response in macrophages. However, the pretreatment of HO-1 inducers or CORM-2, as well as the transfection of HO-1, greatly reversed these alterations [118]. Since several CORMs have been developed [119], many researchers have used different CORMs in their research, and the protective roles of the HO-1/CO pathway in sepsis and endotoxemia have been firmly established. For example, several studies showed that the CO released from CORM-2 or CORM-3 was capable of suppressing the activation of inflammasome related to pyroptosis; as a result, the function of multiple organs and systems (cardiac fibroblasts, intestine and kidney) were preserved, and the elimination of pathogens was promoted in rodents with CLP-induced sepsis [120–123]. Apart from the effects on dampening proinflammatory response and inhibiting pyroptosis, many other mechanisms, such as supporting the energetic metabolism of mitochondrion coupled with activating the biogenesis of mitochondria, reducing the levels of cardiac contractile proteins, inhibiting the activation of NF-κB, downregulating the expression of the TLR4/myeloid differentiation factor-2 (MD2) complex on myeloid cells, suppressing the overactivation of platelets and enhancing autophagy, also contribute to the protective actions of CO in sepsis and endotoxemia [124–129]. Of note, the nuclear factor-erythroid factor 2-related factor 2 (Nrf2) has been shown to be essential to the anti-inflammatory roles of CO released from CORM-2 in LPS-induced endotoxemia, as the deletion of Nrf2 significantly abolished the beneficial effects of CO [130]. Interestingly, exposure of mesenchymal stromal cells to CO enhanced the therapeutic response in mice with CLP-induced sepsis [131]. This study showed that CO exposure greatly improved the treatment efficacy of mesenchymal stromal cells, as these cells have been shown to enhance pathogen elimination, promote inflammation resolution and alleviate organ injury in septic mice. Consistently, increased biosynthesis of CO has also been observed in septic patients; more importantly, survivors had higher levels of CO than nonsurvivors, indicating the beneficial effect of increased CO production in sepsis [64,132].

Only a few investigations have demonstrated that the upregulation of endogenous CO is detrimental to sepsis. For instance, Iwasashi and coworkers found that the active HO-1/CO pathway was associated with liver dysfunction in rats subjected to CLP-induced sepsis. It was reported that HO-1-induced excessive generation of CO led to an immoderate dilation of liver sinusoidal and attendant liver failure, whereas the administration of HO inhibitors (Sn-PP and Zn-PP) significantly alleviated sepsis-induced liver injury, as evidenced by lower plasma aspartate aminotransferase and lower liver cyclic guanosine monophosphate, as well as promoted the survival of rats (61.5% and 66.7% vs. 26.7%) [133].

4.3. H_2S in Sepsis/Septic Shock

The biosynthesis of H_2S is significantly upregulated in sepsis. Accumulating evidence has shown proinflammatory effects of H_2S on sepsis and endotoxemia. Induction of septic shock and endotoxic shock has been reported to greatly increase the arterial level of H_2S in rats, and the elevated H_2S has a negative correlation with the hemodynamic parameters, including the heart rate, the mean arterial pressure and the $+dP/dt$ max in rats [134]. This investigation aroused great attention in exploring the roles of H_2S in sepsis and the mechanisms of action. As showed in a landmark study that detailed the significant role of endogenous H_2S in CLP-induced sepsis, the expression (both mRNA and protein) of CSE and the level of endogenous H_2S were greatly upregulated after the induction of sepsis [135]. The administration of DL-propargylglycine (PAG, 50 mg/kg, intraperitoneal injection), an irreversible inhibitor of CSE, significantly attenuated sepsis-induced neutrophil accumulation, as indicated by tissue myeloperoxidase activity and histological alterations in the liver and lungs, whereas the treatment of sodium hydrosulfide (NaHS, 10 mg/kg, intraperitoneal injection), a fast-releasing H_2S donor, further exacerbated sepsis-associated systemic inflammation and organ injury [135].

Since then, we have conducted a series of studies to further explore the role played by H_2S in sepsis-induced multiple organ dysfunction and to elucidate the underlying mechanism. For example, in mice subjected to CLP-induced sepsis, NF-κB was activated by the elevated H_2S, leading to an upregulation of the production of proinflammatory cytokines (IL-1β, IL-6 and TNF-α) and chemokines (monocyte chemotactic protein-1, and macrophage inflammatory protein-2), the rolling and adherence of leukocytes, the expressions of various adhesion molecules (intercellular adhesion molecule-1, P-selectin and E-selectin) and eventually exaggerated lung injury and liver injury [136,137]. Thereafter, we found that the extracellular signal-related kinase (ERK) pathway was involved in the activation of NF-κB by H_2S following sepsis, as the treatment of the ERK kinase inhibitor significantly abolished H_2S-mediated NF-κB activation and consequently attenuated sepsis-associated systemic inflammation and organ injury [138]. Subsequently, taking the advantage of using the tachykinin precursor 1 gene (the gene that encodes substance P)-deficient mice, as well as using inhibitors of tachykinin receptor 1, the functional receptor of substance P and transient receptor potential vanilloid type 1 (TRPV1), we found that in mice with CLP-induced polymicrobial sepsis, TRPV1-mediated priming of the substance P-tachykinin receptor 1 axis was involved in H_2S-induced activation of the ERK/NF-κB pathway and further resulted in sepsis-associated alterations, including systemic inflammation and organ injury [139–141]. We also found that the involvement of TRPV1-mediated an increase in cyclooxygenase-2 and prostaglandin E metabolite production in H_2S-induced sepsis-associated alterations in mice [142]. Furthermore, the proinflammatory effect of H_2S in polymicrobial sepsis was confirmed as the treatment of small interference RNA that targets the CSE gene reduces the accumulation of leukocytes and the levels of proinflammatory mediators in the liver and lungs [143].

More recently, sepsis was induced in genetically deficient CSE-knockout mice and wild-type mice, by which we further shed light on the proinflammatory action of H_2S in sepsis. The liver sinusoid plays a significant role in maintaining the hepatic function; however, its homeostasis is frequently disrupted in sepsis and endotoxemia. Building on the finding that H_2S serves as a vasoconstrictor in the liver sinusoid in endotoxemia [144,145], we further explored the effects of H_2S on liver sinusoid in sepsis. In wild-type mice, sepsis-induced elevated H_2S caused several alterations in the liver sinusoidal endothelial cells (LSECs), including the defenestration and gaps formation, suggesting that the liver sinusoidal function was impaired by H_2S in sepsis. However, these alterations were significantly reversed in mice genetically deficient in CSE [146,147]. Furthermore, the underlying mechanisms were investigated. We found that the activation of the ERK1/2-NF-κB p65 pathway and the substance P-tachykinin receptor 1 axis plays a central role in H_2S-induced liver sinusoidal dysfunction [146,147]. Similarly, aberrant metabolism of H_2S has been observed in patients with sepsis and animals with LPS-induced endotoxemia [148–150].

A potential beneficial effect of a very low dose of H_2S or slow-release H_2S donors on sepsis and endotoxemia has also been reported. In mice with CLP-induced sepsis, either prophylactic or therapeutic treatment of H_2S donors (NaHS or Lawesson's reagent) promoted the rolling and adhesion of leukocytes and the migration of neutrophils, thereby reducing the bacteremia levels, as well as alleviating hypotension and lung lesions, eventually leading to increased survival. Conversely, the administration of PAG significantly aggravated CLP-induced alterations in mice [151]. It is noteworthy that there were many differences between this study and those discussed above. While this study mainly focused on the role of exogenous H_2S in sepsis, the research discussed above mainly focused on endogenous H_2S. In addition, the way in which mice received the treatment of CSE inhibitors and H_2S donors was also different. This may differ the pharmacological effects of these compounds (subcutaneous injection vs. intraperitoneal injection). Furthermore, the dose of PAG (about 4.5–4.9 g/kg vs. 50 mg/kg) and NaHS (0.56–5.6 mg/kg vs. 10 mg/kg) used in this study and those discussed above was significantly different. These differences may underlie the opposite effects of H_2S in sepsis reported in these investigations. The protective role of exogenous H_2S was further confirmed by Ahmad and colleagues as they found that delayed treatment of NaHS was favorable to rats subjected to CLP-induced sepsis [152]. In addition, it is reported that the preservation of mitochondrial function by NaHS treatment resulted in the improvement of diaphragm weakness and the decline of mortality rate in CLP-induced sepsis rats [153]. More recently, two studies have shown that the pretreatment of GYY4137 (25 mg/kg and 50 mg/kg intraperitoneal injection), a novel slow-releasing H_2S donor, protected against acute lung injury caused by CLP-induced sepsis in mice [154,155]. A similar salutary effect of H_2S was also reported in urinary-derived sepsis, pseudomonas aeruginosa sepsis and pneumococcal pneumosepsis, together with LPS-induced endotoxemia [21,156–159].

5. Gaseous-Mediator-Based Therapeutic Strategy for Sepsis/Septic Shock

The recognition of the involvement of NO, CO and H_2S in the physiopathology of sepsis/septic shock and endotoxemia has led to the development of therapeutic approaches targeting these gaseous mediators for sepsis (Figure 3). To date, in addition to NO inhalation, many other strategies targeting the regulation of the activity of NOS, the clearance of NO and the bioavailability of the substrate have been widely investigated [105,106,111,160]. Inhaled NO possesses promise in sepsis treatment, since it was reported that systemic oxygenation was improved after NO inhalation [161]; however, inhaled NO (40 ppm) failed to augment microcirculatory perfusion and improve organ function in patients with sepsis [162]. As mentioned by the researchers, one possible reason for the failure of inhaled NO in sepsis is that the macrocirculatory hemodynamics of these patients had been optimized prior to the treatment of inhaled NO. Considering the deleterious effects of overproduction of endogenous NO in sepsis, several NOS inhibitors, including selective and nonselective, have been developed, and the therapeutic efficacy and safety of these compounds have been widely investigated. Unfortunately, these studies have produced mixed results; thus, whether to treat sepsis by NOS inhibitors is still a matter of debate [111,163]. While a phase II trial showed that L-NMMA, a nonselective NOS inhibitor, improved systemic vascular response in patients with septic shock, a phase III trial conducted subsequently was terminated. as L-NMMA increased mortality in patients with septic shock [164,165]. The larger phase III investigation was thought to reveal an adverse outcome of L-NMMA treatment that was not detected by the smaller phase II study [165]. In the reactions leading to the NOS-catalyzed production of NO, BH4 does not only act as a crucial cofactor, but it also serves as an endogenous regulator of NOS activity. Overproduction of BH4 and an attendant increase in conversion of BH4 to BH2 result in a high level of BH2 in sepsis. BH2 can bind to NOS and consequently suppress the activity of NOS, leading to a decrease in NO biosynthesis [166]. Promisingly, the administration of the BH4 analog attenuated hemodynamic instability, organ dysfunction and declined mortality in animals with sepsis or endotoxemia [111,163]. Similarly, several novel NO-based therapeutic strategies targeting the enhancement of NO clearance or the improvement of NO bioavailability have been developed for sepsis [111].

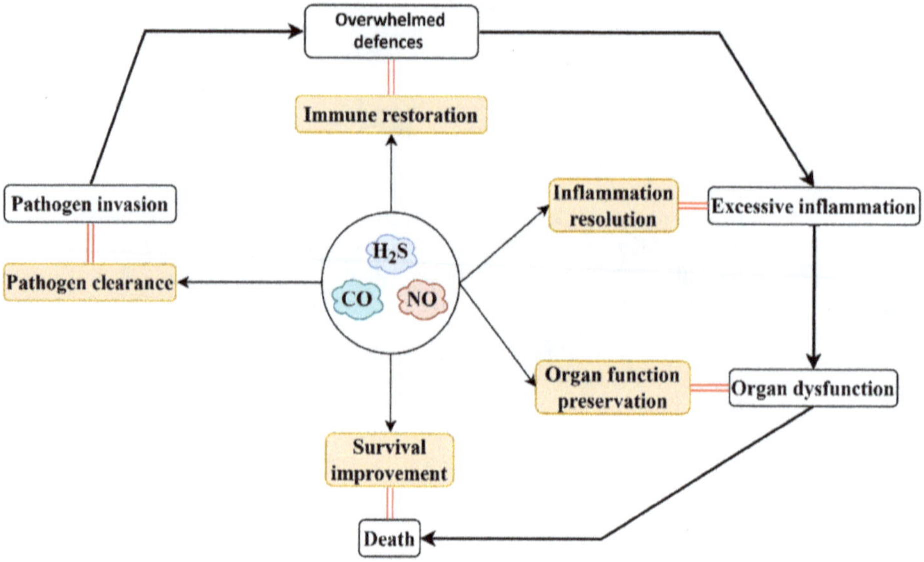

Figure 3. Therapeutic perspective of NO, CO and H_2S in sepsis. Multiple organ dysfunction resulting from excessive inflammation caused by an aberrant host defense response directed at invasive pathogens and their toxins is one hallmark of sepsis, which can eventually lead to death (as shown in white boxes). To promote the survival of sepsis, several medical countermeasures targeting sepsis pathophysiology, including pathogen clearance, immune restoration, inflammation resolution and organ function preservation, have been developed (as shown in light orange boxes). While the feasibility, safety and efficacy of NO-, CO- and H_2S-based treatments should be further assessed, the therapeutic promise of gaseous-mediator-based treatments has been observed in many investigations.

The recognition of the beneficial effects of CO in sepsis in animal models has prompted the development of CO-based therapy for sepsis. The administration of low-dose inhaled CO showed a protective effect in sepsis, as it has been shown to rescue mice from severe sepsis induced by *Staphylococcus aureus* infection [125]. More recently, the feasibility and safety of low-dose inhaled CO administration in patients with sepsis-induced acute respiratory distress syndrome were established in a phase 1 trial [167]. In addition, this study also demonstrated that the administration of low-dose inhaled CO significantly reduced the concentrations of mitochondrial DNA in the plasma, indicating the potential effect of inhaled CO treatment in preserving mitochondrial function in sepsis [167]. Taking advantage of the knowledge that exogenous CO releases from CORMs, including CORM-1, CORM-2 and CORM-3, have potent protective roles in sepsis and endotoxemia, increasing attention has been paid to treating animals with sepsis using these novel compounds [120,121].

Accumulating evidence has revealed the complicated actions played by H_2S in sepsis, leading to the increasing attention being paid to develop therapeutic approaches targeting H_2S for sepsis. Although several preclinical animal studies have shown the protective role of using PAG in sepsis [136,141], many other investigations have also indicated that the administration of inhaled H_2S or H_2S donors, either fast-releasing donors or slow-releasing donors, such as NaHS and GYY4137, is beneficial to sepsis and endotoxemia, resulting in uncertainties in the possible role of H_2S-based treatment for sepsis [20,152–155]. Given the promise of the countermeasures targeting gaseous mediators in sepsis therapy, the establishment of an optimal therapeutic protocol, including the dose and delivery, for gaseous-mediator-based therapy for sepsis will be meaningful and another step forward.

6. Conclusions

In sepsis, a dysregulated host defense response directed at invasive pathogens and their toxins can cause excessive systemic inflammation, consequently leading to multiple organ dysfunction and death. Accumulating evidence, from our laboratory and others, has revealed the important roles of NO, CO and H_2S as novel mediators in regulating the onset, development, progression and outcome of sepsis. Furthermore, the understanding of the significant roles of NO, CO and H_2S in the pathophysiology of sepsis, including their effects on host immune response, pathogen elimination, systemic inflammation and organ dysfunction, has led to the development of several novel therapeutic strategies targeting NO, CO and H_2S for sepsis, such as inhaled NO, CO and H_2S, the inhibitors of NOS, HO and CSE and the slow-releasing donors of NO, CO and H_2S. Although more research is needed to evaluate the feasibility, safety and efficacy of these gaseous-mediator-based treatments, early results have shown the promise of these novel therapeutic strategies. Thus, it is important to put more effort and resources in order to investigate the therapeutic prospects of NO, CO, and H_2S in sepsis.

Author Contributions: Conceptualization, M.B.; supervision, M.B., S.C., Y.Z.; writing—original draft preparation, Z.Z.; writing—review and editing, Z.Z., M.B., S.C., Y.Z.; funding acquisition, M.B. All authors have read and agreed to the published version of the manuscript.

Funding: The research group of Madhav Bhatia is funded by the University of Otago Vice-Chancellor's Strategic Development Fund and Maurice and Phyllis Paykel Trust.

Institutional Review Board Statement: Not applicable.

Informed Consent Statement: Not applicable.

Data Availability Statement: Not applicable.

Conflicts of Interest: The authors declare no conflict of interest.

References

1. Cecconi, M.; Evans, L.; Levy, M.; Rhodes, A. Sepsis and septic shock. *Lancet* **2018**, *392*, 75–87. [CrossRef]
2. Singer, M.; Deutschman, C.S.; Seymour, C.W.; Shankar-Hari, M.; Annane, D.; Bauer, M.; Bellomo, R.; Bernard, G.R.; Chiche, J.-D.; Coopersmith, C.M.; et al. The third international consensus definitions for sepsis and septic shock (Sepsis-3). *JAMA* **2016**, *315*, 801–810. [CrossRef] [PubMed]
3. Rudd, K.E.; Kissoon, N.; Limmathurotsakul, D.; Bory, S.; Mutahunga, B.; Seymour, C.W.; Angus, D.C.; West, T.E. The global burden of sepsis: Barriers and potential solutions. *Crit. Care* **2018**, *22*, 1–11. [CrossRef]
4. Rudd, K.E.; Johnson, S.C.; Agesa, K.M.; Shackelford, K.A.; Tsoi, D.; Kievlan, D.R.; Colombara, D.V.; Ikuta, K.S.; Kissoon, N.; Finfer, S.; et al. Global, regional, and national sepsis incidence and mortality, 1990–2017: Analysis for the global burden of disease study. *Lancet* **2020**, *395*, 200–211. [CrossRef]
5. Reinhart, K.; Daniels, R.; Kissoon, N.; Machado, F.R.; Schachter, R.D.; Finfer, S. Recognizing sepsis as a global health priority—A WHO resolution. *New Engl. J. Med.* **2017**, *377*, 414–417. [CrossRef] [PubMed]
6. Evans, L.; Rhodes, A.; Alhazzani, W.; Antonelli, M.; Coopersmith, C.M.; French, C.; Machado, F.R.; McIntyre, L.; Ostermann, M.; Prescott, H.C.; et al. Surviving sepsis campaign: International guidelines for management of sepsis and septic shock 2021. *Intensive Care Med.* **2021**, *47*, 1181–1247. [CrossRef] [PubMed]
7. Mustafa, A.K.; Gadalla, M.M.; Snyder, S.H. Signaling by Gasotransmitters. *Sci. Signal.* **2009**, *2*, re2. [CrossRef]
8. Wang, R. Gasotransmitters: Growing pains and joys. *Trends Biochem. Sci.* **2018**, *39*, 283–295. [CrossRef]
9. Althaus, M.; Clauss, W.G. Gasotransmitters: Novel regulators of ion channels and transporters. *Front. Physiol.* **2013**, *4*, 27. [CrossRef]

10. Wallace, J.L.; Ianaro, A.; Flannigan, K.L.; Cirino, G. Gaseous mediators in resolution of inflammation. *Semin. Immunol.* **2015**, *27*, 227–233. [CrossRef]
11. Hartmann, C.; Nussbaum, B.; Calzia, E.; Radermacher, P.; Wepler, M. Gaseous mediators and mitochondrial function: The future of pharmacologically induced suspended animation? *Front. Physiol.* **2017**, *8*, 691. [CrossRef] [PubMed]
12. Szabo, C. Gasotransmitters in cancer: From pathophysiology to experimental therapy. *Nat. Rev. Drug Discov.* **2016**, *15*, 185–203. [CrossRef]
13. Kolluru, G.K.; Shen, X.; Yuan, S.; Kevil, C.G. Gasotransmitter heterocellular signaling. *Antioxid. Redox Signal.* **2017**, *26*, 936–960. [CrossRef] [PubMed]
14. Huang, Y.; Zhang, H.; Lv, M.B.; Tang, C.; Du, J.; Jin, H. Sulfur dioxide: Endogenous generation, biological effects, detection, and therapeutic potential. *Antioxid. Redox Signal.* **2022**, *36*, 256–274. [CrossRef] [PubMed]
15. Zuhra, K.; Szabo, C. The two faces of cyanide: An environmental toxin and a potential novel mammalian gasotransmitter. *FEBS J.* **2021**. [CrossRef]
16. Boros, M.; Keppler, F. Methane production and bioactivity-a link to oxido-reductive stress. *Front. Physiol.* **2019**, *10*, 1244. [CrossRef]
17. Yang, G.; Sener, A.; Ji, Y.; Pei, Y.; Pluth, M.D. Gasotransmitters in biology and medicine: Molecular mechanisms and drug targets. *Oxid. Med. Cell. Longev.* **2016**, *2016*, 1–2. [CrossRef]
18. Hendriks, K.; Maassen, H.; van Dijk, P.R.; Henning, R.; van Goor, H.; Hillebrands, J.-L. Gasotransmitters in health and disease: A mitochondria-centered view. *Curr. Opin. Pharmacol.* **2019**, *45*, 87–93. [CrossRef]
19. Winkler, M.S.; Kluge, S.; Holzmann, M.; Moritz, E.; Robbe, L.; Bauer, A.; Zahrte, C.; Priefler, M.; Schwedhelm, E.; Böger, R.H.; et al. Markers of nitric oxide are associated with sepsis severity: An observational study. *Crit. Care* **2017**, *21*, 1–9. [CrossRef]
20. Nakahira, K.; Choi, A.M.K. Carbon monoxide in the treatment of sepsis. *Am. J. Physiol. Cell. Mol. Physiol.* **2015**, *309*, L1387–L1393. [CrossRef]
21. Chen, Y.-H.; Teng, X.; Hu, Z.-J.; Tian, D.-Y.; Jin, S.; Wu, Y.-M. Hydrogen sulfide attenuated sepsis-induced myocardial dysfunction through TLR4 pathway and endoplasmic reticulum stress. *Front. Physiol.* **2021**, *12*, 653601. [CrossRef] [PubMed]
22. Caraballo, C.; Jaimes, F. Organ dysfunction in sepsis: An ominous trajectory from infection to death. *Yale J. Biol Med.* **2019**, *92*, 629–640. [PubMed]
23. Pool, R.; Gomez, H.; Kellum, J.A. Mechanisms of organ dysfunction in sepsis. *Crit. Care Clin.* **2018**, *34*, 63–80. [CrossRef] [PubMed]
24. Lelubre, C.; Vincent, J.-L. Mechanisms and treatment of organ failure in sepsis. *Nat. Rev. Nephrol.* **2018**, *14*, 417–427. [CrossRef]
25. Huang, M.; Cai, S.; Su, J. The pathogenesis of sepsis and potential therapeutic targets. *Int. J. Mol. Sci.* **2019**, *20*, 5376. [CrossRef] [PubMed]
26. Delano, M.J.; Ward, P.A. The immune system's role in sepsis progression, resolution, and long-term outcome. *Immunol. Rev.* **2016**, *274*, 330–353. [CrossRef] [PubMed]
27. Conway-Morris, A.; Wilson, J.; Shankar-Hari, M. Immune activation in sepsis. *Crit. Care Clin.* **2018**, *34*, 29–42. [CrossRef]
28. Lawal, B.; Wang, Y.-C.; Wu, A.T.H.; Huang, H.-S. Pro-Oncogenic c-Met/EGFR, biomarker signatures of the tumor microenvironment are clinical and therapy response prognosticators in colorectal cancer, and therapeutic targets of 3-Phenyl-2H-benzo[e][1,3]-Oxazine-2,4(3H)-Dione derivatives. *Front. Pharmacol.* **2021**, *12*, 691234. [CrossRef] [PubMed]
29. Chousterman, B.G.; Swirski, F.; Weber, G.F. Cytokine storm and sepsis disease pathogenesis. *Semin. Immunopathol.* **2017**, *39*, 517–528. [CrossRef]
30. Jarczak, D.; Kluge, S.; Nierhaus, A. Sepsis—Pathophysiology and therapeutic concepts. *Front. Med.* **2021**, *8*, 628302. [CrossRef] [PubMed]
31. Gotts, J.E.; Matthay, M.A. Sepsis: Pathophysiology and clinical management. *BMJ* **2016**, *353*, i1585. [CrossRef] [PubMed]
32. Liu, D.; Cao, S.; Zhou, Y.; Xiong, Y. Recent advances in endotoxin tolerance. *J. Cell Biochem.* **2019**, *120*, 56–70. [CrossRef] [PubMed]
33. Hamers, L.; Kox, M.; Pickkers, P. Sepsis-induced immunoparalysis: Mechanisms, markers, and treatment options. *Minerva Anestesiol* **2014**, *81*, 426–439. [PubMed]
34. Venet, F.; Monneret, G. Advances in the understanding and treatment of sepsis-induced immunosuppression. *Nat. Rev. Nephrol.* **2018**, *14*, 121–137. [CrossRef]
35. Chen, J.; Wei, H. Immune intervention in sepsis. *Front. Pharmacol.* **2021**, *12*, 718089. [CrossRef] [PubMed]
36. Gantner, B.N.; LaFond, K.M.; Bonini, M.G. Nitric oxide in cellular adaptation and disease. *Redox Biol.* **2020**, *34*, 101550. [CrossRef] [PubMed]
37. Król, M.; Kepinska, M. Human nitric oxide Synthase—Its functions, polymorphisms, and inhibitors in the context of inflammation, diabetes and cardiovascular diseases. *Int. J. Mol. Sci.* **2020**, *22*, 56. [CrossRef] [PubMed]
38. Cinelli, M.A.; Do, H.T.; Miley, G.P.; Silverman, R.B. Inducible nitric oxide synthase: Regulation, structure, and inhibition. *Med. Res. Rev.* **2020**, *40*, 158–189. [CrossRef]
39. Ghafourifar, P.; Cadenas, E. Mitochondrial nitric oxide synthase. *Trends Pharmacol. Sci.* **2005**, *26*, 190–195. [CrossRef]
40. Förstermann, U.; Sessa, W.C. Nitric oxide synthases: Regulation and function. *Eur. Heart J.* **2012**, *33*, 829–837. [CrossRef]
41. Dynnik, V.V.; Grishina, E.V.; Fedotcheva, N.I. The mitochondrial NO-synthase/guanylate cyclase/protein kinase G signaling system underpins the dual effects of nitric oxide on mitochondrial respiration and opening of the permeability transition pore. *FEBS J.* **2019**, *287*, 1525–1536. [CrossRef]

42. DeMartino, A.W.; Kim-Shapiro, D.B.; Patel, R.P.; Gladwin, M.T. Nitrite and nitrate chemical biology and signalling. *Br. J. Pharmacol.* **2019**, *176*, 228–245. [CrossRef] [PubMed]
43. Kapil, V.; Khambata, R.S.; Jones, D.A.; Rathod, K.; Primus, C.; Massimo, G.; Fukuto, J.M.; Ahluwalia, A. The noncanonical pathway for in vivo nitric oxide generation: The nitrate-nitrite-nitric oxide pathway. *Pharmacol. Rev.* **2020**, *72*, 692–766. [CrossRef] [PubMed]
44. Nowaczyk, A.; Kowalska, M.; Nowaczyk, J.; Grześk, G. Carbon monoxide and nitric oxide as examples of the youngest class of transmitters. *Int. J. Mol. Sci.* **2021**, *22*, 6029. [CrossRef] [PubMed]
45. Loscalzo, J. The identification of nitric oxide as endothelium-derived relaxing factor. *Circ. Res.* **2013**, *113*, 100–103. [CrossRef]
46. Cirino, G.; Vellecco, V.; Bucci, M. Nitric oxide and hydrogen sulfide: The gasotransmitter paradigm of the vascular system. *J. Cereb. Blood Flow Metab.* **2017**, *174*, 4021–4031. [CrossRef]
47. Farah, C.; Michel, L.Y.; Balligand, J.-L. Nitric oxide signalling in cardiovascular health and disease. *Nat. Rev. Cardiol.* **2018**, *15*, 292–316. [CrossRef]
48. Ghimire, K.; Altmann, H.M.; Straub, A.C.; Isenberg, J.S. Nitric oxide: What's new to NO? *Am. J. Physiol. Cell Physiol.* **2017**, *312*, C254–C262. [CrossRef]
49. Shah, S.; Karathanasi, A.; Revythis, A.; Ioannidou, E.; Boussios, S. Cancer-Associated thrombosis: A new light on an old story. *Diseases* **2021**, *9*, 34. [CrossRef]
50. Fagone, P.; Mazzon, E.; Bramanti, P.; Bendtzen, K.; Nicoletti, F. Gasotransmitters and the immune system: Mode of action and novel therapeutic targets. *Eur. J. Pharmacol.* **2018**, *834*, 92–102. [CrossRef]
51. Kuschman, H.P.; Palczewski, M.B.; Thomas, D.D. Nitric oxide and hydrogen sulfide: Sibling rivalry in the family of epigenetic regulators. *Free Radic. Biol. Med.* **2021**, *170*, 34–43. [CrossRef]
52. McGinity, C.; Palmieri, E.; Somasundaram, V.; Bhattacharyya, D.; Ridnour, L.; Cheng, R.; Ryan, A.; Glynn, S.; Thomas, D.; Miranda, K.; et al. Nitric oxide modulates metabolic processes in the tumor immune microenvironment. *Int. J. Mol. Sci.* **2021**, *22*, 7068. [CrossRef] [PubMed]
53. Opatrilova, R.; Kubatka, P.; Caprnda, M.; Büsselberg, D.; Krasnik, V.; Veselý, P.; Saxena, S.; Ruia, S.; Mozos, I.; Rodrigo, L.; et al. Nitric oxide in the pathophysiology of retinopathy: Evidences from preclinical and clinical researches. *Acta Ophthalmol.* **2018**, *96*, 222–231. [CrossRef]
54. Hopper, C.P.; Zambrana, P.N.; Goebel, U.; Wollborn, J. A brief history of carbon monoxide and its therapeutic origins. *Nitric. Oxide* **2021**, *111-112*, 45–63. [CrossRef] [PubMed]
55. Wu, L.; Wang, R. Carbon monoxide: Endogenous production, physiological functions, and pharmacological applications. *Pharmacol. Rev.* **2005**, *57*, 585–630. [CrossRef] [PubMed]
56. Duvigneau, J.C.; Esterbauer, H.; Kozlov, A.V. Role of heme oxygenase as a modulator of heme-mediated pathways. *Antioxidants* **2019**, *8*, 475. [CrossRef] [PubMed]
57. Olas, B. Carbon monoxide is not always a poison gas for human organism: Physiological and pharmacological features of CO. *Chem. Interact.* **2014**, *222*, 37–43. [CrossRef]
58. Sethi, J.M. Carbon monoxide. *Crit. Care Med.* **2005**, *33*, S496–S497. [CrossRef]
59. Gullotta, F.; di Masi, A.; Coletta, M.; Ascenzi, P. CO metabolism, sensing, and signaling. *BioFactors* **2011**, *38*, 1–13. [CrossRef]
60. Li, L.; Hsu, A.; Moore, P.K. Actions and interactions of nitric oxide, carbon monoxide and hydrogen sulphide in the cardiovascular system and in inflammation—a tale of three gases! *Pharmacol. Ther.* **2009**, *123*, 386–400. [CrossRef]
61. Ameredes, B.T.; Otterbein, L.E.; Kohut, L.K.; Gligonic, A.L.; Calhoun, W.J.; Choi, A.M.K. Low-dose carbon monoxide reduces airway hyperresponsiveness in mice. *Am. J. Physiol. Cell. Mol. Physiol.* **2003**, *285*, L1270–L1276. [CrossRef] [PubMed]
62. Głowacka, U.; Brzozowski, T.; Magierowski, M. Synergisms, discrepancies and interactions between hydrogen sulfide and carbon monoxide in the gastrointestinal and digestive system physiology, pathophysiology and pharmacology. *Biomolecules* **2020**, *10*, 445. [CrossRef] [PubMed]
63. Motterlini, R.; Foresti, R. Biological signaling by carbon monoxide and carbon monoxide-releasing molecules. *Am. J. Physiol. Physiol.* **2017**, *312*, C302–C313. [CrossRef]
64. Takaki, S.; Takeyama, N.; Kajita, Y.; Yabuki, T.; Noguchi, H.; Miki, Y.; Inoue, Y.; Nakagawa, T.; Noguchi, H. Beneficial effects of the heme oxygenase-1/carbon monoxide system in patients with severe sepsis/septic shock. *Intensive Care Med.* **2010**, *36*, 42–48. [CrossRef] [PubMed]
65. Ryter, S.W.; Ma, K.C.; Choi, A.M.K. Carbon monoxide in lung cell physiology and disease. *Am. J. Physiol. Physiol.* **2018**, *314*, C211–C227. [CrossRef] [PubMed]
66. Owens, E.O. Endogenous carbon monoxide production in disease. *Clin. Biochem.* **2010**, *43*, 1183–1188. [CrossRef] [PubMed]
67. Zuhra, K.; Augsburger, F.; Majtan, T.; Szabo, C. Cystathionine-β-synthase: Molecular regulation and pharmacological inhibition. *Biomolecules* **2020**, *10*, 697. [CrossRef]
68. Szabo, C. A timeline of hydrogen sulfide (H2S) research: From environmental toxin to biological mediator. *Biochem. Pharmacol.* **2018**, *149*, 5–19. [CrossRef] [PubMed]
69. Olson, K.R. H2S and polysulfide metabolism: Conventional and unconventional pathways. *Biochem. Pharmacol.* **2018**, *149*, 77–90. [CrossRef]
70. Kabil, O.; Banerjee, R. Enzymology of H2S Biogenesis, Decay and Signaling. *Antioxidants Redox Signal.* **2014**, *20*, 770–782. [CrossRef]

71. Yang, G.; Wu, L.; Jiang, B.; Yang, W.; Qi, J.; Cao, K.; Meng, Q.; Mustafa, A.K.; Mu, W.; Zhang, S.; et al. H2S as a physiologic vasorelaxant: Hypertension in mice with deletion of cystathionine gamma-lyase. *Science* **2008**, *322*, 587–590. [CrossRef]
72. Kimura, H. Hydrogen sulfide: Its production and functions. *Exp. Physiol.* **2011**, *96*, 833–835. [CrossRef] [PubMed]
73. Kimura, H. The physiological role of hydrogen sulfide and beyond. *Nitric Oxide* **2014**, *41*, 4–10. [CrossRef]
74. Rose, P.; Moore, P.K.; Zhu, Y.Z. H2S biosynthesis and catabolism: New insights from molecular studies. *Cell. Mol. Life Sci.* **2017**, *74*, 1391–1412. [CrossRef] [PubMed]
75. Bhatia, M. Hydrogen sulfide and substance P in inflammation. *Antioxid. Redox Signal.* **2010**, *12*, 1191–1202. [CrossRef] [PubMed]
76. Kimura, H. Hydrogen sulfide (H(2)S) and polysulfide (H(2)S(n)) signaling: The first 25 years. *Biomolecules* **2021**, *11*, 896. [CrossRef] [PubMed]
77. Wang, R. Physiological implications of hydrogen sulfide: A whiff exploration that blossomed. *Physiol. Rev.* **2012**, *92*, 791–896. [CrossRef]
78. Wang, R. Hydrogen sulfide: A new EDRF. *Kidney Int.* **2009**, *76*, 700–704. [CrossRef]
79. Ping, N.-N.; Li, S.; Mi, Y.-N.; Cao, L.; Cao, Y.-X. Hydrogen sulphide induces vasoconstriction of rat coronary artery via activation of Ca 2+ influx. *Acta Physiol.* **2015**, *214*, 88–96. [CrossRef]
80. Ali, M.Y.; Ping, C.Y.; Mok, Y.Y.; Ling, L.; Whiteman, M.; Bhatia, M.; Moore, P.K. Regulation of vascular nitric oxide in vitro and in vivo; a new role for endogenous hydrogen sulphide? *Br. J. Pharmacol.* **2006**, *149*, 625–634. [CrossRef]
81. Lv, B.; Chen, S.; Tang, C.; Jin, H.; Du, J.; Huang, Y. Hydrogen sulfide and vascular regulation–An update. *J. Adv. Res.* **2021**, *27*, 85–97. [CrossRef]
82. Khattak, S.; Zhang, Q.-Q.; Sarfraz, M.; Muhammad, P.; Ngowi, E.; Khan, N.; Rauf, S.; Wang, Y.-Z.; Qi, H.-W.; Wang, D.; et al. The role of hydrogen sulfide in respiratory diseases. *Biomolecules* **2021**, *11*, 682. [CrossRef]
83. Madurga, A.; Golec, A.; Pozarska, A.; Ishii, I.; Mižíková, I.; Nardiello, C.; Vadász, I.; Herold, S.; Mayer, K.; Reichenberger, F.; et al. The H2S-generating enzymes cystathionine β-synthase and cystathionine γ-lyase play a role in vascular development during normal lung alveolarization. *Am. J. Physiol. Cell. Mol. Physiol.* **2015**, *309*, L710–L724. [CrossRef] [PubMed]
84. Dilek, N.; Papapetropoulos, A.; Toliver-Kinsky, T.; Szabo, C. Hydrogen sulfide: An endogenous regulator of the immune system. *Pharmacol. Res.* **2020**, *161*, 105119. [CrossRef]
85. Shefa, U.; Kim, M.-S.; Jeong, N.Y.; Jung, J. Antioxidant and cell-signaling functions of hydrogen sulfide in the central nervous system. *Oxidative Med. Cell. Longev.* **2018**, *2018*, 1–17. [CrossRef] [PubMed]
86. Manandhar, S.; Sinha, P.; Ejiwale, G.; Bhatia, M. Hydrogen sulfide and its interaction with other players in inflammation. *Adv. Exp. Med. Biol.* **2021**, *1315*, 129–159. [CrossRef]
87. Sun, F.; Luo, J.; Yue, T.; Wang, F.; Yang, C.; Zhang, J.; Wang, X.; Wang, C. The role of hydrogen sulphide signalling in macrophage activation. *Immunology* **2021**, *162*, 3–10. [CrossRef] [PubMed]
88. Du, J.; Jin, H.; Yang, L. Role of hydrogen sulfide in retinal diseases. *Front. Pharmacol.* **2017**, *8*, 588. [CrossRef] [PubMed]
89. Dai, J.; Teng, X.; Jin, S.; Wu, Y. The antiviral roles of hydrogen sulfide by blocking the interaction between SARS-CoV-2 and its potential cell surface receptors. *Oxid. Med. Cell Longev.* **2021**, *2021*, 7866992. [CrossRef] [PubMed]
90. Pae, H.-O.; Lee, Y.C.; Jo, E.-K.; Chung, H.-T. Subtle interplay of endogenous bioactive gases (NO, CO and H2S) in inflammation. *Arch. Pharmacal Res.* **2009**, *32*, 1155–1162. [CrossRef] [PubMed]
91. Lee, S.R.; Nilius, B.; Han, J. Gaseous signaling molecules in cardiovascular function: From mechanisms to clinical translation. *Rev. Physiol. Biochem. Pharmacol.* **2018**, *174*, 81–156. [CrossRef]
92. Liew, H.; Khoo, H.; Moore, P.; Bhatia, M.; Lu, J.; Moochhala, S. Synergism between hydrogen sulfide (H2S) and nitric oxide (NO) in vasorelaxation induced by stonustoxin (SNTX), a lethal and hypotensive protein factor isolated from stonefish Synanceja horrida venom. *Life Sci.* **2007**, *80*, 1664–1668. [CrossRef] [PubMed]
93. Whiteman, M.; Li, L.; Kostetski, I.; Chu, S.H.; Siau, J.L.; Bhatia, M.; Moore, P.K. Evidence for the formation of a novel nitrosothiol from the gaseous mediators nitric oxide and hydrogen sulphide. *Biochem. Biophys. Res. Commun.* **2006**, *343*, 303–310. [CrossRef] [PubMed]
94. Chung, H.T.; Choi, B.M.; Kwon, Y.G.; Kim, Y.M. Interactive relations between nitric oxide (NO) and carbon monoxide (CO): Heme oxygenase-1/CO pathway is a key modulator in NO-mediated antiapoptosis and anti-inflammation. *Methods Enzym.* **2008**, *441*, 329–338.
95. Muñoz-Sánchez, J.; Chánez-Cárdenas, M.E. A review on Hemeoxygenase-2: Focus on cellular protection and oxygen response. *Oxidative Med. Cell. Longev.* **2014**, *2014*, 1–16. [CrossRef]
96. Vicente, J.B.; Colaço, H.G.; Mendes, M.I.; Sarti, P.; Leandro, P.; Giuffrè, A. NO* binds human cystathionine β-synthase quickly and tightly. *J. Biol. Chem.* **2014**, *289*, 8579–8587. [CrossRef] [PubMed]
97. Asimakopoulou, A.; Panopoulos, P.; Chasapis, C.T.; Coletta, C.; Zhou, Z.; Cirino, G.; Giannis, A.; Szabo, C.; Spyroulias, G.A.; Papapetropoulos, A. Selectivity of commonly used pharmacological inhibitors for cystathionine β synthase (CBS) and cystathionine γ lyase (CSE). *Br. J. Pharmacol.* **2013**, *169*, 922–932. [CrossRef] [PubMed]
98. Zhao, W.; Zhang, J.; Lu, Y.; Wang, R. The vasorelaxant effect of H(2)S as a novel endogenous gaseous K(ATP) channel opener. *Embo. J.* **2001**, *20*, 6008–6016. [CrossRef]
99. Wesseling, S.; Fledderus, J.O.; Verhaar, M.; Joles, J.A. Beneficial effects of diminished production of hydrogen sulfide or carbon monoxide on hypertension and renal injury induced by NO withdrawal. *Br. J. Pharmacol.* **2015**, *172*, 1607–1619. [CrossRef]

100. Coletta, C.; Papapetropoulos, A.; Erdelyi, K.; Olah, G.; Modis, K.; Panopoulos, P.; Asimakopoulou, A.; Gero, D.; Sharina, I.; Martin, E.; et al. Hydrogen sulfide and nitric oxide are mutually dependent in the regulation of angiogenesis and endothelium-dependent vasorelaxation. *Proc. Natl. Acad. Sci. USA* **2012**, *109*, 9161–9166. [CrossRef]
101. Li, X.-H.; Xue, W.-L.; Wang, M.-J.; Zhou, Y.; Zhang, C.-C.; Sun, C.; Zhu, L.; Liang, K.; Chen, Y.; Tao, B.-B.; et al. H2S regulates endothelial nitric oxide synthase protein stability by promoting microRNA-455-3p expression. *Sci. Rep.* **2017**, *7*, 44807. [CrossRef] [PubMed]
102. Heine, C.L.; Schmidt, R.; Geckl, K.; Schrammel, A.; Gesslbauer, B.; Schmidt, K.; Mayer, B.; Gorren, A.C.F. Selective irreversible inhibition of neuronal and inducible nitric-oxide synthase in the combined presence of hydrogen sulfide and nitric oxide. *J. Biol. Chem.* **2015**, *290*, 24932–24944. [CrossRef] [PubMed]
103. Hua, W.; Chen, Q.; Gong, F.; Xie, C.; Zhou, S.; Gao, L. Cardioprotection of H2S by downregulating iNOS and upregulating HO-1 expression in mice with CVB3-induced myocarditis. *Life Sci.* **2013**, *93*, 949–954. [CrossRef] [PubMed]
104. Jeong, S.O.; Pae, H.O.; Oh, G.S.; Jeong, G.S.; Lee, B.S.; Lee, S.; Kim, D.Y.; Rhew, H.Y.; Lee, K.M.; Chung, H.T. Hydrogen sulfide potentiates interleukin-1beta-induced nitric oxide production via enhancement of extracellular signal-regulated kinase activation in rat vascular smooth muscle cells. *Biochem. Biophys Res. Commun.* **2006**, *345*, 938–944. [CrossRef] [PubMed]
105. De Cruz, S.J.; Kenyon, N.J.; Sandrock, C.E. Bench-to-bedside review: The role of nitric oxide in sepsis. *Expert Rev. Respir. Med.* **2009**, *3*, 511–521. [CrossRef] [PubMed]
106. Hollenberg, S.M.; Cinel, I. Bench-to-bedside review: Nitric oxide in critical illness–update 2008. *Crit. Care* **2009**, *13*, 218–219. [CrossRef] [PubMed]
107. Lin, M.-C.; Ebihara, S.; EL Dwairi, Q.; Hussain, S.N.A.; Yang, L.; Gottfried, S.B.; Comtois, A.; Petrof, B.J. Diaphragm Sarcolemmal injury is induced by sepsis and alleviated by nitric oxide synthase inhibition. *Am. J. Respir. Crit. Care Med.* **1998**, *158*, 1656–1663. [CrossRef] [PubMed]
108. Alkharfy, K.M.; Ahmad, A.; Raish, M.; Vanhoutte, P.M. Thymoquinone modulates nitric oxide production and improves organ dysfunction of sepsis. *Life Sci.* **2015**, *143*, 131–138. [CrossRef]
109. Luo, M.; Luo, S.; Cheng, Z.; Yang, X.; Lv, D.; Li, X.; Guo, Y.; Li, C.; Yan, J. Tubeimoside I improves survival of mice in sepsis by inhibiting inducible nitric oxide synthase expression. *Biomed. Pharmacother.* **2020**, *126*, 110083. [CrossRef]
110. Yu, M.-H.; Chen, M.-H.; Han, F.; Li, Q.; Sun, R.-H.; Tu, Y.-X. Prognostic value of the biomarkers serum amyloid A and nitric oxide in patients with sepsis. *Int. Immunopharmacol.* **2018**, *62*, 287–292. [CrossRef]
111. Lambden, S. Bench to bedside review: Therapeutic modulation of nitric oxide in sepsis—an update. *Intensiv. Care Med. Exp.* **2019**, *7*, 1–14. [CrossRef] [PubMed]
112. Cauwels, A. Nitric oxide in shock. *Kidney Int.* **2007**, *72*, 557–565. [CrossRef] [PubMed]
113. Cui, X.; Besch, V.; Khaibullina, A.; Hergen, A.; Quezado, M.; Eichacker, P.; Quezado, Z.M.N. Neuronal nitric oxide synthase deficiency decreases survival in bacterial peritonitis and sepsis. *Intensiv. Care Med.* **2007**, *33*, 1993–2003. [CrossRef]
114. Yadav, S.; Pathak, S.; Sarikhani, M.; Majumdar, S.; Ray, S.; Chandrasekar, B.S.; Adiga, V.; Sundaresan, N.R.; Nandi, D. Nitric oxide synthase 2 enhances the survival of mice during Salmonella Typhimurium infection-induced sepsis by increasing reactive oxygen species, inflammatory cytokines and recruitment of neutrophils to the peritoneal cavity. *Free Radic. Biol. Med.* **2018**, *116*, 73–87. [CrossRef]
115. Hoetzel, A.; Dolinay, T.; Schmidt, R.; Choi, A.M.K.; Ryter, S.W. Carbon monoxide in sepsis. *Antioxid. Redox Signal.* **2007**, *9*, 2013–2026. [CrossRef] [PubMed]
116. Kyokane, T.; Norimizu, S.; Taniai, H.; Yamaguchi, T.; Takeoka, S.; Tsuchida, E.; Naito, M.; Nimura, Y.; Ishimura, Y.; Suematsu, M. Carbon monoxide from heme catabolism protects against hepatobiliary dysfunction in endotoxin-treated rat liver. *Gastroenterology* **2001**, *120*, 1227–1240. [CrossRef] [PubMed]
117. Chung, S.W.; Liu, X.; Macias, A.A.; Baron, R.M.; Perrella, M.A. Heme oxygenase-1–derived carbon monoxide enhances the host defense response to microbial sepsis in mice. *J. Clin. Investig.* **2008**, *118*, 239–247. [CrossRef] [PubMed]
118. Tsoyi, K.; Lee, T.Y.; Lee, Y.S.; Kim, H.J.; Seo, H.G.; Lee, J.H.; Chang, K.C. Heme-Oxygenase-1 Induction and Carbon Monoxide-Releasing Molecule Inhibit Lipopolysaccharide (LPS)-Induced High-Mobility Group Box 1 Release in Vitro and Improve Survival of Mice in LPS- and Cecal Ligation and Puncture-Induced Sepsis Model in Vivo. *Mol. Pharmacol.* **2009**, *76*, 173–182. [CrossRef] [PubMed]
119. Adach, W.; Błaszczyk, M.; Olas, B. Carbon monoxide and its donors-Chemical and biological properties. *Chem. Interact.* **2020**, *318*, 108973. [CrossRef]
120. Zhang, W.; Tao, A.; Lan, T.; Cepinskas, G.; Kao, R.; Martin, C.M.; Rui, T. Carbon monoxide releasing molecule-3 improves myocardial function in mice with sepsis by inhibiting NLRP3 inflammasome activation in cardiac fibroblasts. *Basic Res. Cardiol.* **2017**, *112*, 16. [CrossRef] [PubMed]
121. Wang, H.; Zhang, S.; Zhao, H.; Qin, H.; Zhang, J.; Dong, J.; Zhang, H.; Liu, X.; Zhao, Z.; Zhao, Y.; et al. Carbon monoxide inhibits the expression of proteins associated with intestinal mucosal Pyroptosis in a rat model of sepsis induced by Cecal ligation and puncture. *Med. Sci. Monit.* **2020**, *26*, e920668-1. [CrossRef] [PubMed]
122. Wegiel, B.; Larsen, R.; Gallo, D.; Chin, B.Y.; Harris, C.; Mannam, P.; Kaczmarek, E.; Lee, P.J.; Zuckerbraun, B.S.; Flavell, R.; et al. Macrophages sense and kill bacteria through carbon monoxide–dependent inflammasome activation. *J. Clin. Investig.* **2014**, *124*, 4926–4940. [CrossRef] [PubMed]

123. Wang, P.; Huang, J.; Li, Y.; Chang, R.; Wu, H.; Lin, J.; Huang, Z. Exogenous carbon monoxide decreases sepsis-induced acute kidney injury and inhibits NLRP3 inflammasome activation in rats. *Int. J. Mol. Sci.* **2015**, *16*, 20595–20608. [CrossRef] [PubMed]
124. Lancel, S.; Hassoun, S.M.; Favory, R.; Decoster, B.; Motterlini, R.; Neviere, R. Carbon monoxide rescues mice from lethal sepsis by supporting mitochondrial energetic metabolism and activating mitochondrial biogenesis. *J. Pharmacol. Exp. Ther.* **2009**, *329*, 641–648. [CrossRef] [PubMed]
125. MacGarvey, N.C.; Suliman, H.B.; Bartz, R.R.; Fu, P.; Withers, C.M.; Welty-Wolf, K.E.; Piantadosi, C.A. Activation of mitochondrial biogenesis by heme oxygenase-1-mediated NF-E2-related factor-2 induction rescues mice from lethal Staphylococcus aureus sepsis. *Am. J. Respir. Crit. Care Med.* **2012**, *185*, 851–861. [CrossRef] [PubMed]
126. Unuma, K.; Aki, T.; Nagano, S.; Watanabe, R.; Uemura, K. The down-regulation of cardiac contractile proteins underlies myocardial depression during sepsis and is mitigated by carbon monoxide. *Biochem. Biophys. Res. Commun.* **2018**, *495*, 1668–1674. [CrossRef] [PubMed]
127. Qin, W.; Zhang, J.; Lv, W.; Wang, X.; Sun, B. Effect of carbon monoxide-releasing molecules II-liberated CO on suppressing inflammatory response in sepsis by interfering with nuclear factor kappa B activation. *PLoS ONE* **2013**, *8*, e75840. [CrossRef]
128. Riquelme, S.A.; Bueno, S.M.; Kalergis, A.M. Carbon monoxide down-modulates Toll-like receptor 4/MD2 expression on innate immune cells and reduces endotoxic shock susceptibility. *Immunology* **2015**, *144*, 321–332. [CrossRef]
129. Liu, D.; Wang, X.; Qin, W.; Chen, J.; Wang, Y.; Zhuang, M.; Sun, B. Suppressive effect of exogenous carbon monoxide on endotoxin-stimulated platelet over-activation via the glycoprotein-mediated PI3K-Akt-GSK3β pathway. *Sci. Rep.* **2016**, *6*, 23653. [CrossRef]
130. Qin, S.; Du, R.; Yin, S.; Liu, X.; Xu, G.; Cao, W. Nrf2 is essential for the anti-inflammatory effect of carbon monoxide in LPS-induced inflammation. *Inflamm. Res.* **2015**, *64*, 537–548. [CrossRef]
131. Tsoyi, K.; Hall, S.R.R.; Dalli, J.; Colas, R.A.; Ghanta, S.; Ith, B.; Coronata, A.; Fredenburgh, L.E.; Baron, R.M.; Choi, A.M.K.; et al. Carbon monoxide improves efficacy of mesenchymal stromal cells during sepsis by production of specialized proresolving lipid mediators*. *Crit. Care Med.* **2016**, *44*, e1236–e1245. [CrossRef] [PubMed]
132. Zegdi, R.; Perrin, D.; Burdin, M.; Boiteau, R.; Tenaillon, A. Increased endogenous carbon monoxide production in severe sepsis. *Intensiv. Care Med.* **2002**, *28*, 793–796. [CrossRef]
133. Iwasashi, H.; Suzuki, M.; Unno, M.; Utiyama, T.; Oikawa, M.; Kondo, N.; Matsuno, S. Inhibition of heme oxygenase ameliorates sepsis-induced liver dysfunction in rats. *Surg. Today* **2003**, *33*, 30–38. [CrossRef] [PubMed]
134. Hui, Y.; Du, J.; Tang, C.; Bin, G.; Jiang, H. Changes in arterial hydrogen sulfide (H2S) content during septic shock and endotoxin shock in rats. *J. Infect.* **2003**, *47*, 155–160. [CrossRef]
135. Zhang, H.; Zhi, L.; Moore, P.K.; Bhatia, M. Role of hydrogen sulfide in cecal ligation and puncture-induced sepsis in the mouse. *Am. J. Physiol. Cell. Mol. Physiol.* **2006**, *290*, L1193–L1201. [CrossRef] [PubMed]
136. Zhang, H.; Zhi, L.; Moochhala, S.; Moore, P.K.; Bhatia, M. Hydrogen sulfide acts as an inflammatory mediator in cecal ligation and puncture-induced sepsis in mice by upregulating the production of cytokines and chemokines via NF-kappaB. *Am. J. Physiol. Lung Cell Mol. Physiol.* **2007**, *292*, L960–L971. [CrossRef] [PubMed]
137. Zhang, H.; Zhi, L.; Moochhala, S.M.; Moore, P.K.; Bhatia, M. Endogenous hydrogen sulfide regulates leukocyte trafficking in cecal ligation and puncture-induced sepsis. *J. Leukoc. Biol.* **2007**, *82*, 894–905. [CrossRef] [PubMed]
138. Zhang, H.; Moochhala, S.M.; Bhatia, M. Endogenous hydrogen sulfide regulates inflammatory response by activating the ERK pathway in polymicrobial sepsis. *J. Immunol.* **2008**, *181*, 4320–4331. [CrossRef]
139. Zhang, H.; Hegde, A.; Ng, S.W.; Adhikari, S.; Moochhala, S.M.; Bhatia, M. Hydrogen sulfide up-regulates substance P in polymicrobial sepsis-associated lung injury. *J. Immunol.* **2007**, *179*, 4153–4160. [CrossRef]
140. Ang, S.-F.; Moochhala, S.M.; Bhatia, M. Hydrogen sulfide promotes transient receptor potential vanilloid 1-mediated neurogenic inflammation in polymicrobial sepsis*. *Crit. Care Med.* **2010**, *38*, 619–628. [CrossRef] [PubMed]
141. Ang, S.F.; Moochhala, S.M.; MacAry, P.A.; Bhatia, M. Hydrogen sulfide and neurogenic inflammation in polymicrobial sepsis: Involvement of substance P and ERK-NF-κB signaling. *PLoS ONE* **2011**, *6*, e24535. [CrossRef] [PubMed]
142. Ang, S.-F.; Sio, S.W.S.; Moochhala, S.M.; Macary, P.A.; Bhatia, M. Hydrogen Sulfide Upregulates Cyclooxygenase-2 and Prostaglandin E Metabolite in Sepsis-Evoked Acute Lung Injury via Transient Receptor Potential Vanilloid Type 1 Channel Activation. *J. Immunol.* **2011**, *187*, 4778–4787. [CrossRef] [PubMed]
143. Badiei, A.; Chambers, S.; Gaddam, R.R.; Bhatia, M. Cystathionine-γ-lyase gene silencing with siRNA in monocytes/macrophages attenuates inflammation in cecal ligation and puncture-induced sepsis in the mouse. *J. Biosci.* **2016**, *41*, 87–95. [CrossRef] [PubMed]
144. Norris, E.J.; Feilen, N.; Nguyen, N.H.; Culberson, C.R.; Shin, M.C.; Fish, M.; Clemens, M.G. Hydrogen sulfide modulates sinusoidal constriction and contributes to hepatic microcirculatory dysfunction during endotoxemia. *Am. J. Physiol. Gastrointest. Liver Physiol.* **2013**, *304*, G1070–G1078. [CrossRef]
145. Poisson, J.; Lemoinne, S.; Boulanger, C.M.; Durand, F.; Moreau, R.; Valla, D.; Rautou, P.-E. Liver sinusoidal endothelial cells: Physiology and role in liver diseases. *J. Hepatol.* **2017**, *66*, 212–227. [CrossRef]
146. Gaddam, R.R.; Fraser, R.; Badiei, A.; Chambers, S.; Cogger, V.C.; Le Couteur, D.G.; Ishii, I.; Bhatia, M. Cystathionine-Gamma-Lyase Gene Deletion Protects Mice against Inflammation and Liver Sieve Injury following Polymicrobial Sepsis. *PLoS ONE* **2016**, *11*, e0160521. [CrossRef]

147. Gaddam, R.R.; Chambers, S.; Fraser, R.; Cogger, V.C.; Le Couteur, D.G.; Ishii, I.; Bhatia, M. Cystathionine-Gamma-Lyase-Derived hydrogen sulfide-regulated substance P modulates liver sieve fenestrations in Caecal ligation and puncture-induced sepsis. *Int. J. Mol. Sci.* **2019**, *20*, 3191. [CrossRef]
148. Liu, S.; Wang, X.; Pan, L.; Wu, W.; Yang, D.; Qin, M.; Jia, W.; Xiao, C.; Long, F.; Ge, J.; et al. Endogenous hydrogen sulfide regulates histone demethylase JMJD3-mediated inflammatory response in LPS-stimulated macrophages and in a mouse model of LPS-induced septic shock. *Biochem. Pharmacol.* **2018**, *149*, 153–162. [CrossRef] [PubMed]
149. Gaddam, R.R.; Chambers, S.; Murdoch, D.; Shaw, G.; Bhatia, M. Circulating levels of hydrogen sulfide and substance P in patients with sepsis. *J. Infect.* **2017**, *75*, 293–300. [CrossRef] [PubMed]
150. Bee, N.; White, R.; Petros, A.J. Hydrogen sulfide in exhaled gases from ventilated septic neonates and children: A preliminary report. *Pediatr. Crit. Care Med.* **2017**, *18*, e327–e332. [CrossRef]
151. Spiller, F.; Orrico, M.I.; Nascimento, D.C.; Czaikoski, P.G.; Souto, F.O.; Alves-Filho, J.C.; Freitas, A.; Carlos, D.; Montenegro, M.F.; Neto, A.F.; et al. Hydrogen sulfide improves neutrophil migration and survival in sepsis via K+ATP channel activation. *Am. J. Respir. Crit. Care Med.* **2010**, *182*, 360–368. [CrossRef] [PubMed]
152. Ahmad, A.; Druzhyna, N.; Szabo, C. Delayed Treatment with sodium hydrosulfide improves regional blood flow and alleviates Cecal ligation and puncture (CLP)-Induced septic shock. *Shock* **2016**, *46*, 183–193. [CrossRef] [PubMed]
153. Zhang, H.-X.; Du, J.-M.; Ding, Z.-N.; Zhu, X.-Y.; Jiang, L.; Liu, Y.-J. Hydrogen sulfide prevents diaphragm weakness in Cecal ligation puncture-induced sepsis by preservation of mitochondrial function. *Am. J. Transl. Res.* **2017**, *9*, 3270–3281. [PubMed]
154. Li, J.; Ma, J.; Li, M.; Tao, J.; Chen, J.; Yao, C.; Yao, S. GYY4137 alleviates sepsis-induced acute lung injury in mice by inhibiting the PDGFRβ/Akt/NF-κB/NLRP3 pathway. *Life Sci.* **2021**, *271*, 119192. [CrossRef]
155. Li, J.; Li, M.; Li, L.; Ma, J.; Yao, C.; Yao, S. Hydrogen sulfide attenuates ferroptosis and stimulates autophagy by blocking mTOR signaling in sepsis-induced acute lung injury. *Mol. Immunol.* **2021**, *141*, 318–327. [CrossRef] [PubMed]
156. Qiu, H.; Chen, X.; Luo, Z.; Zhao, L.; Zhang, T.; Yang, N.; Long, X.; Xie, H.; Liu, J.; Xu, W. Inhibition of endogenous hydrogen sulfide production exacerbates the inflammatory response during urine-derived sepsis-induced kidney injury. *Exp. Ther. Med.* **2018**, *16*, 2851–2858. [CrossRef]
157. Renieris, G.; Droggiti, D.-E.; Katrini, K.; Koufargyris, P.; Gkavogianni, T.; Karakike, E.; Antonakos, N.; Damoraki, G.; Karageorgos, A.; Sabracos, L.; et al. Host cystathionine-γ lyase derived hydrogen sulfide protects against Pseudomonas aeruginosa sepsis. *PLOS Pathog.* **2021**, *17*, e1009473. [CrossRef] [PubMed]
158. Aslami, H.; Pulskens, W.P.; Kuipers, M.T.; Bos, A.P.; Van Kuilenburg, A.B.P.; Wanders, R.J.A.; Roelofsen, J.; Roelofs, J.; Kerindongo, R.P.; Beurskens, C.J.P.; et al. Hydrogen sulfide donor NaHS reduces organ injury in a rat model of pneumococcal pneumosepsis, associated with improved bio-energetic status. *PLoS ONE* **2013**, *8*, e63497. [CrossRef] [PubMed]
159. Cui, W.; Chen, J.; Yu, F.; Liu, W.; He, M. GYY4137 protected the integrity of the blood-brain barrier via activation of the Nrf2/ARE pathway in mice with sepsis. *FASEB J.* **2021**, *35*, e21710. [CrossRef] [PubMed]
160. Harbrecht, B.G. Therapeutic use of nitric oxide scavengers in shock and sepsis. *Curr. Pharm. Des.* **2006**, *12*, 3543–3549. [CrossRef]
161. Teman, N.R.; Thomas, J.; Bryner, B.S.; Haas, C.F.; Haft, J.W.; Park, P.K.; Lowell, M.J.; Napolitano, L.M. Inhaled nitric oxide to improve oxygenation for safe critical care transport of adults with severe hypoxemia. *Am. J. Crit. Care* **2015**, *24*, 110–117. [CrossRef] [PubMed]
162. Trzeciak, S.; Glaspey, L.J.; Dellinger, R.P.; Durflinger, P.; Anderson, K.; Dezfulian, C.; Roberts, B.W.; Chansky, M.E.; Parrillo, J.E.; Hollenberg, S.M. Randomized Controlled trial of inhaled nitric oxide for the treatment of microcirculatory dysfunction in patients with sepsis*. *Crit. Care Med.* **2014**, *42*, 2482–2492. [CrossRef] [PubMed]
163. Dumbarton, T.C.; Maxan, A.; Farah, N.; Sharawy, N.; Zhou, J.; Nantais, J.; Lehmann, C. Tetrahydrobiopterin improves microcirculation in experimental sepsis. *Clin. Hemorheol. Microcirc.* **2017**, *67*, 15–24. [CrossRef] [PubMed]
164. López, A.; Lorente, J.A.; Steingrub, J.; Bakker, J.; McLuckie, A.; Willatts, S.; Brockway, M.; Anzueto, A.; Holzapfel, L.; Breen, D.; et al. Multiple-center, randomized, placebo-controlled, double-blind study of the nitric oxide synthase inhibitor 546C88: Effect on survival in patients with septic shock*. *Crit. Care Med.* **2004**, *32*, 21–30. [CrossRef] [PubMed]
165. Bakker, J.; Grover, R.; McLuckie, A.; Holzapfel, L.; Andersson, J.; Lodato, R.; Watson, D.; Grossman, S.; Donaldson, J.; Takala, J. Administration of the nitric oxide synthase inhibitor NG-methyl-l-arginine hydrochloride (546C88) by intravenous infusion for up to 72 hours can promote the resolution of shock in patients with severe sepsis: Results of a randomized, double-blind, placebo-controlled multicenter study (study no. 144-002)*. *Crit. Care Med.* **2004**, *32*, 1–12. [CrossRef] [PubMed]
166. Kim, H.L.; Park, Y.S. Maintenance of cellular tetrahydrobiopterin homeostasis. *BMB Rep.* **2010**, *43*, 584–592. [CrossRef]
167. Fredenburgh, L.E.; Perrella, M.A.; Barragan-Bradford, D.; Hess, D.R.; Peters, E.; Welty-Wolf, K.E.; Kraft, B.D.; Harris, R.S.; Maurer, R.; Nakahira, K.; et al. A phase I trial of low-dose inhaled carbon monoxide in sepsis-induced ARDS. *JCI Insight* **2018**, *3*, 23. [CrossRef] [PubMed]

International Journal of
Molecular Sciences

MDPI

Review

Nutraceutical Concepts and Dextrin-Based Delivery Systems

Gjylije Hoti, Adrián Matencio, Alberto Rubin Pedrazzo, Claudio Cecone, Silvia Lucia Appleton, Yousef Khazaei Monfared, Fabrizio Caldera and Francesco Trotta *

Department of Chemistry, University of Torino, Via P. Giuria 7, 10125 Torino, Italy; gjylije.hoti@unito.it (G.H.); adrian.matencioduran@unito.it (A.M.); alberto.rubinpedrazzo@unito.it (A.R.P.); claudio.cecone@unito.it (C.C.); silvialucia.appleton@unito.it (S.L.A.); yousef.khazaeimonfared@unito.it (Y.K.M.); fabrizio.caldera@unito.it (F.C.)
* Correspondence: francesco.trotta@unito.it

Abstract: Nutraceuticals are bioactive or chemical compounds acclaimed for their valuable biological activities and health-promoting effects. The global community is faced with many health concerns such as cancers, cardiovascular and neurodegenerative diseases, diabetes, arthritis, osteoporosis, etc. The effect of nutraceuticals is similar to pharmaceuticals, even though the term nutraceutical has no regulatory definition. The usage of nutraceuticals, to prevent and treat the aforementioned diseases, is limited by several features such as poor water solubility, low bioavailability, low stability, low permeability, low efficacy, etc. These downsides can be overcome by the application of the field of nanotechnology manipulating the properties and structures of materials at the nanometer scale. In this review, the linear and cyclic dextrin, formed during the enzymatic degradation of starch, are highlighted as highly promising nanomaterials- based drug delivery systems. The modified cyclic dextrin, cyclodextrin (CD)-based nanosponges (NSs), are well-known delivery systems of several nutraceuticals such as quercetin, curcumin, resveratrol, thyme essential oil, melatonin, and appear as a more advanced drug delivery system than modified linear dextrin. CD-based NSs prolong and control the nutraceuticals release, and display higher biocompatibility, stability, and solubility of poorly water-soluble nutraceuticals than the CD-inclusion complexes, or uncomplexed nutraceuticals. In addition, the well-explored CD-based NSs pathways, as drug delivery systems, are described. Although important progress is made in drug delivery, all the findings will serve as a source for the use of CD-based nanosystems for nutraceutical delivery. To sum up, our review introduces the extensive literature about the nutraceutical concepts, synthesis, characterization, and applications of the CD-based nano delivery systems that will further contribute to the nutraceutical delivery with more potent nanosystems based on linear dextrins.

Keywords: disease; nutraceuticals; nano-carrier; starch; linear dextrin; cyclic dextrin; nanosponges; drug delivery; nutraceutical delivery

Citation: Hoti, G.; Matencio, A.; Rubin Pedrazzo, A.; Cecone, C.; Appleton, S.L.; Khazaei Monfared, Y.; Caldera, F.; Trotta, F. Nutraceutical Concepts and Dextrin-Based Delivery Systems. *Int. J. Mol. Sci.* **2022**, *23*, 4102. https://doi.org/10.3390/ijms23084102

Academic Editor: Maurizio Battino

Received: 19 January 2022
Accepted: 2 April 2022
Published: 7 April 2022

1. Introduction

Industrialization and changing work cultures have caused numerous air and water pollutions, soil, and food contamination because of the extensive use of various harmful man-made items such as chemicals, heavy metals, electromagnetic waves, etc. At the same time, economic development has simultaneously drastically changed human lifestyles, which have become fast-eating cultures with decreasing nutrient quality. Therefore, due to nutritional deficiencies, there is an increase in the development of several diseases such as diabetes, obesity, various cancers, neurodegenerative diseases, heart disease, physiological problems, hypertension and dyslipidemia, chronic and vascular diseases, osteoporosis, arthritis, and many others.

Based on the fact that the raised demands for health care have dramatically increased, more and more people eat minimally processed foods such as vegetables, fruits, and other plant foods, taking dietary supplements or nutraceuticals instead [1–6]. The maintenance of the normal functioning of human body, recognized as a global issue, is reached by obtaining appropriate nutrients from various foods [1,6,7].

For centuries, the major concern of humankind around the world has been the research, development, and commercialization of nutraceuticals, functional food ingredients, and dietary supplements. There are a few challenges to defining the health benefits of certain foods, improving immune function, preventing specific diseases, and reducing side effects and health care costs [3,5,8,9]. The estimation of the mechanism of action and the efficacy of nutraceuticals have been encouraged as a consequence of the challenges of nutraceuticals with safety and health claim trials. Nutraceuticals comprise prebiotics, probiotics, polyunsaturated fatty acids, antioxidants, herbal products, etc. [10,11]. As more consumers use nutraceuticals for disease prevention [12], their efficacy as therapeutic agents is determined by different pathways. Based on drug studies and knowledge, the requirements to achieve the therapeutic purpose comprise the improvement of bioavailability, biocompatibility, solubility, loading efficacy, and toxicity as well as controlling the release, broadening the activity, adjusting the pharmacokinetics (PKs) of the drugs, etc. [13]. Therefore, the therapeutic efficacy of a drug can be improved, and toxic effects can be reduced by developing a drug delivery system. The drug delivery system can be controlled and targeted. Among various drug delivery systems, the molecular nano-carrier has produced great interest within the scientific world [14]. This review will explicitly focus on the application of dextrins in the drug delivery industry because of their non-toxic, biodegradable, and biocompatible nature, water solubility, or high encapsulation for swelling caused by simple chemical modifications [15,16]. These findings on well-explored drug delivery systems will enable the successful development of nutraceutical delivery systems, which are lacking due to the existing contradictory information regarding the nutraceutical term [17].

Dextrins, starch derivatives [18], are well-known for their great potential to develop hydrogels because of their efficient absorption related to degradation by amylases, and their proven clinical tolerability [19]. According to the molecular structure, dextrin can be divided into linear, branched dextrin, and cyclodextrin [20]. Dextrins consist of D-glucose units linked primarily by α (1,4)-glycosidic bonds [21], and branched segments linked by α (1,6)-glycosidic bonds [22]. Cyclodextrins (CDs) and linear dextrins have the same physicochemical, and biological characteristics, but CDs due to their cyclic structure are more resistant to non-enzymatic hydrolysis [23]. CDs are characterized by a typical toroidal cone shape with a lipophilic interior and hydrophilic exterior. Therefore, this peculiar structure enables CDs to form inclusion complexes with the compounds that have the size and polarity compatible with CDs structure [24]. Starch hydrolyzates, with the values of dextrose equivalent (DE, 1–20), are known as maltodextrins. Maltodextrin is a linear dextrin consisting of linear (amylase) and branched (amylopectin) carbohydrates [22,25–27]. Whether the linear dextrin acts comparably as CDs, with one side of the molecule being hydrophilic and the other hydrophobic, is a question that has been addressed over recent decades. A study suggested that the dextrin chains are amphiphilic ribbons, and under certain conditions, the hydrophobic surfaces are noticed by either hydrophobic or amphiphilic molecules [28]. However, viscosity drop during storage, poor solubility, uncontrolled hydration rate, and microbial contamination are some limitations of dextrin [29]. One approach that was applied to address the limitations issue is the use of chemical modification [30,31]. Chemical modification of dextrin can lead to the formation of dextrin polymers and nanosponges (NSs). Nanosponges (NSs) are hyper-cross-linked polymers that can be obtained by reacting CDs or maltodextrins with an appropriate cross-linking agent [32].

As explored from the historical evolution of NSs, cyclodextrin (CD) polymers have found their applications as food component carriers in the 1990s. Whereas CD-NSs are widely explored as drug delivery systems in the new millennium [33]. CD-NSs have shown more advantages compared to the native CDs in entrapping guest molecules, reducing their side effects, improving their stability, extending their release, etc. [34]. This is because CD-NSs, as chemically three-dimensional nanoporous polymeric networks, have various attractive features for use as hydrogels [35]. This strategy facilitates the slower drug elution and maintains a high concentration of drugs in the surrounding tissues over an extended period [36,37]. The CD: cross-linking agent molar ratio affects the nanochannels produced, the swelling, and therefore the loading capacity, and the drug release [35]. The diffusion process in the cross-linked polymer or dense macromolecular networks is slow, and thus the drug release is prolonged and controlled [38]. A free hydrophilic drug, that can freely spread in the aqueous medium, interacts with the hydrophilic zone of the biological membrane model but is unable to pass through the lipophilic layer of the same membrane. For instance, in the case of ester-bridged NSs based on β-CD, the electrostatic interactions of the carboxylic groups of dianhydride bridges with polar moieties of hydrophilic drugs can provide interaction with the hydrophilic layer of the membrane, while the inclusion complex formation with the lipophilic part can provide close interaction with the hydrophobic layer of the membrane, providing a high permeation of the drug [14,35]. Due to electrostatic repulsion, the polymer swells, and the volume of its network increases. Consequently, this increases the pore size of the NSs, and the drug is released [39]. CD-NSs have evolved alongside generations, from the plain NSs to modified NSs, to stimuli-responsive CD polymers, and to molecularly imprinted CD (MIPs-CD) polymers. Delivery challenges for each native CD have been addressed through the CD-NSs generations which have improved the delivery kinetics for most of the therapeutic agents. CD-NSs are well-known delivery systems of several nutraceuticals such as quercetin [40], curcumin [41], resveratrol [42], thyme essential oil [43], melatonin [44], etc.

The potential of any innovative or discovery process is greater when the obstacles between two of its basic ingredients, such as having an idea and testing it, are fewer [45]. Thus, our idea to review the nutraceuticals concept, the CD-NSs for their delivery, along with this entry, will support the use of maltodextrins modification for nutraceuticals delivery. This review attempts to summarize the recent headway on this new health care concept. As Hulda Regehr Clark quoted in her book, "The Cure for All Advanced Cancers": "*. . . never take defeat. When all is lost, try something new. Life is too precious to let it slip away from lack of initiative or plain inertia*".

2. Nutraceuticals

2.1. What Are Nutraceuticals?

Nutraceuticals are natural bioactive or chemical compounds that possess valuable biological activities and demonstrated physiological benefits. Therefore, they offer promotion of the body's natural healing, prevention, and treatment of the disease [5,46,47]. The introduction of many nutraceuticals, as presented in Figure 1, has undoubtedly caused an increasing challenge for nutritionists, physicians, food technologists, and chemists [46] based on the goal of providing a positive impact on human health [48].

The fascinating topic of the food supply has existed throughout history [46]. The connection between the use of relevant foods and health was established by the father of modern medicine, Hippocrates (460–377 BC), more than 2500 years ago, who made the statement "*Let food be thy medicine and medicine be thy food*" [1,3,5]. The term "Nutraceutical" originates from two broad terms such as "Nutrition" and "Pharmaceutical" [4,6,9,17,49–52], and was coined in 1989 by Stephen DeFelice, MD, founder, and chairman of the Foundation for Innovation in Medicine (FIM), Cranford, New Jersey. According to him, the term nutraceutical is defined as "*a food or a part of a food that provides medical or health benefits, including the prevention and the treatment of a disease*", because in his opinion "*the nutraceutical revolution will lead us into a new era of medicine and health, in which the food industry by the year 2000 will become a research-oriented one similar to the pharmaceutical industry*" [45]. This concept has been presented as a modern approach to food science. However, the definition of nutraceuticals and a legitimate assessment of their potential in medicine are still in opposition [17].

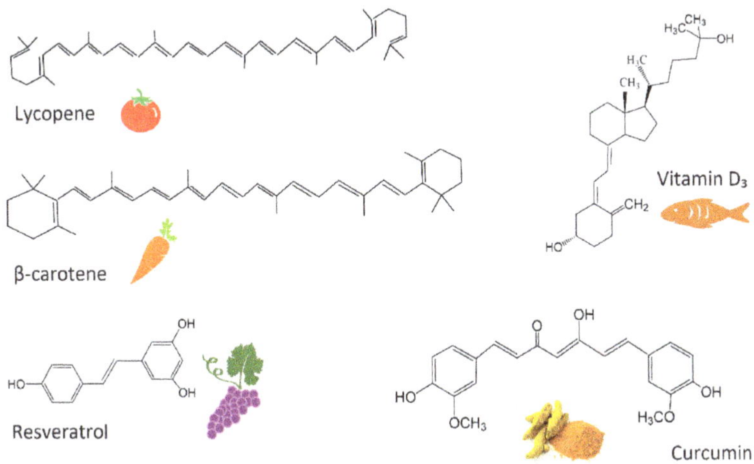

Figure 1. Nutraceuticals.

Stephen L. DeFelice, in a 2014 lecture, admitted that the clinical trials, which have proven the malfunctioning of the dietary supplements and diets, may not have been designed properly. From his standpoint, the reason why dietary supplements and diets do not work is that the cell is not deficient in them and does not need them. Further, he revealed his theory called, "*the cell-nutraceutical acceptance-rejection theory*" related to the lack of efficacy and toxicity. It is a self-explanatory theory, probably explained by the fact that any medication is unlikely to produce benefits if it does not cause harm. In the end, DeFelice stated that the nutrition area will be ongoing, and there is just a need for some new creative thinkers [53,54]. The research in the nutraceuticals area, judging from the number of journal articles indexed in PubMed, has increased steeply since 2000 and is continuously developed [17,53], as presented in Figure 2.

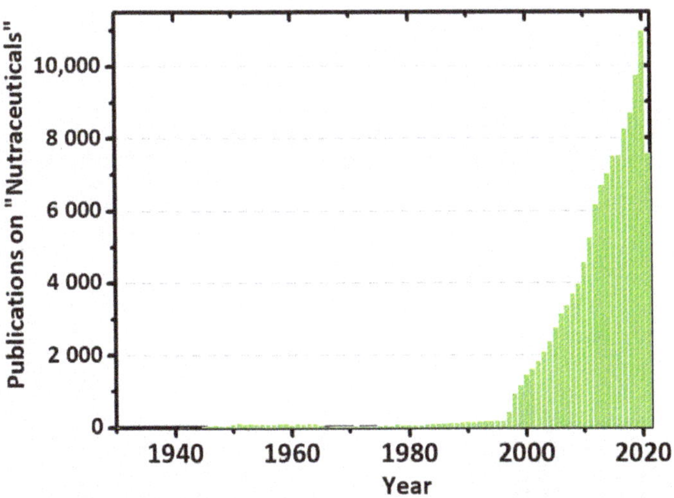

Figure 2. Graph representing the number of research papers (found in PubMed) published on nutraceuticals per year.

2.2. Nutraceuticals vs. Other Definitions/Regulations

The term nutraceutical has no regulatory definition, and the existing contradictory information is generating confusion about the possible effective use of these products. This may be due to a lack of studies on possible mechanisms of action and in vivo research confirming the declared beneficial health effects on specific pathological conditions, as mentioned in the lecture of DeFelice in 2014 [17,49,51]. Therefore, this situation has encouraged the utmost interest for the need of assessing the safety, mechanism of action, and efficacy of nutraceuticals with clinical data. There has been a lot of confusion between the term "nutraceuticals" and the others such as "functional foods", dietary supplements", "designer foods", "medical foods", "pharmafoods", "phytochemicals" etc. (Figure 3). Nutraceuticals, standing between pharmaceuticals and foods, have experienced challenges with safety and health claim trials [11,55]. In comparison to the pharmaceuticals, uni-targeted pure compounds with high-dose use, nutraceuticals are multi-targeted mixtures existing at low concentrations [47]. While the concepts of nutraceuticals, medical or functional foods, and dietary supplements do not have a clear accepted definition, they can most often be used in an exchangeable way [5]. Certain organizations proposed several definitions for nutraceuticals as one of the most active areas of research with a deficiency of a favorable regulatory environment. The understanding of the modern concept of functional food related to the maintenance of health in the general population was proposed by the Japanese academic society in the early 1980s, which implemented the legislation "Foods for Specified Health Use (FOSHU)" [56].

The definition for "functional food" describes, "*food products fortified with special constituents that possess advantageous physiological effects*", whereas, "*the approved health claim; recommended daily intake of the food; nutrition information; guidance on healthy eating; a warning against excessive intake, if necessary; any other special precautions relating to intake, preparation or storage; and other information*", has been the completed FOSHU label. The other legislations that have influenced the dissemination of information to consumers about the relationship between the diet and health in food regulations are the Functional Food Science in Europe (FUFOSE) Concerted Action, NLEA in 1990 [57], the Dietary Supplement Health and Education Act (DSHEA) in 1994 [58], FDA Modernization Act, in 1997 [59], and Functional Food Center (FFC) [56,60–68].

Functional foods provide the required amounts of vitamins, fats, proteins, carbohydrates, etc., that the human body needs for healthy survival. The functional foods are called a Nutraceutical when they assist in the precaution and the treatment of any disease or disorder [69]. As dietary supplements are not considered to treat, cure, or prevent disease [66], the definition of nutraceuticals as, "*those diet supplements that deliver a concentrated form of a presumed bioactive agent from a food, presented in a non-food matrix, and used to enhance health in dosage that exceeds those that could be obtained from normal food*" is used to distinguish whole foods from the isolated components derived from them and to create the category of nutraceuticals for dietary supplements that can obtain pharmacological effects [65,70].

Medical foods are a specific category of therapeutic agents that are intended for the specific dietary management of the disease [66,69]. The term "phytochemicals" refers to a group of plant secondary metabolites that may account for numerous beneficial health effects [71,72] and have the potential of being incorporated into foods, nutraceuticals, or pharmaceuticals [73]. The United Kingdom, Germany, and France are the first countries considering that diet is a more important factor than exercise in achieving good health.

The health ministry of Canada, the Ministry of Agriculture, Fisheries and Food of Britain, the Merriam Webster Dictionary, the dietary supplement industry, the North American Veterinary Nutraceutical Council Inc. presented various definitions that modified the meaning of the term "nutraceutical" [2,53,74,75].

"A substance that is cultivated/produced/extracted or synthesized under optimal and reproducible conditions and when administered orally to patients, would provide the nutrient required for bringing altered body structure and function back to normal, thus improving the health and wellbeing of the patients", is a more appropriate definition for nutraceuticals based on the abovementioned definitions [9].

Nutraceuticals have the advantage over foods and drugs because they may comprise more than a single food or plant component, that may be a contributing active ingredient, but their regulation varies widely around the world [2,9,19]. Nutraceuticals comprised herbal products, isolated nutrients, dietary supplements, diets, genetically engineered foods and processed products such as soups, cereals, and beverages [46] but then, with the passage of the *Dietary Supplement Health* and *Education Act of 1994*, was expanded to include minerals, herbs, vitamins, and other botanicals, aminoacids and any dietary substance [76]. As evidenced by the number of existing nutritional terms that are mentioned above and published elsewhere in the literature, the presence of a consistent definition, to what "functional foods", "nutraceuticals", and other terms mean is essential to properly educate the public about these products which are adequate to deal with future chronic disease prevention and care [56].

2.3. Global Market of Nutraceuticals

Even though there is a disagreement among experts as to what a nutraceutical is, the movement regarding nutraceuticals in the global market is "unstoppable" [77]. The lack of strict regulations controlling nutraceuticals is the main factor that leads to inflating the market share of these products [78]. The beneficial role of nutraceuticals and healthy foods in wellness promotion and disease prevention has been demonstrated by nutritionists and health professionals and has led to an increase in the number of nutraceuticals worldwide [79]. Consumers, thanks to the information available on health-related foods and supplements, spend billions of dollars each year purchasing them. It is significant to evaluate the global nutraceutical market that engenders constant controversy [73]. The industry of nutraceuticals is broken down into functional foods, natural or herbal products, and dietary supplements.

An USD 80 billion nutraceuticals market was identified by the Nutrition Business Journal (NBJ) in 1995 and as USD 91.7 billion in 1997. In 1996, more than USD 6.5 billion were invested in dietary supplements by U.S. consumers, almost doubling this market to USD 12 billion in 1998. Dietary supplements (19.5% per year) and natural or herbal products (11.6% per year) were the most rapidly growing segments of the industry. According to BCC Research, in 2007, the global nutraceuticals market was worth USD 117.3 billion, whereas, in 2013, it was USD 176.7 billion. Nutraceutical foods were the largest market segment in 2007, worth USD 39.9 billion [2,50,70,75]. In 2004, the global nutraceuticals market was estimated to be approximately $106 billion [80], USD 142.1 billion in 2011 [9], and USD 117 billion in 2017 [78]. The global market for food supplements was estimated to be worth between EUR 45 billion and EUR 50 billion in 2009, whereas the EU market was between EUR 8.2 billion and EUR 8.6 billion [81]. Emerging nutraceuticals technology has created a global market with impressive growth rates, with the United States followed by Japan and the European Union as major markets. In 2016, the global nutraceutical industry has experienced an increment of USD 198.7 billion [78] and, in 2018, USD 231 billion is projected to grow at a compound annual growth rate (CAGR) of 7.8% from 2018 to 2023 [82]. Figure 3 presents the global nutraceutical market by region (%) in 2015.

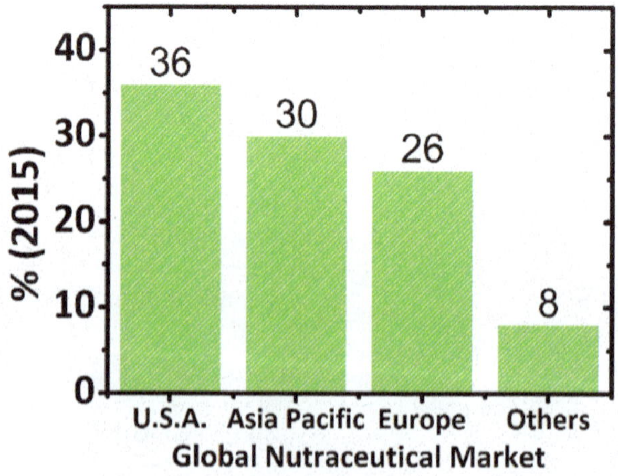

Figure 3. Global Nutraceutical Market by Region (%) in 2015: U.S.A. (36%), Asia Pacific (30%), Europe (26%), and others include the rest of the world (8%) [83].

2.4. Classification of Nutraceuticals

Nutraceuticals can be classified into broad classes based on food bioavailability, mechanism of action of active component, and chemical nature, as presented in Figure 4. Further, they are classified into several sub-classes as follows [84,85]. More comprehensively, nutraceuticals, regarding their promise, can be classified in two ways such as potential nutraceuticals including the majority of nutraceuticals that maintain a promise of a particular health or medical benefit, and established nutraceuticals related to the attainment of the efficient clinical data that prove these benefits [2,6,86]. For instance, folic acid was first considered a potential nutraceutical. Subsequently, it was altered to an established nutraceutical after the release of sufficient clinical evidence that confirmed neural tube defects prevention [45].

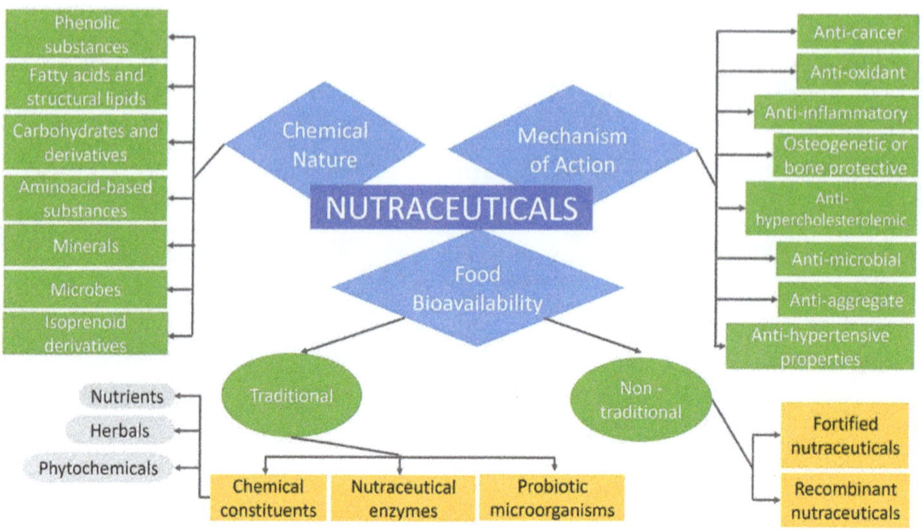

Figure 4. Classification of Nutraceuticals.

2.4.1. Nutraceuticals Based on Food Bioavailability

Regarding the food source, nutraceuticals can be divided into traditional and non-traditional.

Traditional nutraceuticals comprise food that is sourced directly from nature without any further modification. This group, for example, many fruits, fish, grains, tomatoes, salmon or soy, tea, chocolate, which contain various constituents such as fatty acids, lycopene, omega-3, saponins, etc., deliver benefits beyond basic nutrition. Chemical constituents, probiotic microorganisms, and nutraceutical enzymes are several types of traditional nutraceuticals. Nutraceutical enzymes are proteinous in nature, specific in action, and are produced by cells of the body. They can also increase the rate of metabolic activity occurring inside the cells [54,87]. The symptoms of medical conditions such as hypoglycemia, blood sugar disorders, digestive problems, and obesity can be eliminated by adding enzyme supplements to the diet. These enzymes are derived from animal, plant, and microbial sources [88]. Probiotic refers to viable microorganisms that have a vital position in the medical field by making the gastrointestinal tract (GT) more favorable to processes such as metabolism and absorption. Probiotics are counted as an impressive number of microbial species that eradicate toxic flora inside the intestine because of their tolerance to acid and bile salts. With respect to foods, probiotics are considered as, *"viable preparations in foods or dietary supplements to improve the health of humans and animals"* [54,89,90]. In addition, chemical constituents incorporate nutrients, herbals, and phytochemicals. Nutrients are substances with established nutritional functions to sustain the life or health of a person, animal, or part of the body. These substances are antioxidants, minerals, vitamins, amino acids, and fatty acids. Herbals are herbs or botanical products subjected to treatments such as distillation, extraction, fractionation, purification, concentration, etc. They can be found in berries, leaves, roots, and flowers as various parts of plants that are used for medicinal purposes. The combination of herbal products with nutraceuticals can treat many chronic disorders [52–54,84]. Phytochemicals have become more popular as a result of the increment of studies on nutrients. Based on a wide variety of chemical compounds that plants carry, phytochemicals include phenolics, nitrogen-containing compounds, alkaloids, and terpenoids. When phytochemicals are present in plant-rich diets, they lower morbidity and mortality in adult life [71,91].

Further, agricultural or food engineering and product development are the main factors of the appearance of non-traditional nutraceuticals on the market. They are foods enriched with supplements or biotechnologically designed crops to raise the nutrients and ingredients, comprising orange juice fortified with calcium, cereals with added vitamins or minerals, β-carotene-enriched rice, soybeans, and flour with added folic acid [76,92]. They are arranged into fortified and recombinant nutraceuticals. Fortified nutraceuticals comprise fortified foodstuff from agricultural production or the addition of the compatible nutrients to the main ingredients such as flour fortified with calcium, minerals added to cereals, milk fortified with cholecalciferol to treat deficiency of vitamin D, etc. Apart from these, recombinant nutraceuticals involve foodstuffs which are a source of energy. These nutraceuticals, produced using various biotechnological processes, comprise cheese, bread yogurt, vinegar, fermented starch, etc. The cheese and bread, through a fermentation process, extract the enzyme useful for providing necessary nutrients at an optimum level [54,88,93].

2.4.2. Nutraceuticals Based on Chemical Nature

These types are classified based upon nutraceutical chemical nature, more specifically upon functional groups. Based on their primary and secondary metabolite sources, there include several large groups such as isoprenoid derivatives, phenolic substances, fatty acids, structural lipids, carbohydrate derivatives, amino acid derivatives, microbes, and minerals, which provide a basis for subclassification [71,84,94]. Justus von Liebig, the German chemist, proposed that the nutritive value of food and feed can be predicted from the knowledge of the chemical composition of energy-yielding substances such as carbohydrates, fats, proteins, and a few minerals. These substances represent the essentials of a nutritionally adequate diet. The basic structure of proteins, carbohydrates, lipids, vitamins is made up largely of six elements such as hydrogen, oxygen, carbon, nitrogen, phosphorus, and sulfur. The atoms of the aforementioned elements, in organic molecules, are held together by covalent bonds. These bonds are formed when two atoms share a pair of outer orbital electrons and each covalent bond allows the organic molecule to serve as the energy source of the body [95].

Isoprenoid Derivatives

Isoprenoids, also known as terpenoids, are synthesized from a universal compound isopentenyl diphosphate (IPP) and belong to a vast group of secondary metabolites such as carotenoids, polyprenyl alcohols, sterols, ubiquinone (coenzyme Q), prenylated proteins, and heme A [96]. The evidence that isoprenoids are extremely diverse in chemical structure is demonstrated by the characterization of over 23,000 individual isoprenoid compounds and the announcement of hundreds of new structures each year [97].

Phenolic Substances

Phenolic compounds or polyphenols, referring to more than 8000 compounds found in the plant kingdom, are plant secondary metabolites that possess at least an aromatic ring with one or more hydroxyl functional (-OH) groups. They are essential for the growth, development, and reproduction of plants. Their classification can be based on the source of origin, biological function, and chemical structure. To render comprehensible, the classification according to their chemical structure is taken into account. Polyphenols can be divided into several sub-groups such as simple phenols and phenolic acids (hydroxybenzoic and hydroxycinnamic acids), flavonoids (flavones, flavonols, flavanones, isoflavones, flavanonols, anthocyanidins, tannins), stilbenes (resveratrol), and lignans found in plants and foods of plant origin [98–100]. Rich sources of phenolic compounds are fruits, vegetables, whole grains, tea, chocolate, wine, herbs, spices, cereals, oils, seeds, legumes, and others [101–103]. As there are 100 glucosinolates, 200 phytoestrogens, 700 carotenoids, and 4000 mono- and polyphenolics, it is impossible to cover all the information about their mode of action and clinical activity [104].

Fatty Acids and Structural Lipids

Lipids are a heterogeneous group of molecules that are insoluble in water but soluble in organic solvents. They are structurally quite diverse, ranging from simple short hydrocarbon chains to more complex molecules. Their classification includes eight categories such as fatty acyls, glycerophospholipids, sphingolipids, glycerolipids, saccharolipids, sterol lipids, prenol lipids, and polyketides [105–107]. One of the most fundamental categories of biological lipids is the fatty acyl structure representing the major lipid building block of complex lipids [108]. The category of fatty acyls contains fatty acids, aldehydes, alcohols, esters, and amines [107]. Fatty acids are the main constituents of the human cell [109]. They are hydrocarbon chains of varying lengths and degrees of unsaturation, with a methyl group (-CH_3) at one end and a carboxyl group (-COOH) at the other end. The α carbon is the carbon atom next to the carboxyl group, whereas the β carbon is the subsequent one. The last position of fatty acids (-CH_3 group) is designated as omega (ω) carbon. The first step in the synthesis of fatty acids involves the conversion of acetyl-CoA to malonyl-CoA by the enzyme acetyl-CoA carboxylase. Fatty acids can be classified, according to the presence or absence of double bonds, as saturated without double bonds (the most common ones contain 12 and 22 carbon atoms), monounsaturated fatty acids with one C = C located in different positions, and polyunsaturated fatty acids (PUFAa) with more than one double bond. The unsaturated fatty acids can be classified based on the configuration of the double bonds as cis or trans. Further, they can be categorized, based on the first double bond position from the fatty acid methyl-end, as ω-3 PUFAs including primarily α-linolenic acid (ALA) and its metabolic products such as eicosapentaenoic acid (EPA) and docosahexaenoic acid (DHA), as ω-6 PUFAs including primarily linoleic acid (LA) and its derivative arachidonic acid (AA), and as ω-9 monounsaturated fatty acids including primarily oleic acid. These are considered major fatty acids among others. PUFAs, with the first double bond on C3 (α-linolenic acid) and C6 (linoleic acid) from the methyl end, are intended essential because the human body cannot synthesize them, therefore, they have to be taken from a diet [110–112]. The predominant PUFAs, in all diets, are the ω-6 fatty acids with the linoleic acid as their representative whereas α-linolenic fatty acid is the precursor of other ω-3 long-chain PUFAs [109]. Vegetable oils, dairy products, meat products, eggs, soybean, certain seaweeds, grains, and fatty fish or fish oils are the most important dietary sources of fatty acids [110,113].

Carbohydrate Derivatives

Carbohydrates, derived from plant sources, are the most abundant class of organic compounds found in living organisms. They are divided into sugars with a degree of polymerization (DP) 1–2, oligosaccharides with a DP 3–9, and polysaccharides with a DP $\geq$ 10. Sugars include monosaccharides, disaccharides, and sugar alcohols, whereas oligosaccharides include α-glucans and non-α-glucan. As for polysaccharides, they are classified into starch (α-1:4 and 1:6 glucans), and non-starch, or the cell-wall polysaccharides (NSPs). The storage carbohydrates (starch, oligosaccharides and sugars), and the cell-wall polysaccharides (derived from plants, fungi and algae) are two important classes of plant carbohydrates with a contrasting but an important impact on human health [114–116]. The major storage carbohydrate in plants is starch. Starch is a mixture of two glucose polymers such as unbranched amylose comprising (1$\rightarrow$4) α-linked chains of up to several thousand glucose units and highly branched amylopectin comprising (1$\rightarrow$4) and (1$\rightarrow$6) α-linkages of over 100,000 glucose residues. On the other side, cell-wall polysaccharides may be widely grouped into three major categories such as hemicelluloses, cellulose, and pectic polysaccharides. They are mainly found in the plant cell wall and consist of certain monosaccharides residues joined to each other by glycosidic linkages. Plant cell walls are highly complex structures that determine the quality characteristics of many plant-based foods. As one of the main plant cell wall components, cellulose is a complex polysaccharide consisting of a covalent structure as a β (1-4)-linked D-glucan with a DP of more than 10,000 in secondary walls and 2000–6000 in primary cell walls. The formation of hydrogen bonds during the interactions of parallel glucan chains leads to the synthesis of newly cellulose microfibrils conferring the formation of a strong and extensible three-dimensional network. Further, pectic polysaccharides consist of polysaccharides rich in α-D-galacturonic acid (GalA) residues, in which varying proportions of the acid groups are present as methyl esters. As complex macromolecules, they can be composed of 17 different monosaccharides comprising more than 20 different linkages. The major types of pectic polysaccharides backbone are homogalacturonan (HG) and rhamnogalacturonan I (RGI). They, in the primary cell wall, are characterized by certain amounts of neutral sugars present as side chains. The most abundant neutral sugars are arabinan and galactan [114,116–121]. Hemicelluloses as another category of NSPs, are characterized by β-(1$\rightarrow$4)-linked backbones of sugars with an equatorial configuration, a DP of between 150 and 200, and can be extracted with alkaline treatment. They comprise xyloglucans, xylans, mannans, and glucomannans [116,122,123]. The major sources of carbohydrates in the human diet are cereals, tubers, legumes, pulses, fruits, vegetables, fungi, algae, seaweeds, guar, etc. [115].

Amino Acid Derivatives

Proteins are the essential components of tissues in all organisms [124]. The nutrition chemistry, in its investigations, has emphasized the significance of amino acids as the fundamental factors in all concerns in which the proteins have been involved over the years [125]. The three-dimensional structure of proteins affects their function [126]. François Magendie, French experimental physiologist, in 1816 showed that dogs fed foods, containing protein, remained healthy. Whereas dogs fed, containing only fat or carbohydrates, lost weight and developed a corneal ulcer for two weeks and after a month they died. These observations have identified the protein as a specific essential dietary component [95,127]. The nutritional value of dietary protein has been raised as, during hydrolyzation by proteases and peptidases, it generates amino acids, dipeptides, and tripeptides in the lumen of the gastrointestinal tract. A protein contains various amounts of 20 different amino acids which are linked to each other via amide bonds, the so-called peptide bonds [124]. Amino acids, as organic substances that provide nitrogen, hydrocarbon skeletons, and sulfur, have played a significant role in the nutrition and health maintenance of humans and animals. Almost all amino acids have an asymmetric carbon and show optical activity. Glyceraldehyde has been used as a reference to define the absolute configuration of amino acids such as L- or D-isomers. Even though there are more than 100 amino acids in nature, only 20 of them are considered as building blocks of protein. These amino acids have an amino group ($^{+}NH_3$), a hydrogen atom, a carboxyl group (COO^{-}), and a side chain (R) attached to the central α-carbon. Amino acids are classified, based on nitrogen balance, as nutritionally essential or non-essential for humans and animals. Nutritionally non-essential amino acids are synthesized by animals or humans and are not necessary to be provided from the diet. Contrarily, the essential ones cannot be synthesized by an animal, and therefore must be ingested with feed. There are nine essential amino acids, of the 20 standard protein amino acids, including L-leucine, L-valine, L-threonine, L-isoleucine, L-methionine, L-lysine, L-phenylalanine, L-histidine, and L-tryptophan [128–131]. Amino acids that are impressive regulators of key metabolic pathways to improve health, maintenance, growth, immunity, and reproduction of organisms, have led to the development of the functional amino acids concept. The major sources of amino acids are several natural plant proteins and animal products [130].

Microbes and Minerals

In recent years, the great demands, for augmenting the value of nutraceuticals to cure diseases, have notably affected the signs of the progress of nutraceuticals production via metabolic engineering of microbial-based platforms [132]. Some of the microbes, among trillion others that colonize the human body, can potentially be beneficial or harmful. An imbalance of them may cause several diseases, therefore, probiotic and prebiotic supplements may be effective to prevent such conditions. According to a joint Food and Agriculture Organization of the United Nations (FAO) and World Health Organization (WHO) in 2001, probiotics are defined as, "*Live microorganisms which, when administered in adequate amounts, confer a health benefit on the host.*" Afterwards, the study of beneficial bacteria revealed the nondigestible food ingredients, called prebiotics, which stimulate the activity and the growth of these bacteria in the intestinal tract. Human origin, non-pathogenic quality, stability in acid and bile, resistance to technological processes, production of antimicrobial substances, the modulation on the immune system, the persistence within the GI tract, the influence on metabolic activities are some criteria, among others, that a microbe must accomplish to be classified as probiotic [132–134]. The most known probiotics are the lactic acid bacteria *Lactobacillus acidophilus*, *Lactobacillus casei*, and bifidobacterial types. The main sources of them are yogurts and other dairy products such as buttermilk, frozen desserts, milk powder, and acidophilus milk and some non-dairy products such as fruits, vegetables, legumes, and cereals [135,136]. Furthermore, dietary essential minerals are crucial components to uphold several bodily functions [137]. It has been established that several mineral elements are indispensable for normal nutrition constituting approximately between 4% and 6% of body weight. Mineral nutrition has been more important than vitamin nutrition since the body, using some minerals, can replace the lacking vitamins. Whereas, the opposite is hopeless [138]. There are 20 essential minerals for humans divided up into major minerals and trace minerals. Sodium, chloride, phosphorus, potassium, magnesium, calcium, and sulfur are major minerals. On other hand, the trace minerals comprise iron, zinc, iodine, selenium, copper, manganese, fluoride, chromium, and molybdenum. Milk and dairy products are considered to be significant sources of the daily intake of essential minerals [139–141].

2.4.3. Nutraceuticals Based on Mechanism of Action

Concerning specific therapeutic properties, nutraceuticals are known for anti-inflammatory, anti-microbial, anti-oxidant, anti- hypercholesterolemic, anti-aggregate, anti-hypertensive, anti-carcinogenic, osteogenetic, or bone protective properties, etc. [84,94].

Nutraceuticals and Health Benefits

The establishment of a vibrant nutraceutical research community is necessary to spread scientific knowledge about nutraceuticals. This has enabled the creation of the established nutraceuticals from the potential ones and offered the delivery of their enormous benefits across the globe. The reflection of the continuous research, market expansions, and consumer interest is made by the constant changing list of nutraceuticals being investigated [142]. As a result of the unhealthy diet, tobacco use, harmful use of alcohol, irregular sleeping habits, and a lack of daily physical exercise, there are countless global health problems related to the advancement of diabetes mellitus, cardiovascular morbidity and mortality, chronic respiratory diseases, metabolic syndrome, and cancer. They may be summarized with the term "Chronic Non-Communicable Diseases" (NCDs) as a distinguishing feature of lifestyle diseases. Fighting of the aforementioned has evoked several arguments on nutraceuticals efficiency [92,143–145]. The challenge has been to define the interrelationship between the disease and nutrient [104]. Many industries such as foods, herbals, and pharmaceutical manufacturing have evaluated nutraceuticals as beneficial products related to the cure of many health troubles. In addition, the nutraceutical safety and their side effects such as allergic reactions, cardiac arrhythmias, insomnia, their interactions with other nutraceuticals and therapeutic drugs, etc., are marked [146]. Although the adverse side effects of nutraceuticals are usually minimal compared to synthetic drugs [104], their use must be regulated and controlled with experimental assessment or clinical trials [146]. According to the World Health Organization (WHO), lifestyle diseases are one of the most momentous challenges of twenty-first century medicine. The statistics in 2016 have shown that 40.5 million (71%) deaths are due to NCDs among 56.9 million of total premature deaths, 17.9 million (44%) deaths are due to cardiovascular pathologies, 9 million (22%) to cancer, 3.8 million (9%) to chronic respiratory diseases, and 1.6 million (3%) to type 2 diabetes. The main concern is that the toll of deaths can reach 52 million in 2030 if the growth rate continues so [143].

According to the literature described in this subsubsection (2.4.3.), some prominent evidence in the new era of the twenty-first century have shown the enormous growing of nutraceuticals as potent therapeutic supplements. The preventive therapeutic efficacy of new nutraceuticals can be practically extended if their miraculous health benefits are investigated [147]. Therefore, in this review, various nutraceuticals applications will be considered focusing on the rise of more recent afflictions such as cancer, diabetes, neurological, cardiovascular, and chronic diseases, which have emerged as public health problems in many countries [148].

Anti-Microbial Activity

The role of nutraceuticals in the inhibition of microorganisms and alteration of bacterial populations is still implied despite the incompleteness of information. The products such as aloe, goldenseal, St. John's wort, garlic, zinc oxide, echinacea, and zinc gluconate, are studied for their antibacterial activity. As the gram-positive bacteria, Staphylococcus aureus ATCC 29213 is used whereas Escherichia coli ATCC 25922 is used as the gram-negative bacteria. It has been observed that some products can be selective agents in the development of antibiotic resistance and lose their antibacterial properties quickly [12]. Therefore, their effectiveness needs further investigation. Phenolic compounds can also be used as antibiotics, antidiarrheal, or antiulcer agents [149], being involved in various physiological processes of plants and plant defense mechanisms against microbial infections [150].

Anti-Oxidant Activity

The significant and important application of oxygen in clinical medicine can also bring certain toxic effects [151]. Oxygen is considered a double-edged sword; it has promoted and destroyed life for two centuries. Liebig, in 1842, highlighted that toxic oxygen, capable of burning up all the tissues, can be removed from the organism by carbon and hydrogen, that acts as antioxidants, in food. He believed in this since in starvation there is no food to remove the oxygen, and therefore the particles of the brain begin to undergo the process of oxidation. On the other hand, carbon and hydrogen-rich food, considered as antioxidant, by reacting with oxygen can inhibit the destructive influence of oxygen in the tissue [152]. The development of major diseases is supported by oxidation processes that occur naturally in the human body [153]. Therefore, oxygen toxicity has emerged as one of the most fundamental phenomena in biological sciences. Gerschman, in 1954, formulated a general theory of oxygen toxicity describing the oxygen-induced damage that is caused by free radical intermediates. The oxidizing free radicals are generated in excessive amounts when the living organisms are exposed to the increased pressure of oxygen [152,154]. Free radicals and other reactive oxygen species (ROS) are considered potentially harmful agents but are also known to produce various cellular structures and to fight pathogens [155,156]. The superoxide $\bullet O_2-$, hydroxyl radical •OH, and hydrogen peroxide H_2O_2 are free radicals produced by metabolic reactions in the human body, as shown in Figure 5. These are molecules with one or more unpaired electrons. ROS refer to any free radical containing oxygen but can also include non-free radical species such as hydrogen peroxide H_2O_2, ozone O_3, singlet oxygen $^1O_2$, hypochlorite ^-OCl, and peroxynitrite $ONOO^-$. In comparison with non-radicals, the free radical reactions result in new radicals leading to chain reactions.

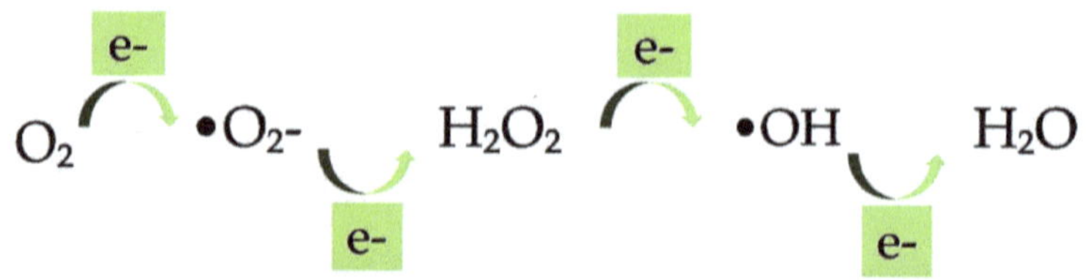

Figure 5. Formation of superoxide radical ($\bullet O_2-$), hydrogen peroxide (H_2O_2), hydroxyl radical (•OH), and water by stepwise, univalent reductions of molecular oxygen.

The $\bullet O_2-$ radical is produced by the first one-electron reduction of molecular oxygen. It can operate as an important second messenger in the cell even though its reactivity and toxicity are low. Further, the H_2O_2 falls as a result of the dismutation of $\bullet O_2-$. Owing to high reactivity, H_2O_2 forms the •OH when it reacts with partially reduced metal ions. The •OH is considered the most important radical, among others, with a high impact on the cell damage as it can directly evoke DNA damage. The general mechanism that the free radicals can be oxidized to oxygen and reduced to water protects the biological systems from the potential hazards of those radicals. The backbone of the cellular antioxidant defense system is composed of the antioxidant enzymes superoxide dismutase (SD), catalase (CAT), and glutathione peroxidase (GPX). The dismutation of $\bullet O_2-$ to H_2O_2 is catalyzed by SD, whereas the detoxification of H_2O_2 is made by CAT. In addition, GPX, using reduced glutathione (GSH) as the electron donor, reduces organic hydroperoxides and H_2O_2. GSH is a tripeptide with a reactive sulfhydryl group, and has multiple effects regarding the antioxidant defense. These effects comprise its action as a scavenger of free radicals such as $\bullet O_2-$, •OH and lipid hydroperoxides, as a substrate for the antioxidant enzyme GPX and in the direct repair of oxidative DNA lesions. As the effects of oxidative stress on human health are considered a serious issue [157], nutraceuticals with antioxidant activities have received attention [96]. The antioxidant activity of nutraceuticals is affected by their chemical structure [158]. Dietary components with important antioxidant functions comprise ascorbate, α-tocopherol, β-carotene, linoleic and linolenic acids, copper, manganese, zinc, selenium, and cysteine [152]. The clinical trials on the role of antioxidants have mainly focused on several compounds, such as carotenoids, vitamins C and E [155]. Carotenoids are efficient antioxidants involved in the scavenging of singlet molecular oxygen and peroxyl radicals. The physical quenching enables the direct transfer of energy between carotenoids and $^1O_2$. The ground state oxygen and a triplet excited carotene are yielded as a result of the transfer of $^1O_2$ energy to the carotenoid molecule. Further, the carotenoid dissipates its energy returning to the ground state throughout its interaction with the surrounding solvent [159]. β-carotene, among the various carotenoids, is an effective quencher of singlet oxygen preventing lipid oxidation (Table 1).

Table 1. The mechanism of the quenching of singlet oxygen by β-carotene.

$^1O_2$	+	β-carotene	⟶	O_2	+	β-carotene *
	β-carotene *		⟶	β-carotene	+	energy (heat)
	β-carotene *		⟶	all-*trans*-β-carotene		

β-carotene * -triplet excited carotene.

Vitamin E or α-tocopherol prevents membrane-mediated effects of oxygen free radicals because it efficiently protects biological membranes from lipid peroxidation, as it is described by a nonenzymatic reaction (Table 2) [160].

Table 2. Vitamin E reaction with oxygen free radicals.

Vitamin E	+	RO	⟶	Vitamin E	+	ROH

The reduction of vitamin E radical must go further by its interaction with ascorbic acid or vitamin C (Table 3). The ascorbate is considered an antioxidant because of its direct participation in the scavenging of the "activated oxygen".

Table 3. The reaction of vitamin E radical with ascorbic acid.

2 Vitamin E	+	ascorbate	⟶	2 Vitamin E	+	dehydroascorbate

The combination of both β-carotene and α-tocopherol inhibits the lipid peroxidation more efficiently than their individual use [159]. The reaction scheme of α-tocopherol during the autoxidation of polyunsaturated fatty acids is presented in Figure 6. Initiation, propagation, and termination, are three stages that the autoxidation as a chain reaction proceeds. The carbon-centered lipid radical or an alkyl radical is produced in the initiation step by the abstraction from a polyunsaturated fatty acid moiety. The alkyl radical, in the propagation step, reacts with molecular oxygen at a very high rate, giving a peroxyl radical. The peroxyl radical is a chain-carrying radical and, therefore, can attack another polyunsaturated lipid molecule. Although the initial peroxyl radical is converted to a hydroperoxide, a new alkyl radical is produced and rapidly converted to another peroxyl radical. The chain reaction continues until inactive products are formed because of the combination of the chain-carrying peroxyl radical with another radical. α-Tocopherol inhibits the propagation step as a chain-breaking antioxidant. A peroxyl radical, after receiving the phenolic hydrogen atom by α-tocopherol is converted to a hydroperoxide. The tocopherol radical, incapable of continuing the chain, is formed. Further, it is removed from the cycle by the reaction with another peroxyl radical to form an inactive and non-radical product. The measure of the antioxidant efficiency of α-tocopherol is made by the rate at which it reacts with peroxyl radicals [161]. Phenolic compounds are considered as strong antioxidants that complement the functions of enzymes and antioxidant vitamins as a protection against oxidative stress caused by excess reactive oxygen species (ROS) [99]. Their elevated capacity is related to scavenging free radicals [162]. Flavonoids, an important class of phenolic compounds, suppress reactive oxygen formation during the antioxidant mechanism, by scavenging reactive species, inhibiting enzymes, chelating trace elements involved in free-radical production, protecting and up-regulating antioxidant defenses [157]. Therefore, they are among the most efficient antioxidant molecules [162]. Tannins [158], terpenes [163], sterols [153], and fiber [164], have also shown antioxidant activities. Further, the excessive production of ROS can also cause many other disorders such as hypertension, inflammation, cataract, cardiovascular disease, diabetes, cancer and neurodegenerative diseases, and osteoporosis emphasizing the significance of phenolic compounds in the inhibition of the ROS formulation [101,165].

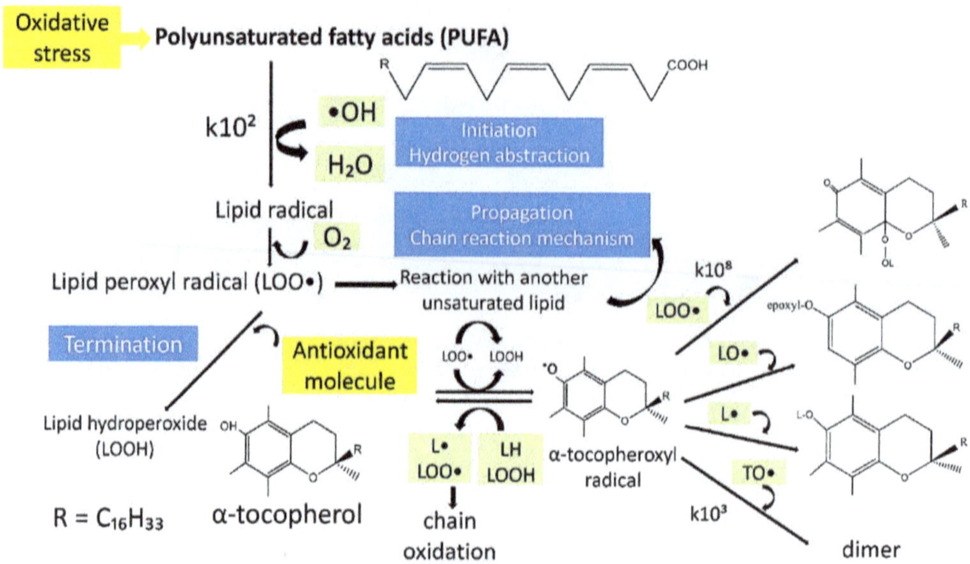

Figure 6. The reaction of α-tocopherol during the autoxidation of unsaturated lipids. LOOH-lipid hydroperoxide; LO•-lipid-alkoxyl radical; L•carbon-centered lipid radical; LOO•-lipid-peroxyl radical; TO•-α tocopheroxyl radical; k-rate constant in M^{-1} s^{-1}.

Anti-Hypertensive Activity

Hypertension is known as one of the most frequent chronic medical conditions in the developed world. It is also considered a major hazard factor for coronary heart disease, stroke, congestive heart failure, and renal disease. Hypertension is a result of the environment-genetics interaction. Inflammation, subsequent gene expression, oxidative stress, nutrient–gene interactions can positively or negatively affect human vascular biology [166]. As hypertension is mainly treated with anti-hypertensive drugs, the use of blood pressure-lowering nutraceuticals is of great interest [167]. Vitamin D3, vitamin C, vitamin B6, amino acids (taurine, arginine, carnitine), chlorogenic acids, melatonin, coenzyme Q10, quercetin, probiotics, and resveratrol are some nutraceutical supplements with an influence in the treatment of hypertension [166].

Chlorogenic acids can be found in fruits, plants, and vegetables such as coffee beans, tomatoes, apples, etc. The mechanism of chlorogenic acids for reducing blood pressure is well-known. Firstly, the consumption of chlorogenic acid is important because it is an antioxidant. The superoxide radical causes hypertension by forming peroxynitrite in vascular walls through the destruction of nitric oxide (NO). NO bioavailability can be increased by inhibiting the reactive oxygen species. This step generates enzymes such as xanthine oxidase and DAD(P)H, and reduces the formation of peroxynitrite. Secondly, the protective role of chlorogenic acid in eNOS causes the induced anti-hypertensive activity.

The antihypertensive response of chlorogenic acid in hypertensive rats (SHR) is blocked by the addition of the N(G)-nitro-L-arginine methyl ester (L-NAME) as a NOS inhibitor. As it is known that the blood pressure is adversely associated with the plasma level of NO metabolites, chlorogenic acid intake increased the urinary NO metabolites in SHR. The effect of blood pressure reduction is mediated by inhibiting angiotensin-converting enzyme (ACE) activity, modulating nitric oxide (NO) production, scavenging free radicals, and improving endothelial function through the products that naturally contain the nutraceuticals [168].

Vitamin D is produced by the non-enzymatic conversion of provitamin D3 to previtamin D3 [169]. Vitamin D receptors are found in the kidney (juxtaglomerular cells), leukocytes, cardiac myocytes, and vascular smooth muscle cells of the human body. A study demonstrated that vitamin D directly suppresses renin synthesis. This is because of the reduction in renin mRNA transcription in the kidney. The plasma renin appears because of vitamin D deficiency. Further, vitamin D alters the epidermal growth factor receptor function and, therefore, inhibits the proliferation of vascular smooth muscle cells. Vitamin D suppresses pro-inflammatory cytokines, reduces asymmetric dimethyl arginine, improves endothelial function and arterial elasticity, increases nitric oxide (NO), and decreases vascular smooth-muscle hypertrophy. For optimal blood pressure lowering effects, it is recommended a vitamin D level of 60 ng/mL [170].

Resveratrol is another nutraceutical that presents an anti-hypertensive effect [171] and can be found in red grapes, and in plants that can survive harsh environmental conditions [172]. Resveratrol improves endothelial dysfunction, prevents the uncoupling of endothelial nitric oxide synthase (eNOS), increases the flow-mediated vasodilation in a dose-related manner, and blocks the effects of angiotensin II [173].

Lycopene can be found in tomatoes, red grapefruits, watermelon, etc. [174]. A study presented a significant blood pressure reduction of 5.4/3 mmHg over six weeks after the administration of standardized tomato lycopene extract [166]. Further, the spontaneously hypertensive rats (SHR) were studied. In this strain, the hypertension was progressively increased over time, and a four-week of lycopene supplemented diet was employed. An effective blood pressure reduction, in both young and adult rats, was observed. This research supported the effectiveness of lycopene in hypertension prevention [175]. This is because of the anti-hypertensive effects of lycopene inhibiting angiotensin-converting enzyme (ACE), reducing oxidative stress that is induced by angiotensin-II, and transversally enhancing the production of nitric oxide in the endothelium [174].

Anti-Inflammatory Activity

The various classes of terpenoids demonstrate health benefits through their connection with key molecular players in animal and human physiology, action as immunostimulants, antioxidant activity booster, blood coagulation hemostasis modulator, related to anti-cancer, anti-malaria, anti-bacterial and anti-viral activities. Terpenoids also modulate transcription factors like the nuclear factor kappa B (NF-κB) related to the regulation of a cascade of events in inflammatory pathways that cause various chronic diseases such as cardiovascular disease, diabetes, Alzheimer's, etc. Scientific studies have shown that several terpene-based volatile compounds occurred in plant essential oils including compounds such as α-pinene, β-limonene, p-cymene, linalool, β-phellandrene, and terpinenes, can have anti-inflammatory, anti-oxidant effects, and can cross the blood-brain barrier and treat the Alzheimer's disease [176].

Fatty acids can affect cellular functions and physiological responses due to their principal roles as energy sources and membrane constituents [105,106]. Fatty acids serve as substrates for the biosynthesis of biologically active lipid mediators and play direct roles in cell signaling that influences gene expression. The responsiveness, and functionality of the cells and tissues can be modified through the mix of complex lipids and their constituent fatty acids. This phenomenon is well-defined for immune, metabolic responses and inflammatory, cardiac and neurological function, platelets, etc. The inflammatory effect of the fatty acids is the most considered. A well-functioning immune system is crucial to human health and serves to protect the host from the effects of infectious agents that exist in the environment. An immune response to host tissue is generated when the immune system recognizes the host antigens as "non-self" rather than as "self", leading to tissue damage as a characteristic of so-called chronic inflammation [177]. Therefore, the inflammatory process appears when the human body tries to fight infection or to repair damaged tissue, leading to the progression of some chronic diseases such as rheumatoid arthritis, inflammatory bowel diseases, asthma, cardiovascular disease, neurological disease, type-1 diabetes, cancer, oncologic or endocrinologic diseases. Fatty acids have a high impact on human health by influencing the appearance or evolution of those diseases [105,109]. Omega-3 and omega-6 PUFAs are the most important fatty acids, and their balance can be important in determining the seriousness and development of the diseases. The inflammatory process can be generated by a high intake of omega-6 PUFAs, particularly arachidonic acid. Conversely, long-chain omega-3 PUFAs are potentially potent anti-inflammatory agents that decrease the expression of adhesion molecules and the production of inflammatory mediators such as eicosanoids, cytokines, and reactive oxygen species. The anti-inflammatory effect of omega-3 PUFAs is related to their direct action attributed to their capability of competing with arachidonic acid or their indirect action of affecting the transcription factors or nuclear receptors responsible for inflammatory gene expression [109,178,179]. The data from experimental and clinical studies [177] have presented the long-chain ω-3 PUFAs as potential therapeutic agents for inflammatory and autoimmune diseases [180]. The most valuable long-chain omega-3 PUFAs are eicosapentaenoic acid (EPA) and docosahexaenoic acid (DHA). Their beneficial effects regarding the cardiovascular diseases interfere with the broad spectrum of anti-arrhythmic, lipid-lowering, anti-thrombotic and anti-inflammatory properties [109,181–183]. Moreover, the direct role of ω-9 fatty acids, in comparison to the ω-3 and ω-6, in inflammatory pathways has been unclear [113] until several in vivo experiments demonstrated it. It has been related to the decreasing production of proinflammatory cytokines such as tumor necrosis factor-alpha (TNF-α) and interleukin 1-beta (IL-1β), and with enhancing production of anti-inflammatory cytokine such as interleukin-10 (IL-10) [184]. Cytokines are a group of cell-derived polypeptides that participate in a complex network of interactions exhibiting both negative and positive regulatory effects in growth, development, or activity of various target cells [177,185]. The ROS-scavenging activity of β-carotene and lycopene has enabled their use as anti-inflammatory substances [186].

Anti-Hypercholesterolemic Activity

The high levels of lipids such as cholesterol, triglycerides, and fat phospholipids in the blood cause the pathological condition called hypercholesterolemia. A prolonged increase in insulin levels, as well as a high level of O-GlcNAc (O-linked β-N-acetylglucosamine), can affect hypercholesterolemia, and lead to dyslipidemia [187]. The development of cardiovascular diseases (CVDs) is more likely to be present in patients with hyperlipidemia [188]. The oxidative stress, induced by reactive oxygen species (ROS) [189], can develop the CVDs and atherosclerosis as previously described in the subsubsubsection of antioxidant activity that oxygen species are likely to be involved in the pathophysiology of many human diseases, such as CVD, cancer, etc. In CVDs, oxidative stress alters the gene expression. Further, the transcription factor activity, particularly NF-κB, activator protein-1 (AP-1), and the peroxisome proliferators-activated receptor (PPAR) family of transcriptional activators are modulated by increased ROS levels. The oxidative modification of low-density lipoprotein (LDL) is one of the first events, in CVDs, that appeared as a consequence of increasing ROS generation [190]. The abnormally low uptake of low-density lipoprotein (LDL) by the liver is caused by a genetic defect of the low-density lipoprotein receptor (LDLR), leading to familial hypercholesterolemia (FH) [191]. It is found that 20% of patients are diagnosed with FH, and only a minimum of them have received the appropriate treatment [192]. Therefore, hypercholesterolemia can be ameliorated by having an appropriate drug treatment, and adequate lifestyle [187]. Red yeast rice, berberine, plant sterols and stanols, dietary fibers, polyphenols, flavonoids, and apple polyphenolic extract are some of the nutraceuticals which claim to have a cholesterol-lowering effect. Plant sterols and stanols are found to inhibit the absorption of cholesterol if they are taken at a dose of g/day. At this dose, they lower LDL-cholesterol (LDL-C) levels by 8 to 10%, and reduce plasma triglycerides between 6 and 9%. However, the effective use of plant sterols and stanols, on total cholesterol and LDL cholesterol, has been observed [193,194]. Further, some bioflavonoids are found to lower cholesterol levels but the information about their bioavailability, presence of contaminants in their original vegetal matrix, and their unwanted side effects, is still lacking [194].

Anti-Aggregate Activity

Another factor that causes the cardiovascular disorders has been the blood platelet activation. The progression of hemostasis, atherosclerosis, and other diseases of the cardiovascular system has been linked to the dysregulation of blood platelet activity. Blood platelet aggregation has been a result of the modulation of platelet function. Platelets, or thrombocytes, are anucleate cells, between 2 and 4 μm in greatest diameter, produced by megakaryocytes. Before being eliminated by the liver and spleen, they circulate in the human bloodstream between seven and ten days [195,196]. Nutraceuticals, as antiplatelet agents, are found to have beneficial effects. As resveratrol has been well-known for its various biological activities, one study [197] investigated the effect of cis-resveratrol on platelet aggregation. The 4-channel aggregometer was used to perform the platelet aggregation. Acid-citrate-dextrose (1:6) was used as an anticoagulant to collect the blood from the abdominal aorta of ether-anesthetized rats. The indicated concentrations of cis- or trans-resveratrol were used to incubate the platelets obtained. Whereas the sub-maximal concentrations of thrombin, collagen, or ADP were used to induce aggregation. This aggregation was further suppressed by both, cis-resveratrol and trans-resveratrol. It is found that 3.6 μg/L resveratrol reduces collagen-induced platelet aggregation by 50.3%. Resveratrol interferes with platelet aggregation by inhibiting Ca^{2+} influx that is essential for platelet aggregation [198]. The modulation of nitric oxide (NO) is another anti-platelet aggregation mechanism of resveratrol. Resveratrol promotes NO production via increasing eNOS expression and activity [199]. This activity is beneficial because NO maintains the vasculature homeostasis, and regulates intracellular signaling pathways. NO limits the thrombotic process by decreasing endothelial cell adhesion and inhibiting platelet aggregation. NO inhibits platelet aggregation by upregulating cyclic guanosine monophosphate, reducing dimerization of integrin $\alpha IIb\beta 3$, and hindering von Willebrand factor (VWF)-mediated platelet adhesion [200].

Anti-Carcinogenic Activity

The treatments of assorted forms of cancer, leading cause of death worldwide, are poorly controlled and have serious side effects. ROS are likely to be involved in the pathophysiology of many human diseases. The imbalance and high level of free radicals such as ROS and reactive nitrogen species (RNS) can affect cancer development. Chemotherapy causes undesired side effects, therefore, diet-related agents are a category of cancer chemopreventive agents that have generated much attention and interest during recent decades [201–203]. However, the physiological relevance of these agents is uncertain [204]. The phenolic compounds are powerful antioxidants that have been used as alternative treatments for cancer [201]. Several studies showed that quercetin, luteolin, kaempferol, apigenin, taxifolin, (-)-epigallocatechin-3-gallate (EGCG) [205], ethanolic extracts of Curcuma rhizome and Zingiber rhizome [206], curcumin [207], are some natural phenolic compounds exhibiting anticancer effects. These compounds affect human cancer cell lines by protecting or reducing the number of tumors and their growth [208]. Their anticancer efficacy may be due to the inhibition of the epithelial-mesenchymal transition (EMT) (Figure 7a), as one of the main pathways employed in cancer development and metastasis, in cancer cells. The phenolic compounds can also prevent cancer initiation, relapse, and metastasis [209].

Further, carotenoids play a significant role in cancer prevention. The effect of a carotenoid on cellular differentiation and proliferation, the prevention of free radical-induced damage to cellular DNA (Figure 7b), and other molecules from the antioxidant function, and the enhancement of immune surveillance in tumorigenesis from the immunomodulatory effects, are several mechanisms related to the cancer prevention of carotenoids. β-carotene, lycopene, lutein and zeaxanthin are the most studied carotenoids [210]. Aside from the beneficial effects, β-carotene and other carotenoids have been found to increase cancer risk but the level of evidence is limited. The effectiveness of chemopreventive agents has been related to the determination of the proper effective dose. For instance, humble levels of folic acid supplementation suppress the development of cancer and vice versa [202]. Furthermore, a high dose of β-carotene can expand the risk of lung cancer [203,210].

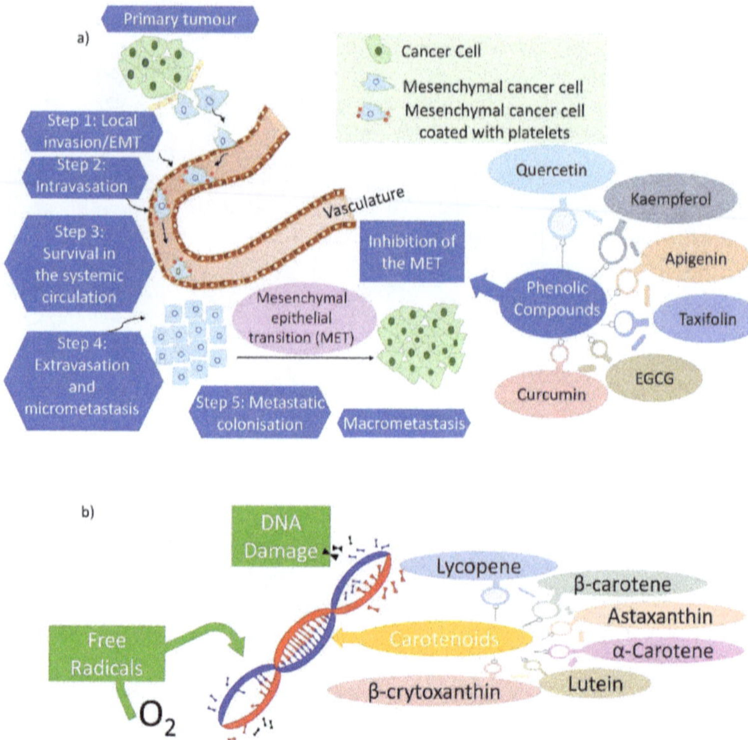

Figure 7. (**a**) The inhibition of the epithelial–mesenchymal transition (EMT) by phenolic compounds. The EMT is a critical part of cancer metastases and consists of the following key steps: Step 1 describes the subjecting of epithelial cancer cells to EMT which cells may then intravasate into the systemic circulation (Step 2). These EMT-induced cells must survive in the circulation (Step 3) before reaching the target. Then, the cells that reach Step 4 must extravasate into the tissue parenchyma upon reaching the target organ site and form micrometastases. At the end (Step 5), the mesenchymal–epithelial transition (MET) activation, another critical event for the metastasis of carcinomas, is required as a subsequent development into potentially life-threatening macrometastases [211,212]. (**b**) The prevention of free radical-induced damage to cellular DNA by carotenoids.

Bone Protective Activity

Bone is a dynamic tissue in a continuous cycle of bone resorption followed by bone formation. Established bone is degraded by osteoclasts through adherence, acidification, and proteolytic digestion. Then, osteoanabolic therapies are used for new bone formation.

A role in promoting bone health is shown by the dietary intake of fruits and vegetables, more specifically by the polyphenols that describe the physiological effects associated with bone material density and bone metabolism [213,214]. Osteoporosis and osteopenia, corresponding with a decrease in bone formation, can be prevented by lifestyle modifications. Adequate nutraceutical supplementation such as calcium intake supplemented by vitamin D, magnesium, potassium, copper, resveratrol, green tea, prebiotics and probiotics, polyunsaturated fatty acids, melatonin, have shown promising results for the management of osteoporosis [215,216].

3. Delivery Systems for Nutraceuticals

A nutraceutical has not always met the requirements to achieve the therapeutic purpose [10]. Therefore, the systems for their delivery must be designated to produce products that have consistent quality attributes. The "biomaterials science" is considered as the main focus toward the development of materials, tailored to a specific application, that can elicit highly precise reactions with proteins and cells. The biomaterials synthesis, characterization, testing, optimization, and the biology of host-material interactions are highlighted during the most intense investigation. The biomaterial must accomplish various requirements such as toxicology, biocompatibility, functional tissue structure, and pathobiology, mechanical and performance requirements, healing, industrial involvement, regulation, etc. (Figure 8) [217]. Nanotechnology is advantageous for manipulating the properties and structures of materials at the nanometer scale, and therefore has opened up new opportunities for numerous applications in biotechnology, molecular biology, medicine, environmental science, etc. [218]. The field of nanotechnology, through the efficacy of nano-drug delivery systems, is contributing to every walk of life improving the bioavailability, biocompatibility, solubility, drug loading efficacy, and surface modifications of bioactive and chemical molecules [219]. The application of nanotechnology in health care is extensively adopted as a robust driver of biomedical novelty [220].

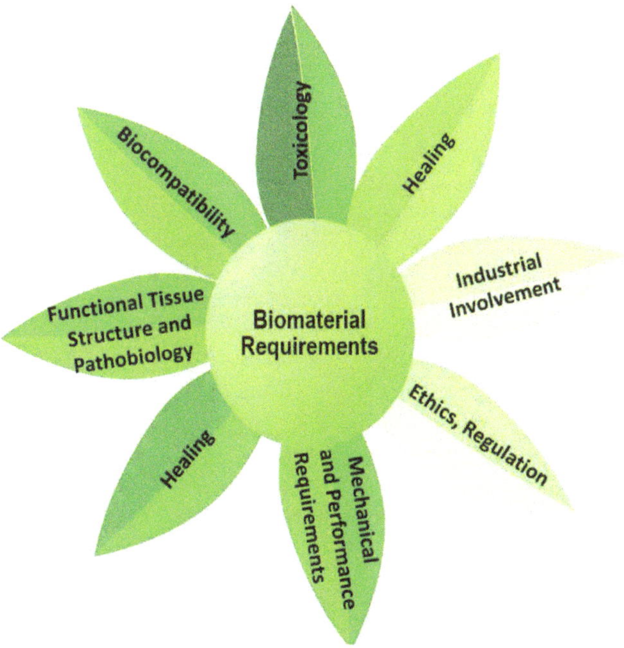

Figure 8. The essential requirements involved in biomaterial design for nutraceuticals delivery.

3.1. Advisable Features of Delivery Systems

3.1.1. Encapsulation and Controlled Release Capacity

To deal with the limitations of the aforementioned nutraceuticals, encapsulation technology has stood out for decades [221]. The encapsulation requires essential considerations such as stability, the inherent physicochemical characteristics, the interactions between the active component and the matrix, etc. [222]. Bioactive molecules such as isoprenoid derivatives, fatty acids, phenolic substances, structural lipids, carbohydrates, aminoacid derivatives, microbes, and minerals, must be encapsulated before their delivery into a system. The incorporation of the bioactive component in a solid or liquid matrix, the dispersion or spraying of liquid in case of a solid matrix solution, and the stabilization of the system through a physical, chemical, or physicochemical process, are the stages that comprise the encapsulation process. It helps that the bioactive substances are protected from adverse environments, thus, improving their bioavailability [223]. The surface release, diffusion through the swollen matrix, and erosion of the matrix are the three steps that result in the release of the bioactive from encapsulants [224]. The release kinetics can be controlled by the diffusion of a drug molecule through the carrier matrix, and it is desirable to develop drug carriers that provide the sustained or controlled release of the drug with a low dosing frequency. Drug release from carriers is affected by various factors including the composition (drug, polymer, and additives), their ratio, physical and/or chemical interactions among the components, and the methods of preparation [225].

3.1.2. Solubility

The foremost issue encountered with the formulation development of new drugs is the low aqueous solubility [226]. Poor water solubility is a significant risk factor in low oral absorption [227] because the molecular dispersion of a drug is necessary for its absorption across biological membranes. The drug, firstly, must be dissolved within the gastrointestinal tract (GIT) [228], and then absorbed. Subsequently, it reaches the systemic circulation that is important for producing the desired pharmacological response after oral administration. The low solubility of many drugs is a major obstacle to the development and the large-scale production of oral solid dosage forms. Accordingly, the major goals of designing and developing new drugs are the improvement of the solubility and the determination of its negative influence on the drug absorption, bioavailability, stability, and therapeutic effect [229,230].

3.1.3. Bioavailability

As previously mentioned, poor bioavailability is the major challenge in designing oral dosage forms. First-pass metabolism, aqueous solubility, dissolution rate, drug permeability, presystemic metabolism, and susceptibility to efflux mechanisms, are various factors on which the oral bioavailability depends [226]. The study of nutraceuticals bioavailability is an important [231], and an urgent necessity, because of the growth of health challenges, and rapid population [232]. β-carotene, vitamin E, various polyphenols such as phenolic acids, stilbenes, flavonoids, lignans, etc., are slowly absorbed, and therefore have a limited bioavailability [233]. Bioavailability is a property of the drug alone and its delivery systems. Low bioavailability of the drug on its therapeutic use can be considered safe for oral administration because it can be administered in excess without any adverse effects. To increase the bioavailability, the development of powerful drug delivery systems, for surviving the harsh acidic environments of the stomach and rising absorption through the intestinal wall, is considered [232,234].

Bioaccessibility (B*), absorption (A*), and transformation (T*), are the three main stages that the studies on the nutraceutical bioavailability climax. Bioaccessibility, the first step is defined as the fraction of ingested nutraceutical that becomes accessible for absorption through the epithelial membrane of the intestine, whereas absorption, the second step, comprises biocomponent absorbed at the level of the gastrointestinal tract (GIT) epithelial cells. Further, transformation, the third step, describes the chemical or biochemical transformations in the GIT fluids during their digestion and metabolism in the liver [235].

3.2. Delivery Systems Design

As the properties of nutraceuticals such as encapsulation, release capacity, bioavailability and biocompatibility, solubility are challenges still to be overcome, nanotechnology has involved the design and development of organic and inorganic materials at the nano scale, with tailor-made physical, chemical, and biological properties [220]. Many scientists, when designing various delivery systems, utilized approaches such as protection of labile compounds, extension of gastric retention time, controlled/delayed-release, lymphatic uptake facilitation, intestinal permeability enhancement, and modulation of metabolic activities for an optimum nutraceutical delivery system [236]. Lipid, surfactant, and biopolymer-based delivery systems are widely explored as carriers for drug delivery.

The strategies needed for the nutraceutical delivery systems are obtained including the strategies based on drug delivery [237]. Lipid-based delivery systems include liposomes, nanoemulsions, solid lipid nanoparticles (SLNs), niosomes, nanostructured lipid carriers (NLCs), and self-emulsifying drug delivery systems (SEDDSs) [238]. In contrast, surfactants tend to self-assemble in aqueous solutions into micelles, bilayers, vesicles, liquid crystals, and reverse micelles. Some of the lipid-based delivery systems are unstable systems over time or when they are exposed to environmental stresses, they are optically opaque, not easily prepared, and they have limitations to encapsulate, deliver, and protect certain substances. Additionally, surfactant-based delivery systems are optically transparent and thermodynamically stable, but their relatively low loading capacity, and taste are considered as drawbacks. Therefore, biopolymer-based delivery systems are mostly preferred because the fat and surfactant level is reduced [239].

3.2.1. Biopolymer-Based Delivery Systems

Biopolymers are biodegradable, biocompatible, and biofunctional materials with a structural backbone made of carbon, oxygen, and nitrogen atoms. Biodegradation enables them to break into carbon dioxide, water, organic macromolecular material, biomass, etc. They are divided into four major categories such as: extracted from microorganisms, biomass, petrochemical, and biotechnological products. Polysaccharides (starches, celluloses, alginates, pectins, gums, and chitosan), proteins of animal origin (whey, collagen, and gelatin), proteins of vegetal origin (zein, soya, and wheat gluten), and lipids (bees wax, carnauba wax, and free fatty acids), are biopolymers from biomass products. Most of the work has been conducted based on polysaccharides because of their better properties compared to proteins or lipids [240,241]. Polysaccharides are non-toxic, stable, biocompatible, cheap, hydrophilic, and chemically modified because of their reactive sites. The most used polysaccharides incorporate carbohydrates derived from animals (such as chitosan, and chondroitin sulfate), carbohydrates derived from plants (such as starch, pectin, and guar gum), and carbohydrates derived from other sources (such as alginate derived from algae) [241]. For this review, we discuss dextrins which are synthetic substances obtained from enzymatic degradation of starch and have been considerably used for drug delivery applications [242].

Starch

Starch is one of the most promising natural, biodegradable, abundant, and renewable biopolymers on earth. It is composed of α-amylose (20–30%), and amylopectin (70–80%) (Figure 9).

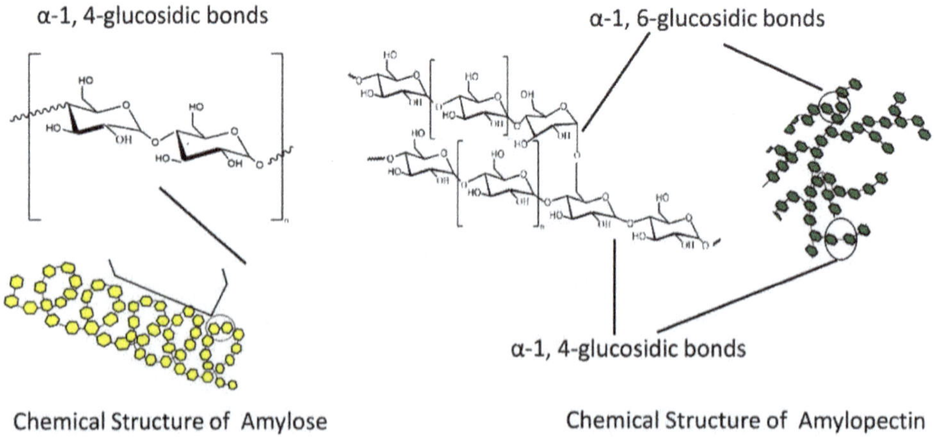

Figure 9. Chemical structure of starch with amylose, and amylopectin.

α-Amylose is a linear macromolecule of several thousands of glucose residues linked by α-(1→4) bonds in a helix conformation. Amylopectin is a branched molecule with α-(1→6) branch points every 24 to 30 glucose residues on average, and consists mainly of α-(1→4)-linked glucose residues. Amylopectin molecules contain up to 106 glucose residues, making them some of the largest molecules in nature [155,240,242,243]. Starch can be isolated from natural resources such as rice, wheat, corn, and potato. Because of the unique physiochemical and functional characteristics of the starch, various studies suggest that starch-based nano and micro-materials can be utilized for a wide range of applications in pharmaceutical and biomedical research [244,245]. However, the native starch is well-known for the substantial swelling and rapid enzymatic degradation causing the fast drug-release, and limitations in the controlled release of drug delivery systems. To enhance the functionalities and new applications, native starch is modified chemically, physically, genetically, and enzymatically [246]. The modification process alters the physicochemical properties and structural attributes and increases the technological value of the native starch [247]. This has led to the use of starch derivatives that are more resistant to enzymatic degradation, the process of cross-linking, and the formation of co-polymers [248]. Since the mid-1900s, a large-scale starch processing industry has emerged. An excellent review published by Van der Maarel et al. [248] detailed the starch-processing industry. The first step is the liquefaction into soluble and short-chain dextrins, and then the saccharification of the starch-hydrolysate syrup to a high concentration glucose syrup. The mobility of the starch chain is caused by the α-(1→4) glucose units. Due to the various hydroxyl groups starch can be easily functionalized [249]. The starch is chemically or enzymatically processed into diverse products such as starch hydrolysates, starch or maltodextrin derivatives, cyclodextrins, glucose syrups, or fructose [250]. The development of modified dextrin-based chemically cross-linked hydrogels for drug delivery applications is the main focus in this review [16].

Dextrin

Dextrin is one of the most noteworthy polymers because of its various features such as hygroscopicity, fermentability, sweetness, stability, gelation, solubility, bioavailability, and molecular compositions. Dextrin is a low-molecular-weight carbohydrate produced by enzymatic and/or acid partial hydrolysis of starch, with the same general formula as starch, but smaller and less complex. It contains α-(1→4) D-glucose units of amylose and the α-(1→4) and α-(1→4,6)-D-glucose units of amylopectin with lower polymerization. During the enzymatic degradation of starch, linear and cyclic dextrins are formed [16,251–254].

Dextrins are widely used in a variety of applications such as adhesive in the manufacture of textiles, in cosmetics, gummed tapes, and paper, biomedical, and pharmaceutical applications [252]. Because dextrins are easily degraded by α-amylase, the chemical modification can tailor the dextrin structure for satisfying a variety of drug delivery objectives [21]. In general, native starches are often modified as a consequence of their unfavorable properties such as high hydrophilicity, and poor solubility, and herein enhance their application. Although there are several methods of starch modification, as described in Figure 10, chemical modification is the most commonly used. The chemical modification means the introduction of functional groups to the molecule of starch giving characteristic properties [247,255,256]. Among dextrins, cyclodextrin (CD) modifications have become a major area of interest for numerous investigations and, therefore they are widely explored as drug delivery systems.

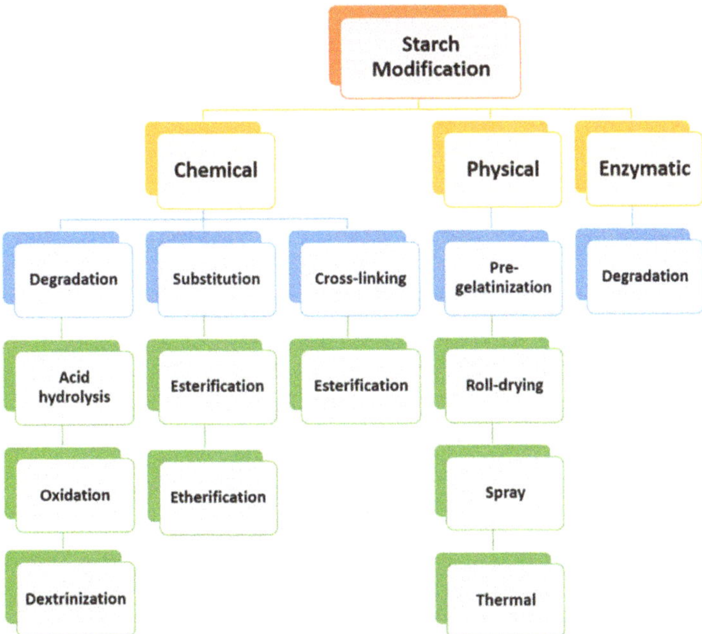

Figure 10. Methods of starch modification.

Nanoparticle-Cell Interactions In Vitro, and In Vivo

Throughout the years of investigating the interaction of dextrin-based nanomaterials with the cells, researchers stumbled upon questions of how this nanoparticle-based on polymeric materials influences the cellular uptake. With all these nanoparticles at hand, and the new ones, more in vitro and in vivo experiments will be performed after a clear understanding of intracellular trafficking and the fate of the nanoparticles after the endocytotic process [257]. The higher cell affinity and easier uptake of the nanoparticles are due to interactions of the ligands on their surface with the receptors on the cell membranes [258]. The in vitro and in vivo stability are altered by nanoparticle size, stability, responsivity, composition, and surface charge. As already discussed, the therapeutics can be encapsulated within the nanoparticle core, can be chemically conjugated to the polymer, bound to the surface of nanoparticles, or can be entrapped in the polymer matrix. Nanocapsules or cavities surrounded by a polymeric membrane, nanospheres, or solid matrix systems, are the most common forms of polymeric nanoparticles (Figure 11a).

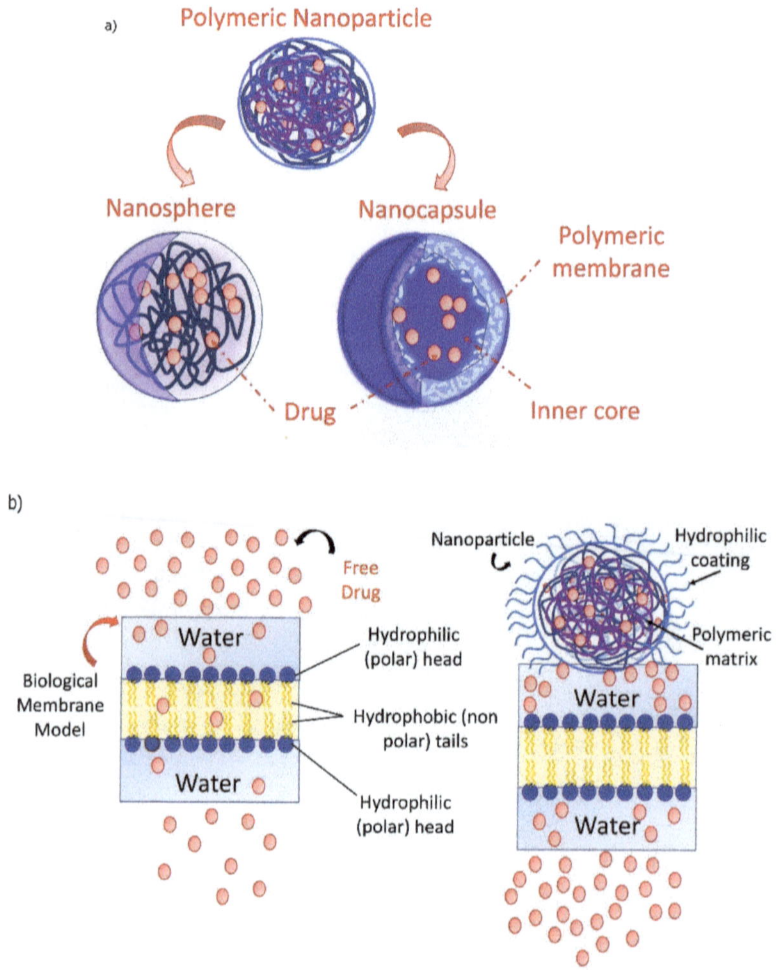

Figure 11. (**a**) Schematic representation of polymeric nanoparticles and (**b**) a model of the interaction between the aqueous phase containing a free hydrophilic drug or drug-loaded nanoparticle and the biological membrane model.

The nanoparticles encounter diverse interactions with the surface of the cell membrane. Cationic nanoparticles can damage the cell membrane and cause cytotoxicity if are too positively charged, whereas anionic nanoparticles, due to repulsive forces, can fight to contact the cell surface. Therefore, the first contact between a cell and a nanoparticle, dependent on nanoparticle and cell properties, can determine the prospect of the nanoparticle, and its therapeutic potential [259]. A huge nanoparticle with hydrophobic moieties is attached to the membrane because the chemical barrier is one of the fundamental elements that regulate the interaction of nanoparticles with the cell membrane. The strength of the chemical barriers is influenced by membrane compositions. When there are enthalpic interactions between nanoparticle hydrophobic ligands and the membrane interior, caused by lipids with longer tails, an increase in barrier strength is observed [260]. Figure 11b) presents the use of nanoparticles as drug delivery systems to prevail the barriers of drug penetration into cells. An appropriate nanoparticle coating can encourage the insight of hydrophilic drugs through the biological membrane. The hydrophilic drug or free drug can freely diffuse in the aqueous medium. However, the diffusion of this drug can be limited because it is incapable of interacting with the outer-hydrophilic zone of the membrane without passing through the lipophilic layer of the same membrane. Therefore, the outer hydrophilic shell of the nanoparticle is introduced to ensure the interaction with the hydrophilic layer of the membrane, whereas the internal lipophilic core of the particle interacts with the hydrophobic layer of the membrane. A high permeation of the drug is provided. Although the major demands for drug delivery devices are still pending, the in vitro and in vivo results showed the constant improvement that has been made starting from microtechnology, crossing to nanotechnology, and recently viewing the selective drug delivery [10].

3.2.2. General Features of Cyclodextrins and Cyclodextrin-Based Polymers as Delivery Systems Matrices

Cyclodextrins (CDs) are cyclic oligosaccharides produced via cyclodextrin (CD)-glycosyltransferase from starch, by certain microbes such as Bacillus macerans. CDs contain six (αCD), seven (βCD), eight (γCD), or more (α-1,4)-linked α-D-glucopyranose units. The truncated shape of CDs is because of the chair conformation of glucopyranose units with the hydroxyl groups orientated to the cone exterior. The primary hydroxyl groups of the sugar residues are at the narrow edge of the cone, and the secondary hydroxyl groups are at the wider edge [261]. CDs tend to form inclusion complexes because of their lipophilic interior and hydrophilic exterior. The mechanism of the complexation includes the absence of the covalent bonds and the presence of the driving force releasing enthalpy-rich water molecules from the cavity part. CDs can include molecules of size and polarity compatible with their lipophilic inner cavity [262,263]. The formation of hydrogen bridges, between the polar hosts and the primary hydroxyls, establishes simultaneously polar interactions, whereas hydrophobic hosts will be housed inside the CD cavity because of the hydrophobic Van der Waals type interactions [23,264,265]. The versatility of CDs is wide and promising in the pharmaceutical and nutraceutical industries, so far, 26,895 research articles about CDs have been published in PubMed, by typing the word "cyclodextrins" [266].

Even though the unique structure of CDs has fascinated scientists around the world, native CDs are appropriate only for the molecule recognition of a wide range of substrates. Moreover, they have various limitations such as the inability of including certain hydrophilic compounds or high molecular-weight drugs, low aqueous solubility, and toxicity in case of β-CD when administered intravenously [267–269]. Therefore, specific applications require overcoming the aforesaid limitations by chemical modifications of CD structures [26,267].

In CDs, hydroxyl groups can be modified by replacing the hydrogen atom or the hydroxyl group with a variety of substituting groups [270]. CD derivatives and CD-based polymers appear as powerful tool. CD derivatives comprise the randomly methylated β-CD, the hydroxypropyl derivatives of β- and γ-CD, sulfobutylether β-CD, and the branched CDs [271], whereas CD-based polymers, containing two or more covalently linked CD-units, can be water-soluble and moderately swelling or insoluble and strongly swelling [272].

CD-based nanosponges (NSs) can easily be obtained by reacting the nucleophilic hydroxyl group of the selected CD with a suitable cross-linking reagent, containing two electrophilic sites, that convert molecular nanocavities into insoluble three-dimensional, nanoporous structures. Widely-used cross-linkers that influence the behavior of the CD units, are epichlorohydrin for hydrophilic NSs synthesis, and diphenyl carbonate (DPC), pyromellitic dianhydride (PMDA), diisocyanates, carbonyldiimidazole (CDI) for hydrophobic NSs synthesis. With a highly porous nanomeric and insoluble nature, CD-NSs are capable of encapsulating a variety of substances, particularly of increasing the solubility of poorly water-soluble drugs, prolonging their release, and improving their bioavailability and stability. Because of these characteristics and their harmlessness, CD-based NSs are used in certain fields such as chemistry, gene delivery, agriculture, cosmetics, food, biomedicine, biotechnology, biocatalysis, etc. In addition, the main area of investigation so far is the pharmacy, in which CD-NSs have been proposed as drug delivery systems [262,263,273–279]. Therefore, it is no wonder that the demand and the need for an explosive scientific and technological revolution have increased over the years.

Historical Developments of Cyclodextrin-Based Nanosponges as Delivery Systems Matrices

Developing effective CD-based systems that can improve the properties of the nutraceutical has been attractive for many applications, particularly the field of pharmaceuticals as delivery systems. Delivery systems with which the CDs are associated comprise nanoparticles, liposomes, microspheres, hydrogels, and NSs [280]. From a historical point of view, CD-based polymers, reviewed by Petitjean et al., can be observed that have progressed as the result of the enormous research conducted over the years. Despite this, the investigations of CD-based polymers application in pharmaceuticals, food chemistry, and biomedicine, are not as abundant as on the parent CDs applications [281]. CD-NSs appear as advanced drug carriers and, therefore, can also contribute as nutraceutical carriers. As with all of the evidence in the history of CD-NSs [33], detailed by Krabicovà et al., over the years significant progress has been made on CD-NSs synthesis and applications in several scientific and technological fields. According to their chemical composition and properties, CD-NSs with particular attention to the pharmaceutical field, are divided into four generations, as overviewed by Caldera et al. [280]. As already described, poor water solubility, low bioavailability, low stability, low permeability, low efficacy, are some drug features that affect the applications of drugs [23,263,280,282,283]. Therefore, to distinguish between the generations of CD-NSs in the improvement of the aforementioned limitations, several experimental results are compared. The first generation of CD-NSs remains among the most commonly explored NSs as drug delivery systems [284].

The First Generation of Cyclodextrin Nanosponges

The first generation comprises urethane (or carbamate), carbonate, ester, and ether CD-NSs [285] (Figure 12).

Cyclodextrin-Based Urethane Nanosponges

Diisocyanates such as hexamethylene diisocyanate (HDI), toluene-2,4-diisocyanate (TDI), are used to synthesize urethane (or carbamate) CD-NSs, and are characterized by a very low surface area (1–2 m^2/g), high resistance to chemical degradation, rigid structure, and a negligible swelling in organic solvent and water [263,285].

A study by Thatiparti and von Recum synthesized CD-based gels by dissolving CD in N, N-dimethylformamide (DMF), and adding 2-isocyanatoethyl 2,6-diisocyanatohexanoate (LTI), and 1,6-diisocyanatohexane (HDI). The antibiotics such as rifampin (RM), novobiocin (NB) sodium salt, and vancomycin (VM) hydrochloride, were loaded to this cross-linked polymer by using a common solvent/solution absorption method. Due to the minimal swelling ability of the synthesized polymer, the percent loading was too low (not more than 3.5%). The antibacterial activity of these antibiotics-loaded gels confirmed a clear zone of inhibition. Additionally, it was observed that the CD-based gels were capable of providing long-term sustained release of these antibiotics. This work developed an anti-infectious drug delivery system [286].

Further, Merritt et al. synthesized CD-based polymers by dissolving γ-CD in N,N dimethylformamide (DMF), or dimethyl sulfoxide (DMSO), and adding hexamethylene diisocyanate (HDI) as a cross-linker. Mitomycin C (MMC), an anti-proliferative drug, was loaded in those polymers. MMC release rates were adjusted to be slower, and more sustained because of the affinity between the MMC and polymer. This therapy, due to less overall exposure to MMC, was suggested to have less clinical risk to the patient and surgical staff compare to traditional treatments [287].

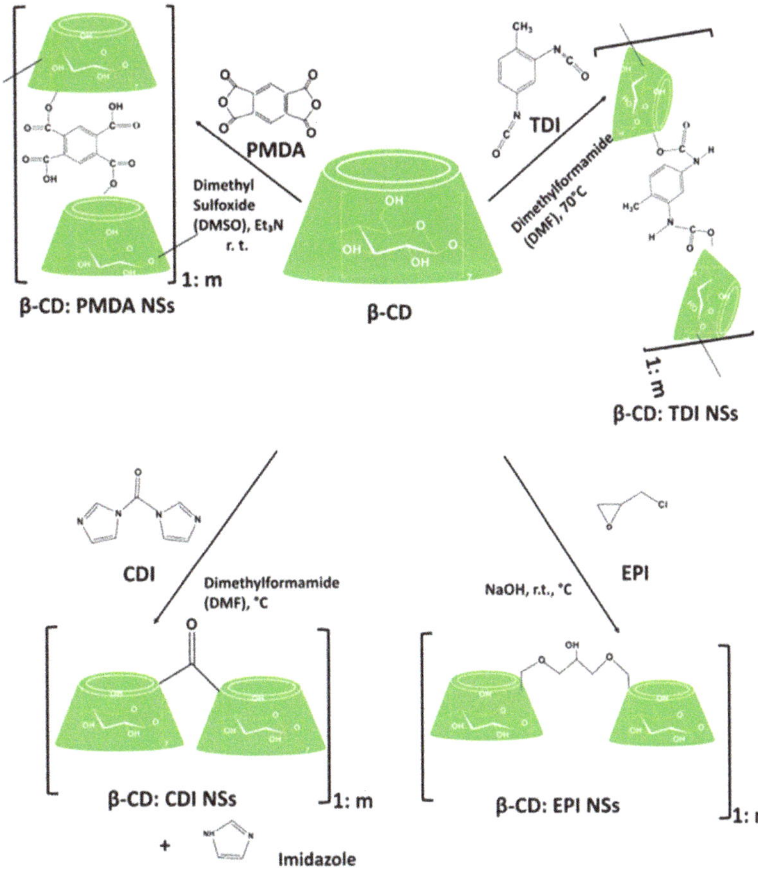

Figure 12. Schematic representation of the synthesis of cyclodextrin (CD)-based ester (PMDA), ether (EPI), carbonate (CDI), and urethane (TDI) nanosponges (NSs).

Cyclodextrin-Based Carbonate Nanosponges

Active carbonyl compounds such as 1,1′-carbonyldiimidazole (CDI), triphosgene, and diphenyl carbonate (DPC), are used to synthesize carbonate CD-NSs, and are characterized by a low surface area (around 2 m^2/g), short cross-linking bridges, good stability to acidic and slightly alkaline solutions, and reduced swelling ability [263,285].

Quercetin, a flavonoid with strong antioxidant activity and many others, has pharmaceutically been challenging because it has a very low solubility of 7.563 μg/mL in water. Therefore, Jullian et al. used β-CD and its derivatives such as hydroxypropyl-β-cyclodextrin (HP-βCD), and sulfobutyl ether-β-cyclodextrin (SBE-βCD) to increase quercetin solubility [288]. Furthermore, Anandam et al. investigated the improvement of the chemical stability, and aqueous solubility of quercetin by utilizing the NSs synthesized from β-CD and DPC in five various molar concentrations (1:2, 1:4, 1:6, 1:8, and 1:10). The solubilization efficiency was increased from 7.563 μg/mL (plain quercetin) to 152.543 μg/mL (quercetin-loaded NSs). As the cross-linking ratio influenced the release of quercetin, it was observed from its rapid burst release of 97% within 60 min. These improvements influenced the increased antioxidant activities of quercetin and offered a potential drug delivery system for oral and topical delivery [40].

Curcumin, another nutraceutical with many potential applications, was loaded into βCD-based NSs synthesized using dimethyl carbonate (DMC) by Darandale and Vavia. The in vitro release study of this formulation showed sustained curcumin release. In comparison with plain curcumin (0.4 μg/mL), and the β-CD complex (5.88 μg/mL), it was also observed higher solubilization of curcumin loaded into CD-NSs (20.89 μg/mL) generating a potential drug delivery system for curcumin in cancer treatment [41].

Ansari et al. presented the possibility to administer resveratrol loaded at two different weight ratios of βCD:CDI (1:2 and 1:4) as buccal delivery and topical application. This is because CD-NSs significantly increased the stability, solubility, and permeation of resveratrol, a polyphenolic phytoalexin with many health benefits [40]. Further, Dhakar et al. demonstrated the high encapsulation efficiency of resveratrol (77.3%) and oxyresveratrol (80.33%) into the β-CD:CDI NSs. In comparison to drug molecules alone, solubilization of resveratrol- and oxyresveratrol-loaded NSs was higher, and therefore, a better antioxidant activity was observed [289]. Matencio et al. studied the complexation of the oxyresveratrol with two different weight ratios of β-CD:CDI NSs (1:4 and 1:8) using a new methodology. The apparent inclusion complex constant (KFapp) between β-CD:CDI NSs, and oxyresveratrol was calculated using the UV-Vis measurement and the Benesi-Hildebrand method with modifications. This study represented the feasibility of the drug-CD-based NSs complexation, and the use of oxyresveratrol, which exhibits a wide range of biological activities, as an ingredient in nutraceutical products [290].

Rezaei et al. incorporated thyme essential oil (TEO), a natural phenolic compound with high antimicrobial and antioxidant activity, into β-CD:DPC (1:4, 1:6, 1:8) NSs. The highest amount of loading capacity, encapsulation efficiency, and solubility were of the ratio of β-CD:DPC (1:4) nanosponge. The solubility of TEO increased from 2.7% (free TEO) up to 41% (TEO encapsulated into the nanosponge). A controlled and slow release of TEO from β-CD:DPC NSs was indicated from the in vitro release. Therefore, TEO-loaded into the β-CD:DPC NSs presented the potential to be used as a natural preservative in the food industry [43].

Norfloxacin [291], kynurenic acid [292], tamoxifen [293], rilpivirine [294], acyclovir [295], camptothecin [296], telmisartan [297], babchi oil [298], chrysin [299], paclitaxel [277], sulfamethoxazole [300], ferulic acid [301], melatonin [44], D-limonene [302], azelaic acid [303], paliperidone [304], griseofulvin [305], flutamide [306], econazole nitrate [307], piperine [308], etc., are several other drugs loaded in carbonate CD-NSs. In all these findings it was observed an enhancement of the biocompatibility and aqueous solubility of the aforementioned drugs when they are loaded in NSs compare to CD-inclusion complexes or uncomplexed drugs. The indicated ones present CD-NSs a promising nanocarrier system.

Cyclodextrin-Based Ether Nanosponges

Cross-linking agents with epoxide groups such as epichlorohydrin (EPI), bisphenol A diglycidyl ether, ethylene glycol, and diglycidyl ether, are used to synthesize ether CD-NSs, and are characterized by a tunable swelling capability, and high chemical resistance. Although the EPI toxicity, the EPI-based CD-NSs, are widely studied [285].

Machín et al. synthesized CD-EPI insoluble polymer, a novel polymeric hydrogel for the controlled release of anti-inflammatory drugs such as naproxen (NAP), and nabumetone (NAB), and antifungal drugs such as naftifine (NF), and terbinafine (TB). According to this study, these hydrogel matrices were considered as potentially suitable sustained release systems [309].

Rodriguez-Tenreiro et al. synthesized new biocompatible hydroxypropyl-β-CD-based hydrogels using ethylene glycol diglycidylether (EGDE) as a cross-linker and incorporated other structurally related polymers such as hydroxypropylmethylcellulose (HPMC). Diclofenac, a nonsteroidal anti-inflammatory drug [310], was chosen as a suitable candidate to be loaded into the synthesized hydrogels. The hydrogels were able to load and sustain the release of diclofenac for several hours [311].

Gami et al. synthesized novel hydrogels from xylan and β-CD using ethylene glycol diglycidyl ether, as a crosslinker, in an alkaline medium. Two anticancer drugs loaded in hydrogels were curcumin (26%), and 5-fluorouracil (98%). The highest cumulative release of 56% 5-fluorouracil, and 37% curcumin, from the synthesized gels, was observed after 24 h. This study synthesized, characterized, and explored hydrogels as an in-vitro drug delivery vehicle [312].

Cyclodextrin-Based Ester Nanosponges

Dianhydrides or di/polycarboxylic acids such as pyromellitic dianhydride (PMDA), ethylenediamine-tetraacetic dianhydride (EDTA dianhydride), butane tetracarboxylic dianhydride, citric acid (CA), etc., are used to synthesize ester- based CD-NSs. These NSs are characterized by a huge absorption of water forming, thus, the hydrogels [263].

Appleton et al. proposed β-CD:PMDA NSs as a solution for the oral delivery of insulin. The loading capacity of insulin was 14.41%, and the encapsulation efficiency was 91.40%. In this study it was hypothesized that the β-CD:PMDA nanosponge can be suitable for loading insulin, protecting it from degradation in the stomach, and promoting its intestinal absorption due to controlled release properties [313].

Another study made by Argenziano et al., used β-CD:PMDA for the topical delivery of imiquimod (IMQ, 1-[2-methylpropyl]-1H-imidazo[4,5-c] quinoline-4-amine), an immune response modifier. The loading capacity of the IMQ into β-CD:PMDA nanosponge was 14.2%, and the encapsulation efficiency was 96.5%. In this work, a nanomedicine-based topical formulation for the prolonged and controlled release kinetics of IMQ through the skin was developed [314].

Ester nanosponge based on PMDA was also used to maximize the therapeutic efficacy of acetyl salicylic acid (a non-steroidal anti-inflammatory and antipyretic drug) [315], doxorubicin (an anti-cancer drug) [316], meloxicam (a non-steroidal anti-inflammatory drug) [317], rilpivirine (an antiviral drug) [318]. In all instances, a controlled and prolonged drug release, a remarkable increase in the drug solubility and bioavailability, was observed.

Other Generations of Cyclodextrin Nanosponges

Other generations of CD-NSs comprise the functionalized CD-NSs, stimuli-sensitive NSs, and molecularly imprinted polymers (MIPs) [285].

Singh et al. enhanced the cellular binding efficiency of β-CD nanosponge by functionalizing its surface with cholesterol as an endogenous physiological molecule. The cholesterol grafting enhanced the adsorption of the anticancer drug Dox because of the hydrophobic charge on the surface. Therefore, the cholesterol-modified β-CD nanosponge system was proposed to be a site-specific drug delivery carrier improving the solubility and bioavailability of small drug molecules with low water-solubility [319], (Figure 13).

Another study made by Asela et al. developed a new nanomaterial based on βCD based NSs for the transport of phenylethylamine (PhEA), as an antidepressant, and 2-amino-4-(4-chlorophenyl)-thiazole (AT), as an anti-microbial, and anti-inflammatory agent. These complexes were functionalized with gold nanoparticles (AuNPs). In comparison to the native βCD, the loading capacity of βCD NSs was eight times higher for PhEA (90%), and AT (150%). Additionally, βCDNS presented a higher degree of solubilization and complexation efficiency of PhEA and AT than native βCD. The immobilization percentage of AuNPs reached 85%. This study demonstrated the versatile materials (βCD NSs, PhEA, AT, and AuNPs) with an efficient loading capacity for potential applications in the transport of therapeutic agents [320].

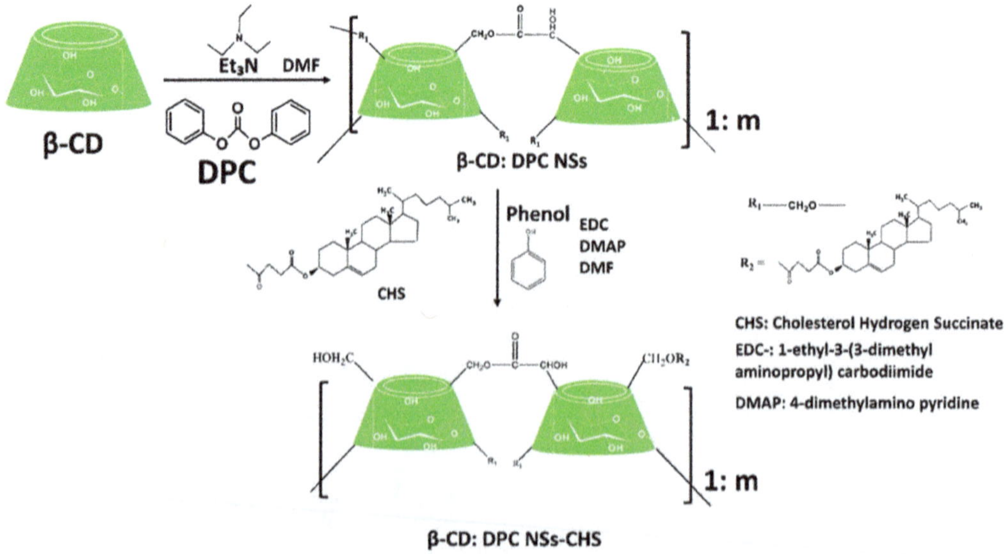

Figure 13. The cholesterol hydrogen succinate (CHS) grafting, in CD-based carbonate NSs (β-CD:DPC NSs), using coupling reaction.

Although CD-NSs are capable of high drug loading, their non-selectivity remains a challenging task regarding pharmacologic efficacy. To offer the advantage of site-specific drug delivery, stimuli-responsive drug delivery was designed by Palminteri et al. In their study, they presented the synthesis of glutathione responsive (GSH) CD-NSs from PMDA and 2-hydroxylethyldisulfide for improving the tumor-specific delivery of resveratrol. GSH-NSs enhanced the aqueous solubility of resveratrol more than four-fold (201 μg/mL) compared to free resveratrol (46 μg/mL). The encapsulation efficiency was 80.64%, whereas the drug loading was 16.12%, representing the GSH-responsive NSs as an effective delivery system for targeting cancer cells. Daga et al. synthesized the GSH-responsive based NSs using β-CD, 2-hydroxyethyl disulfide, and PMDA as cross-linking agents (Figure 14). Doxorubicin, an anticancer drug, was loaded into GSH-NSs. Dox-GSH-NSs showed a good safety profile, and their hepatotoxicity, observed both in vitro and ex vivo, resulted to be comparable with free Dox. It was observed a slowed and prolonged drug release, and no initial burst effect. No cytotoxicity in vitro and ex vivo was caused by GSH-NSs. Therefore, GSH-NSs are considered a suitable carrier of chemotherapeutic drugs [321].

Coviello et al. presented a study of the use of carbonate CD-NSs, based on DPC as a cross-linker, for targeted drug delivery. Into the synthesized NSs, a novel multi-effective heterocyclic compound, 2-(3,4-dimethoxyphenyl)-3-phenyl-4H-pyrido (DB103), was loaded. DB103, as the ideal drug candidate for drug-eluting stents (DES), was efficiently loaded, and slowly released. The novel DB103-NSs system represented an innovative and credible prototype of formulation used for the local therapy of the vessel's wall subjected to percutaneous intervention [322].

Deshmukh, Tanwar et al. functionalized the surface-active CDI cross-linked β-CD NSs by lysozyme. Calcium carbonate, and carboxymethyl cellulose, were added to effectively elevate loading efficiency. Lysozyme was delivered in a conformationally stable structure by damaging bacterial cell walls. This was made by catalyzing the hydrolysis of 1,4-β-linkages between N-acetyl-D-glucosamine and N-acetylmuramic acid residues, present in peptidoglycan layer surrounding the bacterial cell membrane, for controlling the calcium release in hypocalcemia condition for 24 h. The NSs formulation is an encouraging carrier for antibacterial protein [323].

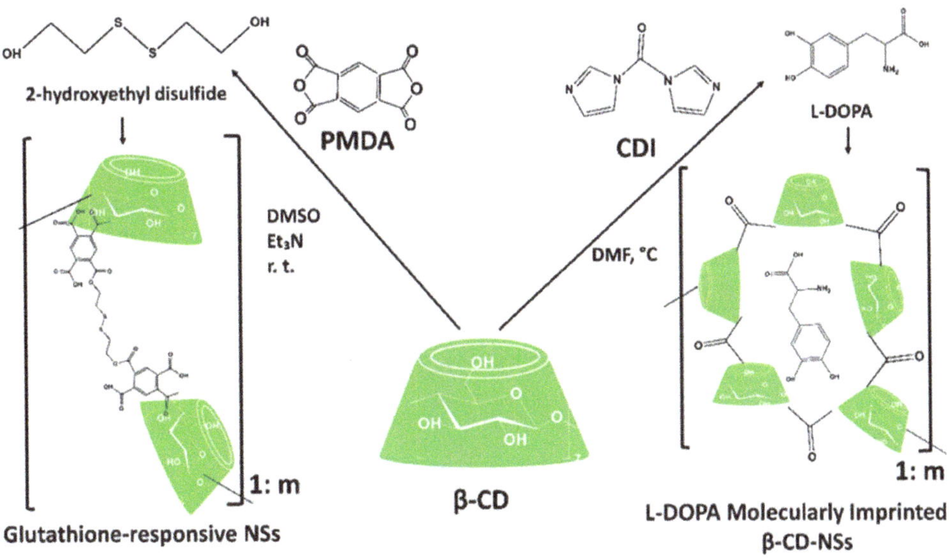

Figure 14. Schematic representation of the synthesis of glutathione-responsive NSs, and molecularly imprinted CD-based carbonate NSs (L-DOPA used as a template).

Deshmukh et al. synthesized molecularly imprinted polymer (MIP) of PMDA cross-linked βCD-based NSs. The work was compared with non-molecularly imprinted polymer (NIP) NSs, and the glucose was used as a template. The synthesized MIP-NSs showed more significant specificity, and binding for glucose compared to their NIP-NSs [324].

Trotta et al. synthesized a MIP NSs of CDI cross-linked β-CD-based NSs. The L-DOPA, as a non-proteinogenic amino acid widely used for the treatment of Parkinson's disease (PD), was used as a template. This study showed that the polycarbonate β-CD-based MIP-NSs, as an oral formulation, can be a promising new drug delivery system for the prolonged release, and the protection of L-DOPA [325] (Figure 14).

The safety issue has invoked important remarks regarding the clinical applications of CD-NSs, as drug delivery systems, as they were predominantly synthesized in organic solvents which may cause cellular damage. To overcome the toxic effects, several surveys comprised the green syntheses of a series of CD-based NSs in natural deep eutectic solvents (NADES) [326], and dextrin-based solvent-free NSs synthesis [327], which need to be further studied for medical, and pharma applications. Another promising platform for therapeutic applications that can be considered is a combination therapy with CD-NSs [328].

3.2.3. General Features of Maltodextrins and Recent Trends in Their Applications as Delivery Systems Matrices

Maltodextrin is defined as a hydrolyzed starch product [329] consisting of D-glucose units linked primarily by α (1,4)-glycosidic bonds [330]. Maltodextrins are classified by their values of dextrose equivalent (DE). The DE ranges up to 20 and expresses the number of reducing sugars present in the polymer. Maltodextrins with different dextrose equivalent (DE) values have different properties. A low fraction of glucose and a high fraction of polysaccharides refers to a low DE, and vice versa [25,331,332]. The different reaction sites in repeating glucose moieties of maltodextrins provide wide alternatives in the chemical conjugation process [333]. The characteristics and the physicochemical, nutritional, functional, and technological properties enable the numerous applications of maltodextrins in the food, medical food products, beverage, and the pharmaceutical industry in tablet and powder applications [334]. Maltodextrin is considered the most used starch hydrolysate by the food industry [335].

Sun et al. developed casein-maltodextrin Maillard conjugates to entrap proanthocyanidins (PAs). PAs are a group of polyphenolic compounds with potent antioxidant capacity. However, the practical application as antioxidants is limited because they are easily affected by environmental stress during the processes and storage, and chemically degraded in the gastrointestinal tract. The in vitro release test presented the exhibition of strong protective effects, by casein-maltodextrin-PAs nanoparticles, during the storage and thermal treatments, and PAs bioaccessibility improvement. Thus, the casein-maltodextrin-PAs nanoparticles have been considered as novel antioxidants for applications in pharmaceutical and nutraceutical products. The release rate of encapsulated PAs and free PAs was 30% and 80% after 120 min [336]. The maltodextrins, through the nanoencapsulation, improved the stability, efficiency, and thermal resistance of ellagitannin extracts from pomegranate peels [337].

Further, Gurturk et al. modified liposomes with maltodextrins, and loaded levodopa as the most effective drug to ameliorate the symptoms of Parkinson disease. Therefore, this study suggested an effective way of targeting the blood-brain barrier (BBB) with controlled and sustained drug release properties, improved cellular binding, and lowered cellular cytotoxicity [338].

Lai et al. used maltodextrins as a film-forming material, and glycerin as a plasticizer to enhance the quercetin oral bioavailability. Maltodextrin films provided long-term storage stability, and fast dissolution rate [339].

Helal et al. chemically conjugated the maltodextrin via the formation of the ester bond with vitamin E succinate. The designed bio-conjugates showed higher aqueous solubility, a slighter toxicological effect on the main body organs, and a higher total antioxidant capacity than the vitamin E succinate [333].

Laurent et al. impregnated the ciprofloxacin (CFX) into an artificial polypropylene (PP) abdominal wall implant that was functionalized with citric acid and hydroxypropyl-Y-cyclodextrin (HPYCD), or with maltodextrin. The use of the HPYCD, and maltodextrins, as two carbohydrate oligomers produced from starch, offered a safe and green process of functionalization, avoiding the common extreme technologies of functionalization such as plasma or radiative technologies. It was observed a higher CFX loading and release over a sustained period in the case of HPYCD with regard to maltodextrins. The HPYCD-finished meshes showed antimicrobial activity, and this not only because of the ionic and hydrogen bonding between CFX and CD-finished textile, but also because of the inclusion in the cavity [18].

Maltodextrins were also used to produce proniosomes formulation with a potential application in the delivery of hydrophobic or amphiphilic drugs [340]. Shruthi et al. prepared the resveratrol-loaded proniosomes using maltodextrin, lactose monohydrate, and pullulan as wall materials. The encapsulated resveratrol presented a more consistent and sustained release, and a higher stability under gastric and intestinal conditions [341].

As CDs, maltodextrins may be chemically modified to improve their physical, and functional characteristics. The maltodextrins are less expensive, and have a higher aqueous solubility than that of comparable cyclic dextrins [25,342]. Compared to modified CDs, modified maltodextrins are less studied in pharmaceutical applications. Therefore, the development of modified CDs, which have been shown for various applications, already discussed in this entry, will fully instruct the linear dextrins to be modified and used for nutraceuticals delivery.

Castro-Cabado et al. synthesized the cross-linked system consisting of maltodextrin with citric acid and tartaric acid [343].

Cecone et al. synthesized the sustainable cationic cross-linked polymers, suitable for eco-friendly scaling-up, based on maize-derived maltodextrin (Glucidex 2®) as a building block, 1,4-butane-diol diglycidyl ether as a cross-linker, and 1,4-diazabicyclo[2.2.2]octane, imidazole, triethylamine, and guanidine hydrochloride were as supplements to obtain the amine-mediated epoxy ring-opening reaction [344]. These materials, considered genuinely good eco-friendly alternatives, can be further studied as green adsorbents for environmental, medical and pharmaceutical applications because of their high adsorbing feature.

Ivàn Melèndez-Ortiz et al. synthesized the hydrogels with antimicrobial activity based on the maltodextrin, previously modified with glycidyl methacrylate (GMA), copolymerized with acrylic acid (AAc) or acrylamide (Aam) to endow mechanical properties and chemical functionality. This step enabled to host bioactive nanoparticles such as zinc oxide nanoparticles (ZnO). ZnO-loaded hydrogels showed antimicrobial activity against Staphylococcus aureus and Escherichia coli. This study has been promising for the possible application of the maltodextrin-based hydrogels as wound dressings [345].

Yan et al. presented a novel design of entrapping simvastatin (SIM), a hypolipidemic drug, to promote bone regeneration in an injectable maltodextrin-based micelle/hydrogel composite system. The SIM-loaded aldehyde-modified micelles were anchored to the hydrogel network by Schiff base linkage. It was observed a great enhancement of the mechanical strength of hydrogels, good biocompatibility, slow and sustained release, and osteogenic capability of SIM [346].

Furthermore, Demasi et al. selected Kleptose Linecaps® DE17 (LC), apart from CD, as a highly soluble maltodextrin. The study synthesized the carbonate and ester NSs, by cross-linking α-, β-, and ϒ-CD, and LC with PMDA and CDI. Dextrin-based NSs encapsulated ailanthone (Ail) as a natural compound with anti-tumor, anti-malarial, anti-inflammatory, anti-parasitic, anti-feedant, and herbicidal activities, but with low insistence and immediate degradation in organic substrates. The encapsulation process increased and extended the phytotoxic activity, preserved the efficacy, and prolonged the release kinetics of Ail [347].

A safe and sustainable synthesis of NSs based on β-CD, and LC with citric acid in water, using sodium hypophosphite monohydrate as catalyst, was performed by Rubin Pedrazzo et al. The synthesized citrate NSs were able to selectively adsorb a significantly higher amount of heavy metals from aqueous solution [348]. Thanks to the helical structure of the amylose chains, maltodextrins can act as complexing agent demonstrating a strategy to integrate drug loaded carriers into hydrogels for drug delivery and tissue engineering applications [347]. We still lack deeper knowledge of the forces involved in the complex formation of maltodextrins with nutraceuticals, and of the pharmaceutical applications.

4. Concluding Remarks

Healthier living, preventive care, and secondary source medication are the focus of the global population since the healthcare costs have risen. The adverse effect of drugs and the risk of toxicity caused a worldwide revolution in nutraceuticals. Nutraceuticals are considered a powerful instrument to facilitate lifespan, optimal health, and quality of life. They act against various nutritionally induced diseases such as obesity, cancers, diabetes, neurodegenerative diseases, cardiovascular disease, allergy, respiratory disorders, arthritis, etc.

In this review, we listed many nutraceuticals classified based on food bioavailability, chemical nature, and mechanism of action, with various health benefits, and summarized common problems that prevent their applications. Similar to the case of any other drug, the health benefits of the nutraceuticals are limited by the low bioavailability, low efficacy, low aqueous solubility, degradation and metabolism, and epithelial permeability.

The polymer-based nanocarriers, with a tripartite structure: monomer, linker, and therapeutic agent, are extensively studied to overcome the aforementioned shortcomings of nutraceuticals. This review shows the use of dextrin-based chemically cross-linked polymer as the most preferred, advanced, biocompatible, and natural system to deliver nutraceuticals. One of the most commonly studied systems is cyclodextrin (CD)-based nanosponges (NSs). CD-NSs are synthesized by reacting CDs with cross-linkers such as carbonyl-diimidazole (CDI), diphenyl carbonate (DPC), pyromellitic anhydride (PMDA), etc. Due to their highly porous nanomeric nature, CD-NSs are found in different applications ranging from chemistry, pharmacy, biomedicine, gene delivery, food and biotechnology, environment, etc.

Nutraceuticals such as curcumin, resveratrol, thyme essential oil, Babchi oil, mitomycin C (MMC), diclofenac, imiquimod, doxorubicin, rilpivirine, melatonin, L-DOPA, etc., are successfully formulated using CD-NSs. CD-NSs are efficient encapsulating agents of delivering nutraceuticals with controlled kinetics through the topical, oral, and parenteral routes. The nanomaterials based on CD-NSs, in contrast to free nutraceuticals and native dextrin units, improved the solubility, biocompatibility, bioavailability, encapsulation, and release capacity of the nutraceuticals, and therefore are considered the most effective delivery systems.

Until now, these nanoparticles have been explored in terms of their physicochemical properties, drug-loading ability, in vitro and in vivo tests. The concerns, such as the specific interaction of the nanoparticles with human organs, tissues, cells, or biomolecules, their effect on human metabolism, and their application in drug delivery, etc., await further intense research, which will be a focus henceforward.

As reviewed above, numerous nanoparticle drug delivery systems, as the most advanced approach contributing to every walk of life, have been prepared. Although maltodextrins are known for a higher aqueous solubility and are less expensive than CDs, their modified forms need to be further studied in pharmaceutical applications. Therefore, the extensive literature about the synthesis, characterization, and applications of the modified CDs, will fully instruct the modified dextrins.

It can be forecasted that the more advanced nanoparticle drug delivery systems will be released and overtake the market.

Author Contributions: Conceptualization, F.T., F.C. and G.H.; investigation, data curation, and writing—original draft preparation, G.H.; supervision, and project administration, F.T.; validation, F.T., G.H., A.M. and A.R.P.; writing—review and editing, and visualization, G.H., A.M., A.R.P., C.C., S.L.A., Y.K.M., F.C. and F.T. All authors have read and agreed to the published version of the manuscript.

Funding: This research received no external funding.

Institutional Review Board Statement: Not applicable.

Informed Consent Statement: Not applicable.

Acknowledgments: This work is the result of an aid to postdoctoral training and improvement abroad (for A.M., number 21229/PD/19) financed by the Consejería de Empleo, Universidades, Empresa y Medio Ambiente of the CARM, through the Fundación Séneca-Agencia de Cienciay Tecnología de la Región de Murcia.

Conflicts of Interest: The authors declare no conflict of interest.

References

1. Prabu, S.; Suriyaprakash, T.K.; Kumar, C.; Kumar, S. Nutraceuticals and their medicinal importance. *Int. J. Health Allied Sci.* **2012**, *1*, 7. [CrossRef]
2. Pandey, M.; Verma, R.K.; Saraf, S.A. Nutraceuticals: New era of medicine and health. *Asian J. Pharm. Clin. Res.* **2010**, *3*, 4.
3. Prabu, S.L.; Suriyaprakash, T.N.K.; Kumar, C.D.; Sureshkumar, S.; Ragavendran, T. Nutraceuticals: A review. *Elixir Pharm.* **2012**, *46*, 8372–8377.
4. Kumari, M.; Jain, S.; Singh, J. Nutraceutical—Medicine of the future. *J. Glob. Biosci.* **2015**, *4*, 2790–2794.
5. Zhao, J. Nutraceuticals, Nutritional Therapy, Phytonutrients, and Phytotherapy for Improvement of Human Health: A Perspective on Plant Biotechnology Application. *Recent Pat. Biotechnol.* **2007**, *1*, 75–97. [CrossRef]
6. Das, L.; Bhaumik, E.; Raychaudhuri, U.; Chakraborty, R. Role of nutraceuticals in human health. *J. Food Sci. Technol.* **2012**, *49*, 173–183. [CrossRef]
7. Agarwal, S.; Hordvik, S.; Morar, S. Nutritional claims for functional foods and supplements. *Toxicology* **2006**, *221*, 44–49. [CrossRef]
8. Shahidi, F. Nutraceuticals and functional foods: Whole versus processed foods. *Trends Food Sci. Technol.* **2009**, *20*, 376–387. [CrossRef]
9. Gupta, R.C. *Nutraceuticals: Efficacy, Safety and Toxicity*; Gupta, R.C., Ed.; Elsevier Inc.: Amsterdam, The Netherlands, 2016; ISBN 9780128021477.

10. Paolino, D.; Mancuso, A.; Cristiano, M.C.; Froiio, F.; Lammari, N.; Celia, C.; Fresta, M. Nanonutraceuticals: The New Frontier of Supplementary Food. *Nanomaterials* **2021**, *11*, 792. [CrossRef]
11. Shekhar, V.; Jha, A.K.; Dangi, J.S. Nutraceuticals: A Re-emerging Health Aid. In Proceedings of the International Conference on Food, Biological and Medical Sciences (FBMS-2014), Bangkok, Thailand, 28–29 January 2014; pp. 105–107. [CrossRef]
12. Ward, P.M.L.; Fasitsas, S.; Katz, S.E. Inhibition, Resistance Development, and Increased Antibiotic and Antimicrobial Resistance Caused by Nutraceuticals. *J. Food Prot.* **2002**, *65*, 528–533. [CrossRef]
13. Vargason, A.M.; Anselmo, A.C.; Mitragotri, S. The evolution of commercial drug delivery technologies. *Nat. Biomed. Eng.* **2021**, *5*, 951–967. [CrossRef] [PubMed]
14. Paolino, D.; Fresta, M.; Sinha, P.; Ferrari, M. Drug Delivery Systems. In *Encyclopedia of Medical Devices and Instrumentation*; John Wiley & Sons, Inc.: Hoboken, NJ, USA, 2006; pp. 437–495, ISBN 0471732877.
15. Sung, Y.K.; Kim, S.W. Recent advances in polymeric drug delivery systems. *Biomater. Res.* **2020**, *24*, 1–12. [CrossRef] [PubMed]
16. Das, D.; Pal, S. Modified biopolymer-dextrin based crosslinked hydrogels: Application in controlled drug delivery. *RSC Adv.* **2015**, *5*, 25014–25050. [CrossRef]
17. Santini, A.; Cammarata, S.M.; Capone, G.; Ianaro, A.; Tenore, G.C.; Pani, L.; Novellino, E. Nutraceuticals: Opening the debate for a regulatory framework. *Br. J. Clin. Pharmacol.* **2018**, *84*, 659–672. [CrossRef] [PubMed]
18. Laurent, T.; Kacem, I.; Blanchemain, N.; Cazaux, F.; Neut, C.; Hildebrand, H.F.; Martel, B. Cyclodextrin and maltodextrin finishing of a polypropylene abdominal wall implant for the prolonged delivery of ciprofloxacin. *Acta Biomater.* **2011**, *7*, 3141–3149. [CrossRef]
19. Molinos, M.; Carvalho, V.; Silva, D.M.; Gama, F.M. Development of a Hybrid Dextrin Hydrogel Encapsulating Dextrin Nanogel As Protein Delivery System. *Biomacromolecules* **2012**, *13*, 517–527. [CrossRef]
20. Xie, H.; Ma, X.; Lin, W.; Dong, S.; Liu, Q.; Chen, Y.; Gao, Q. Linear Dextrin as Potential Insulin Delivery System: Effect of Degree of Polymerization on the Physicochemical Properties of Linear Dextrin–Insulin Inclusion Complexes. *Polymers* **2021**, *13*, 4187. [CrossRef]
21. Hreczuk-Hirst, D.; Chicco, D.; German, L.; Duncan, R. Dextrins as potential carriers for drug targeting: Tailored rates of dextrin degradation by introduction of pendant groups. *Int. J. Pharm.* **2001**, *230*, 57–66. [CrossRef]
22. Saavedra-Leos, Z.; Leyva-Porras, C.; Araujo-Díaz, S.B.; Toxqui-Terán, A.; Borrás-Enríquez, A.J. Technological Application of Maltodextrins According to the Degree of Polymerization. *Molecules* **2015**, *20*, 21067–21081. [CrossRef]
23. Kurkov, S.V.; Loftsson, T. Cyclodextrins. *Int. J. Pharm.* **2013**, *453*, 167–180. [CrossRef]
24. Trotta, F.; Dianzani, C.; Caldera, F.; Mognetti, B.; Cavalli, R. The application of nanosponges to cancer drug delivery. *Expert Opin. Drug Deliv.* **2014**, *11*, 931–941. [CrossRef] [PubMed]
25. Takeiti, C.Y.; Kieckbusch, T.G.; Collares-Queiroz, F.P. Morphological and Physicochemical Characterization of Commercial Maltodextrins with Different Degrees of Dextrose-Equivalent. *Int. J. Food Prop.* **2010**, *13*, 411–425. [CrossRef]
26. Sun, J.; Zhao, R.; Zeng, J.; Li, G.; Li, X. Characterization of Destrins with Different Dextrose Equivalents. *Molecules* **2010**, *15*, 5162–5173. [CrossRef] [PubMed]
27. Marchal, L.M.; Beeftink, H.H.; Tramper, J. Towards a rational design of commercial maltodextrins. *Trends Food Sci. Technol.* **1999**, *10*, 345–355. [CrossRef]
28. Sundari, S.; Raman, B.; Balasubramanian, D. Hydrophobic surfaces in oligosaccharides: Linear dextrins are amphiphilic chains. *Biochim. Biophys. Acta* **1991**, *1065*, 35–41. [CrossRef]
29. Shukla, A.; Singh, A.P.; Maiti, P. Injectable hydrogels of newly designed brush biopolymers as sustained drug-delivery vehicle for melanoma treatment. *Signal Transduct. Target. Ther.* **2021**, *6*, 63. [CrossRef]
30. Brewster, M.E.; Loftsson, T. The use of chemically modified cyclodextrins in the development of formulations for chemical delivery systems. *Pharmazie* **2002**, *57*, 94–101.
31. Cumpstey, I. Chemical Modification of Polysaccharides. *Chem. Funct. Prop. Food Sacch.* **2013**, *2013*, 417672. [CrossRef]
32. Trotta, F.; Zanetti, M.; Cavalli, R. Cyclodextrin-based nanosponges as drug carriers. *Beilstein J. Org. Chem.* **2012**, *8*, 2091–2099. [CrossRef]
33. Krabicová, I.; Appleton, S.L.; Tannous, M.; Hoti, G.; Caldera, F.; Pedrazzo, A.R.; Cecone, C.; Cavalli, R.; Trotta, F. History of cyclodextrin nanosponges. *Polymers* **2020**, *12*, 1122. [CrossRef]
34. Tejashri, G.; Amrita, B.; Darshana, J. Cyclodextrin based nanosponges for pharmaceutical use: A review. *Acta Pharm.* **2013**, *63*, 335–358. [CrossRef] [PubMed]
35. Hoti, G.; Caldera, F.; Cecone, C.; Rubin Pedrazzo, A.; Anceschi, A.; Appleton, S.L.; Monfared, Y.K.; Trotta, F. Effect of the Cross-linking Density on the Swelling and Rheological Behavior of Ester-Bridged β-Cyclodextrin Nanosponges. *Materials* **2021**, *14*, 478. [CrossRef]
36. Hoare, T.R.; Kohane, D.S. Hydrogels in drug delivery: Progress and challenges. *Polymer* **2008**, *49*, 1993–2007. [CrossRef]
37. Matencio, A.; Hoti, G.; Monfared, Y.K.; Rezayat, A.; Pedrazzo, A.R.; Caldera, F.; Trotta, F. Cyclodextrin Monomers and Polymers for Drug Activity Enhancement. *Polymers* **2021**, *13*, 1684. [CrossRef] [PubMed]
38. Gander, B.; Gurny, R.; Doelker, E.; Peppas, N.A. Effect of Polymeric Network Structure on Drug Release from Cross-Linked Poly(Vinyl Alcohol) Micromatrices. *Pharm. Res.* **1989**, *6*, 578–584. [CrossRef] [PubMed]
39. De Stéfano, J.C.Q.; Abundis-Correa, V.; Herrera-Flores, S.D.; Alvarez, A.J. pH-Sensitive Starch-Based Hydrogels: Synthesis and Effect of Molecular Components on Drug Release Behavior. *Polymers* **2020**, *12*, 1974. [CrossRef] [PubMed]

40. Anandam, S.; Selvamuthukumar, S. Fabrication of cyclodextrin nanosponges for quercetin delivery: Physicochemical characterization, photostability, and antioxidant effects. *J. Mater Sci.* **2014**, *49*, 8140–8153. [CrossRef]
41. Darandale, S.S.; Vavia, P.R. Cyclodextrin-based nanosponges of curcumin: Formulation and physicochemical characterization. *J. Incl. Phenom. Macrocycl. Chem.* **2013**, *75*, 315–322. [CrossRef]
42. Ansari, K.A.; Vavia, P.R.; Trotta, F.; Cavalli, R. Cyclodextrin-Based Nanosponges for Delivery of Resveratrol: In vitro Characterisation, Stability, Cytotoxicity and Permeation Study. *AAPS PharmSciTech* **2011**, *12*, 279–286. [CrossRef]
43. Rezaei, A.; Khavari, S.; Sami, M. Incorporation of thyme essential oil into the β-cyclodextrin nanosponges: Preparation, characterization, and antibacterial activitiy. *J. Mol. Liq.* **2021**, *1241*, 1–8. [CrossRef]
44. Mihailiasa, M.; Caldera, F.; Li, J.; Peila, R.; Ferri, A.; Trotta, F. Preparation of functionalized cotton fabrics by means of melatonin loaded β-cyclodextrin nanosponges. *Carbohydr. Polym.* **2016**, *142*, 24–30. [CrossRef] [PubMed]
45. DeFelice, S.L. The nutraceutical revolution: Its impact on food industry R&D. *Trends Food Sci. Technol.* **1995**, *6*, 59–61. [CrossRef]
46. Andlauer, W.; Fürst, P. Nutraceuticals: A piece of history, present status and outlook. *Food Res. Int.* **2002**, *35*, 171–176. [CrossRef]
47. Shahidi, F. Nutraceuticals, functional foods and dietary supplements in health and disease. *J. Food Drug Anal.* **2012**, *20*, 226–230. [CrossRef]
48. Singh, J.; Sinha, S. Classification, Regulatory Acts and Applications of Nutraceuticals for Health. *Int. J. Pharm. Biol. Sci.* **2012**, *2*, 177–187.
49. Kalra, E.K. Nutraceutical—Definition and Introduction. *AAPS J.* **2003**, *5*, 1–2. [CrossRef]
50. Chauhan, B.; Kumar, G.; Kalam, N.; Ansari, S.H. Current concepts and prospects of herbal nutraceutical: A review. *J. Adv. Pharm. Technol. Res.* **2013**, *4*, 4–8. [CrossRef]
51. Cencic, A.; Chingwaru, W. The role of functional foods, nutraceuticals, and food supplements in intestinal health. *Nutrients* **2010**, *2*, 611–625. [CrossRef]
52. Ganapathy, M.; Bhunia, S. Nutraceuticals: The New Generation Therapeutics. *Adv. Tech. Biol. Med.* **2015**, *4*, 2. [CrossRef]
53. Aronson, J.K. Defining 'nutraceuticals': Neither nutritious nor pharmaceutical. *Br. J. Clin. Pharmacol.* **2016**, *83*, 12. [CrossRef]
54. DeFelice, S.L. Nutrition Stymied: The Nutraceutical Solution. XXV National Congress of the Italian Chemical Society-SCI-The University of Calabria. Available online: https://fimdefelice.org/video (accessed on 27 April 2020).
55. Daliu, P.; Santini, A.; Novellino, E. From pharmaceuticals to nutraceuticals: Bridging disease prevention and management. *Expert Rev. Clin. Pharmacol.* **2018**, *12*, 1–7. [CrossRef] [PubMed]
56. Martirosyan, D.M.; Singh, J. A new definition of functional food by FFC: What Makes a New Definition Unique? *Funct. Foods Health Dis.* **2015**, *5*, 209–223. [CrossRef]
57. *Public Law 101–535*; 101st Congress Nutrition Labeling and Education Act of 1990. Library of Congress: Washington, DC, USA, 1990; p. 15. [CrossRef]
58. Young, A.L.; Bass, I.S. The Dietary Supplement Health and Education Act. *Food Drug Law J.* **1995**, *50*, 285–292. [PubMed]
59. Marwick, C. Implementing the FDA modernization act. *J. Am. Med. Assoc.* **1998**, *279*, 815–816. [CrossRef]
60. Glade, M.J. The Dietary Supplement Health and Education Act of 1994—Focus on labeling issues. *Nutrition* **1997**, *13*, 999–1001. [CrossRef]
61. Quinones, R.L.; Winsor, R.D.; Patino, A.; Hoffmann, P. The Regulation of Dietary Supplements Within the United States: Flawed Attempts at Mending a Defective Consumer Safety Mechanism. *J. Consum. Aff.* **2013**, *47*, 328–357. [CrossRef]
62. Bagchi, D. *Nutraceutical and Functional Food Regulations in the United States and around the World*, 2nd ed.; Bagchi, D., Ed.; Elsevier Inc.: Amsterdam, The Netherlands, 2014.
63. Halsted, C.H. Dietary supplements and functional foods: 2 sides of a coin? *Am. J. Clin. Nutr.* **2003**, *77*, 1001S–1007S. [CrossRef]
64. Gupta, S.; Chauhan, D.; Mehla, K.; Sood, P.; Nair, A. An overview of nutraceuticals: Current scenario. *J. Basic Clin. Pharm.* **2010**, *1*, 55–62.
65. Hasler, C.M. Functional Foods: Benefits, Concerns and Challenges—A Position Paper from the American Council on Science and Health. *J. Nutr.* **2002**, *132*, 3772–3781. [CrossRef]
66. Ross, S. Functional foods: The Food and Drug Administration perspective. *Am. J. Clin. Nutr.* **2000**, *71*, 1735S–1738S. [CrossRef]
67. Golodner, L.F. The US Food and Drug Administration Modernization Act of 1997: Impact on consumers. *Clin. Ther.* **1998**, *20*, C20–C25. [CrossRef]
68. Borchers, A.T.; Keen, C.L.; Gershwin, M.E. The Basis of Structure/Function Claims of Nutraceuticals. *Clin. Rev. Allergy Immunol.* **2016**, *51*, 370–382. [CrossRef] [PubMed]
69. Muredzi, P. Food Is Medicine—An Introduction to Nutraceuticals. LAP Lambert Academic Publishing: Saarbrücken, Germany, 2013; ISBN 978-3659437106.
70. Zeisel, S.H. Regulation of " Nutraceuticals". *Sci. Compass* **1999**, *285*, 1853–1855. [CrossRef]
71. Ghosh, D.; Bagchi, D.; Konishi, T. *Clinical Aspects of Functional Foods and Nutraceuticals*; CRC Press, Taylor & Francis Group: Boca Raton, FL, USA, 2015; ISBN 9781466569164.
72. Irene Boye, J. *Nutraceutical and Functional Food Processing Technology*; Wiley-Blackwell: Hoboken, NJ, USA, 2015; ISBN 9781118504949.
73. Dillard, C.J.; Bruce German, J. Phytochemicals: Nutraceuticals and human health. *J. Sci. Food Agric.* **2000**, *80*, 1744–1756. [CrossRef]
74. Valls, J.; Pasamontes, N.; Pantaleon, A.; Vinaixa, S.; Vaque, M.; Soler, A.; Millan, S.; Gomez, X. Prospects of Functional Foods/Nutraceuticals and Markets. *Nat. Prod.* **2013**, 2491–2525. [CrossRef]

75. Ahmad, M.F.; Amir Ashraf, S.; Ali Ahmad, F.; Akhtar Ansari, J.; Ahmad Siddiquee, M.R. Nutraceutical Market and its Regulation. *Am. J. Food Technol.* **2011**, *6*, 342–347. [CrossRef]
76. Singh, U.K.; Deshmukh, S.N. Nutraceuticals. *MIT Int. J. Pharm. Sci.* **2016**, *2*, 43–52.
77. Brower, V. Nutraceuticals: Poised for a healthy slice of the healthcare market? *Nat. Biotechnol.* **1998**, *16*, 728–731. [CrossRef]
78. Helal, N.A.; Eassa, H.A.; Amer, A.M.; Eltokhy, M.A.; Edafiogho, I.; Nounou, M.I. Nutraceuticals' Novel Formulations: The Good, the Bad, the Unknown and Patents Involved. *Recent Pat. Drug Deliv. Formul.* **2019**, *13*, 105–156. [CrossRef]
79. Bagchi, D. Nutraceuticals and functional foods regulations in the United States and around the world. *Toxicology* **2006**, *221*, 1–3. [CrossRef]
80. Dev, R.; Kumar, S.; Singh, J.; Chauhan, B. Potential role of nutraceuticals in present scenerio: A review. *J. Appl. Pharm. Sci.* **2011**, *1*, 26–28.
81. Nicoletti, M. Nutraceuticals and botanicals: Overview and perspectives. *Int. J. Food Sci. Nutr.* **2012**, *63*, 2–6. [CrossRef] [PubMed]
82. Pagliaro, M. Italy's Nutraceutical Industry: A Process and Bioeconomy Perspective into a Key Area of the Global Economy. *Biofuels Bioprod. Biorefining* **2019**, *14*, 1–7. [CrossRef]
83. Lupsea, S. *Nutraceuticals and Food Supplement Sector in Japan—Opportunities for European Produces?* EU-Japan Centre for Industrial Cooperation: Tokyo, Japan, 2016; Available online: https://www.eu-japan.eu/publications/nutraceuticals-and-food-supplements-sector-japan (accessed on 18 January 2022).
84. Chanda, S.; Tiwari, R.K.; Kumar, A.; Singh, K. Nutraceuticals inspiring the current therapy for lifestyle diseases. *Adv. Pharmacol. Sci.* **2019**, *2019*, 6908716. [CrossRef]
85. Ali, A.; Ahmad, U.; Akhtar, J.; Badruddeen; Khan, M.M. Engineered nano scale formulation strategies to augment efficiency of nutraceuticals. *J. Funct. Foods* **2019**, *62*, 16. [CrossRef]
86. Kumar, C.G.; Sripada, S.; Poornachandra, Y. Status and Future Prospects of Fructooligosaccharides as Nutraceuticals. In *Role of Materials Science in Food Bioengineering*; Elsevier Inc.: Amsterdam, The Netherlands, 2018; pp. 451–503, ISBN 9780128114483.
87. Bhaskarachary, K.; Vemula, S.R.; Gavaravarapu, S.R.M.; Joshi, A.K.R. Traditional foods, functional foods and nutraceuticals. *Proc. Indian Natl. Sci. Acad.* **2016**, *82*, 1565–1577. [CrossRef]
88. Keservani, R.K.; Sharma, A.K.; Kesharwani, R.K. *Nutraceutical and Functional Foods in Disease Prevention*; IGI Global Book Series; Advances in Human Services and Public Health: Hershey, PA, USA, 2019; ISBN 2475-6571.
89. Holzapfel, W.H.; Haberer, P.; Geisen, R.; Björkroth, J.; Schillinger, U. Taxonomy and important features of probiotic microorganisms in food and nutrition. *Am. J. Clin. Nutr.* **2001**, *73*, 366S–373S. [CrossRef]
90. Martínez Cruz, P.; Ibáñez, A.L.; Monroy Hermosillo, O.A.; Ramírez Saad, H.C. Use of Probiotics in Aquaculture. *ISRN Microbiol.* **2012**, *2021*, 916845. [CrossRef]
91. DellaPenna, D. Nutritional Genomics: Manipulating Plant Micronutrients to Improve Human Health. *Plant Biotechnol. Food Feed* **1991**, *285*, 375–379. [CrossRef]
92. Galanakis, C.M. *Nutraceuticals and Natural Product Pharmaceuticals*; Academic Press: Cambridge, MA, USA, 2019; ISBN 9780128164501.
93. Srivastava, R.K. Need Of Nutraceuticals / Functional Food Products for Health Benefits to World-Wide People. *J. Biotechnol. Biomed. Sci.* **2018**, *1*, 13. [CrossRef]
94. Wildman, R.E.C. Nutraceuticals and Functional Foods. In *Handbook of Nutraceuticals and Functional Foods*; CRC Press, Taylor & Francis Group: Boca Raton, FL, USA, 2020; ISBN 9781498703727.
95. Stipanuk, M.H.; Caudill, M.A. *Biochemical, Physiological, and Molecular Aspects of Human Nutrition*, 3rd ed.; Elsevier Inc.: Amsterdam, The Netherlands, 2013; ISBN 9781437709599.
96. Chemler, J.A.; Yan, Y.; Koffas, M.A.G. Biosynthesis of isoprenoids, polyunsaturated fatty acids and flavonoids in Saccharomyces cerevisiae. *Microb. Cell Fact.* **2006**, *5*, 9. [CrossRef] [PubMed]
97. Sacchettini, J.C.; Poulter, C.D. Creating lsoprenoid Diversity. *Science* **1997**, *277*, 1788–1789. [CrossRef] [PubMed]
98. Bhuyan, D.J.; Basu, A. Phenolic Compounds Potential Health Benefits and Toxicity. In *Utilisation of Bioactive Compounds from Agricultural and Food Production Waste*; Vuong, Q.V., Ed.; CRC Press, Taylor & Francis: Boca Raton, FL, USA, 2017; pp. 27–59, ISBN 97813151540.
99. Tanase, C.; Cosarcă, S.; Muntean, D.-L. A Critical Review of Phenolic Compounds Extracted from the Bark of Woody Vascular Plants and their Potential Biological Activity. *Molecules* **2019**, *24*, 1182. [CrossRef] [PubMed]
100. Tsao, R. Chemistry and Biochemistry of Dietary Polyphenols. *Nutrients* **2010**, *2*, 1231–1246. [CrossRef]
101. Pérez-Jiménez, J.; Neveu, V.; Vos, F.; Scalbert, A. Identification of the 100 richest dietary sources of polyphenols: An application of the Phenol-Explorer database. *Eur. J. Clin. Nutr.* **2010**, *64*, S112–S120. [CrossRef]
102. Ross, J.A.; Kasum, C.M. DIETARY FLAVONOIDS: Bioavailability, Metabolic Effects, and Safety. *Annu. Rev. Nutr.* **2002**, *22*, 19–34. [CrossRef]
103. Giada, M.D.L.R. Food Phenolic Compounds: Main Classes, Sources and Their Antioxidant Power. In *Oxidative Stress and Chronic Degenerative Diseases—A Role for Antioxidants*; IntechOpen: London, UK, 2013; pp. 87–112. [CrossRef]
104. Lachance, P.A.; Das, Y.T. Nutraceuticals. In *Comprehensive Medicinal Chemistry II*; Elsevier Ltd.: Amsterdam, The Netherlands, 2007; pp. 449–461, ISBN 0-08-044514-4.
105. Burdge, G.C.; Calder, P.C. Introduction to Fatty Acids and Lipids. *World Rev. Nutr. Diet.* **2015**, *112*, 1–16. [CrossRef]
106. Kresge, N.; Simoni, R.D.; Hill, R.L. JBC Historical Perspectives: Lipid Biochemistry. 2010. Available online: https://moam.info/lipid-biochemistry-the-journal-of-biological-chemistry_59e02dbf1723dd1a56a74f84.html (accessed on 18 January 2022).

107. Fahy, E.; Cotter, D.; Sud, M.; Subramaniam, S. Lipid classification, structures and tools. *Biochim. Biophys. Acta (BBA)-Mol. Cell Biol. Lipids* **2011**, *1811*, 637–647. [CrossRef]
108. Fahy, E.; Subramaniam, S.; Brown, H.A.; Glass, C.K.; Merrill, A.H.; Murphy, R.C.; Raetz, C.R.H.; Russell, D.W.; Seyama, Y.; Shaw, W.; et al. A comprehensive classification system for lipids. *J. Lipid Res.* **2005**, *46*, 839–861. [CrossRef]
109. Nagy, K.; Tiuca, I.-D. Importance of Fatty Acids in Physiopathology of Human Body. *IntechOpen* **2017**, 3–22. [CrossRef]
110. Rustan, A.C.; Drevon, C.A. Fatty Acids: Structures and Properties. *Encycl. Life Sci.* **2005**, 7. [CrossRef]
111. Orsavova, J.; Misurcova, L.; Vavra Ambrozova, J.; Vicha, R.; Mlcek, J. Fatty Acids Composition of Vegetable Oils and its Contribution to Dietary Energy Intake and Dependence of Cardiovascular Mortality on Dietary Intake of Fatty Acids. *Int. J. Mol. Sci.* **2015**, *16*, 12871–12890. [CrossRef] [PubMed]
112. Tvrzicka, E.; Kremmyda, L.-S.; Stankova, B.; Zak, A. Fatty Acids as Biocompounds: Their Role in Human Metabolism, Health and Disease—A Review. Part 1: Classification, Dietary Sources and Biological Functions. *Biomed. Pap.* **2011**, *155*, 117–130. [CrossRef] [PubMed]
113. Johnson, M.; Bradford, C. Omega-3, Omega-6 and Omega-9 Fatty Acids: Implications for Cardiovascular and Other Diseases. *J. Glycom. Lipidom.* **2014**, *4*, 8. [CrossRef]
114. Gerschenson, L.N.; Rojas, A.M.; Fissore, E.N. Carbohydrates. In *Nutraceutical and Functional Food Components: Effects of Innovative Processing Techniques*; Elsevier Inc.: Amsterdam, The Netherlands, 2017; pp. 39–101, ISBN 9780128052570.
115. Lovegrove, A.; Edwards, C.H.; De Noni, I.; Patel, H.; El, S.N.; Grassby, T.; Zielke, C.; Ulmius, M.; Nilsson, L.; Butterworth, P.J.; et al. Role of polysaccharides in food, digestion, and health. *Crit. Rev. Food Sci. Nutr.* **2017**, *57*, 237–253. [CrossRef]
116. Cummings, J.H.; Stephen, A.M. Carbohydrate terminology and classification. *Eur. J. Clin. Nutr.* **2007**, *61*, S5–S18. [CrossRef]
117. Voragen, A.G.J.; Coenen, G.-J.; Verhoef, R.P.; Schols, H.A. Pectin, a versatile polysaccharide present in plant cell walls. *Struct. Chem.* **2009**, *20*, 263–275. [CrossRef]
118. Wang, T.; Zabotina, O.; Hong, M. Pectin-Cellulose Interactions in the Arabidopsis Primary Cell Wall from Two-Dimensional Magic-Angle-Spinning Solid-State Nuclear Magnetic Resonance. *Biochemistry* **2012**, *51*, 9846–9856. [CrossRef]
119. Aspinall, G.O. Chemistry of Cell Wall Polysaccharides. In *The Biochemistry of Plants*; Preiss, J., Ed.; Academic Press, Inc.: Cambridge, MA, USA, 1980; Volume 3, pp. 473–500.
120. Zykwinska, A.; Thibault, J.F.; Ralet, M.C. Organization of pectic arabinan and galactan side chains in association with cellulose microfibrils in primary cell walls and related models envisaged. *J. Exp. Bot.* **2007**, *58*, 1795–1802. [CrossRef]
121. Sinha, A.K.; Kumar, V.; Makkar, H.P.S.; De Boeck, G.; Becker, K. Non-starch polysaccharides and their role in fish nutrition—A review. *Food Chem.* **2011**, *127*, 1409–1426. [CrossRef]
122. Waldron, K.W.; Parker, M.L.; Smith, A.C. Plant Cell Walls and Food Quality. *Compr. Rev. Food Sci. Food Saf.* **2003**, *2*, 128–146. [CrossRef] [PubMed]
123. Scheller, H.V.; Ulvskov, P. Hemicelluloses. *Annu. Rev. Plant Biol.* **2010**, *61*, 263–289. [CrossRef] [PubMed]
124. Wu, G. Dietary protein intake and human health. *Food Funct.* **2016**, *7*, 1251–1265. [CrossRef] [PubMed]
125. Osborne, T.B.; Mendel, L.B. Amino-Acids in nutrition and growth. *J. Biol. Chem.* **1914**, *17*, 325–349. [CrossRef]
126. Ray, P.D.; Fry, R.C. The Cell: The Fundamental Unit in Systems Biology. In *Systems Biology in Toxicology and Environmental Health*; Elsevier Inc.: Amsterdam, The Netherlands, 2015; pp. 11–42, ISBN 9780128015681.
127. Carpenter, K.J. A Short History of Nutritional Science: Part 1 (1785–1885). *J. Nutr.* **2003**, *133*, 638–645. [CrossRef]
128. Leuchtenberger, W.; Huthmacher, K.; Drauz, K. Biotechnological production of amino acids and derivatives: Current status and prospects. *Appl. Microbiol. Biotechnol.* **2005**, *69*, 1–8. [CrossRef]
129. Wu, G. Amino acids: Metabolism, functions, and nutrition. *Amino Acids* **2009**, *37*, 1–17. [CrossRef]
130. Wu, G. Functional Amino Acids in Growth, Reproduction and Health. *Adv. Nutr.* **2010**, *1*, 31–37. [CrossRef]
131. Bhagavan, N.V.; Ha, C.-E. Amino Acids. In *Essentials of Medical Biochemistry*; Academic Press: Cambridge, MA, USA, 2015; pp. 21–29. [CrossRef]
132. Wang, J.; Guleria, S.; Koffas, M.A.G.; Yan, Y. Microbial production of value-added nutraceuticals. *Curr. Opin. Biotechnol.* **2016**, *37*, 97–104. [CrossRef]
133. Liong, M.-T. *Beneficial Microorganisms in Food and Nutraceuticals*; Springer: Berlin, Germany, 2015; Volume 27. [CrossRef]
134. Joint FAO/WHO Expert Consultation. *Probiotics in Food; Health and Nutritional Properties and Guidelines for Evaluation*; FAO/WHO: Geneva, Switzerland, 2001. [CrossRef]
135. Soccol, C.R.; de Souza Vandenberghe, L.P.; Spier, M.R.; Medeiros, A.B.P.; Yamaguishi, C.T.; De Dea Lindner, J.; Pandey, A.; Thomaz-Soccol, V. The Potential of Probiotics: A Review. *Food Technol. Biotechnol.* **2010**, *48*, 413–434.
136. Sharma, S.; Agarwal, N.; Verma, P. Probiotics: The Emissaries of Health from Microbial World. *J. Appl. Pharm. Sci.* **2012**, *2*, 138–143.
137. Zhang, Y.Y.; Panozzo, J.; Hall, M.S.; Ajlouni, S. Bioaccessibility of Some Essential Minerals in Three Selected Australian Pulse Varieties Using an In Vitro Gastrointestinal Digestion Model. *J. Food Sci.* **2018**, *83*, 2873–2881. [CrossRef] [PubMed]
138. Gupta, U.C.; Gupta, S.C. Sources and Deficiency Diseases of Mineral Nutrients in Human Health and Nutrition: A Review. *Pedosphere* **2014**, *24*, 13–38. [CrossRef]
139. Park, Y.W. *Bioactive Components in Milk and Dairy Products*; Park, Y.W., Ed.; John Wiley & Sons, Ltd.: Hoboken, NJ, USA, 2009; Volume 66, ISBN 9780813819822.

140. Anderson, J.J.B.; Allen, J.C. Nutrition of Macrominerals and Trace Minerals. In *Functional Foods: Designer Foods, Pharmafoods, Nutraceuticals*; Goldberg, I., Ed.; Springer: New York, NY, USA, 1994; pp. 323–354, ISBN 9781461358619.
141. Górska-Warsewicz, H.; Rejman, K.; Laskowski, W.; Czeczotko, M. Milk and Dairy Products and Their Nutritional Contribution to the Average Polish Diet. *Nutrients* **2019**, *11*, 1771. [CrossRef]
142. Mahantesh, P.; Patil, C.S. Nutraceuticals and Functional Foods in Health Promotion and Disease Risk Management. In Proceedings of the International Conference on Biomedical Engineering and Technology, Kuala Lumpur, Malaysia, 4–5 June 2011; Volume 11, pp. 7–14.
143. Galasso, C.; Gentile, A.; Orefice, I.; Ianora, A.; Bruno, A.; Noonan, D.M.; Sansone, C.; Albini, A.; Brunet, C. Microalgal Derivatives as Potential Nutraceutical and Food Supplements for Human Health: A Focus on Cancer Prevention and Interception. *Nutrients* **2019**, *11*, 1226. [CrossRef]
144. Diem, G.; Brownson, R.C.; Grabauskas, V.; Shatchkute, A.; Stachenko, S. Prevention and control of noncommunicable diseases through evidence-based public health: Implementing the NCD 2020 action plan. *Glob. Health Promot.* **2014**, *23*, 5–13. [CrossRef]
145. Probst-Hensch, N.; Tanner, M.; Kessler, C.; Burri, C.; Künzli, N. Prevention—A cost-effective way to fight the non-communicable disease epidemic. *Eur. J. Med. Sci.* **2011**, *141*, 1–8. [CrossRef]
146. Anjali; Garg, V.; Dhiman, A.; Dutt, R.; Ranga, S. Health benefits of nutraceuticals. *Pharma Innov. J.* **2018**, *7*, 178–181.
147. Sharma, R.; Singh, R.B. Bioactive Foods and Nutraceutical Supplementation Criteria in Cardiovascular Protection. *Open Nutraceuticals J.* **2010**, *3*, 141–153. [CrossRef]
148. Pellett, P.L. World essential amino acid supply with special attention to South-East Asia. *Food Nutr. Bull.* **1996**, *17*, 31. [CrossRef]
149. Bravo, L. Polyphenols: Chemistry, Dietary Sources, Metabolism, and Nutritional Significance. *Nutr. Rev.* **2009**, *56*, 317–333. [CrossRef] [PubMed]
150. Adeboye, P.T.; Bettiga, M.; Olsson, L. The chemical nature of phenolic compounds determines their toxicity and induces distinct physiological responses in Saccharomyces cerevisiae in lignocellulose hydrolysates. *AMB Express* **2014**, *4*, 1–10. [CrossRef] [PubMed]
151. Halliwell, B.; Gutteridge, J.M.C. Oxygen toxicity, oxygen radicals, transition metals and disease. *Biochem. J.* **1984**, *219*, 1–14. [CrossRef] [PubMed]
152. Gilbert, D.L. *Oxygen and Living Processes*, 1st ed.; Springer: New York, NY, USA, 1981; ISBN 9781461258926.
153. Bartosz, G. *Food Oxidants and Antioxidants Chemical, Biological and Functional Properties*; CRC Press, Taylor & Francis: Boca Raton, FL, USA, 2014; Volume 236, ISBN 0021701319376.
154. Gerschman, R.; Gilbert, D.L.; Nye, S.W.; Dwyer, P.; Fenn, W.O. Oxygen Poisoning and X-irradiation: A Mechanism in Common. *Science* **1954**, *119*, 623–626. [CrossRef] [PubMed]
155. Palmer, H.J.; Paulson, K.E. Reactive Oxygen Species and Antioxidants in Signal Transduction and Gene Expression. *Nutr. Rev.* **1997**, *55*, 353–361. [CrossRef]
156. Pizzino, G.; Irrera, N.; Cucinotta, M.; Pallio, G.; Mannino, F.; Arcoraci, V.; Squadrito, F.; Altavilla, D.; Bitto, A. Oxidative Stress: Harms and Benefits for Human Health. *Oxid. Med. Cell. Longev.* **2017**, 1–13. [CrossRef]
157. Nunes, X.P.; Silva, F.S.; da Souza Almeida, J.R.G.; de Lima, J.T.; de Araújo Ribeiro, L.A.; Junior, L.J.Q.; Filho, J.M.B. Biological Oxidations and Antioxidant Activity of Natural Products. In *Phytochemicals as Nutraceuticals—Global Approaches to Their Role in Nutrition and Health*; IntechOpen: London, UK, 2011. [CrossRef]
158. Tian, Y.; Zou, B.; Li, C.; Yang, J.; Xu, S.; Hagerman, A.E. High molecular weight persimmon tannin is a potent antioxidant both ex vivo and in vivo. *Food Res. Int.* **2012**, *45*, 26–30. [CrossRef]
159. Stahl, W.; Sies, H. Antioxidant activity of carotenoids. *Mol. Aspects Med.* **2003**, *24*, 345–351. [CrossRef]
160. Dreher, D.; Junod, A.F. Role of Oxygen Free Radicals in Cancer Development. *Eur. J. Cancer* **1996**, *32A*, 30–38. [CrossRef]
161. Yamauchi, R. Vitamin E: Mechanism of its antioxidant activity. *Food Sci. Technol.* **1997**, *3*, 301–309. [CrossRef]
162. Silva, E.M.; Souza, J.N.S.; Rogez, H.; Rees, J.F.; Larondelle, Y. Antioxidant activities and polyphenolic contents of fifteen selected plant species from the Amazonian region. *Food Chem.* **2007**, *101*, 1012–1018. [CrossRef]
163. Gonzalez-Burgos, E.; Gomez-Serranillos, M.P. Terpene Compounds in Nature: A Review of Their Potential Antioxidant Activity. *Curr. Med. Chem.* **2012**, *19*, 5319–5341. [CrossRef] [PubMed]
164. Saura-Calixto, F. Antioxidant Dietary Fiber Product: A New Concept and a Potential Food Ingredient. *J. Agric. Food Chem.* **1998**, *46*, 4303–4306. [CrossRef]
165. Działo, M.; Mierziak, J.; Korzun, U.; Preisner, M.; Szopa, J.; Kulma, A. The Potential of Plant Phenolics in Prevention and Therapy of Skin Disorders. *Int. J. Mol. Sci.* **2016**, *17*, 160. [CrossRef] [PubMed]
166. Houston, M.C. The role of cellular micronutrient analysis, nutraceuticals, vitamins, antioxidants and minerals in the prevention and treatment of hypertension and cardiovascular disease. *Ther. Adv. Cardiovasc. Dis.* **2010**, *4*, 165–183. [CrossRef]
167. Vercruysse, L.; Smagghe, G.; Herregods, G.; Van Camp, J. ACE Inhibitory Activity in Enzymatic Hydrolysates of Insect Protein. *J. Agric. Food Chem.* **2005**, *53*, 5207–5211. [CrossRef]
168. Chen, Z.-Y.; Peng, C.; Jiao, R.; Wong, Y.-M.; Yang, N.; Huang, Y. Anti-hypertensive Nutraceuticals and Functional Foods. *J. Agric. Food Chem.* **2009**, *57*, 4485–4499. [CrossRef]
169. Rosen, C.J. Vitamin D Insufficiency. *N. Engl. J. Med.* **2011**, *364*, 248–254. [CrossRef]
170. Boldo, A.; Campbell, P.; Luthra, P.; White, W.B. Should the Concentration of Vitamin D be Measured in All Patients with Hypertension? *J. Clin. Hypertens.* **2010**, *12*, 149–152. [CrossRef]

171. Houston, M. The role of nutrition and nutraceutical supplements in the treatment of hypertension. *World J. Cardiol.* **2014**, *6*, 38–66. [CrossRef]
172. Theodotou, M.; Fokianos, K.; Mouzouridou, A.; Konstantinou, C.; Aristotelous, A.; Prodromou, D.; Chrysikou, A. The effect of resveratrol on hypertension: A clinical trial. *Exp. Ther. Med.* **2017**, *13*, 295–301. [CrossRef] [PubMed]
173. Houston, M.C. Treatment of Hypertension with Nutrition and Nutraceutical Supplements: Part 2. *Altern. Complement. Ther.* **2019**, *25*, 23–36. [CrossRef]
174. Mozos, I.; Stoian, D.; Caraba, A.; Malainer, C.; Horbanczuk, J.O.; Atanasov, A.G. Lycopene and Vascular Health. *Front. Pharmacol.* **2018**, *9*, 521. [CrossRef] [PubMed]
175. Ferreira-Santos, P.; Carrón, R.; Montero, M.J.; Sevilla, M.Á. The antihypertensive and antihypertrophic effect of lycopene is not affected by and is independent of age. *J. Funct. Foods* **2021**, *85*, 104656. [CrossRef]
176. Tetali, S.D. Terpenes and isoprenoids: A wealth of compounds for global use. *Planta* **2018**, *249*, 8. [CrossRef]
177. Calder, P.C. Polyunsaturated Fatty Acids, Inflammation, and Immunity. *Lipids* **2001**, *36*, 1007–1024. [CrossRef]
178. Ahmad, T.B.; Rudd, D.; Kotiw, M.; Liu, L.; Benkendorff, K. Correlation between Fatty Acid Profile and Anti-Inflammatory Activity in Common Australian Seafood by-Products. *Mar. Drugs* **2019**, *17*, 155. [CrossRef]
179. Calder, P.C. n-3 Polyunsaturated fatty acids, inflammation, and inflammatory diseases. *Am. J. Clin. Nutr.* **2006**, *83*, 1505S–1519S. [CrossRef]
180. Mori, T.A.; Beilin, L.J. Omega-3 Fatty Acids and Inflammation. *Curr. Atheroscler. Rep.* **2004**, *6*, 461–467. [CrossRef]
181. Endo, J.; Arita, M. Cardioprotective mechanism of omega-3 polyunsaturated fatty acids. *J. Cardiol.* **2015**, *67*, 22–27. [CrossRef]
182. Kinsella, J.E.; Lokesh, B.; Stone, R.A. Dietary n-3 polyunsaturated fatty acids and amelioration of cardiovascular disease: Possible mechanisms. *Am. J. Clin. Nutr.* **1990**, *52*, 1–28. [CrossRef] [PubMed]
183. Sokoła-Wysoczańska, E.; Wysoczański, T.; Wagner, J.; Czyż, K.; Bodkowski, R.; Lochyński, S.; Patkowska-Sokoła, B. Polyunsaturated fatty acids and their potential therapeutic role in cardiovascular system disorders—A review. *Nutrients* **2018**, *10*, 1561. [CrossRef] [PubMed]
184. Medeiros-De-Moraes, I.M.; Gonçalves-De-Albuquerque, C.F.; Kurz, A.R.M.; De Jesus Oliveira, F.M.; Pereira de Abreu, V.H.; Torres, R.C.; Carvalho, V.F.; Estato, V.; Bozza, P.T.; Sperandio, M.; et al. Omega-9 Oleic Acid, the Main Compound of Olive oil, Mitigates Inflammation during Experimental Sepsis. *Oxid. Med. Cell. Longev.* **2018**, *2018*, 13. [CrossRef] [PubMed]
185. Feghali, C.A.; Wright, T.M. Cytokines in acute and chronic inflammation. *Front. Biosci.* **1997**, *2*, 12–26.
186. Kawata, A.; Murakami, Y.; Suzuki, S.; Fujisawa, S. Anti-inflammatory Activity of β-Carotene, Lycopene and Tri-n-butylborane, a Scavenger of Reactive Oxygen Species. *In Vivo* **2018**, *32*, 255–264. [CrossRef]
187. Magno, S.; Ceccarini, G.; Pelosini, C.; Jaccheri, R.; Vitti, J.; Fierabracci, P.; Salvetti, G.; Airoldi, G.; Minale, M.; Saponati, G.; et al. LDL-cholesterol lowering effect of a new dietary supplement: An open label, controlled, randomized, cross-over clinical trial in patients with mild-to-moderate hypercholesterolemia. *Lipids Health Dis.* **2018**, *17*, 8. [CrossRef]
188. Ahangari, N.; Ghayour Mobarhan, M.; Sahebkar, A.; Pasdar, A. Molecular aspects of hypercholesterolemia treatment: Current perspectives and hopes. *Ann. Med.* **2018**, *50*, 303–311. [CrossRef]
189. Venkadeswaran, K.; Muralidharan, A.R.; Annadurai, T.; Ruban, V.V.; Sundararajan, M.; Anandhi, R.; Thomas, P.A.; Geraldine, P. Antihypercholesterolemic and Antioxidative Potential of an Extract of the Plant, Piper betle, and its Active Constituent, Eugenol, in Triton WR-1339-Induced Hypercholesterolemia in Experimental Rats. *Evid.-Based Complement. Altern. Med.* **2014**, *2014*, 478973. [CrossRef]
190. Sharifi-Rad, M.; Anil Kumar, N.V.; Zucca, P.; Varoni, E.M.; Dini, L.; Panzarini, E.; Rajkovic, J.; Tsouh Fokou, P.V.; Azzini, E.; Peluso, I.; et al. Lifestyle, Oxidative Stress, and Antioxidants: Back and Forth in the Pathophysiology of Chronic Diseases. *Front. Physiol.* **2020**, *11*, 21. [CrossRef]
191. Mytilinaiou, M.; Kyrou, I.; Khan, M.; Grammatopoulos, D.K.; Randeva, H.S. Familial hypercholesterolemia: New horizons for diagnosis and effective management. *Front. Pharmacol.* **2018**, *9*, 29. [CrossRef]
192. Goldberg, A.C.; Hopkins, P.N.; Toth, P.P.; Ballantyne, C.M.; Rader, D.J.; Robinson, J.G.; Daniels, S.R.; Gidding, S.S.; de Ferranti, S.D.; Ito, M.K.; et al. Familial Hypercholesterolemia: Screening, diagnosis and management of pediatric and adult patients. *J. Clin. Lipidol.* **2011**, *5*, S1–S8. [CrossRef] [PubMed]
193. Scholle, J.M.; Baker, W.L.; Talati, R.; Coleman, C.I. The Effect of Adding Plant Sterols or Stanols to Statin Therapy in Hypercholesterolemic Patients: Systematic Review and Meta-Analysis. *J. Am. Coll. Nutr.* **2009**, *28*, 517–524. [CrossRef] [PubMed]
194. Santini, A.; Novellino, E. Nutraceuticals in hypercholesterolaemia: An overview. *Br. J. Pharmacol.* **2017**, *174*, 1450–1463. [CrossRef] [PubMed]
195. Rana, A.; Westein, E.; Niego, B.; Hagemeyer, C.E. Shear-Dependent Platelet Aggregation: Mechanisms and Therapeutic Opportunities. *Front. Cardiovasc. Med.* **2019**, *6*, 141. [CrossRef] [PubMed]
196. Olas, B. Dietary Supplements with Antiplatelet Activity: A Solution for Everyone? *Adv. Nutr.* **2018**, *9*, 51–57. [CrossRef] [PubMed]
197. Kim, H.; Oh, S.J.; Liu, Y.; Lee, M.-Y. A Comparative Study of the Anti-Platelet Effects of cis- and trans-Resveratrol. *Biomol. Ther.* **2011**, *19*, 201–205. [CrossRef]
198. Olas, B.; Wachowicz, B. Resveratrol, a phenolic antioxidant with effects on blood platelet functions. *Platelets* **2005**, *16*, 251–260. [CrossRef]
199. Giordo, R.; Zinellu, A.; Hussein Eid, A.; Pintus, G. Therapeutic Potential of Resveratrol in Covid-19-Associated Hemostatic Disorders. *Molecules* **2021**, *26*, 856. [CrossRef]

200. Dutra, L.A.; Guanaes, J.F.O.; Johmann, N.; Lopes Pires, M.E.; Chin, C.M.; Marcondes, S.; Dos Santos, J.L. Synthesis, antiplatelet and antithrombotic activities of resveratrol derivatives with NO-donor properties. *Bioorg. Med. Chem. Lett.* **2017**, *27*, 2450–2453. [CrossRef]
201. Tungmunnithum, D.; Thongboonyou, A.; Pholboon, A.; Yangsabai, A. Flavonoids and Other Phenolic Compounds from Medicinal Plants for Pharmaceutical and Medical Aspects: An Overview. *Medicines* **2018**, *5*, 93. [CrossRef]
202. Benetou, V.; Lagiou, A.; Lagiou, P. Chemoprevention of cancer: Current evidence and future prospects. *F1000Research* **2015**, *4*, 10. [CrossRef] [PubMed]
203. Key, T.J.; Bradbury, K.E.; Perez-Cornago, A.; Sinha, R.; Tsilidis, K.K.; Tsugane, S. Diet, nutrition, and cancer risk: What do we know and what is the way forward? *BMJ* **2020**, *368*, m511. [CrossRef] [PubMed]
204. Aruoma, O.I. Free radicals, oxidative stress, and antioxidants in human health and disease. *J. Am. Oil Chem. Soc.* **1998**, *75*, 199–212. [CrossRef] [PubMed]
205. Brusselmans, K.; Vrolix, R.; Verhoeven, G.; Swinnen, J.V. Induction of Cancer Cell Apoptosis by Flavonoids Is Associated with Their Ability to Inhibit Fatty Acid Synthase Activity. *J. Biol. Chem.* **2005**, *280*, 5636–5645. [CrossRef]
206. Danciu, C.; Vlaia, L.; Fetea, F.; Hancianu, M.; Coricovac, D.E.; Ciurlea, S.A.; Şoica, C.M.; Marincu, I.; Vlaia, V.; Dehelean, C.A.; et al. Evaluation of phenolic profile, antioxidant and anticancer potential of two main representants of Zingiberaceae family against B164A5 murine melanoma cells. *Biol. Res.* **2015**, *48*, 1. [CrossRef]
207. Abusnina, A.; Keravis, T.; Yougbaré, I.; Bronner, C.; Lugnier, C. Anti-proliferative effect of curcumin on melanoma cells is mediated by PDE1A inhibition that regulates the epigenetic integrator UHRF1. *Mol. Nutr. Food Res.* **2011**, *55*, 1677–1689. [CrossRef]
208. Basli, A.; Belkacem, N.; Amrani, I. Health Benefits of Phenolic Compounds Against Cancers. In *Phenolic Compounds*; Soto-Hernandez, M., Palma-Tenango, M., Garcia-Mateos, M.D.R., Eds.; IntechOpen: London, UK, 2017; pp. 193–210. [CrossRef]
209. Amawi, H.; Ashby, C.R., Jr.; Samuel, T.; Peraman, R.; Tiwari, A.K. Polyphenolic Nutrients in Cancer Chemoprevention and Metastasis: Role of the Epithelial-to-Mesenchymal (EMT) Pathway. *Nutrients* **2017**, *9*, 911. [CrossRef]
210. Johnson, E.J. The Role of Carotenoids in Human Health. *Nutr. Clin. Care* **2002**, *5*, 56–65. [CrossRef]
211. Yao, D.; Dai, C.; Peng, S. Mechanism of the Mesenchymal-Epithelial Transition and Its Relationship with Metastatic Tumor Formation. *Mol Cancer Res.* **2011**, *9*, 1608–1620. [CrossRef]
212. Datta, A.; Deng, S.; Gopal, V.; Yap, K.C.-H.; Halim, C.E.; Lye, M.L.; Ong, M.S.; Tan, T.Z.; Sethi, G.; Hooi, S.C.; et al. Cytoskeletal Dynamics in Epithelial-Mesenchymal Transition: Insights into Therapeutic Targets for Cancer Metastasis. *Cancers* **2021**, *13*, 1882. [CrossRef]
213. Lin, Y.; Kazlova, V.; Ramakrishnan, S.; Murray, M.A.; Fast, D.; Chandra, A.; Gellenbeck, K.W. Bone health nutraceuticals alter microarray mRNA gene expression: A randomized, parallel, open-label clinical study. *Phytomedicine* **2016**, *23*, 18–26. [CrossRef] [PubMed]
214. Bhattacharyya, S.; Pal, S.; Mohamed, R.; Singh, P.; Chattopadhyay, S.; Pal China, S.; Porwal, K.; Sanyal, S.; Gayen, J.R.; Chattopadhyay, N. A nutraceutical composition containing diosmin and hesperidin has osteogenic and anti-resorptive effects and expands the anabolic window of teriparatide. *Biomed. Pharmacother.* **2019**, *118*, 109207. [CrossRef] [PubMed]
215. Seely, K.D.; Kotelko, C.A.; Douglas, H.; Bealer, B.; Brooks, A.E. The Human Gut Microbiota: A Key Mediator of Osteoporosis and Osteogenesis. *Int. J. Mol. Sci.* **2021**, *22*, 9452. [CrossRef] [PubMed]
216. Rajput, R.; Wairkar, S.; Gaud, R. Nutraceuticals for better management of osteoporosis: An overview. *J. Funct. Foods* **2018**, *47*, 480–490. [CrossRef]
217. Ratner, B.D.; Hoffman, A.S.; Schoen, F.J.; Lemons, J. Introduction—Biomaterials Science: A Multidisciplinary Endeavor. *Biomater. Sci.* **2004**, 20. [CrossRef]
218. Kumar, S.; Nehra, M.; Kedia, D.; Dilbaghi, N.; Tankeshwar, K.; Kim, K.-H. Nanotechnology-based biomaterials for orthopaedic applications: Recent advances and future prospects. *Mater. Sci. Eng. C* **2020**, *106*, 1–25. [CrossRef]
219. Pillai, S.C.; Borah, A.; Jacob, E.M.; Kumar, D.S. Nanotechnological approach to delivering nutraceuticals as promising drug candidates for the treatment of atherosclerosis. *Drug Deliv.* **2021**, *28*, 550–568. [CrossRef]
220. Lidia, A.-V.; Carlos, Z.-M.; Alicia, R.-M.; Amalia, V.; Jose, V.-B. Nutraceuticals: Definition, applied nanoengineering in their production and applications. *Int. J. Biosens. Bioelectron.* **2019**, *5*, 56–61. [CrossRef]
221. Favaro-Trindade, C.S.; de Matos Junior, F.E.; Okuro, P.K.; Dias-Ferreira, J.; Cano, A.; Severino, P.; Zielińska, A.; Souto, E.B. Encapsulation of Active Pharmaceutical Ingredients in Lipid Micro/Nanoparticles for Oral Administration by Spray-Cooling. *Pharmaceutics* **2021**, *13*, 1186. [CrossRef]
222. Augustin, M.A.; Sanguansri, L. Challenges and Solutions to Incorporation of Nutraceuticals in Foods. *Annu. Rev. Food Sci. Technol.* **2015**, *6*, 463–477. [CrossRef]
223. Reque, P.M.; Brandelli, A. Encapsulation of probiotics and nutraceuticals: Applications in functional food industry. *Trends Food Sci. Technol.* **2021**, *114*, 1–10. [CrossRef]
224. Flores, F.P.; Kong, F. In Vitro Release Kinetics of Microencapsulated Materials and the Effect of the Food Matrix. *Annu. Rev. Food Sci. Technol.* **2017**, *8*, 237–259. [CrossRef] [PubMed]
225. Lee, J.H.; Yeo, Y. Controlled drug release from pharmaceutical nanocarriers. *Chem. Eng. Sci.* **2015**, *125*, 75–84. [CrossRef]
226. Savjani, K.T.; Gajjar, A.K.; Savjani, J.K. Drug Solubility: Importance and Enhancement Techniques. *ISRN Pharm.* **2012**, *2012*, 195727. [CrossRef] [PubMed]

227. Williams, H.D.; Trevaskis, N.L.; Charman, S.A.; Shanker, R.M.; Charman, W.N.; Pouton, C.W.; Porter, C.J.H. Strategies to address low drug solubility in discovery and development. *Pharmacol. Rev.* **2013**, *65*, 315–499. [CrossRef]
228. Porter, C.J.H.; Trevaskis, N.L.; Charman, W.N. Lipids and lipid-based formulations: Optimizing the oral delivery of lipophilic drugs. *Nat. Rev. Drug Discov.* **2007**, *6*, 231–249. [CrossRef]
229. Tambosi, G.; Coelho, P.F.; Soares, L.; Lenschow, I.C.S.; Zétola, M.; Stulzer, H.K.; Pezzini, B.R. Challenges to improve the biopharmaceutical properties of poorly water-soluble drugs and the application of the solid dispersion technology. *Rev. Mater.* **2018**, *23*. [CrossRef]
230. Coltescu, A.-R.; Butnariu, M.; Sarac, I. The Importance of Solubility for New Drug Molecules. *Biomed. Pharmacol. J.* **2020**, *13*, 577–583. [CrossRef]
231. Dima, C.; Assadpour, E.; Dima, S.; Jafari, S.M. Bioavailability and bioaccessibility of food bioactive compounds; overview and assessment by in vitro methods. *Compr. Rev. Food Sci. Food Saf.* **2020**, *19*, 2862–2884. [CrossRef]
232. Al-Obaidi, J.R.; Alobaidi, K.H.; Al-Taie, B.S.; Wee, D.H.S.; Hussain, H.; Jambari, N.N.; Ahmad-Kamil, E.I.; Ariffin, N.S. Uncovering Prospective Role and Applications of Existing and New Nutraceuticals from Bacterial, Fungal, Algal and Cyanobacterial, and Plant Sources. *Sustainability* **2021**, *13*, 3671. [CrossRef]
233. Pressman, P.; Clemens, R.A.; Hayes, A.W. Bioavailability of micronutrients obtained from supplements and food: A survey and case study of the polyphenols. *Toxicol. Res. Appl.* **2017**, *1*, 1–7. [CrossRef]
234. Sietsema, W.K. The absolute oral bioavailability of selected drugs. *Int. J. Clin. Pharmacol. Ther. Toxicol.* **1989**, *27*, 179–211. [PubMed]
235. Dima, C.; Assadpour, E.; Dima, S.; Jafari, S.M. Bioavailability of nutraceuticals: Role of the food matrix, processing conditions, the gastrointestinal tract, and nanodelivery systems. *Compr. Rev. Food Sci. Food Saf.* **2020**, *19*, 954–994. [CrossRef] [PubMed]
236. Ting, Y.; Jiang, Y.; Ho, C.-T.; Huang, Q. Common delivery systems for enhancing in vivo bioavailability and biological efficacy of nutraceuticals. *J. Funct. Foods* **2014**, *7*, 112–128. [CrossRef]
237. Zare, M.; Dziemidowicz, K.; Williams, G.R.; Ramakrishna, S. Encapsulation of Pharmaceutical and Nutraceutical Active Ingredients Using Electrospinning Processes. *Nanomaterials* **2021**, *11*, 1968. [CrossRef]
238. Subramanian, P. Lipid-Based Nanocarrier System for the Effective Delivery of Nutraceuticals. *Molecules* **2021**, *26*, 5510. [CrossRef]
239. McClements, D.J.; Decker, E.A.; Park, Y.; Weiss, J. Structural Design Principles for Delivery of Bioactive Components in Nutraceuticals and Functional Foods. *Crit. Rev. Food Sci. Nutr.* **2009**, *49*, 577–606. [CrossRef]
240. Gheorghita, R.; Anchidin-Norocel, L.; Filip, R.; Dimian, M.; Covasa, M. Applications of Biopolymers for Drugs and Probiotics Delivery. *Polymers* **2021**, *13*, 2729. [CrossRef]
241. Hu, B.; Huang, Q.-R. Biopolymer Based Nano-Delivery Systems for Enhancing Bioavailability of Nutraceuticals. *Chinese J. Polym. Sci.* **2013**, *31*, 1190–1203. [CrossRef]
242. Malafaya, P.B.; Silva, G.A.; Reis, R.L. Natural-origin polymers as carriers and scaffolds for biomolecules and cell delivery in tissue engineering applications. *Adv. Drug Deliv. Rev.* **2007**, *59*, 207–233. [CrossRef]
243. Torres, F.G.; Troncoso, O.P.; Pisani, A.; Gatto, F.; Bardi, G. Natural Polysaccharide Nanomaterials: An Overview of Their Immunological Properties. *Int. J. Mol. Sci.* **2019**, *20*, 5092. [CrossRef] [PubMed]
244. Jacob, J.; Haponiuk, J.T.; Thomas, S.; Gopi, S. Biopolymer based nanomaterials in drug delivery systems: A review. *Mater. Today Chem.* **2018**, *9*, 43–55. [CrossRef]
245. Gopinath, V.; Saravanan, S.; Al-Maleki, A.R.; Ramesh, M.; Vadivelu, J. A review of natural polysaccharides for drug delivery applications: Special focus on cellulose, starch and glycogen. *Biomed. Pharmacother.* **2018**, *107*, 96–108. [CrossRef] [PubMed]
246. Hosseinpourpia, R.; Echart, A.S.; Adamopoulos, S.; Gabilondo, N.; Eceiza, A. Modification of Pea Starch and Dextrin Polymers with Isocyanate Functional Groups. *Polymers* **2018**, *10*, 939. [CrossRef]
247. Alcázar-Alay, S.C.; Meireles, M.A.A. Physicochemical properties, modifications and applications of starches from different botanical sources. *Food Sci. Technol.* **2015**, *35*, 215–236. [CrossRef]
248. Beneke, C.E.; Viljoen, A.M.; Hamman, J.H. Polymeric Plant-derived Excipients in Drug Delivery. *Molecules* **2009**, *14*, 2602–2620. [CrossRef]
249. Serrero, A.; Trombotto, S.; Cassagnau, P.; Bayon, Y.; Gravagna, P.; Montanari, S.; David, L. Polysaccharide Gels Based on Chitosan and Modified Starch: Structural Characterization and Linear Viscoelastic Behavior. *Biomacromolecules* **2010**, *11*, 1534–1543. [CrossRef]
250. Van Der Maarel, M.J.E.C.; Van Der Veen, B.; Uitdehaag, J.C.M.; Leemhuis, H.; Dijkhuizen, L. Properties and applications of starch-converting enzymes of the α-amylase family. *J. Biotechnol.* **2002**, *94*, 137–155. [CrossRef]
251. Carvalho, J.; Gonçalves, C.; Gil, A.M.; Gama, F.M. Production and characterization of a new dextrin based hydrogel. *Eur. Polym. J.* **2007**, *43*, 3050–3059. [CrossRef]
252. Gonçalves, C.; Moreira, S.M.; Carvalho, V.; Silva, D.M.; Gama, M. Dextrin. *Encycl. Biomed. Polym. Polym. Biomater.* **2016**, 2634–2649. [CrossRef]
253. Silva, D.M.; Nunes, C.; Pereira, I.; Moreira, A.S.P.; Domingues, M.R.M.; Coimbra, M.A.; Gama, F.M. Structural analysis of dextrins and characterization of dextrin-based biomedical hydrogels. *Carbohydr. Polym.* **2014**, *114*, 458–466. [CrossRef] [PubMed]
254. Klein, S. Polysaccharides in Oral Drug Delivery—Recent Applications and Future Perspectives. In *ACS Symposium Series*; American Chemical Society: Washington, DC, USA, 2009; pp. 13–30, ISBN 9780841269866.
255. Lewicka, K.; Siemion, P.; Kurcok, P. Chemical Modifications of Starch: Microwave Effect. *Int. J. Polym. Sci.* **2015**, *2015*, 867697. [CrossRef]

256. Haq, F.; Yu, H.; Wang, L.; Teng, L.; Haroon, M.; Khan, R.U.; Mehmood, S.; Bilal-Ul-Amin; Ullah, R.S.; Khan, A.; et al. Advances in chemical modifications of starches and their applications. *Carbohydr. Res.* **2019**, *476*, 12–35. [CrossRef] [PubMed]
257. Mailänder, V.; Landfester, K. Interaction of Nanoparticles with Cells. *Biomacromolecules* **2009**, *10*, 2379–2400. [CrossRef] [PubMed]
258. Li, B.; Li, Q.; Mo, J.; Dai, H. Drug-Loaded Polymeric Nanoparticles for Cancer Stem Cell Targeting. *Front. Pharmacol.* **2017**, *8*, 1–12. [CrossRef] [PubMed]
259. Mitchell, M.J.; Billingsley, M.M.; Haley, R.M.; Wechsler, M.E.; Peppas, N.A.; Langer, R. Engineering precision nanoparticles for drug delivery. *Nat. Rev. Drug Discov.* **2021**, *20*, 1. [CrossRef]
260. Lin, J.; Miao, L.; Zhong, G.; Lin, C.H.; Dargazangy, R.; Alexander-Katz, A. Understanding the synergistic effect of physicochemical properties of nanoparticles and their cellular entry pathways. *Commun. Biol.* **2020**, *3*, 205. [CrossRef]
261. Brewster, M.E.; Loftsson, T. Cyclodextrins as pharmaceutical solubilizers. *Adv. Drug Deliv. Rev.* **2007**, *59*, 645–666. [CrossRef]
262. Osmani, R.A.; Kulkarni, P.; Manjunatha, S.; Gowda, V.; Hani, U.; Vaghela, R.; Bhosale, R. Chapter 9 Cyclodextrin Nanosponges in Drug Delivery and Nanotherapeutics. In *Environmental Nanotechnology*; Dasgupta, N., Ranjan, S., Lichtfouse, E., Eds.; Springer: Berlin/Heidelberg, Germany, 2018; p. 405, ISBN 978-3-319-76090-2.
263. Sherje, A.P.; Dravyakar, B.R. Cyclodextrin-based nanosponges: A critical review. *Carbohydr. Polym.* **2017**, *173*, 37–49. [CrossRef]
264. Martin, J.; Díaz-Montaña, E.J.; Asuero, A.G. *Cyclodextrins: Past and Present*; IntechOpen: London, UK, 2008. [CrossRef]
265. Liu, L.; Guo, Q.X. The driving forces in the inclusion complexation of cyclodextrins. *J. Incl. Phenom.* **2002**, *42*, 1–14. [CrossRef]
266. Matencio, A.; Navarro-Orcajada, S.; García-Carmona, F.; López-Nicolás, J.M. Applications of cyclodextrins in food science. A review. *Trends Food Sci. Technol.* **2020**, *104*, 132–143. [CrossRef]
267. Stella, V.J.; Rajewski, R.A. Cyclodextrins: Their Future in Drug Formulation and Delivery. *Pharm. Res.* **1997**, *14N*, 11. [CrossRef]
268. Rousseau, J.; Menuel, S.; Rousseau, C.; Hapiot, F.; Monflier, E. *Cyclodextrins as Porous Material for Catalysis*; Elsevier Inc.: Amsterdam, The Netherlands, 2016; ISBN 9780128018101.
269. Hedges, A.R. Industrial applications of cyclodextrins. *Chem. Rev.* **1998**, *98*, 2035–2044. [CrossRef] [PubMed]
270. Szejtli, J. Past, present, and future of cyclodextrin research. *Pure Appl. Chem.* **2004**, *76*, 1825–1845. [CrossRef]
271. Jambhekar, S.S.; Breen, P. Cyclodextrins in pharmaceutical formulations I: Structure and physicochemical properties, formation of complexes, and types of complex. *Drug Discov. Today* **2016**, *21*, 356–362. [CrossRef] [PubMed]
272. Bekers, O.; Uijtendaal, E.V.; Beijnen, J.H.; Bult, A.; Underberg, W.J.M. Cyclodextrins in the pharmaceutical Field. *Drug Dev. Ind. Pharm.* **1991**, *17*, 1503–1549. [CrossRef]
273. Chilajwar, S.V.; Pednekar, P.P.; Jadhav, K.R.; Gupta, G.J.; Kadam, V.J. Cyclodextrin-based nanosponges: A propitious platform for enhancing drug delivery. *Expert Opin. Drug Deliv.* **2014**, *11*, 111–120. [CrossRef]
274. Berto, S.; Bruzzoniti, M.C.; Cavalli, R.; Perrachon, D.; Prenesti, E.; Sarzanini, C.; Trotta, F.; Tumiatti, W. Synthesis of new ionic β-cyclodextrin polymers and characterization of their heavy metals retention. *J. Incl. Phenom. Macrocycl. Chem.* **2007**, *57*, 631–636. [CrossRef]
275. Zhang, D.; Lv, P.; Zhou, C.; Zhao, Y.; Liao, X.; Yang, B. Cyclodextrin-based delivery systems for cancer treatment. *Mater. Sci. Eng. C* **2019**, *96*, 872–886. [CrossRef]
276. Swaminathan, S.; Vavia, P.R.; Trotta, F.; Cavalli, R. Nanosponges encapsulating dexamethasone for ocular delivery: Formulation design, physicochemical characterization, safety and corneal permeability assessment. *J. Biomed. Nanotechnol.* **2013**, *9*, 998–1007. [CrossRef]
277. Mognetti, B.; Barberis, A.; Marino, S.; Berta, G.; De Francia, S.; Trotta, F.; Cavalli, R. In vitro enhancement of anticancer activity of paclitaxel by a Cremophor free cyclodextrin-based nanosponge formulation. *J. Incl. Phenom. Macrocycl. Chem.* **2012**, *74*, 201–210. [CrossRef]
278. Ferro, M.; Castiglione, F.; Punta, C.; Melone, L.; Panzeri, W.; Rossi, B.; Trotta, F.; Mele, A. Anomalous diffusion of ibuprofen in cyclodextrin nanosponge hydrogels: An HRMAS NMR study. *Beilstein J. Org. Chem.* **2014**, *10*, 2715–2723. [CrossRef] [PubMed]
279. Trotta, F.; Cavalli, R. Characterization and applications of new hyper-cross-linked cyclodextrins. *Compos. Interfaces* **2009**, *16*, 39–48. [CrossRef]
280. Santos, A.C.; Costa, D.; Ferreira, L.; Guerra, C.; Pereira-Silva, M.; Pereira, I.; Peixoto, D.; Ferreira, N.R.; Veiga, F. Cyclodextrin-based delivery systems for in vivo-tested anticancer therapies. *Drug Deliv. Transl. Res.* **2021**, *11*, 49–71. [CrossRef]
281. Petitjean, M.; García-Zubiri, I.X.; Isasi, J.R. History of cyclodextrin-based polymers in food and pharmacy: A review. *Environ. Chem. Lett.* **2021**, *19*, 3465–3476. [CrossRef]
282. Gambini, J.; Inglès, M.; Olaso, G.; Lopez-Grueso, R.; Bonet-Costa, V.; Gimeno-Mallench, L.; Mas-Bargues, C.; Abdelaziz, K.M.; Gomez-Cabrera, M.C.; Vina, J.; et al. Properties of Resveratrol: In Vitro and In Vivo Studies about Metabolism, Bioavailability, and Biological Effects in Animal Models and Humans. *Oxid. Med. Cell. Longev.* **2015**, *2015*, 837042. [CrossRef]
283. Gandhi, S.R.; Quintans, J.D.S.S.; Gandhi, G.R.; Araujo, A.A.D.S.; Junior, L.J.Q. The use of cyclodextrin inclusion complexes to improve anticancer drug profiles: A systematic review. *Expert Opin. Drug Deliv.* **2020**, *17*, 1069–1080. [CrossRef]
284. Hoti, G.; Appleton, S.L.; Rubin Pedrazzo, A.; Cecone, C.; Matencio, A.; Trotta, F.; Caldera, F. Strategies to Develop Cyclodextrin-Based Nanosponges for Smart Drug Delivery. In *Smart Drug Delivery*; IntechOpen: London, UK, 2021; pp. 1–22. [CrossRef]
285. Caldera, F.; Tannous, M.; Cavalli, R.; Zanetti, M.; Trotta, F. Evolution of Cyclodextrin Nanosponges. *Int. J. Pharm.* **2017**, *531*, 470–479. [CrossRef]
286. Thatiparti, T.R.; Von Recum, H.A. Cyclodextrin Complexation for Affinity-Based Antibiotic Delivery. *Macromol. Biosci.* **2010**, *10*, 82–90. [CrossRef]

287. Merritt, S.R.; Velasquez, G.; Von Recum, H.A. Adjustable release of mitomycin C for inhibition of scar tissue formation after filtration surgery. *Exp. Eye Res.* **2013**, *116*, 9–16. [CrossRef]
288. Jullian, C.; Moyano, L.; Yañez, C.; Olea-Azar, C. Complexation of quercetin with three kinds of cyclodextrins: An antioxidant study. *Spectrochim. Acta-Part A* **2007**, *67*, 230–234. [CrossRef] [PubMed]
289. Dhakar, N.K.; Matencio, A.; Caldera, F.; Argenziano, M.; Cavalli, R.; Dianzani, C.; Zanetti, M.; López-Nicolás, J.M.; Trotta, F. Comparative Evaluation of Solubility, Cytotoxicity and Photostability Studies of Resveratrol and Oxyresveratrol Loaded Nanosponges. *Pharmaceutics* **2019**, *11*, 545. [CrossRef] [PubMed]
290. Matencio, A.; Dhakar, N.K.; Bessone, F.; Musso, G.; Cavalli, R.; Dianzani, C.; García-Carmona, F.; López-Nicolás, J.M.; Trotta, F. Study of oxyresveratrol complexes with insoluble cyclodextrin based nanosponges: Developing a novel way to obtain their complexation constants and application in an anticancer study. *Carbohydr. Polym.* **2020**, *231*, 1–7. [CrossRef] [PubMed]
291. Mendes, C.; Meirelles, G.C.; Barp, C.G.; Assreuy, J.; Silva, M.A.S.; Ponchel, G. Cyclodextrin based nanosponge of norfloxacin: Intestinal permeation enhancement and improved antibacterial activity. *Carbohydr. Polym.* **2018**, *195*, 586–592. [CrossRef]
292. Dhakar, N.K.; Caldera, F.; Bessone, F.; Cecone, C.; Pedrazzo, A.R.; Cavalli, R.; Dianzani, C.; Trotta, F. Evaluation of solubility enhancement, antioxidant activity, and cytotoxicity studies of kynurenic acid loaded cyclodextrin nanosponge. *Carbohydr. Polym.* **2019**, *224*, 1–9. [CrossRef]
293. Torne, S.; Darandale, S.; Vavia, P.; Trotta, F.; Cavalli, R. Cyclodextrin-based nanosponges: Effective nanocarrier for Tamoxifen delivery. *Pharm. Dev. Technol.* **2013**, *18*, 619–625. [CrossRef]
294. Zainuddin, R.; Zaheer, Z.; Sangshetti, J.N.; Momin, M. Enhancement of oral bioavailability of anti- HIV drug rilpivirine HCl through nanosponge formulation. *Drug Dev. Ind. Pharm.* **2017**, *43*, 2076–2084. [CrossRef]
295. Lembo, D.; Swaminathan, S.; Donalisio, M.; Civra, A.; Pastero, L.; Aquilano, D.; Vavia, P.; Trotta, F.; Cavalli, R. Encapsulation of Acyclovir in new carboxylated cyclodextrin-based nanosponges improves the agent's antiviral efficacy. *Int. J. Pharm.* **2013**, *443*, 262–272. [CrossRef]
296. Swaminathan, S.; Pastero, L.; Serpe, L.; Trotta, F.; Vavia, P.; Aquilano, D.; Trotta, M.; Zara, G.; Cavalli, R. Cyclodextrin-based nanosponges encapsulating camptothecin: Physicochemical characterization, stability and cytotoxicity. *Eur. J. Pharm. Biopharm.* **2010**, *74*, 193–201. [CrossRef]
297. Rao, M.; Bajaj, A.; Khole, I.; Munjapara, G.; Trotta, F. In vitro and in vivo evaluation of β-cyclodextrin-based nanosponges of telmisartan. *J. Incl. Phenom. Macrocycl. Chem.* **2013**, *77*, 135–145. [CrossRef]
298. Kumar, S.; Pooja; Trotta, F.; Rao, R. Encapsulation of Babchi Oil in Cyclodextrin-Based Nanosponges: Physicochemical Characterization, Photodegradation, and In Vitro Cytotoxicity Studies. *Pharmaceutics* **2018**, *10*, 169. [CrossRef] [PubMed]
299. Sundararajan, M.; Thomas, P.A.; Venkadeswaran, K.; Jeganathan, K.; Geraldine, P. Synthesis and characterization of chrysin-loaded β-cyclodextrin-based nanosponges to enhance in-vitro solubility, photostability, drug release, antioxidant effects and antitumorous efficacy. *J. Nanosci. Nanotechnol.* **2017**, *17*, 8742–8751. [CrossRef]
300. Yaşayan, G.; Şatıroğlu Sert, B.; Tatar, E.; Küçükgüzel, İ. Fabrication and characterisation studies of cyclodextrin-based nanosponges for sulfamethoxazole delivery. *J. Incl. Phenom. Macrocycl. Chem.* **2020**, *97*, 175–186. [CrossRef]
301. Rezaei, A.; Varshosaz, J.; Fesharaki, M.; Farhang, A.; Jafari, S.M. Improving the solubility and in vitro cytotoxicity (anticancer activity) of ferulic acid by loading it into cyclodextrin nanosponges. *Int. J. Nanomed.* **2019**, *14*, 4589–4599. [CrossRef]
302. Salehi, O.; Masoud, S.; Rezaei, A. Limonene loaded cyclodextrin nanosponge: Preparation, characterization, antibacterial activity and controlled release. *Food Biosci.* **2021**, *42*, 1–9. [CrossRef]
303. Kumar, A.; Rao, R. Enhancing efficacy and safety of azelaic acid via encapsulation in cyclodextrin nanosponges: Development, characterization and evaluation. *Polym. Bull.* **2021**, *78*, 5275–5302. [CrossRef]
304. Sherje, A.P.; Surve, A.; Shende, P. CDI cross-linked β-cyclodextrin nanosponges of paliperidone: Synthesis and physicochemical characterization. *J. Mater. Sci. Mater. Med.* **2019**, *30*, 1–7. [CrossRef]
305. Omar, S.M.; Ibrahim, F.; Ismail, A. Formulation and evaluation of cyclodextrin-based nanosponges of griseofulvin as pediatric oral liquid dosage form for enhancing bioavailability and masking bitter taste. *Saudi Pharm. J.* **2020**, *28*, 349–361. [CrossRef]
306. Allahyari, S.; Esmailnezhad, N.; Valizadeh, H.; Ghorbani, M.; Jelvehgari, M.; Ghazi, F.; Zakeri-Milani, P. In-vitro characterization and cytotoxicity study of flutamide loaded cyclodextrin nanosponges. *J. Drug Deliv. Sci. Technol.* **2021**, *61*, 1–7. [CrossRef]
307. Srivastava, S.; Mahor, A.; Singh, G.; Bansal, K.; Singh, P.P.; Gupta, R.; Dutt, R.; Alanazi, A.M.; Khan, A.A.; Kesharwani, P. Formulation Development, In Vitro and In Vivo Evaluation of Topical Hydrogel Formulation of Econazole Nitrate-Loaded β-Cyclodextrin Nanosponges. *J. Pharm. Sci.* **2021**, *110*, 3702–3714. [CrossRef]
308. Guineo-Alvarado, J.; Quilaqueo, M.; Hermosilla, J.; González, S.; Medina, C.; Rolleri, A.; Lim, L.-T.; Rubilar, M. Degree of crosslinking in β-cyclodextrin-based nanosponges and their effect on piperine encapsulation. *Food Chem.* **2021**, *340*, 128132. [CrossRef] [PubMed]
309. Machín, R.; Isasi, J.R.; Vélaz, I. β-Cyclodextrin hydrogels as potential drug delivery systems. *Carbohydr. Polym.* **2012**, *87*, 2024–2030. [CrossRef]
310. Cassidy, J.; Berner, B.; Chan, K.; John, V.; Toon, S.; Holt, B.; Rowland, M. Human Transbuccal Absorption of Diclofenac Sodium from a Prototype Hydrogel Delivery Device. *Pharm. Res.* **1993**, *10*, 126–129. [CrossRef]
311. Rodriguez-Tenreiro, C.; Alvarez-Lorenzo, C.; Rodriguez-Perez, A.; Concheiro, A.; Torres-Labandeira, J.J. New Cyclodextrin Hydrogels Cross-Linked with Diglycidylethers with a High Drug Loading and Controlled Release Ability. *Pharm. Res.* **2006**, *23*, 121–130. [CrossRef] [PubMed]

312. Gami, P.; Kundu, D.; Seera, S.D.K.; Banerjee, T. Chemically crosslinked xylan–β-Cyclodextrin hydrogel for the in vitro delivery of curcumin and 5-Fluorouracil. *Int. J. Biol. Macromol.* **2020**, *158*, 18–31. [CrossRef] [PubMed]
313. Lucia Appleton, S.; Tannous, M.; Argenziano, M.; Muntoni, E.; Carolina Rosa, A.; Rossi, D.; Caldera, F.; Scomparin, A.; Trotta, F.; Cavalli, R. Nanosponges as protein delivery systems: Insulin, a case study. *Int. J. Pharm.* **2020**, *590*, 1–11. [CrossRef]
314. Argenziano, M.; Haimhoffer, A.; Bastiancich, C.; Jicsinszky, L.; Caldera, F.; Trotta, F.; Scutera, S.; Alotto, D.; Fumagalli, M.; Musso, T.; et al. In Vitro Enhanced Skin Permeation and Retention of Imiquimod Loaded in β-Cyclodextrin Nanosponge Hydrogel. *Pharmaceutics* **2019**, *11*, 138. [CrossRef]
315. Shende, P.K.; Trotta, F.; Gaud, R.S.; Deshmukh, K.; Cavalli, R.; Biasizzo, M. Influence of different techniques on formulation and comparative characterization of inclusion complexes of ASA with β-cyclodextrin and inclusion complexes of ASA with PMDA cross-linked β-cyclodextrin nanosponges. *J. Incl. Phenom. Macrocycl. Chem.* **2012**, *74*, 447–454. [CrossRef]
316. Argenziano, M.; Gigliotti, C.L.; Clemente, N.; Boggio, E.; Ferrara, B.; Trotta, F.; Pizzimenti, S.; Barrera, G.; Boldorini, R.; Bessone, F.; et al. Improvement in the Anti-Tumor Efficacy of Doxorubicin Nanosponges in In Vitro and in Mice Bearing Breast Tumor Models. *Cancers* **2020**, *12*, 162. [CrossRef]
317. Shende, P.K.; Gaud, R.S.; Bakal, R.; Patil, D. Effect of inclusion complexation of meloxicam with β-cyclodextrin- and β-cyclodextrin-based nanosponges on solubility, in vitro release and stability studies. *Colloids Surfaces B Biointerfaces* **2015**, *136*, 105–110. [CrossRef]
318. Rao, M.R.P.; Chaudhari, J.; Trotta, F.; Caldera, F. Investigation of Cyclodextrin-Based Nanosponges for Solubility and Bioavailability Enhancement of Rilpivirine. *AAPS PharmSciTech* **2018**, *19*, 2358–2368. [CrossRef] [PubMed]
319. Singh, P.; Ren, X.; Guo, T.; Wu, L.; Shakkya, S.; He, Y.; Wang, C.; Maharjan, A.; Singh, V.; Zhang, J. Biofunctionalization of β-cyclodextrin nanosponges using cholesterol. *Carbohydr. Polym.* **2018**, *190*, 23–30. [CrossRef] [PubMed]
320. Asela, I.; Donoso-González, O.; Yutronic, N.; Sierpe, R. β-Cyclodextrin-Based Nanosponges Functionalized With Drugs and Gold Nanoparticles. *Pharmaceutics* **2021**, *13*, 513. [CrossRef] [PubMed]
321. Daga, M.; de Graaf, I.A.M.; Argenziano, M.; Martinez Barranco, A.S.; Loeck, M.; Al-Adwi, Y.; Angele Cucci, M.; Caldera, F.; Trotta, F.; Barrera, G.; et al. Glutathione-responsive cyclodextrin-nanosponges as drug delivery systems for doxorubicin: Evaluation of toxicity and transport mechanisms in the liver. *Toxicol. Vitr.* **2020**, *65*, 1–10. [CrossRef]
322. Coviello, V.; Sartini, S.; Quattrini, L.; Baraldi, C.; Gamberini, M.C.; La Motta, C. Cyclodextrin-based nanosponges for the targeted delivery of the anti-restenotic agent DB103: A novel opportunity for the local therapy of vessels wall subjected to percutaneous intervention. *Eur. J. Pharm. Biopharm.* **2017**, *117*, 276–285. [CrossRef]
323. Deshmukh, K.; Tanwar, Y.S.; Sharma, S.; Shende, P.; Cavalli, R. Functionalized nanosponges for controlled antibacterial and antihypocalcemic actions. *Biomed. Pharmacother.* **2016**, *84*, 485–494. [CrossRef]
324. Deshmukh, K.; Tanwar, Y.S.; Shende, P.; Cavalli, R. Biomimetic estimation of glucose using non-molecular and molecular imprinted polymer nanosponges. *Int. J. Pharm.* **2015**, *494*, 244–248. [CrossRef]
325. Trotta, F.; Caldera, F.; Cavalli, R.; Soster, M.; Riedo, C.; Biasizzo, M.; Uccello Barretta, G.; Balzano, F.; Brunella, V. Molecularly imprinted cyclodextrin nanosponges for the controlled delivery of L-DOPA: Perspectives for the treatment of Parkinson's disease. *Expert Opin. Drug Deliv.* **2016**, *13*, 1671–1680. [CrossRef]
326. Cecone, C.; Hoti, G.; Krabicova, I.; Appleton, S.L.; Caldera, F.; Bracco, P.; Zanetti, M.; Trotta, F. Sustainable synthesis of cyclodextrin-based polymers exploiting natural deep eutectic solvents. *Green Chem.* **2020**, *22*, 5806–5814. [CrossRef]
327. Pedrazzo, A.R.; Caldera, F.; Zanetti, M.; Appleton, S.L.; Dahkar, N.K.; Trotta, F. Mechanochemical green synthesis of hyper-crosslinked cyclodextrin polymers. *Beilstein J. Org. Chem.* **2020**, *16*, 1554–1563. [CrossRef]
328. Tannous, M.; Trotta, F.; Cavalli, R. Nanosponges for combination drug therapy: State-of-the-art and future directions. *Nanomedicine* **2020**, *15*, 643–646. [CrossRef] [PubMed]
329. Wang, Y.-J.; Wang, L. Structures and Properties of Commercial Maltodextrins from Corn, Potato, and Rice Starches. *Starch/Stärke* **2000**, *52*, 296–304. [CrossRef]
330. Guntero, V.A.; Peralta, M.; Noriega, P.; Kneeteman, M.N.; Ferretti, C.A. One-Pot Selective Functionalization of Polysaccharides with Urea. *Chem. Proc.* **2021**, *3*, 2–6.
331. Siemons, I.; Politiek, R.G.A.; Boom, R.M.; Van der Sman, R.G.M.; Schutyser, M.A.I. Dextrose equivalence of maltodextrins determines particle morphology development during single sessile droplet drying. *Food Res. Int.* **2020**, *131*, 1–10. [CrossRef] [PubMed]
332. Barthold, S.; Hittinger, M.; Primavessy, D.; Zapp, A.; Groß, H.; Schneider, M. Preparation of maltodextrin nanoparticles and encapsulation of bovine serum albumin—Influence of formulation parameters. *Eur. J. Pharm. Biopharm.* **2019**, *142*, 405–410. [CrossRef] [PubMed]
333. Helal, H.M.; Samy, W.M.; El-Fakharany, E.M.; Kamoun, E.; Mortada, S.M.; Sallam, M.A. Maltodextrin-α-tocopherol conjugates of vitamin E: Influence of degree of derivatization on physicochemical properties and biological evaluation. *J. Drug Deliv. Sci. Technol.* **2020**, *60*, 1–12. [CrossRef]
334. Hofman, D.L.; van Buul, V.J.; Brouns, F.J.P.H. Nutrition, Health, and Regulatory Aspects of Digestible Maltodextrins. *Crit. Rev. Food Sci. Nutr.* **2016**, *56*, 2091–2100. [CrossRef]
335. Rezende, G.; Hashizume, L.N. Maltodextrin and dental caries: A literature review. *Rev. Gaúch Odontol.* **2018**, *66*, 257–262. [CrossRef]

336. Sun, X.; Wu, X.; Chen, X.; Guo, R.; Kou, Y.; Li, X.; Sheng, Y.; Wu, Y. Casein-maltodextrin Maillard conjugates encapsulation enhances the antioxidative potential of proanthocyanidins: An in vitro and in vivo evaluation. *Food Chem.* **2021**, *346*, 1–8. [CrossRef]
337. Okumuş, E.; Bakkalbaşı, E.; Javidipour, I.; Meral, R.; Ceylan, Z. A novel coating material: Ellagitannins-loaded maltodextrin and lecithin-based nanomaterials. *Food Biosci.* **2021**, *42*, 1–7. [CrossRef]
338. Gurturk, Z.; Tezcaner, A.; Dalgic, A.D.; Korkmaz, S.; Keskin, D. Maltodextrin modified liposome for drug delivery through blood-brain barrier. *Medchemcomm* **2017**, *8*, 1337–1345. [CrossRef] [PubMed]
339. Lai, F.; Franceschini, I.; Corrias, F.; Sala, M.C.; Cilurzo, F.; Sinico, C.; Pini, E. Maltodextrin fast dissolving films for quercetin nanocrystal delivery. A feasibility study. *Carbohydr. Polym.* **2015**, *121*, 217–223. [CrossRef] [PubMed]
340. Blazek-Welsh, A.I.; Rhodes, D.G. SEM Imaging Predicts Quality of Niosomes from Maltodextrin-Based Proniosomes. *Pharm. Res.* **2001**, *18*, 656–661. [CrossRef] [PubMed]
341. Shruthi, P.A.; Pushpadass, H.A.; Franklin, M.E.E.; Battula, S.N.; Naik, N.L. Resveratrol-loaded proniosomes: Formulation, characterization and fortification. *LWT-Food Sci. Technol.* **2020**, *134*, 1–12. [CrossRef]
342. Loftsson, T.; Duchêne, D. Cyclodextrins and their pharmaceutical applications. *Int. J. Pharm.* **2007**, *329*, 1–11. [CrossRef]
343. Castro-Cabado, M.; Casado, A.L.; San Román, J. Bio-based thermosets: Effect of the structure of polycarboxylic acids on the thermal crosslinking of maltodextrins. *Eur. Polym. J.* **2016**, *78*, 91–105. [CrossRef]
344. Cecone, C.; Costamagna, G.; Ginepro, M.; Trotta, F. One-step sustainable synthesis of cationic high-swelling polymers obtained from starch-derived maltodextrins. *RSC Adv.* **2021**, *11*, 7653–7662. [CrossRef]
345. Meléndez-Ortiz, H.I.; Betancourt-Galindo, R.; Puente-Urbina, B.; Ledezma, A.; Rodríguez-Fernández, O. Synthesis and characterization of hydrogels based on maltodextrins with antimicrobial properties. *Int. J. Polym. Mater. Polym. Biomater.* **2021**, 1–10. [CrossRef]
346. Yan, S.; Ren, J.; Jian, Y.; Wang, W.; Yun, W.; Yin, J. Injectable Maltodextrin-Based Micelle/Hydrogel Composites for Simvastatin-Controlled Release. *Biomacromolecules* **2018**, *19*, 4554–4564. [CrossRef]
347. Demasi, S.; Caser, M.; Caldera, F.; Dhakar, N.K.; Vidotto, F.; Trotta, F.; Scariot, V. Functionalized dextrin-based nanosponges as effective carriers for the herbicide ailanthone. *Ind. Crops Prod.* **2021**, *164*, 113346. [CrossRef]
348. Pedrazzo, A.R.; Smarra, A.; Caldera, F.; Musso, G.; Dhakar, N.K.; Cecone, C.; Hamedi, A.; Corsi, I.; Trotta, F. Eco-Friendly β-cyclodextrin and Linecaps Polymers for the Removal of Heavy Metals. *Polymers* **2019**, *11*, 1658. [CrossRef] [PubMed]

International Journal of *Molecular Sciences*

MDPI

Article

Neferine, an Alkaloid from Lotus Seed Embryos, Exerts Antiseizure and Neuroprotective Effects in a Kainic Acid-Induced Seizure Model in Rats

Tzu-Yu Lin [1,2], Chih-Yu Hung [3], Kuan-Ming Chiu [4,5], Ming-Yi Lee [4], Cheng-Wei Lu [1,2,*] and Su-Jane Wang [3,6,*]

1 Department of Anesthesiology, Far-Eastern Memorial Hospital, New Taipei City 22060, Taiwan; drlin1971@gmail.com
2 Department of Mechanical Engineering, Yuan Ze University, Taoyuan 32003, Taiwan
3 School of Medicine, Fu Jen Catholic University, New Taipei City 24205, Taiwan; h.chihyu@gmail.com
4 Cardiovascular Center, Division of Cardiovascular Surgery, Far-Eastern Memorial Hospital, New Taipei 22060, Taiwan; chiu9101018@gmail.com (K.-M.C.); mingyi.lee@gmail.com (M.-Y.L.)
5 Department of Electrical Engineering, Yuan Ze University, Taoyuan 32003, Taiwan
6 Research Center for Chinese Herbal Medicine, College of Human Ecology, Chang Gung University of Science and Technology, Taoyuan 33303, Taiwan
* Correspondence: drluchengwei@gmail.com (C.-W.L.); med0003@mail.fju.edu.tw (S.-J.W.)

Citation: Lin, T.-Y.; Hung, C.-Y.; Chiu, K.-M.; Lee, M.-Y.; Lu, C.-W.; Wang, S.-J. Neferine, an Alkaloid from Lotus Seed Embryos, Exerts Antiseizure and Neuroprotective Effects in a Kainic Acid-Induced Seizure Model in Rats. *Int. J. Mol. Sci.* **2022**, *23*, 4130. https://doi.org/10.3390/ijms23084130

Academic Editors: Patrizia Hrelia and Ali Gorji

Received: 25 February 2022
Accepted: 7 April 2022
Published: 8 April 2022

Abstract: Current anti-seizure drugs fail to control approximately 30% of epilepsies. Therefore, there is a need to develop more effective anti-seizure drugs, and medicinal plants provide an attractive source for new compounds. This study aimed to evaluate the possible anti-seizure and neuroprotective effects of neferine, an alkaloid from the lotus seed embryos of *Nelumbo nucifera*, in a kainic acid (KA)-induced seizure rat model and its underlying mechanisms. Rats were intraperitoneally (i.p.) administrated neferine (10 and 50 mg/kg) 30 min before KA injection (15 mg/kg, i.p.). Neferine pretreatment increased seizure latency and reduced seizure scores, prevented glutamate elevation and neuronal loss, and increased presynaptic protein synaptophysin and postsynaptic density protein 95 expression in the hippocampi of rats with KA. Neferine pretreatment also decreased glial cell activation and proinflammatory cytokine (interleukin-1β, interleukin-6, tumor necrosis factor-α) expression in the hippocampi of rats with KA. In addition, NOD-like receptor 3 (NLRP3) inflammasome, caspase-1, and interleukin-18 expression levels were decreased in the hippocampi of seizure rats pretreated with neferine. These results indicated that neferine reduced seizure severity, exerted neuroprotective effects, and ameliorated neuroinflammation in the hippocampi of KA-treated rats, possibly by inhibiting NLRP3 inflammasome activation and decreasing inflammatory cytokine secretion. Our findings highlight the potential of neferine as a therapeutic option in the treatment of epilepsy.

Keywords: neferine; anti-seizure; neuroprotection; antiinflammation; NLRP3 inflammasome; kainic acid; hippocampus

1. Introduction

Epilepsy is a chronic neurological disorder characterized by recurrent, spontaneous, and unpredictable seizures and affects up to 70 million people worldwide [1]. Currently available anti-seizure drugs (ASDs), which are the main treatment for epilepsy, mainly act by blocking Na^+ channels, inhibiting glutamatergic transmission, or enhancing GABAergic transmission [2]. However, long-term treatment with these ASDs is often accompanied by many side effects, and approximately 30% of patients with epilepsy do not respond to these drugs [3,4]. Therefore, there is still a need to search for new, more effective and safer anti-seizure medications. In this context, medicinal plants can potentially play an important role in the development of new ASDs since a diverse group of plant-derived compounds have shown promising anticonvulsant effects in different seizure models [5–9].

Neferine is an alkaloid extracted from the seeds of lotus, *Nelumbo nucifera* Gaertn [10,11], and has been reported to exhibit antioxidant, anti-inflammatory, antithrombotic, antidiabetic, cardioprotective, and antitumor properties [12–17]. In addition, its neuroprotective activity in animal models was also noted. It attenuates brain damage, improves memory and learning abilities, and has antidepressant action [18–23]. However, there is no scientific evidence regarding the anti-seizure effect of neferine. The general objective was therefore to evaluate the action of neferine on kainic acid (KA)-induced seizures in rats and its underlying mechanisms. KA is an analog of glutamate, and the systemic administration of kA to animals causes behavioral seizures and neuropathological lesions which are similar to human epilepsy [24,25]. Therefore, the KA-induced seizure model has been widely recognised as an important experimental model in the research of epilepsy and drug discovery. Our results indicate that pretreatment with neferine significantly attenuated seizure activity, neurotoxicity, and neuroinflammation induced by KA. These beneficial effects are likely mediated by downregulating glutamatergic hyperactivity and the NOD-like receptor 3 (NLRP3)-mediated inflammatory signaling pathway. In fact, experimental models have demonstrated the implication of increased glutamatergic activity and NLRP3 inflammasome activation in the mechanism of epileptogenesis [26–28]. Our findings reveal for the first time that neferine possesses potential as an antiepileptogenic agent as well as efficacy in the management of epilepsy.

2. Results

2.1. Pretreatment with Neferine Attenuates Seizure Activity in KA-Treated Rats

The experimental process is shown in Figure 1A. We first evaluated the anti-seizure effect of neferine. For this, the effect of neferine on seizure activity induced by KA (15 mg/kg, intraperitoneally (i.p.)) was investigated by administering neferine (10 or 50 mg/kg, i.p.) 30 min before KA injection. The dose and schedule of administration were chosen based on our pilot study and others [29,30]. Data analysis by Kruskal–Wallis test yielded statistically significant differences between the tested groups of animals (seizure latency, statistic = 21.9, eta squared = 0.42, $p < 0.001$, n = 9–13/group, Figure 1B; seizure score, statistic = 15.6, eta squared = 0.46, $p < 0.001$, n = 11–13/group, Figure 1C). Dunn's post-hoc test showed that pre-exposure to neferine in KA-treated rats significantly delayed seizure onset and decreased seizure severity compared with the KA alone group ($p < 0.01$; Figure 1B,C). In addition, KA caused 32% mortality in injected rats, and this phenomenon was decreased by 15–18% by neferine pretreatment.

2.2. Pretreatment with Neferine Alleviates Neuronal Damage in the Hippocampi of KA-Treated Rats

Morphological changes in the hippocampus, particularly the CA1 and CA3 regions, have been shown in KA-induced seizure rats [8,31]. To evaluate whether neferine has a neuroprotective effect in KA-treated rats, we performed Nissl staining at 72 h after KA injection to observe the neuronal morphology in the hippocampus. As shown in Figure 2A, neurons in the hippocampi of rats in the control group (rats that received dimethylsulfoxide (DMSO)) were regularly arranged. However, rats in the KA group exhibited disorganized neurons and a considerable loss of neurons in the CA1 and CA3 regions. Neferine pretreatment alleviated these phenomena. One-way ANOVA yielded statistically significant differences between the tested groups of animals (CA1, $F(3, 19) = 48.4$, eta squared = 0.88, $p < 0.001$; CA3, $F(3, 19) = 70.6$, eta squared = 0.91, $p < 0.001$; n = 5–6/group). Tukey post hoc test showed that the number of live neurons in the CA1 and CA3 regions decreased significantly in the KA-treated group compared with the control group ($p < 0.001$). Pre-exposure to neferine in KA-treated rats significantly prevented this reduction compared to the KA alone group ($p < 0.001$) (Figure 2B). In addition, the number of neuronal nuclei (NeuN)-positive cells, which was measured by using NeuN immunohistochemistry in the CA1 and CA3 regions, was counted to investigate neuronal death in the hippocampus (Figure 2A,C). One-way ANOVA yielded statistically significant differences between the tested groups of animals (CA1, $F(3, 21) = 12.9$, eta squared = 0.65,

$p < 0.01$; CA3, F(3, 21) = 25.9, eta squared = 0.79, $p < 0.001$; n = 5–7/group; Figure 2C). Tukey post hoc test showed that the number of NeuN-positive cells in the CA1 and CA3 regions decreased significantly in the KA-treated group compared with the control group ($p < 0.01$). Pre-exposure to neferine in KA-treated rats increased the number of NeuN-positive cells compared to the KA alone group ($p < 0.001$) (Figure 2C). The expression levels of NeuN in the hippocampus were also examined using Western blotting (Figure 2D). One-way ANOVA yielded statistically significant differences between the tested groups of animals (F(3, 16) = 58.8, eta squared = 0.92, $p < 0.001$; n = 5/group). Tukey post hoc test indicated that the expression of NeuN in the hippocampi of KA-treated rats was significantly lower than that in the control group ($p < 0.001$), whereas neferine pretreatment significantly increased the expression of NeuN compared with the KA group ($p < 0.001$) (Figure 2D).

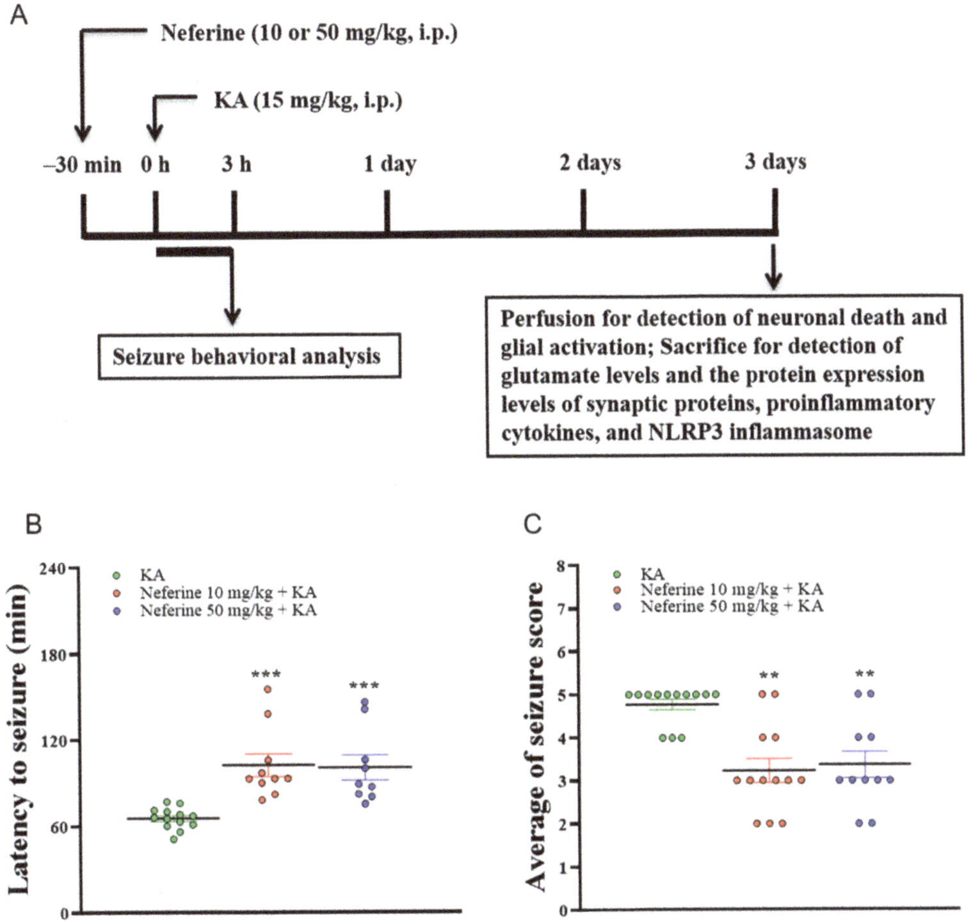

Figure 1. Neferine reduces KA-induced seizure activity. (**A**) Experimental design. (**B**,**C**) Seizure behavior analysis in the different groups (n = 9–13 rats/group). Statistical results showed that neferine increased the latency of seizures (**B**) and decreased seizure score (**C**) (one-way ANOVA). ** $p < 0.01$, *** $p < 0.001$ vs. KA-treated group.

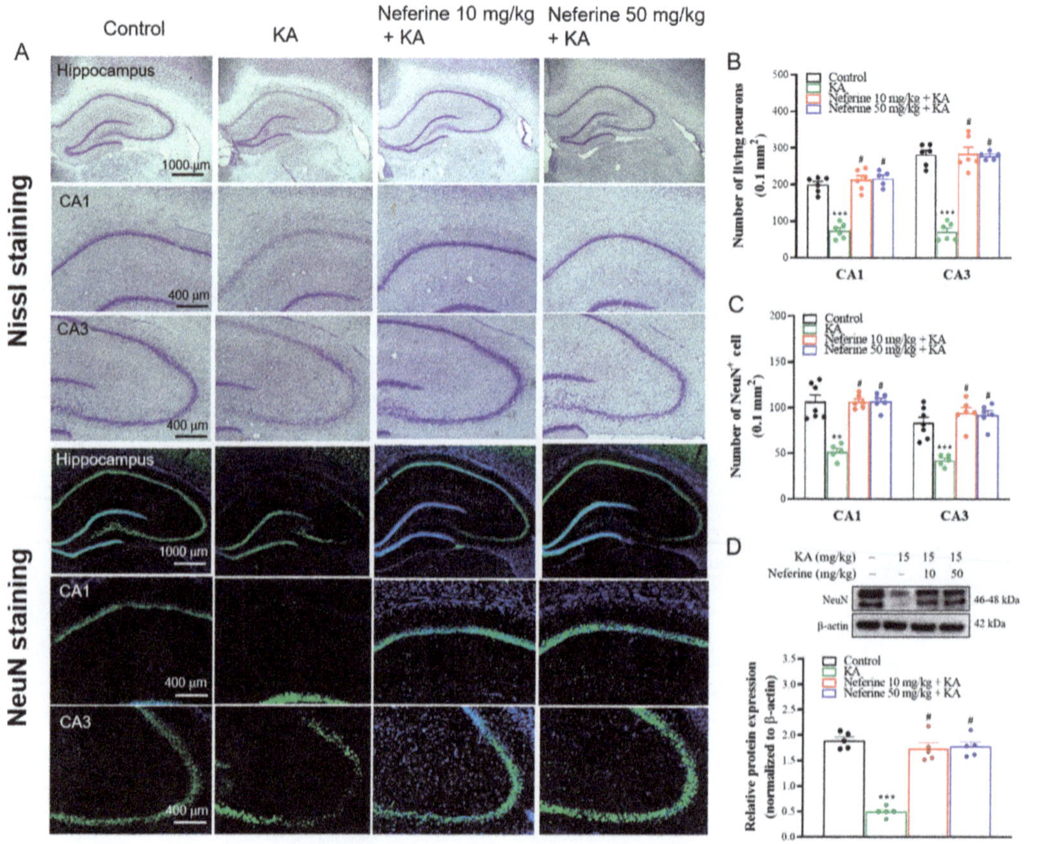

Figure 2. Results of Nissl and NeuN staining in the rat hippocampus. (**A**) Representative images and (**B**,**C**) quantitative data for the number of Nissl- or NeuN-positive hippocampal neurons (n = 5–7 rats/group). Neferine increased the numbers of Nissl- or NeuN-positive hippocampal neurons in the CA1 and CA3 areas (one-way ANOVA). (**D**) Representative Western blot images in the different groups and densitometric values for NeuN were normalized to β-actin levels (n = 5 rats/group). Statistical results of immunoblot analysis show that neferine increased the band intensities of NeuN (one-way ANOVA). ** $p < 0.01$. *** $p < 0.05$ vs. control group. # $p < 0.05$ vs. KA-treated group.

2.3. Pretreatment with Neferine Decreases Glutamate Elevation and Increases the Expression of Synaptic Proteins in the Hippocampi of KA-Treated Rats

Excess glutamate and synaptic dysfunction are observed in the hippocampi of KA-treated rats and are associated with neuronal death [32,33]. Accordingly, we performed high-performance liquid chromatography (HPLC) and Western blot analysis at 72 h after KA injection to evaluate the effect of neferine pretreatment on the concentration of glutamate and the expression of presynaptic protein synaptophysin and postsynaptic density protein 95 (PSD95) in the hippocampus (Figure 3). One-way ANOVA revealed statistically significant differences between the tested groups of animals (glutamate level, $F(3, 16) = 33.5$, eta squared = 0.68, $p < 0.001$, n = 5/group, Figure 3A; synaptophysin, $F(3, 16) = 32.8$, eta squared = 0.86, $p < 0.001$; n = 5/group; PSD-95, $F(3, 16) = 24.4$, eta squared = 0.82, $p < 0.001$; $n = 5$/group; Figure 3B). Tukey post hoc test showed that the concentration of glutamate was increased ($p < 0.001$), whereas the expression levels of synaptophysin ($p < 0.001$) and PSD95 ($p < 0.001$) were decreased in the hippocampi of rats in the KA-treated group compared with the control group. Pre-exposure to neferine in KA-treated rats decreased glutamate levels ($p < 0.001$) but increased synaptophysin and PSD95 expression in comparison with the KA alone group ($p < 0.001$) (Figure 3A,B).

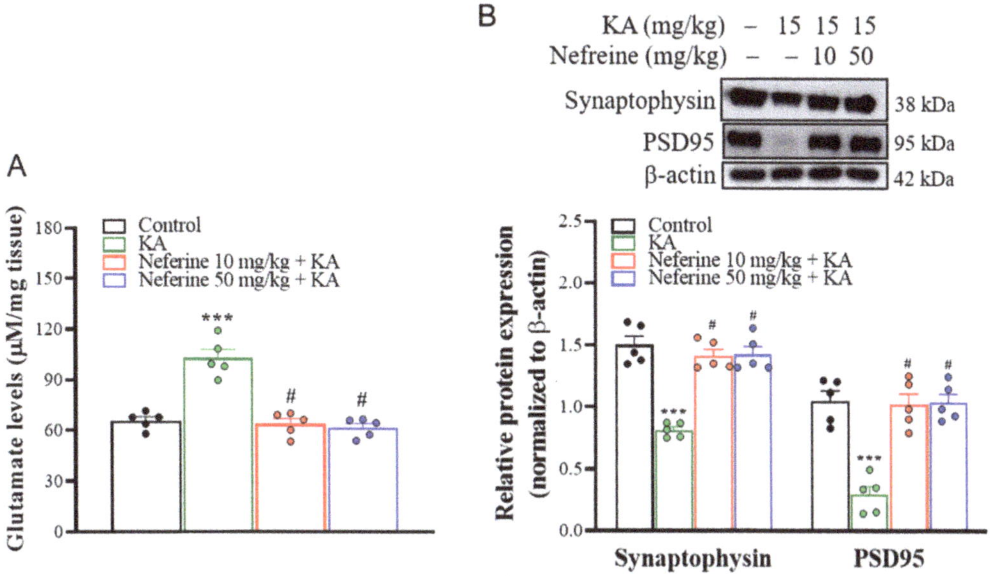

Figure 3. (**A**) HPLC analysis of hippocampal glutamate level in the different groups ($n = 5$ rats/group). Results show that neferine reduced the levels of glutamate (one-way ANOVA). (**B**) Western blot analyses of synaptophysin and PSD95 proteins in rat hippocampal tissue. Representative Western blot images in the different groups and densitometric values for synaptophysin and PSD95 were normalized to β-actin levels ($n = 5$ rats/group). Statistical results of the immunoblot analysis showed that neferine increased the band intensity of synaptophysin and PSD95 (one-way ANOVA). *** $p < 0.05$ vs. control group. # $p < 0.05$ vs. KA-treated group.

2.4. Pretreatment with Neferine Suppresses the Activation of Glial Cells in the Hippocampi of KA-Treated Rats

To gain further insight into the anticonvulsant activity of neferine, we analyzed the activation of glial cells, including astrocytes and microglia, which are known to be involved in the mechanism of epileptogenesis [34]. We assessed the effects 72 h after KA injection by measuring the protein expression level of glial fibrillary acidic protein (GFAP) for astrogliosis and CD11b for microgliosis within the hippocampus (Figure 4). One-way ANOVA yielded statistically significant differences between the tested groups of animals (GFAP, $F(3, 16) = 121.1$, eta squared = 0.96, $p < 0.001$; $n = 5$/group; CD11b, $F(3, 16) = 17.8$, eta squared = 0.77, $p < 0.001$; $n = 5$/group; Figure 4). Tukey post hoc test indicated that KA induced an increase in the expression levels of the GFAP and CD11b proteins in the hippocampus compared with the control group ($p < 0.001$). Pre-exposure to neferine in

KA-treated rats significantly decreased the expression of these two reactive gliosis biomarkers in the hippocampus compared with the KA-treated group ($p < 0.001$; Figure 4). Immunohistochemistry results further revealed that KA-induced gliosis occurred in the hippocampus, as illustrated by the higher numbers of GFAP- and CD11b-positive cells in the CA1 and CA3 regions than in the control group (GFAP: CA1, $F(3, 20) = 27.1$, eta squared = 0.8, $p < 0.001$; CA3, $F(3, 20) = 16.9$, eta squared = 0.72, $p < 0.001$; n = 4–8/group, Figure 5A,B; CD11b: CA1, $F(3, 16) = 123.1$, eta squared = 0.96, $p < 0.001$; CA3, $F(3, 16) = 130.7$, eta squared = 0.96, $p < 0.001$; n = 5/group; Figure 5C,D). Compared with the KA group, KA-treated rats pre-exposed to neferine had reduced numbers of GFAP- and CD11b-positive cells in the CA1 and CA3 regions ($p < 0.001$; Figure 5B,D).

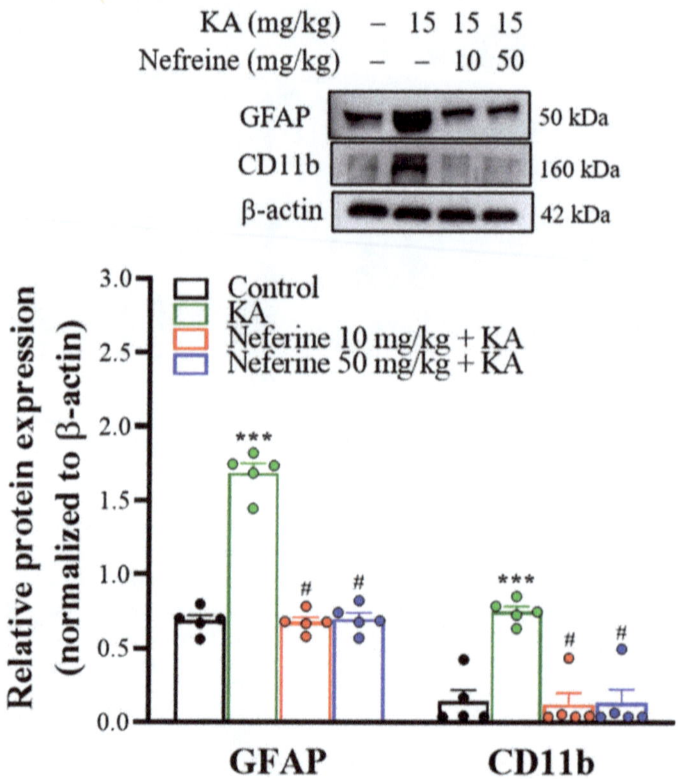

Figure 4. Western blot analyses of GFAP and CD11b proteins in rat hippocampal tissue. Representative images in the different groups and densitometric values for GFAP and CD11b were normalized to β-actin levels (n = 5 rats/group). Statistical results of the immunoblot analysis show that neferine reduced the band intensity of GFAP and CD11b (one-way ANOVA). *** $p < 0.05$ vs. control group. # $p < 0.05$ vs. KA-treated group.

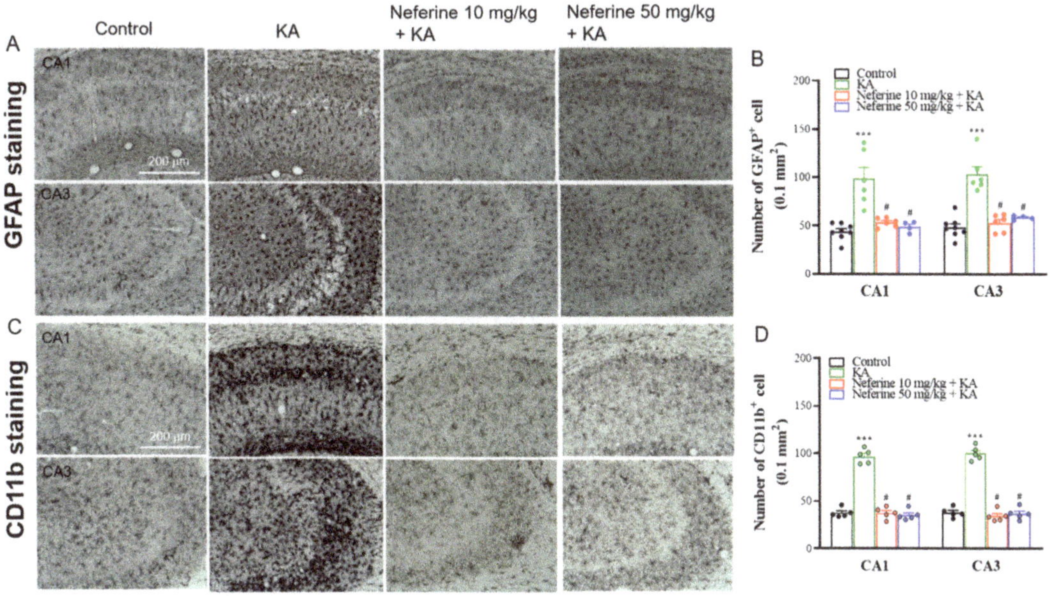

Figure 5. Results of GFAP and CD11b staining in the rat hippocampus. (**A**,**C**) Representative images in the different groups and (**B**,**D**) quantitative data for the number of GFAP- or CD11b-positive hippocampal neurons (n = 4–8 rats/group). Neferine decreased the numbers of GFAP- or CD11b-positive hippocampal neurons in the CA1 and CA3 areas (one-way ANOVA). *** $p < 0.05$ vs. control group. # $p < 0.05$ vs. KA-treated group.

2.5. Pretreatment with Neferine Decreases the Expression of Proinflammatory Cytokines in the Hippocampi of KA-Treated Rats

As KA-induced gliosis is closely associated with the production of proinflammatory cytokines [35,36], we analyzed the protein expression levels of the proinflammatory cytokines interleukin-1β (IL-1β), interleukin-6 (IL-6), and tumor necrosis factor-α (TNF-α) at 72 h after KA injection in the rat hippocampus using Western blot analysis. One-way ANOVA revealed statistically significant differences between the tested groups of animals (IL-1β, $F(3, 16) = 187.1$, eta squared = 0.97, $p < 0.001$; n = 5/group; IL-6, $F(3, 16) = 1651.8$, eta squared = 0.99, $p < 0.001$; n = 5/group; TNF-α, $F(3, 16) = 634.1$, eta squared = 0.99, $p < 0.001$; n = 5/group; Figure 6). Tukey's post hoc test indicated that KA-induced an increase in the expression levels of IL-1β, IL-6, and TNF-α proteins in the hippocampus compared with the control group ($p < 0.001$). Pre-exposure to neferine in KA-treated rats significantly decreased the protein expression levels of these three proinflammatory cytokines in the hippocampus compared with the KA-treated group ($p < 0.001$).

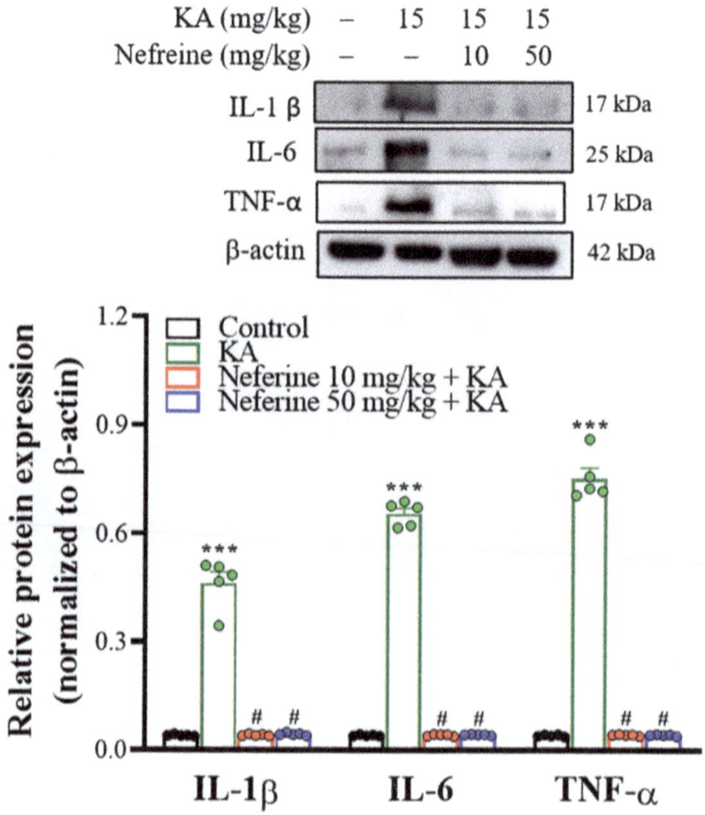

Figure 6. Neferine affects proinflammatory cytokine production in rat hippocampal tissue. Representative Western blot images in the different groups and densitometric values for IL-1β, IL-6, and TNF-α were normalized to β-actin levels (n = 5 rats/group). Statistical results of immunoblot analysis showed that neferine reduced the band intensities of IL-1β, IL-6, and TNF-α (one-way ANOVA). *** $p < 0.05$ vs. control group. # $p < 0.05$ vs. KA-treated group.

2.6. Pretreatment with Neferine Decreases the Activation of the NLRP3 Inflammasome Pathway in the Hippocampi of KA-Treated Rats

To further explore whether the NLRP3 inflammasome pathway is involved in the anti-inflammatory action of neferine, we assessed the expression levels of the NLRP3 inflammasome signaling-related proteins—NLRP3, caspase-1, and interleukin-18 (IL-18) in the hippocampus at 72 h after KA injection. Western blotting analysis with one-way ANOVA revealed statistically significant differences in the expression levels of NLRP3 (F(3, 16) = 463.9, eta squared = 0.99, $p < 0.001$; n = 5/group), caspase-1 (F(3, 16) = 1651.8, eta squared = 0.99, $p < 0.01$; n = 5/group), and IL-18 (F(3, 16) = 634.1, eta squared = 0.93, $p < 0.001$; n = 5/group; Figure 7). Tukey post hoc test indicated that KA induced substantial increase in the protein expression levels of NLRP3, caspase-1, and IL-18 in the hippocampus compared with the control group ($p < 0.001$). Pre-exposure to neferine in KA-treated rats significantly decreased NLRP3, caspase-1, and IL-18 expression in the hippocampus compared with only KA-treated group ($p < 0.001$).

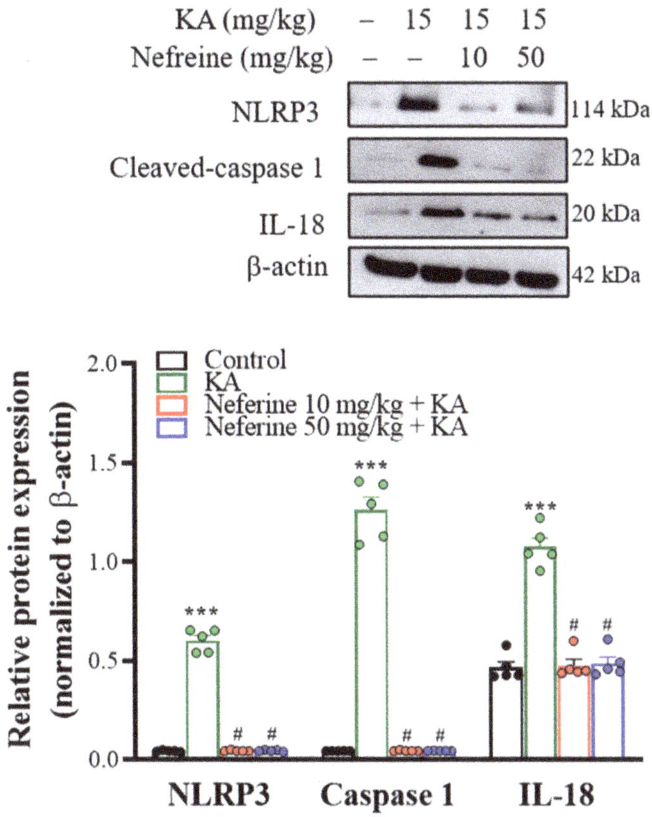

Figure 7. Western blot analyses of NLRP3 inflammasome-related proteins in rat hippocampal tissue. Representative images in the different groups and densitometric values for NLRP3, active caspase-1, and IL-18 were normalized to β-actin levels (n = 5 rats/group). Statistical results of the immunoblot analysis show that neferine reduced the band intensity of NLRP3, active caspase-1, and IL-18 (one-way ANOVA). *** $p < 0.05$ vs. control group. # $p < 0.05$ vs. KA-treated group.

3. Discussion

Epilepsy is a chronic neurological disease that affects millions of people. Current therapy suffers from various limitations, including low efficacy and serious side effects [1,3]. A solution to this problem is to seek out a novel anti-seizure drug from medicinal plants. In fact, plant-derived compounds are known to exhibit anti-seizure activity in animal models with fewer side effects [37]. Therefore, this study aimed to evaluate the effect of neferine pretreatment on seizures induced by the glutamate analog KA in rats. In this model, which mimics the main features of human epilepsy [24], the systemic administration of KA to animals induces tonic-clonic seizures and causes severe loss of neurons and synaptic dysfunction in the central nervous system, particularly in the hippocampus [28,31,32,38]. In the present study, epileptic animals were pretreated with neferine. The results showed that neferine significantly increased seizure latency and decreased seizure severity, demonstrating its anti-seizure activity. Neferine pretreatment also decreased neuronal loss in the CA1 and CA3 regions of the hippocampus, which is prone to damage by KA [39,40]. Furthermore, the decreases in the expression levels of the synaptic proteins synaptophysin and PSD95 in the hippocampi of KA-treated rats were reversed by neferine pretreatment. In fact, downregulation of synaptophysin and PSD95 in the hippocampi of rats treated with KA was associated with the observed hippocampal neuronal loss [31,32]. Therefore, we suggest that neferine can prevent KA-induced seizures and have preventive effects on nerve cell damage and synaptic dysfunction in the hippocampus. Although the ability of neferine to pass the blood-brain barrier remains to be elucidated, its neuroprotective effect on the central nervous system (CNS) has been reported in several animal models of neurological disorders, including cerebral ischemia, Alzheimer's disease, and depression [18,19,21–23].

Various mechanisms have been investigated to understand the etiopathology of KA-seizures; excessive glutamate—a key neurotransmitter in the CNS—has been proposed as one of the main underlying mechanisms [33,41]. Accordingly, the concentration of glutamate in the hippocampus was determined to elucidate the possible mechanism of the anti-seizure and neuroprotective actions of neferine in our study. Consistent with previous studies, KA-treated rats showed a significant increase in hippocampal glutamate concentrations. Interestingly, in KA-treated rats, neferine pretreatment significantly reduced glutamate levels in the hippocampus, suggesting that its anti-seizure and neuroprotective effects might result from the attenuation of glutamatergic hyperactivity. This suggestion was supported by our previous study showing the ability of neferine to inhibit glutamate release from rat cortical nerve terminals [42]. In addition to glutamate, however, further evaluation of the influence of neferine on other neurotransmitter systems, such as γ-aminobutyric acid (GABA) and serotonin, which have been suggested to be involved in epilepsy disorders [43,44], is still warranted.

Suppression of neuroinflammation in the hippocampus of KA-treated rats was another beneficial effect exerted by neferine. Numerous clinical studies and animal experiments have confirmed that the inflammatory response is linked to neuronal death and epileptogenesis [30,42,43]. Glial cells, including astrocytes and microglia, in the CNS, play a key role in the inflammatory reaction. Activation of these cells has been observed in the hippocampi of rats with epilepsy [8,45]. In vivo studies also proved that activated glial cells produce large amounts of proinflammatory cytokines such as TNF-α, IL-1β, and IL-6, which, in turn, cause neuronal damage in the hippocampus of rats with epilepsy [46]. In line with previous studies [8,27,32], we found that treatment of the rats with KA caused significant brain neuroinflammation, as evidenced by increased expression levels of GFAP (an astrocytic marker) and CD11b (a microglial marker), glial cell activation, and proinflammatory cytokine overproduction in the hippocampi of KA-treated rats. These findings suggest the important role of neuroinflammation in the pathophysiology of KA-induced seizures. Additionally, in line with previous studies that have demonstrated that neferine has anti-inflammatory activity [15,16], we found that neferine pretreatment caused significant suppression of neuroinflammation in KA-induced seizure rats, as evidenced by reduced KA-induced gliosis and proinflammatory cytokine production in the hippocampus. Therefore, the ameliorative effect of neferine on inflammation in the hippocampus might participate in its anti-seizure and neuroprotective effects in KA-treated rats.

Previous studies have highlighted the importance of the NLRP3 inflammasome in neuroinflammatory processes during epileptogenesis [27,47–49]. The NLRP3 inflammasome is a multiprotein complex that results from the assembly of NLRP3, apoptosis-associated speck-like protein (ASC), and pro-caspase-1. Activation of the NLRP3 inflammasome causes the activation of caspase-1, which cleaves the inactive proinflammatory cytokines pro-IL-1β and pro-IL-18 into mature IL-1β and IL-18 [50,51]. In animal and clinical epilepsy-related research, increased levels of the NLRP3 inflammasome and neuroinflammatory cytokines have been detected in hippocampal tissues [24,27,46]. Moreover, numerous studies have reported that inhibiting NLRP3 activation and inflammatory cytokine secretion decreases the loss of neurons and the severity of seizures [52,53]. In the present study, the results revealed increases in NLRP3, caspase-1, IL-1β, and IL-18 expression levels in the hippocampi of rats with KA, phenomena that were significantly reversed by neferine pretreatment. Therefore, the suppression of NLRP3 inflammasome activation and downstream inflammatory cytokine secretion by neferine may play a role, at least in part, in the amelioration of seizures and neuronal death in KA-treated rats. Our findings are similar to previous results in which neferine was found to exert neuroprotective effects and effectively suppress NLRP3 inflammatory activation [23,48]. However, how neferine affects NLRP3 inflammasome signaling is still unclear. Further research is necessary to clarify the precise mechanisms.

In the present study, neferine, at 10 and 50 mg/kg, exerts similar effects in all analyzed parameters. We infer that the preventive effects of neferine in KA-induced seizure rats might also occur in a lower dose range. Although we did not examine the effect of neferine at the relatively low dose in this study, our finding is consistent with previous studies, which have reported that neferine at a dose of 12–50 mg/kg attenuates ischemia-induced brain damage in rats [18,19,23].

4. Materials and Methods

4.1. Chemicals and Their Sources

Neferine (purity > 98%, CAS No. 2292-16-2) was purchased from ChemFaces (Wuhan, Hubei, China). KA, DMSO, and other general reagents were purchased from Sigma-Aldrich (St. Louis, MO, USA).

4.2. Animals and Ethics

Male Sprague-Dawley rats weighing 150–200 g were purchased from BioLASCO (Taipei, Taiwan) and were housed under controlled conditions (18–25 °C; 50–60% humidity; 12 h light/dark cycle) with food pellets and water available ad libitum. The animals were used after a one-week period of quarantine and acclimatization. All experimental protocols in this study were approved by the Institutional Animal Care and Use Committee of Fu Jen Catholic University (A10911, 26 August 2020), and the animals were cared for in accordance with the National Institutes of Health Guide for the Care and Use of Laboratory Animals.

4.3. Seizure Induction and Animal Grouping

Seizures were induced by i.p. injection of rats with KA (15 mg/kg), which was dissolved in distilled water. The dose of KA was based on previous studies in KA-treated seizure rats [9,54]. After KA injection, the rats were placed individually in cages and observed for 3 h for the development of seizures. The latency to and intensity of seizures were recorded. Seizure stages were scored according to the scale devised by Racine et al., (1972) [55]: stage 1, facial clonus; stage 2, head nodding; stage 3, forelimb clonus; stage 4, forelimb clonus with rearing; and stage 5, rearing, jumping, and falling. The rats were divided into four groups, including the control (rats-received DMSO), KA, neferine 10 mg/kg + KA, and neferine 50 mg/kg + KA. Neferine was dissolved in a saline solution containing DMSO 1% and administrated (30 μL, i.p.) 30 min before KA injection. 72 h after KA injection, the rats were euthanized using CO_2 or cervical dislocation to obtain the brain tissue samples. Totally 52 rats were used in this study. The seizure behavior was evaluated in all rats. Immunohistochemistry was performed on the fixed brain tissue of 32 rats in four groups (n = 5–8 rats/group). Fresh brain tissue of 20 rats was used for HPLC and Western blotting in four groups (n = 5 rats/group).

4.4. Immunohistochemistry

Seventy-two hours after KA injection, rats were euthanized with CO_2 then perfused intracardially with ice-cold phosphate-buffered saline (PBS) followed by 4% paraformaldehyde. Brains were removed, fixed in 4% paraformaldehyde for 24 h, then underwent dehydration using 30% sucrose, then finally were cut into 25 μm coronal sections (frozen cryosections). For Nissl staining, the sections were mounted on gelatin-coated slides and air-dried overnight. The slides were rehydrated in distilled water and stained in a 0.1% cresyl violet solution (Sigma-Aldrich, St. Louis, MO, USA) for 20 min. After rinsing with distilled water, the slides were gradually dehydrated with a series of alcohols, cleared in xylene, and cover-slipped. Each stained section was observed with a light microscope BX-50 (Olympus, Tokyo, Japan) to assess the degree of nerve cell loss within the hippocampus.

NeuN immunofluorescence staining was performed as described previously [8]. The sections were incubated in phosphate-buffered saline containing 10% normal goat serum for 1 h to block nonspecific antibody binding. The sections were then incubated with NeuN antibody (1:500, Merck Millipore, Billerica, MA, USA) at 4 °C overnight. The next day, the IgG-DyLight 594 (1:1000, Vector Laboratories, Burlingame, CA, USA) was applied for 1 h at room temperature. Finally, 4,6-diamino-2-phenylindole (DAPI) was used for nucleus staining, and sections were mounted onto gelatin-coated glass slides and coverslipped with a mounting medium.

For GFAP and CD11b staining, the sections were treated with primary antibodies (GFAP, 1:1000; Cell Signaling, Beverly, MA, USA; CD11b, 1:500; Abcam, Cambridge, UK) at 4 °C overnight. The sections were then incubated with goat biotinylated anti-mouse secondary antibodies (1:200; Vector Laboratories, Burlingame, CA, USA) for 2 h and subsequently incubated with ExtrAvidin peroxidase (1:1000) for 1h at room temperature. After rinsing in 0.1 M phosphate-buffered saline for 20 min, the sections were reacted with 0.025% 3,3-diaminobenzidine tetrahydrochloride (DAB) solution in phosphate-buffered saline containing 0.0025% hydrogen peroxide for 6 min. The sections were then mounted onto gelatin-coated glass slides and coverslipped with a mounting medium.

Images were captured at 4×, 10×, and 20× magnification using a fluorescence microscope (Zeiss Axioskop 40, Göttingen, Lower Saxony, Germany) under identical settings for each channel. The number of surviving neurons and NeuN-, GFAP-, or CD11b-positive cells were counted in an area of 255 μm × 255 μm of the hippocampal CA1/CA3 in 6 to 8 randomly chosen sections from each animal and averaged for each animal using ImageJ software (Synoptics, Cambridge, Cambridgeshire, UK).

4.5. Quantification of Brain Glutamate Content

The measurement of glutamate levels in the hippocampus is based on a previously described method [54]. In brief, rats were euthanized by cervical dislocation 72 h after KA injection. The hippocampi were rapidly dissected, blotted, weighed, put into 1 mL ice-cold perchloric acid, homogenized, and centrifuged at 10,000× g for 10 min. Subsequently, the supernatant (10 μL) was filtered and injected directly into an HPLC system with electrochemical detection (HTEC500, Eicom, Kyoto, Japan). The relative free glutamate concentration was determined based on peak areas by an external standard method. Glutamate content was expressed as μM/mg of brain tissue.

4.6. Western Blotting

At 72 h after KA injection, the rats were euthanized by cervical dislocation, and the hippocampus was rapidly dissected. Dissected hippocampi were immediately immersed individually in a lysis buffer and sonicated for 10 s. Immunoblotting was performed as described previously [8,54]. Briefly, equal amounts of protein were separated with sodium dodecyl sulfate (SDS)-polyacrylamide gel electrophoresis (Bio-Rad, Hercules, CA, USA) and transferred onto polyvinylidene fluoride membrane by using Trans-Blot Turbo Transfer System (Bio-Rad). The membranes were then incubated overnight at 4 °C with anti-β-actin (1:5000; Cell Signaling, Beverly, MA, USA), anti-NeuN (1:3000; Merck Millipore, Billerica, MA, USA), anti-GFAP (1:5000; Cell Signaling, Beverly, MA, USA), anti-synaptophysin (1:5000; Cell Signaling, Beverly, MA, USA), anti-PSD95 (1:1000; Abcam, Cambridge, Cambridgeshire, UK), anti-CD11b (1:8000; Abcam, Cambridge, Cambridgeshire, UK), anti-IL-1β (1:500; Cell Signaling, Beverly, MA, USA), anti-IL-6 (1:100; Abcam, Cambridge, Cambridgeshire, UK), anti-TNF-α (1:200; Abcam, Cambridge, Cambridgeshire, UK), anti-IL-18 (1:2000; Abcam, Cambridge, Cambridgeshire, UK), anti-NLRP3 (1:3000; Abcam, Cambridge, Cambridgeshire, UK), or anti-cleaved caspase-1 (1:2000; Cell Signaling, Beverly, MA, USA) antibodies. The membrane was then incubated with a horseradish peroxidase-conjugated secondary antibody (1:5000; Genetex, Zeeland, MI, USA) for 2 h at room temperature. After extensive washing, the bands were developed using enhanced chemiluminescence (Amersham ECL Detection Reagents; Cytiva, Little Chalfont, Buckinghamshire, UK). The densities of the bands were measured and analyzed using ImageJ software (Synoptics, Cambridge, Cambridgeshire, UK). To determine the relative band density ratio, all values were normalized against β-actin.

4.7. Data Analysis and Statistics

The data were presented as means ± standard error of the mean (SEM) per group. Data were checked for normal distribution using the Kolmogorov-Smirnov test. Data for Figure 1 were analyzed using Kruskal-Wallis with Dunn post hoc test. Data for Figures 2–7 were analyzed using a one-way analysis of variance (ANOVA) with Tukey's post hoc test. Eta squared was used to assess the effect size [56]. All analyses were conducted using GraphPad Prism 8 (version 8.4.3, GraphPad Software, San Diego, CA, USA). $p < 0.05$ was considered significant.

5. Conclusions

This study introduces new information on the ameliorative effect of neferine on KA-induced seizures in rats. Neferine reduced seizure severity, attenuated neuronal loss, inhibited excess glutamate, and suppressed inflammation in the hippocampi of seizure rats. Suppression of NLRP3 inflammasome activation and inflammatory cytokine secretion is the main mechanism underlying the protective effect of neferine. Therefore, our results demonstrated that neferine might be a potentially valuable resource for the prevention and therapy of epilepsy.

Author Contributions: Conceptualization, T.-Y.L. and C.-W.L.; Data curation, C.-Y.H. and K.-M.C.; Formal analysis, T.-Y.L. and C.-W.L.; Funding acquisition, T.-Y.L., C.-W.L. and S.-J.W.; Investigation, T.-Y.L., C.-W.L. and S.-J.W.; Project administration, K.-M.C. and M.-Y.L.; Resources, C.-Y.H. and K.-M.C.; Supervision, and S.-J.W.; Writing—original draft, S.-J.W.; Writing—review & editing, M.-Y.L. and S.-J.W. All authors have read and agreed to the published version of the manuscript.

Funding: This work was supported by the Ministry of Science and Technology (Grant number MOST 110-2320-B030-005, MOST 110-2314-B-418-008) and Far Eastern Memorial Hospital (Grant number 110-FEMH-FJU-01), Taiwan.

Institutional Review Board Statement: Ethical approval was granted by the Fu Jen Catholic University (A10911, 26 August 2020) and, therefore, experiments were performed in accordance with the ethical standards laid out by the IACUC.

Informed Consent Statement: Not applicable.

Data Availability Statement: The data presented in this study are available on request from the corresponding author.

Conflicts of Interest: The authors declare no conflict of interest.

References

1. Singh, A.; Trevick, S. The Epidemiology of Global Epilepsy. *Neurol. Clin.* **2016**, *34*, 837–847. [CrossRef] [PubMed]
2. Löscher, W.; Klitgaard, H.; Twyman, R.E.; Schmidt, D. New avenues for anti-epileptic drug discovery and development. *Nat. Rev. Drug Discov.* **2013**, *12*, 757–776. [CrossRef] [PubMed]
3. Perucca, P.; Gilliam, F.G. Adverse effects of antiepileptic drugs. *Lancet Neurol.* **2012**, *11*, 792–802. [CrossRef]
4. Tang, F.; Hartz, A.M.S.; Bauer, B. Drug-Resistant Epilepsy: Multiple Hypotheses, Few Answers. *Front. Neurol.* **2017**, *8*, 301. [CrossRef]
5. Auditeau, E.; Chassagne, F.; Bourdy, G.; Bounlu, M.; Jost, J.; Luna, J.; Ratsimbazafy, V.; Preux, P.-M.; Boumediene, F. Herbal medicine for epilepsy seizures in Asia, Africa and Latin America: A systematic review. *J. Ethnopharmacol.* **2019**, *234*, 119–153. [CrossRef]
6. Sharifi-Rad, J.; Quispe, C.; Herrera-Bravo, J.; Martorell, M.; Sharopov, F.; Tumer, T.B.; Kurt, B.; Lankatillake, C.; Docea, A.O.; Moreira, A.C.; et al. A Pharmacological Perspective on Plant-derived Bioactive Molecules for Epilepsy. *Neurochem. Res.* **2021**, *46*, 2205–2225. [CrossRef]
7. He, L.-Y.; Hu, M.-B.; Li, R.-L.; Zhao, R.; Fan, L.-H.; He, L.; Lu, F.; Ye, X.; Huang, Y.-L.; Wu, C.-J. Natural Medicines for the Treatment of Epilepsy: Bioactive Components, Pharmacology and Mechanism. *Front. Pharmacol.* **2021**, *12*, 604040. [CrossRef]

8. Lu, C.-W.; Huang, Y.-C.; Chiu, K.-M.; Lee, M.-Y.; Lin, T.-Y.; Wang, S.-J. Enmein Decreases Synaptic Glutamate Release and Protects against Kainic Acid-Induced Brain Injury in Rats. *Int. J. Mol. Sci.* **2021**, *22*, 12966. [CrossRef]
9. Lu, C.-W.; Lin, T.-Y.; Pan, T.-L.; Wang, P.-W.; Chiu, K.-M.; Lee, M.-Y.; Wang, S.-J. Asiatic Acid Prevents Cognitive Deficits by Inhibiting Calpain Activation and Preserving Synaptic and Mitochondrial Function in Rats with Kainic Acid-Induced Seizure. *Biomedicines* **2021**, *9*, 284. [CrossRef]
10. Huang, Y.; Bai, Y.; Zhao, L.; Hu, T.; Hu, B.; Wang, J.; Xiang, J. Pharmacokinetics and metabolism of neferine in rats after a single oral administration. *Biopharm. Drug Dispos.* **2007**, *28*, 361–372. [CrossRef]
11. Itoh, A.; Saitoh, T.; Tani, K.; Uchigaki, M.; Sugimoto, Y.; Yamada, J.; Nakajima, H.; Ohshiro, H.; Sun, S.; Tanahashi, T. Bisbenzylisoquinoline Alkaloids from Nelumbo nucifera. *Chem. Pharm. Bull.* **2011**, *59*, 947–951. [CrossRef] [PubMed]
12. Pan, Y.; Cai, B.; Wang, K.; Wang, S.; Zhou, S.; Yu, X.; Xu, B.; Chen, L. Neferine enhances insulin sensitivity in insulin resistant rats. *J. Ethnopharmacol.* **2009**, *124*, 98–102. [CrossRef] [PubMed]
13. Zhou, Y.-J.; Xiang, J.-Z.; Yuan, H.; Liu, H.; Tang, Q.; Hao, H.-Z.; Yin, Z.; Wang, J.; Ming, Z. Neferine exerts its antithrombotic effect by inhibiting platelet aggregation and promoting dissociation of platelet aggregates. *Thromb. Res.* **2013**, *132*, 202–210. [CrossRef] [PubMed]
14. Asokan, S.M.; Mariappan, R.; Muthusamy, S.; Velmurugan, B.K. Pharmacological benefits of neferine—A comprehensive review. *Life Sci.* **2018**, *199*, 60–70. [CrossRef]
15. Chiu, K.-M.; Hung, Y.-L.; Wang, S.-J.; Tsai, Y.-J.; Wu, N.-L.; Liang, C.-W.; Chang, D.-C.; Hung, C.-F. Anti-Allergic and Anti-Inflammatory Effects of Neferine on RBL-2H3 Cells. *Int. J. Mol. Sci.* **2021**, *22*, 10994. [CrossRef]
16. Priya, L.B.; Huang, C.; Hu, R.; Balasubramanian, B.; Baskaran, R. An updated review on pharmacological properties of neferine—A bisbenzylisoquinoline alkaloid from Nelumbo nucifera. *J. Food Biochem.* **2021**, *45*, e13986. [CrossRef]
17. Wang, Y.; Wang, S.; Wang, R.; Li, S.; Yuan, Y. Neferine Exerts Antioxidant and Anti-Inflammatory Effects on Carbon Tetrachloride-Induced Liver Fibrosis by Inhibiting the MAPK and NF-κB/IκBα Pathways. *Evid.-Based Complement. Altern. Med.* **2021**, *2021*, 4136019. [CrossRef]
18. Wu, C.; Chen, J.; Yang, R.; Duan, F.; Li, S.; Chen, X. Mitochondrial protective effect of neferine through the modulation of nuclear factor erythroid 2-related factor 2 signalling in ischaemic stroke. *J. Cereb. Blood Flow Metab.* **2019**, *176*, 400–415. [CrossRef]
19. Sengking, J.; Oka, C.; Wicha, P.; Yawoot, N.; Tocharus, J.; Chaichompoo, W.; Suksamrarn, A.; Tocharus, C. Neferine Protects Against Brain Damage in Permanent Cerebral Ischemic Rat Associated with Autophagy Suppression and AMPK/mTOR Regulation. *Mol. Neurobiol.* **2021**, *58*, 6304–6315. [CrossRef]
20. Jung, H.A.; Jin, S.E.; Choi, R.J.; Kim, D.H.; Kim, Y.S.; Ryu, J.H.; Son, Y.K.; Park, J.J.; Choi, J.S. Anti-amnesic activity of neferine with antioxidant and anti-inflammatory capacities, as well as inhibition of ChEs and BACE1. *Life Sci.* **2010**, *87*, 420–430. [CrossRef]
21. Sugimoto, Y.; Furutani, S.; Nishimura, K.; Itoh, A.; Tanahashi, T.; Nakajima, H.; Oshiro, H.; Sun, S.; Yamada, J. Antidepressant-like effects of neferine in the forced swimming test involve the serotonin1A (5-HT1A) receptor in mice. *Eur. J. Pharmacol.* **2010**, *634*, 62–67. [CrossRef] [PubMed]
22. Yin, S.; Ran, Q.; Yang, J.; Zhao, Y.; Li, C. Nootropic effect of neferine on aluminium chloride–induced Alzheimer's disease in experimental models. *J. Biochem. Mol. Toxicol.* **2019**, *34*, e22429. [CrossRef] [PubMed]
23. Zhu, J.-J.; Yu, B.-Y.; Huang, X.-K.; He, M.-Z.; Chen, B.-W.; Chen, T.-T.; Fang, H.-Y.; Chen, S.-Q.; Fu, X.-Q.; Li, P.-J.; et al. Neferine Protects against Hypoxic-Ischemic Brain Damage in Neonatal Rats by Suppressing NLRP3-Mediated Inflammasome Activation. *Oxidative Med. Cell. Longev.* **2021**, *2021*, 6654954. [CrossRef] [PubMed]
24. Upadhya, D.; Kodali, M.; Gitai, D.; Castro, O.W.; Zanirati, G.; Upadhya, R.; Attaluri, S.; Mitra, E.; Shuai, B.; Hattiangady, B.; et al. A Model of Chronic Temporal Lobe Epilepsy Presenting Constantly Rhythmic and Robust Spontaneous Seizures, Co-morbidities and Hippocampal Neuropathology. *Aging Dis.* **2019**, *10*, 915–936. [CrossRef]
25. Costa, A.M.; Lucchi, C.; Simonini, C.; Lustosa, Í.R.; Biagini, G. Status Epilepticus Dynamics Predicts Latency to Spontaneous Seizures in the Kainic Acid Model. *Cell. Physiol. Biochem.* **2020**, *54*, 493–507. [CrossRef]
26. Cristina de Brito Toscano, E.C.; Vieira, É.L.; Dias, B.B.R.; Caliari, M.V.; Gonçalves, A.P.; Giannetti, A.V.; Siqueira, J.M.; Suemoto, C.K.; Leite, R.E.P.; Nitrini, R.; et al. NLRP3 and NLRP1 inflammasomes are up-regulated in patients with mesial temporal lobe epilepsy and may contribute to overexpression of caspase-1 and IL-β in sclerotic hippocampi. *Brain Res.* **2021**, *1752*, 147230. [CrossRef]
27. Mohseni-Moghaddam, P.; Roghani, M.; Khaleghzadeh-Ahangar, H.; Sadr, S.S.; Sala, C. A literature overview on epilepsy and inflammasome activation. *Brain Res. Bull.* **2021**, *172*, 229–235. [CrossRef]
28. Wang, Q.; Yu, S.; Simonyi, A.; Sun, G.Y.; Sun, A.Y. Kainic Acid-Mediated Excitotoxicity as a Model for Neurodegeneration. *Mol. Neurobiol.* **2005**, *31*, 3–16. [CrossRef]
29. Friedman, L.K.; Pellegrini-Giampietro, D.E.; Sperber, E.F.; Bennett, M.V.; Moshe, S.L.; Zukin, R.S. Kainate-induced status epilepticus alters glutamate and GABAA receptor gene expression in adult rat hippocampus: An in situ hybridization study. *J. Neurosci.* **1994**, *14*, 2697–2707. [CrossRef]
30. Spigolon, G.; Veronesi, C.; Bonny, C.; Vercelli, A. c-Jun N-terminal kinase signaling pathway in excitotoxic cell death following kainic acid-induced status epilepticus. *Eur. J. Neurosci.* **2010**, *31*, 1261–1272. [CrossRef]
31. Zhu, X.; Liu, J.; Huang, S.; Zhu, W.; Wang, Y.; Chen, O.; Xue, J. Neuroprotective effects of isoliquiritigenin against cognitive impairment via suppression of synaptic dysfunction, neuronal injury, and neuroinflammation in rats with kainic acid-induced seizures. *Int. Immunopharmacol.* **2019**, *72*, 358–366. [CrossRef] [PubMed]

32. Zhang, F.-X.; Sun, Q.-J.; Zheng, X.-Y.; Lin, Y.-T.; Shang, W.; Wang, A.-H.; Duan, R.-S.; Chi, Z.-F. Abnormal Expression of Synaptophysin, SNAP-25, and Synaptotagmin 1 in the Hippocampus of Kainic Acid-Exposed Rats with Behavioral Deficits. *Cell. Mol. Neurobiol.* **2014**, *34*, 813–824. [CrossRef] [PubMed]
33. Chapman, A.G.; Elwes, R.D.; Millan, M.H.; Polkey, C.E.; Meldrum, B.S. Role of glutamate and aspartate in epileptogenesis; contribution of microdialysis studies in animal and man. *Epilepsy Res. Suppl.* **1996**, *12*, 239–246. [PubMed]
34. Vezzani, A.; Granata, T. Brain Inflammation in Epilepsy: Experimental and Clinical Evidence. *Epilepsia* **2005**, *46*, 1724–1743. [CrossRef]
35. Kim, D.H.; Yoon, B.H.; Jung, W.Y.; Kim, J.M.; Park, S.J.; Park, D.H.; Huh, Y.; Park, C.; Cheong, J.H.; Lee, K.-T.; et al. Sinapic acid attenuates kainic acid-induced hippocampal neuronal damage in mice. *Neuropharmacology* **2010**, *59*, 20–30. [CrossRef]
36. Lu, C.W.; Hsieh, H.L.; Lin, T.Y.; Hsieh, T.Y.; Huang, S.K.; Wang, S.J. Echinacoside, an Active Constituent of Cistanche Herba, Exerts a Neuroprotective Effect in a Kainic Acid Rat Model by Inhibiting Inflammatory Processes and Activating the Akt/GSK3β Pathway. *Biol. Pharm. Bull.* **2018**, *41*, 1685–1693. [CrossRef]
37. Liu, Y.; Wang, S.; Kan, J.; Zhang, J.; Zhou, L.; Huang, Y.; Zhang, Y. Chinese Herbal Medicine Interventions in Neurological Disorder Therapeutics by Regulating Glutamate Signaling. *Curr. Neuropharmacol.* **2020**, *18*, 260–276. [CrossRef]
38. Rusina, E.; Bernard, C.; Williamson, A. The Kainic Acid Models of Temporal Lobe Epilepsy. *eNeuro* **2021**, *8*, 0337-20.2021. [CrossRef]
39. Bahn, S.; Volk, B.; Wisden, W. Kainate receptor gene expression in the developing rat brain. *J. Neurosci.* **1994**, *14*, 5525–5547. [CrossRef]
40. Ben-Ari, Y. Limbic seizure and brain damage produced by kainic acid: Mechanisms and relevance to human temporal lobe epilepsy. *Neuroscience* **1985**, *14*, 375–403. [CrossRef]
41. Chittajallu, R.; Vignes, M.; Dev, K.K.; Barnes, J.M.; Collingridge, G.L.; Henley, J.M. Regulation of glutamate release by presynaptic kainate receptors in the hippocampus. *Nature* **1996**, *379*, 78–81. [CrossRef] [PubMed]
42. Yeh, K.C.; Hung, C.F.; Lin, Y.F.; Chang, D.C.; Pai, M.S.; Wang, S.J. Neferine, a bisbenzylisoquinoline alkaloid of Nelumbo nucifera, inhibits glutamate release in rat cerebrocortical nerve terminals through 5-HT1A receptors. *Eur. J. Pharmacol.* **2020**, *889*, 173589. [CrossRef] [PubMed]
43. Murugesan, A.; Rani, M.R.S.; Hampson, J.; Zonjy, B.; Lacuey, N.; Faingold, C.L.; Friedman, D.; Devinsky, O.; Sainju, R.K.; Schuele, S.; et al. Serum serotonin levels in patients with epileptic seizures. *Epilepsia* **2018**, *59*, e91–e97. [CrossRef] [PubMed]
44. Akyuz, E.; Polat, A.K.; Eroglu, E.; Kullu, I.; Angelopoulou, E.; Paudel, Y.N. Revisiting the role of neurotransmitters in epilepsy: An updated review. *Life Sci.* **2020**, *265*, 118826. [CrossRef] [PubMed]
45. Penkowa, M.; Florit, S.; Giralt, M.; Quintana, A.; Molinero, A.; Carrasco, J.; Hidalgo, J. Metallothionein reduces central nervous system inflammation, neurodegeneration, and cell death following kainic acid-induced epileptic seizures. *J. Neurosci. Res.* **2005**, *79*, 522–534. [CrossRef] [PubMed]
46. Vezzani, A.; Friedman, A.; Dingledine, R. The role of inflammation in epileptogenesis. *Neuropharmacology* **2013**, *69*, 16–24. [CrossRef]
47. Swanson, K.V.; Deng, M.; Ting, J.P.-Y. The NLRP3 inflammasome: Molecular activation and regulation to therapeutics. *Nat. Rev. Immunol.* **2019**, *19*, 477–489. [CrossRef]
48. Wu, C.; Zhang, G.; Chen, L.; Kim, S.; Yu, J.; Hu, G.; Chen, J.; Huang, Y.; Zheng, G.; Huang, S. The Role of NLRP3 and IL-1β; in Refractory Epilepsy Brain Injury. *Front. Neurol.* **2020**, *10*, 1418. [CrossRef]
49. Yue, J.; Wei, Y.J.; Yang, X.; Liu, S.; Yang, H.; Zhang, C. NLRP3 inflammasome and endoplasmic reticulum stress in the epileptogenic zone in temporal lobe epilepsy: Molecular insights into their interdependence. *Neuropathol. Appl. Neurobiol.* **2020**, *46*, 770–785. [CrossRef]
50. Schroder, K.; Tschopp, J. The Inflammasomes. *Cell* **2010**, *140*, 821–832. [CrossRef]
51. Shao, B.-Z.; Xu, Z.-Q.; Han, B.-Z.; Su, D.-F.; Liu, C. NLRP3 inflammasome and its inhibitors: A review. *Front. Pharmacol.* **2015**, *6*, 262. [CrossRef] [PubMed]
52. Meng, X.-F.; Tan, L.; Tan, M.-S.; Jiang, T.; Tan, C.-C.; Li, M.-M.; Wang, H.-F.; Yu, J.-T. Inhibition of the NLRP3 inflammasome provides neuroprotection in rats following amygdala kindling-induced status epilepticus. *J. Neuroinflamm.* **2014**, *11*, 212. [CrossRef] [PubMed]
53. Rong, S.; Wan, D.; Fan, Y.; Liu, S.; Sun, K.; Huo, J.; Zhang, P.; Li, X.; Xie, X.; Wang, F.; et al. Amentoflavone Affects Epileptogenesis and Exerts Neuroprotective Effects by Inhibiting NLRP3 Inflammasome. *Front. Pharmacol.* **2019**, *10*, 856. [CrossRef]
54. Lu, C.-W.; Lin, T.-Y.; Chiu, K.-M.; Lee, M.-Y.; Huang, J.-H.; Wang, S.-J. Silymarin Inhibits Glutamate Release and Prevents against Kainic Acid-Induced Excitotoxic Injury in Rats. *Biomedicines* **2020**, *8*, 486. [CrossRef] [PubMed]
55. Racine, R.J. Modification of seizure activity by electrical stimulation: II. Motor seizure. *Electroencephalogr. Clin. Neurophysiol.* **1972**, *32*, 281–294. [CrossRef]
56. Lakens, D. Calculating and reporting effect sizes to facilitate cumulative science: A practical primer for *t*-tests and ANOVAs. *Front. Psychol.* **2013**, *4*, 863. [CrossRef]

International Journal of
Molecular Sciences

MDPI

Review

COVID-19 and One-Carbon Metabolism

Joanna Perła-Kaján [1] and Hieronim Jakubowski [1,2,*]

[1] Department of Biochemistry and Biotechnology, University of Life Sciences, 60-632 Poznań, Poland; kajan@up.poznan.pl
[2] Department of Microbiology, Biochemistry and Molecular Genetics, Rutgers-New Jersey Medical School, Newark, NJ 07103, USA
* Correspondence: jakubows@rutgers.edu

Abstract: Dysregulation of one-carbon metabolism affects a wide range of biological processes and is associated with a number of diseases, including cardiovascular disease, dementia, neural tube defects, and cancer. Accumulating evidence suggests that one-carbon metabolism plays an important role in COVID-19. The symptoms of long COVID-19 are similar to those presented by subjects suffering from vitamin B_{12} deficiency (pernicious anemia). The metabolism of a cell infected by the SARS-CoV-2 virus is reshaped to fulfill the need for massive viral RNA synthesis, which requires de novo purine biosynthesis involving folate and one-carbon metabolism. Many aspects of host sulfur amino acid metabolism, particularly glutathione metabolism underlying antioxidant defenses, are also taken over by the SARS-CoV-2 virus. The purpose of this review is to summarize recent findings related to one-carbon metabolism and sulfur metabolites in COVID-19 and discuss how they inform strategies to combat the disease.

Keywords: folate; purine biosynthesis; methionine; S-adenosylmethionine; S-adenosylhomocysteine; homocysteine; cysteine; glutathione; choline; methionine sulfoxide

Citation: Perła-Kaján, J.; Jakubowski, H. COVID-19 and One-Carbon Metabolism. *Int. J. Mol. Sci.* **2022**, *23*, 4181. https://doi.org/10.3390/ijms23084181

Academic Editor: María A. Pajares

Received: 11 March 2022
Accepted: 7 April 2022
Published: 10 April 2022

1. Introduction

Severe acute respiratory syndrome coronavirus 2 (SARS-CoV-2) is an enveloped (+)ss-RNA virus, in which the infecting RNA acts as a messenger RNA (mRNA). After entering the host cell, SARS-CoV-2 is replicated. This process involves the translation of viral mRNA by cellular ribosomes to produce the viral replicative enzymes, which generate new RNA genomes and the mRNAs for the synthesis of the components necessary to assemble the new viral particles [1].

SARS-CoV-2 is the cause of a global pandemic of coronavirus disease of 2019 (COVID-19). On 31 December 2019, the World Health Organization's (WHO) country office in China registered cases of 'viral pneumonia' in Wuhan. A month later, on 30 January 2020, WHO's Director-General declared the novel coronavirus outbreak a public health emergency of international concern. On 11 March 2020, WHO made the assessment that COVID-19 could be characterized as a pandemic. Globally, as of 4 April 2022, there have been 494 million confirmed cases of COVID-19, including 6.15 million deaths. The highest number of COVID-19 cases, some 80.1 million, were in the United States, and included 0.98 million deaths. The scale of health and economical threats caused by the pandemic outbreak urged many scientific groups to research the mechanisms triggered by the virus to allow treatment and vaccination. Several SARS-CoV-2 variants have emerged since the first identified strain, apparently with higher transmissibility/virulence and immune escape capabilities.

Interestingly, COVID-19 patients present a diverse severity of clinical manifestations, ranging from no symptoms to death. Of the total COVID-19 cases, about 80% are either asymptomatic or experience a mild course of the disease, while about 14% develop severe symptoms, such as pneumonia, and about 5% present critical symptoms, such as septic shock, respiratory failure, or multi organ failure, and finally about 2% of the subjects die.

In general, the worse course of the disease is associated with old age and comorbidities, especially chronic obstructive lung disease, obesity, diabetes mellitus, cardiovascular disease and hypertension [2].

To complicate matters, in a considerable fraction of patients, SARS-CoV-2 infection is followed by a complication called long COVID-19, which can last for months and has diverse symptoms such as fatigue, headache, 'brain fog', anosmia, myalgia, dizziness, breathlessness, palpitations, and gastrointestinal problems. The prevalence of long COVID-19 is based on ten reporting studies, and ranged from 4.7% to 80%. The frequency of most prevalent long COVID-19 symptoms that may last from weeks to months after acute infection was as follows: chest pain—up to 89%, fatigue—up to 65%, dyspnea—up to 61%, cough and sputum production—up to 59%, cognitive and memory impairment—up to 57.1%, arthralgia—up to 54.7%, sleep disorders—up to 53%, and myalgia—up to 50.6%. The list of other signs and symptoms of long COVID-19 with lower frequency contains over thirty records, all of which are listed with reporting studies [3].

The progression of COVID-19 can be divided into three overlapping phases: early infection, pulmonary phase and hyperinflammation [4]. As the lung parenchyma is targeted by the virus, the organism activates innate immune response, and the following effects may be triggered: inflammation, damage to the vessel walls, vasodilation and endothelial permeability, pulmonary restriction, hypoxemia, and increased cardiovascular stress. Respiratory failure, if present, together with viral infiltration into myocardial tissue and cardiac inflammation leads to cardiac injury [4]. Kumar et al. [2] have described the COVID-19 mechanisms in the human body, including symptomatology, virus–host interactions, and host factors affecting transmissibility, severity and outcomes (age, sex and comorbidities) as well as organ-specific pathologies ongoing in the respiratory, cardiovascular, renal, digestive, and nervous systems during SARS-CoV-2 infection [2].

A summary of over twenty proteomic studies on plasma and serum of COVID-19 patients revealed three deregulated KEGG pathways: complement and coagulation cascades, cytokine-cytokine receptor interaction and cholesterol metabolism [5]. Elevations of inflammation biomarkers such as IL (interleukin)-6, IL-2, IL-7, TNF (tumor necrosis factor), MCP (monocyte chemoattractant protein)-1, MIP (macrophage inflammatory protein)-1, G-CSF (granulocyte-colony stimulating factor), CRP (C-reactive protein), procalcitonin and ferritin are associated with increased mortality [4] and higher disease severity [6]. The results of a cohort study with 84 patients diagnosed with COVID-19 from Wuhan, China, demonstrated that the level of cardiac enzymes, as well as the abnormalities in the ECG, correlate positively with the level of inflammation values, in particular CRP and procalcitonin [7]. As shown recently, multi organ failure in patients with severe COVID-19 complication is caused by systemic vasculitis and cytokine mediated coagulation. Other identified biomarkers are hematological (lymphocyte count, neutrophil count, neutrophil-lymphocyte ratio), erythrocyte sedimentation rate, D-dimer, troponin, creatine kinase, and aspartate aminotransferase. Homocysteine (Hcy) and angiotensin II were also suggested to play significant roles [8].

The symptoms of long COVID-19 are similar to those presented by subjects suffering from pernicious anemia (a condition caused by vitamin B_{12} deficiency), where methylation status is compromised [9,10]. Vitamin B_{12} is a cofactor of the key one-carbon metabolism enzyme—vitamin B_{12}-dependent methionine (Met) synthase (MS) that remethylates Hcy to Met and links Met and folate cycles (Figure 1). MS generates Met, which then is used for the production of S-adenosylmethionine (SAM), a universal methyl donor, for a variety of acceptors, many of which participate in epigenetic regulation of gene expression. Moreover, one-carbon metabolism supports multiple physiological processes, such as biosynthesis of purines and thymidine, amino acid homeostasis of glycine (Gly), serine, and Met, and underlies antioxidant defense via glutathione (GSH) (Figure 1). Additionally, one-carbon metabolism is also important in the generation of energy via adenosine triphosphate (ATP) production in the mitochondria [9–11].

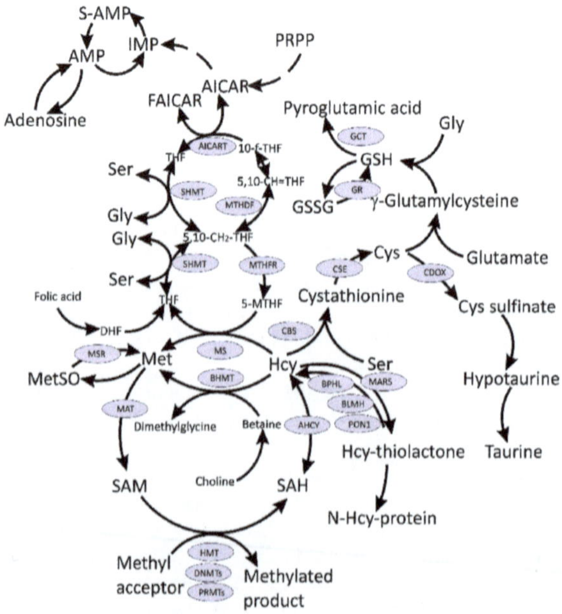

Figure 1. Interconnections between metabolism of folate, one-carbon, and sulfur compounds. Indicated metabolites are discussed in the text. CysGly, a product of GSH catabolism, affected in COVID-19 and discussed in the text, is not shown.

Accumulating evidence suggests that one-carbon metabolism plays an important role in COVID-19. The purpose of this review is to summarize recent findings related to sulfur amino acids and one-carbon metabolism in COVID-19 and discuss how they inform strategies to combat the disease.

2. SARS-CoV-2 Hijacks Host Folate and One-Carbon Metabolism

Metabolism of a cell infected by a virus is reshaped to fulfill the viral needs for its successful replication. Under viral infection, anabolic reactions are dominant and there is an upregulation of the ingestion of an extracellular carbon source (e.g., glucose or glutamine), which is used for lipogenesis and nucleotide synthesis, both of which are crucial for viral replication [1].

Several amino acids and their derivatives play important roles in one-carbon metabolism (Figure 1). Activated Met in the form of SAM is a universal methyl donor, Hcy is an intermediate in Cys synthesis, which in turn is needed for glutathione synthesis responsible for redox homeostasis. Ser is the precursor for the biosynthesis of several amino acids including Gly and Cys and participates in folate metabolism by donating one-carbon units for the biosynthesis of purines and pyrimidines.

Insight into how SARS-CoV-2 remodels the host metabolism to support the virus replication has been provided by ex vivo studies using the African Green Monkey Vero E6 cells [12]. Upon infection with SARS-CoV-2 isolate, the cells rapidly produced viral genomic RNA and nucleocapsid protein by 8 h post-infection, at which time the induction of antiviral genes, NF-kB targets, and ER stress response was observed. These changes were accompanied by a global decrease in host mRNAs with no changes in levels of mRNAs encoding metabolic enzymes. The changes in the patterns of mitochondrial DNA expression indicated ATP depletion. One of the most striking changes was the elevation of *de novo* purine synthesis intermediates (PPRP, FGAR, AIR, SAICAR) (Figure 1) in the SARS-CoV-2-infected cells and the reduction of intracellular folate levels (Table 1). Other metabolites participating in one-carbon metabolism and sulfur amino acid metabolism were also affected by the SARS-CoV-2 infection (Supplementary Data 4 in Ref. [12]). For example, intracellular Met, cystathionine, pyridoxine, betaine, serine, Gly, 5-oxoproline (pyroglutamate), and cysteine-glutathione disulfide levels were attenuated, while reduced glutathione levels were elevated. Intracellular SAM, SAH, cysteine (Cys), oxidized glutathione (GSSG) levels were not affected in the SARS-CoV-2-infected cells. These findings suggest that SARS-CoV-2 hijacks folate and one-carbon metabolism to meet the demands for viral replication [12]. As shown in Table S1 and discussed below, these metabolites were also affected by SARS-CoV-2 infection in vivo in COVID-19 patients.

Table 1. Intracellular metabolites affected by the SARS-CoV-2 infection of Vero E6 cells.

Metabolite	Fold Change [1]	p Value [2]
Purine biosynthesis		
Folate	0.62	0.0020
5-Formimino-tetrahydrofolate	0.18	0.0018
Serine	0.87	0.0029
Glycine	0.71	0.0025
Ribose-5-Phosphate/Xylulose-5-phosphate	0.91	0.405
5-Phosphoribos-1-pyrophosphate (PRPP)	1.44	0.005
Formylglycinamide ribonucleotide (FGAR)	2.38	2×10^{-6}
Aminoimidazole ribonucleotide (AIR)	3.10	0.0113
Succinylaminoimidazolecarboxyamide (SAICAR)	1.24	0.0218
Methionine cycle		
Methionine	0.68	0.0020
S-Adenosylmethionine (SAM)	1.01	NS
S-Adenosylhomocysteine (SAH)	1.19	NS
Trans-sulfuration pathway		
Cystathionine	0.70	0.0507
Cysteine	0.80	NS
Glutathione biosynthesis		
Pyroglutamate/5-Oxoproline	0.73	0.0028
Glutathione, reduced (GSH)	1.71	0.0012
Glutathione, oxidized (GSSG)	0.97	NS
Cysteine-glutathione disulfide	0.34	0.0025
Taurine biosynthesis		
Cysteinesulfinic acid	3.10	0.0291
Choline	1.33	0.0014
Betaine	0.78	0.0497

[1] Recalculated from Supplementary Data 4 in Ref. [12]. [2] t-test, 2-sided, unequal variance.

3. S-Adenosylmethionine and Methylation Index

The SAM/SAH ratio, known as the methylation index, may be affected by SARS-CoV-2 infection. As mentioned earlier, SAM is required for capping of the viral RNA. The RNA cap (m7GpppN-RNA) is composed of a 7-methylguanosine (m7G) linked to the 5′-nucleoside (N) of the RNA chain through a triphosphate bridge (ppp). The cap structure is methylated at the N7 position of the guanosine by the C-terminal (guanine-N7)-methyltransferase (N7-MTase) domain of nonstructural protein 14 (Nsp14), forming cap-0 (m7GpppN-RNA), using SAM as a methyl donor [13,14]. The second methylation reaction during cap synthesis is catalyzed by SAM-dependent Nsp16 methyltransferase, which adds the methyl group on the ribose 2′-O position of the first transcribed nucleotide to form cap-1 (m7GpppNm-RNA). The RNA final cap has several important biological roles in viruses as it is critical for the stability of mRNAs, both for their translation and to evade the host immune response [14].

It has been hypothesized that SARS-CoV-2 infection may lead to SAM depletion in patients suffering long-term consequences of COVID-19. However, although SAM has not been quantified in long COVID-19, this hypothesis doesn't seem to hold much water because several studies showed significant increases or no changes in plasma SAM levels in COVID-19 cases (Table S1). Moreover, SAH levels are either elevated [15,16], attenuated [17] or do not change in COVID-19 [18]. For example, a study on fifty-six COVID-19 patients admitted to the hospital between September and December 2020 in Moscow, Russia, has shown that an elevated SAM level and SAM/SAH and SAM/glutathione ratios have been associated with an increased risk of severe lung injury. Furthermore, an elevated SAM concentration and SAM/SAH and SAM/GSF ratios have been associated with an increased risk of lung damage [18]. Metabolomic analyses have revealed that SAM was significantly elevated in critical cases of COVID-19 [15] and those with a fatal outcome [19] as compared to control, mild and moderate cases of COVID-19 [15] (Table S1). Even though SAM was highest among severe COVID-19 patients, it was associated with a favorable prognosis. On the other hand, while there was no association between dimethylglycine, a by-product of Hcy remethylation to Met by a betaine-dependent enzyme BHMT (Figure 1), and COVID-19 stage, dimethylglycine was significantly lower in patients with an unfavorable progression of COVID-19 [15].

4. Methionine and Methionine Sulfoxide

The results of metabolomic studies on Met are contradictory, showing upregulation [16], downregulation [20,21], or no change in Met levels [17,22] in COVID-19 cases vs. healthy controls (Table S1). The direction of changes in Met level depends on compared groups, i.e., there is a tendency to higher Met levels in critical COVID-19 patients vs. healthy controls, but Met was lower in mild COVID-19 patients vs. healthy controls and there was no change in Met levels in patients with moderate COVID-19 vs. controls [15].

Metabolomic analyses of blood samples from COVID-19 patients and COVID-19-negative subjects revealed the significant impact of SARS-CoV-2 infection on serum Met sulfoxide, which consistently showed increased levels in four independent studies, suggesting increased oxidant stress [16,17,23,24] (Table S1).

5. Glutathione and Related Metabolites

COVID-19 is associated with disrupted redox homeostasis and reactive oxygen species (ROS) accumulation. In May 2020, Polonikov published a hypothesis which stated that [25]: "glutathione deficiency is the most plausible explanation for serious manifestation and death in COVID-19 patients". Glutathione (GSH) depletion has been observed in diseases that increase the risk of COVID-19 [26]. GSH, being the main antioxidant agent, was suggested to be essential for counterbalancing the inflammation observed in SARS-CoV-2 infected patients (reviewed in Ref. [27]).

The glutathione hypothesis of COVID-19 appears to be supported by available data (Table S1). Indeed, GSH levels are consistently decreased in COVID-19 patients [17,18,28,29]. For example, a study on fifty-nine COVID-19 patients admitted to the hospital between August and November 2020 in Moscow, Russia [29], found that the levels of total GSH (tGSH) were significantly lower in moderate and severe COVID-19 patients compared with mildly affected subjects, while reduced CysGly (rCG) was significantly decreased in patients with higher degrees of lung damage based on percentage of lobar involvement (>26%) as compared to subjects with a lower degree of lung damage (0–25% of lobar involvement) (Table 2). tGSH and rCG were suggested to be risk markers for the severity of COVID-19 and lung damage in patients [29]. In addition, a negative correlation between rGSH and advance oxidation protein products in patients with high lung damage was observed [29]. A similar study involving fifty-six COVID-19 patients admitted to the hospital between September and December 2020 in Moscow, Russia [18], found lower GSH concentration in patients with a higher degree of lung damage (>50% of lobar involvement) as compared to patients with a lower degree of lung injury (<25% of lobar involvement). There also has been a significant increase in SAM level and SAM/GSH ratios, and a tendency to higher Hcy levels in subjects with more injured lungs [18] (Table 2).

Table 2. Sulphur metabolites in COVID-19 patients stratified by a degree of lung damage.

Metabolite	Degree of Lung Damage			References
	CT0–1, <5–25% (n = 26)	CT2, 26–49% (n = 16–18)	CT3–4, 50–75% (n = 14–15)	
tGSH, μM	1.81	1.15	1.22 *	[29]
rCG, μM	1.59	1.30	1.29 *	
GSH, μM	1.81	1.15	1.22 #	[18,29]
Hcy, μM	7.4	8.3	9.1	[18]
SAM, nM	59	57	84 #	
SAM/GSH, nM/μM	3.6	7.2 &	5.5	
SAM/GSH, nM/μM	32	57	60 #	

* $p < 0.05$ CT0,1 vs. CT2–4; # $p < 0.05$ CT3,4 vs. CT0,1; & $p < 0.05$ CT2 vs. CT0,1. CT, computer tomography.

Another study involving sixty COVID-19 patients hospitalized in Houston, TX and twenty-four uninfected controls, found that total and reduced red blood cells GSH were significantly lower in COVID-19 patients then in controls. At the same time, measures of lipid peroxidation, indicating oxidative stress (TBARS and F2-isoprostanes), were significantly elevated in COVID-19 patients and increased with age [28]. Severe GSH deficiency and oxidative damage also occur in young COVID-19 patients, and the magnitude of these defects in COVID-19 increased with age (Figure 2).

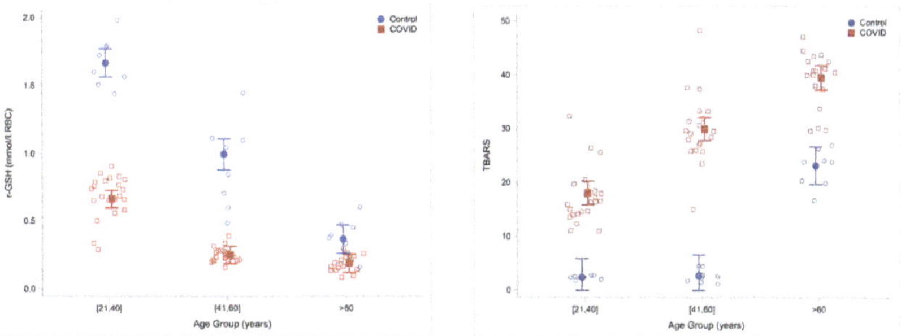

(A) (B)

Figure 2. Reduced glutathione concentrations (**A**) and increased oxidative stress (TBARS) (**B**) in COVID-19 patients and uninfected controls stratified by age. Reproduced from Ref. [28] with permission.

Cys levels were elevated in plasma [24], decreased in serum [21] and plasma [20] or unchanged in serum [22] and plasma [15]. Cystine, the oxidized disulfide form of Cys, was elevated in serum and plasma in several studies [16,21,22] and downregulated with IL-6 increase [23]. In another study, Cys was elevated in moderate COVID-19 patients vs. controls; however, analysis of critical and mild COVID-19 patients showed no changes in Cys levels vs. control group [15]. Cystathionine, an intermediate in the transsulfuration pathway (Figure 1), was either upregulated in plasma [30] and serum [17] or downregulated in the plasma of COVID-19 patients [24]. CysGly, a product of GSH metabolism, was shown to decrease in patients with higher degree of lung damage as diagnosed by computer tomography [29] or increase in plasma of COVID-19 patients as compared with controls [16].

Gly, which participates in GSH biosynthesis, was found to either increase [22], decrease [15,23], or was unchanged [16,17,21,31] in COVID-19 patients. In addition, pyroglutamate, a metabolite that forms from γ-glutamyl-Cys (glutathione precursor) (Figure 1) when Gly is limiting, was shown to be downregulated in COVID-19 patients [17] (Table S1).

6. Homocysteine

Hyperhomocysteinemia (HHcy) is linked to more than a hundred diseases and outcomes and has numerous detrimental effects, including neurotoxic, neuroinflammatory, neurodegenerative, proatherogenic, prothrombotic, and prooxidative. HHcy, a risk factor for cardiovascular disease, may be caused by the C667 > T MTHFR mutation, but also by reduced levels of folic acid and other B-vitamins [32]. The association of cardiovascular damage with COVID-19 and the fact that ischemic heart disease and hypertension are among the most frequent pre-existing comorbidities in COVID-19 patients, led to a suggestion that plasma Hcy can be "a potential marker for severe disease in SARS-CoV-2 patients" [33].

A recent study found that Hcy was significantly elevated in mild (12.73 μM, $n = 74$) and severe (15.62 μM, $n = 43$) COVID-19 cases compared with healthy controls (8.17 μM, $n = 34$). In ROC analysis, Hcy cutoff values of 9.18 μM (sensitivity 76.7%, specificity 76.5%, AUC = 0.951) and 10.3 μM (sensitivity 81.8%, specificity 82.4%, AUC = 1.000) identified mild and severe COVID-19 cases, respectively [34]. In multivariate logistic regression analysis, Hcy and two other variables (monocyte/lymphocyte ratio and fibrin D-dimer) were associated with mild and severe COVID-10. Other relevant variables that are known to be associated with Hcy levels, such as age and sex, have not been included in these analyses. Because there were significant differences in age and sex between COVID-19 cases (mild: 57.96 years, 52.5% male; severe: 71.22 years, 62.8% male) and healthy controls (32.65 years, 20.6% [34], it is not clear whether the differences in Hcy levels were due to COVID-19 or age/gender differences between the groups.

It has also been suggested that *MTHFR C677T* polymorphism may provide explanation for differences in geographical and gender distribution in COVID-19 severity [35] (November 2020). This suggestion was supported by a study [36] of genomic data available in the Genome Aggregation Database (genomAD), and the COVID-19 prevalence and mortality data (as of 27 August 2020), which identified a strong correlation between the prevalence of the *MTHFR C677T* polymorphism and COVID-19 incidence and mortality worldwide. The prevalence of *MTHFR 677T* allele in the Latino and European (non-Finnish) populations, and the incidence and mortality for COVID-19 were higher than reported for most other ethnic groups globally [37]. It has also been suggested that B-vitamins should be used to lower HHcy coexisting with COVID-19 [36]. Consequently to the proposed involvement of HHcy in COVID-19, it was suggested that Hcy may contribute to severe COVID-19 by interfering with G-protein-coupled receptors (GPCRs) (AT1R, B2 and CXCR6), by their upregulation, being an alternative agonist and inducing their heteromerization [38]. However, most of these proposals lack an experimental support.

In a multicenter retrospective study on 313 COVID-19 patients hospitalized between April and September 2020, Hcy was significantly elevated in non-survivors compared with survivors and the authors stated that Hcy was a predictor of severe disease progression; however, this is not shown in the results [39,40]. In another study of hospitalized patients with mild COVID-19 ($n = 273$), Hcy among other variables (age, monocyte-lymphocyte ratio, and period from onset to admission) could predict imaging progression on chest CT at first week from COVID-19 patients [41].

Hcy may be linked to COVID-19 by contributing to coagulopathy and thrombosis, conditions that often develop in SARS-CoV-2 infected patients. Examination of genome-wide associations and tissue-specific gene expression, aimed at elucidating the genetic basis of thrombosis in COVID-19 has led to annotation of various SNPs with five ancestral terms: pulmonary embolism, venous thromboembolism, vascular diseases, cerebrovascular disorders, and stroke. The gene-gene interaction network revealed three clusters that contained hallmark genes for D-dimer/fibrinogen levels, Hcy levels, and arterial/venous thromboembolism with F2 and F5 acting as connecting nodes. Based on these analyses it was suggested that genotyping COVID-19 patients for SNPs examined in this study would help to identify individuals at the greatest risk of complications linked to thrombosis [42].

However, examination of data from several studies shows inconsistent relationships between Hcy and COVID-19 (Table S1). Specifically, Hcy has been found to be *higher* in COVID-19 cases vs. healthy controls in three studies [33,39,41] and *lower* in COVID-19 cases vs. healthy controls in two other studies [17,23]. Other studies reported *no change* in Hcy levels between COVID-19 cases and controls [18,22,24].

7. Other One-Carbon Metabolites

Metabolomic analyses showed that serum serine was enriched in COVID-19 and COVID-19-like patients vs. healthy controls [31], while in other studies serine was found to decrease in more severe COVID-19 cases [20,23] or did not change between patients with different disease severity [16,17,21]. Serine was decreased in critical COVID-19 patients vs. controls [15]. Choline and its derivatives were downregulated in COVID-19 patients [43] (Table S1), suggesting that they might benefit from choline supplementation.

In COVID-19 children with mild symptoms vs. healthy children, methylmalonic acid MMA), Met sulfoxide and Cys increased, while choline, dimethylglycine, and methylcysteine decreased [24] (Table S1). MMA was upregulated 3.2-fold compared to healthy children. In contrast to COVID-19 children, MMA was reduced in COVID-19 adults to 3.1% of its levels in healthy adults. MMA is produced during catabolism of some amino acids (e.g., valine, isoleucine) and lipids (cholesterol, fatty acids), and is further metabolized by a vitamin B_{12}-dependent enzyme malonyl-CoA mutase to succinic acid, which is a substrate for the TCA cycle. Interestingly, MMA was shown to inhibit replication of the mouse hepatitis virus (MHV), a well-known surrogate for SARS-CoV-2 in rat lung epithelial cells L2. Moreover, MMA reduced expression of inflammatory cytokines IL-6, TNF-α, and TGF-β in MHV-infected L2 cells [24]. Taken together, these findings suggest an antiviral and anti-inflammatory role of MMA in COVID-19-children, which can contribute to a mild course of SARS-CoV-2 infection in children compared with adults.

AICAR (5-Aminoimidazole-4-carboxamide ribonucleotide) was shown to be upregulated in COVID-19 patients [17], consistent with a suggestion that SARS-CoV-2 hijacks the host's folate and one-carbon metabolism for viral RNA synthesis [12]. Adenosine, another metabolite generated during hydrolysis of SAH (Figure 1), was shown to be upregulated in COVID-19 patients [16,23], or not affected as a result of SARS-CoV-2 infection [17]. Taurine was either down-regulated [17,23] or did not change between studied groups [16]. Hypotaurine was upregulated in COVID-19 cases, as was cysteinyl-*S*-sulphate [16] (Table S1).

8. SARS-CoV-2 in Relation to RAS

Apart from basic functions such as regulation of blood pressure and vasoconstriction, renin-angiotensin system (RAS) may have pro-inflammatory and pro-fibrotic effects. RAS is implicated in the pathogenesis of hypertension, diabetes mellitus and obesity, which all increase the risk of a severe course of COVID-19. The binding of angiotensin II to its receptors mediates the generation of free radicals and causes mitochondrial dysfunction and tissue damage [44]. As recently discovered, RAS has a direct link to one-carbon metabolism, through Hcy (discussed below), which was found to activate one of the RAS receptors [22]. SARS-CoV-2 infection perturbs RAS and energy metabolism [45].

One of the important RAS players is angiotensin-converting enzyme 2 (ACE2) which converts angiotensin I (Ang I) to angiotensin 1-9 and angiotensin II (Ang II) to angiotensin 1-7. Membrane bound ACE2 (mACE2) is a zinc-containing a single-pass type I membrane protein located on the surface of intestinal enterocytes, renal tubular cells and other cells. The extracellular domain of mACE2 can be cleaved from the transmembrane domain by an enzyme referred to us ADAM17, during the protective phase of RAS. The resulting cleaved protein is known as soluble ACE2 (sACE2). sACE is released into the bloodstream where it cleaves Ang II into angiotensin 1-7 which binds to MasR receptors creating localized vasodilation and hence decreasing blood pressure.

On the other hand, angiotensin-converting enzyme (ACE), cleaves Ang I into the vasoconstricting Ang II that causes a cascade of hormonal reactions, which is part of the body's harmful phase of RAS, leading to an increase in the blood pressure. Hence, sACE2 acts as a counterbalance to the ACE, degrading Ang II into angiotensin 1-7, and lowering blood pressure. A balance between ACE and ACE2 is curtailed for Ang II levels.

Another possible link between RAS and Hcy may involve homocysteinylation of ACE. Modification with Hcy and/or Hcy-thiolactone increases ACE activity leading to endothelial dysfunction [46].

The proper function of RAS depends on a balance between the two axis: the ACE-Ang II-AT1R axis, which has numerous detrimental effects, like vasoconstriction, inflammation, oxidative stress and fibrosis, and the ACE2-Ang 1-7-Mas receptor axis, which displays protective functions, like vasodilation, decreased fibrosis, and decreased inflammation. One of the cellular functions of ACE2 is cleavage of angiotensin II (Ang II) to angiotensin 1-7, and hence protecting against pathogenic effects of ACE-Ang II-AT1R axis of RAS. Obese subjects have increased levels of Ang II and proinflammatory cytokines (TNFα, IL-6, MCP-1) [47]. Tumor necrosis factor-alpha convertase (ADAM17) regulates Ang II and proinflammatory cytokines and mediates regulated ectodomain shedding of ACE2 [48]. The endocytosis of SARS-CoV-2 increases the activity of ADAM17, which in turn leads to the ACE2 shedding [49].

Also, spike protein of SARS-CoV was found to down-regulate ACE2 ectodomain expression [30]. These effects lead to disruption of balance between detrimental ACE-Ang II-AT1R axis and protective ACE2-Ang 1-7-Mas receptor of RAS and shift the RAS effects towards induction of inflammation and ROS production [50].

Patients with comorbidities (hypertension, cardiovascular disease, renal insufficiency, autoimmune disease) associated with severe COVID-19 have increased expression of ACE2 is their lungs [51]. mACE2 also serves as the entry point into cells for some coronaviruses, including HCoV-NL63, SARS-CoV, and SARS-CoV-2 [5]. It has been suggested that subjects with comorbidities may have higher chances of developing severe COVID-19, since ACE2 facilitates SARS-CoV-2 entry into the lung cells [51].

ACE2 is a functional receptor for coronaviruses [52], including SARS-CoV-2, which engages ACE2 for host cell entry through membrane fusion and endocytosis [53]. ACE2 binds to the receptor-binding domain (RBD) of SARS-CoV-2 spike (S) protein. Furthermore, to enter into the host cell, the priming of the viral spike protein (S) is considered essential for its fusion to the host cell membrane, which involves cleavage of the "S" protein by serine proteases called transmembrane serine protease 2 (TMPRSS2) or by cathepsin B or L (CTS-B or -L) and furin present in the host cell membrane [2]. The entry of SARS-CoV-2 to the cell is facilitated by a host factor neuropilin-1 facilitates cell entry and infectivity [54,55].

A molecular dynamic study has suggested that binding of the COVID-19 spike protein to ACE2 is impaired by reduction of the proteins' disulfide bonds [56]. This was confirmed experimentally by showing that the substitution of Cys488 with alanine in SARS-CoV-2 spike protein impaired pseudotyped SARS-CoV-2 infection, syncytium formation, and cell-cell fusion triggered by SARS-CoV-2 spike expression [57]. Consistently, in vitro binding of ACE2 and RBD, spike-mediated cell-cell fusion, and pseudotyped viral infection of VeroE6/TMPRSS2 cells were inhibited by the thiol-reactive compounds *n*-acetyl-Cys (NAC) and a reduced form of glutathione (GSH), which disrupted the Cys488-S-S-Cys480 disulfide bond in the spike protein.

ACE2 is a tissue enzyme and thus circulating levels are low, however, elevated circulating ACE2 is observed in patients with active COVID-19 disease and in the period after infection [58,59] and has been associated with increased risk of major cardiovascular events [60]. Also, ACE2 serum levels were shown to be significantly elevated in smokers, obese and diabetic individuals [61]. In a study of 306 COVID-19 positive subjects, it has been found that high plasma ACE2 during admission was associated with increased maximal illness severity within 28 days (OR = 1.8, 95%-CI: 1.4–2.3, $p < 0.0001$). Additionally, plasma ACE2 was significantly higher in COVID-19 patients with hypertension compared with patients without hypertension and with pre-existing heart conditions and kidney disease compared with patients without these conditions [62].

Ang II plays its detrimental effect through interaction with angiotensin type 1 receptor (AT1R). AT1R is linked to one-carbon metabolism through Hcy, which has been recently discovered to directly interact and activate AT1R, which aggravates vascular injury. It has been shown that the aggravation of abdominal aortic aneurysm by HHcy is abolished with genetic deletion of AT1 receptor and by blocking of AT1 receptor with telmisartan in animal model. Hcy displaces and limits angiotensin II binding to AT1 receptor. There are distinct conformational changes of AT1 receptor upon binding to angiotensin II and Hcy. It has been suggested that Hcy regulates the conformation of the AT1 receptor by forming a salt bridge and a disulfide bond with its Arg167 and Cys289 residues, respectively. Cys289 of AT1 mediates Hcy-induced AT1 receptor activation. Hcy and angiotensin II synergistically activate the AT1 receptor [63]. It remains to be determined if ACE2 can also be modified by homocysteinylation, leading to the change of its structure/function.

9. COVID-19 and Folate Cycle

Apart from activating the glucose metabolism, the SARS-CoV-2 infection activates the folate metabolism. The folate cycle is crucial for the transfer of one-carbon units for nucleotide synthesis. The demand for nucleotide synthesis is increased to match the viral replication needs. Metabolomic studies on Vero E6 cells infected with SARS-CoV-2 have shown that the infection has opposite effects on folate and glutathione abundance, and causes depletion of folate and increase of glutathione level. And interestingly, the elevated glutathione was not crucial for the replication of the virus. However, a drug that is a competitive inhibitor of dihydrofolate reductase and other steps in one-carbon metabolism and nucleotide synthesis, methotrexate, blocks replication and secretion of infectious virions. Experiments with inhibitors of cytosolic and mitochondrial forms of serine hydroxymethyltransferase SHMT1 and SHMT2, respectively), have shown that for virion production particularly important in the cytosolic branch of host one-carbon metabolism, especially for viral subgenomic RNA expression [12].

In one study, plasma folate was significantly reduced in mild (4.7 mg/L, n = 74) and severe (4.6 mg/L, n = 43) COVID-19 cases compared with healthy controls (12.5 mg/L, n = 34) [34]. However, because there were significant differences age and sex between the groups (mild COVID-19: 57.96 years, 52.5% male; severe COVID-19: 71.22 years, 62.8% male; healthy controls (32.65 years, 20.6%) [34], it is not clear whether these differences in folate levels were due to COVID-19 or age/gender differences between the groups.

Some studies have proposed that folic acid might inhibit the binding of the SARS-CoV-2 spike proteins, which blocks the entry of the virus into the cell. One study suggested that vitamin B9 acted as an inhibitor of the furin enzyme, and thus prevented the virus from entering the cell [64]. Another study suggested that folic acid, and its derivatives, 5-methyl tetrahydrofolic acid and tetrahydrofolic acid, have a strong binding affinity against the SARS-CoV-2 [65].

Furin belongs to proprotein convertases family, which cleaves its substrates at Arg-X-X-Arg↓ sites and its impaired activity has been associated with atherosclerosis, cancer and viral infectious diseases. Furin is a ubiquitous endopeptidase, which facilitates SARS-CoV-2 infection by proteolytic cleavage of the spike protein at the S1/S2 cleavage site. This cleavage is essential for entry into human lung cells [66]. Folic acid was tested for the inhibition of furin activity. Docking study results show that folic acid could be an inhibitor of furin and it has been suggested that folic acid could be used in prevention or management of COVID-19-associated respiratory disease in the early stages of the disease [64].

10. COVID-19 and Vitamin B_{12}

In a small prospective study on forty-nine COVID-19 patients, subjects that had worse condition (subjects admitted to ICU or those that have died, n = 9) had significantly higher levels of vitamin B_{12} that those in a better state (n = 40) but in a multivariate regression analysis only age was associated with a worse outcome. Folates and Hcy did not differ significantly between the two groups [67]. In another study, levels of vitamin B_{12} did not differ between COVID-19 children and healthy children [24].

11. Treatment Strategies Targeting One-Carbon Metabolism

Several possible treatment methods for COVID-19 have been recently discussed [68]. The strategies for therapeutics included polymerase inhibitors, protease inhibitors, interferons, and statins. A variety of nutrients and minerals, such as vitamins A, vitamin B_2, B_3, vitamin C, vitamin D, zinc, selenium, and pyrithione have been suggested to be useful in the management of the disease [68]. Additionally, the use of vitamins B_9, B_{12}, probiotics, and magnesium, may also have a positive impact on the prognosis of the infection. Vitamin B_{12} in combination with magnesium and vitamin D has been shown to decrease the severity of COVID-19 [69,70].

Screening hundreds of nutraceuticals compounds against known therapeutic targets of SARS-CoV-2 by molecular docking and the analysis of binding energy have predicted the therapeutic potential of folic acid and its derivatives such as tetrahydrofolic acid and 5-methyl tetrahydrofolic acid against SARS-CoV-2 [65]. Specifically, this computational study found that folic acid was the top nutraceutical predicted to inhibit Spike-ACE-2 interaction, 5-methyltetrahudrofolate bound to PL^{pro}, while folic acid derivatives bound to the NSP15 protein. Notably, folates had binding energies that were similar or better than those for known drugs targeting these SARS-CoV-2 proteins.

However, a suggestion that folates could be valuable drugs against COVID-19 [65] appears to be in conflict with findings in Vero E6 cells infected with SARS-CoV-2 showing that antifoliate drugs targeting purine biosynthesis inhibit the virus propagation via antiviral and anti-inflammatory activity [12]. Methotrexate blocks the replication and secretion of infectious virions in the infected Vero cells and may act synergistically with the antiviral nucleotide analog remdesivir, with competes with ATP for incorporation by the viral RNA polymerase [12].

12. Conclusions

Multi-omics studies revealed that SARS-CoV-2 infection leads to significant changes in numerous metabolites, including those involved in one-carbon metabolism that impact the virus's ability to propagate. However, with the exception of three metabolites (glutathione, Met sulfoxide, and choline) that were consistently affected by COVID-19 in various studies, comparative analyses of COVID-19 vs. control samples, both in metabolomics and single compound studies, have often lead to inconsistent results regarding the direction of the change in a particular differentiating metabolite. This could be due to ethnicity differences between various studies as well as age/sex differences between COVID-19 and control groups, which in some studies differed two- to three-fold [21–23], (Table S1). This may also be caused by differences in COVID-19 cases classification systems and study design (Table S2), including different sampling times in various studies. In some of the studies, COVID-19 patients are compared to healthy controls; however, taking into account the very diverse symptoms and courses of the illness, this may be an oversimplification. On the other hand, other studies compared groups of subjects manifesting different disease intensity, but are using diverse classification systems, making it difficult to compare the results between different studies. Just how important is the classification system of the COVID-19 course is a study [18], in which, depending on the COVID-19 classification system used, there were or were not differences in SAM levels (Table S2). Finally, data discussed in this review suggest that therapeutic interventions aimed at normalizing glutathione, Met sulfoxide, and choline might provide a promising approach to combat the COVID-19 pandemic. Glutathione can be normalized by supplementation with N-acetyl-Cys [71], or more effectively with Gly + N-acetyl-Cys [28] which, by reducing oxidative stress [28] should also normalize Met sulfoxide. Supplementation with choline can restore its normal levels [72]. Further studies are required to determine the therapeutic potential of targeting these metabolic areas.

Supplementary Materials: The following supporting information can be downloaded at: https://www.mdpi.com/article/10.3390/ijms23084181/s1.

Author Contributions: Conceptualization, H.J. and J.P.-K.; methodology, H.J. and J.P.-K.; software, H.J.; validation, H.J. and J.P.-K.; formal analysis, H.J.; investigation, H.J. and J.P.-K.; resources, H.J.; data curation, H.J.; writing—original draft preparation, J.P.-K. and H.J.; writing—review and editing, H.J. and J.P.-K.; visualization, H.J. and J.P.-K.; supervision, H.J.; project administration, H.J.; funding acquisition, H.J. All authors have read and agreed to the published version of the manuscript.

Funding: This research was funded by National Science Foundation, grant numbers 2016/23/B/NZ5/00573, 2018/29/B/NZ4/00771, 2019/33/B/NZ4/01760.

Institutional Review Board Statement: Not applicable.

Data Availability Statement: Not applicable.

Conflicts of Interest: The authors declare that they have no conflict of interest. The funders had no role in the design of the study; in the collection, analyses, or interpretation of data; in the writing of the manuscript, or in the decision to publish the results.

References

1. Mayer, K.A.; Stockl, J.; Zlabinger, G.J.; Gualdoni, G.A. Hijacking the Supplies: Metabolism as a Novel Facet of Virus-Host Interaction. *Front. Immunol.* **2019**, *10*, 1533. [CrossRef] [PubMed]
2. Kumar, A.; Narayan, R.K.; Prasoon, P.; Kumari, C.; Kaur, G.; Kumar, S.; Kulandhasamy, M.; Sesham, K.; Pareek, V.; Faiq, M.A.; et al. COVID-19 Mechanisms in the Human Body-What We Know So Far. *Front. Immunol.* **2021**, *12*, 693938. [CrossRef]
3. Cabrera Martimbianco, A.L.; Pacheco, R.L.; Bagattini, Â.; Riera, R. Frequency, signs and symptoms, and criteria adopted for long COVID-19: A systematic review. *Int. J. Clin. Pract.* **2021**, *75*, e14357. [CrossRef] [PubMed]
4. Akhmerov, A.; Marbán, E. COVID-19 and the Heart. *Circ. Res.* **2020**, *126*, 1443–1455. [CrossRef]
5. Costanzo, M.; Caterino, M.; Fedele, R.; Cevenini, A.; Pontillo, M.; Barra, L.; Ruoppolo, M. COVIDomics: The Proteomic and Metabolomic Signatures of COVID-19. *Int. J. Mol. Sci.* **2022**, *23*, 2414. [CrossRef]
6. Sun, Y.; Dong, Y.; Wang, L.; Xie, H.; Li, B.; Chang, C.; Wang, F.S. Characteristics and prognostic factors of disease severity in patients with COVID-19: The Beijing experience. *J. Autoimmun.* **2020**, *112*, 102473. [CrossRef] [PubMed]
7. Ma, K.-L.; Liu, Z.-h.; Cao, C.-F.; Liu, M.-K.; Liao, J.; Zou, J.B.; Kong, L.-X.; Wan, K.-Q.; Zhang, J.; Wang, Q.-B.; et al. COVID-19 Myocarditis and Severity Factors: An Adult Cohort Study. *medRxiv* **2020**. [CrossRef]
8. Ponti, G.; Maccaferri, M.; Ruini, C.; Tomasi, A.; Ozben, T. Biomarkers associated with COVID-19 disease progression. *Crit. Rev. Clin. Lab. Sci.* **2020**, *57*, 389–399. [CrossRef]
9. Hayden, M.R.; Tyagi, S.C. Impaired Folate-Mediated One-Carbon Metabolism in Type 2 Diabetes, Late-Onset Alzheimer's Disease and Long COVID. *Medicina* **2021**, *58*, 16. [CrossRef]
10. McCaddon, A.; Regland, B. COVID-19: A methyl-group assault? *Med. Hypotheses* **2021**, *149*, 110543. [CrossRef]
11. Ducker, G.S.; Rabinowitz, J.D. One-Carbon Metabolism in Health and Disease. *Cell Metab.* **2017**, *25*, 27–42. [CrossRef]
12. Zhang, Y.; Guo, R.; Kim, S.H.; Shah, H.; Zhang, S.; Liang, J.H.; Fang, Y.; Gentili, M.; Leary, C.N.O.; Elledge, S.J. SARS-CoV-2 hijacks folate and one-carbon metabolism for viral replication. *Nat. Commun.* **2021**, *12*, 1676. [CrossRef] [PubMed]
13. Singh, Y.; Gupta, G.; Kazmi, I.; Al-Abbasi, F.A.; Negi, P.; Chellappan, D.K.; Dua, K. SARS CoV-2 aggravates cellular metabolism mediated complications in COVID-19 infection. *Dermatol. Ther.* **2020**, *33*, e13871. [CrossRef] [PubMed]
14. Romano, M.; Ruggiero, A.; Squeglia, F.; Maga, G.; Berisio, R. A Structural View of SARS-CoV-2 RNA Replication Machinery: RNA Synthesis, Proofreading and Final Capping. *Cells* **2020**, *9*, 1267. [CrossRef] [PubMed]
15. Danlos, F.X.; Grajeda-Iglesias, C.; Durand, S.; Sauvat, A.; Roumier, M.; Cantin, D.; Colomba, E.; Rohmer, J.; Pommeret, F.; Baciarello, G.; et al. Metabolomic analyses of COVID-19 patients unravel stage-dependent and prognostic biomarkers. *Cell Death Dis.* **2021**, *12*, 258. [CrossRef] [PubMed]
16. Su, Y.; Chen, D.; Yuan, D.; Lausted, C.; Choi, J.; Dai, C.L.; Voillet, V.; Duvvuri, V.R.; Scherler, K.; Troisch, P.; et al. Multi-Omics Resolves a Sharp Disease-State Shift between Mild and Moderate COVID-19. *Cell* **2020**, *183*, 1479–1495.e20. [CrossRef] [PubMed]
17. Xiao, N.; Nie, M.; Pang, H.; Wang, B.; Hu, J.; Meng, X.; Li, K.; Ran, X.; Long, Q.; Deng, H.; et al. Integrated cytokine and metabolite analysis reveals immunometabolic reprogramming in COVID-19 patients with therapeutic implications. *Nat. Commun.* **2021**, *12*, 1618. [CrossRef]
18. Kryukov, E.V.; Ivanov, A.V.; Karpov, V.O.; Vasil'evich Aleksandrin, V.; Dygai, A.M.; Kruglova, M.P.; Kostiuchenko, G.I.; Kazakov, S.P.; Kubatiev, A.A. Plasma S-Adenosylmethionine Is Associated with Lung Injury in COVID-19. *Dis. Markers* **2021**, *2021*, 7686374. [CrossRef]
19. Roberts, I.; Wright Muelas, M.; Taylor, J.M.; Davison, A.S.; Xu, Y.; Grixti, J.M.; Gotts, N.; Sorokin, A.; Goodacre, R.; Kell, D.B. Untargeted metabolomics of COVID-19 patient serum reveals potential prognostic markers of both severity and outcome. *Metabolomics* **2021**, *18*, 6. [CrossRef]

20. Li, Y.; Hou, G.; Zhou, H.; Wang, Y.; Tun, H.M.; Zhu, A.; Zhao, J.; Xiao, F.; Lin, S.; Liu, D.; et al. Multi-platform omics analysis reveals molecular signature for COVID-19 pathogenesis, prognosis and drug target discovery. *Signal Transduct. Target. Ther.* **2021**, *6*, 155. [CrossRef]
21. Páez-Franco, J.C.; Torres-Ruiz, J.; Sosa-Hernández, V.A.; Cervantes-Díaz, R.; Romero-Ramírez, S.; Pérez-Fragoso, A.; Meza-Sánchez, D.E.; Germán-Acacio, J.M.; Maravillas-Montero, J.L.; Mejía-Domínguez, N.R.; et al. Metabolomics analysis reveals a modified amino acid metabolism that correlates with altered oxygen homeostasis in COVID-19 patients. *Sci. Rep.* **2021**, *11*, 6350. [CrossRef] [PubMed]
22. Caterino, M.; Costanzo, M.; Fedele, R.; Cevenini, A.; Gelzo, M.; Di Minno, A.; Andolfo, I.; Capasso, M.; Russo, R.; Annunziata, A.; et al. The Serum Metabolome of Moderate and Severe COVID-19 Patients Reflects Possible Liver Alterations Involving Carbon and Nitrogen Metabolism. *Int. J. Mol. Sci.* **2021**, *22*, 9548. [CrossRef] [PubMed]
23. Thomas, T.; Stefanoni, D.; Reisz, J.A.; Nemkov, T.; Bertolone, L.; Francis, R.O.; Hudson, K.E.; Zimring, J.C.; Hansen, K.C.; Hod, E.A.; et al. COVID-19 infection alters kynurenine and fatty acid metabolism, correlating with IL-6 levels and renal status. *JCI Insight* **2020**, *5*, e140327. [CrossRef] [PubMed]
24. Wang, C.; Li, X.; Ning, W.; Gong, S.; Yang, F.; Fang, C.; Gong, Y.; Wu, D.; Huang, M.; Gou, Y.; et al. Multi-omic profiling of plasma reveals molecular alterations in children with COVID-19. *Theranostics* **2021**, *11*, 8008–8026. [CrossRef]
25. Polonikov, A. Endogenous Deficiency of Glutathione as the Most Likely Cause of Serious Manifestations and Death in COVID-19 Patients. *ACS Infect. Dis.* **2020**, *6*, 1558–1562. [CrossRef]
26. Khanfar, A.; Al Qaroot, B. Could glutathione depletion be the Trojan horse of COVID-19 mortality? *Eur. Rev. Med. Pharmacol. Sci.* **2020**, *24*, 12500–12509.
27. Silvagno, F.; Vernone, A.; Pescarmona, G.P. The Role of Glutathione in Protecting against the Severe Inflammatory Response Triggered by COVID-19. *Antioxidants* **2020**, *9*, 624. [CrossRef]
28. Kumar, P.; Osahon, O.; Vides, D.B.; Hanania, N.; Minard, C.G.; Sekhar, R.V. Severe Glutathione Deficiency, Oxidative Stress and Oxidant Damage in Adults Hospitalized with COVID-19: Implications for GlyNAC (Glycine and N-Acetylcysteine) Supplementation. *Antioxidants* **2021**, *11*, 50. [CrossRef]
29. Kryukov, E.V.; Ivanov, A.V.; Karpov, V.O.; Vasil'evich Alexandrin, V.; Dygai, A.M.; Kruglova, M.P.; Kostiuchenko, G.I.; Kazakov, S.P.; Kubatiev, A.A. Association of Low Molecular Weight Plasma Aminothiols with the Severity of Coronavirus Disease 2019. *Oxid. Med. Cell. Longev.* **2021**, *2021*, 9221693. [CrossRef]
30. Kuba, K.; Imai, Y.; Rao, S.; Gao, H.; Guo, F.; Guan, B.; Huan, Y.; Yang, P.; Zhang, Y.; Deng, W.; et al. A crucial role of angiotensin converting enzyme 2 (ACE2) in SARS coronavirus-induced lung injury. *Nat. Med.* **2005**, *11*, 875–879. [CrossRef]
31. Shi, D.; Yan, R.; Lv, L.; Jiang, H.; Lu, Y.; Sheng, J.; Xie, J.; Wu, W.; Xia, J.; Xu, K.; et al. The serum metabolome of COVID-19 patients is distinctive and predictive. *Metabolism* **2021**, *118*, 154739. [CrossRef] [PubMed]
32. Smith, A.D.; Refsum, H. Homocysteine—From disease biomarker to disease prevention. *J. Intern. Med.* **2021**, *290*, 826–854. [CrossRef]
33. Ponti, G.; Ruini, C.; Tomasi, A. Homocysteine as a potential predictor of cardiovascular risk in patients with COVID-19. *Med. Hypotheses* **2020**, *143*, 109859. [CrossRef]
34. Keskin, A.; Ustun, G.U.; Aci, R.; Duran, U. Homocysteine as a marker for predicting disease severity in patients with COVID-19. *Biomark. Med.* 2022, *Epub ahead of print*. [CrossRef] [PubMed]
35. Karst, M.; Hollenhorst, J.; Achenbach, J. Life-threatening course in coronavirus disease 2019 (COVID-19): Is there a link to methylenetetrahydrofolic acid reductase (MTHFR) polymorphism and hyperhomocysteinemia? *Med. Hypotheses* **2020**, *144*, 110234. [CrossRef]
36. Ibrahimagić, O.; Smajlović, D.; Dostović, Z.; Vidović, M.; Tupković, E.; Kunić, S. Comment on an Article: "Homocysteine as a potential predictor of cardiovascular risk in patients with COVID-19". *Med. Hypotheses* **2020**, *143*, 110107. [CrossRef] [PubMed]
37. Ponti, G.; Pastorino, L.; Manfredini, M.; Ozben, T.; Oliva, G.; Kaleci, S.; Iannella, R.; Tomasi, A. COVID-19 spreading across world correlates with C677T allele of the methylenetetrahydrofolate reductase (MTHFR) gene prevalence. *J. Clin. Lab. Anal.* **2021**, *35*, e23798. [CrossRef] [PubMed]
38. Berbert, A. Further comment on articles pertaining to: "Homocysteine as a potential predictor of cardiovascular risk in patients with COVID-19". *Med. Hypotheses* **2021**, *155*, 110676. [CrossRef]
39. Ponti, G.; Roli, L.; Oliva, G.; Manfredini, M.; Trenti, T.; Kaleci, S.; Iannella, R.; Balzano, B.; Coppola, A.; Fiorentino, G.; et al. Homocysteine (Hcy) assessment to predict outcomes of hospitalized Covid-19 patients: A multicenter study on 313 Covid-19 patients. *Clin. Chem. Lab. Med.* **2021**, *59*, e354–e357. [CrossRef]
40. Giovanni, P.; Marco, M.; Gabriella, O.; Tomris, O.; Caterina, F.; Tomasi, A. Predicting COVID-19 Hospitalized Patients' Outcome with Homocysteine. *J. Clin. Cardiol.* **2021**, *2*, 19–22.
41. Yang, Z.; Shi, J.; He, Z.; Lü, Y.; Xu, Q.; Ye, C.; Chen, S.; Tang, B.; Yin, K.; Lu, Y. Predictors for imaging progression on chest CT from coronavirus disease 2019 (COVID-19) patients. *Aging* **2020**, *12*, 6037–6048. [CrossRef] [PubMed]
42. Abu-Farha, M.; Al-Sabah, S.; Hammad, M.M.; Hebbar, P.; Channanath, A.M.; John, S.E.; Taher, I.; Almaeen, A.; Ghazy, A.; Mohammad, A.; et al. Prognostic Genetic Markers for Thrombosis in COVID-19 Patients: A Focused Analysis on D-Dimer, Homocysteine and Thromboembolism. *Front. Pharmacol.* **2020**, *11*, 587451. [CrossRef] [PubMed]
43. Shen, B.; Yi, X.; Sun, Y.; Bi, X.; Du, J.; Zhang, C.; Quan, S.; Zhang, F.; Sun, R.; Qian, L.; et al. Proteomic and Metabolomic Characterization of COVID-19 Patient Sera. *Cell* **2020**, *182*, 59–72.e15. [CrossRef] [PubMed]

44. de Kloet, A.D.; Krause, E.G.; Woods, S.C. The renin angiotensin system and the metabolic syndrome. *Physiol. Behav.* **2010**, *100*, 525–534. [CrossRef]
45. Mori, J.; Oudit, G.Y.; Lopaschuk, G.D. SARS-CoV-2 perturbs the renin-angiotensin system and energy metabolism. *Am. J. Physiol. Endocrinol. Metab.* **2020**, *319*, E43–E47. [CrossRef]
46. Huang, A.; Pinto, J.T.; Froogh, G.; Kandhi, S.; Qin, J.; Wolin, M.S.; Hintze, T.H.; Sun, D. Role of homocysteinylation of ACE in endothelial dysfunction of arteries. *Am. J. Physiol. Heart Circ. Physiol.* **2015**, *308*, H92–H100. [CrossRef]
47. Richard, C.; Wadowski, M.; Goruk, S.; Cameron, L.; Sharma, A.M.; Field, C.J. Individuals with obesity and type 2 diabetes have additional immune dysfunction compared with obese individuals who are metabolically healthy. *BMJ Open Diabetes Res. Care* **2017**, *5*, e000379. [CrossRef]
48. Lambert, D.W.; Yarski, M.; Warner, F.J.; Thornhill, P.; Parkin, E.T.; Smith, A.I.; Hooper, N.M.; Turner, A.J. Tumor necrosis factor-alpha convertase (ADAM17) mediates regulated ectodomain shedding of the severe-acute respiratory syndrome-coronavirus (SARS-CoV) receptor, angiotensin-converting enzyme-2 (ACE2). *J. Biol. Chem.* **2005**, *280*, 30113–30119. [CrossRef]
49. Haga, S.; Yamamoto, N.; Nakai-Murakami, C.; Osawa, Y.; Tokunaga, K.; Sata, T.; Sasazuki, T.; Ishizaka, Y. Modulation of TNF-alpha-converting enzyme by the spike protein of SARS-CoV and ACE2 induces TNF-alpha production and facilitates viral entry. *Proc. Natl. Acad. Sci. USA* **2008**, *105*, 7809–7814. [CrossRef]
50. Wang, K.; Gheblawi, M.; Oudit, G.Y. Angiotensin Converting Enzyme 2: A Double-Edged Sword. *Circulation* **2020**, *142*, 426–428. [CrossRef]
51. Pinto, B.G.G.; Oliveira, A.E.R.; Singh, Y.; Jimenez, L.; Gonçalves, A.N.A.; Ogava, R.L.T.; Creighton, R.; Schatzmann Peron, J.P.; Nakaya, H.I. ACE2 Expression Is Increased in the Lungs of Patients with Comorbidities Associated with Severe COVID-19. *J. Infect. Dis.* **2020**, *222*, 556–563. [CrossRef] [PubMed]
52. Li, W.; Moore, M.J.; Vasilieva, N.; Sui, J.; Wong, S.K.; Berne, M.A.; Somasundaran, M.; Sullivan, J.L.; Luzuriaga, K.; Greenough, T.C.; et al. Angiotensin-converting enzyme 2 is a functional receptor for the SARS coronavirus. *Nature* **2003**, *426*, 450–454. [CrossRef] [PubMed]
53. Hoffmann, M.; Kleine-Weber, H.; Schroeder, S.; Krüger, N.; Herrler, T.; Erichsen, S.; Schiergens, T.S.; Herrler, G.; Wu, N.H.; Nitsche, A.; et al. SARS-CoV-2 Cell Entry Depends on ACE2 and TMPRSS2 and Is Blocked by a Clinically Proven Protease Inhibitor. *Cell* **2020**, *181*, 271–280.e8. [CrossRef] [PubMed]
54. Cantuti-Castelvetri, L.; Ojha, R.; Pedro, L.D.; Djannatian, M.; Franz, J.; Kuivanen, S.; van der Meer, F.; Kallio, K.; Kaya, T.; Anastasina, M.; et al. Neuropilin-1 facilitates SARS-CoV-2 cell entry and infectivity. *Science* **2020**, *370*, 856–860. [CrossRef]
55. Daly, J.L.; Simonetti, B.; Klein, K.; Chen, K.E.; Williamson, M.K.; Antón-Plágaro, C.; Shoemark, D.K.; Simón-Gracia, L.; Bauer, M.; Hollandi, R.; et al. Neuropilin-1 is a host factor for SARS-CoV-2 infection. *Science* **2020**, *370*, 861–865. [CrossRef]
56. Hati, S.; Bhattacharyya, S. Impact of Thiol-Disulfide Balance on the Binding of Covid-19 Spike Protein with Angiotensin-Converting Enzyme 2 Receptor. *ACS Omega* **2020**, *5*, 16292–16298. [CrossRef]
57. Murae, M.; Shimizu, Y.; Yamamoto, Y.; Kobayashi, A.; Houri, M.; Inoue, T.; Irie, T.; Gemba, R.; Kodo, Y.; Nakano, Y.; et al. The function of SARS-CoV-2 spike protein is impaired by disulfide-bond disruption with mutation at cysteine-488 and by thiol-reactive N-acetyl-cysteine and glutathione. *Biochem. Biophys. Res. Commun.* **2022**, *597*, 30–36. [CrossRef]
58. Patel, S.K.; Juno, J.A.; Lee, W.S.; Wragg, K.M.; Hogarth, P.M.; Kent, S.J.; Burrell, L.M. Plasma ACE2 activity is persistently elevated following SARS-CoV-2 infection: Implications for COVID-19 pathogenesis and consequences. *Eur. Respir. J.* **2021**, *57*, 2003730. [CrossRef]
59. Nagy, B.; Fejes, Z.; Szentkereszty, Z.; Sütő, R.; Várkonyi, I.; Ajzner, É.; Kappelmayer, J.; Papp, Z.; Tóth, A.; Fagyas, M. A dramatic rise in serum ACE2 activity in a critically ill COVID-19 patient. *Int. J. Infect. Dis.* **2021**, *103*, 412–414. [CrossRef]
60. Narula, S.; Yusuf, S.; Chong, M.; Ramasundarahettige, C.; Rangarajan, S.; Bangdiwala, S.I.; van Eikels, M.; Leineweber, K.; Wu, A.; Pigeyre, M.; et al. Plasma ACE2 and risk of death or cardiometabolic diseases: A case-cohort analysis. *Lancet* **2020**, *396*, 968–976. [CrossRef]
61. Emilsson, V.; Gudmundsson, E.F.; Aspelund, T.; Jonsson, B.G.; Gudjonsson, A.; Launer, L.J.; Lamb, J.R.; Gudmundsdottir, V.; Jennings, L.L.; Gudnason, V. ACE2 levels are altered in comorbidities linked to severe outcome in COVID-19. *medRxiv* **2020**. [CrossRef]
62. Kragstrup, T.W.; Singh, H.S.; Grundberg, I.; Nielsen, A.L.; Rivellese, F.; Mehta, A.; Goldberg, M.B.; Filbin, M.R.; Qvist, P.; Bibby, B.M. Plasma ACE2 predicts outcome of COVID-19 in hospitalized patients. *PLoS ONE* **2021**, *16*, e0252799. [CrossRef] [PubMed]
63. Li, T.; Yu, B.; Liu, Z.; Li, J.; Ma, M.; Wang, Y.; Zhu, M.; Yin, H.; Wang, X.; Fu, Y. Homocysteine directly interacts and activates the angiotensin II type I receptor to aggravate vascular injury. *Nat. Commun.* **2018**, *9*, 11. [CrossRef]
64. Sheybani, Z.; Heydari, M.; Negahdaripour, M.; Dehdashti, M.; Zolghadr, H.; Moghadami, M.; Masoompour, S.M.; Zolghadr, A.R. The Role of Folic Acid in the Management of Respiratory Disease Caused by COVID-19. *ChemRxiv* **2020**. Available online: https://chemrxiv.org/engage/chemrxiv/article-details/60c749414c8919197ead3059 (accessed on 28 March 2022).
65. Kumar, V.; Kancharla, S.; Jena, M.K. In silico virtual screening-based study of nutraceuticals predicts the therapeutic potentials of folic acid and its derivatives against COVID-19. *VirusDisease* **2021**, *32*, 29–37. [CrossRef] [PubMed]
66. Hoffmann, M.; Kleine-Weber, H.; Pöhlmann, S. A Multibasic Cleavage Site in the Spike Protein of SARS-CoV-2 Is Essential for Infection of Human Lung Cells. *Mol. Cell* **2020**, *78*, 779–784.e5. [CrossRef] [PubMed]

67. Dalbeni, A.; Bevilacqua, M.; Teani, I.; Normelli, I.; Mazzaferri, F.; Chiarioni, G. Excessive vitamin B12 and poor outcome in COVID-19 pneumonia. *Nutr. Metab. Cardiovasc. Dis.* **2021**, *31*, 774–775. [CrossRef] [PubMed]
68. Abd El Hadi, S.R.; Zien El-Deen, E.E.; Bahaa, M.M.; Sadakah, A.A.; Yassin, H.A. COVID-19: Vaccine Delivery System, Drug Repurposing and Application of Molecular Modeling Approach. *Drug Des. Devel. Ther.* **2021**, *15*, 3313–3330. [CrossRef]
69. Tan, C.W.; Ho, L.P.; Kalimuddin, S.; Cherng, B.P.Z.; Teh, Y.E.; Thien, S.Y.; Wong, H.M.; Tern, P.J.W.; Chandran, M.; Chay, J.W.M.; et al. Cohort study to evaluate the effect of vitamin D, magnesium, and vitamin B. *Nutrition* **2020**, *79–80*, 111017. [CrossRef]
70. Asad, D.; Shuja, S.H. Role of Folate, Cobalamin, and Probiotics in COVID-19 Disease Management [Letter]. *Drug Des. Dev. Ther.* **2021**, *15*, 3709–3710. [CrossRef]
71. Di Marco, S.; Foti, G.; Corsico, A.G. Where we are with the use of N-acetylcysteine as a preventive and adjuvant treatment for COVID-19. *Eur. Rev. Med. Pharmacol. Sci.* **2022**, *26*, 715–721. [PubMed]
72. Li, Y.; Freedman, R. Prospects for improving future mental health of children through prenatal micronutrient supplementation in China. *Pediatr. Investig.* **2020**, *4*, 118–126. [CrossRef] [PubMed]

International Journal of *Molecular Sciences*

MDPI

Article

S-Acetyl-Glutathione Attenuates Carbon Tetrachloride-Induced Liver Injury by Modulating Oxidative Imbalance and Inflammation

Rosanna Di Paola [1,†], Sergio Modafferi [2,†], Rosalba Siracusa [3], Marika Cordaro [4], Ramona D'Amico [3], Maria Laura Ontario [2], Livia Interdonato [3], Angela Trovato Salinaro [2,*], Roberta Fusco [5,*], Daniela Impellizzeri [3], Vittorio Calabrese [2,‡] and Salvatore Cuzzocrea [3,‡]

1 Department of Veterinary Sciences, University of Messina, 98168 Messina, Italy; dipaolar@unime.it
2 Department of Biomedical and Biotechnological Sciences, University of Catania, 95124 Catania, Italy; sergio.modafferi@studium.unict.it (S.M.); marialaura.ontario@ontariosrl.it (M.L.O.); calabres@unict.it (V.C.)
3 Department of Chemical, Biological, Pharmaceutical and Environmental Sciences, University of Messina, Viale Ferdinando Stagno D'Alcontres 31, 98166 Messina, Italy; rsiracusa@unime.it (R.S.); rdamico@unime.it (R.D.); linterdonato@unime.it (L.I.); dimpellizzeri@unime.it (D.I.); salvator@unime.it (S.C.)
4 Department of Biomedical, Dental and Morphological and Functional Imaging, University of Messina, Via Consolare Valeria, 98125 Messina, Italy; marika.cordaro@unime.it
5 Department of Clinical and Experimental Medicine, University of Messina, Via Consolare Valeria, 98125 Messina, Italy
* Correspondence: trovato@unict.it (A.T.S.); rfusco@unime.it (R.F.)
† These authors equally contributed to the work.
‡ These authors contributed equally to this work.

Citation: Di Paola, R.; Modafferi, S.; Siracusa, R.; Cordaro, M.; D'Amico, R.; Ontario, M.L.; Interdonato, L.; Salinaro, A.T.; Fusco, R.; Impellizzeri, D.; et al. S-Acetyl-Glutathione Attenuates Carbon Tetrachloride-Induced Liver Injury by Modulating Oxidative Imbalance and Inflammation. *Int. J. Mol. Sci.* **2022**, 23, 4429. https://doi.org/10.3390/ijms23084429

Academic Editor: Stefania Bruno

Received: 8 March 2022
Accepted: 15 April 2022
Published: 17 April 2022

Abstract: Liver fibrosis, depending on the stage of the disease, could lead to organ dysfunction and cirrhosis, and no effective treatment is actually available. Emergent proof supports a link between oxidative stress, liver fibrogenesis and mitochondrial dysfunction as molecular bases of the pathology. A valid approach to protect against the disease would be to replenish the endogenous antioxidants; thus, we investigated the protective mechanisms of the S-acetyl-glutathione (SAG), a glutathione (GSH) prodrug. Preliminary in vitro analyses were conducted on primary hepatic cells. SAG pretreatment significantly protected against cytotoxicity induced by CCl4. Additionally, CCl4 induced a marked increase in AST and ALT levels, whereas SAG significantly reduced these levels, reaching values found in the control group. For the in vivo analyses, mice were administered twice a week with eight consecutive intraperitoneal injections of 1 mL/kg CCl4 (diluted at 1:10 in olive oil) to induce oxidative imbalance and liver inflammation. SAG (30 mg/kg) was administered orally for 8 weeks. SAG significantly restored SOD activity, GSH levels and GPx activity, while it strongly reduced GSSG levels, lipid peroxidation and H_2O_2 and ROS levels in the liver. Additionally, CCl4 induced a decrease in anti-oxidants, including Nrf2, HO-1 and NQO-1, which were restored by treatment with SAG. The increased oxidative stress characteristic on liver disfunction causes the impairment of mitophagy and accumulation of dysfunctional and damaged mitochondria. Our results showed the protective effect of SAG administration in restoring mitophagy, as shown by the increased PINK1 and Parkin expressions in livers exposed to CCl4 intoxication. Thus, the SAG administration showed anti-inflammatory effects decreasing pro-inflammatory cytokines TNF-α, IL-6, MCP-1 and IL-1β in both serum and liver, and suppressing the TLR4/NFkB pathway. SAG attenuated reduced fibrosis, collagen deposition, hepatocellular damage and organ dysfunction. In conclusion, our results suggest that SAG administration protects the liver from CCl4 intoxication by restoring the oxidative balance, ameliorating the impairment of mitophagy and leading to reduced inflammation.

Keywords: liver fibrosis; antioxidant; inflammation

1. Introduction

Liver fibrosis is a characteristic result of chronic liver injury. It is caused by numerous insults, including toxic damage, alcohol abuse, viral infection and metabolic disorders. Additionally, it is well described that a substantial increase in fibrosis and steatosis frequently leads to fatal cirrhosis and even hepatocellular carcinoma in humans [1]. The high incidence of these hepatopathies places them as one of the most severe diseases. Although the mechanism of some etiologies of liver fibrosis is not fully described, it is clear that reactive oxygen species (ROS) play a key role in the pathogenesis of liver diseases [2]. Increased ROS levels are involved in many pathological conditions and diseases, including cardiovascular disease, diabetes, aging and cancer [3–5]. The repeated administration of carbontetrachloride (CCl4) has become one of the most commonly used experimental models for inducing toxin-mediated liver fibrosis [6]. In many aspects, it mimics human chronic disease associated with toxic damage. CCl4 induces severe liver cell damage through the elevation of ROS, resulting in apoptosis, fibrosis and liver injury that contribute to an acute phase reaction characterized by the necrosis of centrilobular hepatocytes, the activation of Kupffer cells and the induction of an inflammatory response [7]. This sequence is associated with the production of several cytokines that cause liver fibrosis [8]. CCl4 is metabolized in the liver by the cytochrome P450 superfamily of monooxygenases (CYP family) to the trichloromethyl radical (CCl*3). Subsequently, this radical reacts with nucleic acids, proteins and lipids, thereby impairing key cellular processes, resulting in an altered lipid metabolism (fatty degeneration and steatosis). The formation of trichloromethylperoxy radicals (CCl_3OO^*) resulting from the oxygenation of $CCl_3{}^*$ further initiates lipid peroxidation and the destruction of polyunsaturated fatty acids. Consequently, the membrane permeability in all cellular compartments (mitochondria, endoplasmic reticulum and plasma membrane) is lowered and generalized hepatic damage occurs that is characterized by inflammation, fibrosis and cirrhosis. Biological membranes are particularly prone to the ROS injuries. The peroxidation of fatty acids in cellular membranes induces a decrease in membrane fluidity and disruption of membrane function and integrity, which is involved in severe pathological changes [9]. It is clear that the direct diminution of ROS levels and inhibition of the oxidative chain reaction induced by CCl4 administration may be critical for the treatment and prevention of CCl4-induced liver damage [10]. In fact, it has been previously described that CCl4 injection increased pro-inflammatory mediators and decreased the antioxidant proteins during liver injury and fibrosis [6,11]. Many endogenous protective mechanisms have been characterized to limit ROS-induced damage [12]. However, additional protective mechanisms of exogenous antioxidants may be important. Thus, many artificial and natural agents possessing antioxidative effects have been suggested to treat and prevent hepatopathies induced by excessive oxidative stress [13,14]. Therefore, supplementation with anti-oxidants is beneficial for human health. Glutathione (γ-L-glutamyl-L-cysteinylglycine, GSH) is an endogenous tripeptide and liver is one of the tissues with the highest content of it. In particular, it is the principal tissue involved in its biosynthesis [15]. Within the cell, GSH is maintained in its thiol-reduced form by an NADPH-dependent enzyme (glutathione disulfide (GSSG) reductase). An additional amount of GSH is present as GSSG and as glutathione conjugates (GS-R). Keeping optimal GSH:GSSG ratios is fundamental for cell survival, since GSH is one of the most important endogenous antioxidant defense systems, removing lipid- and hydrogen-peroxides [16]. A valid approach to replenish the endogenous GSH is using S-acyl prodrugs such as S-acetyl-glutathione (SAG). It is able to cross the cell membrane and increase intracellular SH groups [17]. Differently from GSH, which enters the cells directly, SAG is more stable in blood plasma and, once entered into cell cytoplasm, is converted by cytoplasmatic thioesterases to GSH. In particular, the acetylation of the sulfur atom avoids the GSH decomposition and simplifies its absorption via the intestinal wall [18].

Therefore, in this study, we evaluated the effects of the SAG administration in a mouse model of liver injury, focusing the attention on the CCl4-induced fibrosis, oxidative stress, mitophagy and inflammation.

2. Results

2.1. Effect of SAG on Cytotoxicity and Hepatoprotective Activity in Cells: Preliminary In Vitro Data

Primary hepatic cells only exposed to SAG (0.25–2.00 mM) or the CCl4 vehicle (DMSO 0.5%) showed no changes in cell viability (Figure 1A). On the other hand, primary hepatic cells exposed to CCl4 (4 mM) presented a significant reduction in cell viability when compared to the control group. In turn, pre-treatment with all concentrations of SAG followed by exposure to CCl4 significantly protected against cytotoxicity (Figure 1B).

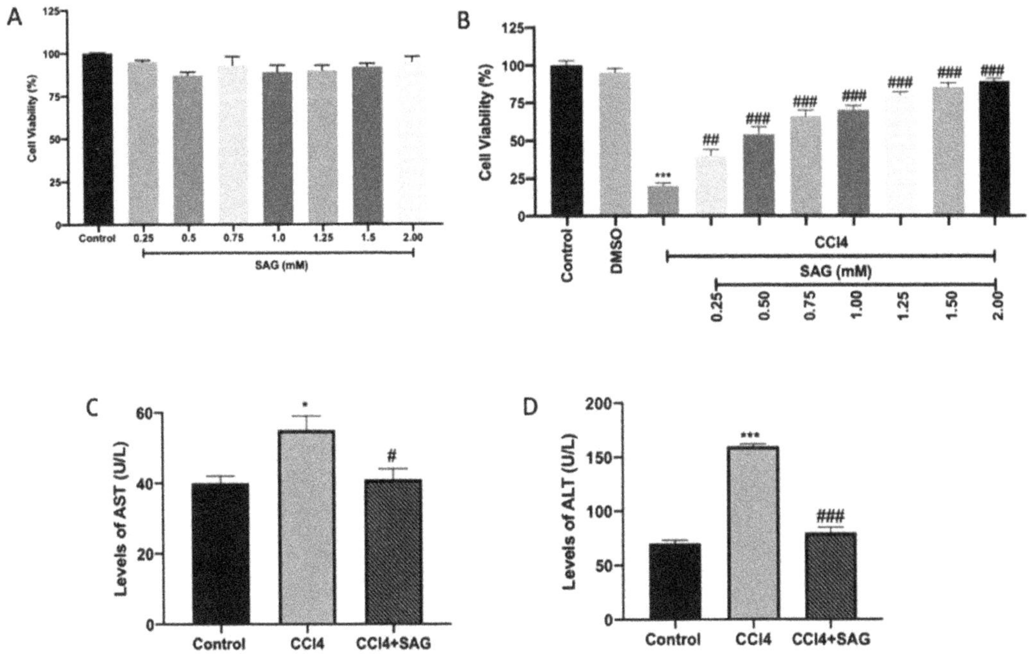

Figure 1. Preliminary data on SAG effect on primary hepatic cells: Viability of cells treated with SAG (**A**); Viability of cells pre-treated for 1 h with SAG followed by exposure to CCl4 (4 mM) for 6 h (**B**); Enzymatic activities of AST (**C**) and ALT (**D**) in the supernatant. The results were analyzed by one-way ANOVA, followed by a Bonferroni post hoc test for multiple comparisons. A p-value of less than 0.05 was considered significant. A p-value of less than 0.05 was considered significant. $^{\#}$ $p < 0.05$ vs. control, * $p < 0.05$ vs. CCL4, $^{\#\#}$ $p < 0.01$ vs. control, $^{\#\#\#}$ $p < 0.001$ vs. control, *** $p < 0.001$ vs. CCL4.

In primary hepatic cells exposed to CCl4, both AST and ALT levels presented a marked increase, respectively (Figure 1C,D, respectively), in relation to the control group. By contrast, pre-treatment with SAG at 2.00 mM significantly reduced both levels (Figure 1C,D).

2.2. Experimental Timeline

CCl4 is well-known to cause hepatic injury, apoptosis and necrosis [6]. In order to investigate the effects of SAG on hepatic damage, mice were intraperitoneally injected with CCl4 1 mL/kg twice a week for 8 consecutive weeks to induce liver fibrosis and were treated orally for 8 weeks with SAG (30 mg/kg) dissolved in saline (Figure 2).

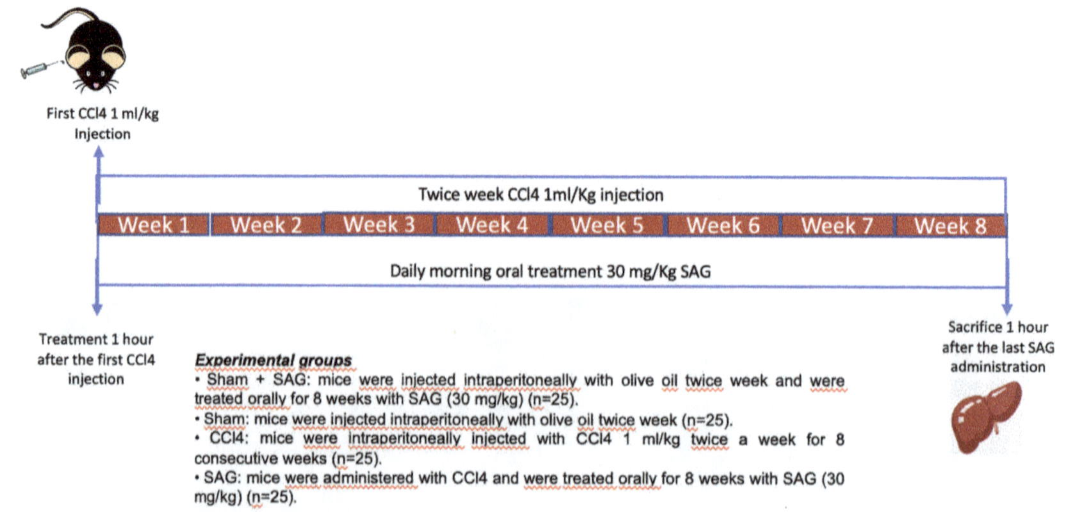

Figure 2. Schematic of study design.

2.3. *Effects of SAG on Oxidative Stress Induced by CCl4 Chronic Exposure*

CCl4 intoxication downregulated SOD (Figure 3A) and GSH (Figure 3B) compared to the sham groups. SAG administration significantly restored SOD activity (Figure 3A) and GSH levels (Figure 3B) in the liver. Additionally, CCl4 administration increased GSSG levels, as compared to the sham groups, whereas it was significantly decreased by SAG administration (Figure 3C). Glutathione peroxidase (GPx) activity was impaired after CCl4 chronic exposure, as compared to the sham groups, whereas SAG administration significantly restored it (Figure 3D). On the same line, SAG supplementation reduced lipid peroxidation (Figure 3E), H_2O_2 levels (Figure 3F) and ROS levels (Figure 3G) in the samples.

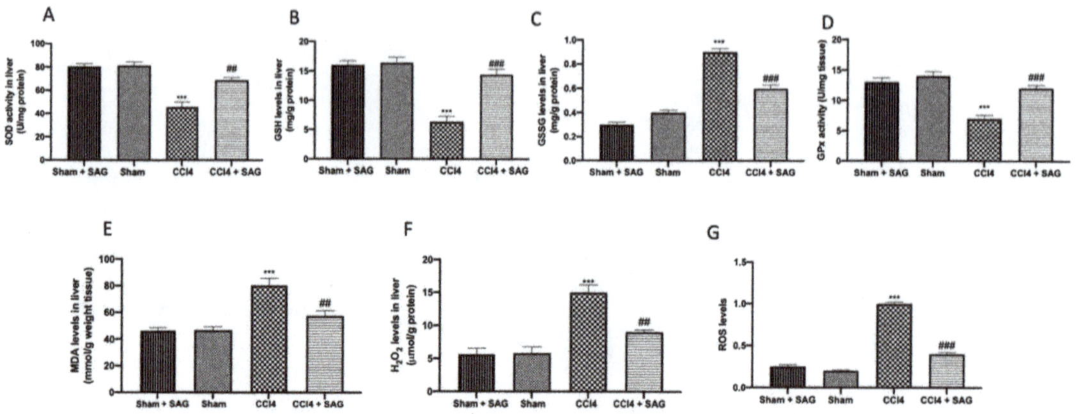

Figure 3. Administration of SAG reduced oxidative stress induced by CCl4 injections: SOD activity (**A**); GSH levels (**B**); GSSG levels (**C**); GPx activity (**D**); MDA levels (**E**); H_2O_2 levels (**F**); ROS levels (**G**). For each analysis, $n = 5$ animals for each group were employed. The results were analyzed by one-way ANOVA, followed by a Bonferroni post hoc test for multiple comparisons. A p-value of less than 0.05 was considered significant. A p-value of less than 0.05 was considered significant. $^{\#\#}$ $p < 0.01$ vs. sham, $^{\#\#\#}$ $p < 0.001$ vs. sham, *** $p < 0.001$ vs. CCL4.

2.4. Effects of SAG on Mitophagy Impairments Induced by CCl4 Chronic Exposure

Western blot analysis displayed that the CCl4 induced a decrease in anti-oxidants, including Nrf2 (Figure 4A), HO-1 (Figure 4B) and NQO-1 (Figure 4C), which were restored by treatment with SAG. Western blot analysis was also employed to evaluate the effect of SAG on the mitochondrial homeostasis. Chronic CCl4 exposure impaired mitophagy and mitochondrial biogenesis. PINK1 (Figure 4D) and Parkin (Figure 4E) levels were decreased following CCl4 intoxication, whereas SAG administration significantly increased their levels.

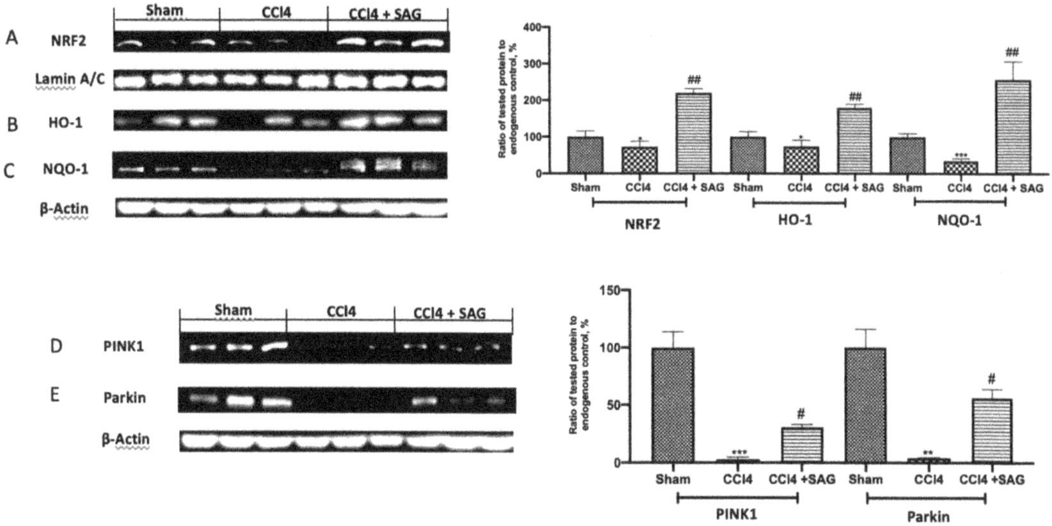

Figure 4. Administration of SAG ameliorated the impaired mitophagy induced by CCl4 injections: Western blot analysis of: NRF2 (**A**), HO-1 (**B**), NQO-1 (**C**), PINK1 (**D**) and Parkin (**E**) levels. For each analysis, $n = 5$ animals for each group were employed. The results were analyzed by one-way ANOVA, followed by a Bonferroni post hoc test for multiple comparisons. A p-value of less than 0.05 was considered significant. A p-value of less than 0.05 was considered significant. $^{\#}$ $p < 0.05$ vs. sham, * $p < 0.05$ vs. CCL4, $^{\#\#}$ $p < 0.01$ vs. sham, ** $p < 0.01$ vs. CCL4, *** $p < 0.001$ vs. CCL4.

2.5. Effects of SAG on Pro-Inflammatory Mediator Secretion Induced by CCl4 Chronic Exposure

ELISA analysis showed elevated levels of TNF-α (Figure 5A), IL-6 (Figure 5B), IL-1β (Figure 5C) and MCP-1 (Figure 5D) in the liver after CCl4 intoxication, which were reduced by SAG administration. In addition, the circulating pro-inflammatory cytokines (Figure 5E, 5F and 5G, respectively), as well as the chemokine (Figure 5H), were significantly increased in the CCl4-treated mice compared to the sham group. However, mice treated with SAG exhibited a significant downregulation of pro-inflammatory mediators in the serum.

2.6. Effects of SAG on TL4/NFkB Signaling Activation Induced by CCl4 Chronic Exposure

In order to confirm SAG anti-inflammatory effects, Western blot analyses were conducted. TLR4 (Figure 6A) and MyD88 (Figure 6B) were significantly increased in CCl4-treated mice, as compared to the sham group. SAG administration significantly reduced their expressions. Next, the NF-kB pathway was examined. The sham group showed basal IkB-α cytosolic expression (Figure 6C) and poor nuclear NF-kB p-65 levels (Figure 6D). CCl4 intoxication importantly degraded cytosolic IkB-α and increased nuclear NF-kB. SAG administration restored IkB-α in cytosol (Figure 6C) and reduced NF-kB nuclear levels (Figure 6D).

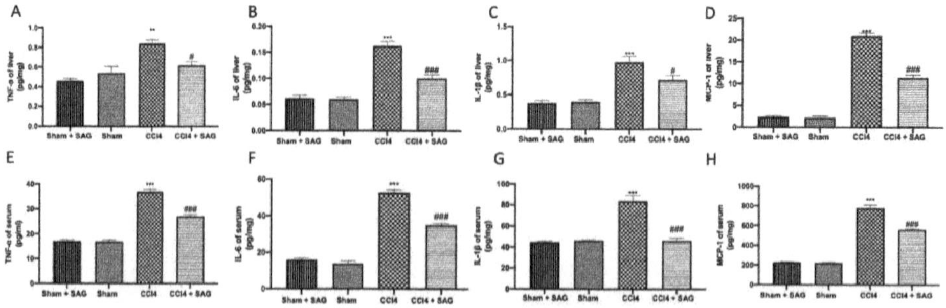

Figure 5. Administration of SAG reduced pro-inflammatory mediator secretion induced by CCl4 injections: Liver levels of TNF-α (**A**), IL-6 (**B**), IL-1β (**C**) and MCP-1 (**D**); Serum levels of TNF-α (**E**), IL-6 (**F**), IL-1β (**G**) and MCP-1 (**H**). For each analysis, $n = 5$ animals for each group were employed. The results were analyzed by one-way ANOVA, followed by a Bonferroni post hoc test for multiple comparisons. A p-value of less than 0.05 was considered significant. A p-value of less than 0.05 was considered significant. # $p < 0.05$ vs. sham, ** $p < 0.01$ vs. CCl4, ### $p < 0.001$ vs. sham, *** $p < 0.001$ vs. CCl4.

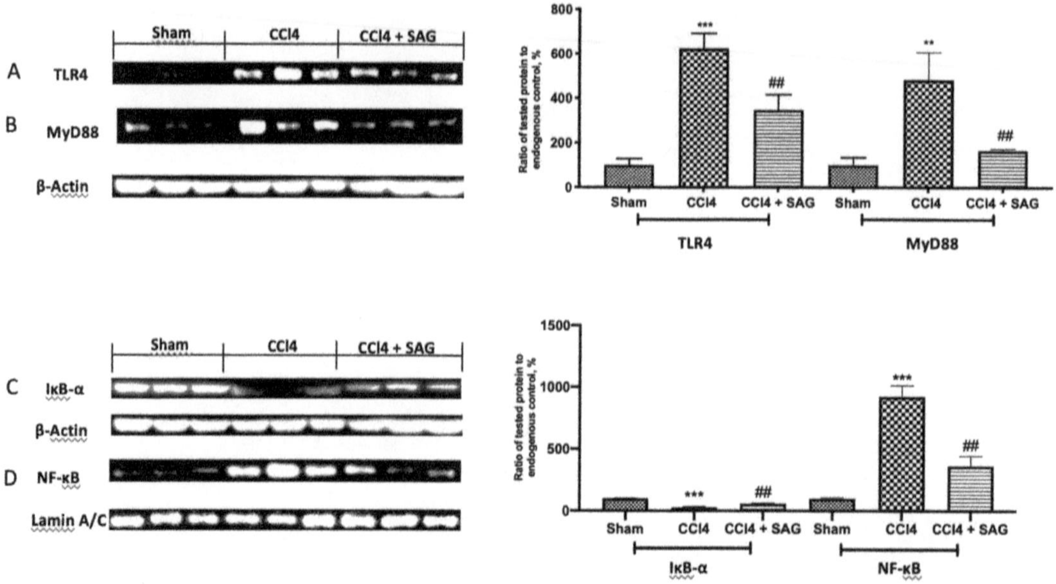

Figure 6. Administration of SAG reduced the activation of the TLR4/NF-kB pathway induced by CCl4 injections: Western blot analysis of: TLR4 (**A**), MyD88 (**B**), IkB-α (**C**) and NF-kB (**D**) levels. For each analysis, $n = 5$ animals for each group were employed. The results were analyzed by one-way ANOVA, followed by a Bonferroni post hoc test for multiple comparisons. A p-value of less than 0.05 was considered significant. A p-value of less than 0.05 was considered significant. ## $p < 0.01$ vs. sham, ** $p < 0.01$ vs. CCl4, *** $p < 0.001$ vs. CCl4.

2.7. Effects of SAG on Liver Fibrosis Induced by CCl4 Chronic Exposure

To evaluate fibrosis, CCl4-induced Masson trichrome staining was performed. CCl4-treated animals showed an altered lobule structure through paraplastic connective tissue, and mild to serious fibrosis was detected (Figure 7C,E) compared to the sham group (Figure 7B,E). No differences were assessed between the sham and sham + SAG groups

(Figure 7A,E). An assessment of hydroxyproline was used to test the collagen content in the liver tissue. The hydroxyproline content was also increased in CCl4-treated mice (Figure 7F), which was well in line with the upregulated levels of α-SMA (Figure 7G) and TGF-β (Figure 7H), whereas IL10 levels were decreased (Figure 7I) in the liver tissue. SAG administration significantly reduced the collagen deposition (Figure 7D,E), hydroxyproline content (Figure 7F) and α-SMA (Figure 7G) and TGF-β (Figure 7H) levels, whereas it increased IL10 expression (Figure 7I) in the liver tissue. Well in line with the tissue results, TGF-β levels (Figure 7J) increased in the serum with CCl4, whereas IL10 levels (Figure 7K) decreased. Both levels in the serum were normalized by SAG administration.

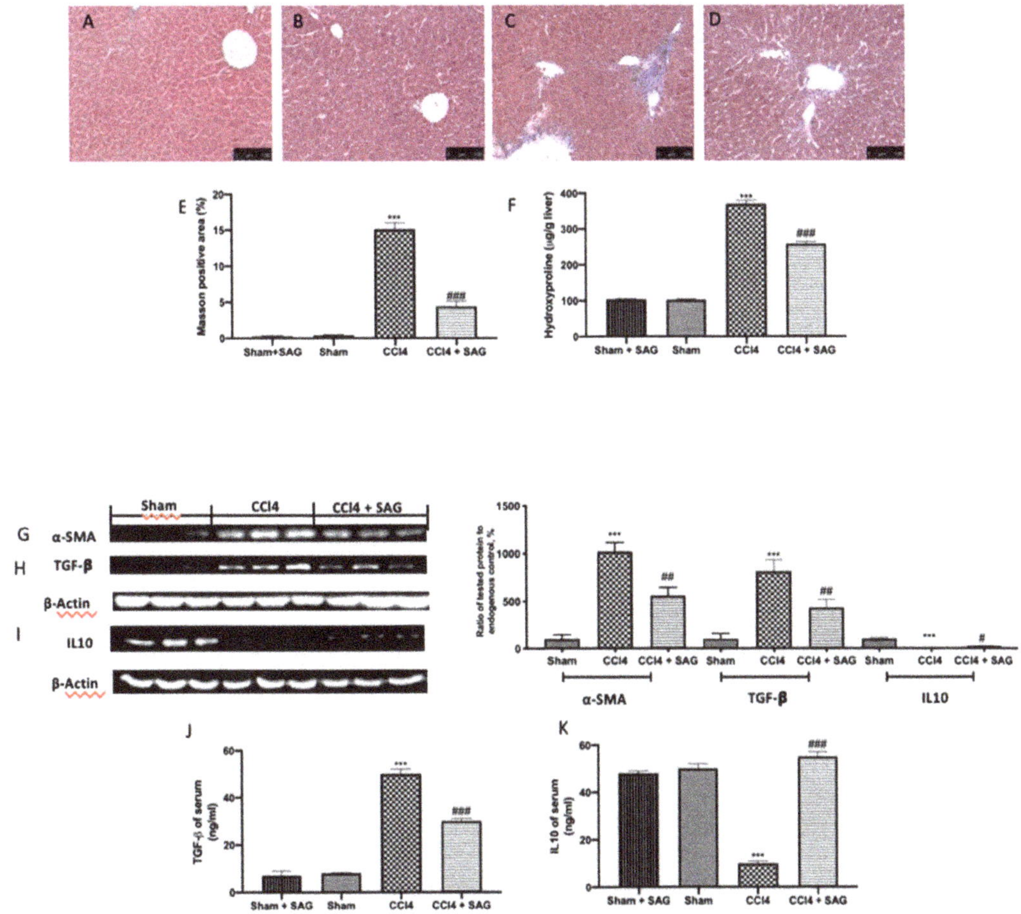

Figure 7. Administration of SAG reduced liver fibrosis induced by CCl4 injections: Masson trichrome staining: Sham + SAG (**A**), Sham (**B**), CCl4 (**C**), CCl4 + SAG (**D**), graphical quantification of Masson-positive area (**E**); Hydroxyproline content (**F**); Western blot analysis of: α-SMA (**G**), TGF-β (**H**) and IL10 (**I**) levels in the liver tissue; Serum levels of TGF-β (**J**) and IL10 (**K**). For each analysis, $n = 5$ animals for each group were employed. The results were analyzed by one-way ANOVA, followed by a Bonferroni post hoc test for multiple comparisons. A *p*-value of less than 0.05 was considered significant. A *p*-value of less than 0.05 was considered significant. $^{\#}$ $p < 0.05$ vs. sham, $^{\#\#}$ $p < 0.01$ vs. sham, $^{\#\#\#}$ $p < 0.001$ vs. sham, *** $p < 0.001$ vs. CCl4.

2.8. Effects of SAG on Histopathological Alterations and Liver Function Induced by CCl4 Chronic Exposure

In liver, CCl4 administration increased the myeloperoxidase (MPO) activity, which is used as an indicator of polymorphonuclear (PMN) cell infiltration (Figure 8F), liver cell damage and necrosis (Figure 8C,E), whereas the sham (Figure 8B,E) and sham + SAG (Figure 8A,E) groups showed normal histological architecture. CCl4 intoxication also compromised liver function, as shown by ALT (Figure 8G) and AST (Figure 8H) levels. SAG administration significantly ameliorated histological damage (Figure 8D,E), MPO activity (Figure 8F) and liver function (Figure 8G,H).

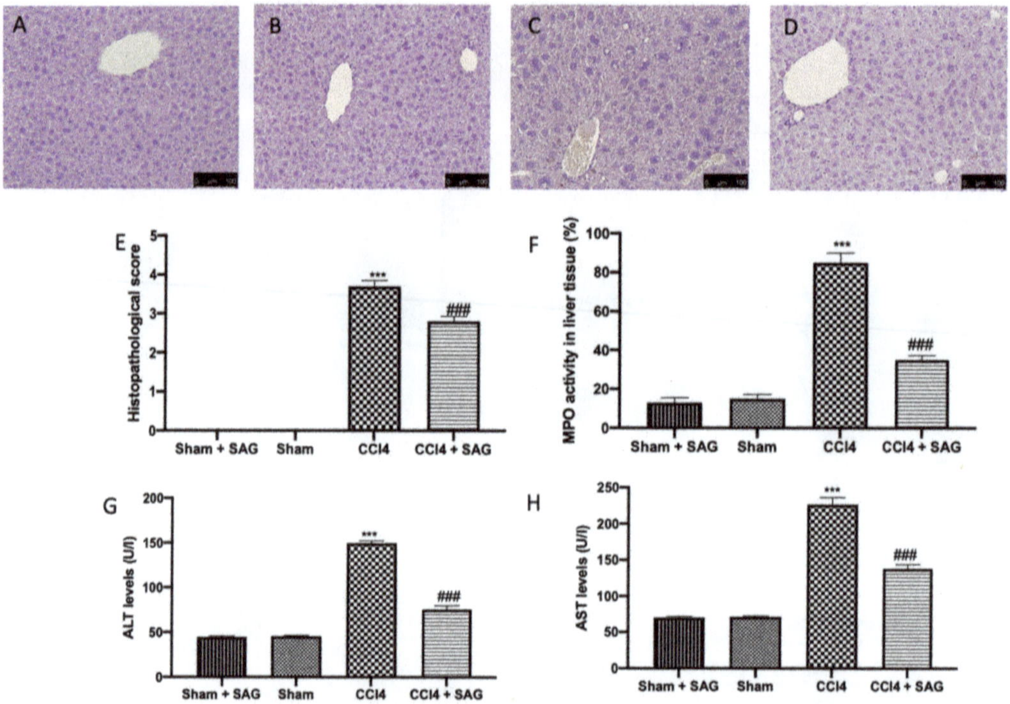

Figure 8. Administration of SAG reduced histological changes and liver dysfunction induced by CCl4 injections: Histological analysis: Sham + SAG (**A**), Sham (**B**), CCl4 (**C**), CCl4 + SAG (**D**), Histological score (**E**); MPO activity (**F**); ALT (**G**) and AST (**H**) levels. For each analysis, $n = 5$ animals for each group were employed. The results were analyzed by one-way ANOVA, followed by a Bonferroni post hoc test for multiple comparisons. A p-value of less than 0.05 was considered significant. A p-value of less than 0.05 was considered significant. ### $p < 0.001$ vs. sham, *** $p < 0.001$ vs. CCl4.

3. Discussion

Liver is an important organ involved in several activities, including bile acid secretion, the generation of blood clotting factors and detoxification. Liver injury may be induced by a variety of factors, including drugs, microbes, xenobiotics and several metabolites [19–22]. Previous studies have indicated that increasing GSH plasma levels has beneficial systemic effects [18], well in line with the described GSH depletion in non-alcoholic fatty liver disease patients compared with controls [23]. However, the oral administration of GSH does not significantly enhance GSH in plasma, while GSH derivatives have been described as able to cross the cell membranes and enhance the oral availability. SAG, a GSH precursor, is more stable in plasma, uptaken by cells and later converted to GSH. In this paper, we evaluated the hepatoprotective effects of SAG in a CCl4-model of liver injury, in which, the liver microsomal oxidizing systems related to cytochrome P-450 produce reactive metabolites of CCl4, including trichloroperoxyl radicals (CCl3O3·) or trichloromethyl radicals (CCl3). These reactive radicals induce lipid peroxidation, causing inflammation and hepatocellular damage and enhancing the production of fibrotic. The pre-treatment of primary hepatic cells with SAG followed by exposure to CCl4 protected cells against the cytotoxicity, indicating that the extracts counteracted the toxicity of products generated by the metabolism of CCl4. This result was corroborated by the analysis of ALT and AST liver enzymes, which overflow to the extracellular medium due to membrane permeability alterations after cellular injury. As expected, exposure to CCl4 induced a marked release of ALT and AST in primary hepatic cells, while the pre-treatment with SAG promoted the normalization of liver enzyme levels, indicating the protection of the cell membrane. SAG administration, thanks to its ability to maintain a cellular reductive state, enhanced antioxidants, including Nrf2, HO-1 and NQO-1, and restored SOD activity and GSH levels in the liver as well. SAG also reduced GSSG levels, lipid peroxidation and H_2O_2 liver levels. Accordingly, recent evidence focused the attention on the importance of the mitochondrial dysfunction induced by the sustained ROS in liver fibrosis [24]. Mitophagy is a useful mechanism that aims to remove impaired mitochondria, and is activated by organelle membrane depolarization [25]. Once activated, this mechanism leads to PINK1 stabilization on the mitochondrial outer membrane and to Parkin recruitment from cytosol. The increased oxidative stress characteristic on liver disfunction causes the impairment of mitophagy and accumulation of dysfunctional and damaged mitochondria. A failure of mitophagy or mitochondrial biogenesis affects hepatocellular function during ischemia/reperfusion [26] and cholestasis [27]. Moreover, interplay between these processes and oxidative metabolism, which may contribute to hepatic cell damage, has been suggested to occur in the steatotic liver [28,29].

Our results showed the protective effect of SAG administration in restoring mitophagy, as shown by the increased PINK1 and Parkin expressions in livers exposed to CCl4 intoxication. As already shown [30,31], CCl4 damage is closely associated with inflammation [32]. SAG administration strongly decreased pro-inflammatory cytokines TNF-α, IL-6, MCP-1 and IL-1β in both the serum and liver. This cytokine release is induced by several pathways, including the TLR4/MyD88, a well-known signaling involved in the CCl4-induced liver injury. In particular, TLR4 recruits the specific mediator MyD88, which triggers downstreaming signaling events for the NFkB phosphorylation and a consequent release of pro-inflammatory cytokines [33,34]. ROS interacts with NF-κB signaling pathways in many ways. The transcription of NF-κB-dependent genes influences the levels of ROS in the cell, and, in turn, the levels of NF-κB activity are also regulated by the levels of ROS [35]. NFkB is a transcription factor normally located in the cytoplasm and bound to the IKK complex. Thanks to the activation of the TLR4 pathway, Ikb-α is phosphorylated and degraded by proteasome, leading to NFkB release and translocation into the nucleus to increase the expression of targeting genes implicated in the inflammatory response [36,37]. Reducing inflammation and oxidative stress, SAG administration importantly reduced liver fibrosis, as confirmed by the decreased collagen deposition, assessed by Masson trichrome staining, and the reduced hydroxyproline contents and α-sma and TGF-β expressions, while it restored IL10 levels. Consequently, SAG administration attenuated histological liver damage, assessed by hematoxylin and eosin staining; MPO activity, used as an indicator of PMN cell infiltration [38–40]; and improved hepatic function, as shown by ALT and AST levels.

In conclusion, the present study indicated the potential protective effects of SAG against CCl4-induced liver damage. The hepatoprotective effects of SAG depend on its ability to reduce the generation of ROS, as well as pro-inflammatory signaling through the de-activation of TLR4/NF-κB signaling (Figure 9). Overall, the present study provides evidence for the protective effects of SAG against CCl4-induced liver injury and suggests SAG as a potential hepatoprotective agent used to prevent oxidative liver damage.

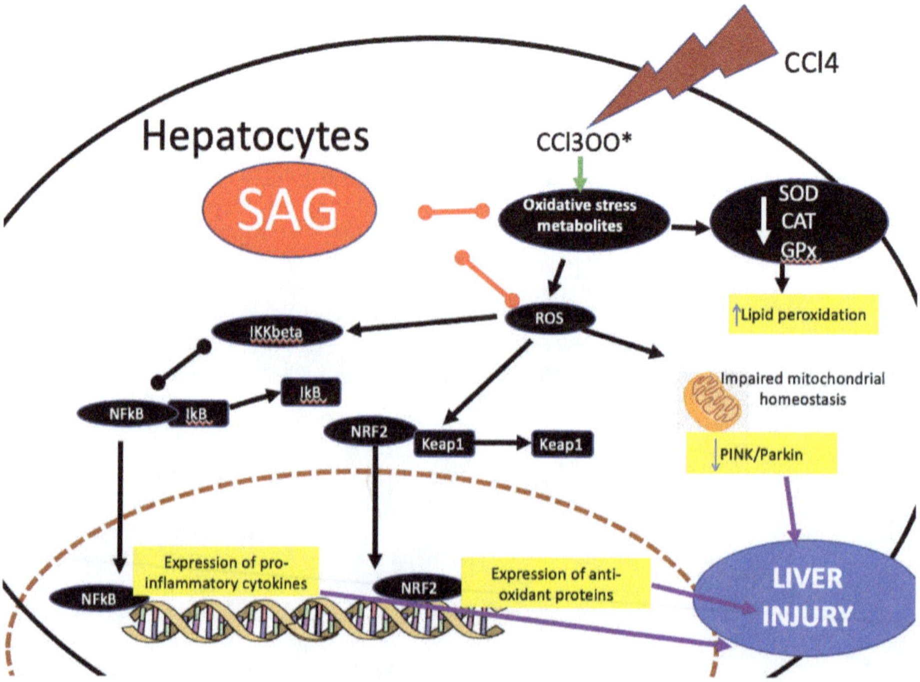

Figure 9. Graphical summary of SAG mechanism.

4. Materials and Methods

4.1. Culture of Hepatocytes

Isolated hepatocytes were prepared from male C57BL/6 mice by the collagenase digestion [41]. Cells were purified by several centrifugations and cultured in DMEM medium supplemented with 10% FBS, HEPES (4.5 mM), penicillin (100 U/mL), L-glutamine (2 mM) and sodium bicarbonate (0.17 M). Cells were maintained under standard culture conditions at 37 °C and 5% CO_2 in a humid environment and trypsinized for experiments whenever the cellular confluence was 70%.

4.2. Cytotoxicity Assay

The assessment of the hepatoprotective effect of SAG was performed using the 3-[4,5-dimethylthiazol-2-yl]-2,5-diphenyl tetrazolium bromide (MTT) reduction assay [42].

Cells (1 × 105 cells/well) were seeded in 96-well plates and maintained for 24 h in culture conditions in order to adhere to them. The experimental design involved pre-treatment of primary hepatic cells with SAG (at 0.25, 0.5, 0.75, 1.0, 1.25, 1.5 or 2.00 mM, dissolved in saline as stock solution and diluted in DMEM) for 1 h followed by exposure to CCl4 (4 mM diluted in 0.5% DMSO) for 6 h. Control group was carried out without CCl4 exposure. At the exposure time, the medium was exchanged and, after 21 h, 10 µL of MTT (5 mg/mL) was added to each well, and the plates were incubated for 3 h in culture conditions. Supernatants were removed and 100 µL of DMSO was added to each well to solubilize the formazan crystals. The absorbance was measured on a microplate reader at 570 nm.

4.3. Activities of ALT and AST in Cell Supernatants

An evaluation of the activities of the liver-specific enzymes alanine aminotransferase (ALT) and aspartate aminotransferase (AST) is recommended for the assessment of hepatocellular function [43]. For this purpose, primary hepatic cells (1×10^6 cells/well) were seeded in 24-well plates, incubated for 24 h in culture conditions and subsequently pre-treated with SAG (at 2.00 mM) followed by exposure to CCl4. Afterwards, supernatants were collected and centrifuged at 5000 rpm for 5 min using a microtube centrifuge. Analyses of the enzymatic activities of AST and ALT were performed according to the manufacturer's instructions (Sigma-Aldrich, Milan, Italy).

4.4. Animals

Sprague Dawley male rats (250 gr, Envigo, Milan, Italy) and C57BL/6 mice (male 20–22 g; age 6–8 weeks) were purchased from Envigo (Milan, Italy) and employed for this study. Messina University Review Board for the carefulness of animals permitted the research (211/2021-PR). All animal experiments agree with the new regulations in Italy (D.Lgs 2014/26), EU regulations (EU Directive 2010/63).

4.5. Carrageenan-Induced Paw Edema (Preliminary Data)

Carrageenan-induced paw edema was performed as previously indicated by a sub-plantar injection of carrageenan (0.1 mL/rat of a 1% suspension in saline) (Sigma-Aldrich, Milan, Italy) into the right hind paw [44]. Increase in paw volume (mL) was measured using a plethysmometer (Ugo Basile, Varese, Italy) immediately prior to the carrageenan injection and every hour for 6 h (Supplemental Figure S1).

4.6. Experimental Groups

Respectively, mice were randomly assigned to different groups, as described below:

- Sham + SAG: mice were injected intraperitoneally with olive oil twice a week and were treated orally for 8 weeks with SAG (30 mg/kg) (Merk, CAS n 3054-47-5, AMBH95E07091) dissolved in saline.
- Sham: mice were injected intraperitoneally with olive oil twice a week.
- CCl4: mice were intraperitoneally injected with CCl4 1 mL/kg (diluted at 1:10 in olive oil) twice a week for 8 consecutive weeks to induce liver fibrosis [11].
- SAG: mice were administered with CCl4 to induce liver fibrosis as vehicle group and were treated orally for 8 weeks with SAG (30 mg/kg) dissolved in saline.

At the end of the experiment, blood samples and liver tissues were collected from the mice for further assays. The dose of SAG has been chosen based on experiments previously conducted in our laboratory.

4.7. Analysis of Biochemical Indicators

Serum alanine aminotransferase (ALT) and aspartate aminotransferase (AST) levels were determined according to manufacturer's instructions (Sigma-Aldrich). As for the hepatic hydroxyproline content, snap-frozen liver specimens were collected and the hydroxyproline content was quantified following the manufacturer's instructions (Sigma-Aldrich) [11]. Determination of SOD activity was performed as already described [45]. SOD activity (U/μg protein) was determined using a microplate reader at 560 nm [46]. GSH levels were determined using a microplate reader at 412 nm and expressed as ng/mg wet tissue [45]. GSSG levels were determined using a microplate reader at 450 nm and expressed as ng/mg wet tissue (MyBiosource MBS749109). Glutathione peroxidase activity was estimated by measuring the oxidation of guaiacol in the liver of treated mice according to a standard method [47]. Lipoperoxidation was estimated using the thiobarbituric acid reactive substances (TBARS) test [48]. Briefly, liver tissue was weighed and homogenized in a 1.15% (*w*/*v*) KCl solution. A 100 mL aliquot of homogenate was then removed and added to a reaction mixture containing 200 mL 8.1% (*w*/*v*) lauryl sulfate, 1.5 mL 20% (*v*/*v*) acetic acid (pH 3.5), 1.5 mL 0.8% (*w*/*v*) thiobarbituric acid and 700 mL distilled water. Samples were then boiled for one hour at 95 °C and centrifuged at 3000× *g* for 10 min. The absorbance of the supernatant was measured spectrophotometrically at 532 nm. MDA levels were expressed as nmol/g wet tissue weight. Whole liver-derived lysates were diluted according to the manufacturer instruction (E-BC-K102-S) and incubated with ammonium molybdate reagent. H_2O_2 content can be calculated by measuring the absorbance value at 405 nm. Relative level of ROS was detected centrifuging liver tissue in the appropriate buffer (20 mmol/L Tir-HCl (pH 7.4), 20 mmol/L NaH_2PO_4, 5 mmol/L $MgCl_2$, 130 mmol/L KCl and 30 mmol/L glucose) and incubating the supernatant with DCFH-DA (2′, 7′-dichlorodihydrofluorescein diacetate) for 15 min at 37 °CC. Then, the reaction was terminated by adding 1 micromol/L of H_2O_2. The absorbance value was determined by fluorescence spectrophotometer [49].

4.8. *Determination of Myeloperoxidase Activity*

MPO activity in liver tissue was used as an indicator of polymorphonuclear (PMN) cell infiltration using a method previously described [50]. Briefly, tissue was weighed and homogenized in a solution containing 0.5% (*w*/*v*) hexadecyltrimethylammonium bromide dissolved in 10 mmol/L potassium phosphate buffer (pH 7.4) and centrifuged for 30 min at 20,000× *g* at 4 °C. An aliquot of supernatant was then removed and added to a reaction mixture containing 1.6 mmol/L tetramethylbenzidine and 0.1 mmol/L hydrogen peroxide (H_2O_2). The rate of change in absorbance was measured spectrophotometrically at 650 nm.

4.9. *ELISA*

The levels of tumor necrosis factor (TNF)-α, interleukin (IL)-1β, IL-6 and monocyte chemoattractant protein (MCP)-1 in serum and liver tissue samples were determined by ELISA kits (R&D systems) [51], The levels of TGF-β and IL10 in serum were determined by ELISA kits (R&D systems) [52].

4.10. *Histopathological Examination*

For histopathological investigations, liver tissues were fixed in formaldehyde solution (10% in PBS); histological sections were stained with hematoxylin and eosin (H&E) and evaluated using a Leica DM6 microscope (Leica Microsystems SpA, Milan, Italy) equipped with a motorized stage and associated with Leica LAS X Navigator software (Leica Microsystems SpA, Milan, Italy) [53]. Histopathologic scores were graded following the Ishak scoring system as follows: 0, no fibrosis; 1, fibrosis expansion of some portal areas ± short fibrous septa; 2, fibrosis expansion of portal areas ± short fibrous septa; 3, fibrosis expansion of most portal areas with occasional portal to portal bridging; 4, fibrosis expansion of portal areas with marked portal to portal bridging, as well as portal to central; 5, marked bridging with occasional nodules (incomplete cirrhosis); 6, cirrhosis, probable or definite [54–56]. Collagen deposition was evaluated by Masson trichrome staining performed according to the manufacturer's protocol (Bio-Optica, Milan, Italy) [57].

4.11. *Western Blot Analysis*

Western blot analyses were made as previously described [58]. Filters were blocked with 1× PBS, 5% (*w*/*v*) no-fat dried milk (PM) for 40 min at room temperature and then probed with one of the next primary antibodies: anti-TGF-β (Santa Cruz Biotechnology, sc-130348), anti-IL-10 (Santa Cruz Biotechnology, sc-8438) or anti-IkB-α (Santa Cruz Biotechnology, sc-1643), or anti- NF-kB p-65 (Santa Cruz Biotechnology, sc-8008), or anti-TLR4 (Santa Cruz Biotechnology, sc-293072), anti-MyD88 (Santa Cruz Biotechnology, sc-74532), or anti-α-SMA (Santa Cruz Biotechnology, sc-53015), or anti-Nrf-2 (Santa Cruz Biotechnology, sc-365949), or anti-HO-1 (Santa Cruz Biotechnology, sc-136960), or NQO-1 (Abcam), or anti-PINK1 (Santa Cruz Biotechnology, sc-517353), or anti-Parkin (Santa Cruz Biotechnology, sc-32282) in 1× PBS, 0.1% Tween-20, 5% *w*/*v* no-fat dried milk (PMT) at 4 °C overnight. Membranes were incubated with peroxidase-conjugated bovine anti-mouse IgG secondary antibody or peroxidase-conjugated goat anti-rabbit IgG (1:2000, Jackson ImmunoResearch, West Grove, PA, USA) [59]. Blots were also incubated with primary antibody against β-actin protein (1:10,000; Sigma-Aldrich Corp.) or lamin (1:10,000; Sigma-Aldrich Corp.), used as internal standards [60]. The relative expressions of the protein bands were detected and quantified by densitometry, as previously explained [61]. In the experiments including Western blot, a representative blot is displayed and densitometric analysis is related in each figure.

4.12. Statistical Evaluation

All values in the images and text are expressed as mean ± standard error of the mean (SEM) of N observations. For in vivo studies, N represents the number of animals. In experiments involving histology and immunohistochemistry, the illustrations represent the outcomes of at least three independent experiments. The results were analyzed by one-way ANOVA, followed by a Bonferroni post hoc test for multiple comparisons. A p-value of less than 0.05 was considered significant. A p-value of less than 0.05 was considered significant. # $p < 0.05$ vs. sham, * $p < 0.05$ vs. vehicle, ## $p < 0.01$ vs. sham, ** $p < 0.01$ vs. vehicle, ### $p < 0.001$ vs. sham, *** $p < 0.001$ vs. vehicle.

Supplementary Materials: The following supporting information can be downloaded at: https://www.mdpi.com/article/10.3390/ijms23084429/s1.

Author Contributions: Conceptualization, R.F. and S.C.; methodology, A.T.S.; software, M.C.; validation, R.S.; formal analysis, R.D.; investigation, D.I.; resources, M.L.O.; data curation, S.M. and L.I.; writing—original draft preparation, R.F.; writing—review and editing, R.D.P.; supervision, V.C.; project administration, R.D.P.; funding acquisition, S.C. All authors have read and agreed to the published version of the manuscript.

Funding: This research received no external funding.

Institutional Review Board Statement: The University of Messina Review Board for animal care. (OPBA) approved the study. All animal experiments agreed with the new Italian regulations (D.Lgs 2014/26), EU regulations (EU Directive 2010/63), and the ARRIVE guidelines.

Informed Consent Statement: Not applicable.

Data Availability Statement: The data used to support the findings of this study are available from the corresponding author upon request.

Conflicts of Interest: The authors declare no conflict of interest.

References

1. Bataller, R.; Brenner, D.A. Liver fibrosis. *J. Clin. Investig.* **2005**, *115*, 209–218. [CrossRef] [PubMed]
2. Poli, G.; Parola, M. Oxidative damage and fibrogenesis. *Free Radic. Biol. Med.* **1997**, *22*, 287–305. [CrossRef]
3. Harrison, D.G.; Cai, H.; Landmesser, U.; Griendling, K.K. Interactions of angiotensin II with NAD (P) H oxidase, oxidant stress and cardiovascular disease. *J. Renin-Angiotensin-Aldosterone Syst.* **2003**, *4*, 51–61. [CrossRef] [PubMed]
4. Inoguchi, T.; Sonta, T.; Tsubouchi, H.; Etoh, T.; Kakimoto, M.; Sonoda, N.; Sato, N.; Sekiguchi, N.; Kobayashi, K.; Sumimoto, H. Protein kinase C–dependent increase in reactive oxygen species (ROS) production in vascular tissues of diabetes: Role of vascular NAD (P) H oxidase. *J. Am. Soc. Nephrol.* **2003**, *14*, S227–S232. [CrossRef] [PubMed]
5. Singh, K.K. Mitochondrial dysfunction is a common phenotype in aging and cancer. *Ann. N. Y. Acad. Sci.* **2004**, *1019*, 260–264. [CrossRef] [PubMed]
6. Scholten, D.; Trebicka, J.; Liedtke, C.; Weiskirchen, R. The carbon tetrachloride model in mice. *Lab. Anim.* **2015**, *49*, 4–11. [CrossRef] [PubMed]
7. Heindryckx, F.; Colle, I.; Van Vlierberghe, H. Experimental mouse models for hepatocellular carcinoma research. *Int. J. Exp. Pathol.* **2009**, *90*, 367–386. [CrossRef]
8. Iwaisako, K.; Jiang, C.; Zhang, M.; Cong, M.; Moore-Morris, T.J.; Park, T.J.; Liu, X.; Xu, J.; Wang, P.; Paik, Y.H.; et al. Origin of myofibroblasts in the fibrotic liver in mice. *Proc. Natl. Acad. Sci. USA* **2014**, *111*, E3297–E3305. [CrossRef]
9. Halliwell, B. Oxidants and human disease: Some new concepts. *FASEB J.* **1987**, *1*, 358–364. [CrossRef]
10. Bosek, P.; Nakano, M. Hepatoprotective effect of rooibos tea (*Aspalathus linearis*) on CCl 4-induced liver damage in rats. *Physiol. Res.* **2003**, *52*, 461–466.
11. Liu, Y.; Wen, P.H.; Zhang, X.X.; Dai, Y.; He, Q. Breviscapine ameliorates CCl4induced liver injury in mice through inhibiting inflammatory apoptotic response and ROS generation. *Int. J. Mol. Med.* **2018**, *42*, 755–768. [PubMed]
12. Sies, H. Strategies of antioxidant defense. *Eur. J. Biochem.* **1993**, *215*, 213–219. [CrossRef] [PubMed]

13. Lieber, C.S. Role of oxidative stress and antioxidant therapy in alcoholic and nonalcoholic liver diseases. *Adv. Pharmacol.* **1997**, *38*, 601–628. [PubMed]
14. Cervinkova, Z.; Drahota, Z. Enteral administration of lipid emulsions protects liver cytochrome c oxidase from hepatotoxic action of thioacetamide. *Physiol. Res.* **1998**, *47*, 151–154.
15. Lu, S.C. Regulation of hepatic glutathione synthesis: Current concepts and controversies. *FASEB J.* **1999**, *13*, 1169–1183. [CrossRef]
16. Meister, A.; Anderson, M.E. Glutathione. *Annu. Rev. Biochem.* **1983**, *52*, 711–760. [CrossRef]
17. Vogel, J.-U.; Cinatl, J.; Dauletbaev, N.; Buxbaum, S.; Treusch, G.; Cinatl, J.; Gerein, V.; Doerr, H.W. Effects of S-acetylglutathione in cell and animal model of herpes simplex virus type 1 infection. *Med. Microbiol. Immunol.* **2005**, *194*, 55–59. [CrossRef]
18. Fanelli, S.; Francioso, A.; Cavallaro, R.A.; d'Erme, M.; Putignano, P.; Miraglia, N.; Mosca, L. Clinical Nutrition & Dietetics. *Clin. Nutr. Diet.* **2018**, *4*, 134.
19. Rowland, A.; Miners, J.O.; Mackenzie, P.I. The UDP-glucuronosyltransferases: Their role in drug metabolism and detoxification. *Int. J. Biochem. Cell Biol.* **2013**, *45*, 1121–1132. [CrossRef]
20. Mao, S.; Gao, D.; Liu, W.; Wei, H.; Lin, J.M. Imitation of drug metabolism in human liver and cytotoxicity assay using a microfluidic device coupled to mass spectrometric detection. *Lab Chip* **2012**, *12*, 219–226. [CrossRef]
21. Dawson, S.; Stahl, S.; Paul, N.; Barber, J.; Kenna, J.G. In Vitro inhibition of the bile salt export pump correlates with risk of cholestatic drug-induced liver injury in humans. *Drug Metab. Dispos.* **2012**, *40*, 130–138. [CrossRef] [PubMed]
22. Luyendyk, J.P.; Kassel, K.M.; Allen, K.; Guo, G.L.; Li, G.; Cantor, G.H.; Copple, B.L. Fibrinogen deficiency increases liver injury and early growth response-1 (Egr-1) expression in a model of chronic xenobiotic-induced cholestasis. *Am. J. Pathol.* **2012**, *178*, 1117–1125. [CrossRef] [PubMed]
23. Videla, L.A.; Rodrigo, R.; Orellana, M.; Fernandez, V.; Tapia, G.; Quinones, L.; Varela, N.; Contreras, J.; Lazarte, R.; Csendes, A.; et al. Oxidative stress-related parameters in the liver of non-alcoholic fatty liver disease patients. *Clin. Sci.* **2004**, *106*, 261–268. [CrossRef] [PubMed]
24. Casalena, G.; Daehn, I.; Bottinger, E. Transforming growth factor-beta, bioenergetics, and mitochondria in renal disease. *Semin. Nephrol.* **2012**, *32*, 295–303. [CrossRef] [PubMed]
25. Kubli, D.A.; Gustafsson, A.B. Mitochondria and mitophagy: The yin and yang of cell death control. *Circ. Res.* **2012**, *111*, 1208–1221. [CrossRef]
26. Czaja, M.J.; Ding, W.X.; Donohue, T.M., Jr.; Friedman, S.L.; Kim, J.S.; Komatsu, M.; Lemasters, J.J.; Lemoine, A.; Lin, J.D.; Ou, J.H.; et al. Functions of autophagy in normal and diseased liver. *Autophagy* **2013**, *9*, 1131–1158. [CrossRef]
27. Serviddio, G.; Bellanti, F.; Stanca, E.; Lunetti, P.; Blonda, M.; Tamborra, R.; Siculella, L.; Vendemiale, G.; Capobianco, L.; Giudetti, A.M. Silybin exerts antioxidant effects and induces mitochondrial biogenesis in liver of rat with secondary biliary cirrhosis. *Free Radic. Biol. Med.* **2014**, *73*, 117–126. [CrossRef]
28. Kim, C.S.; Kwon, Y.; Choe, S.Y.; Hong, S.M.; Yoo, H.; Goto, T.; Kawada, T.; Choi, H.S.; Joe, Y.; Chung, H.T.; et al. Quercetin reduces obesity-induced hepatosteatosis by enhancing mitochondrial oxidative metabolism via heme oxygenase-1. *Nutr. Metab.* **2015**, *12*, 33. [CrossRef]
29. Cavalheri, H.; Both, C.; Martins, M. The Interplay between Environmental Filtering and Spatial Processes in Structuring Communities: The Case of Neotropical Snake Communities. *PLoS ONE* **2015**, *10*, e0127959. [CrossRef]
30. Poling, J.; Gajawada, P.; Richter, M.; Lorchner, H.; Polyakova, V.; Kostin, S.; Shin, J.; Boettger, T.; Walther, T.; Rees, W.; et al. Therapeutic targeting of the oncostatin M receptor-beta prevents inflammatory heart failure. *Basic Res. Cardiol.* **2014**, *109*, 396. [CrossRef]
31. Szpechcinski, A.; Chorostowska-Wynimko, J.; Struniawski, R.; Kupis, W.; Rudzinski, P.; Langfort, R.; Puscinska, E.; Bielen, P.; Sliwinski, P.; Orlowski, T. Cell-free DNA levels in plasma of patients with non-small-cell lung cancer and inflammatory lung disease. *Br. J. Cancer* **2015**, *113*, 476–483. [CrossRef] [PubMed]
32. Shi, H.; Dong, L.; Jiang, J.; Zhao, J.; Zhao, G.; Dang, X.; Lu, X.; Jia, M. Chlorogenic acid reduces liver inflammation and fibrosis through inhibition of toll-like receptor 4 signaling pathway. *Toxicology* **2013**, *303*, 107–114. [CrossRef] [PubMed]
33. Zhu, H.T.; Bian, C.; Yuan, J.C.; Chu, W.H.; Xiang, X.; Chen, F.; Wang, C.S.; Feng, H.; Lin, J.K. Curcumin attenuates acute inflammatory injury by inhibiting the TLR4/MyD88/NF-kappaB signaling pathway in experimental traumatic brain injury. *J. Neuroinflammation* **2014**, *11*, 59. [CrossRef]
34. Ma, Y.; He, M.; Qiang, L. Exercise Therapy Downregulates the Overexpression of TLR4, TLR2, MyD88 and NF-kappaB after Cerebral Ischemia in Rats. *Int. J. Mol. Sci.* **2013**, *4*, 3718–3733. [CrossRef] [PubMed]
35. Morgan, M.J.; Liu, Z.G. Crosstalk of reactive oxygen species and NF-kappaB signaling. *Cell Res.* **2011**, *21*, 103–115. [CrossRef]
36. Wei, H.Y.; Ma, X. Tamoxifen reduces infiltration of inflammatory cells, apoptosis and inhibits IKK/NF-kB pathway after spinal cord injury in rats. *Neurol. Sci.* **2014**, *35*, 1763–1768. [CrossRef]
37. Baldwin, A.S. Regulation of cell death and autophagy by IKK and NF-kappaB: Critical mechanisms in immune function and cancer. *Immunol. Rev.* **2012**, *246*, 327–345. [CrossRef]
38. Di Paola, R.; Cordaro, M.; Crupi, R.; Siracusa, R.; Campolo, M.; Bruschetta, G.; Fusco, R.; Pugliatti, P.; Esposito, E.; Cuzzocrea, S. Protective Effects of Ultramicronized Palmitoylethanolamide (PEA-um) in Myocardial Ischaemia and Reperfusion Injury in vivo. *Shock* **2016**, *46*, 202–213. [CrossRef]

39. Di Paola, R.; Impellizzeri, D.; Fusco, R.; Cordaro, M.; Siracusa, R.; Crupi, R.; Esposito, E.; Cuzzocrea, S. Ultramicronized palmitoylethanolamide (PEA-um((R))) in the treatment of idiopathic pulmonary fibrosis. *Pharmacol. Res.* **2016**, *111*, 405–412. [CrossRef]
40. Fusco, R.; Cirmi, S.; Gugliandolo, E.; Di Paola, R.; Cuzzocrea, S.; Navarra, M. A flavonoid-rich extract of orange juice reduced oxidative stress in an experimental model of inflammatory bowel disease. *J. Funct. Foods* **2017**, *30*, 168–178. [CrossRef]
41. Lee, M.K.; Yeo, H.; Kim, J.; Kim, Y.C. Protection of rat hepatocytes exposed to CCl4 in-vitro by cynandione A, a biacetophenone from *Cynanchum wilfordii. J. Pharm. Pharmacol.* **2000**, *52*, 341–345. [CrossRef] [PubMed]
42. de Oliveira Fernandes, T.; de Avila, R.I.; de Moura, S.S.; de Almeida Ribeiro, G.; Naves, M.M.V.; Valadares, M.C. *Campomanesia adamantium* (Myrtaceae) fruits protect HEPG2 cells against carbon tetrachloride-induced toxicity. *Toxicol. Rep.* **2015**, *2*, 184–193. [CrossRef] [PubMed]
43. Boone, L.; Meyer, D.; Cusick, P.; Ennulat, D.; Bolliger, A.P.; Everds, N.; Meador, V.; Elliott, G.; Honor, D.; Bounous, D. Selection and interpretation of clinical pathology indicators of hepatic injury in preclinical studies. *Vet. Clin. Pathol.* **2005**, *34*, 182–188. [CrossRef] [PubMed]
44. Cordaro, M.; Siracusa, R.; Fusco, R.; D'Amico, R.; Peritore, A.F.; Gugliandolo, E.; Genovese, T.; Scuto, M.; Crupi, R.; Mandalari, G.; et al. Cashew (*Anacardium occidentale* L.) Nuts Counteract Oxidative Stress and Inflammation in an Acute Experimental Model of Carrageenan-Induced Paw Edema. *Antioxidants* **2020**, *9*, 660. [CrossRef] [PubMed]
45. Siracusa, R.; D'Amico, R.; Cordaro, M.; Peritore, A.F.; Genovese, T.; Gugliandolo, E.; Crupi, R.; Impellizzeri, D.; Cuzzocrea, S.; Fusco, R.; et al. The Methyl Ester of 2-Cyano-3,12-Dioxooleana-1,9-Dien-28-Oic Acid Reduces Endometrial Lesions Development by Modulating the NFkB and Nrf2 Pathways. *Int. J. Mol. Sci.* **2021**, *22*, 3991. [CrossRef]
46. Fusco, R.; Cordaro, M.; Siracusa, R.; D'Amico, R.; Genovese, T.; Gugliandolo, E.; Peritore, A.F.; Crupi, R.; Impellizzeri, D.; Cuzzocrea, S.; et al. Biochemical Evaluation of the Antioxidant Effects of Hydroxytyrosol on Pancreatitis-Associated Gut Injury. *Antioxidants* **2020**, *9*, 781. [CrossRef]
47. Dutta, S.; Chakraborty, A.K.; Dey, P.; Kar, P.; Guha, P.; Sen, S.; Kumar, A.; Sen, A.; Chaudhuri, T.K. Amelioration of CCl4 induced liver injury in swiss albino mice by antioxidant rich leaf extract of Croton bonplandianus Baill. *PLoS ONE* **2018**, *13*, e0196411. [CrossRef]
48. Siracusa, R.; Fusco, R.; Peritore, A.F.; Cordaro, M.; D'Amico, R.; Genovese, T.; Gugliandolo, E.; Crupi, R.; Smeriglio, A.; Mandalari, G.; et al. The Antioxidant and Anti-Inflammatory Properties of *Anacardium occidentale* L. Cashew Nuts in a Mouse Model of Colitis. *Nutrients* **2020**, *12*, 834. [CrossRef]
49. Liu, B.; Zhang, J.; Sun, P.; Yi, R.; Han, X.; Zhao, X. Raw Bowl Tea (Tuocha) Polyphenol Prevention of Nonalcoholic Fatty Liver Disease by Regulating Intestinal Function in Mice. *Biomolecules* **2019**, *9*, 435. [CrossRef]
50. Peritore, A.F.; D'Amico, R.; Cordaro, M.; Siracusa, R.; Fusco, R.; Gugliandolo, E.; Genovese, T.; Crupi, R.; Di Paola, R.; Cuzzocrea, S.; et al. PEA/Polydatin: Anti-Inflammatory and Antioxidant Approach to Counteract DNBS-Induced Colitis. *Antioxidants* **2021**, *10*, 463. [CrossRef]
51. Cordaro, M.; Siracusa, R.; Impellizzeri, D.; D'Amico, R.; Peritore, A.F.; Crupi, R.; Gugliandolo, E.; Fusco, R.; Di Paola, R.; Schievano, C.; et al. Safety and efficacy of a new micronized formulation of the ALIAmide palmitoylglucosamine in preclinical models of inflammation and osteoarthritis pain. *Arthritis Res. Ther.* **2019**, *21*, 254. [CrossRef] [PubMed]
52. Saber, S.; Mahmoud, A.A.A.; Helal, N.S.; El-Ahwany, E.; Abdelghany, R.H. Renin-angiotensin system inhibition ameliorates CCl4-induced liver fibrosis in mice through the inactivation of nuclear transcription factor kappa B. *Can. J. Physiol. Pharmacol.* **2018**, *96*, 569–576. [CrossRef] [PubMed]
53. Peritore, A.F.; Siracusa, R.; Fusco, R.; Gugliandolo, E.; D'Amico, R.; Cordaro, M.; Crupi, R.; Genovese, T.; Impellizzeri, D.; Cuzzocrea, S.; et al. Ultramicronized Palmitoylethanolamide and Paracetamol, a New Association to Relieve Hyperalgesia and Pain in a Sciatic Nerve Injury Model in Rat. *Int. J. Mol. Sci.* **2020**, *21*, 3509. [CrossRef] [PubMed]
54. Uesugi, T.; Froh, M.; Arteel, G.E.; Bradford, B.U.; Thurman, R.G. Toll-like receptor 4 is involved in the mechanism of early alcohol-induced liver injury in mice. *Hepatology* **2011**, *34*, 101–108. [CrossRef]
55. Nallagangula, K.S.; Nagaraj, S.K.; Venkataswamy, L.; Chandrappa, M. Liver fibrosis: A compilation on the biomarkers status and their significance during disease progression. *Future Sci. OA* **2018**, *4*, FSO250. [CrossRef]
56. Su, T.H.; Shiau, C.W.; Jao, P.; Yang, N.J.; Tai, W.T.; Liu, C.J.; Tseng, T.C.; Yang, H.C.; Liu, C.H.; Huang, K.W.; et al. Src-homology protein tyrosine phosphatase-1 agonist, SC-43, reduces liver fibrosis. *Sci. Rep.* **2017**, *7*, 1728. [CrossRef]
57. Fusco, R.; Siracusa, R.; Peritore, A.F.; Gugliandolo, E.; Genovese, T.; D'Amico, R.; Cordaro, M.; Crupi, R.; Mandalari, G.; Impellizzeri, D.; et al. The Role of Cashew (*Anacardium occidentale* L.) Nuts on an Experimental Model of Painful Degenerative Joint Disease. *Antioxidants* **2020**, *9*, 511. [CrossRef]
58. Fusco, R.; Gugliandolo, E.; Siracusa, R.; Scuto, M.; Cordaro, M.; D'Amico, R.; Evangelista, M.; Peli, A.; Peritore, A.F.; Impellizzeri, D.; et al. Formyl Peptide Receptor 1 Signaling in Acute Inflammation and Neural Differentiation Induced by Traumatic Brain Injury. *Biology* **2020**, *9*, 238. [CrossRef]
59. D'Amico, R.; Fusco, R.; Cordaro, M.; Siracusa, R.; Peritore, A.F.; Gugliandolo, E.; Crupi, R.; Scuto, M.; Cuzzocrea, S.; Di Paola, R.; et al. Modulation of NLRP3 Inflammasome through Formyl Peptide Receptor 1 (Fpr-1) Pathway as a New Therapeutic Target in Bronchiolitis Obliterans Syndrome. *Int. J. Mol. Sci.* **2020**, *21*, 2144. [CrossRef]

60. Esposito, E.; Campolo, M.; Casili, G.; Lanza, M.; Franco, D.; Filippone, A.; Peritore, A.F.; Cuzzocrea, S. Protective Effects of Xyloglucan in Association with the Polysaccharide Gelose in an Experimental Model of Gastroenteritis and Urinary Tract Infection. *Int. J. Mol. Sci.* **2018**, *19*, 1844. [CrossRef]
61. Fusco, R.; Cordaro, M.; Genovese, T.; Impellizzeri, D.; Siracusa, R.; Gugliandolo, E.; Peritore, A.F.; D'Amico, R.; Crupi, R.; Cuzzocrea, S.; et al. Adelmidrol: A New Promising Antioxidant and Anti-Inflammatory Therapeutic Tool in Pulmonary Fibrosis. *Antioxidants* **2020**, *9*, 601. [CrossRef] [PubMed]

International Journal of
Molecular Sciences

MDPI

Article

Overexpression of Neuregulin-1 Type III Has Impact on Visual Function in Mice

Nan Su [1,†], Weiqi Zhang [2], Nicole Eter [1], Peter Heiduschka [1,‡] and Mingyue Zhang [2,*,‡]

1 Research Laboratory, Department of Ophthalmology, University Hospital Muenster, 48149 Muenster, Germany; yanshengsu266@gmail.com (N.S.); nicole.eter@ukmuenster.de (N.E.); peter.heiduschka@ukmuenster.de (P.H.)
2 Laboratory for Moleculare Neuroscience, Clinic for mental Health, University Hospital Muenster, 48149 Muenster, Germany; wzhang@uni-muenster.de
* Correspondence: mingyue.zhang@ukmuenster.de; Tel.: +49-251-8351824; Fax: +49-251-8357123
† Current address: Su Nan Department of Ophthalmology, The First Affiliated Hospital of Zhengzhou University, Erqi District, Zhengzhou 450052, China.
‡ These authors contributed equally to this work.

Abstract: Schizophrenia is associated with several brain deficits, including abnormalities in visual processes. Neuregulin-1 (Nrg1) is a family of trophic factors containing an epidermal growth factor (EGF)-like domain. It is thought to play a role in neural development and has been linked to neuropsychiatric disorders. Abnormal Nrg1 expression has been observed in schizophrenia in clinical studies. Moreover, in schizophrenia, there is more and more evidence found about pathological changes of the retina regarding structural, neurochemical and physiological parameters. However, mechanisms of these changes are not well known. To investigate this, we analysed the function of the visual system using electroretinography (ERG) and the measurement of visual evoked potentials (VEP) in transgenic mice overexpressing Nrg1 type III of three different ages (12 weeks, 24 weeks and 55 weeks). ERG amplitudes tended to be higher in transgenic mice than in control mice in 12-week old mice, whereas the amplitudes were almost similar in older mice. VEP amplitudes were larger in transgenic mice at all ages, with significant differences at 12 and 55 weeks (p values between 0.003 and 0.036). Latencies in ERG and VEP measurements did not differ considerably between control mice and transgenic mice at any age. Our data show for the first time that overexpression of Nrg1 type III changed visual function in transgenic mice. Overall, this investigation of visual function in transgenic mice may be helpful to understand corresponding changes that occur in schizophrenia, as they may find use as biomarkers for psychiatric disorders as well as a potential tool for diagnosis in psychiatry.

Keywords: electroretinography; visual evoked potential; neuregulin 1; schizophrenia

Citation: Su, N.; Zhang, W.; Eter, N.; Heiduschka, P.; Zhang, M. Overexpression of Neuregulin-1 Type III Has Impact on Visual Function in Mice. *Int. J. Mol. Sci.* **2022**, *23*, 4489. https://doi.org/10.3390/ijms23094489

Academic Editor: Raffaele Capasso

Received: 15 March 2022
Accepted: 17 April 2022
Published: 19 April 2022

1. Introduction

Schizophrenia is a complex disorder that affects 0.5–1% of the adult population throughout different ethnicities in the world. According to findings reported in the literature, several genes have been associated with the neuropathology in diverse populations [1]. Among these candidates, genes encoding the proteins neuregulin (Nrg1) and its receptor ErbB4 have been shown to be promising susceptibility genes of schizophrenia [1]. Nrg1 is a pleiotropic growth and differentiation factor, which can be classified in six major isoforms, (Nrg1 type I–VI). Types I, II, IV, V and VI are single transmembrane proteins, whereas type III contains a cysteine-rich domain that loops back intracellularly along with its N-terminal sequence [2]. Moreover, Nrg1 types1, II and III are best characterized in the peripheral nervous system [3]. Clinical studies show that some patients show abnormal levels of expression of Nrg1 and ErbB4 isoforms in different brain regions [4]. Schizophrenic patients are impaired in cognitive abilities, including executive control and working memory [5]. In addition, elevated levels of Nrg1 and ErbB4 proteins have been found in studies on post-mortem schizophrenic patients [6]. Furthermore, our early work on mice has shown that elevated Nrg1 expression (Nrg1-III-tg) showed ventricular enlargement and symptoms similar to schizophrenia [7]. Similarly, Olaya and colleagues showed in 2018 that overexpression of Nrg1 type III in mice confers schizophrenia-like behavior [8]. Importantly, another study confirmed that Nrg1/ErbB4 regulates visual cortical plasticity [9].

Researchers have long been aware of the link between schizophrenia and visual processing impairments, which are accompanied by multiple structural and functional disturbances in patients. Furthermore, the retina may be particularly affected, as it belongs to the central nervous system and shows similarities to the brain and spinal cord in terms of structure, functionality, response to insult, and immunology [10]. In addition, from an embryonic point of view, the retina and optic nerve, which have a neuroectodermal origin, emerge from diencephalon and can be seen with the naked eye in its natural state in the living organism [11]. Impairments of visual processing are well established in schizophrenia, including multiple structural and functional disturbances in patients. In addition, in studies in which patients were checked for factors such as psychotic symptoms and auditory distortions, visual distortions (including those of the retina) have been associated with suicidal ideation [12]. These alterations of the visual system include dopaminergic abnormalities, abnormal output, maculopathies and retinopathies, cataracts, poor visual acuity, and thinning of the retinal fibre layer (RNFL) [11]. Looking at this from another angle, some classic ocular pathologies have been found to occur in the context of several major neurodegenerative disorders, and RNFL thinning is related to brain volume loss in aging and illness progression as well as cognitive decline in multiple sclerosis and Alzheimer's disease [13].

The purpose of this study was to check changes in the function of the visual system of transgenic mice overexpressing Nrg1 type III, as this protein is linked with both schizophrenia and visual processing. Visual function was assessed by the flash electroretinography (ERG) and measurement of visual evoked potentials (VEP). While several ERG anomalies have been identified in patients with psychiatric disorders [14], the underlying mechanisms and visual processing abnormalities in schizophrenia are still unknown. Moreover, we checked retinal morphology by optical coherence tomography (OCT). This method is non-invasive and rapid, providing the quantitative measurement of RNFL thickness and retinal volume (e.g., macula volume/MV) [15].

There are various proteins in the retina that could probably be influenced by overexpression of Nrg1 type III. We checked frozen sections of the eyes of control mice and transgenic mice for the immunoreactivity (IR) for an isoform of glutamic acid decarboxylase (GAD65), the voltage-gated K^+ channel Kv2.1, and the postsynaptic density protein 95 (PSD-95).

2. Materials and Methods

2.1. Animals

The experiments were performed in accordance with European Communities Council Directive (86/EEC) and were approved by the Federal State Office for Consumer Protection and Food Safety of North Rhine-Westphalia (LANUV), Germany (file no. 84-02.04.2016.A417). All efforts were made to minimize animal suffering and to reduce the number of animals used in the experiments to the minimum necessary for reliable statistical analyses.

Animals of three different ages were investigated, young (12 weeks old), midlife (24 weeks old) and older (55 weeks old) mice. In each age group, transgenic and control mice were compared. Generation and genotyping of transgenic Nrg1-III-tg mice have been described in detail in [16].

2.2. *Visual Electrophysiology*

Electroretinography was performed as described previously in Schubert et al., 2015 [17]. Briefly, the mice were anaesthetised using a standard intraperitoneal ketamine/xylazine injection. Sleeping animals were placed on a heating pad to dilate the pupils by tropicamide and neosynephrine eye drops. Desensitisation of the cornea was achieved by a drop of proparacaine.

For the ERG and VEP measurement, the commercial measuring device RetiPort from Roland Consult (Brandenburg, Germany) was used. During the measurement, the animals were placed on a heated plate at 37 °C to prevent cooling of the animals. For the ERG measurement, a gold ring electrode was placed on the cornea of the left eye without damaging the cornea. VEP was recorded simultaneously by inserting a stainless-steel needle electrode subcutaneously above the visual centre of the mice on top of the skull between the ears. Another gold electrode that was moistened with saline and placed into the mouth of the animals served as the reference.

Measurement was performed in the scotopic mode, with animals that were dark-adapted for at least 12 h. Visual stimulation was performed by the application of flashes of six different light intensities, ranging from 0.0003 to 30 cd·s/m^2. Responses of the visual system were recorded, averaged, and stored by the RetiPort device. After the measurement, the still sleeping animals were kept in a separate box and given back into the cage after awakening.

2.3. *OCT*

To measure the thickness of the retina, in vivo imaging was performed. Mice were anaesthetised and the pupils dilated as described above. The mice were put in front of the "Spectralis" device by Heidelberg Engineering and were examined by optical coherence tomography (OCT). Formation of cataract was delayed by dropping distilled water onto the eyes.

Images were centered with the optic nerve head in the centre of the image. Retinal thickness was determined using the built-in routines in the nasal, superior, temporal and inferior quadrants of the inner circle and the outer circle of the ETDRS grid.

2.4. *Immunohistochemistry*

Eyes of euthanised mice were isolated and fixed in 4% paraformaldehyde for 1 h, washed 2× in PBS pH 7.4 for 5 min and frozen in NEG-50™. Cryo sections (thickness 10 µm) were cut using a Cryostar NX70 cryostat (Thermo Fisher Scientific, Waltham, MA, USA), placed on Starfrost Advanced Adhesive glass slides (Engelbrecht) and stored at −20 °C until used for immunohistochemistry.

For immunohistochemistry, sections were blocked with Power Block™ reagent (HK085-5K, BioGenex) at room temperature for 6 min, then washed 3× with 0.1 M PBS and incubated overnight with primary antibodies at 4 °C. The sections were then washed 3× with 0.1 M PBS and incubated with appropriate secondary antibodies for 1 h at room temperature. The nuclei were counterstained with DAPI (4′6′-diamidino-2-phenylindole dihydrochloride), diluted with pure water 1:300, for 7 min at room temperature. The primary antibody was diluted in 1% bovine serum albumin containing 0.1% Triton X-100, and the secondary antibody goat anti rabbit were diluted with 1% bovine serum albumin (Table 1). Finally, sections were washed 3× with 0.1M PBS and mounted under glass coverslips using mounting medium (ImmuMount™, Thermo Scientific).

Table 1. Antibodies used in this study.

Primary Antibodies					
Specificity	**Host**		**Supplier**	**Catalogue No.**	**Dilution**
Kv-2.1	Rabbit		Synaptic Systems	231 002	1:500
GAD 2/GAD 65	Guinea pig		Synaptic Systems	198104	1:500
Neuregulin-1	Rabbit		GeneTex	GTX133355	1:1000
PSD-95	Rat		Santa Cruz Biotechnology	sc-32290	1:300
Secondary Antibodies					
Specificity	**Host**	**dye**	**Supplier**	**Catalogue No.**	**Dilution**
Anti-Rabbit	goat	Texas Red	Abcam	ab150080	1:800
Anti-Guinea pig	goat	Alexa Fluor 488	Abcam	ab150185	1:600

2.5. *Quantitative Analysis of Fluorescence in the Images of Histological Staining*

Overview images of the mice retina sections were taken with 40× magnification (Axio Imager M2, Carl Zeiss, Jena, Germany). To ensure comparability of the multiple specimens, all procedures of staining and image acquisition were performed in an identical way. Digital images were processed using ImageJ (version 1.46r, NIH, Bethesda, MD, USA). Regions of interest (ROIs) were set manually in the ganglion cell, amacrine cell, and inner and outer plexiform layers using the implemented ROI manager, and the intensity of staining against KV, GAD, and PSD within the ROIs was measured to determine the amount of staining. To achieve reliable data, a uniform background in all areas of the image was necessary. The threshold tool settings were the same in every image.

2.6. *Data Analysis*

Data are presented as mean ± SEM. Evaluation of the data was performed by separately comparing the means of each parameter obtained in the control mice and the Nrg1-III-tg mice. After testing for normal distribution with a Shapiro-Wilk test, a parametric unpaired Student's t-test or a non-parametric Mann–Whitney test wa used to determine the difference between control mice and Nrg1-III-tg mice. The level of statistical significance was set as $p = 0.05$, statistical significance is indicated as an * $p < 0.05$, ** $p < 0.01$, *** $p < 0.001$.

3. Results

3.1. *Investigation of the Function of the Visual System Measured by ERG and VEP*

ERG and VEP measurements were carried out in younger (12 weeks old), midlife (24 weeks old) and older (55 weeks old) mice; in these age groups, transgenic mice and control mice were compared.

In each age group, amplitudes of scotopic a-waves and b-waves displayed obvious differences. In younger mice, these amplitudes showed a trend to be smaller in control mice than in transgenic mice. However, amplitudes were higher in control mice than in transgenic mice in midlife mice and older mice, and the difference was statistically significant in midlife animals (Figure 1A). The latencies in scotopic ERG were in the same range in the three groups, and no relevant differences were seen (Figure 1B).

Amplitudes of scotopic oscillatory potentials were significantly higher in transgenic mice than in control mice in younger mice, however, not in midlife and older mice. No significant differences in the latencies of oscillatory potentials were found between the groups (Figure 1C). The b/a ratio tended to be slightly higher in control mice compared to transgenic mice (Figure 1D).

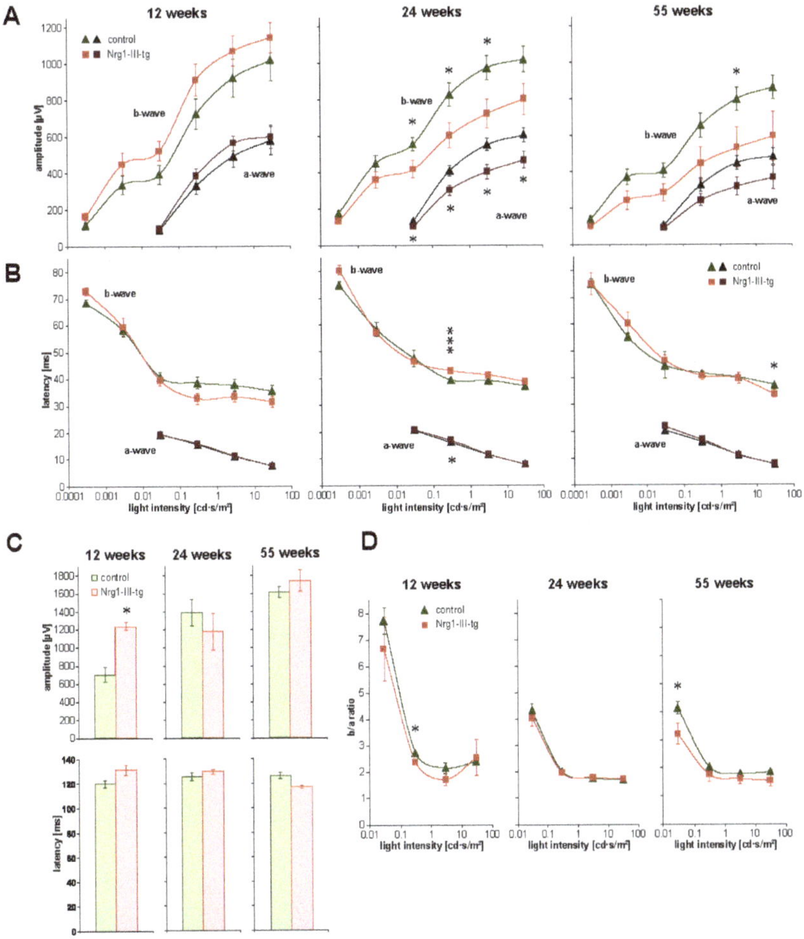

Figure 1. ERG Parameters in 12-, 24 and 55-week control and transgenic mice. Scotopic amplitudes (**A**) and latencies (**B**) of a-waves and b-waves at different light intensities as indicated: (**C**): amplitudes and latencies of scotopic oscillatory potentials; (**D**): the ratio of amplitudes of b-waves and a-waves. Number of animals at the age of 12, 24 and 55 weeks: $n = 8$, $n = 13$ and $n = 9$ control mice and $n = 6$, $n = 5$ and $n = 8$ Nrg1-III-tg age-matched mice, respectively. Data are shown as mean values. Error bars show standard error of means. Statistical significance of differences is indicated as follows: * $p < 0.05$, *** $p < 0.001$.

The amplitudes of scotopic VEP were significantly higher in the younger and older transgenic mice than in control mice at most light intensities (Figure 2A). Moreover, amplitudes of VEP were higher in transgenic mice than in control mice in the midlife mice, though without significance of the difference. Despite some deviations in the younger mice and the midlife mice, latencies did not show relevant differences between transgenic mice and control mice (Figure 2B).

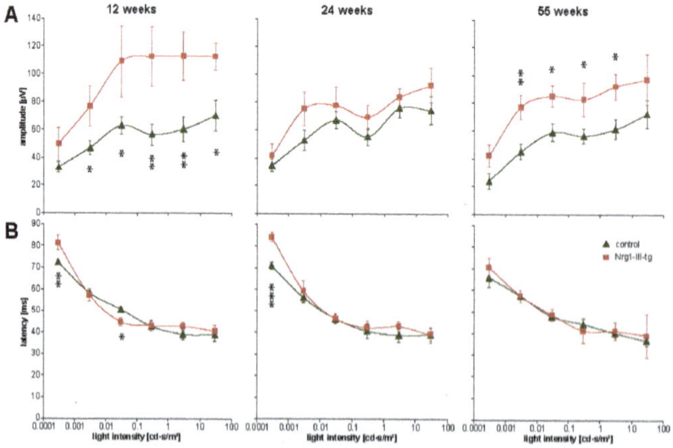

Figure 2. Scotopic VEP amplitudes and latencies in 12-, 24- and 55-week-old mice. Scotopic amplitudes (**A**) and latencies (**B**) of visual evoked potentials at different light intensities as indicated. Data are shown as mean values. Error bars show standard error of means. Number of animals at the age of 12, 24 and 55 weeks: $n = 8$, $n = 13$ and $n = 9$ control mice and $n = 6$, $n = 5$ and $n = 8$ Nrg1-III-tg age-matched mice, respectively. Statistical significance of differences is indicated as follows: * $p < 0.05$, ** $p < 0.01$, *** $p < 0.001$.

Examples of waveforms of ERG and VEP measurements in young animals are shown in Figure 3. As a general observation, amplitudes of electrophysiological parameters decreased with increasing age. It appeared that amplitudes decreased more clearly in transgenic mice than in control mice.

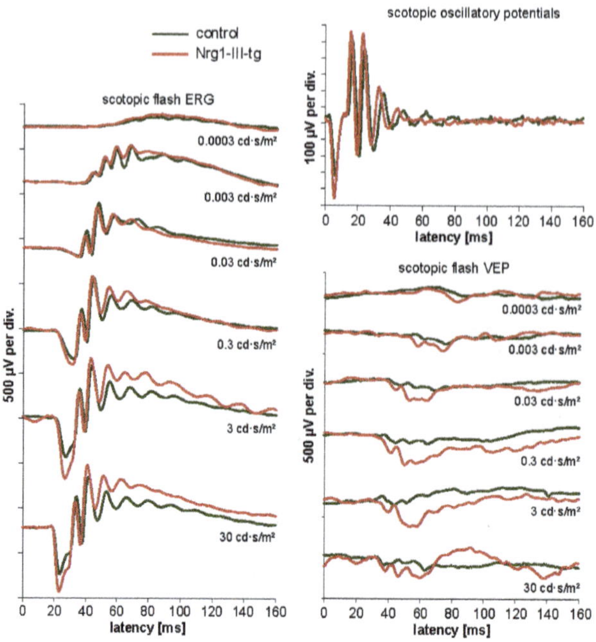

Figure 3. Typical waveforms of ERG and VEP measurements in 12-week-old control and transgenic mice.

3.2. Investigation of the Thickness and Volume of the Mouse Retina Measured by OCT

OCT measurements were carried out in the younger, midlife and elder mice, and thickness and volume of the retina were measured.

We did not find any significant differences in the thickness (data not shown) between transgenic mice and control mice at any age. Furthermore, there was no significant difference between transgenic mice and control mice when the volumes were compared (Figure 4).

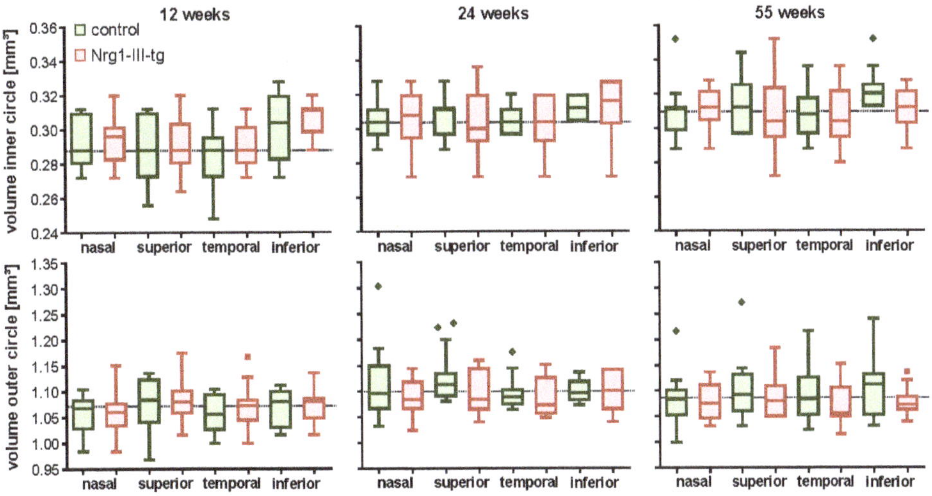

Figure 4. The Volume of retina in 12-, 24- and 55-week-old mice. Data are shown as mean values. Error bars show standard error of means. Number of animals at the age of 12, 24 and 55 weeks: $n = 12$, $n = 13$ and $n = 14$ control mice and $n = 16$, $n = 8$ and $n = 10$ Nrg1-III-tg age-matched mice, respectively.

3.3. Semi-Quantitative Analysis of Histological Staining

We first performed immunohistochemical staining against Nrg1. Immunoreactivity (IR) for Nrg1 was found in the ganglion cell layer, the inner nuclear layer, outer plexiform layer and in the photoreceptor inner segments (Figure 5). It was clearly higher in transgenic mice than in control mice in midlife animals. Conversely, the IR for Nrg1 appeared to be lower in transgenic mice than in control mice in the older animals (Figure 5).

To determine the effect of overexpression of Neuregulin on the GABAergic system in the retina, we checked retinal sections for GAD65 IR (Figure 6). GAD65 IR was seen in the inner plexiform layer, and it was higher in transgenic mice than in control mice in midlife animals. In older mice, no difference between the two groups was seen.

We observed changes of ERG amplitudes in transgenic mice compared to control mice (Figure 1). Because voltage-gated K^+ channels play a role in the visual process, we checked IR for Kv2.1 in the retina. Kv2.1 IR was seen in the ganglion cell layer, in traces in the inner plexiform layer and amacrine cells, and particularly strongly in the photoreceptor inner segments. The intensity of Kv2.1 IR was higher in transgenic mice than in control mice in midlife animals. In older mice, the difference between the two groups was much weaker (Figure 6).

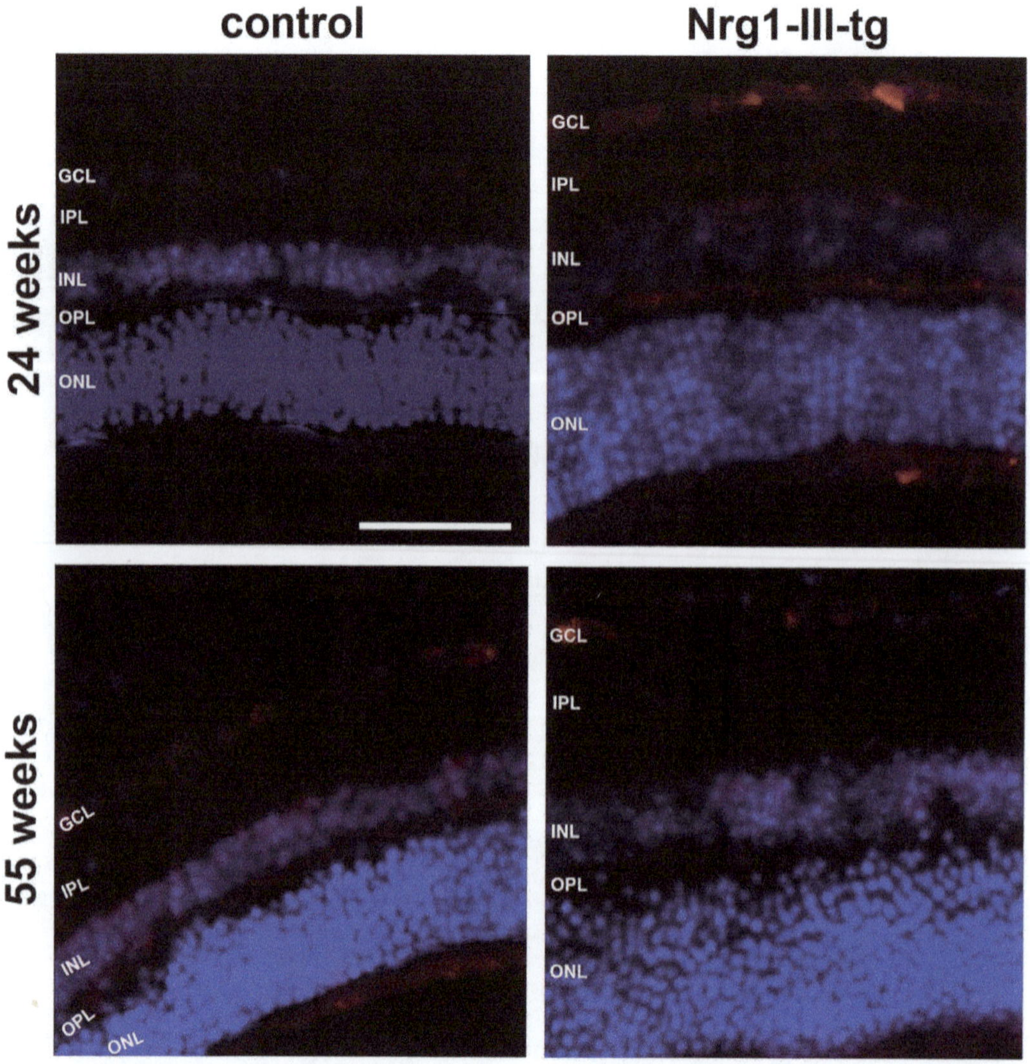

Figure 5. Immunohistochemical staining against neuregulin. The trophic factor neuregulin (red) in frozen sections of the retina of the mouse eye at 24 weeks and 55 weeks of control and Nrg1-III-tg as indicated. The cell nuclei were stained with DAPI (blue); GCL-ganglion cell layer; IPL-inner plexiform layer; INL-inner nuclear layer; OPL-outer plexiform layer; ONL-outer nuclear layer. Scale bar: 50 μm.

IR for PSD-95 was found on the outer rim of the outer plexiform layer, i.e., most probably in the region of the synapses of the photoreceptors. The intensity of PSD-95 IR was similar in both groups in midlife animals. It decreased in older animals, and this decrease appeared to be more obvious in transgenic mice than in control mice (Figure 7).

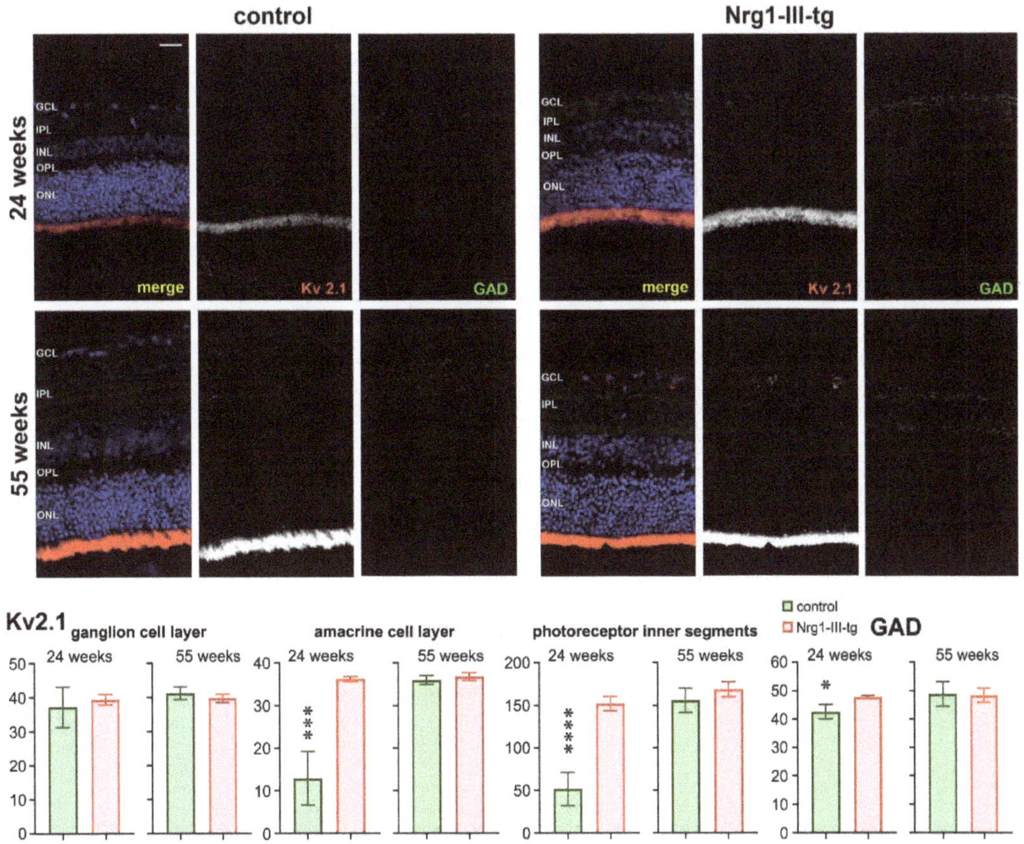

Figure 6. Immunohistochemical staining against Kv2.1 and GAD. Kv2.1 (red) and GAD (green) in frozen sections of the retina of the mouse eye at 24 weeks and 55 weeks of control and Nrg1-III-tg as indicated. The cell nuclei were stained with DAPI (blue); GCL-ganglion cell layer; IPL-inner plexiform layer; INL-inner nuclear layer; OPL-outer plexiform layer; ONL-outer nuclear layer. Scale bar: 50 µm. * $p < 0.05$, *** $p < 0.001$, **** $p < 0.0001$.

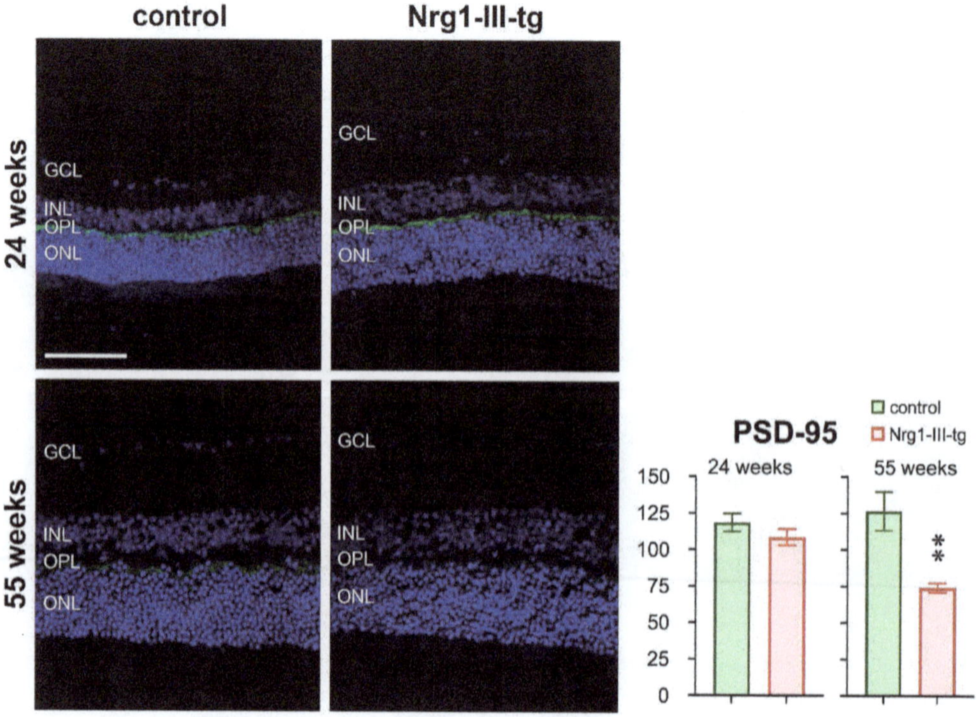

Figure 7. Immunohistochemical staining against PSD-95. PSD-95 (green) in frozen sections of the retina of the mouse eye at 24 weeks and 55 weeks of control and Nrg1-III-tg as indicated. The cell nuclei were stained with DAPI (blue); GCL-ganglion cell layer; IPL-inner plexiform layer; INL-inner nuclear layer; OPL-outer plexiform layer; ONL-outer nuclear layer. Scale bar: 50 μm. ** $p < 0.01$.

4. Discussion

In the current study, we investigated the influence of neuronal overexpression of Nrg1-Type III on the visual function in mice by electroretinography and VEP measurement. The above data demonstrate that ERG responses had a significantly reduced scotopic responses in transgenic mice compared to control mice in midlife and older, but not in the younger mice, where amplitudes were larger in transgenic mice. VEP amplitudes were enhanced in Nrg1 III-tg mice at all ages, with significance in younger and older mice. Latencies did not show any relevant differences between the groups at any age, either in ERG, or in VEP measurements.

The human neuregulin1 (Nrg1) gene is a major schizophrenia susceptibility gene, and its association with the illness has been found in different populations [18]. In addition, dysregulated expression of Nrg1, including elevation of expression of Nrg1, increases disease susceptibility and has been found in studies of post-mortem brain tissue from schizophrenia patients [19]. In our previous mouse studies, gene disruption that increased expression of Nrg1-Type III can confer distinct schizophrenia-like phenotypes in behaviour and brain biology [7].

More and more studies are showing that schizophrenia is associated with several brain deficits, including visual processing deficits [20,21]. As key player in visual processing, the retina is part of the central nervous system and is composed of several layers. To evaluate the function of specific layers of neurons of the retina, ERG can be used [22]. The cornea-negative a-wave indicates the electrical activity of the photoreceptors, and ON-bipolar cells are the source of the b-wave in the retina of mice [23,24]. Our results show that electroretinographic a-waves and b-waves were smaller in transgenic mice compared to control mice in the midlife and elder mice, but not in the younger mice, suggesting that Nrg1-III-tg mice had impaired retinal function at higher ages. Considering the origin of the a-waves and b-waves and the transmission and processing of nerve signals in the retina as well as to better understand the function and mechanism of Nrg1 in the retina, we then performed a series of immunostaining. As has been found in the brain, the Nrg1 signalling pathway has an effect on neurotransmission and synaptic plasticity [7]. Hence, further studies are required to determine which changes in the neural signalling occur in the retina of transgenic mice. Nrg1 has been implicated in neural excitatory synapse development and regulated neuronal migration in the neuronal system [25,26]. The inhibitory neurotransmitter GABA is synthesized by two isoforms of glutamic acid decarboxylase (GAD65, GAD67) [27], and mice lacking GAD 65 show an impaired visual cortical plasticity [28]. The 65-KD isoform (GAD65) is found primarily in the axon terminals in the visual cortex and reversibly bound to the membrane of synaptic vesicles, playing a role in the control of the synaptic release of GABA [29,30]. We found an enhanced IR of GAD65 in midlife transgenic mice, but not in older mice, implying increased GABA synthase and release from the synaptic vesicle in retina. On the other hand, we have previously shown that electrophysiological recordings in an acute cortical slice revealed a moderate increase in frequency [7]. Whether neuregulin has a common mechanism of affecting synaptic transmission and synaptic plasticity in the retina and the cortex, requires further studies. The amount of neurotransmitter release depends on the neuronal membrane excitability, shape of the action potential, or probability of vesicular release [31,32]. In addition, potassium channels regulate neuronal excitability, setting resting membrane potentials and decreasing and increasing excitability [33]. The voltage-gated Kv2.1subunit is nearly ubiquitously expressed in the mammalian brain and is present in most neuronal cell types, including pyramidal cells and interneurons and plays an important role in its ability to dynamically contribute to neuronal excitation [34,35]. The results show that overexpression of NRG1 has an impact on IR of KV2.1 both in the amacrine cell layer and in the photoreceptor inner segments layer in midlife mice. Meanwhile, these data are consistent with ERG data that overexpression of Nrg1 affects the a-wave mainly in midlife mice, as the a-wave is generated mainly by the rod and cone receptor photocurrent [36]. However, there were no differences in older mice (Figure 6), suggesting an interference of development of Kv2.1 in the overexpression of Nrg1transgenic mice. Moreover, Nrg1 regulates the stability of PSD-95 in GABAergic neurons in a way that requires its receptor ErbB4 [37]. Reduced PSD-95 IR in older mice suggest that NRG1 regulates the PSD-95 in the synaptic terminal in the retina of transgenic mice.

The visual evoked potential (VEP) is an electrical potential recorded from the visual cortex in response to a visual stimulus. Therefore, the function of the optic nerve and the quality of signal processing in the visual cortex can be evaluated by measuring the VEP [38]. We observed an increase in the VEP amplitudes in transgenic mice. It may be caused by the presence of abnormal ganglion cells or an abnormal number of ganglion cells. A changed signal processing in the visual cortex is also possible. This must be evaluated in further studies.

We observed that there was more IR for Nrg1 in the retina of young transgenic mice than in control mice (Figure 5). As overexpression of Nrg1 is performed under the control of the Thy-1 promoter, it could be anticipated that mainly retinal ganglion cells showed immunoreactivity for Nrg1, because these cells especially produce Thy-1 at young ages [39]. Given that the expression of Nrg1 seems to change with age, this may suggest that NRG/ErbB signalling plays a role during the early development of the retina.

Our previous work showed that dysregulation of Nrg1 by cortical pyramidal neurons disrupts GABAergic and glutamatergic neurotransmission in cortex as well as synaptic plasticity [7]. All these changes in transgenic mice have the potential to alter the visual system and to further eventually impact ERG and VEP.

Given the findings of the current study, further investigations should more deeply explore the mechanisms by which these visual anomalies occur, as they will be helpful for understanding the biological basis of the psychiatric disorder. Taken together, these results indicate a critical role of Nrg1 on retinal synaptic system.

Author Contributions: Conceptuation, M.Z. and P.H.; Methodology, N.S., M.Z. and P.H.; data curation and analysis, N.S., M.Z. and P.H.; Writing—original draft preparation M.Z.; writing—review editing, N.S. and P.H.; M.Z., N.S. and P.H., Founding, W.Z. and P.H.; W.Z. and N.E. provided study materials, instrumentation and analyis tools. All authors have read and agreed to the published version of the manuscript.

Funding: This work was supported by the Deutsche Forschungsgemeinschaft (ZH34/3-1) and supported in part by Henan Province Health Ministry, PR China, project no. 2018061 (NS).

Institutional Review Board Statement: The animal study protocol was by the Federal State Office for Consumer Protection and Food Safety of North Rhine-Westphalia (LANUV), Germany (file no. 84-02.04.2016.A417 and date of approval 3 April 2017).

Informed Consent Statement: Not applicable.

Data Availability Statement: Not applicable.

Acknowledgments: The authors would like to thank Christiane Schettler, Kathrin Schwarte, Helen Haupt and Mechthild Wissing for their support to this study.

Conflicts of Interest: The authors declare that have no competing interest.

References

1. Harrison, P.J.; Law, A.J. Neuregulin 1 and Schizophrenia: Genetics, Gene Expression, and Neurobiology. *Biol. Psychiatry* **2006**, *60*, 132–140. [CrossRef] [PubMed]
2. Falls, D.L. Neuregulins: Functions, forms, and signaling strategies. *Exp. Cell Res.* **2003**, *284*, 14–30. [CrossRef]
3. Pan, P.; Dobrowsky, R.T. Differential expression of neuregulin-1 isoforms and downregulation of erbin are associated with Erb B2 receptor activation in diabetic peripheral neuropathy. *Acta Neuropathol. Commun.* **2014**, *1*, 39. [CrossRef] [PubMed]
4. Law, A.J.; Kleinman, J.E.; Weinberger, D.R.; Weickert, C.S. Disease-associated intronic variants in the ErbB4 gene are related to altered ErbB4 splice-variant expression in the brain in schizophrenia. *Hum. Mol. Genet.* **2007**, *16*, 129–141. [CrossRef] [PubMed]
5. Lewis, D.A.; Sweet, R.A. Schizophrenia from a neural circuitry perspective: Advancing toward rational pharmacological therapies. *J. Clin. Investig.* **2009**, *119*, 706–716. [CrossRef] [PubMed]
6. Chong, V.Z.; Thompson, M.; Beltaifa, S.; Webster, M.J.; Law, A.J.; Weickert, C.S. Elevated neuregulin-1 and ErbB4 protein in the prefrontal cortex of schizophrenic patients. *Schizophr. Res.* **2008**, *100*, 270–280. [CrossRef]
7. Agarwal, A.; Zhang, M.; Trembak-Duff, I.; Unterbarnscheidt, T.; Radyushkin, K.; Dibaj, P.; Martins-De-Souza, D.; Boretius, S.; Brzózka, M.M.; Steffens, H.; et al. Dysregulated Expression of Neuregulin-1 by Cortical Pyramidal Neurons Disrupts Synaptic Plasticity. *Cell Rep.* **2014**, *8*, 1130–1145. [CrossRef]
8. Olaya, J.C.; Heusner, C.L.; Matsumoto, M.; Sinclair, D.; Kondo, M.; Karl, T.; Weickert, C.S. Overexpression of Neuregulin 1 Type III Confers Hippocampal mRNA Alterations and Schizophrenia-Like Behaviors in Mice. *Schizophr. Bull.* **2018**, *44*, 865–875. [CrossRef]
9. Sun, Y.; Ikrar, T.; Davis, M.F.; Gong, N.; Zheng, X.; Luo, Z.D.; Lai, C.; Mei, L.; Holmes, T.; Gandhi, S.P.; et al. Neuregulin-1/ErbB4 Signaling Regulates Visual Cortical Plasticity. *Neuron* **2016**, *92*, 160–173. [CrossRef]
10. London, A.; Benhar, I.; Schwartz, M. The retina as a window to the brain—From eye research to CNS disorders. *Nat. Rev. Neurol.* **2013**, *9*, 44–53. [CrossRef]
11. Silverstein, S.M.; Rosen, R. Schizophrenia and the eye. *Schizophr. Res. Cogn.* **2015**, *2*, 46–55. [CrossRef] [PubMed]
12. Granö, N.; Salmijärvi, L.; Karjalainen, M.; Kallionpää, S.; Roine, M.; Taylor, P. Early signs of worry: Psychosis risk symptom visual distortions are independently associated with suicidal ideation. *Psychiatry Res.* **2015**, *225*, 263–267. [CrossRef] [PubMed]
13. Gordon-Lipkin, E.; Chodkowski, B.; Reich, D.S.; Smith, S.A.; Pulicken, M.; Balcer, L.J.; Frohman, E.M.; Cutter, G.; Calabresi, P.A. Retinal nerve fiber layer is associated with brain atrophy in multiple sclerosis. *Neurology* **2007**, *69*, 1603–1609. [CrossRef] [PubMed]
14. Ong, Y.-T.; Hilal, S.; Cheung, C.Y.; Venketasubramanian, N.; Niessen, W.J.; Vrooman, H.; Anuar, A.R.; Chew, M.; Chen, C.; Wong, T.Y.; et al. Retinal neurodegeneration on optical coherence tomography and cerebral atrophy. *Neurosci. Lett.* **2015**, *584*, 12–16. [CrossRef]
15. Ferrari, L.; Huang, S.-C.; Magnani, G.; Ambrosi, A.; Comi, G.; Leocani, L. Optical Coherence Tomography Reveals Retinal Neuroaxonal Thinning in Frontotemporal Dementia as in Alzheimer's Disease. *J. Alzheimer's Dis.* **2017**, *56*, 1101–1107. [CrossRef]

16. Hébert, M.; Mérette, C.; Gagné, A.-M.; Paccalet, T.; Moreau, I.; Lavoie, J.; Maziade, M. The Electroretinogram May Differentiate Schizophrenia From Bipolar Disorder. *Biol. Psychiatry* **2020**, *87*, 263–270. [CrossRef]
17. Schönfeldt-Lecuona, C.; Kregel, T.; Schmidt, A.; Kassubek, J.; Dreyhaupt, J.; Freudenmann, R.W.; Connemann, B.J.; Gahr, M.; Pinkhardt, E.H. Retinal single-layer analysis with optical coherence tomography (OCT) in schizophrenia spectrum disorder. *Schizophr. Res.* **2020**, *219*, 5–12. [CrossRef]
18. Velanac, V.; Unterbarnscheidt, T.; Hinrichs, W.; Gummert, M.N.; Fischer, T.M.; Rossner, M.; Trimarco, A.; Brivio, V.; Taveggia, C.; Willem, M.; et al. Bace1 processing of NRG1 type III produces a myelin-inducing signal but is not essential for the stimulation of myelination. *Glia* **2012**, *60*, 203–217. [CrossRef]
19. Schubert, T.; Gleiser, C.; Heiduschka, P.; Franz, C.; Nagel-Wolfrum, K.; Sahaboglu, A.; Weisschuh, N.; Eske, G.; Rohbock, K.; Rieger, N.; et al. Deletion of myosin VI causes slow retinal optic neuropathy and age-related macular degeneration (AMD)-relevant retinal phenotype. *Cell. Mol. Life Sci.* **2015**, *72*, 3953–3969. [CrossRef]
20. Stefansson, H.; Steinthorsdottir, V.; Thorgeirsson, T.; Gulcher, J.; Stefansson, K. Neuregulin 1 and schizophrenia. *Ann. Med.* **2004**, *36*, 62–71. [CrossRef]
21. Weickert, C.S.; Tiwari, Y.; Schofield, P.; Mowry, B.J.; Fullerton, J.M. Schizophrenia-associated HapICE haplotype is associated with increased NRG1 type III expression and high nucleotide diversity. *Transl. Psychiatry* **2012**, *2*, e104. [CrossRef] [PubMed]
22. Samani, N.N.; Proudlock, F.A.; Siram, V.; Suraweera, C.; Hutchinson, C.; Nelson, C.P.; Al-Uzri, M.; Gottlob, I. Retinal Layer Abnormalities as Biomarkers of Schizophrenia. *Schizophr. Bull.* **2018**, *44*, 876–885. [CrossRef] [PubMed]
23. Adams, S.A.; Nasrallah, H.A. Multiple retinal anomalies in schizophrenia. *Schizophr. Res.* **2018**, *195*, 3–12. [CrossRef] [PubMed]
24. Pinto, L.H.; Invergo, B.; Shimomura, K.; Takahashi, J.S.; Troy, J.B. Interpretation of the mouse electroretinogram. *Doc. Ophthalmol.* **2007**, *115*, 127–136. [CrossRef]
25. Penn, R.D.; Hagins, W.A. Signal Transmission along Retinal Rods and the Origin of the Electroretinographic a-Wave. *Nature* **1969**, *223*, 201–205. [CrossRef]
26. Robson, J.; Maeda, H.; Saszik, S.; Frishman, L. In vivo studies of signaling in rod pathways of the mouse using the electroretinogram. *Vis. Res.* **2004**, *44*, 3253–3268. [CrossRef]
27. Flames, N.; Long, J.E.; Garratt, A.; Fischer, T.M.; Gassmann, M.; Birchmeier, C.; Lai, C.; Rubenstein, J.L.; Marin, O. Short- and Long-Range Attraction of Cortical GABAergic Interneurons by Neuregulin-1. *Neuron* **2004**, *44*, 251–261. [CrossRef]
28. Ting, A.K.; Chen, Y.; Wen, L.; Yin, N.-M.; Shen, C.; Tao, Y.; Liu, X.; Xiong, W.-C.; Mei, L. Neuregulin 1 Promotes Excitatory Synapse Development and Function in GABAergic Interneurons. *J. Neurosci.* **2011**, *31*, 15–25. [CrossRef]
29. Soghomonian, J.-J.; Martin, D.L. Two isoforms of glutamate decarboxylase: Why? *Trends Pharmacol. Sci.* **1998**, *19*, 500–505. [CrossRef]
30. Hensch, T.K.; Fagiolini, M.; Mataga, N.; Stryker, M.P.; Baekkeskov, S.; Kash, S.F. Local GABA Circuit Control of Experience-Dependent Plasticity in Developing Visual Cortex. *Science* **1998**, *282*, 1504–1508. [CrossRef]
31. Guo, Y.; Kaplan, I.V.; Cooper, N.G.; Mower, G.D. Expression of two forms of glutamic acid decarboxylase (GAD67 and GAD65) during postnatal development of cat visual cortex. *Dev. Brain Res.* **1997**, *103*, 127–141. [CrossRef]
32. Christgau, S.; Aanstoot, H.J.; Schierbeck, H.; Begley, K.; Tullin, S.; Hejnaes, K.; Baekkeskov, S. Membrane anchoring of the autoantigen GAD65 to microvesicles in pancreatic beta-cells by palmitoylation in the NH_2-terminal domain. *J. Cell Biol.* **1992**, *118*, 309–320. [CrossRef] [PubMed]
33. Rutecki, P.A. Neuronal excitability: Voltage-dependent currents and synaptic transmission. *J. Clin. Neurophysiol.* **1992**, *9*, 195–211. [CrossRef] [PubMed]
34. Lanore, F.; Silver, R.A. Extracting Quantal Properties of Transmission at Central Synapses. *Neuromethods* **2016**, *113*, 193–211. [CrossRef]
35. Dodson, P.; Forsythe, I. Presynaptic K+ channels: Electrifying regulators of synaptic terminal excitability. *Trends Neurosci.* **2004**, *27*, 210–217. [CrossRef]
36. Ikematsu, N.; Dallas, M.L.; Ross, F.A.; Lewis, R.W.; Rafferty, J.N.; David, J.A.; Suman, R.; Peers, C.; Hardie, D.G.; Evans, A.M. Phosphorylation of the voltage-gated potassium channel Kv2.1 by AMP-activated protein kinase regulates membrane excitability. *Proc. Natl. Acad. Sci. USA* **2011**, *108*, 18132–18137. [CrossRef]
37. Asi, H.; Perlman, I. Relationships between the electroretinogram a-wave, b-wave and oscillatory potentials and their application to clinical diagnosis. *Doc. Ophthalmol.* **1992**, *79*, 125–139. [CrossRef]
38. Ridder, W.; Nusinowitz, S. The visual evoked potential in the mouse—Origins and response characteristics. *Vis. Res.* **2006**, *46*, 902–913. [CrossRef]
39. Barnstable, C.; Dräger, U.C. Thy-1 antigen: A ganglion cell specific marker in rodent retina. *Neuroscience* **1984**, *11*, 847–855. [CrossRef]

International Journal of
Molecular Sciences

MDPI

Article

Urolithin A Inactivation of TLR3/TRIF Signaling to Block the NF-κB/STAT1 Axis Reduces Inflammation and Enhances Antioxidant Defense in Poly(I:C)-Induced RAW264.7 Cells

Wen-Chung Huang [1,2,3,†], Chian-Jiun Liou [2,4,†], Szu-Chuan Shen [5], Sindy Hu [6,7], Jane C-J Chao [8], Chien-Yu Hsiao [7,9,*] and Shu-Ju Wu [7,9,*]

1 Research Center for Food and Cosmetic Safety, Graduate Institute of Health Industry Technology, College of Human Ecology, Chang Gung University of Science and Technology, Taoyuan City 33303, Taiwan; wchuang@mail.cgust.edu.tw
2 Division of Allergy, Asthma, and Rheumatology, Department of Pediatrics, Chang Gung Memorial Hospital, Linkou, Taoyuan City 33303, Taiwan; ccliu@mail.cgust.edu.tw
3 Department of Pediatrics, New Taipei Municipal Tu Cheng Hospital, Chang Gung Memorial Hospital, and Chang Gung University, New Taipei City 23678, Taiwan
4 Department of Nursing, Division of Basic Medical Sciences, Research Center for Chinese Herbal Medicine, and Graduate Institute of Health Industry Technology, Chang Gung University of Science and Technology, Taoyuan 33303, Taiwan
5 Graduate Program of Nutrition Science, National Taiwan Normal University, 88 Ting-Chow Rd, Sec 4, Taipei 11677, Taiwan; scs@ntnu.edu.tw
6 Department of Cosmetic Science, College of Human Ecology, Chang Gung University of Science and Technology, Guishan Dist., Taoyuan City 33303, Taiwan; sindyhu@hotmail.com
7 Department of Dermatology, Aesthetic Medical Center, Chang Gung Memorial Hospital, Linkou, Taoyuan City 33303, Taiwan
8 School of Nutrition and Health Sciences, College of Nutrition, Taipei Medical University, 250 Wu-Hsing Street, Taipei 11031, Taiwan; chenjui@tmu.edu.tw
9 Department of Nutrition and Health Sciences, Research Center for Chinese Herbal Medicine, College of Human Ecology, Chang Gung University of Science and Technology, Taoyuan City 33303, Taiwan
* Correspondence: nulycopene@gmail.com (C.-Y.H.); sjwu@mail.cgust.edu.tw (S.-J.W.); Tel.: +886-3-2118999 (ext. 5494) (C.-Y.H.); +886-3-2118999 (ext. 593) (S.-J.W.)
† These authors contributed equally to this work.

Citation: Huang, W.-C.; Liou, C.-J.; Shen, S.-C.; Hu, S.; Chao, J.C.-J.; Hsiao, C.-Y.; Wu, S.-J. Urolithin A Inactivation of TLR3/TRIF Signaling to Block the NF-κB/STAT1 Axis Reduces Inflammation and Enhances Antioxidant Defense in Poly(I:C)-Induced RAW264.7 Cells. *Int. J. Mol. Sci.* **2022**, 23, 4697. https://doi.org/10.3390/ijms23094697

Academic Editor: Malgorzata Kloc

Received: 25 March 2022
Accepted: 20 April 2022
Published: 23 April 2022

Abstract: Urolithin A is an active compound of gut-microbiota-derived metabolites of polyphenol ellagic acid that has anti-aging, antioxidative, and anti-inflammatory effects. However, the effects of urolithin A on polyinosinic acid-polycytidylic acid (poly(I:C))-induced inflammation remain unclear. Poly(I:C) is a double-stranded RNA (dsRNA) similar to a virus and is recognized by Toll-like receptor-3 (TLR3), inducing an inflammatory response in immune cells, such as macrophages. Inflammation is a natural defense process of the innate immune system. Therefore, we used poly(I:C)-induced RAW264.7 cells and attenuated the inflammation induced by urolithin A. First, our data suggested that 1–30 μM urolithin A does not reduce RAW264.7 cell viability, whereas 1 μM urolithin A is sufficient for antioxidation and the decreased production of tumor necrosis factor-α (TNF-α), monocyte chemoattractant protein-1 (MCP-1), and C-C chemokine ligand 5. The inflammation-related proteins cyclooxygenase-2 and inducible nitric oxide synthase were also downregulated by urolithin A. Next, 1 μM urolithin A inhibited the levels of interferon (INF)-α and INF-β. Urolithin A was applied to investigate the blockade of the TLR3 signaling pathway in poly(I:C)-induced RAW264.7 cells. Moreover, the TLR3 signaling pathway, subsequent inflammatory-related pathways, and antioxidation pathways showed changes in nuclear factor-κB (NF-κB) signaling and blocked ERK/mitogen-activated protein kinase (MAPK) signaling. Urolithin A enhanced catalase (CAT) and superoxide dismutase (SOD) activities, but decreased malondialdehyde (MDA) levels in poly(I:C)-induced RAW264.7 cells. Thus, our results suggest that urolithin A inhibits TLR3-activated inflammatory and oxidative-associated pathways in macrophages, and that this inhibition is induced by poly(I:C). Therefore, urolithin A may have antiviral effects and could be used to treat viral-infection-related diseases.

Keywords: urolithin A; poly(I:C); TLR3; NF-κB; MAPK

1. Introduction

Mammalian Toll-like receptors (TLRs) recognize various pathogen-associated molecular patterns (PAMPs) and play crucial roles in the innate immune system [1]. Polyinosinic-polycytidylic acid (poly(I:C)) is a double-stranded RNA (dsRNA), and the specific ligand of TLR3 recognizes viral dsRNA [2]. Recently, poly(I:C) was shown to be able to activate the TLR3/TRIF pathway in the systemic inflammatory response [3]. The TLR family interacts with PAMPs and recruits a TIR domain-containing adapter inducing interferon (IFN)-β (TRIF), which initiates TLR signaling to activate the transcription factors nuclear factor-κB (NF-κB) and interferon-regulatory factor 3 (IRF3) to induce innate immune responses and inflammatory mediators [4]. The activation of TLRs immediately regulates cytokines, chemokines, and type I IFN production with the subsequent induction of IFN-responsive genes, such as antiviral genes and other inflammatory mediators, to protect the host from microbial infection [5,6]. This process initiates the host's defensive responses, such as inflammation and antioxidants, to protect the host from microbial infection [1].

Normally, NF-κB is localized in the cytoplasm and binds to the inhibitory κB (IκB) proteins. A variety of inflammatory stimuli induce IκB phosphorylation by kinases, leading to the degradation of IκB. The phosphorylation of IκBα leads to NF-κB activation, and the active NF-κB translocates into the nucleus and activates the target genes. Through TLR3 signaling, poly(I:C) mediates the phosphorylation of IκB and NF-κB activation [7], which then modulates the innate immune and oxidative stress responses. Studies have indicated that the activation of NF-κB increases cellular oxidative stress [8,9]. In addition, studies have indicated that interactions between the nuclear factor erythroid 2-related factor 2 (Nrf2) and NF-κB pathways in cells cause oxidative stress. Thus, NF-κB modulates Nrf2 transcription and activity, and the NF-κB pathway can have both anti- and pro-oxidant roles in the setting of oxidative stress [8,10]. On the other hand, Kelch-like ECH-associating protein 1 (KEAP1) is an IκB kinase (IKK)β E3 ubiquitin ligase that can prevent NF-κB pathway activation. As such, NF-κB can have different roles in Nrf2 expression and activity [11]. The TLR3-induced generation of intracellular reactive oxygen species (ROS) is required for the activation of NF-κB, and the phosphorylation of the IRF3-induced signal transducer and the activator of transcription 1 (STAT1) contributes to the generation of inflammatory mediators and innate immune responses in macrophages. Alternatively, TLR3-ROS signaling plays a role in innate immune responses by activating STAT1 [12]. The ROS molecules induce proteins and lipid peroxidation, as well as DNA damage. NF-κB activation has been shown to increase manganese superoxide dismutase (Mn-SOD) levels in TNF-α-treated Ewing's sarcoma cells. Thus, the activation of the NF-κB pathway may also induce antioxidant activity [13]. In addition, STAT1 and NF-κB signaling may induce the transcriptional activation of type I IFNs (IFN-α/β) in activated macrophages [14,15], as well as many antiviral genes for survival [16].

In particular, TRIF by poly(I:C), which phosphorylates IRF3 to induce the transcription of pro-inflammatory cytokines and type I IFNs, increases the expression of cytokines, such as IFN-β [17,18]. Studies have indicated that poly(I:C)-induced IRF3 phosphorylation immediately activates mitogen-activated protein kinase (MAPK) signaling. Simultaneously, TRIF-dependent signaling can activate MAPK signaling [19–21]. There are three MAPKs in mammalian cells: c-Jun N-terminal kinase (JNK), extracellular signal-regulated kinase (ERK), and p38 MAPK. A variety of transduced signals induce MAPK phosphorylation, activating MAPK in the nucleus to mediate cellular processes, including stress responses and cancer progression. In addition, studies have shown that MAPK mediates important physiological and pathological functions in innate immunity [22,23]. TRIF induces IRF3, NF-κB, and MAPK kinase activation, affecting viral infections [24]. Moreover, poly(I:C) could activate TLR3-induced cyclooxygenase-2 (COX-2) protein expression and increase

JNK, ERK, and p38 MAPK phosphorylation to promote an inflammatory process in the brain [25].

Pomegranate (*Punica granatum* L.) contains polyphenols, including punicalagin, punicalin, ellagic acid, and gallic acid, which are rich in the fruit, leaves, and peel [26]. However, the bioavailability of dietary polyphenol is low, and ellagic acid and related polyphenols from the consumption of pomegranates are metabolized to urolithin A by gut microbes [27,28]. Thus, urolithin A is a microflora-derived metabolite. Studies have found that a mitophagy activator improves muscle mitochondrial and cellular health in older humans [29]. By improving mitochondrial functioning and mitigating MAPK/NF-κB/Akt signaling to decrease ROS and pro-inflammatory cytokine levels, urolithin A attenuates metabolic diseases in multiple tissues, preventing Alzheimer's disease, type 2 diabetes mellitus, non-alcoholic fatty liver disease, and attenuated ox-LDL-induced cholesterol accumulation [30,31]. Interestingly, urolithin A is a potential neuroprotective agent, decreasing oxidative stress and increasing antioxidant enzymes in H_2O_2-actived Neuro-2a cells [32]. Through the modulation of the gut microbiota, urolithin A also decreases body weight gain and improves inflammation and dysfunctional lipid metabolism in high-fat-diet-induced obese mice [33].

The antiviral mechanisms of urolithin A remain unclear because its therapeutic effects are not fully explained by its interaction with TLR3. Previous studies have shown antioxidant, anti-inflammation, and anti-aging effects, but the direct action of urolithin A inhibition by poly(I:C) and its attenuated potential on innate immunity remains unexplored. The present study aimed to explore the effects of urolithin A on the poly(I:C)-induced activation of TLR3's antiviral effects, counterbalancing NF-κB/MAPK's inflammatory responses and enhancing the NRF2 antioxidation pathway. We found that urolithin A inhibits poly(I:C)-induced IFN-α and IFN-β levels and induces phosphorylated STAT1 (pSTAT1) downstream of these cytokines. In addition, urolithin A, accompanied by the ERK/MAPK inhibitor (PD98059), reduces the TNF-α, MCP-1, and INF-β levels induced by poly(I:C) in RAW264.7 macrophages. Therefore, the decrease in poly(I:C)-induced IFN-α and IFN-β levels after urolithin A treatment could depend on the inhibited phosphorylation of IRF3.

2. Results

2.1. Urolithin A Inhibited Inflammatory Cytokines and Inactived TLR3 Pathway Protein Expression in Poly(I:C)-Stimulated RAW264.7 Cells

Urolithin A at concentrations of ≥30 μM showed no significant cytotoxicity in RAW264.7 cells, but concentrations of ≥60 μM significantly reduced cell numbers (Figure 1B). Therefore, all experiments used 1–30 μM urolithin A. When RAW264.7 cells were treated with urolithin A in six-well plates and stimulated with poly(I:C) (1 μg/mL) at concentrations of ≥1 μM, this significantly reduced inflammatory cytokines, including TNF-α, MCP-1, and CCL-5, compared to poly(I:C) alone ($p < 0.01$; Figure 1C–E). Interestingly, ≥1 μM urolithin A significantly inhibited TLR3 protein expression ($p < 0.05$), and ≥10 μM urolithin A significantly decreased TRIF and pIRF protein expression compared to poly(I:C) alone ($p < 0.05$; Figure 2A–F). A previous study found that a ligand of TLR3 contributes to the viral-infection-induction of chemokine CCL5/RANTES secretion by airway epithelial cells to attract inflammatory cells [34]. In addition, cytokines TNF-α and MCP-1, produced by macrophages and dendritic cells, are important for innate immunity against viruses. TRIF and IRF3 are activated in poly(I:C)-exposed macrophages, leading to the production of inflammatory mediators [34,35]. Our data showed that TLR3, TRIF, and pIRF3 are activated in poly(I:C)-exposed macrophages, leading to the production of TNF-α, MCP-1, and CCL-5. Urolithin A significantly inhibited TLR3, TRIF, and pIRF3 protein expression in poly(I:C)-stimulated RAW264.7 cells compared to poly(I:C) alone.

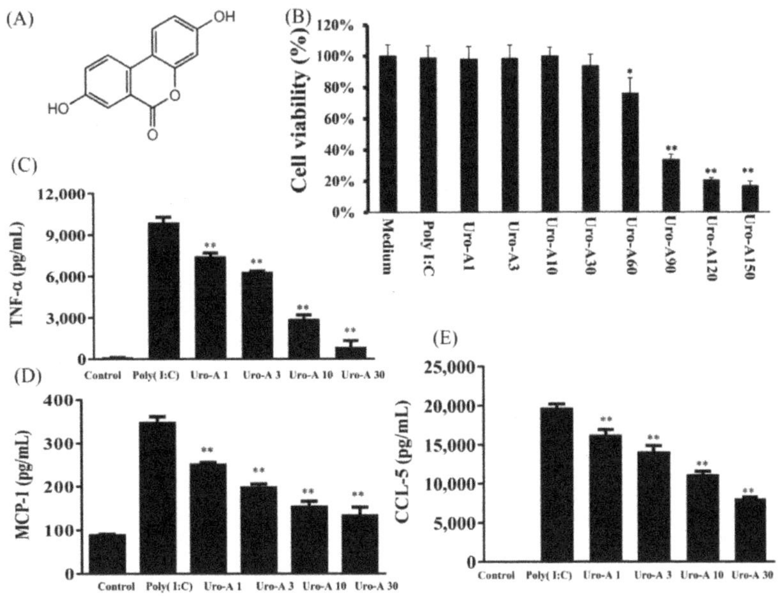

Figure 1. Decreased inflammation-related cytokine secretion. (**A**) Structure of Uro-A. (**B**) Cell viability of MTT. (**C**) TNF-α, (**D**) MCP-1, and (**E**) CCL-5 concentrations. Cells were pre-treated with Uro-A for 1 h, and then with added poly(I:C) (1 µg/mL) for 24 h. Uro-A significantly decreased the inflammatory cytokine concentration in poly(I:C)-stimulated RAW264.7 cells. Data are presented as mean ± SD. * $p < 0.05$, ** $p < 0.01$ compared to RAW264.7 cells stimulated with poly(I:C) alone.

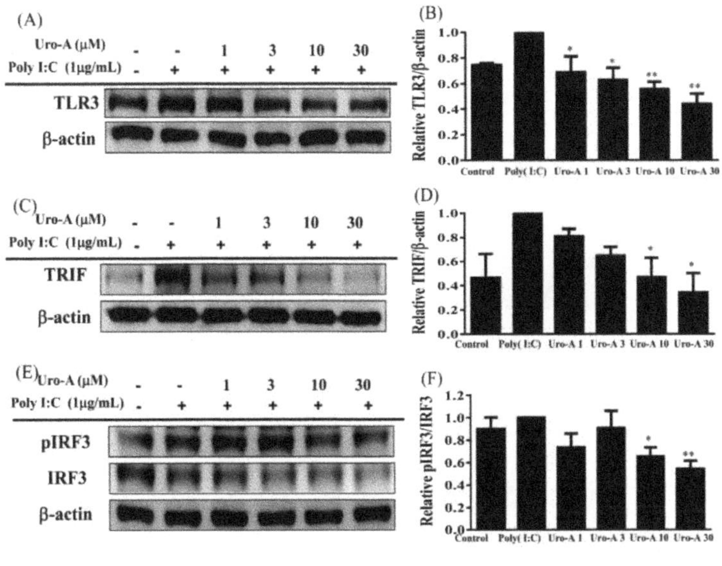

Figure 2. Urolithin A (Uro-A) suppressed TLR3 pathway protein expression in poly(I:C)-stimulated RAW264.7 cells. (**A**) TLR3 protein expression. (**B**) Fold-change in TLR3 expression relative to β-actin expression. (**C**) TRIF protein expression. (**D**) Fold–change in TRIF expression relative to β-actin expression. (**E**) pIRF3 and IRF3 protein expression. (**F**) Fold-change in pIRF3 expression relative to IRF3 expression. RAW264.7 cells were precultured with various concentrations of Uro-A, and then

incubated with poly(I:C) (1 μg/mL) for 30 min or 24 h. Data are presented as mean ± SD. * $p < 0.05$, ** $p < 0.01$ compared to RAW264.7 cells stimulated with poly(I:C) alone.

2.2. Urolithin A Blocked the NF-κB/STAT1 Pathway to Inhibit the Expression of Inflammatory Mediators in Poly(I:C)-Induced RAW264.7 Cells

Urolithin A concentrations of ≥3 μM significantly blocked NF-κB activation (Figure 3A–C) and concentrations of ≥10 μM inhibited pIκB expression (Figure 3D,E) compared to poly(I:C) alone. Furthermore, urolithin A concentrations of ≥3 μM significantly suppressed STAT1 phosphorylation compared to poly(I:C) alone ($p < 0.01$). Previous studies have indicated that poly(I:C)-induced NF-κB and STAT1 activation promotes IFN-β secretion in aortic valve interstitial cells [35]. In addition, poly(I:C)-induced TLR3 expression and the activation of the STAT1 pathway regulates the innate immune response, resulting in the production of IFN-α/β in epidermal keratinocytes [36]. Therefore, we evaluated the ability of urolithin A to inhibit NF-κB and STAT1 in relation to the suppression of IFN-α/β secretion and decreased inflammatory mediators in poly(I:C)-stimulated RAW264.7 cells. Immunofluorescent staining to investigate urolithin A showed that it could inhibit the translocation of NF-κB p65 from the cytoplasm into the nucleus when cells were pre-cultured with or without urolithin A (1, 3, 10, 30 μM) for 1 h and added poly(I:C) (1 μg/mL) for 30 min. Urolithin A increased the p65 subunit retention in the cytoplasm in poly(I:C)-stimulated RAW264.7 cells (Figure 3A). Moreover, we used immune blots to analyze NF-κB and IκB protein expression and found that urolithin A decreased NF-κB and IκB phosphorylation compared to poly(I:C) alone (Figure 3B–E). Importantly, urolithin A at concentrations of ≥1 μM strikingly decreased STAT1 phosphorylation compared to poly(I:C) alone (Figure 4A,B). We also found that urolithin A at ≥10 μM decreased the expression of the inflammatory mediators COX-2 and iNOS compared to poly(I:C) alone (Figure 4C–F). Our results indicate that urolithin A inhibits inflammatory mediator expression in poly(I:C)-induced RAW264.7 cells via the blockade of the NF-κB and STAT1 pathways.

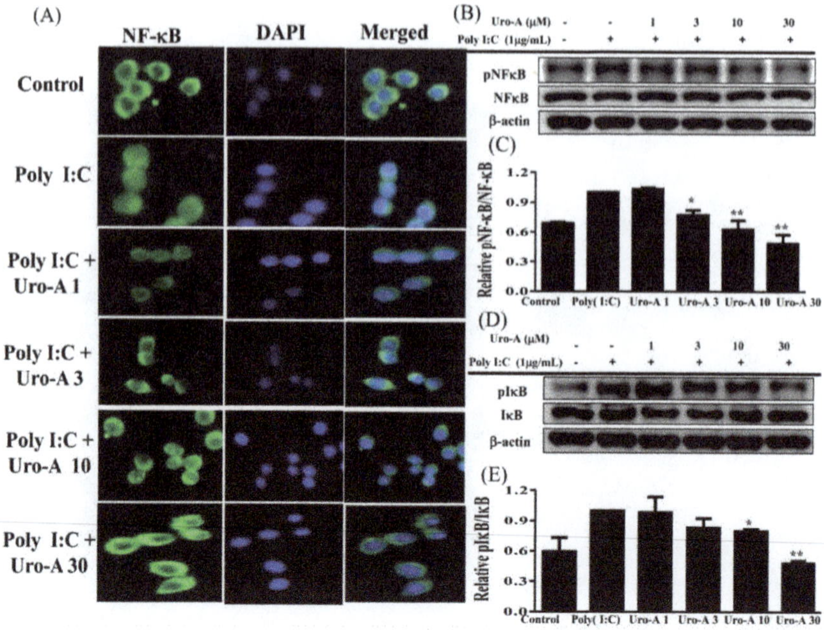

Figure 3. Urolithin A (Uro-A) blocked the NF-κB activation in poly(I:C)-induced RAW264.7 cells.

(**A**) DAPI staining showed that Uro-A inhibited NF-κB p65 from translocating from the cytoplasm into the nucleus. Cells were pre-cultured with Uro-A, and then incubated with poly(I:C) (1 μg/mL) for 30 min. Immunofluorescent staining was performed to evaluate NF-κB p65 (green) translocation. Blue is nuclear staining with DAPI. (**B**) pNF-κB and NF-κB protein expression. (**C**) Fold-change in μpNF-κB expression relative to NF-κB expression. (**D**) pI-κB and I-κB protein expression. (**E**) Fold-change in pI-κB expression relative to I-κB expression. RAW264.7 cells were precultured with various concentrations of Uro-A, and then incubated with poly(I:C) (1 μg/mL) for 30 min or 24 h. Data are presented as mean ± SD. * $p < 0.05$, ** $p < 0.01$ compared to RAW264.7 cells stimulated with poly(I:C) alone.

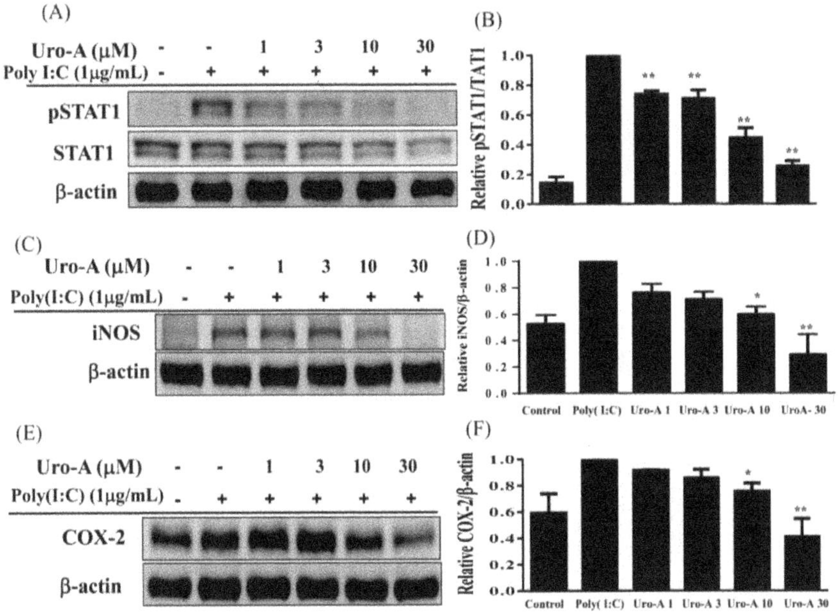

Figure 4. Urolithin A (Uro-A) inhibited the expression of inflammatory mediators in poly(I:C)-induced RAW264.7 cells. (**A**) pSTAT1 and STAT1 protein expression. (**B**) Fold-change in pSTAT1 expression relative to STAT1 expression. (**C**) iNOS protein expression. (**D**) Fold-change in iNOS expression relative to β-actin expression. (**E**) COX-2 protein expression. (**F**) Fold-change in COX-2 expression relative to β-actin expression. RAW264.7 cells were precultured with various concentrations of Uro-A, and then incubated with poly(I:C) (1 μg/mL) for 30 min or 24 h. Data are presented as mean ± SD. * $p < 0.05$, ** $p < 0.01$ compared to RAW264.7 cells stimulated with poly(I:C) alone.

2.3. Urolithin A Suppressed the ERK/MAPK Pathway and the ERK Inhibitor Decreased Cytokine Secretion in Poly(I:C)-Stimulated RAW264.7 Cells

Next, we evaluated whether urolithin A inhibits MAPK activation and MAPK inhibitors decrease inflammation in poly(I:C)-induced RAW264.7 cells. First, to evaluate MAPK signaling protein expression, cells were pre-cultured with or without urolithin A (1, 3, 10, 30 μM) for 1 h and added poly(I:C) (1 μg/mL) for 30 min or 24 h. The collected protein samples were analyzed by immunoblot. Urolithin A concentrations of ≥3 μM significantly decreased pERK1/2 expression compared to poly(I:C) alone ($p < 0.05$; Figure 5A,B). Nevertheless, urolithin A did not affect pp38 or pJNK protein expression (Figure 5C–F). Next, we evaluated the effect of the ERK/MAPK inhibitor on urolithin A-induced changes to inflammatory cytokine concentrations in poly(I:C)-induced RAW264.7 cells. RAW264.7 cells were pre-cultured with urolithin A (3 and 10 μM) and/or 10 μM ERK1/2 inhibitor PD98059 before incubation with poly(I:C) (1 μg/mL). After 24 h, the collected supernatant was evaluated by ELISA for cytokine concentrations. Although urolithin A at 10 μM + poly(I:C) significantly decreased MCP-1 and IFN-β ($p < 0.01$ and $p < 0.05$, respectively), that urolithin A at 10 μM decreased TNF-α utility is equivalent to the ERK inhibitor + poly(I:C) alone. Urolithin A at 3 μM + poly(I:C) cannot achieve the same effect as the ERK1/2 inhibitor + poly(I:C) alone. In addition, the ERK1/2 inhibitor + poly(I:C) + urolithin A at concentrations of ≥3 μM markedly decreased the concentrations of TNF-α and MCP-1 levels, and urolithin A ≥10 μM significantly decreased IFN-β levels compared to the poly(I:C) + ERK1/2 inhibitor ($p < 0.01$). Moreover, the ERK1/2 inhibitor + poly(I:C) + urolithin A at concentrations of ≥3 μM significantly decreased TNF-α, MCP-1, and IFN-β levels compared to poly(I:C) + urolithin A alone ($p < 0.01$). These results showed there is an additive effect with a combination of urolithin A and the ERK1/2 inhibitor. Furthermore, Uro-A at ≥1 μM remarkably inhibited IFN-α and IFN-β concentrations compared to poly(I:C) alone in poly(I:C)-induced RAW264.7 cells ($p < 0.01$) (Figure 6D,E). Therefore, we suggest that, in poly(I:C)-stimulated RAW264.7 cells, urolithin A attenuated the NF-κB/STAT1 and ERK/MAPK pathways to inhibit the expression of inflammatory mediators (COX-2 and iNOS) and decrease inflammatory cytokine (TNF-α and MCP-1) and IFN-α/β concentrations.

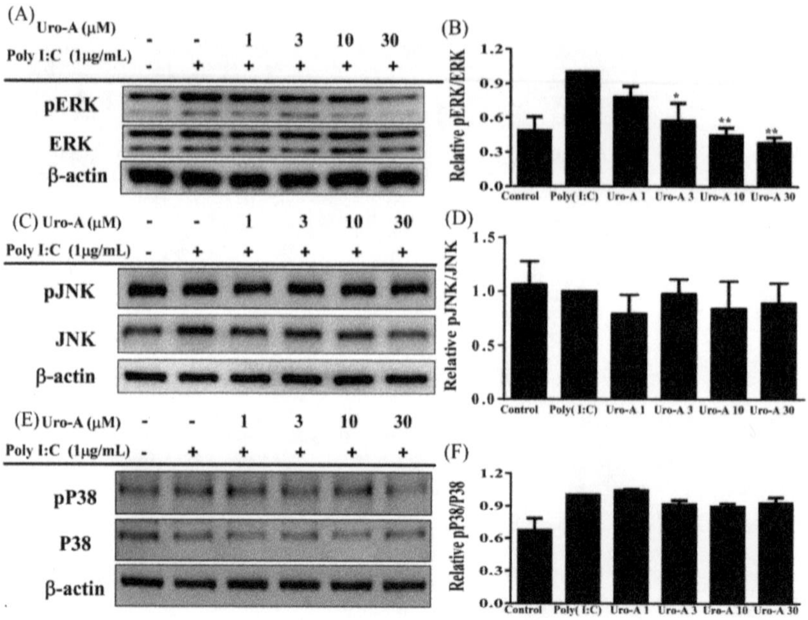

Figure 5. Urolithin A (Uro-A) inhibited inflammatory pERK signaling in poly(I:C)-induced RAW264.7 cells. (**A**) pERK and ERK protein expression. (**B**) Fold-change in pERK expression relative to ERK expression. (**C**) pJNK and JNK protein expression. (**D**) Fold-change in pJNK expression relative to JNK expression. (**E**) pP38 and P38 protein expression. (**F**) Fold-change in pP38 expression relative to P38 expression. RAW264.7 cells were precultured with various concentrations of Uro-A, and then incubated with poly(I:C) (1 μg/mL) for 30 min or 24 h. Data are presented as mean ± SD. * $p < 0.05$, ** $p < 0.01$ compared to RAW264.7 cells stimulated with poly(I:C) alone.

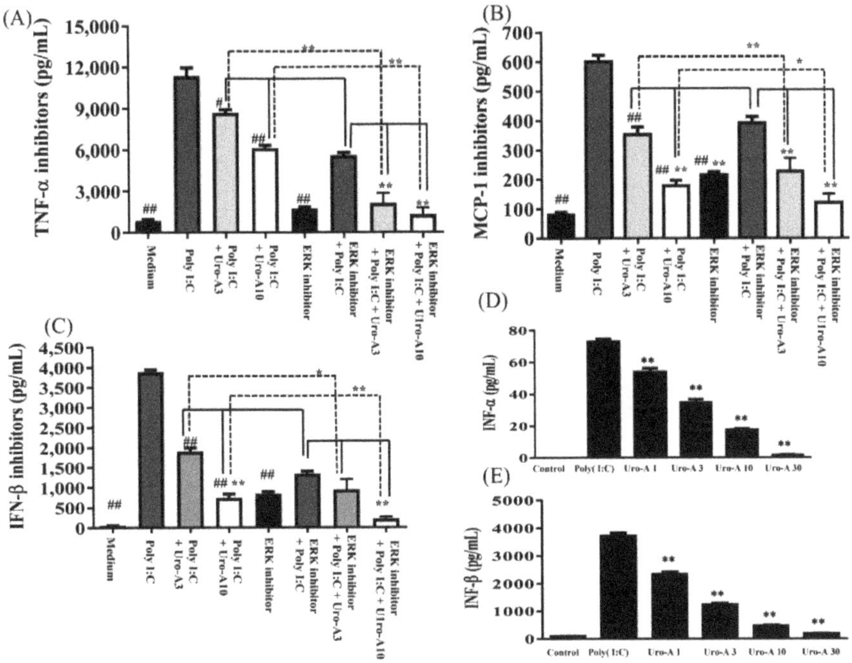

Figure 6. Urolithin A (Uro-A) and the ERK inhibitor suppressed inflammation-related cytokine secretion in poly(I:C)-induced RAW264.7 cells. RAW264.7 cells were pre-cultured with various concentrations of Uro-A, and then incubated with poly(I:C) (1 μg/mL) for 24 h. (**A**) ELISA showed inflammation-related cytokine concentrations of TNF-α, (**B**) MCP-1, and (**C**) IFN-β. (**D**) Uro-A inhibited IFN-α and (**E**) IFN-β concentrations. Data are presented as mean ± SD. # $p < 0.05$, ## $p < 0.01$ compared to poly(I:C) alone in RAW264.7 cells stimulated with poly(I:C). * $p < 0.05$, ** $p < 0.01$ compared to RAW264.7 cells stimulated with the ERK inhibitor + poly(I:C) alone and the ERK inhibitor + poly(I:C) + Uro-A, respectively.

2.4. *Urolithin A Elevated Nrf2 Transcriptional Regulation and Enhanced Antioxidant Cytoprotective Defense in Poly(I:C)-Stimulated RAW264.7 Cells*

Immunofluorescent staining was performed to evaluate urolithin A's regulation of the antioxidant transcription factor Nrf2. RAW264.7 cells were pre-cultured with urolithin A (1–30 μM) for 1 h and then with added poly(I:C) (1 μg/mL) for 30 min. Nrf2 was retained in the cytoplasm of poly(I:C)-stimulated RAW264.7 cells, whereas urolithin A at concentrations of ≥3 μM affected Nrf2 translocation from the cytoplasm into the nucleus (Figure 7A). Furthermore, urolithin A at concentrations of ≥1 μM significantly enhanced the CAT and Mn-SOD activities compared to poly(I:C) alone (Figure 7B,C). MDA was also significantly reduced by urolithin A at concentrations of ≥1 μM compared to poly(I:C) alone (Figure 7D). According to the above results, we suggest that urolithin A has a Nrf2-mediated dose response to enhanced antioxidant defense in poly(I:C)-stimulated RAW264.7 cells.

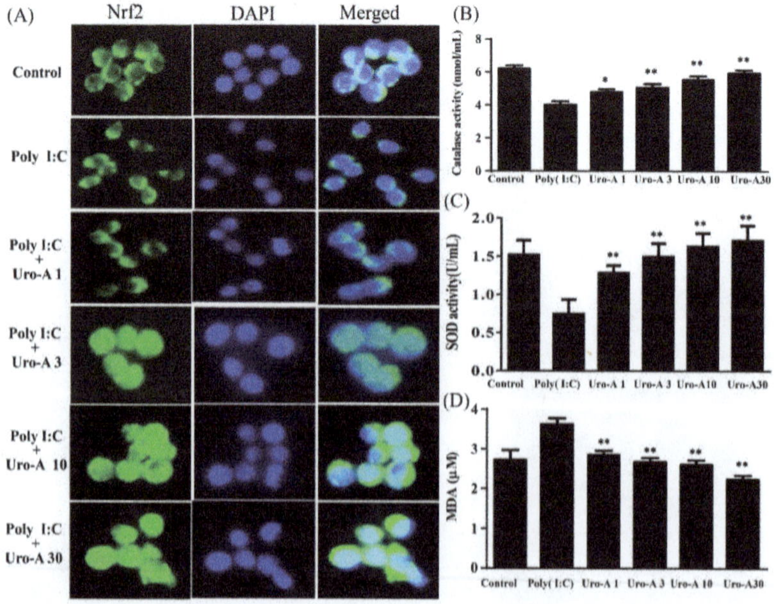

Figure 7. Urolithin A (Uro-A) activated Nrf2 expression and enhanced antioxidant defense in poly(I:C)-stimulated RAW264.7 cells. (**A**) Uro-A promoted Nrf2 (green) translocation from the cytoplasm into the nucleus (DAPI, blue). (**B**) Uro-A increased the activities of antioxidant enzymes SOD and (**C**) CAT, and (**D**) attenuated MDA production. RAW264.7 cells were pre-cultured with different concentrations of Uro-A for 1 h, and then incubated with poly(I:C) (1 ng/mL) for 24 h. Data are presented as mean ± SD. * $p < 0.05$, ** $p < 0.01$ compared to RAW264.7 cells stimulated with poly(I:C) alone.

3. Discussion

Human immune cells, including macrophages, elicit IFN-α/β production to activate natural killer (NK) cells via TLR3 receptors on the endosomal membrane, which are sensors of viral dsRNA and poly(I:C) [37]. In viral infection, TLR3 recognizes dsRNA and, through TRIF, transmits signals to activate the transcription factor IRF3. The phosphorylation of IRF3 results in its translocation into the nucleus, leading to IFN-α/β and inflammatory cytokine production and antiviral immune responses [38]. Therefore, it is worth exploring the TLR3-mediated type I IFN signaling pathway. Here, we demonstrated that urolithin A can prevent TLR3-mediated IFN-α/β antiviral responses (Figures 2 and 6D,E).

Previous studies have shown that poly(I:C), through the activation of the NF-κB pathway, induces iNOS and COX-2 overexpression, subsequently producing proinflammatory cytokines [39,40]. Urolithin A is an excellent inhibitor, inactivating NF-κB (Figure 3) and attenuating the expression of proinflammatory mediators (e.g., iNOS, COX-2) (Figure 4C,D) to achieve decreased cytokine (TNF-α, MCP-1, and CCL-5) secretion (Figure 1C–E). Furthermore, NF-κB has crosstalk with the STAT and IRF3 pathways in the antiviral response. Studies have indicated that NF-κB and IRF3 translocate into the nucleus in poly(I:C)-induced MEF cells [41]. Poly(I:C)-activated NF-κB simultaneously induces IRF3 and STAT1, leading to INF-α/β secretion, and IKKα is activated by dsRNA-mediated STAT1 phosphorylation in antiviral signaling by the innate immune system [7]. In the inflammatory response, transcription factors NF-κB and STAT1 are important regulators of inflammatory cytokine production and inflammatory cell infiltration. During the infection, STAT1 is a TLR-mediated antibody response and B cell differentiation [42]. STAT1 plays an essential role in the TLR-mediated antibody response of the marginal zone during inflammation and infection [43]. Thus, STAT1 is an important target in preventing or treating viral-infection-induced inflammation. In the present study, we demonstrated that poly(I:C)-induced NF-κB and STAT1 phosphorylation, and pre-treatment with urolithin A efficiently inhibited pNF-κB and pSTAT1 protein expression (Figures 3 and 4A). We suggest that urolithin A's blockade of NF-κB and STAT pathways is mediated by TLR3 in poly(I:C)-induced RAW264.7 cells.

Macrophages play important roles in immunity and various inflammatory diseases via the production of cytokines and the inflammatory mediators iNOS and COX-2. Many studies have elucidated the relationship between cytokines and inflammatory mediators involved in the MAPK inflammatory pathway [40]. Previous studies have indicated that poly(I:C) binds to endosomes in the cells and induces transcription factor activation, resulting in the secretion of antiviral IFN-I, which feeds back to antiviral gene replication. Furthermore, MAPK/p38 and ERK are required for poly(I:C)/TLR3-mediated cytokine production [44,45]. This result indicates that poly(I:C) increases iNOS and COX-2 overexpression and induces cytokine production. In the present study, we investigated the mechanisms underlying poly(I:C)-induced inflammatory and antiviral MAPK signaling. First, urolithin A significantly attenuated pERK/MAPK expression, but we did not observe the same expression in pP38 and pJNK/MAPK (Figure 5). In addition, when we blocked the ERK/MAPK pathway, we observed anti-inflammatory and antiviral effects. As stated above, reducing ERK/MAPK protein expression may involve the expression of the downstream inflammatory mediator iNOS and COX-2. Urolithin A and the ERK inhibitor perfectly decreased cytokines and IFN-I concentration (Figure 6). These findings demonstrate that urolithin A has antiviral immunity and anti-inflammation effects via the modulation of the ERK/MAPK pathway.

In addition, poly(I:C) has been indicated as activating TLR3, subsequently causing inflammation and IFN-β expression and inducing ROS production [46]. Mounting evidence suggests that transcription factor Nrf2 regulates cellular-defense oxidative stress and prevents inflammation-related diseases. On the other hand, studies have shown that inflammatory responses are the second line in innate immunity, and the activation of TLR3 signaling and Nrf2 expression work together to regulate the innate immune system [47]. The interplay between TLR signaling and the Nrf2 pathway may be beneficial for the prevention or treatment of inflammation. Nrf2 signaling is a major modulator of antioxidant enzymes and attenuates inflammation. Nrf2 signaling can be activated by TLR agonists and enhance antioxidant defense [48]. Previous studies have implicated TLR ligands as inflammation mediators and immuno-modulators through the activation of Nrf2 to regulate downstream antioxidant enzymes [49]. Importantly, activated Nrf2 limits IκB phosphorylation and keeps NF-κB in the cytoplasm, reducing inflammatory mediators, such as iNOS, COX-2, and cytokines [50]. However, whether the interaction between Nrf2 and TLR signaling regulates the antioxidant and anti-inflammation activity of macrophages infected by poly(I:C) remains unknown. The oxidative-stress-related enzymes SOD and CAT and the lipid peroxidation product MDA were detected in this study. Furthermore, urolithin A promoted Nrf2 translocation from the cytoplasm into the nucleus, enhancing SOD and CAT activities to reduce MDA levels in poly(I:C)-induced macrophages (Figure 7). We also observed that NF-κB, IκB phosphorylation, and inflammatory mediators were inhibited by urolithin A. These results suggest that urolithin A depends on the activation of Nrf2 to enhance antioxidative defense and anti-inflammatory actions.

Taken together, the data indicate that urolithin A can inhibit TLR3 expression and reduce IFN-α/β secretion by suppressing IRF3 phosphorylation. Moreover, urolithin A suppressed the activation of NF-κB/STAT1 and ERK/MAPK signaling, reducing inflammatory mediators (iNOS, TNF-α, MCP-1, and CCL-5) in poly(I:C)-stimulated macrophages. Importantly, urolithin A significantly modulated Nrf2 to increase antioxidant enzymes (SOD and CAT) and decrease the inflammatory responses in poly(I:C)-stimulated macrophages (Figure 8).

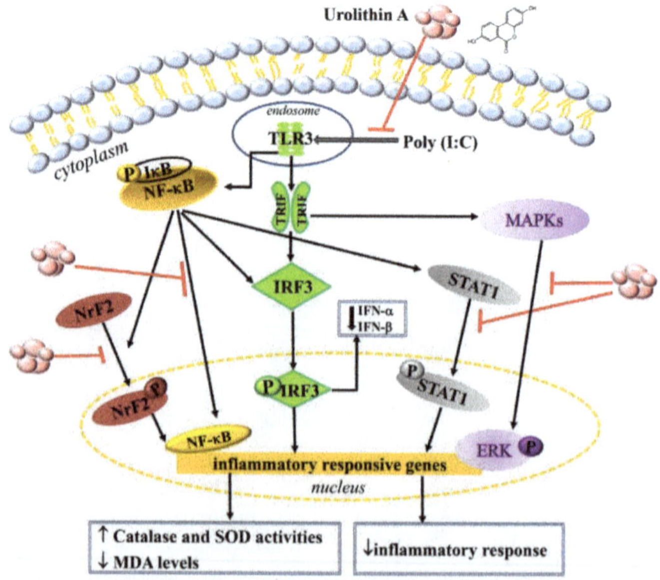

Figure 8. Model explaining the mechanisms of urolithin A's attenuation of viral-induced inflammation and oxidative stress in macrophages stimulated by poly(I:C).

4. Materials and Methods

4.1. Materials

Urolithin A (≥97% purity by HPLC, Figure 1A) was purchased from Sigma-Aldrich (St. Louis, MO, USA), cell viability assays (MTT), and poly(I:C) were purchased from InvivoGen (San Diego, CA, USA). Enzyme-linked immunosorbent assay (ELISA) kits were purchased from R&D Systems (Minneapolis, MN, USA). DAPI solution was purchased from Sigma-Aldrich (St. Louis, MO, USA). The inhibitor PD98059 was purchased from Enzo Life Sciences (Farmingdale, NY, USA). Antibodies against β-actin, TLR3, TRIF, IRF, STAT1, NF-κB, I-κB, COX-2, iNOS, Nrf2, and phosphorylated IRF (pIRF), STAT1, NF-κB (pNF-κB), and IκB (pIκB) were purchased from Santa Cruz Biotechnology (Santa Cruz, CA, USA). Antibodies against JNK, ERK, p38, and phosphorylated JNK (pJNK), ERK (pERK), and p38 (pp38) were purchased from Millipore (Billerica, MA, USA).

4.2. Preparation of Urolithin A and Cell Culture

The urolithin A stock solution was prepared at 100 mM and stored at −20 °C. DMSO was added to the medium at a final concentration of ≤0.1% as described previously [33]. The mouse macrophage RAW264.7 cell line was purchased from Bioresource Collection and Research Center (BCRC, Taiwan). Cells were cultured in Dulbecco's Modified Eagle Medium (Invitrogen-Gibco, Paisley, Scotland) with 10% heat-inactivated fetal bovine serum (FBS; Invitrogen-Gibco, Paisley, Scotland) and 100 U/mL penicillin G, 100 μg/mL streptomycin, and 50 ng/mL gentamycin at 37 °C in a humidified atmosphere of 5% CO_2. RAW264.7 cells were seeded in serum-free medium for 8 h. Subsequently, RAW264.7 cells were cultured in DMEM medium with 10% FBS. RAW264.7 cells were pre-cultured with or without urolithin A at various concentrations (1, 3, 10, 30 μM) for 1 h, and then with added poly(I:C) (1 μg/mL). After 24 h, the RAW264.7 cells were lysed for Western blot analysis and the supernatant used for ELISA.

4.3. Cell Viability Assay

We used MTT kits (Sigma-Aldrich) to assess the inhibitory effect of urolithin A on RAW264.7 cell viability. Cells were seeded in 96-well plates at a concentration of 10^5 cells/well and urolithin A was added at a concentration of between 1–30 μM for 24 h before culturing at 37 °C for 2 h. After 2 h, MTT solution was added to the plate and a microplate reader (Multiskan FC; Thermo, Waltham, MA, USA) was used at 450 nm to assay cell viability. Each concentration was repeated three times in independent measurements and the inhibitory effect on RAW264.7 cells was reported as a percentage relative to the cells without urolithin A treatment.

4.4. ELISA Assay

RAW264.7 cells (10^5 cells/mL) were pre-cultured with or without urolithin A (1–30 μM) in 24-well plates for 1 h, and then poly(I:C) (1 μg/mL) was added before continuing the culture for 24 h. The levels of TNF-α, MCP-1, CCL-5, INF-α, and INF-β were measured in the supernatant using specific ELISA kits following the manufacturers' instructions. A microplate reader (Multiskan FC; Thermo) was used at 450 nm to determine the optical density (OD) value.

4.5. Preparation of Total Proteins

RAW264.7 cells (2×10^5 cells/mL) were pre-cultured with or without urolithin A (1–30 μM) and stimulated with or without poly(I:C) (1 μg/mL) for 24 h or 30 min in six-well plates to evaluate the total protein and phosphorylated protein content, respectively. We used 300 mL of lysis buffer (Tris–HCl (pH 7, 450 mM); EDTA (1 mM); NaCl (150 mM); DTT (1 mM); NP40 (0.5%); and sodium dodecyl sulfate (SDS, 0.1%)) containing a protease inhibitor cocktail and phosphatase inhibitors (Sigma, St. Louis, MO, USA) to harvest cells. The BCA assay kit (Pierce) was used to quantitate all protein concentrations.

4.6. Western Blot Analysis

Proteins were separated on 10% SDS polyacrylamide gels and transferred to polyvinylidene fluoride (PVDF) membranes (Millipore, Billerica, MA, USA). The membranes were incubated overnight at 4 °C with a primary antibody against actin (Sigma) and the protein of interest. The membranes were then washed three times with Tris-buffered saline with Tween 20 (TBST) and incubated at room temperature for 1 h with a secondary antibody. Finally, the proteins were detected using Luminol/Enhancer solution (Millipore) and the base quantitated by the Bio Spectrum 600 system (UVP, Upland, CA, USA).

4.7. Immunofluorescence

RAW264.7 cells (2×10^5 cells/mL) were seeded in six-well plates until they reached 50–60% confluence and pre-cultured with or without urolithin A (1–30 μM) for 1 h. Poly(I:C) (1 μg/mL) was added for 30 min before suctioning out the medium. The cells were washed with PBS and fixed with paraformaldehyde (4%, *w*/*v*). The fixed cells were incubated with anti-NF-kB (1:100; Cell Signaling Technology, MA, USA) and NrF2 (1:100; Santa Cruz, CA, USA) antibody overnight at 4 °C. The next day, the medium was removed, and the cells washed with PBS before incubating with secondary antibodies at room temperature for 1 h. The cells were washed with PBS again to remove the dye, and DAPI solution (4′,6-diamidino-2-phenylindole, Sigma) was added to stain the nucleus. A fluorescence microscope (Olympus, Tokyo, Japan) was used to acquire the images. The control groups were treated with poly(I:C) alone, and all experiments were repeated three times.

4.8. Antioxidant Defense

Lipid peroxidation produced malondialdehyde (MDA), and the antioxidant enzyme superoxide dismutase (SOD) and catalase (CAT) activities were analyzed as an indicator of the antioxidant defense. First, RAW264.7 cells were treated with urolithin A (1–30 μM) for 1 h, and then stimulated with poly(I:C) (1 μg/mL) for 24 h. Using commercial kits, we analyzed the MDA, SOD, and CAT according to the manufacturer's instructions (Sigma-Aldrich). A spectrophotometer was used to measure the absorbance at 532 nm for MDA, 450 nm for SOD, and 585 nm for CAT.

4.9. Statistical Analysis

Image Lab software (Bio-Rad) was used to quantify the intensity of Western blot bands. Data are presented as the mean ± standard deviation (SD) of at least three independent experiments. We used one-way analysis of variance (ANOVA) followed by Tukey's post-hoc test. A p-value of < 0.05 was considered significant.

5. Conclusions

This study first demonstrated that urolithin A may be a TLR3 inhibitor, blocking the NF-κB/STAT1 axis and modulating the Nrf2/NF-κB pathway to enhance antioxidant defense and attenuate inflammatory responses in poly(I:C)-stimulated macrophages.

Author Contributions: Investigation, W.-C.H., C.-J.L. and S.-C.S.; data analysis, S.H. and J.C.-J.C.; methodology, W.-C.H., C.-J.L. and C.-Y.H.; writing and editing, S.-J.W. All authors have read and agreed to the published version of the manuscript.

Funding: This research was supported by grants from Chang Gung Memorial Hospital (CMRPF1K0082, CMRPF1L0011, CMRPF1J0042), the Ministry of Science and Technology (109-2320-B-255-006-MY3), and Chang Gung University of Science and Technology (ZRRPF3K0111 ZRRPF3L0091) of Taiwan.

Institutional Review Board Statement: Not applicable.

Informed Consent Statement: Not applicable.

Data Availability Statement: The data that support the findings of this study are available from the corresponding author upon reasonable request.

Conflicts of Interest: The authors declare no potential conflict of interest.

References

1. Takumi, K.; Taro, K. Toll-Like Receptor Signaling Pathways. *Front. Immunol.* **2014**, *5*, 461.
2. Takemura, N.; Kawasaki, T.; Kunisawa, J.; Sato, S.; Lamichhane, A.; Kobiyama, K.; Aoshi, T.; Ito, J.; Mizuguchi, K.; Karuppuchamy, T.; et al. Blockade of TLR3 protects mice from lethal radiation-induced gastrointestinal syndrome. *Nat. Commun.* **2014**, *5*, 3492. [CrossRef]
3. Chen, E.; Chen, C.; Niu, Z.; Gan, L.; Wang, Q.; Li, M.; Cai, X.W.; Gao, R.; Katakam, S.; Chen, H.; et al. Poly(I:C) preconditioning protects the heart against myocardial ischemia/reperfusion injury through TLR3/PI3K/Akt-dependent pathway. *Signal Transduct. Target. Ther.* **2020**, *5*, 216. [CrossRef]
4. Chen, C.; Gao, R.; Li, M.; Wang, Q.; Chen, H.; Zhang, S.; Mao, X.; Behensky, A.; Zhang, Z.; Gan, L.; et al. Extracellular RNAs-TLR3 signaling contributes to cognitive decline in a mouse model of postoperative cognitive dysfunction. *Brain Behav. Immun.* **2019**, *80*, 439–451. [CrossRef]
5. Janeway, C.A.; Medzhitov Jr, R. Innate immune recognition. *Annu. Rev. Immunol.* **2002**, *20*, 197–216. [CrossRef]
6. Kawai, T.; Akira, S. The role of pattern-recognition receptors in innate immunity: Update on toll-like receptors. *Nat. Immunol.* **2010**, *11*, 373–384. [CrossRef]
7. Jiang, Z.; Mak, T.W.; Sen, G.; Li, X. Toll-like receptor 3-mediated activation of NF-κB and IRF3 diverges at Toll-IL-1 receptor domain-containing adapter inducing IFN-beta. *Proc. Natl. Acad. Sci. USA* **2004**, *9*, 3533–3538. [CrossRef]
8. Lingappan, B. NF-κB in Oxidative Stress. *Curr. Opin. Toxicol.* **2018**, *7*, 81–86. [CrossRef]
9. Toledano, M.B.; Leonard, W.J. Modulation of transcription factor NF-κB binding activity by oxidation-reduction in vitro. *Proc. Natl. Acad. Sci. USA* **1991**, *88*, 4328–4332. [CrossRef]
10. Farzane, S.; Shikh, P.; Adity, B.; Luca, C. NRF2 and NF-κB interplay in cerebrovascular and neurodegenerative disorders: Molecular mechanisms and possible therapeutic approaches. *Redox Biol.* **2019**, *21*, 101059.
11. Rushworth, S.A.; Zaitseva, L.; Murray, M.Y.; Shah, N.M.; Bowles, K.M.; MacEwan, D.J. The high Nrf2 expression in human acute myeloid leukemia is driven by NF-κB and underlies its chemo-resistance. *Blood* **2012**, *26*, 5188–5198. [CrossRef] [PubMed]
12. Chul-Su, Y.; Jwa-Jin, K.; Lee, S.J.; Hwang, J.H.; Lee, C.H.; Lee, M.S.; Jo, E.K. TLR3-Triggered Reactive Oxygen Species Contribute to Inflammatory Responses by Activating Signal Transducer and Activator of Transcription-1. *J. Immunol.* **2013**, *15*, 6368–6377.

13. Djavaheri-Mergny, M.; Javelaud, D.; Wietzerbin, J.; Besançon, F. NF-κB activation prevents apoptotic oxidative stress via an increase of both thioredoxin and Mn-SOD levels in TNF alpha-treated Ewing sarcoma cells. *FEBS Lett.* **2004**, *578*, 111–115. [CrossRef] [PubMed]
14. Gao, J.J.; Filla, M.B.; Fultz, M.J. Autocrine/paracrine IFN-αβ mediates the lipopolysaccharide-induced activation of transcription factor Stat1α in mouse macrophages: Pivotal role of Stat1α in induction of the inducible nitric oxide synthase gene. *J. Immunol.* **1998**, *161*, 4803–4810.
15. Kim, H.S.; Lee, M.S. Essential role of STAT1 in caspase-independent cell death of activated macrophages through the p38 mitogen-activated protein kinase/STAT1/reactive oxygen species pathway. *Mol. Cell. Biol.* **2005**, *25*, 6821–6833. [CrossRef]
16. Yang, C.S.; Shin, D.M.; Kim, K.H. NADPH oxidase 2 interaction with TLR2 is required for efficient innate immune responses to mycobacteria via cathelicidin expression. *J. Immunol.* **2009**, *182*, 3696–3705. [CrossRef]
17. Moore, T.C.; Petro, T.M. IRF3 and ERK MAP-kinases control nitric oxide production from macrophages in response to poly-I:C. *FEBS Lett.* **2013**, *587*, 3014–3020. [CrossRef]
18. Sharma, S.; TenOever, B.R.; Grandvaux, N. Triggering the interferon antiviral response through an IKK-related pathway. *Science* **2003**, *30*, 1148–1151. [CrossRef]
19. Navarro, L.; David, M. p38-dependent activation of interferon regulatory factor 3 by lipopolysaccharide. *J. Biol. Chem.* **1999**, *274*, 35535–35538. [CrossRef]
20. Nociari, M.; Ocheretina, O.; Murphy, M. Adenovirus induction of IRF3 occurs through a binary trigger targeting Jun N-terminal kinase and TBK1 kinase cascades and type I interferon autocrine signaling. *J. Virol.* **2009**, *83*, 4081–4091. [CrossRef]
21. Zhang, B.; Li, M.; Chen, L.; Yang, K.; Shan, Y.; Zhu, L.; Sun, S.; Li, L.; Wang, C. The TAK1-JNK cascade is required for IRF3 function in the innate immune response. *Cell Res.* **2009**, *19*, 412–428. [CrossRef] [PubMed]
22. Anupama, M.; Rajagopal, R. Mitogen-Activated Protein Kinases and Their Role in Radiation Response. *Genes Cancer* **2013**, *4*, 401–408.
23. Arthur, J.S.C.; Ley, S.C. Mitogen-activated protein kinases in innate immunity. *Nat. Rev. Immunol.* **2013**, *13*, 679–692. [CrossRef] [PubMed]
24. Hu, W.; Jain, A.; Gao, Y.; Dozmorov, I.M.; Mandraju, R.; Wakeland, E.K.; Pasare, C. Differential outcome of TRIF-mediated signaling in TLR4 and TLR3 induced DC maturation. *Proc. Natl. Acad. Sci. USA* **2015**, *10*, 13994–13999. [CrossRef]
25. Yousif, N.M.; de Oliveira, A.C.P.; Brioschi, S.; Huell, M.; Biber, K.; Fiebich, B.L. Activation of EP2 receptor suppresses poly(I: C) and LPS-mediated inflammation in primary microglia and organotypic hippocampal slice cultures: Contributing role for MAPKs. *Glia* **2018**, *66*, 708–724. [CrossRef]
26. Vučić, V. Composition and Potential Health Benefits of Pomegranate: A Review. *Curr. Pharm. Des.* **2019**, *25*, 1817–1827. [CrossRef]
27. Zhao, R. Pomegranate peel polyphenols reduce chronic low-grade inflammatory responses by modulating gut microbiota and decreasing colonic tissue damage in rats fed a high-fat diet. *Food Funct.* **2019**, *13*, 8273–8285. [CrossRef]
28. Begoña, C.; Paula, P.; Juan, C.E. Identification of urolithin a as a metabolite produced by human colon microflora from ellagic acid and related compounds. *J. Agric. Food. Chem.* **2005**, *53*, 5571–5576.
29. Andreux, P.A.; William, B.B.; Ryu, D.; Burdet, F.; Ibberson, M.; Aebischer, P.; Auwerx, J.; Singh, A.; Rinsch, C. The mitophagy activator urolithin A is safe and induces a molecular signature of improved mitochondrial and cellular health in humans. *Nat. Metab.* **2019**, *1*, 595–603. [CrossRef]
30. Ashley, M.T.; Darius, F.; Virginia, C.; Ramer-Tait, A.E.; Chung, S. Immunomodulatory Role of Urolithin A on Metabolic Diseases. *Biomedicines* **2021**, *9*, 192.
31. Han, Q.A.; Su, D.; Shi, C.; Liu, P.; Wang, Y.; Zhuab, B.; Xia, X. Urolithin A attenuated ox-LDL-induced cholesterol accumulation in macrophages partly through regulating miR-33a and ERK/AMPK/SREBP1 signaling pathways. *Food Funct.* **2020**, *1*, 3432–3440. [CrossRef] [PubMed]
32. Guillermo, C.; Francisco, L.; Carmen, C.F.; Hugo, M.; López, V. The Metabolite Urolithin-A Ameliorates Oxidative Stress in Neuro-2a Cells, Becoming a Potential Neuroprotective Agent. *Antioxidants* **2020**, *2*, 177.
33. Abdulrahman, A.O.; Alzubaidi, M.Y.; Nadeem, M.S.; Khan, J.A.; Rather, I.A.; Khan, M.I. The Utilization of Urolithin A—A Natural Polyphenol Metabolite of Ellagitannins as a Modulator of the Gut Microbiota for Its Potential Use in Obesity Therapy. *Proceedings* **2021**, *79*, 12.
34. Tetsuya, H.; Satoshi, M.; Takashi, H.; Ohnishi, T.; Kimura, T.; Kurokawa, M.; Ieki, K.; Odaka, M.; Suzuki, S.; Watanabe, S. Cooperative Activation of CCL5 Expression by TLR3 and Tumor Necrosis Factor-α or Interferon-γ through Nuclear Factor-κB or STAT-1 in Airway Epithelial Cells. *Int. Arch. Allergy Immunol.* **2010**, *152*, 9–17.
35. Laura, D.R.; Paola, G.; Francesca, P.; Soldati, L. COVID-19: Is there a role for immunonutrition in obese patient? *J. Transl. Med.* **2020**, *18*, 415.
36. Dai, X.; Sayama, K.; Yamasaki, K.; Tohyama, M.; Shirakata, Y.; Hanakawa, Y.; Tokumaru, S.; Yahata, Y.; Yang, L.; Yoshimura, A. SOCS1-Negative Feedback of STAT1 Activation Is a Key Pathway in the dsRNA-Induced Innate Immune Response of Human Keratinocytes. *J. Investig. Dermatol.* **2006**, *126*, 1574–1581. [CrossRef]
37. Field, A.K.; Tytell, A.A.; Lampson, G.P. Inducers of interferon and host resistance, II. Multistranded synthetic polynucleotide complexes. *Proc. Natl. Acad. Sci. USA* **1967**, *58*, 1004–1010. [CrossRef]
38. Matsumoto, M.; Seya, T. TLR3: Interferon induction by double-stranded RNA including poly(I:C). *Adv. Drug Deliv. Rev.* **2008**, *7*, 805–812. [CrossRef]

39. Makoto, A.; Matsumoto, T.; Taguchi, K.; Kobayashi, T. Poly (I:C) impairs NO donor-induced relaxation by overexposure to NO via the NF-κB/iNOS pathway in rat superior mesenteric arteries. *Free Radic. Biol. Med.* **2017**, *112*, 553–566.
40. Pinheiro de Oliveira, A.C.; Yousif, N.M.; Bhatia, H.S.; Hermanek, J.; Huell, M.; Fiebich, B.L. Poly(I:C) increases the expression of mPGES-1 and COX-2 in rat primary microglia. *J. Neuroinflamm.* **2016**, *13*, 11. [CrossRef]
41. Maciej, C.; Zbigniew, K.; Wiktor, P.; Kochańczyk, M.; Jaruszewicz-Błońska, J.; Tudelska, K.; Błoński, S.; Kimmel, M.; Brasier, A.R.; Lipniacki, T. Cell fate in antiviral response arises in the crosstalk of IRF, NF-κB and JAK/STAT pathways. *Nat. Commun.* **2018**, *9*, 493.
42. Liu, T.; Zhang, L.; Joo, D. NF-κB signaling in inflammation. *Signal Transduct. Target Ther.* **2017**, *2*, 17023. [CrossRef] [PubMed]
43. Lei, D.; Yi, L.; Lin, C.; Shi, G.; Dong, Z.; Li, J.; Fan, P.; Wang, Q.; Su, X.; Zhang, S. SARI attenuates colon inflammation by promoting STAT1 degradation in intestinal epithelial cells. *Mucosal Immunol.* **2019**, *12*, 1130–1140.
44. Shawn, M.F.; Danielle, S.B.; Jason, A.S.; Athen, S.R.; Guinn, Z.P.; Pinkerton, T.S.; Petro, T.M.; Moore, T.C. MEK/ERK MAP kinase limits poly I:C-induced antiviral gene expression in RAW264.7 macrophages by reducing interferon-beta expression. *FEBS Lett.* **2021**, *30*, 2665–2674.
45. Kim, H.J.; Khan, I.; Shahidullah, A.; Halimi, S.M.A.; Rauf, A.; Lee, J.Y.; Kim, Y.J.; Kim, B.Y.; Park, W. Diospyrin Modulates Inflammation in Poly I:C-Induced Macrophages via ER Stress-Induced Calcium-CHOP Pathway. *Processes* **2020**, *8*, 1050. [CrossRef]
46. Nanae, H.; Takafumi, M.; Hirotsugu, U.; Harada, M. Transfection of poly(I:C) can induce reactive oxygen species-triggered apoptosis and interferon-β-mediated growth arrest in human renal cell carcinoma cells via innate adjuvant receptors and the 2-5A system. *Mol. Cancer* **2014**, *13*, 217.
47. Yin, S.; Cao, W. Toll-Like Receptor Signaling Induces Nrf2 Pathway Activation through p62-Triggered Keap1 Degradation. *Mol. Cell. Biol.* **2015**, *35*, 2673–2683. [CrossRef]
48. Shikha, M.; Damodar, G. Crosstalk of toll-like receptors signaling and Nrf2 pathway for regulation of inflammation. *Biomed. Pharm.* **2018**, *108*, 1866–1878.
49. Ahmed, N.; Nahid, S.; Naif, O.; Al-Harbi, M.M.; Ahmad, S.F. TLR-7 agonist attenuates airway reactivity and inflammation throughNrf2-mediated antioxidant protection in a murine model of allergic asthma. *Int. J. Biochem. Cell. Biol.* **2016**, *73*, 53–56.
50. Marta, R.; Remzi, O.E.; Baijayanti, J.; Desponds, C.; Snäkä, T.; Prevel, F.; Isorce, N.; Lye, L.F.; Owens, K.L.; Lopes, U.G. The antioxidant response favors Leishmania parasites survival, limits inflammation and reprograms the host cell metabolism. *PLoS Pathog.* **2021**, *25*, e1009422.

International Journal of
Molecular Sciences

MDPI

Article

Effects of Hypocalcemic Vitamin D Analogs in the Expression of DNA Damage Induced in Minilungs from hESCs: Implications for Lung Fibrosis

Esmeralda Magro-Lopez †, Irene Chamorro-Herrero † and Alberto Zambrano *

Chronic Diseases Program, Institute of Health Carlos III, 28220 Madrid, Spain; esme-1988@live.com (E.M.-L.); irenechamorrrohrro@gmail.com (I.C.-H.)
* Correspondence: azambra@isciii.es
† These authors contributed equally to this work.

Abstract: In our previous work, we evaluated the therapeutic effects of 1α,25-Dihydroxyvitamin D_3, the biologically active form of vitamin D, in the context of bleomycin-induced lung fibrosis. Contrary to the expected, vitamin D supplementation increased the DNA damage expression and cellular senescence in alveolar epithelial type II cells and aggravated the overall lung pathology induced in mice by bleomycin. These effects were probably due to an alteration in the cellular DNA double-strand breaks' repair capability. In the present work, we have evaluated the effects of two hypocalcemic vitamin D analogs (calcipotriol and paricalcitol) in the expression of DNA damage in the context of minilungs derived from human embryonic stem cells and in the cell line A549.

Keywords: human pluripotent stem cells; hESCs; minilungs; vitamin D; vitamin D analogs; paricalcitol; calcipotriol; lung fibrosis

Citation: Magro-Lopez, E.; Chamorro-Herrero, I.; Zambrano, A. Effects of Hypocalcemic Vitamin D Analogs in the Expression of DNA Damage Induced in Minilungs from hESCs: Implications for Lung Fibrosis. *Int. J. Mol. Sci.* **2022**, *23*, 4921. https://doi.org/10.3390/ijms23094921

Academic Editor: Melpo Christofidou-Solomidou

Received: 6 April 2022
Accepted: 27 April 2022
Published: 28 April 2022

1. Introduction

DNA damage and cellular senescence underlie the physiopathology associated with idiopathic pulmonary fibrosis (IPF) and other chronic conditions that can evolve with fibrosis. IPF is a form of progressive interstitial pneumonia of unknown etiology with an estimated survival of 3 to 4 years [1]. IPF pathogenesis is the consequence of an excessive matrix deposition leading to tissue scarring and irreversible organ injury, probably due to a persistent input of damage and tissue repair response. It has been reported that cellular senescence is implicated in the tissue repair program, and its occurrence in IPF, unfortunately, has a detrimental role in contrast to other fibrogenic conditions [2–4]. Vitamin D and its analogs have been demonstrated to be active in the regulation of fibrosis that characterizes multiple chronic diseases, including pulmonary fibrosis [5–8]. For instance, the preventive use of vitamin D supplementation was associated with a general improvement in the lung fibrosis symptomatology induced in mice, probably due to its anti-inflammatory effects [8,9]. However, as we have reported in our previous work, the "therapeutic" treatment of mice having bleomycin-induced fibrosis seemed to worsen the pathology: the mice treated with vitamin D showed increased architectural distortion, subpleural scarring and more areas of aberrant reepithelization compared to the controls. These areas were defined by the accumulation of alveolar epithelial type II (ATII) cells harboring high levels of DNA damage in the form of DNA double-strand breaks (DSBs). DSBs were also observed in cells throughout respiratory bronchioles or immersed in alveolar fields. The bulk of the DNA damage was preferably associated with epithelial cells; fibroblasts, however, seemed to be more resistant to DNA damage than epithelial cells [10]. Senescence can be induced prematurely as a result of a persistent DNA damage response (DDR) secondary to oxidative stress that induces DNA double-strand breaks [11]. Indeed, DSBs are potent inducers of cell arrest and a typical hallmark of cell senescence [12]. Our results also showed significantly greater levels of DSBs and cell senescence in epithelial cells than in fibroblasts, and they were consistent with the central hypothesis underlying IPF indicating that epithelial injury and impaired regeneration activate fibroblasts and that cellular senescence induced by persistent epithelial damage may be the origin of aberrant epithelial regeneration and the promotion of fibrosis [13–16].

Although bleomycin reproduces well many aspects of the general pulmonary fibrosis and some lesions present in IPF, it has never been promoted as an experimental equivalent of IPF. The strength of the bleomycin model consists of the reproducibility and versatility as a model of general fibrosis. In addition, its high efficiency levels inducing DSBs make bleomycin a very interesting model to analyze DNA damage [17].

In the present study, we have evaluated various vitamin D analogs in the context of DNA damage induced by bleomycin. A large amount of vitamin D analogs have been synthesized throughout the years, and the clinical use for secondary hyperparathyroidism, osteoporosis or psoriasis has been approved for many of them [18,19]. The potent effect of vitamin D on intestinal calcium and phosphorus absorption and bone mineral mobilization, often leading to the development of hypercalcemia and hyperphosphatemia, has precluded its therapeutic use for many conditions. The ideal analog would retain vitamin D receptor-binding capacities and have minimal effects on calcium and phosphorus metabolism. Our working hypothesis is that hypocalcemic vitamin D analogs could show a lower incidence in the expression of DNA damage upon a bleomycin insult than the active form of vitamin D.

2. Results

We have tested our postulated hypothesis in the cell line A549, an immortalized counterpart of ATII cells, in 2D minilungs (lung and alveolar differentiated cells from hESCs arranged in bidimensional cultures) and in 3D minilungs from hESCs (human lung bud organoids embedded in Matrigel^{TM} sandwiches). The lung organoids generated from hESCs have enormous advantages over cell lines or simple primary cultures as they offer an unlimited availability of primary cells, show the complete lung epithelial spectrum and emulate structural and functional features of the original organ.

The exposure of A549 cells to a sublethal bleomycin shock (12 μg/mL for 6 h) induces the expression of DNA damage (DD) foci containing TP53BP1, a reliable marker of DSBs [20–24]. These conditions allow the accurate quantification of DSBs. To avoid a potential effect of bleomycin on the beginning of the transcription triggered by vitamin D or its analogs, we have performed short pre-treatments with these ligands before the addition of bleomycin. In addition, as vitamin D and its analogs show differential binding affinities to their receptor, we assessed increasing amounts of ligands in the dose–response assay using the expression of the gene *CYP24A1*, a vitamin D target gene, as the response. The results showed that, in the case of A549 cells or 2D minilungs, the amount of 5 nmol/L is adequate as we found no significant differences between vitamin D and the analogs evaluated in terms of *CYP24A1* expression. However, in the case of the minilungs embedded in Matrigel^{TM} sandwiches (3D organoids), this low concentration of ligand gave rise to significant differences between vitamin D and the analogs evaluated, probably due to the limited access of the ligands to the organoid cells. The amount of 50 nmol/L, however, resulted in similar responses of vitamin D and the hypocalcemic analogs (Supplementary Figure S1).

After bleomycin treatments, DD foci can be rapidly visualized as discrete foci in a pan-nuclear pattern (Figure 1A). As previously reported [10], the exposure of A549 cells to vitamin D, in the presence of bleomycin, increased the levels of DD foci, both the percentage of damaged cells and the levels of severely damaged cells harboring more than 20 DD foci per nucleus (Figure 1B–D) (n = 3; >150 cells were analyzed; $p < 0.001$). However, the two hypocalcemic vitamin D analogs tested (paricalcitol and calcipotriol) were able to drastically reduce the bulk of the DD expression compared to vitamin D in the presence of bleomycin (Figure 2C,D; n = 3; >150 cells were analyzed; ANOVA $p < 0.001$). Figure 2A,B shows that the treatment of A549 cells with vitamin D or its analogs, in the absence of bleomycin, did not alter the low basal level of damage of the cell population as previously described for vitamin D [10] ($n = 3$; >150 cells were analyzed; ANOVA $p < 0.001$). In order to reproduce these results in much more reliable models of lung structure and function, we generated minilungs from hESCs as previously described [25,26]. On the one hand, we generated lung airway and epithelial cells arranged in bidimensional cultures (2D minilungs) from the hESC line AND-2 as previously described [25]. Briefly, good hESCs colonies are grown along inactivated MEFs (iMEFs), picked up and passaged to new plates with iMEFs in order to accumulate material for lung differentiation. Figure 3A shows the expression of pluripotency marker SOX-2 in a good AND-2 colony and representative micrographs at various times of the differentiation process: embryoid bodies (EBs), anterior foregut endoderm (AFE), cultures at day 23 (lung progenitors) and at day 60 (differentiated lung airway and alveolar cells). The cultures from day 50 onwards show the expression of representative markers of the lung airway and alveolar cells, illustrating the heterogeneity in cell shape, including the presence of flat cells with a crescent shape morphology and granular and roughly cuboidal-shaped cells, likely corresponding to ATI and ATII cells, respectively (Figure 3A; d60). Although from day 50 they can be considered mature, the cultures were used for the desired experimentation from day 60 on. Figure 3B shows a RT-qPCR result illustrating the complexity of these cultures ($n = 3$; >4 organoids per experimental replicate were used; ANOVA $p < 0.001$). As previously described by us and others [25–29], the differentiation protocol applied here yields cultures enriched in alveolar epithelial cells (ATI and ATII cells). On the other hand, the generation of 3D minilungs implies the formation of nascent organoids in suspension at certain times of the protocol (see Figure 3C and Methods for details) and their final embedding in MatrigelTM sandwiches to reach the desirable state of differentiation characterized by the presence of lung buds more or less branched [lung buds organoids (LBOs)], as previously described [26,29] (Figure 3D). Figure 3E shows representative micrographs of histochemical analysis (H&E staining) and immunohistochemical analysis with surfactant antibodies performed on LBOs sections. In order to analyze the expression of DNA damage and the effect of vitamin D and its analogs, 2D minilungs were treated with 12.5 μg/mL of bleomycin for 72 h. All the cell types of these complex cultures seemed to be affected equally by the bleomycin treatment. As in the case of the A549 cells, neither vitamin D nor its hypocalcemic analogs altered significantly the basal levels of DD in the absence of bleomycin (Figure 4A,B) ($n = 3$; >150 cells were analyzed; ANOVA $p < 0.001$). As expected, the exposure of these cultures to bleomycin and vitamin D increased the levels of DD foci reached by bleomycin itself (Figure 4C,D). As in the case of the A549 cells, the treatment with paricalcitol and calcipotriol did not further increase the DD levels reached by bleomycin and seemed to reduce significantly the DD expression induced by bleomycin (Figure 4C,D, ($n = 3$; >4 organoids per condition were used and >150 cells were analyzed; ANOVA $p < 0.001$). Equivalent assays were performed on 3D minilungs embedded into matrigelTM sandwiches. As previously reported, lung buds minilungs are mainly constituted by ATII cells [26,29]. Although to a lesser extent, we found similar results to those obtained in the case of 2D minilungs (Figure 4E–H) [$n = 3$; >4 organoids per condition were used and >150 cells were analyzed; ANOVA $p > 0.05$ (panels 4E,F) and ANOVA $p < 0.001$ for data represented in panels 4G,H]. This reduction in the extension of damage was probably because the organoids are embedded into matrigel^{TM} sandwiches and the access of bleomycin and the ligands to them is more restricted. Finally, we evaluated, in A549 cells, a continuous cell line counterpart of ATII cells, the expression of γH2AFX marker, which is a reliable marker of DD, cell senescence and aging, as previously described [20,30]. Figure 4I shows the significant increase in the γH2AFX expression levels in the presence of bleomycin and vitamin D compared to the controls and the drastic reduction induced by paricalcitol. Equivalent assays were performed using a battery of commercially available less-hypercalcemic vitamin D analogs, including 22-oxacalcitriol, tacalcitol and vitamin D2. All the vitamin D analogs seemed not to further increase the expression levels of γH2AFX reached in bleomycin-treated cells (Figure 4J). Moreover, in the case of paricalcitol, a significant reduction in the expression of γH2AFX compared to the bleomycin-treated cells was observed.

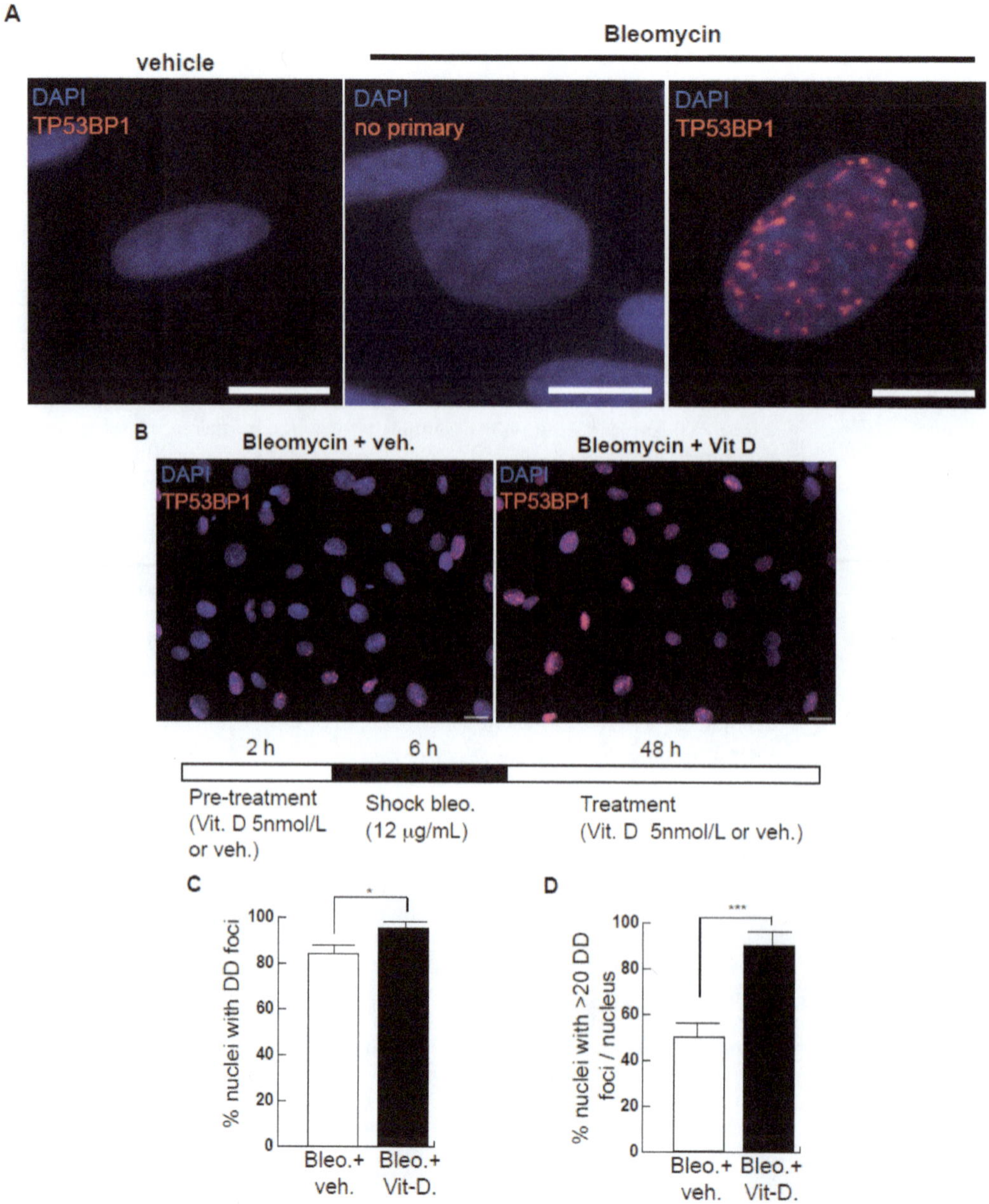

Figure 1. Effects of bleomycin and vitamin D in A549 cells. (**A**) A549 cells were treated with bleomycin with a sublethal bleomycin shock (12 μg/mL for 6 h) incubated for 48 h and then processed for indirect immunofluorescence to detect DNA damage foci containing TP53BP1 (red dots), a reliable marker of DNA double-strand breaks (DSBs). Veh.: bleomycin vehicle (PBS). No primary: negative control of the immunofluorescence assay consisting of the absence of primary antibody. Scale bar: 5 μm. (**B**) Expression of DSBs (TP53BP1 foci (red dots)) induced by vitamin D in the presence of bleomycin. The cells were pre-treated with 5 nmol/L vitamin D for 2 h, subjected to a bleomycin shock (12 μg/mL for 6 h) and then treated with 5 nmol/L vitamin D or its vehicle. Representative micrographs

taken at 48 h post-shock are shown. Scale bar: 20 µm. (**C**,**D**) Quantification of damaged cell: % of nuclei with TP53BP1 foci (**C**) or % of severely damaged cells (more than 20 TP53BP1 foci per nucleus (**D**)). Data from three experiments are represented; more than 150 cells per condition were analyzed ($p < 0.001$). *: $p < 0.05$, ***: $p < 0.001$.

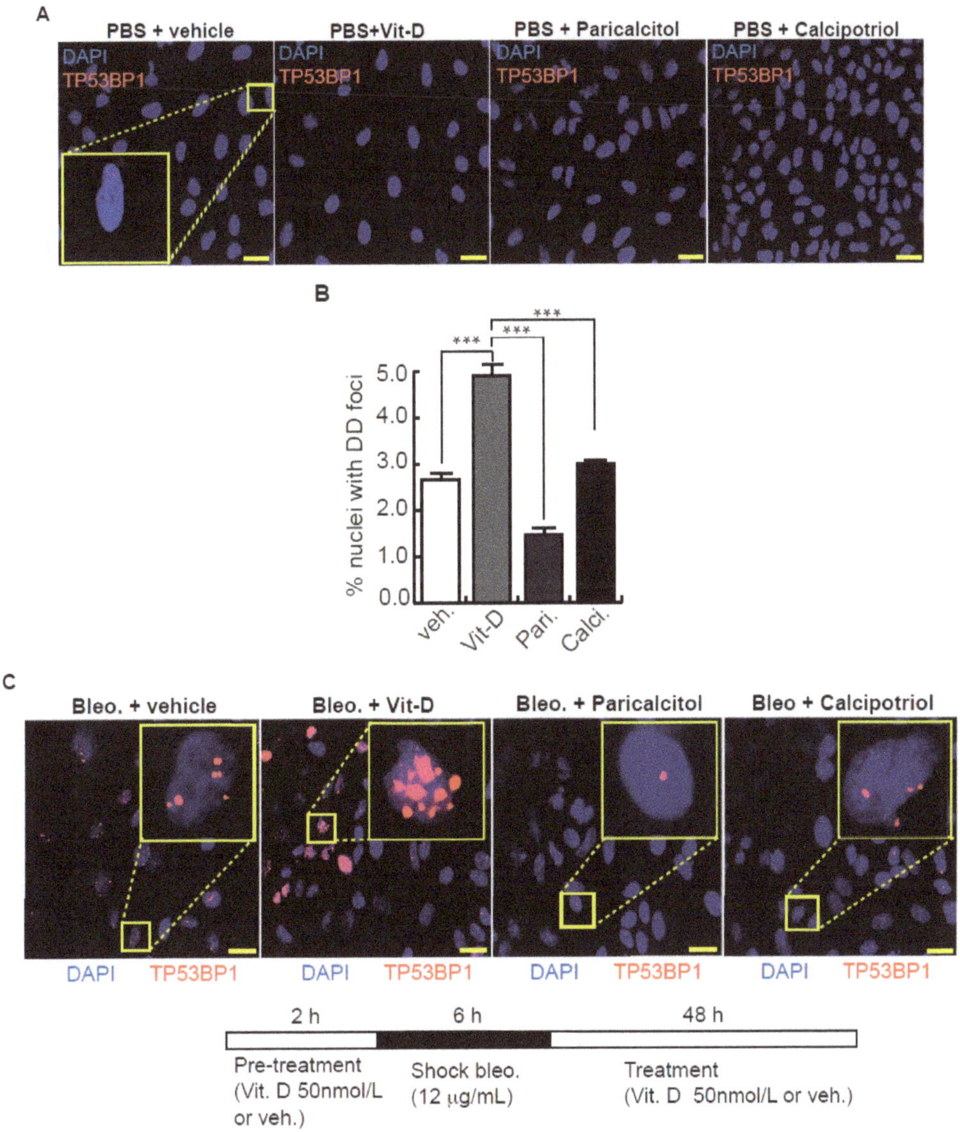

Figure 2. *Cont.*

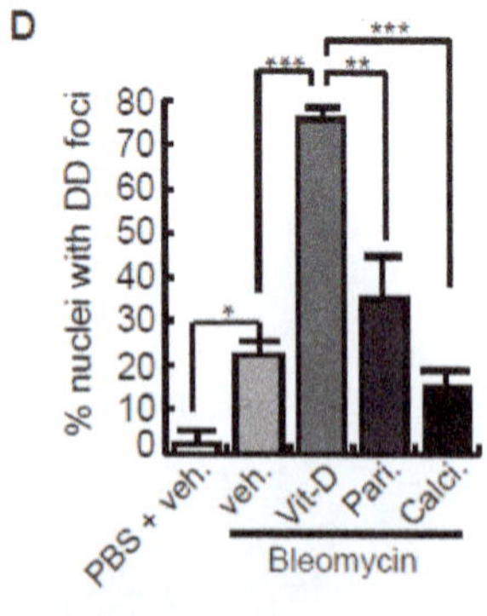

Figure 2. Effects of vitamin D and two hypocalcemic analogs in the expression of DNA damage induced by bleomycin in A549 cells. (**A**) Basal expression of DSBs (TP53BP1 foci (red dots)) induced by vitamin D and the analogs paricalcitol and calcipotriol in the absence of bleomycin. A549 cells were treated with 50 nmol/L vitamin D (or the corresponding analog) for 48 h and then subjected to immunofluorescence to detect DD foci. Representative micrographs are shown. PBS is the bleomycin vehicle; vehicle: vitamin D or analogs vehicle; scale bar: 10 µm. (**B**) Quantification of damaged cells (nuclei with TP53BP1 foci). Data from three experiments are represented; more than 150 cells per condition were analyzed. ANOVA $p < 0.001$. (**C**) Expression of TP53BP1 foci (red dots) in cultures of A549 cells pre-treated with 50 nmol/L vitamin D, analogs or its vehicle for 2 h and subjected to a bleomycin shock (12 µg/mL) for 6 h. After that, the cultures were treated with 50 nmol/L vitamin D and the analogs paricalcitol and calcipotriol (or vehicle) for 48 h. Representative micrographs taken at 48 h post-shock are shown. Scale bar: 10 µm. (**D**) Quantification of damaged cells; veh.: vitamin D or analogs vehicle. Data from three experiments are represented; more than 150 cells per condition were analyzed; ANOVA $p < 0.001$. The results presented in the figures are means ± SEM. Significance of the analysis is indicated as *: $p < 0.05$, **: $p < 0.01$, ***: $p < 0.001$.

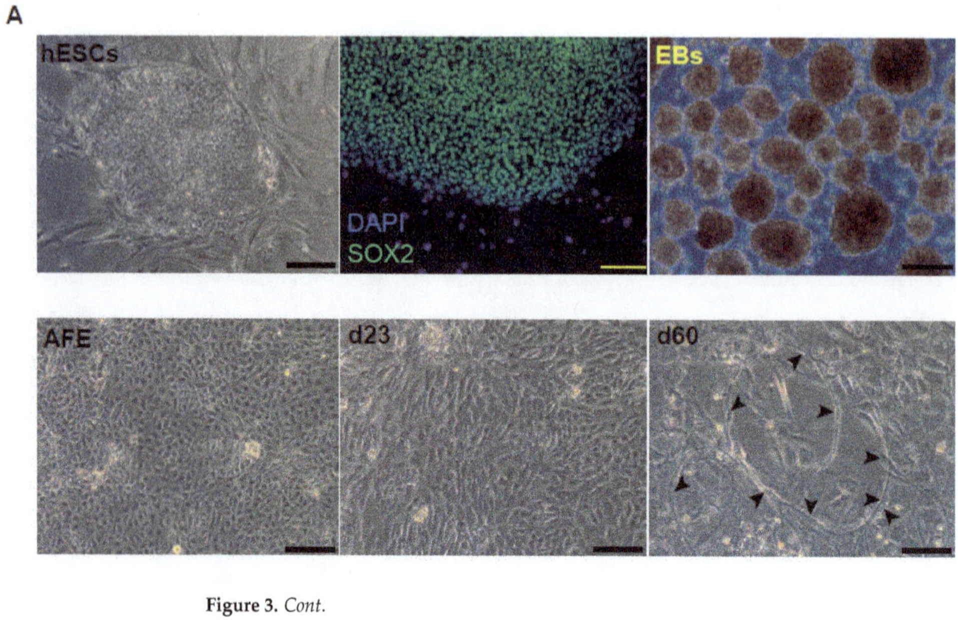

Figure 3. *Cont.*

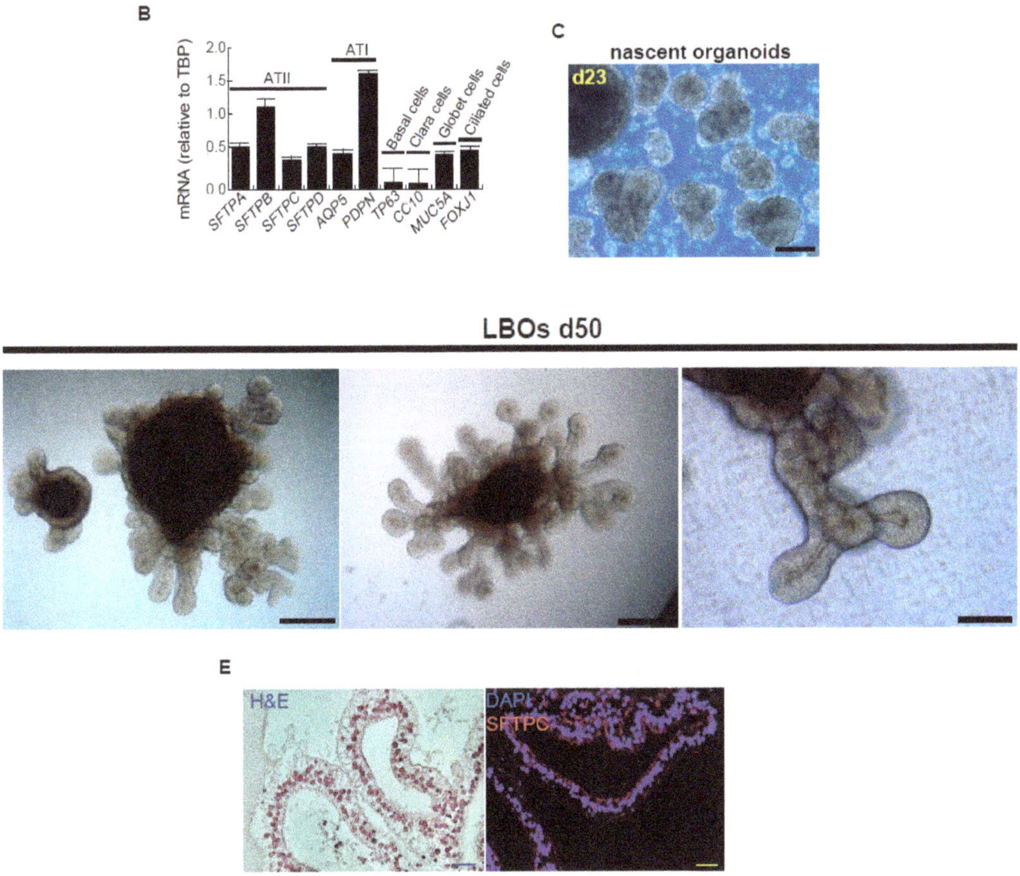

Figure 3. Representative micrographs of the sequential differentiation processes and expression markers. (**A**) Left upper micrograph: AND-2 colony growing along with feeder cells (inactivated MEFs (iMEFs)); scale bar: 100 µm); central upper micrograph: expression of SOX2 (SRY (sex-determining region Y)-box 2) in an undifferentiated colony of AND-2; scale bar: 100 µm. Right upper micrograph: representative micrograph of EBs (embryoid bodies). Bottom panels: AFE (anterior foregut endoderm) and representative micrographs of cultures at day 23 of differentiation (lung progenitors) and at day 60 (differentiated lung and airway cells); black arrowheads signal cells with a typical flat and crescent shape morphology denoting alveolar type I cells (ATI cells); scale bar: 100 µm. (**B**) Levels of expression (relative to TBP (TATA box binding protein)) of lung and airway epithelial cells markers at day 60 (n = 3; > 4 organoids per condition were used). (**C**) Representative micrograph of nascent organoids growing in suspension at day 23. (**D**) Representative micrograph of LBOs at day 50 embedded in Matrigel^{TM} sandwiches; scale bar: 100 µm and 50 µm (micrograph on the right). (**E**) Micrograph on the left: histochemical analysis of LBO sections (H&E staining); representative micrograph of an immunohistochemical staining of LBO sections (micrograph on the right) with an SFTP-C antibody. Scale bar: 100 µm.

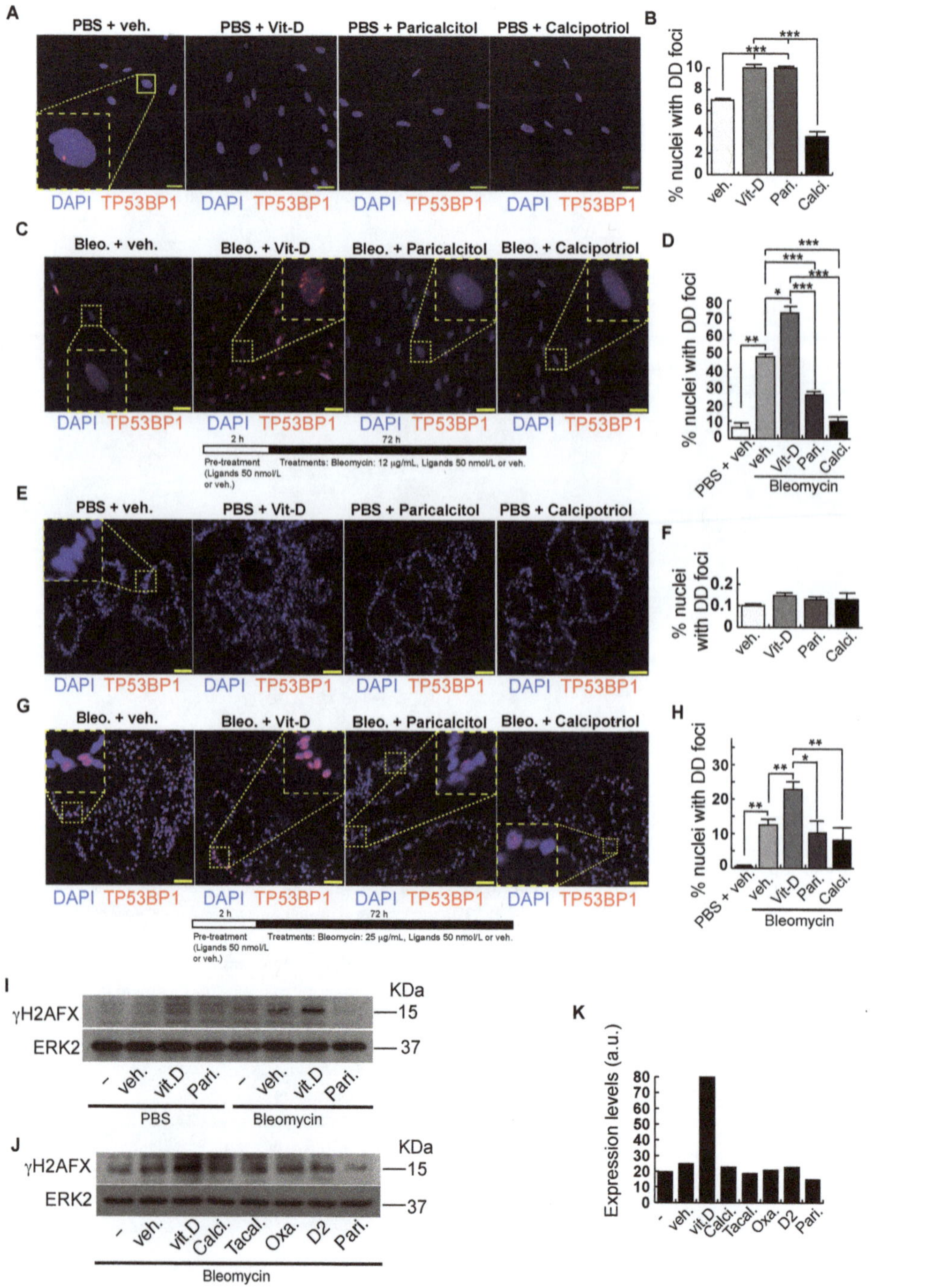

Figure 4. Effects of vitamin D and hypocalcemic analogs in the expression of DNA damage induced by bleomycin in the minilungs generated from hESCs. (**A**) Basal expression of DNA damage (TP53BP1 foci (red dots)) induced by vitamin D and the analogs paricalcitol and calcipotriol in the absence of

bleomycin. 2D minilungs were treated with 50 nmol/L vitamin D (or the corresponding analog) for 72 h and then subjected to immunofluorescence to detect DD foci. Representative micrographs are shown. PBS is the bleomycin vehicle; veh.: vitamin D or analogs vehicle; scale bar: 10 μm. (**B**) Quantification of damaged cells (nuclei harboring TP53BP1 foci) corresponding to conditions A; (n = 3; >150 cells were analyzed; ANOVA $p < 0.001$). (**C**) Expression of TP53BP1 foci (red dots) in 2D minilungs cells pre-treated with 50 nmol/L vitamin D, analogs or its vehicle for 2 h and then with bleomycin (12.5 μg/mL), vitamin D or analogs (50 nmol/L) for 72 h. After that, the cultures were processed for indirect immunofluorescence to detect DNA damage foci containing TP53BP1 (red dots); representative micrographs are shown. Scale bar: 10 μm. (**D**) Quantification of damaged cells corresponding to conditions C. veh.: vitamin D or analogs vehicle; (n = 3; >150 cells were analyzed; ANOVA $p < 0.001$). (**E**) Basal expression of DNA damage in 3D minilungs (LBOs): TP53BP1 foci (red dots) induced by vitamin D and the analogs paricalcitol and calcipotriol in the absence of bleomycin. 3D minilungs (LBOs) were treated with 50 nmol/L vitamin D (or the corresponding analog) for 72 h and then subjected to immunofluorescence to detect DD foci. Representative micrographs are shown. PBS is the bleomycin vehicle; veh.: vitamin D or analogs vehicle; scale bar: 10 μm. (**F**) Quantification of damaged cells corresponding to condition E (nuclei harboring TP53BP1 foci); (n = 3; >4 organoids per condition were used and >150 cells were analyzed; ANOVA $p > 0.05$). (**G**) Expression of TP53BP1 foci (red dots) in 3D minilungs (LBOs) cells pre-treated with 50 nmol/L vitamin D, analogs or its vehicle for 2 h and then with bleomycin (25 μg/mL), vitamin D or analogs (50 nmol/L) for 72 h. After that, the cultures were processed for indirect immunofluorescence to detect DNA damage foci containing TP53BP1 (red dots); representative micrographs are shown. Scale bar: 10 μm. (**H**) Quantification of damaged cells corresponding to condition G (nuclei harboring TP53BP1 foci); (n = 3; >4 organoids per condition were used and >150 cells were analyzed; ANOVA $p < 0.001$). The results presented in the figures are means ± SEM. Significance of the analysis is indicated as *: $p < 0.05$, **: $p < 0.01$, ***: $p < 0.001$. (**I**) Detection of γH2AFX in A549 cell extracts. Cells were pre-treated with 50 nmol/L vitamin D for 2 h, subjected to a bleomycin shock (12 μg/mL for 6 h) and then treated with 50 nmol/L vitamin D or its vehicle. Cell extracts were obtained at 48 h post-shock; Pari: paricalcitol; ERK2 was used as loading control. KDa: kilodaltons. (**J**) Detection of γH2AFX in A549 cell extracts. Cells were treated as in J. Cell extracts were obtained at 48 h post-shock; Calci: calcipotriol; Tacal: tacalcitol; Oxa.: 22-oxacalcitriol; D2: vitamin D2; Pari: paricalcitol; ERK2 was used as loading control. KDa: kilodaltons. (**K**): densitometry analysis of western J; a.u: arbitrary units.

3. Discussion

Besides its function in the mineral homeostasis and immune system, vitamin D plays a role in multiple chronic diseases involving the respiratory system. Epidemiological studies have suggested a link between vitamin D deficiency and the risk of development of chronic lung diseases, such as asthma, chronic obstructive pulmonary disease (COPD), cystic fibrosis and respiratory infections [31]. This association has led to the notion that vitamin D supplementation might ameliorate the progress of these diseases. Vitamin D supplementation, however, needs to be evaluated carefully as it can be a factor contributing to vitamin-D-mediated hypercalcemia and hypercalciuria [32]. In addition, cases of vitamin D toxicity associated with overdoses due to manufacturing or intake errors have been reported [33]. Moreover, we have reported a detrimental role of vitamin D supplementation in a therapeutic experimental system, very likely associated with an impairment in the cellular DSBs repair capabilities and cell senescence [10].

Vitamin D may affect the progression of fibrosis at different stages: anti-fibrinolytic coagulation cascade, inflammation, fibroblasts activation and on the negative regulation of the renin–angiotensin system. Vitamin D seemed to prevent the experimental lung fibrosis induced by bleomycin [8,9,34–36]. However, in these experimental studies, vitamin D is administered either before or very early after the bleomycin insult, so the effects observed were very likely due to the inherent anti-inflammatory properties of vitamin D. Thus, these studies can be defined as preventive. In addition, various hypocalcemic analogs, such as paricalcitol, calcipotriol and 22-oxacalcitriol, have been demonstrated to be active as anti-fibrotic agents in different experimental systems and types of fibrosis [37–49]. Vitamin

D less-hypercalcemic analogs might provide an alternative to vitamin D supplementation to treat many conditions related to fibrosis. The ideal analog would retain vitamin D receptor-binding capacities and have minimal effects on mineral metabolism.

Bleomycin can induce pulmonary fibrosis and fibrogenic cytokine release by oxidant-mediated DNA scission in a variety of animal models. The principal drawbacks of the bleomycin model relate to the rapid lung remodeling and the emphysema-like changes induced [17]. However, it reproduces well many aspects of the general pulmonary fibrosis and some lesions present in IPF, although it should be stated here that bleomycin is not a reliable model of IPF [17]. However, the potential of bleomycin in the induction of DSBs and senescence in many cell types is extraordinary [10,13]. We have made use of this advantage in the present study to explore the influence of various vitamin D hypocalcemic analogs in the context of A549 cells and minilungs generated from hESCs. Our experiment approaches the initial steps of the fibrogenic conditions, i.e., the expression of DNA damage underlying many conditions evolving towards fibrosis.

The generation of human minilungs that share the structural features and some extent of the functionality of the native organ may serve as a system model to emulate the DNA damage inflicted during the course of fibrogenic conditions, such as IPF. Currently, the more efficient protocols to generate airway and alveolar epithelial cells from the direct differentiation of hPSCs are biased to the production of alveolar cells [26,27,29]. We have employed either shocks or continuous exposures of bleomycin. The sublethal bleomycin shocks employed here allow the accurate quantification of DNA damage in the form of DSBs and the observation of subtle differences between the experimental conditions that might otherwise be masked by the extraordinary potential of bleomycin. Bleomycin seems to inflict DNA damage in the form of DSBs in all the epithelial cells equally, even when the cell organization is the form of lung buds embedded in Matrigel^{TM} sandwiches. However, the assembly of organoids into Matrigel^{TM} sandwiches can make difficult the access of bleomycin and ligands to the cells. The reduction in the extent of DNA damage inflicted by bleomycin compared to the 2D minilungs or A549 cells might reflect this fact.

As a continuation of our earlier work [10], our current hypothesis states that hypocalcemic vitamin D analogs could show a lower incidence in the expression of DNA damage upon a bleomycin insult than the active form of vitamin D. The initial results presented here suggest that less-hypercalcemic analogs do not show the deleterious effects observed by vitamin D treatment in the presence of bleomycin and could be an alternative to vitamin D supplementation. In addition, the treatment with such vitamin D analogs could be tested as efficient agents to reduce the bulk of the DD expression underlying multiple diseases that can evolve with DNA damage, fibrosis and aging, such as IPF and other lung interstitial conditions. Future in vivo work in this direction will be necessary.

4. Materials and Methods

4.1. Cell Culture

Alveolar epithelial cells type II (A549, ATCC) were maintained in DMEM medium supplemented with 10% FBS (Sigma-Aldrich, Burlington, MA, USA), 2 mM glutamine and 100 U/mL of penicillin and streptomycin (Lonza, Basel, Switzerland). We used the active form of vitamin D (1α,25-Dihydroxyvitamin D_3 or calcitriol) (cat.#D1530; Sigma-Aldrich; Vitamin D stock was 10 μM in ethanol) and the following vitamin D analogs (stocks were 50 μM in ethanol): calcipotriol (cat.#203537; Santa Cruz Biotechnology, Dallas, TX, USA), paricalcitol (cat.#477938; Santa Cruz Biotechnology), tacalcitol (cat.#sc-361371a; Santa Cruz Biotechnology, Dallas, TX, USA), 22-Oxacalcitriol (cat.#sc-361076; Santa Cruz Biotechnology, Dallas, TX, USA) and vitamin D2 (cat.#sc- sc-205988; Santa Cruz Biotechnology, Dallas, TX, USA). Treatments were performed in cells maintained in DMEM supplemented with 10% hormone-depleted serum. This serum was prepared by using the anion exchange resin AGR1-X8 from BIO-RAD (cat.#1401441; BIO-RAD, Hercules, CA, USA) as previously described [20]. Bleomycin sulfate (cat.# CAYM13877–50) was purchased to VWR [Radnor, PA, USA] (bleomycin stock: 50 mM in PBS).

4.2. Maintenance of hESCs

The hESCs line AND-2 was obtained from the "Biobanco de células madre de Granada" (ISCIII, Spain); passages 26–40. Mouse embryonic fibroblasts (MEFs) were obtained at 13.5 days post-coitum from C57BL/6 mice as described previously [20]. MEFs were mitotically inactivated by an overnight treatment with 2 μg/mL of mitomycin C (cat.#M4287; Sigma-Aldrich, Darmstatdt, Germany) and plated at a density of approximately 16,000 cells/cm^2. hESCs were cultured along with MEFs under standard conditions (http://www.stembook.org (accessed on 5 April 2022)). The maintenance medium was composed of KO-DMEM (cat.#10829-018 Gibco; Life Technologies, Carlsbad, CA, USA), 20% KO serum replacement (cat.#10828010 Gibco; Life Technologies, Carlsbad, CA, USA), 0.1 mM β-mercaptoethanol (cat.#21985-023 Gibco; Life Technologies, Carlsbad, CA, USA), 2 mM Glutamax (cat.#35050-061,Gibco; Life Technologies, Carlsbad, CA, USA), nonessential aminoacids (cat.#11140-050 Gibco; Life Technologies (Carlsbad, CA, USA) and primocin (cat.#12I05MM; InvivoGen, Toulouse, France). The medium was filtered by using 0.22-μ pore filter systems (cat.#431097; Corning, NY, USA); 10 ng/mL recombinant human basic fibroblast growth factor (hbFGF) (cat.#PHG6015; Invitrogen, Waltham, MA, USA) and 10 μM Y-27632 (cat.#1254; Tocris, R&D Systems, Bristol UK) were added before use. The medium was changed on a daily basis and cells were passaged either by enzymatic (collagenase IV method) (collagenase IV: cat.#11140050; Gibco; Life Technologies, Carlsbad, CA, USA) or mechanical procedures (http://stembook.org (accessed on 5 April 2022)). Cells were maintained in an undifferentiated state in a 5% CO_2/air environment. The differentiation process was carried out under hypoxic conditions in a 5% CO_2/5% O_2/95% N_2 environment (Galaxy 48R incubator (New Brunswick); Eppendorf, Hamburg, Germany) or in normoxy, as indicated in the corresponding differentiation step.

4.3. Primitive Streak Formation and Induction of Definitive Endoderm (DE)

Induction of endoderm was performed as previously described [25,26]. Primitive streak formation (day 0; 24 h) and endoderm induction (days 1–4) were performed in serum-free differentiation (SFD) medium. SFD medium was composed of a mix of IMDM:F12 (3:1) media (cats.#B12-722F and 10-080 CVR; Corning, Corning, NY, USA), supplemented with N2 (cat.#17502-048, Gibco; Life Technologies, Carlsbad, CA, USA), B27 (cat.#17504-044, Gibco; Life Technologies, Carlsbad, CA, USA), 2 mM Glutamax (cat.#35050-061 Gibco; Life Technologies, Carlsbad, CA, USA), 1% penicillin-streptomycin (DE17-602E; Lonza, Basel, Switzerland) and 0.05% bovine serum albumin (BSA) (cat.#A7906; Sigma-Aldrich, Burlington, MA, USA). The medium was filtered using a 0.22 μ-pore filter system (cat.#431097; Corning, Corning, NY, USA); 50 μg/mL ascorbic acid (cat.#A4554; Sigma-Aldrich, Burlington, MA, USA) and 0.04 μL/mL monothioglycerol (stock > 97%) (cat.#M6145; Sigma-Aldrich, Burlington, MA, USA) were added before use. MEFs were depleted by passaging hESCs lines onto Matrigel^{TM}-coated (cat.#354230; Life Technologies, Carlsbad, CA, USA) plates for at least 48 h. Cells were briefly trypsinized into small 3–10 cell clumps and the reaction was halted with stop medium (IMDM medium (BE12722F) supplemented with 50% foetal bovine serum (F7524, Sigma-Aldrich, Burlington, MA, USA), 2 mM Glutamax, 1% penicillin-streptomycin and 30 ng/mL DNase I (cat.#260913-10MU; Calbiochem, San Diego, CA, USA). Cells were then centrifuged 5 min at 850 rpm and washed carefully two times with an excess of SFD medium. To form embryoid bodies (EBs), the clumps were plated onto low-attachment 6-well plates (cat.#3471; Corning, Corning, NY, USA) and maintained in SFD medium in a 5% CO_2/5% O2/95% N_2 environment (Galaxy 48R incubator; New Brunswick).

For primitive streak formation, 10 μM Y-27632, 10 ng/mL Wnt3a (cat.#5036-WN; R&D Systems, Minneapolis, MN, USA) and 3 ng/mL human BMP4 (cat.#314-BP; R&D Systems, Minneapolis, MN, USA) were used. EBs were collected, resuspended carefully in endoderm induction medium containing 10 μM Y-27632, 0.5 ng/mL human BMP4, 2.5 ng/mL hbFGF and 100 ng/mL human Activin (cat.# 338-AC; R&D Systems, Minneapolis, MN, USA).

Cells were fed after 36–48 h, depending on cell density, by removing half the old medium and adding half fresh medium.

4.4. Induction of Anterior Foregut Endoderm (AFE)

AFE (days 4, 5 or 5) was induced as previously described [25,26]. EBs were dissociated into single cells with trypsin. Dissociated cells were transferred to a conical tube containing stop medium to neutralize trypsin. Cells were centrifuged for 5 min at 850 rpm, washed carefully twice with SFD medium and counted. For AFE induction, 25,000–30,000 cells/cm^2 were plated on fibronectin-coated (F0895; Sigma-Aldrich, Burlington, MA, USA, USA) 12-well tissue culture plates in AFE induction medium 1 (SFD medium supplemented with 10 mM SB-431542 (cat.#1614; Tocris, Bristol, UK) and 100 ng/mL of NOGGIN (cat.#6057; R&D Systems, Minneapolis, MN, USA). After 24 h of incubation, the medium was aspirated and AFE induction medium 2 (SFD medium supplemented with 1 μM IWP2 (cat.#3533; Tocris, Bristol, UK) and 10 μM of SB-431542) was added to the cultures. This process was carried out under hypoxic conditions only for the bidimensional cultures.

4.5. Lung Progenitors Induction and Expansion

Lung progenitor induction and expansion was carried out as previously described [25,26]. On day 6,5–7, AFE cultures treated for 20 days with the ventralization medium consisting of SFD medium supplemented with 3 μM CHIR99021 (cat.#04; Tocris, Bristol, UK), 10 ng/mL human FGF10 (cat.#345-FG; R&D Systems, Minneapolis, MN, USA), 10 ng/mL human KGF (cat.#251KG-010; R&D Systems, Minneapolis, MN, USA), 10 ng/mL human BMP4 (cat.#314-BP; R&D Systems, Minneapolis, MN, USA), 10 ng/mL murine EGF (cat.#2028-EG-200; R&D Systems, Minneapolis, MN, USA) and 50 nM all-trans retinoic acid (cat.#R2625; Sigma-Aldrich, Burlington, MA, USA). The culture medium was changed every two days. At a time point between days 8 and 12, cultures were incubated under normoxic conditions. At day 16, cultures were briefly digested with trypsin in order to remove potential nonectodermal contaminating cells. Supernatant of this brief digestion containing single cells and small clumps were removed. The remaining cell clumps were replated onto fibronectin-coated MW12 plates at 1:3 dilutions in fresh medium after trypsin neutralization and careful washing. Plates were returned to the hypoxic conditions (5% CO_2/5%O_2/95%N_2 environment).

4.6. Lung and Airway Epithelial Cells Maturation

At day 26, cultures were incubated with SFD medium supplemented with 3 μM CHIR99021, 10 ng/mL human FGF10, 10 ng/mL human FGF10, 0.1 mM 8-bromocAMP (cat.# B5386; Sigma-Aldrich, Burlington, MA, USA), 0.1 mM IBMX (3,7-dihydro-1-methyl-3-(2methylpropyl)-1H-purine-2,6-dione; cat.# I5879; Sigma-Aldrich, Burlington, MA, USA) and 60 nM dexamethasone (cat.#D5902; Sigma-Aldrich, Burlington, MA, USA). The medium was changed every two days and plates were maintained under hypoxic conditions (5%CO_2/5%O_2/95%N_2 environment). Cultures were carried further under these conditions until their experimental use at day 50. Treatments were performed in minilungs maintained in day 26 medium as indicated in the corresponding experiments.

4.7. Formation of Lung Bud Organoids

In this case, the differentiation process was performed under normoxic conditions from the anteriorization stage on. At day 8, cells were briefly trypsinized into small 3–10 cell clumps and the reaction was halted with stop medium (IMDM medium (BE12-722F) supplemented with 50% fetal bovine serum (FBS; F7524; Sigma-Aldrich, Burlington, MA, USA), 2 mM Glutamax, 1% penicillin-streptomycin). Cells were then centrifuged for 5 min at 850 rpm and washed carefully twice with an excess of SFD medium. The clumps were plated onto low-attachment six-well plates (cat.#3471; Corning, Corning, NY, USA) in branching medium (SFD medium containing 3 μM CHIR99021, 10 ng/mL FGF10, 10 ng/mL KGF, 10 ng/mL BMP4, 50 nM all-trans retinoic acid). These three-dimensional clumps (nascent lung bud organoids) were incubated and fed every other day for approximately 20–25 days. After that, these nascent organoids were embedded into a Matrigel^{TM} sandwich assembled on MW96 wells. 50 µL of Matrigel^{TM} were loaded on the MW96 well and allowed to gel. Nascent organoids were picked up with a wide mouth plastic Pasteur pipette, divided into MW96 wells containing 50% Matrigel^{TM}, diluted in branching media and immediately transferred onto the first layer of Matrigel^{TM}. After solidification of this intermediate layer containing the nascent organoids, 50 µL of Matrigel^{TM} were added on top. Finally, each sandwich containing various organoids was incubated with 50 µL branching media. Medium was changed every 2–3 days. Growing branching structures were easily visualized under the microscope after 1 or 2 weeks. Treatments were performed in minilungs maintained in branching medium as indicated in the corresponding experiments.

4.8. Indirect Immunofluorescence of A549 Cells and 2D Minilungs

Cells were seeded in 8-well chambers (cat.#154,534; Thermofisher Scientific, Waltham, MA, USA) at a density of 20,000 cells/well. The following day, the cells were treated as indicated in the corresponding experiments. Immunofluorescence was performed as previously described [20]. Basically, the cells were fixed in 2% PFA in PBS for 10 min at RT and permeabilized with 0.1% Triton X-100 and 0.1% sodium citrate for 5 min at RT. Preparations were washed with PBS and washing solution (PBS/0.25% BSA/0.1% Tween 20), blocked for 30 min with blocking solution (washing solution + 2.5% BSA) and incubated overnight with antibodies against TP53BP1 (1:500; sc-16565; Invitrogen, Waltham, MA, USA). Preparations were then washed with washing solution and incubated with secondary antibodies conjugated with Alexa fluor dyes (488, 546) from Life Technologies (cat.#A-11029, cat.#A-11035) for 1 h at RT. Nuclei were counterstained with DAPI, and samples were mounted with ProLong Diamond (cat.#P36961; Life Technologies, Waltham, MA, USA). Cell images were captured with fluorescence microscope (Zeiss Axio) equipped with a camera (AxiocamMRm) and AxioVision software. DNA damage foci were quantified by counting from >150 cells for each experimental condition. For 2D minilungs, the glass chamber slides were incubated overnight at 4 °C with human fibronectin in order to plate the differentiated cells. Cultures from day 50 were digested with trypsin, neutralized with stop medium and washed with SFD medium. Approximately 40,000 differentiated epithelial cells per well were plated in the epithelial maturation medium. Cultures were maintained under normoxic conditions for one day before treatments.

4.9. Indirect Immunofluorescence of Lung Bud Organoids

Organoids were picked up from the MW96 wells, transferred into a well of a MW12 and fixed with 4% paraformaldehyde (PFA) for 15 min at RT. After that, the organoids were washed three times with PBS for 10 min and incubated overnight at 4 °C with 30% sucrose. The sucrose was exchanged for a solution of 7.5% gelatin/15% sucrose and incubated for 15 min at 37 °C. The organoids were carefully transferred to cryomolds and progressively embedded in various layers of solidified 7.5% gelatin/15% sucrose. These preparations were cut into 10-µm sections in a Leica CM3050 cryostat. The mounted sections were washed with PBS and permeabilized with PBS/1% BSA/0.25% Triton X-100 for 5 min at RT. After that, the sections were washed and blocked for 30 min at RT with blocking solution (PBS-BSA 1%). The sections were incubated for 2 h with antibodies against TP53BP1 (1:500; sc-16565; Invitrogen, Waltham, MA, USA) or the pro surfactant protein C (1:200; ab3785, Merck). Preparations were washed with washing solution and incubated with a secondary antibody conjugated with Alexa fluor dye (546) from Life Technologies (cat.#A-11035, Waltham, MA, USA) for 1 h at room temperature. Nuclei were counterstained with DAPI and samples were mounted with ProLong Diamond (cat.#P36961; Life Technologies, Waltham, MA, USA). Cell images were captured with a fluorescence microscope (Zeiss Axio) equipped with a camera (AxiocamMRm) and AxioVision software. DNA damage foci were counted from >150 cells for each experimental condition.

4.10. Analysis of Proteins by Western Blot

Cell monolayers were washed with ice-cold PBS and lysed in triple-detergent lysis buffer (50 mM Tris-HCl pH 8.0, 150 mM NaCl, 0.02% sodium azide, 0.1% SDS, 1% NP-40, 0.5% sodium deoxycholate, 100 µg/mL PMSF, 2 µg/mL pepstatin, 2 µg/mL aprotinin, 2 µg/mL leupeptin, and phosphatase inhibitors cocktail 2 or 3 (cat.#P5726, P0044, Sigma-Aldrich, St. Louis, MO, USA)). SDS-PAGE and immunoblotting were performed under standard conditions. Basically, samples in Laemmli buffer (30 µg/lane) were separated through 12% gels and transferred to nitrocellulose membranes for 90 min at RT in the presence of 20% methanol and 0.1% SDS. Membranes were blocked with 3% BSA in PBS-Tween 0.05% (PBST-BSA) and incubated O/N at 4 °C with a γH2AFX antibody (cat.#05–636, Millipore, Burlington, MA, USA) diluted 1:1000 in PBST-BSA. Densitometry analysis of bands was performed by using Image J software (https://imageJ.nih.gov/ (accessed on 22 April 2020)).

4.11. Quantitative real-time RT-PCR (RT-qPCR) of minilungs

Total RNA was extracted using Trizol (cat.#15596026; Ambion) following manufacturer's instructions. cDNA was generated using the High-Capacity cDNA kit (cat.#4387406; Applied Biosystems, Waltham, MA, USA). Real-time qPCR was performed by using the powerUpSYBR Green mix (cat.#A25742) on the Quantstudio-3 system (Applied Biosystems, Waltham, MA, USA) following manufacturer's instructions. Absolute quantification of each gene was obtained using a standard curve of serial diluted genomic DNA (cat.#11807720, Roche) and normalized to housekeeping gene TBP (TATA box binding protein).

The genes analyzed and the sequences of the oligonucleotides employed in this study were the following: TBP [Tata-Box Binding Protein; Forward: 5′-TGAGTTGCTCATACCGTGCTGCTA, Reverse: 5′-CCCTCAAACCAACTTGTCAACAGC]; TP63 (Tumor Protein P63, marker of basal cells) [Forward: 5′-CCTATAACACAGACCACGCGCAGA, Reverse: 5′-GTGATGGAGAGAGA GCATCGAAG]; MUCIN5AC (Mucin 5AC, marker of globet cells) [Forward: 5′GCACCAACGA CAGGAAGGATGAG, Reverse: 5′-CACGTTCCAGAGCCGGACAT]; SCGB1A1 (Secretoglobin Family1A Member1 or CC10, marker of clara cells) [Forward: 5′-TCATGGACACACCCTCCAG TTATGAG.

Reverse: 5′-TGAGCTTAATGATGCTTTCTCTGGGC]; *PDPN* (Podoplanin, marker of AT-I cells) [Forward: 5′- AGGAGAGCAACAACTCAACGGGA, Reverse: 5′- TTCT-GCCAGGACCCAGAGC]; *AQP5* (Aquaporin 5, marker of AT-I cells) [Forward: 5′- GC-CATCCTTTACTTCTACCTGCTC, Reverse: 5′- GCTCATACGTGCCTTTGATGATGG]; *SFTPA* (Surfactant Protein A, marker of AT-II cells) [Forward: 5′-GTGCGAAGTGAAGGACGTTTGTG, Reverse: 5′-TTTGAGACCATCTCTCCCGTCCC]; *SFTPB* (Surfactant Protein B, marker of AT-II cells) [Forward: 5′-TCTGAGTGCCACCTCTGCATGT, Reverse: 5′-TGGAGCATTGCC TGTGGTATGG]; *SFTPC* (Surfactant Protein C, marker of AT-II cells) [Forward: 5′-CCTTCTT ATCGTGGTGGTGGTGGT, Reverse: 5′-TCTCCGTGTGTTTCTGGCTCATGT]; *SFTPD* (Surfactant Protein D, marker of AT-II cells) [Forward: 5′-TGACTGATTCCAAGACAGAGGGCA, Reverse: 5′-TCCACAAGCCCTGTCATTCCACTT]; *FOXJ1* (Forkhead Box J1, marker of ciliated cells) [Forward: 5′-GGCATAAGCGCAAACAGCCG, Reverse: 5′-TCGAAGATGGCCT CCCAGTCAAA]; *CYP24A1* (cytochrome P450 family 24 subfamily A member 1) [Forward: 5′-GGTGACATCTACGGCGTAC, Reverse: 5′-CTTGAGACCCCCTTTCCAGAG].

4.12. Statistical Analysis

Data were subjected to the Shapiro–Wilk test and D'Agostino and Pearson omnibus test to verify their normality. Statistical significance of data was determined by applying a two-tailed Student's *t*-test or analysis of variance followed by the Newman–Keuls or Bonferroni post hoc tests for experiments with more than two experimental groups; $p < 0.05$ is considered significant. Significance of analysis of variance post hoc test or the Student's *t*-test is indicated in the figures as *, $p < 0.05$; **, $p < 0.01$ and ***, $p < 0.001$. Statistics were calculated with the Prism 9 software (GraphPad Software). The results presented in the figures are means ±SEM. Experiments were repeated three times.

5. Conclusions

The bleomycin treatment of cells and organoids might be used to reproduce the bulk of the DNA damage expression underlying multiple conditions that evolve with fibrosis. The treatment with vitamin D less-hypercalcemic analogs does not increase the DNA damage expression reached by the sole treatment with bleomycin. In terms of DNA damage expression, these initial results indicate that the vitamin D analog paricalcitol behaves as a potential DNA damage eraser. This fact might be exploited to set up a therapy for fibrogenic diseases.

Supplementary Materials: Supplementary materials can be found at https://www.mdpi.com/article/10.3390/ijms23094921/s1.

Author Contributions: E.M.-L. performed and designed experiments, analyzed data. I.C.-H. performed and designed experiments, analyzed data experiments. A.Z. performed and designed experiments, analyzed data, wrote paper and conceived the project. All authors have read and agreed to the published version of the manuscript.

Funding: This work was supported by Grant PI19CIII/00003 from the Institute of Health Carlos III (ISCIII) to Alberto Zambrano.

Institutional Review Board Statement: The use of the hESC line AND-2 and the experimental procedures of this study were approved by the ISCIII Ethics Committee (ref. no. CEI PI 10_2015-v2) and the National Committee of Guarantees for the Use and Donation of Human Cells and Tissues (ref. no. 345 288 1 and 436 351 1).

Informed Consent Statement: Not applicable.

Data Availability Statement: Please contact the corresponding author for data requests.

Acknowledgments: We thank the histology facility (Manolo and Marta) of the ISCIII for technical help.

Conflicts of Interest: The authors have no conflict of interest to declare.

Abbreviations

AFE	Anterior foregut endoderm
ATI	Alveolar type I cells
ATII	Alveolar type II cells
BMP4	Bone Morphogenic Protein 4
BSA	Bovine serum albumin
DD	DNA damage
DD foci	DNA damage foci
DSBs	DNA double-strand breaks
EBs	Embryoid bodies
FBS	Foetal bovine serum
FGF	Fibroblast growth factor
FOXJ1	Forkhead Box J1
H2AFX	H2A histone family member X
hbFGF	Human basic fibroblast growth factor
hESCs	Human embryonic stem cells
hPSCs	Human pluripotent stem cells
IBMX	Isobutylmethylxanthine
IPF	Idiopathic pulmonary fibrosis
KGF	Keratinocyte growth factor
LBOs	Lung bud organoids
MEFs	Mouse embryonic fibroblasts
MUCIN5AC	Mucin 5AC, Oligomeric Mucus/Gel-Forming
PBS	Phosphate-buffered saline
PDPN	Podoplanin
RT-qPCR	Quantitative real-time RT-PCR (reverse transcription polymerase chain reaction)
SCGB1A1	Secretoglobin Family 1A Member 1; CC10

SEM	Standard error of the mean
SFD	Serum-free differentiation
SFTPA	SurfactantProtein A
SFTPB	SurfactantProtein B
SFTPC	SurfactantProtein C
SFTPD	SurfactantProtein D
SOX2	SRY (sex-determining region Y)-box 2
TBP	TATA box binding protein
TP53BP1	Tumor Protein P53 Binding Protein 1
TP63	Tumor Protein P63
µm	Micrometer

References

1. King, T.E., Jr.; Pardo, A.M. Idiopathic pulmonary fibrosis. *Lancet* **2011**, *378*, 1949–1961. [CrossRef]
2. Hecker, L.; Logsdon, N.J.; Kurundkar, D.; Kurundkar, A.; Bernard, K.; Hock, T.; Meldrum, E.; Sanders, Y.Y.; Thannickal, V.J. Reversal of persistent fibrosis in aging by targeting Nox4-Nrf2 redox imbalance. *Transl. Med.* **2014**, *6*, 231ra47. [CrossRef] [PubMed]
3. Lv, X.X.; Wang, X.X.; Li, K.; Wang, Z.Y.; Li, Z.; Lv, Q.; Fu, X.M.; Hu, Z.W. Rupatadine protects against pulmonary fibrosis by attenuating PAF-mediated senescence in rodents. *PLoS ONE* **2013**, *8*, e68631. [CrossRef] [PubMed]
4. Shivshankar, P.; Brampton, C.; Miyasato, S.; Kasper, M.; Thannickal, V.J.; Le Saux, C.J. Caveolin-1 deficiency protects from pulmonary fibrosis by modulating epithelial cell senescence in mice. *Am. J. Respir. Cell Mol. Biol.* **2012**, *47*, 28–36. [CrossRef] [PubMed]
5. Ding, N.; Yu, R.T.; Subramaniam, N.; Sherman, M.H.; Wilson, C.; Rao, R.; Leblanc, M.; Coulter, S.; He, M.; Scott, C.; et al. A vitamin D receptor/SMAD genomic circuit gates hepatic fibrotic response. *Cell* **2013**, *153*, 601–613. [CrossRef] [PubMed]
6. Ito, I.; Waku, T.; Aoki, M.; Abe, R.; Nagai, Y.; Watanabe, T.; Nakajima, Y.; Ohkido, I.; Yokoyama, K.; Miyachi, H.; et al. A nonclassical vitamin D receptor pathway suppresses renal fibrosis. *J. Clin. Investig.* **2013**, *123*, 4579–4594. [CrossRef] [PubMed]
7. Meredith, A.; Boroomand, S.; Carthy, J.; Luo, Z.; McManus, B. 1,25 Dihydroxyvitamin D3 Inhibits TGFβ1-Mediated Primary Human Cardiac Myofibroblast Activation. *PLoS ONE* **2015**, *10*, e0128655. [CrossRef]
8. Zhang, Z.; Yu, X.; Fang, X.; Liang, A.; Yu, Z.; Gu, P.; Zeng, Y.; He, J.; Zhu, H.; Li, S.; et al. Preventive effects of vitamin D treatment on bleomycin-induced pulmonary fibrosis. *Sci. Rep.* **2015**, *5*, 17638. [CrossRef]
9. Tan, Z.X.; Chen, Y.H.; Xu, S.; Qin, H.Y.; Zhang, C.; Zhao, H.; Xu, D.X. Calcitriol inhibits bleomycin-induced early pulmonary inflammatory response and epithelial-mesenchymal transition in mice. *Toxicol. Lett.* **2016**, *240*, 161–171. [CrossRef]
10. Guijarro, T.; Magro-Lopez, E.; Manso, J.; Garcia-Martinez, R.; Fernandez-Aceñero, M.J.; Liste, I.; Zambrano, A. Detrimental pro-senescence effects of vitamin D on lung fibrosis. *Mol. Med.* **2018**, *24*, 64. [CrossRef]
11. Wyman, C.; Kanaar, R. DNA double-strand break repair, all's well that ends well. *Rev. Genet.* **2006**, *40*, 363–383. [CrossRef] [PubMed]
12. Campisi, J.; d'Adda di Fagagna, F. Cellular senescence, When bad things happen to good cells. *Nat. Rev. Mol. Cell Biol.* **2007**, *8*, 729–740. [CrossRef] [PubMed]
13. Aoshiba, K.; Tsuji, T.; Kameyama, S.; Itoh, M.; Semba, S.; Yamaguchi, K.; Nakamura, H. Senescence-associated secretory phenotype in a mouse model of bleomycin-induced lung injury. *Exp. Toxicol. Pathol.* **2013**, *65*, 1053–1062. [CrossRef] [PubMed]
14. Aoshiba, K.; Zhou, F.; Tsuji, T.; Nagai, A. DNA damage as a molecular link in the pathogenesis of COPD in smokers. *Eur. Respir. J.* **2012**, *39*, 1368–1376. [CrossRef]
15. Chilosi, M.; Carloni, A.; Rossi, A.; Poletti, V. Premature lung aging and cellular senescence in the pathogenesis of idiopathic pulmonary fibrosis and COPD/emphysema. *Transl. Res.* **2013**, *162*, 156–173. [CrossRef] [PubMed]
16. Kuwano, K.; Araya, J.; Hara, H.; Minagawa, S.; Takasaka, N.; Ito, S.; Kobayashi, K.; Nakayama, K. Cellular senescence and autophagy in the pathogenesis of chronic obstructive pulmonary disease (COPD) and idiopathic pulmonary fibrosis (IPF). *Respir. Investig.* **2016**, *54*, 397–406. [CrossRef]
17. Chua, F.; Gauldie, J.; Laurent, G.J. Pulmonary fibrosis, Searching for model answers. *Am. J. Respir. Cell Mol. Biol.* **2005**, *33*, 9–13. [CrossRef]
18. Leyssens, C.; Verlinden, L.; Verstuyf, A. The future of vitamin D analogs. *Front. Physiol.* **2014**, *5*, 122. [CrossRef]
19. Slatopolsky, E.; Finch, J.; Brown, A. New vitamin D analogs. *Kidney Int. Suppl.* **2003**, *85*, S83–S87. [CrossRef]
20. Zambrano, A.; García-Carpizo, V.; Gallardo, M.E.; Villamuera, R.; Gómez-Ferrería, M.A.; Pascual, A.; Buisine, N.; Sachs, L.M.; Garesse, R.; Aranda, A. The thyroid hormone receptor β induces DNA damage and premature senescence. *J. Cell Biol.* **2014**, *204*, 129–146. [CrossRef]
21. Rogakou, E.P.; Boon, C.; Redon, C.; Bonner, W.M. Megabase chromatin domains involved in DNA double-strand breaks in vivo. *J. Cell Biol.* **1999**, *146*, 905–916. [CrossRef] [PubMed]
22. Schultz, L.B.; Chehab, N.H.; Malikzay, A.; Halazonetis, T.D. p53 binding protein 1 (53BP1) is an early participant in the cellular response to DNA double-strand breaks. *J. Cell Biol.* **2000**, *151*, 1381–1390. [CrossRef] [PubMed]
23. Abraham, R.T. Checkpoint signalling, Focusing on 53BP1. *Nat. Cell Biol.* **2002**, *12*, E277–E279. [CrossRef]

24. Ward, I.M.; Minn, K.; Jorda, K.G.; Chen, J. Accumulation of checkpoint protein 53BP1 at DNA breaks involves its binding to phosphorylated histone H2AX. *J. Biol. Chem.* **2003**, *278*, 19579–19582. [CrossRef] [PubMed]
25. Magro-Lopez, E.; Guijarro, T.; Martinez, I.; Martin-Vicente, M.; Liste, I.; Zambrano, A. A Two-Dimensional Human Minilung System (Model) for Respiratory Syncytial Virus Infections. *Viruses* **2017**, *9*, 379. [CrossRef]
26. Magro-Lopez, E.; Palmer, C.; Manso, J.; Liste, I.; Zambrano, A. Effects of lung and airway epithelial maturation cocktail on the structure of lung bud organoids. *Stem Cell Res. Ther.* **2018**, *9*, 186. [CrossRef]
27. Huang, S.X.; Green, M.D.; de Carvalho, A.T.; Mumau, M.; Chen, Y.W.; D'Souza, S.L.; Snoeck, H.W. The in vitro generation of lung and airway progenitor cells from human pluripotent stem cells. *Nat. Protoc.* **2015**, *10*, 413–425. [CrossRef]
28. Huang, S.X.; Islam, M.N.; O'Neill, J.; Hu, Z.; Yang, Y.G.; Chen, Y.W.; Mumau, M.; Green, M.D.; Vunjak-Novakovic, G.; Bhattacharya, J.; et al. Efficient generation of lung and airway epithelial cells from human pluripotent stem cells. *Nat. Biotechnol.* **2014**, *32*, 84–91. [CrossRef]
29. Chen, Y.W.; Huang, S.X.; de Carvalho, A.L.R.T.; Ho, S.H.; Islam, M.N.; Volpi, S.; Notarangelo, L.D.; Ciancanelli, M.; Casanova, J.L.; Bhattacharya, J.; et al. A three-dimensional model of human lung development and disease from pluripotent stem cells. *Nat. Cell Biol.* **2017**, *19*, 542–549. [CrossRef]
30. Martínez, I.; García-Carpizo, V.; Guijarro, T.; García-Gomez, A.; Navarro, D.; Aranda, A.; Zambrano, A. Induction of DNA double-strand breaks and cellular senescence by human respiratory syncytial virus. *Virulence* **2016**, *7*, 427–442. [CrossRef]
31. Finklea, J.D.; Grossmann, R.E.; Tangpricha, V. Vitamin D and chronic lung disease, A review of molecular mechanisms and clinical studies. *Adv. Nutr.* **2011**, *2*, 244–253. [CrossRef] [PubMed]
32. Tebben, P.J.; Singh, R.J.; Kumar, R. Vitamin D-Mediated Hypercalcemia, Mechanisms, Diagnosis, and Treatment. *Endocr. Rev.* **2016**, *37*, 521–547. [CrossRef] [PubMed]
33. Planté-Bordeneuve, T.; Berardis, S.; Bastin, P.; Gruson, D.; Henri, L.; Gohy, S. Vitamin D intoxication in patients with cystic fibrosis, Report of a single-center cohort. *Sci. Rep.* **2021**, *11*, 7719. [CrossRef] [PubMed]
34. Tzilas, V.; Bouros, E.; Barbayianni, I.; Karampitsakos, T.; Kourtidou, S.; Ntassiou, M.; Ninou, I.; Aidinis, V.; Bouros, D.; Tzouvelekis, A. Vitamin D prevents experimental lung fibrosis and predicts survival in patients with idiopathic pulmonary fibrosis. *Pulm. Pharmacol. Ther.* **2019**, *55*, 17–24. [CrossRef] [PubMed]
35. Chang, J.; Nie, H.; Ge, X.; Du, J.; Liu, W.; Li, X.; Sun, Y.; Wei, X.; Xun, Z.; Li, Y.C. Vitamin D suppresses bleomycin-induced pulmonary fibrosis by targeting the local renin-angiotensin system in the lung. *Sci. Rep.* **2021**, *11*, 16525. [CrossRef]
36. Elwakeel, E.E.; Mohamed, A.Z.; Shaalan, W.M. Therapeutic effects of mesenchymal stem cells and vitamin D on Bleomycin triggered lung damage in male adult albino rats. *Ultrastruct. Pathol.* **2022**, 1–14. [CrossRef]
37. Tan, X.; Li, Y.; Liu, Y. Paricalcitol attenuates renal interstitial fibrosis in obstructive nephropathy. *J. Am. Soc. Nephrol.* **2006**, *17*, 3382–3393. [CrossRef]
38. Green, D.M.; Leonard, A.R.; Paranjape, S.M.; Rosenstein, B.J.; Zeitlin, P.L.; Mogayzel, P.J., Jr. Transient effectiveness of vitamin D2 therapy in pediatric cystic fibrosis patients. *J. Cyst. Fibros.* **2010**, *9*, 143–149. [CrossRef]
39. Inoue, K.; Matsui, I.; Hamano, T.; Fujii, N.; Shimomura, A.; Nakano, C.; Kusunoki, Y.; Takabatake, Y.; Hirata, M.; Nishiyama, A.; et al. Maxacalcitol ameliorates tubulointerstitial fibrosis in obstructed kidneys by recruiting PPM1A/VDR complex to pSmad3. *Lab Investig.* **2012**, *92*, 1686–1697. [CrossRef]
40. Meems, L.M.; Cannon, M.V.; Mahmud, H.; Voors, A.A.; van Gilst, W.H.; Silljé, H.H.; Ruifrok, W.P.; de Boer, R.A. The vitamin D receptor activator paricalcitol prevents fibrosis and diastolic dysfunction in a murine model of pressure overload. *J. Steroid Biochem. Mol. Biol.* **2012**, *132*, 282–289. [CrossRef]
41. Hirose, M.; Nishino, T.; Obata, Y.; Nakazawa, M.; Nakazawa, Y.; Furusu, A.; Abe, K.; Miyazaki, M.; Koji, T.; Kohno, S. 22-Oxacalcitriol prevents progression of peritoneal fibrosis in a mouse model. *Perit. Dial. Int.* **2013**, *33*, 132–142. [CrossRef] [PubMed]
42. González-Mateo, G.T.; Fernández-Míllara, V.; Bellón, T.; Liappas, G.; Ruiz-Ortega, M.; López-Cabrera, M.; Selgas, R.; Aroeira, L.S. Paricalcitol reduces peritoneal fibrosis in mice through the activation of regulatory T cells and reduction in IL-17 production. *PLoS ONE* **2014**, *9*, e108477.
43. Lai, C.C.; Liu, C.P.; Cheng, P.W.; Lu, P.J.; Hsiao, M.; Lu, W.H.; Sun, G.C.; Liou, J.C.; Tseng, C.J. Paricalcitol Attenuates Cardiac Fibrosis and Expression of Endothelial Cell Transition Markers in Isoproterenol-Induced Cardiomyopathic Rats. *Crit. Care Med.* **2016**, *44*, e866–e874. [CrossRef] [PubMed]
44. Lee, A.S.; Jung, Y.J.; Thanh, T.N.; Lee, S.; Kim, W.; Kang, K.P.; Park, S.K. Paricalcitol attenuates lipopolysaccharide-induced myocardial inflammation by regulating the NF-κB signaling pathway. *Int. J. Mol. Med.* **2016**, *37*, 1023–1029. [CrossRef]
45. Wahsh, E.; Abu-Elsaad, N.; El-Karef, A.; Ibrahim, T. The vitamin D receptor agonist, calcipotriol, modulates fibrogenic pathways mitigating liver fibrosis in-vivo, An experimental study. *Eur. J. Pharmacol.* **2016**, *789*, 362–369. [CrossRef]
46. Gözel, N.; Duran, F.; Yildirim, A.; Yolbaş, S.; Önalan, E.; Özercan, İ.H.; Koca, S.S. Paricalcitol Inhibits Wnt/β-Catenin Signaling Pathway and Ameliorates Dermal Fibrosis in Bleomycin Induced Scleroderma Model. *Arch. Rheumatol.* **2017**, *33*, 288–294. [CrossRef] [PubMed]
47. Wang, X.; Wang, G.; Qu, J.; Yuan, Z.; Pan, R.; Li, K. Calcipotriol Inhibits NLRP3 Signal Through YAP1 Activation to Alleviate Cholestatic Liver Injury and Fibrosis. *Front. Pharmacol.* **2020**, *11*, 200. [CrossRef] [PubMed]

48. Martínez-Arias, L.; Panizo, S.; Alonso-Montes, C.; Martín-Vírgala, J.; Martín-Carro, B.; Fernández-Villabrille, S.; García Gil-Albert, C.; Palomo-Antequera, C.; Fernández-Martín, J.L.; Ruiz-Torres, M.P.; et al. Effects of calcitriol and paricalcitol on renal fibrosis in CKD. *Nephrol. Dial. Transplant.* **2021**, *36*, 793–803. [CrossRef] [PubMed]
49. Gong, J.; Gong, H.; Liu, Y.; Tao, X.; Zhang, H. Calcipotriol attenuates liver fibrosis through the inhibition of vitamin D receptor-mediated NF-κB signaling pathway. *Bioengineered* **2022**, *13*, 2658–2672. [CrossRef]

International Journal of *Molecular Sciences*

MDPI

Article

Sulfated Phenolic Substances: Preparation and Optimized HPLC Analysis

Lucie Petrásková [1], Kristýna Káňová [1,2], Katerina Brodsky [1,2], Anastasiia Hetman [1,3], Barbora Petránková [1,4], Helena Pelantová [1], Vladimír Křen [1] and Kateřina Valentová [1,*]

1 Institute of Microbiology of the Czech Academy of Sciences, Vídeňská 1083, 142 20 Prague, Czech Republic; petraskova@biomed.cas.cz (L.P.); astriik@gmail.com (K.K.); katerina.brodsky@biomed.cas.cz (K.B.); hetmananastasiia@seznam.cz (A.H.); barbora.petrankova@natur.cuni.cz (B.P.); pelantova@biomed.cas.cz (H.P.); kren@biomed.cas.cz (V.K.)
2 Department of Biochemistry and Microbiology, University of Chemistry and Technology Prague, Technická 3, 166 28 Prague, Czech Republic
3 Department of Pharmacology and Toxicology, Faculty of Pharmacy in Hradec Králové, Charles University, Heyrovského 1203, 500 05 Hradec Králové, Czech Republic
4 Department of Analytical Chemistry, Faculty of Science, Charles University, Albertov 6, 128 43 Prague, Czech Republic
* Correspondence: kata.valentova@email.cz

Abstract: Sulfation is an important reaction in nature, and sulfated phenolic compounds are of interest as standards of mammalian phase II metabolites or pro-drugs. Such standards can be prepared using chemoenzymatic methods with aryl sulfotransferases. The aim of the present work was to obtain a large library of sulfated phenols, phenolic acids, flavonoids, and flavonolignans and optimize their HPLC (high performance liquid chromatography) analysis. Four new sulfates of 2,3,4-trihydroxybenzoic acid, catechol, 4-methylcatechol, and phloroglucinol were prepared and fully characterized using MS (mass spectrometry), ^{1}H, and ^{13}C NMR. The separation was investigated using HPLC with PDA (photodiode-array) detection and a total of 38 standards of phenolics and their sulfates. Different stationary (monolithic C18, C18 Polar, pentafluorophenyl, ZICpHILIC) and mobile phases with or without ammonium acetate buffer were compared. The separation results were strongly dependent on the pH and buffer capacity of the mobile phase. The developed robust HPLC method is suitable for the separation of enzymatic sulfation reaction mixtures of flavonoids, flavonolignans, 2,3-dehydroflavonolignans, phenolic acids, and phenols with PDA detection. Moreover, the method is directly applicable in conjunction with mass detection due to the low flow rate and the absence of phosphate buffer and/or ion-pairing reagents in the mobile phase.

Keywords: aryl sulfotransferase; *Desulfitobacterium hafniense*; HPLC analysis; sulfates; flavonoids; polyphenols; phenolic acid

Citation: Petrásková, L.; Káňová, K.; Brodsky, K.; Hetman, A.; Petránková, B.; Pelantová, H.; Křen, V.; Valentová, K. Sulfated Phenolic Substances: Preparation and Optimized HPLC Analysis. *Int. J. Mol. Sci.* **2022**, *23*, 5743. https://doi.org/10.3390/ijms23105743

Academic Editor: David Arráez-Román

Received: 2 May 2022
Accepted: 19 May 2022
Published: 20 May 2022

1. Introduction

Plant constituents such as flavonoids and phenolic acids are an essential part of a so-called healthy diet. These polyphenolic substances, once considered antioxidants and protective agents against ROS, act more at the receptor level [1]. However, their function is still not fully understood, and it is necessary to thoroughly study their toxicology, metabolism, and possible interactions with dietary supplements and drugs.

Flavonoids and phenolic acids are preferentially metabolized (sulfated, methylated, or glucuronidated) via the II biotransformation phase [2,3]. Sulfated flavonoids and other polyphenols are therefore of great pharmacological interest. They can serve as potential (pro)drugs and have antiviral, antitumor, anticoagulant, and anti-inflammatory activities [4]. Sulfation is also associated with molecular recognition, cell signaling, and hormone regulation [5]. A detailed study of all these properties requires substantial amounts, which usually cannot be isolated from the biological material. On the other hand, nature-inspired sulfated conjugates can be synthesized chemically or enzymatically to provide well-characterized standards.

The enzymatic sulfation of (poly)phenolic compounds has recently been preferred to the chemical one [6–13]. Bacterial aryl sulfotransferases such as the one from *Desulfitobacterium hafniense* use *p*-nitrophenyl sulfate (*p*-NP-S) as a sulfate donor. When monitoring the reaction progress, we solve the analytical separation of polar (phenols, *p*-nitrophenol, *p*-NP) and highly polar substances (phenolic sulfates, *p*-NP-S). Besides TLC, the HPLC method is the analytical method of the first choice for monitoring sulfation reactions due to its robustness and general availability. Sulfates of various polyphenols carrying one or more highly charged sulfate group(s) are inherently very polar compounds. If the sulfate group is not modified, or if the effect of the modification on the overall polarity is small, conventional reversed-phase liquid chromatography will not provide adequate retention. This shows up on the chromatogram as peak fronting, tailing, or very broad peaks.

One way to solve this problem is to modify the sulfate group using ion-pairing liquid chromatography (IPLC). IPLC uses a conventional reversed stationary phase in combination with a mobile phase enriched in an ion-pairing reagent. The ion-pairing reagent (generally an amine) forms a hydrophobic pair with the highly charged molecule of opposite charge. This dramatically increases the retention of the analytes and thus the sharpness of the peaks. IPLC has been used, for example, for the analysis of oligonucleotides [14], for the separation of reaction mixtures with sulfates of 2-phenylethyl alcohol, *p*-NP-glycerol, *p*-NP-glucose, phenyl-glucose, *p*-NP-*N*-acetyl-D-glucosamine (GlcNAc) [6], and for 3-sulfo-17-beta-estradiol-4,4′-disulfate, and for monitoring the sulfation of resveratrol and phloretin [7,15]. However, IPLC applications may also have disadvantages, such as artifacts when using gradient elution, incompatibility with mass spectrometry (MS) and preparative chromatography, or more complex mobile phase preparations.

Therefore, in some cases, a better solution is required. Resolution and sensitivity can be greatly improved by a careful combination of stationary and mobile phases, flow rate, temperature, and the type of detection. Hereunder we show the advantages and disadvantages of analytical methods developed for the separation of sulfated substances.

C18 columns in combination with phosphate-containing mobile phases and an ion-pairing reagent required over 20 min for the separation of indoxyl sulfate, sulfated primary aliphatic alcohols, estradiol, or bisphenol-A [6,7,16]. Significantly shorter times (2.5–6 min) were reached when separating silybin or isosilybin sulfates. However, the width of the peaks in 5% of their height was greater than one minute (1–4.5 min) in all cases [9]. Polyamine II column with a linear gradient of phosphate buffers was used to separate fluorescently labeled *N*-sulfated polysaccharides [17]. However, phosphate is incompatible with the LC-MS interface.

For monitoring the metabolic fate of (poly)phenolic substances in complex experiments on animals, MS/MS detection can preferably be used. However, these devices are expensive and not generally available. The sulfated metabolites of naringin and hesperidin were monitored in in vivo experiments using the UHPLC-ESI-MS/MS (ultra-high performance liquid chromatography coupled with electrospray ionization tandem mass spectrometry) analysis (Waters Acquity BEH C18 column, gradient elution). The metabolites were identified based on chemical composition, retention time, MS/MS fragmentation pattern, and comparison with available standards and references [2]. The sulfated product of chlorocatechol was identified using the NUCLEODUR HILIC (hydrophilic interaction chromatography) column with an ion trap mass analyzer. The peak of chlorocatechol sulfate was broad and the other components of the reaction mixture (*p*-NP, 3-chlorocatechol) were not separated from the baseline [18]. The biotransformation products of quercetin, isoquercitrin, and taxifolin sulfates formed by HepG2 cells were monitored by UHPLC (XDB phenyl column) and ESI MS/MS detection [19]. The sulfation reaction (catalyzed by aryl sulfotransferase from *Clostridium innocuum*) of various phenols, flavonoids, quinones, primary alcohols, and sugars was analyzed using a C18 column (mobile phase acetonitrile, water, trifluoroacetic acid) and the identity of the reaction products was confirmed by MS [20].

The analyses of hydroxytyrosol and acetyltyrosol sulfates [21], luteolin, myricetin, and ampelopsin sulfates [8], quercetin mono- and disulfates [12] or quercetin, isoquercitrin, and taxifolin sulfates [11] on the pentafluorophenyl column (mobile phase water, methanol, trifluoroacetic acid) suffered from tailing peaks. Tailing peaks were also observed during the separation of sulfated polysaccharides (cellobiose and GlcNAc-linker-tBoc) on the Multospher APS-HP-5 μm HILIC column (mobile phase acetonitrile, ammonium acetate) [22]. HPLC chromatograms of 2,3-dehydrosilybin sulfation obtained by separation on a monolithic Chromolith C18 column (mobile phase acetonitrile, water, formic acid) also showed tailing [13].

Separation of a chemically prepared mixture of mono- and disulfates of quercetin on a BDS -Hypersil C18 column (mobile phase ammonium acetate with methanol) gave good analytical separation (sharp peaks, baseline separation); up to 20 min was achieved in the separation [23]. The sulfation reagent sulfur trioxide *N*-triethylamine probably also served as an ion-pairing reagent and allowed sharp peaks. Analyses of the sulfates of quercetin and epicatechin on the AQUA reversed-phase C18 column and the Spherisorb S3OD-2 C18 column (mobile phase trifluoroacetic acid in water and methanol, respectively) took over 30 min [24].

The following parameters can be considered as disadvantages for the separation of sulfated polyphenols: Phosphate in the mobile phase, use of ion-pairing reagent (both incompatible with LC-MS), MS/MS detection (not a common detector), separation time over 20 min, and tailing and/or coelution of the peaks (poorer separation of the peaks in the reaction mixture). The disadvantages of the published methods with at least one undesirable parameter are summarized in Table 1.

Table 1. An overview of HPLC methods for sulfate separation published to date and their parameters.

Column	C18									C18 [a]	HILIC [b]		Phenyl [a]	PFP [c]			Polyamine
Phosphate in mobile phase	+	+	+	-	-	-	-	+	+	-	-	-	-	-	-	-	+
Ion-pairing reagent	-	+	+	-	-	-	-	+	+	-	-	-	-	-	-	-	-
MS/MS, QTof detection	-	-	-	-	-	+	-	+	-	+	+	-	+	-	-	-	Fluorescence
Long separation time (≥ 20 min)	? [d]	+	+	+	+	-	-	+	-	+	+	-	-	+	+	-	-
Tailing peaks, coelution	?	-	+	?	-	?	+	-	+	-	+	+	+	+	+	+	+
Reference	[16]	[6]	[7]	[24]	[23]	[20]	[11]	[15]	[9]	[2]	[18]	[22]	[19]	[21]	[8]	[12,13]	[17]

[a] Ultra-high performance liquid chromatography (UHPLC), [b] hydrophilic interaction chromatography, [c] pentafluorophenyl, [d] information not provided in the reference.

The aim of this work was to obtain a large library of sulfated phenols, phenolic acids, flavonoids, and (2,3-dehydro)flavonolignans and to develop a robust and reliable HPLC analytical method suitable for the separation of enzymatic sulfation reaction mixtures of these phenolics. The authentic standards of sulfates of 2,3,4-trihydroxybenzoic acid, catechol, 4-methylcatechol, and phloroglucinol were prepared in the frame of this work. Four types of stationary phases (pentafluorophenyl (PFP), C18, C18 Polar, and HILIC) were compared using mobile phases without or with buffer and with PDA detection. The optimal method is also suitable for mass detection because it does not use phosphate buffers and ion-paired reagents in the mobile phase. The duration of separation was up to 20 min in most cases without peak tailing and coelution.

2. Results and Discussion

2.1. Synthesis of Sulfated Phenolics

A large number of metabolic standards used in this work were prepared earlier using the bacterial aryl sulfotransferase of *Desulfitobacterium hafniense* [8,10–13,19]. Besides previously published products, new authentic standards were prepared in the present work with the same very effective enzymatic method for selective sulfation of polyphenols (Figure 1) and fully characterized with HPLC, MS, and NMR (see Supplementary Material).

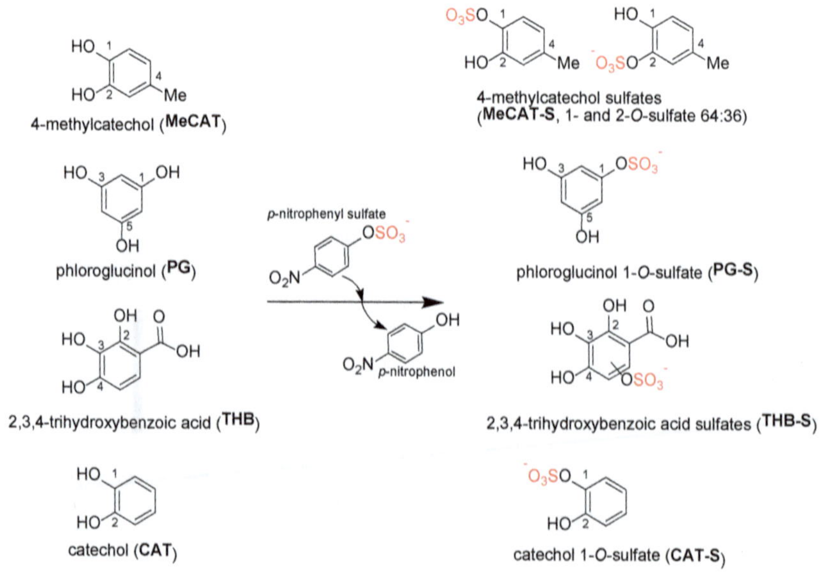

Figure 1. Sulfation of 4-methylcatechol (MeCAT), phloroglucinol (PG), 2,3,4-trihydroxybenzoic acid (THB), and catechol (CAT) using aryl sulfotransferase from *Desulfitobacterium hafniense* and *p*-nitrophenyl sulfate as sulfate donor.

The structures of all phenolic compounds and their respective sulfated derivatives used in the present study are shown in Figures 2–4. Altogether, these compounds form a unique library of sulfated simple phenols (Figure 2a), phenolic acids (Figure 2b), flavonoids (Figure 3), and flavonolignans (Figure 4).

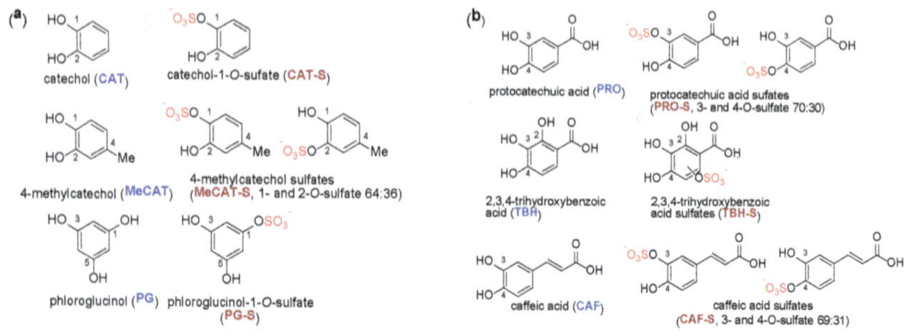

Figure 2. Structures of simple phenols (**a**) and phenolic acids (**b**, both blue) and their sulfates (crimson) used in this study. The sulfate groups are highlighted in red.

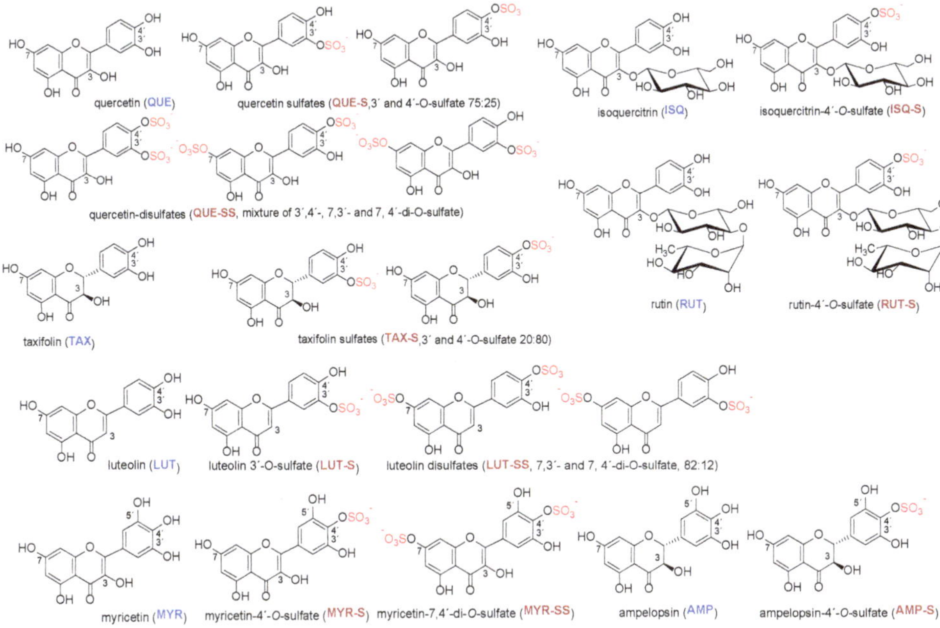

Figure 3. Structures of flavonoids (blue) and their sulfates (crimson) used in this study. The sulfate groups are highlighted in red.

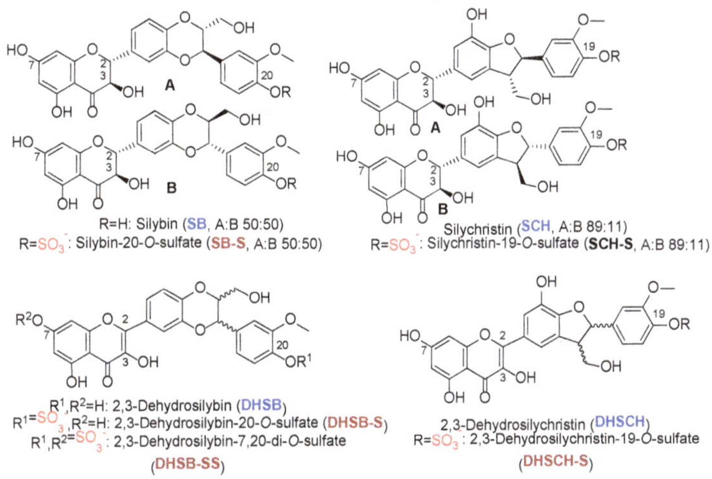

Figure 4. Structures of flavonolignans (blue) and their sulfates (crimson) used in this study. The sulfate groups are highlighted in red.

2.1.1. Sulfation of 4-Methylcatechol (MeCAT)

Sulfation of 4-methylcatechol yielded a product (MeCAT-S) that appeared as one spot on TLC and one peak on HPLC using Method M2. However, NMR analysis revealed a mixture of two isomers: 4-methylcatechol-1-*O*-sulfate and 4-methylcatechol-2-*O*-sulfate in a ratio of 64:36 (Table S8a,b). The reaction (Figure 1) was rapid, with the entire amount of MeCAT-S synthesized within the first 20 min. After purification by gel chromatography, 180 mg of the product was obtained from 200 mg of starting material (isolated yield 58 mol%) as a white powder with an overall purity of 99% (HPLC, Figure S1) and corresponding *m/z* (Figure S2).

2.1.2. Sulfation of Phloroglucinol (PG)

In the case of PG, the formation of at least two products was observed (TLC, HPLC-Method M2, Figure 1), but only one of them was successfully isolated and characterized. According to NMR and MS analyses, it was confirmed to be a monosulfate. The product was isolated as a brownish powder (73 mg, 44% yield, 90% purity, Figures S11 and S12, Table S13).

2.1.3. Sulfation of Protocatechuic Acid (PRO)

When PRO was sulfated, only one sulfated product was observed (TLC, HPLC-Method M2). According to the NMR analysis, it was a mixture of two regioisomers: protocatechuic acid-3-*O*-sulfate and protocatechuic acid-4-*O*-sulfate in the ratio of 7:3. The product tended to associate with solvents and other chemicals used during the purification process (H_2O, MeOH, HCOOH); however, after repeated purification, it was isolated as a white powder (12 mg, 4% yield, 99% HPLC purity, Figure S3, Tables S9a,b).

2.1.4. Sulfation of 2,3,4-Trihydroxybenzoic Acid (THB)

The gradual formation of one product was observed (TLC, HPLC-Method M2, Figure 1). Furthermore, this product tended to associate with solvents. The purification on gel chromatography gave a brown oil product (34 mg, 8% yield, 96% HPLC purity, Figures S5 and S6, Tables S10a,b).

2.1.5. Sulfation of Caffeic Acid (CAF)

The sulfation of caffeic acid yielded a product that behaved as one spot on TLC and one peak on HPLC. Again, NMR analysis revealed a mixture of 3- and 4-*O*-sulfate in the ratio of 69:31. After purification, we obtained 301 mg of white-brownish oily solid (100% yield, 99% HPLC purity, Figures S7 and S8, Table S11a,b).

2.1.6. Sulfation of Catechol (CAT)

The sulfation of catechol (Figure 1) led to surprisingly only one regioisomer, catechol-1-*O*-sulfate, as confirmed by MS and NMR. After purification, 116 mg of the product was obtained (34% yield, 98% HPLC purity, Figures S9 and S10, Table S12) as a white powder.

2.2. *Development and Optimization of Analytical Methods for the Separation of Sulfated Polyphenols*

The typical composition of the reaction mixture to be analyzed was as follows: *p*-NP, *p*-NP-S, the parent compound, and its sulfate (see Figure 1; in some cases, isomers and/or disulfate(s) were also present). The main focus was on monitoring the separation of all components in the mixture, the duration of the separation, and the width of the peaks at five percent of their height. The following limits were established for evaluating the width of the peaks: $w_{0.05} < 0.300$, very good; $0.300 < w_{0.05} < 0.500$, good; and $w_{0.05} > 0.500$, poor.

2.3. *Separations without Buffer in the Mobile Phase*

Based on our previous experience and the articles published so far (Table 1), we decided to test these four columns: Kinetex PFP, ZICpHILIC, Chromolith RP 18e, and Luna Omega Polar C18. Initially, the Kinetex PFP, Chromolith RP 18e, and Luna Omega Polar C18 columns were each tested without the use of buffers in the mobile phase (Table 2). The use of buffers in the mobile phase is always associated with an increased washing effort of the analytical system and thus a higher time requirement, so we tried to avoid this in the first step. Since the samples contain ionizable compounds, formic acid or TFA (0.1%) was added to all mobile phases to achieve a pH of about 3. The non-ionized form is less polar and therefore more strongly retained in a reverse phase system.

Table 2. An overview of the columns used and separation conditions.

Stationary Phase	Method Number	Mobile Phase A	Mobile Phase B	Flow Rate [mL/min]	T [°C]	Gradient
PFP	M1	10 mM CH_3COONH_4/HCOOH (100/0.1, *v*/*v*)	MeOH	0.6	45	0 min 40% B, 0–20 min 40–72% B, 20–21 min 72–40% B, 21–24 min 40%
	M2	10 mM CH_3COONH_4/HCOOH (100/0.1, *v*/*v*)	MeOH	0.6	45	0 min 40% B, 0–20 min 20–50% B, 20–21 min 50–20% B, 21–24 min 20%
	M3	H_2O/CH_3COOF_3 (100/0.1, *v*/*v*)	MeOH	0.6	45	0 min 40% B, 0–25 min 40–80% B, 25–26 min 80–40% B, 26–28 min 40%
ZICpHILIC	M4	AcCN/HCOOH (100/0.1, *v*/*v*)	10 mM CH_3COONH_4/HCOOH (100/0.1, *v*/*v*)	0.4	25	0 min 5% B, 0–7.5 min 5–20% B, 7.5–10 min 20% B, 10–12 min 20–5% B, 12–15 min 5% B, 15–17 5% B
C18	M5	10 mM CH_3COONH_4/HCOOH (100/0.1, *v*/*v*)	MeOH	1	25	0–2 min 5% B, 2–7 min 5–90% B, 7–8 min 90%B, 8–11 min 90–5% B, 11–14 min 5% B
	M6	AcCN/H_2O/HCOOH(5/95/0.1, *v*/*v*)	AcCN/H_2O/HCOOH (80/20/0.1)	1	25	0–5 min 0–30% B, 5–7 min 30–0% B, 7–9 min 0% B
C18 Polar	M7	H_2O, HCOOH (100/0.1, *v*/*v*)	AcCN/H_2O/HCOOH (80/20/0.1)	0.4	25	0–7 min 0–90% B, 7–8 min 90% B, 8–11 min 90–0% B, 11–14 min 0% B

The pentafluorophenyl column offers unique polar and aromatic selectivity thanks to the fluorine atoms at the periphery of the phenyl units anchored to the core-sell silica support. The Kinetex PFP column was tested using Method M3 (0.1% TFA, MeOH as mobile phases) for the separation of quercetin (QUE), isoquercitrin (ISQ), rutin (RUT), taxifolin (TAX), luteolin (LUT), myricetin (MYR), and ampelopsin (AMP) and their sulfates. Although the pH of the mobile phase was very low (pH = 2.0, column limit is 1.5) and the ionization of compounds (sulfates and their parent compounds) should be suppressed, the separation was not satisfactory in many cases. The width of the peak in the 0.5% of its height ($w_{0.05}$) ranged from 0.295 to 1.703 (Table 3). The widest peaks were observed in quercetin and quercetin sulfates ($w_{0.05}$ 0.554–1.703), luteolin sulfates ($w_{0.05}$ 0.655–1.703), and myricetin sulfates ($w_{0.05}$ 0.542–1.398). The values of $w_{0.05}$ for the other tested substances were up to 0.500. No coelution of peaks was observed in the reaction mixtures. The longest retention time was 20.010 min (luteolin), but for most of the analyzed substances, it ranged from 6 to 16 min.

Table 3. Comparison of retention times and peak widths of selected analytes using different columns and methods.

Stationary Phase	PFP [a]				ZICpHILIC [b]		C18				C18-Polar	
Method [c]	M1		M2		M4		M5		M6		M7	
Analyte [d]	t_R [e]	$w_{0.05}$ [f]	t_R [e]	$w_{0.05}$ [f]	t_R [e]	$w_{0.05}$ [f] [min]	t_R [e]	$w_{0.05}$ [f]	t_R [e]	$w_{0.05}$ [f]	t_R [e]	$w_{0.05}$ [f]
DHSB	23.101	0.329	-	-	1.160	0.251	7.293	0.146	7.039	0.177	9.032	0.143
DHSB-S	17.385	0.347	-	-	3.636	0.849	6.914	0.189	6.771	0.608	7.904	0.234
DHSB-SS	8.280	0.525	-	-	8.853	0.588	6.097	0.392	3.180	1.701	8.490	0.146
DHSCH	17.010	0.425	-	-	2.025	1.209	6.691	0.108	6.957	0.101	7.573	0.125
DHSCH-S	12.060	0.576	-	-	5.448	0.465	6.258	0.177	6.771	0.478	6.768	0.322
SCH	12.449, 13.285 [i]	0.299, 0.312	-	-	3.047	1.210	6.002	n.d. [g]	5.753	0.156	7.413	0.132
SCH-S	6.176, 6.654 [h]	0.338	-	-	6.332	0.540	5.552	n.d. [g]	4.985	0.393	6.313	0.395
SB	15.202, 15.521 [i]	n.d. [g]	-	-	1.094	0.647 [i]	6.480, 6.514 [i]	n.d. [g]	4.141, 4.225 [i]	n.d. [g]	7.410	0.186
SB-S	13.728, 14.479 [h]	0.287 0.415	-	-	2.470, 3.007, 4.365 [j]	n.d. [g]	6.487, 6.579 [h]	n.d. [g]	3.165	0.981	6.555	0.186
CAF	5.150	0.238	12.527	0.356	2.540	0.660	4.954	0.102	3.087	0.211	5.987	0.270
CAF-S	3.720	0.267	9.233	0.671	4.194	0.528	4.326, 4.494 [j]	n.d. [g]	5.847	2.555	5.366	1.040
PRO	3.721	0.270	7.213	0.337	3.184	0.787	2.816	0.410	1.822	0.237	3.653	0.471
PRO-S	2.990	0.225	5.541	0.403	6.255	0.558	2.104	0.276	4.697	2.579	2.500	0.920
THB	3.568	0.246	6.240	0.368	4.456	0.638	1.944	0.175	2.058	0.259	4.122	0.383
THB-S	2.693, 3.033 [j]	0.329, 0.266	4.949	0.317	n.d. [k]	-	n.d. [l]	-	0.718	0.050	2.966	0.582
CAT	4.578	0.256	7.453	0.350	1.257	0.270	2.920	0.356	2.317	0.246	4.331	0.403
CAT-S	3.240	0.207	4.938	0.386	n.d. [k]		4.290	n.d. [g]	5.109	2.564	7.627	0.305
MeCAT	5.978	0.335	11.944	0.508	0.963	n.d. [g]	4.946	0.226	3.769	0.300	6.302	0.261
MeCAT-S	4.112	0.249	7.910	0.524	1.049	0.314	4.362 [j]	n.d. [g]	6.578	2.546	5.536	0.507
PG	2.906	0.205	4.004	0.235	4.641	0.430	1.180	0.288	1.039	0.197	1.641	0.310
PG-S	2.503	0.177	3.339	0.266	9.500	0.732	1.465	0.567	2.683	1.534	1.265	0.308
*p*NP	11.528	0.338	20.542	0.385	0.984	0.273	5.609	0.169	5.031	0.221	7.274	0.278
*p*NP-S	4.930	0.457	10.401	0.498	1.494	0.600	5.216	0.900	0.796	0.140	5.567	0.621

[a] Kinetex pentafluorophenyl, [b] hydrophilic interaction chromatography, [c] for details on the individual methods, see Table 2, [d] full names and structures of the analytes are shown at Figures 1–3, [e] retention time, [f] the width of the peak in 5% of its height, [g] the peak shape did not allow the determination of $w_{0.05}$, [h] separation of sulfated stereoisomers A and B, [i] partial separation of stereoisomers A and B, [j] partial separation of sulfated regioisomers, [k] the compound was decomposed during the analysis, only the parent compound without sulfate was detected, [l] the compound was not caught on the column and eluted with a dead volume. Dark green means $w_{0.05} < 0.300$, light green means $0.300 < w_{0.05} < 0.500$, and red means $w_{0.05} > 0.500$.

We also tested two C18 columns, one monolithic (Chromolith RP18e, Method M6), and one specially designed for increased retention of polar compounds (Luna Omega Polar C18, Method M7, Table 2). The mobile phase was the same in both cases (acetonitrile 5% and 80%, water, 0.1% formic acid, pH 2.7; the pH limits of the columns are 2.0 and 1.5, respectively). The Luna Omega Polar C18 stationary phase is a combination of a universal C18 ligand and a polar modified surface that provides improved polar retention and aqueous stability. Due to the smaller particle size of the stationary phase (3 μm) and the associated higher backpressure, the flow rate was only 0.4 mL/min. In contrast, the monolithic C18 column has much higher permeability and porosity, so the flow rate was 1 mL/min.

Nevertheless, for most analytes, we observed a lower $w_{0.05}$ of the peaks using this C18 Polar column compared to the monolithic column, or the widths of the peaks were very similar on both columns (e.g., 2,3-dehydrosilychristin DHSCH, caffeic acid CAF or *p*-NP). The width of the peaks was very good for 2,3-dehydrosilybin DHSB, 2,3-dehydrosilybin-20-*O*-sulfate DHSB-S, and 2,3-dehydrosilybin-7,20-di-*O*-sulfate DHSB-SS, further for DHSCH, silychristin SCH, silybin SB, caffeic acid CAF, 4-methylcatechol MeCAT, and *p*-NP; good for 2,3-dehydrosilychristin-19-*O*-sulfate DHSCH-S, silychristin-19-*O*-sulfate SCH-S, protocatechuic acid PRO, 2,3,4-trihydroxybenzoic acid THB, catechol CAT, catechol-1-*O*-sulfate CAT-S, phloroglucinol PG, and its sulfate. Broader peaks were detected for five compounds, namely CAF-S, PRO-S, THB-S, MeCAT-S, and *p*-NP-S. In contrast, eight broad peaks were detected in the monolithic C18 column, most of which were sulfates (DHSB-S, DHSB-SS, SB-S, CAF-S, PRO-S, CAT-S, MeCAT-S, and PG-S). With the monolithic column, coelution of the peaks was observed in the case of *p*-NP-S and THB-S in the respective reaction mixtures.

The retention times of the eluted compounds were up to 7 min. With the Polar C18 column, coelution of the peaks was observed only in the case of *p*-NP-S and MeCAT-S. The retention times of the eluted compounds were up to 9 min. Thus, when we compare the two C18 columns, the separation on the C18 Polar column shows narrower peaks and also a lower consumption of the mobile phase due to the lower flow rate (0.4 mL/min). In addition, this method is directly applicable in conjunction with a mass detector. The method with the monolithic column (M6) would require further modification in case of connection with the mass detector (flow rate 1 mL/min). The comparison of HPLC chromatograms of individual compounds is in Supplementary Material (Figures S13–S51).

2.4. *Separation with Buffer in the Mobile Phase*

As can be seen in the example of the separations of polyphenol sulfates on the PFP column and the C18 columns mentioned above, it is clear that in some cases, the mere acidification of the mobile phase is not sufficient to deionize ionizable compounds (broad peaks). Fine-tuning of the separation of highly polar sulfated polyphenols can be achieved by changes in the mobile phase, e.g., by choice of buffer, and thus consistent control of pH. The most commonly used buffers for HPLC with UV detection are acetate and phosphate. Since phosphate buffer is not compatible with MS detection, we chose ammonium acetate buffer with a pH of 3.8, and the buffer strength was set at 10 mM. All methods tested with 10 mM buffer were also tested with 5 mM concentration on several samples, but the width of the peaks was broader than with 10 mM buffer in all cases (data not shown).

Separation of quercetin (QUE), ampelopsin (AMP), luteolin (LUT), myricetin (MYR), isoquercitrin (ISQ), rutin (RUT), taxifolin (TAX), and their respective sulfates on PFP column using 10 mM ammonium acetate buffer and methanol as mobile phases (Method M1) showed good separation of most compounds ($0.300 < w_{0.05} < 0.500$). Only in two cases (QSS and AMP) was this range slightly exceeded (0.642 and 0.519, respectively). Moreover, the retention time of most compounds was shorter than in Method M3 (Table 3). No coelution of peaks was observed in the reaction mixtures. When we compare the separation of these compounds on the same PFP column, it is clear that we obtained better or equal separation with the acetate buffer method (Method M1) than with the method without acetate buffer (Method M3) in all cases except for ampelopsin ($w_{0.05}$ 0.519 versus 0.314, Table 4).

Table 4. Comparison of retention times and peak widths of selected analytes using PFP column with (M1) and without buffer (M3).

Method	M1 [a]		M3 [b]	
Analyte [c]	t_R [d] [min]	$w_{0.05}$ [e] [min]	t_R [d] [min]	$w_{0.05}$ [e] [min]
QUE	16.127	0.385	17.470	0.554
QUE-S	12.139	0.397	15.084	0.632
QUE-SS	6.014	0.642	11.084	1.703
AMP	5.480	0.519	6.062	0.314
AMP-S	6.007	0.352	7.449	0.437
LUT	17.916	0.427	20.010	0.422
LUT-S	13.340	0.377	16.741	0.501
LUT-SS	12.830	0.362	13.595	1.426
MYR	12.716	0.390	14.014	0.410
M-S	8.566	0.478	15.877	0.542
M-SS	4.602	0.400	12.497	1.398
ISQ	9.397	0.344	9.615	0.312
ISQ-S	6.681	0.311	8.247	0.381
RUT	8.869	0.342	9.156	0.322
RUT-S	6.055	0.261	7.175	0.287
TAX	7.380	0.285	7.767	0.466
TAX-S	5.582	0.321	7.557	0.368
*p*NP	11.528	0.338	13.414	0.367
*p*NP-S	4.930	0.457	6.751	0.295

[a] With ammonium acetate buffer, [b] with 0.1% TFA (for details, see Table 2), [c] full names and structures of the analytes are shown in Figures 1–3, [d] retention times, [e] width of the peak in 5% of its height; dark green means $w_{0.05} < 0.300$, light green means $0.300 < w_{0.05} < 0.500$, and red means $w_{0.05} > 0.500$.

In addition, we tested the separation of other polyphenolic compounds and their sulfates on this column so that we can compare it with other stationary phases used in this work. The separation of CAF, CAF-S, PRO, PRO-S, THB, THB-S, CAT, CAT-S, MeCAT-S, and PG-S was very good. The separation of DHSB-S, DHSCH, SCH, SCH-S, SB-S, MeCAT, *p*-NP, and *p*-NP-S was good; only the peaks of DHSB-SS and DHSCH-S were slightly broader ($w_{0.05}$ was 0.525 and 0.576, respectively). Retention times of analytes ranged from 2 to 23 min; more than half of them were less than 10 min. No coelution of peaks in reaction mixtures was observed. From our previous unpublished experiments, we know that this gradient in method M1 is too fast for many phenolic acids, benzoic acids, and their sulfates. Therefore, we tried to modify this method. We reduced the initial concentration of mobile phase B from 40 to 20% and the final concentration to 50% (instead of the original 72%), and the length of the gradient was maintained (Method M2). However, we did not observe any significant improvement in $w_{0.05}$, only the retention times were longer compared with Method 1.

Separation of phenolic compounds and their sulfates on the monolithic C18 column was also performed in 10 mM ammonium acetate buffer as the mobile phase (Method M5). The separation of many compounds was very good or good using this column (DHSB and its sulfates, DHSCH, DHSCH-S, CAF, PRO, PRO-S, THB, CAT, MeCAT, PG, and *p*-NP). However, for several compounds (SCH, SCH-S, SB, SB-S, CAF-S, CAT-S, MeCAt-S), the width of the peaks could not be determined because of their unusual shape (double hunch, see Section 2.5). The peak of THB-S was not caught on the stationary phase and eluted with the dead volume. The retention times of all compounds were up to 7 min. Coelution of *p*-NP and SCH-S was observed in the reaction mixture.

The Zic-pHILIC column was designed by the manufacturer for difficult separations of polar hydrophilic compounds. It is a polymer-based column with densely bound zwitterionic functional groups with a charged equilibrium of 1:1 and the lowest pH stability of 2. The mobile phases were acetonitrile and 10 mM ammonium acetate, both acidified with formic acid (pH 3.8), and the compounds were separated by gradient elution (Method M4). Almost all substances analyzed showed broad peaks except for DHSB, DHSCH-S, CAT-S, MeCAT-S, PG, and *p*-NP. Retention times were not measured for THB-S and CAT-S because the sulfated substance is likely degraded to the parent substance on the column. Retention times ranged from 0.9 to 9.5 min. Coelution of *p*-NP and SB, *p*-NP-S, and CAT was observed in the reaction mixture. The comparison of HPLC chromatograms of individual compounds is in Supplementary Material (Figures S19–S51).

2.5. *Separation of Regioisomers and Stereoisomers*

NMR analyses have shown that in some cases, multiple regioisomers are formed in enzymatic sulfation reactions, as in the case of QUE-S, QUE-SS, TAX-S, CAF-S, MeCAT-S, PRO-S, and THB-S. Separation of these regioisomers was achieved on a monolithic C18 column (Method M5) in CAF-S (peaks at 4.326 min and 4.494 min, resolution 1.251) and in MeCAT-S (double hunch peak at 4.362 min, resolution 0.540). The separation of two THB-S regioisomers was observed in Method M1 (peaks at 2.827 min and 3.033 min, resolution 1.944). The separation of two sulfated regioisomers was previously published on a C18 UHPLC column with MS/MS detection (3-phenylpropionic acid 4′-*O*-sulfate (RT 4.92 min) and 3-phenylpropionic acid 3′-*O*-sulfate (RT = 5.22 min) or caffeic acid-4′-*O*-sulfate (RT = 3.97 min) and caffeic acid-3′-*O*-sulfate) [2]. Baseline separation of disulfate regioisomers of myricetin was reported on the PFP column (mobile phase 0.1% trifluoroacetic acid in water and methanol) [8].

Since silybin is a mixture of two diastereoisomers, silybin A and B (in a ratio of approximately 1:1), we observed the separation of these two diastereoisomers in methods M1, M5, and M6. Sulfated diastereoisomers of silybin were separated by Methods M1 and M5. The resolution of the peaks was 1.365 (M1), 0.2 (M5), and 0.02 (M6). In the case of Method M4, we observed even three peaks (partially separated), and all of them have the same absorption maximum (286 nm). In this case, it is probably peak cleavage due to the unsuitability of the stationary phase for the separation of polyphenolic substances, as can be seen from the large peak width of most of the compounds examined. Although silychristin is also a mixture of two diastereoisomers A and B, we observed their separation only on the PFP column (M1). The separation of sulfated silychristin diastereoisomers was observed with the PFP column and Method M1 (Table 3). The separation of diastereoisomers of silychristin or silybin at the reverse phase is described in many papers, e.g., [25–27], but the separation of sulfated diastereoisomers have not been described so far.

2.6. Selection of the Best Method

In this study, we compared four types of stationary phases, namely pentafluorophenyl, ZICpHILIC, monolithic C18, and C18 stationary phases, with treatment for better polar retention. We combined these stationary phases with either a mobile phase without a buffer or with a buffer. The combination of pentafluorophenyl stationary phase and 10 mM acetate buffer/methanol in gradient elution proved to be the best. This resulted in sharp peaks for almost all test compounds without tailing and very good separation of all components in the mixture for up to 20 min (Figure 5). The other columns/mobile phases tested were only suitable for some of the compounds analyzed. The retention times and peak widths of all compounds tested are shown in Tables 3 and 4. The peak shapes and comparisons of HPLC chromatograms of individual compounds in all methods tested can be found in Supplementary Materials (Figures S13–S51).

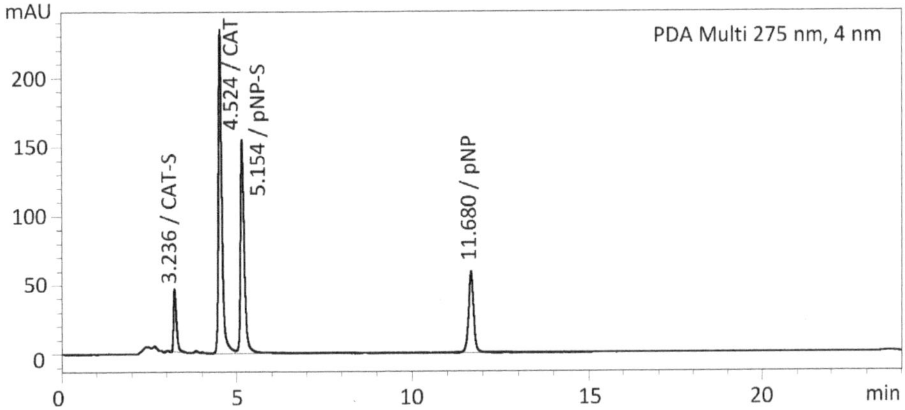

Figure 5. HPLC chromatogram—an example of typical composition of the enzymatic reaction mixture to be analyzed: *p*-NP, *p*-NP-S, parent compound (CAT), and its sulfate (CAT-S), Method M1.

2.7. Method Validation

Due to the large number of samples measured and columns tested, one representative from the group of sulfated phenols, phenolic acids, flavonoids, and flavonolignans, namely 4-methylcatechol sulfate (MeCAT-S), caffeic acid sulfate (CAF-S), ampelopsin sulfate (AMP-S), and silychristin sulfate (SCH-S), was selected for the validation of Method M1 (the most universal PFP column and mobile phase for all samples tested). The linearity, limit of detection, the limit of quantification, precision, accuracy, recovery and repeatability are summarized in Table 5.

Table 5. The linearity, limit of detection (LOD), limit of quantification (LOQ), intermediate precision, repeatability, accuracy, and recovery for four representatives of sulfated phenol (MeCAT-S), phenolic acid (CAF-S), flavonoids (AMP-S), and flavonolignans (SCH-S).

Sample	Regression Equation	R^2	LOD [mM]	LOQ [mM]	Repeatability [%] [a]	Intermediate Precision [%] [a]	Accuracy [%] [b]	Recovery [%]
MeCAT-S	y = 161,221 × c + 17,188	0.9999	0.032	0.108	1.55	2.18	1.3	104
CAF-S	y = 26,266 × c + 4120	0.9998	0.560	1.680	3.63	8.13	2.2	104
AMP-S	y = 96,081 × c	0.9977	0.061	0.202	1.74	6.30	2.4	103
SCH-S	y = 127,596 × c + 27,799	0.9999	0.340	1.020	1.78	9.04	3.3	105

R^2 correlation coefficient; [a] expressed as relative standard deviation, n = 6; [b] expressed as relative standard deviation, n = 3.

The calibration curves were linear in the range from 0.625 to 50 mM for all analytes. The correlation coefficients were greater than 0.9997 in all cases (except AMP-S, 0.9977), demonstrating a high degree of correlation and good linearity of the method. LOD and LOQ ranged from 0.032 to 1.680 mM. This indicates that our method has adequate sensitivity. The ranges of %RSD parameters for repeatability (intra-day precision) and intermediate (inter-day) precision were 1.55 to 3.63 and 2.18 to 9.04, respectively. The accuracy ranges were from 1.3 (MeCAT-S) to 3.3% (SCH-S), and recoveries were in the range of 103–104% for all tested samples. Sample stability was evaluated by storing unprocessed samples at ambient temperature up to 24 h and freeze/thaw cycles after three cycles at −18 °C. The experiments indicated that all four analytes tested were stable in the period of 24 h, as the recoveries ranged between 97–102% for SCH-S, 94–104% for CAF-S, 98–103% for AMP-S, and 97–102% for MeCAT-S.

3. Materials and Methods

3.1. Material

Acetonitrile, methanol, formic acid (all VWR chemicals, Stříbrná Skalice, Czech Republic analytical grade), deionized water (Ultrapure, Watrex, Prague, Czech Republic), ammonium acetate (Lach-Ner, Neratovice, Czech Republic), ampelopsin (Herb Nutritionals, Shanghai, China), 3,4-dihydroxycinamic acid (caffeic acid), *p*-nitrophenol sulfate, 2,3,4-trihydroxybenzoic acid, 3,4-dihydroxybenzoic (protocatechuic) acid, catechol (all Acros Organics, Thermo Fisher Scientific, Waltham, MA, USA), 4-methylcatechol (Aldrich, Merck KGaA, Darmstadt, Germany), quercetin (Sigma, Merck KGaA, Darmstadt, Germany), phloroglucinol (Alfa Aesar, Haverhill, MA, USA), *p*-nitrophenol, luteolin, and myricetin (abcr GmbH, Karlsruhe, Germany). Standards of silybin, 2,3-dehydrosilybin, and silychristin were prepared and fully characterized in the Laboratory of Biotransformation, Institute of Microbiology, Prague, Czech Republic [28–30].

The description of the preparation and full characterization (NMR, HRMS, and HPLC) of the following sulfates have already been published: silybin A-20-*O*-sulfate and silybin B-20-*O*-sulfate (SB-S, 50:50) [10], silychristin-19-*O*-sulfate (SCH-S), 2,3-dehydrosilybin-20-*O*-sulfate (DHSB-S), 2,3-dehydrosilychristin-19-*O*-sulfate (DHSCH-S) [13], quercetin-3′-*O*-sulfate and quercetin-4′-*O*-sulfate (QUE-S, 75:25) [11,12,19], rutin-4′-*O*-sulfate (RUT-S), taxifolin-4′-*O*-sulfate and taxifolin-3′-*O*-sulfate (TAX-S, 80:20), isoquercitrin-4′-*O*-sulfate (ISQ-S) [11,19], ampelopsin-4′-*O*-sulfate (AMP-S), luteolin-3′-*O*-sulfate (LUT-S), luteolin-7, 3′- and 7, 4′-di-*O*-sulfates (LUT-SS), myricetin-4′-O-sulfate (MYR-S), myricetin-di-7, and 4′-*O*-sulfate (MYR-SS) [8].

3.2. Preparation of the Enzyme

The aryl sulfotransferase from *Desulfitobacterium hafniense* used for the sulfation was prepared as described previously in our works [11,12].

3.3. General Method for the Preparation of Sulfates

The substrate (MeCAT, PRO, THB, CAF, CAT, PG; 200 mg of each, 1 eq) was dissolved in 5 mL of acetone in a flask and then *p*-NPS (25 mg/mL, 1 or 2 eq in 100 mM Tris-glycine buffer pH 8.9), 24 mL of Tris-glycine buffer, and 2 mL of AST enzyme (360 U/mL) were added. The reaction mixture was then incubated under an inert atmosphere (Ar) at 30 °C for 5 h. The monitoring of the reaction was performed using TLC (mobile phase ethyl acetate/chloroform/trifluoroacetic acid, 16/1/0.01). After incubation, the reaction mixture was heated up to 95 °C to terminate the enzymatic reaction and stored at −20 °C until purification.

3.4. Purification of Sulfates

In the case of MeCAT, PRO, THB, CAF, and CAT, the reaction mixture was partially evaporated in a rotary evaporator to remove acetone from the mixture. The pH was adjusted to 7.5–7.7 using formic acid, and then the mixture was extracted with ethyl acetate (3 × 50 mL) to remove *p*-nitrophenol (control by TLC, mobile phase ethyl acetate/chloroform/trifluoroacetic acid, 16/1/0.01). The aqueous phase was then evaporated, dissolved in 2–5 mL of 80% methanol, centrifuged (5000 × rpm, 20 min), and loaded onto a Sephadex LH-20 column (GE Healthcare Bio-Sciences, Uppsala, Sweden; 30 g of dry weight, 3 cm i.d.) with 80% methanol as a mobile phase (0.2 mL/min). The elution time was usually 2 days, and the fraction detection was performed by TLC (ethyl acetate/chloroform/trifluoroacetic acid, 16/1/0.01). Purification of the phloroglucinol reaction mixture was performed using preparative HPLC (Section 3.5) using an ASAHIPAK GS-310 20F column (Shodex, Munich, Germany). The reaction mixture (100 mg) was dissolved in 1 mL of 50% methanol, filtered, and injected onto the column (5 mL/min, 25 °C, detection at 254 and 369 nm). The fractions containing the desired product were then joined, fully evaporated, and lyophilized. Low purity fractions were re-purified using the same methodology.

3.5. Preparative HPLC

The preparative HPLC (Shimadzu, Kyoto, Japan) system consisted of an LC-8A high-pressure pump with an SPD-20A dual-wavelength detector (with preparative cell), FRC-10A, and fraction collector. The system was connected to a PC using a CBM20A command module and controlled by the LabSolution 1.24 SPI software suite supplied with the instrument.

3.6. Mass Spectrometry (MS)

The samples were dissolved in MeOH and introduced into the mobile phase flow (MeOH/H2O 4:1; 100 μL/min) using a 2 μL loop. Spray voltage, capillary voltage, tube lens voltage, and capillary temperature were 4.0 kV, −16 V, −120 V, and 275 °C, respectively.

3.7. Nuclear Magnetic Resonance (NMR)

NMR spectra were recorded on a Bruker Avance III 600 MHz and 400 MHz spectrometers at 30 °C in dimethylsulfoxide (DMSO-d_6); residual solvent signal (δ_H 2.499 ppm, δ_C 39.46 ppm) served as an internal standard. NMR experiments: ^{1}H NMR, ^{13}C NMR, gCOSY, gHSQC, and gHMBC were performed using the standard manufacturer's software. The position of sulfate attachment was determined using typical changes in chemical shifts of the attached and adjacent carbons (compared with starting acceptors) as described in [11].

3.8. Analytical HPLC System

The Shimadzu Prominence LC analytical system comprised Shimadzu CBM-20A system controller, Shimadzu LC-20AD binary HPLC pump, Shimadzu CTO-10AS column oven, Shimadzu SIL-20ACHT cooling autosampler, and Shimadzu SPD-20MA diode array detector (Shimadzu, Kyoto, Japan).

3.9. Analytical Columns and Mobile Phases

Kinetex PFP column (150 × 4.6 mm, 5 μm), guard column PFP (4 × 3 mm, 5 μm), both Phenomenex (USA); Method M1: mobile phase A = 10 mM ammonium acetate, 0.1% HCOOH, pH 3.8; mobile phase B = 100% MeOH; gradient elution: 0 min 40% B, 0–20 min 40–72% B, 20–21 min 72–40% B, 21–24 min 40% for equilibration of the column; flow rate 0.6 mL/min, 45 °C, PDA detection (200–400 nm), the wavelength of the absorption maximum of the respective compound was extracted. The method with the same mobile phases, flow rate, temperature, and detection but a different gradient was used for the analysis of phenolic acids PRO, THB, CAF, and also for CAT, MeCAT, PG, respectively. (Method M2): 0 min 20% B, 0–20 min 20–50% B, 20–21 min 50–20% B, 21–24 min 20% B for equilibration of the column. Another method (Method M3) that has also been tested with this column: mobile phase A = 0.1% TFA (pH = 2.0), B = 100% MeOH; gradient elution 0 min 40% B, 0–25 min 40–80% B, 25–26 min 80–40% B, 26–28 min 40% for equilibration of the column; flow rate 0.6 mL/min, 45 °C.

ZicpHILIC column (100 × 2.1 mm, 5 μm), guard column ZicHILIC (20 × 2.1 mm, 5 μm), both Merck (DE); Method M4: mobile phase A = 100% acetonitrile, 0.1% HCOOH; mobile phase B = 10 mM ammonium acetate, 0.1% HCOOH, pH 3.8; gradient elution: 0 min 5% B, 0–7.5 min 5–20% B, 7.5–10 min 20% B, 10–12 min 20–5% B, 12–15 min 5% B, 15–17 5% B for equilibration of the column; flow rate 0.4 mL/min, 25 °C, PDA detection (200–400 nm), the wavelength of the absorption maximum of the respective compound was extracted.

Chromolith RP 18e column (100 × 3 mm, monolith), guard column Chromolith RP 18-e (5 × 4.6 mm, monolitith), both Merck (DE); Method M5: mobile phase A = 10 mM ammonium acetate, 0.1% HCOOH, pH 3.8, mobile phase B = 100% MeOH; gradient elution: 0–2 min 5% B, 2–7 min 5–90% B, 7–8 min 90%B, 8–11 min 90–5% B, 11–14 min 5% B for equilibration of the column; flow rate 1 mL/min, 25 °C, PDA detection (200–400 nm), the wavelength of the absorption maximum of the respective compound was extracted. Another method (Method M6) has also been tested with this column: mobile phase A = 5% acetonitrile, 0.1% HCOOH, B = 80% acetonitrile, 0.1% HCOOH; gradient elution 0–5 min 0–30% B, 5–7 min 30–0% B, 7–9 min 0% B for equilibration of the column; flow rate 1.0 mL/min, 25 °C.

Luna Omega Polar C18 column (100 × 2.1 mm, 3 μm), guard column Polar C18 for 2.1 mm ID, both Phenomenex (USA); Method M7: mobile phase A = 5% acetonitrile, 0.1% HCOOH, mobile phase B = 80% acetonitrile, 0.1% HCOOH; gradient elution: 0–7 min 0–90% B, 7–8 min 90% B, 8–11 min 90–0% B, 11–14 min 0% B for equilibration of the column; flow rate 0.4 mL/min, 45 °C, PDA detection (200–400 nm), the wavelength of the absorption maximum of the respective compound was extracted.

All of the above methods, where 10 mM ammonium acetate was used in the mobile phase, were also tested with 5 mM ammonium acetate under the same conditions. An overview of the methods used is in Table 2.

3.10. Sample Preparation

Samples of individual sulfates were dissolved in water or parent compounds in MeOH/water (1/1, *v/v*) (1 mg/mL), centrifugated, and injected (1 μL). The reaction mixtures (10 μL) were diluted by water (50 μL), centrifuged, and injected (1 μL).

3.11. Method Validation

Four representatives from the group of phenol, phenolic acid, flavonoid, and flavonolignan sulfates, namely 4-methylcatechol sulfate (MeCAT-S), caffeic acid sulfate (CAF-S), ampelopsin sulfate (AMP-S), and silychristin sulfate (SCH-S), were selected for validation on the PFP column (the most universal column for all samples tested, Method M1, see Table 2).

3.11.1. Linearity

Linearity was evaluated by measuring five concentrations of each analyte in duplicates. The concentrations of the analytes prepared in the volumetric flasks were 50, 25, 12.5, 6.25, and 0.625 mM. The results were examined for a linear relationship by plotting the peak area and the corresponding concentrations of the analyte, followed by linear least squares regression and calculation of the slope and correlation coefficient.

3.11.2. Limit of Detection (LOD) and Limit of Quantification (LOQ)

The values of LOQ and LOD were determined according to the following equations: LOD = 3 × noise/slope of the corresponding calibration curve, LOQ = 10 × noise/slope of the corresponding calibration curve.

3.11.3. Precision and Repeatability

Intra-day precision (repeatability) and inter-day precision (intermediate precision) were calculated by analyzing four selected samples at the concentration of 12.5 mM. Results were expressed as the relative standard derivation (RSD). Repeatability measurement was performed on the same day in six replicates; the intermediate precision was measured on six different days.

4. Conclusions

We have developed a robust HPLC analytical method suitable for the separation of enzymatic sulfation reaction mixtures of flavonoids, dehydroflavonoids, phenolic acids, and catechols with PDA detection. This method is based on the combination of pentafluorophenyl stationary phase and 10 mM acetate buffer/methanol in gradient elution. Moreover, the low flow rate (0.6 mL/min) and the absence of phosphate buffer and/or ion-pairing reagents in the mobile phase make the method directly applicable in combination with mass detection. Last but not least, four authentic standards of 2,3,4-trihydroxybenzoic acid sulfates, catechol sulfate, 4-methylcatechol sulfate, and phloroglucinol sulfate were prepared in this work.

Supplementary Materials: The following are available online at https://www.mdpi.com/article/10.3390/ijms23105743/s1.

Author Contributions: Conceptualization, L.P., V.K. and K.V.; data curation, K.K.; formal analysis, L.P., A.H. and H.P.; funding acquisition, L.P. and K.V.; investigation, L.P., K.K., K.B., A.H., B.P. and H.P.; methodology, L.P.; project administration, K.V.; resources, L.P.; supervision, L.P., V.K. and K.V.; validation, L.P.; visualization, A.H.; writing—original draft, L.P.; writing—review and editing, L.P., K.K., K.B., A.H., B.P., H.P., V.K. and K.V. All authors have read and agreed to the published version of the manuscript.

Funding: This research was funded by the Czech Science Foundation, grant number 19-00043S, and Czech Health Research Council, grant number NU21-02-00135.

Institutional Review Board Statement: Not applicable.

Informed Consent Statement: Not applicable.

Data Availability Statement: The data presented in this study are available in the article or supplementary material.

Conflicts of Interest: The authors declare no conflict of interest.

Abbreviations

AMP	ampelopsin
AMP-S	ampelopsin-4′-O-sulfate
CAF	caffeic acid
CAF-S	caffeic acid 3- and 4-O-sulfate (69:31)
CAT	catechol
CAT-S	catechol-O-sulfate
DHSB	2,3-dehydrosilybin
DHSB-S	2,3-dehydrosilybin-20-O-sulfate
DHSB-SS	2,3-dehydrosilybin-7,20-di-O-sulfate
DHSCH	2,3-dehydrosilychristin
DHSCH-S	2,3-dehydrosilychristin-19-O-sulfate
ISQ	isoquercitrin
ISQ-S	isoquercitrin-4′-O-sulfate
LUT	luteolin
LUT-S	luteolin-3′-O-sulfate
LUT-SS	luteolin-7,3′- and 7, 4′-di-O-sulfates (82:12)
MeCAT	4-methylcatechol
MeCAT-S	4-methylcatechol-1- and 2-O-sulfate (64:36)
MYR	myricetin
MYR-S	myricetin-4′-O-sulfate
MYR-SS	myricetin-7,4′-di-O-sulfate
p-NP	*p*-nitrophenol
p-NPS	*p*-nitrophenyl sulfate
PG	phloroglucinol
PG-S	phloroglucinol-O-sulfate
PRO	protocatechuic acid
PRO-S	protocatechuic acid 3- and 4-O-sulfates (70:30)
QUE	quercetin
QUE-S	quercetin-3′- and 4′-O-sulfate (75:25)
QUE-SS	quercetin- 3′,4′-, 7,3′- and 7, 4′-di-O-sulfate
RUT	rutin
RUT-S	rutin-4′-O-sulfate
SB	silybin A and B (50:50)
SB-S	silybin A-20-O-sulfate and silybin B-20-O-sulfate (50:50)
SCH	silychristin A and B (90:10)
SCH-S	silychristin-19-O-sulfate
THB	2,3,4-trihydroxybenzoic acid
TX	taxifolin
TX-S	taxifolin-4′- and 3′-O-sulfate (80:20)

References

1. Kerimi, A.; Williamson, G. At the interface of antioxidant signalling and cellular function: Key polyphenol effects. *Mol. Nutr. Food Res.* **2016**, *60*, 1770–1788. [CrossRef] [PubMed]
2. Guo, X.; Li, K.; Guo, A.; Li, E. Intestinal absorption and distribution of naringin, hesperidin, and their metabolites in mice. *J. Funct. Foods* **2020**, *74*, 104158. [CrossRef]
3. Křen, V.; Marhol, P.; Purchartová, K.; Gabrielová, E.; Modriansky, M. Biotransformation of Silybin and its Congeners. *Curr. Drug Metab.* **2013**, *14*, 1009–1021. [CrossRef] [PubMed]
4. Correia-da-Silva, M.; Sousa, E.; Pinto, M.M.M. Emerging sulfated flavonoids and other polyphenols as drugs: Nature as an inspiration. *Med. Res. Rev.* **2014**, *34*, 223–279. [CrossRef] [PubMed]
5. Chapman, E.; Best, M.D.; Hanson, S.R.; Wong, C.-H. Sulfotransferases: Structure, mechanism, biological activity, inhibition, and synthetic utility. *Angew. Chem. Int. Ed.* **2004**, *43*, 3526–3548. [CrossRef] [PubMed]
6. Hartog, A.F.; Wever, R. Substrate engineering and its synthetic utility in the sulfation of primary aliphatic alcohol groups by a bacterial arylsulfotransferase. *Adv. Synth. Catal.* **2015**, *357*, 2629–2632. [CrossRef]
7. Hartog, A.F.; Wever, R. Sulfation made easy: A new versatile donor for enzymatic sulfation by a bacterial arylsulfotransferase. *J. Mol. Catal. B-Enzym.* **2016**, *129*, 43–46. [CrossRef]

8. Káňová, K.; Petrásková, L.; Pelantová, H.; Rybková, Z.; Malachová, K.; Cvačka, J.; Křen, V.; Valentová, K. Sulfated metabolites of luteolin, myricetin, and ampelopsin: Chemoenzymatic preparation and biophysical properties. *J. Agric. Food Chem.* **2020**, *68*, 11197–11206. [CrossRef]
9. Marhol, P.; Hartog, A.F.; van der Horst, M.A.; Wever, R.; Purchartová, K.; Fuksová, K.; Kuzma, M.; Cvačka, J.; Křen, V. Preparation of silybin and isosilybin sulfates by sulfotransferase from *Desulfitobacterium hafniense*. *J. Mol. Catal. B-Enzym.* **2013**, *89*, 24–27. [CrossRef]
10. Purchartová, K.; Engels, L.; Marhol, P.; Šulc, M.; Kuzma, M.; Slámová, K.; Elling, L.; Křen, V. Enzymatic preparation of silybin phase II metabolites: Sulfation using aryl sulfotransferase from rat liver. *Appl. Microbiol. Biotechnol.* **2013**, *97*, 10391–10398. [CrossRef]
11. Purchartová, K.; Valentová, K.; Pelantová, H.; Marhol, P.; Cvačka, J.; Havlíček, L.; Křenková, A.; Vavříková, E.; Biedermann, D.; Chambers, C.S.; et al. Prokaryotic and eukaryotic aryl sulfotransferases: Sulfation of quercetin and its derivatives. *ChemCatChem* **2015**, *7*, 3152–3162. [CrossRef]
12. Valentová, K.; Káňová, K.; Di Meo, F.; Pelantová, H.; Chambers, C.S.; Rydlová, L.; Petrásková, L.; Křenková, A.; Cvačka, J.; Trouillas, P.; et al. Chemoenzymatic preparation and biophysical properties of sulfated quercetin metabolites. *Int. J. Mol. Sci.* **2017**, *18*, 2231. [CrossRef] [PubMed]
13. Valentová, K.; Purchartová, K.; Rydlová, L.; Roubalová, L.; Biedermann, D.; Petrásková, L.; Křenková, A.; Pelantová, H.; Holečková-Moravcová, V.; Tesařová, E.; et al. Sulfated metabolites of flavonolignans and 2,3-dehydroflavonolignans: Preparation and properties. *Int. J. Mol. Sci.* **2018**, *19*, 2349. [CrossRef] [PubMed]
14. Goyon, A.; Yehl, P.; Zhang, K. Characterization of therapeutic oligonucleotides by liquid chromatography. *J. Pharm. Biomed. Anal.* **2020**, *182*, 113105. [CrossRef]
15. Van der Horst, M.A.; Hartog, A.F.; El Morabet, R.; Marais, A.; Kircz, M.; Wever, R. Enzymatic sulfation of phenolic hydroxy groups of various plant metabolites by an arylsulfotransferase. *Eur. J. Org. Chem.* **2015**, *2015*, 534–541. [CrossRef]
16. Banoglu, E.; King, R.S. Sulfation of indoxyl by human and rat aryl (phenol) sulfotransferases to form indoxyl sulfate. *Eur. J. Drug. Metab. Pharmacokinet.* **2002**, *27*, 135–140. [CrossRef]
17. Sarıbaş, A.S.; Mobasseri, A.; Pristatsky, P.; Chen, X.; Barthelson, R.; Hakes, D.; Wang, J. Production of *N*-sulfated polysaccharides using yeast-expressed *N*-deacetylase/*N*-sulfotransferase-1 (NDST-1). *Glycobiology* **2004**, *14*, 1217–1228. [CrossRef]
18. Ji, Y.; Islam, S.; Mertens, A.M.; Sauer, D.F.; Dhoke, G.V.; Jakob, F.; Schwaneberg, U. Directed aryl sulfotransferase evolution toward improved sulfation stoichiometry on the example of catechols. *Appl. Microbiol. Biotechnol.* **2019**, *103*, 3761–3771. [CrossRef]
19. Roubalová, L.; Purchartová, K.; Papoušková, B.; Vacek, J.; Křen, V.; Ulrichová, J.; Vrba, J. Sulfation modulates the cell uptake, antiradical activity and biological effects of flavonoids in vitro: An examination of quercetin, isoquercitrin and taxifolin. *Bioorg. Med. Chem.* **2015**, *23*, 5402–5409. [CrossRef]
20. Mozhaev, V.V.; Khmelnitsky, Y.L.; Sanchez-Riera, F.; Maurina-Brunker, J.; Rosson, R.A.; Grund, A.D. Arylsulfotransferase from *Clostridium innocuum*—A new enzyme catalyst for sulfation of phenol-containing compounds. *Biotechnol. Bioeng.* **2002**, *78*, 567–575. [CrossRef]
21. Begines, P.; Biedermann, D.; Valentová, K.; Petrásková, L.; Pelantová, H.; Maya, I.; Fernández-Bolaños, J.G.; Křen, V. Chemoenzymatic synthesis and radical scavenging of sulfated hydroxytyrosol, tyrosol, and acetylated derivatives. *J. Agric. Food Chem.* **2019**, *67*, 7281–7288. [CrossRef] [PubMed]
22. Islam, S.; Laaf, D.; Infanzón, B.; Pelantová, H.; Davari, M.D.; Jakob, F.; Křen, V.; Elling, L.; Schwaneberg, U. KnowVolution campaign of an aryl sulfotransferase increases activity toward cellobiose. *Chem. Eur. J.* **2018**, *24*, 17117–17124. [CrossRef] [PubMed]
23. Jones, D.J.; Jukes-Jones, R.; Verschoyle, R.D.; Farmer, P.B.; Gescher, A. A synthetic approach to the generation of quercetin sulfates and the detection of quercetin 3′-*O*-sulfate as a urinary metabolite in the rat. *Bioorg. Med. Chem.* **2005**, *13*, 6727–6731. [CrossRef]
24. Dueñas, M.; González-Manzano, S.; Surco-Laos, F.; González-Paramas, A.; Santos-Buelga, C. Characterization of sulfated quercetin and epicatechin metabolites. *J. Agric. Food Chem.* **2012**, *60*, 3592–3598. [CrossRef] [PubMed]
25. Fenclova, M.; Novakova, A.; Viktorova, J.; Jonatova, P.; Dzuman, Z.; Ruml, T.; Kren, V.; Hajslova, J.; Vitek, L.; Stranska-Zachariasova, M. Poor chemical and microbiological quality of the commercial milk thistle-based dietary supplements may account for their reported unsatisfactory and non-reproducible clinical outcomes. *Sci. Rep.* **2019**, *9*, 11118. [CrossRef] [PubMed]
26. Graf, T.N.; Cech, N.B.; Polyak, S.J.; Oberlies, N.H. A validated UHPLC-tandem mass spectrometry method for quantitative analysis of flavonolignans in milk thistle (*Silybum marianum*) extracts. *J. Pharm. Biomed. Anal.* **2016**, *126*, 26–33. [CrossRef] [PubMed]
27. Petrásková, L.; Káňová, K.; Biedermann, D.; Křen, V.; Valentová, K. Simple and rapid HPLC separation and quantification of flavonoid, flavonolignans, and 2,3-dehydroflavonolignans in silymarin. *Foods* **2020**, *9*, 116. [CrossRef] [PubMed]
28. Biedermann, D.; Buchta, M.; Holečková, V.; Sedlák, D.; Valentová, K.; Cvačka, J.; Bednárová, L.; Křenková, A.; Kuzma, M.; Škuta, C.; et al. Silychristin: Skeletal alterations and biological activities. *J. Nat. Prod.* **2016**, *79*, 3086–3092. [CrossRef]
29. Křenek, K.; Marhol, P.; Peikerová, Ž.; Křen, V.; Biedermann, D. Preparatory separation of the silymarin flavonolignans by Sephadex LH-20 gel. *Food Res. Int.* **2014**, *65*, 115–120. [CrossRef]
30. Pyszková, M.; Biler, M.; Biedermann, D.; Valentová, K.; Kuzma, M.; Vrba, J.; Ulrichová, J.; Sokolová, R.; Mojović, M.; Popović-Bijelić, A.; et al. Flavonolignan 2,3-dehydroderivatives: Preparation, antiradical and cytoprotective activity. *Free Radic. Biol. Med.* **2016**, *90*, 114–125. [CrossRef]

International Journal of
Molecular Sciences

MDPI

Article

Viability of Glioblastoma Cells and Fibroblasts in the Presence of Imidazole-Containing Compounds

Elisabeth Christiane Seidel [1,2], Claudia Birkemeyer [3], Rainer Baran-Schmidt [1], Jürgen Meixensberger [1], Henry Oppermann [1,4,†] and Frank Gaunitz [1,*,†]

1 Klinik und Poliklinik für Neurochirurgie, Universitätsklinikum Leipzig, 04103 Leipzig, Germany; christiane.seidel@medizin.uni-leipzig.de (C.E.S.); rainer.baran-schmidt@medizin.uni-leipzig.de (R.B.-S.); juergen.meixensberger@medizin.uni-leipzig.de (J.M.); henry.oppermann@medizin.uni-leipzig.de (H.O.)
2 Klinik und Poliklinik für Neurologie, Universitätsklinikum Leipzig, 04103 Leipzig, Germany
3 Institut für Analytische Chemie, Universität Leipzig, 04103 Leipzig, Germany; birkemeyer@chemie.uni-leipzig.de
4 Institut für Humangenetik, Universitätsklinikum Leipzig, 04103 Leipzig, Germany
* Correspondence: frank.gaunitz@medizin.uni-leipzig.de
† These authors contributed equally to this work.

Abstract: The naturally occurring dipeptide carnosine (β-alanyl-*L*-histidine) specifically attenuates tumor growth. Here, we ask whether other small imidazole-containing compounds also affect the viability of tumor cells without affecting non-malignant cells and whether the formation of histamine is involved. Patient-derived fibroblasts and glioblastoma cells were treated with carnosine, *L*-alanyl-*L*-histidine (LA-LH), β-alanyl-*L*-alanine, *L*-histidine, histamine, imidazole, β-alanine, and *L*-alanine. Cell viability was assessed by cell-based assays and microscopy. The intracellular release of *L*-histidine and formation of histamine was investigated by high-performance liquid chromatography coupled to mass spectrometry. Carnosine and LA-LH inhibited tumor cell growth with minor effects on fibroblasts, and *L*-histidine, histamine, and imidazole affected viability in both cell types. Compounds without the imidazole moiety did not diminish viability. In the presence of LA-LH but not in the presence of carnosine, a significant rise in intracellular amounts of histidine was detected in all cells. The formation of histamine was not detectable in the presence of carnosine, LA-LH, or histidine. In conclusion, the imidazole moiety of carnosine contributes to its anti-neoplastic effect, which is also seen in the presence of histidine and LA-LH. Despite the fact that histamine has a strong effect on cell viability, the formation of histamine is not responsible for the effects on the cell viability of carnosine, LA-LH, and histidine.

Keywords: carnosine; glioblastoma; fibroblasts; imidazole-containing compounds; cell viability; high-performance liquid chromatography coupled to mass spectrometry

Citation: Seidel, E.C.; Birkemeyer, C.; Baran-Schmidt, R.; Meixensberger, J.; Oppermann, H.; Gaunitz, F. Viability of Glioblastoma Cells and Fibroblasts in the Presence of Imidazole-Containing Compounds. *Int. J. Mol. Sci.* **2022**, *23*, 5834. https://doi.org/10.3390/ijms23105834

Academic Editor: David Arráez-Román

Received: 19 April 2022
Accepted: 19 May 2022
Published: 23 May 2022

1. Introduction

With 3.23 new cases per 100,000 inhabitants in the United States, glioblastoma (GBM) is the most frequent malignant tumor of the human brain [1]. GBM is an astrocytic tumor that is classified, according to the World Health Organization (WHO), with the highest WHO grade IV. Despite the best possible treatment, which consists of maximal safe resection of the tumor, radiotherapy, and adjuvant chemotherapy with temozolomide, the 5-year overall survival of GBM patients is only 7.2% [1]. Furthermore, the effectiveness of this therapy is highly dependent on the genetic properties of the tumor, namely, the methylation status of the *O*-6-methylguanine-DNA methyltransferase (MGMT) promotor [2]. In view of the poor prognosis and missing alternatives to standard therapy, there is ongoing research for new treatment strategies and drugs that could improve the corresponding outcome.

In recent years, we and others have demonstrated that the naturally occurring dipeptide *L*-carnosine (β-alanyl-*L*-histidine; note that throughout the text, "carnosine" refers to

"*L*-carnosine"), which was originally discovered more than 120 years ago [3], may be a potential anti-neoplastic drug for different types of cancer in general and also for glioblastoma in particular (for reviews, see [4,5]). As carnosine is rapidly degraded in human plasma due to the presence of serum carnosinase, it has long been thought that its use as a systemically administered drug may be limited. Although there is now evidence that carnosine can escape from degradation by its uptake into erythrocytes [6], other compounds with higher stability may be considered as alternatives. Therefore, it is reasonable to study and better understand carnosine's mode of action on tumor cell viability, which also requires an understanding of the bioactive characteristics of the molecule. In view of the observation that *L*-histidine is able to mimic carnosine's anti-neoplastic effect [7], we asked whether other small imidazole-containing compounds are also able to mimic carnosine's effect, whether the β-alanyl moiety is required, and whether the effects of other compounds are indeed comparable to those of carnosine. In addition, we also wanted to know whether the formation of histamine from *L*-histidine is involved in the anti-neoplastic effect.

2. Results

2.1. *Viability of Glioblastoma Cells and Patient-Derived Fibroblasts in the Presence of Imidazole-Containing Compounds*

In the first series of experiments, we investigated the effect of imidazole-containing compounds on cell viability using five patient-derived fibroblast cell cultures (13/16, 90/15, 69/15, 60/15, 52/15) and four glioblastoma cell lines (U87, T98G, U87, G55T2). The cells were incubated for 48 h in the presence of carnosine, *L*-alanyl-*L*-histidine (LA-LH), β-alanyl-*L*-alanyl (βA-LA), *L*-histidine, histamine, imidazole, *L*-alanine, and β-alanine (all 50 mM), and cell viability was determined measuring the amount of ATP in cell lysates and dehydrogenase (DH) activity in living cells. The data were compared to untreated control cells. The results of experiments with fibroblasts from culture 52/15 and cells from the glioblastoma cell line U343 are presented in Figure 1 (all other experiments are presented in detail in Supplement S1 with Figure S1a for ATP in cell lysates and Figure S1b for dehydrogenase activity).

The summary of all data obtained is shown in Table 1.

In summary, as revealed by both assays, carnosine significantly reduced viability in all glioblastoma cells to at least 80% (with the exception of measuring DH in T98G). As shown in Table 1, fibroblasts also responded with a small reduction of viability in the presence of carnosine in almost all cells (maximal reduction to 90%), which was, in all cases, lower than that observed in GBM cell lines. Using LA-LH, we could also detect a significant reduction of viability in glioblastoma cells, as determined by both assays, whereas the viability of fibroblasts was not affected by the compound. Overall, carnosine and LA-LH exerted a comparable impact on glioblastoma cells, with differences regarding the various cell lines. Histidine reduced glioblastoma cell viability in a stronger manner than carnosine and LA-LH and significantly affected the viability of fibroblasts. In contrast, the non-imidazolyl-containing dipeptide βA-LA did not reduce GBM or fibroblast viability. Histamine strongly reduced fibroblast viability and glioblastoma cell viability without a significant difference between both cell types. Comparing the effect of histamine with histidine, a significantly stronger reduction by histamine was detected in fibroblasts but not in glioblastoma cells. Imidazole also strongly reduced the viability of fibroblasts and glioblastoma cells without discriminating between the cell types. In comparison to imidazole, histamine strongly reduced the cell viability of fibroblasts and exerted a comparable effect on glioblastoma cells.

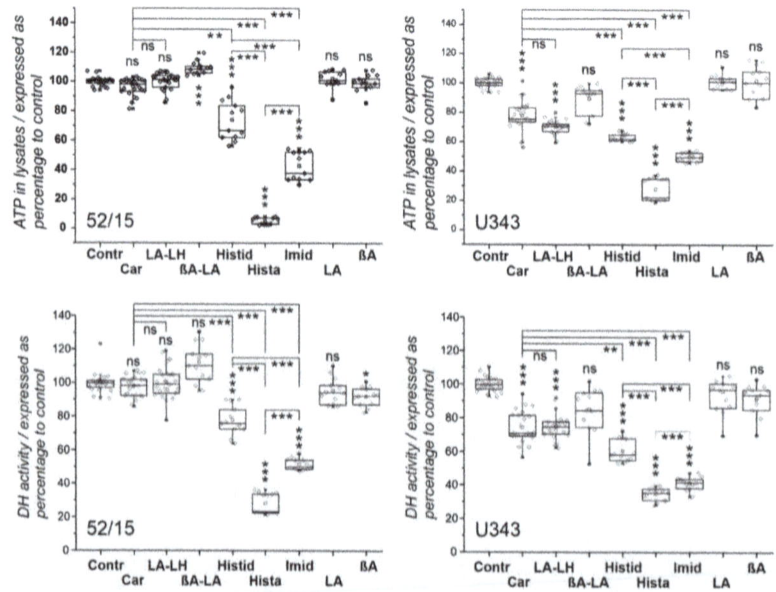

Figure 1. Viability of fibroblasts from culture 52/15 (**left**) and glioblastoma cells from the U343 line (**right**) after treatment with different compounds. Cells were treated for 48 h with carnosine (Car), *L*-alanyl-*L*-histidine (LA-LH), β-alanyl-*L*-alanine (βA-LA), *L*-histidine (Histid), histamine (Hista), imidazole (Imid), *L*-alanine (LA), and β-alanine (βA) (all 50 mM) or vehicle control (Contr) for 48 h. Cell viability was measured by determining ATP in cell lysates (upper panels) and dehydrogenase activity (DH) in living cells (lower panels). Results are presented as box plots. Statistical analysis was performed using a one-way ANOVA. The level of significance between different compounds is indicated by horizontal lines and compared to Contr above the boxes: *: $p < 0.05$; **: $p < 0.005$; ***: $p < 0.0005$; ns: not significant.

Table 1. Effect of different compounds on viability of fibroblasts and glioblastoma cells.

	Fibroblasts										Glioblastoma Cells							
Culture	13/16		90/15		69/15		60/15		52/15		U343		T98G		U87		G55T2	
Assay	ATP	DH	ATP	DH	ATP	DH	ATP	DH	ATP	DH	ATP	DH	ATP	DH	ATP	DH	ATP	DH
Carnosine																		
LA-LH																		
Histidine																		
Imidazole																		
Histamine																		
bAla-L-Ala																		
L-Ala																		
b-Ala																		
Reduction to control (%):					100–90		90–80		80–70		70–60		60–50		50–40		<40	

The reduction of ATP in cell lysates and dehydrogenase activity (DH) in living cells is color-indicated compared to untreated control cells after 48 h exposure to the compounds. Imidazole-containing compounds are shown in bold. Note: only statistically significant effects are indicated, and white fields indicate no significance.

2.2. *Necrosis and Morphological Changes of GBM Cells and Patient-Derived Fibroblasts after Treatment with Different Imidazole-Containing Compounds*

In the preceding section, we analyzed the viability of GBM cells and fibroblasts by measuring ATP in cell lysates and dehydrogenase activity in living cells after treatment with different compounds. These experiments indicated that carnosine and LA-LH more clearly reduced the investigated parameters in GBM cells than in fibroblasts; histidine and imidazole also appeared to affect GBM cells more severely than fibroblasts, and histamine similarly affected fibroblasts and tumor cells. Nonetheless, only carnosine and LA-LH appeared to have no effect on the viability of fibroblasts, although they obviously inhibited the production of ATP and dehydrogenase activity in GBM cells in general to below 80%.

In order to get a more detailed picture of the effect of the compounds on GBM cells and fibroblasts, we investigated the subsequent morphological changes after treatment with the selected compounds, and we determined cell numbers and necrotic cells by propidium iodide staining. In addition, we also stained living cells using Calcein-AM. For this experiment, cells from the four glioblastoma cell lines (G55T2, T98G, U87, and U343) and from fibroblast cultures (90/15, 69/15, 60/15, and 52/15) were exposed to our compounds (each 50 mM) for 48 h. Then, staining with Hoechst 33343, Calcein-AM, and propidium iodide was performed, and the effects were monitored by phase contrast and fluorescence microscopy. Figure 2 presents the result of the experiment and its analysis, along with representative pictures of cells from the U343 line and from fibroblast culture 52/15. A summary of all data obtained is presented in Table 2.

Pictures from experiments with other cell lines and fibroblast cultures are presented in Supplement S2 with microscopic images for 53/15 (Figure S2a); 60/15 (Figure S2b); 69/15 (Figure S2c); 90/15 (Figure S2d); G55T2 (Figure S2e); T98G (Figure S2f); U87 (Figure S2g); U343 (Figure S2h) and the statistical analysis (Figure S2i). In addition, a summary of the observations from all cells is given in Table 2.

In summary, the strongest reduction of the number of living cells was achieved in the presence of imidazole, which did not significantly discriminate between fibroblasts or glioblastoma cells, indicating the generally high toxicity of this compound for both cell types. In some cases (e.g., U87), the loss of cells was so high that the calculation of the ratio between dead and living cells was not useful ("nd" in Table 2). Histidine and histamine, on the other hand, affected fibroblasts and glioblastoma cells significantly differently, with many more dead cells in the latter. Compared to histidine and histamine, carnosine and LA-LH seemed to be less toxic but, in most cases, exhibited a stronger effect on glioblastoma cells than on fibroblasts. Comparing the cell counts presented in Table 2 to the effects on physiological parameters (Table 1), it seems to be likely that the effects of the different compounds on cells may differ with regard to their mechanisms of action on energy metabolism.

Table 2. Comparison of living cells and the ratio between dead and living cells in fibroblasts and glioblastoma cell cultures under the influence of different compounds.

	Fibroblasts								**Glioblastoma**							
culture	90/15		69/15		60/15		52/15		U343		T98G		G55T2		U87	
	%	ratio	%	ratio	%	ratio	%	ratio	%	ratio	%	ratio	%	ratio	%	ratio
Car																
LALH																
Histid																
Imid																nd
Hista																
living cells to control (%)						**≥ 100**	**<100–90**	**<90–80**	**<80–70**	**<70–60**	**<60–50**	**<50–40**	**<40–30**	**<30–20**	**<20–10**	**<10**
ratio dead/living							**≤0.5**	**>0.5–1**	**>1–2**	**>2–4**	**>4–8**	**>8–16**	**>16–32**	**>32–64**	**>65**	

The total number of living cells in four fibroblast cultures and four glioblastoma cell lines after treatment with different compounds compared to untreated control cells (in %); the ratios between dead cells and living cells are color-indicated.

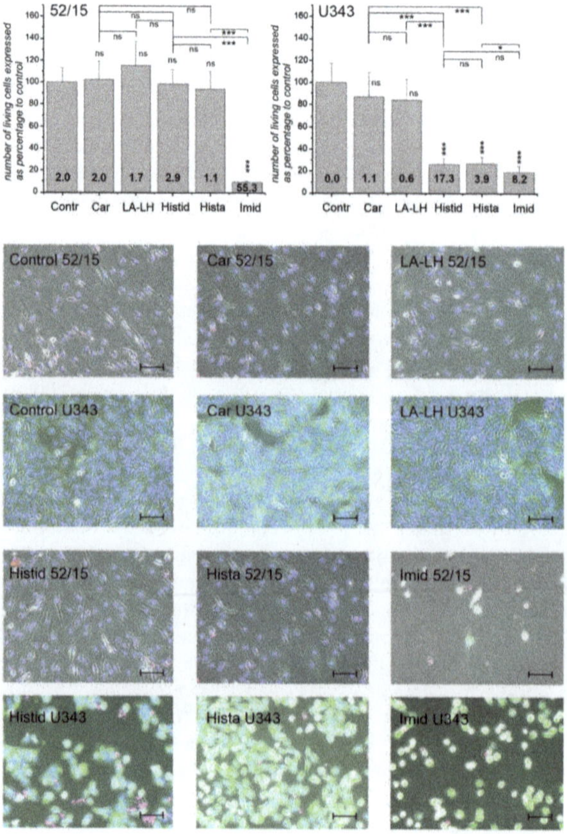

Figure 2. Microscopic analysis of fibroblasts (52/15) and glioblastoma cells (U343) in the presence of different compounds. Cells were treated for 48 h with carnosine (Car), *L*-alanyl-*L*-histidine (LA-LH), β-alanyl-*L*-alanine (βA-LA), *L*-histidine (Histid), histamine (Hista), imidazole (Imid), *L*-alanine (LA), and β-alanine (βA) (all 50 mM) or vehicle control (Contr) for 48 h. Then, microscopic images after staining with Hoechst 33343 (nuclei, blue), propidium iodide (dead cells, red), and Calcein-AM (living cells, green) were compared (representative images are presented as overlays in the lower panels, which also include an image obtained by phase contrast). In order to determine the number of living cells using ImageJ, the total number of nuclei was determined, subtracting the nuclei of dead cells (bars in upper panel; note: determination of living cells from images of cells positive for Calcein-AM staining was not performed because of high errors due to technical reasons). The ratio of dead cells to living cells is presented by bold numbers in the bars in the upper panel. Statistical analysis was performed using a one-way ANOVA. The level of significance between different compounds is indicated by horizontal lines and compared to Contr above the bars: *: $p < 0.05$; ***: $p < 0.0005$; ns: not significant. (Size bars: 100 μm).

2.3. The Release of L-Histidine from L-Alanyl-L-Histidine Does Not Result in the Formation of Histamine

Next, we investigated whether different amounts of *L*-histidine are released from carnosine and LA-LH and whether significant intracellular amounts of histamine can be formed from intracellular *L*-histidine. Figure 3a presents the intracellular amounts of *L*-histidine when cells are incubated in the presence of carnosine or LA-LH, respectively (both 50 mM), as determined in fibroblast cell cultures ($n = 5$) and glioblastoma cell lines ($n = 4$). As can be seen, no significant increase of intracellular *L*-histidine compared to untreated control cells was observed in cells incubated in the presence of carnosine. In contrast, incubation in the presence of LA-LH resulted in a significant increase in intracellular *L*-histidine. In order to detect a possible formation of histamine, we incubated cells from the GBM line T98G directly in the presence of *L*-histidine (25 mM) and histamine (25 μM) and determined the amount of both compounds in the cells. As can be seen in Figure 3b, we could not detect histamine in cells incubated in the presence of *L*-histidine, although we observed a steep rise in the intracellular amount of histamine when added to the medium. In addition, we did not detect histamine or its degradation products in cells exposed to carnosine, LA-LH, or *L*-histidine, whereas we observed the formation of *N*-methylhistamine in cells cultivated in the presence of histamine.

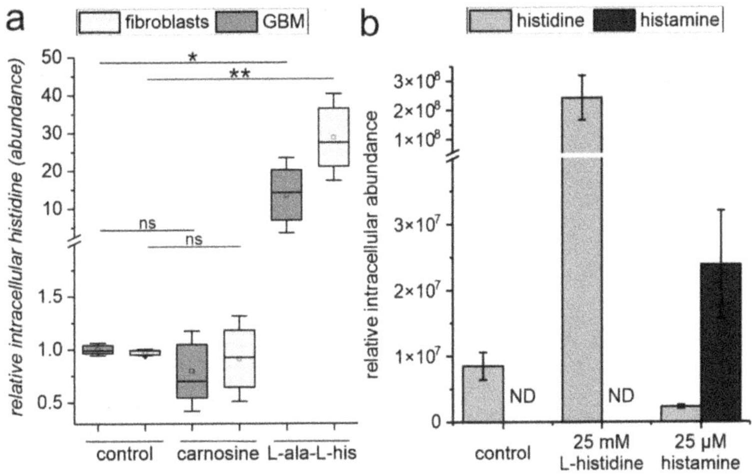

Figure 3. Release of histidine from carnosine and LA-LH and formation of histamine from histidine in fibroblasts and glioblastoma cells. (**a**) Four different glioblastoma cell lines and five different fibroblast cultures were treated with carnosine or *L*-alanyl-*L*-histidine (LA-LH) (both 50 mM) or vehicle control for 48 h. Afterwards, metabolites were extracted and intracellular histidine was determined by LC–MS. Results are presented as box plots that were obtained from the median of the replicates of each cell culture. (**b**) T98G cells were treated with 25 mM L-histidine, 25 μM histamine, or vehicle control for 48 h. Afterwards, metabolites were extracted, and intracellular histamine and *L*-histidine were determined by LC–MS. Statistical analysis was performed using a one-way ANOVA with the Games–Howell post hoc test. The level of significance is indicated as: *: $p < 0.05$; **: $p < 0.005$; not significant (ns): $p > 0.05$. ND: not detected.

3. Discussion

Several decades ago, the anti-neoplastic effect of carnosine was first described in vivo by Nagai and Suda [8]. Later, this observation was confirmed in vivo and in vitro by several groups and for different types of cancer, such as gastric carcinoma [9], colon carcinoma [10], cervical carcinoma [11], and glioblastoma [12]. Just recently, we demonstrated that L-histidine, one amino acid of the dipeptide, reduces GBM cell viability even more potently than carnosine [13]. Therefore, we wondered whether L-histidine itself or other histidine-containing compounds would also discriminate between malignant GBM cells and non-malignant fibroblasts, as demonstrated for carnosine [14,15]. Here, we observed that carnosine significantly reduced cell viability in GBM cells, whereas there was no significant effect observed in fibroblasts with regard to the amount of ATP in cell lysates and only a small but significant effect with regard to dehydrogenase activity. A comparable discrimination between tumor and non-tumor cells was also observed after the application of LA-LH. Although L-histidine did significantly reduce the viability of fibroblasts, the effect was stronger in GBM cells but significantly different between GBM cells and fibroblasts only with regard to dehydrogenase activity.

As L-histidine also has anti-neoplastic effects on GBM cells, this observation raises the question of whether *L*-histidine could be used as a therapeutic agent instead of carnosine. First of all, at the concentrations employed in our experiments (50 mM), the amino acid obviously also affects fibroblasts. Comparable observations have been made by Rauen et al., who detected that in cultivated liver cells, *L*-histidine at a concentration of 76 mM had a ~2.5 higher toxicity compared to carnosine at a concentration of 198 mM [16]. As *L*-histidine is a proteinogenic amino acid, it could also be assumed that orally ingested *L*-histidine is taken up by other cells and may not reach the tumor. Orally applied carnosine, on the other hand, is rapidly degraded by serum carnosinase [17] and may, therefore, be limited in delivering its histidine moiety to cancer cells. In this regard, it has to be noted that we recently demonstrated that carnosine can escape from degradation by uptake into erythrocytes [6], explaining the observation that the dipeptide can be detected in the urine of volunteers up to 5 h after oral ingestion [18]. In addition, there are several reports of the therapeutic effects of orally ingested carnosine that point towards the delivery of intact carnosine, especially to the brain [19–21].

Given the fact that LA-LH has an anti-neoplastic effect comparable to carnosine, being able to discriminate between tumor and non-tumor cells, the other question is whether this compound or other imidazole-containing dipeptides could be a useful alternative to carnosine. This question cannot be answered yet, but it is interesting to note that our experiments demonstrate that LA-LH is intracellularly more rapidly degraded to its amino acid constituents than carnosine (Figure 3). Therefore, future experiments should investigate whether the release of *L*-histidine results in a more rapid loss of the bioactive imidazolyl moiety in the cells and whether more stable *L*-histidine-containing dipeptides could be an alternative to carnosine or LA-LH. More complex synthetic compounds derived from imidazole have already been discussed as potential anti-cancer drugs (for a review, see [22]). In fact, some of the more complex imidazole-derived compounds have already entered the clinics with regard to a number of diseases (for a review, see [23]), but one has to take into account that synthetic drugs require intensive testing before being used in therapy. On the other hand, carnosine is a naturally occurring compound that has already been used in a number of studies with human patients, and, together with its constituent β-alanine, it has a high acceptance as a food supplement for athletes [24]. At this point, it is also interesting to note that carnosine, instead of negatively affecting non-malignant cells, has protective effects on normal cells and has been discussed as a neuroprotector, especially in various pathological brain conditions [25]. Imidazole itself has also been discussed as a potential drug for the treatment of colon cancer [26]. However, although Long and Wang used low concentrations in their culture experiments with colon carcinoma cells (up to 36 μM), the high toxicity towards fibroblasts, as seen in our experiments, should be taken into account when considering the use of imidazole as an anti-cancer drug.

Another question addressed by the presented experiments is whether histamine obtained by decarboxylation of *L*-histidine could be responsible for carnosine's and L-histidine's anti-neoplastic effect. This notion has been discussed by others [27] and could have been deduced by the fact that histamine has a very strong effect on viability (Figure 1). However, we now rule out this possibility by the observation that we could not detect the formation of histamine after the exposure of GBM cells to carnosine, LA-LH, or *L*-histidine. In addition, we also did not detect *N*-methylhistamine, a degradation product of histamine, in cells exposed to carnosine, LA-LH, or L-histidine, which were detectable in cells exposed to histamine. At this point, it should also be noted that *AOC1* (gene encoding diamine oxidase (EC 1.4.3.22), which is responsible for the conversion of *L*-histidine to histamine) is almost never present in GBM cells and normal brain tissue (transcripts per million transcripts ~2), as revealed by in silico analysis using data from the TCGA Research Networks (https://www.cancer.gov/tcga; accessed on 24 December 2022) and the GTex database (https://gtexportal.org/home/; accessed on 24 December 2022) using GEPIA (Gene Expression Profiling Interactive Analysis; http://gepia.cancer-pku.cn/; accessed on 24 December 2022 [28]). In addition, it should be noted that the neuroprotective properties of carnosine are also independent of its metabolization via the *L*-histidine-histamine pathway [29].

In conclusion, our experiments demonstrate the importance of the *L*-histidine moiety of carnosine for its anti-neoplastic effect. Although the molecular mechanisms by which this moiety exerts its anti-neoplastic effect have to be revealed in detail, there is evidence that imidazolyl-containing compounds are able to inhibit mitochondrial ATP production [30]. In addition, it has been shown that they can induce cell cycle arrest [31], most likely by binding the imidazolyl moiety to DNA [32]. Despite the observation that the imidazolyl moiety contributes to the anti-neoplastic effect, it is important to note that we recently demonstrated that carnosine's influence on tumor cell viability is accompanied by an influence on the pentose phosphate pathway through its interaction with the glycolytic intermediates glyceraldehyde-3-phosphate and dihydroxyacetone phosphate. Therefore, carnosine may have a broader influence on tumor cells than *L*-histidine alone or other imidazolyl-containing dipeptides [13]. Finally, it would also be interesting to test whether the other compounds used in our study also affect the migration of GBM cells, as demonstrated for carnosine [15]. On the other hand, at least for compounds with high toxicity, such as *L*-histidine, imidazole, or histamine, one certainly needs to test lower concentrations than those used in our present investigation.

4. Material and Methods

4.1. Reagents

If not stated otherwise, all chemicals were purchased from Sigma-Aldrich (Taufkirchen, Germany), Merck (Darmstadt, Germany), or Carl Roth (Karlsruhe, Germany). Carnosine was kindly provided by Flamma s.p.a. (Chignolo d'Isola, Italy); *L*-alanyl-*L*-histidine and β-alanyl-*L*-alanine were purchased from Bachem (Bubendorf, Switzerland).

4.2. Cell Lines and Fibroblast Cultures

The glioblastoma cell line G55T2 was obtained from Sigma (Taufkirchen, Germany), the cell lines U87 and T98G from the ATCC (Manassas, VA, USA), and the U343 line from the German Collection of Microorganisms and Cell Cultures (Braunschweig, Germany). All cells were genotyped (Genolytic GmbH, Leipzig, Germany), and their identities were confirmed.

For cultivation, cells were propagated in 250 mL culture flasks (Sarstedt AG & Co., Nümbrecht, Germany) using 10 mL of standard culture medium (DMEM/4.5 g/L glucose, without pyruvate (Life Technologies, Darmstadt, Germany) supplemented with 10% fetal bovine serum (FBS superior, Biochrom, Berlin, Germany), 2 mM GlutaMax (Life Technologies), and penicillin–streptomycin (Life Technologies)) at 37 °C and 5% CO_2 in humidified air in an incubator.

Fibroblast cultures were established as described previously [15]. Briefly, freshly removed galea tissue was washed with PBS (phosphate-buffered saline) and minced with a scalpel blade. After mincing, small tissue pieces were transferred to a 25 cm^2 culture flask (TPP, Trasadingen, Switzerland) sprinkled with AmnioMax complete medium (Gibco, Darmstadt, Germany). Tissue pieces were incubated for 30 min at room temperature, and finally, 1 mL AmnioMax complete medium was added. Incubation was then performed at 37 °C, with 5% CO_2 and humidified air in an incubator. Medium was changed after 72 h. As soon as a confluent layer was obtained, cells were removed from culture flasks by the use of Accutase (PAA, Pasching, Austria) and transferred to 75 cm^2 culture flasks (TPP). AmnioMax medium with AmnioMax supplement was used for the first 2–3 weeks of cultivation. Thereafter, fibroblasts were cultivated under the same conditions as glioblastoma cells. Galea tissue was obtained during standard surgery performed at the Neurosurgery Department of the University Hospital Leipzig in 2015 and 2016. All patients provided written informed consent according to German law, as confirmed by the local committee (144/08-ek).

4.3. Cell Viability Assays

For cell viability assays, cells were counted and seeded into sterile 96-well plates (μClear, Greiner Bio One, Frickenhausen, Germany) at a density of 5000 cells/well in 200 μL standard medium. After 24 h of cultivation (37 °C, 5% CO_2/95% air), the medium was aspirated and fresh medium containing supplements was added, as indicated in each experiment (100 μL/well), and the cells were incubated for an additional 48 h. Then, the CellTiter-Glo Luminescent Cell Viability Assay (Promega, Mannheim, Germany) was employed to determine viable cells by measuring ATP in cell lysates, and the CellTiter-Blue Cell Viability Assay (Promega) was used to quantify the cell's dehydrogenase activity in living cells. All assays were carried out according to the manufacturer's protocols. Luminescence and fluorescence were measured using a SpectraMax M5 multilabel reader (Molecular Devices, Biberach, Germany).

4.4. Staining and Determination of Live and Dead Cells

The number of live and dead cells after treatment with different compounds was determined in 12-well plates (TPP). Cells were seeded at a density of 80,000 cells per well in 1 mL of medium. After 24 h, cells received fresh medium with the test compounds. After 48 h in the presence of the compounds, cells were washed with Hanks balanced salt solution (calcium, magnesium, 1 g/L glucose, pH 7.4; Thermo Fisher Scientific, Darmstadt, Germany) before DMEM containing Calcein-AM (2 μM), propidium iodide (1.5 μM), and Hoechst 33343 (2 μM) was added for 1 h. Microscopic pictures were taken using a BZ-X800 microscope (Keyence, Neu-Isenburg, Germany) using phase contrast and fluorescence to identify nuclei (360/460 nm), living cells (470/525 nm), and dead cells (560/630 nm). For the determination of the number of nuclei and of dead cells, 9 images (at 4× magnification) from each well were taken, and ImageJ was used to determine the number of dead cells and nuclei [33]. (Note: we used this approach instead of a FACS analysis, as the detachment of fibroblasts, especially under high toxicity conditions, contributes to additional toxicity.)

4.5. Determination of Intracellular L-Histidine

Intracellular amounts of histidine and histamine were determined as described previously [13]. Briefly, cells were seeded at a density of 300,000 cells per well into a 6-well plate in 2 mL of culture medium. After 24-h cultivation, the culture medium was removed and replaced with fresh medium containing specific compounds for each experiment, and cells were incubated for an additional 48 h. Then, cells were washed thrice with 1 mL of ice-cold washing buffer, followed by extraction and by the addition of 400 μL of ice-cold methanol. After 10 min of gentle shaking on ice, extracts were collected in 1.5 mL Eppendorf tubes and wells were rinsed twice in 400 μL distilled high-quality water (Milli-Q). Samples were evaporated to dryness by lyophilization (Martin Christ Gefriertrocknungsanlagen, Osterode, Germany). For derivatization, the freeze-dried extracts were redissolved in 100 μL high-quality water (Milli-Q), and 100 μL 0.5% *ortho*-phthalaldehyde (dissolved in methanol) was added. Derivatization was carried out at 37 °C in a thermomixer for 45 min, followed by the addition of 800 μL 0.1% formic acid in HPLC grade water. The obtained solution (200 μL) was transferred into 250 μL conic glass inserts of 2 mL ND10 vials, followed by high-performance liquid chromatography coupled to mass spectrometry (HPLC–MS). After extraction, the protein of the remaining layer of fixed cells was extracted by the addition of 200 μL lysis buffer (77 mM K_2HPO_4, 23 mM KH_2PO_4, 0.2% TritonX-100, pH 7.8). Then, the total protein was determined by using the Pierce 660 nm Protein Assay (Thermo Scientific, Braunschweig, Germany).

4.6. HPLC–MS Set Up and Data Analysis

An Agilent 1100 series HPLC consisting of a variable wavelength detector, a well plate autosampler, and a binary pump, coupled with a Bruker Esquire 3000 plus electrospray ionization mass spectrometer, was used. The column was a Phenomenex Gemini 5 μ C18 110 Å 150 mm × 2 mm column with a precolumn. The eluent system consisted of two solvents, with eluent A: 0.1% formic acid in acetonitrile and eluent B: 0.1% formic acid in HPLC grade water. Mobile phase flow rate was 0.5 mL/min with the following gradient for separation: 0–10 min 90% B, 90% to 0% B within 15 min, 25–35 min 0% B, 0% to 90% B within 5 min, and 40–47 min 90% B for column equilibration. The mass spectrometer operated in positive mode (target mass: m/z 300; mass range: m/z 70–400), and the dry gas temperature was set to 360 °C (flow rate: 11 L/min; 70 psi). Data were analyzed using OpenChrom version 2.0.103.v20150204-1700 [34]. Histidine and histamine were identified by standards, and target masses m/z 272 (histidine) and 228 (histamine) were used for quantification. If not stated otherwise, the abundance of a metabolite is defined by the peak area determined from the selected ion chromatogram of an experiment, normalized to the total cellular protein (µg). It should be noted that this method does not allow discrimination between *D* and *L* stereoisomers. Thus, when referring to signals obtained by HPLC–MS, only histidine (and not *L*-histidine) is mentioned.

4.7. Statistical Analysis and Graphical Representation

Statistical analysis was carried out using SPSS (IBM, Armonk, NY, USA; version: 28.0.0.0 (190)). For multiple comparisons, a one-way ANOVA after testing for normality of distribution (Kolmogorov–Smirnov test) was employed, using a Games–Howell or Bonferroni post hoc test after testing for equality of variances (Levene's test). Results were considered to be statistically significant at a value of $p < 0.05$. Graphical representations were prepared using OriginPro (2021b; OriginLab Corporation, Northampton, MA, USA) and CorelDraw Graphics Suite 2020 (Corel Corporation, Ottawa, ON, Canada).

Supplementary Materials: The following are available online at https://www.mdpi.com/article/10.3390/ijms23105834/s1 Supplement S1: Viability of cells from four glioblastomas and from five fibroblast cultures derived from patients, cultivated in the presence of imidazole-containing compounds. Supplement S2: Microscopic analysis of fibroblast cell cultures and glioblastoma cells in the presence of different compounds.

Author Contributions: Conceptualization, H.O. and F.G.; Methodology, H.O., C.B. and F.G.; Formal Analysis, H.O., C.B., C.E.S. and F.G.; Investigation, H.O., C.B., C.E.S., R.B.-S. and F.G.; Resources, F.G. and J.M.; Data Curation, H.O. and F.G.; Writing—Original Draft Preparation, H.O., C.E.S., and F.G.; Writing—Review and Editing, H.O., C.B., C.E.S., J.M. and F.G.; Visualization, H.O., C.E.S. and F.G.; Supervision, H.O. and F.G.; Project Administration, F.G. and J.M. All authors have read and agreed to the published version of the manuscript.

Funding: Funded by the Open Access Publishing Fund of Leipzig University supported by the German Research Foundation within the program Open Access Publication Funding.

Institutional Review Board Statement: The study was conducted according to the guidelines of the Declaration of Helsinki and approved by the Ethics Committee of the Medical Faculty of the University of Leipzig (#144/08-ek) (approval date: 5 June 2008).

Informed Consent Statement: Informed consent was obtained from all subjects involved in the study.

Data Availability Statement: All data used in this study are presented in the manuscript and Supplementary Materials.

Acknowledgments: LC–MS analyses were carried out at the mass spectrometry core facility of the Faculty of Chemistry and Mineralogy at Leipzig University, MS-UL; we thank Susan Billig for technical assistance. We like to thank Flamma (Flamma s.p.a. Chignolo d'Isola, Italy (http://www.flammagroup.com); accessed on 18 May 2022) for the generous supply of very high-quality carnosine for all our experiments. In addition, we like to thank Hans-Heinrich Foerster from Genolytic GmbH (Leipzig, Germany) for genotyping and confirmation of cell identity.

Conflicts of Interest: The authors declare that they have no potential conflicts of interest.

References

1. Ostrom, Q.T.; Patil, N.; Cioffi, G.; Waite, K.; Kruchko, C.; Barnholtz-Sloan, J.S. CBTRUS Statistical Report: Primary Brain and Other Central Nervous System Tumors Diagnosed in the United States in 2013–2017. *Neuro-Oncology* **2020**, *22*, iv1–iv96. [CrossRef] [PubMed]
2. Hegi, M.E.; Diserens, A.-C.; Gorlia, T.; Hamou, M.-F.; de Tribolet, N.; Weller, M.; Kros, J.M.; Hainfellner, J.A.; Mason, W.; Mariani, L.; et al. MGMT gene silencing and benefit from temozolomide in glioblastoma. *N. Engl. J. Med.* **2005**, *352*, 997–1003. [CrossRef] [PubMed]
3. Gulewitsch, W.; Amiradzibi, S. Ueber das Carnosin, eine neue organische Base des Fleischextraktes. *Ber. Dtsch. Chem. Ges.* **1900**, *33*, 1902–1903. [CrossRef]
4. Turner, M.D.; Sale, C.; Garner, A.C.; Hipkiss, A.R. Anti-cancer actions of carnosine and the restoration of normal cellular homeostasis. *Biochim. Biophys. Acta Mol. Cell Res.* **2021**, *1868*, 119117. [CrossRef]
5. Hipkiss, A.R.; Gaunitz, F. Inhibition of tumour cell growth by carnosine: Some possible mechanisms. *Amino Acids* **2014**, *46*, 327–337. [CrossRef]
6. Oppermann, H.; Elsel, S.; Birkemeyer, C.; Meixensberger, J.; Gaunitz, F. Erythrocytes Prevent Degradation of Carnosine by Human Serum Carnosinase. *Int. J. Mol. Sci.* **2021**, *22*, 12802. [CrossRef]
7. Letzien, U.; Oppermann, H.; Meixensberger, J.; Gaunitz, F. The antineoplastic effect of carnosine is accompanied by induction of PDK4 and can be mimicked by L-histidine. *Amino Acids* **2014**, *46*, 1009–1019. [CrossRef]
8. Nagai, K.; Suda, T. Antineoplastic effects of carnosine and beta-alanine—Physiological considerations of its antineoplastic effects. *J. Physiol. Soc. Jpn.* **1986**, *48*, 741–747.
9. Shen, Y.; Yang, J.; Li, J.; Shi, X.; Ouyang, L.; Tian, Y.; Lu, J. Carnosine inhibits the proliferation of human gastric cancer SGC-7901 cells through both of the mitochondrial respiration and glycolysis pathways. *PLoS ONE* **2014**, *9*, e104632. [CrossRef]
10. Iovine, B.; Iannella, M.L.; Nocella, F.; Pricolo, M.R.; Baldi, M.R.; Bevilacqua, M.A. Carnosine inhibits KRas-mediated HCT-116 proliferation by affecting ATP and ROS production. *Cancer Lett.* **2012**, *315*, 122–128. [CrossRef]
11. Ditte, Z.; Ditte, P.; Labudova, M.; Simko, V.; Iuliano, F.; Zatovicova, M.; Csaderova, L.; Pastorekova, S.; Pastorek, J. Carnosine inhibits carbonic anhydrase IX-mediated extracellular acidosis and suppresses growth of HeLa tumor xenografts. *BMC Cancer* **2014**, *14*, 358. [CrossRef] [PubMed]
12. Renner, C.; Seyffarth, A.; de Arriba, S.; Meixensberger, J.; Gebhardt, R.; Gaunitz, F. Carnosine Inhibits Growth of Cells Isolated from Human Glioblastoma Multiforme. *Int. J. Pept. Res. Ther.* **2008**, *14*, 127–135. [CrossRef]
13. Oppermann, H.; Purcz, K.; Birkemeyer, C.; Baran-Schmidt, R.; Meixensberger, J.; Gaunitz, F. Carnosine's inhibitory effect on glioblastoma cell growth is independent of its cleavage. *Amino Acids* **2019**, *51*, 761–772. [CrossRef] [PubMed]
14. Holliday, R.; McFarland, G.A. Inhibition of the growth of transformed and neoplastic cells by the dipeptide carnosine. *Br. J. Cancer* **1996**, *73*, 966–971. [CrossRef] [PubMed]
15. Oppermann, H.; Dietterle, J.; Purcz, K.; Morawski, M.; Eisenlöffel, C.; Müller, W.; Meixensberger, J.; Gaunitz, F. Carnosine selectively inhibits migration of IDH-wildtype glioblastoma cells in a co-culture model with fibroblasts. *Cancer Cell Int.* **2018**, *18*, 111. [CrossRef]
16. Rauen, U.; Klempt, S.; de Groot, H. Histidine-induced injury to cultured liver cells, effects of histidine derivatives and of iron chelators. *Cell. Mol. Life Sci.* **2007**, *64*, 192–205. [CrossRef]
17. Qiu, J.; Hauske, S.J.; Zhang, S.; Rodriguez-Niño, A.; Albrecht, T.; Pastene, D.O.; van den Born, J.; van Goor, H.; Ruf, S.; Kohlmann, M.; et al. Identification and characterisation of carnostatine (SAN9812), a potent and selective carnosinase (CN1) inhibitor with in vivo activity. *Amino Acids* **2019**, *51*, 7–16. [CrossRef]
18. Gardner, M.L.G.; Illingworth, K.M.; Kelleher, J.; Wood, D. Intestinal-Absorption of the Intact Peptide Carnosine in Man, and Comparison with Intestinal Permeability to Lactulose. *J. Physiol.* **1991**, *439*, 411–422. [CrossRef]
19. Baraniuk, J.N.; El-Amin, S.; Corey, R.; Rayhan, R.; Timbol, C. Carnosine Treatment for Gulf War Illness: A Randomized Controlled Trial. *GJHS* **2013**, *5*, 69–81. [CrossRef]
20. Hajizadeh-Zaker, R.; Ghajar, A.; Mesgarpour, B.; Afarideh, M.; Mohammadi, M.-R.; Akhondzadeh, S. l-Carnosine as an Adjunctive Therapy to Risperidone in Children with Autistic Disorder: A Randomized, Double-Blind, Placebo-Controlled Trial. *J. Child Adolesc. Psychopharmacol.* **2018**, *28*, 74–81. [CrossRef]
21. Mehrazad-Saber, Z.; Kheirouri, S.; Noorazar, S.-G. Effects of l-Carnosine Supplementation on Sleep Disorders and Disease Severity in Autistic Children: A Randomized, Controlled Clinical Trial. *Basic Clin. Pharmacol. Toxicol.* **2018**, *123*, 72–77. [CrossRef] [PubMed]
22. Akhtar, J.; Khan, A.A.; Ali, Z.; Haider, R.; Shahar Yar, M. Structure-activity relationship (SAR) study and design strategies of nitrogen-containing heterocyclic moieties for their anticancer activities. *Eur. J. Med. Chem.* **2017**, *125*, 143–189. [CrossRef] [PubMed]
23. Siwach, A.; Verma, P.K. Synthesis and therapeutic potential of imidazole containing compounds. *BMC Chem.* **2021**, *15*, 12. [CrossRef] [PubMed]
24. Saunders, B.; Elliott-Sale, K.; Artioli, G.G.; Swinton, P.A.; Dolan, E.; Roschel, H.; Sale, C.; Gualano, B. β-alanine supplementation to improve exercise capacity and performance: A systematic review and meta-analysis. *Br. J. Sports Med.* **2017**, *51*, 658–669. [CrossRef]

25. Lopachev, A.V.; Abaimov, D.A.; Filimonov, I.S.; Kulichenkova, K.N.; Fedorova, T.N. An assessment of the transport mechanism and intraneuronal stability of L-carnosine. *Amino Acids* **2021**, 1–8. [CrossRef]
26. Long, Y.; Wang, D. Inhibition of Colon Cancer Cell Growth by Imidazole Through Activation of Apoptotic Pathway. *Med. Sci. Monit.* **2019**, *25*, 7597–7604. [CrossRef]
27. Zhang, L.; Yao, K.; Fan, Y.; He, P.; Wang, X.; Hu, W.; Chen, Z. Carnosine protects brain microvascular endothelial cells against rotenone-induced oxidative stress injury through histamine H_1 and H_2 receptors in vitro. *Clin. Exp. Pharmacol. Physiol.* **2012**, *39*, 1019–1025. [CrossRef]
28. Tang, Z.; Li, C.; Kang, B.; Gao, G.; Li, C.; Zhang, Z. GEPIA: A web server for cancer and normal gene expression profiling and interactive analyses. *Nucleic Acids Res.* **2017**, *45*, W98–W102. [CrossRef]
29. Bae, O.-N.; Majid, A. Role of histidine/histamine in carnosine-induced neuroprotection during ischemic brain damage. *Brain Res.* **2013**, *1527*, 246–254. [CrossRef]
30. Shi, Y.; Lim, S.K.; Liang, Q.; Iyer, S.V.; Wang, H.-Y.; Wang, Z.; Xie, X.; Sun, D.; Chen, Y.-J.; Tabar, V.; et al. Gboxin is an oxidative phosphorylation inhibitor that targets glioblastoma. *Nature* **2019**, *567*, 341–346. [CrossRef]
31. Morelli, M.B.; Amantini, C.; Nabissi, M.; Cardinali, C.; Santoni, M.; Bernardini, G.; Santoni, A.; Santoni, G. Axitinib induces senescence-associated cell death and necrosis in glioma cell lines: The proteasome inhibitor, bortezomib, potentiates axitinib-induced cytotoxicity in a p21(Waf/Cip1) dependent manner. *Oncotarget* **2017**, *8*, 3380–3395. [CrossRef] [PubMed]
32. Uzlikova, M.; Nohynkova, E. The effect of metronidazole on the cell cycle and DNA in metronidazole-susceptible and -resistant Giardia cell lines. *Mol. Biochem. Parasitol.* **2014**, *198*, 75–81. [CrossRef] [PubMed]
33. Schneider, C.A.; Rasband, W.S.; Eliceiri, K.W. NIH Image to ImageJ: 25 years of image analysis. *Nat. Methods* **2012**, *9*, 671–675. [CrossRef] [PubMed]
34. Wenig, P.; Odermatt, J. OpenChrom: A cross-platform open source software for the mass spectrometric analysis of chromatographic data. *BMC Bioinform.* **2010**, *11*, 405. [CrossRef] [PubMed]

International Journal of *Molecular Sciences*

MDPI

Article

Synthesis and NLRP3-Inflammasome Inhibitory Activity of the Naturally Occurring Velutone F and of Its Non-Natural Regioisomeric Chalconoids

Tiziano De Ventura [1,†], Mariasole Perrone [2,3,†], Sonia Missiroli [2,3], Paolo Pinton [2,3,4], Paolo Marchetti [1], Giovanni Strazzabosco [1], Giulia Turrin [1], Davide Illuminati [1], Virginia Cristofori [1], Anna Fantinati [5], Martina Fabbri [1], Carlotta Giorgi [2,3,*], Claudio Trapella [1,3,*] and Vinicio Zanirato [1]

1 Department of Chemistry, Pharmaceutical and Agricultural Sciences, University of Ferrara, Via Luigi Borsari 46, 44121 Ferrara, Italy
2 Department of Medical Sciences, Section of Experimental Medicine, University of Ferrara, Via Fossato di Mortara, 64/b, 44121 Ferrara, Italy
3 Laboratory for Technologies of Advanced Therapies (LTTA), Via Fossato di Mortara, 70, 44121 Ferrara, Italy
4 Maria Cecilia Hospital, GVM Care & Research, 48033 Cotignola, Italy
5 Department of Environmental and Prevention Sciences, University of Ferrara, Via Fossato di Mortara 17, 44121 Ferrara, Italy
* Correspondence: carlotta.giorgi@unife.it (C.G.); trap@unife.it (C.T.); Tel.: +39-0532-455802 (C.G.); +39-0532-455924 (C.T.)
† These authors contributed equally to this work.

Abstract: Plant-derived remedies rich in chalcone-based compounds have been known for centuries in the treatment of specific diseases, and nowadays, the fascinating chalcone framework is considered a useful and, above all, abundant natural chemotype. Velutone F, a new chalconoid from *Millettia velutina*, exhibits a potent effect as an NLRP3-inflammasome inhibitor; the search for new natural/non-natural lead compounds as NLRP3 inhibitors is a current topical subject in medicinal chemistry. The details of our work toward the synthesis of velutone F and the unknown non-natural regioisomers are herein reported. We used different synthetic strategies both for the construction of the distinctive benzofuran nucleus (**BF**) and for the key phenylpropenone system (**PhP**). Importantly, we have disclosed a facile entry to the velutone F via synthetic routes that can also be useful for preparing non-natural analogs, a prerequisite for extensive SAR studies on the new flavonoid class of NLRP3-inhibitors.

Keywords: flavonoid; chalcone-based compounds; NLRP3-inflammasome inhibitors

Citation: De Ventura, T.; Perrone, M.; Missiroli, S.; Pinton, P.; Marchetti, P.; Strazzabosco, G.; Turrin, G.; Illuminati, D.; Cristofori, V.; Fantinati, A.; et al. Synthesis and NLRP3-Inflammasome Inhibitory Activity of the Naturally Occurring Velutone F and of Its Non-Natural Regioisomeric Chalconoids. *Int. J. Mol. Sci.* **2022**, *23*, 8957. https://doi.org/10.3390/ijms23168957

Academic Editor: Atsushi Matsuzawa

Received: 13 July 2022
Accepted: 9 August 2022
Published: 11 August 2022

1. Introduction

Chalcone-based compounds can act as photoinitiators of polymerization under visible light with an excellent profile for (i) free radical polymerization, (ii) cationic polymerization, (iii) synthesis of interpenetrating polymer networks (IPNs), and (iv) thiol-ene reactions [1]. Additionally, fluorescent chalcone derivatives have been used for the development of a mouse embryonic stem cell probe [2]. On the other hand, in-depth pharmacological studies concluded that natural extracts containing chalcone-based compounds exhibit an impressive array of biological activities, including anti-inflammatory/anticancer effects [3–5]. Actually, chalcones may inhibit specific enzymes such as different kinases [6–8], the aldose reductase [9,10], cyclooxygenases [11,12], and inducible nitric oxide synthase [13,14]. Moreover, it has been shown that both synthetic and natural chalconoids play a healthy role in several diseases by inhibiting the NLRP3-inflammasome formation [15–19]. The NLRP3 inflammasome is a large protein complex controlling the production of caspase-1 and ultimately of pro-inflammatory cytokines (IL-1 and IL-18). In this context, it was reported that velutone F (**1**) (Figure 1), a retrochalcone [20–22] recently identified in the ethanolic extract of the leguminous plant *Millettia velutina* [23,24], inhibits the formation of the NLRP3 active complex. Among the eight new flavonoids identified in the lipophilic crude residue derived from 10 kg of dry vine stems of *Millettia velutina*, compound **1** exhibited the most potent inhibitory effect against nigericin-induced IL-1 release in THP-1 cells. Velutone F, featuring the 1-phenyl-2-propen-1-one moiety (**PhP**) and a substituted benzofuran core (**BF**), can be classified as a hybrid chalcone. The development of hybrid molecules incorporating different pharmacophores, each with its own molecular target, is an important area of research in medicinal chemistry [25]. Actually, both natural and synthetic benzofuran-derived compounds have potential therapeutic interests ranging from antibacterial, antifungal, anti-inflammatory, analgesic, antidepressant, anticonvulsant, anticancer, anti-HIV, antidiabetic, antituberculosis, and antioxidant [26–29].

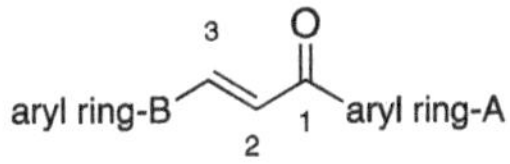

Chalcone compounds

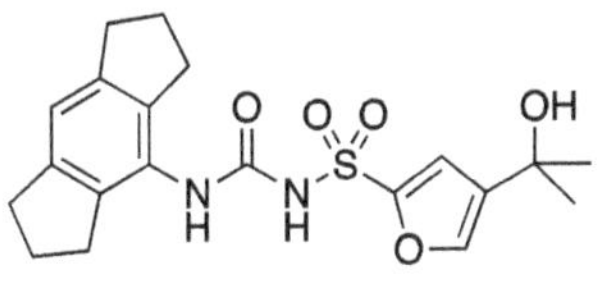

MCC950
o
CP-456,773

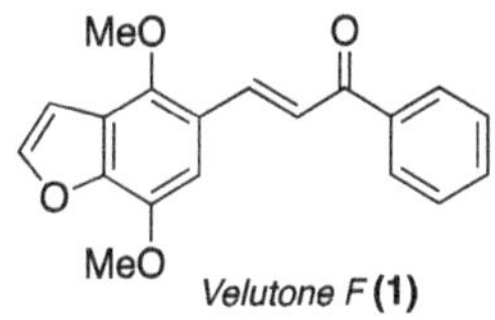

Velutone F (**1**)

Figure 1. Natural and non-natural inhibitors of the NLRP3 inflammasome activation.

We are currently involved in a multidisciplinary study aimed at identifying new anti-inflammatory/anticancer compounds that mimic the MCC950 molecular structure (Figure 1). It has been demonstrated that the diarylsulfonylurea MCC950 powerfully inhibits the NLRP3 activation selectively [30,31]. In detail, MCC950 would seem to reversibly bind the NLRP3 multi-protein complex making it unable to generate the active complex, namely the NLRP3-inflammasome [32,33]. Because of the profoundly different chemical structures of the synthetic MCC950 compared to the one of velutone F (**1**), it is reasonable to assume that they can block the activation of the NLRP3-inflammasome by interfering with different bio-chemical targets. We were intrigued by the possibility of disposing of multi-milligrams amount of the natural substance **1** for the purpose of undertaking pharmacochemical investigations and mostly to shed some light on the mechanism by which compound **1** inhibits the nigericin-induced IL-1 release. Actually, when we started the search, the chemical identity of **1** was ascertained by spectroscopic means exclusively, primarily 2D NMR [23]. However, very recently [24], the same research teams got proof of the chalconoid structure of compound **1** by its semi-synthesis from Khellin.

The reasons mentioned above prompted us to design/develop synthetic routes for preparing both the natural substance originally extracted from the tropical plant *Millettia velutina* and analogs to be studied by the biologist team as promising anti-inflammatory agents.

2. Results and Discussion

A host of synthetic strategies has been developed to create the trans-carbon-carbon double bond of the 1,3-diaryl-2-propen-1-one moiety featuring chalcone compounds, including Claisen-Schmidt's condensations, Wittig, and Julia-Kocienski olefinations. Instead, palladium-catalyzed reactions, namely Heck, Sonogashira, and Suzuki-Miyaura cross couplings, have been used to establish the 1,3-diaryl-enone framework through A-ring/C-1 and/or C1/C2 bond formation [34].

Initially, we planned a synthetic strategy to compound **1** based on the introduction of the required alkene by Wittig olefination of 5-formyl-4,7-dimethoxy benzofuran **7** (**5-FBF**) with the 1-phenyl-2-(triphenylphosphoranylidene)ethanone counterpart **8** [35]. To this end we elaborated two synthetic approaches for the preparation of the key intermediate

5-FBF differing in the way the benzofuran core (**BF**) could be formed via annellation of the carbocyclic ring system onto a preformed furan, synthetic pathway A (Scheme 1), or alternatively, by creating the furan ring by intramolecular cyclization on a preexistent carbocyclic, synthetic pathway B (Scheme 2).

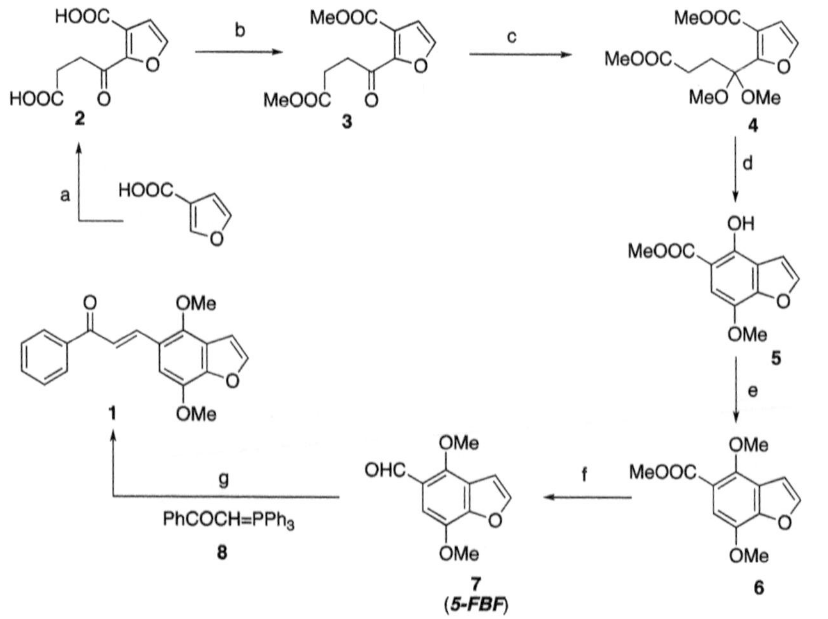

Scheme 1. Synthetic pathway A for target compound **1**. Reagents and conditions: (**a**) i: LDA, THF, −78 °C, succinic anhydride; ii: dil.HCl; 35%; (**b**) MeOH, H_2SO_4, 90%; (**c**) $HC(OMe)_3$, p-TsOH, MeOH reflux, 36 h, 90%; (**d**) i: t-BuOK, THF, −78 °C; ii: dil.HCl, 62%; (**e**) MeI, Cs_2CO_3, DMF, rt, 16 h, 96%; (**f**) i: $LiAlH_4$, Et_2O, 12 h; ii: PCC, CH_2Cl_2, rt, 1.5 h, 60% two steps; (**g**) MeCN, MW, 135 °C, 1.5 h, 70%.

2.1. *Synthetic Pathway A for Target Compound* 1

The first route starts by treating the 3-furoic acid with excess LDA to produce the corresponding C-2 lithiated carboxylate lithium salt, which, by reacting with succinic anhydride, gave the dicarboxylic acid **2**. The desired acyl derivative **3** could be isolated in a modest yield after esterification of **2** [36,37]. The subsequent carbonyl group protection as dimethyl ketal **4** opened the way to the creation of the annellated six-membered carbocyclic ring. Thus, the Dieckmann cyclization, performed with potassium tert-butoxide at −78 °C, occurred with simultaneous elimination of MeOH from the dimethyl ketal group yielding the aromatic derivative **5**, which was taken to the benzofuran derivative **6** by etherification. Subsequently, the methoxycarbonyl group was converted to the required formyl group by a two-step process entailing reduction with $LiAlH_4$ and oxidation of the resulting primary alcohol with pyridinium chlorochromate (PCC). The resulting **5-FBF** was eventually reacted with the stabilized phosphorous ylide **8**, in turn, prepared according to the literature [35]. The Wittig olefination under microwave irradiation provided velutone F (**1**) in 70% yield after chromatographic purification. As expected, NMR spectroscopic data for the synthesized velutone F were superimposable to the ones originally reported for the retrochalcone isolated from *Millettia velutina*.

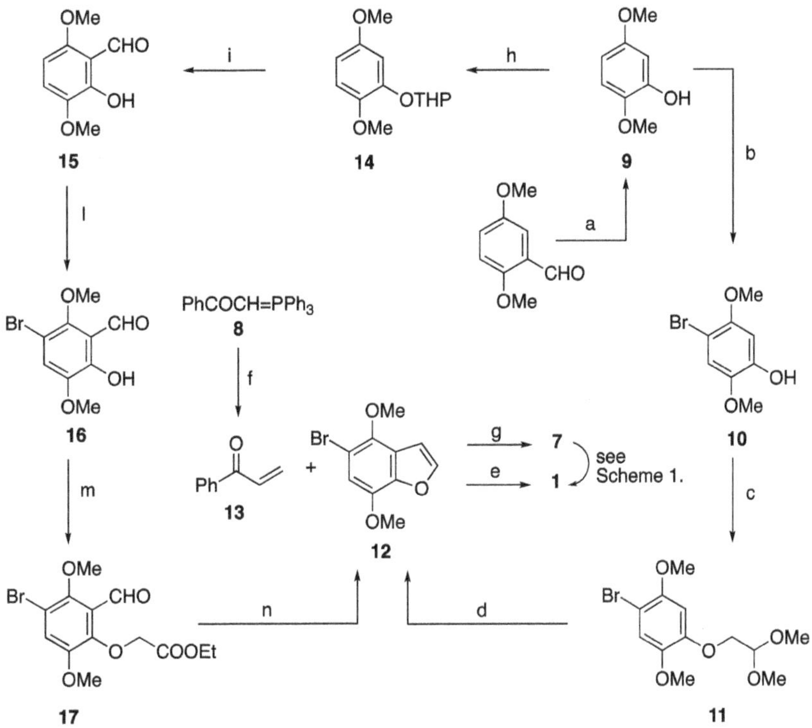

Scheme 2. Synthetic pathway B for target compound **1**. Reagents and conditions: (**a**) i: H_2O_2, t-BuOH, cat.SeO_2, rt, 12 h; ii: MeOH, K_2CO_3, 90% two steps; (**b**) NBS, MeCN, 0 °C, 30 min, 97%; (**c**) $BrCH_2CH(OMe)_2$, KOH, DMA, 140 °C, 1.5 h, 83%; (**d**) PPA, PhCl, reflux, 30 min, 50%; (**e**) $Pd(OAc)_2$, Tri(o-tolyl)phosphine, DIPEA:DMF (1:1), 32%; (**f**) CH_2Cl_2-formalin, reflux, 12 h, 80%; (**g**) i: BuLi, THF, −78 °C; ii: DMF, 50%; (**h**) DHP, PPTS, CH_2Cl_2, rt, 12 h, 91%; (**i**) i: BuLi, Et_2O, 0 °C, 2 h; ii: DMF, −78 to rt, 2 h; iii: HCl 6M, 80% overall; (**l**) Br_2-AcOH, AcONa, rt, 30 min, 80%; (**m**) $BrCH_2COOEt$, DMF, Cs_2CO_3, rt, 1 h, 85%; (**n**) i: LiOH, THF-H_2O, rt, 12 h; ii: AcONa, Ac_2O, 130 °C, 2 h; iii: EtOH, 60 °C, 2 h, 57%.

2.2. *Synthetic Pathway B for Target Compound* 1

The starting move of the alternative synthetic approach to **7** entailing the creation of the furan ring by intramolecular cyclization of the bromobenzene derivative **11** (Scheme 2) was the Dakin-like oxidation of the cheap 2,5-dimethoxybenzaldehyde by using the H_2O_2/cat. SeO_2 system in *tert*-BuOH [38]. Methanolysis of the resulting arylformate promptly furnished the 2,5-dimethoxyphenol **9**, which underwent regioselective bromination with NBS affording **10**. The exclusive C-4 bromination accounted for the marked para orienting effect of the phenolic hydroxyl group [39].

At this stage, with the aim of creating the annellated 2,3-unsubstituted furan ring, we needed to introduce an O-tethered functionalized two carbon fragment. Thus, compound **10** was easily converted to the aryl ether **11** by treatment with bromoacetaldehyde dimethyl acetal and KOH in dimethylacetamide (DMA) [40]. The anticipated intramolecular electrophilic aromatic substitution was carried out with polyphosphoric acid (PPA), providing the benzofuran derivative **12** in a 50% yield. Transformation of the 5-bromo benzofuran derivative **12** into 5-formyl-4,7-dimethoxy benzofuran **7** (**5-FBF**) was achieved through halogen-metal exchange with BuLi followed by reaction with DMF [41].

In order to make the critical furan ring annellation step more efficient, we turned our attention to a recently reported protocol for the synthesis of 6-hydroxybenzofuran based on an unusual cycloaddition-cycloreversion sequence [42]. To this end, we prepared the penta-substituted benzene derivative **16** starting from the tri-substituted phenol derivative **9** (Scheme 2), exploiting the well-known ability of the OTHP as an ortho-directing group for the metalation [43]. Accordingly, compound **9** was promptly transformed into the corresponding tetrahydropyranyl ether **14** that was at first ortho-lithiated by treatment with BuLi and later reacted with DMF to give the tetra-substituted phenyl derivative **15**. The subsequent bromination para to the phenol group [44] afforded the desired penta-substituted benzene **16**, to which the O-ethoxycarbonylmethylene fragment was easily inserted by standard etherification reaction. In previous saponification, compound **17** was treated with the Ac_2O-AcONa system with heating to give the required 5-bromo benzofuran derivative **12** in an appreciable 57% yield. The mechanism proposed for the interesting cyclization reaction entails dehydration of the carboxyl group to give an unstable ketene intermediate that is trapped intramolecularly by the formyl group. The thermal [2 + 2] heterocycloaddition reaction is followed by a cycloreversion with the expulsion of CO_2 and production of the 2,3-unsubstituted benzofuran derivative **12** [42].

Having the suitably derivatized aryl ring-B moiety in hand, we conceived preparing compound **1** by exploiting the Mizoroki-Heck cross-coupling reaction between **12** and 1-phenyl-2-propen-1-one **13**. The latter reagent was, in turn, easily prepared by reacting phosphorous ylide **8** and formaldehyde according to a known Wittig protocol [45] (Scheme 2). Disappointingly, the Pd(0)-catalyzed reaction provided the retrochalcone **1** with a modest 32% yield [46].

2.3. *Synthetic Pathways Providing the Non-Natural Regioisomers of Velutone F*

With the aim of learning about stereo-electronic properties of the hybrid benzofuran-retrochalcone scaffold, we decided to prepare the non-natural compounds **22, 23,** and **28** featuring the **PhP** moiety attached, respectively, at C-2, C-6, and C-3 of the 4,7-dimethoxy benzofuran core (**BF**). The non-natural regioisomers of velutone F are previously unknown compounds.

2.3.1. Synthetic Pathways to the Isomers 22 and 23

The direct formylation of electron-rich arenes can be conveniently accomplished via the Vilsmeier–Haack (V–H) reaction. Indeed, the benzo[b]furan nucleus is reported to yield the 2-formyl derivative by reaction with the V–H electrophilic species [47]. We anticipated that the regioselectivity of the V–H reaction could change if electron-donating groups were present on the phenyl ring of the benzo-fused system. In line with our hypothesis, we decided exploring the behavior of 4,7-dimethoxy benzo[b]furan **19** under V-H reaction conditions (Scheme 3). We planned to build the substituted benzo[b]furan **19** from **9** by creating the annellated 2,3-unsubstituted furan ring according to our previously sound synthetic pathway B for target compound **1** (Scheme 2). Thus, once etherified the phenolic group of **9** with the functionalized two carbon fragment, the resulting compound **18** was cyclized to **19** under the action of PPA (Sn-β zeolite also showed to efficiently promote this transformation [48]). As expected, we found the subsequent electrophilic aromatic substitution reaction was poorly regioselective: all but one of the regioisomeric formyl benzofuran derivatives **2-FBF**, **5-FBF**, and **6-FBF** were formed. In detail, chromatographic purification of the residue from the V-H reaction led us to isolate compounds **20** (**2-FBF**) together with **7** (**5-FBF**) in 37% yield (^{1}H NMR and HPLC analysis showed the isomers were in a 3.5:6.5 ratio), and compound **21** (**6-FBF**) in 30% yield. At this stage, we submitted the separated fractions to the Wittig olefination with the stabilized phosphorous ylide **8**. We obtained chalcone **23** from **6-FBF**, while in the same manner, the inseparable mixture of **2-FBF** and **5-FBF** furnished chalcones **22** and **1**, which, gratifyingly, could be easily separated by column chromatography.

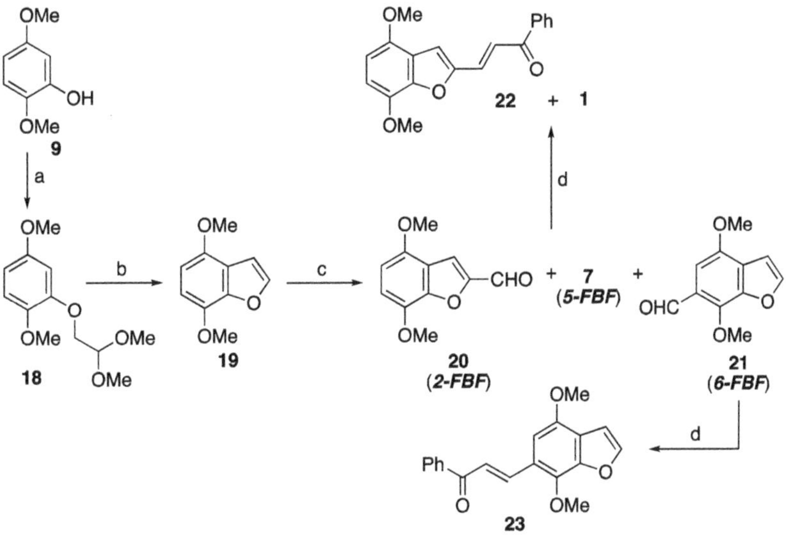

Scheme 3. Synthetic pathways to regioisomers **22** and **23**. Reagents and conditions: (**a**) KOH, DMAC, $BrCH_2CH(OMe)_2$, 140 °C, 2 h, 70%; (**b**) Sn-β zeolite, PhCl, 105 °C, 5 h, 62%; (**c**) $POCl_3$, DMF, 80 °C, 12 h, (**20**, 24%), (**7**, 13%), (**21**, 30%); (**d**) **8**, toluene, reflux, 12 h 50%.

2.3.2. Synthetic Pathway to the Isomer 28

We envisaged setting up the **PhP** moiety of **28** by reaction of 3-formyl benzofuran derivative **27** (**3-FBF**) with the ylide **8** according to the previously tested protocol entailing a classical Wittig olefination. About **3-FBF** preparation, we selected compound **25** as the direct precursor having in mind a controlled oxidation of its C-3 methyl substituent in order to derive the pivotal electrophilic functional group [49,50] (Scheme 4). Thus, we devised preparing compound **25** in two steps from phenol **9**, namely: etherification with bromoacetone followed by acid-promoted cyclization of the resulting aryloxy acetone derivative **24**. The planned SeO_2-oxidation of the C-3 methyl residue of compound **25** furnished a 1:1 mixture of the primary alcohol **26** and the aldehyde **27**, which were separable by chromatography. However, we found it very easy to carry out the oxidation of **26** to **3-FBF** by using the Corey–Suggs reagent (PCC). Eventually, the microwave-promoted Wittig reaction between the aryl aldehyde **27** (**3-FBF**) and the phosphorous ylide **8** provided the aimed chalconoid **28** in good yield.

2.3.3. Inhibition of IL-1β Release In Vitro and In Vivo

Among the inflammasomes, NLRP3 is the most studied and characterized due to its implication in the pathogenesis of different human diseases [51]. The activation of NLRP3 inflammasomes consists of caspase-1 activation, which in turn induces secretion of the inflammatory cytokine IL-1β. Hence, IL-1β release is the most used read-out for NLRP3 inflammasome activation. In this study, IL-1β release was determined by ELISA assay to assess the inhibitory effects of synthesized compounds on NLRP3 inflammasome activation both in mouse bone marrow-derived macrophages (BMDMs) and in human PMA-differentiated lipopolysaccharide (LPS)-primed THP-1 macrophages (Figure 2). Velutone F (**1**) and the non-natural compounds **22**, **23**, and **28** demonstrated a high inhibitory capacity on the release of IL-1β.

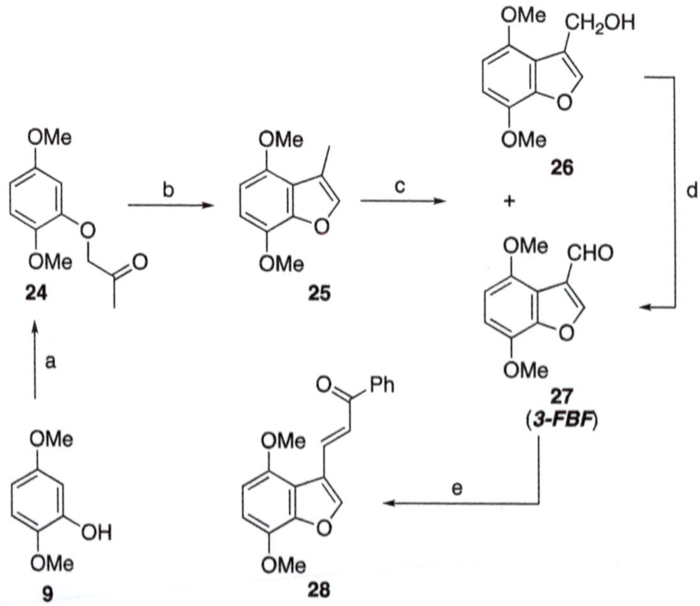

Scheme 4. Synthetic pathway to regioisomer **28**. Reagents and conditions: (**a**) $BrCH_2COCH_3$, Cs_2CO_3, DMF, rt, 12 h, 63%; (**b**) PPA, PhCl, 110 °C, 20 min, 50%; (**c**) SeO_2, dioxane, 75 °C, 96 h, 80%; (**d**) PCC, CH_2Cl_2, rt, 12 h, 95%; (**e**) **8**, MeCN, MW, 135 °C, 2 h, 65%.

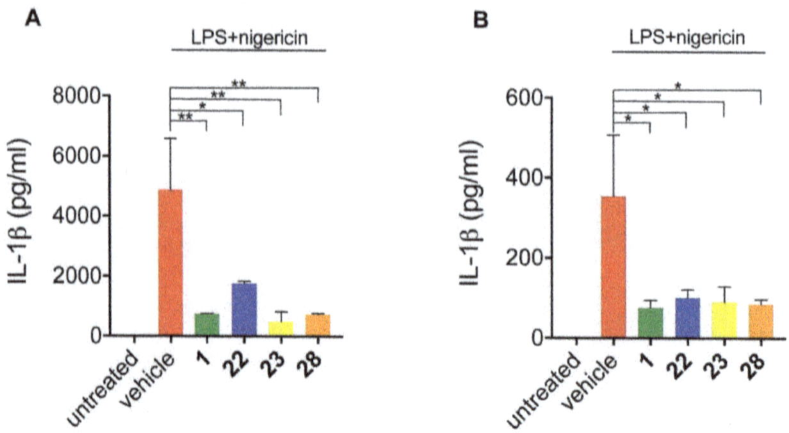

Figure 2. In vitro inhibition of IL-1β release. (**A**) BMDMs were stimulated with LPS (1 μg/mL) for 2 h, then treated with compounds for 30 min, and then activated with nigericin (10 μM) for 1 h. (**B**) LPS-primed PMA-differentiated THP-1 cells were stimulated with LPS (1 μg/mL) for 3 h, then treated with compounds for 30 min, and then activated with nigericin (10 μM) for 1 h. Cell supernatants were analyzed with ELISA test for the evaluation of IL-1β release (n = 3, mean ± SEM). Statistics differences were analyzed using one-way ANOVA: ** $p < 0.01$, * $p < 0.05$.

To confirm the anti-inflammatory activity of these compounds in vivo, we used an LPS-induced inflammation treatment. Mice were intraperitoneally (IP) pre-injected with velutone F (**1**) and the non-natural isomers **22**, **23**, and **28** or vehicle at 25 mg/kg and then were IP injected with LPS (1 mg/kg). After 4 h, plasma and peritoneal exudate were collected from mice and analyzed for evaluation of IL-1β release. The mice that received pre-treatment with our compounds displayed a dramatic reduction in IL-1 production both in peritoneal exudate and blood samples (Figure 3).

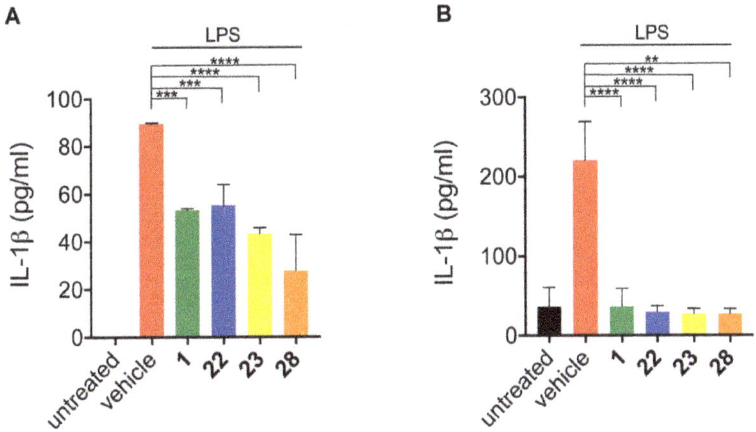

Figure 3. In vivo inhibition of IL-1β release. (**A**) Exudate and (**B**) plasma levels of IL-1β from C57BL/6 mice pretreated with compounds or vehicle control as determined by ELISA 4 h after intraperitoneally IP LPS (1 mg/kg) injection, n = 6/group. Data are shown as mean ± SEM. Statistical differences were analyzed using one-way ANOVA: **** $p < 0.0001$, *** $p < 0.001$, ** $p < 0.01$.

Taken together, these results revealed that synthesized velutone F (**1**), as well as its regioisomers 22, 23, and 28, exerted a strong inhibition on the NLRP3 inflammasome activation both in vitro and in vivo.

3. Materials and Methods

3.1. Chemistry: Materials and General

All reagents and solvents that were commercially purchased were directly used without prior treatment. Reaction temperatures were recorded using a regular thermometer without correction. Melting points were determined on the Reichert Termovar apparatus and are uncorrected. Reactions were monitored by analytical thin-layer chromatography (TLC) on silica gel F254 glass plates (Merck, Darmstad, Germany) and visualized under UV light 254 nm and $KMnO_4$. Mobile phases abbreviations: A = Ethyl acetate, P = petroleum ether, DCM = dichloromethane. ^{1}H NMR spectra were recorded with a Mercury Place Varian 400 MHz NMR spectrometer (Varian Palo Alto, CA, USA) at room temperature. Chemical shifts (in ppm) were recorded as parts per million (ppm) downfield to tetramethylsilane (TMS). The following abbreviations are used for a multiplicity of NMR signals: s, singlet; d, doublet; t, triplet; q, quartet; m, multiplet; dd, double doublet; dt, double triplet; ddd, doublet of doublet of doublets; dtd, doublet of triplet of doublets. MS was carried out using electrospray ionization (ESI MICROMASS ZQ 2000, Waters, Milford, MA, USA). HRMS was acquired by CIGS University of Modena e Reggio Emilia (Modena, Italy) using a Q-Exactive Hybrid Quadrupole Orbitrap (Thermo Scientific, Waltham, MA, USA). IR spectra were taken on a Perkin-Elmer FT-IR spectrum 100 spectrometer (Foster City, CA, USA). Analytical HPLC was performed with Beckman System Gold 168 (Milan, Italy) using a C18 100 A Phenomenex Kinetex (150 × 4.6 mm) and UV-DAD detection. [A]: 0.1% TFA in water and [B]: 0.1% TFA in acetonitrile was used as a binary mobile phase at a flow rate of 0.7 mL/min. Gradient: from 0% to 100% [B] in 25 min. Microwave reactions were performed in a Biotage Initiator Plus reactor (Biotage, Sweden) using 2–5 mL glass vials with a rubber cap. For all experimental spectra see Supplementary Materials.

3.1.1. Methyl 3-(4-Methoxy-4-oxobutanoyl)furan-2-carboxylate (3)

A solution of 3-Furoic acid (1 g, 8.92 mmol, 1 equiv.) in anhydrous THF (6 mL) was added dropwise, under argon atmosphere, to a cooled (−78 °C) solution of LDA, which was previously prepared from *n*-BuLi (12.5 mL, 1.6 M in hexane, 19.96 mmol, 2.2 equiv.) and *i*-Pr_2NH (2.9 mL, 19.96 mmol, 2.2 equiv.) in THF (25 mL). The reaction mixture was stirred at −78 °C for 2 h, then a solution of succinic anhydride (1 g, 9.99 mmol, 1.1 equiv.) in anhydrous THF (8 mL) was added to the reaction flask. The cooling bath was removed, and aqueous 2 M HCl (30 mL) was added at room temperature. The mixture was extracted with chloroform, the organic phase was dried over Na_2SO_4, and solvent was evaporated. The solid residue was dissolved in MeOH (50 mL), cooled at 0 °C, and H_2SO_4 (0.45 mL, 95%) was added dropwise. The mixture was stirred at room temperature for 24 h, then a saturated $NaHCO_3$ solution was added, and the mixture was extracted with AcOEt. The organic phase, dried over Na_2SO_4, was evaporated under reduced pressure to give a crude that was purified by silica gel column chromatography with A/P 1:9, *v*/*v* as a solvent system. Title compound (**3**) was isolated as a colorless oil. Yield 32% over two steps. ^{1}H NMR (400 MHz, Chloroform-*d*) δ 7.51 (d, *J* = 1.7 Hz, 1H), 6.81 (d, *J* = 1.7 Hz, 1H), 3.89 (s, 3H), 3.68 (s, 3H), 3.31 (t, *J* = 6.7 Hz, 2H), 2.73 (t, *J* = 6.6 Hz, 2H). ^{13}C NMR (101 MHz, Chloroform-*d*) δ 187.72, 172.96, 162.89, 150.83, 144.36, 122.29, 113.62, 52.46, 51.83, 34.92, 27.47 [36,37].

3.1.2. Methyl 3-(1,1,4-Trimethoxy-4-oxobutyl)furan-2-carboxylate (4)

A solution of compound **3** (0.9 gr, 3.75 mmol, 1 equiv.), $HC(OMe)_3$ (2.1 mL, 19.48 mmol, 5.2 equiv.), and *p*-TsOH (0.05 gr) in MeOH (8 mL) was refluxed for 48 h. The reaction mixture was cooled to room temperature, pyridine (0.18 mL) was added, and the solvent evaporated.The residue was purified by silica gel column chromatography with A/P 1:4, *v*/*v* as a solvent system. Title compound **4** was obtained as a transparent oil. Yield 90%. ^{1}H NMR (400 MHz, Chloroform-*d*) δ 7.36 (d, *J* = 1.7 Hz, 1H), 6.66 (d, *J* = 1.7 Hz, 1H), 3.81 (s, 3H), 3.58 (s, 3H), 3.20 (s, 6H), 2.47 (t, *J* = 6.7 Hz, 2H), 2.16 (t, *J* = 6.6 Hz, 2H). ^{13}C NMR (101 MHz, Chloroform-*d*) δ 173.51, 163.81, 154.87, 141.84, 117.12, 112.38, 111.59, 101.72, 52.18, 51.97, 49.51, 30.35, 28.92 [36,37].

3.1.3. Methyl 4-Hydroxy-7-methoxybenzofuran-5-carboxylate (5)

tert-BuOK (0.61 g, 5.45 mmol, 2.2 equiv.) was poured into anhydrous THF (65 mL) under an argon atmosphere, and the suspension was cooled to −78 °C. A solution of compound **4** (0.71 g, 2.48 mmol, 1 equiv.) in anhydrous THF (2 mL) was added dropwise, and the mixture turned orange. After 1.5 h of stirring at −78 °C, anhydrous HCl (4 M in 1,4-dioxane) was added. The resulting transparent yellow reaction mixture was allowed to room temperature and stirred for 1 h. The precipitate was filtered through a Gooch filter, and the solvent was evaporated under reduced pressure. Chromatographic purification of the residue with A/P 1:9 *v*/*v* as the eluent phase provided title compound **5** as a white solid. Melting point: 125–127 °C. Yield 62%. ^{1}H NMR (400 MHz, Chloroform-*d*) δ 11.14 (s, 1H), 7.60 (dd, *J* = 2.1 Hz, 1H), 7.17 (s, 1H), 6.98 (d, *J* = 2.1 Hz, 1H), 3.97 (d, *J* = 1.6 Hz, 6H). ^{13}C NMR (101 MHz, Chloroform-*d*) δ 171.09, 152.43, 149.52, 144.76, 139.19, 118.90, 105.66, 105.38, 105.02, 56.51, 52.29 [36,37].

3.1.4. Methyl 4,7-Dimethoxybenzofuran-5-carboxylate (6)

A mixture of compound **5** (0.21 g, 0.95 mmol, 1 equiv.), Cs_2CO_3 (1.55 g, 5.75 mmol, 5 equiv.), and CH_3I (0.29 mL, 4.75 mmol, 5 equiv.) in anhydrous DMF (6 mL) was stirred at room temperature for 16 h under an argon atmosphere. After complete conversion of starting material (checked by TLC A/P9 1:9), the reaction was partitioned between HCl 2 M and AcOEt; the organic phase was washed with HCl 2 M (×2) and brine, dried over Na_2SO_4, and the solvent was evaporated to dryness. Purification of the crude by column chromatography with A/P 1:4 as the eluent phase afforded title compound **6** as a colorless oil. Yield 96%. ^{1}H NMR (300 MHz, Chloroform-*d*) δ 7.63–7.61 (m, 1H), 7.26 (s, 1H), 6.95–6.94

(m, 1H), 4.00 (d, J = 0.8 Hz, 6H), 3.94 (d, J = 0.8 Hz, 3H). ^{13}C NMR (75 MHz, Chloroform-d) δ 167.04, 149.35, 148.04, 145.57, 141.84, 123.31, 117.58, 108.69, 105.80, 62.46, 56.77, 52.52.

3.1.5. 4,7-Dimethoxybenzofuran-5-carbaldehyde (**7**)

A solution of compound **6** (0.2 g, 0.84 mmol, 1 equiv.) in dry Et_2O (4 mL) was added dropwise to a stirred to a cooled (0 °C) suspension of $LiAlH_4$ (0.08 g, 2.1 mmol, 2.5 equiv.) in anhydrous Et_2O (5 mL) under argon atmosphere. The reaction mixture was stirred at room temperature overnight, then it was cooled to 0 °C, and aqueous NaOH solution (5 mL, 2 M) was added. After 30 min, the mixture was filtered on a celite pad, and the phases were separated. The aqueous phase was extracted with Et_2O. The combined organics were dried over Na_2SO_4, and the solvent was evaporated to give (4,7-dimethoxybenzofuran-5-yl)methanol as a white crystalline solid, which was used in the next oxidative step without further purification.

A solution of the primary alcohol (0.14 g, 1.29 mmol, 1 equiv.) in anhydrous DCM (5 mL) was added to a suspension of PCC (0.42 g, 1.93 mmol, 1.5 equiv.) in DCM (5 mL). The dark brown reaction mixture was magnetically stirred at room temperature for 1.5 h then Et_2O was added. The supernatant was decanted from the black insoluble residue, which was further washed with Et_2O (×3). The combined organic phases were filtered through a pad of silica, and the solvent was evaporated. Purification of the crude by column chromatography with A/P 1:4 v/v as a solvent system furnished title compound **7** as a white crystalline solid. Melting point: 142–145 °C. Yield 50%.^{1}H NMR (400 MHz, Chloroform-d) δ 10.46 (s, 1H), 7.67 (dd, J = 2.3, 0.4 Hz, 1H), 7.25 (s, 1H), 7.02 (d, J = 2.3 Hz, 1H), 4.15 (s, 3H), 4.00 (s, 3H). ^{13}C NMR (101 MHz, Chloroform-d) δ 189.09, 149.92, 145.30, 142.04, 122.24, 120.30, 105.88, 103.45, 61.74, 56.29.

3.1.6. 2,5-Dimethoxyphenol (**9**)

H_2O_2 (2.4 mL, 30% w/w, 78.21 mmol, 5.4 equiv.) was added dropwise to a magnetically stirred mixture of 2,5-dimethoxybenzaldehyde (2.42 g, 14.56 mmol, 1 equiv.), SeO_2 (0.038 g, 0.342 mmol, 0.023 equiv.) in *tert*-BuOH (6 mL). The reaction mixture was stirred overnight at room temperature then Et_2O (20 mL) was added. The resulting mixture was washed with brine (×3), the organic phase was dried over Na_2SO_4, and the solvent was evaporated to dryness. The residue was dissolved in MeOH (10 mL), and a solution of K_2CO_3 (2 g) in H_2O (10 mL) was added to the solution. After 1 h, HCl 6M was added to reach pH 1, and the reaction mixture was extracted with DMC (20 mL ×3). The combined organic phases were dried over anhydrous Na_2SO_4, and the solvent was evaporated to dryness. Purification was carried out by silica gel column chromatography with A/P 1.5:8.5 v/v to obtain compound **9** as a transparent oil. Yield 90%. ^{1}H NMR (400 MHz, Chloroform-d) δ 6.77 (d, J = 8.8 Hz, 1H), 6.56 (d, J = 2.9 Hz, 1H), 6.38 (dd, J = 8.8, 2.9 Hz, 1H), 5.66 (s, 1H), 3.84 (s, 3H), 3.75 (s, 3H). ^{13}C NMR (101 MHz, Chloroform-d) δ 154.56, 146.44, 140.96, 111.45, 104.23, 101.73, 56.58, 55.66 [38].

3.1.7. 4-Bromo-2,5-dimethoxyphenol (**10**)

NBS (1.16 g, 6.55 mmol, 1 equiv.) was added to a cooled (0 °C) solution of **9** (1.0 g, 6.55 mmol, 1 equiv.) in acetonitrile (130 mL). Stirring at 0 °C was continued for 30 min. Then, $Na_2S_2O_3$ saturated solution was added, and the mixture was extracted with AcOEt (×3). The combined organic phases were washed with brine, dried over Na_2SO_4, and the solvent was evaporated. Purification of the residue by column chromatography by eluting with A/P 1:4 v/v gave **10** as a transparent oil. Yield 97%. ^{1}H NMR (400 MHz, Chloroform-d) δ 7.02 (s, 1H), 6.61 (s, 1H), 5.65 (s, 1H), 3.83 (s, 3H), 3.81 (s, 3H). ^{13}C NMR (101 MHz, Chloroform-d) δ 150.66, 145.71, 140.94, 115.77, 100.51, 99.81, 56.77, 56.74.

3.1.8. 1-Bromo-4-(2,2-dimethoxyethoxy)-2,5-dimethoxybenzene (**11**)

$BrCH_2CH(OMe)_2$ (1.0 mL, 8.82 mmol, 1.5 equiv.) was added dropwise to a stirred solution of **10** (1.37 g, 5.88 mmol, 1 equiv.) and KOH (0.66 g, 11.756 mmol, 2 equiv.) in

DMAC (12 mL). The reaction mixture was stirred at 140 °C for 1.5 h and then cooled to room temperature. H_2O (20 mL) was added, and the resulting mixture was extracted with Et_2O (40 mL ×3). The combined organic phases were washed with aqueous NaOH (30 mL, 5%) and H_2O (30 mL), dried over Na_2SO_4, and the solvent was evaporated to dryness. Purification of the residue by column chromatography with A/P 1:4 *v*/*v* as the eluent furnished **11** as a white crystalline solid. Melting point: 57–60 °C. Yield 83%. ^{1}H NMR (400 MHz, Chloroform-*d*) δ 7.05 (s, 1H), 6.63 (s, 1H), 4.72 (t, *J* = 5.2 Hz, 1H), 4.04 (d, *J* = 5.2 Hz, 0H), 3.83 (s, 3H), 3.80 (s, 3H), 3.46 (s, 6H). ^{13}C NMR (101 MHz, Chloroform-*d*) δ 150.30, 148.22, 144.66, 117.47, 102.52, 102.07, 69.94, 57.10, 56.87, 54.51.

3.1.9. 5-Bromo-4,7-dimethoxybenzofuran (**12**) by Cyclization of (**11**)

A solution of compound **11** (0.37 g, 1.55 mmol, 1 equiv.) in chlorobenzene (3 mL) was added dropwise to a heated (120 °C) mixture of polyphosphoric acid (0.5 g) in chlorobenzene (15 mL). After 30 min, the reaction mixture was cooled to room temperature, diluted with Et_2O, and washed with H_2O. The removal of volatiles under reduced pressure gave a residue that was purified by column chromatography with A/P 0.5:9.5 *v*/*v*. Compound **12** was isolated as a white crystalline solid. Melting point: 63–65 °C. Yield 50%. ^{1}H NMR (400 MHz, Chloroform-*d*) δ 7.59 (dd, *J* = 2.2, 0.5 Hz, 1H), 6.92 (s, 1H), 6.87 (dd, *J* = 2.1, 0.4 Hz, 1H), 3.97 (d, *J* = 1.7 Hz, 6H). ^{13}C NMR (101 MHz, Chloroform-*d*) δ 145.11, 144.54, 143.95, 142.09, 122.44, 110.55, 107.66, 104.60, 61.06, 56.57.

3.1.10. 4,7-Dimethoxybenzofuran-5-carbaldehyde (**7**) via Halogen-Metal Exchange

n-BuLi (1.32 mL, 1.6 M in hexane, 2.1 equiv.) was added to a cooled (−78 °C) solution of compound **12** (0.26 g, 1.0 mmol, 1.0 equiv) in dry THF (5 mL) under argon atmosphere. The reaction mixture was stirred at −78 °C for 20 min. Then, DMF (0.25 mL, 3.25 equiv.) was added, and the cooling bath was removed. After 12 h, AcOEt (10 mL) was added, and the reaction was quenched with NH_4Cl saturated aqueous solution. The organic layer was separated, washed with brine, dried over Na_2SO_4, and the solvent was evaporated to dryness. Chromatographic purification of the residue with A/P 1:4 *v*/*v* afforded the aldehyde **7** as a white crystalline solid. Melting point: 142–145 °C. Yield 50%. ^{1}H NMR (400 MHz, Chloroform-*d*) δ 10.46 (s, 1H), 7.67 (dd, *J* = 2.3, 0.4 Hz, 1H), 7.25 (s, 1H), 7.02 (d, *J* = 2.3 Hz, 1H), 4.15 (s, 3H), 4.00 (s, 3H). ^{13}C NMR (101 MHz, Chloroform-*d*) δ 189.09, 149.92, 145.30, 142.04, 122.24, 120.30, 105.88, 103.45, 61.74, 56.29.

3.1.11. 1-Phenylprop-2-en-1-one (**13**)

Formalin (0.8 mL, 37%, 10 mmol, 5 equiv.) was added to a solution of the ylide **8** (0.76 g, 2 mmol, 1 equiv.) in DCM (8 mL), and the mixture was heated at reflux overnight. After washing with brine, the organic phase was dried over anhydrous Na_2SO_4 and evaporated to dryness. Purification of the residue by column chromatography with A/P 1:4 *v*/*v* as the eluent phase furnished compound **13** as a clear oil. Yield 80%. ^{1}H NMR (400 MHz, Chloroform-*d*) δ 7.97–7.93 (m, 2H), 7.58–7.55 (m, 1H), 7.51–7.45 (m, 2H), 7.16 (ddd, *J* = 17.2, 10.6, 0.6 Hz, 1H), 6.44 (dd, *J* = 17.2, 1.7 Hz, 1H), 5.95–5.91 (m, 1H). ^{13}C NMR (101 MHz, Chloroform-*d*) δ 191.14, 137.35, 133.06, 132.47, 130.26, 129.00, 128.77, 128.70, 128.07 [45].

3.1.12. 2-(2,5-Dimethoxyphenoxy) tetrahydro-2H-pyran (**14**)

DHP (4.16 mL, 46 mmol, 10 equiv.) was added to a solution of compound **9** (0.71 g, 4.6 mmol, 1 equiv.) and PPTS (0.115 g, 0.46 mmol, 0.1 equiv.) in DCM (10 mL). The reaction mixture was stirred at room temperature overnight, then it was diluted with DCM and washed with NaOH solution and brine. The organic phase was dried over Na_2SO_4 and evaporated. Purification was performed by column chromatography using A/P 1:9 *v*/*v* as the eluent phase. Compound **14** was isolated as a clear oil. Yield 91%. ^{1}H NMR (400 MHz, DMSO-d_6) δ 6.89 (d, *J* = 8.8 Hz, 1H), 6.68 (d, *J* = 2.9 Hz, 1H), 6.50 (dd, *J* = 8.9, 2.9 Hz, 1H), 5.39 (t, *J* = 3.3 Hz, 1H), 3.82 (ddd, *J* = 11.1, 8.9, 3.8 Hz, 1H), 3.71 (s, 3H), 3.67 (s, 3H), 3.53 (dtd, *J* = 11.5, 4.3, 1.2 Hz, 1H), 1.94–1.38 (m, 7H) [44].

3.1.13. 2-Hydroxy-3,6-dimethoxybenzaldehyde (**15**)

n-BuLi (4 mL, 1.6 M in hexane, 6.37 mmol, 1.1 equiv.) was added dropwise to a cooled (0 °C) solution of compound **14** (1.38 g, 5.79 mmol, 1 equiv.) in anhydrous Et_2O (60 mL) under argon atmosphere. The reaction mixture turned purple, then yellow, and stirring was continued at room temperature for 2 h. The reaction flask was cooled to −78 °C, and anhydrous DMF (1.78 mL, 23.16 mmol, 4 equiv.) was added. The reaction mixture was stirred at room temperature for 2 h, quenched with HCl (5 mL, 6 N), and stirred at room temperature for 1 h. The aqueous phase was extracted with AcOEt (×3), the combined organics were dried over Na_2SO_4, and the solvent was evaporated. Purification of the residue was performed by column chromatography with A/P 1:4 v/v as the eluent phase. Compound 15 was isolated as bright yellow solid. Melting point: 67–69 °C. Yield 80%. ^{1}H NMR (400 MHz, Chloroform-*d*) δ 12.17 (s, 1H), 10.31 (s, 1H), 7.02 (d, *J* = 8.9 Hz, 1H), 6.27 (d, *J* = 8.9 Hz, 1H), 3.84 (d, *J* = 0.8 Hz, 6H). ^{13}C NMR (101 MHz, Chloroform-*d*) δ 194.91, 155.87, 153.58, 142.08, 120.27, 111.09, 99.36, 56.90, 55.78 [44].

3.1.14. 3-Bromo-6-hydroxy-2,5-dimethoxybenzaldehyde (**16**)

AcONa (0.091 gr, 1.1 mmol, 1.1 equiv.) and then Br_2 (0.056 mL, 1.1 mmol, 1.1 equiv.) was added to a cooled (0 °C) solution of compound **15** (0.182 g, 1 mmol, 1 eq) in acetic acid (3 mL). The reaction mixture was stirred at room temperature for 30 min. Then, H_2O (10 mL) was added, and the mixture was extracted with DCM (10 mL × 3). The combined organic phases were washed with H_2O (10 mL), dried over Na_2SO_4, and evaporated. Purification of the residue by column chromatography with A/P 1:4 v/v as the eluent phase furnished compound **16** as a light-yellow solid. Melting point: 92–95 °C. Yield 80%. ^{1}H NMR (400 MHz, Chloroform-*d*) δ 11.89 (s, 1H), 10.21 (s, 1H), 7.19 (s, 1H), 3.92 (s, 3H), 3.87 (s, 3H). ^{13}C NMR (101 MHz, Chloroform-*d*) δ 194.91, 152.63, 152.47, 145.51, 122.50, 115.15, 104.06, 63.41, 56.67 [44].

3.1.15. Ethyl 2-(4-bromo-2-formyl-3,6-dimethoxyphenoxy)acetate (**17**)

Cs_2CO_3 (0.36 g, 1.1 mmol, 1.1 equiv.) and $BrCH_2COOEt$ (0.12 mL, 1.1 mmol, 1.1 equiv.) were added to a solution of compound **16** (0.261 g, 1 mmol, 1 equiv.) in DMF (5 mL). The reaction mixture was stirred at room temperature for 1 h. H_2O (10 mL) was added, and the mixture was extracted with AcOEt (20 mL × 3). The combined organic phases were washed with H_2O (20 mL), dried over Na_2SO_4, and the solvent was evaporated. Compound **17** was isolated as a white grainy solid after trituration with petroleum ether. Melting point: 85–86 °C. Yield 85%. ^{1}H NMR (400 MHz, Chloroform-*d*) δ 10.49 (s, 1H), 7.27 (s, 1H), 4.76 (s, 2H), 4.23 (q, *J* = 7.1 Hz, 2H), 3.85 (d, *J* = 2.1 Hz, 6H), 1.28 (t, *J* = 7.1 Hz, 3H). ^{13}C NMR (101 MHz, Chloroform-*d*) δ 189.32, 168.96, 150.74, 148.89, 148.75, 121.27, 112.82, 69.75, 62.64, 61.36, 56.66, 14.24.

3.1.16. 5-Bromo-4,7-dimethoxybenzofuran (**12**) via Cyclization of (**17**)

$LiOH{\cdot}H_2O$ (0.032 g, 0.78 mmol, 3 equiv.) was added to a magnetically stirred solution of compound **17** (0.09 g, 0.26 mmol, 1 equiv.) in a 3:1 THF/H_2O (4 mL) mixture. After stirring at room temperature overnight, the reaction mixture was acidified with 1 M HCl and extracted with AcOEt (10 mL × 3). The combined organics were washed with brine, dried over anhydrous Na_2SO_4, and evaporated to dryness. The white solid residue was dissolved in Ac_2O (3 mL), AcONa (0.085 g, 1.035 mmol, 1.5 equiv.) was added, and the mixture was heated at 130 °C for 2 h. Successively, the temperature was lowered to 60 °C, and EtOH (3 mL) was added dropwise to consume acetic anhydride. The stirring was continued at 60 °C for 2 h. then, the reaction mixture was cooled to room temperature. H_2O (3 mL) was added, and the mixture was extracted with AcOEt (10 mL). The organic phase was washed with NaOH (5 mL 2 N) and with H_2O (5 mL), dried over Na_2SO_4, and evaporated. Chromatographic purification of the residue with A/P 0.5:9.5 provided compound 12 as a white crystalline solid. Melting point: 63–65 °C. Yield 57%.^{1}H NMR (400 MHz, Chloroform-*d*) δ 7.59 (dd, *J* = 2.2, 0.5 Hz, 1H), 6.92 (s, 1H), 6.87 (dd, *J* = 2.1, 0.4 Hz,

1H), 3.97 (d, *J* = 1.7 Hz, 6H).^{13}C NMR (101 MHz, Chloroform-*d*) δ 145.11, 144.54, 143.95, 142.09, 122.44, 110.55, 107.66, 104.60, 61.06, 56.57.

3.1.17. Synthesis of (**1**) from (**12**) via Mizoroki-Heck

A reaction flask containing DMF (1 mL) and DIPEA (1 mL) was charged with phenylpropenone **13** (0.165 g, 1.25 mmol, 1.25 equiv.), aryl bromide **12** (0.257 g, 1 mmol, 1 equiv.), P(o-tol)$_3$ (0.121 gr, 0.4 mmol), and Pd(OAc)$_2$ (0.011 g, 0.05 mmol, 0.05 equiv.). The reaction mixture was stirred under an argon atmosphere at 110 °C for 5 h. Then, AcOEt (10 mL) and H_2O (10 mL) were added to the residue after solvent evaporation. The insoluble solid was filtered on celite, and the organic phase was washed with brine and dried over Na_2SO_4. The solvent was evaporated, and Chromatographic purification of the residue with A/P 1:4 *v*/*v* provided compound **1** [10] as a yellow crystalline solid. Melting point: 98–103 °C. Yield 32%. ^{1}H NMR (400 MHz, Chloroform-*d*) δ 8.22 (d, J = 15.8 Hz, 1H), 8.05–8.01 (m, 2H), 7.62 (d, *J* = 2.2 Hz, 1H), 7.59–7.48 (m, 4H), 7.05 (s, 1H), 6.95 (dd, *J* = 2.2, 0.6 Hz, 1H), 4.05 (s, 3H), 4.04 (s, 3H). ^{13}C NMR (101 MHz, Chloroform-*d*) δ 191.00, 148.07, 144.97, 141.96, 140.32, 138.60, 132.53, 128.54, 128.50, 121.68, 121.47, 120.74, 105.34, 104.97, 61.51, 56.50. HPLC r.t. = 23,233 min. ESI = 309,2086 [M + H]$^+$. HRMS *m*/*z*: [M + H]$^+$ calc for $C_{19}H_{16}O_4$ 309.11214, found 309.1121; [M + Na]$^+$ calc 331.09408 found 331.0937. I.R. (neat) cm^{-1} = 3126, 1654, 1598, 1586, 1572, 1480, 1440, 1354, 1287, 1231, 1202, 1185, 1159, 1072, 1013, 995, 935, 875, 853, 771, 738, 718, 704, 685, 662.

3.1.18. Synthesis of (**1**) from (**7**) via Wittig

A microwave vial was charged with aldehyde **7** (0.065 g, 0.3 mmol, 1 equiv.), ylide **8** (0.125 g, 0.33 mmol, 1.1 equiv.), and acetonitrile (3 mL); microwave was set at 135 °C for 1.5 h. The solvent was evaporated and the residue chromatographed by eluting with A/P 1:4 *v*/*v*. Yield 70%.

3.1.19. (E)-3-(4,7-Dimethoxybenzofuran-2-yl)-1-phenylprop-2-en-1-one (**22**)

A microwave vial was charged with aldehyde (**20**, in mixture ratio 65:35 with 7) (0.08 g, 0.38 mmol, 1 equiv.), ylide **8** (0.144 g, 0.38 mmol, 1 equiv.), and acetonitrile (3 mL); microwave was set at 135 °C for 1.5 h. The solvent was evaporated and the residue chromatographed by eluting with A/P 1:4 *v*/*v*. Yield 70%. Melting Point: 155–160 °C. ^{1}H NMR (400 MHz, Chloroform-*d*) δ 8.12–8.06 (m, 2H), 7.71 (d, *J* = 4.9 Hz, 2H), 7.59 (ddt, *J* = 8.3, 6.6, 1.4 Hz, 1H), 7.55–7.47 (m, 2H), 7.13 (s, 1H), 6.80 (d, *J* = 8.6 Hz, 1H), 6.54 (d, *J* = 8.6 Hz, 1H), 4.01 (s, 3H), 3.91 (s, 3H). ^{13}C NMR (101 MHz, Chloroform-*d*) δ 189.04, 151.81, 147.38, 145.17, 139.38, 137.41, 132.41, 130.13, 128.09, 128.05, 121.06, 120.40, 110.05, 108.56, 102.49, 56.09, 55.25. HPLC r.t. = 22,617 min. ESI = 3,092,086 [M + 1]. HRMS *m*/*z*: [M + H]$^+$ calc for $C_{19}H_{16}O_4$ 309.11214, found 309.1122; [M + Na]$^+$ calc 331.09408 found 331.094.

3.1.20. Synthesis of (**22** and **1**) from (**20** and **7**) via Wittig by Convectional Heating

A round bottom flask vial was charged with aldehydes (**20**, in mixture ratio 65:35 with **7**) (0.08 g, 0.38 mmol, 1 equiv.), ylide 8 (0.144 g, 0.38 mmol, 1 equiv.), and toluene (3 mL); the mixture was refluxed overnight. After complete conversion of the starting material, the solvent was evaporated and the residue chromatographed by eluting with A/P 1:4 *v*/*v*. Yield 70%.

3.1.21. (E)-3-(4,7-Dimethoxybenzofuran-6-yl)-1-phenylprop-2-en-1-one (**23**)

A microwave vial was charged with aldehyde **21** (0,08 mg, 0.38 mmol, 1 equiv.), ylide **8** (144 mg, 0.38 mmol, 1 equiv.), and acetonitrile (3 mL); the microwave was set at 135 °C for 1.5 h. The solvent was evaporated and the residue chromatographed by eluting with A/P 1:4 *v*/*v*. Yield 70%. Melting Point 138–141 °C ^{1}H NMR (400 MHz, Chloroform-*d*) δ 8.23 (d, *J* = 15.9 Hz, 1H), 8.05–8.02 (m, 2H), 7.62 (d, *J* = 2.1 Hz, 1H), 7.61–7.49 (m, 4H), 6.89 (d, *J* = 2.1 Hz, 1H), 6.87 (s, 1H), 4.17 (s, 3H), 3.97 (s, 3H).^{13}C NMR (101 MHz, Chloroform-*d*) δ 191.16, 148.59, 147.46, 145.45, 140.43, 138.72, 132.61, 128.64, 128.61, 122.92, 122.26, 122.11,

105.07, 101.56, 61.41, 55.92.HPLC r.t. = 24,183 min. ESI = 3,092,086 [M + 1]. HRMS *m/z*: $[M + H]^+$ calc for $C_{19}H_{16}O_4$ 309.11214, found 309.1121; $[M + Na]^+$ calc 331.09408 found 331.0937.

3.1.22. 2-(Dimethoxymethoxy)-1,4-dimethoxybenzene (**18**)

Bromoacetaldehyde dimethyl acetale (1.14 mL, 9.73 mmol, 1.5 equiv.) was added dropwise to a solution of **9** (1 g, 6.49 mmol, 1 equiv.), KOH (727 mg, 13 mmol, 2 equiv.) in DMAC (13 mL) at room temperature. The mixture was heated at 140 °C and stirred for 2 h until the starting material was no longer detected by TLC (A/P 1:4). It was allowed to cool down to room temperature and quenched with H_2O (20 mL). The crude was extracted with Et_2O (40 mL × 3), and combined organic phases were washed with an aqueous solution of 5% NaOH (30 mL) and H_2O (30 mL). After drying over anhydrous Na_2SO_4, the ether extraction phase was evaporated in vacuo. The residue was purified by silica gel column chromatography (elution system A/P 1:4 v/v) to obtain **18** as a white crystalline solid. Yield 70%. ^{1}H NMR (400 MHz, Chloroform-*d*) δ 6.80 (d, J = 8.8 Hz, 1H), 6.56 (d, J = 2.9 Hz, 1H), 6.44 (dd, J = 8.8, 2.8 Hz, 1H), 4.77 (t, J = 5.2 Hz, 1H), 4.03 (d, J = 5.2 Hz, 2H), 3.78 (d, J = 21.9 Hz, 6H), 3.46 (s, 6H). ^{13}C NMR (101 MHz, Chloroform-*d*) δ 154.26, 149.08, 144.18, 113.21, 104.69, 102.73, 102.37, 69.16, 56.84, 55.73, 54.35.

3.1.23. 4,7-Dimethoxybenzofuran (**19**)

Compound **18** (1 g, 4.12 mmol, 1 equiv.) and Sn-β zeolite (4.12 g) (kindly prepared by the analytical chemistry division of the University of Ferrara, Prof. Pasti Luisa) were poured in chlorobenzene (42 mL). The heterogeneous mixture was stirred at 105 °C. After 5 h, it was allowed to cool down to room temperature, filtered through a Gooch filter, and evaporated under reduced pressure. The residue was purified by silica gel column chromatography (elution system A/P 1:9 v/v) to obtain a white crystalline solid. Yield 62%. ^{1}H NMR (400 MHz, Chloroform-*d*) δ 7.56 (d, J = 2.1 Hz, 1H), 6.86 (d, J = 2.1 Hz, 1H), 6.70 (d, J = 8.5 Hz, 1H), 6.54 (d, J = 8.5 Hz, 1H), 3.97 (s, 3H), 3.90 (s, 3H). ^{13}C NMR (101 MHz, Chloroform-*d*) δ 147.60, 143.90, 140.38, 119.45, 106.43, 104.43, 102.72, 56.53, 55.80.

3.1.24. 1-(2,5-Dimethoxyphenoxy)propan-2-one (**24**)

Compound **9** (560 mg, 4.21 mmol, 1 equiv.) was dissolved in DMF (0.2 M), then Cs_2CO_3 (1.5 g, 4.631 mmol, 1.1 equiv.) and bromoacetone (0.63 gr, 4.63 mmol, 1.1 equiv.) were added. The mixture was stirred overnight at room temperature and quenched with brine and H_2O. The mixture was extracted with AcOEt, dried over Na_2SO_4, and the solvent was evaporated. The residue was purified by silica gel column chromatography (elution system A/P 1:4 v/v) to obtain 24 as a yellowish oil. Yield 63%.^{1}H NMR (400 MHz, Chloroform-*d*) δ 6.83 (d, J = 8.8 Hz, 1H), 6.47 (dd, J = 8.8, 2.8 Hz, 1H), 6.39 (d, J = 2.8 Hz, 1H), 4.57 (s, 2H), 3.84 (s, 3H), 3.74 (s, 3H), 2.28 (s, 3H).^{13}C NMR (101 MHz, Chloroform-*d*) δ 206.07, 154.22, 148.22, 144.00, 113.22, 105.29, 102.74, 74.46, 56.73, 55.77, 26.56.

3.1.25. 4,7-Dimethoxy-3-methylbenzofuran (**25**)

A mixture of polyphosphoric acid (0.52 g) in chlorobenzene (15 mL) was heated at 120 °C, and a solution of 24 (370 mg, 1.55 mmol, 1 equiv.) in chlorobenzene (3 mL) was added. The mixture turned black, and after 20 min, it was allowed to cool at room temperature. The chlorobenzene phase was decanted, and the residue was rinsed three times with 20 mL of Et_2O. The ether was evaporated, and cyclized product 25 was purified on silica gel column chromatography by eluting with A/P 0.2:9.8, v/v, to obtain a yellowish solid. Yield 50%. ^{1}H NMR (400 MHz, Chloroform-*d*) δ 7.28 (d, J = 1.4 Hz, 1H), 6.66 (d, J = 8.5 Hz, 1H), 6.49 (d, J = 8.5 Hz, 1H), 3.95 (s, 3H), 3.86 (s, 3H), 2.35 (d, J = 1.3 Hz, 3H). ^{13}C NMR (101 MHz, Chloroform-*d*) δ 149.34, 145.93, 140.50, 140.36, 120.11, 116.40, 106.33, 102.40, 56.58, 55.80, 29.78.

3.1.26. (4,7-Dimethoxybenzofuran-3-yl)methanol (**26**), 4,7-Dimethoxybenzofuran-3-carbaldehyde (**27**)

Compound **25** (109 mg, 0.57 mmol, 1 equiv.) and SeO_2 (125 mg, 1.134 mmol, 2 equiv.) in 1,4-dioxane were stirred at 75 °C for 96 h. The cooled mixture was filtered through Gooch, and the solvent evaporated under reduced pressure. Chromatographic purification of the residue using A/P 3:7 v/v as the eluent phase provided title compounds **26** and **27** in 1:1 ratio as brown and red solid, respectively. Yield 80%. ^{1}H NMR (400 MHz, Chloroform-*d*) δ 7.47 (d, *J* = 1.0 Hz, 1H), 6.71 (d, *J* = 8.6 Hz, 1H), 6.56 (d, *J* = 8.6 Hz, 1H), 4.73 (d, *J* = 0.9 Hz, 2H), 3.95 (s, 3H), 3.94 (s, 3H), 2.85 (s, 1H). ^{13}C NMR (101 MHz, Chloroform-*d*) δ 147.06, 146.03, 140.71, 140.55, 120.85, 118.54, 106.82, 102.70, 56.55, 55.94.

3.1.27. 4,7-Dimethoxybenzofuran-3-carbaldehyde (**27**)

A solution of **26** (50 mg, 0,24 mmol, 1 equiv.) in anhydrous DCM (5 mL) was added to a suspension of PCC (129 mg, 0.6 mmol, 2.5 equiv.) in anhydrous DCM (5 mL). The red mixture turned dark brown and was stirred at room temperature overnight. The insoluble cake was rinsed three times with anhydrous Et_2O. The ether portions were combined, filtered through a pad of Florisil, and evaporated. Compound **27** was afforded by chromatographic purification using A/P 3:7 v/v as the eluent phase. Yield 95%. ^{1}H NMR (400 MHz, Chloroform-*d*) δ 10.41–10.41 (m, 1H), 8.21–8.18 (m, 1H), 6.77 (d, *J* = 8.6 Hz, 1H), 6.65 (d, *J* = 8.6 Hz, 1H), 3.96 (s, 3H), 3.92 (s, 3H). ^{13}C NMR (101 MHz, Chloroform-*d*) δ 187.31, 148.52, 148.25, 145.99, 140.31, 123.46, 107.68, 104.36, 56.57, 55.82.

3.1.28. (E)-3-(4,7-Dimethoxybenzofuran-3-yl)-1-phenylprop-2-en-1-one (**28**)

A microwave vial was charged with aldehyde **27** (65 mg, 0.3 mmol, 1 equiv.), ylide **8** (125 mg, 0.33 mmol, 1.1 equiv.), and acetonitrile (3 mL); the microwave was set at 135 °C for 2 h. The solvent was evaporated, and the residue was purified on silica gel chromatography eluting with A/P 1:4 v/v to obtain the desired product **28** as a yellow-ocher solid. Yield 65%. ^{1}H NMR (400 MHz, Chloroform-*d*) δ 8.08–8.02 (m, 3H), 7.97–7.84 (m, 2H), 7.61–7.48 (m, 3H), 6.78 (d, *J* = 8.6 Hz, 1H), 6.65 (d, *J* = 8.7 Hz, 1H), 3.98 (s, 3H), 3.96 (s, 3H). ^{13}C NMR (101 MHz, Chloroform-*d*) δ 190.58, 148.41, 146.56, 145.78, 140.38, 138.38, 134.75, 132.62, 128.56, 128.46, 124.00, 119.85, 116.75, 107.53, 104.04, 56.56, 55.96. HPLC r.t. = 23,017 min. ESI = 3,092,086 [M + 1]. HRMS *m/z*: $[M + H]^+$ calc for $C_{19}H_{16}O_4$ 309.11214, found 309.1121; $[M + Na]^+$ calc 331.09408 found 331.0937.

3.1.29. Cell Cultures and Stimulation

Bone marrow-derived macrophages (BMDMs) were isolated from C57BL/6 mice as described [52] and differentiated for 7 days in Iscove's Modified Dulbecco's Medium supplemented with 15% fetal bovine serum (FBS, Gibco), 1% penicillin/streptomycin (P/S) and 10 ng/mL M-CSF. THP-1 cells were grown in RPMI medium supplemented with 10% FBS, 100 U/mL penicillin, and 100 mg/mL streptomycin. THP-1 cells were stimulated by 100 ng/mL PMA overnight to differentiate into macrophages. All cells were grown in a 5% CO_2 incubator at 37 °C. BMDMs were seeded at 5×10^5 in 24 well plates. After 12 h, the medium was removed, and cells were treated with LPS from *Escherichia coli* 055:B5 (1 μg/mL) in fresh Iscove's Modified Dulbecco's Medium for 2 h. After that, the medium was removed and replaced with a serum-free medium containing DMSO or compounds (10 μM) for 30 min. Cells were then stimulated with Nigericin (10 μM) for 1 h. Human THP-1 cells were seeded at 3×10^5 cells per well in 24 well plates. The following day, the overnight medium was replaced, and cells were stimulated with LPS (1 μg/mL) for 3 h. The medium was removed and replaced with a serum-free medium containing DMSO or compounds (10 μM) for 30 min. Cells were then stimulated with Nigericin (10 μM) for 1 h.

3.1.30. In Vivo LPS Challenge

C57BL/6 mice were IP injected with compounds (25 mg/kg) or vehicle control (DMSO) 30 min before IP injection of LPS 1 mg/kg (4 h) and then were euthanized, and blood and peritoneal exudate were isolated. Mouse plasma was collected after blood centrifugation (1000× *g*, 15 min at 4 °C). ELISA for IL-1β was performed according to the manufacturer's instructions (R&D Systems).

3.1.31. ELISA

Supernatants from BMDMs and THP-1 cell culture were assayed for mouse or human IL-1β, respectively, by ELISA according to the manufacturer's instructions (R&D Systems).

4. Conclusions

We succeeded in synthesizing the bioactive compound velutone F (**1**), the chalconoid contained in *Millettia velutina* stem, together with other 21 flavonoids. Our chemical total synthesis of velutone F is advantageous compare to the recently reported [24] seven steps semi-synthesis requiring the naturally occurring furanochromone derivative Khellin as the costly starting material (25 G > 800 USD). We could develop synthetic routes A and B; the first one establishes the benzofuran nucleus by creating the annellated carbocyclic ring onto a furan ring, while the second one proceeds exactly the other way. Instead, the Wittig olefination and the Heck coupling reaction are the featuring steps allowing for the assemblage of the 1,3-diaryl enone scaffold. Both the multi-step synthetic routes A and B establish the enone moiety at position C-5 of the 4,7-dimethoxybenzofuran thus, slightly modified synthetic approaches were designed in order to achieve the non-natural chalconoids **22**, **23**, and **28**, which formally are the C-2, C-3, and C-6 regioisomers of velutone F (**1**). The anti-inflammatory effects of the newly synthesized compounds are also reported.

Supplementary Materials: The following supporting information can be downloaded at: https://www.mdpi.com/article/10.3390/ijms23168957/s1.

Author Contributions: Conceptualization, C.T., V.Z. and C.G.; methodology, T.D.V., M.P. and S.M.; validation, P.P., C.G. and V.Z.; synthesis of calchonoids, T.D.V., D.I., P.M., G.T., V.C., G.S. and M.F.; writing—original draft preparation, V.Z., A.F., C.T., S.M. and C.G.; writing—review and editing, A.F., C.T. and T.D.V.; supervision, P.P., C.G. and C.T.; funding acquisition, C.T., P.P. and C.G. All authors have read and agreed to the published version of the manuscript.

Funding: The authors are greatly thankful to the University of Ferrara for the support. The Signal Transduction Laboratory is supported by the Italian Association for Cancer Research (IG-23670 to P.P., IG-19803 to C.G.), A-ROSE, Progetti di Rilevante Interesse Nazionale (PRIN2017E5L5P3 to P.P and PRIN20177E9EPY to C.G.), the Italian Ministry of Health (GR-2019-12369646 to S.M.), the European Research Council (853057-InflaPML to C.G.) and local funds from the University of Ferrara to C.G., P.P., V.Z. and C.T. M.P. is supported by AIRC research fellowship (ID: 26665). P.P. is grateful to C. Degli Scrovegni for her continuous support.

Institutional Review Board Statement: Procedures involving animals and their care were in conformity with institutional guidelines, and Animal Ethics Committee approved all experimental protocols (Authorization N° 481/2017-PR and CBCC2.N.BH4 approved by Italian Ministry of Health).

Acknowledgments: The authors thank Giorgia Macedonio for NMR analysis and Erika Marzola for HPLC analysis.

Conflicts of Interest: The authors declare no conflict of interest.

References

1. Tehfe, M.-A.; Dumur, F.; Xiao, P.; Delgove, M.; Graff, B.; Fouassier, J.-P.; Gigmes, D.; Lalevée, J. Chalcone Derivatives as Highly Versatile Photoinitiators for Radical, Cationic, Thiol-ene and IPN Polymerization Reactions Upon Visible Lights. *Polym. Chem.* **2014**, *5*, 382–390. [CrossRef]
2. Lee, S.-C.; Kang, N.-Y.; Park, S.-J.; Yun, S.-W.; Chandran, Y.; Chang, Y.-T. Development of a fluorescent chalcone library and its application in the discovery of a mouse embryonic stem cell probe. *Chem. Commun.* **2012**, *48*, 6681–6683. [CrossRef]
3. Nowakowska, Z. A review of anti-infective and anti-inflammatory chalcones. *Eur. J. Med. Chem.* **2007**, *42*, 125–137. [CrossRef] [PubMed]

4. Zhou, B.; Xing, C. Diverse Molecular Targets for Chalcones with Varied Bioactivities. *Med. Chem.* **2015**, *5*, 388–404. [CrossRef] [PubMed]
5. Zhuang, C.; Zhang, W.; Sheng, C.; Zhang, W.; Xing, C.; Miao, Z. Chalcone: A Privileged Structure in Medicinal Chemistry. *Chem. Rev.* **2017**, *117*, 7762–7810. [CrossRef]
6. Funakoshi-Tago, M.; Tanabe, S.; Tago, K.; Itoh, H.; Mashino, T.; Sonoda, Y.; Kasahara, T. Licochalcone A Potently Inhibits Tumor Necrosis Factor α-Induced Nuclear Factor-*k*B Activation through the Direct Inhibition of I*k*B Kinase Complex Activation. *Mol. Pharmacol.* **2009**, *76*, 745–753. [CrossRef]
7. Pandey, M.K.; Sandur, S.K.; Sung, B.; Sethi, G.; Kunnumakkara, A.B.; Aggarwal, B.B. Butein, a Tetrahydroxychalcone, Inhibits Nuclear Factor (NF)-*k*B and NF-*k*B-regulated Gene Expression through Direct Inhibition of I*k*Bα Kinase β on Cysteine 179 Residue. *J. Biol. Chem.* **2007**, *282*, 17340–17350. [CrossRef]
8. Wang, J.P.; Tsao, L.T.; Raung, S.L.; Lin, C.N. Investigation of the inhibitory effect of broussochalcone A on respiratory burst in neutrophils. *Eur. J. Pharmacol.* **1997**, *320*, 201–208. [CrossRef]
9. Maccari, R.; Ottana, R. Targeting Aldose Reductase for the Treatment of Diabetes Complications and Inflammatory Diseases: New Insights and Future Directions. *J. Med. Chem.* **2015**, *58*, 2047–2067. [CrossRef]
10. Aida, K.; Tawata, M.; Shindo, H.; Onaya, T.; Sasaki, H.; Yamaguchi, T.; Chin, M.; Mitsuhashi, H. Isoliquiritigenin: A New Aldose Reductase Inhibitor from Glycyrrhizae Radix. *Planta Med.* **1990**, *56*, 254–258. [CrossRef]
11. Song, N.R.; Kim, J.E.; Park, J.S.; Kim, J.R.; Kang, H.; Lee, E.; Kang, Y.G.; Son, J.E.; Seo, S.G.; Heo, Y.S.; et al. Licochalcone A, a Polyphenol Present in Licorice, Suppresses UV-Induced COX-2 Expression by Targeting PI3K, MEK1, and B-Raf. *Int. J. Mol. Sci.* **2015**, *16*, 4453–4470. [CrossRef] [PubMed]
12. Nyandoro, S.S.; Nkunya, M.H.H.; Josepha, C.C.; Odalo, J.O.; Sattler, I. New Glucopyranosylglyceryl-N-Octenyl Adipate and Bioactivity of Retro and Condensed Chalcones from *Toussaintia orientalis*. *Tanz. J. Sci.* **2012**, *38*, 108–126, eISSN: 2507–7961. Print ISSN 0856-1761.
13. Lee, M.H.; Kim, J.Y.; Ryu, J.-H. Prenylflavones from *Psoralea corylifolia* Inhibit Nitric Oxide Synthase Expression through the Inhibition of I-*k*B-α Degradation in Activated Microglial Cells. *Biol. Pharm. Bull.* **2005**, *28*, 2253–2257. [CrossRef]
14. Daikonya, A.; Katsuki, S.; Kitanaka, S. Antiallergic Agents from Natural Sources 9. Inhibition of Nitric Oxide Production by Novel Chalcone Derivatives from *Mallotus philippinensis* (Euphorbiaceae). *Chem. Pharm. Bull.* **2004**, *52*, 1326–1329. [CrossRef]
15. Chen, H.; Chen, X.; Sun, P.; Wu, D.; Yue, H.; Pan, J.; Li, X.; Zhang, C.; Wu, X.; Hua, L.; et al. Discovery of dronedarone and its analogues as NLRP3 inflammasome inhibitors with potent anti-inflammation activity. *Bioorg. Med. Chem. Lett.* **2021**, *46*, 128160. [CrossRef]
16. Zhang, C.; Yue, H.; Sun, P.; Hua, L.; Liang, S.; Ou, Y.; Wu, D.; Wu, X.; Chen, H.; Hao, Y.; et al. Discovery of chalcone analogues as novel NLRP3 inflammasome inhibitors with potent anti-inflammation activities. *Eur. J. Med. Chem.* **2021**, *219*, 113417. [CrossRef] [PubMed]
17. Choi, H.R.; Lim, H.; Lee, J.H.; Park, H.; Kim, H.P. Interruption of *Helicobacter pylori*-Induced NLRP3 Inflammasome Activation by Chalcone Derivatives. *Biomol. Ther.* **2021**, *29*, 410–418. [CrossRef]
18. Swanson, K.V.; Deng, M.; Ting, J.P.Y. The NLRP3 inflammasome: Molecular activation and regulation to therapeutics. *Nat. Rev. Immunol.* **2019**, *19*, 477–489. [CrossRef] [PubMed]
19. Wang, Z.; Xu, G.; Gao, Y.; Zhan, X.; Qin, N.; Fu, S.; Li, R.; Niu, M.; Wang, J.; Liu, Y.; et al. Cardamonin from a medicinal herb protects against LPS-induced septic shock by suppressing NLRP3 inflammasome. *Acta Pharm. Sin. B* **2019**, *9*, 734–744. [CrossRef]
20. Aybe, S.I.; Furuya, T. Studies on Plant Tissue Cultures. Part 36. Biosynthesis of a Retrochalcone, Echinatin, and Other Flavonoids in the Cultured Cells of *Glycyrrhiza echinate*. A New Route to a Chalcone with Transposed A- and B-Rings. *J. Chem. Soc. Perkin Trans.* **1982**, *1*, 2725–2734. [CrossRef]
21. Aybe, S.I.; Furuya, T. Biosynthesis of a Retrochalcone, Echinatin: A Feeding Study with Advanced Precursors. *Tetrahedron Lett.* **1981**, *22*, 2097–2098. [CrossRef]
22. Ayabe, S.I.; Yoshikawa, T.; Kobayashi, M.; Furuya, T. Biosynthesis of a Retrochalcone, Echinatin: Involvement of *O*-Methyl Transferase to Licodione. *Phytochemistry* **1980**, *19*, 2331–2336. [CrossRef]
23. Ma, X.; Zhao, M.; Tang, M.-H.; Xue, L.-L.; Zhang, R.-J.; Liu, L.; Ni, H.-F.; Cai, X.-Y.; Kuang, S.; Hong, F.; et al. Flavonoids with Inhibitory Effects on NLRP3 Inflammasome Activation from *Millettia velutina*. *J. Nat. Prod.* **2020**, *83*, 2950–2959. [CrossRef] [PubMed]
24. Zhang, R.; Hong, F.; Zhao, M.; Cai, X.; Jiang, X.; Ye, N.; Su, K.; Li, N.; Tang, M.; Ma, X.; et al. New Highly Potent NLRP3 Inhibitors: Furanochalcone Velutone F Analogues. *ACS Med. Chem. Lett.* **2022**, *13*, 560–569. [CrossRef]
25. Mishra, S.; Singh, P. Hybrid molecules: The privileged scaffolds for various pharmaceuticals. *Eur. J. Med. Chem.* **2016**, *124*, 500–536. [CrossRef]
26. Miao, Y.H.; Hu, Y.H.; Yang, J.; Liu, T.; Sun, J.; Wang, X.J. Natural source, bioactivity and synthesis of benzofuran derivatives. *RSC Adv.* **2019**, *9*, 27510–27540. [CrossRef]
27. Chand, K.; Hiremathad, A.; Singh, M.; Santos, M.A.; Keri, R.S. A review on antioxidant potential of bioactive heterocycle benzofuran: Natural and synthetic derivatives. *Pharm. Rep.* **2017**, *69*, 281–295. [CrossRef]
28. Radadiya, A.; Shah, A. Bioactive benzofuran derivatives: An insight on lead developments, radioligands and advances of the last decade. *Eur. J. Med. Chem.* **2015**, *97*, 356–376. [CrossRef]
29. Khanam, H. Bioactive Benzofuran derivatives: A review. *Eur. J. Med. Chem.* **2015**, *97*, 483–504. [CrossRef]

30. Schwaid, A.G.; Spencer, K.B. Strategies for Targeting the NLRP3 Inflammasome in the Clinical and Preclinical Space. *J. Med. Chem.* **2021**, *64*, 101–122. [CrossRef]
31. Zhang, X.; Xu, A.; Lv, J.; Zhang, Q.; Ran, Y.; Wei, C.; Wu, J. Development of small molecule inhibitors targeting NLRP3 inflammasome pathway for inflammatory diseases. *Eur. J. Med. Chem.* **2020**, *185*, 1118222. [CrossRef]
32. Coll, R.C.; Hill, J.R.; Day, C.J.; Zamoshnikova, A.; Boucher, D.; Massey, N.L.; Chitty, J.L.; Fraser, J.A.; Jennings, M.P.; Robertson, A.A.; et al. MCC950 directly targets the NLRP3 ATP-hydrolysis motif for inflammasome inhibition. *Nat. Chem. Biol.* **2019**, *15*, 556–559. [CrossRef]
33. Tapia-Abellán, A.; Angosto-Bazarra, D.; Martínez-Banaclocha, H.; de Torre-Minguela, C.; Cerón-Carrasco, J.P.; Pérez-Sánchez, H.; Arostegui, J.I.; Pelegrin, P. MCC950 closes the active conformation of NLRP3 to an inactive state. *Nat. Chem. Biol.* **2019**, *15*, 560–564. [CrossRef]
34. Goyal, K.; Kaur, R.; Goyal, A.; Awasthi, R. Chalcones: A review on synthesis and pharmacological activities. *J. Appl. Pharm. Sci.* **2021**, *11* (Suppl. S1), 1–14. [CrossRef]
35. Chen, Z.; Hu, F.; Huang, S.; Zhao, Z.; Mao, H.; Qin, W. Organocatalytic Enantioselective Selenosulfonylation of a C–C Double Bond to Form Two Stereogenic Centers in an Aqueous Medium. *J. Org. Chem.* **2019**, *84*, 8100–8111. [CrossRef]
36. Gammill, R.B. Synthetic Routes to Benzofurans and Benzothiophenes and Intermediates Therefor. U.S. Patent 4,609,739, 2 September 1986.
37. Gammill, R.B. Preparing 4,7-Dialkoxybenzofurans, and Intermediates Used Therein. European Patent 95,835, 7 December 1983.
38. Guzman, J.A.; Mendoza, V.; Garcia, E.; Garibay, C.F.; Olivares, L.Z.; Maldonado, L.A. Baeyer-Villiger Oxidation of β-Aryl Substituted Unsaturated Carbonyl Compounds with Hydrogen Peroxide and Catalytic Selenium Dioxide. *Synth. Commun.* **1995**, *25*, 2121–2133. [CrossRef]
39. Pang, Y.; An, B.; Lou, L.; Zhang, J.; Yan, J.; Huang, L.; Li, X.; Yin, S. Design, Synthesis, and Biological Evaluation of Novel Selenium- Containing Isocombretastatins and Phenstatins as Antitumor Agents. *J. Med. Chem.* **2017**, *60*, 7300–7314. [CrossRef]
40. Ma, W.; Huang, J.; Huang, X.; Meng, S.; Yang, Z.; Li, C.; Wang, Y.; Qi, T.; Li, B. Direct construction of 2,3-unsubstituted benzofurans and benzothiophenes via a metal-free catalyzed intramolecular Friedel-Crafts reaction. *Org. Chem. Front.* **2019**, *6*, 493–497. [CrossRef]
41. Liu, H.; Ge, L.; Wang, D.; Chen, N.; Feng, C. Photoredox-Coupled F-Nucleophilic Addition: Allylation of *gem*-Difluoroalkenes, *Angew. Chem. Int. Ed.* **2019**, *58*, 3918–3922. [CrossRef]
42. Song, S.-Y.; Lu, H.-L.; Wang, G.-F.; Yang, Y.-Q.; Huang, Y.-S. An improved and scale-up synthesis of 6-hydroxybenzofuran. *Res. Chem. Intermed.* **2016**, *42*, 4433–4442. [CrossRef]
43. Bunce, R.A.; Moore, J.D. Tetrahydropyranyloxy-Directed *ortho* Lithiation of Aromatic Systems. Synthesis of *o*-Hydroxycinnamate Esters from Phenols. *Org. Prep. Proc. Int.* **1997**, *29*, 293–299. [CrossRef]
44. Wriede, U.; Fernandez, M.; West, K.F.; Harcour, D.; Moore, H.W. Synthesis of Halodimethoxy-1,2-benzoquinones. *J. Org. Chem.* **1987**, *52*, 4485–4489. [CrossRef]
45. Ma, T.; Fu, X.; Kee, C.W.; Zong, L.; Pan, Y.; Huang, K.-W.; Tan, C.-H. Pentanidium-Catalyzed Enantioselective Phase-Transfer Conjugate Addition Reactions. *J. Am. Chem. Soc.* **2011**, *133*, 2828–2831. [CrossRef]
46. Mikami, S.; Kitamura, S.; Negoro, N.; Sasaki, S.; Suzuki, M.; Tsujihata, Y.; Miyazaki, T.; Ito, R.; Suzuki, N.; Miyazaki, J.; et al. Discovery of Phenylpropanoic Acid Derivatives Containing Polar Functionalities as Potent and Orally Bioavailable G Protein-Coupled Receptor 40 Agonists for the Treatment of Type 2 Diabetes. *J. Med. Chem.* **2012**, *53*, 3756–3776. [CrossRef]
47. Sayson, L.V.; Custodio, R.J.P.; Ortiz, D.M.; Lee, H.J.; Kim, M.; Jeong, Y.; Lee, Y.S.; Kim, H.J.; Cheong, J.H. The potential rewarding and reinforcing effects of the substituted benzofurans 2-EAPB and 5-EAPB in rodents. *Eur. J. Pharmacol.* **2020**, *885*, 173527. [CrossRef]
48. Sun, N.; Huang, P.; Wang, Y.; Mo, W.; Hu, B.; Shen, Z.; Hu, X. Zeolite-catalyzed synthesis of 2,3-unsubstituted benzo[b]furans via the intramolecular cyclization of 2-aryloxyacetaldehyde acetals. *Tetrahedron* **2015**, *71*, 4835–4841. [CrossRef]
49. Zaidlewicz, M.; Chechlowska, A.; Prewysz-Kwinto, A. Enantioselective Synthesis of 2- and 3-benzofuryl β-amino alcohols. *Heterocycles* **2001**, *55*, 569–577. [CrossRef]
50. Ando, K.; Kawamura, Y.; Akai, Y.; Kunitomo, J.I.; Yokomizo, T.; Yamashita, M.; Ohta, S.; Ohishi, T.; Ohishi, Y. Preparation of 2-, 3-, 4- and 7-(2-alkylcarbamoyl-1-alkylvinyl)benzo[*b*]furans and their BLT1 and/or BLT2 inhibitory activities. *Org. Biomol. Chem.* **2008**, *6*, 296–307. [CrossRef]
51. Missiroli, S.; Perrone, M.; Boncompagni, C.; Borghi, C.; Campagnaro, A.; Marchetti, F.; Anania, G.; Greco, P.; Fiorica, F.; Pinton, P.; et al. Targeting the NLRP3 inflamasome as a new therapeutic option for overcoming cancer. *Cancers* **2021**, *13*, 2297. [CrossRef]
52. Ying, W.; Cheruku, P.S.; Bazer, F.W.; Safe, S.H.; Zhou, B. Investigation of macrophage polarization using bone marrow derived macrophages. *J. Vis. Exp.* **2013**, *76*, 50323. [CrossRef]

MDPI
St. Alban-Anlage 66
4052 Basel
Switzerland
www.mdpi.com

International Journal of Molecular Sciences Editorial Office
E-mail: ijms@mdpi.com
www.mdpi.com/journal/ijms

www.ingramcontent.com/pod-product-compliance
Lightning Source LLC
Chambersburg PA
CBHW041931260926
51850CB00018B/44
* 9 7 8 3 7 2 5 8 0 9 9 9 8 *